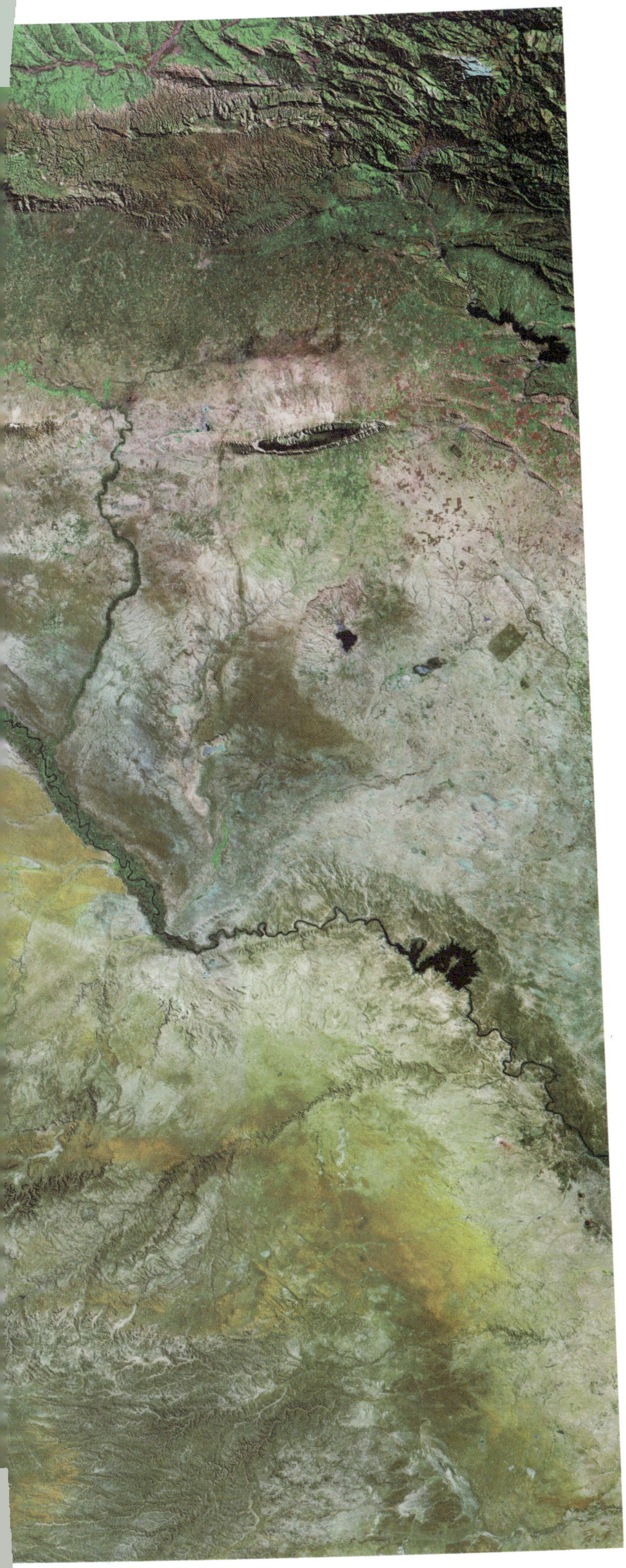

AF846121

LANDSAT satellite imagery (circa 1990-2000) of Syria and surrounding areas with shaded relief texture from the 90 m SRTM topographic dataset. The satellite imagery, prepared under contract by the Earth Satellite Corporation, is provided free by NASA's Applied Sciences Directorate at the John C. Stennis Space Center (https://zulu.ssc.nasa.gov/mrsid/).This 583 Mbyte image was produced by Global Mapper®. The projection is UTM for Zone 37 (36º-42ºE), datum is WGS-84, with 50 m pixels. The image extends from 35º to 43ºE, and from 32º to 38ºN.

Geological Framework of the Levant

Volume I: Cyprus and Syria

Edited by

Valery A. Krasheninnikov
Geological Institute of the Russian Academy of Sciences
Pyzhevsky Pereulok 7,
109017 Moscow, Russia

John K. Hall
Geological Survey of Israel
Marine Geology, Mapping & Tectonics Division
30 Malchei Israel Street,
Jerusalem 95 501, Israel

Francis Hirsch
Geological Survey of Israel
30 Malchei Israel Street,
Jerusalem 95 501, Israel

Chaim Benjamini
Department of Geological and Environmental Sciences
Ben-Gurion University of the Negev
P.O.B. 653, Beersheva 84 105, Israel

Akiva Flexer
Department of Geophysics and Planetary Sciences,
Tel-Aviv University, Tel-Aviv 69978, Israel

Jerusalem, April 2005

Front Cover. Topography of Cyprus. based upon the 25 m digital terrain model (DTM) of Cyprus by Hall (1998). The making of this DTM is discussed by Hall (2005) in Chapter 15 of Volume II. This three-dimensional image was made using Interactive Visualization Systems, Inc. (IVS) Fledermaus® software. The Troodos Complex occupies the center of the mountainous area in south-western Cyprus. The highest peak is Mount Olympus at 1951 m altitude. To the southwest of the Troodos is the Mamonia Complex. These areas are discussed in detail in the text.

Inside Front Cover. Geological Map of Cyprus (Geological Survey Department, Ministry of Agriculture, Natural Resources and Environment, Government of Cyprus, ©1995) with shaded relief texture from the 25 m digital terrain model (DTM) of Cyprus (Hall, 1998; see Chapter 15, Volume II). The texture was applied using Global Mapper® software. Note the change in topographic fabric between the differing geological facies of the Mamonia, Kyrenia, and Troodos Terranes, as well as the Circum Troodos sedimentary and Arakapas transform sequences.

Back Cover. Topography of Syria and vicinity. This image was made using the Shuttle Radar Topographic Mission (SRTM) 3" (≤ 90 m) dataset, obtained in February 2001 by NASA's Space Shuttle Endeavour during mission STS-99. A 60 m long boom was deployed from the shuttle's payload bay and C-band interferometric radar used over a 10 day period to derive an approximate 1" (≤ 30 m) grid of the Earth's land areas between 62°N and 62°S. Following extensive processing by NASA's Jet Propulsion Laboratory the 90 m dataset was released for Eurasia in November 2003, and for Africa in April 2004. The digital data is available for downloading from the U. S. Geological Survey's ftp://edcsg9.cr.usgs.gov/pub/data/srtm/ website. The horizontal datum is the WGS-84 datum that is the basis for positions derived from the Global Positioning System (GPS). The SRTM dataset is still in its infancy, with small holes resulting from areas in radar shadow, water bodies with incoherent returns etc. However this dataset, appearing just as these two volumes are going to press, is a hundredfold improvement over the previous 0.5' (≤ 900 m) GLOBE, GTOPO30, HYDRO1k, and ACE datasets (see Chapter 16 of Volume II) that were previously available from analysis of topographic maps and satellite altimetry. Global Mapper® 5.10 software (www.globalmapper.com) was used to convert from the native 1° by 1° SRTM .hgt tiles in geographic coordinates between 35°E and 43°E and 32°N and 38°N, to a single binary .grd file in Universal Transverse Mercator (UTM) coordinates (Zone 37) for Golden Software Inc. Surfer® 8 for Windows. Surfer® 8 was then used to render the image shown on a reduced grid of 0.1' (≤ 185.2 m).

Inside Back Cover. LANDSAT satellite imagery (circa 1990-2000) of Syria and surrounding areas with shaded relief texture from the 90 m SRTM topographic dataset. The satellite imagery, prepared under contract by the Earth Satellite Corporation, is provided free by NASA's Applied Sciences Directorate at the John C. Stennis Space Center (https://zulu.ssc.nasa.gov/mrsid/). This 583 Mbyte image was produced by Global Mapper®. The projection is UTM for Zone 37 (36°-42°E), datum is WGS-84, with 50 m pixels. The image extends from 35° to 43°E, and from 32° to 38°N.

©2005 Historical Productions-Hall
All rights reserved

All materials subject to this copyright and included
in this volume may be photocopied for the non-commercial
purpose of scientific and educational advancement.

Copyright is not claimed on any material prepared
by government employees within the scope of their employment.

Typeset using Adobe® InDesign® 2.2 CS for Macintosh from original Microsoft® Word 2000 files.

Prepress, layout, color separations, plates and printing by Art-Plus
Harashei Barzel 19, Talpiot Industrial Zone, Jerusalem 93420, Israel. Tel: +972-2-648-2228.

Printed and bound in Israel

ISBN 965-7297-02-8

Contents of Volume I

Plates

(In separate plate box)

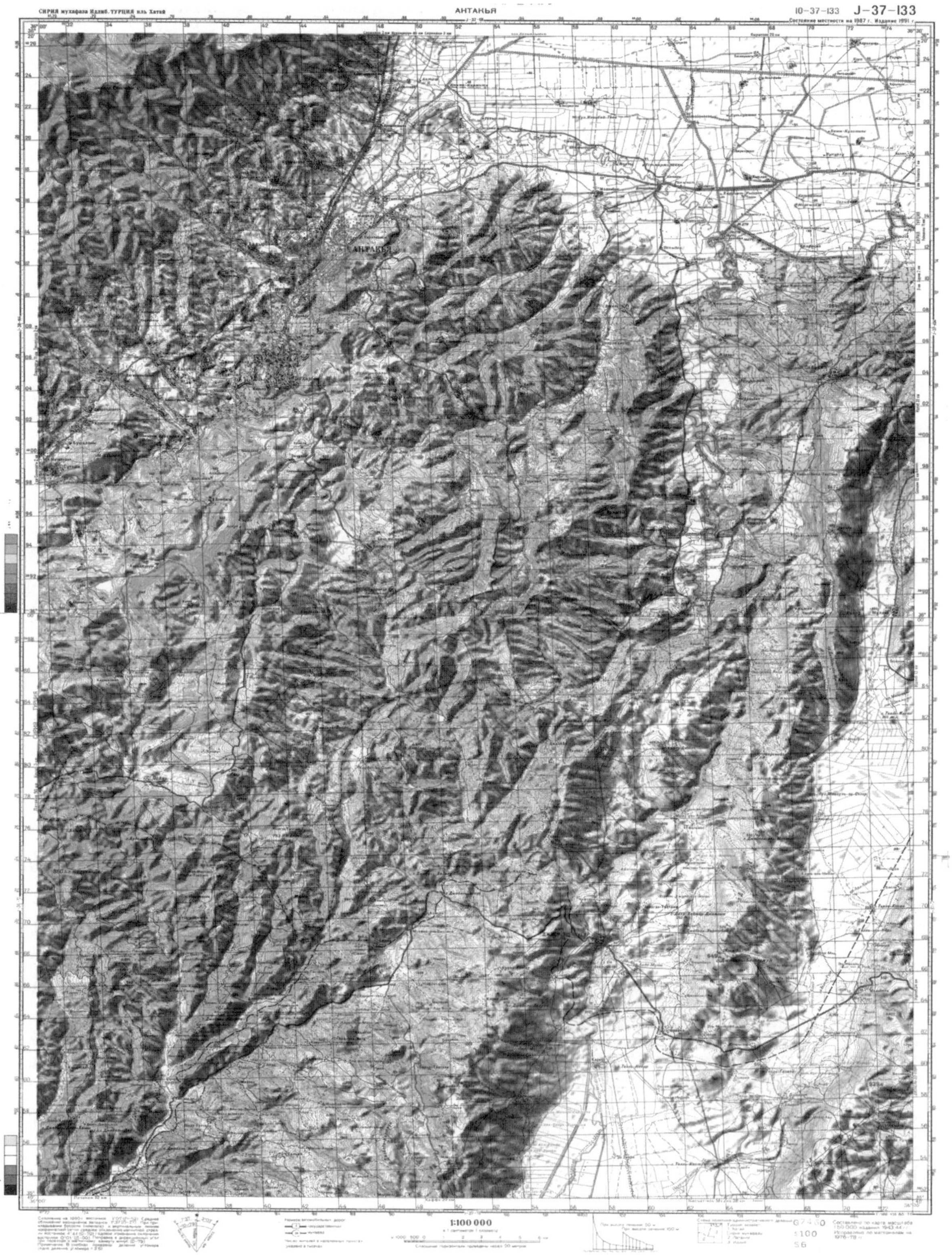

Example of the Russian topographic mapping of Syria (1970-1991), initiated by the geological work of the Russian Academy of Sciences. Two 1:100,000 scale contour sheets (J-37-133 – Antakya and I-37-1) have been analytically hill shaded using the gridded 90 m SRTM topography. The area shown is the northern part of the Al Ghab graben in northwestern Syria (discussion in Chapter 8).
Digital map files downloaded from http://www.lib.berkeley.edu/EART/syria/100k.html, registered as GeoTiffs and shaded by Global Mapper®

Preface and Acknowledgements

Valery A. Krasheninnikov
Geological Institute of the Russian Academy of Sciences, Pyzhevskiy Pereulok 7, 109107 Moscow, Russian
John K. Hall
Geological Survey of Israel, 30 Malchei Israel Street, Jerusalem 95 501, Israel
Francis Hirsch
Geological Survey of Israel, 30 Malchei Israel Street, Jerusalem 95 501, Israel
Chaim Benjamini
Department of Geological and Environmental Sciences, Ben-Gurion University of the Negev, P.O.B. 653, Beer-Sheva 84 105, Israel

The eastern Mediterranean area, embracing the triple junction of the African, Arabian and Anatolian plates, has attracted the interest of many geologists. A great number of publications have attempted to explain the geological features in the nine countries of this area. Unfortunately the political mosaic has often put limits on the free exchange of ideas, which were restricted to international as well as occasional regional meetings. The complicated problems of correlation in the eastern Mediterranean and the Middle East can only be understood and completed by international collaboration, *e.g.* the recent Peri-Tethys Program, in which scientists of all countries involved were invited and did participate in the preparation of the Tethys Atlas (1993) and the Peri-Tethys Atlas (2000).

This has motivated us to bridge these gaps with these two volumes that focus on the most representative areas of the eastern Mediterranean, namely Southwestern Cyprus and Syria in Volume I, and the Levantine Basin and Israel in Volume II.

In our opinion, this task can only be achieved by incorporating data from both the onshore and offshore domains. This has involved the participation of scientists from a broad spectrum of disciplines such as geophysics, oceanography, volcanology, petrology, tectonics, geochemistry, geochronology, paleontology, stratigraphy and others.

The editors are well aware that the region is too large to be covered in great and consistent detail in these two volumes. However the present publication has provided a canvas on which the respective authors, who have devoted much of their professional lives to the investigation of these areas, can summarize and synthesize their mostly original findings.

These two volumes cannot fulfill the job of giving all the details of this sizeable area. Beyond the general knowledge imparted, we see its task as complementing and not supplanting the extensive works of others, often in allied disciplines. Hence the Russian scientific contribution to the geology of south-western Cyprus consists of three main topics.

1) A classic petrological description of the volcanic rocks of the Troodos and Mamonia Complex;

2) A very detailed micropaleontological investigation that has greatly improved the Radiolarian and Conodont biostratigraphy of the Mamonia Complex and a definition of the stratigraphy and paleogeography of the neoautochthon cover through planktonic foraminifer determinations; and

3) models for the tectonic evolution and structure of southwestern Cyprus and the east central Mediterranean Sea from detailed structural mapping and marine geophysical measurements.

In Syria, the pioneering field and laboratory work of scientists of the Russian Academy of Sciences and the Ministry of Geology, together with their Syrian colleagues from the General Establishment of Geology and Mineral Resources and the Syrian Petroleum Company over a period of nearly 40 years is summarized. The exhaustive field mapping included topography, geology, primary geophysics, and other pure and applied aspects of geology. These studies of primarily surface geology have provided the basis for all the following studies, and nicely supplement the modern mostly subsurface geophysical and tectonic compilations of groups such as that led by Prof. Muawia Barazangi at Cornell University, Ithaca, New York, USA.

Caveats - an explanation to prevent misinterpretation

A number of difficulties had to be bridged, which were of a scientific as well as of a technical character. The most difficult concerned the striking difference in the neo-western and classical Russian methodologies of stratigraphic subdivision. The concept and interpretation of the terms 'suite' and 'formation' is different for the two. Since both have their advantages and shortcomings, it is not a matter of validity or superiority. Thus we want the reader to be aware that for the chapters on Southwest Cyprus and Syria, the Russian method was used, whereas in the chapters on Israel, the neo-western approach is applied. It is hoped that some benefit may result from the difference in approaches.

In Israel the subdivisions are based on lithology, using 'group' and 'formation'. This method is based mainly on the proposals of Hollis D. Hedberg, which have been adopted by many national surveys in the world. Stratigraphic boundaries are lithological boundaries. In Southwest Cyprus and Syria, the subdivisions are based primarily on regional biostratigraphic units. While lithologies are recognized, the boundaries are chrono-stratigraphic. In both approaches, biostratigraphy plays an important role. But as described herein, the method used in Southwest Cyprus and Syria dictates the subdivision, whereas in Israel, it dates the unit, allowing the formation to be hetero-chronic.

Since in reality many subdivisions are caused by gaps and unconformities, the final effect may be the same, as groups are often identical with series or stages. However, in several cases, beginning from the Cretaceous, group limits are hetero-chronous and may not coincide with series or stage boundaries.

In the present study, no formation names are given for the Cretaceous-Cenozoic research on Syria. An exception was made for names introduced by Dubertret in Lebanon and subsequently used in Syria. Other formation names, introduced by oil geologists, are not always applicable. Therefore, we introduce no new names since this is a prerogative of the local national surveys. As a result we resorted to the use of mapping symbols, which are based on biostratigraphic data.

Since these data for Syria are highly reliable, and represent the state of the art, they allow correlation of the units they define with the units that yield the same assemblages in neighboring Jordan and Israel.

The scope of Parts II (Syria) and IV (Israel) of these books has been dictated by non-scientific realities. Part II relates to a huge country, with difficult logistics, while Part IV relates to a relatively tiny, intensively-surveyed area. The preparation of two volumes on their respective geologies is therefore difficult, since most of the data presented for Syria are new or for the first time in English, while for Israel the synthesis of the mass of available data within a single volume had never been attempted.

Among the technical problems, language and translation were paramount. The Southwestern Cyprus and Syrian chapters were originally written in Russian, and then translated, or written directly in Russian-English. On the other hand the Israeli chapters were mostly written in English by contributors with a facility for the language, even though it is not their mother tongue.

A Dedication and Appreciation

We are deeply aware that the task of producing our two volumes would have been impossible without the immense contributions of the many geologists who have preceded us and to whom we are indebted for having paved the way in the investigation of the geology of these regions. By way of thanks, these volumes are dedicated to the seven geologists below who pioneered and led the way in the geological study of the Near East.

Max Blanckenhorn - Geologist/Paleontologist (1867-1947)
Leopold Kober - Tectonics (1883-1970)
Louis Dubertret - Geologist (1904-1979)
Victor Ponikarov - Geologist (1917-1996)
Leo Picard - Geologist (1900-1997)
Zeev Reiss - Paleontologist (1917-1996)
Abraham Parnes - Paleontologist (1892-1989)

We acknowledge the following for their assistance in acquiring photographs of these geological greats: the Geologen Archiv (Freiburg/Br), and in particular Prof. Eugen Seibold, for the portrait of Max Blanckenhorn; Prof. Alexander Tollmann and Dr. Fred Roegl (Vienna), for the portrait of Leopold Kober; Ms. N. P. Akhmetieva (Moscow) for the portrait of Louis Dubertret; Ms. T. V. Ponikarova for the portrait of Victor Ponikarov; Eliyahu Wakshal (Rehovoth) for the photograph of Leo Picard and portrait of Avraham Parnes; and Ms. Pnina Merling Reiss for the portrait of her husband Zeev Reiss

Acknowledgements

We thank our respective institutions, the Geological Institute (GIN), the Institute of Geochemistry and Analytical Chemistry, the Institute of Physics of the Earth, and the Institute of Oceanology of the Russian Academy of Sciences, the Geological Survey of Israel (GSI), the Department of Geology at Ben-Gurion University (BGU) of the Negev, and the Department of Geophysics and Planetary Sciences, Tel Aviv University (TAU), for providing us with the necessary research facilities and the time to edit the books. In particular we thank Academician Yuri Leonov, Director of GIN, and Dr. Amos Bein at the GSI and his predecessor, Dr. Gideon Steinitz.

During 1987-1990, scientists of the former USSR Academy of Sciences carried out joint investigations in Cyprus with the Geological Survey Department of Cyprus (Nicosia). We express our deep thanks to the geologists of Cyprus, and in particular to Dr. George Constantinou and Dr. Costas Xenophontos, as well as to Prof. L. V. Dmitriev of the Institute of Geochemistry and Analytical Chemistry in Moscow.

We gratefully acknowledge the individual contributions of many who assisted in particular sections of these books.

Prof. Jörn Thiede, Director of the Alfred Wegener Institute in Bremerhaven, and former director of GEOMAR in Kiel, rendered a particularly valuable service in permitting the extensive reprocessing of the deep seismic reflection data of the Strakhov 5 cruise.

Dr. Valeria Kornetova (Mrs. Krasheninnikov) for her critical reading of the chapter on the Volcanics of Syria, for constant companionship and consolation during the long months of work laying out the text and figures of this volume.

Mr. Donald Cannon in Moscow who corrected the English of the chapter on the Cretaceous of Syria.

Ms. Miriam Lederman, neighbor and good friend, who assisted in the last Sisyphusian stages of preparation of Chapter 22, and for her extensive proof reading skills as the books approached completion.

Our colleagues at the Geological Survey of Israel in Jerusalem: Dr. Ron Bogoch, Dr. Tuvia Weissbrod, and Ms. Bevi Katz for their editing skills on various chapter manuscripts; Dr. Arie Gilat for translation of a particularly difficult chapter, and for several years of assistance in all ways imaginable, and even some unimaginable; our librarians, Ms. Varda Arad and Oksana Zaripov and their colleagues Ms. Zelda Colodner and Debby Magen at the Hebrew University Geology Department library, who at our request sometimes ransacked their treasured holdings for us; and Ms.

Bat-Sheva Cohen and Ms. Chana Netzer-Cohen at the Graphics Department of the GSI for some remarkable figures.

Others helped in inestimable ways to make our life easy. Doron Hirsch for his graphical skills on the computer. Ms. Channa Eisenmann for her professional hospitality to the principal editor during his seven months in Jerusalem. In the run-up to publication, for scanning and duplication of materials, we were greatly assisted by Yaakov Zrubabely and Nissim Sabag at SuperGraph Ltd. And the pre-press and actual production of these volumes was in the capable hands of Haim Gross and Boris Menukhin at Art Plus in Jerusalem.

Addendum

It is not everyday that such volumes are prepared, and it is certainly infrequent that they are published with the objective of making them available at no cost to libraries, institutions, and individuals who need the information. However this is the case with these two books, and perhaps some sort of explanation is in order.

The financing for these volumes comes from the benevolence of Thomas Harper Blodgett (1887-1964), the maternal grandfather of Dr. John K. Hall. Thomas H. Blodgett was the long-time president of the American Chicle Company, the makers of Chiclets and Dentyne chewing gum. Mr. Blodgett was a very successful businessman, and his success has allowed his grandchildren the luxury of helping worthwhile projects get off the ground.

For the implementation of this project, we gratefully acknowledge the assistance of Mr. Lee Metzendorf, Senior Vice-President of Fiduciary Trust Company International and Adv. Dana Hiscock in making available the funds to bring this project to fruition.

These volumes were not created under the best of conditions. The bulk of the work has been carried out in Jerusalem during the second Al Aksa intifada. Prof. Krasheninnikov's two stays of seven months total were punctuated by six nearby bombings which took the lives of over fifty people, and in the case of his first lodging, what were once his apartment windows were blown out by the particularly cruel blast at the Moment Cafe on the Gaza Road. But the carnage was not confined to Israel. On September 11, 2001, the suicide plane that hit the second World Trade Center building obliterated the top floor offices of Fiduciary Trust Company International, killing 93 of their employees.

It is our hope that these volumes will bring a little light to these dark times.

Thomas Harper Blodgett (1887-1964), President of the American Chicle Company, at his desk in the Chrysler Building, New York City. Photo from a 1945 greeting card sent to employees and stockholders. Mr. Blodgett is the maternal grandfather of John K. Hall. His success in business and subsequent beneficence made possible these two new volumes and the previous volume on the Structural Framework of the North Eastern Mediterranean.

REFERENCES CITED

Dercourt, J., M. Gaetani, B. Vrielynck, E. Barrier, B. Biju-Duval, M. F. Brunet, J. P. Cadet, S. Crasquin, and M. Sandulescu (Eds.), 2000. Atlas Peri-Tethys: Paleogeographical Maps, Comm. for the Geological Map of the World, Paris, 268+xx p.

Dercourt, J., L. E. Ricou, and B. Vrielynk (Eds.), 1993. Atlas Tethys Palaeoenvironmental Maps. Gauthier-Villars, Paris, 307 p., 14 maps, 1 pl.

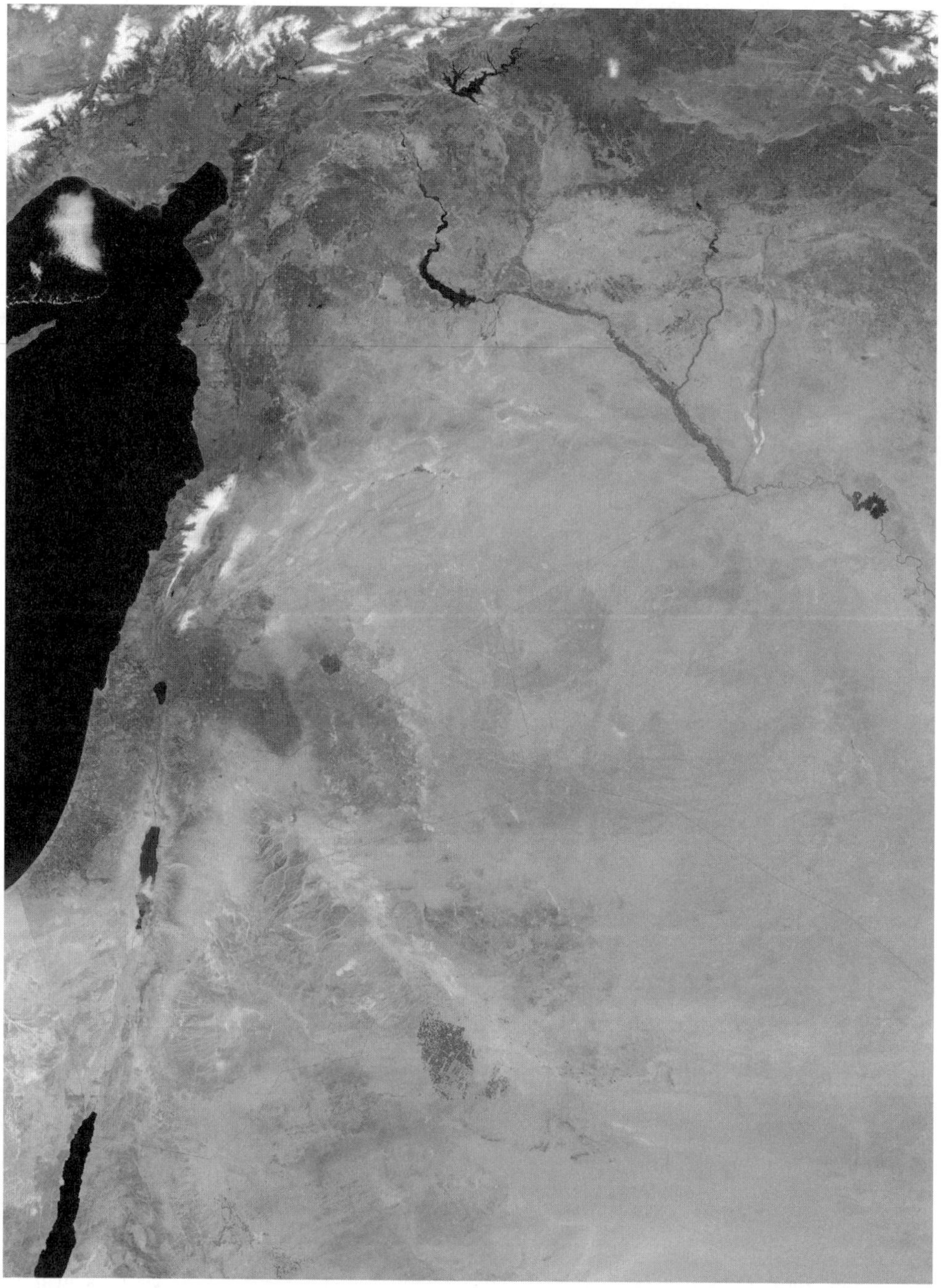

The Levant as of March 22, 2004. Grayscale look at a natural color image from the MODIS camera on NASA's Terra satellite. The **Levant** is a general geographical term for the area east of the Mediterranean Sea, south of the Taurus Mountains, and west of the northern Arabian Desert and Mesopotamia. It does not include any part of the Arabian Peninsula or the Caucasus; Sinai is considered a land bridge from the Levant to Egypt. For the purposes of these books Cyprus and the Levantine Basin are also included.

PART I

South-Western Cyprus
Introduction

Valery A. Krasheninnikov
Geological Institute of the Russian Academy of Sciences, Pyzhevskiy pereulok 7, 109017 Moscow, Russia
Sergey A. Silantyev
Vernadsky Institute of Geochemistry and Analytical Chemistry of the Russian Academy of Sciences, Kosygin Str.19, 117975 Moscow, Russia

The south-western part of Cyprus extends from the Akamas Peninsula in the west to the surroundings of Limassol in the east (**Fig. I.1**). Beyond a narrow strip of the seashore plain, the foothills of the Troodos Massif appear as a dissected relief, that stretches to the high mountains of the Troodos Massif, culminating in the high peak of Mount Olympus (1951 m). On this small piece of land, nature created a unique geological garden. It includes complexes of intrusive, effusive, metamorphic and sedimentary rocks of different genesis and various facies. The very complicated tectonic style juxtaposes elements that are difficult to unravel.

Therefore it is not surprising that geologists paid special attention to the geology and mineral resources of this island. Th. Pantazis, in his excellent article "A revised bibliography of Cyprus geology" (1969), mentions that the history of the geological investigations and the mining and mineral industry had gone back to the prehistoric Greek, Phoenician, and Roman times. He also gave a long list of geologists of the second half of the 19th century and the first half of the 20th century who had published data about the geological structure, volcanic processes, paleontology, and seismicity of Cyprus: A. Gaudry, D. Bate, F. Reed, C. Cullis, A. Edge, F. Henson, J. McGinty, L. Bear, C. Bellamy, R. Browne, D. Bishopp, D. Burdon, A. Cockbain, I. Gass, and C. Renz et al.

Systematic investigation of the regional geology of Cyprus is connected with the work of geologists of the Geological Survey of Cyprus (F. Ingham, Y. Hadjisravrinou, A. Panayiotou, Th. Pantazis, W. Turner, R. Wilson, Ch. Constantinou, C. Xenophontos). Their studies revealed the main features of the geology of this island. Peculiarities of the geological structure of South-western Cyprus which are considered in this part of the monograph were outlined by Kluyver (1969), Lundberg (1969), Bagnall (1960), Zomenis (1972), Mantis (1972, 1977), and Hadjistavrinou and Afrodisis (1977).

Geological investigations in Cyprus have received a strong push since the start of drilling in oceanic basins (1968), when marine geology obtained the status of a special branch in the geological sciences, and mobilism became dominant in the investigation of the lithosphere.

The idea that the Troodos Ophiolite Massif and the complicated association of igneous, metamorphic, and sedimentary rocks of the Mamonia Complex are remnants of the oceanic crust and sedimentary cover of the Mesozoic Tethys, emplaced on a continental margin, was extremely prolific. Moreover, the dismembered fragments of magmatic and sedimentary assemblages belonging to passive continental margins, accreted within the Mamonia complex added puzzling elements. It stimulated the research of geologists of various countries (*e.g.* Cyprus, Great Britain, Canada, France, Germany, Greece, Japan, USA, etc.) in the field of petrology, tectonics, lithology, and stratigraphy, as well as the land and marine geophysics of Cyprus. Publications of A. H. F. Robertson, N. H. Woodcock, U. Bernarz, H. U. Schminke, Y. Dilek, I. G. Gass, H. Lapierre, A. Miyashiro, J. Makris, P. Robinson, J. Malpas, R. E. Swarbrick, P. Thy, E. M. Moores, F. J. Vine, W. R. Baragar, W. E. Cameron, A. Panayotou, J. G. Spray, J. C. Roddick, J. A. Pearce, C. Xenophontos, R. J. Varga, G. Constantinou and others deciphered the general elements of the tectonic structure and many events of the geological evolution of this unique region, but also faced some complicated problems. It is well known today that numerous geochemical data obtained during the last decades convincingly demonstrated that magmatic suites of different subduction origin take part in the construction of the magmatic sequence of famous ophiolite complexes, among which the Troodos ophiolite is one of the best known (Miyashiro, 1975; Cameron, 1985). On other hand, at the same time it was shown that typical spreading centers exist in back-arc and fore-arc basins in which oceanic-type crust is formed (Karig, 1971; Saunders and Tarney, 1979; Taylor and Nesbit, 1995). However magmatic rocks related with this crust of suprasubduction origin are characterized by clear geochemical evidence.

International conferences and symposia devoted to various topics of petrology and tectonics that were held in Cyprus (1979, 1987, 2000) were of great significance. Cyprus as a geological object is a most popular target for geological field excursions and one of the best field schools for the international student community. The progress in the study of the geology of south-western Cyprus is obvious and this circumstance has influenced the whole mobilistic concept.

During 1987-1990 some institutions of the USSR Academy of Sciences (the Geological Institute, Vernadsky Institute of Geochemistry and Analytical Chemistry, Institute of Physics of the Earth) undertook geological investigations in the south-western part of Cyprus and the adjacent area of the Mediterranean Sea (the Levantine Basin). These intensive multidisciplinary studies have been carried out by the above mentioned institutions in close collaboration with the Geological Survey of Cyprus in the framework of the "Tethys" Project supported by the Russian Academy of Sciences (1987-1995). The main goal of these investigations in the Cyprus region was the reconstruction of the geodynamic conditions which are responsible for the formation of the Troodos ophiolite and Mamonia Complex.

In the following chapters we attempt to compile our results and those of studies by scientists of other countries.

REFERENCES CITED

Bagnall, P. S., 1960. The geology and mineral resources of the Pano Lefkara-Larnaca area. Geol. Surv. Dept. Cyprus, Mem. No 5, 116 p.

Cameron, W. E., 1985. Petrology and origin of primitive lavas from the Troodos ophiolite, Cyprus. Contributions to Mineralogy and Petrology, v. 89, p. 239-255.

Hadjistavrinou, Y. and S. Afrodisis, 1977. Geology and hydrogeology of the Paphos region. Geol. Surv. Dept. Cyprus, Bull. No 7, p. 1-44.

Hall, J. K., 1998a. Digital terrain model of the island of Cyprus. In: T. Weissbrod (ed.), Isr. Geol. Surv., Current Research, Vol. 11, p. 45-50.

Hall, J. K., 1998b. Three-dimensional map of the Island of Cyprus. Grayscale sheet of land topography at 1:250,000 scale, Geological Survey of Israel and Cyprus, Jerusalem, one sheet. Sheet prepared for ICGEM-3, Third International Conference on the Geology of the Eastern Mediterranean, Nicosia, Cyprus, September 23-26, 1998.

Karig, D. E., 1971. Origin and development of marginal basin in the Western Pacific. J. Geophys. Res., 76(11), p. 2542-2561.

Kluyver, H. M., 1969. Report on a regional geological mapping in Paphos area. Geol. Surv. Dept. Cyprus, Bull. No 4, p. 21-36.

Lundberg, B. R., 1969. The geology of the Anglisidhes-Lefkara area. Geol. Surv. Dept. Cyprus, Bull. No 2, p. 19-39.

Mantis, M., 1972. Some planktonic foraminifera from the Paleogene of South Cyprus. Geol. Surv. Dept. Cyprus, Bull. No 5, p. 91-125.

Mantis, M., 1977. Upper Eocene Radiolaria in Cyprus. Geol. Surv. Dept. Cyprus, Bull. No 7, p. 75-84.

Miyashiro, A., 1975. Classification, characteristics, and origin of ophiolites. J. Geol., 83(2), p. 249-281.

Panayides, I., C. Xenophontos, and J. Malpas (Editors), 2000. Proceedings, Third International Conference on the Geology of the Eastern Mediterranean. The Geological Survey Department, Nicosia, Cyprus, 376 p.

Pantazis, Th. M., 1969. A revised bibliography of Cyprus geology. Geol. Surv. Dept. Cyprus, Bull. No 2, p. 57-81.

Saunders, A. D., and J. Tarney, 1979. The geochemistry of basalts from a back-arc spreading center in the East Scotia Sea. Geochemica and Cosmochimica Acta, v. 43, p. 555-572.

Taylor, R. N., and R. W. Nesbitt, 1995. Arc volcanism in an extensional regime at the initiation of subduction: a geochemical study of Hahajima, Bonin Islands, Japan. In: J. L. Smellie (ed.), Volcanism associated with extension at consumed plate margins, Geol. Soc. Spec. Publ., London, p. 165-172.

Zomenis, S. L., 1972. Stratigraphy and hydrogeology of the Neogene rocks in the Northern foothills of the Troodos Massif, Cyprus. Geol. Surv. Dept. Cyprus, Bull. No 5, p. 22-90.

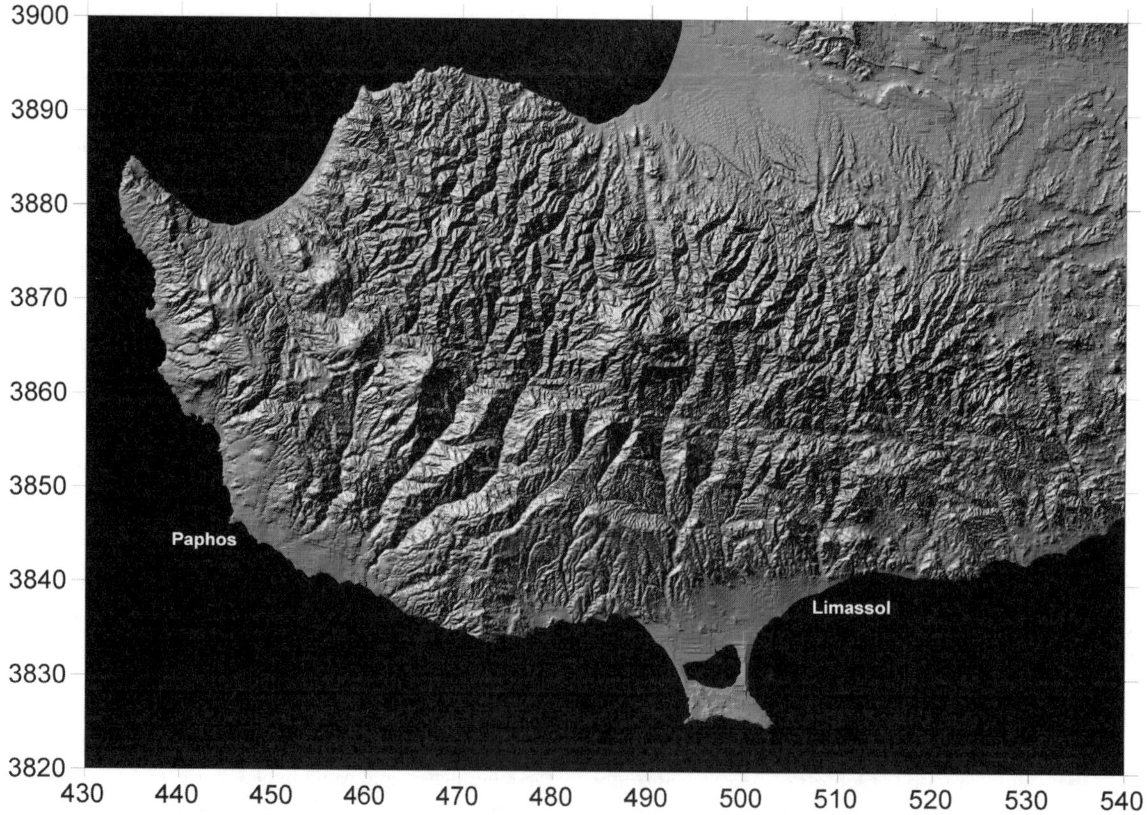

Fig I.1. Shaded relief map of the south-eastern part of the island of Cyprus. Image produced from the 25 m DTM of Cyprus (Hall, 1998a, b; Chapter 15 in Volume II).

Chapter 1

The Troodos Ophiolite Complex (Structural Para-Autochthon)

Sergey A. Silantyev, Maxim V. Portnyagin
Vernadsky Institute of Geochemistry and Analytical Chemistry of the Russian Academy of Sciences, Kosygin Str. 19, 117975 Moscow, Russia
Nikita Yu. Bragin, Kirill G. Kaleda, Kirill A. Krylov, Lyubov G. Bragina
Geological Institute of the Russian Academy of Sciences, Pyzhevskiy Pereulok 7, 109017 Moscow, Russia

A. Introduction

Sergey A. Silantyev
Vernadsky Institute of Geochemistry and Analytical Chemistry of the Russian Academy of Sciences

The ophiolitic succession in the Troodos Massif in Cyprus (**Fig. 1.1**), is one of the better-studied and best-documented examples of an ophiolite complex in the world. Historically, the discussion concerning the origin and development of the Troodos ophiolites demonstrates a history of ideas in the persistent controversy over the petrogenesis of collision-related zones in ancient oceanic basins. The canonic model of ophiolite formation suggests an origin for ophiolitic assemblages in a mid-oceanic ridge followed by their tectonic emplacement on a continental margin by collision with the oceanic basin (Coleman, 1971, 1977; Dewey and Bird, 1971; Moores and Vine, 1971).

Numerous geochemical evidence was obtained during the last few decades which convincingly demonstrated that different magmatic suites took part in constructing ophiolitic complexes of different ages (Miyashiro, 1975; Saunders et al., 1980; Cameron, 1985). It was also shown that typical spreading centers exist in back-arc and fore-arc basins, where oceanic-type crust is formed (Karig, 1971; Saunders and Tarney, 1979; Taylor and Nesbit, 1995). The compositional distinction between ophiolites can generally be found in the relationships and proportions of the distinct magmatic assemblages generated within the subducting slab in the intra-arc and island-arc environments. The causes for this variability could be related to the age and the composition of the downwelling slab and to the maturity of the arc.

The Troodos Massif, consisting of a WNW-ESE trending elongated ophiolite terrain, is located close to the center of the island of Cyprus (Fig. 1.1). Ultramafic, mafic and intermediate to felsic (minor) plutonic rocks are exposed mainly in the core of the Troodos Massif, around the summit of Mount Olympus. Rocks belonging to the lava and sheeted dikes complex are the most widespread igneous rocks making up the periphery of the massif. Ophiolitic pillow lavas are also present within the Larnaca Graben, east of the main ophiolite belt. In the Akamas Forest area of southwest Cyprus, a large structural block of the Troodos Ophiolite is juxtaposed with the rocks of the Mamonia terrain.

In the case of the Troodos ophiolite, Robertson (1990), Robinson and Malpas (1990), Sobolev et al. (1993), and others have suggested an evolution of subduction-related magmatic melts. Robinson and Malpas (1990) and Robertson and Xenophontos (1993) described the evolution and mode of emplacement of the Troodos complex. Robertson and Woodcock (1980) suggested that the history of the Troodos Ophiolite began in Early Cretaceous time with the development of active spreading in the Neo-Tethys oceanic basin. The existence of a series of calc-alkaline igneous rocks and shoshonites along the east-west-trending Kyrenia Ridge in northern Cyprus, point to a north-dipping fossil subduction zone. The early spreading-related magmatic assemblage was formed in an intra-arc basin (Robinson and Malpas, 1990), whereas the large magmatic stage took place in an island arc environment (Pearce, 1975).

The distribution and compositional characteristics of the Eastern Mediterranean ophiolites (Cyprus, Syria, Turkey), and their structural relationships with the African-Arabian continental margins, suggest a system of spreading centers and transform faults separated by a series of microcontinents. Dilek and Moores (1990) postulated that a complicated structure of oceanic basin provenance existed above a south-dipping zone of subduction, where the northern crust of the Paleo-Tethys subducted. This model thus suggested simultaneous opening of the Neo-Tethys and closing of the Paleo-Tethys. Later on, the compressional regime obducted the oceanic crust over a passive continental margin consisting of volcano-sedimentary complexes. This southward obduction and the closing of the Mesozoic Tethys Ocean were linked to the convergence of Africa and Eurasia (Dilek and Moores, 1990; Xenophontos, 1989).

On the basis of paleostress reconstruction, Grand et al. (1993)

suggested a scheme for the tectonic development of the Troodos Massif:

1) Early-Late Cretaceous east-west or ESE-WNW large-magnitude extensional strains, when the Troodos ophiolites were formed, and lateral displacement, striking 120°, taking place along the Arakapas Fault Zone;

2) Late Cretaceous-Eocene regional compression in the 160° direction;

3) Late Neogene N-S extensional strains.

The estimated U-Pb zircon age for plagiogranites from the Troodos Ophiolite is 91.6±1.4 Ma (Mukasa and Ludden, 1987). This age belongs to the Turonian interval (Blome and Irwin, 1985; Staudigel et al., 1986). Earlier radiometric datings on gabbroic and diabase rocks indicated 100-110 Ma (Aptian-Albian) (Desmet et al., 1978; Delaloye et al., 1980). A much younger Senonian age (~80-85 Ma) was suggested for the Troodos ophiolitic lavas, based on the Perapedhi and Kannaviou formations that overlie the ophiolitic lavas. A biostratigraphic (radiolarian) age of circa 89-91 Ma for umbers also points to the age of the lava succession (Blome and Irwin, 1985). About 75 million years ago the whole ophiolite complex was obducted and emplaced above the continental basement (Blome and Irwin, 1985).

B. The Lava Complex

Maxim V. Portnyagin
Vernadsky Institute of Geochemistry and Analytical Chemistry of the Russian Academy of Sciences, Kosygin Str. 19, 117975 Moscow, Russia

General characteristics

The main development of the ophiolite lavas is found around the periphery of the Troodos Massif. The lava pile varies considerably in thickness, from a hundred meters to 1,000-1,500 m in some areas (Taylor, 1990). The overlying rocks are Cretaceous metalliferous sediments, and pelagic and chaotic deposits. Very minor (less than one percent) but diverse sedimentary rocks occur sporadically as thin interbeds among the extrusives in the upper stratigraphic levels. Near the base of the volcanic succession there are transitional strata, the so-called Basal Group, that separate the sheeted dike complex and the overlying lava sequences. This unit consists of highly variable proportions of isolated pillows, pillowed screens, and dikes. The amount of volcanic components increases stepwise with stratigraphic height.

The lavas are located within a series of structural blocks separated by subvertical faults from two different periods.

An earlier generation of faults formed during a period of volcanic activity. They have a low vertical displacement (up to several tens of meters) and are often filled by synvolcanic hydrothermally cemented breccias. The later faults, responsible for the formation of the block structure of the lava complex, are probably associated with the final stage of the ophiolite emplacement, during tectonic accretion and obduction of a continental margin.

Much has been written about the geochemical nature of the Troodos lavas, mostly dealing with deciphering the tectonic settings of their origin and development. The ideas of earlier authors, based mainly on petrographic criteria and field geometrical relationships, have been summarized by Gass and Masson-Smith (1963), who concluded that the extrusion of pillow lavas was greatly affected by an extensional regime in an oceanic basin located between Eurasia and Africa.

In the early stages of plate-tectonic theory, Gass (1968) postulated that the only satisfactory mechanism to account for such an extension is sea-floor spreading at a mid-oceanic ridge. Moores and Vine (1971) have proposed a model of spreading ridge segmentation for the Troodos, describing the volcanically active zone as a system consisting of a number of on-axis cells operated independently, but tapped to provide volcanic material; it was also suggested that the Arakapas Fault Belt in Southern Cyprus might be a fossil oceanic transform fault.

Miyashiro (1973, 1975) was the first to recognize a clear geochemical similarity between the major-element geochemistry of the Troodos lavas and the extrusives from island-arc assemblages, and concluded that the effusion of ophiolite lavas may have resulted from supra-subduction spreading at a plate margin. Similarly, Pearce and his associates (Pearce and Cann, 1973; Pearce, 1975, 1977), based on certain stable element data (*e.g.* Zr, Ti and Y), have interpreted the tectonic setting in terms of some sort of subduction-related zone geometry.

Much work has been done on the petrologic characterization of the Troodos extrusives, and the relevant information can be found in numerous publications from recent decades: *e.g.* Pearce, 1980; Cameron, 1985; McCulloch and Cameron, 1983; Robinson et al., 1983; Mehegan and Robinson, 1991; Mehegan, 1988; Schmincke et al., 1983; Rautenschlein et al., 1985, 1987; Rautenschlein, 1987; Bednarz, 1993; Bednarz and Schmincke, 1994; Malpas and Langdon, 1984; Taylor, 1988, 1990; Taylor and Nesbitt, 1988; Thy and Xenophontos, 1991; Thy and Esbensen, 1993; Malpas, 1990; Malpas et al., 1993; Sobolev et al., 1991, 1993a, b; Portnyagin and Sobolev, 1993, 1995; and Portnyagin et al., 1997.

Main Lava Types and Volcanic Stratigraphy

The Troodos lavas span a significant spectrum of compositions. The most abundant rocks are andesites, basaltic andesites, and basalts. Silica-rich dacites, rhyodacites, basic picrites, and other compositional end members are generally less

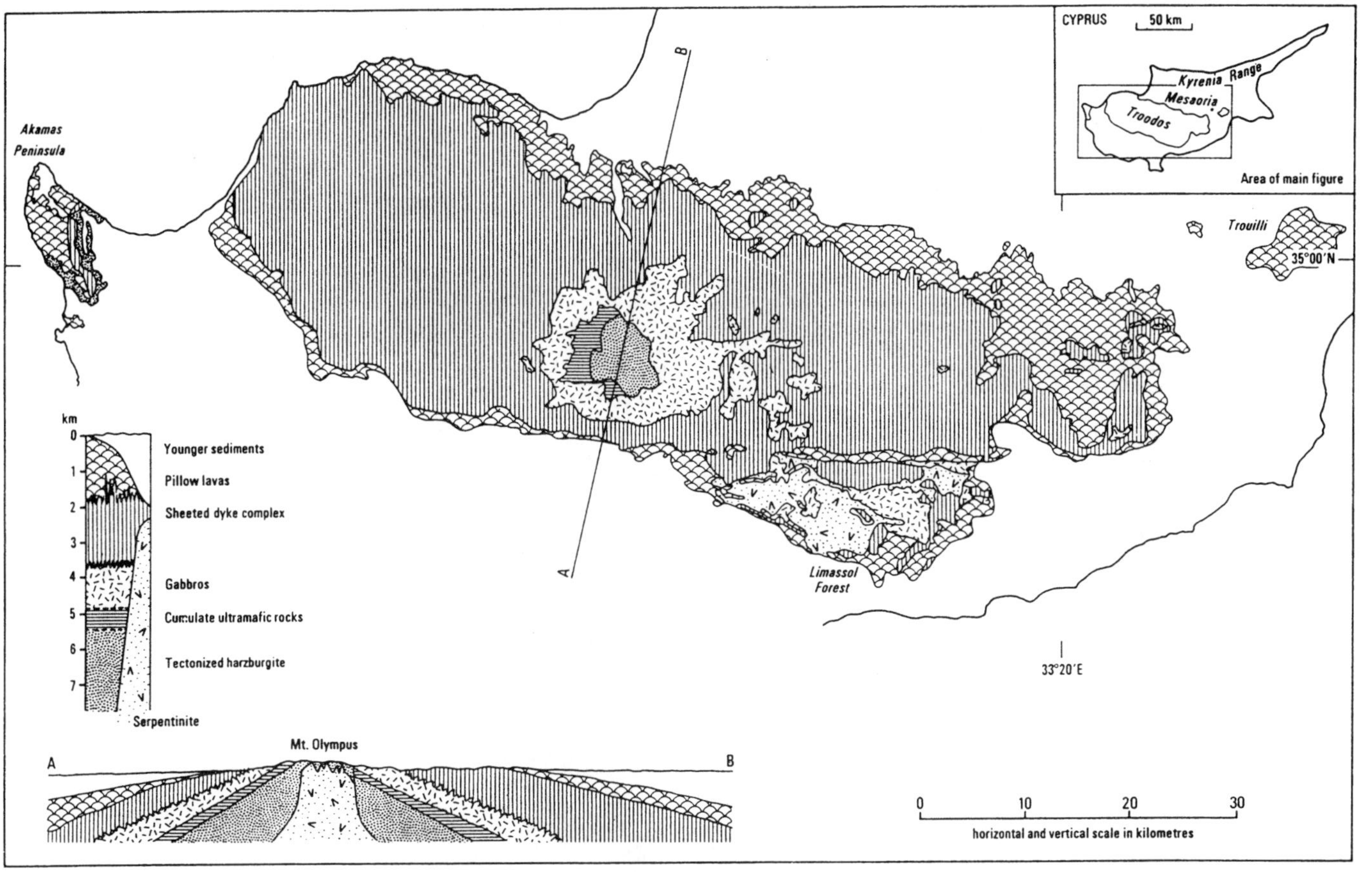

Fig. 1.1. Geological sketch map, section and ophiolite sequence of the Troodos Massif (after Gass, 1980).

common (**Fig. 1.2**). Strong hydrothermal alteration of the lavas is typical for most areas. The alteration was shown not only to have led to replacement of primary minerals but also strongly affected the whole rock composition, especially in alkali and alkaline earth elements (*e.g.* Bednarz and Schmincke, 1989). More reliable information on the primary composition of the rocks and their parental magmas is provided by the data on fresh volcanic glasses and glassy rocks which sporadically occur in the lava sections (*e.g.* Robinson et al., 1983). The data provide the basis of modern views on the compositional peculiarities of the Troodos magmas, their origin, spatial distribution, and evolution in time.

Traditionally, two major lava suites are distinguished within the Troodos volcanic complex. Scientists from the Cyprus Geological Survey (*e.g.* Bear, 1960; Gass, 1960) provided an initial subdivision of the extrusives by identifying the olivine phyric group at the top of the sequence (the Upper-Pillow Lavas or UPL) resting on generally aphyric or rare pyroxene-plagioclase phyric lavas (the Lower Pillow Lavas or LPL). The stratigraphy was modified by Gass and Smewing (1973), who identified lower 'axis' and upper 'off-axis' sequences, which were suggested to be chemically continuous but separated by a partial unconformity and metamorphic discontinuity.

Later, an attempt was made to integrate field volcanic, tectonic and alteration features within the new mineralogical and geochemical data on the lavas (Schmincke et al., 1983; Robinson et al., 1983; Rautenschlein et al., 1985; Schmincke and Bednarz, 1990). It was found that two major compositional suites (**Fig. 1.3**) compose the extrusive section of Troodos. The relatively evolved arc-tholeiite suite of andesites, dacites, and rhyodacites has been recognised in the lower section of the lava pile. According to the elevated content of TiO_2 in the rocks (generally above 1 wt.%), the suite was called 'the High-Ti Series' (HTS). Relatively primitive highly magnesian basalts and andesites displaying a similarity to boninites were found to dominate at the top of lava sequences. This suite, characterized by a TiO_2 content less than 1 wt.%, has been named 'the Low-Ti Series' (LTS). In a few localities the boundary between these suites corresponds with the boundary between the so-called 'Upper' and 'Lower Pillow Lavas' (*e.g.* in the Margi area), but in other cases it occurred at a much deeper stratigraphic level (*e.g.* the Akaki River canyon).

More recently, lavas relatively poor in titanium were also found at the base of the lava pile. These 'Lower Low-Ti Series' lavas (Lo-LTS) (Bednarz and Schmincke, 1994) occur directly above the Sheeted Dyke Complex and thus represent the first eruptive products of the Troodos Ridge. These lavas are relatively fresh and reach their maximum thickness in the Analiondas-Kataliondas area

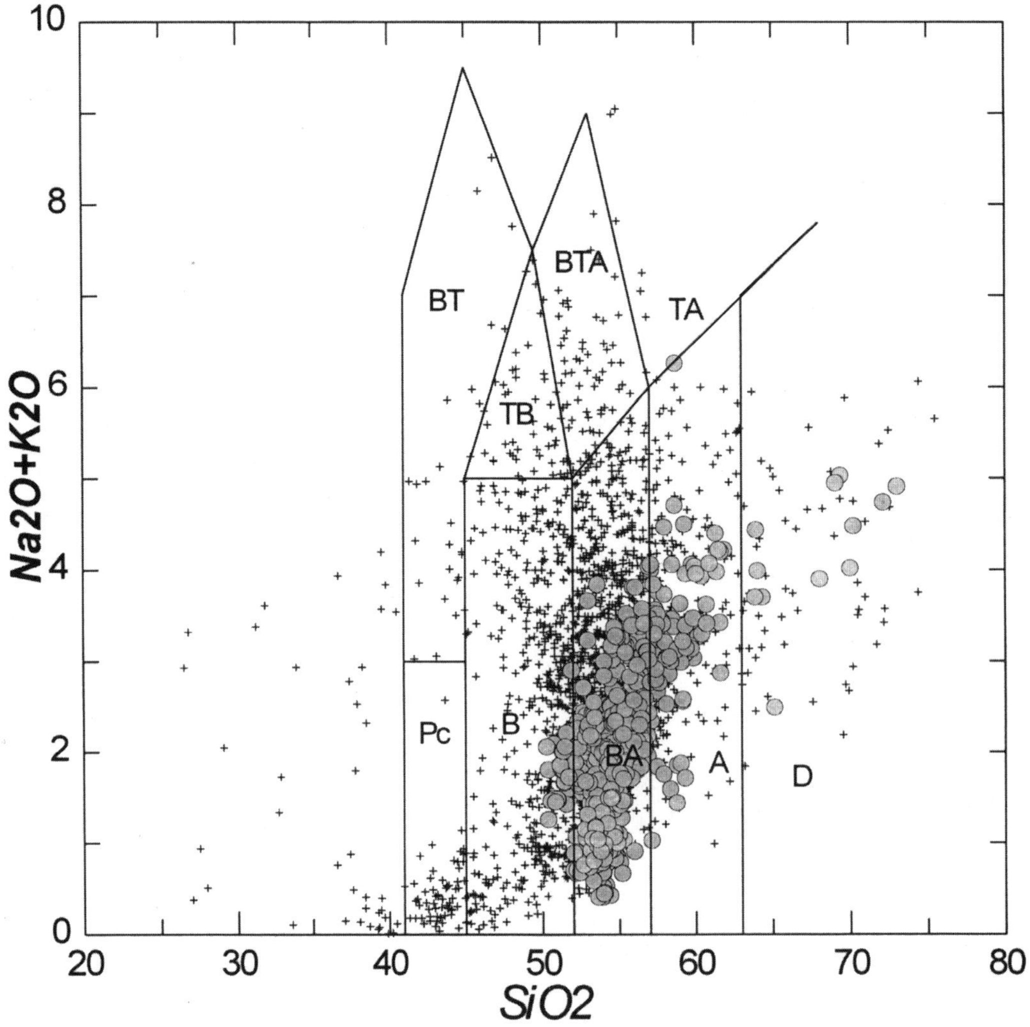

Fig. 1.2. TAS diagram showing compositions of the Troodos lavas (in wt.%). Dots - volcanic glasses, crosses - whole rocks. For data sources see Sobolev et al., (1993a and references therein). Pc - picrobasalt, B - basalt, BA - basaltic andesite, A - andesite, D - dacite, BT - basanite-tephrite, TB - trachybasalt, BTA - basaltic trachyandesite, TA - trachyandesite

(Taylor, 1990; Portnyagin et al., 1997). Elsewhere, the lavas are strongly altered and occur as pillow screens in the Basal Group, or as highly dissected lava units at the base of the lava sequence (Desmet, 1976; Taylor, 1990; Bednarz and Schmincke, 1994).

Detailed volcanologic study in the north-eastern part of the Troodos Ophiolite Massif has led to the recognition of three principal types of submarine volcanoes, namely, pillow volcanoes, sheet flow volcanoes, and breccia flow volcanoes, which were identified at several locations (Schmincke et al., 1983; Schmincke and Bednarz, 1990). Their repeated development is interpreted as being connected with volcano-tectonic-hydrothermal cycles (**Fig. 1.4**), which are recorded as ~6-10 volcanic units within the local stratigraphic sequences. In addition to these intermittent cycles, three main stages were generally recognized as subdivisions of the extrusive crust in NE Troodos (Schmincke and Bednarz, 1990):

a) During the first stage, non-depleted andesitic and dacitic to rhyodacitic lavas, up to tens of meters thick, appear to have erupted from breccia type volcanoes being developed near to active spreading axes or centers. The most common facies variety consists of distal glassy sheet flows, completely brecciated and disintegrated during their flowing phase and then converted almost entirely into breccioid lava strings, due to syn-volcanic and subsequent faulting. This facies association was found in the (~300 m thick) lower part of the extrusive sequence.

b) Extrusives of the second stage are confined largely to sheet flow eruptions. Fragments of sheet flow volcanoes are composed exclusively of extensive sheet flows, most of which are andesitic, although they can also be basaltic. In the lower sections, there are many smooth transitions to the distal facies of breccia flow volcanoes. Sheeted flow volcanoes are the prominent facies component in the central part (~100-300 m) of the extrusive sequence in the Northern Troodos.

c) The third stage is dominated by pillow volcanoes, which are considered to be small loaf-shaped edifices (with summit

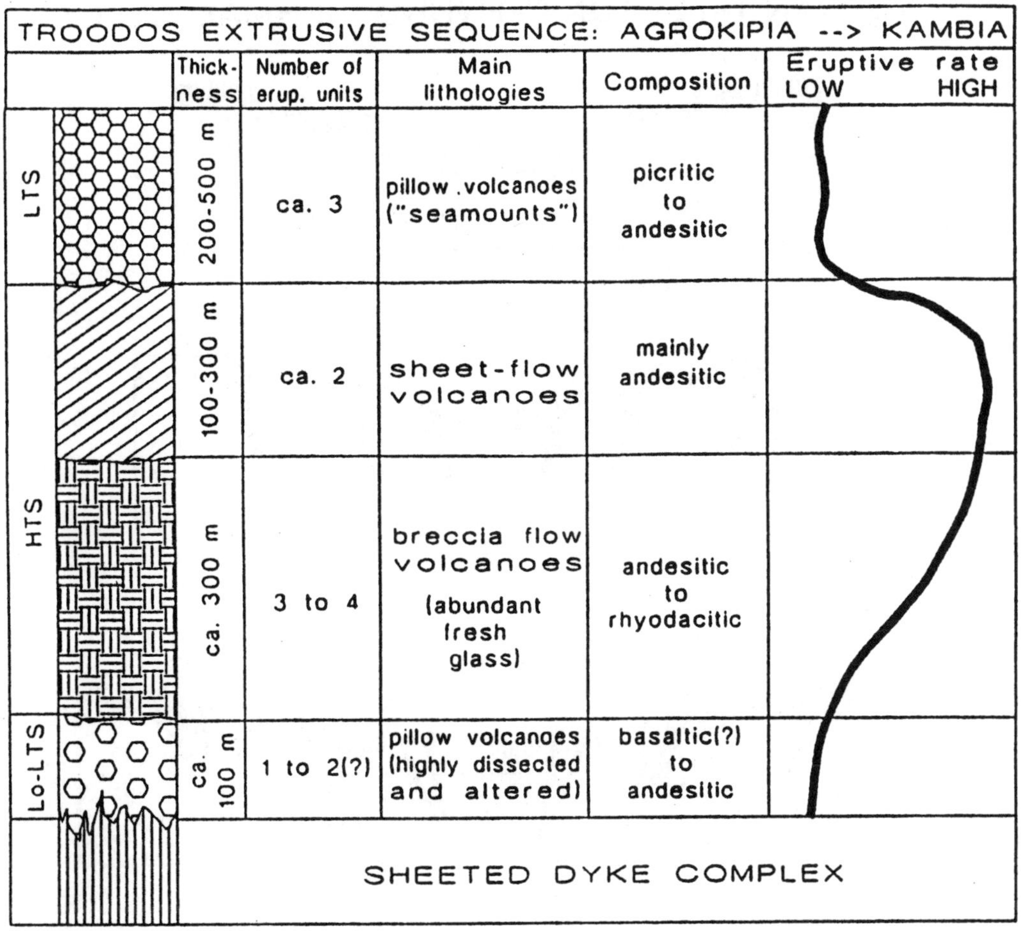

Fig. 1.3. Schematic diagram of succession of volcanic units in the Troodos Extrusive Series from Agrokipia in the west to Kambia in the east as summarized from Holes CY-1, CY-1A, CY-2, CY-2A, the Pediaeos area and re-interpretation of the Akaki sequence (after Bednarz, 1988).

elevations of only 100-200 m), composed of low viscosity magmas (a characteristic typical of basaltic andesites). According to their primary geochemistry, pillow lavas erupted off the on-axis sites or outside the active spreading centers. Despite the fact that the effusion of pillow basalts was documented within distinct stratigraphic intervals, including the oldest portion (~300 m) of the Lava Complex, the pillow lavas typically occur in the uppermost part (~200-300 m) of the extrusive crust in the Northern Troodos, suggesting a waning stage in the development of ophiolite volcanism.

The boundary between the HTS and LTS suites is not always evident. Evidence of a temporary halt in volcanic activity between HTS and LTS eruptions is provided by a thin sediment horizon separating the lava types at Margi and Ayios Onouphrios in the NE Troodos (Boyle and Robertson, 1984; Taylor, 1990). In other places, a careful examination of the HTS-LTS contact reveals a lithologically sharp limit without any of the features of a stratigraphic gap (*e.g.* Koronia Hill) or a gradational transition with distinct petrographic/geochemical characteristics (*e.g.* Taylor, 1990).

From the above, the Troodos extrusives can be subdivided into older on-axis Lo-LTS and HTS sequences (including the Basal Group unit) and more recent off-axis lavas of LTS. The current division of the Lava Complex is based on a number of criteria including petrographic features, chemical differences, lithology, abundance of dikes, intensities of metamorphic alternation, as well as relationship to sulfide mineralization.

The composition and origin of the older lava sequence (Lower Low-Ti series and High-Ti Series volcanics)

The greater part of the Lava Complex is exposed along the northern and southern margins of the Troodos Massif. This complex is subdivided into the HTS and Lo-LTS suites. The Troodos Lavas contain a remarkable variety of volcanic rocks ranging from picrites to rhyolite-dacites (**Table 1.1**). Andesites and andesite-dacites of the HTS are the most abundant rock types. All these rocks typically have a degree of metamorphism ranging from greenschist to zeolite facies.

Most of the HTS rocks are fine-grained or glassy and

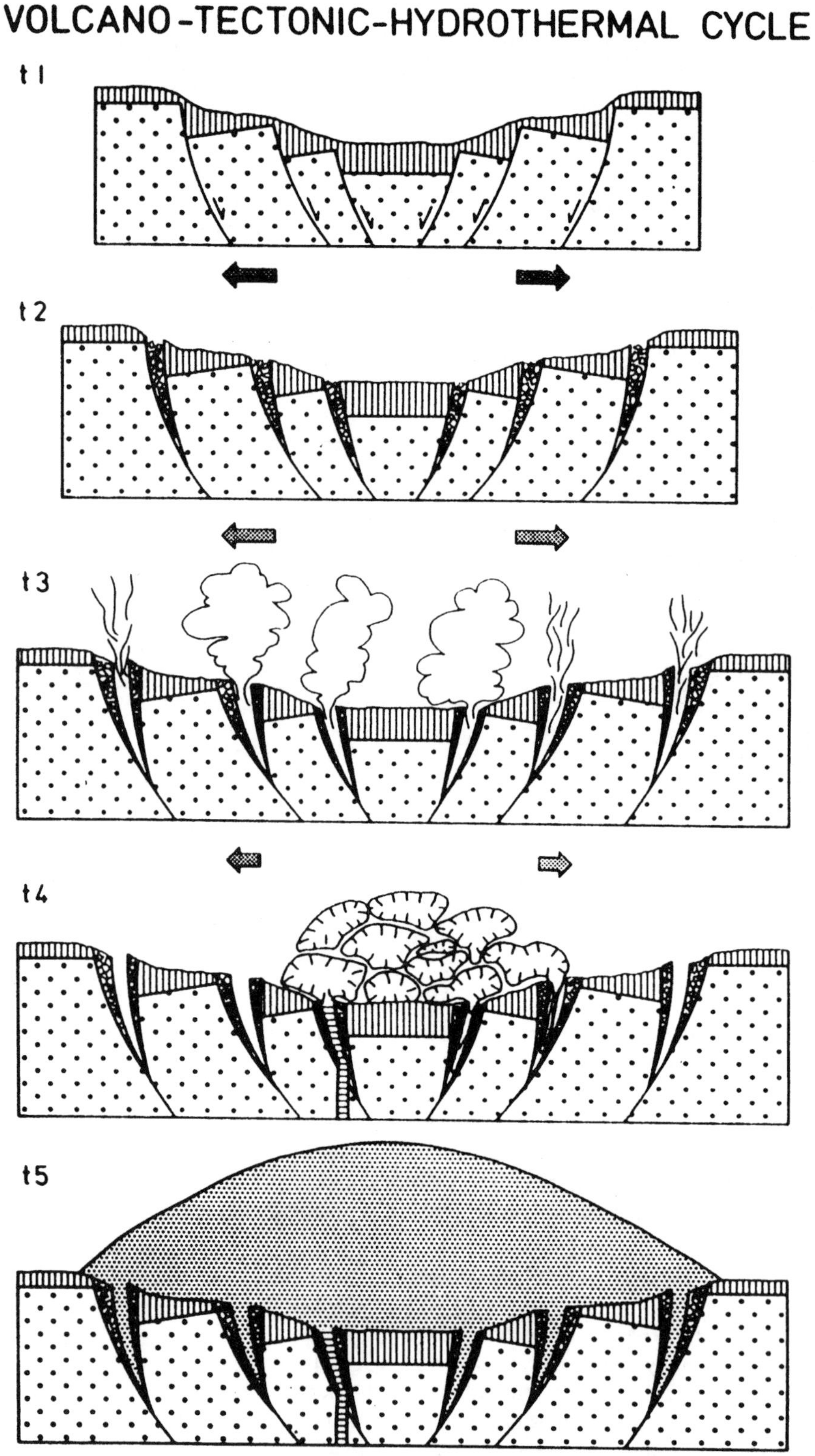

Fig. 1.4. Schematic representation of 5 stages in the evolution of volcano-tectonic-hydrothermal cycles following Schmincke and Bednarz (1990). t1: Graben formation and synchronous sheet-flow extrusion (vertically ruled) ponded at fault scarps; t2: Continued crustal extension resulting in open fractures and collapse of fracture walls; t3: Hydrothermal stage of fluid convection in the crust facilitated by enhanced permeability along extensional fractures; t4: Early stage of formation of graben centered pillow-lava volcano starting with megatubes extrusion; t5: Completion of pillow volcano. Scale of crustal extension (indicated by size of horizontal arrows) decreasing during the volcano-tectonic-hydrothermal cycle.

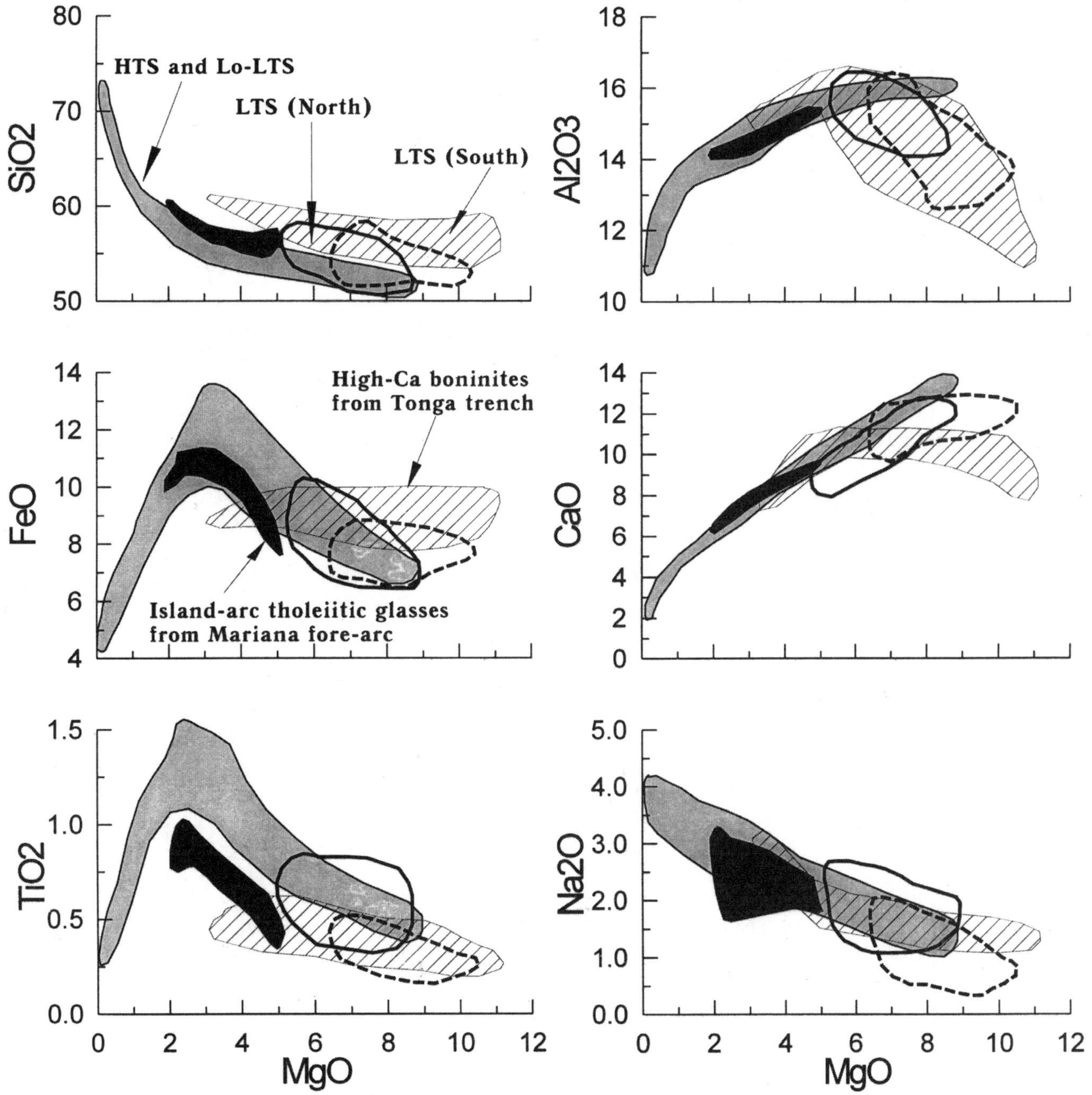

Fig. 1.5. Major element composition of the Troodos volcanic glasses (in wt.%). Data from Robinson et al. (1983); Rautenschlein et al. (1985); Flower and Levine (1987); Mehegan (1988); Thy and Xenophontos (1991); Sobolev et al. (1993a); Portnyagin et al. (1997) and our unpublished materials. Compositional fields of possible modern analogous of the Troodos glasses are shown. Data for Mariana fore-arc glasses are from Meijer (1980); Meijer and Reagan (1981); Tongan high-Ca boninites are by Danyushevsky et al. (1995) and references therein.

very poor in phenocrysts. The most common phenocrysts are plagioclase (An 85-30), clinopyroxene (En 29-46, Fs 10-41, Wo 21-48), orthopyroxene (En 52-69, Fs 24-41, Wo 1-5), and, in places, pigeonite (Bednarz and Schmincke, 1994; Robinson et al., 1983; Schmincke et al., 1983; Thy and Xenophontos, 1991). The phenocrysts of titanomagnetite and sparse ilmenite are characteristic for the dacites (Bednarz and Schmincke, 1994).

All these volcanics have many chemical features in common. Most of them show very little regional variations of major-element and trace-element characteristics. A typical feature is their relatively depletion in MgO. Its content in volcanic glasses is between 3.8 and 0.7 weight percent (**Fig. 1.5**; Mehegan and Robinson, 1991; Robinson et al., 1983). The iron and titanium in the glasses demonstrate a clear arc-tholeiite trend and an increase along the

Table 1.1. Representative compositions of the Troodos High-Ti Series and Lower Low-Ti Series Lavas

NN	1	2	3	4	6	5
Sample number	TRV-359	353	370	86-7	TRD-147	TRD-150
	GL	GL	GL	GL	GL	WR
SiO_2, wt.%	68.08	63.95	59.89	56.92	51.41	44.42
TiO_2	0.50	1.07	1.10	1.21	0.49	0.34
Al_2O_3	12.41	13.57	13.76	15.92	15.91	10.25
FeO*	5.47	8.77	10.79	10.40	6.81	8.16
MnO	0.13	0.17	0.18	0.17	0.14	0.16
MgO	0.49	1.95	3.30	4.45	8.40	22.23
CaO	3.40	5.78	7.73	7.49	13.90	7.96
Na_2O	3.50	3.82	2.79	2.78	1.55	0.51
K_2O	0.40	0.58	0.24	0.24	0.12	0.07
P_2O_5	0.08	0.12	0.03	0.10	0.05	0.05
Cr_2O_3					0.04	0.14
LOI	6.89**	5.82	3.89	5.08	1.53**	5.9
La, ppm	5.35	3.56	2.07	2.4	1.00	0.80
Ce	17.0	10.9	6.14	7.9	3.18	2.00
Nd	13.3	10.4	6.09	6.7	2.49	
Sm	4.53	3.85	2.37	2.52	0.96	0.71
Eu	1.31	1.35	0.89	0.96	0.36	0.25
Dy	8.75	6.98	4.34		1.9	
Er	5.76	4.68	2.92		1.25	
Yb	6.37	4.59	2.84	3.10	1.36	0.83
Lu		0.69	0.42	0.47		0.13
Ta		0.1	0.058	0.073		0.02
Nb					0.34	<0.7
Zr	2.2	85	48	63	20	14
Hf		2.79	1.65			0.38
Y	52.3	47	37	34	11.2	7.7
V		63		346		136
Rb		6.87	5.20	11		0.91
Sr	68.5	97.3	80.6	95	62	38
Ba	57.2	38.4	31.4	61	11.4	3.1
Th				0.37		0.147
U						0.036
Sc		21.6	31.5	28.9		26.7
Ni		8	8	36		605
Cr	3	3	1	10	389	1498
Co		14	30	28		72
$^{87}Sr/^{86}Sr$		0.70391	0.70389			
$^{143}Nd/^{144}Nd$		0.51304	0.51301			

*All iron as FeO; ** H_2O in glass; GL - volcanic glass, WR - whole rock.
1- dacite from Peristerona River canyon (our unpublished);
2- andesite-dacite from Akaki River canyon (Rautenschlein et al., 1985);
3- andesite from Akaki River canyon (Rautenschlein et al., 1985);
4- andesite-basalt from Pediaeous River canyon (Bednarz and Schmincke, 1994);
5-6 - basalt and picrite, respectivelly, from vicinity of Analiondas village (Portnyagin et al., 1997).

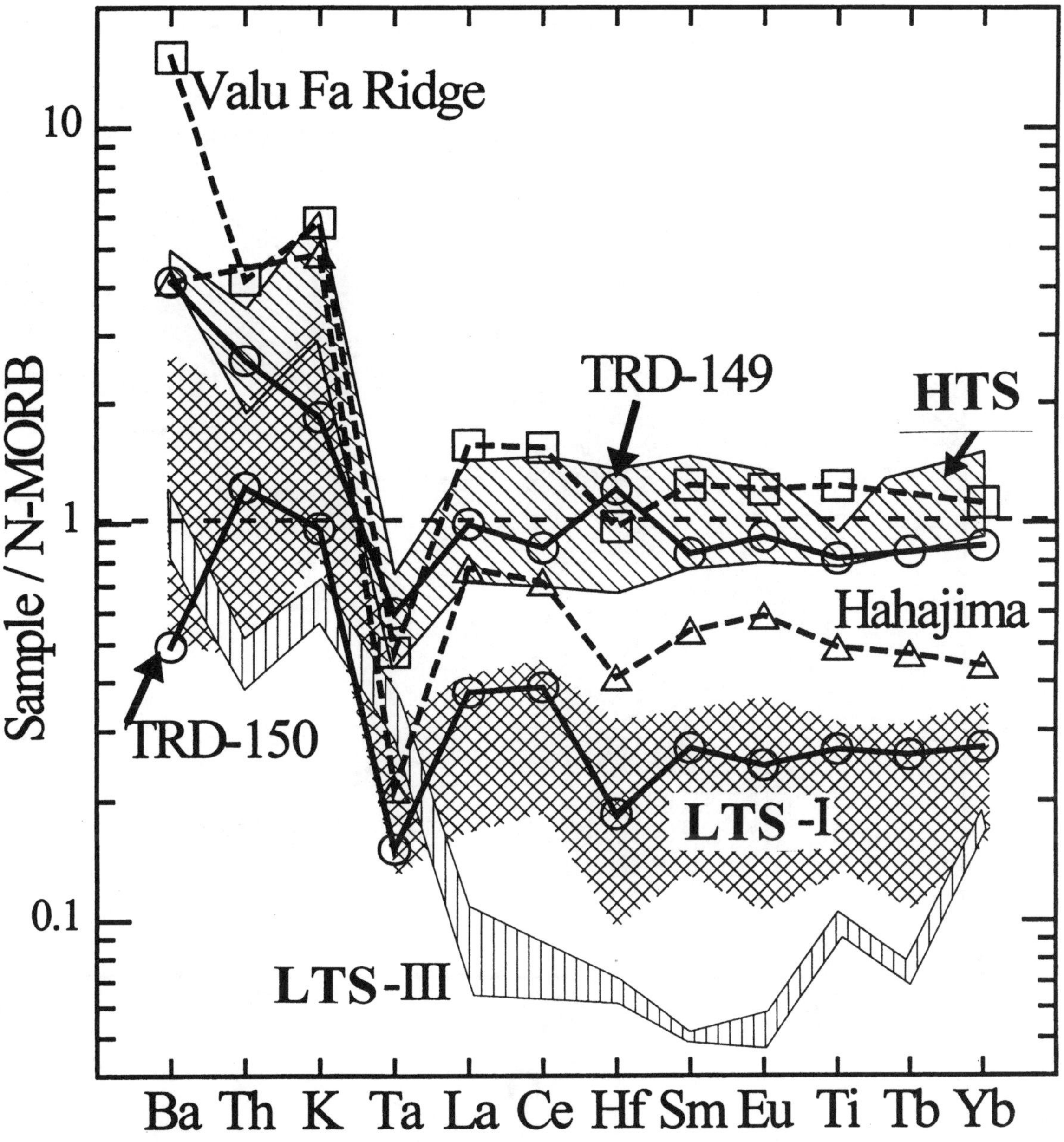

Fig. 1.6. Trace element compositions of the Troodos HTS and Lo-LTS after Portnyagin at al. (1997). Compositions of representative samples from Analiondas area (picrite TRD-150 and andesite TRD-149) are shown. Field of evolved HTS lavas after Rautenschlein et al. (1985). Fields of types I and III LTS picrites (MgO 23.2-35.4 wt.% and 31.9-32.8 wt.% respectively) after Sobolev et al. (1993a). Bonin forearc tholeiite with ~6.5 wt.% MgO (Hahajima Island) after Taylor and Nesbitt (1995). Valu Fa Ridge andesite after Vallier et al. (1991). Normalization to N-MORB (Sun and McDonough 1989).

liquid line of descent and then a sharp drop in most differentiated varieties at an MgO below 3.5 wt.%.

Despite small regional variability, HTS compositions essentially overlap as they demonstrate relatively high concentrations of incompatible elements (REE, Ti, Zr, Y), in contrast to overlaying LTS lavas. A characteristic feature of HTS, observed on mantle normalized plates, are the well pronounced minima in Nb and Ta concentrations, and the maxima in Large Ion Lithophile elements (K, Ba, Rb) relative to neighboring light rare-earth elements. In this regard, the HTS rocks resemble depleted island-arc tholeiites and differ from the N-MORBs (Normal Mid-Oceanic Ridge Basalts) as argued by Robinson et al. (1983), Rautenschlein et al. (1985), and Bednarz and Schmincke (1994).

The most primitive members of the older lava sequence, described in some areas in the NE Troodos, consist of andesite-basalts and basalts with a MgO content above 4 wt.% and belonging to Lo-LTS (Taylor, 1990; Portnyagin et al., 1997; Fig. 1.5 and **Fig. 1.6**, and Table 1.1). These rocks contain rare phenocrysts of olivine, clinopyroxene, and plagioclase (Bednarz and Schmincke, 1994). Very unusual high magnesian mafic to ultramafic lavas,

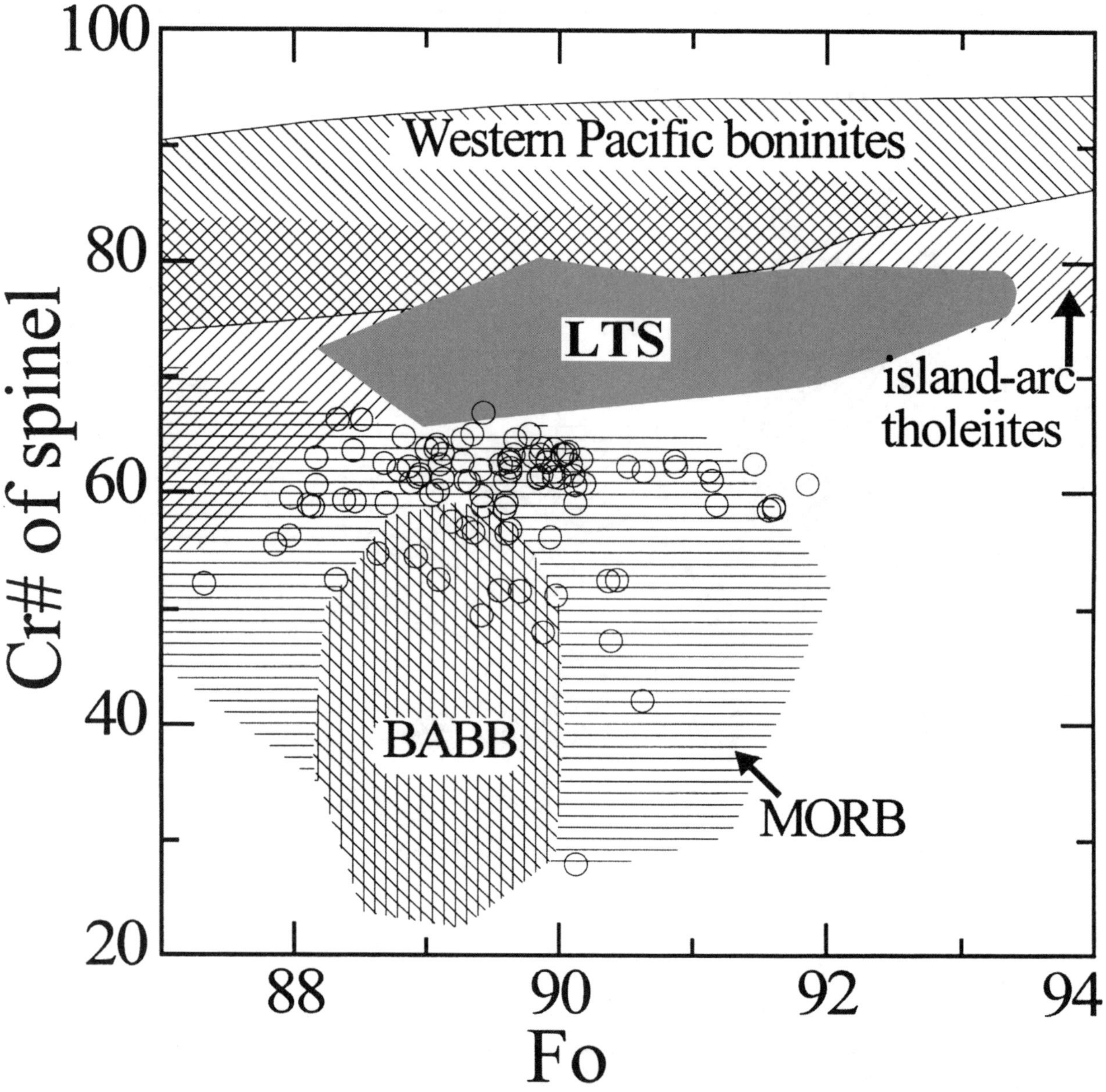

Fig. 1.7. Compositions of spinel inclusions in olivine from high-magnesian Lo-LTS. The compositional fields of spinel inclusions in olivine from Troodos LTS, western Pacific boninites, primitive island-arc tholeiite, primitive mid-oceanic ridge picrites and basalts (MORB), back-arc basin basalts (BABB) are shown. For data sources see Portnyagin et al. (1997) and references therein.

with well preserved quenched glasses, are only present among the early volcanites at the Analiondas section (Portnyagin et al., 1993, 1997). The phenocrysts in these rocks are composed of high-magnesian olivine (Fo 83-91.7), clinopyroxene, and chrome-spinel (Cr/(Cr+Al)=0.27-0.67) (**Fig. 1.7**).

The origin of HTS was a subject of long debate in the literature. Two alternative scenarios were put forward. Robinson et al. (1983) have proposed that the HTS represent an independent geochemical suite that is similar to the evolved arc-tholeiitic lavas. Lo-LTS lavas were suggested to represent the most primitive members of the suite (Bednarz and Schmincke, 1994). Alternatively, Cameron (1985) suggested that the volcanics assigned to the HTS subdivision are the result of deep fractionation of the least depleted varieties of the LTS.

Recent detailed studies of Lo-LTS lavas from the Analiondas area (Portnyagin et al., 1993, 1997; Sobolev et al., 1993a) shed further light on this question and provided the first quantitative estimates of parental magmas for the association of HTS and Lo-LTS. It has been shown that the Lo-LTS lavas belong to an independent geochemical rock type, similar to island arc low-K tholeiites, which cannot be related to typical LTS by late-stage fractionation at low pressures. Parental melts of Lo-LTS were

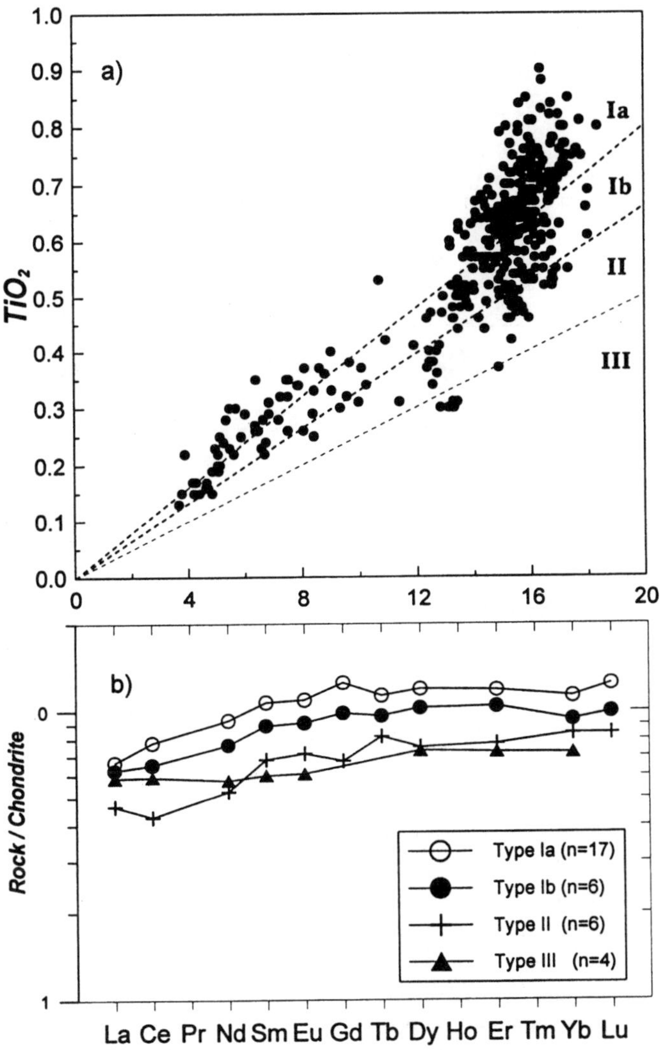

Fig. 1.8. Compositions of the primitive lavas from North Troodos (see Sobolev et al., 1993a for data sources).

a) Al_2O_3 vs. TiO_2 (in wt.%) plot showing major geochemical types of the North Troodos LTS. See Portnyagin et al. (1993); Sobolev et al. (1993a, b); Taylor and Nesbitt (1988) for details of the discrimination and sample choice.

b) Averaged compositions of the different geochemical types. REE contents are normalized to ordinary chondrite (after Anders and Grevesse, 1989) and to 15wt.% of Al_2O_3 in order to eliminate influence of crystal (OL ± CPX ± OPX) fractionation on the rocks composition.

shown to be basalts and picrite-basalts with MgO and water content of 10-15 wt.% and ~1.5 wt.% respectively. The origin of the melts is interpreted as a second stage melting of depleted mantle source residual after several percent of N-MORB extraction. Remelting of the refractory source at SSZ occurred at a temperature of ~1300°°C and a pressure of ~10 kbar and was promoted by the influx of hydrous fluid from altered subducted oceanic crust. Most of the lavas within the lower sequence appear be related to one another and are likely to represent evolved magmas, that have resulted from fractional crystallization of more primitive parents. Andesitic and dacitic lavas, which dominate within the sequence, may have originated from over 50% crystallization of olivine and clinopyroxene, following that of clinopyroxene, orthopyroxene and plagioclase, under a water pressure of about 0.5 kbar (Thy and Xenophontos, 1991; Portnyagin et al., 1997).

The composition and origin of the upper Low-Ti series volcanics

The most typical development of the LTS occurs on the northern flank of the Troodos Ophiolite Massif, where phyric to highly phyric basalts and andesite-basalts with phenocrysts of olivine and clinopyroxene are the typical and dominant rock types. The only characteristic of these rocks are metamorphics in zeolite facies and the rarity of dikes. The MgO content in the volcanic glasses of these rocks is between 8.5 and 3.9 wt.% and the TiO_2 concentration is less than 1 wt.% (Robinson et al., 1983).

Characteristic orthopyroxene-bearing lavas in boninite-type extrusives along the southern flank of the Troodos Massif petrographically closely resemble typical LTS rocks of the northern sections. However they contain phenocrysts of magnesian orthopyroxene (Cameron, 1985). The MgO content in the glasses of these rocks is up to 10.5 wt.% (Fig. 1.5; Flower and Levine, 1987).

The ultramafic lavas, long recognized by Gass (1958), constitute a specific group among the LTS. According to Sobolev et al. (1993), these contain 40-70% of large (up to 10 cm in diameter) phenocrysts of olivine (Fo 87-94) scattered in a vitreous to finely crystalline groundmass; minor phenocrysts of green clinopyroxene, chrome-spinel, and orthopyroxene are also present. Taylor (1990) explains the formation of these ultramafic lavas, which are closely associated with less porphyritic olivine-bearing basalts, in two ways. They may be formed by accumulation *in situ* of the coarse olivine crystals within the lower portions of large lava flows and pillows, or they were precipitated at the time that the magma moved to the sea floor.

A significant distinction between LTS and the most primitive members of Lo-LTS, related to HTS, cannot be made on the basis of the different degree of fractional crystallization, but only by comparison of their volcanic glass compositions and mineralogic features. The primitive glasses of the LTS show a lower content of CaO and Al_2O_3, being enriched with SiO_2 and FeO (Fig. 1.5 and Table 1.1; Portnyagin et al., 1997) at a given MgO content (~8 wt.%). Phenocrysts of high magnesian olivine (Fo up to 94) and chromium spinel (Cr/(Cr+Al) = 0.65-0.8) also characterize these rocks (Fig. 1.7; Sobolev et al., 1993a; Portnyagin et al., 1997). From the above evidence, the LTS of the north Troodos Massif are transitional between island-arc tholeiites and high-calcium boninites.

One of the interesting aspects of the Lava Complex in the Troodos Massif is a general upward increase in geochemical heterogeneity. Laterally, main HTS lithologies remain constant and share many common geochemical features. By contrast, the LTS show complex vertical and lateral variations in the geochemistry of its lavas (McCulloch and Cameron, 1983; Cameron, 1985). With respect to these changes a strict distinction among the different lava

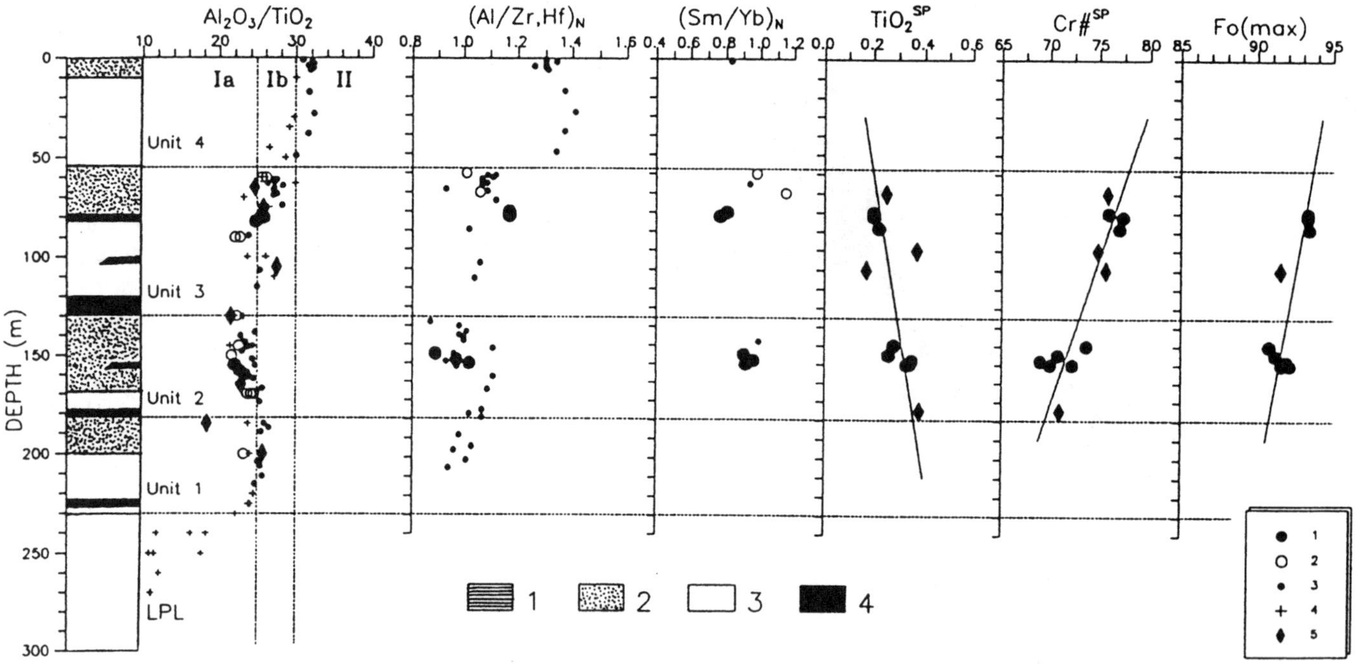

Fig. 1.9. The compositional stratigraphy of the LTS sequence from Margi area (North-Eastern Troodos). Stratigraphic column legend: 1- sediments, 2- olivine-phyric basalts, 3- aphyric and sparsely phyric basalts, 4- ultramafic lavas. Data legend: 1, 5- ultramafic lavas, 2-4 - basalts. 1 and 2 after Sobolev et al. (1993b), 3 after Taylor (1988), 4 and 5 -- after Bailey et al. (1991).

types cannot be explained as the result of fractional crystallization of a single magmatic melt, and this is taken as evidence that these variations must have been related to different compositions of parent magmas.

Recent investigations (see Sobolev et al., 1993a and references therein) not only support this key interpretation but also make it possible to outline the general trends in the development of the Troodos magmatism. One of the main conclusions from these works was the recognition of two regionally distinct trends in the geochemical evolution of the late-stage magmatism of Troodos (Cameron, 1985; Mehegan and Robinson, 1991; Sobolev et al., 1993a; Taylor and Nesbitt, 1988). One of the trends is represented in the LTS sequence of the Northern and Eastern Troodos. The other one is observed in LTS lavas along the southern boundary of the ophiolite massif.

Many details of the LTS rocks of the N and N-E Troodos were studied (Fig. 1.5 and **Fig. 1.8**). Cameron (1985) identified three geochemical lava-types, using Al_2O_3/TiO_2 ratios. Since this ratio is not influenced by crystallization of olivine and clinopyroxene in common liquidus phases of LTS, it demonstrates significant variations in the composition of parental magmas, as it varies in the primitive lavas from 20 to 45. The concentrations of the moderately incompatible elements such as titanium, sodium, zirconium and yttrium and of the heavy and middle REE decrease with increasing Al_2O_3/TiO_2 ratio (Fig. 1.8 and **Tables 1.2** and **1.3**). Comparing degrees of depletion, the resulting proportions of these types is about 80:15:5, indicating that the less depleted lavas are most abundant in the areas studied.

The stratigraphic distribution of the different lava types is not random. Detailed reconstructions of the LTS sections of Northern Troodos reveal that the temporal distribution of different lava types is generally systematic as shown by the evident progressive upward depletion of the incompatible elements (Taylor, 1990; Bednarz and Schmincke, 1994). There is also a clear change of the liquidus mineral compositions, *e.g.* the upward increase of both the maximum olivine Fo content and the Cr/(Al+Cr) ratio in spinel (**Fig. 1.9**; Portnyagin and Sobolev, 1993; Sobolev et al., 1993b).

The geochemical evolution of the North Troodos LTS is interpreted as the result of progressive melting of a single mantle source that led to the general upward depletion of the lavas. The proposed model is explained more fully by Wood (1979), Cameron (1985), Duncan and Green (1987), and Sobolev et al. (1993a). The combination of theoretical modeling data and melt-inclusion studies (Sobolev et al., 1986; 1993a) indicates that the primary melts of LTS were rich in MgO (15-22 wt.%) and in H_2O (up to 2 wt.%). Their origin is suggested to be the result of mantle melting initiated above the suprasubduction zone and proceeding at a pressure of 15-20 kbar and at temperatures above 1400°C. It is important to note that the most plausible source for the older Lo-LTS and HTS was restricted to shallower levels in the mantle and lower temperatures (Portnyagin et al., 1997; Sobolev et al., 1993a). According to modern thermal models for supra-subduction zones (*e.g.*, Davies and Bickle, 1991), the sources of HTS and LTS could not be related to one another. Thus, the results imply the involvement of two independent mantle sources as

Table 1.2. Representative compositions of the North Troodos Low-Ti Series Lavas

NN	1	2	3	4	5	6
Sample number	316	315	345	M.12	CYP-2	CYP-1
Geoch. type	Ia	Ib	II	III	Ia	Ib
	GL	GL	GL	WR	WR	WR
SiO_2, wt.%	53.36	55.84	56.01	55.44	45.79	43.87
TiO_2	0.71	0.57	0.45	0.31	0.35	0.17
Al_2O_3	15.81	15.10	14.20	13.43	7.54	4.65
FeO*	7.49	8.04	7.8	8.57	8.13	8.09
MnO	0.14	0.16	0.16	0.16	0.16	0.13
MgO	7.84	6.69	7.98	9.83	24.03	35.38
CaO	12.42	11.24	11.24	10.83	5.99	3.65
Na_2O	2.05	2.01	1.86	1.15	0.58	0.39
K_2O	0.11	0.23	0.20	0.20	0.08	0.15
P_2O_5		0.03		0.02		
LOI	1.97	2.98	3.15	4.19	5.57	2.44
Al_2O_3/TiO_2	22.3	26.5	31.5	43.3	21.5	27.4
La, ppm	1.41	1.12	0.88	1.74	1.04	0.61
Ce	4.22	3.01	2.39	4.12	3.45	1.43
Nd	4.03	2.86	2.43	2.98		
Sm	1.52	1.15	1.03	1.04	0.84	0.35
Eu	0.62	0.46	0.42	0.43	0.30	0.11
Dy	2.76	2.33	2.16	2.15		
Er	1.85	1.56	1.44	1.42		
Yb	1.79	1.57	1.44	1.41	1.03	0.48
Lu	0.27	0.23	0.22	0.24		
Ta	0.039	0.098	0.071		0.034	0.02
Zr	29	27	22	22		
Hf	0.95	0.88	0.77	0.65	0.597	0.193
Y	22	19	14	11		
V	215		237	237		
Rb	2.17	5.64	5.36	4.9	1.8	2.2
Sr	97.5	87.1	68.4	90		
Ba	16.9	25.5	22.1		9.7	17.4
Th					0.17	0.11
U					0.075	0.055
Sc	34.9	34.0	38.3	42	24.4	10.5
Ni	56	12	62	244	803	1655
Cr	215	67	264	737	1815	3362
Co	35	32	34	33	79	114
$^{87}Sr/^{86}Sr$	0.70333	0.70432	0.70426			
$^{143}Nd/^{144}Nd$	0.51303	0.51301	0.51300			

*All iron as FeO; GL - volcanic glass, WR - whole rock.
1-3- basaltic andesites from Akaki River canyon (Rautenschlein et al., 1985);
4- basaltic andesite from Koronia Hill (Taylor, 1990);
5-6- picrites from vicinity of Margi village (Sobolev et al., 1993b).

Table 1.3. Representative compositions of the South Troodos Low-Ti Series Lavas

NN	1	2	3	4	5	6	7
Sample number	TRV-135	TRDS-27	KA	92	TRV-155	LIM	7PK-30
Geoch. type	I	II	II	III	III	III	III
	GL	GL	GL	WR	GL	GL	WR
SiO_2, wt.%	51.46	53.85	55.01	51.74	54.05	58.35	43.07
TiO_2	0.70	0.41	0.42	0.19	0.18	0.29	0.08
Al_2O_3	15.84	15.52	15.15	12.47	12.93	14.37	5.50
FeO*	7.94	7.3	7.57	7.18	7.91	7.15	8.40
MnO	0.06	0.12	0.15	0.14	0.17	0.11	0.13
MgO	7.17	7.28	8.43	12.74	9.37	7.46	29.87
CaO	12.12	11.56	11.38	9.88	12.19	10.71	4.00
Na_2O	1.97	1.46	1.61	1.10	0.40	1.38	0.11
K_2O	0.09	0.14	0.13	0.39	0.06	0.22	0.04
P_2O_5	0.05	0.03	0.12	0.01	0.03		
LOI	2.5**	2.1**	2.77	4.16	2.1**	3.18	7.08
Al_2O_3/TiO_2	22.6	37.9	36.1	65.6	71.8	49.6	68.8
La, ppm	1.33	1.40	1.15	1.09	0.31	0.69	0.16
Ce	4.32	3.20	2.79	2.19	0.55	1.75	
Nd	3.65	2.14	2.51	1.02	0.28	1.49	
Sm	1.40	0.69	0.84	0.32	0.18	0.63	0.13
Eu	0.54	0.29	0.35	0.11	0.10	0.26	0.05
Dy	2.72	1.15	1.78	0.93	0.78	1.56	
Er	1.72	0.83	1.24	0.75	0.69	1.12	
Yb	1.55	0.82	1.28	0.84	0.83	1.19	0.55
Lu			0.20			0.19	
Ta							0.047
Nb	0.6	1.1	1.2	1	0.6	0.6	
Zr	33.8	21.1	17	16	3.5	33	
Hf				0.3			0.13
Y	16.4	9.3	8	7	5.8	10	
V		212		203			
Rb			2.9			6.16	2.2
Sr	77.9	77.3	83	62	8.9	72.2	
Ba	14.9		26.1	45	7.3	19.5	7.7
Th							0.047
U							0.026
Sc				40			18.7
Ni				339			1334
Cr	254	336		940	515		2948
Co							89.2
$^{87}Sr/^{86}Sr$			0.70355	0.70638		0.70464	
$^{143}Nd/^{144}Nd$			0.51304	0.51193		0.51295	

*All iron as FeO; GL - volcanic glass, WR - whole rock.
1 - basaltic glass from Ayios Mamas area (our unpublished);
2 - basaltic andesite from Asgata area (Sobolev et al., 1993a);
3 - basaltic andesite from Kalavasos area (Rauteschlein, unpubl. Ph. D. thesis);
4 - basalt from Trimiklini area (Cameron, 1985);
5 - andesitic glass from vicinity of Mottikoloni village (our unpublished);
6 - andesitic glass from Limni mine (Rauteschlein, unpubl. Ph. D. thesis);
7 - picrite from Kalavasos area (Sobolev et al., 1993b).

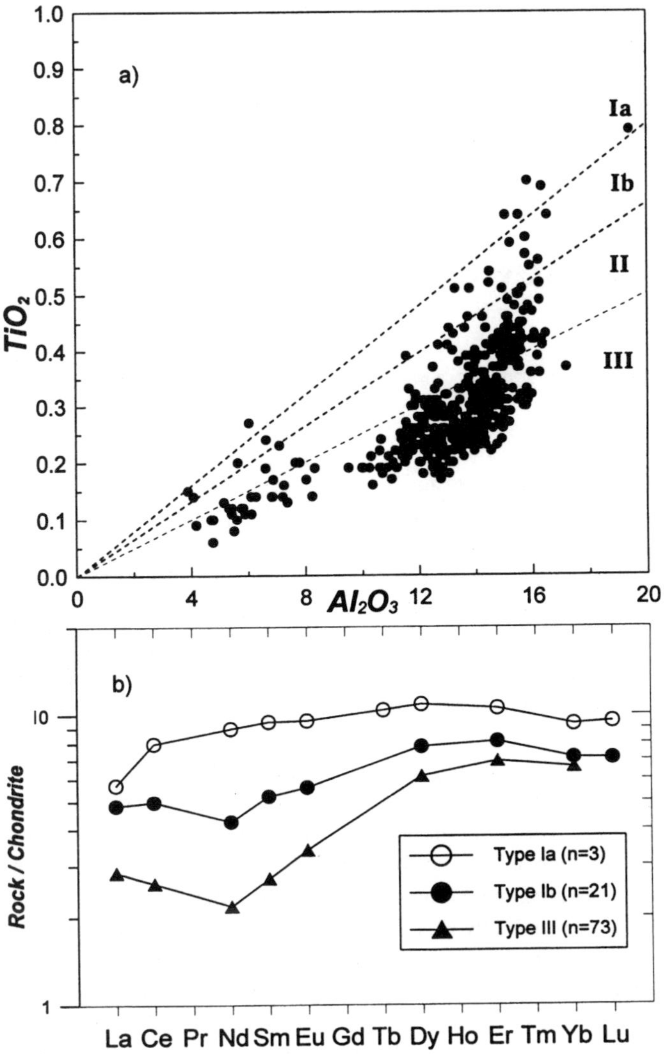

Fig. 1.10. Compositions of the primitive lavas from South Troodos (see Sobolev et al., 1993a for data sources). a) Al_2O_3-TiO_2 (in wt.%) plot showing major geochemical types of the South Troodos LTS. b) Averaged compositions of the different geochemical types of the South Troodos LTS. REE contents are normalized to ordinary chondrite (after Anders and Grevesse, 1989) and to 15wt.% of Al_2O_3 in order to eliminate the influence of crystal fractionation on the rock compositions.

the origin of the whole lava sequence of the north Troodos. Despite fundamental differences, an obvious similarity of the LTS and Lo-LTS mantle sources exists. This is expressed in the depletion by previous melt extraction and remelting above the subduction zone (Sobolev et al., 1993a).

The lava assemblage associated with the southern regions of the Troodos Ophiolite Massif extends to the Akamas Peninsula Ophiolite and to the South Troodos Transform Zone that includes the narrow Arakapas Fault Belt and much of the Limassol Forest Ophiolite. It is also present as blocks of 'Troodos lithologies' occurring in the Mamonia Complex terrain (see Chapter 2 in this book).

As in northern branch volcanites, the southern branch lavas represent three distinct geochemical groups, characterized by varying depletion of incompatible elements. (Fig. 1.5 and **Fig. 1.10**). The most widespread of these lavas represent the third group, which is extremely depleted in incompatible elements, particularly titanium (Al_2O_3/TiO_2 >40), sodium, strontium, zirconium, as well as the middle REE (Fig. 1.3) with resulting U-shaped patterns (Fig. 1.10). The analytical data show that some of the rocks have extremely and abnormally low concentrations of the incompatible elements, which have not been reported elsewhere. The presence of magnesian orthopyroxene phenocrysts is a characteristic feature of these rocks.

These rocks have long been a popular focus for geochemical study. Crawford et al. (1989) used these rocks for the establishment of a specific group of the very primitive high-CaO boninites, which may have formed in a supra-subduction zone in a primitive intra-oceanic arc or incipient backarc basin. On the other hand, many Troodos students agree that the boninitic lavas are closely associated with the development of the Arakapas extensional transform (*e.g.*, Rogers et al., 1989 and Murton, 1990). In addition, the Tonga Trench transform-generated boninites (Danyushevsky et al., 1995) may serve as a plausible model for the formation of the Troodos boninitic lavas.

The mildly depleted lavas of the first and second types are geochemically similar to the LTS lavas of the northern branch (Mehegan, 1988; Taylor and Nesbitt, 1988) but show linear to slightly U-shaped REE patterns (Fig. 1.10).

Because of the structural complexity in Southern Cyprus, the relationships between the different lava types are poorly understood. In some localities, lavas of quite distinct geochemical characteristics are interlayering (Mehegan, 1988; Robinson and Malpas, 1990; Rogers et al., 1989). In other places, the lava sequence resembles that of Northern Troodos (Flower and Levine, 1987).

The origin and development of the southern branch lavas is still debated. A widely held view is that the parent magmas are typically the final result of the melting of a highly depleted shallower mantle source, with pressures less than 10 kbar (Cameron, 1985; Duncan and Green, 1987; Kostopoulos and Murton, 1992). There is evidence to suggest that at least two enriched components were involved in the magma-forming process in the South Troodos regions (Kostopoulos and Murton, 1992; Portnyagin et al., 1996; Rogers et al., 1989; Sobolev et al., 1993a; Taylor and Nebitt, 1988). One of them was a water-rich fluid containing highly variable amounts of LILE and possibly LREEs (Portnyagin et al., 1996; Sobolev and Chaussidon, 1996). Some other, more 'exotic' components were responsible for the elevated contents LREEs, low эNd and high эSr in the lavas (Cameron, 1985). A possible role of interaction of these enriched components with mantle material is still poorly understood (Rogers et al., 1989).

Conclusions

It has been recognised for a long time that the

understanding of the origin of the Troodos lava series is critical for the establishment of a model of the formation of the Troodos ophiolite as a whole. Despite the progress that has been made over the recent decades, many aspects concerning the formation of the Troodos lavas are still debated. Additional investigations, including detailed isotope and trace element studies, together with careful volcanologic and structural examination of the lava sequences, should help to establish a comprehensive model for magma-forming processes, taking into account both the physico-chemical and tectonic controls of the origin of the diverse products of the Troodos ophiolitic volcanism.

C. The Dyke Complex

Maxim V. Portnyagin
Vernadsky Institute of Geochemistry and Analytical Chemistry of the Russian Academy of Sciences, Kosygin Str. 19, 117975 Moscow, Russia

Geologic Characteristics

The dike swarm assemblage of the Troodos Massif is one of the most developed and best-documented of all the eastern Mediterranean ophiolites. It offers unique evidence for extensional stress and fracturing on a variety of scales in a sea-floor spreading environment. The dike sheets cover a wide area on the flanks of the Troodos Range around its plutonic core, and occur over a 1-1.5 km depth interval (Baragar et al., 1989, 1990).

There is a transitional zone between the base of the Lava Complex and a zone of almost 100 percent dikes, which is traditionally referred to as the Basal Group (BG). This mixed zone of dikes, pillowed lava screens, and isolated pillows is, in places, only a few tens of meters thick, and characterized by the continuous decrease downwards of the pillow lava content. In many cases, the dike density increases rapidly from 20 percent at the top of the BG to 70 percent near its base over a stratigraphic interval some 25-30 m thick (Gass, 1980).

The intrusive nature of these dikes, previously designated the Sheeted Diabase Group by Bishopp (1952), was recognized by Wilson (1959) and was interpreted by them in terms of a hypabyssal rock complex. Wilson (1959) pointed out that sheeted dikes were greatly influenced by an extensional regime during their formation and constitute a long-lived system of feeders to the overlying pillow lavas.

Gass (1968) postulated that the origin and development of the Sheeted Intrusive Complex must have taken place in a mid-oceanic ridge environment. Moores and Vine (1971), who identified the Troodos Massif as an ophiolite, argued that the dike complex is structurally similar to the dike swarm assemblages of modern mid-ocean ridges, and assumed that the Troodos dikes and the overlying lavas were derived from a number of on-axis cells operating independently, but tapped to provide volcanic material to the surface. Based on statistical analysis of the chilled directions in the dikes, Kidd (1977) and Kidd and Cann (1974) concluded that the oceanic crust was generated at a narrow accretion zone, located to the west of the island of Cyprus.

Subsequent detailed mapping and various petrologic and structural studies demonstrated the compositional heterogeneity of the sheeted-dikes and revealed structural complexity within the complex not previously recognized. Several episodes of dike injection have been identified in a number of localites on the basis of petrographic features and age relationships of the various dike types (Desmet et al., 1978). Dyke orientation over a wide area shows considerable regional variation in the original strike and dip of the dike sets, representing different generations of dike swarms with distinctive cross-cutting relations (Searle and Panayiotou, 1980).

Detailed mapping and paleomagnetic investigations led to the recognition of the remains of a series of ancient domains and discrete asymmetric graben-type paleostructures of characteristic dike attitude in the northern portion of the Troodos Ophiolite Massif (Varga and Moores, 1985, 1990). After their intrusion, the dikes were probably slightly rotated along listric normal faults that bounded each structure and whose trend is generally parallel to that of the dike swarms **(Fig. 1.11**). These faults probably flatten with depth, forming a wide detachment fault zone near the sole of each graben (Agar and Klitgord, 1995). In some localites, these faults are intersected by undeformed dikes that indicate simultaneous faulting and magmatism. Generally speaking, several spreading-related structures in the Troodos dike complex were interpreted to reflect a spreading axis shifting eastwards due to progressive ridge jumps (Varga and Moores, 1985; Varga et al., 1987). A plausible general model for formation of a dike assemblage structure, after the example of the Larnaca Graben, was introduced by Varga and Moores (1990**) (Fig. 1.12).**

Geochemistry and Mineralogy of the dikes

There is a close similarity in the compositional range between the lavas and the sheeted dikes, and most, if not all, compositional variations of the extrusives are also represented in the dike assemblage. As shown in **Fig. 1.13** and **Table 1.4**, the dike rocks grade from quartz diorites and diorites to dolerites and locally, to picrites forming a complete gradation through the entire spectrum of compositions in the dike complex.

The main rock types are altered dolerites and diabases with sparsely phyric to phyric texture, containing phenocrysts of

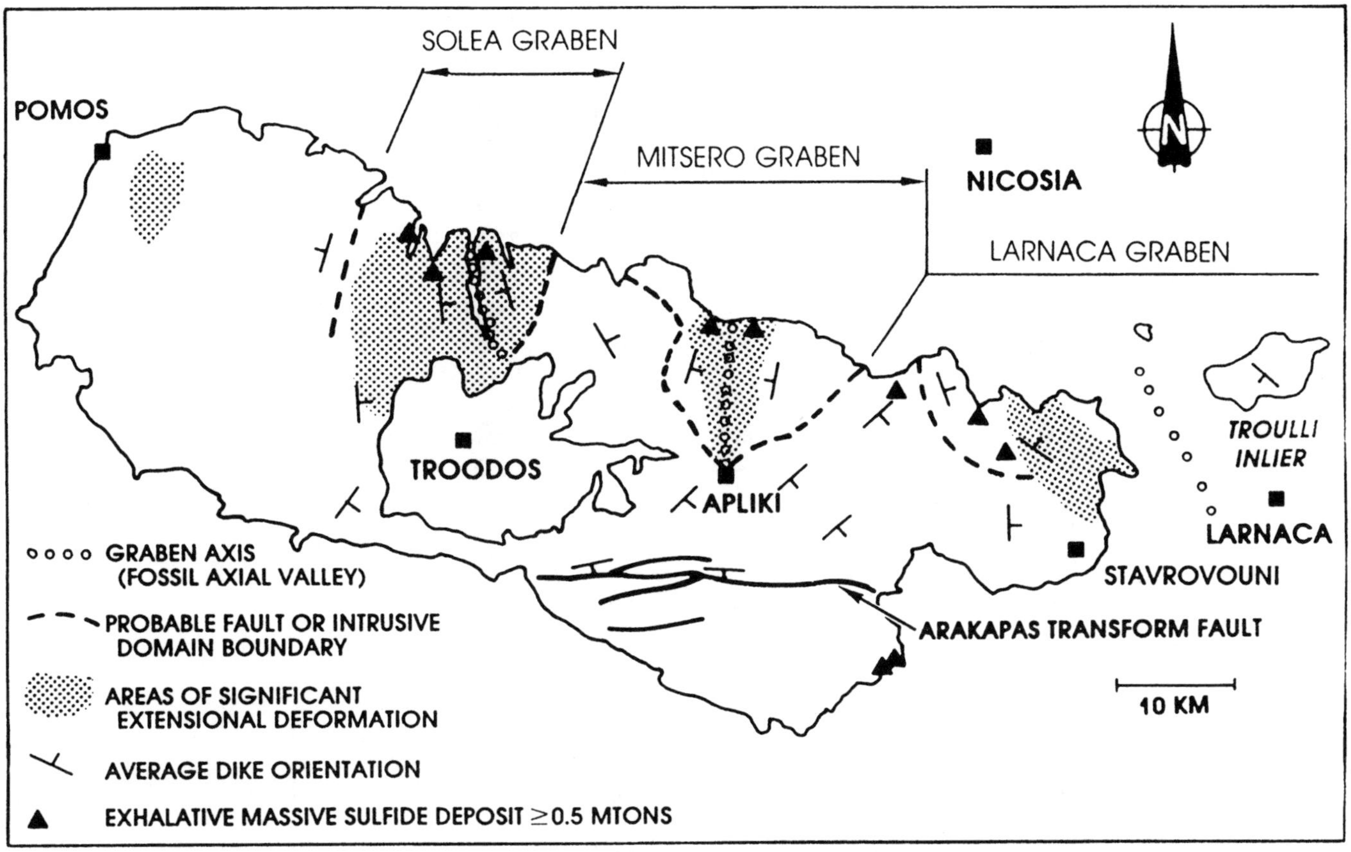

Fig. 1.11. Location of major grabens on the northern flank of the Troodos ophiolite (after Varga and Moores, 1985, 1990).

plagioclase and clinopyroxene, and less commonly olivine. The olivine-phyric dikes are rare, and belong to the latest dike generation. If present, abundant phenocrysts of highly magnesian olivine (up to Fo 94) are concentrated largely in the central part of these dike, whereas their margins are composed of aphyric varieties. In the Limassol Forest area, there are dikes containing phenocrysts of olivine, clinopyroxene, and orthopyroxene in varying proportions, which cross-cut the mantle peridotites (Murton, 1990).

Fig. 1.3 and Table 1.4 demonstrate a continuous compositional spectrum in the dike rocks grading from dacite to picrite. The low magnesian dike rocks are similar in composition to the typical differentiated extrusives of the HTS. The MgO-enriched dike rocks correlate to both the LTS and the primitive varieties of the Lo-LTS. Based on their incompatible-element contents, not only do the dike rocks correlate to the geochemical types of lavas of Cameron (1985), but also contain more depleted varieties unknown among the north Troodos lavas (Baragar et al., 1989, 1990; Tarney and Marsh, 1991).

Relationships between composition of dikes and timing of intrusion

Because the dikes represent feeders along which the different magmas approached the sea floor and extruded onto its surface as lava flows, they must have recorded the same compositional range as the simultaneous lavas. This assumption was challenged by Baragar et al. (1989, 1990) on the basis of their study of the dike 'chilled' margins. They reported that there are no persistent relationships between age and composition of the titanium-rich and titanium-poor dikes, which were suggested to be the geochemical equivalents to the HTS and LTS. The conclusion was reached that the intrusion of dikes of different geochemical types were probably simultaneous, although titanium-enriched rocks are statistically prevalent in the dike complex. As these results were in conflict with the observed stratigraphic relationships of HTS and LTS in the Troodos lava complex, Baragar et al. (1990) suggested that different types of magmas coexisted simultaneously in the separate magma chambers during volcanic activity.

More recently these conclusions were revised. Using discriminant analysis for the geochemical classification of lavas and dikes Thy and Esbensen (1993) have shown that the compositions of about 98% of all of the dikes distributed within the multi-component space closely correspond to those of the lower lavas. This is in keeping with the earlier suggestion by Bear (1960) that the sheeted dikes and the Lower Pillow Lavas are probably cogenetic. The general mineralogical similarity of the dike rocks and the HTS andesitic extrusives also supports this view (Thy, 1990). Another informative feature is the presence of

Table 1.4. Representative rocks compositions of the Sheeted Dyke Complex of the Troodos ophiolite

NN	1	2	3	4	5
Sample number	B94	B178	LQ48	TRDS-50	TRDS-51
$SiO_{2,\ wt.\%}$	50.5	50.59	56.79	46.14	49.04
TiO_2	0.33	0.69	1.13	0.22	0.19
Al_2O_3	12.8	15.69	14.69	7.51	9.99
FeO*	9.88	7.7	10.81	8.00	7.97
MgO	9.35	7.78	3.35	24.18	19.19
CaO	8.00	9.48	5.24	7.1	10.35
Na_2O	2.29	3.09	4.80	0.27	0.13
K_2O	0.50	0.20	0.08	0.04	0.08
P_2O_5	0.03	0.06	0.08	0.04	0.01
LOI	4.6	3.42	2.51	5.62	2.85
La, ppm				0.25	0.29
Ce	2.03	3.94	5.53	0.71	0.84
Nd	1.90	4.07	5.62	0.84	0.6
Sm				0.35	0.18
Eu	0.33	0.63	0.85	0.14	0.09
Yb	1.33	1.98	2.65	0.65	0.83
Lu			0.42	0.10	0.13
Zr				4.5	2.3
Hf	0.42		1.49		
Y	21	10	29	4.6	5.3
V	220	210	30		
Rb				0.9	2.2
Sr	47	140	83	1.7	12
Ba	50	10	11	<10	<10
Th		0.19	0.26		
Sc				28	35
Ni	65	70			
Cr	180	230	10	2400	1900
Co	41	28	22	74	67

*All iron as FeO.
1- 4 - dykes from the northern part of the massif;
5- dyke from the Limassol Forest Complex.
1-3 data from Baragar et al., 1989; 5- our unpublished data.

abundant primitive, titanium-rich and titanium-poor lavas often alternating with the typical evolved lavas within the Lo-LTS and HTS sequence of the eastern part of the Troodos Massif (Taylor, 1990; Bednarz and Schmincke, 1994; Portnyagin et al., 1997). Keeping in mind this important observation, the data of Baragar et al. (1989) can be interpreted in the following way. Feeders for the LTS are probably only those latest magnesian dikes with two chilled margins, such as the olivine-phyric dikes described by Desmet et al. (1988), Portnyagin et al. (1994), and Sobolev et al. (1993a). Whereas the older magnesian low-titanium dikes, with one or no chilled margins, are probably feeders for the Lo-LTS lavas found at the base of lava sequence. Direct evidence for this suggestion comes from the comparison of Lo-LTS thicknesses and the average composition of the dikes in the different Troodos graben structures (**Fig. 1.14**), which demonstrates more magnesian dike composition for the Mitsero graben where Lo-LTS unit is well developed.

Thus, current petrologic thinking is that the main portions of the sheeted dike assemblage and their associated HTS and Lo-LTS volcanics may have formed by repeated injections of magmas of variable composition generated from multiple magma chambers developed near active spreading axes or overlapping/shifting centers above a subduction zone in a marginal basin. By contrast, the formation of the feeders for the LTS type took place either outside the main volcanic zone or away from the on-axis sites or during the waning episodes of volcanic activity on the Troodos.

D. The Plutonic Complex

Sergey A. Silantyev
Vernadsky Institute of Geochemistry and Analytical Chemistry of the Russian Academy of Sciences, Kosygin Str. 19, 117975 Moscow, Russia

The well-studied Troodos mafic-ultramafic plutonic assemblages have been used as a geochemical and petrological type locality for characterization of ophiolite suites (Moores and Wine, 1971; Coleman, 1977; Gass, 1980).

Diverse plutonic rocks are mainly exposed in the central part of the Troodos Ophiolite and in the Limassol Forest Ophiolite Massifs, which are separated by the Arakapas Fault Zone. According to their field relationships, petrologic characteristics, and deformation style, these igneous rocks were divided into two main assemblages:

1) A plutonic core complex, composed entirely of mantle peridotites;

2) Younger ultramafic to mafic cumulates of the layered series with minor granitoid rocks at the top, which are stratigraphically overlying the ophiolite core complex.

The first assemblage of residual peridotites is exposed in two areas of the main ophiolite belt and consists of the following succession (from bottom to top):

a) The ultramafic tectonites are exposed around Mt. Olympus and in the Limassol Forest. These almost certainly represent a refractory residue after fractional melting of the lithospheric mantle peridotite (Allen, 1975; Panayiotou, 1987). The mantle derived spinel-bearing lherzolites were interpreted as a separate block about 3 km long, southeast of the Troodos peridotitic core (Batanova et al., 1994).

b) The harzburgite-dunite transition zone, where mantle peridotite passes upwards into dunite metacumulate that is overlain by wehrlites, lherzolites, websterites, and clinopyroxenites (Panayiotou, 1987).

c) Melanocratic olivine gabbro demarcates(?) the lower boundary of the next gabbroic cumulate complex. It grades up into olivine-free pyroxene gabbro, passing upward into magnetite-hornblende gabbro and diorite. In many cases, the lower boundary of this complex is a tectonic contact with the underlying peridotites formed due to the protrusion of serpentinized harzburgites.

d) The plagiogranites typically occur as isolated igneous bodies at the boundary between the gabbroid complex and the sheeted dike assemblage. Small plagiogranite bodies are also found in the upper part of the gabbroid sequence and within the lower dike complex.

High-angle faults characterize the plutonic part of the ophiolite sequence. The contact between the upper gabbros and the dike complexes has a tectonic character which corresponds to a decoupling zone (Agar and Klitgord, 1995) that formed due to many planar detachment surfaces.

Petrography and Mineralogy

Two main plutonic rock groups with different petrologic characteristics and deformation style were recognized in the

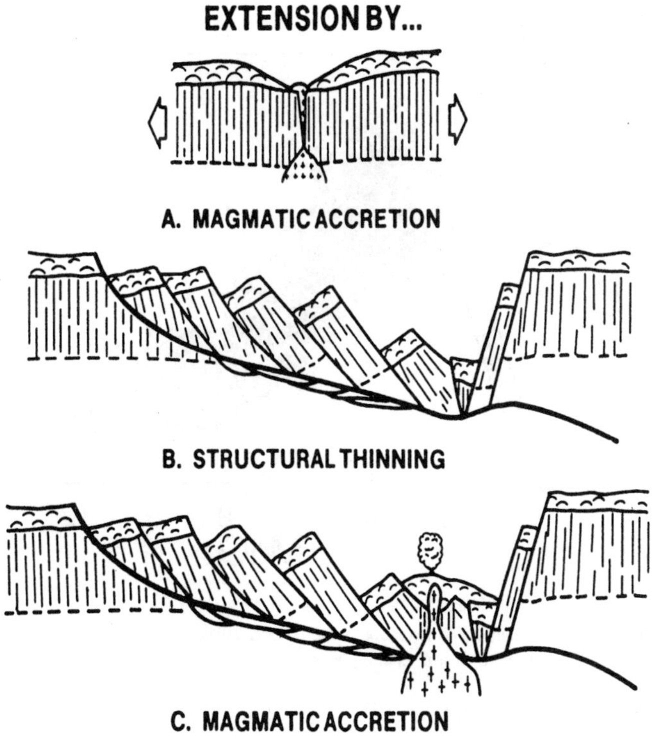

Fig. 1.12. Non-steady state model for magmatism and tectonic extension (after Varga and Moores, 1990). (A) Spreading accommodated by magmatic injection when crustal accretion keeps pace with plate separation. (B) Spreading accommodated by crustal attenuation as magmatism does not keep pace with plate separation. (C) Return to spreading by magmatic accretion. See also discussion in Varga and Moores (1990).

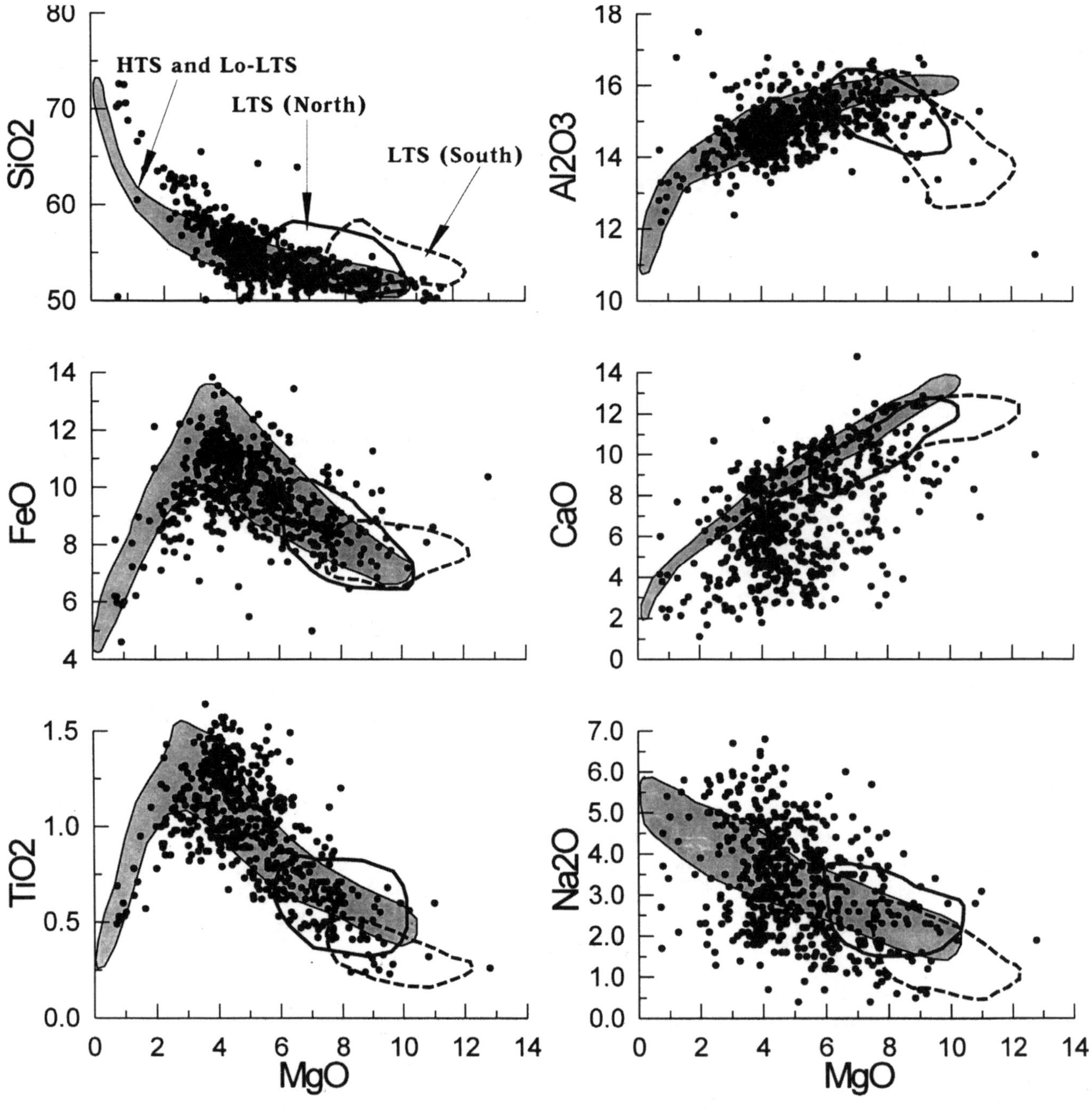

Fig. 1.13. Major element composition of the Troodos sheeted dikes (wt.%). The compositions of rocks from northern flank of the ophiolite are shown. Data from Baragar et al. (1989). Fields are the same as in Figure 1.5.

Troodos Massif (Malpas et al., 1987). They are as follows:

1) Rocks characterized by tectonic deformation under conditions of high temperature. This group includes harzburgite tectonites, dunites with chromite bodies, banded olivine pyroxenite, plagioclase peridotites, pyroxenites, two-pyroxene gabbros, plagiogranites, and dikes comprised of high-magnesium microgabbro. The high-angle contacts between the major lithologies are sharply transected by a gently-dipping (and contradictory?) banding of tectonic origin. All these rocks are intruded by the following second magmatic assemblage.

2) Diverse ultramafic rocks consisting mainly of wehrlites. The gabbroid rocks are mainly found at marginal parts of the ultramafic plutons, where they occur as vertical dike swarms.

More fractionated rocks, such as non-layered gabbros, diorites, and plagioigranites, become dominant in the higher levels of the plutonic complex, where the early and late magmatic groups are structurally indistinguishable (Malpas et al., 1987). In the vicinity of the Caledonian Waterfall, Benn and Laurent (1987) divided the

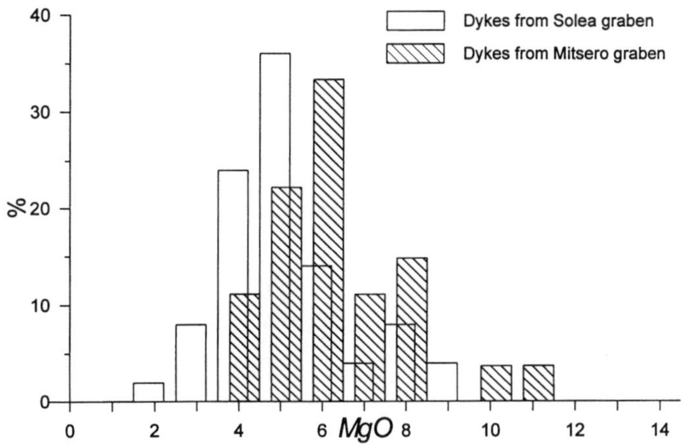

Fig. 1.14. Variations of MgO content in dikes from different Troodos grabens. Note the dikes from the Mitsero graben where magnesian Lo-LTS were found are statistically more magnesian compared to dikes from the Solea graben. Data on dikes composition from Baragar et al. (1989) were used.

plutonic rocks into two distinct magmatic series: a lower (early) concordant body of banded gabbro and pyroxenites, intruded by a later poikilitic wehrlites, and gabbro-norites. The boundary between the two series is marked by a structural discontinuity.

The mantle peridotite around the Troodos summit and in the Limassol Forest is represented by two main groups: harzburgites with various degrees of melting, and spinel lherzolites. Depleted harzburgites and minor spinel-bearing lherzolites represent these rocks. A noticeable feature of these rocks is well developed banding caused by the orientation of the olivine crystals and banded segregations of large orthopyroxene crystals.

Extensive deformation and subsequent alteration has destroyed original important diagnostic fabrics, thus the metamorphic textures and structures are dominant in these rocks. A noticeable fabric is a near-vertical and generally parallel fine-scale banding of segregated orthopyroxene crystals. The main rock type is medium to coarse- grained porphyroclastic harzburgites which contain less than 5% primary clinopyroxene (En 45-51.3, Wo 44.4-46.5, Fs 4.3-8.5) (Hebert and Laurent, 1990). The Troodos harzburgites contain relics of olivine (Fo 89.1-90.4), orthopyroxene (En 88-89.8) and spinel (Cr/(Cr+Al) = 0.41-0.61) (Allen, 1975; Hebert and Laurent, 1990), indicating a very high degree of melting in the mantle substratum. Only pyroxene bearing harzburgites from the Troodos are close to the most depleted residual harzburgites of the mid-oceanic ridges (**Fig. 1.15**). Additional evidence supporting this interpretation is the incompatible elements of the clinopyroxene from the Troodos residual peridotites that show a similar pattern of heavy REE content in comparison with abyssal peridotites (Sobolev and Batanova, 1995).

A complete gradation exists between the clinopyroxene-bearing harzburgites and the spinel-bearing lherzolites consisting of 65-75% of olivine, 20-25% orthopyroxene, 5-7.5% clinopyroxene, and 1-2% accessory chrome spinel (Sobolev and Batanova, 1995). These depleted rocks (Fig. 1.15 and **Fig. 1.16**) are enriched in light REE as compared to residual peridotites from modern mid-oceanic ridges, although they have similar mineralogy.

The boundary between the mantle peridotites and the overlying peridotitic cumulates coincides with a structural discontinuity, where harzburgite mylonites are typically associated with dunites and verlites.

Peridotitic Cumulates

The sequence of ultramafic to mafic cumulates is about 1500 m thick and consists of dunites, chromitites, clinopyroxene dunites, clinopyroxenites, olivine websterites, plagioclase lherzolites, poikilitic wehrlites, and minor troctolites. Their textures vary from adcumulate to heteradcumulate; the banding is better developed in the dunites (Panayiotou, 1987; Hebert and Laurent, 1990).

Coarse-grained wehrlites occur within the harzburgite tectonite in the western part of the Limassol Forest area, where they form large apophysate intrusives with no evidence of chilled contacts. This implies that the unserpentinized harzburgites were in a warmed-up state at the time of the intrusion of the wehrlites (Murton and MacLeod, 1987). Conversely, the olivine-phyric picrite dikes were emplaced into cool harzburgites and wehrlites with clear chilled contacts.

Dunites with granoblastic textures occur as isolated bodies within the harzburgite massifs. These rocks, characterized by banding, are generally concordant with foliation in the harzburgite's host rocks. The disseminated chromite mineralization and chromite podiform bodies are restricted to the dunites adjacent to their contacts with the harzburgites.

The dunites consist of highly magnesian olivine (Fo 89.8-91.2), chromium-rich spinel (Cr/(Cr+Al) = 0.51-0.71), and less than 5% orthopyroxene (En 88.9-90.3) and clinopyroxene (En 48-51.1, Wo 45.1-46.7, Fs 3.8-8.3) (Allen, 1975; Thy et al., 1989; Hebert and Laurent, 1990). According to Hebert and Laurent (1990), the dunites were subjected to stress at high temperature (>1000°C). There is a close similarity in composition between the spinel and clinopyroxene from the Troodos dunites, and the boninitic assemblage from the Gibbs Island peridotites (South Shetland Islands Arc, Antarctica), as well as the boninite-bearing ophiolite from the Bay of Islands, Newfoundland (Fig. 1.16 and **Fig. 1.17**).

Panayiotou (1987) reported that the apodunite serpentinites are characterized by a transition into intensely serpentinized clinopyroxene dunites, which are sharply cut by small dikes of rodingitizated gabbroids having well-developed chilled contacts. The clinopyroxene dunites either grade into the serpentinized wehrlites, or are tectonically juxtaposed with them (Panayiotou, 1987).

Poikilitic peridotites belong to the late plutonic assemblage intruding the harzburgites, and the dunites are typically wehrlites and clinopyroxenites. Banding is well developed in these rocks due to the segregation of olivine in individual bands. The deformed wehrlite bodies are composed of cumulus crystals of olivine and chromite, as well as large (2-3 cm) oikocrysts of pyroxene, and minor plagioclase (Murton and Macleod, 1987; Panayiotou, 1987).

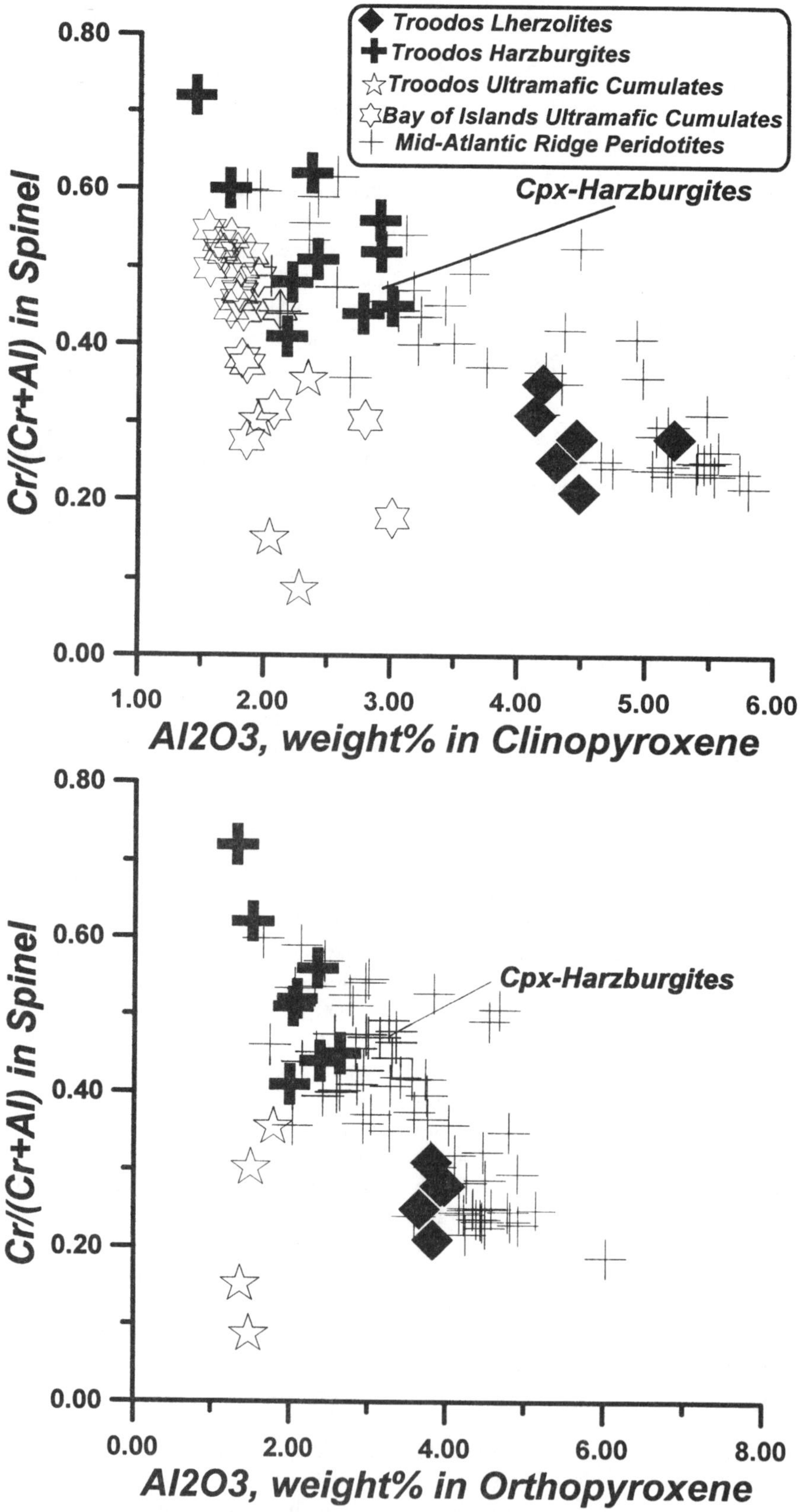

Fig. 1.15. Variations of compositions in Spinel, Clinopyroxene and Orthopyroxene from mantle peridotites of the Troodos Ophiolite (data by Sobolev and Batanova, 1995), Bay of Islands Ultramafic Cumulates - BOIC (data by Elthon et al., 1984 and Komor et al., 1985), and Mid-Atlantic Ridge Peridotites.

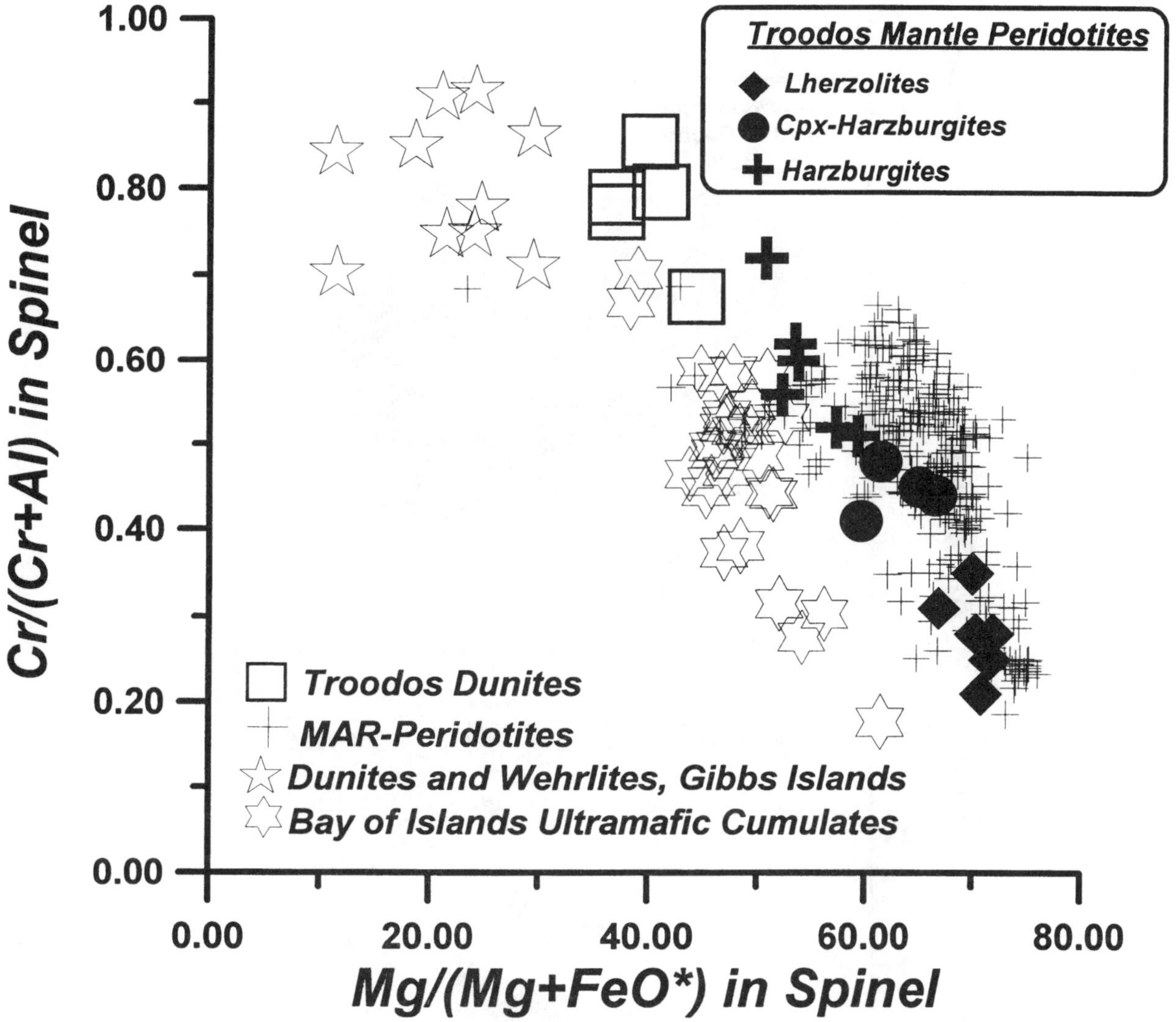

Fig. 1.17. Variations of Al_2O_3 versus Mg# (a) and Al_2O_3 versus Cr_2O_3 (b) in Clinopyroxene from Mantle Peridotites and ultramafic cumulates of the Troodos Ophiolite (data by Allen, 1975; Thy et al., 1989; Sobolev and Batanova, 1995). Black asterisks - different rocks from BOIC suite (data by Elthon et al., 1984 and Komor et al., 1985) and Mid-Atlantic Ridge Peridotites.

Rounded and partly serpentinized inclusions of olivine are present within these oikocrysts (Hebert and Laurent, 1990).

Mafic Cumulates

Various gabbroic rocks occur as large intrusive bodies surrounding the core of the Troodos Massif, or as small intrusive bodies in the Limassol Forest area. Small dike-like bodies of alternating anorthosite and gabbro are present within the gabbroid section of the Limassol Forest Ophiolite. There is evidence of tectonic deformation and the juxtaposition of the layered gabbroids and mantle ultramafics, which is probably due to the post- kinematic intrusion of poiklitic wehrlite (Panayiotou, 1987).

Hole CY-4 of the Cyprus Crustal Study Project was drilled south of Palekhori, and penetrated 2263 m of the lower Sheeted Dyke Complex and the Plutonic Complex. Most of this data was published by Laurent (1990), Malpas (1990), Thy et al. (1989), and Thy (1990). The general stratigraphic relations between the main rock units in drillhole CY-4 are as follows. The ultramafic to mafic cumulates were found to occur at a depth of 2263-1000 m. Between 2263 and 1746 m the cumulate ultramafics grade upward into coarse-grained layered gabbroic cumulates (their top is at about 1300 m depth). The mineral assemblage of the lower sequence is typified by low titaniferous clinopyroxene and orthopyroxene cumulus phases; olivine disappears at approximately 1450 m depth, whereas plagioclase (An>95) occurs above a depth of 1750 m. In contrast, the overlying medium-grained gabbroic cumulates with cryptic layering occur between 1300 and 900 m, and contain plagioclase (An 95-96), high titaniferous clinopyroxene, and orthopyroxene as cumulus phases throughout the sequence. Iron

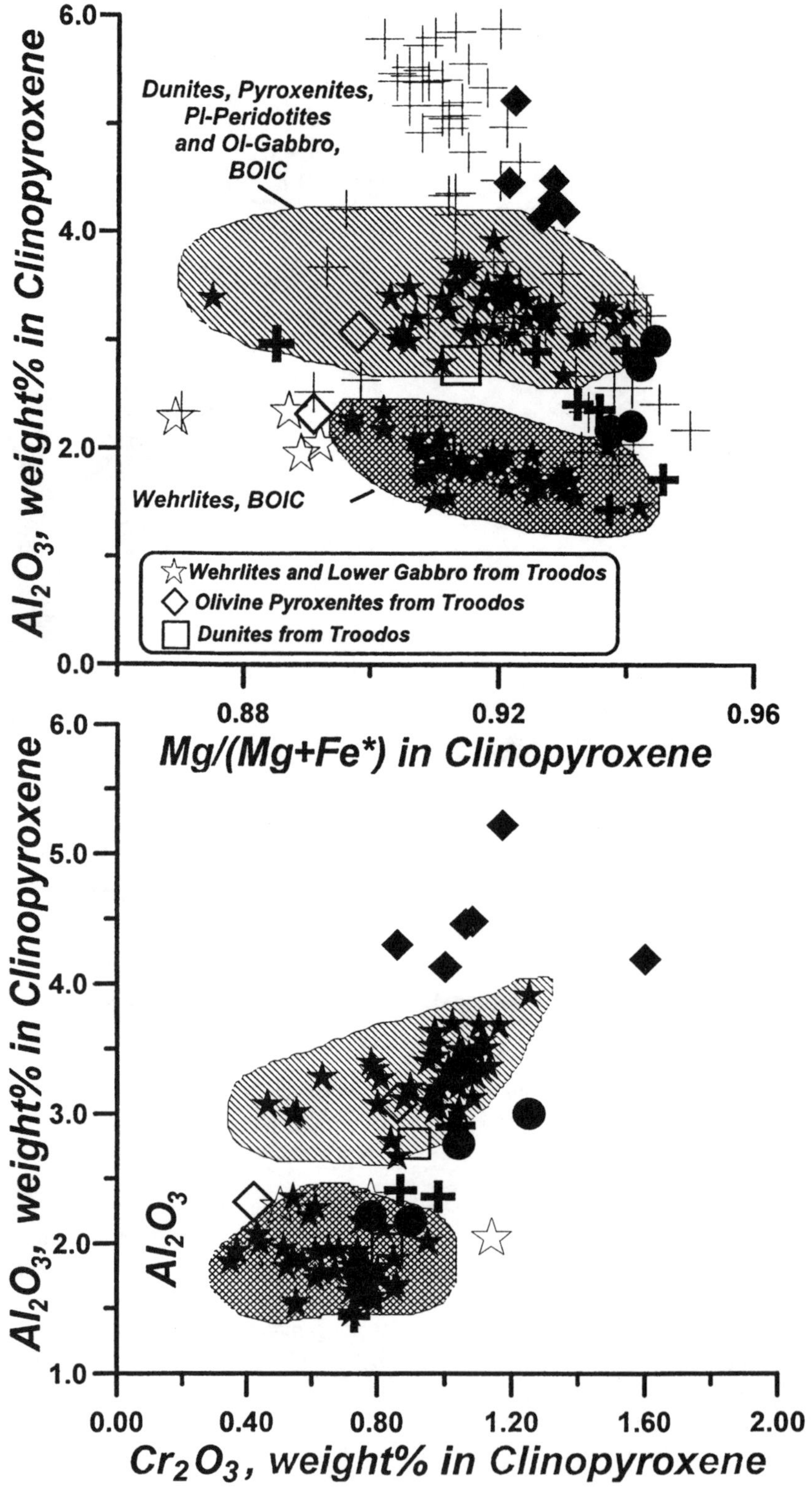

Fig. 1.17. Variations of Al_2O_3 versus Mg# (a) and Al_2O_3 versus Cr_2O_3 (b) in Clinopyroxene from Mantle Peridotites and ultramafic cumulates of the Troodos Ophiolite (data by Allen, 1975; Thy et al., 1989; Sobolev and Batanova, 1995). Black asterisks - different rocks from BOIC suite (data by Elthon et al., 1984 and Komor et al., 1985) and Mid-Atlantic Ridge Peridotites.

and titanium oxides appear locally at about 1200 m depth.

The lower cumulates consist of low titaniferous clinopyroxene and orthopyroxene as cumulus phases, whereas olivine appears at about 1600 m and plagioclase (An>95) occurs above 1750 m (Laurent, 1990; Thy, 1990). There is evidence for an intrusive contact between the lower coarse-grained gabbros and websterites, and the upper (~1200 m) medium-grained gabbroic rocks (Thy, 1990). According to Thy (1990), the general distinction between the lower and upper cumulates is in the different appearance of the cryptic layering. In the upper cumulates, the inconspicuous banding is marked by an alternation of the three main lithologies. These produce cyclical units of ~200 m thickness, each starting with a most primitive rock composition. Nevertheless, the same stratification in the lower cumulate gabbros is determined by latent layering related to only a single boundary which is characterized by an inversely trending change in the cumulus phase (Thy, 1990). The first appearance of precipitated olivine crystals, as cumulus phases, was found in the lower cumulates, whereas crystallization of plagioclase and pyroxenes first appear in the upper cumulate gabbros.

It was therefore concluded that three main lithologies are recognized between 2263 m and 1000 m in the CY-4 drillhole: (a) a lower horizon of websterites; (b) an intermediate level with layered gabbro-norites, and (c) upper gabbro-norites.

Amphibole-bearing gabbros associated with minor plagiogranite bodies occur at the uppermost part of this sequence (between 1000 and 700 m), adjacent to the contact with the dike complex (Laurent, 1990).

The lower assemblage of dikes occurs down to a depth of 640 m. The detailed mapping in the vicinity of the CY-4 drillhole revealed that the plagiogranite bodies cut the dike swarms and contain diabase xenoliths (Malpas, 1990).

A small pluton in the central Troodos Massif, near the Hji Pavlow chromite deposit, is a post-kinematic lower cumulate emplaced within the harzburite tectonites (Malpas et al., 1987). From bottom to top this pluton is comprised of dunites with minor wehrlites, poikilitic plagioclase wehrlites with minor pyroxenites and chromitites, and pyroxenites with minor plagioclase wherlites and lherzolites. Most of the differentiated rocks are magnetite-bearing gabbro, pegmatoid leucogabbro and diorite that occur as dikes, and isolated bodies emplaced close to the tectonized zones. This sequence is a strict analog to the Troodos lower layered cumulates. It was probably an integral part of the post-tectonic magmatic series, intruding both the harzburgite-dunite transition zone and the overlying lower (early) metacumulates.

The latest plutonic rocks in the central part of the Limassol Forest area are the isotropic gabbros intruding the mantle harzburgites. These occur as veins and apophyses of aplite-like leucogabbro (Murton and Macleod, 1987). The magmatic banding in these rocks is always discordant with the orientation of the older rocks, reflecting syntectonic crystallization (Murton and MacLeod, 1987; Malpas, 1990). The plagiogranitic 'chilled' zone facies of the late-magmatic assemblage, which contains gabbroic rock xenoliths, crosscuts the upper gabbroids (Malpas et al., 1987).

Mesocratic and Leucocratic rocks

Small homogeneous intrusions of intermediate and felsic rocks occur mainly at the boundary of and within the diabase dike series of the Troodos plutonic core. Inclusions of both the oldest diabase dike and the younger gabbroids are present within the plagiogranites. The small tongues of microgabbro exhibit a chilled margin against the granitoid rocks. The late dike swarm crosscuts all these rocks.

Although in fieldwork these rocks have long been divided into mappable units, their petrographic classification is still not certain. Moores and Vine (1971) described a quartz-diorite consisting of abundant quartz, zonal plagioclase (andesine in the core and albite on the margins), and secondary chlorite replacing dark-colored minerals. Malpas et al. (1987) termed rocks of similar composition, but probably with less alteration, as being plagiogranite and trondhjemite composed of saussuritized plagioclase and clinopyroxene that was replaced partly by actinolite and magnetite. Furthermore, there is a porphyritic quartz-albite rock, like the quartz diorite, whose primary hornblende crystals occur as elongated needles (Malpas et al., 1987; Malpas, 1990). In some places (*e.g.* near the CY-4 drillhole) these rocks contain abundant xenoliths of diabase. There are also granitoid rocks, associated with gabbro, where the only dark-colored mineral is epidote. The plagiogranites with granophyric texture consist of 70% plagioclase, 20% quartz and 10% amphibole, and in some fissures they contain secondary epidote.

Geochemistry of Plutonic Rocks

Geochemical evidence concerning the Plutonic Complex is mainly given by Moores and Vine (1971), Allen (1975), Coleman (1977), Sobolev and Batanova (1995), and Batanova et al. (1996).

Residual Mantle Peridotites

Major elemental characteristics reflect variations in the modal mineralogy of the tectonized harzburgite, which is typically expressed by a variable proportion of olivine, pyroxene and clinopyroxene. However, the mineral and chemical composition of these rocks depends on the degree of melting. The system of MgO/SiO_2-Al_2O_3/SiO_2 (Jagoutz et al., 1979) seems to offer a general basis for understanding the nature of the Troodos mantle harzburgites. Their small distribution on **Fig. 1.18** is typical of the most depleted mantle peridotites from mid-oceanic ridges (in the lower plot), and both the Troodos cumulated dunites, and the collision-related cumulate boninitic peridotite (upper points).

The abyssal peridotites commonly exhibit the mobility of MgO and SiO_2 which accompanies the serpentinization of harzburgites in oceanic basins. The general MgO and SiO_2 enrichment corresponds to low-and medium-temperature alterations, respectively (Bazylev, 1989). In this regard, the small compositional shifts along the MgO/SiO_2 axis (Fig. 1.18) can be

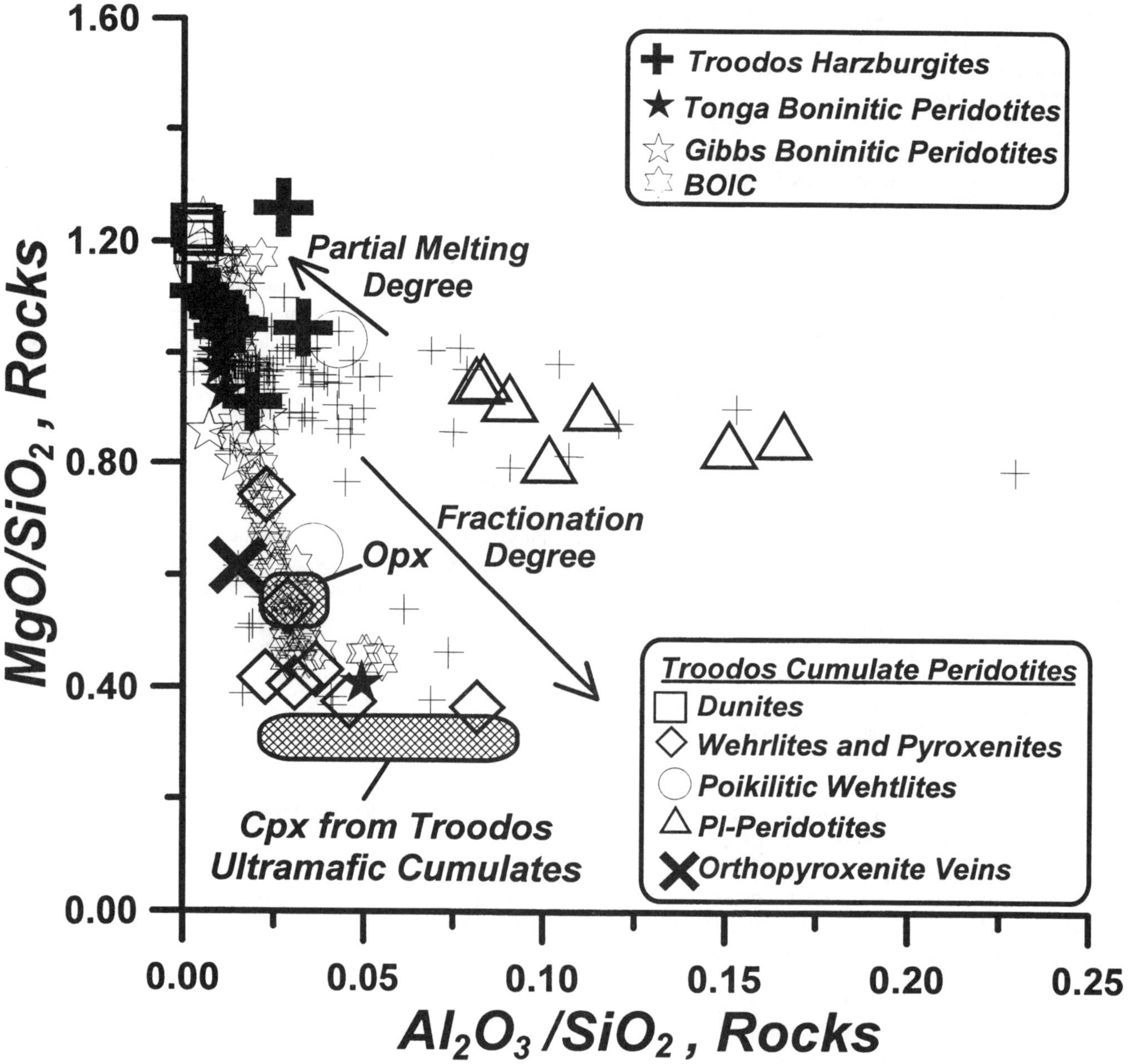

Fig. 1.18. Variations of Al_2O_3/SiO_2 versus MgO/SiO_2 in Mantle Peridotites and ultramafic cumulates of the Troodos Ophiolite (data by Allen, 1975), BOIC (data by Komor et al., 1985), boninitic peridotites from Tonga Trench (data by Sharaskin et al., 1983), boninitic peridotites from Gibbs Island (data by Silantyev et al., 1997) and Mid-Atlantic Ridge Peridotites.

interpreted as a result of low-temperature metamorphism. Because the Al_2O_3 is, in fact, immobile under metamorphic conditions, the observed changes in composition along the Al_2O_3/SiO_2 axis are mainly connected with variations in the pyroxene content. This determines the whole petrographic spectrum of the Troodos harzburgites, ranging from the most depleted harzburgites to clinopyroxene-bearing harzburgites and spinel-bearing lherzolites.

Cumulate Peridotites

The two-fold division of the ultramafic cumulates is supported by geochemical data. Malpas et al. (1987) and Hebert and Laurent (1990) showed that the early harzburgite-dunite-pyroxenite assemblage and island arc tholeiites from the lower pillow lavas are complementary, whereas the late wehrlite-clinopyroxenite assemblage is genetically related to the upper pillow lavas having high MgO and SiO_2 content. Geochemical comparison of the Troodos cumulated peridotites with the boninitic peridotites from the Tonga trench and Gibbs Island (South Shetland Islands,

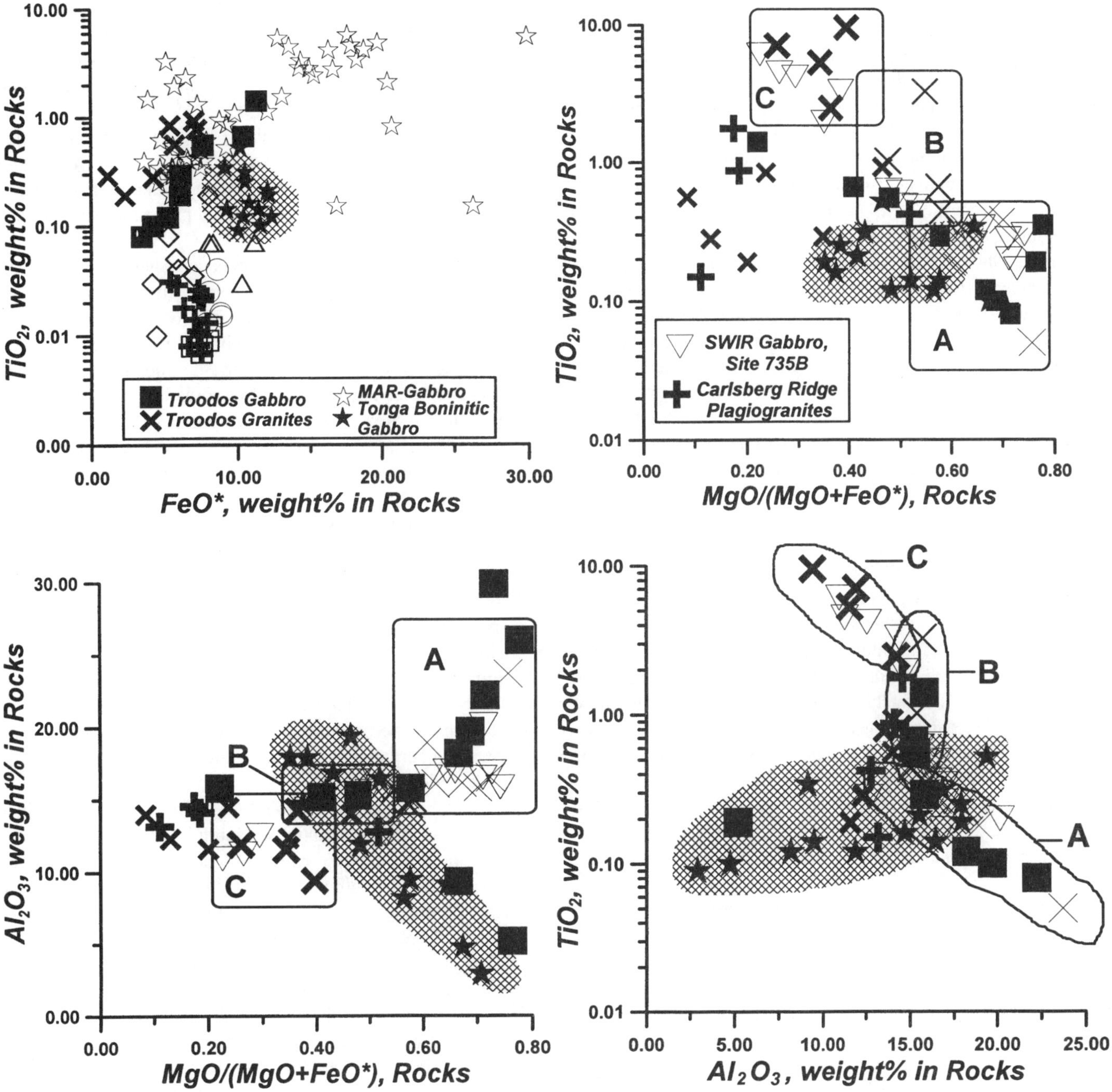

Fig. 1.19. Variations of some major element contents in gabbro and plagiogranites from the Troodos Ophiolite Complex (data by Moores and Vine, 1971 and Coleman, 1977), Mid-Atlantic Ridge gabbro — empty asterisks (big oblique crosses: very pale - gabbro of early stage (A), pale — gabbro of middle stage (B), thick — gabbro of late stage (C) (Miyashiro and Shido, 1980), South-Western Indian Ridge gabbro (data by Dick et al., 1991) and plagiogranites from Carlsberg Ridge, Indian Ocean — thick straight cross (data by Engel and Fisher, 1975). The compositional field of boninitic gabbro from Tonga Trench (black asterisks) is marking (data by Sharaskin et al., 1983).

Antarctic), and ultramafic cumulates from the Bay of Islands ophiolite, indicates a genetic connection between the Troodos late cumulates and the ultramafic boninitic layered series (Fig. 1.18). By contrast, the early plagioclase peridotites display an independent composition reflecting fractionation of the tholeiitic magma.

Mafic Cumulates

The melt inclusions in plagioclases from the olivine gabbro are indications of water-rich andesitic melts that could be generated during fractionation of the lower pillow lava parent magmas

(Batanova et al., 1996). These authors also point to the identical content of incompatible elements in the clinopyroxene from these early-stage gabbroid rocks and the spreading-related cumulated gabbros. By contrast, the latest isotropic gabbros from the Limassol Forest are perhaps genetically related to the boninitic pillow lavas (Murton and MacLeod, 1987).

A general plot of the major element content from the Troodos gabbroids and the main Mid-Atlantic Ridge gabbroids (**Fig. 1.19**) ranges from troctolite and olivine-pyroxene gabbro, through augite gabbro, to iron-titanium oxide-enriched gabbro (Miyashiro and Shido, 1980). An analogous relationship has been found in the sequence of troctolites, olivine gabbros, gabbro-norites, and iron-titanium oxide-rich gabbros from the Southwestern Indian Ridge (Drillhole 735B; Dick et al., 1991), and not analogous with the Tonga Trench boninitic gabbros (Fig. 1.19). In the Limassol Forest however, minor mafic intrusive rocks enriched in MgO and poor in alumina and titanium characterize the boninite series. Cameron (1985) suggests that these rocks are the plutonic analogues of the Troodos type-III boninitic lavas.

Intermediate and Felsic Plutonic Rocks

A close similarity in SiO_2 and K_2O content is found between the earliest differentiated lavas (LPL) and the quartz diorites (76.60-64.50 wt.% and 0.18-0.58 wt.%) respectively, (Moores and Vine, 1971) (Table 1.1). Malpas et al. (1987) considered that the plagiogranites and the trondhjemites represent a final magmatic stage that resulted from fractionation of basaltic magma. Because few of these late felsic plutons intruded the diabase dike complex their petrochemical character reflects assimilation with the diabase. The latter felsic plutonic rocks from the Troodos Massif show complete overlap in major-element chemistry with the typical oceanic plagiogranites from the Carlsberg Ridge, Indian Ocean (Fig. 1.19), indicating their tholeiitic fractionation.

Origin and Development of the Plutonic Complex

All of the Troodos mantle peridotites indicate the geochemical features of supra-subductional restites. One very important component is spinel lherzolite that represents a zone of incipient melting of a suboceanic mantle source. This was also responsible for the generation of the basaltoids belonging to the N-MORB type (Sobolev and Batanova, 1995), but has assimilated a subduction component. As noted above, the typical Troodos harzburgites could not result from greater degrees of melting of a lherzolite protolith in a closed magmatic chamber system, but instead reflect an inter-reaction of the parental lherzolite material with enriched percolating magmatic liquids (Sobolev and Batanova, 1995).

A plausible mechanism for the formation of dunites closely associated with harzburgites in both the modern oceanic basin and the ophiolite complexes is the model of Kelemen et al. (1995) that suggests the existence of dunite feeders within a mantle substratum, which appear to be caused by the inter-reaction of clinopyroxene-bearing mantle rocks with percolating mantle liquids.

In accordance with this model, the Troodos mantle peridotites can be interpreted as evidence of multiple melting of a single mantle substratum. The resulting genetic sequence is considered to be as follows: spinel lherzolite - clinopyroxene harzburgite - harzburgite - dunite (Fig. 1.4).

The overlying layered complex consists of two main magmatic assemblages with distinct genetic origins. The early association corresponds to a series of island arc-evolved tholeiites. Compositionally, these gabbroic rocks are more akin to those of the Mid-Atlantic Ridge and the Indian Ocean Ridge, which characterize formation in a low spreading-rate setting. The late plutonic rocks are associated with primitive to very primitive volcanites from the upper part of the extrusive succession. These are probably identical with the boninitic magmas generated in the Mariana and Tonga island arcs (Hebert and Laurent, 1990).

The formation of the two cumulate series is probably due to two-stage melting of a mantle diapir, but could also be due to difference in the density of a crystal mush zone occurring directly below the ‘petrologic Moho’ (Benn and Laurent, 1987). This mixture, or suspension consisting of crystals and melt, was probably very sensitive to stress field changes (Laurent, 1992) and was formed into a mantle diapir located in the subduction zone. During its uplift this diapir has reached the base of the Troodos crust causing its deformation, during which a crystal mush zone was finally incorporated into the crustal substratum forming the cumulate peridotite bodies.

Within the Arakapas Fault Zone the boninite series intrusives were intruded into the mantle peridotite blocks during the left-lateral slip. The oblique extension along this shear surface coincided with the beginning of an anticlockwise rotation of the Cyprus Microplate that led to the reactivation of the pre-existing transform-related structures (Murton and MacLeod, 1987).

The establishment of an east-west trending, vertically dipping zone of tectonic melange in the harzburgite tectonites reflects the late stages in the tectonic development of the Troodos Massif (Murton and MacLeod, 1987). Within such zones, blocks of various lithologies, ranging from harzburgite and dunite to the typical upper-crustal rock types, were incorporated within a serpentinite matrix. This suggests the protrusion of serpentinite masses into shallower-crustal horizons. It is interesting to note that small deformed serpentinite bodies occur sporadically among the lava-breccias (Murton and MacLeod, 1987). It is evident that a conventional model for the mid-oceanic ridge serpentinite diapirism is favored for the development of diapiric tectonics in the Troodos Massif.

E. Metamorphism of the Troodos Ophiolite Complex

Sergey A. Silantyev
Vernadsky Institute of Geochemistry and Analytical Chemistry of the Russian Academy of Sciences, Kosygin Str. 19, 117975 Moscow, Russia

The general attributes of the so-called 'ophiolite metamorphism' are as follows:

a) Indications of a low-pressure regime and low to medium temperature conditions during metamorphism;

b) A distribution of metamorphic grades (zones) depending on a considerable temperature increase downwards in the ophiolite succession, that varies from the zeolite facies which occurred mainly in the upper portions of the extrusive sequence, to the epidote-amphibolite facies in the plutonic rocks;

c) Evidence of non-isochemical metamorphism that is strongly pronounced in the lavas; and

d) Features of interaction between seawater-derived fluids and igneous rocks, which may explain all the metamorphic changes (Silantyev, 1993).

Metamorphism of the Ultramafic Rocks

Very little is known of the mineralogy of the Troodos serpentinites. According to Coleman (1971), these rocks consist mainly of lizardite and clinochrysotile, with minor amounts of antigorite, brucite, talc, magnetite, and carbonate. Because the peridotites are almost invariably associated with partly or wholly serpentinized primary harzburgites, it is evident that late intra-crustal metamorphism occurred at the temperature regime characteristic of the lower greenschist facies.

Oxygen-isotope studies (Bonatti et al., 1984) indicate that chrysotile and lizardite in abyssal peridotites form at temperatures of 200°-120°C. By contrast, the association chrysotile+antigorite in serpentinites, which appears to develop directly from the lizardite, corresponds to temperatures of about 250°C (O'Hanley and Wicks, 1995).

In the serpentinized ultramafic cumulates penetrated by the CY4 drillhole, there is oxygen-isotope evidence of metamorphic regeneration in the presence of submarine solutions (Vibetti et al., 1989a). The character of the low-temperature recrystallization of the residual and cumulate peridotites is the strongest point favoring the development of serpentinization by oceanic metamorphism. On the other hand the typical presence of brucite in ophiolitic serpentinites suggests another environment of formation in relation to the abyssal peridotites, as argued by Bazylev (1989). There is also stable isotope data (Wenner and Taylor, 1974) suggesting the probability of serpentinization in the presence of meteoric water.

All these data suggest that at least part of the Troodos serpentinites may have formed in a non-oceanic environment. It is possible that early serpentinization of the Troodos ultramafic protolith took place in an oceanic environment under the influence of submarine hydrothermal solutions. Late low-temperature serpentinization associated with the formation of brucite perhaps reflects an uplift of the ultramafic massif to the land surface taking place outside of the oceanic basin. These episodes of Troodos diapir movement are recorded by the presence of serpentinite olistoliths in some olistostrome deposits. Later episodes took place long after the final suturing of the oceanic basin, when low-density serpentinite masses appeared at or near the land surface.

Metamorphism of the Mafic and Felsic Plutonic Rocks

There is evidence of low- to medium-temperature metamorphism in most of the plutonic rocks of the Troodos Massif. Secondary amphibole, epidote, and chlorite appear as a result of metamorphic hydration of the parent rocks of the gabbroid protolith. Plagioclase becomes more enriched in sodium during this process. The assemblage of secondary minerals is characteristic of uralite gabbro. However, Moores and Vine (1971) considered the coexistence of two discrete amphibole generations in the Troodos plutonic rocks: a primary amphibole that could crystallize directly from parent melt, and a secondary one replaced by pyroxene during metamorphism. Hebert and Laurent (1990) suggested that the formation of low alumina-low titanium actinolite is a result of fractional crystalization, whereas the appearance on the liquidus of the same actinolite together with clinopyroxene reflects a geochemical peculiarity of the parent melt and a high content of H_2O in the magmatic system.

However, it is clear from data on amphibole compositions from the Troodos plutonic rocks presented in Thy et al. (1989), as well as from phase relations characteristic of uralitic gabbro, that actinolite and high-alumina actinolite from Troodos gabbroids have a metamorphic affinity.

The data in **Fig. 1.20** support this view. This conclusion is based on the comparison of the compositional characteristics of the amphiboles from the Troodos uralite gabbros with those from the metagabbro of mid-oceanic ridges, and the composition fields of magmatic amphibole from igneous rocks and mantle xenolites. The metamorphism of the Troodos gabbroids is associated with temperature conditions of the upper greenschist facies, judging from variations in the Al_2O_3 content of the amphibole. These metamorphic conditions are common for oceanic metagabbro.

On the western side of the Solea Graben, where the gabbroids and the dike swarm assemblage are juxtaposed along a detachment zone, the level of metamorphism has been estimated as a prehnite-actinolite stage of the greenschist facies, formed at a temperature of 300°-350°C (Schiffman et al., 1987). These rocks, displaying

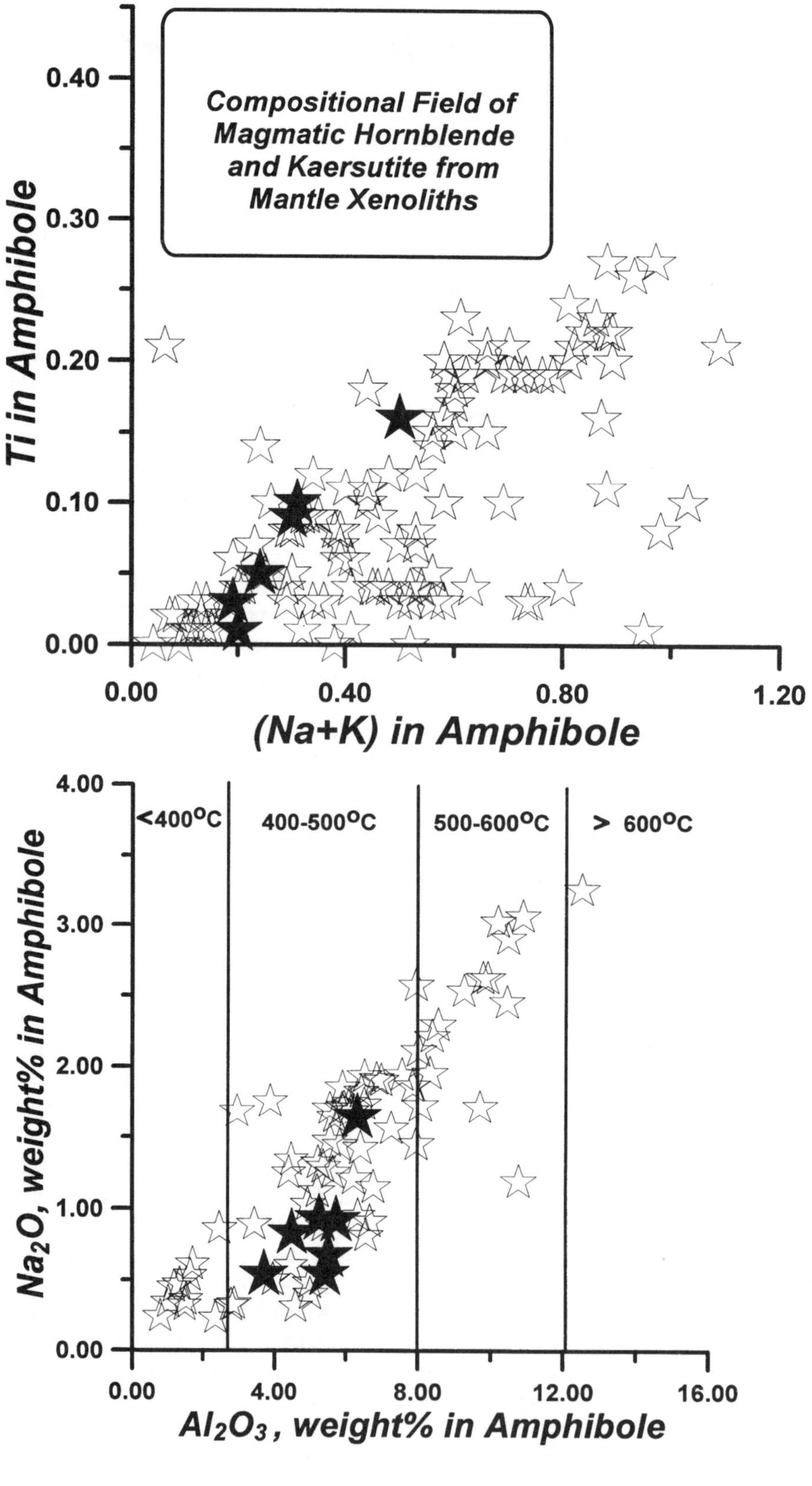

Fig. 1.20. Variations of compositional parameters in amphibole from Troodos gabbro. Field of magmatic hornblende and kaersutite from mantle xenoliths defined by Silantyev (1985); temperature range of oceanic metamorphism determined by Silantyev (1995).

original igneous textures, are crossed by many laumontite veins, which never cross the surface of detachment. This suggests that the gabbroids were metamorphosed prior to the main episodes of deformation at low water/rock (W/R) ratios, versus those for the dike rocks. Most low-temperature metamorphic changes in the gabbroids are reflected by the zeolitization, indicative of later microgabbro dikes crossing upper portions of the gabbroids and plagiogranites (Malpas et al., 1987).

Oxygen-isotope investigations of the layered complex penetrated by the CY4 drillhole indicate evidence for subsolidus isotopic exchange between seawater and the host rocks. Indications of considerable infiltration of seawater commonly occur from the sheeted dike assemblage through upper portions of the gabbro complex, to the banded cumulate gabbro and peridotites (Vibetti et al., 1989a).

There is evidence to suggest that hydrothermal alteration of the gabbroids occurred in the presence of seawater derived solutions having oxygen isotope compositions different from those of pure seawater, and at a lower temperature in comparison to the dike rock metamorphism. Data on the amphibole compositions of dolerites and gabbro also support this view. Alt (1991) suggested that although evolved arc island magmas were enriched in $\delta^{18}S$ the main control of variations in sulfur isotope composition in the gabbroids was the composition of seawater sulfate.

Mineralogical evidence of greenschist to epidote-amphibolite facies of metamorphism were found elsewhere in the quartz diorites, trondhjemites, and plagiogranites. Plagioclase from these rocks is partly replaced by zoisite, epidote, and smectite, and there is replacement of clinopyroxene by chlorite and actinolite with variable amounts of Al_2O_3 (Moores and Vine, 1971; Malpas et al., 1987). Mafic minerals in some granitoids are only represented by epidote. In quartz diorites, primary needle hornblende occurs together with metamorphic amphibole (Malpas, et al., 1987). The formation of the epidote in fissures reflects the latest episodes of recrystallization of the plagiogranites.

Metamorphism of the Dyke and Lava Complexes

The sheeted dike assemblage in the Solea Graben provides a good example of hydrothermal metamorphism in a preserved portion of a former rift valley of the oceanic spreading center. This typical site demonstrates the nature of metamorphic changes in an oceanic substratum up to the paleo-Moho continuity, related closely to hydrothermal activity (Varga et al., 1987). A plausible model for three-dimensional distribution of metamorphic isograds and isopleths of oxygen isotope compositions is given in Schiffman et al. (1987).

The huge northeast-trending zone of altered rocks formed by the ascending hydrothermal solutions is limited to the area of ubiquitous occurrence of epidosites within the dike assemblage. Because of cross-cutting relationships between the epidosite bodies and the main structural elements of the graben, the dominant phase of hydrothermal metamorphism appears to be earlier than this phase of extensional deformation. The mapped localites with prominent hydrothermal activity features (some 10 km wide and 2-3 km thick) are comparable in size to modern-day submarine hydrothermal systems originating at or near long-lived mid-oceanic ridge spreading centers (Schiffman et al., 1987).

Secondary minerals from the upper portions of the dike complex are quartz, celadonite, calcite, smectite, mixed-layer smectite-chlorite, and minor zeolite, whose association is indicative of the zeolite facies of metamorphism (Varga et al., 1987; Schiffman et al., 1987). Mineral assemblages of the greenschist facies characteristically occur downward in the section, suggesting that cool seawater percolated through the dike rock column.

The oxygen isotope composition in the dike rocks indicates that their alteration occurred in the presence of little-changed seawater and at low values of the W/R ratio (Vibetti et al., 1989a). Good evidence for the existence of pronounced vertical temperature gradients during a period of hydrothermal circulation comes from the analysis of the distribution pattern of $\delta^{18}O$ values. Variations in the sulfur isotope composition occurred in the altered feeder dike rocks suggesting interaction of seawater from the Cretaceous oceanic basin and remobilized sulfur derived from the ophiolite substratum (Alt, 1991).

Schiffman et al. (1987) reported that three zones of epidosites have been mapped on the western side of the Solea Graben. These recrystallized dike rocks, consisting of epidote, quartz, and minor chlorite were produced under maximum temperatures and at the highest values of the W/R ratio. In this regard, they have no strict analogs among the hydrothermal metamorphism occurrences in the Troodos Massif. In some places (*e.g.* the Apilki cupriferous sulfide deposit), the epidosites probably underlie the massive sulfide ore bodies. It is evident, however, that the most altered dike rocks are depleted in Cu and Zn.

The temperatures of metamorphism were estimated by the method of homogenization of fluid inclusions in the quartz. This study shows that the temperature reached during the formation of the epidosites was between 270°C and 340°C (Schiffman et al., 1987). The lowest temperatures of metamorphism correspond to formation in dolerites of potassium feldspar and quartz, which are typical products of low-temperature hydrothermal alteration of seafloor basalts.

The most important information on the metamorphism of the lavas has been obtained in the course of the Cyprus Crystal Study Project from the drilling results in the northeastern part of the Troodos Massif (wells CY1, CY1A, CY2, and CY2A). Detailed investigations were made of the composition and zonal structure of the lava alteration in the Akaki River canyon and at the Agrokipia stockwork ore deposit (Gillis and Robinson, 1985; Bednarz et al., 1991; Erzinger et al., 1991).

Three main zones of hydrothermal alteration were recognized. They are as follows:

1) The lower zone, up to 500-600 m thick, occurs at the base of the lava succession. It is represented by island-arc tholeites, andesitic and dacitic hyaloclastites, and minor diabase dikes. According to Bednarz et al. (1991), the level of metamorphism of these rocks has been estimated at greenschist facies, based on the

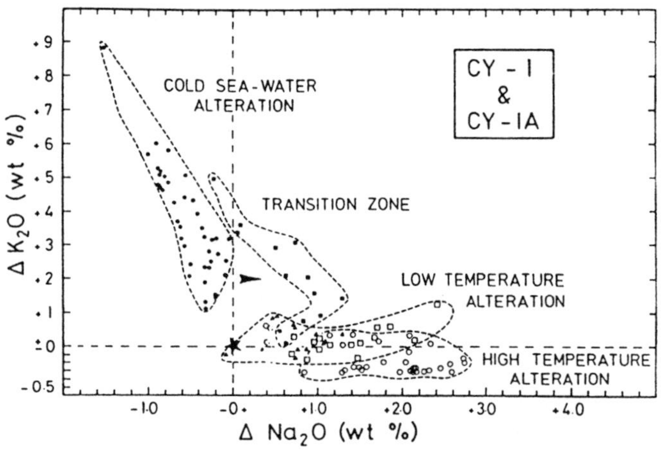

Fig. 1.21. ΔNa_2O vs ΔK_2O variation diagram for drill holes CY-1 and CY-1A discriminating the different alteration zones in the Troodos upper crust. Arrow indicates shift of trend by influence of low temperature alteration. Big star indicates original composition of rocks. After Bednarz and Schmincke (1989).

mineral paragenesis epidote+sphene+chlorite. As shown in Fig. 1.21, K depletion and Na enrichment at a W/R ratio averaging around 15 are characteristic of these rocks. Secondary minerals are smectite, aragonite, laumontite, quartz, and chalcedony. The presence of laumontite indicates a temperature of metamorphism of 150°-250°C.

2) The second (middle) zone up to 500 m is generally composed of high magnesium – high silica basaltic rocks in the form of pillow lavas and massive flows. There are also local horizons in which rocks are enriched in K_2O and Na_2O (**Fig. 1.21**). Secondary minerals are represented by zeolites, including analcime, natrolite, phillipsite, gmelinite, and schabazite, celadonite, clinoptilolite, smectite, and calcite; palygorskite and calcite become more abundant in upper parts of this zone. The temperature of metamorphism may not exceed 170°C, and the W/R ratio is equal to 20 (Bednarz et al., 1991).

3) In the third (upper) zone (upper sequence of the section, *i.e.* 20-200 m), extending from the sediment-lava boundary downwards into the lava pile room, the pillow lavas demonstrate a high degree of oxidation and leaching. This zone has been interpreted as a zone of long-term oxidative sea-floor weathering (Gillis and Robinson, 1985). There is a remarkable lack of low temperature alteration features, where the deep-sea metalliferous sediments (iron-manganese umbers) rest directly upon the lava flows. Geochemically, the mechanism of introduction of cool seawater solutions into newly formed extrusive crust is rather straightforward. These rocks are typically characterized by the strong enrichment in K and loss of Na at W/R ratios of more than 100 (Fig. 1.21). The mineral assemblage contains smectite, zeolites, calcite, iron oxides, and secondary potassium feldspar. This style of alteration is related to the oxidative environment at or near the sea floor, corresponding to temperatures of 4° to 100°C (Gillis and Robertson, 1985; Bednarz et al., 1991).

Many isotopic data have been interpreted to reflect a considerable contribution of seawater-derived hydrothermal fluids to the metamorphic processes throughout the development of the Troodos volcanism over a wide range of temperatures (Schiffman et al., 1987; Vibetti et al., 1989a; Alt, 1991).

Staudigel et al. (1986) reported isotopic age determinations by K-Ar and ^{87}Rb-^{86}Sr isochron methods on vein celadonite from the altered volcanics of the upper part of the lava sequence. For two individual samples, the measured dates are 77-80 and 87-92 Ma. The K-Ar total-rock measurement on this material gives an age of 84.7+2.3 Ma. These data provide a minimum age for the time of the formation of celadonite, and confirm radiolarian fauna dating (Staudigel et al., 1986). These age estimations can be interpreted as evidence of the long-term formation of celadonite, ranging from 7 Ma to 15 Ma. They stress that a similar process of precipitation of vein minerals from hydrothermal solutions has continued over a period of 12-19 million years in modern oceanic basins after the consolidation of the oceanic crust (Staudigel et al., 1986).

Metamorphic History of the Troodos Ophiolites

There is a clear mineralogic and geochemical similarity in the metamorphic features of the Troodos ophiolite rocks and the magmatic rocks in modern oceanic basins. In comparing the conditions of metamorphism in the dikes and extrusives from the Solea Graben and the Akaki River canyon, it is concluded that the temperature of metamorphism increases down the section, whereas the W/R ratio decreases. Conditions estimated for the Upper Pillow Lavas range from 80°C to 150°C at a W/R ratio of more than 100; for Lower Pillow Lavas: 150°C to 250°C at a W/R ratio = 15; and for the Dyke Complex: between 270°C and 340°C at low W/R ratios. In addition, a fluid-inclusion study of ultramafic rocks of the greenschist facies metamorphism (Vibetti et al., 1989b) indicates that the most plausible sources for the NaCl-enriched inclusions are the high-temperature seawater fluids. Their primary compositions were changed considerably when they passed through the host rocks.

Such a style of distribution of metamorphic conditions and trends in the associated geochemistry (K enrichment and Na depletion) is similar to that of the general features of the metamorphism of the highest parts of the sequence of the oceanic crust (Layers 2 and 2B).

The analogy between the Troodos-type sulfide mineralization and that commonly associated with basalts generated in modern spreading-related settings is also evident. In both cases, the sulfides are precipitated from the metalliferous hydrothermal fluids migrating to the seafloor along growth faults at the time of formation into the host rocks' greenschist facies mineral assemblages. The accumulation of ores in the volcanic rocks of the Troodos ophiolite took place via the mixing of ascending reduced fluids of seawater origin and descending low temperature shallow seawater solutions, at the upper parts of the volcanic sequence (Erzinger et al., 1991).

A two-stage history of metamorphism is perhaps recorded in the underlying gabbroids. The early metamorphic stage reflects the temperature conditions estimated for the upper greenschist to epidote-amphibolite facies, that led to the formation of actinolite and high-alumina actinolite. The late stage is associated with extensive zeolitization of the gabbroid substratum. Such trends in the metamorphic transformation are characteristic of the gabbroids generated in the mid-oceanic ridges (Silantyev, 1995).

The intense serpentinization of the peridotites is related to the protrusion of mantle diapirs (tectonite harzburgite) into the shallow-crustal horizons of the ophiolite succession. A characteristic result of the migratory movement of seawater solutions reaching the deep-seated horizons is the hydration of the ultramafic rocks, a process which led to considerable deconsolidation of the existing rocks. This may have made the upward tectonic movement of the mantle and cumulate rocks easier. The same mechanism is responsible for the development of diapiric tectonics in the highest parts of the mid-oceanic ridges (Silantyev, 1995). The final episode of serpentinite diapirism possibly continued after the disappearance of the subduction zone and the closing of the oceanic basin in which all of the igneous units of Troodos ophiolite complex were formed.

F. Brief Review of the Associated Mineralization of the Troodos Ophiolite Complex

Sergey A. Silantyev
Vernadsky Institute of Geochemistry and Analytical Chemistry of the Russian Academy of Sciences, Kosygin Str. 19, 117975 Moscow, Russia

Economic mineralization in the Troodos Massif is confined to chromite and chrysotile asbestos occurrences within the Plutonic Complex, and to massive metallic sulfide deposits in the Troodos lavas, which have been mined since pre-Roman times.

The chromites occur exclusively within the dunite bodies which are either developed among the mantle harzburgites or occur relatively close to their contacts (Greenbaum, 1977). Following Malpas et al. (1987), most of the chromites appear together with the olivine-rich rocks, reflecting high pressure-high temperature conditions of formation, and showing evidence of a cumulate origin. The podiform chromites are always associated with the dunites, and they were deformed contemporaneously, showing a general concordance with the banded structures of the host harzburgite tectonites. The Hagipavlou ore deposit located 1.5 km to the south of Mt. Olympus is the most typical of all the chromite deposits of Cyprus.

The vein asbestos mineralization, confined to a zone of intensively brecciated and serpentinized harzburgite, lacks systematic orientation and crosses the shattered host rocks in random directions (Panayiotou, 1987; Malpas et al., 1987). In the largest Pano Amiandos deposit (on the eastern side of Mt. Olympus), the asbestos veins consist mainly of chrysotile and associated antigorite (Panayiotou, 1987). In our opinion, the Cyprus-type asbestos mineralization can be interpreted as being caused by relatively low-temperature recrystallization of mantle substratum.

The massive metallic sulfide deposits are genetically linked to the volcanic complex. Characteristically, in Cyprus the sulfide deposits are located below the high magnesian olivine-phyric extrusives assigned to the Lower Pillow Lavas (Constantinou, 1987). In the five main ore districts, the copper reserves, as described by Constantinou (1987), are between 50,000 and 20,000,000 tons. The zinc concentration in these ores is largely below 0.2%, with some increase in the oxidized ore bodies.

Most of the ore occurrences show a pronounced vertical zonation that is best studied in the Skouriotissa deposit. Constaninou's (1987) work provides one of the best-documented examples of the zonal arrangement in the vertical distribution of different types of sulfide mineralization in ancient spreading-related volcanic complexes. According to his model, the sulfides were precipitated in an orderly sequence. From bottom to top, this idealized sequence is as follows: The well-developed stockwork zone underlies most of the economic ore bodies. Above this is a thin zone of pyrite-bearing rocks rich in silica. The next zone of massive sulfide ores with conglomeratic structure consists of pyrite, chalcopyrite, and sphalerite. These rocks are directly overlain by black, unconsolidated pyrite-bearing sands containing blocks of compact sulfide material showing coloform banding. These unconsolidated 'conglomeratic ores' pass upward into a oxidized zone where the predominant mineral is goethite. The zone of oxidized ores grades up-section into an iron-rich sediment that is marked as ochre (Constantinou, 1987).

The current thinking is that the formation of the Cyprus-type sulfide is equated with the seafloor hydrothermal metamorphism associated with modern volcanic activity in an oceanic spreading-related setting. An important conclusion following the finding of remains of pyritic smoker vents (Constantinou, 1987; Varga et al., 1987) is that the Troodos massive sulfides are similar in origin to the ore deposits related to oceanic black smokers,.

G. Geodynamic Regime of Formation of the Troodos Ophiolites

Maxim V. Portnyagin and Sergey A. Silantyev
Vernadsky Institute of Geochemistry and Analytical Chemistry of the Russian Academy of Sciences, Kosygin Str. 19, 117975 Moscow, Russia

In summarizing the existing data on the compositional characteristics of the Troodos Ophiolite Complex rocks, one can recognize within this succession two main, broadly-defined magmatic rock assemblages with distinct genetic origins. Each of these assemblages corresponds to the two main stages of tectonic evolution of the Cyprus Region of the eastern Mediterranean.

The lower cumulative sequence of the Plutonic Complex, the Sheeted Dyke Complex, and the Lower Pillow Lavas, are all typical constituents of the lower magmatic rock assemblage and belong to the first stage of formation of the Troodos Complex. Its formation must be related to the consequent extensional condition in a supra-subduction setting. These rocks are derivatives of island arc tholeitic magmas, which are characteristic of intra-arc and backarc basins of the transitional zones. The geochemical signatures indicate that these rocks are probably the direct result of secondary melting of a relatively low-temperature mantle source of lherzolite composition in the shallow lithospheric mantle above the subduction zone. In this regard, the lherzolites occurring at Mt. Olympus are considered to be the restites of such magmas.

The upper magmatic rock assemblage includes the upper portions of the cumulative sequence, the latest mafic dikes, and the Upper Pillow Lavas belong to the second stage of magmatic evolution of the Troodos complex. These rocks are thought to represent the products of evolution of high MgO - high SiO_2 bonitite-type magmas. It is generally agreed that such associations are characteristic of the entire basement of intraoceanic island arcs. The parent magmas of this magmatic assemblage are probably the result of melting of a highly depleted mantle source, similar to the mantle restite sources of mid-oceanic ridges or oceanic islands. On the other hand, these rocks may represent restites extracted from the parent magmas of the early magmatic ophiolites complexes (e.g. an early magmatic assemblage of the Troodos Massif).

Taking into account the spatial relationships between the two above-mentioned magmatic assemblages, we suggest that the following sequence of geodynamic events led to the formation of the complete ophiolite succession in the Troodos. The evolution of this ophiolite began with the establishment of a subduction zone within a marginal basin of the paleo-Tethys ocean about 95 million years ago. This early stage of magmatism was greatly influenced by a spreading-related regime. In this stage, most portions of the Troodos ophiolite formed entirely in a forearc basin (in concordance with the 'Mariana Fore-Arc Model' of Robinson et al., 1983; Portnyagin et al., 1997) or in a backarc basin above an obliquely-dipping subduction zone (the 'Andaman Sea Model' of Moores and Vine, 1971). In either case, they probably formed due to progradation of a backarc spreading center into an island arc lithosphere (the 'Valu Fa Ridge Model' of Portnyagin et al., 1997). The greater part of the Troodos ophiolite complex was formed during this stage of magmatism. It is possible that the Arakapas Fault Zone, interpreted as a transform fault separating the entire Troodos Ophiolite and the Limassol Forest Ophiolite (the so-called Anti-Troodos terrain), existed as early as this stage (Murton and Macleod, 1987).

The paleontological data (Blome and Irwin, 1985) indicate that there is no evidence for a lengthy pause in the formation of the lower and upper magmatic assemblages. In some areas, however, the transition between these two assemblages is marked by a break in volcanic activity, the first appearances of hydrothermal effects, and the tectonic reconstruction of the upper part of ophiolite crust (Varga and Moores, 1990).

The second-stage of magmatism is recorded by extensive eruptions of the Upper Pillow Lavas and by the emplacement of small masses of post-kinematic intrusions, which intrude both the early Plutonic Complex and the dike swarm assemblage. The productivity of this magmatic stage was sufficiently lower as compared with the early magmatic stage. These rocks appear to have been associated with an island arc environment, judging from their geochemical features. However, no volcaniclastic component common to island-arc complexes has been found in this assemblage. This phenomenon allows us to suggest the formation of a second-stage magmatic complex of the Troodos ophiolites within an immature island arc, partly or wholly occurring on the early-formed ophiolitic crust.

The paleobathymetric interpretations of the chilled glass compositions (Schmincke and Bednarz, 1990) and the oceanic metamorphism features indicate that there was a general deep-water environment of formation for the entire magmatic complex of the Troodos ophiolite succession.

Thus, the Troodos Massif, long recognized as a preserved fragment of the oceanic lithosphere of the paleo-Tethys ocean, is unique in the fact that the geologic characteristics and genetic models for this small terrain, in their wider context, may be applicable to other ophiolites over the whole of the eastern Mediterranean, and may therefore be used as a standard for global geodynamic correlation. In this regard, recent models of petrogenesis of the Troodos ophiolitic magmas can serve as a reliable adjunct in the study of the effects of accretion of the oceanic crust on development of the Troodos magmatism, which apparently ranges from generation of mantle melts to the formation of active, ore-forming hydrothermal systems (Batanova et al., 1994; Agar and Klitgord, 1995; Sobolev and Chaussidon, 1996; Portnyagin et al., 1997). In the meantime, the current progress in the study of the Troodos ophiolites is promising to provide new results in the future that will be very fruitful for the geodynamic reconstruction of the geological evolution of the Tethyan Ophiolitic Belt as a whole.

H. The Sedimentary Cover of the Troodos Massif

Nikita Yu. Bragin, Lyubov G. Bragina, Kirill G. Kaleda, and Kirill A. Krylov
Geological Institute of the Russian Academy of Sciences, Pyzhevskiy Pereulok 7, 109017 Moscow, Russia

The sedimentary deposits of the Troodos Massif are represented by:

1) Sediments of the upper part of the ophiolitic sequence that overlie the Lower Pillow Lavas and are covered by the Upper Pillow Lavas;

2) Deposits of the Troodos Ophiolite Complex sedimentary cover, which are subdivided into three units: (a) volcaniclastic breccias and tuffaceous sandstones; (b) metalliferous sediments of the Perapedhi Formation (the Turonian-Lower Campanian) and; (c) bentonite clays of the Kannaviou Formation (the Campanian-Lower Maastrichtian).

Sediments of the upper part of the ophiolitic sequence

The sedimentary deposits which overlie the Lower Pillow Lava unit are represented by commonly metalliferous mudstones and siltstones with occasional horizons of volcaniclastic sandstones. The thickness of this sedimentary unit ranges from 250 m to a few centimeters (MacLeod, 1990). Paleontological remains are very rare. They are represented by poorly preserved benthic foraminifers, radiolarians and ichthyolits that have no stratigraphic value.

The Troodos Ophiolite Complex sedimentary cover

Volcaniclastic breccias

A unit of volcaniclastic breccias and sandstones directly overlies the eroded surface of the Upper Pillow Lavas or forms interbeds in the basalts. This unit has its widest distribution along the Arakapas Fault (Simonian and Gass, 1978), and it can be subdivided in this region into two members: a lower, chaotic, and an upper, with elements of stratification.

The lower member is well exposed near Perapedhi village. It is represented by dark-gray, brownish-gray and black chaotic breccia without any stratification, with rectangular, non-rounded fragments of gabbro-diabases and pillow lavas ranging in size from 1 cm to 0.5 m. These fragments represent more than 50% of the total rock composition. The breccia has a clayey or tuffaceous matrix.

The upper stratified member consists of non-cemented breccias, gravelites, sandstones, siltstones, and mudstones that are distributed along the whole Arakapas Fault. These rocks commonly form layers with thicknesses ranging from 1 to 30 m and with graded bedding. The basal parts of these layers are represented by breccias which have sharp contacts with the underlying rocks. The breccias gradually pass upwards into gravelites and sandstones and then to siltstones and mudstones. Fragments in the breccias are angular and range in size from 40 to 50 cm. They consist mainly of metabasalts, metadolerites, and more rarely of gabbroids and mudstones that are intercalated with breccias. Ultramafic rocks were not found among the breccias. Fine-grained clastics are characterized by grains of clinopyroxene, plagioclase and quartz. These clastics are interbedded with pillow lavas and are sometimes intruded by doleritic sills.

The thickness of this member varies widely over the area, and depends on the surface relief of the underlying volcanics. A maximal total thickness of 500 m (including intercalations of volcanics) is indicated in depressions. The overall thickness of sedimentary rocks in this case is not more than 150 m.

There are very few paleontological data from the breccia unit. Campanian-Early Maastrichtian radiolaria were found in mudstones of the upper member east of Perapedhi village (Robertson and Hudson, 1974). The Late Campanian-Maastrichtian radiolaria *Amphipyndax tylotus* Foreman, *Afens lirioides* Riedel et Sanfilippo, *Archaeodictyomitra lamellicostata* Foreman, *Cornutella californica* Campbell et Clark, *Dictyomitra koslovae* Foreman, *Pseudoaulophacus pargueraensis* Pessagno, *Stichomitra communis* Squinabol, *Theocapsomma comys* Foreman were found in siltstone horizons of the upper member north of Agios Photios village (Osozawa and Okamura, 1993). No fossil data are known from the lower member.

The above-mentioned features of the volcaniclastic breccias allow us to interpret this unit as an ophiolitic olistostrome that was deposited along the transform fault valley. The clastics of the lower member were formed by submarine erosion and collapse of the steep slopes of the fault valley. The upper member was formed by high-density debris flows and turbidites. Volcanic activity continued during deposition of the breccia unit.

The Perapedhi Formation

The Perapedhi Formation (Mantis, 1970) represents the oldest paleontologically dated deposits within the Troodos Ophiolite Complex sedimentary cover, which directly overlies the Upper Pillow Lavas or their associated volcaniclastic breccias. The Perapedhi Formation consists mainly of metalliferous sediments (umbers) that are distributed over the paleodepressions as lenses of various thickness. Stratigraphic information from the Perapedhi Formation is of primary importance in the reconstruction of the geologic history of Cyprus in the Late Cretaceous, and for structural analysis of this area. This is because the deposition of the Perapedhi sediments took place directly after the formation of the Troodos ophiolites, and before their collision with the allochthonous Mamonia Complex (Robertson and Woodcock, 1979). Indeed, the age of the Perapedhi Formation is still in

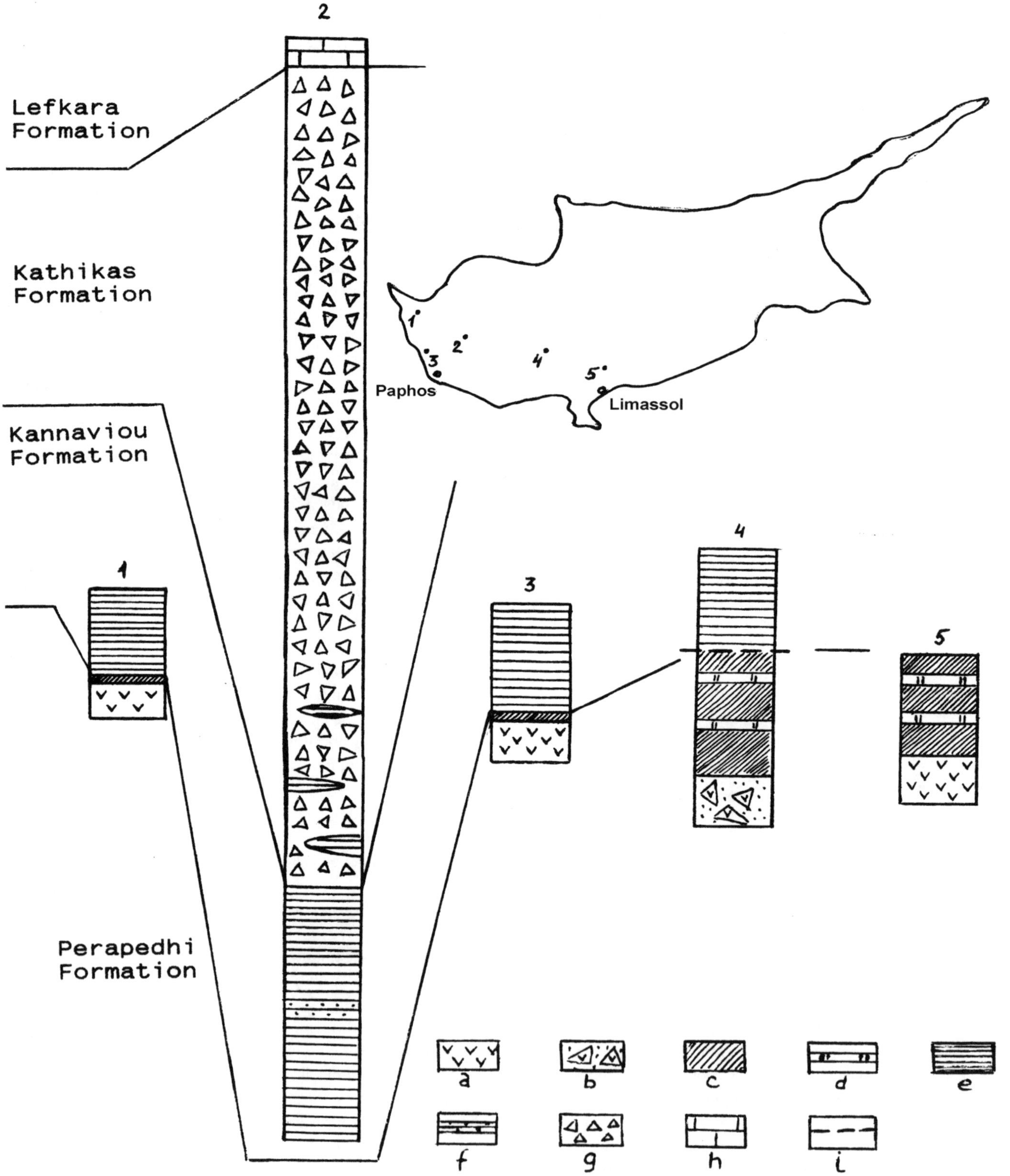

Fig. 1.22. Sections of the Upper Cretaceous cover of the Troodos Massif.
Outcrops are: 1) Akamas Peumsula; 2)Kritou Marattou-Agios Dimitrianos Road; 3) Mavrokolymos River; 4) Peripedi Village; 5) Magdalene Quarry.
Lithologies are: a, b) Pillow basalts and volcanoclastic breccia; c) Umbers; d) Cherts; e) Clays; f) Sandstones; g) Olistostromal deposits; h) Limestones and marls; i) Tectonic contact.

question. Some authors suppose it to be the Campanian (Swarbrick and Robertson, 1980), or Turonian (Blome and Irwin, 1985). Thurow (1991) concluded that it spans a wide stratigraphic range - from the Turonian to Lower Campanian. Paleontologic remnants from the Perapedhi Formation are represented only by radiolarians in need of more detailed studies. The genesis of umbers from the Perapedhi Formation was the subject of much discussion (Wilson, 1959; Morel, 1960; Constantinou and Govett, 1972; Robertson and Hudson, 1973). They were interpreted as resulting from subaqueus weathering of pillow lavas (Constantinou and Govett, 1972), or as ancient analogues of hydrothermal products of mid-oceanic rifts (Robertson, 1975).

In the Magdalene quarry (on the southern flank of the Limassol Forest Massif) the Upper Pillow Lavas are overlaid by massive and platy umbers with rare chert intercalations (total thickness up to 12 m) (Fig. 1.22). Cherts yielded the Turonian radiolaria: *Alievium* aff. *superbum* (Squinabol), *Archaeodicty-omitra squinaboli* Pessagno, *Crucella irwini* Pessagno, *Dictyomitra formosa* Squinabol, *Orbiculiforma monticelloensis* Pessagno, *Praeconocaryomma lipmanae* Pessagno, *Pseudoaulo-phacus lenticulatus* (White), *Pseudodictyomitra pseudo-macrocephala* (Squinabol), *P. nakasekoi* Taketani, and *Pyramispongia glascockensis* Pessagno (Blome and Irwin, 1985). This assemblage represents the oldest paleontological data from umbers of the Perapedhi Formation. Mixtites that are tentatively assigned to the Moni Formation overlie the Perapedhi Formation in this section.

The Perapedhi Formation type section is located in the central part of Cyprus near Perapedhi village. Outcrops of the Perapedhi Formation were studied in a road cut near the Saittas village (150 m from the Perapedhi-Saittas-Kouka road connection). The sequence is represented by the following layers (**Fig. 1.22**) (Bragina and Bragin, 1996):

1) Volcaniclastic breccias with a tuffaceous matrix and fragments of basalt and diabase. The observed thickness in this outcrop is 1 m. The same breccias have a thickness of several dozen meters to the south of the road cut.

2) Red and yellow to rust-colored clays which fill the pockets on the eroded surface of the breccias. The thickness varies from 0 to 1 m.

3) Black and dark-brown loose manganese-bearing mudstones with weakly rounded fragments (up to 2 cm) of yellowish-gray phosphatic clayey limestones (8-10 cm). Pebbles of limestones are parallel to the bedding, and yield poorly preserved radiolarians (probably Late Cretaceous). The thickness is 0.1-0.2 m.

4) Umbers that are dark brown and yellow, platy, sometimes massive, with brownish-black intercalations of massive Fe-Mn-bearing cherts, rare intercalations (1-2 cm) of yellow tripoli-like silty radiolarites, a few intercalations (1 cm) of pinkish-white carbonatized radiolarites, and rare small iron-manganese concretions. The thickness is 10 m.

Radiolarites that were found in the upper part of member 4 are characterized by an abundant, extremely rich and morphologically diverse radiolarian assemblage, which shows a high taxonomic diversity of such genera as *Dictyomitra*, *Stichomitra*, *Spongosaturnalis*, *Crucella*, *Praeconocaryomma*. It is difficult to suggest any quantitative domination by any species. Most common are: *Annikaella omanensis* De Wever, *Dictyomitra densicostata* Pessagno, *Cryptamphorella conara* (Foreman), *Pseudodictyomitra* aff. *nakasekoi* Taketani, *Spongosaturnalis* (?) *ichikawaji* Foreman, *Stichomitra perapedhia* Bragina, and *Triactoma fragilis* Bragina. On the other hand, *Afens lirioides* Riedel et Sanfilippo, *Artostrobium urna* Foreman, *Pseudoaulophacus lenticulatus* (White) are relatively rare.

Many species of this assemblage are extinct in the Campanian, including *Alievium superbum* (Squinabol), *Archaeospongoprunum bipartitum* Pessagno, *Artostrobium urna* Foreman, *Crucella cachensis* Pessagno, and *C. messinae* Pessagno. This indicates that the age of the assemblage cannot be younger then the Campanian. In addition, *Sciadiocapsa cansia* Foreman and *Theocapsomma comys* Foreman first appear in the Campanian.

Taxa such as *Alievium gallowayi* (White), *Patellula verteroensis* (Pessagno), *Pseudoaulophacus floresensis* Pessagno, *P. lenticulatus* (White), *Afens lirioides* Riedel et Sanfilippo, *Amphipyndax pseudoconulus* (Pessagno), *Artostrobium urna* Foreman and *Xitus asymbatos* (Foreman) are typical of the *Amphipyndax pseudoconulus* Zone of the tropical Campanian (Empson-Morin, 1981). This point of view can be supported by data on similar radiolarian assemblages from the vicinities of Perapedhi and Trimiklini villages (Osozawa and Okamura, 1993; Urquhart and Banner, 1995).

Nevertheless, some features of this assemblage allow us to suppose a Late Santonian-Early Campanian age for the studied deposits. First it should be noted that *Theocorys antiqua* (Squinabol) and *Squinabollum fossilis* (Squinabol) are extinct at the Santonian-Campanian boundary. Second, the relatively rare tests of *Amphipyndax pseudoconulus* (Pessagno), which are represented by specimens with a weakly developed nodose wall structure, were recovered together with the typical specimens.

The following species are present in this assemblage together with the above-mentioned taxa: *Acaeniotyle* sp., *Actinomma* sp., *Archaeospongoprunum distributum* Bragina, *Halesium* sp., *Multastrum regalis* Vishnevskaya, *Orbiculiforma vacaensis* Pessagno, *Patulibracchium* aff. *dickinsoni* Pessagno, *P.* aff. *lawsoni* Pessagno, *Phaseliforma carinata* Pessagno, *Praeconocaryomma irwini* Pessagno, *Praeconocaryomma* cf. *lipmanae* Pessagno, *P. universa* Pessagno, *Pseudoaulophacus praefloresensis* Pessagno, *Spongosaturnalis* aff. *horridus* (Squinabol), *S.* ex gr. *hueyi* Pessagno, *Stylosphaera pusilla* Campbell and Clark, *Amphipyndax stocki* (Campbell and Clark), *Archaeodictyomitra regina* (Campbell and Clark), *Bathropyramis* sp., *Cornutella californica* Campbell and Clark, *Dictyomitra* cf. *andersoni* (Campbell and Clark), *D. formosa* Foreman, *D. multicostata* Zittel, *D.* ex gr. *torquata* Foreman, *Neosciadiocapsa jenkinsi* Pessagno, *Stichomitra cechena* Foreman, *S.* aff. *insignis* (Squinabol), *Theocampe* (?) *cypraea* Bragina, *Theocampe* cf. *salillum* Foreman, and *Theocoronium subtriquetrus* Bragina.

Many species of this assemblage are similar to those from the coeval Californian assemblages. Approximately half of the

Table 1.5 Characteristic analyses of umbers and associated rocks from the vicinity of Perapedhi village

Sample	**K-1**	**K-2**	**K-4**	**K-6**	**B-1**
SiO_2	20.61	13.93	84.47	63.77	76.06
TiO_2	0.45	0.49		0.32	0.48
Al_2O_3	1.69	2.35		0.86	5.93
Fe_2O_3	5.08	14.62	11.23	26.85	4.98
FeO	0.63		0.38	0.12	0.14
MnO	3.35	16.25	0.09	0.28	0.09
MgO	0.48	0.94	0.17	0.11	0.47
CaO	2.45	20.56	0.68	1.82	2.17
Na_2O	0.22	0.60	0.07	0.15	0.76
K_2O	0.50	0.29	0.11	0.14	1.26
P_2O_5	0.42	0.11	0.14	0.57	0.20
H_2O^+	12.95	5.52	2.30	3.89	2.72
H_2O^-	1.31	7.20	0.38	0.79	1.26
CO_2		15.70			
Sum	**100.14**	**98.56**	**100.02**	**99.67**	**96.52**
Fe+Mn/Ti	154	78		100	13
Al+Fe+Mn	0.02	0.05		0.02	0.46
Cr	36	26	78	167	
Ni	117	89	30	41	
V	510	205	255	700	
Cu	43	70	32	130	
Pb	68	18	13	15	
Co	34	88	6	26	
Mo	11.2	1.5	7.3	18.5	
Ge	2.2	2.0	1.7	1.6	
Ga	10	10	10	10	
La	54	55	12	34	
Ce	35	20	16	54	
Sm	18	14	3.2	7.2	
Eu	3.8	3.0	0.9	1.4	
Tb	1.8	0.4	0.6	1.1	
Yb	6.4	4.5	1.6	2.6	
Lu	1.0	0.8	0.3	0.5	

Sample K1 — umber, K2 — manganese-bearing umber, K4 and K6 — cherts, B1 — clay from the Kannaviou Formation. Analyses were done in the Chemical laboratory of the Geological Institute of the Russian Academy of Sciences. Contents of oxides are shown in weight percents, contents of rare elements — in grams per ton.

taxonomic composition is represented by species known from the Great Valley Sequence (Pessagno, 1976), and approximately 25% are typical of the Coast Range formations (Foreman, 1968). On the other hand, the taxonomical affinity with tropical oceanic assemblages is not so high (Foreman, 1975; Empson-Morin, 1981, 1984; Sanfilippo and Riedel, 1985).

Other outcrops of the Perapedhi Formation are known westward along the southern flank of the Troodos Massif, near the villages of Pano Panagia and Kannaviou. They are also known within the area of distribution of the Mamonia Allochthonous Complex. In the central part of the Akamas Peninsula, the umbers of the Perapedhi Formation (0.5 m) lie upon the Upper Pillow Lavas (Fig. 1.22). The Perapedhi Formation is overlain in this section by bentonitic clays of the Kannaviou Formation. At the basal horizon of the Perapedhi Formation radiolarians *Amphipyndax tylotus* Foreman and *Patellula verteroensis* (Pessagno) were found. They indicate a considerably younger age, *i.e.* Late Campanian to Maastrichtian (Urquhart and Banner, 1995).

Table 1.6 Characteristic analyses of umbers and associated rocks from the basin of the Mavrokolymbos River

Sample	**51-1**	**51-4**	**51-11**	**51-19**	**51-20**
SiO_2	48.52	30.16	51.04	31.58	15.15
TiO_2	0.55	0.55	0.48	0.50	0.53
Al_2O_3	12.11	6.18	10.75	8.86	5.01
Fe_2O_3	5.82	42.76	13.95	31.99	60.86
FeO	3.52				
MnO	0.17	0.56	0.28	7.34	0.35
MgO	9.17	2.60	3.90	3.66	2.06
CaO	8.49	1.54	1.92	1.32	1.87
Na_2O	0.81	0.67	1.28	1.01	0.61
K_2O	0.89	1.34	2.32	1.51	0.74
P_2O_5	0.00	0.33	0.25	0.11	1.30
H_2O^+	3.48	8.01	6.03	7.41	9.41
H_2O^-	5.52	4.39	7.33	5.37	2.60
CO_2	0.45	0.50			
Sum	**99.50**	**99.59**	**99.53**	**100.66**	**100.49**
Fe+Mn/Ti	21	92	35	94	135
Al+Fe+Mn	0.48	0.10	0.36	0.14	0.06
Cr	100	1000	1000	1000	1000
Ni	220	200	130	220	470
Cu	110	560	73	540	280
Pb	20	210	50	200	340
Zn	50	260	110	15	230
Ba	53	22	320	66	13
As	15	100	10	260	54
Rb	15	48	39	28	55
Sr	84	99	260	200	88
Zr	10	48	17	110	53
Y	5	38	27	5	99
Nb	10	64	5	5	58
Th	30	220	30	210	410
La	10	15	22	36	66
Ce	10	15	10	10	15

Sample 51-1 — argillirated boninite, 51-19 — manganese-bearing umber, 51-4 and 51-20 — umbers, 51-11 — argillirated hyaloclastite. Analyses were done in the Chemical laboratory of the Geological Institute of the Russian Academy of Sciences. Contents of oxides are shown in weight percents, contents of rare elements — in grams per ton

Therefore, the sections of the Perapedhi Formation south of the Troodos Massif are characterized by discordant stratigraphic data based on radiolarians. The age of the Perapedhi Formation ranges from the Turonian (the eastern region, Mangaleni) to the Late Santonian-Early Campanian (the central region, Perapedhi), and to the Late Campanian (the western region, Akamas). This age discordance requires an explanation. It may be caused by the diachronous character of the lower boundary of the Perapedhi Formation. A more ancient (Turonian?) radiolarian assemblage could also be extracted from the pebbles of the basal horizon of the Perapedhi Formation in the type section. On the other hand, the Perapedhi Formation overlies volcaniclastic breccias in the type section, and lies directly on the Upper Pillow Lavas in the other sections. The age of the volcaniclastic breccias from the Perapedhi Section can differ from the age of the pillow lavas. These breccias are insufficiently studied and could be of Turonian to Santonian age. It should also be noted that the volcaniclastic sandstones of the western sections are considerably younger, and are dated as

Late Campanian-Maastrichtian (Osozawa and Okamura, 1993). Despite many unsolved problems, the obtained data show that the development of the Troodos Ophiolite Complex was completed before the end of the Santonian or beginning of the Campanian.

The most common lithologies of the Perapedhi Formation are umbers. They are massive or platy rocks with a variable brownish color. Typically they feature a high content of Fe and Mn and a relatively low content of rare elements (**Tables 1.5 and 1.6**).

Umbers from the southern flank of the Troodos Massif slightly differ in their chemical, mineralogic and petrographic composition from those that are known from the area of distribution of the Mamonia Allochthonous Complex. About 90% of the total volume of the former type of umber is represented by a cryptocrystallic iron-silica mass that is composed of layered silicate (dioctahedric smectite or/and mixed-layered mineral like smectite/illite), goethite, and uncrystallized substance. Cherty umbers yield chalcedony. These umbers are characterized by a high value of the titanium modulus $(Fe+Mn)/Ti > 90$, a relatively low value of the aluminium modulus $Al/(Al+Fe+Mn) < 0.1$, and with $FeO+MnO > 30\%$. Therefore, they can be assigned to ore-bearing deposits according to the classification of Butuzova (1989). The main volume of the second type of umber is composed of uncrystallized substance and mixed-layered mineral like smectite-illite with minor goethite. Nearly 25% of the total volume of these rocks is represented by an argillirated volcanoclastic substance (secondary clayey material). The second type of umbers is characterized by titanium modulus $(Fe+Mn)/Ti = 20\text{-}100$ and by relatively low of aluminium modulus $Al/(Al+Fe+Mn) = 0.1\text{-}0.2$, with $FeO+MnO = 14\text{-}40\%$. Therefore, they represent metalliferous sediments, according to the classification of Butuzova (1989).

Umbers from the southern flank of the Troodos Complex are associated with cherts, sometimes pyrite-bearing, phosphorites, radiolarites and rhodochrosites. Phosphorites (Kaleda and Krylov, 1992) (**Table 1.7**) are commonly represented by intercalations (up to 10 cm in thickness) within rocks with structure. Less common are phosphorites with the collomorph texture that forms crusts on the chert beds, and pebbles of phosporite-bearing limestones. The main minerals of phosphorites are fluorapatite and hydroxylapatite. The wilkeite and collophane are additionally present in the crusts, and francolite is present in the phosphate-bearing limestones. The hyaloclastites and grainy phosphorites are associated with umbers in the area of the Mamonia Allochthonous Complex. Hyaloclastites form intercalations (up to 0.5 m in thickness). They are composed of vitroclasts and fragments of basaltoids (up to 7-10 cm in size) that are practically replaced completely by dioctahedrical smectite like montmorillonite. They are characterized by low values of the titanium modulus (10-40) and relatively high values of the aluminium modulus (0.2-0.6).

The overall results prove that umbers of Cyprus have much in common with the recent hydrothermal sediments of the mid-oceanic ridges. Additionally, they show several important differences. The first important feature is the presence of associated phosphorites that are unknown in the recent mid-oceanic ridges. Phosphorites of the Perapedhi Formation are lithologically and mineralogically similar to the phosphorites of seamounts.

On the other hand, the Upper Pillow Lavas of the Troodos Massif have none of the characteristics of mid-oceanic ridge basalts. They are represented by boninite and calc-alkaline series (Sharaskin et al., 1980; Sobolev at al., 1986) that were formed under a subduction zone (Beccaluva et al., 1979). Some of the geochemical characteristics of the umbers also differ from those of typical hydrothermal deposits of the mid-oceanic ridges (MOR). The umbers are characterized by a low barium content, which is very common for the hydrotherms of MOR. Lead has a high content in umbers, but in the MOR hydrotherms its concentration is less than that of Cu and Zn. Additionally, umbers yield high concentrations of Th, Y, Nb, As, and Sr, which have a direct correlation with the content of Fe and Mn, and are caused by hydrogenic sedimentation of these elements as hydroxides of Fe and Mn.

The sediments of the Perapedhi Formation might also be the resulted of hydrothermal activity of the kind that occurs on seamounts, rather than that found on the typical mid-oceanic ridge. The prominent difference between umbers from the southern flank of the Troodos Massif and those from the area of the Mamonia Allochthonous Complex can be explained by reduced hydrothermal activity during the deposition of umbers from the Mamonia area.

The Kannaviou Formation

The Kannaviou Formation is represented by bentonite clays that lie upon the deposits of the Perapedhi Formation and on the Upper Pillow Lavas. The Kannaviou Formation is widely distributed along the southern flank of the Troodos Massif and in the area of the Mamonia Allochthonous Complex. Clays of the Kannaviou Formation are usually non-carbonate bearing and commonly yield abundant radiolarians that have primary importance for biostratigraphy. The thickness of the Kannaviou Formation is variable. Usually there are several dozen meters of clays, but in the type area near Kannaviou village their thickness can be estimated at one hundred meters.

In the section near Perapedhi village, the Kannaviou Formation has a tectonic contact with the underlying strata of the Perapedhi Formation (Bragina and Bragin, 1996) (Fig. 1.22). The Kannaviou Formation is represented by greenish-gray and brownish-gray silty clays, intercalated with light-gray or greenish-gray cherty siltstones with the radiolarians *Afens lirioides* Riedel and Sanfilippo, *Alievium gallowayi* (White), *Clathropyrgus* sp., *Crucella robusta* Bragina, *Dictyomitra densicostata* Pessagno, *D. formosa* Foreman, *D.* ex gr. *torquata* Foreman, *Patellula planoconvexa* (Pessagno), *Praeconocaryomma universa* Pessagno, *Pseudoaulophacus floresensis* Pessagno, *P. praefloresensis* Pessagno, *Rhopalastrum* cf. *tumidum* Lipman, *Amphipyndax pseudoconulus* (Pessagno), *Stichomitra* cf. *campi* (Campbell and Clark), and *Xitus asymbatos* Foreman, presumably implying the Campanian age. The observed thickness is 12 m.

The deposits of the Kannaviou Formation are widely distributed in the type area near Kannaviou village. The lower part is exposed along the Kannaviou-Pano Panagia road. Green

Table 1.7 Chemical analyses of phosphorites, associated with umbers

Sample	**163-90**	**167-90**	**640-90**	**646-90**	**3-87**	**637-90**
SiO_2	19.22	25.08	9.30	8.68	14.73	6.96
TiO_2	0.02	0.03	0.06	0.04	0.02	0.03
Al_2O_3	0.41	0.51	1.79	1.68	0.01	0.01
Fe_2O_3	3.35	3.47	11.57	13.18	5.85	8.34
MnO	1.35	0.71	1.62	1.26	5.39	2.10
MgO	0.33	1.91	1.12	1.53	0.64	0.46
CaO	36.80	36.79	35.05	35.54	34.35	39.78
Na_2O	0.46	0.58	0.58	0.58	1.04	0.66
K_2O	0.02	0.02	0.04	0.04	0.12	0.02
P_2O_5	24.13	22.66	23.53	23.81	22.27	26.44
H_2O^-	0.86	0.68	3.34	2.06	2.47	1.82
LOI	5.80	5.10	7.94	6.56	8.25	7.08
S=O	0.22	0.30	0.10	0.10	0.10	0.10
Sum	**92.95**	**97.84**	**95.94**	**94.94**	**97.58**	**93.69**
S	0.44	0.60	0.10	0.17	0.10	0.10
CO_2	1.98	1.98	3.52	2.42	4.30	3.74
F	2.67	2.67	0.16	0.33	1.06	1.65

Samples 163/90 and 167/90 are from the vicinity of Kannaviou village; samples 640/90, 646/90, 637/90, and 3/87 are from the vicinity of Perapedhi village. Samples 163/90, 167/90, 640/90, and 646/90 are from layered grain phosphorites. Sample 3/87 — phosphatised envelope on chert, sample 637/90 — pebble of phosphatized limestone. Analyses were done in the Chemical laboratory of the Geological Institute of the Russian Academy of Sciences. Contents are shown in weight percents.

non-layered bentonite clays with intercalation of clayey siltstones directly lie on the Upper Pillow Lavas and are overlain by greenish-gray bentonite clays, tuffaceous sandstones, and thin-layered light-gray cherty mudstones. The total thickness of the basal unit is 25 m.

The upper part of the Kannaviou Formation has been studied along the Kritou Marattou-Agios Dimitrianos road. The section has the following units (Bragina and Bragin, 1995) (Fig. 1.22):

1) The Kannaviou Formation consists of bentonite clays that are greenish-gray and massive, enclosing interlayers of light-gray cherty claystone and greenish-gray tuffaceous siltstones. A 2-m-thick horizon in the lower part of the outcrop consists of white and light-gray massive lithoclastic tuffs. Rocks above this horizon yielded the following radiolarians: *Afens lirioides* Riedel and Sanfilippo, *Amphipyndax pseudoconulus* (Pessagno), *A. stocki* (Campbell and Clark), *A. tylotus* Foreman, *Alievium gallowayi* (White), *A. superbum* (Squinabol), *Artostrobium* sp., *Crucella irwini* Pessagno, *Dictyomitra andersoni* (Campbell and Clark), *D. densicostata* Pessagno, *D. duodecimcostata* (Squinabol), *D. koslovae* Foreman, *Euchitonia santonica* Lipman, *Phaseliforma laxa* Pessagno, *Praeconocaryomma californica* Pessagno, *P. universa* Pessagno, *Pseudoaulophacus pargueraensis* Pessagno, *P. praefloresensis* Pessagno, *Rhopalosyringium* sp., *Stichomitra* cf. *campi* (Campbell and Clark), and *S.* ex gr. *cechena* Foreman. The apparent thickness is 30 m.

2) The Kathikas Formation consists of mixtites with a matrix consisting of brick-red sandy unlaminated clays. Inclusions are unrounded, unsorted and variable in size from dominant small (1-10 cm) blocks to large individual (1-5 m) blocks. They are mainly composed of the rocks of the Mamonia Complex: Vlambouros and Akamas sandstones, siliceous rocks of the Episkopi Formation, and volcanic rocks and limestones of the Dhiarizos Group. The lower boundary of the Kathikas Formation is sharp and irregular and has pockets. In the basal part of the formation, lenses of bentonite clays (up to 5 m in thickness) similar in composition to those from the Kannaviou Formation, bear the following radiolarians: *Acaeniotyle* ex gr. *diaphorogona* Foreman, *Afens lirioides* Riedel and Sanfilippo, *Alievium* sp., *Amphipyndax pseudoconulus* (Pessagno), *A. stocki* (Campbell and Clark), *Clathropyrgus titthium* Riedel and Sanfilippo, *Dictyomitra* ex gr. *duodecimcostata* (Squinabol), *Hemicryptocapsa conara* Foreman, *Histiastrum latum* Lipman, *Phaseliforma laxa* Pessagno, *Pseudoaulophacus floresensis* Pessagno, *P. pargueraensis* Pessagno, *P. praefloresensis* Pessagno, *Rhopalastrum tumidum* Lipman, *Septinastrum dogeli* Gorbovets, *Stylosphaera* ex gr. *pusilla* Campbell and Clark, and *Xitus asymbatos* (Foreman). The thickness is 100-150 m.

3) The Lower Lefkara Formation consists of limestones and marls that are white or light-gray, platy and cloddish, and

which bear foraminifers of the Late Maastrichtian (according to determinations by V. A. Krasheninnikov). The apparent thickness is 5-10 m.

In the lower course of the Mavrokolimbos River, 1 km above its mouth, there are outcrops of pillow-lavas overlain by Upper Cretaceous deposits. From the base upward, this section includes the following beds (Bragina and Bragin, 1995) (Fig. 1.22):

1) Weathered pillow-basalts. The observed thickness is 1 m.

2) Clays that are reddish to brown or dark-brown, laminated, sometimes platy, and which bear small lenses of yellow siliceous umbers. At their base, they fill pockets on the eroded surface of the basalts and include basalt pebbles. The thickness is 0-1 m.

3) Bentonite clays that are greenish-gray, massive, weakly laminated, sometimes platy, with interlayers of gray platy cherty shale (20-30 cm) and rare interlayers of gray siltstone (up to 1 cm). In the basal and higher parts of the bed, the following radiolarians occur: *Afens lirioides* Riedel and Sanfilippo, *Alievium gallowayi* (White), *A.* aff. *riedeli* Pessagno, *A.* sp., *Amphipyndax pseudoconulus* (Pessagno), *A. stocki* (Campbell and Clark), *Archaeodictyomitra regina* (Campbell and Clark), *Crucella* sp., *Dictyomitra duodecimcostata* (Squinabol), *D. formosa* (Squinabol), *D. koslovae* Foreman, *Dictyomitra torquata* Foreman, *Halesium sexangulum* Pessagno, *Patellula euessceei* Empson-Morin, *P. verteroensis* Pessagno, *Patulibracchium* sp., *Praeconocaryomma californica* Pessagno, *P. lipmanae* Pessagno, *P. universa* Pessagno, *Pseudoaulophacus pargueraensis* Pessagno, *P. venadoensis* Pessagno, *Rhopalosyringium* sp., *Stichomitra* ex gr. *alamedaensis* (Campbell and Clark), *S. campi* (Campbell and Clark), *Thanarla veneta* (Squinabol), *Theocampe urna* (Foreman), and *Xitus asymbatos* (Foreman). The first level of appearance of *Alievium superbum* (Squinabol), *Crucella messinae* Pessagno, *Dictyomitra multicostata* Zittel, *Pseudoaulophacus praefloresensis* Pessagno, *Rhopalastrum* cf. *tumidum* Lipman, and *Stichomitra cechena* Foreman is 0.5 m above the base. Another level (1.2 m above the base) is characterized by the first appearance of *Amphipyndax tylotus* Foreman and *Hemicryptocapsa* cf. *conara* Foreman, but taxa such as *Alievium superbum* (Squinabol), *Crucella messinae* Pessagno, and *Theocampe urna* (Foreman) disappear here. The thickness is 1.8 m.

4) Clays that are greenish-gray and massive with unclear lamination, comprising frequent interbeds (20-30 cm) of gray platy cherty claystones and scarce lensed interlayers of white flocculate tripoli. The first appearance of *Phaseliforma laxa* Pessagno and *Pseudoaulophacus floresensis* Pessagno is at a level of 4 m above the base. Higher up (5.5 m above the base), *Crucella irwini* Pessagno and *Rhopalastrum tumidum* Lipman appear, but *Phaseliforma laxa* Pessagno disappears here. Specimens of *Archaeospongoprunum bipartitum* Pessagno, *Dictyomitra densicostata* Pessagno, *Clathropyrgus titthium* Riedel and Sanfilippo, *Crucella* cf. *cachensis* Pessagno, *Histiastrum latum* Lipman, *Prunobracchium longum* Pessagno, and *Rhopalastrum* ex gr. *ingens* Lipman occur 6.5 m above the base, whereas *Dictyomitra multicostata* Zittel disappears here.

The first occurrences of *Acaeniotyle* ex gr. *diaphorogona* Foreman, *A.* ex gr. *umbilicata* (Rust), *Archaeospongoprunum venadoensis* Pessagno were noted 8 m above the base. This level is also marked by the disappearance of many of the species that occurred below: *Archaeospongoprunum bipartitum* Pessagno, *Crucella* cf. *cachensis* Pessagno, *Histiastrum latum* Lipman, *Prunobracchium longum* Pessagno, and *Rhopalastrum* ex gr. *ingens* Lipman.

The next level, 10 m above the base, revealed the first occurrence of *Acidnomelos* sp. and *Phaseliforma carinata* Pessagno, and the absence of *Acaeniotyle* ex gr. *diaphorogona* Foreman and *Rhopalastrum tumidum* Lipman as well as several taxa characteristic of Bed 3, such as *Crucella* cf. *messinae* Pessagno, *Dictyomitra duodecimcostata* (Squinabol), *Halesium sexangulum* Pessagno, and *Xitus asymbatos* (Foreman).

The upper level at 10.5 m marks the first occurrences of *Histiastrum aster* Lipman, *Hemicryptocapsa conara* Foreman, *Theocampe* ex gr. *ascalia* Foreman and the disappearance of *Acaeniotyle* ex gr. *umbilicata* (Rust), *Amphipyndax tylotus* Foreman, *Archaeospongoprunum venadoensis* Pessagno, *Clathropyrgus titthium* Riedel and Sanfilippo, *Phaseliforma carinata* Pessagno, *Pseudoaulophacus praefloresensis* Pessagno, and *Stichomitra* ex gr. *alamedaensis* (Campbell and Clark). The thickness is 13 m.

Thus, the Kannaviou and basal Kathikas deposits are characterized by representative radiolarian assemblages that provide insight into their stratigraphy. In tropical oceanic regions, the Campanian-Maastrichtian interval is currently subdivided into the *Amphipyndax pseudoconulus* (Campanian) and *Amphipyndax tylotus* (Late Campanian-Maastrichtian) Zones of the radiolarian scale (Sanfilippo and Riedel, 1985). These zones are also distinguishable in the Cyprus sections.

We identified the *Amphipyndax pseudoconulus* Zone in the basal part of Bed 3 of the Mavrokolimbos section. Considering the occurrences here of *Afens lirioides*, *Alievium gallowayi*, *Dictyomitra koslovae*, and *Pseudoaulophacus pargueraensis*, one could suggest that, in this particular section, we are dealing with the upper part of the oceanic *Amphipyndax pseudoconulus* Zone, which is characterized by these species (Sanfilippo and Riedel, 1985). However, it also contains *Theocampe urna* and *Alievium superbum*, neither of which are typical of the upper part of this zone (Sanfilippo and Riedel, 1985). In the Mavrokolimbos section, the zone has a very limited thickness (only up to 1.2 m). It may represent a condensed succession, which would explain the co-occurrence of species that are different in age. However, it is not inconceivable that the stratigraphic intervals of the above species in the section under consideration are somewhat different from those in the oceanic successions.

The base of the *Amphipyndax tylotus* Zone is distinctly marked by an index species that appears in Bed 3 of the Mavrokolimbos section at a level 1.2 m above the base. This index species and associated forms occur at higher levels in the Kannaviou Formation succession and within the basal part of the Kathikas Formation, where it encloses lenses of bentonite clays. Therefore, deposits of the *Amphipyndax tylotus* Zone have a considerably greater thickness than beds of the previous radiolarian zone. This may also indicate that only its upper part is present in this section. The first interpretation is inconsistent with the monotonous lithology of the bentonite clays that are observed all through the Kannaviou Formation section, including the *Amphipyndax tylotus* Zone, where these deposits are thick and do not display the lithologic features indicative of breaks in sedimentation, rewashing, and other events that could be responsible for the formation of a condensed succession.

Radiolarians of the oceanic *Amphipyndax tylotus* Zone include some species that are characteristic of its lower part and which became extinct during the Maastrichtian (Sanfilippo and Riedel, 1985). In the studied section, these radiolarians are *Afens lirioides*, *Alievium gallowayi*, *Clathropyrgus titthium*, *Dictyomitra koslovae*, *Pseudoaulophacus floresensis*, and *P. pargueraensis*. They obviously indicate that only the lower part of the zone (the Upper Campanian-Lower Maastrichtian) is represented in the described successions of the Kannaviou and Kathikas Formations. This is consistent with the fact that, near the village of Kannaviou, the Kathikas Formation underlies the Upper Maastrichtian carbonate deposits. It is also interesting that, in Cyprus, *Amphipyndax pseudoconulus* is distributed throughout the entire interval of the *Amphipyndax tylotus* Zone. In oceanic successions, this form became extinct in the very basal part of this zone, presumably before the Maastrichtian (Sanfilippo and Riedel, 1985). If this is true, and if *Amphipyndax pseudoconulus* has the same stratigraphic range in Cyprus, then we have to admit that the deposits with radiolarians of the *Amphipyndax tylotus* Zone correspond to a very narrow stratigraphic interval (the Upper Campanian) despite the considerable thickness of the sediments.

Similar stratigraphic data were obtained by other investigators. The Late Campanian radiolarians of the *Amphipyndax tylotus* Zone were found in the numerous sections of the Kannaviou Formation: in the Akamas Peninsula, Paleomylon Valley, Arkhimandrita village, and Paralimni village (Urquhart and Banner, 1995). In the Anadkhiu-Sarama section these authors also found the foraminifers *Pseudoguembelina costulata* (Cushman), *Pseudotextularia elegans* (Rzehak), *Spiroplecta globulosa* (Ehrenberg), *S. striata* (Ehrenberg), *Tesseraella planata* (Cushman), *Biglobigerina prairiehillensis* (Pessagno), *B. ultramicra* (Subbotina), *Contusotruncana fornicata* (Plummer), *Globotruncanita stuartiformis* (Dalbiez), *Globotruncana linneiana* (d'Orbigny), and *G. stephensoni* Pessagno, which were interpreted as Late Campanian (Urquhart and Banner, 1995).

The data obtained provide more refined age determinations for the Kannaviou and Kathikas Formations. Formerly, the Kannaviou Formation, which also bears *Globotruncana*, was considered as either a Maastrichtian (Mantis, 1970; Ealey and Knox, 1975; Blome and Irwin, 1985), or Campanian unit (Swarbrick and Robertson, 1980). Due to the recognition of the *Amphipyndax pseudoconulus* and *Amphipyndax tylotus* Zones of the radiolarian scale, we suggest that it corresponds to the stratigraphic range of the Campanian-Early Maastrichtian, bearing in mind that *Afens lirioides*, *Amphipyndax pseudoconulus*, *Dictyomitra koslovae*, and other forms that became extinct during the Maastrichtian occur here until the top of the formation. Swarbrick and Robertson (1980) referred the Kathikas Formation to the Maastrichtian. However, in the lower part it bears radiolarians of the Amphipyndax *tylotus* Zone, suggesting that this level is Late Campanian in age. Accordingly, this formation, which underlies the upper Maastrichtian carbonates, can be considered as Late Campanian-Early Maastrichtian in age.

Bentonite clays are the most important sediments of the Kannaviou Formation. Dioctahedric smectite, like montmorillonite but with 10-20% of mica-like layers is the main rock-constituent mineral of these clays. Less common are chlorite, palygorskite, opal-CT, and zeolites like heilandite or clinoptilolite. Tuffaceous sandstones and siltstones that are widely distributed in sections of the Kannaviou Formation near Kannaviou village are commonly massive, but some beds have horizontal or convolute layering, and even cross-bedding is present. These sediments are composed of angular grains of quartz, plagioclases that are usually zonal, and fragments of plagioclase-porphyrite volcanics with microlite and (sometimes) the trachyte structure of the main rock mass. Less common are grains of hornblende, hyperstene, clinopyroxene, biotite, cherts, carbonates, and vitroclasts. There are two type of vitroclasts:

1) non-colored, with numerous mineral inclusions, and with an index of refraction of 1.508-1.510, and

2) light-brown, porous, with an index of refraction of 1.522. The cement is clayey or carbonatic.

Clays of the Kannaviou Formation are characterized by high values of the aluminium modulus $Al/(Al+Fe+Mn) > 0.3$, by low total content of iron and manganese $Fe+Mn < 10\%$, and by low values of the titanium modulus $(Fe+Mn)/Ti < 40$. According to these distinctive features, the clays strongly differ from the underlying sediments of the Perapedhi Formation. We suggest that the bentonite clays were formed by the composite process of terrigenous sedimentation and the submarine weathering of volcanics.

The character of the contact with bentonite clay and the underlying pillow lavas is of primary importance for understanding this process. Usually the 1 m-thick upper part of pillow-lava is strongly weathered and the basalts are completely replaced by montmorillonite that has common characteristics with the montmorillonite from bentonite clays. The primary structures of the volcanics remain in this case. Therefore, we can suggest that a process, similar to that which forms bentonite clays, might occur in the replacing of volcanics by montmorillonite. Additionally, this suggestion gains support from the geochemical characteristics of the Kannaviou Formation, like its relative enrichment in nickel, that can be related to boninites.

On the other hand, the lithologic composition of clastics from the Kannaviou Formation shows an active input of acidic volcanic material. Such volcanic rocks are unknown among Troodos and Mamonia lithologies. This fact allows one to suggest (Robertson, 1977), that the tuffaceous sandstones of the Kannaviou Formation were formed by volcanoclastic material derived from some presently non-existing volcanic upland, probably a volcanic island arc. The wide distribution of volcaniclastic sediments southwest of the Troodos Massif (near Kannaviou village) can be interpreted as indicating the presence of a rift or another depression with composite clayey and clastic sedimentation in the Campanian and Maastrichtian.

REFERENCES CITED

Agar, S. M., and Klitgord, K. D., 1995. A mechanism for decoupling within the oceanic lithosphere revealed in the Troodos ophiolite. Nature, 374(6519), p. 232-238.

Allen, C. R., 1975. The Petrology of a Portion of the Troodos Plutonic Complex, Cyprus. Unpub. Ph.D. thesis, Cambridge University, 465 p.

Alt, J. C., 1991. Sulfur Isotopic Profile Through the Troodos Ophiolite, Cyprus: Preliminary Results and Implications. In: Tj. Peters et al. (eds.), Ophiolite Genesis and Evolution of the Oceanic Lithosphere, Ministry of Petroleum and Minerals, Sultanate of Oman, p. 427-443.

Anders, E., and Grevesse, N., 1989. Abundances of the elements: meteoritic and Solar. Geochimica et Cosmochimica Acta, 53 (1), p.197-214.

Bailey, D. G., Langdon, G. S., Malpas, J., and Robinson, P. T., 1991. Ultramafic and related lavas from the Margi area, Troodos ophiolite. In: I. L. Gibson, J. Malpas, P. T. Robinson and C. Xenophontos (eds.), Cyprus Crustal Study Project: Initial Report, Holes CY-1 and 1A. Geological Survey of Canada, Paper 90-20, p. 187-202.

Baragar, W. R. A., Lambert, M. B., Baglow, N., and Gibson, N., 1989. Sheeted dikes of the Troodos ophiolite. In: I. L. Gibson, J. Malpas, P. T. Robinson and C. Xenophontos (eds.), Cyprus Crustal Study Project: Initial Report, Hole CY-4. Geological Survey of Canada, Paper 99-9, p. 69-106.

Baragar, W. R. A., Lambert, M. B., Baglow, N., and Gibson, N., 1990. The sheeted dike zone in the Troodos ophiolite. In: J. Malpas, E. M. Moores, A. Panayiotou and C. Xenophontos (eds.), Ophiolites; Oceanic Crustal Analogues. Proc. Symp. Troodos 1987. Geological Survey Department, Nicosia, p. 37-51.

Batanova, V. G., Sobolev, A. V., and Schminke, H.-U., 1994. Mantle Lherzolites from Troodos Ophiolites: Mineralogy and ion probe geochemistry of clinopyroxenes. Mineralogical Magazine, 58A(A-K), p. 57-58.

Batanova, V. G., Sobolev, A. V., and Schmincke, H.-U., 1996. Parental melts of the intrusive cumulates of Troodos Massif, Cyprus: A study of clinopyroxenes and melt inclusions in plagioclase. Petrology, 4(3), p.255-264.

Bazylev, B. A., 1989. Metamorphism of the oceanic hyperbasites. Unpublished Ph.D. thesis, Vernadsky Institute of Geochemistry of the Russian Academy of Sciences, Moscow, 213 p. (In Russian).

Bear, L. M., 1960. The geology and mineral resources of the Akaki-Lithrodhonda area. . Cyprus Geological Survey Department, Memoir No. 3, 122 p.

Beccaluva, L., Ohnenstetter, D., and Ohnenstetter, M., 1979. Geochemical discrimination between ocean floor and island arc tholeiites - Application to some phiolites. Can. J. Earth Sci., 16(9), p. 1874-1882.

Bednarz, U., 1988. Volcanology, petrology, geochemistry and hydrothermal alteration in the northeastern Troodos Extrusive Series (Cyprus). Unpublished Ph.D. thesis, Ruhr-Universitaet Bochum, 195 p.

Bednarz, U., Götte, P., and Schmincke, H. -U., 1991. The Petrography of altered submarine lavas and major element mobility in Drillholes CY-1 and CY-1A, Cyprus. In: I. L. Gibson, J. Malpas, P. T. Robinson, and C. Xenophontos (eds.). Cyprus Crustal Study Project: Initial Report, Hole CY-4, Geological Survey of Canada, Paper 90-20, p. 95-116.

Bednarz, U., and Schmincke, H. -U., 1989. Mass transfer during sub-seafloor alteration of the upper Troodos crust (Cyprus). Contributions to Mineralogy and Petrology, 102(1), p. 93-101.

Bednarz, U. and Schmincke, H.-U., 1994. Petrological and geochemical evolution of the Northeastern Troodos Extrusive Series, Cyprus. Journal of Petrology, 35(2), p. 489-523.

Benn, K., and Laurent, R., 1987. Intrusive suite documented in Troodos ophiolite plutonic complex, Cyprus. Geology, 15(9), p. 821-824.

Bishopp, D. W., 1952. The Troodos Massif, Cyprus. Nature, 169(4299), p. 489-490.

Blome, C. D., and Irwin, W. P., 1985. Equivalent radiolarian ages from ophiolitic terranes of Cyprus and Oman. Geology, 13(6), p. 401-404.

Bonatti, E., Lawrence, J. R., and Morandi, N., 1984. Serpentinization of oceanic peridotites: temperature dependence of mineralogy and boron content. Earth Planet. Sci. Lett., 70(1), p. 88-94.

Boyle, J. F., and Robertson, A. H. F., 1984. Evolving metallogenesis at the Troodos spreading axis. In: I. G. Gass, S. J. Lippard and A. W. Shelton (eds.), Ophiolites and Oceanic Lithosphere. Geological Society, London, Special Publication 13, Blackwell Scientific Publications, p. 169-181.

Bragina, L. G., and Bragin, N. Yu., 1995. Radiolarians and stratigraphy of the Campanian-Maastrichtian deposits in South-western Cyprus. Stratigraphy and Geol. Correlation, 3(2), p. 147-155, Moscow. (In Russian).

Bragina, L. G., and Bragin, N. Yu., 1996. Stratigraphy and radiolarians of the type section of the Upper Cretaceous Perapedhi Formation (Cyprus). Stratigraphy and Geol. Correlation, 4(3), p. 38-45, Moscow. (In Russian).

Butuzova G. Yu., 1989. Types of modern hydrothermal and hydrothermal-sedimentary units in the active zone of the World Ocean Lithology and mineral resources, No 2, p. 3-23, Moscow. (In Russian).

Cameron, W. E., 1985. Petrology and origin of primitive lavas from the Troodos ophiolite, Cyprus. Contributions to Mineralogy and Petrology, 89(2-3), p. 239-255.

Coleman, R. G., 1971. Plate tectonic emplacement of upper mantle peridotites along continental edges. J. Geophys. Res., 76(5), p. 1212-1222.

Coleman, R. G., 1977. Ophiolites. Ancient Oceanic Lithosphere? Springer-Verlag, Berlin, Heidelberg, New York, 229 p.

Constantinou, G., 1987. Massive sulfide deposits. In: Field Excursion Guidebook, Symposium Troodos 87, Ophiolites and Oceanic Lithosphere, Nicosia, Cyprus, p. 87-109.

Constantinou, G., and Govett, G. J. S., 1972. Genesis of sulfide deposits, ochre and umbers of Cyprus. Trans. Inst. Mining and Metal., Sect. B, 81(783), p. 834-846.

Crawford, A. J., Falloon, T. J., and Green, D. H., 1989. Classification, petrogenesis and tectonic setting of boninites. In: A. J. Crawford (ed.), Boninites. Unwin Hyman, London, p. 1-49.

Danyushevsky, L. V., Sobolev, A. V., and Falloon, T. J., 1995. North Tongan high-Ca boninite petrogenesis: the role of Samoan plume and subduction zone-transform fault transition. Journal of Geodynamics, 20(3), p. 219-241.

Davies, J. H., and Bickle, M. J., 1991. A physical model for the volume and composition of melt produced by hydrous fluxing above subduction zones. Philosophical Transactions of the Royal Society of London, v. 335 (Ser. A), p. 355-364.

Desmet, A., Lapierre, H., Rocci, G., Gagny, Cl., Parrot, J.-F., and Delaloye, M., 1978. Constitution and significance of the Troodos sheeted complex. Nature, 273(5663), p. 527-530.

Dewey, J. F., and Bird, J. M., 1971. Origin and emplacement of the ophiolite suite: Appalachian ophiolites in Newfoundland. J. Geophys. Res., 76(14), p. 3179-3206.

Dick, H. J. B., Meyer, P. S., Bloomer, S., Kirby, S., Stakes, D., and Mayer, C., 1991. Lithostratigraphic evolution of an in-situ section of oceanic layer 3. In: R. P. Von Herzen, P. T. Robinson et al. (eds.). Proceedings of the Ocean Drilling Program, Scientific Results, v. 118, p. 439-538.

Dilek, J., and Moores, E. M., 1990. Regional tectonics of the eastern Mediterranean ophiolites. In J. Malpas, E. M. Moores, A. Panayiotou, and C. Xenophontos (eds.), Ophiolites. Oceanic Crustal Analogues, Proceedings of the Symposium Troodos 1987, The Geological Survey Department, Ministry of Agriculture and Natural Resources, Nicosia, p. 295-307.

Duncan, R. A., and Green, D. H., 1987. The genesis of refractory melts in the formation of oceanic crust. Contributions to Mineralogy and Petrology, 96(3), p. 326-342.

Ealey, P. J., and Knox, G. J., 1975, The pre-Tertiary rocks of S. W. Cyprus. Geol. en Mijnbouw., 54(1-2), p.85-100.

Elthon, D., Casey, J. F., and Komor, S., 1984. Cryptic mineral-chemistry variations in a detailed traverse through the cumulate ultramafic rocks of the North Arm Mountain Massif of the Bay of Islands ophiolite, Newfoundland. In: I. G. Gass, S. J. Lippard, and A. W. Shelton (eds.), Ophiolites and Oceanic Lithosphere, Geological Society, London, Special Publication 13, Blackwell Scientific Publications, p. 83-97.

Empson-Morin, K. M., 1981. Campanian Radiolaria from DSDP Site 313, Mid-Pacific Mountains. Micropaleontology, 27(3), p. 249-292.

Empson-Morin, K. M., 1984. Depth and latitude distribution of Radiolaria in Campanian (Late Cretaceous of tropical and subtropical oceans. Micropaleontology, 30(1), p. 87-115.

Engel, C. G., and Fisher, R. L., 1975. Granitic to ultramafic rock complexes of the Indian Ocean Ridge System, Western Indian Ocean. Geol. Soc. Amer. Bull., 86(11), p. 1553-1578.

Erzinger, J., Emmermann, R., and Mandler, M., 1991. Geochemistry of Altered Lavas of the Troodos Ophiolite, Cyprus crustal Study Project, Hole CY-2a. In I. L. Gibson, J. Malpas, P. T. Robinson, and C. Xenophontos (eds.). Cyprus Crustal Study Project: Initial Report, Hole CY-4, Geological Survey of Canada, Paper 90-20, p. 81-94.

Flower, M. F. J., and Levine, H. M., 1987. Petrogenesis of a tholeiite-boninite sequence from Ayios Mamas, Troodos ophiolite: evidence for splitting of a volcanic arc? Contributions to Mineralogy and Petrology, 97(4), p. 509-524.

Foreman, H., 1968. Upper Maastrichtian radiolaria of California. Paleontol. Assoc. Spec. Pap. Paleontology, London, No 3, p.1-82.

Foreman, H., 1975. Radiolaria from the North Pacific, Leg 32. In: K. J. Hsu, L. Montadert et al. (eds.), Initial Reports of the Deep Sea Drilling Project, v. 32, p. 579-676.

Gass, I. G., 1958. Ultrabasic pillow lavas from Cyprus. Geological Magazine, XCV(3), p. 241-251.

Gass, I. G., 1960. The geology and mineral resources of the Dhali area. Cyprus Geological Survey Department, Memoir No 3, 116 p.

Gass, I. G., 1968. Is the Troodos Massif of Cyprus a Fragment of Mesozoic Ocean Floor? Nature, 220(5162), p. 39-42.

Gass, I. G., 1980. The Troodos Massif: its role in the unraveling of the ophiolite problem and its significance in the understanding of constructive plate margin processes. In: A. Panayiotou (ed.), Ophiolites. Proc. Intern. Ophiolite Symposium Cyprus 1979. Geological Survey Department, Nicosia, p. 23-35.

Gass, I. G., and Masson-Smith, D., 1963. The geology and gravity anomalies of the Troodos Massif, Cyprus. Trans. of the Royal Soc. of London, v. 222 A, p. 417-467.

Gass, I. G., and Smewing, J. D., 1973. Intrusion, extrusion, and metamorphism at constructive margins: evidence from Troodos Massif, Cyprus. Nature, 242, p. 26-29.

Gillis, K. M., and Robinson, P. T., 1985. Low-Temperature Alteration of the extrusive Sequence, Troodos Ophiolite, Cyprus. Canadian Mineralogist, 23(3), p. 431-441.

Grand, T., Lapierre, H., and Mascle, G. H., 1993. Superimposed tectonics of the Cyprus Ophiolitic Massifs. Tectonics, 12(1), p. 93-101.

Greenbaum, D., 1977. The chromitiferous rocks of the Troodos ophiolite complex, Cyprus. Economical Geology, 72(7), p. 1175-1194.

Hebert, R., and Laurent, R., 1990. Mineral chemistry of the plutonic section of the Troodos ophiolite: New constraints for genesis of arc-related ophiolites. In J. Malpas, E. M. Moores, A. Panayiotou, and C. Xenophontos (eds.), Ophiolites. Oceanic Crustal Analogues, Proceedings of the Symposium Troodos 1987, The Geological Survey Department, Ministry of Agriculture and Natural Resources, Nicosia, p. 149-163.

Jagoutz, E., Palme, H., Baddenhausen, H., Blum, K., Cendales, M., Dreibus, G., Spettel, B., Lorenz, V., and Wanke, H., 1979. The abundances of major, and trace elements in the earth's mantle as derived from primitive ultramafic nodules. In: Abstracts of Papers submitted to the Tenth Lunar and Planet Sciences Conference, Part 2, p. 610-612.

Kaleda, K. G., and Krylov, K. A., 1992. Composition, geodynamic occurrence and genesis of the Cyprus umbers. Trans. (Doklady) of the Russian Academy of Sciences, 326(5), p. 865-870.

Karig, D. E., 1971. Origin and development of marginal basin in the Western Pacific. J. Geophys. Res., 76(11), p. 2542-2561.

Kelemen, P. B., Whitehead, J. A., Aharonov, E., and Jordahl, K. A., 1995. Experiments on flow focusing in soluble porous media, with applications to melt extraction from the mantle. J. Geophys. Res., 100(B1), p. 475-496.

Kidd, R. G. W., 1977. A model for the process of formation of the upper oceanic crust. Geophys. J. Roy. Astronom. Soc., 50(1), p. 149-183.

Kidd, R. G. W., and Cann, J. R., 1974. Chilling statistics indicate an ocean-floor spreading origin for the Troodos complex, Cyprus. Earth Planet. Sci. Lett., 24(1), p. 151-155.

Komor, S. C., Elthon, D., and Casey, J. F., 1985. Mineralogic variation in a layered ultramafic cumulate sequence at the North Arm Mountain Massif, Bay of Islands Ophiolite, Newfoundland. J. Geophys. Res., 90 (B9), p. 7705-7736.

Kostopoulos, D. K., and Murton, B. J., 1992. Origin and distribution of components in boninite genesis: significance of the OIB component. In: L. M. Parson, B. J. Murton and P. Browning (eds.), Ophiolites and their Modern Oceanic Analogues. Geological Society Special Publication, 60, p. 133-154.

Laurent, R., 1990. Parental magma and crystal fractionation modeling of the CY-4 plutonic rocks, Troodos ophiolite, Cyprus. In: J. Malpas, E. M. Moores, A. Panayiotou and C. Xenophontos (eds.), Ophiolites, Oceanic Crustal Analogues, Proceedings of the Symposium Troodos 1987, Geological Survey Department, Ministry of Agriculture and Natural Resources, Nicosia, p. 139-148.

Laurent, R., 1992. Peridotite intrusions emplaced in the fossil suprasubduction zone environment of Cyprus. In L. M. Parson, B. J. Murton and P. Browning (eds.), Ophiolites and their Modern Oceanic Analogues, Geological Society Special Publication, 60, p. 233-239.

MacLeod, C. J., 1990. Role of the Southern Troodos Transform Fault in the rotation of the Cyprus microplate: evidence from the Eastern Limassol Forest Complex. In: J. Malpas, E. M. Moores, A. Panayiotou and C. Xenophontos (eds.), Ophiolites: Oceanic Crustal Analogues, Proceedings of the Symposium Troodos 1987, Nicosia, 1987, Cyprus, p. 75-85.

Malpas, J., 1990. Crustal accretionary processes in the Troodos ophiolite, Cyprus: evidence from field mapping and deep crustal drilling. In: Ophiolites, Oceanic Crustal Analogues, Proceedings of the Symposium Troodos 1987, Geological Survey Department, Ministry of Agriculture and Natural Resources, Nicosia, p. 65-75.

Malpas, J. G., Calon, T., and Xenophontos, C., 1987. Plutonic rocks of the Troodos ophiolite. In: Field Excursion Guidebook, Symposium Troodos 87, Ophiolites and Oceanic Lithosphere, Nicosia, Cyprus, p. 158-181.

Mantis, M., 1970. Upper Cretaceous-Tertiary foraminiferal zones in Cyprus. Cyprus Scientific Researches, v. 3, p. 227-241.

McCulloch, M. T., and Cameron, W. E., 1983. Nd-Sr isotopic study of primitive lavas from the Troodos ophiolite, Cyprus: Evidence for a subduction-related setting. Geology, 11(12), p. 727-731.

Meijer, A., 1980. Primitive arc volcanism and a boninite series: examples from Western Pacific island arcs. In: D. E. Hayes (ed.), The Tectonic and Geologic Evolution of Southwest Asian Seas and Islands. Geophysical Monograph 23, American Geophysical Union, p. 269-282.

Meijer, A., and Reagan, M., 1981. Petrology and geochemistry of the island of Sarigan in the Mariana arc: calc-alkaline volcanism in an oceanic setting. Contributions to Mineralogy and Petrology, 77(4), p. 337-354.

Mehegan, J. M., 1988. Temporal, spatial, and chemical evolution of the Troodos ophiolite lavas, Cyprus: Supra-subduction zone volcanism in the Tethys Sea. Unpublished Ph.D. Thesis, Dalhousie University, Halifax, 700 p.

Mehegan, J. M., and Robinson, P. T., 1991. Lava groups and volcanic stratigraphy of the CCSP Boreholes CY-1 and CY-1A, Troodos ophiolite, Cyprus. In: I. L. Gibson, J. Malpas, P. T. Robinson and C. Xenophontos (eds.), Cyprus Crustal Study Project: Initial Report, Holes CY-1 and 1A. Geological Survey of Canada, Paper 90-20, p. 177-186.

Miyashiro, A., 1973. The Troodos ophiolitic complex was probably formed in an island arc. Earth Planet. Sci. Lett., 19(2), p. 218-224.

Miyashiro, A., 1975. Classification, characteristics, and origin of ophiolites. J. Geology, 83(2), p. 249-281.

Miyashiro, A., and Shido, F., 1980. Differentiation of gabbros in the Mid-Atlantic Ridge near 24°N. Geochemical Journal, 14(4), p. 145-154.

Moores, E. M., and Vine, F. G., 1971. The Troodos Massif, Cyprus and other ophiolites as oceanic crust: evaluation and implications. Phil. Trans. Roy. Soc. London, Ser. A, v. 268, p. 443-466.

Morel, S. W., 1960. The geology and mineral resources of the Apsiou-Akrotiri Area. Mem. Geol. Surv. Dep. Cyprus, v. 7, p. 51-88.

Mukasa, S. B., and Ludden, J. N., 1987. Uranium-lead isotopic ages of plagiogranites from the Troodos ophiolite, Cyprus, and their tectonic significance. Geology, 15(9), p. 825-828.

Murton, B. J., 1990. Was the Southern Troodos Transform Fault a victim of microplate rotation? In: J. Malpas, E. M. Moores, A. Panayiotou and C. Xenophontos (eds.), Ophiolites; Oceanic Crustal Analogues. Proceedings of the Symposium Troodos 1987. Geological Survey Department, Ministry of Agriculture and Natural Resources, Nicosia, p. 87-98.

Murton, B. J., and MacLeod, C. J., 1987. The Limassol Forest and Arakapas Fault Belt. In: Field Excursion Guidebook, Symposium Troodos 87, Ophiolites and Oceanic Lithosphere, Nicosia, Cyprus, p. 214-233.

O'Hanley, D. S., and Wicks, F. J., 1995. Conditions of formation of lizardite, chrysotile and antigorite, Cassiar, British Columbia. Canadian Mineralogist, 33(4), p. 753-773.

Osozava, S., and Okamura, M., 1993. New radiolarian ages from the Troodos ophiolite and their tectonic significance. Island Arc, 2(1), p. 1-6.

Panayiotou, A., 1987. An overview of the geology of Cyprus. In: Field Excursion Guidebook, Symposium Troodos 87, Ophiolites and Oceanic Lithosphere, Nicosia, Cyprus, p .7-22.

Pearce, J. A., 1975. Basalt geochemistry used to investigate past tectonic environments on Cyprus. Tectonophysics, 25(1-2), p. 41-67.

Pearce, J. A., 1980. Geochemical evidence for the genesis and eruptive setting of lavas from Tethyan ophiolites. In: A. Panayiotou (ed.). Ophiolites. Proc. Intern. Ophiolite Symposium Cyprus 1979. Geol. Survey Dep., Nicosia, p. 261-272.

Pearce, J. A., and Cann, J. R., 1973. Tectonic setting of basic volcanic rocks determined using trace element analyses. Earth Planet. Sci. Lett., 19(2), p. 290-300.

Pessagno, E. A., 1976. Radiolarian zonation and stratigraphy of the Upper Cretaceous portion of the Great Valley sequence, California Coast Ranges. Cushman Foundation for Foraminiferal Research, Spec. Publ., No 2, 95 p.

Portnyagin, M. V., Danyushevsky, L. V., and Kamenetsky, V. S., 1997. Coexistence of two distinct mantle sources during formation of ophiolites: a case study of primitive pillow-lavas from the lowest part of the volcanic section of the Troodos Ophiolite, Cyprus. Contributions to Mineralogy and Petrology, 128(2-3), p. 287-301.

Portnyagin, M. V., Kamenetsky, V. S., and Sobolev, A. V., 1993. A possible parental melt of the lower section of Troodos lava sequence. EUG VII Abst., Suppl. No 1. Terra Nova, p. 69.

Portnyagin, M. V., Magakyan, R., and Schmincke, H.-U., 1996. Geochemical Variability of Boninite Magmas: Evidence from Magmatic Inclusions in Highly Magnesian Olivine from Lavas of Southwestern Cyprus. Petrology, 4(3), p. 231-246.

Portnyagin, M. V., and Sobolev, A. V., 1993. Geochemical Zonation of the Lava Complex in the Troodos Ophiolite. Transactions of the Russian Academy of Sciences, 330(4), p. 212-217. (In Russian).

Portnyagin, M. V., and Sobolev, A. V., 1995. Melting beneath immature island-arc: Evidence from melt inclusions study of Cyprus volcanics. EOS Trans. AGU. Spring Meet. Suppl., p. S 267.

Portnyagin, M. V., Sobolev, A. V., Dmitriev, L. V., and Danyushevsky, L. V., 1994. Primitive melts of the Troodos ophiolite (Cyprus) trapped as inclusions by chromium spinels. 16th General Meeting IMA, Pisa, p. 334.

Rautenschlein, M., Jenner, G. A., Hertogen, J., Hofmann, A. W., Kerrich, R., Schmincke, H. -U., White, W. M., 1985. Isotopic and trace element composition of volcanic glasses from the Akaki Canyon, Cyprus: implications for the origin of the Troodos ophiolite. Earth Planet. Sci. Lett., 75(4), p. 369-383.

Robertson, A. H. F., 1975. Cyprus umbers: basalt-sediment relation in the Mesozoic ocean floor. J. Geol. Soc. London, 131(5), p. 511-531.

Robertson, A. H. F., 1977. Tertiary uplift history of the Troodos Massif, Cyprus. Geol. Soc. Am. Bull., 88(12), p. 1763-1772.

Robertson, A. H. F., 1990. Tectonic evolution of Cyprus. In: J. Malpas, E. M. Moores, A. Panayiotou, and C. Xenophontos (eds.), Ophiolites; Oceanic Crustal Analogues. Proc. Symp. Troodos 1987. Geological Survey Department, Ministry of Agriculture and Natural Resources, Nicosia, p. 235-250.

Robertson, A. H. F., and Hudson, J. D., 1973. Cyprus umbers: chemical precipitates on a Tethyan ocean ridge. Earth Planet. Sci. Lett., 18(1), p. 93-101.

Robertson, A. H. F., and Hudson, J. D., 1974. Pelagic sediments in the Cretaceous and Tertiary history of the Troodos Massif, Cyprus. In: K. J. Hsu, and H. C. Jenkins (eds.), Pelagic sediments on land and under the sea, Spec. Publ. Intern. Assoc, Sedimentol., v. 1, p. 403-436.

Robertson, A. H. F., and Woodcock, N. H., 1979. Mamonia Complex, Southwest Cyprus: evolution and emplacement of a Mesozoic continental margin. Geol. Soc. Am. Bull., 90(7), p.651-655.

Robinson, P. T., and Malpas, J., 1990. The Troodos ophiolite of Cyprus: New perspectives on its origin and emplacement. In J. Malpas, E. M. Moores, A. Panayiotou, and C. Xenophontos (eds.), Ophiolites. Oceanic Crustal Analogues, Proceedings of the Symposium Troodos 1987, The Geological Survey Department, Ministry of Agriculture and Natural Resources, Nicosia, p. 13-26.

Robinson, P. T., Melson, W. G., O'Hearn, T., and Schmincke, H.-U., 1983. Volcanic glass compositions of the Troodos ophiolite, Cyprus. Geology, 11(7), p. 400-404.

Rogers, N. W., MacLeod, C. J., and Murton, B. J., 1989. Petrogenesis of boninitic lavas from the Limassol Forest Complex, Cyprus. In: A. J. Crawford (ed.), Boninites. Unwin Hyman, London, p. 288-313.

Sanfilippo, A., and Riedel, W. R., 1985. Cretaceous radiolaria. In: Plankton Stratigraphy. Cambridge Univ. Press, Earth Sci. Ser., p. 573-630.

Saunders, A. D., and Tarney, J., 1979. The geochemistry of basalts from a back-arc spreading center in the East Scotia Sea. Geochimica et Cosmochimica Acta, 43(4), p. 555-572.

Saunders, A. Tarney, D., J., Marsh, N. G., and Wood, D. A., 1980. Ophiolites as ocean crust or marginal basin crust: A geochemical approach. In: A. Panayiotou (ed.), Ophiolites, Proceedings International Ophiolite Symposium Cyprus, 1979, Ministry of Agriculture and Natural Resources Geological Survey Department, p. 193-204.

Schiffman, P., Smith, B. M., Varga, R. J., and Moores, E. M., 1987. Geometry, conditions and timing of off-axis hydrothermal metamorphism and ore-deposition in the Solea graben. Nature, 325(6103), p. 423-425.

Schmincke, H.-U., and Bednarz, U., 1990. Pillow, sheet flow and breccia flow volcanoes and volcano-tectonic hydrothermal cycles in the extrusive series of the northeastern Troodos ophiolites, Cyprus. In: J. Malpas, E. M. Moores, A. Panayiotou, and C. Xenophontos (eds.), Ophiolites; Oceanic Crustal Analogues. Proc. Symp. Troodos 1987. Geological Survey Department, Ministry of Agriculture and Natural Resources, Nicosia, p. 185-206.

Schmincke, H.-U., Rautenschlein, M., Robinson, P. T., and Mehegan, J. M., 1983. Troodos extrusive series of Cyprus: A comparison with oceanic crust. Geology, 11(7), p. 405-409.

Searle, D. L., and Panayiotou, A., 1980. Structural implications in the evolution of Troodos Massif, Cyprus. In: A. Panayiotou (ed.), Ophiolites: Proceedings of the International Ophiolite Symposium, Cyprus, 1979. Cyprus Geological Survey Department, p. 50-60.

Simonian, K. O., and Gass, I. G., 1978. Arakapas fault belt, Cyprus: a fossil transform fault. Geol. Soc. Am. Bull., 89(8), p. 1220-1230.

Sharaskin, A. Ya., Dobretsov, N. J., and Sobolev, A. V., 1980. Marianites: the clinoenstatite-bearing lavas associated with the ophiolite assemblage of Mariana Trench. Ophiolites. Proceedings Intern. Ophiolite Symp. Cyprus, 1979, p. 473-479.

Sharaskin, A. Ya., Pustchin, I. K., Zlobin, S. K., and Kolesov, G. M., 1983. Two ophiolite sequences from the basement of the Northern Tonga Arc. Ofioliti, 8(3), p. 411-430.

Silantyev, S. A., 1985. Genesis of Hornblende Gabbro in Ophiolite Layered Complexes and Oceanic Crust. Geochimia, No 6, p. 761-773.

Silantyev, S. A., 1993. Metamorphic Complexes of the Eastern Mediterranean: Metamorphic Conditions, Protoliths, and Geodynamic Environment. Petrology, 1(4), p. 395-416.

Silantyev, S. A., 1995. Metamorphism of Contemporary Oceanic Basins. Petrology, 3(1), p. 20-30.

Silantyev, S. A., Bazylev, B. A., Udintsev, G. B., and Schenke, H. W., 1997. The Origin and Genetic Conditions of the Hyperbasite Complex of Gibbs Island, South Shetland Islands, West Antarctica. Petrology, 5(3), p. 278-290.

Sobolev, A. V., and Batanova, V. G., 1995. Mantle lherzolites of the Troodos ophiolite complex, Cyprus: Clinopyroxene geochemistry. Petrology, 3(5), p. 440-448.

Sobolev, A. V., and Chaussidon, M., 1996. H_2O concentrations in primary melts from island arcs and mid-ocean ridges: Implications for H_2O storage and recycling in the mantle. Earth Planet. Sci. Lett., 137(1-4), p. 45-55.

Sobolev, A. V., Dmitriev, L. V., Tsameryan, O. P., Kononkova, N. N., and Robinson, P. T., 1991. A possible primary melt composition for the ultramafic lavas of the Margi area, Troodos ophiolite, Cyprus. In: I. L. Gibson, J. Malpas, P. T. Robinson, and C. Xenophontos (eds.), Cyprus Crustal Study Project: Initial Report, Holes CY-1 and 1A. Geological Survey of Canada, p. 203-218.

Sobolev, A. V., Portnyagin, M. V., Dmitriev, L. V., Tsameryan, O. P., Danyushevsky, L. V., Kononkova, N. N., Shimizu, N., and Robinson, P. T., 1993a. Petrology of ultramafic lavas and associated rocks of the Troodos Massif, Cyprus. Petrology, 1(4), p. 331-361.

Sobolev, A. V., Tsameryan, O. P., Dmitriev, L. V., and Kononkova, N. N., 1986. Water-bearing komatite as a new type of komatiite melt and origin of ultramafic lavas of the Troodos Mountains, Cyprus. Transactions (Doklady) of the USSR Academy of Sciences, 286(1), p. 160-163. (In Russian).

Sobolev, A. V., Tsameryan, O. P., Portnyagin, M. V., Kamenetsky, V. S., Danyushevsky, L. V. Dmitriev, L. V., and Joron, Y.-L., 1993b. Geochemistry and mineralogy of the ultramafic magmas of the Troodos ophiolite Massif, Cyprus. Geochemistry International, 30(9), p. 25-43.

Staudigel, H., Gillis, K., and Duncan, R., 1986. K/Ar and Rb/Sr ages of celadonites from Troodos ophiolite, Cyprus. Geology, 14(1), p. 72-75.

Sun, S.-S., and McDonough, W. F., 1989. Chemical and isotopic systematics of oceanic basalts: implications for mantle composition and processes. In: A. D. Saunders and M. J. Norry (eds.), Magmatism in the Ocean Basins. Geological Society Special Publication, London, p. 313-345.

Swarbrick, R. E., and Robertson, A. H. F., 1980. Revised stratigraphy of the Mesozoic rocks of Southern Cyprus. Geol. Mag., 117(5), p. 547-563.

Tarney, J., and Marsh, N. G., 1991. Major and trace element geochemistry of Holes CY-1 and CY-4: implications for petrogenetic models. In: I. L. Gibson, J. Malpas, P. T. Robinson, and C. Xenophontos (eds.), Cyprus Crustal Study Project: Initial Report, Holes CY-1 and 1a. Geological Survey of Canada, Paper 90-20, p. 133-176.

Taylor, R. N., 1988. The stratigraphy, geochemistry and petrogenesis of the Troodos extrusive sequence, Cyprus. Unpublished. Ph.D. Thesis, Southampton Univ., 303 p.

Taylor, R. N., 1990. Geochemical stratigraphy of the Troodos extrusive sequence: temporal developments of a spreading center magma chamber. In J. Malpas, E. M. Moores, A. Panayiotou, and C. Xenophontos (eds.), Ophiolites. Oceanic Crustal Analogues, Proceedings of the Symposium Troodos 1987, The Geological Survey Department, Ministry of Agriculture and Natural Resources, Nicosia, p. 173-183.

Taylor, R. N., and Nesbitt, R. W., 1988. Light rare-earth enrichment of supra subduction-zone mantle: Evidence from the Troodos ophiolite, Cyprus. Geology, 16(5), p. 448-451.

Taylor R. N., and Nesbitt, R. W., 1995. Arc volcanism in an extensional regime at the initiation of subduction: a geochemical study of Hahajima, Bonin Islands, Japan. In: J. L. Smellie (ed.) Volcanism associated with extension at consumed plate margins. Geol. Soc. Spec. Publ., London, p. 115-134.

Thy, P., 1987. Magmas and magma chamber evolution, Troodos ophiolite, Cyprus. Geology, 15(4), p. 316-319.

Thy, P., 1990. Cryptic variation of a cumulate sequence from the plutonic complex of the Troodos ophiolite. In: J. Malpas, E. M. Moores, A. Panayiotou, and C. Xenophontos (eds.), Ophiolites; Oceanic Crustal Analogues. Proc. Symp. Troodos 1987. Geological Survey Department, Ministry of Agriculture and Natural Resources, Nicosia, p. 165-172.

Thy, P., and Esbensen, K. H., 1993. Seafloor spreading and the ophiolitic sequences of the Troodos complex: a principal component analysis of lava and dike compositions. J. Geophys. Res., 98(B7), p. 11799-11805.

Thy, P., Schiffman, P., and Moores, E. M., 1989. Igneous Mineral Stratigraphy and Chemistry of the Cyprus Crustal Study Project Drill Core in the Plutonic Sequences of the Troodos Ophiolite. In I. L. Gibson, J. Malpas, P. T. Robinson, and C. Xenophontos (eds.). Cyprus Crustal Study Project: Initial Report, Hole CY-4. Geological Survey of Canada, Paper 88-9, p. 147-185.

Thy, P., and Xenophontos, C., 1991. Crystallization orders and phase chemistry of glassy lavas from the pillow sequences, Troodos ophiolite, Cyprus. Journal of Petrology, 32(2), p. 403-428

Thurow, J., 1991. Upper Cretaceous radiolarians from Cyprus evidence for increased productivity/preservation during global paleooceanographic events In: Interrad VI. Abstracts. Univ. degli studi Firenze, p. 81.

Urquhart, E., and Banner, F. T., 1995. Biostratigraphy of the supra-ophiolite sediments of the Troodos Massif, Cyprus: the Cretaceous Perapedhi, Kannaviou, Moni and Kathikas Formations. Geol. Mag., 131(4), p. 499-518.

Vallier T. L., Jenner, G. A., Frey, F. A., Gill, J. B., Davis, A. S., Volpe, A. M., Hawkins, J. W., Morris, J. D., Cawood, P. A., Morton, J L., Scholl, D. W., Rautenschlein, M., White, W. M., Williams, R. W., Stevenson, A. J., and White, L. D., 1991. Subalkaline andesite from Valu Fa Ridge, a back-arc spreading center in southern Lau Basin: petrogenesis, comparative chemistry, and tectonic implications. Chemical Geology, 91(3), p. 227-256

Varga, R. J., and Moores, E. M., 1985. Spreading structure of the Troodos ophiolite, Cyprus. Geology, 13(12), p. 846-850.

Varga, R. J., and Moores, E. M., 1990. Intermittent magmatic spreading and tectonic extension in the Troodos Ophiolite: implications for exploration of black smoker-type ore deposits. In: J. Malpas, E. M. Moores, A. Panayiotou, and C. Xenophontos (eds.), Ophiolites; Oceanic Crustal Analogues. Proc. Symp. Troodos 1987. Geological Survey Department, Ministry of Agriculture and Natural Resources, Nicosia, p. 53-64.

Varga, R. J., Schiffman, P., Smith, B. M., Moores, E. M., and Hurst, S., 1987. Structure and hydrothermal metamorphism of the Solea graben, Troodos ophiolite. In: Field Excursion Guidebook, Symposium Troodos 87, Ophiolites and Oceanic Lithosphere, Nicosia, Cyprus, p. 182-209.

Vibetti, N. J., Kerrich, R., and Fyfe, W. S., 1989a. Oxygen and Carbon Isotope Studies of Hydrothermal Alteration in the Troodos Ophiolite Complex, Cyprus. In: I. L. Gibson, J. Malpas, P. T. Robinson, and C. Xenophontos (eds.), Cyprus Crustal Study Project: Initial Report, Hole CY-4, Geological Survey of Canada, Paper 88-9, p. 221-228.

Vibetti, N. J., Kerrich, R., and Fyfe, W. S., 1989b. Hypersaline Fluids Discovered in Troodos Ophiolite. In: I. L. Gibson, J. Malpas, P. T. Robinson, C. Xenophontos (eds.), Cyprus Crustal Study Project: Initial Report, Hole CY-4, Geological Survey of Canada, Paper 88-9, p. 229-234.

Wenner, D. B., and Taylor, H. P., 1974. D/H and O^{18}/O^{16} studies of serpentinization of ultramafic rocks. Geochimica et Cosmochimica Acta, 38(8), p. 1255-1286.

Wilson, R. A. M., 1959. The geology and mineral resources of the Xeros-Troodos area. Cyprus Geol. Surv. Dep. Mem., 1, 135 p.

Wood, D. A., 1979. Dynamic partial meting: its application to the petrogenesis of basalts erupted in Iceland, the Faeroe Islands, the Isle of Skye (Scotland) and the Troodos Massif (Cyprus). Geochimica et Cosmochimica Acta, 43(7), p. 1031-1046.

Xenophontos, C. 1989. Troodos Ophiolite - An Ancient Analogue of a Special Type of Marginal Basin. 28th IGC Abstract, Washington, v. 3, p. 388

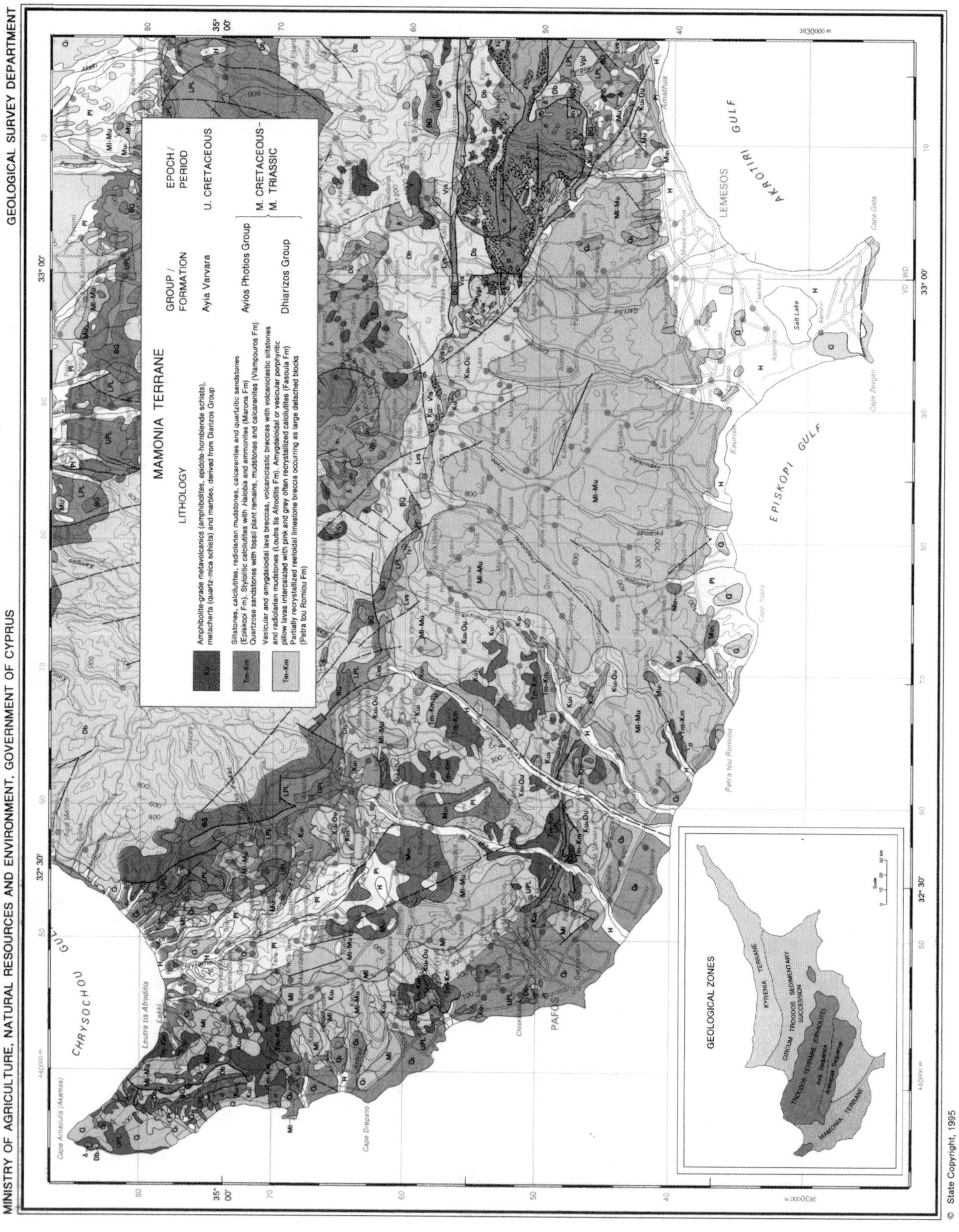

The Mamonia Terrane on the Island of Cyprus. Grayscale image of the colorful Geological Map of Cyprus (1995), modified to just show the Mamonia Terrane in southwestern Cyprus. This image will serve as a guide to the geological locations in Chapter 2.

Chapter 2

Igneous, Metamorphic and Sedimentary Rocks of the Mamonia Complex (Structural Allochthon)

Sergey A. Silantyev, Maxim V. Portnyagin, Boris A. Basylev
Vernadsky Institute of Geochemistry and Analytical Chemistry of the Russian Academy of Sciences, Kosygin Str. 19, 117975 Moscow, Russia
Nikita Yu. Bragin, Kirill A. Krylov, Lyubov G. Bragina
Geological Institute of the Russian Academy of Sciences, Pyzhevskiy per. 7, 109017 Moscow, Russia

A. Introduction

Sergey A. Silantyev
Vernadsky Institute of Geochemistry and Analytical Chemistry of the Russian Academy of Sciences, Kosygin Str. 19, 117975 Moscow, Russia

The allochthonous Mamonia terrain occurs in western and southwestern Cyprus. It occupies the southwestern periphery of the Troodos Massif and consists of igneous, metamorphic, and sedimentary rocks. These include radiolarites, volcanoclastics, bentonites, pelagic limestones, and marls as well as reefal limestones (Malpas and Xenophontos, 1987).

The Mamonia Complex has been subdivided into two main tectonostratigraphic units: the Ayios Photios Group, consisting mainly of Triassic to Mid-Cretaceous sediments and the Dhiarizos Group, consisting largely of pillow basalts, trachybasalts, and trachytes of Triassic-Jurassic age (Malpas and Xenophontos, 1987).

The Ayios Photios Group consists of Late Triassic to Middle Cretaceous quartz sandstones, calcareous-siliceous lutites/pelites and hemipelagic sediments, derived from a continental margin (Lapierre, 1975; Malpas and Xenophontos, 1987).

The Dhiarizos Group encompasses Triassic-Jurassic pillow lavas interbedded in shallow- and deeper-water sediments, among which pyroclastic tuffs and agglomerates are also present. (Malpas and Xenophontos, 1987). The Dhiarizos Group is considered to represent the sole of the stack of nappes. In many cases, the volcanics are alkali basalts, andesites, and trachytes, spatially juxtaposed with the fragmented reefal limestones. It appears, however, that the Dhiarizos volcanic assemblage is far more diverse than that suggested earlier, and consists, at least, of four geochemical rock types (Malpas and Xenophontos, 1987).

Both groups are juxtaposed to large serpentinite sheets and slivers of Troodos-type pillow lavas. The largest block of ophiolites derived from the Troodos Massif is exposed in the Akamas Peninsula, as a tectonic enclave (Lapierre, 1975). Metamorphic blocks incorporated within the serpentinite sheets are characteristic only of SW Cyprus. Amphibolites, hornblende-schists, and metasediments are the main components of these blocks. In many cases, metamorphic blocks are discordantly juxtaposed with the pillow lava packets. There is sharp contrast in the amount of deformation between the Mamonia Complex and the overlying, weakly disrupted sedimentary succession.

In a discontinuous arcuate belt extending from the Phasoula area to the Akamas Peninsula, Troodos serpentinites are closely associated with volcanics of the Dhiarizos Group. These lengthy serpentinite sheets are located along the E-W trending vertical fault zones. Their high-angle north and northeast dips are similar to those of the associated volcanic rocks of different ages.

In SW Cyprus, the serpentinites form two subparallel ridges, separated by a fault zone, where pillow lavas, fragmented volcanic and sepentinite breccias, and a conglomeratic association of volcanites, gabbroids, and serpentinites are all present. This brecciated complex resembles breccias of the Arakapas Fault Zone, which have been assigned to the talus slope deposits of a transform fault. Bastite pseudomorphs replacing original orthopyroxene grains are often present in the serpentinite. This mineralogical peculiarity together with structural features, implies that the protolith likely consisted of mantle harzburgite tectonites (Malpas and Xenophontos, 1987). In the southern ridge, serpentinites are associated with *en-echelon* oriented blocks of diabase and gabbro. Malpas and Xenophontos (1987) reported that the serpentinite massif in the Ezousa River valley is transected by branching and converging gabbroic dikes with chilled margins. They suggested that these serpentinites appeared on the oceanic basin floor in the environment of simultaneous high volcanic activity, vertical tectonic movements, pillow lava eruptions, and intrusion of gabbroic dikes and sills. Such a setting also resembles the scenario suggested for development of the Arakapas Fault Zone.

The Mamonia Complex rocks are generally covered by the neoautochthonous succession of the Late Cretaceous to Cenozoic sediments (calcarenite, limestone, sandstone, clay, gypsum and chert). Erosional windows reveal the pre-Campanian deformed igneous and sedimentary formations of the Mamonia Complex.

The Mamonia Complex tectonically overlies the southern edge of the Troodos Massif, where the Mamonia rocks are juxtaposed with bentonitic clays of the Kannaviou Formation, forming an integral part of the sedimentary surrounding of the Troodos Massif. These strongly deformed Campanian clays, volcanoclastic sandstones, and subordinate pelagic clays have a volcanoclastic origin but are not related genetically to the Troodos low-K volcanics, judging from petrographic and other data (Robertson, 1977).

Fragments of the Troodos-type pillow lavas are tectonically incorporated within the Mamonia Complex (Lapierre (1975). These lavas represent basaltoids highly depleted in HFSE, characteristic of the Arakapas Fault Zone and its surroundings (Malpas and Xenophontos, 1987). As with the Troodos pillow lavas, these are elsewhere interbedded with volcaniclastic deposits. The boninitic olivine phyric pillow lavas are present among the Late Cretaceous Troodos-type volcanics in SW Cyprus (Malpas and Xenophontos, 1987).

It is now proven that the western portion of the Arakapas Fault Zone interacted with the Mamonia Complex. Tectonic mixing of Arakapas Fault Zone rocks and Mamonia Complex rocks probably reflects rotation of the Troodos Microplate in the Late Cretaceous (Clube et al., 1985). In concordance with this concept, many structural elements of the Mamonia Complex, particularly large steeply-dipping faults, are significantly younger than proposed by previous workers.

In some areas of SW Cyprus, near the ophiolitic Troodos Massif, both the Kannaviou Formation deposits and deformed rocks of the Mamonia Complex are conformably overlain by the Kathikas Formation olistostrome. This contains heterolithologic fragments of almost all Mamonia lithologies, including cherts, sandstones and pillow lavas, set in a yellowish-pink argillaceous matrix (Malpas and Xenophontos, 1987). The structure of the Kathikas olistostrome is generally massive. Another variant on the stratification arrangement is observed in some areas, where thin horizons of pelagic limestones dated to the Maastrichtian occur among the olistostrome deposits (Swarbrick and Naylor, 1980). The Kathikas olistostrome appears to have been deposited within fault lineaments active in the Late Cretaceous. These are interpreted as wrench fault zones, along which the Troodos and Mamonia terrains became juxtaposed in pre-Cenozoic time. There is a manifestation of positive crustal vertical movements along these currently active lineaments, which could had led to uplift of the olistostromic deposits formed in a deep-water environment. With cessation of strike-slip faulting, the olistostrome complex was transgressively covered by pelagic carbonate sediments (Malpas and Xenophontos, 1987). In some areas, the presence of significant serpentinite masses makes it clear that extensional tectonism within the strike-slip deformation zones resulted in protrusion of serpentinites to the surface of the seafloor.

Although the allochthony of the Dhiarizos and Ayios Photios Groups is evident, their origin is controversial (Lapierre, 1975; Robertson, 1977; Clube and Robertson, 1986; Malpas and Xenophontos, 1987).

In SW Cyprus, along the strike of the serpentinite belts, small tectonic blocks, commonly less than 100 m long, of schistose amphibolites, with igneous and sedimentary protoliths, are juxtaposed with ultramafic rocks. The largest block of metamorphic rocks (about 1.6 km) belongs to the Ayia Varvara Formation, and is situated on the left bank of the Ezousa River north of Ayia Varvara village. Pervasive schistosity, associated with the juxtapositioning of serpentinites and amphibolites (Malpas and Xenophontos, 1987), is particularly notable in the serpentinite near the contact with these amphibolities.

According to Spray and Roddick (1981), the Ayia Varvara Block metamorphics are the products of low- and medium-pressure dynamothermal metamorphism of alkali and tholeiitic mafic rocks as well as associated sediments. The age of metamorphism of these rocks, determinated by the $^{40}Ar/^{39}Ar$ method, is 83 Ma to 90 Ma. These authors also suggested that the metamorphism of the Ayia Varvara Formation took place in an environment resembling modern oceanic transform fault zones during extensional strike-slip deformation associated with the development of the Arakapas transform structure. In this way, interaction of a newly formed heated oceanic crust, generated at a spreading center, with cool blocks of older rocks, derived from different segments of a mid-ocean ridge and involved by lateral displacements along the extensional transform, is a plausible model for the metamorphism of the Ayia Varvara metamorphics. Also, the frictional heat caused by strike-slip adjustment movements of crustal blocks within the wrench fault zone, influenced the temperature of metamorphism (Spray and Roddick, 1981). A tectonic origin for the alternations between metasediments and orthoamphibolites within the Ayia Varvara Block cannot be excluded (Malpas and Xenophontos, 1987).

According to Malpas and Xenophontos (1987), lithologic characteristics of the protolith of the meta-cherts and marbles reflect its formation in a shallow-water or deep-water marine basin, whereas the protolith of the quartz-feldspatic rocks is interpreted as being connected with a nearby continental provenance.

Perhaps, all these proposed protoliths are present as tectonic blocks within the adjacent melange in the Mamonia Complex. Diagnostic features of the tectonic juxtaposition of the two complexes under low-stress conditions involving highly viscous substrata are as follows: (a) the presence of large blocks of metamorphic schists within the serpentinites, (b) characteristic changes in orientation of schistosity or its smoothing-out at the margins of amphibolite bodies and (c) the occurrence of boudinage in serpentinites near their contacts with the amphibolites. In the opinion of Malpas and Xenophontos (1987), these features indicate the absence of a primary thrust-related contact between the ultramafic rocks and their underlying amphibolites.

In summary, the allochthonous Mamonia Complex includes the oldest sedimentary and volcanic rocks in this part of Cyprus, as well as diverse ultramafic and mafic plutonic rocks, and metamorphic rocks. Analysis of the geology of this complex is extremely important for understanding the Mesozoic geologic history of the whole Mediterranean. The structure of the Mamonia

Complex can be described below as follows: (1) sedimentary rocks of the Ayios Photios Group and the Dhiarizos Group; (2) volcanic rocks; (3) ultramafic rocks (hyperbasites), (4) mafic rocks (gabbroides), and (5) metamorphic (amphibolites, serpentinites) and metasedimentary rocks.

B. Sedimentary rocks of the Mamonia Complex

Nikita Yu. Bragin, Kirill A. Krylov and Lyubov G. Bragina
Geological Institute of the Russian Academy of Sciences, Pyzhevskiy per. 7, 109017 Moscow, Russia

Sedimentary rocks within the Mamonia Complex have been mainly identified in the Ayios Photios Group. In the Dhiarizos Group, consisting mainly of volcanic rocks, sediments are rather subordinate.

The Ayios Photios Group

The Ayios Photios Group consists of the Late Triassic Vlambouros Formation and the Middle Jurassic to Middle Cretaceous Episkopi Formation. These formations are quite distinct in lithology and genesis (**Fig. 2.1**).

The Vlambouros Formation

The Triassic Vlambouros Formation is represented mainly by sandstones and siltstones, commonly with plant remains (Robertson and Woodcock, 1979), interbedded by calcarenites and micritic limestones with remains of *Halobia* and radiolarians. In the south-eastern part of the Mamonia Terrane, on the left bank of the Dhiarizos River, 2.7 km to the east of Mamonia village, the top of the Vlambouros Formation consists locally of monotonous micritic limestones up to 20 m thick. Swarbrick and Robertson (1980) distinguished these limestones as the Marona Formation (**Fig. 2.2**). It is partly covered by recent slumps. In the northern and north-western parts of the region the Marona Formation is not present (Swarbrick and Robertson, 1980), and the Episkopi Formation directly overlies the Vlambouros Formation.

The deposits of the Vlambouros Formation were interpreted as rift-related sediments formed on the southern passive margin of the Tethys (Robertson and Woodcock, 1979; Kazmin et al., 1987). Permanent deep-water conditions of sedimentation during the Jurassic, Early Cretaceous and Cenomanian presumably prevailed after this rifting (Robertson, 1990). Similar sequences occur in adjacent areas – northwestern Syria (Delaune-Mayere et al., 1977) and southwestern Turkey (Robertson, 1990; Robertson and Woodcock, 1981).

Another interpretation of the Vlambouros Formation suggests its formation near the continental rise of the Turkey microcontinent on the northern passive margin of Tethys.

On the left bank of the Mavrokolymbos River, near the Argakin tou Vatkion Spring, the most complete section of the Vlambouros Formation is exposed and paleontologically well dated. It is represented by the following units (from base to top) (Fig. 2.1 and **Fig. 2.3**):

1) Yellowish-gray, coarse- to middle-grained, thick-bedded sandstones with well developed A-C Bouma intervals. The thickness of some beds is 1 m. Sandstones bear abundant plant detritus confined to the bedding surfaces and remains of *Halobia*

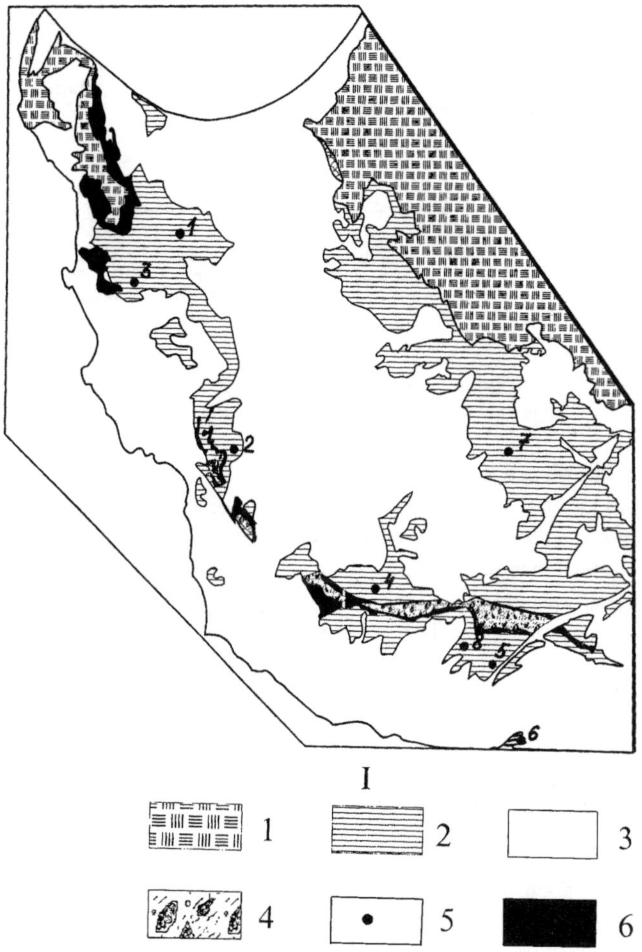

Fig. 2.1. Schematic map of distribution (exposures) of the Troodos Complex, Mamonia Complex and Neoautochthon (Maastrichtian-Cenozoic) rocks and location of the main sections of the Ayios Photios Group. 1- The Troodos Complex (Subautochthon); 2. - The Mamonia (Allochthonous) Complex; 3. - The Neoautochthon; 4. - Serpentinite melange (type 1); 5. - Sections of the Mamonia Complex; 6. - Serpentinite melange (type 2). Sections numbered: 1 - Phasli; 2 - the Mavrokolymbos River; 3 - Lara; 4 - Episkopi; 5 - Souskiou; 6 - Khapotami; 7 - Ayios Photios; 8 - the Xeropotamos River.

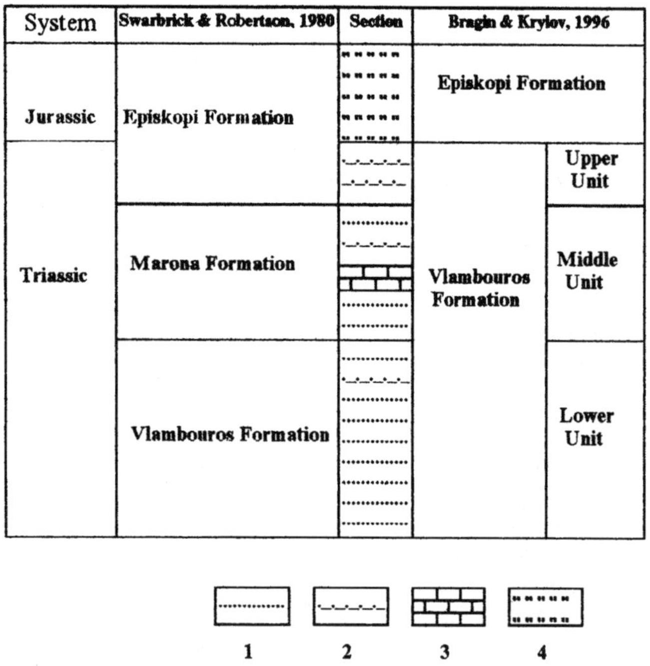

Fig. 2.2. Stratigraphic subdivision of the Vlambouros Formation. 1 - sandstone; 2 - siltstone; 3 - limestone; 4 - chert.

sp. The observed thickness is 2 m.

2) Light-gray middle- to fine-grained, massive and thick-bedded sandstones with numerous intercalations of greenish-gray siltstones and rare beds of light-gray micritic limestones. Sandstones have typical turbidity structures with distinctive B-C and B-C-D-E Bouma intervals. The limestones yielded the Carnian to Early Norian conodont *Neogondolella nodosa* (Hayashi). 10 m.

3) Greenish-gray, fine-grained sandstones and siltstones with intercalations of gray and pink micritic limestones and pink calcarenites. Limestones bear Late Carnian to Norian conodonts such as *Epigondolella* sp. 8 m.

4) Gray and reddish-gray sandstones, intercalated with greenish-gray, or, sometimes, reddish-gray, fine-grained siltstones, red cherts and pink calcarenites. Cherts yielded the Late Norian to Rhaetian radiolarians *Livarella gifuensis* Yoshida, *L.* sp., *Betraccium* sp. 10 m.

5) Red, middle- to fine-grained sandstones with carbonate cement, intercalated with brick-red siltstones. 5 m.

6) Gray, fine-grained sandstones, intercalated with brick-red siltstones and rare beds of red cherts with the Late Norian to Rhaetian radiolarians *Livarella* sp. 10 m.

These sediments are covered without visible discordance by red cherty mudstones with numerous thin (3 cm) intercalations of red and green cherts bearing the Callovian to Early Oxfordian radiolarians *Guexella nudata* (Kocher), *Hsuum brevicostatum* Ozvoldova, *Protunuma ochiensis* Matsuoka, *Zhamoidellum mikamense* Aita. They were considered as basal layers of the Episkopi Formation.

Near Lara Bay on the Akamas Peninsula a 20 m thick section of the Vlambouros Formation consists of gray, yellowish and reddish-gray, fine-grained, cross-bedded, platy sandstones with intercalations of gray, cross-bedded siltstones, pink calcarenites, gray and pink limestones and rare concretions of sandstone with carbonate cement. (Figs. 2.1 and 2.3). This sequence can be correlated with layers 3 and 4 of the Mavrokolymbos River section. The overlying sandstones, intercalated with siltstones (members 5 and 6 of the Mavrokolymbos River section), are absent here, and basal layers of the Episkopi Formation (the Callovian-Lower Oxfordian) are in direct contact with the middle part of the Vlambouros Formation. Unfortunately, there are no fossil findings from the sediments of the Vlambouros Formation at Lara Bay. They can be correlated only lithologically with the layers of other sections.

Upward follow red, brick-red, sometimes light-green, platy cherts of the Episkopi Formation with the Callovian-Early Oxfordian radiolarians *Eucyrtidiellum ptyctum* (Riedel et Sanfilippo), *Guexella nudata* (Kocher), *Haliodictya hojnosi* (Riedel et Sanfilippo), *Higumastra imbricata* (Ozvoldova), *Hsuum brevicostatum* (Ozvoldova), *Mirifusus guadalupensis* (Pessagno), *Parvicingula dhimenaensis* (Baumgartner), *Protunuma ochiensis* (Matsuoka), *Ristola altissima* (Rust), and *Zhamoidellum mikamense* Aita. 12 m.

Opposite the village of Souskiou, on the right side of the Dhiarizos River (Figs. 2.1 and 2.3), in the central part of the Mamonia Complex area, a strongly deformed, partly inverted sequence of the Vlambouros Formation can be partially seen. The section includes the following beds:

1) Gray, middle- to fine-grained, thick-bedded sandstones with plant detritus and rare intercalations of gray micritic limestones 10 m below an unexposed interval of 30-40 m

2) Gray and greenish-gray, fine-grained sandstones with frequent intercalations of pink calcarenites and gray micritic limestones with the Late Norian radiolarians: *Betraccium* cf. *deweveri* Pessagno et Blome, *Livarella* cf. *densiporata* Kozur et Mostler, *L. longus* Yoshida, *L.* sp., *Lysemelas olbia* Sugiyama, *Pentactinocarpus* cf. *sevaticus* Kozur et Mostler, *P. magnus* (Kozur et Mostler), *Praemesosaturnalis* aff. *gracilis* (Kozur et Mostler), and *P. multidentatus* (Kozur et Mostler). The observed thickness is 15 m.

3) Reddish-gray, fine-grained, cross-bedded sandstones, intercalated with reddish-gray siltstones and with rare beds of red cherts. The observed thickness is 5 m.

A similar section is exposed at the right side of the Ezousa River, 2 km downstream from Episkopi Village. Here, it consists of the following units (Figs. 2.1 and 2.3):

1) Gray, medium-grained, massive sandstones with interlayers of greenish-gray, middle- to fine-grained, sometimes silty, platy sandstones, which contain intercalations of greenish-gray and light-gray, micritic limestones. Massive sandstones yield intercalations of pink calcarenites and lense-like concretions of white and pink, micritic limestones with the Late Carnian-Late Norian conodonts *Epigondolella abneptis* (Huckriede). The observed thickness is 20 m.

2) Gray, middle-grained calcarenites (calciturbidites) and

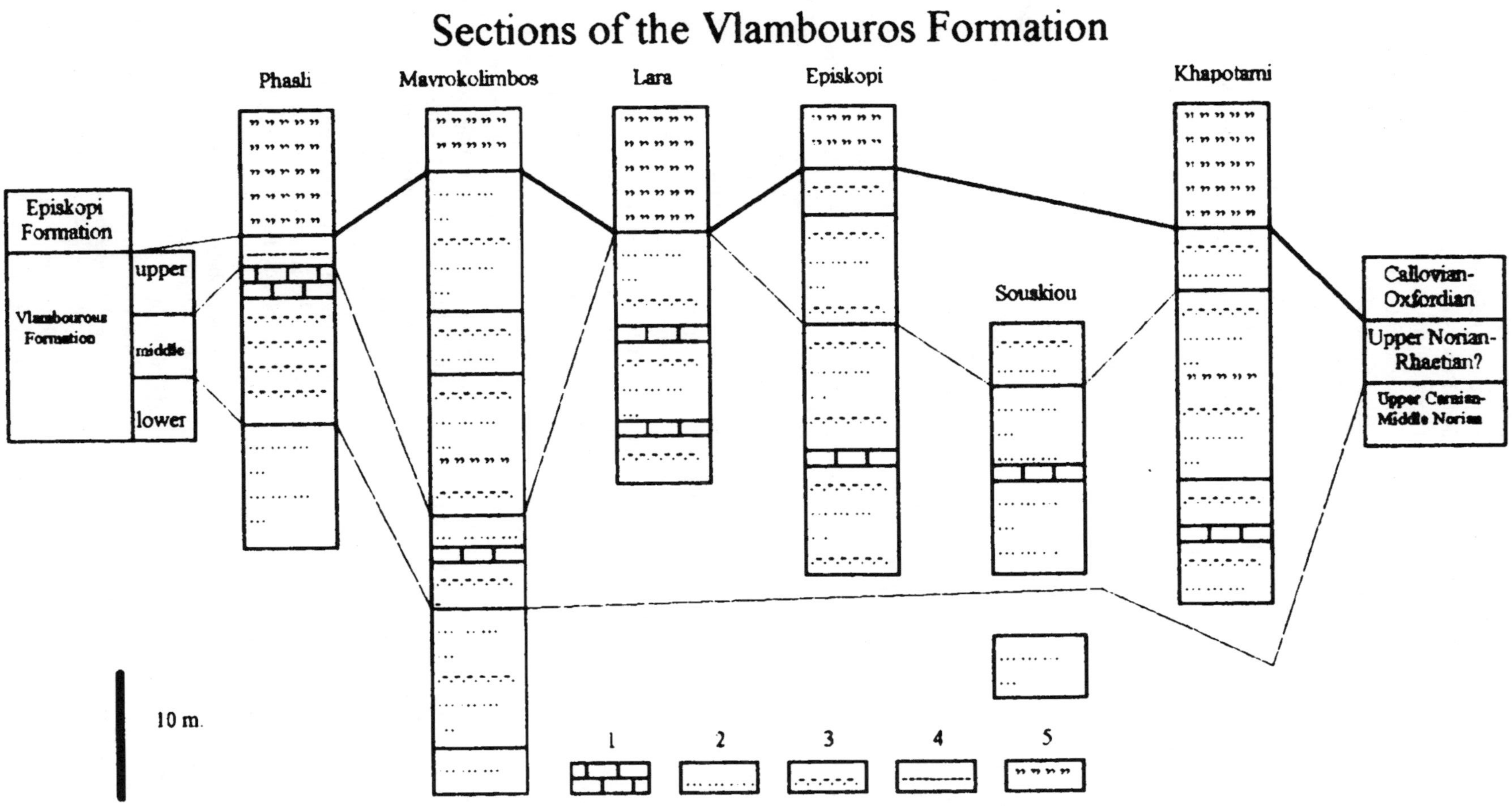

Fig. 2.3. Sections of the Vlambouros Formation and their correlation. 1 - limestone; 2 - sandstone; 3 - siltstone; 4 - mudstone; 5 - chert.

sandstones with carbonate cement and frequent intercalations of brown and brick-red siltstones alternating with rare beds of gray micritic limestones and pink cherts, which yielded the Late Norian-Rhaetian radiolarians: *Betraccium* sp., and *Livarella densiporata* Kozur et Mostler. The observed thickness is 8 m.

3) Gray, fine-grained, silty, cross-bedded, gradational sandstones interbedded with reddish-brown mudstones with thin interlayers of siltstones. The observed thickness is 3 m.

Following this are 3.5 m of greenish, pink, and brick-red, sometimes banded cherts, intercalated with brick-red, cherty mudstones belonging to the Episkopi Formation. Cherts bear the Callovian-Early Oxfordian radiolarians: *Higumastra imbricata* (Ozvoldova), *Hsuum brevicostatum* (Ozvoldova), and *Protunuma ochiensis* Matsuoka.

On the right side of the Khapotami River, 3 km upstream, the Vlambouros Formation displays a slightly different lithology:

1) Reddish-gray and gray siltstones interbedded with gray, micritic limestones, pink, cross-bedded calcarenites and gray, fine-grained sandstones. The observed thickness is 10 m.

2) Gray and light-gray, fine-grained sandstones with sinuous bedding and intercalations of light-gray siltstones, pink and light-red cherts, and pink, micritic limestones with the Late Norian radiolarians: *Betraccium* cf. *deweveri* Pessagno et Blome, *Livarella longus* Yoshida, *Praemesosaturnalis multidentatus* (Kozur et Mostler), and *Pentactinocarpus* cf. *sevaticus* Kozur et Mostler. The thickness is 15 m.

3) Gray and light-gray, fine-grained, thin-bedded sandstones, intercalated with reddish-gray siltstones. The thickness is 5 m.

The overlying Episkopi Formation consists of red and brick-red, calcitized cherts with numerous thin intercalations of red cherty mudstones, with the Callovian-Early Oxfordian radiolarians: *Guexella nudata* (Kocher), *Hsuum brevicostatum* (Ozvoldova), and *Protunuma ochiensis* Matsuoka. 20 m.

The correlation of the studied sequences allows us to subdivide the Vlambouros Formation into three main members (Fig. 2.2):

Member 1. Middle-grained sandstones with plant detritus and rare beds of micritic limestones, which contain the Late Carnian-Early Norian conodonts: *Epigondolella abneptis* (Huckriede) and *Neogondolella nodosa* (Hayashi). The presence of the Middle Norian deposits is not inconceivable. The thickness is 10 m.

Member 2. Fine-grained sandstones and siltstones with numerous intercalations of limestones, calcarenites and (sometimes) cherts with the Late Norian radiolarians *Betraccium* cf. *deweveri* Pessagno et Blome, *Livarella longus* Yoshida and conodont *Epigondolella bidentata* Mosher. The thickness is less than 20 m.

Member 3. Reddish, fine-grained sandstones and siltstones with rare beds of cherts with Late Norian-Rhaetian radiolarians. The thickness varies from 0 to 20 m.

The reddish sandstones and siltstones of member 3 were interpreted by previous investigators as the basal layer of the overlying Episkopi Formation (Robertson and Woodcock, 1979; Swarbrick and Robertson, 1980). Lithological and paleontological evidence allows us to include this member into the Vlambouros Formation. This conclusion also touches upon the problem of the monotonous limestones. In some sections these were distinguished by previous scientists (Robertson and Woodcock, 1979; Swarbrick and Robertson, 1980) as the lithologically distinct Marona Formation, intercalated between the Vlambouros Formation in the definition of Robertson and Woodcock (1979) and the red-colored clastic rocks of our member 3 of the Vlambouros Formation. In our interpretation the Marona Formation is the facial equivalent of member 2 of the Vlambouros Formation (Fig. 2.2). The middle part of the Vlambouros Formation is characterized by an abundance of carbonates. Unfortunately, the type section of the Marona Formation is now covered by a landslide.

In the Akamas Peninsula, 2 km south-southwest of Fasli (Fig. 2.3) a section of the Vlambouros Formation containing a carbonate middle member occurs as follows:

1) Gray and yellowish-gray, middle- to fine-grained, platy sandstones with plant detritus and intercalations of pinkish-gray and greenish-gray thin-bedded siltstones. The thickness is 20 m.

2) Pink and white, platy, micritic limestones with styloliths and thin intercalations of yellowish-gray mudstones. Limestones are characterized by the Late Norian conodonts *Epigondolella bidentata* Mosher, and *E. abneptis* (Huckriede). The thickness is 2 m.

3) Red, thin-bedded mudstones. The observed thickness is 2 m.

This sequence is covered by the Episkopi Formation, that consists of 10 m of red, platy cherts with thin intercalations of red and brick-red, cherty mudstones. Cherts contain the Callovian-Early Oxfordian radiolarians *Guexella nudata* (Kocher), *Hsuum brevicostatum* (Ozvoldova), and *Protunuma ochiensis* Matsuoka.

All three members of the Vlambouros Formation are present in this section.The middle member consists of pure limestones. Their thickness is similar to the total thickness of limestones in the middle member of other sections. It seems more logical to consider the Marona Formation as a facial equivalent of the middle member of the Vlambouros Formation (Fig. 2.2).

Previous investigators pointed out that the amount of coarse-grained clastics in the sections of the Vlambouros Formation is constantly decreasing from north-west to south-east (Swarbrick and Robertson, 1980). This is only correct to some extent. The Khapotami River section is relatively depleted in sandstones, and more abundant in limestones and cherts. Nevertheless, other sections are almost similar in these characteristics, and there are no basis for speculating on the decreasing ratio of coarse-grained clastics from north-west to south-east, only because of the notable exception of the Khapotami section. Moreover, the lower part of the Vlambouros Formation in the Dhiarizos River section is represented by the most coarse-grained facies among all the sections. This probably indicates that the Vlambouros Formation has a more complicated facies development than was previously supposed. Its facies variations fit well the model of turbidite fans. Rare and thin intercalations of radiolarian cherts can be interpreted in this case as background sediments in a relatively deep basin.

The middle and upper members of the Vlambouros Formation yield a radiolarian assemblage typical of the *Betraccium deweveri* Zone (Blome, 1984; Carter, 1993) that defines the

Upper Norian. It includes the following taxa: *Betraccium* cf. *deweveri* Pessagno et Blome, *Neopaurinella* cf. *sevatica* Kozur et Mostler, *Livarella longus* Yoshida, *L.* cf. *densiporata* Kozur et Mostler, *L. gifuensis* Yoshida, *L.* sp., *Lysemelas olbia* Sugiyama, *Pentactinocarpus* cf. *sevaticus* Kozur et Mostler, *P. magnus* (Kozur et Mostler), *Praemesosaturnalis* aff. *gracilis* (Kozur et Mostler), *P. multidentatus* (Kozur et Mostler), *P.* aff. *multidentatus* (Kozur et Mostler), *P. quadriradiatus* (Kozur et Mostler), and *P. zapfei* (Kozur et Mostler). This assemblage can be correlated with coeval Late Norian assemblages from the Alps (Kozur, Mostler, 1972, 1979, 1981), Eastern Russia (Bragin, 1991), Japan (Yao, 1982; Yoshida, 1986), and the western regions of North America (Blome, 1984). The low taxonomic diversity of this assemblage should be noted. Representatives of the family *Pantanelliidae* are very rare, with exception of the genus *Betraccium*. Cyrtoid nassellarians are also rare and the genus *Canoptum*, which is common in this time interval, is absent. There is a certain quantitative predominance of the genera *Livarella* and *Praemesosaturnalis*. This may point to rather semi-pelagic environmental conditions. Sometimes, when the taxonomic diversity was very low, a wider Late Norian-Rhaetian age range could not be excluded, though the presence of Rhaetian deposits could not be established with certainty in the Vlambouros Formation.

Exceptionally interesting and unusual relations between the Vlambouros and overlying Episkopi Formation suggest a very long hiatus, which includes an approximately Hettangian-Bathonian stratigraphic interval. This hiatus can be traced in all studied sections and considered as regional. Two interpretations can be proposed for this phenomenon:

1) The gap in sedimentation during the entire Early and part of the Middle Jurassic resulted from abyssal conditions similar to those in recent depositional areas of red oceanic clays, with a very low sedimentation rate;

2) Erosion of the Lower and Middle Jurassic deposits.

The probability of one of these scenarios occurring can only be ascertained by analysis of the lithologic evolution of the Vlambouros Formation based on new stratigraphic data.

The lower member of the Vlambouros Formation is represented by an alternation of thick-bedded, sandy turbidites. Their individual rhythms can have a thickness of 1 m. In this case, the A-C intervals of the Bouma turbidity cycle (Bouma, 1962) are distinct. More fine-grained sandstones with typical turbidity structures showing elements B-C, B-C-D, and more rarely B-C-D-E, appear in the middle part of the lower member. The micritic limestones, which can be interpreted as the horizon E of a turbidity cycle or the typical background basin sediment (Walker, 1967), appear at the same level. The upper part of the lower member is represented by the alternation of middle- to fine-grained sandstones with calcilutites and calcarenites. The Bouma elements C-D-E are most common here, and the elements B-C-D and A-B-C are rare. Generally, this part of the section shows a gradual decrease of thickness of the separate rhythms (sandstone beds), a gradual change of the grain size of the sandstones from coarse- and middle- to fine-grained (as Bouma cycle horizons A and B become rare), and the appearance of calcarenites in the upper part of the member. It should be noted that some sediments of the lowermost part of the Vlambouros Formation, common in the Akamas Peninsula, were interpreted by several authors as very shallow deltaic deposits, *e.g.* the products of very quick sedimentation (Robertson and Woodcock, 1979).

Petrographic investigation of the sandstones permits their classification into quartz subarkoses, litharenites, and calcarenites. The subarkoses are composed of angular and semi-rounded quartz grains with subordinate grains of plagioclase, K-feldspar, perthite, hornblende, epidote, mica, and rare ore and heavy minerals. Some quartz grains have a Fe-oxide coating that suggests a long history of transportation or multistage redeposition. The cement consists largely of clay minerals (mostly hydromicas) building up to 20-30% of the rock volume. The features allow us to refer these sandstones to the subarkosic wacke type (Middleton, 1970; Middleton and Hampton, 1973). Apart from quartz, feldspar and micas, litharenites contain angular fragments of carbonate rocks, acid volcanics, quartzite and micaceous schists; their Bouma intervals B and C bear fragments of coalified wood. Carbonate fragments often appear in various elements of the turbidity cycles, giving rise to calcarenites and lithocalcarenites.

Oncolitic sandstones were also noted. They form up to 1 m thick beds, composed of several layers, ranging around 10-15 cm in thickness. The oncolite cores are represented by quartz grains or by the afore-mentioned rock fragments. This structure of the oncolites suggests that they were formed in a shallow-water area, and were redeposited by turbidity or autokinetic flows (Walker, 1967). Oncolites constitute a small percentage of all the sandstones in the lower member of the Vlambouros Formation.

The middle member of the Vlambouros Formation is distinguished by an abundance of micritic limestones, calcarenites and cherts, which form numerous intercalations among the sandstones and siltstones. Sandstones of this member are petrographically similar to this rock type in the lower member, and are predominantly of turbiditic origin. In addition to turbidity sandstones, this sequence contains fine-grained varieties with very fine sinuous to horizontal bedding and very sharp contacts between the layers indicating a very high degree of granulometric sorting. All of these features, as well as the basal structures, can be interpreted as the result of the redeposition of these sediments by bottom currents (Curray and Moore, 1974; Hampton, 1972; Hein, 1982; Hollister et al., 1974; Hubert, 1964). Such sandstones can be described as contourites, although the petrographic composition of contourites and turbidites is very similar.

It should also be noted that oncolitic sandstones are very rare at this level, and are completely absent among the contourites. The upper parts of the turbidity cycles often contain micritic limestones, representing the E Bouma interval. These limestones are usually pink in color, and contain numerous remains of radiolarians and scattered shells of *Halobia*. These features, as well as the distribution of the calcareous intercalations relative to turbidity cycles, indicate that the micritic limestones are the product of normal background sedimentation, which was disturbed by the simultaneous formation of turbidites and contourites.

The upper member of the Vlambouros Formation has an essentially different lithology. This member is characterized by thin interbedding of fine-grained sandstones, siltstones, mudstones, calcarenites and calcilutites. Here, two types of litharenites were observed. The first one is represented by thin beds (2-5 cm) with sinuous- to cross-bedded, well sorted materials, while graded bedding is missing. Abrupt changes in bedding with sharp sandstone/siltstone boundaries as well as flow grooves are typical. The second type is represented by thicker (5-10 cm) beds, which may show C-D and B-C-D Bouma intervals. In addition to indistinct graded bedding, there is pendulum, cross, and criss-cross bedding in the uppermost rocks. Sandstone layers with very thin parallel bedding sometimes occur. Very small scale laminations (1-2 mm) can be noticed in such beds from a distance of several dozens of centimeters. The petrographic composition of all these sandstones is identical to that in the underlying clastic rocks.

It should be noted that calcarenites are more widespread in the upper member. These rocks are predominantly fine- and medium-grained. They contain a clastic fraction (0-30%) of quartz, feldspars, acid volcanics, quartz-mica schists, quartzites, microquartzites, and siliceous rocks. Carbonate material is represented by micritic grains, pellets, radiaxial and tangential oolites, recrystallized foraminifera, shell detritus, and sponge spicules. This material is undoubtedly of shallow-water origin. Intraclasts are represented by fragments of organoclastic limestones, marls and micritic limestones.

Calcilutites usually have fine laminar-flow structures due to the changing grain size from micritic to calcarenitic. Graded bedding is typical for thin layers, which sometimes contain coarse-grained carbonate or terrigenous clastic material and shell filaments. Terrigenous material is represented by rare grains of quartz and plagioclase, and by more frequent mica flakes, oriented parallel to bedding planes. Carbonate material is represented mostly by calcite grains. Carbonate micritic grains are rare, scattered, and contain radiolarian remains replaced by calcite. Some varieties contain fine-grained carbonate material composed of shell detritus and scattered calcitized radiolarian fragments. Thin bivalve shells are oriented along bedding planes. Some types of micritic limestones are bioturbated. They show clotty structures and calcitized feeding burrows. The upper parts of calcilutite beds usually exhibit silicification and gradual transitions to the radiolarian cherts. They can be interpreted as background sediments of a basin with normal salinity and aeration.

The petrography of clastic sediments of the Vlambouros Formation suggests that erosion affected the supracrustal metamorphic complex and was accompanied by the redeposition of shallow-water (shelf) carbonate sediments. Erosion of the metamorphic complex decreased with time, while the redeposition of carbonate sediments gained more importance. Background 'pelagic' sedimentation also took place, forming radiolaria-bearing micrite and chert beds. This sedimentation was of secondary importance. All these data allow us to conclude that the Vlambouros Formation was deposited in the marginal part of a basin, relatively close to a continental block or microcontinent, and surrounded by carbonate platforms. These conclusions are in agreement with the suggestion that the Vlambouros Formation and its equivalents in other regions, *e.g.* north-western Syria, were a part of the southern passive margin of the Tethys (Delaune-Mayere et al., 1977). Nevertheless, the exact position of the Vlambouros deposits in the southern areas of the Tethys is still questionable (Kazmin et al., 1987; Robertson, 1990; Robertson and Woodcock, 1981; Krylov et al., 1995).

The gradual transition from proximal to more distal turbidites, and then to contourites ends with the erosion and redeposition of turbiditic sediments by currents. Such changes can be attributed to rifting within the continental margins that accompanied the widening and deepening Neotethys.

The deposition of the Vlambouros Formation is also coeval with a large eustatic transgressive cycle. (Reding, 1990a, b). This is in agreement with the upward decrease in clastic material and the wider development of carbonate shelf areas, which were the provenance for the calciturbidite sediments of the Vlambouros Formation. The general transgression and widening of the basin probably changed the current patterns and caused the erosion of some sediments. This is supported by the observation of vertical changes in lithology and sporatic appearance of the upper member of the Vlambouros Formation, which is completely absent in some localities. It thus seems reasonable to believe that the stratigraphic hiatus between the Vlambouros and Episkopi Formations is a result of bottom current activity. At the same time, it should be noted that a hiatus, when caused only by currents, cannot span the entire Lower and Middle Jurassic everywhere over such a wide area.

Episkopi Formation

The Episkopi Formation overlies the Triassic Vlambouros Formation. Its sediments span the Callovian-Cenomanian interval, representing the most complete section of the Middle Jurassic-Lower Cretaceous deposits in the region.

In Lara Bay on the Akamas Peninsula, the Episkopi Formation overlies the Vlambouros Formation. It consists of the following subunits (Fig. 2.1, **Fig. 2.4**):

1) Red, cherry-red, rarely light-green, platy cherts with radiolarians *Acanthocircus suboblongus* (Yao), *A.* cf. *trizonalis* (Rust), *Amphipyndax* (?) aff. *tsunoensis* Aita, *Angulobracchia purisimaensis* (Pessagno), *Archaeodictyomitra* cf. *rigida* Pessagno, *Guexella nudata* (Kocher), *Higumastra imbricata* (Ozvoldova), *H.* cf. *inflata* Baumgartner, *Hsuum brevicostatum* Ozvoldova, *H. maxwelli* Pessagno, *H.* sp., *Mirifusus* cf. *guadalupensis* Pessagno, *M. dianae* (Karrer), *Palinandromeda* cf. *podbielensis* (Ozvoldova), *Paronaella mulleri* Pessagno, *Parvicingula* cf. *dhimenaensis* Baumgartner, *Protunuma ochiensis* Matsuoka, *P. turbo* Matsuoka, *Ristola altissima* (Rust), *Semihsuum sourdoughense* Pessagno, Blome et Hull, *Stichocapsa robusta* Matsuoka, *S. convexa* Yao, *S.* aff. *himedaruma* Aita, *Theocapsa* cf. *pentagona* Aita, *Tricolocapsa* cf. *parvipora* Tan, and *Tritrabs ewingi* (Pessagno), presumably of the Callovian-Early Oxfordian age. The thickness is 12 m.

2) Red-brown, coarse conglomerates with calcareous cement and pebbles of radiolarian chert, siliceous mudstone and micrite; enclosed are interlayers of cross-laminated calcarenite. Pebbles

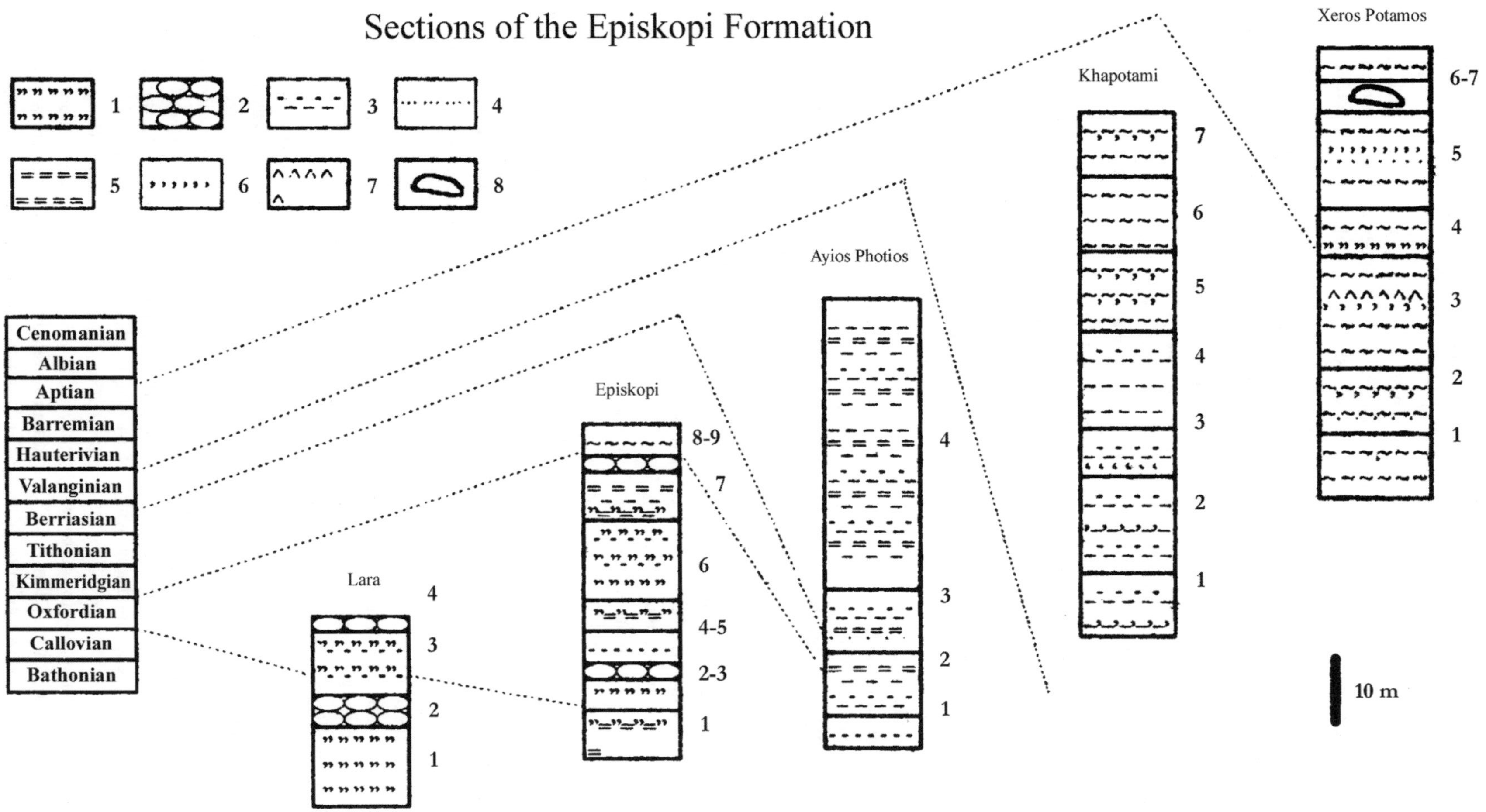

Fig. 2.4. Sections of the Episkopi Formation and their correlation. 1 - chert; 2 - conglomerate; 3 - siltstone; 4 - sandstone; 5 - calcarenite; 6 - cherty mudstone; 7 - tuff; 8 - Triassic limestone block.

have varying degrees of roundness. Moreover, the unit comprises fragments of radiolarite, containing remains of Jurassic radiolarians similar to those from Bed 1. The thickness is 4 m.

3) Light-gray and greenish cherts with interlayers of red cherts and siliceous green mudstone. Cherts yielded the radiolarians *Acanthocircus suboblongus* Yao, *Amphipyndax* (?) aff. *tsunoensis* Aita, *Archaeodictyomitra rigida* Pessagno, *Hsuum brevicostatum* Ozvoldova, *Hsuum* sp., *Stichocapsa robusta* Matsuoka, *S. himedaruma* Aita, *Stichomitra tairai* Aita, *Stylocapsa oblongula* Kocher, and *Theocapsomma cordis* Kocher, probably of Oxfordian age. The thickness is 10 m.

4) Red-brown conglomerates with calcareous cement, grading upward into cross-laminated calcarenites and banded green cherts enclosing the same radiolarian assemblage as below. The observed thickness is 3.5 m.

The lower reaches of the Mavrokolymbos River expose a similar section, only without coarse conglomerates. Calcarenite interlayers contain shallow water benthic foraminifers such as *Ammobaculites* cf. *braunsteini* Cushm. et Applin, *Haplophragmium* sp., *Haurania* cf. *desepta* Hens., *Kurnubia* sp., *Mayncina termieri* Hottin., *Orbitopsella* sp., *Pfenderina* sp., *Protopeneroplis striata* Weyn., and *Valvulina lugeoni* Sept. Radiolarians, presumably of Callovian-Early Oxfordian age similar to those from the section near Lara Bay, are also found in this locality. Foraminifers are certainly redeposited. This assumption is supported by the presence of benthic foraminifers only in calcarenite interlayers with abundant oolites, pellets, shell detritus, *i.e.* carbonate material of shallow-water origin.

The left slope of the Ezousa River valley, opposite the northern outskirts of Episkopi, exposes the type section of the Episkopi Formation, although only the lower part of the formation is exposed (Figs. 2.1 and 2.4). The contact with the Vlambouros Formation is not visible here. The succession of subunits from the base to top is as follows.

1) Red, pink, more rarely greenish and light-gray, platy, in places clayey cherts with interlayers of pink and green, silicified calcarenites include a cherty conglomerate horizon (0.5 m) in its upper part. The thickness is 6 m.

2) Alternation of radiolarian cherts, similar to those from the previous bed, with red and crimson-colored, more rarely greenish-gray pelitic cherts; the subunit is made of two beds (up to 0.5 m) of cherty conglomerates at the base and bears radiolarians similar to those found in the lower bed of the Lara Bay section, presumably of Callovian-Early Oxfordian age. The thickness is 5 m.

3) Coarse conglomerates and brown-gray to brown massive gravelstones with abundant detritus and fragments of redeposited chert and calcarenite beds. The lower contact of the subunit is abrupt and uneven. The calcarenite-calcirudite horizon (1 m) is located in the middle part. The thickness is 4 m.

4) Brick-red and reddish-brown loose siliceous mudstone with red clay interlayers. The thickness is 4 m.

5) Greenish-gray, silicified cherts and calcarenites with interlayers of greenish-gray, siliceous mudstone. The thickness is 3 m.

6) Reddish-brown and greenish-gray, siliceous mudstone with frequent interlayers of pelitic, reddish-brown cherts in the lower part and with red clay interlayers in the upper part. The thickness is 13 m.

7) Alternation of greenish-gray siliceous pelites, silicified greenish-gray and reddish calcarenites, greenish-gray radiolarites, and greenish-gray to reddish-brown mudstones. The thickness is 8 m.

Subunits 4-7 are characterized by the radiolarian assemblage of *Acanthocircus suboblongus* Yao, *Hsuum brevicostatum* Ozvoldova, *Protunuma ochiensis* Matsuoka, *Stylocapsa oblongula* (Kocher), *Stichocapsa himedaruma* Aita, and *Theocapsomma cordis* Kocher, presumably of the Oxfordian age.

8) Grayish-brown, cherty and carbonate-detrital lumpy gravelstones with fragments of redeposited chert and calcarenite beds. The lower contact is very uneven, truncating underlying beds and filled pockets. The thickness is 2 m.

9) Reddish-brown clay with interlayers of greenish-gray and white cherts bearing the radiolarians *Podocapsa amphitreptera* Foreman and *Spongocapsula perampla* (Rust), presumably Kimmeridgian -Tithonian. The observed thickness is 3 m.

Subunits 1-3 correspond to the interval exposed near Lara Bay. These beds represent the lower, mainly siliceous member of the Episkopi Formation. The intraformational conglomerates are found in other areas as well, except for the Khapotami River basin. Subunits 4-8 of the type section belong to the middle, predominantly clayey member of the formation. Subunit 9 of the type section is attributed to the upper part of the Episkopi Formation near Ayios Photios village (Figs. 2.1 and 2.4).

Along the road from Ayios Photios to Phalia, the section exposes the following subunits.

1) Alternation of light green-gray and pinkish-green calcarenites, in places banded and often silicified, with greenish-gray and red radiolarian cherts and brown-crimson siliceous mudstone. The subunit yields the Kimmeridgian-Tithonian radiolarians *Acaeniotyle diaphorogona* Foreman, *Alievium* sp., *Archaeospongoprunum* cf. *patricki* Jud, *Emiluvia orea* Baumgartner, *E. pessagnoi* Foreman, *E. premyogii* Baumgartner, *Lychnocanomma* cf. *xiphophora* (Rust), *Mirifusus dianae* (Karrer), *Paronaella* aff. *kotura* Baumgartner, *P.* ex gr. *pristidentata* Baumgartner, *Podobursa triacantha* (Fischli), *Podocapsa amphitreptera* Foreman, *Pseudocrucella adriani* Baumgartner, *Pseudoeucyrtis* sp., *Ristola altissima* (Rust), *Spongocapsula perampla* (Rust), and *S p.* sp. The radiolarians *Obesacapsula verbana* (Parona) and *Sethocapsa cetia* Foreman appear in the upper part of the bed. The thickness is 4 m.

2) Siliceous, brown-crimson thin-platy mudstone and siltstone with horizons (less than 0.05 m) of light greenish-gray banded calcarenites; radiolarians are *Archaeodictyomitra apiara* (Rust), *Alievium helenae* Schaaf, *Ditrabs sansalvadorensis* (Pessagno), *Emiluvia pessagnoi* Foreman, *Mirifusus dianae* (Karrer), *Parvicingula boesii* (Parona), *Podobursa polylophia* Foreman, *Podocapsa amphitreptera* Foreman, *Pseudodictyomitra carpatica* (Lozyniak), and *Xitus* aff. *alievi* (Foreman), most likely of Tithonian-Valanginian age. The thickness is 8 m.

3) Alternation of siliceous brown-crimson thin-platy conchoidal

mudstone and siltstone with gray and greenish-gray medium-grained massive quartz-feldspathic sandstone displaying graded bedding (the thickness of interlayers is up to 0.2 m), and with light greenish-gray banded calcarenites and pink-green compact conchoidal marls. Quartz-feldspathic sandstone differs in composition from that of the Triassic Vlambouros Formation and is similar to massive sandstone found in tectonic blocks in the Akamas Peninsula. The radiolarian assemblage is similar to that from the previous subunit. The thickness is 10 m.

4) Brown-crimson and reddish-brown cherty mudstones with thin (a few centimeters) interlayers of greenish-gray calcarenites that are repeated at intervals of approximately 1-1.5 m, and with rare intercalations of greenish-white and pink cherty marls. The radiolarian assemblage is represented by *Cecrops septemporatus* (Parona), *Mirifusus dianae* (Karrer), *Podobursa polylophia* Foreman, *Parvicingula cosmoconica* (Foreman), *Sethocapsa pseudouterculus* Aita, and *Xitus spicularius* (Aliev), indicating a Valanginian-Hauterivian age. The observed thickness is 45 m.

The upper part of the Episkopi Formation is exposed on the left slope of the Khapotami River, four km upstream from the previous exposure. It comprises from bottom to top:

1) Red and violet-gray cherty mudstones with rare interlayers of white radiolarite-spongolite cherts, sometimes silicified to flints. The thickness is 10 m.

2) Light-gray and greenish-gray cherty mudstones and clays with common interbeds of white spongolitic cherts and light-gray cherty siltstones. The thickness is 15 m.

3) Red, greenish-gray and violet-gray cherty mudstones and clays with frequent interbeds of white spongolitic cherts. The thickness is 7 m.

4) Red, violet-gray and gray cherty mudstones, with rare interbeds of black manganese-bearing cherts and occasional interlayers of white spongolitic cherts. The thickness is 15 m.

Subunits 1-4 yield the radiolarians *Holocryptocanium barbui* Dumitrica, *Thanarla conica* (Aliev), *T. pulchra* (Squinabol), *Pseudodictyomitra carpatica* (Lozyniak), and *Crolanium* sp. of Early Cretaceous age, presumably Barremian(?)-Albian.

5) Light-gray and greenish-gray clays and mudstones with interbeds of black manganese-bearing cherts. A 3 m thick layer of red and violet-gray cherty mudstones was observed in the upper part of the member. The thickness is 15 m.

6) Red, brick-red and violet-gray clays and mudstones with frequent interbeds of black manganese-bearing cherts and occasional interlayers of dark-gray bituminous thin-laminated mudstones. The observed thickness is 15 m.

7) Red, brick-red, sometimes greenish-gray and violet-gray clays and mudstones with occassional beds of black manganese-bearing cherts and light-gray spongolitic pelites with rare radiolarians *Holocryptocanium barbui* Dumitrica and *Thanarla* cf. *elegantissima* (Cita), that indicate late Early Cretaceous to Cenomanian age. The thickness is 15 m.

8) Red and brick-red clays with interbeds of black manganese-bearing cherts. The observed thickness is 10 m.

Swarbrick and Robertson (1980) hinted at the close similarity of these fine clastic and red-colored clayey deposits with the Mavrokolympos Formation. The Mavrokolymbos Formation, whose type section was described in the Xeros Potamos River valley by Swarbrick and Robertson (1980), was interpreted as a sedimentary sequence overlying the sheets of Triassic to Lower Jurassic volcanics of the Dhiarizos Group (Robertson and Woodcock, 1979). This section was restudied in 1990 during our field work in Cyprus. We noted that the sedimentary rocks of the type-section clearly have tectonic relationships with the underlying volcanics of the Dhiarizos Group. These volcanics contain numerous lenses, intercalations and interstitial fillings of limestones. These limestones yield the Upper Triassic conodonts *Epigondolella abneptis* (Huckriede) and *E. spatulata* (Hayashi). Thus, the Cretaceous sediments described here can more likely be interpreted as being the upper part of the Ayos Photios sedimentary sequence that has no sedimentary relationship with the Dhiarizos Group. Taking this point of view we can use the term 'Mavrokolymbos Formation' for the Middle Cretaceous argillaceous units of the Mamonia Complex.

The Episkopi Formation, exposed in the section of the Xeros Potamos valley 2 km north from Nikokleia and 3.5 km south-south-west of Choletria, is separated by a tectonic contact from the Upper Triassic volcanics (Fig. 2.2) and includes the following subunits in ascending order:

1) Red and green clays with rare intercalations of black manganese-bearing spongolites composed of more than 50% of sponge debris. The thickness is 10 m.

2) Red, light-red and green argillaceous clays with horizons of grayish-green, fine-grained sandstones and black spongolites in the upper part of the member. The thickness is 10 m.

3) Red clays with intercalations of greenish-gray and light-gray clays, white thin-bedded lithoclastic tuffs and black spongolites. The thickness is 20 m.

4) Greenish-gray and brick-red clays and mudstones with small lenses and intercalations of dark-gray spongolites, light greenish-gray siltstones, light-gray fine-grained sandstones and calcarenites, light-gray micritic limestones and pink radiolarian cherts with the following radiolarians: *Acaeniotyle* cf. *A. tribulosa* Foreman, *A. umbilicata* (Rust), *Archaeodictyomitra* ex gr. *A. vulgaris* Pessagno, *Dictyomitra gracilis* (Squinabol), *Holocryptocanium barbui* Dumitrica, *H. geysersensis* Pessagno, *Mita magnifica* Pessagno, *M.* sp. *A* (Pessagno, 1977), *Obeliscoides vinassai* (Squinabol), *Obesacapsula somphedia* (Foreman), *Pseudodictyomitra pseudomacrocephala* (Squinabol), *P. nakasekoi* Taketani, *P. paronai* (Aliev), *Spongosaturnalis horridus* (Squinabol), *S.* sp. ex gr. *S. horridus* (Squinabol), *Stichomitra communis* Squinabol, *Thanarla veneta* (Squinabol), *Torculum coronatum* (Squinabol), and *Xitus clava* (Parona), of Albian-Cenomanian age. Radiolarian cherts are pink to yellowish-pink, clayey, thin-bedded, with very thin lamination, and with visible radiolarians on the weathered surface. Their intercalations are less then 0.3 m thick. The thickness is 7 m.

5) Red and green clays with thin layers of black spongolites, with blocks of dark-gray spongolitic chert at the base of the member and

with horizons of brownish-gray fine-grained sandstone in the upper part of the member. The thickness is 15 m.

6) Reddish-brown mudstones with limestone blocks and with intercalations of gray, greenish-gray and pink marly radiolarian cherts containing Albian to Cenomanian radiolarians: *Acaeniotyle umbilicata* (Rust), *Dictyomitra gracilis* (Squinabol), *Holocryptocanium barbui* Dumitrica, *H. geysersensis* Pessagno, *Mita* sp., *Obeliscoides vinassai* (Squinabol), *Obesacapsula somphedia* (Foreman), *Pseudodictyomitra pseudomacrocephala* (Squinabol), *P. nakasekoi* Taketani, *Spongosaturnalis horridus* (Squinabol), *Stichomitra communis* Squinabol, *Thanarla veneta* (Squinabol), *Torculum coronatum* (Squinabol), and *Xitus clava* (Parona).

Small (0.5-2 m) blocks of light-gray micritic limestone yield Late Carnian-Early Norian (Late Triassic) radiolarians *Capnodoce anapetes* De Wever, *Capnuchosphaera tricornis* De Wever, *Triassocampe* sp., and *Xiphotheca* sp. These limestone blocks have an elongated shape and parallel the dip of the mudstones. The contacts between the mudstones and limestones are sharp and thin, without milonite or tectonic clay. The surfaces of the limestone blocks sometimes exhibit pockets filled by mudstone. Thus, these limestones show no clear evidence of tectonic contacts and cannot be the result of shearing. They can be interpreted as the result of gravity emplacement into the basin where the mudstones were accumulated. The thickness is 5 m.

7) Red clays and mudstones. The observed thickness is 5 m.

Summing up: The Episkopi Formation is represented by a relatively thick, up to 300 m, and complicated assemblage of sediments that represent different facies. The sequence can be subdivided to the following members:

1) Radiolarian cherts of the Callovian-Lower Oxfordian (up to 30 m).

2) Cherty mudstones of the Oxfordian (up to 30 m).

3) Alternating cherty mudstones, sandstones, calcarenites and radiolarites of the Kimmeridgian-Hauterivian (up to 70 m).

4) Cherty mudstones and clays with interbeds of spongolites of the Barremian-Albian (more then 50 m).

5) Clays and cherty mudstones with interbeds of spongolites, sandstones and radiolarites of the Albian-Cenomanian (up to 100 m).

Member 1 shows a fine rhythmic alternation of extended radiolarite beds up to 5-7 cm thick with siliceous mudstones (discontinuous beds up to 2-3 cm thick). Calcilutite interlayers are few and far between. Horizons of coarse, calcareous-siliceous conglomerates are also found. They are formed due to destruction and washout of accumulating sediments. Diverse radiolarian assemblages occur everywhere. The sediments appear to be accumulated in a deep basin with calm sedimentation conditions, *i.e.* below the carbonate compensation depth (CCD), only rarely disturbed and perturbated by distal calciturbidites but sometimes interrupted by redeposition, with formation of intraformational conglomerates. During the formation of the conglomerates, the radiolarites were slightly lithified resulting in the appearance of deformed fragments of radiolarite beds among pebbles of siliceous rocks. The carbonate matrix did not seem to be at all lithified. These structural peculiarities could reflect formation on a slope and respective detachment, sliding and redeposition of the fragments.

The deposits are similar in type with synchronous sequences of the western Mediterranean region (Baumgartner, 1984; Aita, 1987; Baumgartner et al., 1995). The presumably Callovian-Lower Oxfordian radiolarian assemblage is characterized by the presence of *Guexella nudata*, *Protunuma turbo*, and *Theocapsomma* cf. *pentagona*, together with many transitional species. The composition of the assemblage indicates open marine basin conditions. However, shallow-water areas of the carbonate shelf were not far from the region of siliceous mud accumulation. The areas served as a source of carbonate material for calciturbidites and conglomerates, and redeposition of benthic foraminifers took place as well. All this indicates the washout of shelf deposits. Calciturbidites also include terrigenous clastics, though in lesser amounts than found in the Triassic deposits. This material is usually well-rounded and suggests a supracrustal rock complex subjected to erosion. In some sections the presence of horizons with signs of sliding of unconsolidated sediments shows that sediments accumulating on the bottom had a low slope gradient of less than 1°.

Siliceous mudstones are dominant in Member 2. They are usually red in color and represent pelitic background sediments bearing rare and isolated radiolarian remains and lacking carbonate or terrigenous sandy-silty admixture. Calcarenite and calcilutite interlayers containing foraminifers locally occur among mudstones. Interlayers of radiolarian cherts are less frequent, and radiolarian assemblages are impoverished as compared to those from Member 1. An increase in sedimentation rate and the higher proportion of fine clastic calciturbidites is apparent here and indicates a greater influx of clastic material from the shelf. This resulted in the impoverishment of radiolarian assemblages, which, nevertheless, can be correlated with the Oxfordian radiolarians of the western Mediterranean (Aita, 1987; Baumgartner, 1984; Baumgartner et al., 1995). The general trend of sedimentation in the basin did not change; it still developed below the CCD under conditions of good aeration and enhanced oxygen potential.

Member 3 has very diversified lithologies. Here the key rocks are siliceous mudstones similar to those from the previous level. The member also contains calcarenites and calcilutites in abundance. Some of them are rich in radiolarians of varying abundance and diversity. Radiolarians correlate with assemblages of the Kimmeridgian-Valanginian zones of the western Mediterranean region and the Atlantic Ocean (Baumgartner, 1984; Baumgartner et al., 1995). The quartz sandstones distinguished earlier as the Akamas beds (Robertson, Woodcock, 1979; Swarbrick, Robertson, 1979) and slightly silicified marl interbeds are typical of the member as well.

The further enhancement of the influx of terrigenous material is obvious for this stage. The quartz grains in the sandstone are very well rounded and have ferruginous envelopes indicating their prolonged transportation and derivation from sands, which experienced several cycles of sedimentation and redeposition as well as being subjected to aeolian erosion. In the Akamas Peninsula, quartz sandstones are represented by sporatic lumps and blocks 10-20 m thick irregularly distributed along the strike. Structural peculiarities suggest that sandstones represent the channel facies of turbidity and density currents. In most of the rocks, radiolarians are less frequent than in Member 2, being abundant only in calcarenites and calcilutites. Such a phenomenon may indicate the rewashing and subsequent concentration of radiolarians. The small thickness of the member is likely due to the rewashing and partial erosion of the sediment. The radiolarian assemblages themselves are typical of an open tropical basin with normal salinity, which had a good connection with other oceanic basins.

Member 4 is characterized by a predominance of clays and clayey mudstones with subordinate beds of spongolites that are sometimes enriched by iron-manganese hydroxides. The coarse-grained sediments are practically absent, but the thickness and rate of sedimentation are higher than for the lower members. Radiolarites were not found, and radiolarians are rare and are represented by assemblages of low diversity. In contrast to the underlying member, there is a lack of significant concentrations of resedimented radiolarian debris in the turbidites (the spongolites may belong to the turbidites). This is probably related to the lesser productivity of radiolarian plankton caused by either the partial isolation of the basin or by small deviations in salinity. The spongolites consist of well-sorted sponge skeletal elements that were probably transported from the shelf edge area. Both the reduction in turbiditic sedimentation and the increase in background argillaceous sedimentation are characteristic of the fourth member.

Member 5 closely resembles the underlying member, and consists largely of clays and mudstones, with common spongolite interbeds, often mineralized. Unlike Member 4, it contains cross-bedded sandstone interbeds and rare horizons of radiolarian cherts. The radiolarian fauna again becomes abundant and diverse suggesting improvements in the links with an open marine basin. The presence of rich radiolarian fauna is coeval with the appearance of blocky horizons containing detached blocks of Triassic limestones with Mamonia Complex lithologies. There are two possible source units: the Vlambouros Formation of the Ayos Photios Group, and the Choletria Member of the Dhiarizos Group. The studied limestones did not contain clastic material and they have affinity with sediments of the Choletria Member. The presence of such olistoliths indicates certain tectonic activity during Albian-Cenomanian time that caused erosion of the Triassic strata. The displacement, uplift, and partial erosion of the Dhiarizos Group probably occurred during an early phase of the accretion of the Mamonia Complex. The other possible explanations are extensional tectonism or rifting.

From the above evidence, the development of the Late Mesozoic basin can be summarized as follows. The general regressive character of the sedimentation is recorded by the upwards gradual decrease in the amount of cherts and the increase of clastics. In the course of this development the style of clastic sedimentation changed considerably from the active deposition of turbidites during the Late Jurassic, to a background clayey sedimentation with minor turbidites in Albian-Cenomanian times. The basin, either a marginal sea or an immature ocean devoid of the constant pelagic conditions of strict oceanic sedimentation, was nevertheless open to the World Ocean during most of the time. In general, sedimentation during this period of time can be defined as hemipelagic. Only in the Callovian-Oxfordian, was it eupelagic, exhibiting the prevalence of biogenic sedimentation and a low sedimentation rate.

The Dhiarizos Group

The Dhiarizos Group is represented mainly by volcanics: pillow lavas, volcanic breccias, and other igneous rocks. Sedimentary rocks are present as lenses, interbeds and inclusions of limestones and cherts. These sediments are sometimes characterized by bivalves, corals, conodonts, and radiolarians that provide stratigraphic data.

Previous investigators (Robertson and Woodcock, 1979; Swarbrick and Robertson, 1980) subdivided the Dhiarizos Group into several formations:

1) The Phasoula Formation, consisting of pillow lavas overlain by micritic limestones and cherts;

2) The Petra tou Romiou Formation, made up of pillow lavas with blocks of reefal limestones;

3) The Loutra tis Aphroditis Formation, consisting of volcanic breccia with lava flows and interbeds of micritic limestones and cherts.

These three formations are strongly deformed, and the various melanges sometimes incorporate large blocks from these formations.

Phasoula Formation

The Phasoula Formation is widely distributed over the whole area of the Mamonia Complex. The most complete section of the Phasoula Formation can be observed in the central part of the Akamas Peninsula, 2.5 km east of Lara Bay. This section is represented by:

1) Brown to greenish-brown strongly fractured basalts, sometimes displaying pillow structures. The thickness is over 100 m.

2) White platy recrystallized limestones with interlayers of pink limestones and red cherts; the thickness is 8 to 10 m.

3) Light-red and pink thin-bedded cherts with rare interlayers of pink limestones, with Upper Norian to Rhaetian radiolarians: *Betraccium sp.*, *Livarella densiporata* Kozur et Mostler, *L. longus* Yoshida, and *Pentactinocarpus sevaticus* (Kozur et Mostler); approximately 5 m. Tectonic contact, subparallel to bedding surfaces.

4) Red thin-bedded cherts, intercalated by thin (0.1-0.2 cm) layers of red cherty mudstones, with occasional interlayers of greenish-gray cherts. The Middle Jurassic radiolarians *Guexella* cf. *nudata* (Kocher), *Stichocapsa robusta* Matsuoka, *Tricolocapsa plicarum* Yao, *T.* cf. *tetragona* Matsuoka, and *Williriedellum* sp. were found in this member. The thickness is 10 m.

The thick basalts at the base are similar to the volcanics 1-1.5 km east of Choletria. There, basaltic pillow lavas with common inclusions of pink and sometimes greenish-gray limestones bear volcaniclastic material. These limestones are characterized by occasional filaments, probably juvenile *Halobia* and abundant recrystallized radiolarians. The Late Triassic (Late Carnian-Early Norian) conodonts *Epigondolella spatulata* (Hayashi) and *E. abneptis* (Huckriede) were found here. The total thickness of this lava unit may be more then 100 m.

The type section of the Phasoula Formation is located directly north of the village of Phasoula (Swarbrick and Robertson, 1980). The section is represented by a 100-200 m thick succession of pillow lavas with rare thin interbeds of red clayey cherts with the Late Carnian-Early Norian conodont *Epigondolella abneptis* (Huckriede).

The limestones and cherts that overlie the basalts of the Phasoula Formation are known as the Choletria Member (Swarbrick and Robertson, 1980). Blocks of these limestones are common in the melanges of the Mamonia Allochthonous Complex. One such block was studied along the right bank of the Ezousa River 1 km north of Agia Varvara village (Bragin and Krylov, 1999). This 14 m thick block consists of white and pink platy micritic limestones with intercalations of gray and yellowish-pink chert. This limestone block is interpreted as a sedimentary lense derived from the Triassic part of the Dhiarizos Group. The fact that these limestones do not contain the clastic intercalations common for the carbonate sediments of the Ayos Photios Group adds to this conclusion (Swarbrick and Robertson 1980; Bragin and Krylov 1996).

The limestones and cherts contain abundant recrystallized radiolarians, replaced by calcite. Only one sample yielded pyritized radiolarians, with well-preserved morphological features. The radiolarian assemblage is characterized by high taxonomic diversity, being represented by the following taxa: *Annulotriassocampe* cf. *sulovensis* (Kozur et Mock), *Archaeocenosphaera* sp., *Bulbocyrtium latum* Bragin, *Caphtorocyrtium tenerum* Bragin, *Capnodoce ruesti* Kozur et Mock, *Capnuchosphaera deweveri* Kozur et Mostler, *C. theloides minor* Bragin, *C.* cf. *theloides* De Wever, *C.* aff. *carpathica* Kozur et Mock, *Carinaheliosoma carinata* (Kozur et Mostler), *Entactinosphaera* aff. *simoni* Kozur et Mostler, *Ferresium* aff. *conclusum* Carter, *Foremanellina* (?) sp., *Haeckelicyrtium carterae* Bragin, *Heliosoma* (?) *riedeli* Kozur et Mostler, *Icrioma tetrancistrum* De Wever, *Kahlerosphaera aspinosa* Kozur et Mock, *K. norica* Kozur et Mostler, *Karnospongella bispinosa* Kozur et Mostler, *Kinyrosphaera trispinosa* Bragin, *K. helicata* Bragin, *Laxtorum* (?) sp., *Liassosaturnalis parvus* Kozur et Mostler, *Loffa* (?) sp., *Multimonilis pulcher* Yeh, *Nabolella trispinosa* Bragin, *Neopylentonema* aff. *procera* Sugiyama, *Palaeosaturnalis triassicus* (Kozur et Mostler), *P. latiannulatus* Kozur et Mostler, *P. mocki* Kozur et Mostler, *Paronaella norica* Kozur et Mock, *Pentactinocarpus* aff. *tetracanthus* Dumitrica, *Praemesosaturnalis* cf. *multidentatus* (Kozur et Mostler), *Poulpus piabyx* De Wever, *Pseudosaturniforma carnica* Kozur et Mostler, *Praenanina veghae* Kozur, *Praeorbiculiformella goestlingensis* Kozur et Mostler, *Pseudostylosphaera* (?) sp., *Sarla* (?) sp., *Sepsagon* sp., *Sethocapsa*

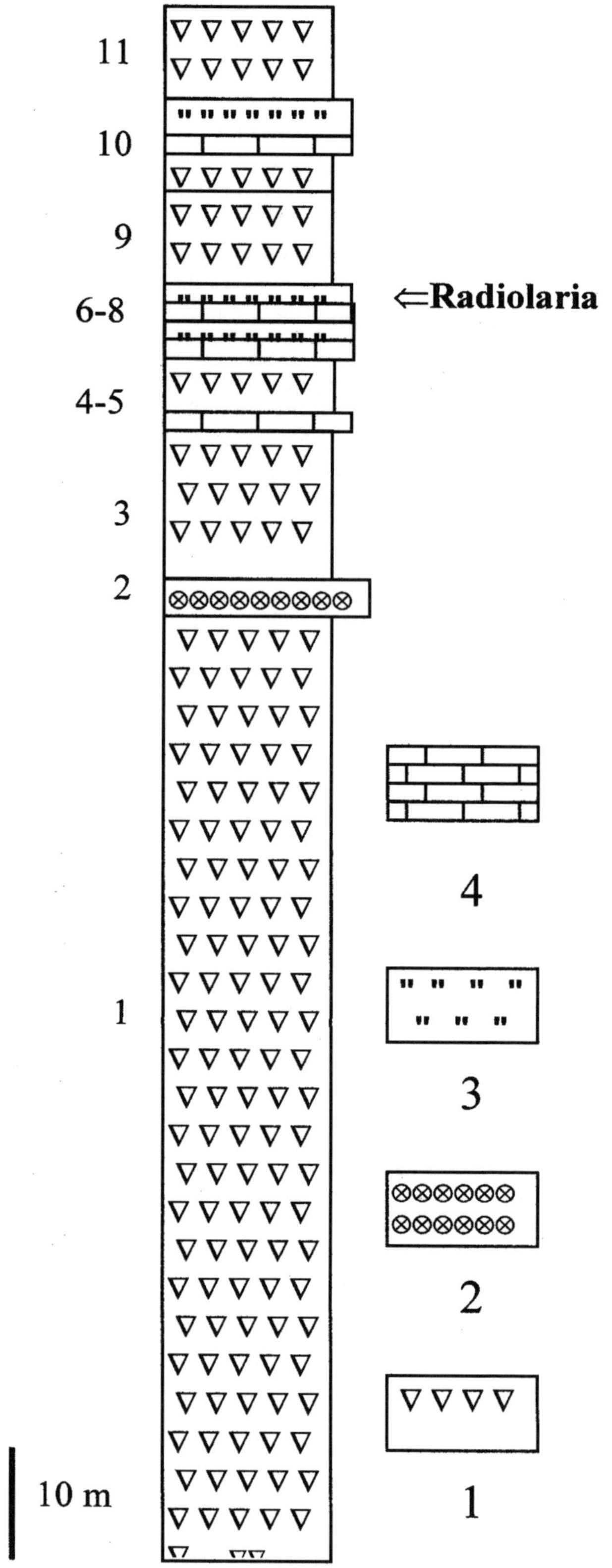

Fig. 2.5. Sections of the Loutra tis Afroditis Formation in the central part of the Akamas Peninsula. 1 - ophiolite xlasts breccia; 2 - pillow lavas; 3 - radiolarites; 4 - micritic limestones.

sp., *Spongostylus carnicus* Kozur et Mostler, *S. tortilis* Kozur et Mostler, *Sulovella constricta* Kozur et Mock, *Syringocapsa batodes* De Wever, *Trialatus robustus* (Nakaseko et Nishimura), *Triassobipedis* (?) sp., *Triassocrucella triassica* (Kozur et Mostler), *Veghicyclia* cf. *robusta* Kozur et Mostler, *Vinassaspongus transitus* Kozur et Mock, *Whalenella* aff. *perfecta* (Blome), *W. robusta* Bragin, *Xiphotheca rugosa* Bragin, *X. longa* Kozur et Mock, *X.* (?) *spinellifera* Bragin, and *Zhamojdasphaera proceruspinosa* Lahm. All these radiolarians come from a single sample.

The Early Norian age of this assemblage is confirmed by the conodont *Epigondolella spatulata* (Hayashi) and such radiolarian taxa as *Capnodoce ruesti*, *Sulovella constricta,* and *Vinassaspongus transitus*. This radiolarian assemblage is typical for the *Capnodoce ruesti* Zone (Kozur, Mostler, 1994) with some exceptions. Some species *i.e. Karnospongella bispinosa*, *Praenanina veghae*, and *Spongostylus carnicus* were known only from Carnian or older deposits. The Cyprus locality may represent the latest occurrence of these taxa.

Therefore, a composite section of the Phasoula Formation can be represented as follows:

1) Pillow lavas, Carnian-Early Norian. More then 100 m.

2) Limestones, Norian, 10-20 m.

3) Cherts. Norian to Middle Jurassic. 15-20 m, but still no Early Jurassic fossils.

Petra tou Romiou Formation

The Petra tou Romiou Formation is characterized by abundant limestone blocks of shallow-water origin, yielding Late Triassic bivalves and corals (Lapierre, 1975; Swarbrick and Robertson, 1980). The type locality of the formation is at the Petra tou Romiou beach, where large blocks of volcanics and adjacent limestones are incorporated into a melange with clayey matrix (see Chapter 4 of this volume).

Loutra tis Aphroditis Formation

The Loutra tis Aphroditis Formation is located along the north-east shore of the Akamas Peninsula. The type section is represented by alternating thick (more than 50 m) beds of volcanic breccia composed of angular fragments of basalts, diabases and amphibolites, and thinner beds of pillow lavas, micritic limestones and cherts of low metamorphic grade. Cherts from this locality yield only poorly preserved radiolarians.

A similar section, 2 km north-east of Lara Bay, was described by the authors (**Fig. 2.5**):

1) Green, greenish-brown and red ophiolitic breccia composed predominantly of basalts and diabases, with minor blocks of gabbroids; 100 m thick.

2) Pink, gray and light-red ophicalcites with common inclusions of basalts and serpentinized gabbroics; 4 m thick.

3) Breccia similar to that of layer 1, sometimes fine-grained, with inclusions of hyaloclastites; 15 m thick.

4) Thin-bedded pink micritic limestones with stylolites and with intercalations of red cherty mudstones near the top and bottom of unit; 2 m in thickness.

5) Breccia similar to that of layer 1; 1-3 m thick.

6) Thin-bedded dark-red and crimson radiolarian cherts; 2-4 m in thickness.

7) Pink micritic limestones; 1 m thick.

8) Radiolarian cherts similar to layer 6; 4 m in thickness.

9) Green pillow basalts with transition upwards to basaltic breccia; 10 m thick.

10) Basaltic breccia with lenses of pink and red micritic limestones (the lower part of this unit) and intercalations of red radiolarian cherts (the upper part of this unit); 10 m thick.

11) Basaltic breccia; 10 m thick.

The abundant and moderately well-preserved radiolarians were found in subunit 6. Other cherts of this section are characterized by similar but poorly preserved radiolarian remains. The assemblage from subunit 6 consists of the following taxa: *Ares* sp., *Archaeocenosphaera laseekensis* Pessagno et Yang, *Bipedis patricki* Whalen et Carter, *Broctus* aff. *kuensis* Pessagno et Whalen, *Canoptum* sp., *Canutus* sp., *Crucella* sp., *Gigi* aff. *fistis* De Wever, *Gorgansium gongyloideum* Kishida et Hisada, *G.* sp., *Katroma* aff. *westermanni* Whalen et Carter, *Laxtorum* sp., *Pantanellium sixi* Whalen et Carter, *Parahsuum* cf. *simplum* Yao, *P.* aff. *ovale* Hori et Yao, *Paronaella* spp., *Pobum* cf. *infinitum* (Pessagno et Poisson), *Praeconocaryomma* cf. *media* Pessagno et Poisson, *Protokatroma* sp., *Saitoum* cf. *keki* De Wever, and *Trillus* sp., of Early Jurassic age, probably Late Sinemurian to Pliensbachian.

The presumably younger, Late Jurassic-Cretaceous volcanics and sediments of the Dhiarizos Group remain undated. Some melanges of southwestern Cyprus contain blocks of Late Mesozoic cherts and adjacent lavas, but their origin is unclear. It is actually not possible to exclude a derivation of these cherts from the Ayos Photios Group. The description of some of these localities is given below.

Along the left bank of the Ezousa River, 1.5 km NNE of Agia Varvara, strongly altered volcanics with blocks of white recrystallized limestones and red cherts can be observed. These cherts bear the Early Cretaceous radiolarians: *Alievium helenae* Schaaf, *Archaeodictyomitra apiara* (Rust), *A. excellens* Baumgartner, *Cecrops septemporatus* (Parona), *Pantanellium lanceola* (Parona), *Parvicingula boesii* (Parona), *Podocapsa amphitreptera* Foreman, and *Xitus alievi* (Foreman).

In the roadcut between Nea Choletria and Stavrokono altered volcanics are in direct (but clearly tectonic) contact with intensively folded thin-bedded red cherts with these late Middle Jurassic radiolarians: *Acanthocircus suboblongus* (Yao), *Archaeodictyomitra* cf. *rigida* Pessagno, *Guexella nudata* (Kocher), *Hsuum brevicostatum* Ozvoldova, *H. maxwelli* Pessagno, *Mirifusus* cf. *guadalupensis* Pessagno, *Protunuma ochiensis* Matsuoka, and *Tricolocapsa* cf. *parvipora* Tan, that confirm the Bajocian-Callovian age. All these localities need more precise study.

Therefore, the Dhiarizos Group consists of three volcanogenic formations of distinct composition and age: the Upper Triassic-Middle Jurassic Phasoula Formation, the Upper Triassic Petra tou Romiou Formation, and the Lower Jurassic Loutra tis Aphroditis Formation.

C. Volcanic Rocks of the Mamonia Complex

Maxim V. Portnyagin
Vernadsky Insititute of Geochemistry and Analytical Chemistry of the Russian Academy of Sciences, Kosygin Str. 19, 117975 Moscow, Russia

General Characteristics

The Dhiarizos Group of Triassic-Jurassic age occurs at the base of the allochthonous Mamonia terrain (Swarbrick and Robertson, 1980). They consist mainly of pillow lavas and subordinate massive lavas and lava-breccias. These rocks are exposed within the erosional windows which together form a discontinuous arcuate belt along the southern and south-eastern parts of the Mamonia terrain (**Fig. 2.6**) (Malpas and Xenophontos, 1987).

According to the stratigraphic classification of Swarbrick and Robertson (1980), the Dhiarizos Group includes two main types of volcano-sedimentary formations. The Phasoula and Loutra tis Aphroditis Formations consist of mafic lavas associated with carbonate sediments. The volcanic rocks associated with hemipelagic mudstones are characteristic of the Mavrokolymbos and Petra tou Romiou Formations. According to radiolarian fauna, determined by N. Bragin and L. Bragina (see this Chapter), the Phasoula Formation belongs to the Upper Triassic-Middle Jurassic, while the Petra tou Romiou Formation has a Late Triassic age, and the Loutra tis Aphrodites Formation is attributed to the Lower Jurassic. The Dhiarizos Group rocks are tectonically overlain by sedimentary rocks of the Triassic-Early Cretaceous Ayios Photios Group. The Late Cretaceous volcanites of the 'Troodos type' (Lapierre, 1975) are closely associated with the Dhiarizos Group rocks. These volcanites are genetically related to the establishment of the ophiolitic crust of the Troodos Complex which forms the core of the island of Cyprus, to the north of the collisional zone of the Mamonia Complex.

Main Petrographical and Geochemical Lava Types

Early workers (*e.g.* Lapierre, 1975; Pearce, 1975) designated the Dhiarizos Group volcanites as variably fractionated alkali basalts of intra-plate origin. It is now known, however, that this group includes at least two main types of volcanics with different morphologic appearance in the field. They are also quite distinct in their content of major and trace elements, which are inert under the low-temperature conditions of post-magmatic alteration.

According to Malpas and Xenophontos (1987) and Malpas et al. (1992, 1993), more than 80 per cent of the Dhiarizos volcanic rocks are olivine-phyric variolitic basalts forming pillow lava flows with individual pillows 20-50 cm in diameter. They are orange-brown or red when weathered. In some cases, individual pillows are characterized by interpillow voids or have interpillow layering of alternating bands of chloritized and zeolitized basalts. Shrinkage voids as well as those between adjacent pillows are usually filled with pink and gray calcilutite. Minor red, green and brown cherts and black iron-manganese-bearing cherts occur as void infillings or discontinuous interbeds within the lava flows.

The pillow margins contain olivine phenocrysts set in a crystalline matrix. In the central parts of the pillows the amount of olivine phenocrysts increases up to 15%. The olivine is completely altered in the rocks. The groundmass, with a variolitic or interseptal texture, is composed of skeletal plagioclase, as well as clinopyroxene, olivine, and ore minerals which occur in the interstices.

The composition of the Dhiarizos Group olivine basalts from the Ayia Varvara area is shown in **Table 2.1**. This rock (sample 1) indicates a mildy evolved composition with Mg/(Mg+Fe) = 0.56. Based on the major element content, this can be classified as olivine tholeiite (Yoder and Tilley, 1962) closely resembling typical MORB. Malpas and Xenophontos (1987) and Malpas et al. (1992) reported that the Dhiarizos Group tholeiites are generally poor in TiO_2 and Zr (0.7-1 wt.% and about 25-40 ppm, respectively). Plots in Malpas et al. (1993) show varieties enriched in TiO_2 (up to 2 wt.%) and other incompatible elements, although the relevant analytical data are not given. On the Zr/TiO_2-Nb/Y diagram (**Fig. 2.7**), the olivine basalts fall in a discrete field, having a tendency to some Nb-enrichment, and an increase of the Nb/Y ratio resembling moderately incompatible element-enriched oceanic tholeiites of E-MORB type (LeRoex, 1987). The Dhiarizos Group tholeiites display a typically flat REE-pattern, slightly depleted in LREE and HREE with respect to the MREE (**Fig. 2.8;** Malpas et al., 1993).

Following Malpas et al. (1993), the remaining 20% of the Dhiarizos Group volcanics are alkali basalts. Two main petrographic varieties are typical and dominant among these rocks:

a) pillow aphyric basalts with individual pillows of about 50 cm, with red-violet weathering, and calcite and zeolite amygdales, and

b) plagioclase-phyric basalts, light-gray when weathered, consisting of 50% or more of large (0.5-2 cm) plagioclase phenocrysts completely replaced by prehnite and set in an augite-plagioclase groundmass with a sub-ophitic texture. More commonly, these rocks form pillow lava flows; massive sheet flows and dikes of similar composition are subordinate (*e.g.* in the Kholetria-Ayia Varvara area). The alkali lavas are closely associated with both olivine basalt lavas of the same age and younger sedimentary rocks.

Compositions of the typical alkali basalts are shown in Table 2.1 (samples 2 and 3) and Figs. 2.7 and 2.8. These rocks are poorer in silica (lesser than 48 wt.%) and richer in TiO_2 (more than 1.5 wt.%), and are also slightly enriched in alkalis compared with the

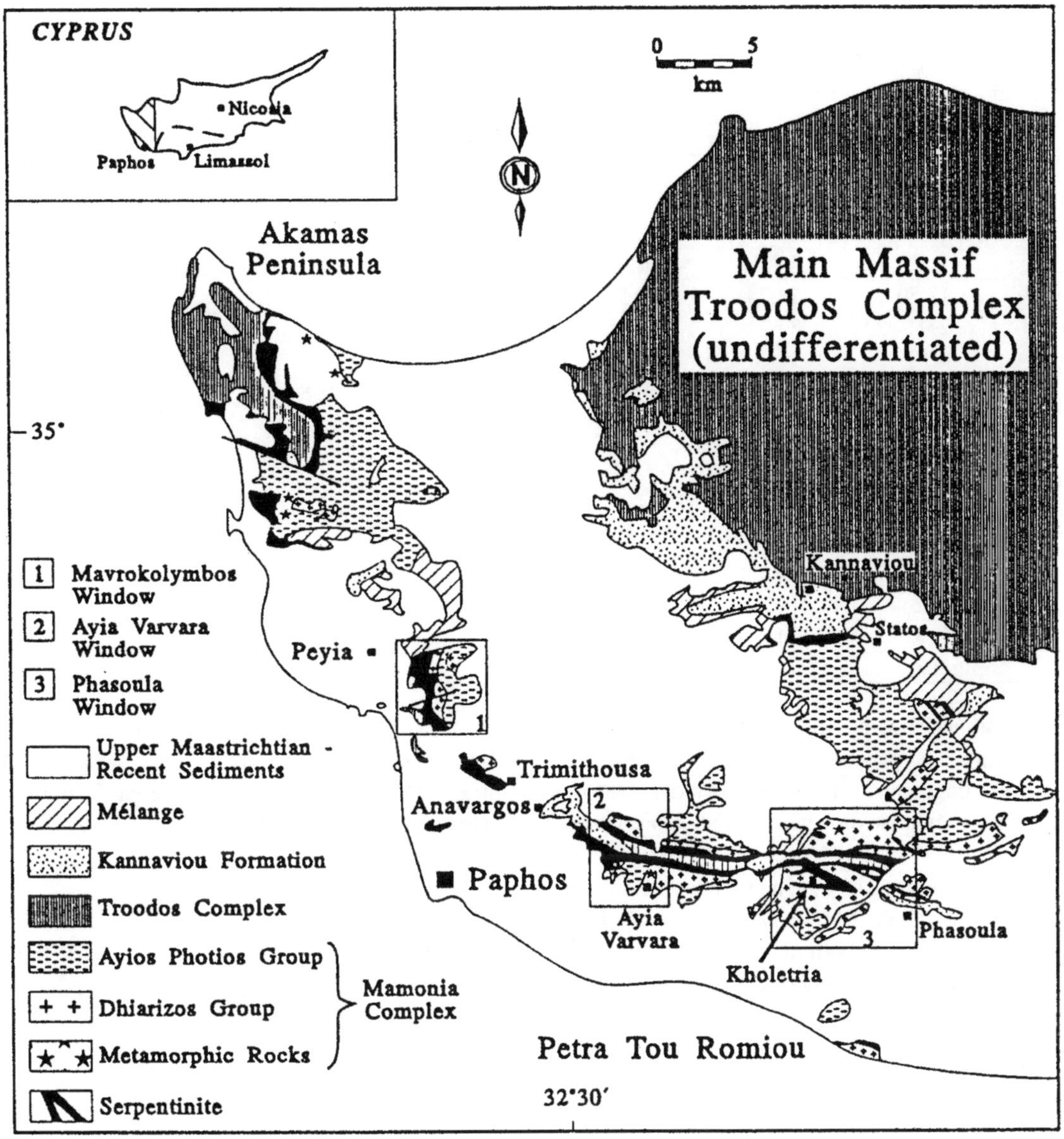

Fig. 2.6. Geological sketch map of SW Cyprus showing the main erosional windows where volcanic rocks of the Dhiarizos Group and the Troodos Complex are exposed (after Malpas et al., 1993).

Dhiarizos Group tholeiities. In Fig. 2.7, the basalts form a discrete group with Zr/TiO_2 values from 60 to 80 and the Nb/Y ratio reaching 1; the latter is the result of significant Nb enrichment of up to 20-60 ppm.

The considerable difference between the alkali basalts from the Kholetria area (Malpas et al., 1993) and the Dhiarizos Group tholeiites and all Troodos volcanics illustrated in Fig. 2.8 is that the Kholetria basalts display REE-enriched spectra with a chondrite normalized La/Yb ratio of up to 10. The whole-rock chemical compositions of the aphyric and plagiolase-phyric alkali basalts are generally similar. According to previously published data (Malpas and Xenophontos, 1987; Malpas et al., 1992), the slight difference between the two types of rocks is that the plagioclase-phyric basalts are richer in Zr and to a lesser degree in TiO_2, reflected in higher Zr/TiO_2 values when compared to the aphyric basalts (Fig. 2.7). The concentrations of other elements in both rocks are similar.

The ankeramites containing phenocrysts of titan-augite, olivine and minor plagioclase form a small volumetric proportion among the Dhiarizos Group volcanics. In the Petra tou Romiou area, these rocks occur as pillow lava flows closely associated with iron-manganese-bearing cherts. The ankermite fragments have also been found in the deformed argillaceous shales of the Mavrokolymbos Formation.

The chemical composition of the ankeramite from the Petra tou Romiou area is given in Table 2.1 (sample 4), and demonstrates the evident alkali nature that is reflected in the low content of SiO_2

Table 2.1. Representative compositions of the volcanic rocks from Southwestern Cyprus

wt.%	1	2	3	4	5	6	7
SiO_2	47.8	38.0	45.7	45.8	60.23	66.34	51.6
TiO_2	0.86	1.32	2.52	2.68	0.47	0.44	0.18
Al_2O_3	14.4	12.0	16.4	15.3	13.5	15.41	10.6
FeO	10.48	6.40	10.02	9.88	5.02	4.94	8.15
MnO	0.19	0.17	0.27	6.00	0.52	0.21	0.15
MgO	7.52	5.73	4.00	7.12	5.67	0.61	14.53
CaO	11.50	18.24	9.56	3.19	4.72	6.09	8.42
Na_2O	2.93	3.63	3.98	3.24	4.22	5.07	1.33
K_2O	0.13	1.04	1.10	0.13	0.06	0.06	0.40
P_2O_5	0.04	0.16	0.41	0.67	0.08	0.06	0.01
LOI	3.74	13.10	4.05	6.11	5.02	0.91	5.22
Ppm							
Y	34	26	40	35	63	48	14
Zr	31	72	198	239	645	621	<10
Nb	7	24	63	59	162	124	<5
V	192	145	301	286	52	25	282
Ni	168	34	36	-	20	4	246
Cr	508	74	64	-	5	7	823

1 - pillow-lava of olivine tholeiite from the Ayia Varvara area; 2 - pillow-lava of alkaline aphyric basalt from the Ayia Varvara area; 3 - pillow-lava of alkaline plagioclase-phyric basalt from the Ayia Varvara area; 4 - ankaramite pillow-lava from the Petra tou Romiou area; 5-6 - fragments of travhyte lava's enclosed in limestones from the Petra tou Romiou area; 7 - pillow-lava of the Troodos type from the Ayia Varvara area. All data represented after Malpas and Xenophontos (1987).

and the higher concentrations of TiO_2, Zr, and Nb compared with the olivine tholeiites, although the MgO content (6-7 wt.% in the case of the ankeramite) is similar in both rock types. In general, the ankeramites closely resemble the plagioclase-phyric alkali basalts from the Ayia Varvara area.

Other fractionated products of the alkali magmas are present in very small amounts. The trachytes and trachybasalts occur either as single thin lava flows (*e.g.* in the Kholetria area) or as rock fragments in the limestones of the Petra tou Romiou area (Malpas and Xenophontos, 1987). The chemical compositions shown in Table 2.1 (samples 5 and 6) indicate that these rocks are characterized by a higher SiO_2 content (~60 wt.%), low levels of MgO and FeO, and (Figs. 2.7 and 2.8) are strongly enriched in the most incompatible elements, particularly Nb, Zr, and LREE. Such geochemical signatures are indicative of highly fractionated mafic parent magmas closely resembling the ankeramatic lavas.

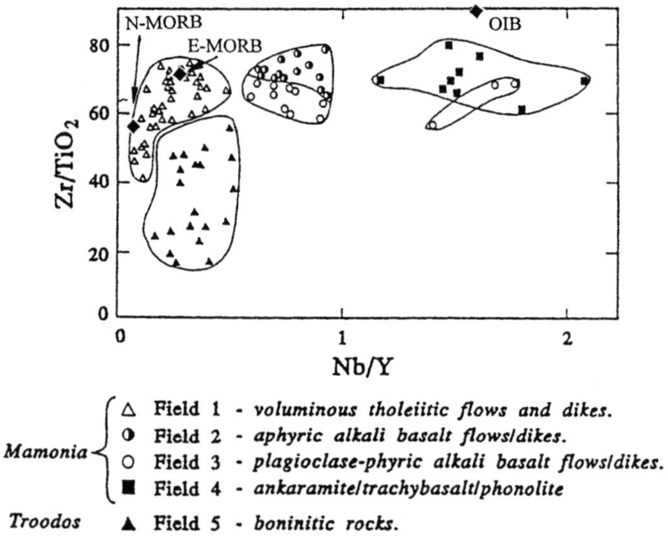

Fig. 2.7. Zr/TiO_2 versus Nb/Y for rocks of the Dhiarizos Group and the Troodos Complex (after Malpas et al., 1993). The average compositions of N-MORB (Normal-MORB), E-MORB (Enriched MORB) and OIB (Oceanic Island Basalts) are from Sun and McDonough (1989).

Troodos Complex Lavas

In many areas of the Mamonia terrain, the Dhiarizos Group volcanics are associated with Late Cretaceous volcanics originating from the Troodos Complex. The minimum Late Cretaceous age of the Complex is based on dating of the overlying volcaniclastic sediments of the Kannaviou Formation (Mantis, 1970). According to Malpas and Xenophontos (1987) and Malpas et al. (1993), the Troodos lavas occurring within the Mamonia Complex are

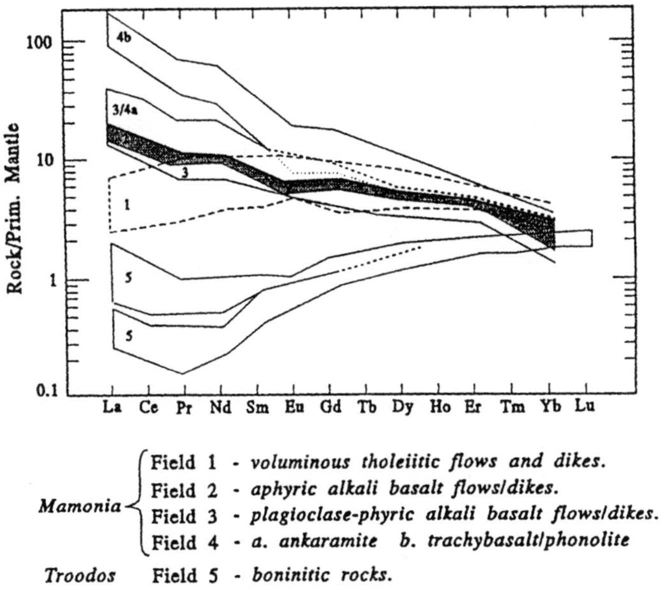

Fig. 2.8. Normalized REE concentrations in volcanic rocks from the Mamonia Complex. Fields are after Malpas et al. (1993).

divisible into two units. The lower volcanics are pillow olivine-phyric lavas consisting of small (15-30 cm) individual pillows with glassy chilled margins; the content of olivine in these rocks is always between 5 and 15%. The upper unit is formed of pillow lava flows with individual pillows up to 1 m in diameter. Olivine-rich subvolcanic bodies, which are probably feeders and sills, are frequently associated with these pillow lavas (*e.g.* in the Mavrokolymbos area; Portnyagin et al., 1996). The magmatic contact between the Troodos lavas and the gabbroic rocks and serpentinites has been documented in some exposures (Malpas et al., 1993). Both the lower boninite series lavas belonging to the Troodos Complex are overlain by onlapping bentonitic clays, radiolarian mudstones, and volcanically derived siltstones and sandstones of Maastrichtian age (Kannaviou Formation; (Malpas et al., 1993).

The compositional characteristics of typical Troodos lavas are given in Table 2.1 and Figs. 2.7 and 2.8. Whole rock and volcanic glass compositions indicate significant MgO enrichment, high silica content, and extremely low concentrations of Zr, Ti, and REE (Malpas et al., 1992; Magakyan, 1994; Portnyagin et al., 1996). In their appearance, chemical and petrographic features, these volcanics are similar to the boninite-type lavas from the Arakapas Fault Belt and Limassol Forest Complex (see Chapter 1). Significantly, they all display the U-shaped REE pattern characteristic of Troodos-type lavas from the Mamonia terrain (Fig. 2.8).

Geochemical Classification

A geochemical classification of its volcanic rocks provides significant data for determining the original tectonic position of the Mamonia allochthon. The Dhiarizos Group lavas are intensely altered. This excludes the use of their major-element contents. However, there is evidence based on immobile incompatible elements such as HFSE (High Field Strength Elements), including Ti, Zr, Hf, Nb, Ta, and Th, and to a lesser degree of REE and Y.

Pearce (1975) first classified the Dhiarizos Group lavas of the Petra tou Romiou and Kholetria areas as 'within-plate basalts', using the Ti-Zr-Y diagram and analytical data from his small sample collection. Although these criteria cannot distinguish between continental and oceanic basalts, Pearce's analysis of the Mamonia Complex sedimentary and magmatic rocks led to the conclusion that Dhiarizos Group lavas reflect ocean island magmatism from a near continental margin. This conclusion remains valid to the present.

Studies of the Dhiarizos Group lavas were recently resumed. Malpas and Xenophontos (1987) and Malpas et al. (1992, 1993) took the same general approach to lava classification. However, new important conclusions were reached due to the new representative collections. As shown in **Fig. 2.9** (Malpas et al., 1993), the Mamonia Complex volcanics constitute two discrete groups. The

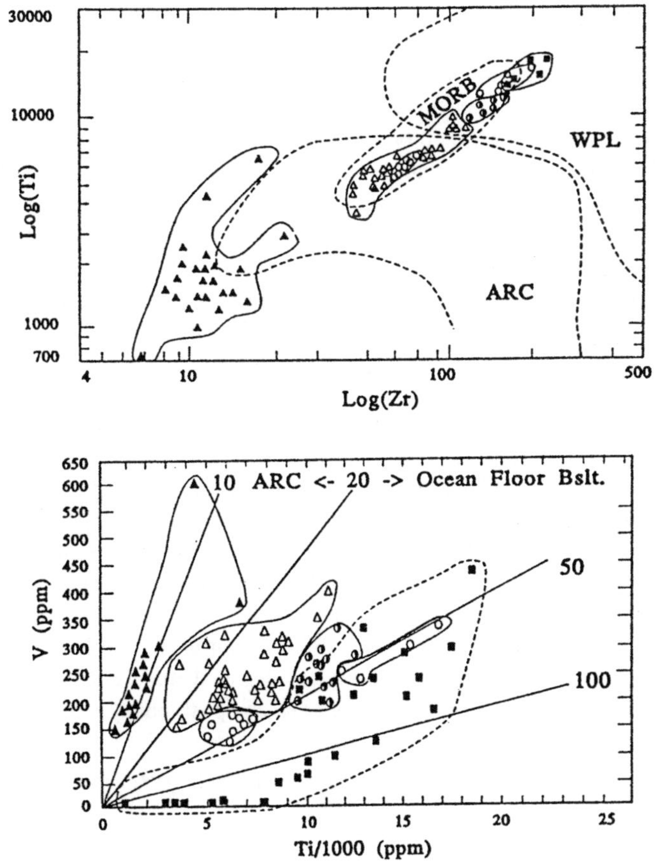

Fig. 2.9. Ti vs. Zr and V vs. Ti plots for rocks of the Dhiarizos Group and the Troodos Complex (from Malpas et al., 1993). Symbols as on Fig. 2.8. WPL - Within Plates Basalts; ARC - Arc Basalts.

first group includes the Troodos-type lavas indicating minimum concentrations of Ti, Zr, and Y, resembling highly incompatible element-depleted island arc volcanics. The classification of these rocks, in which fresh volcanic glasses are preserved, is accurate. They are very similar to the boninite-type lavas of the Arakapas Fault Zone and the Limassol Forest Complex, which are the most strongly depleted in incompatible elements among the other Troodos volcanics (Figs. 2.7 and 2.8).

The second group, the Dhiarizos lavas, have a compositional field similar to that of MORBs, with an enrichment in incompatible elements, although there is some similarity to the intra-plate alkali basalt field. The relationship of Ti and V contents in the Dhiarizos lavas (**Fig. 2.10**) shows an inclination to oceanic basalt fields, since the Dhiarizos lavas are less depleted in Ti (at similar V contents) compared with both Troodos lavas and the majority of island arc volcanics. In this regard, the main conclusion that emerges from the analytical data published by Malpas et al. (1993) is that the Dhiarizos olivine basalts depleted in incompatible elements correspond to the N-MORBs and thus represent Tethyan oceanic crust basement. The alkali lavas are considered to be products of seamount volcanism, in accordance with Pearce (1975). The basalts enriched in incompatible elements are also compositionally similar to the most enriched varieties of the MORBs (Le Roex, 1987). In fact, part of these rocks, together with the depleted tholeiites, could represent oceanic crust formed at a spreading center; the petrographic features do not contradict this suggestion.

In this regard, an important aspect is the origin of this oceanic basin. One of the alternative hypotheses suggests the break-up of Gondwana and the occurrence of intial oceanic rifting leading to the formation of a small oceanic basin, similar to the modern Red Sea (Robertson and Dixon, 1984; Robertson, 1990). An alternative hypothesis suggests the formation of a back-arc type basin above a south-dipping subduction zone (Dilek and Moores, 1990).

Evidence for the concept of the formation of the Dhiarizos lavas in a basin resembling the Red Sea comes from an environmental interpretation of the associated sedimentary rocks (**Fig. 2.11**; Robertson, 1990), and a comparison of the compositions of the Dhiarizos lavas and the Red Sea basalts which are similar in their mineralogical and trace-element features (Fig. 2.10). The general similarity between the two types of volcanics is stressed by the chemical variability of the Red Sea basalts, which include both analogs of depleted oceanic tholeiites and alkali basalts (Schilling, 1969; Eissen et al., 1989; Barrat et al., 1990, 1993). In addition, the presence of slightly depleted mantle lherzolites in the Mamonia Complex, which are compositionally similar to those of continental margins, and particularly closely resembling the lherzolites associated with the transitional stage of continental to oceanic rifting (Island of Zabargad in the Red Sea), is additional evidence for the model of Robertson and Dixon (1984).

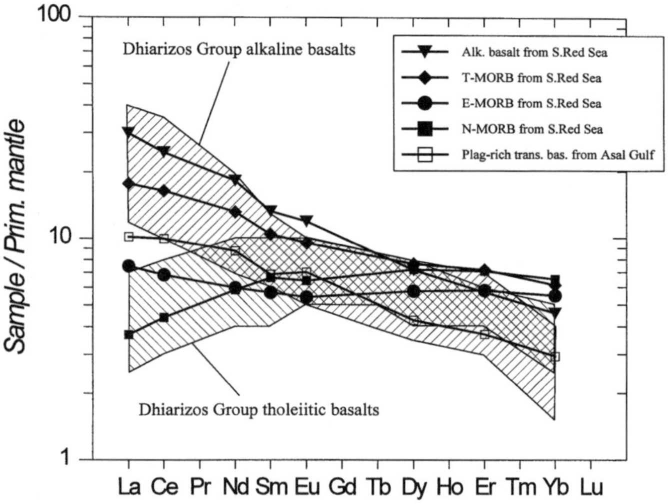

Fig. 2.10. Normalized REE concentrations in volcanic rocks of the Dhiarizos Group compared to compositions of selected samples from the southern Red See and Asal Rift regions. Fields of the Dhiarizos Group volcanics are from Malpas et al. (1993), compositions of basalts from the Red Sea and Asal Rift are from Barrot et al. (1990). Normalization to primitive mantle values after Sun and McDonough (1989).

Because of its proximity to the continental crust, the Red Sea alkali basalts have geochemical features consistent with the magma-forming process in the lithospheric mantle of the African craton (Barrat et al., 1990, 1993). If the Dhiarizos Group lavas were generated in a similar geodynamic setting in which magma-forming processes were similar to that for the Red Sea basalts, their geochemical features might be inherited from the lithospheric mantle source. Additional investigations should help to test this suggestion. Although Red Sea volcanism is a most plausible model for the formation of the Dhiarizos Group lavas, the back-arc basin model (Dilek and Moores, 1990) may also be applicable to the Mamonia terrain, although geochemical data is scanty. Further detailed geochemical studies of the lavas in SW Cyprus, aimed at deciphering their geodynamic nature, should serve as a basis for a valid model of the tectonic evolution of Cyprus and contribute to the knowledge of the whole eastern Mediterranean region.

D. Ultramafic rocks of the Mamonia Complex

Boris A. Bazylev
Vernadsky Institute of Geochemistry and Analytical Chemistry of the Russian Academy of Sciences
Kosygin Str. 19, 117975 Moscow, Russia

Ultramafic rock outcrops are mainly restricted to the arcuate belt extending from the Akamas Peninsula in the north-west to the Arkhimandrita in the south-east. In the Anavargos-Marathounda area, this serpentinite belt changes strike from nearly N-S to sub-

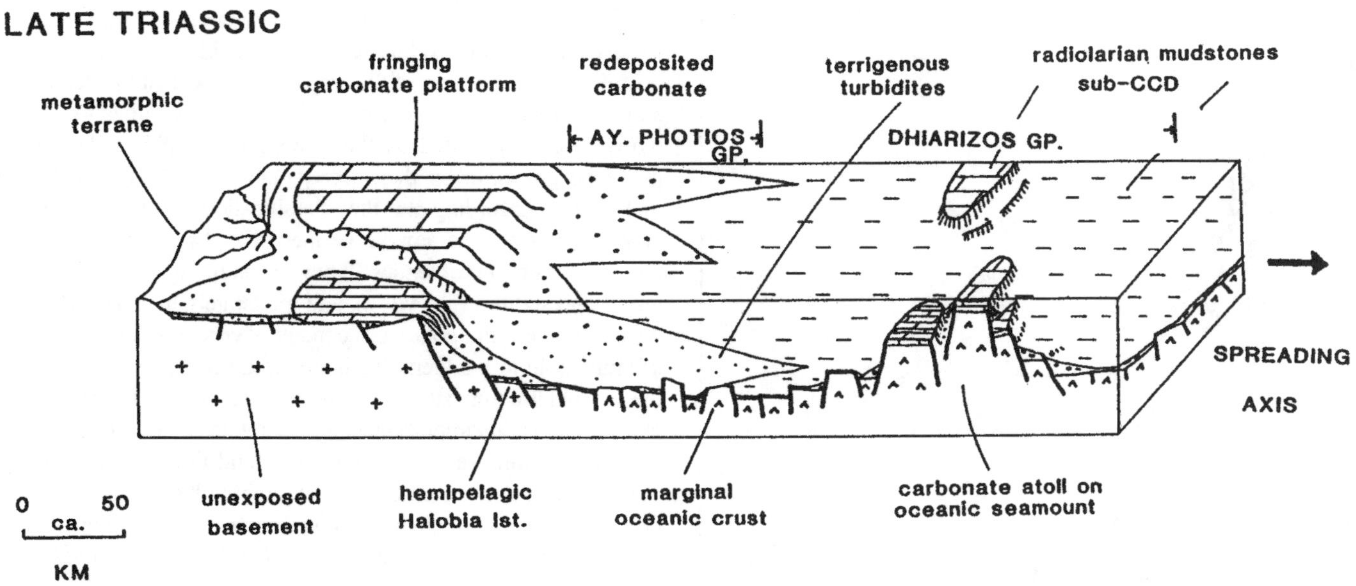

Fig. 2.11. Reconstruction of the Mamonia Complex (Ayios Photios and Dhiarizos Groups) as a passive margin of the Red Sea-type small ocean basin (from Robertson, 1990). Olivine tholeiites of the Dhiarizos Group can be viewed as remnants of oceanic floor, alkaline basalts with associated reefal limestones as remnants of seamounts.

E-W, and then extends eastward to the middle part of the Dhiarizos River. A few separate, relatively large ultramafic bodies occur outside this belt near the villages of Kanaviou and Galataria and the Petra tou Romiou cliffs (**Fig. 2.12**). It is now thought that the slivers of serpentinized ultramafics are tectonic emplacements connected with the Trodoos Ophiolite (Lapierre, 1975; Malpas and Xenophontos, 1987).

The ultramafics form separate tectonic blocks (slivers), up to several hundred meters thick, sometimes coupled together or separated by zones of tectonic breccia, schist, and boudinage. In most cases, their contacts with the Dhiarizos Group basaltoids and the Ayia Varvara Formation amphibolites and schists are tectonic, steeply dipping to the north and north-east. Primary, non-tectonic contacts of serpentinites and pillow lavas were also documented (Malpas and Xenophontos, 1987).

During the course of mapping (Lapierre, 1975; Turner, 1969; Malpas and Xenophontos, 1987), Mamonia Complex ultramafic rocks were not petrologically classified, but were broadly designated as serpentinites, bastite serpentinites, and minor spinel harzburgites. Later, petrographic and mineralogic studies of these rocks (Bazylev et al., 1993) revealed significant compositional heterogeneity, and three main rock types were established:

1) residual spinel harzburgites,
2) residual spinel lherzolites, and
3) cumulate dunites and pyroxenites.

In the investigated exposures throughout the Mamonia Complex, their distribution is as follows: harzburgite comprises approximately 60%, lherzolite about 40%, and cumulate rocks less than 1%. The harzburgite and lherzolite tend to occur together with no systematic arrangement even within the smaller blocks (less than 100 m long) or groups of such blocks, but no gradations between the two rock types have been yet observed. There is, however, a tendency for localization of the harzburgite, in the form of relatively consolidated boudins set in an intensely sheared serpentinized matrix of the original lherzolitic composition. The cumulate rocks form dikes and small irregular bodies, up to a few meters in thickness, among the harzburgites.

In the Akamas Forest, a large (~8 km long) nearly N-S trending block is entirely composed of harzburgite with subordinate cumulate ultramafics. By contrast, the adjacent nearly E-W trending Loutra tis Aphroditis block, ~1.5 km long, is located just to the north of the harzburgite and is composed exclusively of lherzolite (exluding the sheared contact zones). In the Ayia Varvara area, ultramafic slivers show an *en echelon* distribution: harzburgite slivers form a sublatitudinal zone about 500 m wide, north and south of which lherzolites are exposed.

Petrography and Mineralogy

High-temperature plastic flow and associated recrystallization typical of mantle peridotites are common in Mamonia Complex residual peridotites. When these rocks are fresh, they are pan-allotriomorphic granular or inequigranular with a massive or weakly banded structure. Near shear zones, these rocks are fine-grained and schistose. Large (1-2 cm) pyroxene porphyroblasts contain abundant solid solution lamellae which are usually deformed (*e.g.* kink banding and curved cleavage), whereas small (up to 0.5 mm) grains (neoblasts) lack these features. Clino-pyroxene is present in the harzburgites in the form of rare small neoblasts or individual lamellae within orthopyroxene porphyroblasts. The pyroxenes

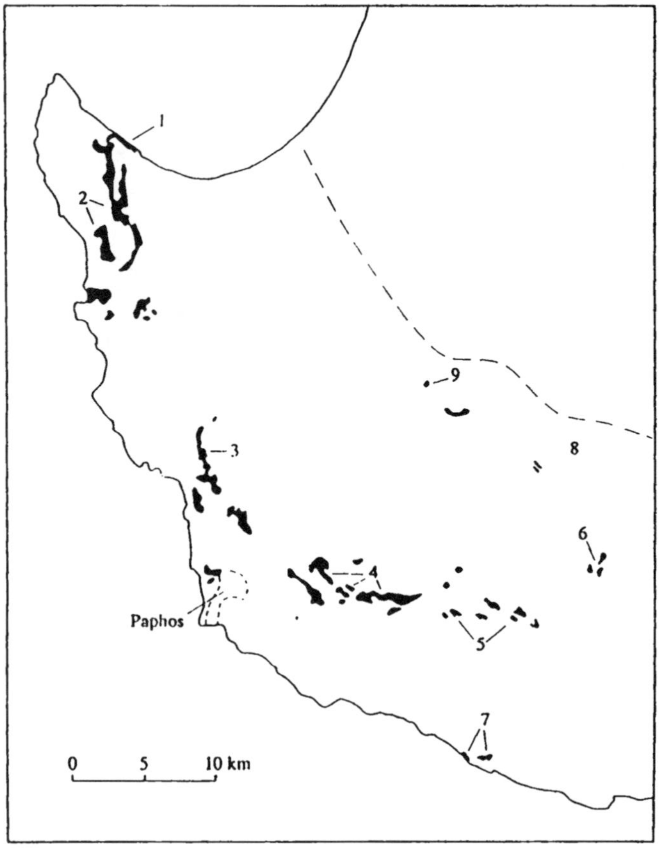

Fig. 2.12. Map of ultramafic outcrops in ranges of the Mamonia Complex (Bazylev et al., 1993). 1 – Loutra tis Aphroditis, 2 – Akamas Forest, 3 – Mavrokolymbos, 4 – Ayia Varvara, 5 – Phasoula, 6 – Prastion (Potima), 7 – Petra tou Romiou, 8 – Galataria, 9 – Kannaviou.

from the lherzolites are enriched in alumina (4.1-7.2 wt.% and 3.5-6.4 wt.% in clinopyroxene and orthopyroxene, respectively) with relatively high titanium (0.05-0.44 wt.%) and sodium (0.22-1.29 wt.%) in the clinopyroxene. The pyroxenes from the harzburgites are poor in alumina (1.8-4.1 wt.%), titanium (0.02-0.22 wt.%) and sodium (0.02-0.35 wt.%).

Grains of olivine, up to 5-8 mm in diameter, show wavy and mosaic extinction and are elongated parallel to the axes of pyroxene porphyroblasts. The 100*Mg/(Mg+Fe) ratios of the olivines from the lherzolite and harzburgite are 89.1-91.1 wt.% and 91.0- 91.7 wt.%, respectively, although the nickel content in both is similar (0.37-0.43 wt.% and 0.38-0.43 wt.%, respectively).

Euhedral xenomorphic grains (up to 1 mm) of yellowish chrome-spinel are dominant in the lherzolites. Both idiomorphic and xenomorphic grains of reddish-brown chrome-spinel are present in the harzburgites. No segregation features are seen. Cr/(Cr+Al) ratios of the spinel from the lherzolite and harzburgite are 0.10-0.31 and 0.33-0.71 respectively, and the titanium contents are up to 0.07 and 0.10 wt.%, respectively.

In some lherzolites, a colorless monoclinic amphibole occurs as short prismatic or oval grains up to 0.3-0.5 mm long. Intense amphibolization is almost invariably associated with sheared rocks having a porphyroblastic texture and foliation structure. The post-deformation origin of the amphibole is evident in such cases, and its formation was accompanied by the recrystallization of the pyroxenes, as indicated by the disappearance of lamellae in the pyroxene grains, or by their segregation in the form of euhedral inclusions. According to the classification of Leake (1978), amphiboles from the amphibole lherzolites belong to magnesian and paragasite hornblendes. They contain high alumina (>10 wt.%) and sodium (2-4 wt.%), and moderately high TiO_2 (0.5-1.0 wt.%), whereas the potassium content is very low. The formation of amphibole in the lherzolite is not accompanied by a change in the Cr-spinel composition. The recrystallization of the pyroxenes in the course of amphibolitzation is reflected in a decrease of both the alumina (up to 1.9-4.8 wt.%) and titanium and sodium concentrations. In the case of the amphibole lherzolites, relicts of original pyroxene remain in the central part of large porphyro-blasts, and are identical to those from the amphibole-free lherzolites.

The pyroxenites and dunites have a hypidiomorphic granular and coarse granular to pegmatoid texture and a massive structure. The chrome-spinels are represented in these rocks by abundant small (up to 0.5 mm) opaque euhedral grains. The Cr/(Cr+Al) ratio in the spinels is 0.69-0.80 and the titanium content is 0.10-0.20 wt.%. Olivine forms rounded grains, relatively idiomorphic compared to the orthopyroxene. The 100*Mg/(Mg+Fe) ratio of the olivine from the pyroxenites is 89.8-90.1, and the NiO content is rather low (0.26-0.34 wt.%?). Orthopyroxene dominates in the pyroxenites and occurs as large hypidiomorphic short prismatic crystals. Xenomorphic grains of clinopyroxene and plagioclase (anorthite) are rare. The absence of deformation and solid solution lamellae is noted in the minerals from the pyroxenites. The pyroxenes in these rocks have a low Al_2O_3 content (0.7-2.1 wt.%) and are depleted in TiO_2 and Na_2O_3 (in the clinopyroxene these are 0.03-0.06 wt.% and 0.05-0.27 wt.%, respectively).

Geochemistry of the Mamonia Complex Ultramafic Rocks

Some samples of relatively weakly serpentinized harzburgite and lherzolite show evidence of isochemical metamorphism. Based on a recalculation of water-free compositions, the lherzolites, which contain 5-15% normative clinopyroxene in amphibole-free varieties, and up to 15% amphibole in the amphibole-bearing varieties, have an Al_2O_3 content varying between 1.5 and 4.05 wt.%. Most fresh lherzolites are characterized by almost constant values of the Al_2O_3:CaO ratio that are close to 1.17. From this recalculation, it can also be seen that the harzburgites with 0.7-2.3% normative clinopyroxene contain 0.5-0.8 wt.% Al_2O_3. For fresh harzburgites and lherzolites, the bulk rock Mg number (100*Mg/(Mg+Fe)) is nearly the same as that for olivine and orthopyroxene in these rocks.

The formation of the hornblende is in some cases attributable to isochemical reactions, although some Na influx is apparent. The loss of Ca (related to serpentinization) is evident in well-serpentinized rocks only where the LOI exceeds 10.7 wt.% (for lherzolites) and 12-13 wt.% (for harzburgites). The Mg-loss in a significant part of the lherzolites and harzburgites may be related either to serpentinization or to submarine weathering of peridotites (Snow and Dick (1995).

A comparision with other typical lherzolites with a similar depletion in the basaltic component (Al_2O_3 averaging 3.0 wt.% in a water-free composition) indicates that the Mamonia Complex lherzolites are poorer in sodium (about 0.10 wt.% in the amphibole-free varieties). Lherzolites originating from subcon-tinental mantle material typically have sodium contents varying between 0.20-0.23 wt.%.

Metamorphism of the Mamonia Complex Ultramafic Rocks

Metamorphism of Mamonia lherzolites and amphibole lherzolites occurred at temperatures of 880±50°C and pressures of 11-16 kbars; based on the amphibole stability field, the water pressure under these conditions was not less than 5 kbar (Bazylev et al., 1993). Higher temperatures (830 to 1090°C) and lower pressures (8 to 13 kbars), characteristic of metamorphism of some of the Mamonia harzburgites, were probably in response to their location near active channelways marked by an accumulation of dunite and pyroxenite on their walls.

The composition of high-temperature pyroxene generation corresponds to temperatures of about 1040-1070°C at an accompanying pressure of about 8-13 kbar. These estimates conform probably to the final stage of pyroxene crystallization, and are close to the equilibrium parameters calculated for harzburgites localized near these pyroxenites.

The subsequent medium-temperature metamorphism involves the replacement of spinel by ferrite-chromites and of primary pyroxenes by tremolite and chlorite. The composition of metamorphic spinels requires a temperature of crystallization of 450-580°C (except for very rare estimated values of up to 620°C), judging from the solvus estimation in the Mg-Fe-Cr-Al system (Sack and Ghiorso, 1991). Such metamorphic occurrences are mainly restricted to zones of contemporaneous faulting and rock-fracturing within harzburgite sheets and at the contacts of diverse peridotitic rocks with basalts and metamorphics, where metamorphic silicate minerals are wholly replaced by the late serpentine.

Serpentinization of Mamonia peridotites is the rule rather than the exception. The lherzolites and harzburgites are in all cases partially or wholly replaced by serpentine. Large volumes (80 to 100%) of these rock have been converted to serpentine and their freshest varieties contain about 50% serpentine. In the dunites, the serpentinization is complete. Pyroxenites show much greater resistance to serpentinization. In some of the orthopyroxenites, only relatively little serpentinization (up to 10-15%) is apparent. In peridotites with a massive structure and inconspicuous banding, olivine, the predominant mineral, is replaced by greenish mesh serpentine, and pyroxenes by greenish bastites; colorless serpentines with an hourglass structure are also present is some rocks. In foliated peridotites, the development of fine-grained colorless serpentine accompanied by magnetite is usual. Petro-graphically, there is no evident distinction between the serpentinization of the Mamonia peridotites and those of ophiolitic ultramafic rocks (Wicks and Whittacker, 1977) or of oceanic peridotites (Bazylev, 1989).

Carbonatization effects are not very common in Mamonia Complex peridotites, being mainly restricted to the sheared ultramafics. A common carbonate is MnO-enriched (up to 1.6%) calcite. SrO-rich (0.70 wt.%) carbonate, which is believed to be the result of aragonite recrystallization, is also present. Such Sr-enriched aragonites characteristically occur in abyssal peridotites exposed on the modern-day ocean floor (Bonatti et al., 1980; Bazylev, 1995b).

Petrogenesis of the Mamonia Complex Peridotites

When compared with oceanic peridotites (Dick and Bullen, 1984), Mamonia Complex spinel lherzolites and harzburgites show a wide range of Cr# (Cr/(Cr+Al) values in the spinels which overlap those of oceanic sources. Ophiolitic complexes showing these features are interpreted as having a polygenetic igneous origin that reflects their multistage development (Dick and Bullen, 1984). As shown in **Fig. 2.13**, the Mamonia spinels have a clear bimodal distribution of the Cr-content, with two distinct peaks: one (Cr# ~0.2) is characteristic of the lherzolites and the other (Cr# ~0.6) indicative of the harzburgites. Hence, at least a two-stage development of the Mamonia Complex ultrabasites may be assumed.

Cr# variations in the spinels from the Mamonia harzburgites are generally comparable with those of island-arc basement harzburgites (Dick and Bullen, 1984; Arai, 1987). This similarity suggests that the Mamonia harzburgite may have been formed in a similar geodynamic setting and can represent a product of mantle melting above a subduction zone. Mineralogically, the Mamonia harzburgites closely resemble those from the Troodos Massif (Allen, 1975; Sobolev and Batanova, 1995). The composition of the clinopyroxenes reflect the influence of a subduction component in the formation of the Troodos harzburgites **(Fig. 2.14**; Sobolev and Batanova, 1995). The original alumina content in their fractionated products is between 10.2 and 14.5 mass.% (Bazylev et al., 1993). These values accord well with compositional characteristics of primary magmas of the Troodos Complex extrusives (Sobolev et al., 1993), considered to be parent melts for both the boninitic-type magmatic series and the island-arc tholeiites. In the author's opinion, partial melting (about 20% at a pressure of 15-20 kbar, in the presence of an independent source of Na and K) of mantle material corresponding to average Mamonia lherzolite, could result in the generation of melts similar to those estimated for Troodos (Sobolev et al., 1993), and for residuals similar to the spinel harzburgites of the Troodos Massif and of the Mamonia Complex (dotted trend in Fig. 2.14; Bazylev et al., 1993).The cumulate peridotite bodies occur as magma channelways in the Mamonia harzburgites. These cumulates represent products of crystallization of melts ultimately leaving the harzburgite at depths corresponding to pressures of not less than 7-8 kbar.

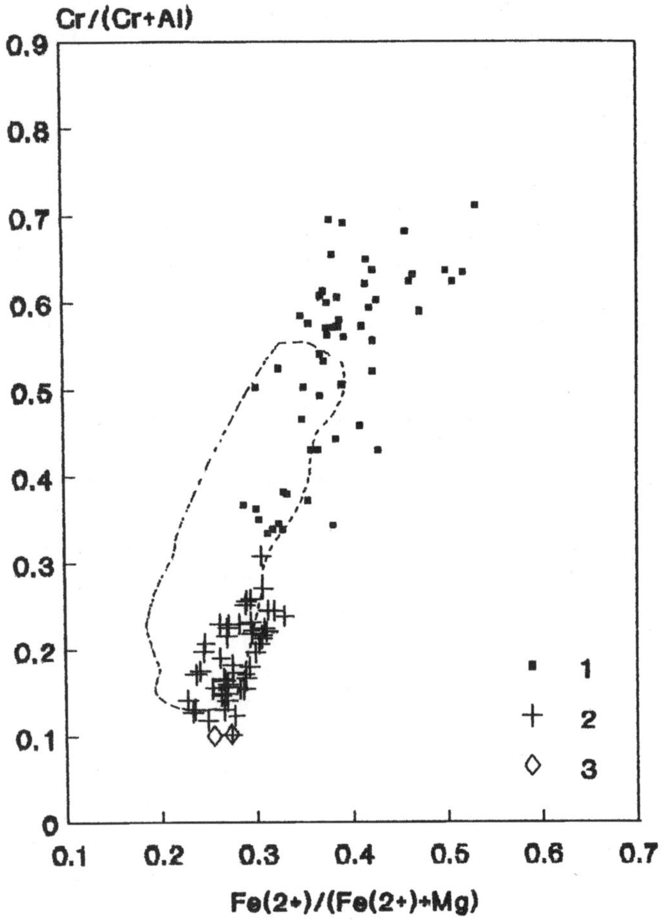

Fig. 2.13. Chrome spinel compositions in the Mamonia Complex ultramafics.1 – harzburgites, 2 – lherzolites, 3 – the least depleted oceanic spinel lherzolites (Bonatti et al., 1993). The field of spinel compositions in abyssal spinel peridotites shown as in Dick and Bullen (1984).

The important aspects of petrogenesis of the Mamonia Complex lherzolites are the limited ranges of Cr# in the spinels (0.10-0.31), the significant variations of the Na-content in the primary clinopyroxenes, and the local occurrence of pargasite amphiboles. It is clear from Fig. 2.14 and **Fig. 2.15** that the most primitive lherzolite of the Mamonia Complex (sample 257M; Loutra tis Aphroditis) closely resembles the most primitive lherzolites of mid-ocean ridges (Bonatti et al., 1993). These mid-ocean lherzolites are associated, however, with alkali olivine basalts (generally containing normative nepheline), which differ considerably from typical mid-ocean ridge magmatic rocks. Moreover, these most primitive oceanic lherzolites are very similar in mineralogy to subcontinental mantle lherzolites: for both, the low-Cr# spinels and clinopyroxenes with stable high Na content relative to oceanic peridotites are typical (Bonatti and Michael, 1988; Bazylev, 1995a). The Cr# level in the spinels from the Mamonia lherzolites resembles that of spinel lherzolites generated at slowly spreading mid-oceanic ridges (Dick, 1989), although the spinels of oceanic lherzolites are generally more enriched in Cr, and the Na content of the pyroxenes from some of the Mamonia lherzolites are higher.

Significant variations in Na concentration in the primary clinopyroxenes from the Mamonia lherzolites were interpreted as evidence for the presence of analogs of both oceanic-type and subcontinental-type lherzolite residues derived from a melt yielding both MORB-like tholeiitic and intraplate-type magmas, respectively (Bazylev et al., 1993). Later, the author showed that the lherzolites with both the low-Na pyroxene and the high-Na pyroxene do not form discrete units but occur as alternations in the continuous section (Loutra tis Aphroditis). From this distribution one can conclude that the Mamonia Complex lherzolites might have formed in a geodynamic setting representing a transitional stage from the continental to oceanic rifting.

In accordance with the petrogenetic model of rift genesis (Bonatti and Seyler, 1987; Bonatti and Michael, 1987), the transition from alkaline to tholeiitic magmatism takes place during the initial spreading stages, and the mineralogical features of the mantle residual peridotites can be attributed to the transition from typical subcontinental to typical oceanic peridotites. This type of peridotite exposures occur within areas of transition from continental to oceanic crust which mark recent or ancient passive continental margins. In fact, the primary mineral compositions of the Mamonia lherzolites closely resemble those of the Iberian (Galician) continental margin peridotites (Evans and Girardeau, 1988), which contain spinels with Cr# = 0.10-0.35, olivines with Mg# = 89.4-90.6, and primary clinopyroxenes (not equilibrated

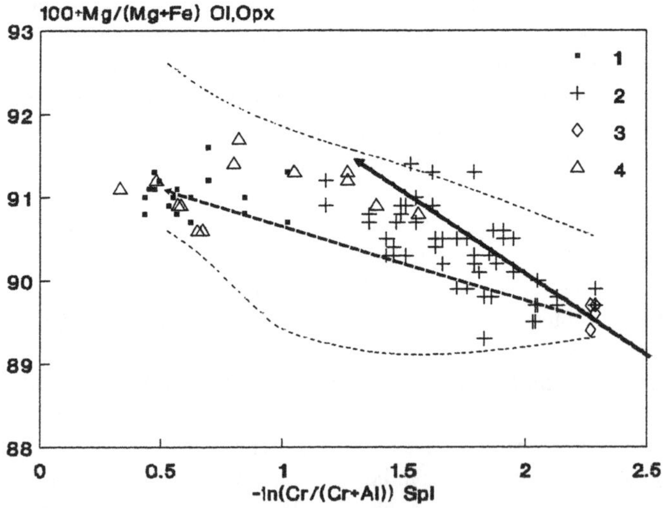

Fig. 2.14. Variations of Mg# (100*Mg/(Mg+Fe)) in olivine and orthopyroxene from the Mamonia Complex ultramafics. Continuous arrow indicates the residual mineral chemistry changes during lherzolite formation; discontinuous arrow – those during harzburgite formation. Discontinuos lines limit the Olivine-Spinel Mantle Array (Arai, 1987). 4 – spinel harzburgites, clinopyroxene harzburgites and spinel lherzolites of the Troodos Massif (Sobolev, Batanova, 1995). Other indexes are same as in Fig. 2.13.

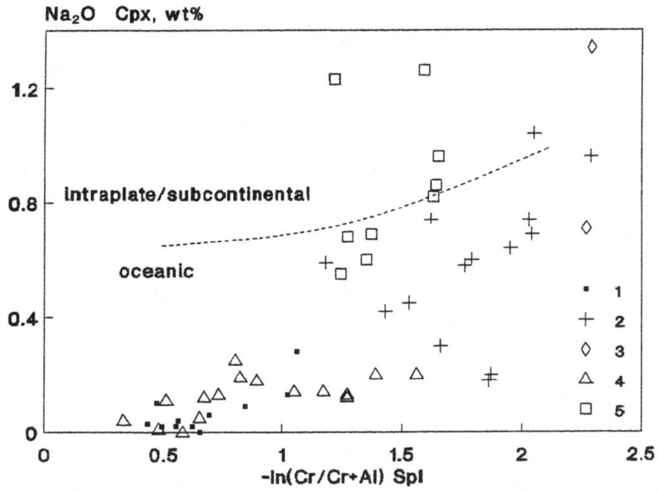

Fig. 2.15. The dependence of the clinopyroxene sodium content on the spinel Cr # in the Mamonia Complex Ultramafics. The discrimination line between the oceanic-type and intraplate (subcontinental)-type spinel peridotites is shown as in Bazylev, (1995). 5 – Lherzolites of the Iberian passive continetal margin (the relict mineral compositions which are not equilibrated with plagioclase are plotted only) (Evans and Girardeau, 1988; Kornprobst and Tabit, 1988; Agrinier, Mevel and Girardeau, 1988). Other indexes (1 - 4) are the same as in Figs. 2.13 and 2.14.

with the plagioclase) with highly variable sodium concentrations (Fig. 2.15).

Quantitative temperature and pressure estimates for the Mamonia lherzolite subsolidus metamorphism are in good accord with Ringwood's suboceanic temperature interval (Ringwood, 1975). However, the high water pressure at such deep levels is uncharacteristic of the lithosphere in the axial zone of oceanic rifts, and pargasite hornblendes are atypical for mid-oceanic ridge peridotites. By contrast, the pargasite mineralization considered to be caused by the occurrence of pre-rifting water metasomatism (Agrinier et al., 1988) is common in the peridotites formed during the stages of transition from continental to oceanic rifting (*e.g.* Zabargad, St. Paul, Iberian margin). On the other hand, there are cases of close spatial association of pargasite-bearing spinel lherzolites with proven island-arc rocks, suggesting the existence of subduction zones which served as sources of water, localized below the amphibolized mantle rocks (Bonatti et al., 1990). In the case of the Mamonia lherzolites, in accordance with the appropriate geodynamic formation setting of these rocks proposed above, the source of water during lherzolite amphibolization can hardly be related to the subducted slab. This problem needs further investigation.

Judging from the primary mineral compositions of the Mamonia lherzolites, the segregated melts span a diverse compositional range from MORB-type tholeiites to intraplate-type alkali olivine basalts. If it is assumed that these escape from the residuals at a depth corresponding to 20 kbar (which is equivalent to the upper pressure limit of the lherzolite subsolidus metamorphism), the alumina content estimates made for these melts (Bazylev et al., 1993) indicate values between 13.7-15.3 wt.%. In accordance with the proposed structural model of the lithosphere (Bonatti and Seyler, 1987), at the stage of formation of the Mamonia lherzolite, the escaped melts could undergo both high-level differentiation and contamination due to interaction with continental crust material. From the above evidence, the Dhiarizos Group lavas are probably extrusive expressions of the same magma generation events.

In contrast to the Mamonia lherzolites, the Troodos Massif spinel lherzolites investigated by Sobolev and Batanova (1995) display a significantly greater degree of partial melting relative to the Mamonia lherzolites: spinel is characterized by Cr# = 0.21-0.35, and olivine and orthopyroxene have Mg# = 89.9-91.5. The stable low levels of sodium in the clinopyroxenes is also a diagnostic feature of these rocks (Fig. 2.15). High temperature amphibolization does not occur in the Troodos lherzolites.

The clinopyroxenes from both the Troodos spinel lherzolites and harzburgites are anomalously enriched in LREE, and Sr and Zr which seem to be derived from a subducting slab, suggesting a genetic affinity with the Troodos spinel lherzolites and harzburgites (Sobolev and Batanova, 1995).

In fact, some of the Mamonia lherzolites might be genetic analogs of the Troodos-type spinel lherzolites rather than of the more typical and dominant slightly depleted Mamonia lherzolites. The presence of both moderately chromian spinel and low sodium chlinopyroxene in part of the Mamonia lherzolites does not eliminate the possibility of their being of island-arc origin.

Geological Evolution of the Mamonia Complex Ultramafics

From the evidence above, the Mamonia Complex ultramafics include polygenetic rocks of different ages. The formation of most of the lherzolites reflects the earliest stage in the occurrence of oceanic spreading in the Mesozoic Tethys, which took place in the Triassic (Malpas and Xenophontos, 1987). After melt extraction, the lherzolites could be emplaced into the overlying crust (Dhiarizos group volcanics), as were peridotites of passive continental margins. The high temperature amphibolization affecting the lherzolites probably took place at the same stage. However, the Mamonia lherzolites do not contain plagioclase which is characteristic of the peridotites occurring on the ocean floor of continental margins or reflecting pre-existing continental rifting (*e.g.* Zabargad and Iberian margin).

All these features, plus the absence of metamorphism, correspond to a temperature range of 600-830°C, suggesting that when tectonic emplacement of the Mamonia lherzolites took place, it was characterized by a small amplitude zone restricted to the lowest crustal levels. In other words, after the segregation of melts, the Mamonia lherzolites cooled at deep levels outside the plagioclase stability field (at an assumed pressure of more than 6 kbar).

The next stage in the geologic evolution of the Mamonia

peridotites is marked by tectonization, metamorphic recrystallization under the conditions of upper greenschist-amphibolite grade (450-600°C), and by tectonic juxtaposition of the lherzolites and harzburgites, suggesting a closing of the Mesozoic Tethys and a collision of the African plate with an island arc during the Late Campanian-Maastrichtian (Malpas and Xenophontos, 1987).

If the generally accepted model of northward-dipping subduction of the Tethys oceanic lithosphere is used, this subducted lithospheric plate must have bent adjacent to the trench causing both the extension of upper lithosphere levels (and probable formation of normal tension fractures) and the compression of lithospheric mantle. These strains progressively increased during the stage of convergence of the trench with the thicker lithosphere of the northern passive margin of the African Plate. These strain-related openings could allow penetration of water into the deep levels of the passive margin lithosphere, causing hydration of the lherzolites, and their extrusion into the upper crustal levels.

Because evidence is lacking of a simultaneous metamorphic recrystallization outside the schistosity zones, the uplift of the lherzolite blocks must have been rapid. As with the lherzolites, the overlying lower-crustal gabbroids (in agreement with the proposed structural model of continental margin lithosphere of Bonatti and Seyler, 1987) probably also experienced metamorphism, uplift, and schistosity. The separate metabasite blocks of the Ayia Varvara Formation, with no field evidence for temperature zoning, can be interpreted as just such gabbroids, judging from their mineralogical and chemical features, and the P-T conditions of metamorphism estimated by Silantyev (1993).

The established close association of lherzolites with harzburgites within a number of exposures throughout the Mamonia Complex (Bazylev et al., 1993) seems to reflect a complex juxtaposition of tectonic blocks or slivers representing both passive margin and island arc lithospheric fragments. Malpas and Xenophontos (1987) proposed just such a juxtaposition of crustal blocks, formed in different geodynamic settings, based upon mapping of the volcanic rocks of the Mamonia Complex. In fact, the ultramafics also seem to provide an opportunity to identify the lithosphere blocks from the different geodynamic settings.

Finally, it should be noted that sporadic findings of spinel lherzolites are known in other parts of the southern margin of the Tethys, including the Bassit Massif in Syria (Silantyev, 1993) and the ophiolites of Oman (Browning, 1982). Some mineralogical features of these rocks (spinel and clinopyroxene compositions) allow us to interpret them as analogs of the Mamonia Complex spinel lherzolites. Therefore, SW Cyprus is evidently not the only region in the Eastern Mediterranean where exposures of peridotite are preserved that were formed during the earliest stage of spreading of the Mesozoic Tethys.

E. Gabbroids of the Mamonia Complex

Sergey A. Silantyev
Vernadsky Institute of Geochemistry and Analytical Chemistry of the Russian Academy of Sciences,
Kosygin Str. 19, 117975 Moscow, Russia

Introduction

Gabbroids are the least studied igneous rocks in SW Cyprus. Their exposures are scattered throughout the area of the Mamonia Complex, from the Xeros Potamos River in the east to the Akamas Peninsula in the west. Typically, their outcrops are closely associated with tectonized serpentinite sheets, deformed pillow lavas, and dolerite dikes and sills. Because of the sporadic exposure of relatively large gabbroid bodies, their compositional characteristics and relationships with associated rock units may be illustrated from the more extensive and better-studied exposures.

At the middle reach of the Xeros Potamos River, south-east of the village of Galataria, sporadic outcrops of isolated gabbroid bodies are closely associated with dolerite dikes in the narrow gorge on the west bank of the river. Both sides of this gorge are mainly composed of intensely deformed serpentinite and subordinate pillow lavas and lithified (mainly arkosic) sediments. Tectonized serpentinite, which is tectonically juxtaposed with the gabbroids at the mouth of the gorge, is also exposed across the river from this locality.

The most extensive gabbroid outcrops are in the lower reaches of the Ezousa River, where some of them reach 180 x 250 m in size. On the left side of the valley, these rocks are exposed to the north of the Ayia Varvara metamorphic block, which is the largest exposure of metamorphic rocks within the Mamonia Complex.

A large serpentinite sheet, whose southern and northern boundaries are marked by fault zones filled with serpentinite-tectonite, separates these gabbroids from the metamorphics. Gabbro and associated diabase bodies occur as separate small blocks within the lava succession whose northern edge is in faulted contact with the serpentinite-tectonite. On the right bank, across the river from these gabbroid outcrops, occurs the most extensive gabbro exposure in SW Cyprus. Its northern edge is covered by alluvium, and the southern edge is in faulted contact with serpentinite which is tectonically juxtaposed with a thick lava succession, including a long diabase sill. As in the west bank exposures, the southern boundary of this succession of pillow lavas coincides with a fault zone filled with tectonized serpentinite.

To the south of this zone, there is a tectonic block composed of gabbroids and diabases enclosed in serpentinite. In addition, small gabbroid and diabase fragments and blocks are incorporated within the serpentinite melange. According to Malpas and Xenophontos

(1987), the lava successions occurring on both sides of the Ezousa River valley belong to both the Dhiarizos Group volcanics and the Troodos-type high-magnesian lavas which are in close association with fault-bounded serpentinite sheets. In the Ezousa River valley it appears that the E-W and ESE-WNW trending fault zones bound the main lithologies within the magmatic complex.

The greater part of the Akamas Peninsula is composed of Troodos ophiolite-related hyperbasites and pillow lavas (Lapierre, 1975), whose outcrops form a SSE-NNW trending belt. Sporadic outcrops of gabbroids associated with diabases occur in the central part of the peninsula, at the Kefalovrysia area, where the tectonized serpentinite sheets alternate with bodies of highly altered diabases, gabbroids, and pillow lavas. Evidence is found of intense hydrothermal activity, which is characteristic of the pillow lavas that show abundant sulfide mineralization. Some pillow lavas are akin to the Troodos boninitic lavas, with a light gray color and the presence of abundant olivine phenocrysts. In some cases, the relationships between the diabase and microgabbro suggest their genetic affinity with the same subvolcanic series. To the north-west of the village of Neon Chorion, the diabases form outcrops up to 300 m long with morphologic features conformable with the Troodos sheet dikes. In this dike complex, the transition from typical dolerite to microgabbro with an ophitic texture is continuous.

Mineralogy and Geochemistry of the Mamonia Complex Gabbroids

Xeros Potamos River Valley Gabbros.

Texturally diverse gabbroids outcropping within this area are represented by microgabbro, ophitic gabbro, gabbro with distinctive cumulate textures, and pegmatitic gabbro. Their closely associated rocks are diabases. In some cases, the transition from diabase through gabbro-dolerite to medium crystalline gabbro is continuous. Shattered contacts of gabbroids with the surrounding rocks (typically serpentinites) mask the original relationship of the gabbroids with their host rocks, thus reconstruction of the history of the gabbroid bodies in this area is always difficult.

The cumulate gabbro consists of clinopyroxene, plagioclase, titano-magnetite, and secondary actinolite and chlorite. Spinel is present in more leucocratic varieties. The highly deformed pegmatoid gabbro consists predominantly of pyroxene (diallage) and plagioclase. The associated diabases have the same mineralogical composition but also contain secondary low-temperature carbonate. The clinopyroxene from the cumulate gabbro is ferrous augite ($En_{38}Wo_{42}Fs_{20}$) or augite ($En_{45}Wo_{43}Fs_{12}$), and is similar in alumina and titanium content to augites from the alkali igneous rock group of monozonitoids (**Fig. 2.16A, B**). The plagioclase in these rocks is completely albititized (An_{1-3}). Actinolite and chlorite replace clinopyroxene, probably as an equilibrium association with the albite.

The major element chemistry of the gabbroids occurring at the middle reach of the Xeros Potamos River corresponds to composition fields of alkaline igneous rocks: Na_2O = 5.01-5.28, K_2O = 0.76-0.59, TiO_2 = 1.64-1.41, MgO = 7.65-7.55, and SiO_2 = 49.66-50.39 (**Fig. 2.17**).

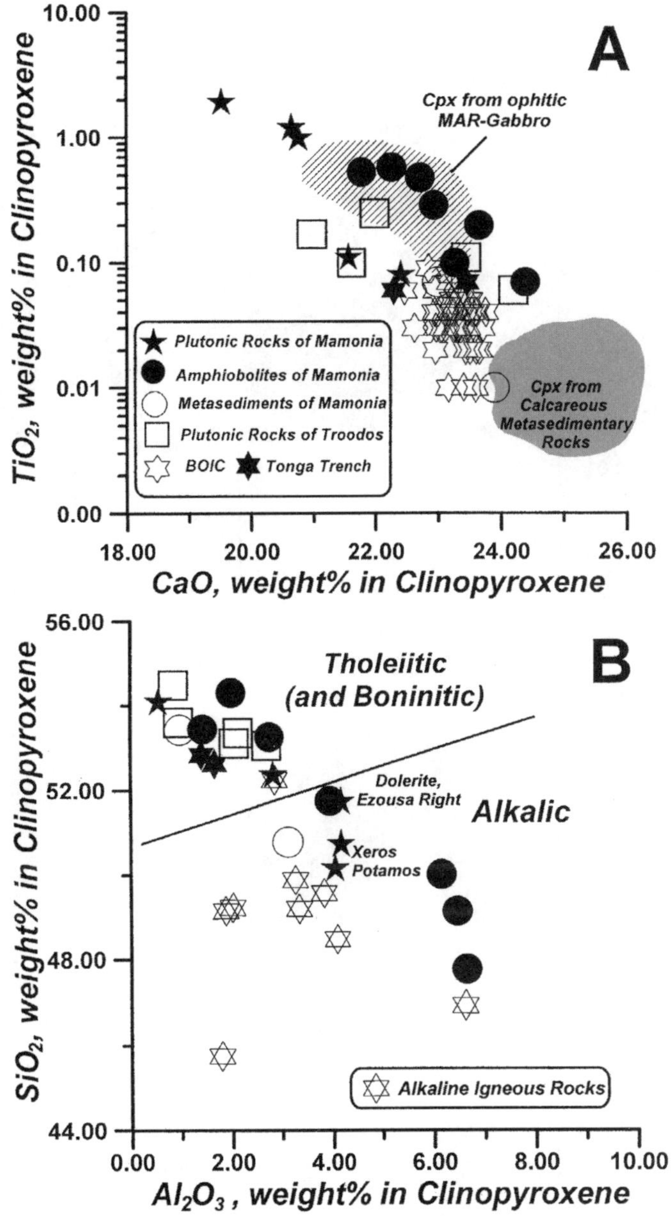

Fig. 2.16. Variations of compositions of Clinopyroxene in plutonic and metamorphic rocks from the Mamonia Complex. A - TiO_2 vs Cao; B - SiO_2 vs Al_2O_3. Data for the Troodos Ophiolite are presented by Thy (1990), for the Bay of Islands Ultramafic Cumulates (BOIC) by Elthon et al. (1984); Komor et al. (1985), and for the Tonga Trench by Bloomerand Fisher (1987). The compositional fields for pyroxene from tholeiitic and alkaline magmatic rocks shown as in Natland 1980). The compositions of clinopyroxene from alkaline igneous rocks (shown on Fig. 1.16B as open stars) were presented by Deer, Howie and Zussman (1963).

The pattern distribution of the rare earth elements (**Fig. 2.18A** and Fig. 2.20) is also supported by these ratios: (La/

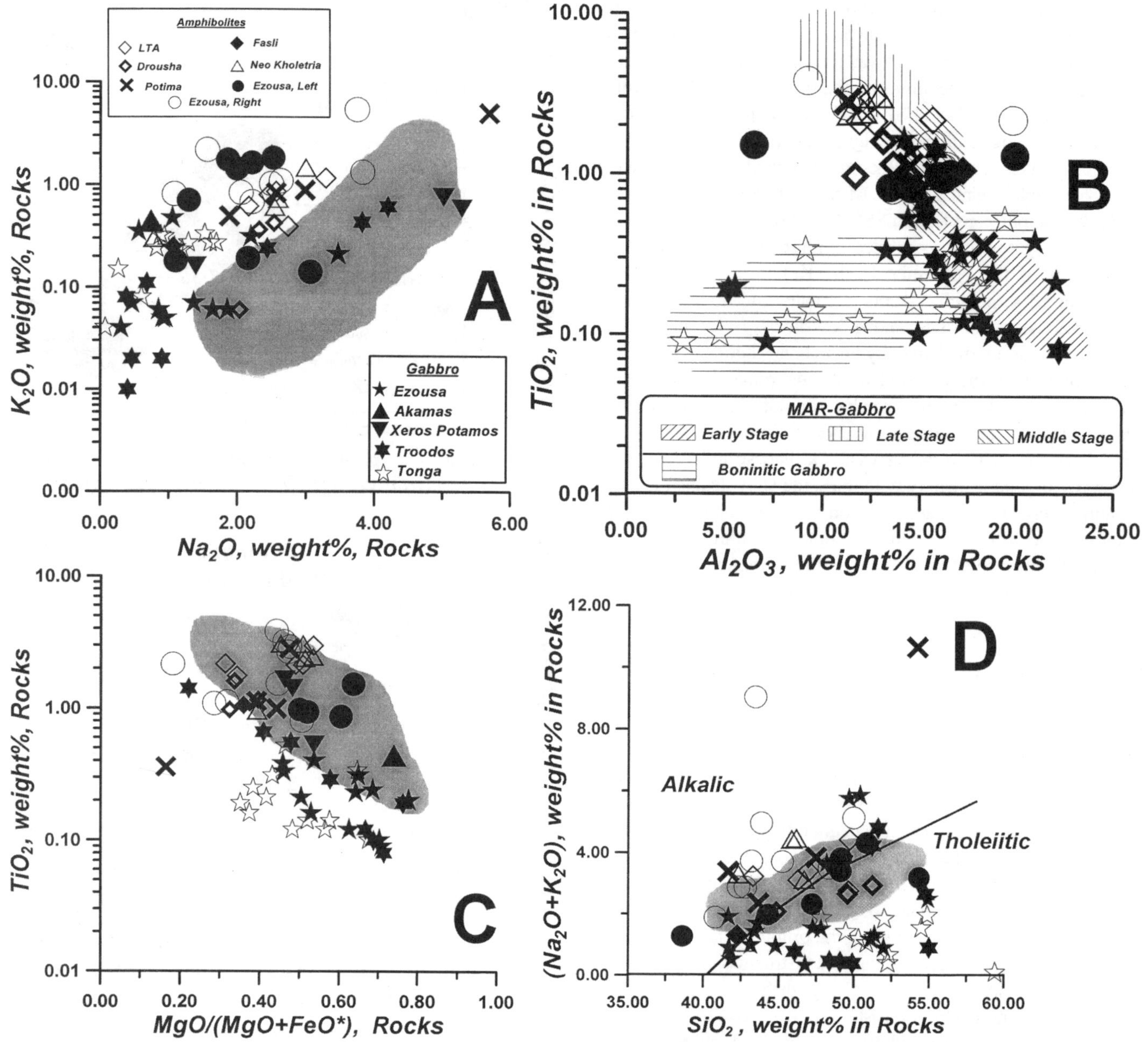

Fig. 2.17. Variations of major elements content in plutonic and metamorphic rocks from the Mamonia Comlex. Compositions of gabbro from the Tonga trench are presented by Sharaskin et al. (1983), from the Troodos ophiolite by Moores and Vine (1971) and Coleman (1977). The compositional fields for MAR-Gabbro of the early, middle, and late stages are presented after Miyashiro and Shido (1980). The compositional fields of alkalic and tholeiitic magmatic series shown as Wilson (1989).

$Sm)_{cn}$=9.94 and $(Yb/Sm)_{cn}$=1.09. The major element covariance indicates immobility during the metamorphism of the gabbroids; components such as CaO, Na_2O, SiO_2, FeO, and MgO reflect the lack of significant shifts in the primary composition of the gabbro from the Xeros Potamos River exposures during the course of their alteration. On the other hand, the carbonatization of their associated diabases must have had an influence on the primary composition of these rocks.

Ezousa River Valley Gabbros.

These are largely cumulate gabbros which vary considerably in their plagioclase content, from melanocratic two-pyroxene-olivine gabbro-cumulate to clinopyroxene-plagioclase pegmatoid and plagioclase-enriched leucocratic gabbro. Their closely associated diabases are typical greenstone-altered dolerites exhibiting either an intersertal texture or a more coarse-grained, micro-ophitic texture. Almost all these gabbroids and diabases contain a secondary mineralogical association including amphibole (aluminous actinolite), chlorite, epidote, and albite. The gabbroids and dolerites occur as isolated bodies mostly located close to the shatter zone and faults.

The pyroxene from the clinopyroxene-plagioclase gabbro is augite ($En_{44}Wo_{46}Fs_{10}$) and is similar in alumina and titanium content to that in igneous rocks with normal alkalinity (Fig. 2.16A, B). The plagioclase in these rocks is partially albitized and represented by bytownite (An_{89}). The melanocratic gabbro is composed of augite ($En_{46}Wo_{46}Fs_8$), orthopyroxene (bronzite: $En_{79}Wo_3Fs_{18}$), associated olivine ($Fo_{80\text{-}81}$) and subordinate plagioclase; the secondary amphibole is tremolite, and the epidote is close to clinozoisite ($Al/(Al+Fe^3)$=0.99). The diabase consists mainly of clinopyroxene and plagioclase, similar in composition to those from the gabbro: the clinopyroxene is $En_{40}Wo_{44}Fs_{16}$ and the plagioclase ranges from An_{85} to An_{89}. The secondary mineral association characteristic of the greenstone-altered basaltoids includes amphibole (actinolite-aluminous actinolite), chlorite, and albite, and is identical to that of their associated altered gabbroids.

There is no correlation between the major element content of the Ezousa River gabbroids and their degree of alteration. This implies that the differences in chemical composition between the protolith of these rocks and products of their alteration are comparatively small. On the other hand, the albitization of the plagioclase evident in some diabase specimens should have accompanied the enrichment in Na and loss of Ca, in accordance with the compositional change during spilitization of the basaltoids. Petrochemically, the melanocratic cumulates outcrop-ping on both river banks (including the large gabbro block on the right bank of the Ezousa River valley) are identical to the ultramafic cumulates of the boninitic series: (values in mass.%) SiO_2=46.76, TiO_2=0.09, MgO=25.9 and SiO_2=41.88, TiO_2=0.20, and MgO=30.10 (Fig. 2.17).

Moderately magnesian-enriched gabbroids from the same area are attributable to more leucocratic and felsic derivatives of the same magmatic series: SiO_2=43.13-51.37, TiO_2=0.10-0.30, MgO=9.42-14.75 (Fig. 2.17). Direct evidence for the boninite nature of the Ezousa River gabbroids comes from their chondrite-normalized REE-element abundances with well-developed V-shaped patterns (**Fig. 2.19**). **Fig. 2.20** shows the variation of $(La/Sm)_{cn}$ and $(Yb/Sm)_{cn}$ values in these rocks, which range from 1.57 to 4.18 and from 1.47 to 5.43, respectively. It is clear that the above data on the clinopyroxene compositions are in agreement with the assumption of the boninitic nature of the Ezousa River gabbroids, because the compositional parameters of the mineral data (Fig. 2.20B) do not allow separation with confidence between the tholeitic and boninitic rock groups.

The associated diabases have identical mineralogical and geochemical signatures. Their resemblance to the gabbroids is supported by the following geochemical data: $(La/Sm)_{cn}$=1.57 and $(Yb/Sm)_{cn}$=1.47 in rocks from the left bank of the Ezousa River, and $(La/Sm)_{cn}$=6.89 and $(Yb/Sm)_{cn}$= 2.71 in those from the right bank (Fig. 2.20).

From the above evidence, the Ezousa River gabbroids and

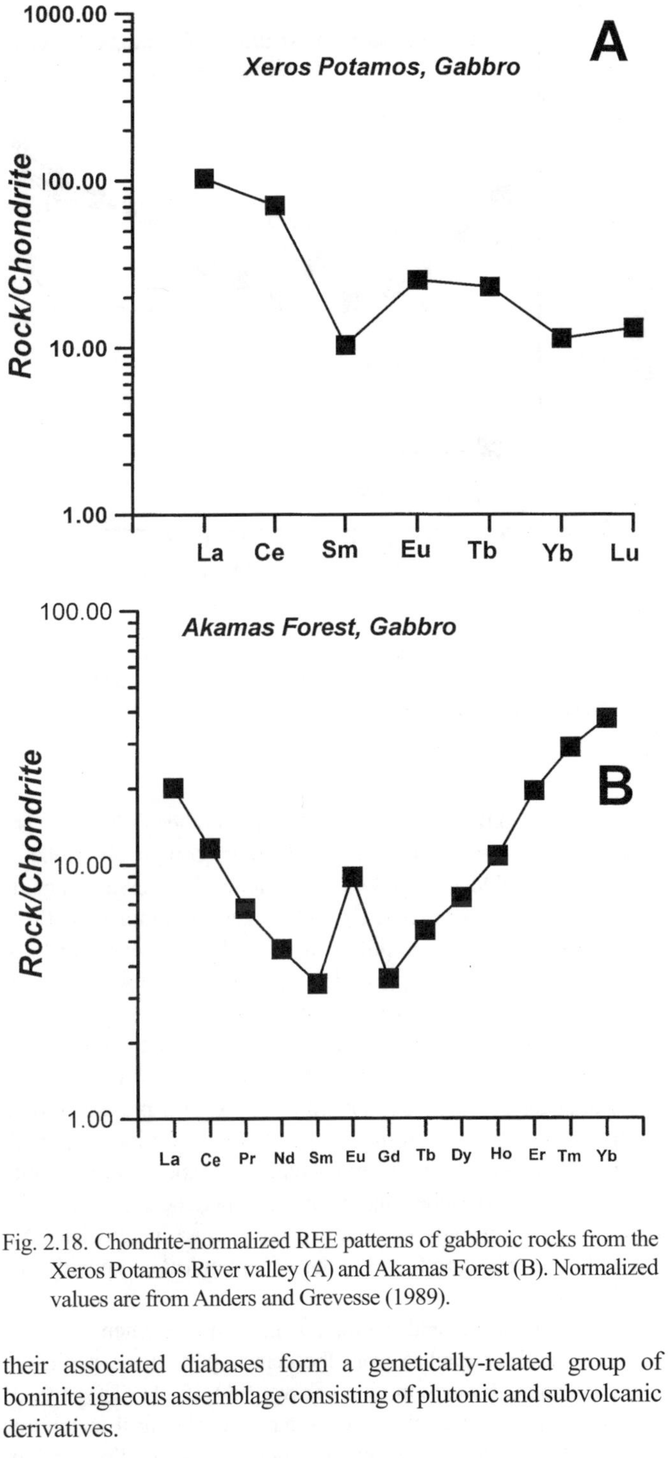

Fig. 2.18. Chondrite-normalized REE patterns of gabbroic rocks from the Xeros Potamos River valley (A) and Akamas Forest (B). Normalized values are from Anders and Grevesse (1989).

their associated diabases form a genetically-related group of boninite igneous assemblage consisting of plutonic and subvolcanic derivatives.

Akamas Peninsula Gabbros.

The rocks are typically microgabbros. Their relationships with the hosted diabases and spilitizated pillow lavas suggest that all these rocks are an integral part of a single magmatic series, where they form either a sheeted dike complex or subvolcanic sills. In the case of the extensive exposures (not less than 300 m in length) at the middle part of the peninsula, the diabase forming the

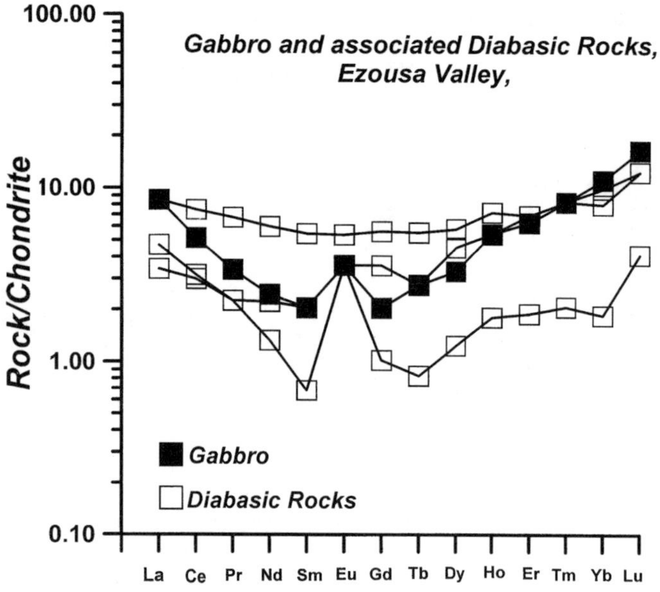

Fig. 2.19. Chondrite-normalized REE patterns of gabbro and associated diabasic rocks from Ezousa River valley. Normalized values are from Anders and Grevesse (1989).

Conditions and Geodynamic Nature of Metamorphism of the Mamonia Complex Gabbroids

Elsewhere in SW Cyprus, the gabbroids contain a secondary mineral association of constant phase composition, usually including actinolite, aluminous actinolite, albite, epidote, and chlorite. Their associated diabases have the same metamorphic parageneses but in some cases show evidence of carbonatization. An identical association is typical of modern oceanic metagabbros and metadolerites, and it is also characteristic of the products of metamorphism of the Troodos gabbros and dolerites.

The variation in alumina content in the amphibole from the gabbros (0.34-7.46 wt.%) and diabases (2.82-6.75 wt.%) of SW Cyprus emphasizes the greenschist grade of metamorphism (400-500°C). In addition, the chlorine content of the amphibole from the Mamonia Complex plutonic rocks is identical to low-alumina amphiboles from oceanic greenschist metabasites. Significantly, the Mamonia gabbroid amphibole is very similar in composition to the metamorphic amphibole of the boninitic gabbros of the Tonga Trench, metamorphosed under the same conditions (Silantyev and Zlobin, 1986). The association of albite (An_{1-3}) with chlorite and zoisite in the Mamonia's gabbros and diabases reflects a low-temperature (300-350°C) stage of metamorphism, which has

sheeted dike complex and (rarer) bedded bodies (sills) consists of clinopyroxene and plagioclase, which are completely replaced by actinolite, chlorite, epidote, and albite. This assemblage is typical of spilite metabasites. Another feature common to these rocks and their associated pillow lavas is clear evidence for local hydrothermal activity, expressed in the abundant sulfide mineralization, and the concomitant formation of calcite, quartz, and chlorite. Within thick (not less than 300 m) dike swarms, the transition from diabase through gabbro-dolerite to microgabbro is continuous. Typical microgabbro of this kind is present in dike complex outcrops west of the village of Neon Chorion. This rock exhibits a subophitic texture and is mainly composed of actinolite replacing clinopyroxene, and albitizated plagioclase; bytownite (An_{90-92}) is present only as relicts of the primary plagioclase. No marked differences are found in the mineralogical composition of the dolerites.

There is a general similarity in major element content between the microgabbro from the central part of the peninsula and the Ezousa River boninitic gabbro: SiO_2=51.11, TiO_2= 0.33, and MgO=11.23. Both these rocks have geochemical signatures associated with boninite magmatism derivatives: $(La/Sm)_{cn}$ =5.89 and $(Yb/Sm)_{cn}$=11.04 (Figs. 2.17, 2.18 and 2.20).

The gradual transition between the diabases forming the sheeted dike complex and the microgabbro implies the genetic affinity of these rocks in an association of boninite-like gabbroids and diabases for the Ezousa River valley and the central portion of the Akamas Peninsula.

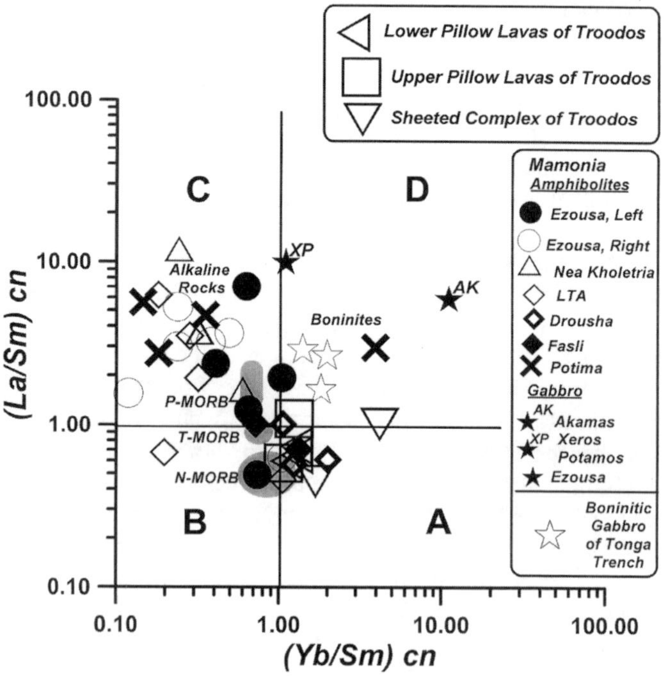

Fig. 2.20. Variations of $(La/Sm)_{cn}$ versus $(Yb/Sm)_{cn}$ in plutonic and metamorphic rocks from the Mamonia Complex. Volcanic rocks of the Troodos are presented by Portnyagin et al. (1996); boninitic gabbro of the Tonga Trench by Sharaskin et al. (1983); composition fields of MORB by Schilling et al. (1983) and Wilson (1989). Symbols as on Fig. 2.17.

a regressive relationship with a higher-temperature para-genesis including the aluminous actinolite. These relationships between secondary minerals are typical of the third layer of the oceanic crust, and are diagnostic features of oceanic meta-morphism.

The following geochemical tendencies are characteristic of modern oceanic crust metamorphism:

1) CaO loss and Na_2O gain ('spilitization trend');

2) FeO increases with decreasing SiO_2, CaO and Na_2O;

3) Enrichment in CaO and loss of SiO_2 and Na_2O ('carbonatization trend'); and

4) Inert behaviour of major elements (Silantyev, 1995).

In this regard, the low mobility of major elements (or even immobility) during metamorphism of the gabbros in SW Cyprus, as well as the tendency toward splititization and carbonatization of some diabases, is an additional argument for an oceanic metamorphic origin for their alteration. It is important to stress that hydrothermal features, including sulfide mineralization, and pyrite+calcite and quartz+chlorite associations, have been found in the microgabbro and diabases of the central Akamas Peninsula, and they are also indicative of lavas and dikes in the Solea Graben within the Troodos Massif. These lavas and dikes are usually interpreted (*e.g.* Varga et al., 1987) as being formed within a fossil spreading zone with clear evidence for black smoker activity.

From the above petrological and geochemical features, the major agent of oceanic metamorphism affecting the plutonic rocks of SW Cyprus is considered to be a fluid of marine origin. Also, taking into account the boninitic nature of the gabbroids from the Ezousa River valley and the central part of the Akamas Peninsula, it may be suggested that their metamorphic alteration took place within a back-arc or fore-arc basin. Because the Xeros Potamos River alkali gabbros also suffered greenschist facies meta-morphism under the probable influence of submarine fluids, the geodynamic setting accompanying this alteration could also be characterized by subaqueous conditions.

Origin of the Mamonia Complex Gabbroids and their Relationship to the Troodos Plutonic Complex

The gabbroids occurring at the southwestern edge of Cyprus within the Mamonia Complex have some features in common, in their composition, and their relationship with the host rocks. In all studied sections, the gabbroids are closely associated with dike complex rocks, and the gradual transition between these rocks is evident in the Akamas Peninsula area. Most of the gabbroids and their associated diabases belong to the boninite series. Only the Xeros Potamos River alkali gabbroids are the products of intra-plate alkaline magmatism. The origin of their associated diabases and their genetic relationships to the gabbros is still debated because of the lack of petrological and geochemical data. Field observations in the Xeros Potamos River valley suggest, however, that the diabase and gabbroids form a single magmatic complex, observed in the more western areas. However, in contrast with the boninite-type dolerite-gabbroic association exposed in the Ezousa River valley and to the south-west, this complex was formed during the magmatic evolution of melts of the intra-plate type.

Because of the tectonically disturbed contacts between the rocks of individual gabbro plutons elsewhere in SW Cyprus, their original relationships remain unknown. The similarity between these gabbroids and the plutonic complex sections of the Troodos Massif immediately comes to mind. In the case of the Troodos sections, this succession starts with the cumulate melanocratic olivine gabbro passing upward into the olivine-free pyroxene gabbro. Significantly, diorites and plagiogranites associated with gabbro and diabases at the top of the Troodos layered succession have not been found in the Mamonia Complex. Because of the lack of plagiogranite bodies formed during an early stage of formation of the Troodos plutonic complex (including the harzburgite tectonites, dunites, plagioclase peridotites, two-pyroxene gabbros, plagiogranites, and microgabbro dikes), the Mamonia gabbroids are probably equivalent to the late-stage gabbroids of the second, plagiogranite-free magmatic rock group (wherlites, poikilitic wherlites, and cumulate gabbro including gabbro-norites), which are typically localized in the marginal parts of large intrusives where they occur as veins or swarms of vertical dikes intruding the early-stage plutonic rocks.

This assumption is based on both observation of the spatial relationships and on the geochemical similarity between the diabases and the gabbros from SW Cyprus. As noted above, in many cases the Mamonia Complex gabbros are closely associated with dike complex rocks, or, in the case of the central Akamas Peninsula, they form dike bodies. Petrological and geochemical data providing additional arguments regarding the genetic affinity of these rocks are as follows

1) The composition of the rock-forming minerals in the Mamonia gabbro and the late-stage Troodos gabbro;

2) Their common geochemistry is characteristic of boninite series plutonic rocks;

3) The similarity in petrological and geochemical features of the metamorphism of the boninitic gabbro from the Mamonia and Troodos complexes; and

4) The characteristic absence of representatives of tholeiitic magmatic products among plutonic assemblages in SW Cyprus.

Thus, in SW Cyprus, the Mamonia Complex rocks are juxtaposed with the gabbroids, most of which are of the boninite-type and identical to those of the late-stage magmatic assemblage from the Troodos Complex. These gabbros are associated with geochemically similar diabases and pillow lavas belonging to the same magmatic series. It seems reasonable to suggest that most of the Mamonia gabbroids were formed within the same geodynamic setting as the plutonic complex of Troodos, or, at least that of its late-stage boninitic rocks. The present geological position of the boninitic gabbros in SW Cyprus reflects events which took place after the formation of the Troodos Complex. The absence of the representative island-arc tholeiite series rocks which characterize the late stage of development of Troodos magmatism is probably the result of large-scale lateral movements along a wrench fault. It coincides with the surface of a structural discontinuity within the main gabbro succession of the Troodos, where it separates the lower and upper gabbroic horizons. In accordance with this tectonic scenario, the youngest Troodos plutonic rocks, *i.e.* the Limassol Forest boninitic gabbros exposed along the Arakapas Fault Zone, could be tectonically incorporated within the volcanic and sedimentary rocks of the Mamonia Complex.

A distinct kind of gabbroic rock that has no analogy among the gabbroids of the Troodos ophiolite is only found in the Xeros Potamos River valley. These gabbros, with typical alkaline characteristics, are considered to be plutonic derivatives of the Dhiarizos Group magmatic complex, consisting mainly of pillow lavas with alkali basaltic, trachybasaltic, and trachytic compositions.

Because of the sporadic exposure of the plutonic rocks which display the proposed genetic association with the Dhiarizos Group, it is reasonable to assume that the autochthonous structural units composed of magmatic Triassic rocks in SW Cyprus have undergone a lesser degree of tectonic disintegration as compared with the magmatic rocks within the Troodos Massif.

F. Amphibolites and Metasediments from Tectonic Blocks within the Serpentinite Melange

Sergey A. Silantyev
Vernadsky Institute of Geochemistry and Analytical Chemistry of the Russian Academy of Sciences, Kosygin Str. 19, 117975 Moscow, Russia

Sporadic outcrops of schistose and massive varieties of mafic amphibolites and metasediments are present elsewhere in SW Cyprus. These rocks, which differ in composition and texture from the associated slightly metamorphosed igneous and sedimentary rocks of both the Mamonia Complex (Dhiarizos and Ayios Photios Groups) and the Troodos Complex, are known only in the southwestern part of the island where they occur as small (usually tens of meters) tectonic blocks always associated with tectonized serpentinites (**Fig. 2.21**).

The amphibolites and associated metasediments described below have been studied in eight isolated tectonic blocks within the main SE-NW trending serpentinite belt of SW Cyprus, where they are irregularly scattered along its strike from the Dhiarizos River valley to the northeastern coastal margin of the Akamas Peninsula (**Fig. 2.22**).

Petrographic and Geochemical Features of the Metamorphic Rocks from Tectonic Blocks

Potima Block.

This small (25x10m) outcrop of interbedded meta-sediments and amphibolites is located on the left bank of the Dhiarizos River near the village of Potima, south-west of Prastion village. The southern and northern margins of the block are bounded by siliceous sedimentary rocks and schistose serpentinites, respectively. The amphibolites are represented by both massive melanocratic varieties similar to hornblendite, and typical banded hornblende-schist. The metasediments are mainly metamorphic arkoses.

The most common rock type is massive amphibolite with a grano-neoblastic texture, consisting of amphibole and minor plagioclase, sphene, chlorite, and relict apatite. In more leucocratic banded varieties, the plagioclase occurs equally with amphibole. Banded garnet-amphibolite consisting of plagioclase, potassium feldspar, biotite, amphibole, and garnet are subordinate. Arkosic schists with a relict psammitic structure are composed of plagioclase, quartz, potassium feldspar, biotite, chlorite, calcite, and rare clasts of ilmenite. In the massive amphibolite (hornblendite) the amphibole is edenite-pargasite hornblende (Al_2O_3=12-13 wt.%) and is practically chlorine-free (<0.01 wt.%). The amphibole from the leucocratic banded amphibolite is very similar in composition to that in the massive amphibolite (Al_2O_3=12-14 wt.% and Cl<0.01 wt.%) and is associated with albite (An= 3-8%). The plagioclase from the meta-sandstones is albite (An=3-7%) or oligoclase (An=27%), probably relict in nature. The biotite from this rock is compositionally eastonite-siderophyllite (TiO_2=2.31-2.96 wt.%; Mg# =0.38-0.40). Its compositional parameters differ from those in the biotite from alkali igneous rocks (**Fig. 2.23**), but are characteristic of metasedimetary rocks. In the leucocratic garnet-amphibole the hornblende is associated with ferriferous garnet (Mg# =0.10-0.12; $Alm_{56\text{-}58}Spess_{18}Pyr_{6\text{-}8}Gross_{18}$), ferriferous biotite (Mg# = 0.36; TiO_2 1.30-1.81wt.%), oligoclase-albite (An=0.16%), and potassium feldspar. Relicts of magmatogenic silicates have not been found in the Potima Block ampibolites. Despite this, their massive variety, similar to the hornblendite, is probably orthoamphibolite. By contrast, the leucocratic banded varieties, (at least, the biotite-bearing amphibolites) may be assigned to metasedimentary or metasedimentary-volcanic rocks.

The major-element characteristics of the Potima Block melanocratic amphibolites (orthoamphibolites) are clearly comparable to those of the potassium-enriched basic igneous rocks (up to 1.0 wt.%) belonging to the alkali magmatic series (Fig. 2.17D and Fig. 2.25). By contrast, it seems unlikely that a hypothesis of original magmatic nature can by itself account for the leucocratic banded garnet-amphibolites in the Potima Block, taking in account their geochemical features, phase composition and texture (Fig. 2.17A-C and 2.25). Evidence for a proposed sedimentary nature of their protolith comes from the REE distribution patterns (**Fig. 2.24**). In contrast to these rocks, the REE distribution in the massive amphibolites conforms to that for the typical intra-plate alkali

Fig. 2.21. Outcrops of serpentinite breccia in the Mavrokolymbos River valley (photo of K. Krylov).

igneous rocks.

The sole exception (sample MAM-153) is a melanocratic amphibolite with V-shaped chondrite-normalized REE pattern (Fig. 2.24). If its protolith had a magmatic origin, this REE distribution pattern might reflect the connection with boninite series rocks. However, such an assertion is untenable on the basis of the bulk composition of this rock which differs from boninitic rocks and conforms to that of the tholeiite group rocks: (in wt.%) SiO_2=43.63, TiO_2=0.99, MgO=8.19, FeO*=10.35, and CaO=16.69. See Fig. 2.17B and **Fig. 2.25**).

This contradiction in the geochemical data disappears if we asume a sedimentary protolith for this amphibolite. Enrichment in heavy REE is, in fact, characteristic of marine carbonate oozes (Balashov, 1976). Their admixture with a volcaniclastic material derived from a basaltoid source could lead to the observed distribution of REE in the protolith of metasedimentary amphibolite, in a manner resembling that of the boninite spectrum. Despite the observed similarity in shape of the two spectra, the total concentration of REEs in the amphibolite is much higher than that in boninite.

The Nea Choletria Block

The Nea Choletria Block is a small (50x10m) tectonic block exposed at the watershed of the Dhiarizos and Xeros Potamos Rivers to the north of Nea Choletria village. Immediately near the northern contact with serpentinites, the block is composed of thinly banded siliceous-carbonate metasediments consisting of quartz and calcite, with minor chlorite. The amphibolites are found

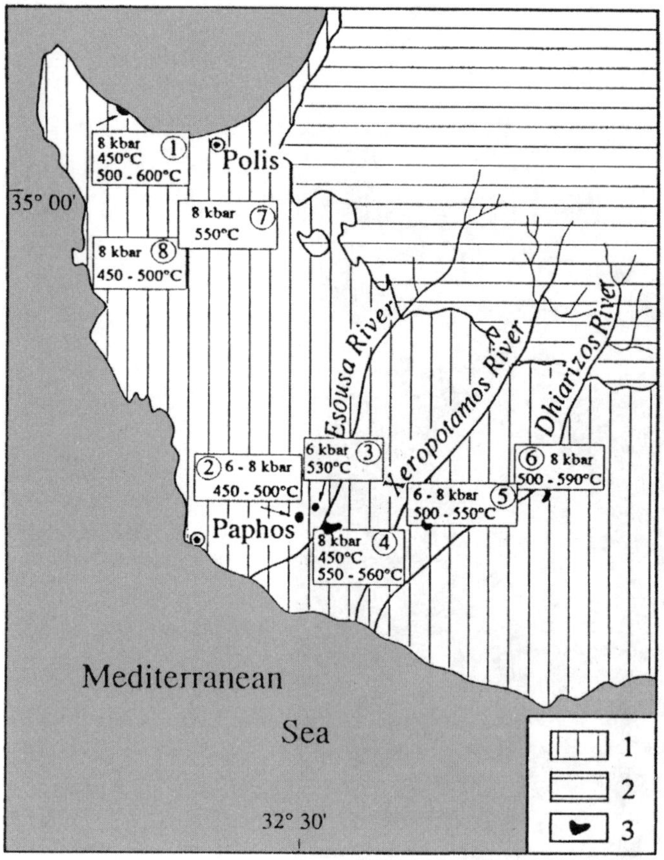

Fig. 2.22. Distributions of metamorphic blocks and related metamorphic conditions in SW Cyprus. In cubicks: 1 - area occupied by the Mamonia Complex; 2 - the Troodos ophiolite; 3 - outcrops of metamorphic rocks. In circles: 1 - Loutra tis Aphroditis Block, 2 - Mersinion Block, 3 - Kakorkakas Block, 4 - Ayia Varvara Block, 5 - Neo Kholetria Block, 6 - Potima Block, 7 - Drousha Block, 8 - Fasli Block. Presented after Silantyev (1993).

to the south and extend continuously to a southern contact with pillow lava packs. Massive melanocratic amphibolite similar to the hornblendite and consisting of amphibole and minor plagioclase and sphene, and associated banded hornblende-schist composed of amphibole, plagioclase, epidote, and minor muscovite, are the dominant rock types in the block.

The amphibole in the amphibolites has a constant composition (Al_2O_3=11-13 wt.% and Cl<0.02 wt.%) and belongs to the pargasite-edenite series. The plagioclase is albite (An=6-10%). Relicts of magmatogenic minerals are absent in these rocks.

The massive amphibolites are similar in major-element variations to intra-plate basic magmatic rocks: (in wt.%) SiO_2=42-46, TiO_2=2.5-3.1, MgO=9.2-12.7, (Na_2O+K_2O)=3.7-4.4 (Figs. 2.17D and 2.25). On the other hand, some varieties of the banded hornblende-schist have petrochemical characteristics suggesting their sedimentary nature: for example, the high values of CaO (more than 20 wt.%) do not correlate to values of the other major elements characteristic of magmatic rocks. The REE distribution in the massive amphibolites is identical to that in the alkali plutonic rocks, implying a magmatic protolith of the intraplate type (Fig. 2.20 and **Fig. 2.26**).

The Ayia Varvara Block

The Ayia Varvara Block is the largest (~1.6x0.5 km) within the Mamonia Complex. It is exposed along the left side of the Ezousa River valley to the north of Ayia Varvara village. A prominent southeast-northwest fault zone filled with serpentinites separates the Ayia Varvara metamorphics from a thick serpentinite sheet outcropping farther to the north. Near the contact with the amphibolites, the serpentinites are intensely schistosed. This is interpreted (*e.g.* Malpas and Xenophontos, 1987) as being caused by later deformation associated with the tectonic juxtaposition of serpentinites with the amphibolites. The rocks on the southern and eastern sides of this metamorphic block are in tectonic contact with pillow lavas (**Fig. 2.27**).

Thinly banded melanocratic amphibolites and siliceous-carbonate metasediments are the dominant rocks within the sequence of the Ayia Varvara Block. The melanocratic amphibolite consists mainly of hornblende and plagioclase, with accessory apatite; relicts of clinopyroxene ($En_{39}Wo_{49}Fs_{12}$) are rare. Massive amphibolite forms small isolated bodies resembling boudins. This rock consists of hornblende, plagioclase, potassium feldspar, sphene, and epidote, and exhibits a relict gabbro-ophitic texture indicating a magmatic protolith. The differences in lithologic composition of the metasediments reflect the variability of their mineralogical composition. The siliceous and siliceous-carbonate rocks are composed of quartz, hornblende, calcite, actinolite, and potassium feldspar. The carbonate rocks consist of calcite,

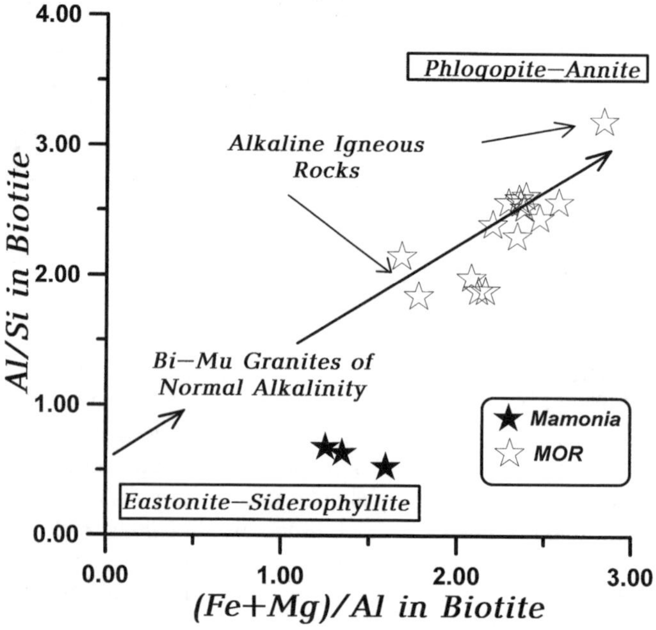

Fig. 2.23. Variations of Al/Si versus (Fe+Mg)/Al in biotite from metamorphic rocks of the Mamonia Complex. Compositional field of biotite from magmatic rocks of different alkalinity is shown as in Marakushev and Tararin (1965).

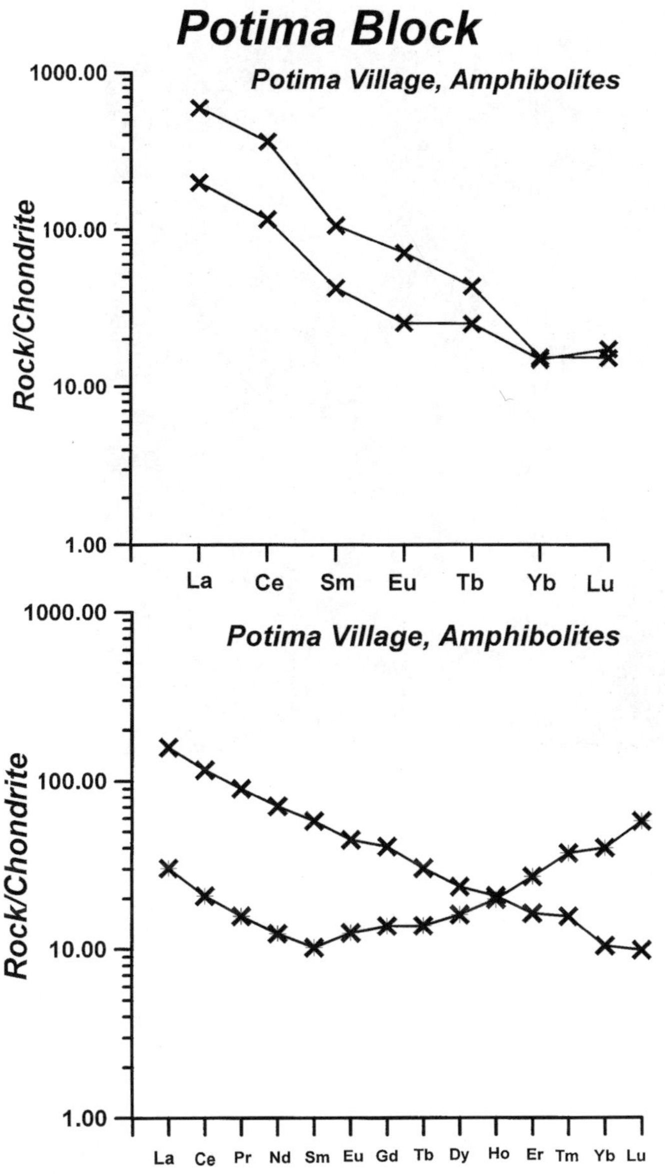

Fig. 2.24. Chondrite-normalized REE patterns of amphibolites from the Potima Block. Normalized values are from Anders and Grevesse (1989).

dolomite, hornblende, chlorite, sphene, biotite, and epidote.

The main amphibole in the Ayia Varvara metamorphic rocks is pargasite hornblende; on the other hand, a high-Fe high-K hornblende (enriched in hastingsite) is present in some amphibolites, in many cases in association with potash feldspar. From the correlation between the petrochemical feautures of the Ayia Varvara amphibolites and the compositional characteristics of the amphibole from these rocks, we can conclude that the potassium content in the hornblende has no influence on the bulk composition of the host rocks. However, the high-potassium pargasite common in the meta-carbonate rock contains significant MgO relative to some metasedimentary rocks of the Ayia Varvara

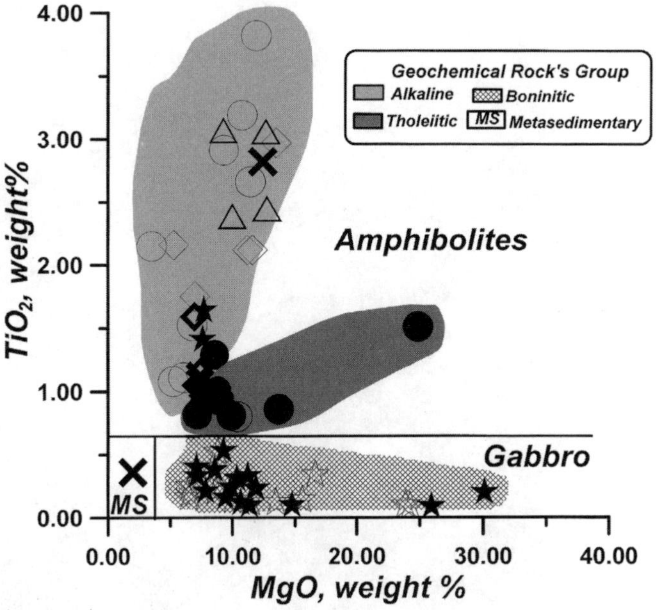

Fig. 2.25. Variations of TiO_2 versus MgO in plutonic (black stars) and metamorphic (symbols as on Fig. 2.17) rocks from the Mamonia Complex.

Formation, whereas the hornblendes in these samples are compositionally similar to amphiboles from such alkali magmatic rocks as trachybasalt, shoshonitic basalt, syenite, and monzodiorite (**Fig. 2.28**).

This implies that three generations of hornblende are present in the Ayia Varvara Block metamorphics:

1) Metamorphic hornblende;

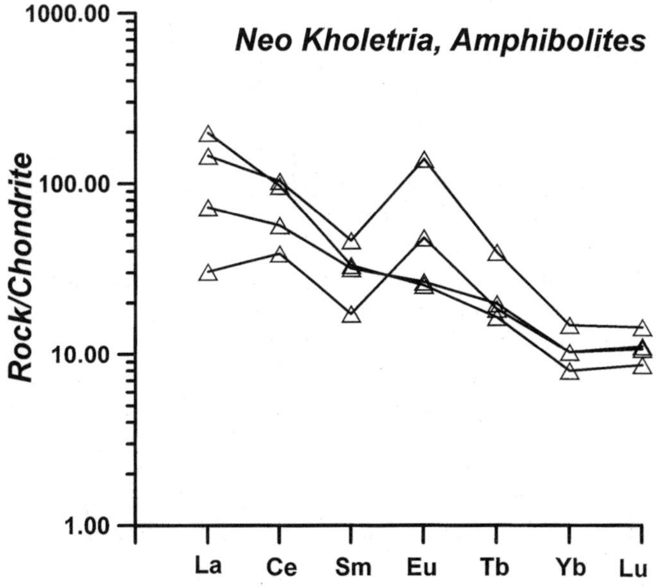

Fig. 2.26. Chondrite-normalized REE patterns of amphibolites from the Potima Block. Normalized values are from Anders and Grevesse (1989).

Fig. 2.27. Contacts of serpentinites (gray and blue tops of slopes) and pillow-lavas (pink) at right slope of the Ezousa River valley to the west of Ayia Varvara village (photo of K. Krylov)

2) Relict magmatic hornblende; and
3) Clastic magmatic hornblende in the metasediments.

A comparision of amphiboles from the Ayia Varvara blocks with those in oceanic metagabbros and amphibolites is given in **Fig. 2.29**. It is evident that the hornblendes from the main rock type in the Ayia Varvara Block differs from the oceanic metabasite amphiboles by its enrichment in potassium and by the absence of chlorine. Chrome occurs in high concentrations in the pargasite hornblende from the most common amphibolite rock type. This is probably the result of replacement of clinopyroxene.

The plagioclase in the Ayia Varvara Block metamorphics is either oligoclase-andesine or albite. Potassium feldspar occurs in amphibolites with both igneous and metasedimentary protoliths and is characterized by a relatively constant composition: 32% Ab-27% An-40% Or to 29% Ab-28% An-43% Or. This may imply that the potassium feldspar in these amphibolites belongs to the relict phase associated with the primary hastingsite hornblende, which is common in the metamorphic rocks examined. The close compositional similarity between potassium feldspar from the orthoamphibolites and the reflects the clastic nature of the metasediments.

The proportion of orthoamphibolites and metasediments (including para-amphibolites) within the Ayia Varvara Block is nearly equal. The metasediments have significant compositional variations in the major elements, particularly SiO_2, CaO, and TiO_2. In contrast, the orthoamphibolites show a certain compositional monotony, with only variable amounts of potassium (0.19 and 1.81 wt.%). The diagnostic feature of metasediments is the markedly lower concentrations of chrome compared with the orthoamphibolites.

Taking into account both the mobility of some major elements during metamorphism and the mostly inert behaviour of REE under medium-temperature metamorphism (Silantyev, 1995), it is interesting to compare these constituents in the studied amphibolites. This comparison is not meant to imply a significant change in major-element distribution in the course of metamorphism, and, therefore, it can be used for the geochemical identification of the pre-metamorphic protolith. Some of the Ayia Varvara orthoamphibolites are enriched in light REEs (Fig. 2.20 and **Fig. 2.30A**) probably reflecting the geochemical nature of their protoliths. The highest enrichment is characteristic of the amphibolites with plutonic protoliths: $(La/Sm)_{cn}$=7.06 and $(Ce/Yb)_{cn}$=4.43. In their chondrite-normalized REE distribution and total REE concentrations, they resemble alkali intrusive rocks similar to syenite and monozonite (Gullers and Graf, 1984); the petrochemical data (Fig. 2.17D) supports this view.

On the other hand, the less enriched amphibolites are identical in their geochemical and petrochemical features to LREE-depleted igneous rocks similar to tholeiites belonging to T-MORBs or N-MORB (Figs. 2.20 and 2.30A). The composition of relict

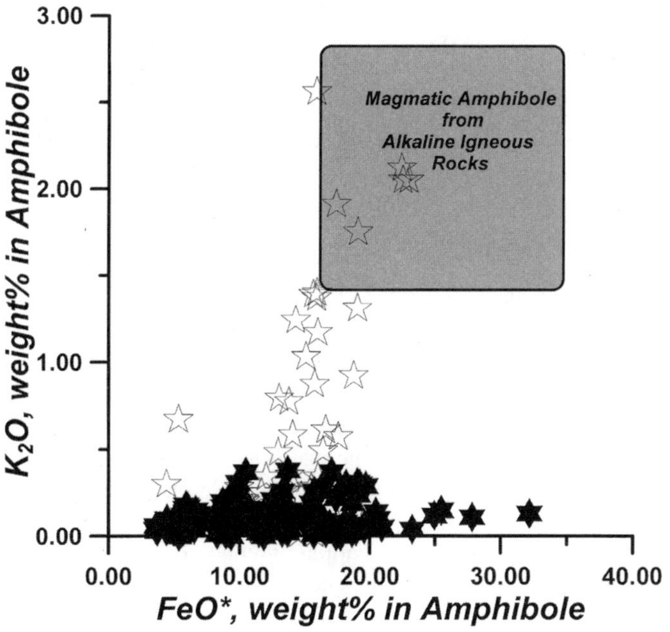

Fig. 2.28. Variations of K_2O versus FeO* in amphibole from amphibolites of the Mamonia Complex (open stars) and oceanic metaplutonic rocks and amphibolites (black stars).

clinopyroxene from these rocks confirms their derivation from tholeiite series rocks (Fig. 2.16). The level of concentrations and distribution of REE in the metasedimentary rocks allow us to distinguish different protolith lithologies corresponding to carbonate, carbonate-siliceous, and siliceous sediments.

The Kakorkakas Block

The Kakorkakas Block is located 700 m northwest of the Ayia Varvara Block, on the west bank of the Ezousa River near the mouth of the Argakin tis Kakorkakas brook. It is an allochthonous 120x100 m block of interbedded amphibolite and metasediment incorporated within the serpentinites. They form the western extension of the serpenitinite sheet outcropping further towards the east across the river; to the south, this sheet is bounded by a disturbed lava succession containing small gabbro and diabase blocks (**Fig. 2.31**).

As in all the other Mamonia metamorphic blocks, two main types of amphibolitic rocks are present. One is a massive amphibolite (hornblendite) with a grano-neoblastic texture, containing pargasite hornblende (Al_2O_3=11-12 wt.% and Cl<0.02 wt.%); no relicts of magmatogenic minerals have been found. The other type, a more leucocratic banded amphibolite, exhibits relict cumulate textures and contains in addition to the metamorphic minerals - amphibole, epidote, sphene and relicts of hastingsite hornblende (Al_2O_3=14-15 wt.% and K_2O= 2.6 wt.%) which is associated with clinopyroxene Na-augite) from alkali igneous rocks: (mass.%) Al_2O_3=5.80-6.15, SiO_2=0.58-0.66 and Na_2O=1.23-1.53 (Fig. 2.16).

Siliceous-carbonate rocks consisting of quartz, calcite, garnet, clinopyroxene, epidote, and sphene, are the dominant metasediments. Clinopyroxene in these rocks is a salite-ferrosalite ($En_{31}Wo_{51}Fs_{18}$; Fig. 2.16) and the garnet represents andradite, which is common in the meta-carbonate rocks. Biotite-feldspar crystalline schists are subordinate, and are typically composed of oligoclase-andesine (33% An) of assumed clastic origin, albite (7% An), and biotite identical to that in the metasedimentary rocks of the Potima Block (Fig. 2.23).

The variation in major element content of the massive orthoamphibolites and amphibolites with a relict cumulate texture demonstrates chemical similarity with typical monzonitoids (Figs. 2.17D and 2.25), in good agreement with the data on the composition of relict clinopyroxene from these rocks. The alkaline nature of the protoliths of the orthoamphibolites (with signs of a cumulate origin), is stressed by both the REE distribution in these rocks (Figs. 2.20 and 2.30B) and the level of their light REE enrichment: $(La/Sm)_{cn}$=3.19-5.24 and $(Ce/Yb)_{cn}$=4.68-11.91. Some varieties of massive amphibolite with no evidence for relict magmatic textures or mineral phases have parameters comparable with tholeiite series magmatic rocks, *i.e.* $(La/Sm)_{cn}$=0.63 and $(Ce/Yb)_{cn}$=0.72 (Figs. 2.20 and 2.30B).

The Mersinion Block

The Mersinion Block is exposed on the east bank of the Argakin tis Mersinion brook, 1 km west of the Ezousa River. It is a 60x45 m metamorphic block incorporated within tectonized serpentinite containing small blocks of pillow lavas. The melanocratic amphibolite, consisting of amphibole, plagioclase, and sphene, is the dominant rock type within this block. Gneissic banded amphibolite, composed of amphibole, plagioclase, epidote, and clinopyroxene, and banded siliceous-carbonate metasediments

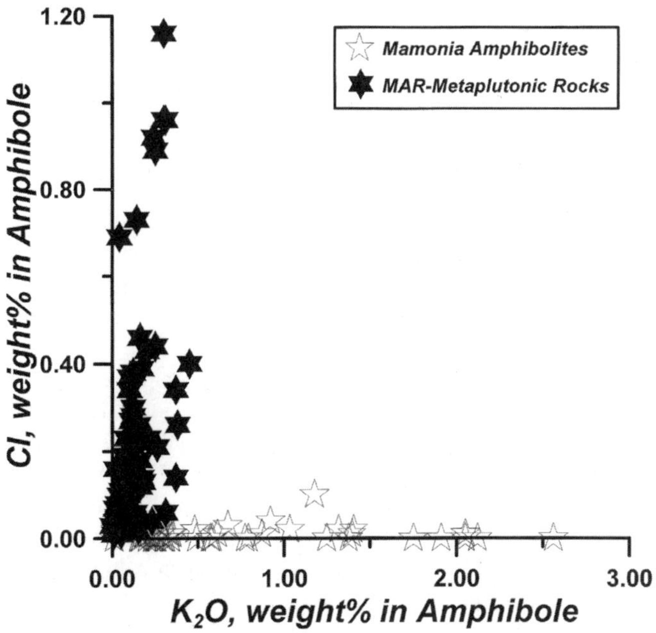

Fig. 2.29. Variations of Cl versus K_2O in amphibole from amphibolites of the Mamonia Complex and oceanic metaplutonic rocks and amphibolites (symbols as on Fig. 2.17.).

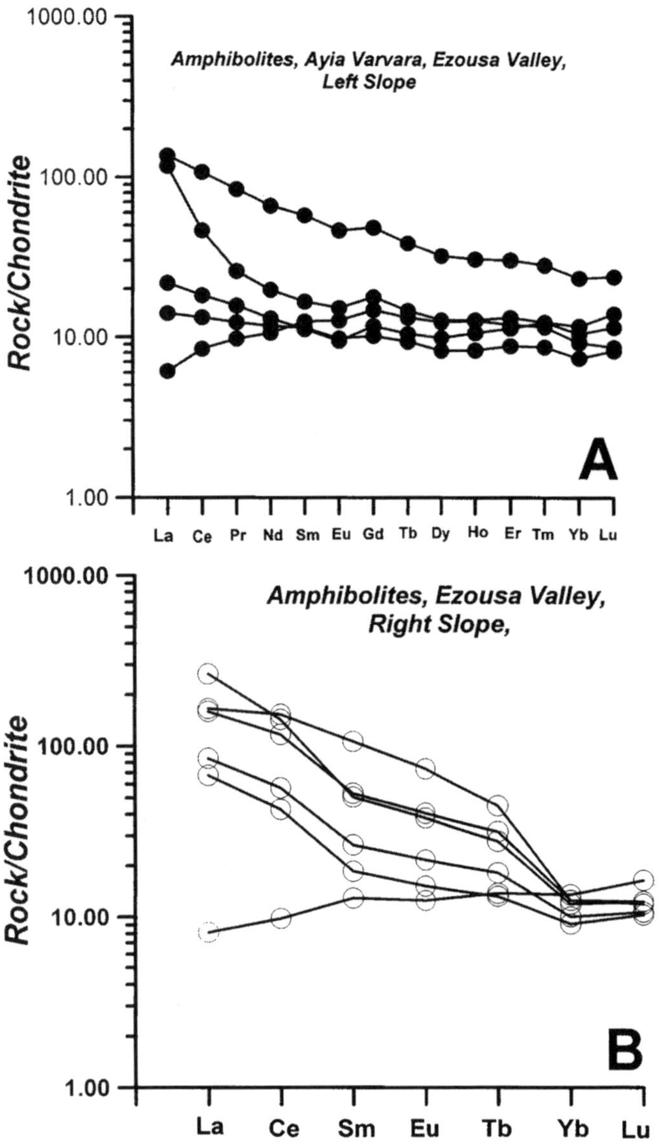

Fig. 2.30. Chondrite-normalized REE patterns of amphibolites from the Ayia Varvara Block (A), and Kakorkakas and Mersinion Blocks (B). Normalized values are from (Anders and Grevesse, 1989).

are subordinate.

In all textural varieties, the amphibole is exclusively represented by pargasite hornblende (Al_2O_3 = 10-10.8 wt.% and Cl< 0.04 wt.%) associated with albite (1-8% An). The relict magmatogenic hornblende (Al_2O_3 = 13 wt.% and K_2O=1.85 wt.%) and associated relict plagioclase (48% An) and clinopyroxene have been found in the gneissic leucocratic amphibolites. The clinopyroxenes in these rocks are very similar in composition to relict magmatogenic pyroxenes from the orthoamphibolites of the Kakorkakas Block: Al_2O_3 = 6.46 wt.% and TiO_2 = 0.53 wt.% (Fig. 2.16).

Petrochemically, the Mersinion Block amphibolites are congeneric with those of the Kakorkakas Block: (in wt.%) SiO_2 42-45, TiO_2 3.2-3.8, MgO 10.8-11.9, CaO 11-12, and (Na_2O+K_2O) 3.0-3.7 (Figs. 2.17D and 2.25). As in the Kakorkakas orthoamphibolites, these rocks are probably the result of metamorphism of monozonite-type alkali gabbroids. Also, as is shown in Figs. 2.20 and 2.30B, the distribution pattern of the REEs in both the massive and banded amphibolites and their enrichment in the lanthanoides $(La/Sm)_{cn}$=1.56-2.98 and $(Ce/Yb)_{cn}$=9.27-12.15) support this assumption.

The Fasli Block

The Fasli Block is from the central part of the Akamas Peninsula, where metamorphic rocks occur as small allochthonous blocks incorporated within serpentinite breccia together with rock fragments of various lithologies: *e.g.* carbonate rocks, cherts, diabases. The Fasli Block, that lies half way between the villages of Lara and Fasli is the smallest (5x8 m) among the eight metamorphic blocks that were investigated.

This block is almost entirely composed of a schistose amphibolite consisting of hornblende, plagioclase, epidote, sphene, and accessory ilmenite. The amphibole (Al_2O_3=12 wt.%; Cl<0.01 wt.%) is associated with albite (1-4% An) or oligoclase (21% An). Relict andesine-labradorite (43-54% An) and ilmenite are present in some samples. Quartz-mica schist composed of biotite, quartz, and calcite only occurs as a thin bed within the banded melanocratic amphibolites.

It is clear from Figs. 2.17D and 2.25) that the major element variations of most of these rocks fall in the compositional fields of differently fractionated tholeiites of the MORB family: SiO_2 = 44.82-51.23 wt.%, TiO_2 = 0.97-1.59 wt.%, MgO = 6.91-7.60 wt.%, FeO* = 11.27-15.84 wt.%, CaO = 14.36-9.36 wt.%, and (Na_2O+K_2O) = 2.09-2.95 wt.%. Their affinity with the metamorphic derivatives of the tholeiitic magmatism products is also supported by the REE distribution in these amphibolites (Fig. 2.20 and **Fig. 2.32A**). The schistose amphibolites with $(La/Sm)_{cn}$ = 0.57-1.01 and $(Ce/Yb)_{cn}$ = 0.35-0.96 are considered to be formed by metamorphism of N-MORB or T-MORB types of tholeiites.

The Drousha Block

The Drousha Block is a small (10x15 m) block of strongly weathered, rubbly amphibolitic rocks outcropping in the up-slope cliff along the Pegeia-Polis road north-east of Drousha village and south of Polis town. The host rocks are tectonized serpentinites containing disintegrated packages of pillow lavas.

The mineralogic composition and textural fabric of the Drousha amphibolite are similar to the Fasli Block. Both are composed of amphibole, plagioclase, relict clinopyroxene, and rare rutile. The amphibole in the Drousha amphibolite is a high alumina pargasite-edenite hornblende (Al_2O_3 = 12-14 wt.% and Cl< 0.01 wt.%) associated with albite (1-4% An) and sphene. The relict clinopyroxene is compositionally similar to augite from tholeiites: Al_2O_3 = 3.94 wt.%, TiO_2 = 0.29 wt.%, Na_2O = 0.99 wt.%, CaO 22.93 wt.%, and Mg# =0.67 (Fig. 2.16).

The major-element characteristics of the Drousha amphibolite, as well as some of the Fasli Block rocks, indicate that these

Fig. 2.31. The right slope of the Ezousa River valley opposite to the metamorphic Ayia Varvara Block. Dark hill to the right consists of serpentinities and its top is composed of metamorphic rocks from the Kakorkakas Block. There is a gabbro outcrop at the left, down corner of photo (photo of K. Krylov)

rocks can be a result of metamorphism of plagioclase-enriched fractionated products of the MORB tholeiite series: SiO_2 = 42 wt.%, TiO_2 = 1.5 wt.%, Al_2O_3 = 17.23 wt.%, MgO = 17.04 wt.%, FeO* = 12.55, CaO = 15.85 wt.%, and (Na_2O+K_2O) = 1.31 wt.% (Figs. 2.17D and 2.25). Chondrite-normalized REE abundance diagrams for these rocks (Figs. 2.20 and 2.32B) are also interpreted as reinforcing the above indication.

The Loutra tis Aphroditis Block

The Loutra tis Aphroditis Block is exposed along some 60 m of the shore cliff in the northeastern part of the Akamas Peninsula. It consists of alternating metasediments, which constitute more than 50% of the rocks in the block and are mainly quartz-mica schists and quartzites with subordinate meta-cherts, and massive melanocratic and banded leucocratic amphibolites. All these rocks are characterized by an almost vertical dip in accordance with their host's steeply-dipping serpentinite sheets. In the west the block is tectonically juxtaposed with a zone of sheeted diabase dikes.

The metasedimentary rocks are represented mostly by quartz-mica schists and quartzites. As with the other blocks, these amphibolites can be divided into massive mafic and banded leucocratic varieties. The massive mafic variety consists of hornblende, plagioclase, and epidote. In addition there is garnet occurring in the banded amphibolites. Both varieties of amphibolite contain a high alumina pargasite hornblende that is almost uniform in composition: *i.e.* Al_2O_3 = 11.5-13.5 wt.%, TiO_2 = 0.46-0.96, K_2O = 0.16-1.06 wt.%, and Cl< 0.01 wt.% . Its associated plagioclase has the composition of albite (2% An). The garnet was only found in banded amphibolites and is represented by almandine (Alm_{60} $Spess_6Pyr_{15}Gross_{19}$; Mg# = 0.21-0.27).

The major- and rare earth element abundances in the Loutra tis Aphroditis amphibolites resemble those of the ortho-amphibolites from the Nea Choletria Block (Figs. 2.17D, 2.20, 2.25 and 2.32C). On the other hand, significant differences in the REE chemistry exist between the Nea Choletria orthoamphibolites and the garnet-bearing amphibolites from Loutra tis Aphroditis; the former have unusually high total REE content (300 ppm) and higher $(Ce/Yb)_{cn}$ value: *i.e.* 16 (Fig. 2.32C). On the contrary, these parameters in the massive amphibolites from Loutra tis Aphroditis are clearly compatible with those of the Nea Choletria orthoamphibolites. The total REE concentrations range from 72 to 140 ppm and from 60 to 115 ppm, respectively. The $(Ce/Yb)_{cn}$ ratio is 5 for the Loutra tis Aphroditis amphibolites, whereas the ratios of the orthoamphibolites of the Nea Choletria Block range from 5 to 7. These facts suggest that the Loutra tis Aphroditis amphibolitic rocks now represented by the massive amphibolites were comagmatic in origin with the alkali intrusive rocks. Their mineralogy is typically reflected in the cumulation of plagioclase, judging from the characteristic

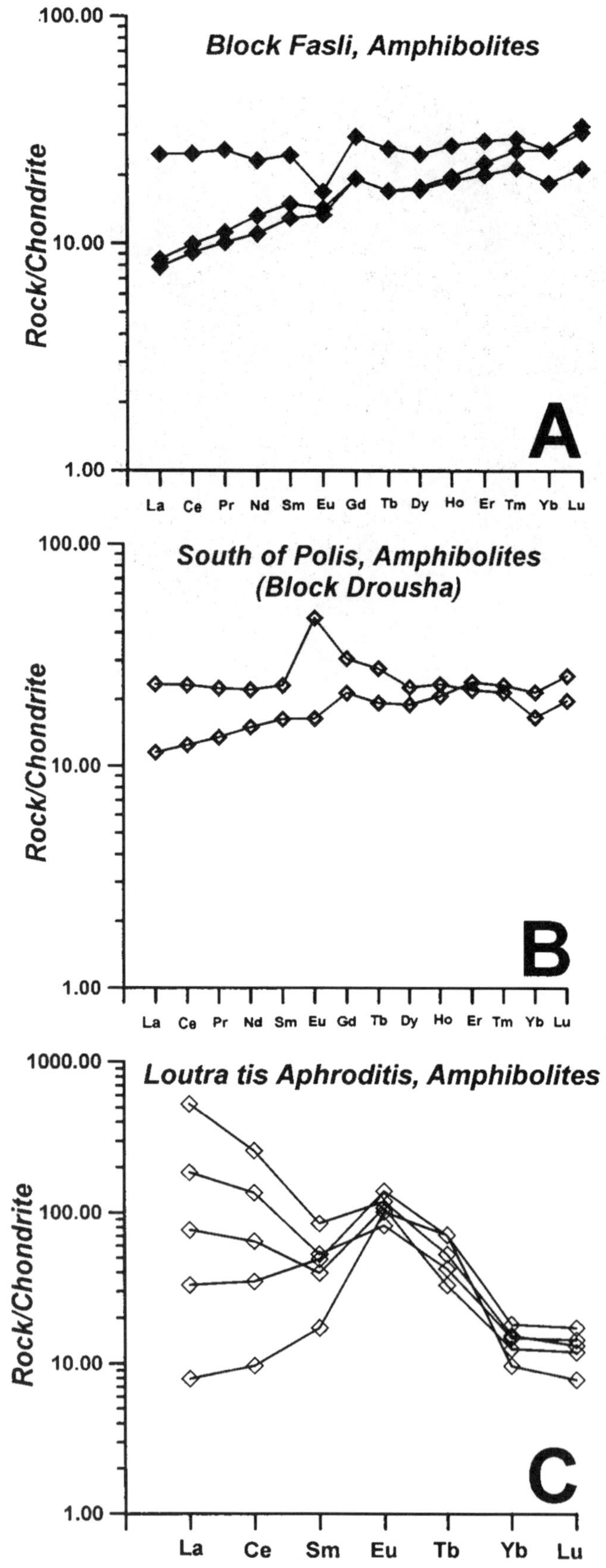

Europium extremum on the distribution patterns of REE's for these rocks (Fig. 2.32C). In this way, the formation of the garnet-bearing amphibolite is explained by metamorphism of highly-fractionated cumulates of the same alkali intrusive series.

Conditions of Metamorphism and Proposed Protoliths

The typical and dominant occurrence of epidote-albite amphibolites containing indicative albite+hornblende paragenesis serves as a very useful and basic criteria for reconstruction of the P-T conditions of metamorphism of the Mamonia metamorphic rocks. As noted by Miyashiro (1973), the albite-hornblende metabasites commonly occur as an integral part of the high-pressure metamorphic belts. The above-mentioned compositions of the coexisting mineral phases allow quantitative estimation of the formation pressure and temperature of the Mamonia metamorphics. Our quantitative P-T estimates have been made by using the experimental amphibole-plagioclase geothermobarometer of Plyusnina (1983), and the experimental data of Spear (1981) and Plyusnina (*op. cit.*) on the pressure and temperature dependence of the alumina content in the calcic-amphibole of metabasites. In this way, the formation conditions of the massive amphibolites and hornblende-schists from all the metamorphic blocks that were studied are considered to have been as follows:

Potima Block: T=500°C and P=8 kbar
Nea Choletria Block: T=550-500°C and P=6-8 kbar
Ayia Varvara Block: T=450°C and T=550-560°C and P=8 kbar
Kakorkakas Block: T=530°C and P=6 kbar
Argakin tis Mersinion Block: T=450-500°C and 6-8 kbar
Fasli Block: T=450-500°C and P=8 kbar
Drousha Block: T=550°C and P=8 kbar
Loutra tis Aphroditis Block: T=450°C and P=8 kbar.

In addition, evidence for high-pressure metamorphism of the Mamonia amphibolites comes from the low mean (<2) of the Al^4/Al^6 ratio for the amphiboles from these rocks. The temperature of metamorphism of the metasediments was determined using the equilibrium coexistence of dolomite and calcite in the carbonate-silicate metasediments, following the technique of Goldsmith and Newton (1969) and Rice (1977). The distribution of magnesium between equilibrium calcite and dolomite from these rocks indicates temperatures of about 450°C.

The formation temperatures of the garnet-bearing amphibolites and biotite-garnet metasedimentary rocks have also been determinated using garnet-amphibole and garnet-biotite geothermometers (Perchuk, 1970; Perchuk and Ryabchikov, 1976; Ferry and Spear, 1978). From our calculations, the reaction that led to the appearance of these parageneses in the metasediments of the Potima and Loutra tis Aphroditis Blocks occurred at temperatures of 550°C (garnet+biotite) and 500° to 600°C (garnet+amphibole),

Fig. 2.32. Chondrite-normalized REE patterns of amphibolites from the Akamas Forest area. A - Fasli Block, B - Drousha Block, and C - Loutra tis Aphroditis Block. Normalized values are from Anders and Grevesse, (1989).

respectively.

The results of all these calculations indicate that the sedimentary and igneous rocks from which the Mamonia Complex metamorphics were formed were metamorphosed under uniform temperature (450° to 550°C) and pressure (6-8 kbar) conditions, and no relationship exists between the intensity or character of metamorphism and the location and size of the individual metamorphic blocks (Fig. 2.22).

As noted above, one of the most distinctive features of the amphibole from the Mamonia Complex metamorphic rocks is the low mean concentration of chlorine in its anionic group. This indicates a marked difference in amphibole chemistry between the high-pressure metabasites of SW Cyprus and the amphibolitized gabbros and amphibolites of the modern ocean basins and ophiolitic complexes of different ages, including those of the Troodos Massif (Fig. 2.29). Another common feature of the Mamonia amphiboles is their relative enrichment in potassium compared with oceanic metabasites. It follows that these high-pressure orthoamphibolites correspond to a protolith that includes alkali igneous rocks, which may have been metamorphosed under a fluid regime distinctly different from that of the oceanic-type metamorphism. In addition, the isotopic compositions of carbon and sulfur in the investigated orthoamphibolites from tectonic blocks in the Ezousa, Xeros Potamos, and Dhiarizos river valleys suggest the absence of fluids of marine origin within the metamorphic system (Silantyev, 1993).

The geochemical particularities of the Mamonia orthoamphibolites reflect the original igneous rocks from which they were derived. Their proposed protoliths include rocks with intra-plate alkalic magmatic, and tholeiite series affinities. This agrees with the idea advanced by Spray and Roddick (1981). The meta-morphic rocks making up the Ayia Varvara Block are geochemically similar to the Dhiarizos Group volcanic rocks (tholeiitic basalts, trachybasalts and trachytes) and the siliceous-carbonate sedimentary rocks of the Ayia Photios Group. The close similarity in REE distribution of the Xeros Potamos River gabbroids and of orthoamphibolites from the Nea Choletria and Loutra tis Aphroditis Blocks strongly supports this origin (Figs. 2.18A, 2.26 and 2.32C). On the other hand, the original sedimentary lithologies characteristic of the Ayia Varvara Block protolith are practically identical to those of the lithologically heterogeneous deposits forming the non-metamorphic sedimentary complex in the same area.

Judging by the geochemical data given above, the protholith of the rocks from which the Mamonia orthoamphibolites were formed consisted of the products of intraplate magmatism. Within the Ayia Varvara, Kakorkakas, Argakin tis Mersinion, and Potima Blocks these are probably metamorphosed trachybasalts and trachytes. Their metamorphic plutonic equivalents are present within the Ayia Varvara, Nea Choletria, and Loutra tis Aphroditis Blocks. A second, subordinate type of pre-metamorphic igneous rock is MORB typically found only in the central Akamas Peninsula (within the Fasli and Drousha Blocks). In the Ezousa River area these highly metamorphosed tholeiite series rocks are associated with the intra-plate magmatic rocks of the Kakorkakas and Ayia Varvara Blocks. In all the investigated Mamonia metamorphic blocks however, there is a notable absence of metamorphosed equivalents of those boninite series rocks characteristic of the Troodos Ophiolite Complex. The above-mentioned case of an uncharacteristic, boninite-like distribution of REE's in amphibolites from the Potima Block can be attributed solely to mixing of the sediment-derived carbonate matter with basic volcanogenic material.

Such a sedimentary protolith would mainly include siliceous-carbonates and meta-arkoses, which are typical constituents of continental-margin sedimentary sequences. It is significant that the Ayios Photios Group rocks, including quartz sandstones and carbonate-siliceous lutites, belong to such depositional sequences.

Geodynamic Regime of Metamorphism of the Mamonia Complex Amphibolites and Metasediments

Metamorphic blocks within the Mamonia Complex are similar in lithology and textural fabric to specific ophiolite-related metamorphics, connected with the so-called 'subophiolitic soles' but differ in both P-T conditions and structural features from the products of metamorphism of the different igneous rocks belonging to the ophiolite sequence. These rocks, showing intense schistosity and plication, occur as aureoles or separate tectonic blocks within a serpentinite melange at the sole of ophiolitic plates. Such rocks are products of high-pressure metamorphism at a wide range of temperature. The inverted zonation of temperature grade, from highest at the ophiolite plate contact to lower grades outward, is characteristic of many such metamorphic aureoles. More typically this type of metamorphism reflects conditions close to an isochemical system, and the metamorphism-forming fluids were probably significantly different from those of sea-floor metamorphism of oceanic rocks and ophiolites. In contrast to the 'ophiolite-type' metamorphics, diverse sedimentary rocks metamorphosed under isofacial conditions (together with their associated igneous rocks), occur widely among the subophiolitic high-pressure complexes. All of the above-mentioned features are common for the high-grade metamorphic rocks from the Mamonia Complex.

Based upon the tectonic and geophysical characteristics of modern transform faults, Spray and Roddick (1981) proposed that the Mamonia Complex amphibolites were formed in an oceanic transform setting. They believed that these rocks were generated in a low-pressure environment and that they are transitional between greenschist and amphibolite facies metamorphism. Petrologically and geochemically this model is untenable, because all existing data on metamorphism within modern oceanic transform faults contradict what is known about the thermodynamic regime of metamorphism of the Mamonia amphibolites. The hornblende compositions of orthoamphibolites from SW Cyprus are also distinct from those of oceanic amphibolites, including those from the transform faults. In any case, fluids of marine origin were absent in the metamorphic system which formed these rocks.

Based on data obtained from geothermobarometry, the P-T regime characterizing the metamorphism of the Mamonia

Complex amphibolites ranged relatively widely from 450°C to 550°C. These temperature variations during metamorphism are probably evidence of temperature zonation which produced the dynamothermal metamorphic aureoles at the base of the ophiolitic plates (*e.g.* Malpas, 1979; Ghent and Stout, 1981). The geochemical signature of the metamorphosed igneous and sedimentary rocks from the Mamonia tectonic blocks indicate that their protoliths consisted of igneous and sedimentary rocks, typical of the marginal areas of oceanic basins, including the passive continental margin. Modern transform faults have a completely different compositional spectrum. Thus the tectonic scenario for the Mamonia metamorphics implies a geodynamic setting different from that of the modern ocean basins, including their transform faults.

The P-T conditions and geochemistry of the metamorphic rocks in the Mamonia metamorphic blocks are similar to those of the 'metamorphic sole' rocks at the base of the ophiolites of the Dinarides, Turkey, Iran, and Oman (Pamic, 1977; Karamata et al., 1980; Okay, 1982; Piskin et al., 1983; Alavi-Tehrani, 1980; Pamic and Adib, 1982; Ghent and Stout, 1981). Most of these metamorphic complexes were formed at pressures of 5-9 kbar and temperatures of 300°-800°C (Silantyev, 1993). Where the high-pressure metamorphic complexes did not undergo tectonic disruption, they were restricted to basal hyperbasite horizons within thick ophiolite nappes. The resulting temperature zonation is represented by the inverted aureole. For example, at the base of the Semail ophiolite complex (Ghent and Stout, 1981) the zonation corresponds to the granulite to greenschist facies metamorphism.

All of these subophiolitic metamorphic complexes share one notable common aspect – that both the high-pressure metamorphics formed in the absence of marine fluids which develop along the base of the ophiolite plate, and the low-pressure ophiolitic oceanic metamorphics formed in the presence of such fluids, have the same or very close ages (Silantyev, 1993). The case of SW Cyprus nicely illustrates this regularity: the metamorphism of the high-pressure amphibolites has an age of 83 and 90 Ma according to $^{40}Ar/^{39}Ar$ age determinations (Spray and Roddick, 1981), while the K-Ar and Rb-Sr age determinations of the low-pressure metamorphism affecting the pillow lavas from the Troodos Massif is between 80 and 92 Ma (Studigel, Gillis, and Dunkan, 1986; **Fig. 2.33**). This similarity in age of metamorphism of the two contrasting pressure types of metamorphism which are now occurring in tectonically juxtaposed complexes, is evidence for their formation in different geodynamic settings.

As noted above, most subophiolitic high-pressure metamorphics in the eastern Mediterranean, including the Mamonia Complex, may have been formed by progressive obduction of ophiolite masses onto the continental margin (volcano-sedimentary successions) from about 95 Ma until about 66 Ma (Silantyev, 1993). The proposed formation of high-pressure complexes outside the subduction zones is based on their close spatial association with ophiolitic complexes of the same age, despite a lack of high-pressure metamorphism features in the ophiolites themselves (Silantyev, 1993). Such a tectonic scenario has the following weak points:

1) If the protolith was metamorphosed outside a subduction zone, then the thickness of the obducted ophiolitic nappe (or nappes) must be no less than 20 km;

2) Taking into account the probable genetic affinity between the protoliths of the Mamonia high-pressure metamorphics and the Dhiarizos and Ayia Photios Group rocks, it is difficult to explain the close spatial association of the two groups within the context of the proposed mechanism; and

3) The simultaneity of the obduction-type high-pressure metamorphism affecting the Mamonia Complex metamorphics and of the oceanic-type low-pressure metamorphism of the Troodos ophiolitic rocks, suggests that low-temperature alteration in the presence of submarine hydrothermal solutions continued during the obduction of the ophiolitic masses upon the continental margin complex.

For the obduction model, there are the following contrary arguments:

1) The proposed thickness of the ophiolite succession at the Troodos Massif is not more than 6 km (Coleman, 1984), and the estimated depth of the Moho discontinuity in the central part of the massif (in the Agros area) is not more than 14 km (Agar and Klitgord, 1995).

2) The high-pressure metamorphic tectonic blocks show sharp tectonic contacts everywhere with both the Dhiarizos and Ayia Photios Group rocks and the surrounding serpentinite plates.

3) Finally, because the presumed major sources of heat are active magma chambers, whose geometry and distribution control the circulation pattern of marine hydrothermal solutions in the spreading zones, magmatic activity and oceanic metamorphism appeared to be practically simultaneous at the Troodos. This suggests that the obductional interaction of Troodos ophiolitic rocks with Mamonia high-presure metamorphics took place at a certain deep level within a back-arc or fore-arc basin during an active stage in the development of the Troodos Massif magmatic complex.

Existing geodynamic models for the formation of collision-related high-pressure metamorphic complexes focus on the assumption of change from subductional to obductional regimes, causing transport of high-pressure metamorphics to the surface (Schreyer, 1985). This tectonic scenario defines the characteristic loop-like P-T trajectory of metamorphism within the collisional zones. In the case of a 'cool' subduction zone, it is suggested that subduction of a thick, cool oceanic plate from about 100 Ma, occurred beneath a hanging lithospheric wedge. In such a subduction zone, temperatures of about 500°C correspond to pressures of 20 kbar and depths of 60 km (Delany and Helgeson, 1978). According to Hacker's calculations (Hacker, 1991), in the case of fast (5 cm/year) subduction of young oceanic lithosphere having the age of 2 Ma, the 450°C isotherm will cross the boundary between the hanging block and the surface of the downgoing slab at the depth of about 7 km (2.2 kbar) some 3 Ma after the beginning of the subduction; in 6 and 9 Ma this isotherm will reach the deep levels of 7.5 km (2.4 kbar) and 9 km (2.9 kbar), respectively.

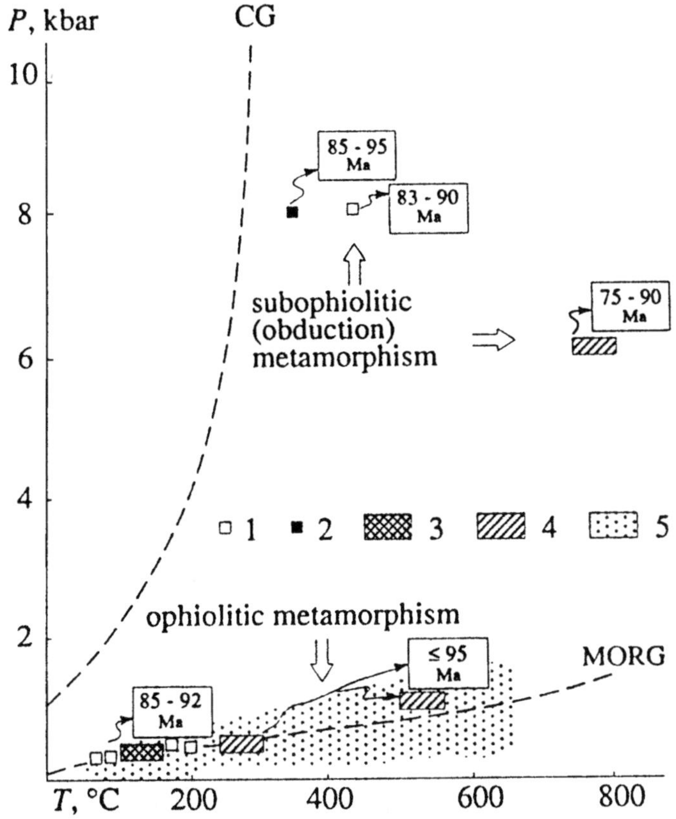

Fig. 2.33. Comparison of metamorphic P-T conditions and ages of high pressure amphibolites and associated ophiolitic rocks from the Eastern Mediterranean. 1 - Mamonia and Troodos, 2 and 3 - NW Syria (Tourkmannli Block and Baer Bassit ophiolite), 4 - Oman (data by Chen and Pallister, (1981) and Lippard (1983), 5 - metamorphism of the axial zone of the Mid-Oceanic Ridge. MORG - geotherm beneath the Mid-Oceanic Ridge, CG - continental geotherm. After Silantyev (1993).

From these calculations, the P-T regime for formation of the Mamonia high-pressure metamorphics is akin to 'warm' (relatively young) subduction of oceanic lithosphere (**Fig. 2.34**). According to Cloos (1993), the temperature gradient under this kind of subduction has values of 10°C/km to 25°C/km, and the subducting oceanic lithosphere has the age of about 10-20 Ma. Based on the assumption that the Mamonia protholith was formed in a continental margin and metamorphism of these rocks took place in a subduction zone, it is reasonable to suggest that the ocean basin related to this continental margin would be small and short-lived.

One of the principal problems associated with the present tectonic position of the high-pressure metamorphics is to identify the mechanism bringing them to the surface. To build a reasonable model for this transport, it is necessary to interpret the conditions of formation of the inverted thermal aureole at a subduction zone. Peacock (1987) suggested the possibility that fast subduction beneath young (less than 10 Ma) oceanic lithosphere would be supplied with efficient heat transfer from the hanging lithospheric block, leading to an inverted thermal gradient in the upper portions of the subducting plate. He also suggested that the preservation of this gradient in metamorphic rocks (in the case of the Mamonia Complex, 450°-550°C) is caused by the accretion of these rocks at the base of the hanging lithosphere block. Because of the generally homogeneous phase indicative of high-pressure amphibolites, their transportation toward the surface was characterized by high rates of movement.

The case of the shallower subduction model for the formation of the Mamonia Complex high-pressure metamorphics raises the question about the emplacement mechanism which caused the large-scale transport of these rocks to the surface. In this connection it is necessary to remember that in SW Cyprus all the tectonic metamorphic blocks are closely associated with thick high-angle serpentinite slivers. It is enticing to consider these pure ultramafic rocks as the high-pressure metamorphics transported from the subduction zone to upper crustal levels. Evidence for such an exumation mechanism for the high-grade amphibolites is probably to be found within the petrologic heterogeneity of the ultramafics containing the Mamonia metamorphic blocks. If these high-pressure metamorphics were formed in the upper parts of a subducting lithospheric block, the mantle residual ultramafic rocks of non-ophiolitic origin were then originally located beneath them, at the base of the subducting lithosphere. When these ultramafics moved upward by plastic flow, piercing the overlying layers, they

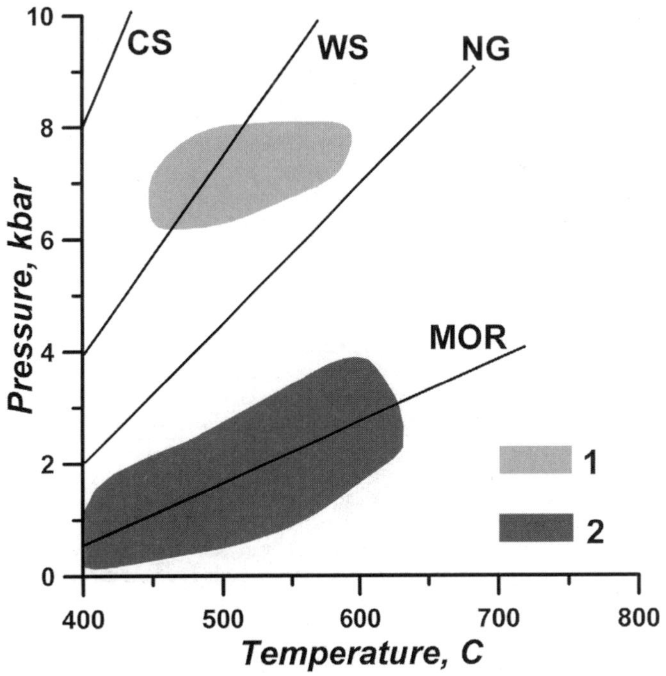

Fig. 2.34. P-T grid for geodynamic conditions related to different subduction styles. CS - 'cold' subduction, WS - 'warm' subduction, NG - normal geothermal gradients, presented by Cloos (1993)). MOR - Mid-Oceanic Ridge geotherm. 1 - high pressure metamorphic rocks from the Mamonia Complex, 2 – meta-igneous rocks from contemporary oceanic basins.

took the metamorphics with them and transported them towards the surface. Because the Troodos ophiolitic peridotites belong to the supra-subductional complex, we must assume that they have distinctly different petrologic and geochemical features from those ultramafic rocks derived from the base of the subducted lithospheric plate.

Thus, ultramafic rock diapirism is probably responsible for the tectonic disruption of the inverted thermal aureole within the subduction zone, the transportation to the surface of its fragments in the form of tectonic blocks, and their tectonic incorporation into the non-metamorphosed continental margin rocks and Troodos ophiolitic rocks. On the basis of this interpretation we should re-evaluate the tectonic nature of the Arakapas Fault Zone, as the suture line marking a fossil subduction zone.

REFERENCES CITED

Agar, S. M., and Klitgord, K. D., 1995. A mechanism for decoupling within the oceanic lithosphere revealed in the Troodos ophiolite. Nature, 374(6519), p. 232-238.

Agrinier, P., Mevel, C., and Girardeau, J., 1988. Hydrothermal alteration of the peridotites cored at the ocean/continent boundary of the Iberian margin: Petrologic and stable isotope evidence. In: Boillot, G., Winterer, E. L. et al. Proc. ODP, Sci. Results, v. 103. College Station, TX (Ocean Drilling Program), p. 225-234.

Aita, Y., 1987. Middle Jurassic to Lower Cretaceous radiolarian biostratigraphy of Shikoku with reference to selected sections in Lombardy Basin and Sicily. Sci. rep. of the Tohoku Univ. Sendai. Second ser. (Geology), 58(1), p. 1-91.

Alavi-Tehrani, N., 1980. The distribution of ophiolites in Iran and their significance. Ofioliti, v. 2, Special Issue, p. 315-334.

Allen, C. R., 1975. The petrology of a portion of the Troodos plutonic complex, Cyprus. Ph D Thesis. Cambridge University.

Anders, E., and Grevesse, N., 1989. Abundances of the elements: meteoritic and solar. Geochimica et Cosmochimica Acta, 53(1), p. 197-214.

Arai, S., 1987. An estimation of the less depleted spinel peridotite on the basis of olivine-spinel mantle array. Neues. Jahrb. Mineral. Monatsh., No. 8, p. 347-354.

Balashov, Yu. A., 1976. Geochemistry of the Rare Earth Elements. Moscow, Nauka, 267 p.

Barrat, J. -A., Jahn, B. M., Joron, J. L., Auvray, B., and Hamdi, H., 1990. Mantle heterogeneity in northeastern Africa: evidence from Nd isotopic compositions and hygromagmaphile element geochemistry of basaltic rocks from the Gulf of Tadjoura and southern Red Sea regions. Earth and Planetary Science Letters, 101(2-4), p. 233-247.

Barrat, J. -A., B. Jahn, M., Fourcade, S., and Joron, J. L., 1993. Magma genesis in an ongoing rifting zone: The Tadjoura Gulf (Afar area). Geochimica et Cosmochimica Acta, 57(10), p. 2291-2302.

Baumgartner, P.O., 1984. A Middle Jurassic-Early Cretaceous low-latitude radiolarian zonation based on Unitary Associations and age of Tethyan radiolarites. Eclogae geol. Helv, 77(3), p. 729-837.

Baumgartner, P. O., O'Dogherty, L., Gorican, S., et al., 1995. Middle Jurassic to Lower Cretaceous Radiolaria of Tethys: Occurences, Systematics, Biochronology. Mem. Geol. (Lausanne), No 23, 1172 p.

Bazylev, B. A., Magakyan, R., Silantyev, S. A., Ignatenko, K. I., Romashova, T. V., and Xenophontos, C., 1993. Petrology of peridotites from the Mamonia complex, southwestern Cyprus. Petrology, 1(4), p. 302-330.

Bazylev, B. A., 1989. Metamorphism of oceanic hyperbasites. Ph D Thesis. Moscow. Vernadsky Institute of Geochemistry. 213 p.

Bazylev, B. A., 1995a. Clinopyroxene and spinellide composition of restite hyperbasites as indicators of origin and composition of associated primary mantle magmas. Geokhimiya, 1995(7), p. 915-923. (In Russian).

Bazylev, B. A., 1995b. Geochemical effect of carbonatization of peridotites in oceanic crust. In 5th Zonenshain conference of plate tectonics. Abstracts. Moscow. November 22-25, p. 4-5.

Blome, Ch. D., 1984. Upper Triassic Radiolaria and radiolarian zonation from Western North America. Bull. Amer. Paleontol., 85(1), p. 1-88.

Bloomer, S. H., and Fisher, R. L., 1987. Petrology and geochemistry of igneous rocks from the Tonga Trench; A non-accreting plate boundary. J. Geology, 95(4), p. 469-495.

Bonatti, E., Lawrence, J. R., Hamlin, P. R., and Breger, D., 1980. Aragonite from deep sea ultramafic rocks. Geochimica et Cosmochimica Acta, 44(8), p. 1207-1214.

Bonatti, E., and Michael, P. J., 1989. Mantle peridotites from continental rifts to ocean basins to subduction zones. Earth Planet. Sci. Letters, 91(3-4), p. 297-311.

Bonatti, E., and Seyler, M., 1987. Crustal underplating and evolution in the Red Sea rift: Uplifted gabbro/gneiss crustal complexes on Zabargad and Brothers Islands. J. Geophys. Res., 92(B12), p. 12,803-12,821.

Bonatti, E., Seyler, M., Channell, J., Girardeau, J., and Mascle, G., 1990. Peridotites drilled from the Tyrrhenian sea, ODP Leg 107. In: K. A. Kastens, J. Mascle et al. Proc. ODP, Sci. Results, v. 107: College Station, TX (Ocean Drilling Program), p. 37-47.

Bonatti, E., Seyler, M., and Sushevskaya, N., 1993. A cold suboceanic mantle belt at the Earth's equator. Science, 261(5119), p. 315-320.

Bouma, A. H., 1962. Sedimentology of some flysh deposits. A graphic approach to facies interpretation. Amsterdam: Elsevier, 168 p.

Bragin, N. Yu., 1991. Radiolaria and Lower Mesozoic units of the USSR eastern regions. Moscow. Nauka, 125 p. (In Russian).

Bragin, N. Yu., and Krylov, K. A., 1999. Early Norian Radiolaria from Cyprus. Geodiversitas, 21(4), p. 539-569.

Browning, P., 1982. The petrology, geochemistry, and structure of the plutonic rocks of the Oman ophiolite. Ph D Thesis, Dept. of Earth Sciences, The Open University.

Cameron, W. E., 1985. Petrology and origin of primitive lavas from the Troodos ophiolite, Cyprus. Contributions to Mineralogy and Petrology, 89(2-3). p. 239-255.

Carter, E. S., 1993. Biochronology and Paleontology of the uppermost Triassic (Rhaetian) radiolarians, Queen Charlotte Islands, British Columbia, Canada. Mem. Geol. (Lausanne). No 11. 175 p.

Chen, J. H., and Pallister, J. S., 1981. Lead Isotopic Studies of the Samail Ophiolite, Oman. J. Geophys. Res, 86(B4), p. 2699-2709.

Coleman, R. G., 1977. Ophiolites. Ancient Oceanic Lithosphere? Springer-Verlag, Berlin-Heidelberg-New York, 229 p.

Coleman, R. G., 1984. Preaccretion tectonics and metamorphism of ophiolites. Ofioliti, 9(3), p. 205-222.

Cloos, M., 1993. Lithospheric buoyancy and collisional orogenesis: subduction of oceanic plateaus, continental margins, island arcs, spreading ridges, and seamounts. Bull. Geol. Soc. Am., 105(6), p. 715-737.

Clube, T. M. M., Creer, K. M., and Robertson, A. H. F., 1985. Palaeorotation of the Troodos microplate. Nature, 317(6037),. p. 552-525.

Curray, J. R., and Moore, D. G., 1974. Sedimentary and tectonic processes in the Bengal deep-sea fan and geosyncline. Geol. continental margins. New York: Springer-Verlag, p. 617-627.

Deer, W. A., Howie, R. A., and Zussman, J., 1963. Chain Silicates .Rock-Forming Minerals. V. 2. Longmans. London. 400 p.

Delaney, J. M., and Helgeson, H. C., 1978. Calculation of the thermodynamic consequences of dehydratation in subducting oceanic crust to 100 kb and >800°C. Amer. J. Sci., 278(5), p. 638-687.

Delaune-Mayere, Marcoux, M., J., Parrot, J. -F., and Poisson, A., 1977. Modele d'Evolution Mesozoique de la paleo-marge tethysienne au niveau des nappes radiolaritiques et ophiolitiques du Taures Lycien, d'Antalya et du Baer-Bassit. - Structural History of the Mediterranean basins. In B. Biju-Duval, and L. Montadert, (eds.). Paris: Technip, p. 79-94.

Dick, H. J. B., and Bullen, T., 1984. Chromian spinel as a petrogenetic indicator in abyssal and alpine-type peridotites and spatially associated lavas. Contrib. Mineral. Petrol., 86(1), p. 54-76.

Dick, H. J. B., 1989. Abyssal peridotites, very slow spreading ridges, and ocean ridge magmatism. Magmatism in the ocean basins. Geol. Soc. Spec. Publ., No 42, p. 71-105.

Dilek, Y., and Moores, E. M., 1990. Regional tectonics of the eastern Miditerranean ophiloites. In: J. Malpas, E. M. Moores, A. Panayiotou and C. Xenophontos (eds.), Ophiolites - Oceanic Crustal Analogues. Proc. Symp. Troodos 1987. Geological Survey Department, Nicosia, p. 295-309.

Eissen, J. -P., Juteau, T., Joron, J. -L., Dupre, B., Humler, E., Al'mukhamedov, A., 1989. Petrology and geochemistry of basalts from the Red Sea axial rift at 18° north. J. Petrology, 30(4), p. 791-839.

Elthon, D., Casey, J. F., and Komor, S., 1984.Cryptic mineral-chemistry variations in a detailed traverse through the cumulate ultramafic rocks of the North Arm Mountain Massif of the Bay of Islands ophiolite, Newfoundland. In I. G. Gass, S. J. Lippard, A. W. Shelton (eds.), Ophiolites and Oceanic Lithosphere, Geol. Soc. London, Spec. Publ. 13, Blackwell Scientific Publishers, p. 83-97.

Evans, C. A., and Girardeau, J., 1988. Galicia margin peridotites: undepleted abyssal peridotites from the North Atlantic. In: G. Boillot, E. L. Winterer et al., (eds.), Proc. ODP, Sci. Results, v. 103: College Station, TX, p. 195-208.

Ferry, J. M., and Spear, F. S., 1978. Experimental colibration of the partitioning of Fe and Mg between bitotite and garnet. Contrib. Mineral. Petrol., 66(2), p. 113-117.

Ghent, E. D., and Stout, M. Z., 1981. Metamorphism at the Base of the Samail Ophiolite, Southeastern Oman Mountains. J. Geophys. Res., 86(B4), p. 2557-2571.

Goldsmith, J. R., and Newton, R. C., 1969. P-T-X relations in the system $CaCO_3$-$MgCO_3$ at the high temperature and pressure. Amer. J. Sci., 267-A, p. 160-190.

Granath, V. C. H., Papike, J. J., and Labotka, T. C., 1983. The Notch Peak contact metamorphic aureole, Utah: petrology of the Big Horse limestone member of the Orr formation. Geol. Soc. Amer. Bull., 94(7), p. 889-906.

Gullers, R. L., and Graf, J. L., 1984. Rare Earth Elements igneous rocks of the continental crust: predominantly basic and ultrabasic rocks. In: P. Henderson, (ed.), Rare earth element geochemistry, Amsterdam-Oxford-New York-Tokyo, Elsevier Pub. Co., p. 237-268.

Hacker, B. A., 1991. The role of deformation in the formation of metamorphic gradients: Ridge subduction beneath the Oman ophiolite. Tectonics, 10(2), p. 455-473.

Hampton, M. A., 1972. The role of subaqueous debris flow in generating turbidity currents. J. Sedim. Petrol., 42(4), p. 775-793.

Hein, F. J., 1982. Depositional mechanisms of deep sea coarse clastic sediments, Cap Enrage Formation, Quebec. Can. J. Earth Sci., 19(2), p. 267-287. Also: Erratum, Can. J. Earth Sci., 19(8), p. 1726.

Hofmann, A. W., 1988. Chemical differentiation of the Earth: the relationship between mantle, continental crust and oceanic crust. Earth and Planetary Science Letters, 90(3), p. 297-314.

Hollister, C. D., Johnson, D. A., and Lonsdale, P. F., 1974. Current-controlled abyssal sedimentation Samoan Passage, Equatorial West Pacific. J. Geology, 82(3), p. 275-300.

Hubert, J. E., 1964. Textural evidence for deposition of many western North Atlantic deep-sea sands by oceanobottom currents rather than turbidity currents. J. Geology, 72(6), p. 757-785.

Karamata, S., Majer, V., and Pamic, J., 1980. Ophiolites of Yugoslavia. Ofioliti. Tethyan Ophiolites, v. 1, p. 105-125.

Kazmin, V. G., Ricou, L. E., and Sborschikov, I. M., 1987. Structure and development of the passive margin of eastern Tethys. In: History of the Tethys. Moscow, Nauka, p. 39-57.

Komor, S. C., Elthon, D., and Casey, J. F., 1985. Mineralogic variation in a layered ultramafic cumulate sequence at the North Arm Mountain Massif, Bay of Islands Ophiolite, Newfoundland. J. Geophys. Res., 90(B9), p. 7705-7736.

Kornprobst, J., and Tabit, A., 1988. Plagioclase-bearing ultramafic tectonites from the Galicia margin (Leg 103, Site 637): Comparison of their origin and evolution with low-pressure ultramafic bodies in Western Europe. In: G. Boillot, E. L. Winterer et al., (eds.), Proc. ODP, Sci. Results, v. 103: College Station, TX (Ocean Drilling Program), p. 253-268.

Kozur, H., and Mostler, H., 1972. Beitrage zur Erforschung der Mesozoischen Radiolaria. T. 1. Geol. Palaontol. Mitt. Innsbruck, Bd. 2, No 819, S. 1-60.

Kozur, H., and Mostler, H., 1979. Beitrage zur Erforschung der Mesozoichen Radiolaria. T. III. Geol. Palaontol. Mitt. Innsbruck, Bd. 9, No 112, S. 1-132.

Kozur, H., and Mostler, H., 1981. Beitrage zur Erforschung der Mesozoichen Radiolaria. T. IY. Geol. Palaontol. Mitt. Innsbruck, S.-Bd., S. 1-208.

Krylov, K. A., K. G. Kaleda, and N. Yu. Bragin, 1995a. Terrigeneous chaotic complexes of south-western Cyprus. In: 5th Zonenschain conference on plate tectonics. Moscow, November 22-25. Abstr., p. 77.

Krylov, K. A., Kaleda, K. G., and Bragin, N. Yu., 1995. Tectonic evolution south-western of Cyprus in Mesozoic. In: 5th Zonenschain conference on plate tectonics. Moscow. November 22-25. Abstr., p. 77.

Lapierre, H., 1975. Les formations sedimentaire et eruptive des nappes de Mamonia et leurs relation aves le massif de Troodos (Chypre occidentale). Soc. Geol. France Mem., No 123, 132 p.

Leake, B. E., 1978. Nomenclature of amphiboles. Can. Mineralogist, 16(4), p. 501-520.

Le Roex, A. P., 1987. Source regions of mid-ocean ridge basalts: evidence for enrichment processes. In: M. A. Menzies (ed.) Mantle metasomatism. London. Academic Press, p. 389-421.

Lippard, S. J., 1983. Cretaceous high pressure metamorphism in NE Oman and its relationship to subduction and ophiolite nappe emplacement. J. of Geol. Soc. London, 140(1), p. 97-104.

MacLeod, C. J., 1990. Role of the Southern Troodos Transform Fault in the rotation of the Cyprus microplate: evidence from the Eastern Limassol Forest Complex. In: J. Malpas, E. M. Moores, A. Panayiotou and C. Xenophontos (eds.), Ophiolites - Oceanic Crustal Analogues. Proc. Symp. Troodos 1987. Geological Survey Department, Nicosia, p. 75-85.

Magakyan, R., 1993. Primitive boninites from Mamonia, SW Cyprus. Eos Trans. AGU. Fall Meet. Suppl., 44 (75), p. 746-747.

Malpas, J. G., 1979. The dynamothermal aureole of the Bay of Islands ophiolite suite. Canad. J. Sci., 16(11), p. 2086-2102.

Malpas, J. G., and Xenophontos, C., 1987. Mamonia Complex and its relation to the Troodos Ophiolites. In: Field Excursion Guidebook. Symposium Troodos 87. Ophiolites and Oceanic Lithosphere, Nicosia, Cyprus, p. 234-259.

Malpas, J. G., Xenophontos, C., and Williams, D., 1992. The Ayia Varvara Formation of S.W. Cyprus, a product of complex collisional tectonics. Tectonophysics, 212(3-4), p. 193-211.

Malpas, J. G., Calon, T., and Squires, G., 1993. The development of a late Cretaceous microplate suture zone in SW Cyprus. In: H. M. Prichard et al., (eds.) Magmatic processes and plate tectonics. Geological Society Special Publication, No 76, p. 177-195.

Mantis, M., 1970. Upper Cretaceous-Tertiary foraminiferal zones in Cyprus. Scientific Research Centre Cyprus, p. 227-241.

Marakushev, A. A., and Tararin, A. A., 1965. Mineralogical criteria of alkalinity of granitoids. Academy of Sciences of USSR News, Geological Series, No 3, p. 20-37. (In Russian).

Middleton, G. V., 1970. Experimental studies related to problems of flysh sedimentation. Geol. Soc. Can. Spec. Pap., 7, p. 253-272.

Middleton, G. V., and Hampton, M. A., 1973. Sedimentary gravity flows: mechanisms of flow deposition. Turbidites and deep water sedimentation. Soc.Econ. Paleontol.Miner. Pacific Short course. Anaheim. p. 1-38.

Miyashiro, A., 1973. Metamorphism and Metamorphic Belts. London George Allen and Unwin Ltd., 520 p.

Miyashiro, A., and Shido, F., 1980. Differentiation of gabbros in the Mid-Atlantic Ridge near 24°N. Geochemical Journal, 14(4), p. 145-154.

Moores, E. M., and Vine, F. J., 1971. The Troodos massif, Cyprus and other ophiolites as oceanic crust: evaluation and implications. In: Philosophical Transactions of the Royal Society of London, v. A268, p. 443-466.

Natland, J. H., 1980. The Progression of Volcanism in the Samoan Linear Volcanic Chain. Am. J. Sci., v. 280-A, p. 709-735.

Okay, A. I., 1982. Incipient Blueschist Metamorphism and Metasomatism in Tavsanli Region, North-West Turkey. Contrib. Mineral. Petrol., 79(4), p. 361-368.

Pamic, J., 1977. Variation in geothermometry and geobarometry of peridotite intrusion in the Dinaride Central Ophiolite Zone, Yugoslavia. Amer. Mineral., 62(9-10), p. 874-886.

Pamic, J., and Adib, D., 1982. High grade amphibolites and granulites at the base of the Neiriz Peridotites in Southern Iran. N. Jb. Miner. Abh., B.143, No 2, p. 113-121.

Peacock, S. M., 1987. Creation and Preservation of subduction-related inverted metamorphic gradients. J. Geophys. Res., 92(B12), p. 12,763-12,781.

Pearce, J. A., 1975. Basalt geochemistry used to investigate past tectonic environments on Cyprus. Tectonophysics, 25(1-2), p. 41-67.

Perchuk, L. L., 1970. Equilibria of the Rock-Forming Minerals. Nauka. Moscow. 390 p. (In Russian).

Perchuk, L. L., and Ryabchikov, I. D., 1976. Phase co-existence in mineral systems. Nedra. Moscow. 286 p. (In Russian).

Piskin, O., Chessex, R., Delaloye, M., Moritz, R., and Wagner, J. J., 1983. A tectonic approach of the Kizil Dag Ophiolite. Ofioliti. Oceanic tectonics and metamorphism, v .8. p. 50.

Plank, T., and Langmuir, C. H., 1988. An evaluation of the global variations in the major element chemistry of arc basalts. Earth Planet. Sci. Lett., 90(4), p. 349-370.

Plyusnina, L. P., 1983. Experimental investigation of metamorphism of basites. Moscow. Nauka. 159 p. (In Russian).

Portnyagin, M. V., Magakyan, R. G., and Schminke, H. -U., 1996. Geochemical variability of Boninite Magmas: Evidence from Magmatic Inclusions in Highly Magnesian Olivine from lavas of Southwestern Cyprus. Petrology, 4(3), p. 231-246.

Reding, H., 1990a. Facies and sedimentation. V. 1. Moscow. Mir. 351 p.

Reding, H., 1990b. Facies and sedimentation. V. 2. Moscow. Mir. 382 p.

Rice, J. M., 1977. Progressive metamorphism of impure dolomitic limestone in the Marysville aureole, Montana. Amer. J. of Sci., 227(1), p. 1-24.

Ringwood, A. E., 1975. Composition and petrology of the earth's mantle. New York: McGraw -Hill. 618 p.

Robertson, A. H. F., 1977. The Kannaviou Formation, Cyprus, volcanoclastic sedimentation of a probable late Cretaceous volcanic arc. J. Geol. Soc. London, 134(3), p. 269-292.

Robertson, A. H. F., 1990a. Tectono-sedimentary evolution of the Eastern Mediterranean Neotethys: Summaries, questions and answers. Intern. Earth Sci. Cong. on Aegean Reg. 1-6 Oct.1990, Izmir, Turkey. IESCA Publ., v. 2, p. 236-278.

Robertson, A. H. F., 1990b. Tectonic evolution of Cyprus. In: J. Malpas, E. M. Moores, A. Panayiotou and C. Xenophontos (eds.), Ophiolites - Oceanic Crustal Analogues. Proc. Symp. Troodos 1987. Geological Survey Department, Nicosia, p. 232-252.

Robertson, A. H. F., and Dixon, J. E., 1984. Introduction: Aspects of the geological evolution of the Eastern Mediterranean. In: J. E. Dixon, and A. H. F. Robertson, (eds.) The geological evolution of the Eastern Miditerranean. Geological Society Special Publication. No 17, p. 1-74.

Robertson, A. H. F., and Woodcock, N. H., 1979. Mamonia Complex, Southwest Cyprus: evolution and emplacement of a Mesozoic continental margin. Bull. Geol. Soc. Am., 90(1), p. 651-665.

Robertson, A. H. F., and Woodcock, N. H., 1981. Goden Zone, Antalya Complex: volcanism and sedimentation along Mesozoic continental margin. Geol. Rundschau, 70(3), p. 1177-1214.

Robertson, A. H. F., and Xenophontos, C., 1993. Development of concepts concerning the Troodos ophiolite and adjacent units in Cyprus. In: Prichard, H.M. et al. (eds.) Magmatic processes and plate tectonics. Geological Society Special Publication. No 76, p. 89-119.

Sack, R. O., and Ghiorso, M. S., 1991. Chromian spinels as petrogenetic indicators: thermodynamics and petrological applications. Am. Mineral., 76(5-6), p. 827-847.

Sharaskin, A. Ya., Pustchin, I. K., Zlobin, S. K., and Kolesov, G. M., 1983. Two ophiolite sequences from the basement of the Northern Tonga Arc. Ofioliti, 8(3), p. 411-430.

Schilling, J. -G., 1969. Red Sea floor origin: rare-earth evidence. Science, 165(3900), p. 1357-1360.

Schilling, J. -G., Zajac, M., Evans, R., Johnston, T., White, W., Devine, J. D., and Kingsley, R., 1983. Petrological and geochemical variations along the Mid-Atlantic Ridge from 27° to 73°N. Am. J. Sci., 283(6), p. 510-586.

Schreyer, W., 1985. Metamorphism of Crustal rocks at Mantle Depths: High-Pressure Minerals and Mineral Assemblages in Metapelites. Fortschr. Miner., 63(2), p. 227-261.

Silantyev, S. A., 1993. Metamorphic Complexes of the Eastern Mediterranean: Metamorphic Conditions, Protoliths, and Geodynamic Environment. Petrology, 1(4), p. 395-416.

Silantyev, S. A., 1995. Metamorphism of Contemporary Oceanic Basins. Petrology, 3(1), p. 20-30.

Silantyev, S. A., and Zlobin, S. K., 1986. Metamorphism in the rocks from inner slope of the northern part of Tonga Trench. In: O.A.Bogatikov, (ed.), Oceanic magmatism. Moscow. Nauka, p. .250-262.

Snow, J. E., and Dick, H. J. B., 1995. Pervasive magnesium loss by marine weathering of perodotite. Geochim. et Cosmochim. Acta, 59(20), p. 4219-4235.

Sobolev, A. V., and Batanova, V. G., 1995. Mantle lherzolites of the Troodos ophiolite complex, Cyprus: Clinopyroxene geochemistry. Petrology, 3(5), p. 440-448.

Sobolev, A. V. Portnyagin, M. V., Dmitriev, L. V., Tsameryan, O. P., Danyushevsky, L. V., Kononkova, N. N., Shimizu, N., and Robinson, P., 1993. Petrology of ultramafic lavas and associated rocks of the Troodos massif, Cyprus. Petrology, 1(4), p. 331-361.

Spear, F. S., 1981. An experimental study of hornblende stability and compositional variability in amphibolite. Am. J. Science, 281(6), p. 697-734.

Spray, J. G., and Roddick, J. C., 1981. Evidence for Upper Cretaceous transform fault metamorphism in West Cyprus. Earth Planet. Sci. Lett., 55(2), p. 273-291.

Staudigel, H., Gillis, K., and Duncan, R. E., 1986. K/Ar and Rb/Sr ages of celadonites from the Troodos ophiolite, Cyprus. Geology, 14(1), p. 72-75.

Stolper, E., and Newman, S., 1994. The role of water in the petrogenesis of Mariana Trough magmas. Earth Planet. Sci. Lett., 121(3-4), p. 293-325.

Sun, S. -S., and McDonough, W. F., 1989. Chemical and isotopic systematics of oceanic basalts: implications for mantle composition and processes. Magmatism In: A. D. Saunders, and M. J. Norry, (eds.). The Ocean Basins. London. Geol. Soc. Spec. Publ., No 42, p. 313-345.

Swarbrick, R. E., and Naylor, M. A., 1980. The Kathikas melange SW Cyprus: late Cretaceous submarine debris flows. Sedimentology, 27(1), p. 63-78.

Swarbrick, R. E., and Robertson, A. H. F., 1980. Revised stratigraphy of the Mesozoic rocks of southern Cyprus. Geol. Mag., 117(5), p. 547-563.

Thy, P., 1990. Cryptic variation of a cumulate sequence from the plutonic complex of the Troodos ophiolite. In: J. Malpas, E. M. Moores, A. Panayiotou and C. Xenophontos (eds.), Ophiolites - Oceanic Crustal Analogues. Proc. Symp. Troodos 1987. Geological Survey Department, Nicosia, p. 165-172.

Turner W. M., 1969. Geology of the Polis-Kathikas area, Cyprus. Geological Survey Department, Cyprus.

Varga, R. J., Schiffman, P., Smith, B. M., Moores, E. M., and Hurst, S., 1987. Structure and hydrothermal metamorphism of the Solea graben, Troodos ophiolite. In: Field Excursion Guidbook. Symposium Troodos 87. Ophiolites and Oceanic Lithsphere, Nicosia, Cyprus, p. 182-209.

Walker, R. G., 1967. Turbidite sedimentary structures and their relationship to proximal and distal depositional environments. J. Sedim. Petrol., 37(1), p. 25-43.

Wicks F J., and Whittaker, E. J. W., 1977. Serpentine textures and serpentinization. Can. Mineralogist, 15(4), p. 459-488.

Wilson, M., 1989. Igneous Petrology. London. Ulwin, Hyman. Boston-Sydney-Wellington. 465 p.

Yao, A., 1982. Middle Triassic to Early Jurassic radiolarians from the Inuyama Area, Central Japan. J. Geosci. Osaka City Univ., v. 25, p. 53-70.

Yoder, H. S., and Tilley, C. E., 1962 Origin of basaltic magmas: an experimental study of natural and synthetic rock systems. J. Petrology, 3(3), p .342-529.

Yoshida, H., 1986. Upper Triassic to Lower Jurassic radiolarian biostratigraphy in Kagamigahara City, Gifu Prefecture, Central Japan. J. Earth Sci. Nagoya Univ., v. 34, p. 1-21.

Chapter 3

Stratigraphy and Lithology of the Neoautochthon

Valery A. Krasheninnikov and Kirill G. Kaleda
Geological Institute of the Russian Academy of Sciences, Pyzhevskiy pereulok 7, 109017 Moscow, Russia

The present chapter outlines the stratigraphy of the formations of the neoauthochthonous cover in south-western Cyprus, taking into account their lithologic characteristics and lateral changes.

Late Cretaceous (Campanian to Maastrichtian) and Cenozoic marine sediments form the neoautochthonous sequence that overlies the para-autochthonous Troodos and allochthonous Mamonia complexes. These rocks have been studied by Ovey (1937), Henson et al. (1949), and Cockbain (1961). Over the last three decades, they have been intensively investigated and described in the numerous publications of Mantis, 1970, 1972, 1977; Pantazis, 1967, 1978; Robertson, 1976, 1977a-d; Robertson and Hudson, 1974; Swarbrick and Robertson, 1980; Swarbrick and Naylor, 1980; Robertson et al., 1991, 1995; Kluyver, 1969; Robertson and Grasso, 1995a, b; Payne and Robertson, 1995; Robertson and Dixon, 1984; McCallum and Robertson, 1995; Baroz and Bizon, 1974, 1977; Orszag-Sperber et al., 1989; Orszag-Sperber and Rouchy, 1979; Turner, 1968, 1971; and others.

Much of the work on the sedimentology, stratigraphy and micropaleontology of the neoautochthonous sequence, done by Ph.D. students from the University of Edinburgh, University College of London, and the Université de Paris-Sud (S. Eaton, 1987; P. Elion, 1983; E. Follows, 1990; M. Hasket, 1984; J. Herguera, 1982; and P. Ortiz, 1982), regretfully remains unpublished.

The subdivision of the neoautochthonous sedimentary sequence into lithostratigraphic units is summarized in **Table 3.1**.

In 1987 and 1990, the authors studied the Upper Cretaceous and Cenozoic of the southern margin of the Troodos Massif, between Akamas Peninsula and Limassol Forest. This area includes the structural depression between the Mamonia uplift and the Limassol Forest Massif (**Fig. 3.1**), The most important sections include:

1) the sequence near Perapedhi, extending southward as far as Polemidhia;
2) the sequence near Ayios Nikolaos, extending in the direction of Evdhimou;
3) the sequence near Pano Panayia; and
4) the sequence in the area of Petra tou Romiou.

These investigations included the relationships of neoautochthonous and underlying Mammonia and Troodos complexes in the surroundings of Akrounda, Kannaviou, Phiti, Kathikas, Simou, Kallepia, Ayios Dhimitrianos, Dhrymou, Pissouri, Polemi, Letimbou, Strombi, and in the upper part of the Mavrokolymbos Valley.

The aim of this study was to:

a) provide a valid stratigraphic framework for the Late Cretaceous, Paleogene, and Neogene of southern Cyprus,
b) analyze the history of basin development, with respect to its paleogeographical, paleoecological, and paleotectonic characteristics and
c) recognize the importance of South-West Cyprus in the overall framework of the eastern Mediterranean region.

The careful examination of the sections presents the authors' interpretation of the regional stratigraphy in this area. The succession consists of the Lower Lefkara Formation (Maastrichtian to Early Eocene), the Middle Lefkara Formation (Early-Middle Eocene), the Upper Lefkara Formation (Middle-Late Eocene), the Kilani Formation (Oligocene-earliest Early Miocene), the Arsos Formation (Early Miocene), the Pakhna Formation (Early to Late Miocene), the Kalavasos Formation (Messinian gypsums, Late Miocene), and the Pissouri Formation (Pliocene). More accurate ages were determined for the traditional formations, as well as for the Kilani and Arsos formations, which were previously studied by the authors (Krasheninnikov and Kaleda, 1994).

The stratigraphic range and age of the above-mentioned rock units correspond to those in other eastern Mediterranean countries (Syria, Israel, Egypt), thus facilitating correlation of the Cenozoic deposits and helping to decipher the geological history of this realm. This is the first peculiarity of our scientific approach. The second peculiarity is connected with our understanding of the term formation. In the Russian stratigraphic nomenclature it genetically designates a complex of rocks of similar provenance (*e.g.* an iron ore formation, phosphatic formation, etc.). In Russia, for the purpose of geological mapping or applied geology the term 'suite' is used. A suite is a regional stratigraphic unit which is characterized by identical lithology and with lower and upper lithological boundaries coinciding with the biostratigraphical ones. Consequently, suite boundary surfaces are isochronous and this characteristic indicates the principal difference between a Russian suite and a formation as accepted in the West.

In the study area, the formations under consideration are usually laterally persistent units which can be traced along the south-western margin of the Troodos Massif. We mapped their boundaries with the help of the zonal (foraminiferal and radiolarian) stratigraphy and these lithological levels coincide with the chronostratigraphical ones. In some cases, however, there are distinct lithologic variations controlled by tectono-environmental factors. These led to the recognition of a few local formational units (*e.g.* the Kathikas and Terra formations) as integral parts of the regional succession.

Table 3.1. Formations of the Upper Cretaceous and Cenozoic deposits of Cyprus according to different authors (after Krasheninnikov and Kaleda, 1994).

AGE			GROUPS AND FORMATIONS						
			PANTAZIS (1967)		TURNER (1971)		MANTIS (1970-75)		KRASHENINNIKOV and KALEDA, (1994)
PLIOCENE	PIACENZIAN	N21 N20		ATHALASSA		PISSOURI		NICOSIA / MARLS OF MIRTOU	PISSOURI
PLIOCENE	ZANCLEAN	N19 N18		ATHALASSA		PISSOURI		NICOSIA / MARLS OF MIRTOU	PISSOURI
MIOCENE	MESSINIAN	N17	DHALI GROUP	KORONIA / KALA-VASOS	DHALI GROUP	YIOLOU		KORONIA / GYPSUM	KALAVASOS
MIOCENE	TORTONIAN	N16 N15	DHALI GROUP	PAKHNA	DHALI GROUP	PAKHNA		PAKHNA	PAKHNA
MIOCENE	SERRAVALLIAN - UPPER LANGHIAN	N14 N13 N12 N11 N10 N9	DHALI GROUP	PAKHNA	DHALI GROUP	PAKHNA	LEFKARA GROUP	PAKHNA / TERRA	PAKHNA
MIOCENE	BURDIGALIAN	N8 N7	LEFKARA GROUP	UPPER CHALKS: BEDDED		TERRA	LEFKARA GROUP	TERRA	PAKHNA / ARSOS
MIOCENE	AQUITANIAN	N6 N5 N4	LEFKARA GROUP	UPPER CHALKS: BEDDED	LAPITHOS GROUP	MAVROLI	LEFKARA GROUP	LEFKARA TERMINAL	ARSOS / KILANI
OLIGOCENE	CHATTIAN	P22 P21	LEFKARA GROUP	UPPER CHALKS: BEDDED	LAPITHOS GROUP	MAVROLI	LEFKARA GROUP	LEFKARA TERMINAL	KILANI
OLIGOCENE	RUPELIAN	P20 P18-19	LEFKARA GROUP	UPPER CHALKS: BEDDED	LAPITHOS GROUP	MAVROLI	LEFKARA GROUP	LEFKARA TERMINAL	KILANI
EOCENE	PRIABONIAN	P17 P16 P15	LEFKARA GROUP	UPPER CHALKS: MASSIVE	LAPITHOS GROUP	MAVROLI	LEFKARA GROUP	UPPER LEFKARA	UPPER LEFKARA
EOCENE	BARTONIAN	P14 P13	LEFKARA GROUP	UPPER CHALKS: MASSIVE	LAPITHOS GROUP	PANO-LEFKARA	LEFKARA GROUP	UPPER LEFKARA	UPPER LEFKARA
EOCENE	LUTETIAN	P12 P11 P10	LEFKARA GROUP	MASSIVE / CHALKS WITH CHERTS	LAPITHOS GROUP	PANO-LEFKARA	LEFKARA GROUP	UPPER LEFKARA	MIDDLE LEFKARA
EOCENE	YPRESIAN	P9 P8 P7 P6	LEFKARA GROUP	CHALKS WITH CHERTS	LAPITHOS GROUP	PANO-LEFKARA	LEFKARA GROUP	PANO-LEFKARA	LOWER LEFKARA
PALEOCENE	THANETIAN	P5 P4	LEFKARA GROUP	CHALKS WITH CHERTS	LAPITHOS GROUP	PANO-LEFKARA	LEFKARA GROUP	PANO-LEFKARA	LOWER LEFKARA
PALEOCENE	SELANDIAN	P3	LEFKARA GROUP	CHALKS WITH CHERTS	LAPITHOS GROUP	PANO-LEFKARA		HIATUS	LOWER LEFKARA
PALEOCENE	DANIAN	P2 P1	LEFKARA GROUP	CHALKS WITH CHERTS	LAPITHOS GROUP	PANO-LEFKARA		HIATUS	LOWER LEFKARA
CRETACEOUS	MAASTRICHTIAN		LEFKARA GROUP	LOWER CHALKS	LAPITHOS GROUP	SIMOU		LOWER CHALKS	LOWER LEFKARA
CRETACEOUS	CAMPANIAN			PERAPEDHI	LAPITHOS GROUP	SIMOU		PERAPEDHI	LOWER LEFKARA / KANNAVIOU

Thus, during our investigations in Cyprus, we used suites in practice, as regional stratigraphic subdivisions, but this term is unknown to a non-Russian reader so we continued using the formational names.

At this point it is necessary to give a short explanation of the Cenozoic stratigraphic scheme we applied to the subdivision of the Paleogene and Neogene sediments of Cyprus (and Syria).

The deep-sea drilling in the World Ocean which started in 1968 showed by means of orthostratigraphic groups of planktonic microfossils (foraminifera and nannoplankton) that zonal schemes are the basis of the Cenozoic subglobal (latitudinal) stratigraphy. These zones were identified within the entire warm-water belt of the Earth (40ºS–50ºN) and provide correlation of the Cenozoic marine sediments in oceans, seas, and on continents. Taking into account principal levels of change of planktonic foraminifera, zones can be united into subdivisions of higher rank – stages, subseries, and series. In our opinion, the results of the deep-sea drilling definitely define 'the face' of Cenozoic stratigraphy for the 21st century. The next step in the scientific development of Cenozoic stratigraphy, will be the correlation of low-latitude zonal sequences, based on the calcareous plankton, with high-latitude ones, based on siliceous plankton (radiolarians, diatoms), thus converting the subglobal Cenozoic stratigraphy into a global one.

In accordance with this approach, the Paleogene succession in Cyprus (and Syria) consists of:

Paleocene: Danian Stage (the *Turborotalia pseudobulloides* Zone, P1b to the *Acarinina uncinata* Zone, P2); Selandian Stage (the *Morozovella angulata* Zone s.l., P3 which is sometimes subdivided into the *Morozovella angulata* Zone s.str., P3a and the *Igorina pusilla* Zone, P3b); Thanetian Stage (the *Planorotalites pseudomenardii* Zone, P4 and *Morozovella velascoensis* Zone, P5).

Lower Eocene: Ypresian Stage (the *Morozovella subbotinae* Zone, P6 to the *Acarinina pentacamerata* Zone, P9);

Middle Eocene: Lutetian Stage (the *Hantkenina aragonensis* Zone, P10 and *Globigerapsis kugleri* Zone, P11 which sometimes are combined in the *Acarinina bullbroki* Zone, P10-P11, as well as the *Morozovella lehneri* Zone, P12) and Bartonian Stage (the *Orbulinoides beckmanni* Zone, P13 and the *Truncorotaloides rohri* Zone, P14);

Upper Eocene: Priabonian Stage (the *Globigerapsis semiinvoluta* Zone, P15 to the *Turborotalia centralis– Globigerina gortanii* Zone, P17);

Lower Oligocene: Rupelian Stage (the *Globigerina tapuriensis* Zone, P18 to the *Globigerina ampliapertura* Zone, P20); Chattian Stage (the *Paragloborotalia opima* Zone, P21 to the *Globorotalia kugleri* Zone, N4a).

The Neogene succession includes the following subdivisions:

Lower Miocene: Aquitanian Stage (the *Globigerinoides primordius-Globorotalia kugleri* Zone, N4b to the *Catapsydrax stainforthi* Zone, N6); Burdigalian Stage (the *Globigerinatella insueta–Globigerinoides trilobus* Zone, N7 and the *Praeorbulina glomerosa* Zone, N8);

Middle Miocene: Serravallian Stage s.l. (the *Orbulina suturalis–Globorotalia peripheroronda* Zone, N9, the *Globorotalia peripheroacuta* Zone, P10, undifferentiated *Globorotalia fohsi fohsi* to *Sphaeroidinellopsis subdehiscens– Globigerina druryi* Zones, N11–N13 and the *Globigerina nepenthes - Globorotalia siakensis* Zone, N14);

Upper Miocene: Tortonian Stage (the *Globorotalia continuosa* Zone, N15 and the *Globorotalia acostaensis* Zone, N16).

Sediments of the Uppermost Miocene (Messinian Stage) do not contain informative assemblages of planktonic foraminifera. But for the Pliocene the regional (Mediterranean) zones can be applied: the *Sphaeroidinellopsis seminulina* Acme-Zone and the *Globorotalia margaritae* Zone (Zanclean Stage) and the *Globorotalia puncticulata–Globorotalia inflata* Zone (Piacenzian Stage).

We placed the Lower/Middle Miocene boundary at the base of the *Orbulina suturalis–Globorotalia peripheroronda* Zone, N9 where planktonic foraminifera underwent significant evolutional change. This level was recommended by the Regional Committee on the Mediterranean Neogene Stratigraphy (Excursion 'A', 1975) as the Lower/Middle Miocene boundary. Transfer of this boundary to the base of the *Praeorbulina glomerosa* Zone, N8, indicates the preservation of the Langhian Stage which was proposed in 1865, based on lithological features in the Chessole Formation, Italy.

The serious limitations of the stratotypes of the Paleogene and Neogene stages, established in the second half of the 18th century, have been considered in numerous publications and are well known. These stratotypes reflect local or regional biological and geological events, and represent the scientific state-of-the-art at that time. The results of drilling in the oceanic basins have since provided us with the tools to create a subglobal/global Cenozoic stratigraphy for the 21st century. This trend has guided us in applying a subglobal stratigraphy to the Cenozoic subdivision of Cyprus (and Syria).

Kathikas Formation

A description of the Kathikas Formation is given in Chapters 1 and 4 of this volume. Below we will emphasize its main features.

In the area of the Mamonia Complex and along the southwest margin of the Troodos Massif, the chaotic mixtites of the Kathikas Formation represent the oldest part of the neoautochthonous sediments of South-West Cyprus. It is dated by radiolarians of the *Amphipyndax tylotus* Zone as Late Campanian to Early Maastrichtian. The Kathikas Formation is believed to be equivalent to the marls and marly limestones of the basal part of the Lower Lefkara Formation in the southeastern part of the Troodos Massif.

Outcrops of the Kathikas Formation are concentrated in three separate areas (**Fig. 3.2**):

1) in the vicinity of Kathikas, Arodhes, Peya and the Mavrokolymbos Valley, north of Paphos;

2) near Episkopi, Ayia Varvara, Kholetria, and Mamonia, east of Paphos; and

3) in the vicinity of Simou, Phiti, Dhrymou, Kannaviou, Statos, Ayios Photios, Galataria, Pendalia, Ayios Dhimitrianos and

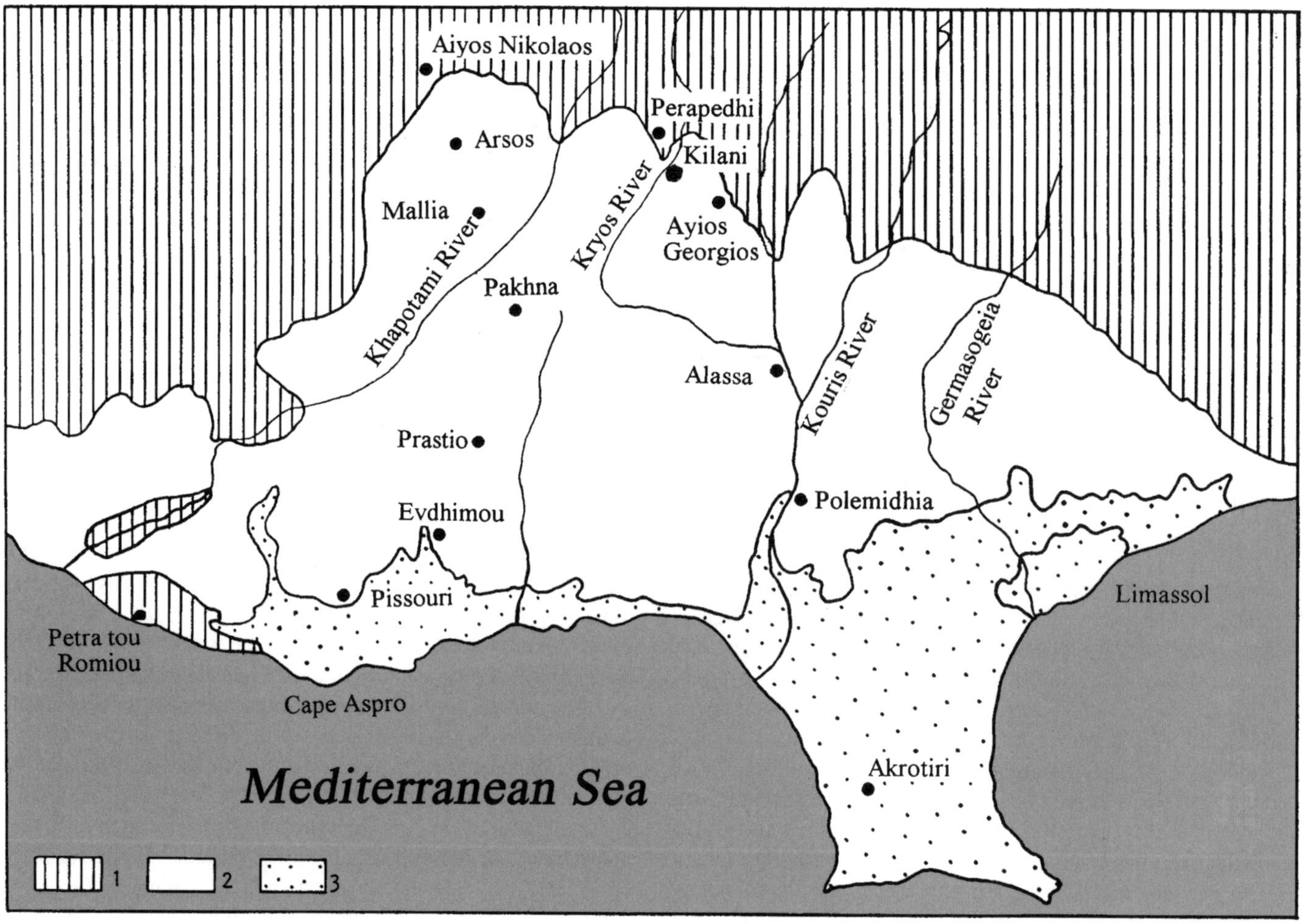

Figure 3.1. The Upper Cretaceous-Cenozoic depression of South-Western Cyprus between the Khapotami River and the Germasogeia River and position of the reference sections of the Upper Cretaceous, Paleogene and Neogene deposits. The western reference section is located along the line: Ayios Nikolaos - Arsos - Mallia - Pakhna - Prastio - Evdhimou - Pissouri. The eastern reference section stretches along the line: Perapedhi - Kilani - Ayios-Georgios - Alassa - Polemidhia. 1 - pre-Campanian-Maastrichtian rocks; 2 - outcrops of Campanian-Maastrichtian, Paleogene and Neogene deposits; 3 - Quaternary sediments.

Kelokendhara, along the south-western margin of the Troodos Massif.

The Kathikas Formation was described by Lapierre (1975), Robertson and Woodcock (1979), Swarbrick and Robertson (1980) and Swarbrick and Naylor (1980). Near Kathikas, the formation reaches a maximum thickness of up to 250-270 m. It is represented by mixtites in a clayey matrix. In most localities, the Kathikas Formation is separated from the underlying bentonitic clays of the Kannaviou Formation by an erosional surface. Nevertheless, the upper layers of the Kannaviou Formation and the whole Kathikas Formation both belong within the *Amphipyndax tylotus* Zone (Bragina and Bragin, 1995.

The angular fragments of the mixtites, derived from the Mamonia Complex, are found in a matrix of reddish brown clays, redeposited from the Mavrokolymbos Formation, the Ayios Photios Group of Mamonia Complex, and bentonitic clays of the Kannaviou Formation, which cover the Troodos Complex.

An important aspect regarding the origin and development of the mixtites is the gradual lateral transition from the relatively undeformed Mamonia Complex into the Kathikas mixtites. This is observed upstream on the Xeros Potamos River (Swarbrick and Naylor, 1980), in the so-called 'blocky formation', that consists of brecciated fragments of Mamonia lithologies. These are tectonized blocks with no clayey matrix, indicating fracturation and rotation. Above this is found the 'immature melange unit', a transitional unit whose fragments, consisting of large blocks of argillites and bedded cherts in an argilliceous matrix, differ completely in their composition from the typical Kathikas clasts. Upward these rocks pass gradually into the typical chaotic Kathikas lithologies.

The Kathikas Formation is therefore considered as an olistostrome, representing debris flows and subaqueous gravity-slide processes, for which the term 'tectono-gravitative mixtites' was coined by M. Leonov (1981), nicely depicting the true nature of the Kathikas olistostrome.

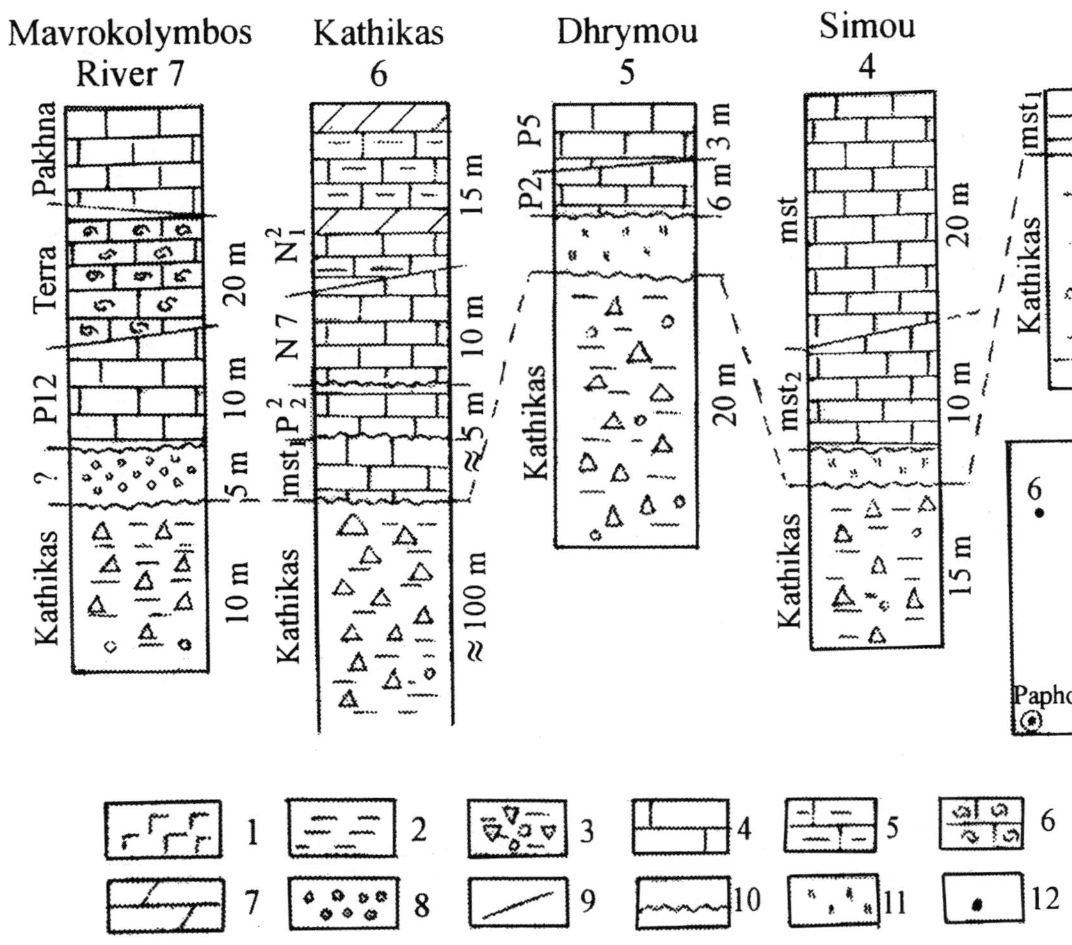

Figure 3.2. Some sections of the Upper Cretaceous, Paleogene and Neogene deposits in limits of the Mamonia Complex rocks and their location. 1 - lavas; 2 - non-calcareous clays of the Kannaviou Formation; 3 - mixtites of the Kathikas Formation; 4 - limestones; 5 - clayey limestones; 6 - detrital organogenous limestones of the Terra Formation; 7 - marls; 8 - small-pebble conglomerate and gravelite; 9 - faults observed in sections; 10 - erosional contacts; 11 - no outcrops; 12 - position of sections. Sections: 1 - Kannaviou; 2 - Ayios Dhimitrianos; 3 - Phiti; 4 - Simou; 5 - Dhrymou; 6 - Kathikas; 7 - the Mavrokolymbos River.

The relationship between the Kathikas Formation and the underlying Kannaviou Formation is explained in the contribution of Krylov et al. in Chapter 4 of this volume. These authors describe the sharp and irregular lower contact of the chaotic deposits within erosion pockets of the underlying Kannaviou Formation. The matrix of the mixtites closely resembles the bentonites of the Kannaviou Formation, which near Kannaviou yields the radiolarians: *Acaeniotyle* ex gr. *diaphorogona* Foreman, *Afens lirioides* Riedel et Sanfilippo, *Alievium* sp., *Amphipyndax pseudoconulus* (Pessagno), *A. stocki* (Campbell et Clark), *A. tylotus* Foreman, *Clathropyrgus titthium* Riedel et Sanfilippo, *Dictyomitra* ex gr. *duodecimcostata* (Squinabol), *Hemicryptocapsa conara* Foreman, *Crucella latum* Lipman, *Phaseliforma laxa* Pessagno, *Pseudoaulophacus floresensis* Pessagno, *P. pargueraensis* Pessagno, *P. praefloresensis* Pessagno, *Rhopalastrum tumidum* Lipman, *Septinastrum dogeli* Gorbovets, *Stylosphaera* ex gr. *pusilla* Campbell et Clark, *Xitus asymbatos* (Foreman) (Bragina and Bragin, 1995). This assemblage corresponds to the early part of the *Amphipyndax tylotus* Zone, which suggests a Late Campanian to Early Maastrichtian age for the Kathikas olistostrome.

Lower Lefkara Formation

The carbonate deposits of the Lower Lefkara Formation represent a long time interval, from Late Cretaceous (Late Campanian-Maastrichtian) at its base, to Paleogene (Paleocene and most of Early Eocene). The latter ends at the top of the *Morozovella aragonensis* Zone. Laterally this unit varies considerably in stratigraphic volume, thickness, facies, and relationship to substratum.

The Cretaceous portion of the formation is intermittently exposed along the southern edge of the Troodos Massif. In the east, near Perapedhi, its lowest member consists of pink marly limestones and marls, up to 12 m thick, which in the process of weathering were rounded, exposing layers resembling sheets of paper (**Fig. 3.3**). The cryptocrystalline calcitic matrix contains highly variable proportions of clayey and organogenic siliceous materials. Insolubles in the unsilicified carbonates vary from 18 to 46 percent. The clay fraction is represented by smectite, a mixed-layer mineral of the smectite-mica type, polygorskite and associated opal-CT and quartz (?chalcedony).

The Lower Lefkara Formation rests directly on the Kannaviou Formation without clear evidence of any angular unconformity or a stratigraphic gap. It must nevertheless be emphasized that the minimal amount of exposure does not permit observation of the contact between the Kannaviou and Lower Lefkara Formations. The basal marls in this section contain a rich fauna of radiolarians. The most common forms are *Amphipyndax tylotus* Foreman, *A. stocki* (Campbell et Clark), *Dictyomitra multicostata* Zittel, *D. formosa* Squinabol, *D. kozlovae* Foreman, and *Alievium gallowayi* (White). This characteristic assemblage of the *Amphipyndax tylotus* Zone clearly indicates a Late Campanian to Early Maastrichtian age (Khokhlova et al., 1994).

Rare benthic agglutinated and calcareous foraminifera belong to the genera *Glomospira, Valvulina, Bolivinopsis, Eponides, Cristellaria, Osangularia* and *Anomalina*, whereas planktonic foraminifera are completely absent (destroyed by selective dissolution). This suggests that the carbonate rocks in this section represent deep-water sediments which were deposited below the level of the planktonic foraminifera lysocline but above the benthic calcareous foraminifera lysocline.

In the area of Ayios Nikolaos, 10 km to the west of Perapedhi, the Lower Lefkara Formation rests with angular disconformity on the Kannaviou Formation. The Late Cretaceous and a part of the Paleocene sequence are missing and the marls at the base of the Lower Lefkara Formation belong to the *Morozovella velascoensis* Zone of Late Paleocene age.

East of Pano Panayia, the Lower Lefkara Formation overlies the Troodos Complex. Its Upper Cretaceous part is well exposed, and overlies the pillow lavas and brick brown argillaceous umbers, up to 25-30 cm thick, of the Perapedhi Formation.

Above an unexposed interval of 11 m, a 92 m thick sequence of the Lower Lefkara Formation outcrops. It can be subdivided into four members (**Fig. 3.4**):

Member 1: l6 m of alternating (20-60 cm) light gray marls and argillaceous limestones showing the concentric and rounded weathering with paper sheet layering of the very thinly bedded (1-2 cm) biomicrites. Calcarenites are present as thin (up to 20-40 cm) interlayers. In places these are silicified and become more abundant in the higher beds of the member. Abundant radiolaria are characteristic for the *Amphipyndax tylotus* Zone of Late Campanian to Maastrichtian age. Rare casts of planktonic foraminifera appear at the top of the member. We conclude that the sediments of this member were deposited below the level of the planktonic foraminifera lysocline.

Member 2: the 31 m thick sequence is largely composed of marls and argillaceous limestones. The beds are thin to thick-bedded, ranging from 0.2 to 1.2 m in thickness. They are interlayered with subordinate 30-50 cm thick beds of biomicritic limestones and occasionally calcarenite layers a few centimeters thick.

Member 3: a 33 m thick sequence of marls and argillaceous limestones with thin (to 15 cm) layers of silicified, biomicritic limestones with horizontal bedding. The upper 5 m thickness of this member is predominantly composed of essentially horizontal biomicrites with wavy bedding, current marks, and signs of bioturbation.

Members 2 and 3 were deposited somewhat above the level of the planktonic foraminifera lysocline. They are almost totally devoid of radiolaria. On the other hand planktonic foraminifera become rather numerous upwards in the section, although poorly preserved, and apparently subjected to dissolution. Their identification is possible only at the generic level: *Globotruncana, Hedbergella, Globigerinelloides*, and *Heterohelix*.

Member 4: Its base consists of an intraformational scoured surface. It is 12 m thick and consists of concentrically weathered gray limestones. Abundant and rather well-preserved planktonic foraminifera suggest that the uppermost part of the Cretaceous sequence consists of typical pelagic foraminiferal oozes. These

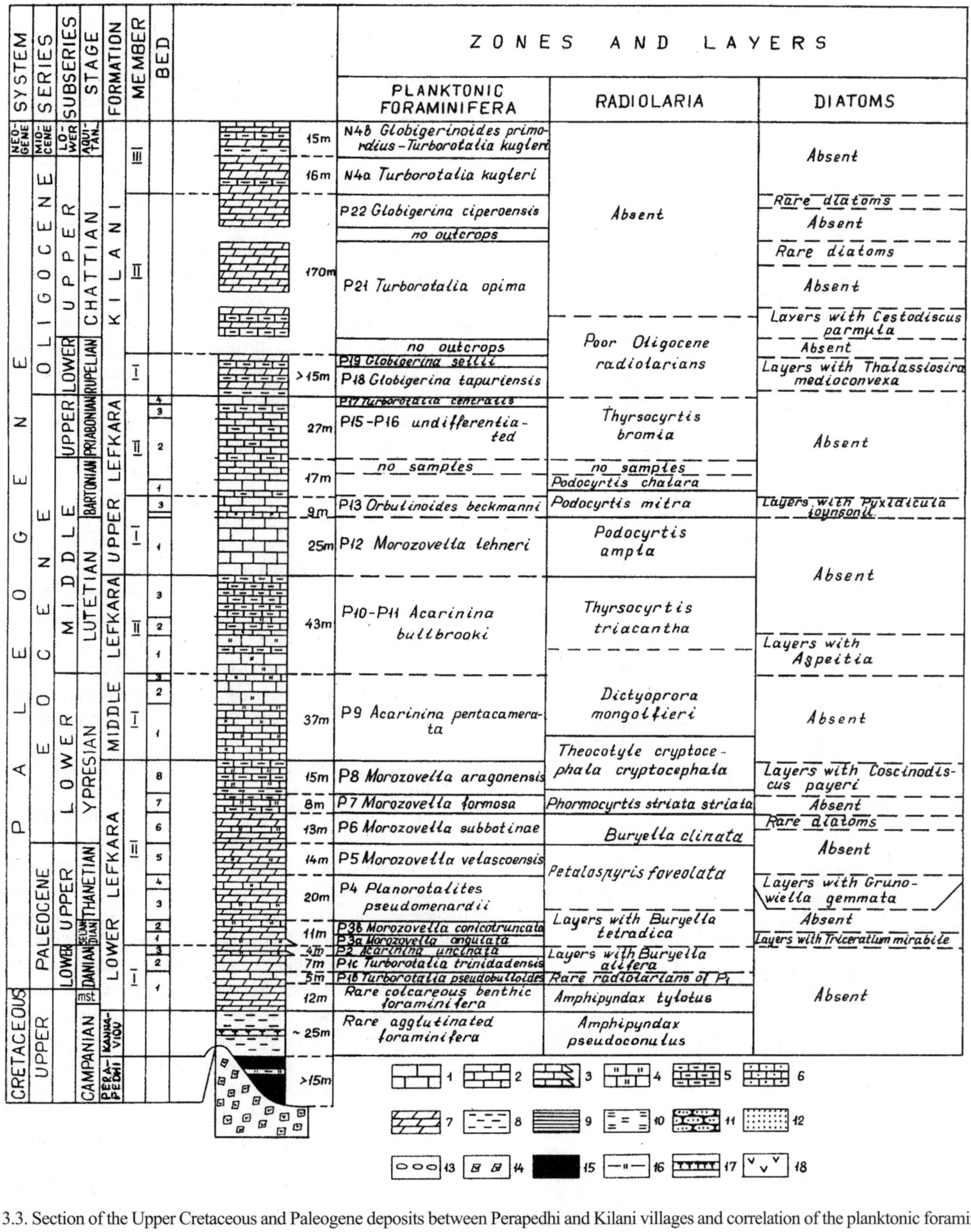

Figure 3.3. Section of the Upper Cretaceous and Paleogene deposits between Perapedhi and Kilani villages and correlation of the planktonic foraminifera, radiolaria and diatom zonal scales (after V.A. Krasheninnikov and K.G. Kaleda, 1994). Lithological symbols: 1 - thick-bedded limestones; 2 - thin-bedded limestones; 3 - calcarenites; 4 - silicified limestones; 5 - clayey limestones; 6 - sandy limestones; 7 - marls; 8 - clays; 9 - argillites; 10 - clays enriched by organic matter; 11 - sandstones; 12 - sands; 13 - conglo-breccia; 14 - volcano-clastic breccia; 15 - umbers; 16 - intercalations of flints; 17 - tripoli-earth; 18 - gypsum.

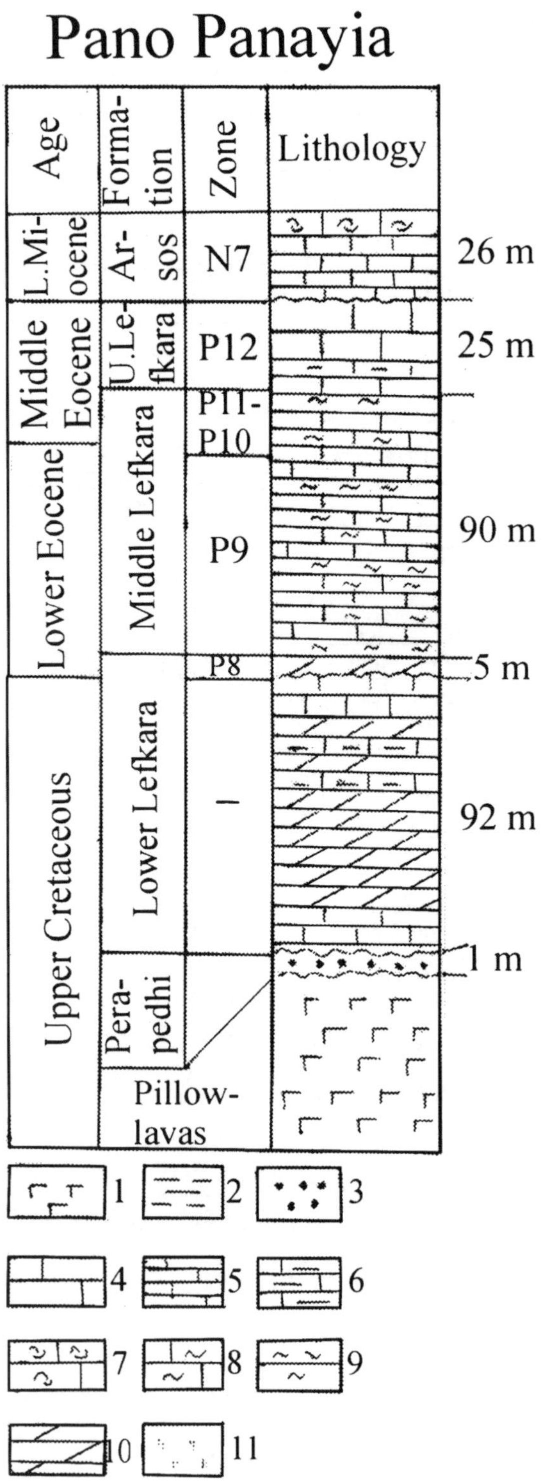

Figure 3.4. Section of the Upper Cretaceous and Paleogene deposits of Pano Panayia village. Lithological symbols (for Figs. 3.4 and 3.5): 1 - pillow lavas; 2 - non-calcareous clays of the Kannaviou Formation; 3 - umbers of the Perapedhi Formation; 4 - thick-bedded limestones; 5 - thin-bedded limestones; 6 - clayey limestones; 7 - detrital organogenous limestones; 8 - silicified limestones; 9 - flints; 10 - marls; 11 - no outcrops.

include *Abathomphalus mayaroensis, Gansserina gansseri, Rosita contusa, Globotruncana aegyptiaca, Globotruncanella havanensis, Hedbergella holmdelensis, Globigerinelloides subcarinata, Heterohelix striata, Pseudotextularia elegans, Race--miguembelina fructicosa*, and *Rugoglobigerina hexacamerata*. They belong to the top of the Maastrichtian, *i.e.* the lower part of the *Abathomphalus mayaroensis* Zone, as indicated by the presence of *Gansserina gansseri*.

The top of the member is eroded and unconformably overlain by layers of the Early Eocene *Morozovella aragonensis* Zone.

In the vicinity of Phity village (Fig. 3.2), above the red-coloured mixtites of the Kathikas Formation, follows a unit of up to 5-6 m white and light-gray, coarse thick-bedded, soft to hard and dense fine-grained chalky limestones with a characteristic blocky parting. They contain a diverse assemblage of well-preserved planktonic foraminifera, including *Globotruncana aegyptiaca, G. ventricosa, G. arca, G. falsostuarti, Globotruncanella havanensis, G. petaloidea, Globotruncanita stuarti, G. stuartiformis, Rosita fornicata, Rugoglobigerina rugosa, Rugotruncana subcircumnodifer, Hedbergella holmdelensis, Pseudoguembelina costulata, P. excolata*, and *Pseudotextularia elegans*. This assemblage evidently belongs to the upper part of the *Globotruncana aegyptiaca* Zone of Early Maastrichtian age, as indicated by the rare *Racemiguembelina fructicosa*.

The westernmost Late Cretaceous exposure can be observed near Simou village (Fig. 3.2). Here, north and immediately south of the road to Polis, small outcrops of brownish red mixtites of the Kathikas Formation and Maastrichtian limestones are exposed respectively. The latter are separated by a gently dipping fault (with a zone of fracturing) that divides this exposure into a lower and an upper block.

The stratigraphically 10 m thick lower block is composed of white chalky limestones showing a rough bedding and blocky partings. It contains thin lenses of brownish silicified limestones 5-10 cm thick. Abundant planktonic foraminifera are found in these deposits, including *Gansserina gansseri, Abathomphalus intermedius, Globotruncana aegyptiaca, G. ventricosa, Globotruncanita stuarti, Rosita contusa*, and *Pseudoguembelina costulata* of the Late Maastrichtian *Gansserina gansseri* Zone.

The 20 m thick upper block consists of pale brownish shaded chalky limestones and frequent interbeds and lenses of firm silicified limestones. The planktonic foraminifera assemblages found in several samples from this block are very unusual. They consist of very small shells of *Heterohelix, Guembelitria, Hedbergella, Pseudoguembelina,* and *Rugoglobigerina,* in combination with individual, normal-sized shells of *Globotrucana*. This sorting, according to shell-size, may be attributed to distal turbidite interbeds.

In the area of the Mamonia Complex of South-West Cyprus, near Kathikas village (Fig. 3.2), pink pelitomorphic limestones up to 5-6 m thick form a recumbent slump fold that rests directly on mixtites of the Kathikas Formation. It is dated by the Maastrichtian *Globotruncana aegyptiaca* Zone. They are unconformably overlain by Middle Eocene white limestones of the Upper Lefkara

Formation.

As mentioned above, the direct contact between the Kathikas Formation and the Lower Lefkara Formation cannot be seen in all the sections. It is therefore not beyond reason that the Kathikas Formation mixtites may be covered by different zones of the Maastrichtian (*Globotruncana aegyptiaca* Zone and the *Gansserina gansseri* Zone). One can assume that in some sections the mixtites replace different stratigraphic intervals of the Upper Cretaceous part of the Lower Lefkara Formation.

The Paleogene portion of the Lower Lefkara Formation is well exposed along the Kryos River near Perapedhi. There, the continuous stratigraphic sequence up to 110 m thick subdivides into the Danian, Selandian, Thanetian, and early part of the Ypresian Stages, based upon planktonic foraminifera and siliceous microfossils (radiolaria and diatoms). The Paleocene deposits can add up to 61 m and the Lower Eocene part to 36 m (Fig. 3.3).

The mainly gray marls and marly limestones of the Paleogene remain similar to the underlying Upper Cretaceous rocks. Thin, up to 15 cm, interbeds of calcarenites and dense biomicrites with nodular silicification occur among the marls. An upward increase in the rate of silicification and decrease of argillaceous content characterize these sediments, which are rich in radiolaria and foraminifera. Commonly, about 15% of the matrix consists of shells, but in some beds the content of the microscopically identifiable biogenic constituents reaches over 50%. Foraminifera of the Danian deposits are mostly recrystallized and replaced by silica. Well preserved foraminifera become more common in the higher strata.

Based on planktonic foraminifera the biostratigraphic zonation of the Paleocene and Early Eocene sediments is as follows.

Danian: about 16 m, consists in three zones:

1) *Turborotalia pseudobulloides* Zone, characterized by the presence of the index-species and associated forms: *Turborotalia varianta, Subbotina triloculinoides, Eoglobigerina eobulloides, E. tetragona, E. fringa, Chiloguembelina midwayensis*, and *Guembelitria irregularis*;

2) *Turborotalia trinidadensis* Zone with the index-species, *Turborotalia varianta, T. pseudobulloides, Subbotina triloculinoides, S. trivialis, Planorotalites compressa, Globoconusa daubjergensis*, and *Chiloguembelina midwayensis*;

3) *Acarinina uncinata* Zone with the index-species, *Acarinina inconstans, A. praecursoria, Turborotalia pseudobulloides, T. varianta, T. quadrata, Subbotina triloculinoides, S. trivialis*, and *Planorotalites compressa*.

Selandian, consists of two zones:

1) *Morozovella angulata* Zone, contains abundant specimens of the index-species and associated forms: *Planorotalites ehrenbergi, P. compressa, Turborotalia varianta, T. pseudobulloides, Subbotina triloculinoides*, and *S. trivialis*;

2) *Morozovella conicotruncata* Zone, indicated by the first appearance of the index-species and *Igorina pusilla*, as well as the continuous development of *Morozovella angulata, Turborotalia pseudobulloides, T. varianta, Subbotina triloculinoides*, and *S. trivialis*.

Thanetian, consists of two zones:

1) The *Planorotalites pseudomenardii* Zone, characterized by the index-species, *Igorina laevigata, Morozovella occlusa, M. velascoensis, Acarinina subsphaerica, A. intermedia, Subbotina velascoensis, S. nana, S. quadritriloculinoides*;

2) *Morozovella velascoensis* Zone, includes abundant specimens of the index-species accompanied by *Morozovella acuta, M. aequa, Acarinina acarinata, A. subsphaerica, Muricoglobigerina soldadoensis, Igorina convexa, Subbotina nana, S. velascoensis*, and *S. compressaformis*.

The overlying Lower Eocene deposits (Ypresian) are represented by three foraminiferal zones:

1) *Morozovella subbotinae* Zone, including the index-species, *Morozovella aequa, M. formosa gracilis, Acarinina pseudotopilensis, A. subsphaerica, A. acarinata, Muricoglobigerina soldadoensis, Subbotina compressaformis*, and *S. eocaenica*;

2) *Morozovella formosa* Zone, consisting of the assemblage of *Morozovella formosa formosa, M. formosa gracilis, M. marginodentata, M. lensiformis, M. wilcoxensis, M. subbotinae, Acarinina camerata, A. pseudotopilensis, A. acarinata, Muricoglobigerina soldadoensis, Subbotina compressaformis*, and *S. eocaenica*;

3) *Morozovella aragonensis* Zone, defined by the first appearance of the index-species associated with *Acarinina interposita, A. triplex, A. pentacamerata, A. pseudotopilensis, Muricoglobigerina soldadoensis, Subbotina pseudoeocaena, S. eocaenica*, and *S. inaequispira*.

The Lower Lefkara Formation is rich in radiolaria, whose diverse assemblages have also been valuable in providing a biostratigraphic subdivision for the Paleocene to Lower Eocene succession. Following Khokhlova et al. (1994), the Paleocene part of the sequence can be subdivided into the *Buryella alifera* beds (Danian), the *Buryella tetradica* beds (Selandian) and the *Petalospyris foveolata* Zone (Thanetian). The Lower Eocene interval is subdivided into the *Buryella clinata* Zone, the *Phormocyrtis striata striata* Zone and the lower part of the *Theocotyle cryptocephala cryptocephala* Zone.

Diatoms are also present in some layers. According to Radionova (1994), the Paleogene sequence near Perapedhi is subdivided into the *Triceratium mirabile* beds (Selandian), *Grunowiella gemmata* beds (within the *Planorotalites pseudomenardii* Zone, Thanetian), and the *Coscinodiscus payeri* beds (within the *Morozovella aragonensis* Zone, Ypresian).

North of Ayios Nikolaos, the Lower Lefkara Formation starts with the *Morozovella velascoensis* Zone of the Upper Paleocene and, as mentioned before, rests unconformably on bentonitic clays of the Kannaviou Formation (**Fig. 3.5**). The 7 m thick basal unit of the carbonate sequence consists of alternating pale cream colored, fine-grained biomicritic limestones, brownish marly limestones and marls, and shows concentric weathering.

Overlying the basal unit is a 20-25 m thick section of similar soft marls and argillaceous limestones, with thin (10-15 cm) silicified limestone interbeds which represent the Lower Eocene. They contain the diverse planktonic foraminifera assemblages of the *Morozovella subbotinae*, *Morozovella formosa*, and *Morozovella aragonensis* Zones.

In the section near Pano Panayia, the 3 m thick uppermost layers of the Lower Lefkara Formation are represented by spheroidally weathered gray argillaceous limestones. These are dated as Lower Eocene (*Morozovella aragonensis* Zone). An obscure unconformity separates these deposits from the underlying Maastrichtian limestones. Consequently, a considerable part of the Lower Lefkara Formation (the Paleocene and the main part of the Lower Eocene) is missing here (Fig. 3.4).

In the vicinity of Dhrymou, the contact of the Lower Lefkara Formation with the Kathikas Formation mixtites is not exposed. The first unit of the Lower Lefkara Formation has an apparent thickness of about 10 m and consists of massively bedded, white chalky limestones containing the diverse and well preserved planktonic foraminifera asemblages of the Danian *Acarinina uncinata* Zone (Fig. 3.2). Above these, a fault-bounded unit (over 15 m thick) is found, composed of light-pink chalky limestones with brownish chert lenses, containing the characteristic assemblage of the Thanetian *Morozovella velascoensis* Zone. It must be emphasized that these layers are lithologically different from the stratigraphically equivalent deposits near Perapedhi, and more closely resemble Middle-Upper Eocene carbonates of the Upper Lefkara Formation.

The Lower Lefkara Formation thus appears to represent a continuous sequence only in one section, near Perapedhi. Moving westward, significant variations in the stratigraphic completeness of the formation can be observed with the disappearance of some of the Paleocene and Lower Eocene units. There are also lateral facies changes in the same direction, from deep-water sediments with planktonic foraminifera and siliceous fossils into less deep-water, pelagic micrites containing abundant planktonic foraminifera with no signs of dissolution of their shells.

Middle Lefkara Formation

In the Perapedhi section, the up to 80 m thick Middle Lefkara Formation rests conformably on the Lower Lefkara Formation. The gradual passage to the mainly light-gray and white chalky biomicrites of the Middle Lefkara Formation, with a variable amount of calcarenitic interbeds, includes extensive silicification, particularly in the lower part of the sequence. Diagenetic cherts are responsible for the development of the silicified limestones, which in some places represent a third of the sequence. The chalky biomicrites, that represent the bulk of the formation, are characterized by strong recrystallization of its cryptocrystalline groundmass and skeletal biomicrites. It is therefore difficult to estimate the initial matrix/bioclast ratio, though the latter evidently formed up to 50% of the rock. Pure chalky limestones of the Middle Lefkara Formation, showing no signs of silicification, contain a low content of insoluble residue which varys between 6.70 and 15.06%. The clay fraction is represented by smectite and palygorskite, with subsidiary chlorite, opal-CT, quartz (?chalcedony), and heulandite (Fig. 3.3).

The Early to Middle Eocene age of this sequence is supported by distinctive planktonic foraminifer assemblages. The lower part of the succession, where intercalation of cherts is most typical, has yielded fauna of the *Acarinina pentacamerata* Zone (Ypresian) including *Acarinina pentacamerata, A. interposita, A. triplex, A. pseudotopilensis, Muricoglobigerina soldadoensis, M. senni, Morozovella aragonensis, M. caucasica, Subbotina pseudoeocaena, S. eocaena,* and *S. inaequispira.*

The Middle Eocene upper part of the section, in which the amount of cherty layers diminishes, belongs to the *Acarinina bullbrooki* Zone (Lutetian), yielding the index-species and *Acarinina aspensis, A. triplex, Muricoglobigerina senni, Globigerapsis index, Subbotina boweri, S. pseudoeocaena,* and *S. eocaena.*

The radiolarian fauna allows subdivision of the Middle Lefkara Formation into three zones ranging from the upper part of the *Theocotyle cryptocephalia cryptocephalia* Zone through the *Dictyoprora mongolfieri* Zone into the *Thyrsocyrtis triacantha* Zone. The boundary between the Lower and Middle Eocene is considered to be within the *Dictyoprora mongolfieri* Zone. Only one diatom assemblage, found at the base of the Middle Eocene, permits its correlation with the *Aspeitia* beds.

Near Ayios Nikolaos (Fig. 3.5), a similar ~70 m thick section of the Middle Lefkara Formation consists mainly of moderately to well-bedded, white and light-gray, dense, fine-grained limestones. Many individual layers from 5 to 20-30 cm thick and lenses of silicified, gray, aphanitic limestones are intercalated within these rocks. Softer marl interbeds are present in places, but usually they are very rare.

In the section near Pano Panayia, the Middle Lefkara Formation attains a thickness of 90 m and consists largely of white, dense and hard limestones (Fig. 3.4). Silicifications occur either as brownish and black chert concretions and lenticules or as interbeds and lenses of gray silicified limestones. Carbonate turbidites, consisting of 1-3 cm clasts derived from the Mamonia Complex volcanics, as well as calcarenites with graded lamination, are often interbedded within the main limestone lithofacies.

The upper part of the succession contains frequent greenish gray marl interbeds and irregular, yellowish and pinkish chert concretions, up to 0.6 m in diameter. Calciturbidites are also well represented, and contain clastic debris (1-5 cm) derived from both the Kannaviou Formation and Mamonia Complex.

Despite the fact that planktonic foraminifera are difficult to extract from these dense limestones, the following Early Eocene microfauna was recognized in the lower part of the sequence: *Acarinina pentacamerata, A. interposita, Morozovella caucasica, M. aragonensis,* and *Subbotina pseudoeocaena*. It is interesting to note that redeposited Late Cretaceous foraminifera may sometimes be found together with these microfossils. By contrast, the soft marl interbeds from the upper part of the section yielded a relatively diverse fauna of planktonic foraminifera, including *Acarinina bullbrooki, A. aspensis, A. pentacamerata, Subbotina boweri, Morozovella spinulosa, M. caucasica, Globigerapsis sp., Truncorotaloides rohri, Hantkenina nuttalli, H.* aff. *liebusi,*

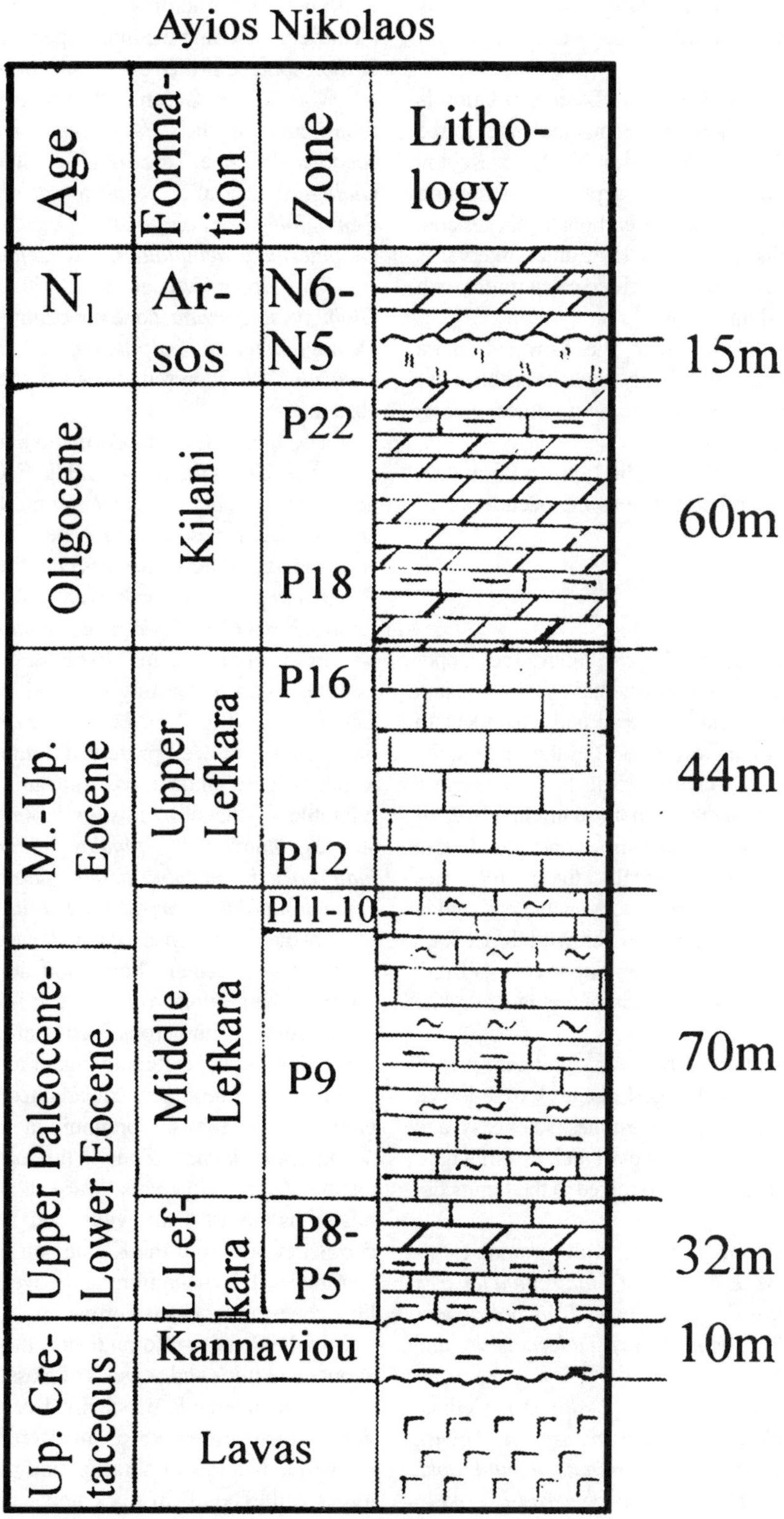

Figure 3.5. Section of the Upper Cretaceous and Paleogene deposits of Ayios-Nikolaos village. Lithological symbols: see Fig. 3.4.

Guembelitrioides higginsi, and *Muricoglobigerina senni*. This assemblage strongly suggests a Middle Eocene age (*Acarinina bullbrooki* Zone).

In the south of the island, the Middle Lefkara Formation is exposed above the Petra tou Romiou terrigenous melange, in the coastal cliff near Petra tou Romiou (Krylov et al., 1995; Krylov et al., and this volume, Chapter 4). Here, a prominent horizon of slump breccia forms the base of the carbonate succession. Stratigraphically overlying this thin unit, a 15 m thick interval is composed of alternating biomicrites and silicified calcarenites with many penecontemporaneous slump folds.

In the section near Kathikas, where the Lower Lefkara Formation is missing, Maastrichtian slightly pink limestones are immediately overlain by the Middle Eocene white limestones of the Upper Lefkara Formation.

In the section near Akrounda, the Middle Lefkara Formation is also absent, a fact which is probably due to the occurrence of a large fault.

Upper Lefkara Formation

In the Perapedhi section of the Kryos Valley, the Upper Lefkara Formation is 70 m thick and consists mainly of white and light-gray, firm and massive limestones, externally identical to chalky limestones (Fig. 3.3). In the lower 30 m of the section, the limestones are thick-bedded with characteristic large concentric and blocky fracture surfaces. In the upper part, many interbeds of small, concentrically weathering silicified limestones emphasize the well-bedded stratification. More typically, these limestones are composed of 50% or more of foraminifera shells, embedded in a micritic matrix. In contrast to those of the Middle Lefkara Formation, these limestones are almost non-recrystallized. However, the insoluble residue and clay fraction are generally the same in both formations.

The Upper Lefkara Formation embraces a significant part of the Middle Eocene and the whole Upper Eocene. Extraction of planktonic foraminifera from the dense limestones is not easy, and therefore, may not represent the true diversity of their assemblages. Despite this, the following zones were recognized in the Perapedhi section:

The *Morozovella lehneri* Zone (Lutetian) with common forms as *Acarinina rotundimarginata, A. bullbrooki, Hantkenina lehneri, Globigerapsis index, Morozovella spinulosa, Truncorotaloides topilensis, T. rohri, Subbotina frontosa, S. pseudoeocaena*, and *Muricoglobigerina senni*;

The *Orbulinoides beckmanni* Zone (Bartonian), defined by *Turborotalia centralis*, *Hantkenina alabamensis*, *H. liebusi*, *Morozovella spinulosa*, *Acarinina rotundimarginata*, *A. bullbrooki*, *Truncorotaloides topilensis*, *T. rohri*, *Globigerapsis index*, *G. kugleri*, *Dentoglobigerina pseudovenezuelana*, *Subbotina frontosa*, and *S. pseudoeocaena*;

The top of the Middle Eocene is determined by the last occurrence of *Acarinina* sp., but the *Truncorotaloides rohri* (Bartonian) remains poorly defined because of the lack of well-preserved microfossils, though the following forms were identified, including the index-species, *Globigerina praebulloides, Turborotalia centralis*, and *Hantkenina alabamensis,*;

The Upper Eocene is only partly defined by planktonic foraminifers, as the *Globigerapsis semiinvoluta* and *Turborotalia cocoaensis* Zones are undifferentiated. Common forms are *Globigerapsis* aff. *semiinvoluta, G. index*, *G. tropicalis, Dentoglobigerina corpulenta, D. galavisi, D. pseudovenezuelana, Globigerina praebulloides*, and *Pseudohastigerina micra*;

The Upper Eocene ends with the *Turborotalia centralis-Globigerina gortanii* Zone containing *Globigerina praebulloides, Dentoglobigerina corpulenta, D. galavisi*, and *Turborotalia centralis*; the representatives of the genus *Globigerapsis* are absent.

The Upper Lefkara Formation contains rich assemblages of radiolaria. Using them, the Middle Eocene is subdivided into three zones: the *Podocyrtis ampla* Zone, the *Podocyrtis mitra* Zone, and the *Podocyrtis chalara* Zone. The *Thyrsocyrtis bromia* Zone corresponds to the Upper Eocene. Diatoms are very rare and the *Pyxidicula joynsonii* beds were only found in sediments of the foraminiferal *Orbulinoides beckmanni* Zone.

In the area of Ayios Nikolaos, the Upper Lefkara Formation thins to 44 m, consisting of dense limestones of the Middle to Late Eocene (Fig. 3.5). Fossils are difficult to extract from these limestones and well-preserved foraminifera are rare. However, a relatively abundant assemblage of planktonic foraminifera, referable to the *Morozovella lehneri* Zone, contain the index-species, *Morozovella spinulosa, Acarinina rotundimarginata, A. bullbrooki, Hantkenina liebusi*, *Globigerapsis index, G. kugleri, Truncorotaloides rohri,* and *Subbotina pseudoeocaena*.

In the section near Pano Panayia (Fig. 3.4), the basal part of the Upper Lefkara Formation lithologically differs from that found in the sections to the east. It is represented by a 25 m thick sequence with numerous intercalations of thick-bedded, dense, white limestones, externally again resembling chalky limestones, thin-bedded limestones, calcarenites, soft white limestones, and greenish-gray marls. Foraminifera assign this sequence to the *Morozovella lehneri* Zone. The succession is characterized by distinct slump structures, warped layers, large synsedimentary olistoliths up to 1 m wide, and breccia lenses composed of debris derived from the Kannaviou Formation and the Mamonia Complex. It is unconformably overlain by the Early Miocene limestones of the Arsos Formation.

In the Petra tou Romiou area, the Upper Lefkara Formation is about 40 m thick and consists of massively bedded limestones with occasional interbeds of silicified calcarenites and concentrically weathering argillaceous limestones. Slump structures and numerous features of abrasion and redeposition are evident in its lower half. The 20 m thick upper half consists of thick bedded limestones unconformably covered by Middle Miocene deposits, although a tectonic contact separating these rocks cannot be excluded.

In South-West Cyprus, in the Mavrokolymbos Valley (Fig. 3.2), the Upper Lefkara Formation directly overlies the Kathikas

Formation, starting with a basal puddingstone conglomerate derived from the Kathikas Formation. It consists of about 70 m of alternating argillaceous and chalky limestones, calcarenites and cherts, whose upper 50 m is comprised of thick-bedded so-called chalky limestones with calcarenite horizons. Planktonic foraminifera refer the section to the Middle Eocene *Morozovella lehneri* and *Orbulinoides beckmanni* Zones, both index-species being present. The Middle Eocene is unconformably overlain by Middle Miocene limestones and there is no faunal evidence for the presence of the Upper Eocene in this section.

Near Kathikas, the Upper Lefkara Formation is reduced to a thin, nearly 5 m thick sliver of white dense limestones, unconformably overlying Maastrichtian deposits of the Lower Lefkara Formation and is itself unconformably overlain by Lower Miocene Arsos Formation (Fig. 3.2).

In the vicinity of Akrounda, north of the storage reservoir in the riverbed of a tributary of the Germasogeia River, the Upper Lefkara Formation reaches a thickness of over 100 m. Here it ranges from the Middle to the Upper Eocene. Alluvium covers the base of the section. The exposed Middle Eocene strata are composed almost completely of roughly bedded, very dense, white to light-gray, fine-grained limestones, split by blocky cleavage. Other notable features include occasional interbeds of soft marly limestones, one of which contains planktonic foraminifera of the *Orbulinioides beckmanni* Zone, i.e. *Turborotalia centralis, Hantkenina alabamensis, Truncorotaloides topilensis, Globigerapsis index, Dentoglobigerina pseudovenezuelana* and *Acarinina* sp.

The Upper Eocene begins with frequent alternations of white to light-gray dense limestones and gray argillaceous limestones, with intense cleavage and containing rare microfossils. Layers up to 6 m in thickness, consisting of argillaceous limestones and greenish-gray soft marls, appear up-section. The latter yield abundant planktonic foraminifera, characteristic of the *Turborotalia cocoaensis* Zone: *Turborotalia cocoaensis, Globigerapsis tropicalis, Cribrohantkenina bermudezi, Hantkenina suprasuturalis, Dentoglobigerina corpulenta, D. galavisi,* and *Pseudohastigerina micra*. The uppermost part of the section, about 20 m thick, consists of interbedded gray argillaceous limestones and concentrically weathered green marls. The latter contain *Turborotalia centralis, Dentoglobigerina corpulenta, D. galavisi, Globigerina officinalis, G. praebulloides,* and *G. ouachitaensis*, which indicate the *Turborotalia centralis-Globigerina gortanii* Zone. The top of this zone has been taken as the boundary between the Eocene and Oligocene.

In lithologic terms, the Middle Lefkara and Upper Lefkara Formations have many sedimentological features in common, with the general presence of silicified limestones and cherts being a diagnostic feature for the deposits of the Lower Eocene and the basal Middle Eocene. Nevertheless silicification is not equally occurring over the whole sequence, its major concentration being in the Middle Lefkara Formation, where to a considerable degree cherts determine the general appearance of the formation. More typically, the silicification is associated with calcarenite interbeds. Where these occur widely, the layers of homogenous cherts are of various shades of gray and sometimes brown or black. They usually show sharp contacts with adjacent beds. Where the calcarenites are rare or absent, biogenic constituents are subjected to silicification that tends to form nodular cherts and isolated lenses of silicified limestones. These have gradational contacts and a complete gradation exists between the non-silicified chalky limestones and the cherty bodies.

By contrast, the proportion of silica is always low in the upper part of the Midlle Lefkara Formation and throughout the Upper Lefkara Formation. Silicification has affected the skeletal material and is responsible for the replacement of some casts of foraminifers by opal. When the degree of silicification increases, the amount of foraminiferal shells replaced by silica also increases. Replacement and secondary growth of microcrystals appear to have nucleated around skeletal remains. The recrystallization gradually, spreads outwards, leading to a generalized development of amoeba-shaped nodules in a cryptograined matrix. In most silicified rocks, silification attains over 90%, leaving less than 10% relict calcite. It should be noted that silicified rocks preserve the macroscopical and microscopical features of the initial calcareous rock.

Based on X-ray diffraction analysis, we related the main mineralogical characteristics of the silica to the degree of silicification. Opal-CT dominates in the less-silicified carbonate rocks, whereas chalcedony is characteristic of the strongly silicified rocks. The maximum chert occurrence is in the Middle Lefkara Formation and differs from the other formations by its intense recrystallization. Recrystallization and silicification are probably related to post-sedimentary redistribution of substances within the sediments. In some places (*e.g.*, Petra tou Romiou and Pano Panayia) the cherts display intricately folded beds and/or blocky separation, developed during subaqueous sliding. This suggests that the diagenetic replacement by siliceous material took place shortly after deposition (the stage of diagenesis) but before thorough lithification (consolidation) of the sediment.

One distinctive feature for the whole Eocene sequence is the presence of calcarenites, especially in its middle portion. These are well represented near Pano Panayia, where in places they form up to 50 percent of the sequence. There is little or no evidence of calcarenites in the sections of the Kryos and Mavrokolymbos valleys.

The calcarenites are exclusively made up of foraminiferal remains. In most cases, these thin, laterally persistent layers are calciturbidites. Individual beds, generally ranging from 5 to 15 cm and occasionally 20 cm, correspond to a Bouma-sequence. The lower part, with abrupt and sometimes wavy base, is a well-graded unit. It passes via a massive, ungraded part, into a horizontally-laminated interval. The uppermost part of the beds shows signs of cross-lamination. In some beds near Ayios Nikolaos and Perapedhi this sequence is often reduced because the lower subdivisions are missing, suggesting that only the distal portion of the turbidite is represented. A third relatively minor type of calcarenite is an ungraded bioclast bed. These calcerenites have sharp bases and tops, which contain 'floating', flattened clasts of chalky biomicritic limestones and horizontally laminated and cross-bedded layering.

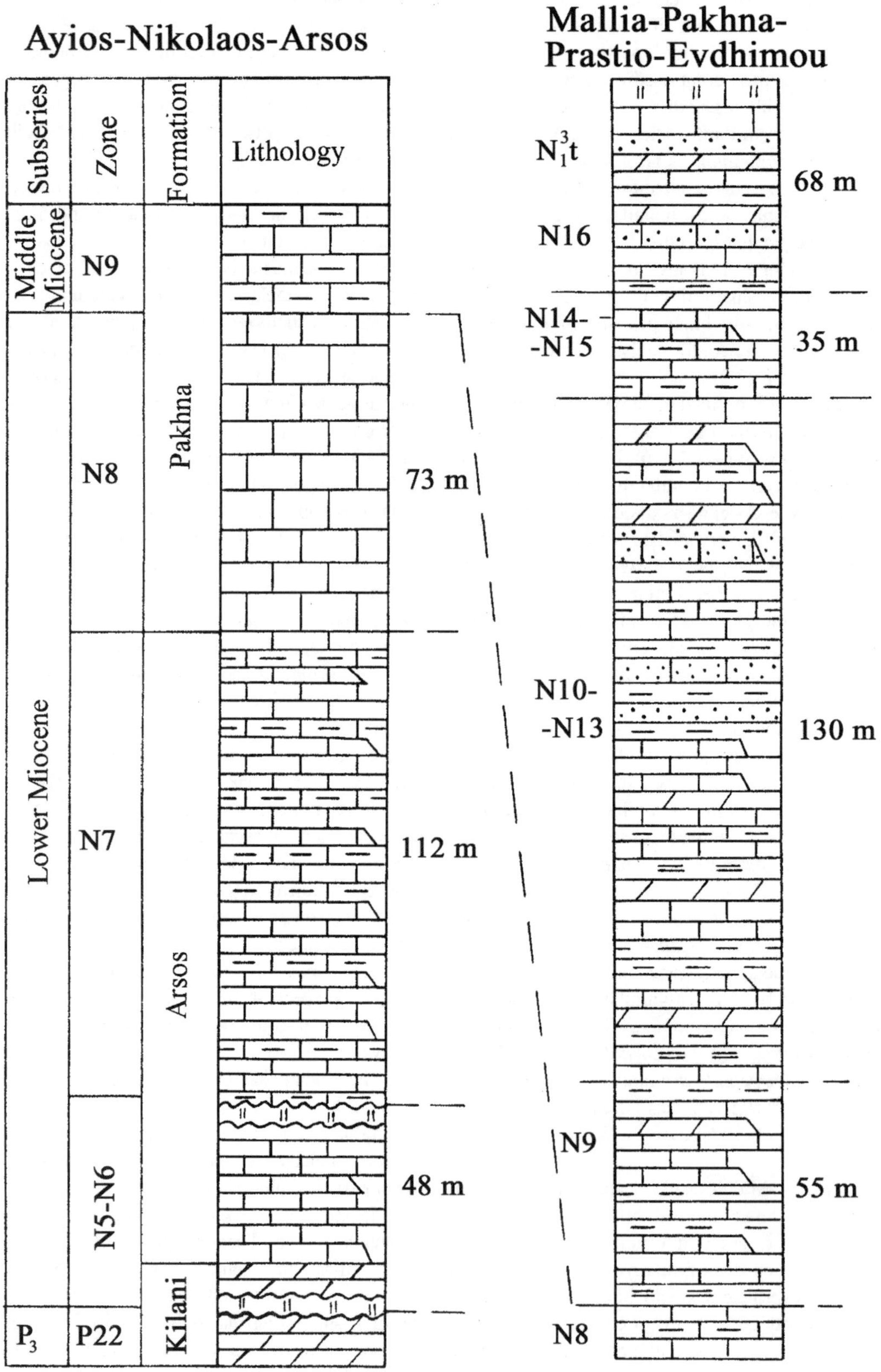

Figure 3.6. Section of the Miocene deposits from Ayios-Nikolaos village to Arsos, Mallia, Pakhna, Prastio, and Evdhimou villages. Lithological symbols: see Fig. 3.3.

Such calcarenites occur widely in the vicinity of Ayios Nikolaos and are interpreted as being connected with bottom-current action

No terrigenous material is present in the Eocene carbonates with the exception of the deposits near Pano Panayia, where the Middle Lefkara and Upper Lefkara formations contain clastic debris. These are either in the form of lithoclast admixtures in carbonate beds or separate exotic polymictic interbeds of red siliceous argillites and cherts, gray sandstones and limestones, as well as volcanics. These are derived from the Mamonia Complex and the gray-green clays of the Kannaviou Formation. Sand-sized material characteristically occurs as an admixture at the base of the calciturbidites and in the chalky limestones. Puddingstone interbeds consisting of angular 3 cm clasts are present at the base of some calciturbidites. Interfingering with the background chalky limestones are isolated angular fragments, up to 30-40 cm wide, as well as beds composed of sand- to fine gravel-sized lithoclasts from 10 cm to a few meters thick. Another important feature is the presence of synsedimentary slump–structures near the top of the Middle Lefkara Formation and the base of the Upper Lefkara Formation.

Kilani Formation

The Kilani Formation consists of greenish-gray marly limestones with minor biomicrite interbeds. As defined by Krasheninnikov and Kaleda (1994), the Kilani Formation represents the Oligocene. These deposits are only developed in the basin bounded by the Mamonia Complex on the west, the Limassol Forest on the east, and probably extending south into the Mediterranean. In the northern part of the basin, by the Troodos Massif, it is exposed in the upper reaches of the Kouris, Kryos, and Dhyarrizos valleys. Sections were measured in the vicinity of the Ayios Yeoryios, Kilani, and Ayios Nikolaos villages (Figs. 3.3 and 3.5). The formation varies in thickness from a minimum of about 60 m near Ayios Nikolaos to a maximum of 200-230 m in its type locality at Kilani. It is primarily built up of weakly recrystallized marly limestones and marls, in which foraminiferal shells may represent up to 30%. The insoluble residue, consists mainly of smectite with subordinate amounts of hydromica, palygorskite, chlorite, quartz, and heulandite, representing 18-21% of the rock.

In the Kryos Valley, between the Ayia Mavra Chapel and the village of Kilani, an Oligocene sequence is well defined by planktonic foraminifera:

The *Globigerina tapuriensis* Zone, with the index-species, *Globigerina ampliapertura, G. prasaepis, G. ouachitaensis, G. officinalis, G. praebulloides, Dentoglobigerina galavisi, D. pseudovenezuelana, Cassigerinella chipolensis, Pseudohastigerina micra, Turborotalia gemma, Globigerinita unicava,* and *Chiloguembelina cubensis*;

The *Globigerina sellii* Zone, with an assemblage identical to that above, with the addition of abundant *Globigerina sellii*;

The *Turborotalia opima* Zone is characterized by the index-species, *Turborotalia nana, T. gemma, Globigerina angulisuturalis, G. praebulloides; G. ouachitaensis, G. offi-cinalis, G. prasaepis, Dentoglobigerina rohri, Globigerinita unicava,* and *Cassigerinella chipolensis*;

The *Globigerina ciperoensis* Zone containing abundant *Globigerina ciperoensis, G. angulisuturalis, G. angustiumbilicata, G. praebulloides,* and *G. ouachitaensis*, with subordinate specimens of *Turborotalia nana, T. brevispira, Globigerinita unicava, Cassigerinella chipolensis, Dentoglobigerina rohri,* and *Globorotaloides suteri*;

The *Turborotalia kugleri* Zone, where the index-species is rare and is accompanied by *Turborotalia brevispira, Globigerina praebulloides, G. juvenilis, G. angustiumbilicata, Catapsydrax unicava, C. dissimilis* and *Globorotaloides suteri*.

The *Globigerina ampliapertura* Zone, which normally lies between the *Globigerina sellii* Zone and the *Turborotalia opima* Zone was not identified in this sections because of lack of exposure.

At the very top of the formation, greenish-gray marly limestones and marls with subordinate biomicritic limestone belong to the base of the Lower Miocene. This is indicated by an assemblage of *Turborotalia kugleri, T. nana, T. siakensis, Globigerina venezuelana, G. juvenilis, G. woodi, G. praebulloides, Globoquadrina praedehiscens, Catapsydrax dissimilis,* and *C. unicava*, together with rare *Globigerinoides primordius* of the *Globigerinoides primordius-Turborotalia kugleri* Zone. The Lower Miocene interval reaches a maximium thickness of 15 m near Kilani. In the Kouris Valley the unit is again concealed.

The distribution of Oligocene, reduced to a basin between the Mamonia Complex and the Limassol Forest and totally absent further west, as in Pano Panaya, suggests that a regressive episode marks the geologic development of the region at that time.

Arsos Formation

The Arsos Formation (Krasheninnikov and Kaleda, 1994) consists predominantly of distinctive hard white platy limestones. It is particularly well exposed in three areas:

i) Pano Panayia (Fig. 3.4),

ii) along the Ayios Nikolaos- Arsos road (**Fig. 3.6**), and

iii) the right bank of the Kouris River, near Ayios Yeoryios (**Fig. 3.7**). There the full sequence of the Arsos Formation, between 100 and 140 m thick, lies conformably on the earliest Lower Miocene *Globigerinoides primordius-Turborotalia kugleri* Zone. It is divisible into two planktonic foraminiferal biostratigraphic units:

1) The undifferentiated *Catapsydrax dissimilis* Zone and *Catapsydrax stainforthi* Zone (Aquitanian) with *Catapsydrax dissimilis, C. unicava, Globigerina venezuelana, G. woodi, G. praebulloides, G. juvenilis, G. angustiumbilicata, Globoquadrina praedehiscens,* and *Turborotalia siakensis*, associated with rare *Globigerinoides trilobus*;

2) The *Globigerinatella insueta-Globigerinoides trilobus* Zone (Burdigalian), containing abundant specimens of *Globigerinoides trilobus* together with *G. diminutus, G. subquadratus, Globigerina foliata, G. bollli, G. falconensis, G. angustiumbilicata, Globoquadrina dehiscens, G. altispira,* and *Turborotalia*

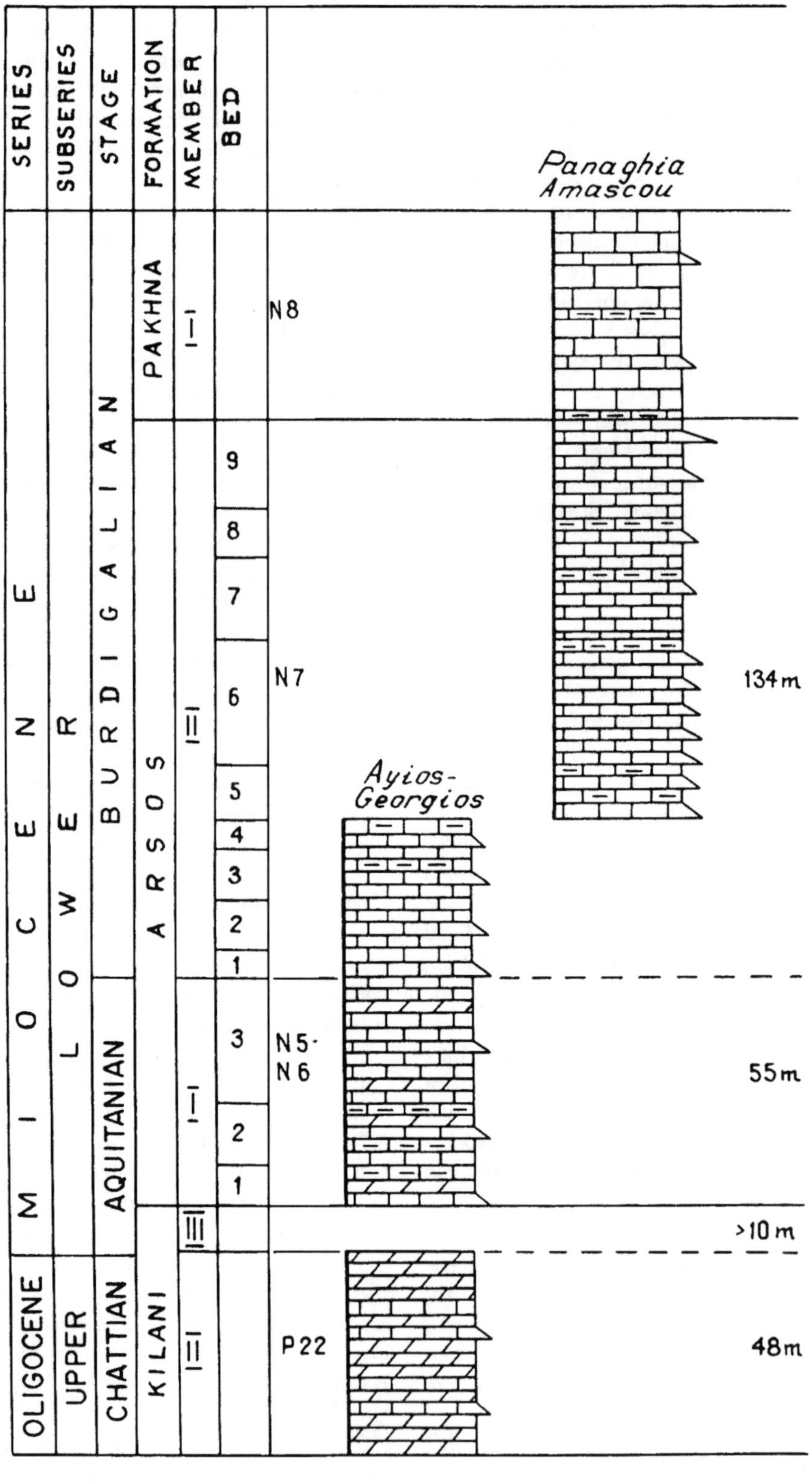

Figure 3.7. Section of the Lower Miocene deposits between the village of Ayios-Georgios and the Panaghia Amascou Monastery. Lithological symbols: see Fig. 3.3.

siakensis.

Near Arsos and Ayios Yeoryios, the Arsos Formation rests conformably on the lowermost Miocene marly limestones that form the top of the Kilani Formation. Near Pano Panayia, it paraconformably covers Middle Eocene deposits, leaving a huge hiatus of the uppermost Middle Eocene, Late Eocene, Oligocene, and lowermost Miocene.

The white marly limestones are composed of homogenous micrites with few (~10%) small foraminiferal shells. Platy partings and diagenetic solution structures occur on the bed surfaces. Concentrically weathered, light-gray biomicrites and light-pink calcarenites are intercalated within the marly limestones. The up to 10 cm thick biomicrite interbeds are usually composed of 50% foraminifera shells and they display bioturbation structures. In the 2 cm to 1 m thick calcarenite interbeds, the percentage of shelly material is even higher. The sedimentary structures of most calcarenites suggest an origin as calciturbidites, but some of them may also represent sediments formed by currents and turbidity flows. The distribution of the calcarenites is not uniform. Near Ayios Nikolaos and Ayios Yeoryios calcarenites constitute up to 25% of the sequence, whereas near Pano Panayia they are rare.

Three horizons of sub-rounded conglomerates appear in the upper part of the section near Pano Panayia. Up to 3 m thick each, they are considered to represent debris flows. They are composed of algal limestone, marly limestone and chert clasts, derived from both the Middle and Upper Lefkara formations, as well as sandstones, siliceous argillites and cherts from the Mamonia Complex, and of syngenetic debris with coral fragments and echinid spines. The dominant clasts are angular and attain lengths of 3 to 10 cm. Large limestone bed fragments up to 1 m long from the Upper Lefkara Formation also occur in places.

Despite the presence of terrigenous rock fragments in the debris flows, both the calcarenites and marly host limestones are characteristically poor in terrigenous components. In the Kryos Valley the amount of the terrigenous component in calcarenites and marly limestones ranges between 1.8% and 5.4%, respectively.

The sections in the Kouris and Kryos valleys exhibit a continuous process of sedimentation during the Oligocene and Early Miocene.

By contrast, near Pano Panayia, an obscure unconformity separates the underlying Middle Eocene from the overlying Early Miocene *Globigerinatella insueta-Globigerinoides trilobus* Zone of the Arsos Formation. A big hiatus is quite evident here.

A similar situation may be found near Akrounda, in the Limassol Forest, where the Arsos Formation also consists of sediments of the Early Miocene *Globigerinatella insueta-Globigerinoides trilobus* Zone, directly overlying Upper Eocene limestones of the Upper Lefkara Formation. It is still unclear if this boundary is of a stratigraphic or tectonic nature, since the lower contact of the thin platy limestones of the Arsos Formation is covered by coarse Quaternary conglomerates. Therefore it is difficult to conclude whether the absence of the Oligocene and lowermost Miocene (Aquitanian) is a stratigraphic hiatus, or connected with tectonic activity on the Arakapas fault.

Terra Formation

The Terra Formation was only investigated in the Mavrokolymbos Valley. It consists of a large slump block of roughly bedded bioclastic limestones. These are composed of skeletal detritus, mainly consisting of large foraminifera, algal fragments and echinid spines with an admixture of terrigenous material. Some beds contain angular clasts and pebbles of carbonate, siliceous and volcanic rocks from the Mamonia Complex, and Paleogene limestone and chert fragments. The Early Miocene age of the formation is based on the presence of *Miogypsina*.

In the area of the Mamonia Complex, where we only observed the Terra Formation in a (tectonically?) slumped block, so far there is only meager evidence for suggesting a lateral facies change from the shallow-marine Terra Formation into the well-bedded pelagic limestones of the Arsos Formation.

Pakhna Formation

The Pakhna Formation has a wide distribution in southern Cyprus, between Limassol Forest in the east and the Dhiarrizos River in the west. Occasional Pakhna Formation outcrops are also present in the area of the Mamonia Complex.

Continuous sections were studied along the Alassa-Polemidhia (**Fig. 3.8**) and Arsos-Mallia-Evdhimou highways (Fig. 3.6), where a continuous transition from the Arsos Formation to the Pakhna Formation is observed. The same relationship can also be seen near Akrounda. However, in other places, we found various stratigraphic intervals of the Pakhna Formation unconformably overlying chaotic rocks of the Kathikas Formation, Maastrichtian limestones of the Lower Lefkara Formation, and Middle to Upper Eocene limestones of the Upper Lefkara Formation. Such relations have been observed around Petra tou Romiou, Kili and Stroumbi, and near Kathikas, and in some other places.

South of the Troodos Massif, the complete section of the Pakhna Formation reaches a thickness of 350-370 m. To the west, near Kili and Stroumbi, the thickness decreases to about 150 m. and in the vicinity of Petra tou Romiou to as little as 100 m.

The entire Pakhna Formation can be subdivided into four members: (from bottom to top):

Member 1 consists of about 100 m of mainly thick bedded (0.5 to 1.5 m), light-gray, chalky limestones with generally about 10% of insoluble residue. They are interlayered with greenish-gray marls, thick-bedded (1.0-1.2 m), light-gray and yellow, residue-poor (3.4%) calcarenites with structures indicating a turbidity current origin. Violet-gray and gray, thick-bedded marly carbonates are rich in organic matter. Bioturbation characterizes all these rocks. In contrast to the underlying formations, the sediments of this member are light and plastic and their clay fraction includes kaolinite, mica and feldspars, together with smectite, palygorskite and quartz.

Member 2, varying from 80 m near Pakhna to about 135 m

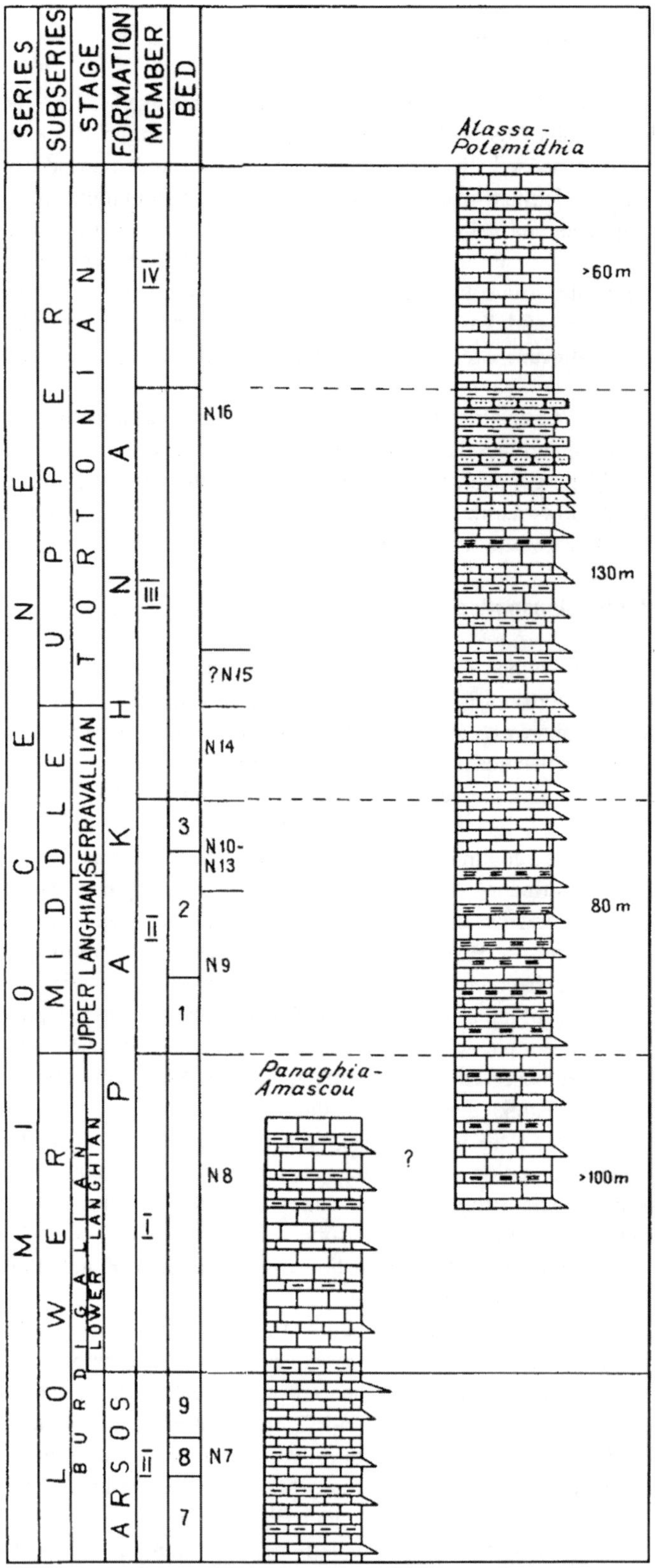

Figure 3.8. Section of the Lower - Upper Miocene deposits from the Panaghia Amascou Monastery to Alassa and Polemidhia villages. Lithological symbols: see Fig. 3.3.

along the Alassa-Polemidhia highway, is made up of the same rock types as below, but most layers are thin bedded and between 10 and 20 cm thick. Another persistent feature is the presence of discrete, lithologically homogenous subunits, some of which are essentially marly, calcarenitic or of other types. For example, distinct subunits of carbonates rich in organic matter occupy different stratigraphic intervals. Pyrite impregnations and carbonised plant remains often occur. When freshly exposed, the predominant color is dark-gray to blue-gray. Thin intercalations (up to 2 m) occur, consisting of either very dark-gray to black marls containing coalified plant debris or massively bedded, poorly cemented calcarenites, which form individual layers 0.5 and 5 cm thick. A typical marl from the organically rich horizons consists of approximately 50% calcite and 24% of insoluble residue. The matrix is composed of micrite with foraminiferal shells, scattered, angular, silt-sized grains of quartz, feldspars, mica, cherts and redeposited carbonates, alternating with layers of high or low carbon content and fine terrigenous material.

Member 3 ranges in thickness from 45 m near Pakhna to 125-130 m along the road from Alassa to Polemidhia. It largely consists of calcarenites, up to 0.7 m thick, with minor interbeds of strongly bioturbated chalky limestones. The subaqueous slumping structures and their associated syngenetic breccia bodies are well exposed along this road. The calcarenites display graded bedding, with characteristic concentrations of gastropod and pelecypod shells at the base of some layers. Some calcarenites are formed by redeposited clasts of biomicritic limestones, and small amounts (up to 15-20%) of quartz, feldspar, chert, and argillaceous-carbonates with grain sizes ranging from sand to sometimes even gravel.

Member 4 is well-bedded. It is similar to the underlying member in its lower part, but in its upper part it shows signs of brief episodes of local marine erosion, often accompanied by the formation of syngenetic breccias up to 1.5 m thick. Calcarenites dominate here and contain considerable amounts of terrigenous, sand to sometimes gravel-sized particles, together with algal and coral fragments, and shelly debris. Some calcarenite layers, up to 1.5 m thick, and their associated chalky limestone beds are made up of dense, strongly lithified carbonates, distinguishing them from the usually incompletely lithified and poorly cemented carbonates of the member. The general upward increase in both the amount of clastic material of different origins and the frequency of breccia occurrence are characteristic for the upper part of the member. The top of the Pakhna Formation consists of distinctive 1.5 to 8 m thick brecciated limestone layers in the Mavrokolymbos Valley, of bioclastic limestones near Petra tou Romiou, and of gravelstones in the vicinity of Pakhna, where this member attains a thickness of more than 80 m.

Overall, the Pakhna Formation encompasses a considerable stratigraphic interval, extending from the late Early Miocene *Praeorbulina glomerosa* Zone to the lower part of the Upper Miocene (Tortonian).

The Lower-Middle Miocene boundary is well defined by the evolutionary appearance of the first representatives of the genus *Orbulina*. It corresponds to the base of the Middle Miocene *Orbulina suturalis-Globorotalia peripheroronda* Zone. While the Italian nomenclature would say that the Pakhna Formation encompasses the Langhian and Serravaglian stages of the Middle Miocene, and the Tortonian stage of the Late Miocene, the following succession of zones is observed based on planktonic foraminiferal assemblages.

In the lowest thick-bedded member of the Pakhna Formation, the Early Miocene *Praeorbulina glomerosa* Zone is characterized by the association of *Praeorbulina glomerosa glomerosa, P. glomerosa circularis, P. glomerosa curva, P. transitoria, Globigerinoides bisphaerica, G. trilobus, G. subquadratus, G. diminutus, Globoquadrina dehiscens, G. altispira, Turborotalia siakensis, Globorotalia periphero-ronda, G. praescitula, Globigerina bollii, G. praebulloides, G. woodi, G. druryi, G. falconensis,* and *G. angustiumbilicata*.

In the second member of the Pakhna Formation, two Middle Miocene zones were recognized. The lower, *Orbulina suturalis-Globorotalia peripheroronda* Zone, is marked by the common occurrence of the index-species and *Biorbulina bilobata,* together with *Globorotalia peripheroronda, Turborotalia siakensis, T. obesa, Praeorbulina glomerosa, P. transitoria, Globigerinoides bisphaerica, G. trilobus, G. obliquus, Globoquadrina altispira, G. dehiscens, Globigerina druryi, G. bulloides, G. woodi,* and *G. falconensis*.

The younger subdivision consists of the undifferentiated *Globorotalia peripheroacuta, Globorotalia fohsi lobata, Globorotalia fohsi robusta*, and *Sphaeroidinellopsis subdehiscens-Globigerina druryi* Zones. A more precise subdivision of these zones is difficult because of the absence of the genus *Sphaeroidinellopsis* and important elements of the *Globorotalia fohsi* and *Globorotalia praemenardii* groups. The total assemblage is, however, quantitatively rich and includes such species as *Orbulina suturalis, Biorbulina bilobata, Globigerinoides obliquus, G. bollii, G. trilobus, G. subquadratus, Globigerina druryi, G. bulloides, G. decoraperta, G. falconensis, Globoquadrina dehiscens, G. altispira, Turborotalia siakensis, T. obesa, Globorotalia praescitula,* and *G. scitula*.

In the lower part of the third member, the uppermost Middle Miocene *Globigerina nepenthes-Turborotalia siakensis* Zone is recognized. It is characterized by the first appearance of *Globigerina nepenthes* and *Orbulina universa* and by very rare occurrence of *Turborotalia siakensis*. Among others, it contains *Globigerinoides obliquus, G. bollii, G. trilobus, Biorbulina bilobata, Globoquadrina altispira, G. dehiscens, Globigerina bulloides, G. druryi, G. decoraperta, G. falconensis, Globorotalia scitula,* and *Orbulina suturalis*.

The upper part of the third member belongs to the Tortonian *Globorotalia acostaensis* Zone of the Upper Miocene. This zone is marked by the dominant occurrence of *Orbulina universa, Globigerinoides extremus, Globigerinella aequilateralis, Globigerina nepenthes,* and *G. bulloides*. The other very characteristic species are rare, *i.e. Globorotalia merotumida, G. acostaensis, G. menardii,* and *G. scitula*. Associated species are *Biorbulina bilobata, Orbulina suturalis, Globoquadrina dehiscens, G. altispira, Globigerinoides bollii, G. trilobus, G. obliquus, Globigerina decoraperta, G. quinqueloba,* and *G. falconensis*.

The fourth member, consisting of more shallow-water sediments, still contains rare and poorly preserved planktonic foraminifera of the same Tortonian assemblage.

Kalavasos Formation

Messinian evaporites are widely distributed over the whole Mediterranean region (Hsü et al., 1973). In southern Cyprus, such deposits are represented in four synclinal structures by the Kalavasos Formation:

i) South-west of the Troodos Massif, around Polemi;

ii) south of the Troodos Massif, near Pissouri and Alassa;

iii) and in south-eastern Cyprus, near Psematismenos and Kalavasos.

The various aspects of the Kalavasos Formation have been well-examined by several authors (Orszag-Sperber and Rouchy, 1979; Orszag-Sperber et al., 1980, 1989; Pantazis, 1967, 1978; Baroz et al., 1978; Cita, 1982; Kluyver, 1969; Rouchy et al., 1980; and others). Robertson et al. (1995) discussed the lithological as well as environmental characteristics of the Messinian evaporites of southern Cyprus. This study showed the existence of a number of discrete lagoonal basins such as the Polemi, Pissouri, Alassa, Maroni (Kalavasos), and Mesaoria basins. The evaporites must have formed in response to a global Late Miocene eustatic fall in sea level, desiccation, and tectonic movements.

As regards depositional conditions, two major facies assemblages have been defined in the gypsiferous sequence. An association of coarse-crystalline (**Fig. 3.9**), sugary, botryoidal, banded gypsums (selenite) (**Fig. 3.10**) and stromatolites occurs in the shallow-water (up to 10 m deep) facies that coincides with the periphery of the basins. In the relatively deeper facies, reaching depths up to 70 m, fine-grained gypsums (alabaster) were chemically precipitated. In the deeper parts of the basins, considerable amounts of material derived from the shallower parts were redeposited by bottom and turbidity currents. Terrigenous materials, transported by debris flows, are also common, and sometimes abundant. Local and regional tectonic movements largely controlled the paleogeographical and paleohydrological conditions. These range from tectonic instability to extensional deformation and rifting.

According to our observations in the vicinity of Pissouri, Evdhimou, Kallepia, Stroumbi, Polemidhia and Letimbou, the Kalavasos Formation can be subdivided into three members.

North of Evdhimou, the lower member of the Kalavasos Formation rests on the Pakhna Formation and consists of alternating light-yellow, dense, bioclastic limestones and coquina, with no evidence of a turbiditic origin. Near Kallepia, the lower part of the Kalavasos Formation is made up of alternating gray, pelitomorphic limestones and gray, thin bedded, friable, tripoli-like rocks.

The middle member of the Kalavasos Formation is composed of several varieties of gypsum. These are finely- to coarsely crystalline, and sometimes brecciated. This gypsum member

Figure 3.9. Coarse-crystalline gypsum ('swallow's tail') from the Messinian deposits of the vicinity of Polemi village (photograph by K. Krylov).

varies in thickness from 30-40 m near Pissouri to up to 80 m in the vicinity of Stroumbi.

Near Kallepia, the upper member of the Kalavasos Formation which overlies the gypsum consists of white, massive diatomites with minor mudstone interbeds. A more complicated sequence of this upper member is observed in the vicinity of Pissouri, where it starts with a horizon of soft white marls, up to 1.5 m thick, followed by characteristically alternating conglomeratic breccias, pelitimorphic and biogenic grainstones, as well as minor interbeds of calcarenites and mudstones. The conglomeratic breccias consist of unsorted rounded fragments, up to 20 cm in diameter, composed of pelitomorphic and bioclastic limestones, coquinas, and calcarenites. By contrast, pelmicrites are poor in residue (4.28%). The grainstone interbeds consist of resedimented particles of pelitomorphic limestone and usually contain small amounts of quartz and clay minerals, including glauconite. Marls with sand and clay admixtures are also common. The thickness of the post-gypsum upper member of the Kalavasos Formation varies from 20 m near Kallepia to 30 m in the vicinity of Pissouri.

Pissouri Formation

The Pissouri Formation (= Nicosia Formation) is found in south-western Cyprus (Pantazis, 1967). It is well exposed in the

Figure 3.10. Sugary banded gypsums (selenite) from the Messinian deposits of the vicinity of Polemi village (photograph by K. Krylov).

Polis Graben on the south-western edge of the island (Payne and Robertson, 1995), in the central section of the Ezousa River, and near Letimbou and Kallepia, as well as in the southern part of the island, and near Pissouri in the vicinity of Cape Aspro.

Near Pissouri the type section of the Pissouri Formation (**Fig. 3.11**), overlies the carbonates and evaporites of the Kalavasos Formation and consists of about 120 m of alternating light gray, concentrically weathering marls and grayish-green or brownish clays. These contain a dispersed admixture of fine terrigenous material. In the lower part of the section 5 to 25 cm thick interbeds of very fine- to fine-grained sands and poorly cemented sandstones are present. They appear as beds or lenses, consisting of quartz, feldspar and glauconite grains, chert and limestone fragments, particles of clay, and hornblende and clinopyroxene.

Marly limestone interbeds, up to 10 cm thick, contain considerable amounts of chlorite and mixed-layer clay minerals of the chlorite-smectite type in their clay fraction. The latter differs in composition from those in the underlying formations. Layers of fine-grained, horizontally bedded calcarenites with plant detritus, up to 1.5 m thick, appear in the middle part of the section. In the upper part, yellow, roughly cross-bedded, fine-grained sandstones up to 30-40 cm thick occur as individual beds.

The Pissouri Formation can be subdivided into three biostratigraphic units using planktonic foraminiferal assemblages which assign it to the Pliocene. The sequence begins with the *Sphaeroidinellopsis* Acme-Zone, where *Sph. seminulina* and *Sph. subdehiscens* are the dominant forms. The list of other species includes *Orbulina universa, Globigerinoides sacculifer, G. trilobus, G. extremus, G. obliquus, Globoquadrina altispira, Globorotlia acostaensis, G. scitula,* and *Globigerina bulloides*. This zone was also recognized at the base of the Pissouri Formation in the Polis graben (Payne and Robertson, 1995).

The *Globorotalia margaritae evoluta* Zone contains a very diverse planktonic foraminifera assemblage, including *G. margaritae evoluta*, *Globigerinoides conglobatus, G. trilobus, G. ruber, G. sacculifer, G.extremus, G. obliquus, G. elongatus, Orbulina universa, Globoquadrina altispira, Globorotalia crassaformis, Globigerina bulloides,* and *G. quinqueloba*; speciments of *Sphaeroidinellopsis subdehiscens* are very rare.

Therefore the planktonic foraminifera suggest an Early

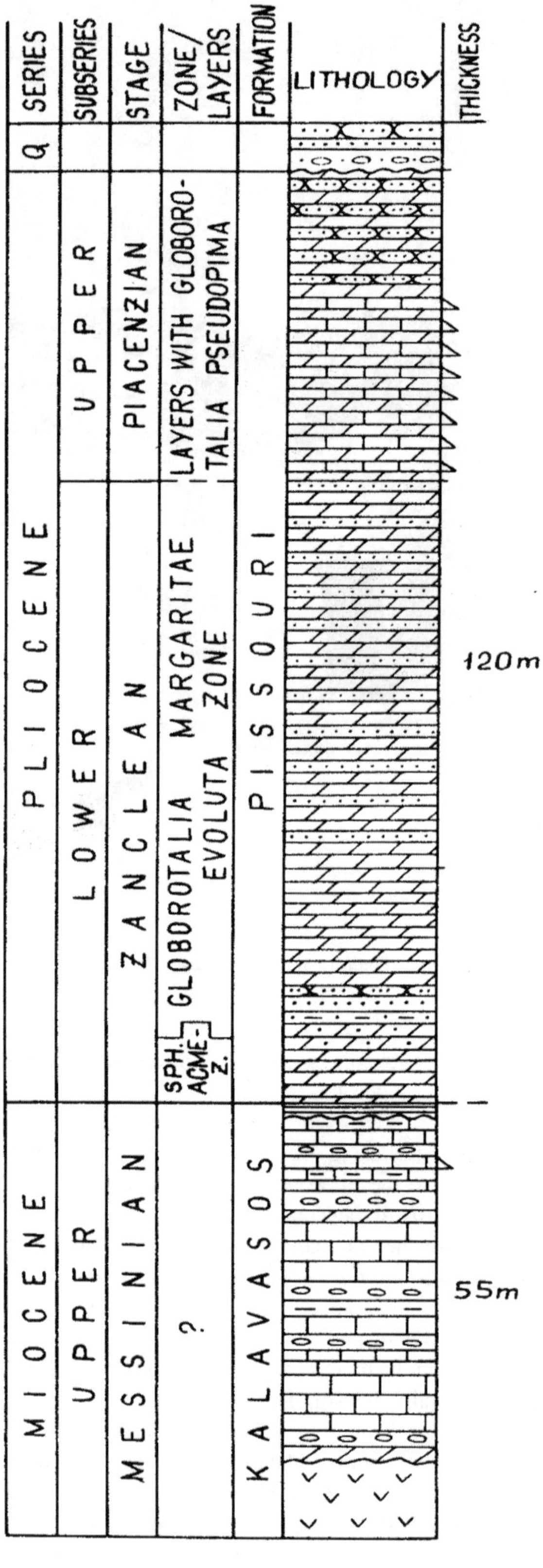

Figure 3.11. Section of the Pliocene deposits near the village of Pissouri (after V.A. Krasheninnikov and K.G. Kaleda, 1994). Lithological symbols: see Fig. 3.3.

Pliocene age (Zanclean) for both zones.

The planktonic foraminifera from the higher part of the Pissouri section are rather numerous but their assemblage is of low diversity. This assemblage includes *Globorotalia pseudopima, Orbulina universa, Globigerinoides ruber, G. sacculifer, G. trilobus, Globigerina bulloides, G. quinqueloba,* and is probably Late Pliocene (Piacenzian).

A Quaternary breccia rests disconformably on the Pliocene Pissouri Formation. Its clasts, up to 0.5 m in size, are composed of carbonates derived from the Pakhna and Upper Lefkara Formations. Ophiolite-derived greenish-gray and black sands occur as separate lenses at the base of the breccia.

DEPOSITIONAL PATTERNS AND GEOLOGIC DEVELOPMENT OF THE NEOAUTOCHTHON IN SOUTHERN CYPRUS

The peculiarities of the structure and geological development of the Southern Cyprus neoautochthon have been deciphered through analysis of the litho-stratigraphic units (formations) in combination with detailed biostratigraphic investigations (using zonal schemes based on planktonic foraminifera and radiolaria).

As has already been shown in Table 3.1, the stratigraphic intervals of the Upper Cretaceous and Cenozoic formations of southern Cyprus had been interpreted in different ways by various authors. It should be added that in publications of the last years *e.g.* Payne and Robertson (1995; p. 617, Fig. 4) and Robertson et al. (1995; p. 236, Table 1), the interpretation of the stratigraphic intervals of formations differs from those in papers by Pantazis (1967), Turner (1971), and Mantis (1970, 1972). Our approach to understanding the formations of southern Cyprus is connected not only with the local geological settings but also with the regional geological conditions for the eastern Mediterranean as a whole. Therefore the Lower, Middle and Upper Lefkara, Kilani, Arsos, Pakhna, Kalavasos and Pissouri Formations are for us stratigraphically analogous with corresponding units in other countries of the Near East (*viz.* Syria, Israel, Jordan and Egypt).

The major conclusions that can be drawn from our study are summarized in the following points.

1) Ancient structural features appear to have significantly controlled the pattern of deposition and stratigraphic completeness of the neoautochthonous succession. In south-western Cyprus, these essential elements are the structural depression between the Dhiarrizos and Germasogeia Rivers and the two relative highs which now appear as the Mamonia Complex in the west and the Limassol Forest Ophiolite Massif in the east.

A continuous Campanian to Pliocene succession is only present in the above-mentionned depression. The Late Cretaceous and Paleogene pelagic sediments, documented by planktonic foraminifera, radiolaria and diatoms, were deposited in relatively deep-water environment. In Early and Middle Miocene times pelagic sediments continued to accumulate in this area, in progressively shallower depths, and consisted exclusively of calcareous plankton. The overall shallowing of the basin is expressed by the shallow-water deposits characteristic for the Late Miocene and Pliocene formations.

By contrast, reduced sections of Upper Cretaceous and Cenozoic deposits are associated with elevated synsedimentary elements of the seafloor, bounding the deeper-water regions. Many local scour surfaces and stratigraphic gaps are associated with the relatively large amounts of chaotic sedimentation in the Kathikas Formation in the west, and the Moni Formation in the east, as well as the general shallow-water facies pattern in the younger formations. These features are characteristic of the neoautochthonous cover of the Mamonia and Limassol Forest uplifts.

2) In most cases, the above-mentioned formations extend from the depression to the two uplifts. However, distinct variations occur and formations sometimes replace each other laterally. These variations include the following peculiarities.

a) Lithostratigraphic boundaries, which in general are clearly recognizable over the entire investigated areas, do not always coincide with the chronostratigraphic datum-levels. As an example, the upper boundary of the Kilani Formation, in its type section near Kilani, coincides with the top of the *Globigerinoides primordius - Globorotalia kugleri* Zone (earliest Aquitanian), whereas near Ayios Nikolaos, this formation also includes the basal part of the overlying *Catapsydrax dissimilis* Zone (Aquitanian). The section of the Pakhna Formation near Alassa starts with the *Praeorbulina glomerosa* Zone, whereas in the sections near Arsos and Kathikas, it also includes the uppermost part of the underlying *Globigerinatella insueta - Globigerinoides trilobus* Zone.

b) As to the completeness of the stratigraphic succession, significant gaps have been have been found in several formations in different structural environments. In the section near Perapedhi, the Lower Lefkara Formation ranges in age from Maastrichtian to Early Eocene (the *Morozovella aragonensis* Zone), whereas near Pano Panayia this formation is reduced to the Maastrichtian only. By contrast, Late Cretaceous deposits are missing in the section of Ayios Nikolaos, where it is only represented by Late Paleocene (the *Morozovella velascoensis* Zone) to Lower Eocene (up to the *Morozovella aragonensis* Zone).

The Lower and Middle Lefkara formations are completely absent in the Mavrokolymbos Valley.

In the continuous section near Perapedhi, the Upper Lefkara Formation ranges from the Middle Eocene (the *Morozovella lehneri* Zone) to the Late Eocene. Near Pano Panayia this formation represents only the *Morozovella lehneri* Zone and the younger deposits of the Middle and Upper Eocene are absent.

The Arsos Formation includes three Early Miocene zones-the *Catapsydrax dissimilis*, *Catapsydrax stainforthi* and *Globigerinatella insueta-Globigerinoides trilobus* Zones in the representative sections near Arsos and Ayios Nikolaos. Only the youngest zone is present in the reduced sections near Pano Panayia and Akrounda.

c) In general, a clear distinction occurs between the deeper-water lithofacies of the depression and the more shallow water one on the Mamonia and Limassol Forest uplifts. In the depression, at Perapedhi, the Campanian-Maastrichtian limestones contain radiolarians and benthic calcareous foraminifera but no planktonic foraminifera. This suggests deposition above the carbonate compensation depth but somewhat below the level of the planktonic foraminifera lysocline. Closer to the Mamonia Massif, near Pano Panayia, the lower part of Lower Lefkara Formation already contains radiolaria and planktonic foraminifera, but the latter are poorly preserved and apparently subjected to strong dissolution. Higher up, chalky limestones from the top of the Maastrichtian contain abundant and well-preserved planktonic foraminifera and no radiolarians at all. Well into the Mamonia massif, near Simou, Phiti, and Kathikas, genuine foraminiferal chalky limestones rest directly on red-colored mixtites.

The same pattern is characteristic for Paleocene deposits. Planktonic foraminifera are abundant in the Danian at Perapedhi, but their shells were dissolved and replaced by silica. In the depressional facies the Selandian and Thanetian microfossil assemblages consist mainly of mixed planktonic foraminifera and radiolaria, whereas in the Mamonia uplift, near Dhrymou, only foraminifera occur, with no trace of radiolarians.

During the field work the Maastrichtian and Paleocene marly limestones near Simou, Kathikas, Phity, and Dhrymou were often confused and assigned to the Upper Lefkara Formation.

Whereas the pelagic foraminiferal-radiolarian oozes that form the limestones and marls of the Lower Lefkara Formation at Perapedhi were accumulated under lower bathyal conditions, the chalky limestones in the Mamonia Massif represent upper bathyal foraminiferal oozes. This suggests that the water depths of the pelagic Late Cretaceous and Paleogene ranged at least from 1000 to 1500 m.

At Perapedhi and Ayios Nikolaos, the deep-water pelagic carbonates of the Middle Lefkara and Upper Lefkara Formations consist mainly of limestones with siliceous concretions, as well as thick-bedded massive limestones. To the west, on the Mamonia Massif at Pano Panayia, this lithofacies is replaced by limestones with abundant calcarenites and calciturbidites with individual 'floating' pebbles. Such differences characterize these two structural regions, which were subjected to different erosional synsedimentary regimes.

The platy limestones of the Arsos Formation generally show no lithological variation between Arsos and the Kouris and Kryos valleys. In contrast, up to 3 m thick horizons of coarse-clastic debris flows appear only in the west, at Pano Panayia.

The most pronounced facies changes, caused by a combination of tectonic and erosional factors, took place in Late Cretaceous and Early Miocene times.

The Maastrichtian limestones at the base of the Lower Lefkara Formation represent the base of the neoautochthon and were deposited in the central part of the structural depression, where the seafloor was the deepest, as at Perapedhi and Pano Panayia. In the areas of the Mamonia Complex and the Limassol Forest Ophiolitic Massif, the deep water limestones are partly replaced by the chaotic mixtites of the Kathikas and Paleo-Moni Formations respectively.

In the case of the Arsos Formation, a clear facies change, from relative deep-water platy limestone in the center of the depression into shallow-water bioclastic limestone on the Mamonia uplift, reflects the distinct paleobathymetric conditions of the Early Miocene basin.

Within the deep-water deposits near Perapedhi, calcarenites are very rare. They essentially occur only in the Lower Eocene, and lack the basal graded elements of the Bouma succession. Calcarenites of such type were deposited in the distal parts of turbidity currents. Evidently, at Perapedhi the Maastrichtian-Paleogene section was remote from any source area, and is thus characterized by typical pelagic sediments. In the vicinity of Ayios Nikolaos, the number of turbidite beds appreciably increases, with the lower elements of the cycles also missing. Calcarenites attain their fullest development near Pano Panayia in the Maastrichtian and Eocene intervals of the sequence. These represent the proximal facies of high-density turbidite flows, as indicated by a well-graded bed at their base, often with terrigenous gravel-size clasts. Middle Eocene gravity-slide debris flows, derived from a submarine high, constitute still another diagnostic feature of slope environment.

d) The thickness of the neoautochthonous cover varies considerably between the depression, located around the Kryos and Kouris valleys, and the uplift of the Mamonia Complex. Between Perapedhi and Polemidhia, the continuous Late Cretaceous-Cenozoic sequence approaches 1400 m. On the Mamonia uplift, a continuous stratigraphic sequence is absent and a composite column of Late Cretaceous and Cenozoic strata does not exceed 700 m.

e) It is evident that the rate of sedimentation can only be estimated for the continuous basinal sequences. As a result, Krasheninnikov and Kaleda (1994) recognized two main stages in the development of the neoautochthonous basin. The Lefkara Group was deposited during the Late Campanian to Eocene first stage. Very slow accumulation, gradually increasing upwards from 1 to 13 mm/1000 y, is characteristic for the lower bathyal foraminiferal-radiolarian oozes. Such rates of sedimentation match those in oceanic basins. During the second stage, higher rates between 20 and 40 mm/1000 y characterize the Oligocene to Pliocene carbonates and clastics of the Kilani, Arsos, Pakhna, and Pissouri formations.

These higher rates of sedimentation are found on the Mamonia uplift. For example, in the case of the Lower Lefkara Formation, the sequence thickens as one moves away from the deep-water area towards the Mamonia uplift. The 12 m of Maastrichtian in the deep-water section of Perapedhi thickens to 92 m in the shallow-water section of Pano Panayia, implying a nearly 8 fold increase in rate of sedimentation. This is partly due to the effect of dissolution on the calcareous nannofossils and foraminifera in the deeper-water environment of the Perapedhi section.

3) The most important aspect regarding the nature of the neoautochthon is the change in sedimentary types, from the Campanian, Maastrichtian and Paleocene abyssal non-carbonatic radiolarian oozes and lower-bathyal carbonate foraminiferal-radiolarian oozes of the Kannaviou and Lower Lefkara formations, to the Late Miocene-Pliocene shallow-water carbonate-terrigenous and evaporite deposits of the Arsos, Pakhna and Kalavasos formations. This trend in basinal development was described by Robertson (1976), Robertson and Hudson, (1974), Baroz and Bizon (1977) and fully confirmed by subsequent investigations.

Vertical replacement of deep-water sediments by more shallow-water ones is also observed on the Mamonia uplift at Pano Panayia, Phiti, Simou, and Dhrymou. The bathymetric environment on this uplift was shallower than in the depression and appeared earlier than in the deeper-water sections of the depression, as derived from paleontological factors such as diversity, quantity, state of preservation and dissolution of planktonic foraminifera.

These distinctions in depositional depths are well illustrated in the basal part of the Lower Lefkara Formation. At Perapedhi, as mentioned above, the Maastrichtian sediments contain only radiolarians, whereas planktonic foraminifera are totally absent. The Danian is rich in planktonic foraminifera, but their shells underwent dissolution and were replaced by silica. Well-preserved planktonic foraminifera associated with radiolarians appear only in Selandian deposits.

By contrast, at Pano Panayia, the Maastrichtian sequence also begins with radiolarian bearing strata, underlying layers with semi-dissolved planktonic foraminifera, with the youngest Maastrichtian *Abathomphalus mayaroensis* Zone containing well preserved planktonic forminifera. Abundant and well-preserved planktonic foraminifera of the Early Maastrichtian *Globotruncana aegyptiaca* and Late Maastrichtian *Gansserina gansseri* Zones are found in marls at Phiti and Simou. The Danian *Acarinina uncinata* Zone in marly limestones at Dhrymou contains abundant and excellently preserved planktonic foraminifera and no radiolarians. We may therefore conclude that the Maastrichtian and Danian pure carbonate pelagic foraminiferal oozes are typical constituents of the shallower sequence of the Mamonia uplift.

The geodynamic events that produced the neoautochthonous basin and determined the subsequent development of Southern Cyprus area are still poorly understood. They consist of the following:

a) Creation of the sub-basement of the Troodos, Mamonia and Limassol Forest elements, which were then covered by the mostly deep water sediments of the neoautochton during an early phase of abyssal sedimentation;

b) Formation of the Mamonia and Limassol Forest uplifts, separated by the Perapedhi depression;

c) Steady decrease in the depth of water over the Cyprus block, ending in its emergence as an island.

The vertical movements must have been related to the subduction of the African Plate, whereas the horizontal movements may be connected to the rotation of Troodos.

4) Study of both the Cretaceous-Cenozoic lithology and the stratigraphy of the neoautochthon allow us to recognize several stages in the geologic evolution of South-West Cyprus. Each stage reflects a combination of tectonic, paleogeographic and bathymetric factors of local (Cyprus), regional (eastern Mediterranean) and subglobal significance.

The evolution of the neoautochthon began with the

establishment of a deep-water marine basin during Maastrichtian, Paleocene, and Early and Middle Eocene times, when pelagic foraminiferal-radiolarian oozes were deposited. During this period pelagic sediments were deposited over all of Southern Cyprus. The transgressive maximum coincided with the latest Early Eocene (the *Acarinina pentacamerata* Zone) and the Middle Eocene, as seen from the maximum extent of sediments and the presence of chert and silicified limestones. Pelagic sediments continued to accumulate in the Late Eocene, but their restricted occurrence implies the beginning of a regression. This regression culminated in the Oligocene, and it was the most severe regression in the Paleogene. Sediments of this age are known only in the depression of South-West Cyprus, and their facies shows an abundant influx of terrigenous clay particles.

The beginning of the Neogene transgression is recorded by the Early Miocene deposits, mainly consisting of pelagic carbonate oozes of the Arsos Formation, which contain no siliceous plankton. These platy limestones were deposited in the upper bathyal zone. On the structural highs, these deposits are replaced by the bioclastic limestones of the Terra Formation.

The late Early Miocene (the *Praeorbulina glomerosa* Zone) and Middle Miocene reflect the maximum of the Neogene marine transgression. This is evidenced by the pelagic carbonates of the lower half of the Pakhna Formation, which are the most widespread transgressive sediments in south-western Cyprus.

The Late Miocene echoes a general marine regression. The terrigenous influx of silt- to sand-size particles is a common feature of the pelagic carbonates of the Tortonian, and are especially abundant at its uppermost part. The final phase of regression culminated in the deposition of the Messinian evaporites, which filled the semi-restricted basins of south-western Cyprus.

The Pliocene carbonatic-terrigenous sediments are related to a regional marine transgression. In fact the terrigenous clastics show a variety of fragments, but do not contain any ophiolite clasts. The latter are only present in the Quaternary sediments.

It may therefore be concluded that most neoautochthon deposits are carbonates and that despite the variation in paleogeographic environments, the influx of terrigenous material into the basin was generally insignificant. Evidence for the beginning of subaerial denudation of the Ophiolite Complex is not found before the Quaternary.

The same stage by stage evolution, each with some locally controlled modifications, can be distinguished in the Maastrichtian-Cenozoic basins of the entire Near Eastern realm.

REFERENCES CITED

Baroz, F., and Bizon, G., 1974. Le Neogène de la chaine du Pentadactylos et de la partie nord de la Mésaoria (Chypre). Étude stratigraphique et micropaléontologique. Rev. Inst. France Pétrol., 29(3), p. 327-359.

Baroz, F., and Bizon, G., 1977. La couverture Tertiaire du flanc nord du massif du Troodos et de la partie méridionale de la Mésaoria; étude stratigraphique et micropaléontologique. Rev. Inst. France Pétrol., 32(5), p. 719-760.

Baroz, F., Bernoulli, D., Biju-Duval, B., Bizon, G., Bizon, J. J., and Letouzey, J., 1978. Correlation of the Neogene formations of the Florence Rise and of Northern Cyprus: paleogeographic and structural implications. Init. Rep. of the DSDP, v. 42, part 1. Washington (U. S. Gov. Print. Off.), p. 903-926.

Bragina, L. G., and Bragin, N. Yu., 1995. Radiolarians and stratigraphy of the Campanian-Maastrichtian deposits in South-western Cyprus. Stratigraphy and Geol. Correlation, 3(2), p. 147-155, Moscow. (In Russian).

Bragin, N. Yu., Krylov, K. A., and Bragina, L. G., 2002. Sedimentary rocks of the Mamonia Complex. This volume, Chapter 2.

Cita, M. B., 1982. The Messinian salinity crisis in the Mediterranean: a review. Am. Geophys. Un. Geodynam. Series, v. 7, p. 113-140.

Cockbain, A. E., 1961. Foraminiferal faunas from the Lapithos Group of Cyprus. Geol. Mag., 98(3), p. 177-194.

Eaton, S., 1987. The sedimentology of mid to late Miocene carbonates and evaporites in southern Cyprus. Unpublished Ph.D. Thesis, Univ. of Edinburgh, --- p.

Excursion 'A'. Marine Neogene in Austria and Czechoslovakia, 1975. Regional Committee on Mediterranean Neogene Stratigraphy. VIth Congress, Bratislava.

Fabricius, F. H., Heimann, K. O., and Braune, K., 1978. Comparison of Site 374 with Circum-Ionian land sections; implications for the Messinian «salinity crisis» on the basis of a «dynamic model». Init. Rep. of the DSDP, v. 42, part 1. Washington (U.S. Gov. Print. Off.), p. 927-942.

Henson, F. R., Browne, R. V., and McGinty, J., 1949. A synopsis of the stratigraphy and geological history of Cyprus. Quart. J. Geol. Soc., London, v. 105, pt. 1, no. 417, p. 1-41.

Hsü, K. J., Ryan, W. B. F., and Cita, M. B., 1973. Late Miocene dessication of the Mediterranean. Nature, 242(5395), p. 240-244.

Khokhlova, I. E., Bragina, L. G., and Krasheninnikov, V. A., 1994. Zonal stratigraphy of the Upper Cretaceous and Paleogene deposits of the key Perapedhi section (Southern Cyprus) by means of radiolarians and correlation with the foraminiferal zones. In: V. A. Krasheninnikov and J. K. Hall, (eds.), Geological structure of the Northeastern Mediterranean, Historical Productions-Hall, Jerusalem, p. 219-250.

Kluyver, H. M., 1969. Report on a regional geological mapping in Paphos. Bull. Cyprus Geol. Surv. Dept., v. 4, p. 21-36.

Krasheninnikov, V. A., and Kaleda, K. G., 1994. Stratigraphy and lithology of Upper Cretaceous and Cenozoic deposits of the key Perapedhi section (neoautochthon of Southern Cyprus). In: V. A. Krasheninnikov and J. K. Hall, (eds.), Geological structure of the Northeastern Mediterranean, Historical Productions-Hall, Jerusalem, p. 195-218.

Krylov, K. A., Kaleda, K. G., and Bragin, N. Yu., 1995. Terrigeneous chaotic complexes of south-western Cyprus. In: 5th Zonenshain conference on plate tectonics, Moscow, p. 77.

Lapierre, H., 1975. Les formations sédimentaires et eruptives des nappes de Mamonia et leures relations avec le Massif du Troodos (Chypre occidentale). [The sedimentary and volcanic formations of the Mamonia Nappe and their relation to the Troodos Massif, western Cyprus]. Mem. Soc. Geol. France, 54(123), 132 p.

Leonov, M. G., 1981. Olistostromes in the structure of folded areas. Trans. of the Geological Institute of the USSR Acad. Sci., v. 344, Moscow, 175 p. (in Russian).

Mantis, M., 1970. Upper Cretaceous-Tertiary foraminiferal zones in Cyprus. Scientific Research Centre of Cyprus, p. 227-241.

Mantis, M., 1972. Some planktonic foraminifera from the Paleogene of the South Cyprus. Bull. Geol. Surv. Dept., Cyprus, No. 5, p. 91-126.

Mantis, M. 1977. Upper Eocene Radiolaria in Cyprus. Bull. Geol. Surv. Dept., Cyprus, No. 7, p. 75-80.

McCallum, J. E., and Robertson, A. H. F., 1995. Late Pliocene-Early Pleistocene Athalassa formation, north central Cyprus: carbonate sand bodies in a shallow seaway between two emerging landmasses. Terra Nova, 7(2), p. 265-277.

Orszag-Sperber, F., and Rouchy, J. M., 1979. Le Miocène terminal et le Pliocène inférieur au Sud de Chypre. V Séminaire sur le messinien. IGCP, Project 117 (Evènements géologiques a la limite. Miocène-Pliocène). Paris, 60 p.

Orszag-Sperber, F., Rouchy, J. M., and Elion, P., 1989. The sedimentary expression of regional tectonic events during the Miocene-Pliocene transition in the southern Cyprus basins. Geol. Mag., 126(3), p. 291-299.

Orszag-Sperber, F., Rouchy, J. M., Bizon, G., Bizon, J. J., Cravatte, J., and Müller, C., 1980. La sédimentation messinienne dans le bassin de Polemi (Chypre). Géologie Méditerranéenne, 7(1), p. 91-102.

Ovey, C. D., 1937. Some Tertiary foraminifera from Cyprus. J. Roy. Microscop. Soc., 57(---), p. ---.

Pantazis, T. M., 1967. The geology and mineral resources of the Pharmakas - Kalavasos area. Mem. Cyprus Geol. Surv. Dept., v. 8, 190 p.

Pantazis, T. M., 1978. Cyprus evaporites. Init. Rep. of the DSDP, v. 42, part 2. Washington (U. S. Gov. Print. Off.), p. 1185-1194.

Payne, A. S., and Robertson, A. H. F., 1995. Neogene supra-subduction zone extension in the Polis graben system, west Cyprus. J. Geol. Soc., London, 152(4), p. 613-628.

Radionova, E. P., 1994. Diatoms in the Upper Cretaceous and Paleogene sediments of the Perapedhi section (Southern Cyprus). In: V. A. Krasheninnikov and J. K. Hall, (eds.), Geological structure of the Northeastern Mediterranean, Historical Productions-Hall, Jerusalem, p. 251-264.

Robertson, A. H. F., 1976. Pelagic chalk and calciturbidites from the Lower Tertiary of the Troodos Massif, Cyprus. J. Sedim. Petrology, 46(---), p. 1007-1016.

Robertson, A. H. F., 1977a. The Kannaviou formation, Cyprus: volcaniclastic sedimentation of a probable late Cretaceous volcanic arc. J. Geol. Soc. London, 134(3), p. 269-292.

Robertson, A. H. F., 1977b. Tertiary uplift history of the Troodos Massif, Cyprus. Geol. Soc. Am. Bull., 88(12), p. 1763-1772.

Robertson, A. H. F., 1977c. The origin and diagenesis of cherts from Cyprus. Sedimentology, 24(1), p.11-30.

Robertson, A. H. F., 1977d. The Moni melange, Cyprus: An olistostrome formed at a destructive plate margin. J. Geol. Soc. London, 133(5), p. 447-466.

Robertson, A. H. F., and Dixon, J. E., 1984. Introduction: aspects of the geological evolution of the Eastern Mediterranean. In: The geological evolution of the Eastern Mediterranean. Spec. Publ. Geol. Soc. London, 17, p. 1-74.

Robertson, A. H. F., and Grasso, M., 1995a. Later Tertiary-Quaternary Mediterranean tectonics and paleo-environments - an introduction. Terra Nova, 7(2), p. 112-113.

Robertson, A. H. F., and Grasso, M., 1995b. Overview of the Late Tertiary-Recent tectonic and palaeo-environmental development of the Mediterranean region. Terra Nova, 7(2), p. 114-127.

Robertson, A. H. F., and Hudson, J. D., 1974. Pelagic sediments in the Cretaceous and Tertiary history of Cyprus. In: Pelagic sediments: on land and under the sea. Intern. Assoc. Sedimentol. Spec. Publ., v. 1, p. 403-436.

Robertson, A. H. F., and Woodcock, N.H. 1979. The Mamonia complex, SW Cyprus: the evolution and emplacement of a Mesozoic continental margin. Geol. Soc. Amer. Bull., 90(7), p. 651-665.

Robertson, A. H. F., Eaton, S., Follows, E. J., and McCallum, J. E., 1991. The role of tectonics versus global sea-level changes in the Neogene evolution of the Cyprus active margin. Intern. Assoc. Sedimentol., Spec. Publ., 12, p. 331-369.

Robertson, A. H. F., Eaton, S., Follows, E. J., and Payne, A. S., 1995. Depositional processes and basin analysis of Messinian evaporites in Cyprus. Terra Nova, 7(2), p. 233-253.

Rouchy, J. M., Orszag-Sperber, F., Bizon, G., and Bizon, J. J., 1980. Mise en évidence d'une phase émersion fini messinienne dans le bassin de Pissouri (Chypre): une modalité de passage Miocène / Pliocène en Méditerranée orientale. C. R. Acad. Sci. Paris, D 291(9), p. 729-732.

Swarbrick, R. E., and Naylor, M. A., 1980. The Kathikas melange, SW Cyprus: late Cretaceous submarine debris flow. Sedimentology, 27(1), p. 63-78.

Swarbrick, R. E., and Robertson, A. H. F., 1980. Revised stratigraphy of the Mesozoic rocks of southern Cyprus. Geol. Mag., 117(6), p. 547-563.

Turner, W. M., 1968. A progress report on the geology of western Cyprus including the Akamas Peninsula: the general stratigraphy of western Cyprus including the Akamas Peninsula. Univ. New Mexico, Albuquerque, New Mexico, v. 1, p. 1-41, v. 2, p. 1-14.

Turner, W. M., 1971. Geology of the Polis - Khatikas area, Cyprus. Doctoral dissertation. Geol. Dept. of the Univ. of New Mexico, Albuquerque, New Mexico, 430 p.

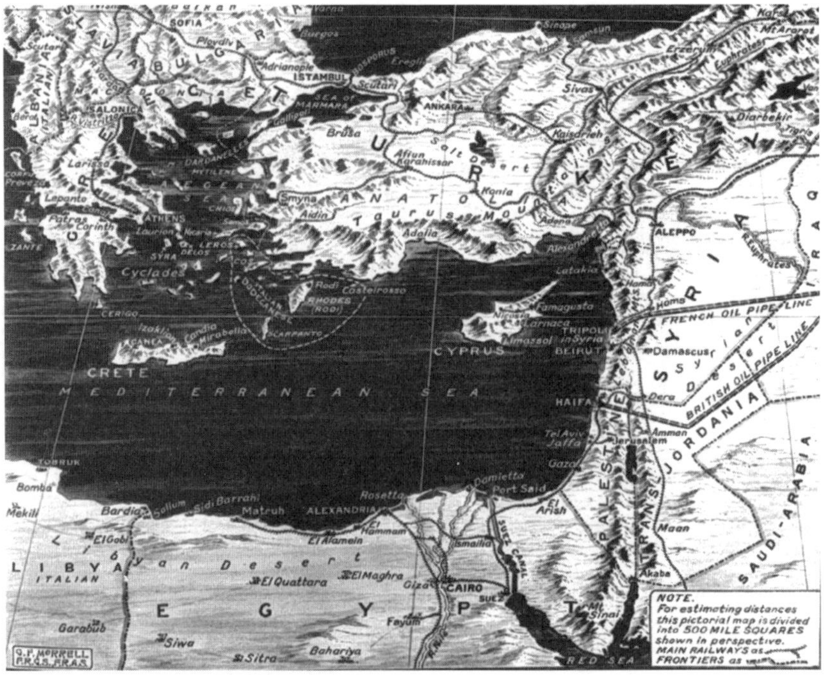

The Levant: Battleground of the Empire Today. World War II map from the Illustrated London News, 16 November 1940, p. 621. Map image (No. 81814) courtesy of the University of San Diego History Department website at: http://history.sandiego.edu/gen/WW2Index/picindexmaps.html.

Chapter 4

Chaotic Rock Complexes in South-Western and Central Cyprus

Kirill A. Krylov, Kirill G. Kaleda, and Nikita Yu. Bragin
Geological Institute of the Russian Academy of Sciences, Pyzhevsky pereulok, 7, 109017 Moscow, Russia

INTRODUCTION

Detailed mapping in southwestern Cyprus has led to the recognition of several distinct chaotic rock complexes which are considered to reflect the main stages of the geodynamic development of the region. Some of these assemblages are described here for the first time, as *e.g.* the Arakapas Complex. Others were reinterpreted, as was the Moni Complex or redefined, as was the Petra tou Romiou Complex.

Discussions of stratigraphy and sedimentology in some of these chaotic rocks *e.g.*, the Kathikas Formation, are presented in Chapters 2 and 3 of this volume. Despite this, the following outline reviews important aspects regarding the nature of the main chaotic rock formations in the region.

The development of chaotic rocks of various origin in Southern Cyprus is evident in the stratigraphic levels from Late Cretaceous to the Eocene and Miocene. Such mixtites are sporadically distributed in the south-west of the island and in the Limassol Forest. Understanding the link between the plate-tectonic setting and the environment of chaotic sedimentation is crucial to the general interpretation of the geologic history of the Eastern Mediterranean.

The literature abundantly discusses the distinctions between terrigenic melanges and olistostromes, *e.g.*, Hsü (1974), Sokolov (1979), and M. Leonov (1981). Regretfully, the term 'melange' is often used for the Cyprean mixtites, without a well-defined classification (Pantazis, 1967; Turner, 1973; Robertson, 1977; Simonian and Gass, 1978; Swarbrick and Naylor, 1980).

The classification proposed herein attempts to address the chaotic rocks in Cyprus, via their stratigraphic position, lithological characteristics, and the events controlling their formation.

Arakapas chaotic rock complex

This new term applies to the Late Cretaceous mixtites of the Limassol Forest. This olistostrome has been studied in detail by Simonian and Gass (1978), but was never described as a separate rock unit. This complex is found in a narrow belt, some 50 km long and 5 km wide. Structurally, its occurrence coincides with the east-west-trending Arakapas Fault Belt that marks the boundary between the Troodos and Limassol Forest ophiolite masses. Numerous workers, starting with Moores and Vine (1971), have interpreted this suture zone as a fossil transform structure.

The Arakapas Complex can be divided into two major units of different origin. In the vicinity of Perapedhi, the lower unit characteristically represents a chaotic breccia with no signs of stratification. By contrast, the upper unit consists of a stratified gravitational deposit, alternating with lava flows and deep-water background sediments, which was developed almost continuously along the whole extent of the Arakapas Fault zone.

The lower breccia consists of more than 50 percent chaotic angular fragments of metamorphosed gabbro-diabases and pillow lavas (between 1 cm and 50 cm across) derived from the Troodos Complex, set in a non-metamorphosed argillaceous matrix. With regard to the source of the olistoliths, following the terminology of Sokolov (1979), the lower breccias represent ophiolite-derived olistostrome deposits.

The chaotic breccias are in many places clearly adjacent to ancient escarpments and composed of blocks of the same lava flows and sheeted dikes. These fault-scarp walls have in many cases undergone intense tectonic brecciation and slickensiding. The densely packed fragments, minor fine-clastics, and matrix show the same metamorphism. This suggests their formation by gravity-driven collapse of the submarine escarpments bounding fault valleys associated with the Arakapas transform zone, as already proposed by Simonian and Gass (1978). In regard to the inferred mode of origin, the breccias of the lower unit belong to the class of tectono-gravitational mixtites, in concordance with the terminology used by M. Leonov (1981).

The upper unit is a 1-30 m thick, layered, graded and upwards fining chaotic sequence. It displays sharp basal contacts. More typically, each bed consists of uncemented breccias at its base, gravels and sandstones in its middle part, and siltstones as well as mudstones at the top. The breccias 40-50 cm sized fragments consist mainly of metabasalts, metadolerites, and rare gabbro clasts. The petrology of these fragments closely resemble that of the Troodos Complex (Robertson and Hudson, 1974; Simonian and Gass, 1978). In some places, locally derived argillite fragments are intercalated within the ophiolite-derived breccias. The fine clastic rocks are formed mainly of clinopyroxene, plagioclase and quartz grains.

The formation of graded beds must be related to repeated high-density debris flows and associated turbidity-currents, called 'incoherent slides' by Hoedemaeker (1973). They were accompanied by volcanic activity, as indicated by the presence of pillow lava and dolerite sills between gravity-slides (Robertson and Hudson, 1974). The association of multiple debris flows and interstratified tongues of pelagic metalliferous and siliceous argillites, containing Campanian-Early Maastrichtian radiolaria (Robertson and Hudson, 1974), indicates deposition in a deep-water environment.

Paleo-topography largely controls the thickness of the whole Arakapas Complex, which varies considerably over the area. The succession is thickest in structural depressions, where mixtites and volcanics together may reach a thickness approaching 500 m. Their thickness is only about 150 m on structural highs, where contemporaneous extrusive and dike rocks are absent.

The age of the Arakapas olistostrome is Cenomanian–Santonian, as derived from its stratigraphic relations. The Arakapas mixtites overlie the Cenomanian-Turonian Troodos Complex (Robertson and Hudson, 1974), They alternate with Troodos pillow lavas, and the lower chaotic breccias are locally overlain by the Santonian-Early Campanian pelagic Perapedhi Formation (Bragin and Bragina, 1991; Bragina and Bragin, 1996). The age of the Arakapas Complex is further constrained by the Campanian-Early Maastrichtian age of the overlying Kannaviou Formation (Mantis, 1970; Robertson and Hudson, 1974; Bragina and Bragin, 1995).

Kathikas chaotic rock melange

The name Kathikas Melange was coined by Lapierre (1975) for the sedimentary melange or olistostrome, locally deposited in southwestern Cyprus. More precisely, it is found between the villages of Kathikas and Ayia Varvara, where it attains a maximum thickness of over 200 m.

The Kathikas mixtites were further investigated by Robertson and Woodcock (1979) and by Swarbrick and Naylor (1980). Their observations are summarized below together with our own comments.

Stratigrapically, the Kathikas Complex is a chaotic layered sequence between the Kannaviou Formation and the Lower Lefkara Formation. The lower contact is erosional. Bentonitic clays, dated as Late Campanian to Early Maastrichtian, form tongues within the mixtites of the lower part of the Kathikas Formation (Bragina and Bragin, 1995; Krasheninnikov and Kaleda, this volume, Chapter 3). Careful examination of the upper boundary reveals that it is either transitional or disconformable. Gradational contacts are found in most localities where the Kathikas rocks are overlain by, or pass laterally into, pelagic carbonates of the Lower Lefkara Formation. Near the villages of Kathikas, Phity, and Simou, the basal horizons of the Lower Lefkara Formation are Maastrichtian. Carbonatic interbeds, lithologically resembling the rocks of the Lower Lefkara Formation, are present in the upper part of the Kathikas sequence. Where an erosional scoured surface demarcates the base of the neoautochthonous complex, the age of the base of the latter grades from Danian, near Drimou, to Thanetian, at Pano Panayia, or Lutetian at Milos tou Mavrokolymbos.

Structurally, the Kathikas Complex is restricted to the boundary between the Troodos and Mamonia terrains. Near Ayia Varvara, these rocks are integrated into the tectonized serpentinite melange. Clasts within the melange are interpreted as due to post-magmatic alteration of Troodos ultramafic and mafic rocks. This monogenic melange is directly overlain by the imbricated thrust plates associated with the Mamonia nappe (Malpas et al., 1992; Krylov et al., 1993).

The fundamental relationship of the Mamonia allochthon to the Kathikas Complex has been shown in the upper part of the Xeros Potamos River (Swarbrick and Naylor, 1980). Here, untectonized Mamonia Complex rocks pass laterally into the tectonic breccia horizon, the so-called Blocky Formation. Here, blocks of Mamonia lithologies are intensely fractured and rotated and no matrix is present. The overlying 'immature melange' consists of large blocks of siliceous mudstones and bedded cherts, set in an argillaceous matrix. This transitional unit grades upwards into the chaotic rocks with all the characteristics of the Kathikas olistostrome.

Petrographically, the Kathikas olistoliths are distinctly different from those in the Arakapas olistostrome by being polymictic and entirely composed of Mamonia Complex rocks (**Fig. 4.1**). Important aspects with regard to the formation of the Kathikas olistostrome are the presence of blocks of metamorphites from the Ayia Varvara Formation, and inclusions of serpentinized ultramafic and mafic igneous rocks. In terms of clast-size, the angular mixtites range from cm-sized clasts to huge blocks up to 200-500 m across, but on average are about 1m.

On a local scale and within the individual exposures, the structure of the Kathikas mixtites is chaotic. Despite this, many olistolith horizons show some degree of systematic arrangement. For instance, most large blocks are composed of Late Triassic limestone of the Dhiarizos Group. Of these, the largest is the angular exotic block, up to 1.5 km long, on which the picturesque village of Episkopi is situated.

The matrix is represented by prevalently argillaceous, heterogeneous material derived from the erosion of pelitic rocks of the Albian to Cenomanian Mavrokolymbos Formation (Ayios Photios Group of the Mamonia Complex) and the underlying Kannaviou Formation (Lapierre, 1975; Bragina and Bragin, 1995). Direct evidence for the general ascendant change in matrix composition has been obtained from the continuous Kathikas sequence near Kannaviou village. In the lower part of this section, grayish-green, weakly lithified bentonitic clays, locally derived from the Kannaviou Formation, constitute the most persistent component in a relictic non-homogenous matrix. Moving upwards, reddish-brown, previously consolidated siltstones become more prominent within the binding material of the higher olistolith horizons, thus making up the dominant matrix.

It may be concluded that the origin and development of the Kathikas olistostrome is related to the Maastrichtian thrusting of the Mamonia allochthone over the Troodos ophiolite. Good evidence for the pre-existing tectonization of the parent rocks may be provided, both by the presence of abundant slip-scratches and slickensided surfaces within the olistoliths, and by their absence in the matrix of olistostromal deposits. The term 'tectono-gravitative mixtites' (Leonov, 1981) is appropriate for describing these types of chaotic rocks. The Kathikas olistostrome is probably the response to debris flows and associated subaqueous gravity-driven processes (Fig. 4.2). In the definitions of Hoedemaeker (1973) and Sokolov (1979), the Kathikas Complex can be classified as ophiolite-clast

Fig. 4.1. Outcrop of the Kathikas Formation near Kannaviou village. Blocks of various size and composition in the greenish-gray clayey matrix (photograph by K. Krylov).

olistostrome, formed mostly by repeated incoherent slides.

Moni chaotic rock complex

The name Moni Melange was introduced by Pantazis (1967) to describe all chaotically mixed rocks in southwestern Cyprus. The term is now restricted to the polygenetic tectono-gravitational mixtites along the WNW-trending Yerasa Fault Zone. Developed only here, along a strike length of more than 30 km, the Moni Melange stratigraphically separates the Late Cretaceous Kannaviou Formation from the Miocene Koronia Formation. It attains an apparent thickness of at least 200 m, as argued by Robertson (1977) and Robertson et al. (1991). In our opinion, the Moni Complex subdivides into two chaotic rock units of distinct genetic origin which possess a two stage history. Thus the 'Paleo-

Moni ophiolite-clast olistostrome' and the 'Neo-Moni terrigenous tectonic melange', never formally defined, are used as informal lithostratigraphic units.

The Paleo-Moni unit is a chaotic layered sequence dated as Maastrichtian. Along its northern boundary, this unit depositionally overlies ophiolite volcanics of the Troodos Massif or passes laterally into bentonitic clays and radiolarites of the Kannaviou Formation, and is also underlain by ophiolitic pillow lavas (Robertson, 1977). The bentonitic clays, which are similar in age and composition to those from the Kannaviou Formation, constitute the prevalent matrix of the whole Moni Complex mixtites (Mantis, 1970).

The clasts and blocks composing the olistostrome are represented mainly by Mamonia Complex lithologies. Exotic constituents in the common clast assemblage include the Parekklisha Sandstone and the Monagroulli Argillite units, unknown outside the Moni Complex. More typically, the olistoliths range in size from several meters to 1.0-1.5 km across.

Blocks belonging to the Mamonia nappe have a general NNW-SSE trend and SSW inclination. Large, east-dipping serpentinites, some of which are about 5 km long, are more or less centrally located. The nature of these serpentinites is still debated. They resemble some of the highly altered ultramafic and mafic igneous rocks occurring within the Troodos Massif. The uncemented Parekklisha sandstone blocks are always oriented in the same way as the surrounding blocks. Characteristically, they occur along the boundary with the Troodos terrain and trend roughly parallel to the strike of the rocks at the top of the ophiolite succession. Such circumstances are not to be expected in association with a tectonic melange (Robertson, 1977).

In contrast with the Kathikas multiple debris flows, the Paleo-Moni mixtites indicate an accumulation due to sliding, the 'coherent slides' of Hoedemaeker (1973), and they may have been deposited in a deep-water trough filled with Kannoviou pelagic clays. Because of the general similarity in their age and main sources, the olistoliths of the Kathikas and Paleo-Moni olistostromes are considered to have formed during a major phase of collision between the Troodos and Mamonia terrains, during Maastrichtian time.

The formation of the Neo-Moni melange must be related to repeated post-Late Eocene overthrusting along the Yerasa Fault Zone. Despite the general lithological similarity between the Paleo-Moni gravity-slide deposits and the Neo-Moni mixtites, the outstanding feature of the tectono-stratigraphic nature of the latter is the presence of abundant limestone blocks. These are derived from the neoautochthonous Lower Lefkara Formation, depositionally overlying the Paleo-Moni olistostrome. These blocks appear to be detached from their original position and to have been included as tectonic enclaves in the older bentonic clay matrix (**Fig. 4.2**). Obvious deformation by fracturing and rotation is characteristic of these tectonic inclusions, in contrast with the pre-existing Paleo-Moni olistoliths. Another notable feature is the presence in the binding material of admixtures of mylonitized material derived from the Ayios Photios Group.

The relationship between the whole Moni Complex and the

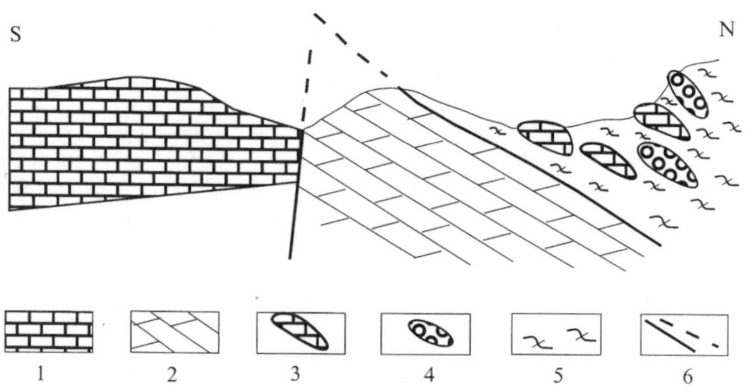

Fig. 4.2. Neo Moni Melange and its relationship with neoautochthonous deposits near Akrounta village. 1 - the Upper Lefkara Formation, Middle to Upper Eocene, limestones, marls and calcarenites; 2 - Middle Eocene chalky limestones; 3 - blocks of various rocks including the Upper Lefkara Formation; 4 - blocks of sandstones; 5 - matrix represented by gray mylonitized bentonitic clays (the Kannaviou Formation) or red mylonitized mudstones (the Ayos Photios Group); 6- thrusts and faults (dotted line - hypothetical).

adjacent portions of the pre-Miocene neoautochthonous cover may be observed along the south-western edge of its exposures (*e.g.* near the village of Akrounda; Fig. 4.3). All these rocks were subjected to considerable tectonic deformation. This is indicated by the presence of a series of imbricated north-northeast thrust faults, along which the Paleo-Moni olistostrome deposits have been thrust into place over the Middle Eocene carbonate rocks. Strong slickensiding occurs in the Moni mixtite matrix, and intense fracture cleavage is observed in the Upper Eocene deposits.

Petra tou Romiou chaotic rock complex

We apply this new term to a separate tectonic melange, occurring in southwestern Cyprus between Petra tou Romiou and Ayia Varvara (Krylov et al., 1993). Exposures of this complex extend along the WNW-trending Petra tou Romiou Lineament, which extends nearly parallel to the Yerasa Fault Zone.

Though devoid of direct evidence as to its age, the Petra tou Romiou melange is, according to Krasheninnikov and Kaleda (this volume, Chapter 3), unconformably overlain by the Middle Lefkara Formation (**Fig. 4.3**). It contains neoautochthon-derived limestone blocks of the Lower Lefkara Formation. The youngest fragments may possibly indicate Paleocene-Early Eocene ages. Within the type locality, the complex has an exposed thickness of at least 70 m.

Structurally, the complex represents the lowest tectono-stratigraphic unit of the Mamonia Nappe system. It is particularly evident near Ayia Varvara. In the north, the Petra tou Romiou rocks are overlain by a series of imbricated thrust plates of the Mamonia allochthon. In other places, Troodos ophiolite rocks appear to have been thrusted over Petra tou Romiou melange rocks. By contrast, along the southern edge of the exposures, the boundaries of the melange unit are either hidden under the

Fig. 4.3. Outcrop of the Petra tou Romiou Melange in the type area. Large blocks of alkaline basalts and Triassic limestones are surrounded by reddish-brown mylonites. Lower right corner of the photo - neoautochthonous limestones of the Lower Eocene (Krasheninnikov and Kaleda, this volume, Chapter 3) (photograph by K. Krylov).

Fig. 4.4. Small blocks of limestones derived from the Dhiarizos Group surrounded by matrix. Petra tou Romiou area (photograph by K. Krylov).

Fig. 4.5. Matrix represented by mylonitized red siltstones of the Ayos Photios Group (dark) and gray-greenish clays that are probably derived from the Kannaviou Formation (light) (photograph by K. Krylov).

Fig. 4.6. Matrix represented by reddish-brown mylonites (the Ayos Photios Group) (dark) and gray-greenish mylonites (the Dhiarizos Group) (light) (photograph by K. Krylov

Table 4.1. Chaotic rock complexes of Southwestern and Central Cyprus and their comparison.

Name of complex	Occurrence	Lithology		Contacts	Time of formation	Specific features	Type of formation
		Blocks	Matrix				
Arakapas	Limassol Forest	Gabbro-diabases and pillow lavas of theTroodos Ophiolite Complex	Argillaceous	Commonly stratigraphic	Cenomanian? - Maastrichtian	Presence of stratification. Extreme domination of the Troodos-type rocks	Ophiolite-derived olistostrome
Kathikas	Southwestern Cyprus	All lithological varieties of the Mamonia Allochthonous Complex	Composite argillaceous	Commonly stratigraphic	Early Maastrichtian	Presence of stratification. Presence of background sediments	Ophiolite-bearing polymictic olistostrome formed by incoherent slides
Moni: Paleo-Moni	Limassol Forest	Rocks derived from the Troodos Complex and Mamonia Complex	Benthonitic clays from the Kannaviou Formation (Campanian - Maastrichtian)	Tectonic and relict stratigraphic	Maastrichtian	Presence of exotic rocks of the Parekklisha Formation. Lateral passing to the Kannaviou Formation	Ophiolite-bearing polymictic olistostrome formed by coherent slides
Moni: Neo-Moni	Limassol Forest	Same varieties together with limestones from the Lower Lefkara Formation	Same clays together with mylonitized mudstones of the Ayos Photios Group	Tectonic	Pre-Late Miocene	High tectonic influence. Presence of blocks that have younger age then matrix.	Tectonic melange
Petra tou Romiou	Southwestern Cyprus	Rocks derived from the Troodos Complex and Mamonia Complex	Mylonitized clays and mudstones from the Troodos Complex and Mamonia Complex	Lower contact is tectonic, upper is stratigraphic	Paleocene	High mylonitization of matrix. Common slickensiding.	Tectonic melange

undisturbed neoauthochthonous sequence or cut by the Coastal Boundary Fault.

In lithological terms, the Petra tou Romiou Complex represents a typical terrigenous-volcaniclastic melange. It is formed of lithologically heterogeneous blocks, up to several tens of meters in size. They are composed of rock types of Mamonia Complex lithologies, extrusives and serpentinites derived from the Troodos Massif, and limestones from the Lower Lefkara Formation. The spatial distribution of blocks of various lithologies is not uniform and their pattern is rather complicated. It is observed that the same kind of rock fragments tend to be concentrated in places.

The relationship of the melange blocks to the matrix is best seen in the many small quarries (**Fig. 4.4**) The binder consists mainly of mylonites with the compositional characteristics of the adjacent blocks. The most common varieties are ophiolite-derived boninitic lavas and red brown siltstones and mudstones derived from the Ayios Photios Group (Mamonia Complex), which are usually mixed together with strongly tectonized bentonitic clays (**Fig. 4.5**). In places with fewer melange blocks, the spaces between the dispersed rock fragments are entirely filled with mylonites, associated with the Ayios Photious Group rocks (**Fig. 4.6**).

In many places, the formation of the binder is clearly accompanied by intense foliation and mylonitization, closely associated with slickensiding. These features, indicating severe deformation and shearing, are interpreted as being caused by regional thrusting in Paleocene-Early Eocene times. It is also evident that the youngest limestone incursions were subjected to strong brecciation, earlier than the fault rocks (mylonites). Thus, the Petra tou Romiou mixtites are considered to be a typical tectonic melange.

CONCLUSIONS

The formation of the chaotic rock complexes reflects major geodynamic episodes in the structural history of Cyprus. They represent an environmental response to tectonic events. The general trend in the progressive change and development of the mixtites is apparent (**Table 4.1**). This rather genetic sequence is considered to be typical for the whole Alpine Belt. The only exception is the absence of typical olistostromes indicating the occurrence of incipient rifting in the Mesozoic Tethys.

The onset of chaotic sedimentation in the Cenomanian-Campanian took place in a spreading-related setting. The Arakapas ophiolitic olistostrome assemblage could have formed through collapse of the fault scarps bounding the active faults marking the fossil transform zone in a fore-arc basin (the Troodos Complex).

The formation of the Kathikas Complex and Paleo-Moni olistostromes must be related to the Maastrichtian thrusting of the Mamonia terrain over the Troodos subautochthon. Much of the ophiolitic material has been derived from the frontal portions of the nappes, which underwent intense destruction during the thrusting.

By contrast, the tectonic melange units closely associated with areas of intense deformation and shearing are believed to be characteristic of the collision zone (suture) segments. The Petra tou Romiou terrigenous-volcaniclastic melange is interpreted as the remote result of underthrusting of the African plate towards the 'Cyprus Microplate' during a major collision phase in the Paleocene-Early Eocene. The generation of the Neo-Moni terrigenous melange may have been caused by tectonic movements along large, northwest-trending, transpressional wrench faults which occurred in the Eocene to Miocene.

All these chaotic rock complexes separate geologic bodies and their occurrence has to be represented on geologic maps.

REFERENCES CITED

Bragin, N. Yu., and Bragina, L. G., 1991. Radiolarian biostratigraphy of Troodos Upper Cretaceous sediments (Cyprus). Interrad VI (VI Meeting Int. Assoc. Radiolarian Paleontologists), Universita Degli Studi di Firenze, p. 17.

Bragina, L. G., and Bragin, N. Yu., 1995. Radiolarians and stratigraphy of the Campanian-Maestrichtian deposits in Southwestern Cyprus. Stratigraphy and Geological Correlation, 3(2), p. 147-155. (In Russian with English translation).

Bragina, L. G., and Bragin, N. Yu., 1996. Stratigraphy and Radiolaria of the type-section of Upper Cretaceous Perapedhi Formation (Cyprus). Stratigraphy and geological correlation. Moscow, 4(3), p. 38-45. (In Russian with English translation).

Hoedemaeker, Ph. J., 1973. Olistostromes and other delapsional deposits and their occurrence in the region Moratalla (Province of Murcia, Spain). Scripta. geol., v. 19, 207 p.

Hsü, K. J., 1974. Melanges and their distinction from olistostromes. Modern and ancient geosynclinal sedimentation, R. H. Dott and R. H. Shaver (eds.). Spec Publ. Soc. Econ. Paleont. Miner., Tulsa, v. 19., p. 321-333.

Krylov, K. A., Kaleda, K. G., and Bragin, N. Yu., 1993. Structure of Southwestern Cyprus. Abstr. L. P. Zonenshain Mem. Conf. on Plate Tectonics, Moscow, November 17-20, 1993. Inst. Oceanology Russ. Acad. Sci., GEOMAR, Kiel, Germany, p. 91-92.

Lapierre, H., 1975. Les formations sédimentaires et éruptives des nappes de Mamonia et leur relations avec le Massif du Troodos. Mém. Soc. Géol. Fr., 123 p.

Leonov, M. G., 1981. Olistostromes and their position in the structures of folding belts. Moscow, Nauka, Trans. Geol. Inst. Acad. Sci. USSR, v. 344, 175 p. (In Russian).

Malpas, J., Xenophonthos, C., and Williams, D., 1992. The Ayia Varvara Formation of SW Cyprus: a product of complex collisional tectonics. Tectonophysics, 212(3-4), p. 193-213.

Mantis, M., 1970. Upper Cretaceous-Tertiary foraminiferal zones in Cyprus. Epithris, Cyprus, Scientific Research Centre, v. 3, p. 227-241.

Moores, E. M., and Vine, F. J., 1971. The Troodos Massif, Cyprus, and other ophiolites as oceanic crust: evaluation and implications. Phil. Trans. of the Royal Society, London, A.268(1192), p. 433-466.

Pantazis, Th. M., 1967. The geology and mineral resources of the Phamakios - Kalavasos area. Mem. Geol. Surv. Dept. Cyprus, v. 8, 190 p.

Robertson, A. H. F., 1977. The Kannaviou Formation, Cyprus: volcanoclastic sedimentation of a probable late Cretaceous volcanic arc. J. Geol. Soc. London, 133(5), p. 447 - 467.

Robertson, A. H. F., Eaton, S., Follows, E. J., and McCallum, J. E., 1991. The role of local tectonic versus global sea-level change in the Neogene evolution of the Cyprus active margin. Spec. Publ. Int. Assoc. Sedimentology, v. 12, p. 331-369.

Robertson, A. H. F., and Hudson, J. D., 1974. Pelagic sediments in the Cretaceous and Tertiary history of the Troodos Massif, Cyprus. In: Pelagic sediments: On Land and Under the Sea, K. J. Hsü and H. C. Jenkyns (eds.). Spec. Publ. Int. Assoc. Sedimentolology, v. 1, p. 403-436.

Robertson, A. H. F., and Woodcock, N. H., 1979. Mamonia Complex, Southwest Cyprus: evolution and emplacement of a Mesozoic continental margin. Geol. Soc. Am. Bull., 90(7), p. 651-665.

Simonian, K. O., and Gass, I. G., 1978. Arakapas fault belt, Cyprus: A fossil transform fault. Geol. Soc. Am. Bull., 89(8), p. 1220-1230.

Sokolov, S. D., 1979. Ophiolitic and ophiolitoclastic olistostromes of the fold belts. Geotectonics, Moscow, No 3, p. 76-88. (In Russian).

Swarbrick, R. E., and Naylor, M. A., 1980. The Kathikas Melange, southwest Cyprus: late Cretaceous submarine debris flows. Sedimentology, 27(1), p. 63-78.

Turner, W. M., 1973. The Cyprean gravity nappe and the autochthonous basement of Cyprus. In: K. A. De Jong and R. Scholten (eds.), Gravity and Tectonics, John Wiley & Sons, New York, p. 287-301.

Chapter 5

The Tectonic Structure and Evolution of South-Western Cyprus

Kirill A. Krylov
Geological Institute of the Russian Academy of Sciences, Pyzhevskiy pereulok 7, 109017 Moscow, Russia
Sergey A. Silantyev
Vernadsky Institute of Geochemistry and Analytical Chemistry of the Russian Academy of Sciences, Kosygib Str. 19, 117975 Moscow, Russia
Valery A. Krasheninnikov
Geological Institute of the Russian Academy of Sciences, Pyzhevskiy pereulok 7, 109017 Moscow, Russia

INTRODUCTION

Within the framework of the Tethys Project, between 1987 and 1993, researchers of the Geological Institute, Vernadsky Institute of Geochemistry and Analytical Chemistry, and Schmidt Institute of Physics of the Earth of the Russian Academy of Sciences, performed multidisciplinary geological studies in the south-western part of Cyprus. This area displays (1) rocks of the Mamonia fold-and-thrust pile, (2) the Troodos ophiolite rocks, and (3) the overlying sedimentary cover. These circumstances render south-western Cyprus an appealing proving ground on which to unravel the relations between these rock complexes. A thorough and massive sampling program, covering all the rock varieties exposed there (for the purposes of geochemical and other studies), made it possible to spatially delineate individual rock assemblages and to establish new ones. This chapter discusses the materials thus obtained in light of the problem addressed here and with due regard to previous research.

Initially, by analogy with other ophiolitic terranes, the Troodos Massif was believed to make up the uppermost unit in the tectonic stratigraphy. This viewpoint was later challenged by Lapierre (1975), whose study of the fold-and-thrust pile of southwestern Cyprus led her to conclude that the Troodos ophiolite occurs structurally below the nappes composed of Mamonia rocks. This corollary automatically redefines the Troodos ophiolite as a structurally unique, or at least atypical rather than 'classical' ophiolite, as it was viewed previously. Detailed 1:50,000 scale mapping (Lapierre, 1975) furnished further support for the concept that viewed the Mamonia fold-and-thrust complex as an independent structural unit (Bear, 1960), distinct from the Troodos ophiolite, and for the presence of complex melanges collectively named Moni.

Robertson and Woodcock (1979) further refined this stratigraphic scheme, subdividing the Mamonia complex into two groups: the Dhiarizos and Ayios Photios. The former comprises alkali basaltic sequences with volcanic breccias overlain by, or interbedded with, pelagic sediments-limestones and radiolarites. Robertson and Woodcock (1979) attributed to the same group the reef-related limestones, the Petra tou Romiou carbonate breccias and high-Ti alkaline basalts, and the Louthra tis Aphroditis 'volcanic breccias'. The Ayios Photios Group, by contrast, was defined as a purely sedimentary sequence comprising terrigenous, carbonate, and siliceous deposits subdivided into the Vlambouros and Episkopi Formations (Swarbrick, Robertson, 1980; Bragin, Krylov, 1995, 1996). The thick sequence of the Upper Cretaceous to Quaternary, mostly carbonate rocks, occurring along the periphery of the Troodos Massif, was customarily described as a sedimentary cover (Robertson, 1977; Clube and Robertson, 1986). This is how, by the 1980's, the main concepts concerning the architecture of south-western Cyprus had taken shape, then remaining virtually unrevised through the later publications (Lapierre, 1975; Robertson and Woodcock, 1979, 1981; Swarbrick, 1980, 1993; Swarbrick and Naylor, 1980; Robertson, 1998a, b; Kazmin, 1991; Kazmin et al., 1987, 1989).

Previous studies, however, left the inner structure of the Mamonia pile poorly understood, although certain structural issues referring to the Mamonia pile were addressed in the latest works (Swarbrick, 1993; Malpas et al., 1993, 1998; Robertson, 1998a, b). The following issues remained unresolved: (1) how many nappes occur in the pile, (2) which lithologies make up which nappe, and (3) what setting the metamorphic rocks occupy in the inner structure of the Mamonia pile. The topics of the accretionary vs. collisional origin of the Mamonia pile, age of formation of its structure, and the character of the metamorphic rocks occurring within the pile remained open to debate, as can be seen from a number of works (Silantyev, 1993; Swarbrick, 1993; Malpas et al., 1992, 1998). In practice, the entire geological structure was subdivided following the geographical principle based upon the areal distribution of the rocks of the Troodos ophiolite and of those attributed to the Mamonia complex. The tectonic subdivision of Cyprus was conducted in a rather sketchy manner. Thus, Robertson (1990) recognized the Kyrenia, Troodos, and Mamonia Massifs as component parts of the structural assemblage of Cyprus. The mode of occurrence of the Troodos ophiolites with respect to the Mamonia complex remained obscure, although Lapierre (1975) interpreted them to be in thrust relations, the Mamonia complex being allochthonous. Note that none of the workers have ever defined the neoautochthonous complex as a separate structural unit.

METHODS USED

Our studies were based on (a) large-scale (1:5,000) geological mapping of key localities in which rocks of various lithotectonic complexes are exposed, (b) construction of individual profiles, (c) documentation of reference sections, and (d) revisiting of previously documented sections including detailed stratigraphic studies. This kind of approach enabled us to identify the mappable units, each corresponding to a single lithotectonic complex. Subsequently, the relations among the mappable units thus recognized were studied, and their structural settings established. In this manner, the overall structural organization of the study area was unraveled. Using detailed stratigraphic scales (a radiolarian scale for the Mesozoic deposits and a foraminiferal scale for the Upper Mesozoic-Cenozoic strata) afforded precise datings for the mappable units, further refined the schemes of subdivision of the deposits, and enabled us to unravel the scenario of formation of the structural ensemble of south-western Cyprus. A detailed geochemical sampling program across all the major exposures of igneous and metamorphic rocks contributed significantly to the understanding of the architecture of the Mamonia complex.

The main locality for studying the relationships between the Troodos ophiolite and the Mamonia fold-and-thrust pile as well as the latter's inner structure was chosen near Ayia Varvara village. Additional localities were picked in the valley of the Mavrokolymbos Rivar, on the Akamas Peninsula, and in the interfluve between the Dhiarizos and Xeropothamos Rivers. The relationships between the nappe systems and overlying carbonate deposits of the Upper Cretaceous to Cenozoic age were studied along the southern margin of the Troodos Massif near Perapedhi village and near the villages of Kannaviou, Simou, Kathikas, Ayios Nikolaos, Akrounda, and along the Mavrokolymbos River.

PRESENT DAY STRUCTURAL ORGANIZATION OF SOUTH-WESTERN CYPRUS (A BRIEF GEOLOGICAL DESCRIPTION)

Currently, the Island of Cyprus makes up the principal segment of the Cyprus island arc, which is being underthrust by the margin of the African continent (**Figs. 5.1** and **5.2**). The present-day subduction zone dips northward and northeastward and is characterized by vigorous seismicity. It is represented by a slab clearly delineated by earthquake hypocenters. From the south, the Cyprean island arc is bounded by a well-defined trench, along which the main plate boundary runs, whereas in the north Cyprus is separated from the continental block of Turkey by a backarc basin. The modern setting of Cyprus is determined by dynamic equilibrium between the uplifting of the Troodos Massif due to pressure exerted by the subducting African margin and the compensatory action of erosion.

In the structural grain of south-western Cyprus, three major structural units are recognized:

1) The Troodos ophiolite together with its overlying sedimentary formations (Perapedhi, Kannaviou, and Kathikas) –the **subautochthon** (or **para-autochthon**);

2) The Mamonia fold-and-thrust pile (or Mamonia thrust system) – the **allochthon**, which overrides the Troodos ophiolite tectonically;

3) The Upper Cretaceous to Quaternary sedimentary deposits – the **neoautochthon**, which overlaps the entire tectonic pile.

Below is a brief description of the inner structure characteristics of these structural complexes and the relations among them based on our analysis of key localities and sections. These complexes are described in greater detail in other chapters of this monograph.

Subautochthon – The Troodos ophiolite complex

The geology of the Troodos complex is addressed in a multitude of publications because it is one of the world's best understood ophiolites (Miyashiro, 1973; Pearce, 1975; Pearce et al, 1984; Gass, 1968, 1980; Thy and Moores, 1988; Robinson and Malpas, 1990, 1998; Sobolev et al., 1993; and others). The ophiolite comprises all the basic constituents of a typical ophiolite section, the relationships among them being only slightly upset by faulting. **Fig. 5.3** depicts a schematic section of the ophiolite. From bottom to top, it consists of:

1) Tectonized mantle peridotites – mainly depleted harzburgite

2) Intrusive rocks, represented by ultramafic cumulates, isotropic gabbro, and plagiogranite,

3) The sheeted dike complex,

4) Pillow lavas, and

5) Pelagic sediments of the Perapedhi and Kannaviou Formations (Constantinou and Govet, 1972) and the olistostrome of the Kathikas Formation. Based on radiolarians, the Perapedhi Formation has been dated to the Turonian-Santonian (Blome and Irwin, 1985; Bragin and Bragina, 1993).

The pillow lavas make up a sequence a few hundred meters to 1.0-1.5 km thick, exposed along the periphery of the Troodos Massif and showing a bipartite structure (Robinson et al., 1983; Bernarz et al., 1991; Sobolev et al., 1993). The lower group is composed chiefly of low-Mg, high-Ti aphyric lavas with subordinate high-Mg, low-Ti aphyric and olivine-clinopyroxene-phyric lavas. It is interpreted as an island-arc tholeitic series ranging compositionally from picrite and basalt through to dacite. The upper group is composed mainly of a low-Ti series of picrite, basalt, and basaltic andesite corresponding in mineral composition and geochemical fingerprints to boninites. Petrologic studies of the rocks of this complex point to a cogenetic character of its constituent members, suggesting that they make up a continuous section formed in one and the same geodynamic setting (Thy and Moores, 1988; Gass, 1980; Sobolev et al., 1993). The magmatic portion of the section, based on U-Pb dates from the plagiogranites, was formed at 90-92.4 Ma (Mukasa and Ludden, 1987).

Direct relationships are commonly observable between (1) the allochthonous Mamonia fold-and-thrust pile plus the neoautochthon and (2) the upper portions of the above-described succession (pillow lavas and sedimentary deposits). South-west of the Troodos Massif, the Perapedhi Formation is overlain by

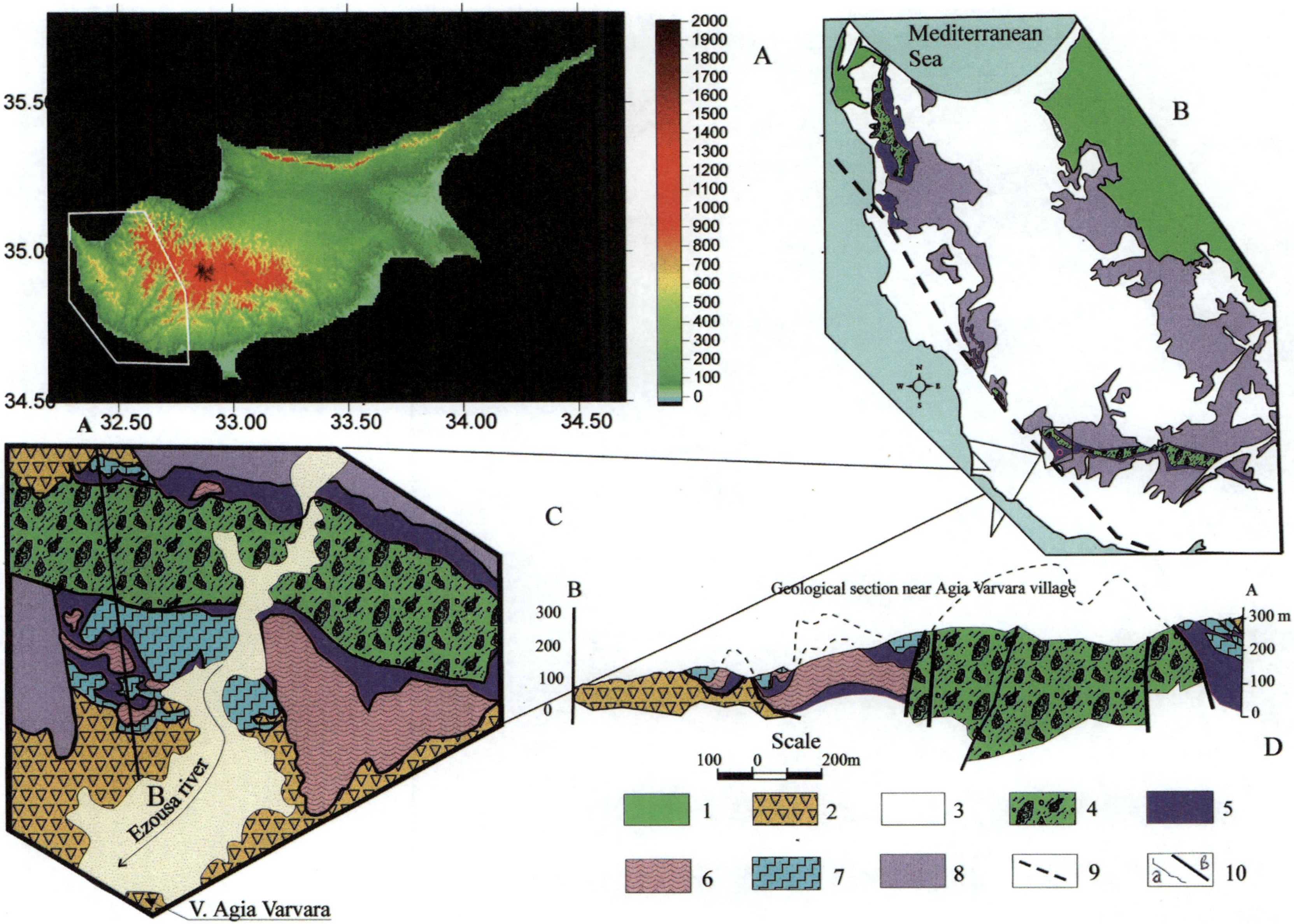

Fig. 5.1. Geological structural scheme of south-western Cyprus. Letters label: A - Numerical model for Cyprus topography; B - Tectonic scheme of south-western Cyprus; C - Geological map of the key Ayia Varvara village locality; D - Geological profile through the key Ayia Varvara village locality. Units of the Troodos ophiolitic terrane: 1 - Relative structural subautochthon; 2 - Terrigenous tectonic melanges; 3 - Neoautochthonous cover deposits; 4 - Relative subautochthon-monomictic melange after the Troodos ophiolitic rocks; 5-8 - Mamonia composite terrane: 5 - Polymictic serpentinite melange; 6 - Nappe composed of the Ayia Varvara metamorphic rocks; 7 - Nappe composed of volcanogenic, chert, and carbonate deposits of the Dhiarizos Group; 8 - Nappe composed of terrigenous and chert deposits of the Ayios Photios complex; 9 - Faults, inferred; 10 - Faults, authentic.

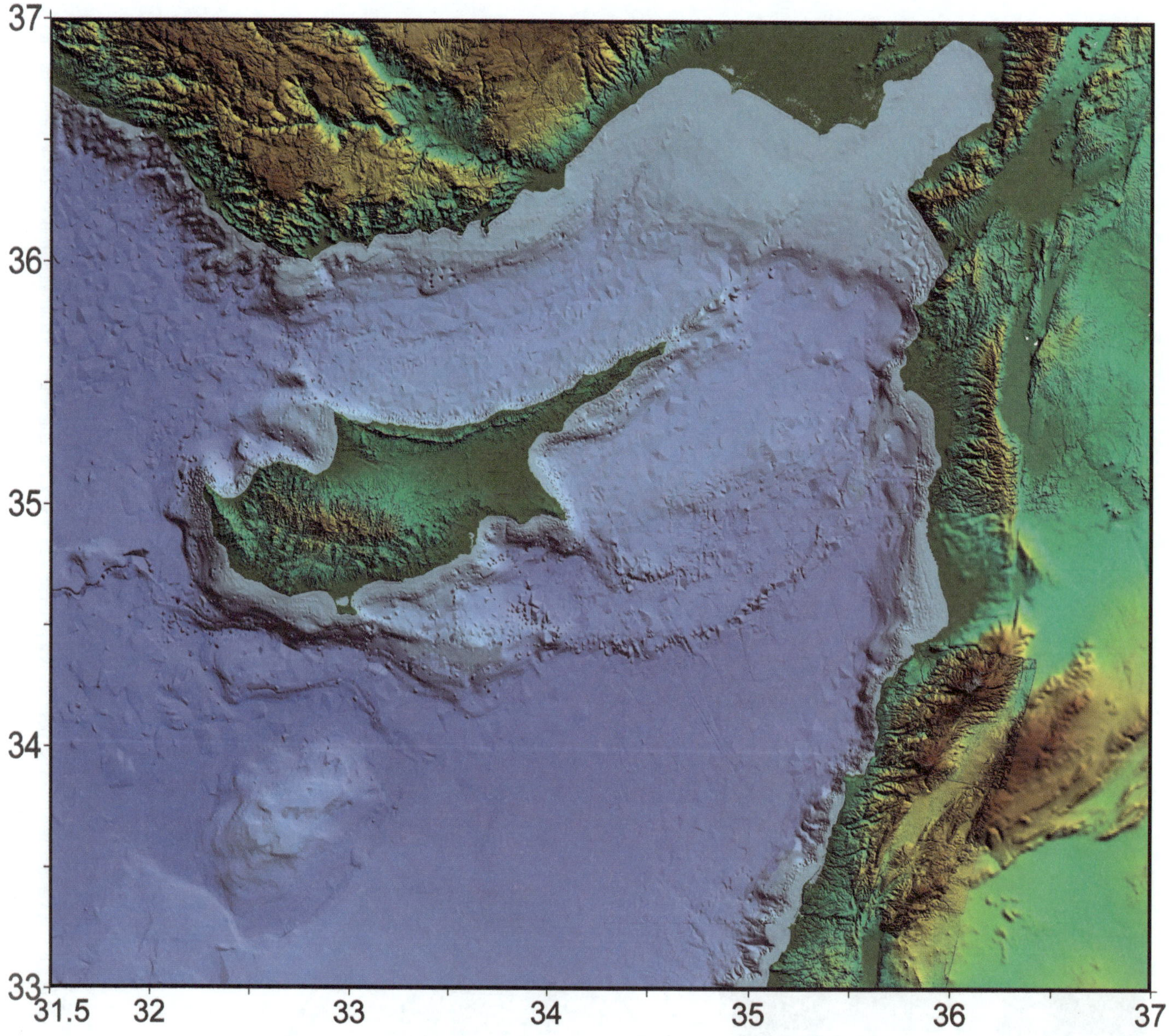

Fig. 5.2. Shaded relief of the Northeastern Mediterranean based on latest available grid models, Mercator projection, 250 m pixels. The bathymetry is based upon the digitized contours of the 1:625,000 scale chart of the Eastern Mediterranean with 50 m contours (Hall, 1994), interpolated by TIN together with all available spot soundings and data from the Strakhov cruise. Land coverage is as follows: Eurasian land areas north of 35°N are based on 90 m Shuttle Radar Terrain Mission (SRTM - Bamler, 1999) data; Israel is 25 m DTM of Hall (1997); Cyprus is 25 m DTM of Hall (1998); Lebanon is from 30 m ASTER data (Miyazaki et al., 1996), while ACE 0.5' (~900 m - Berry et al., 2002) data covers Jordan and a small part of coastal Syria.

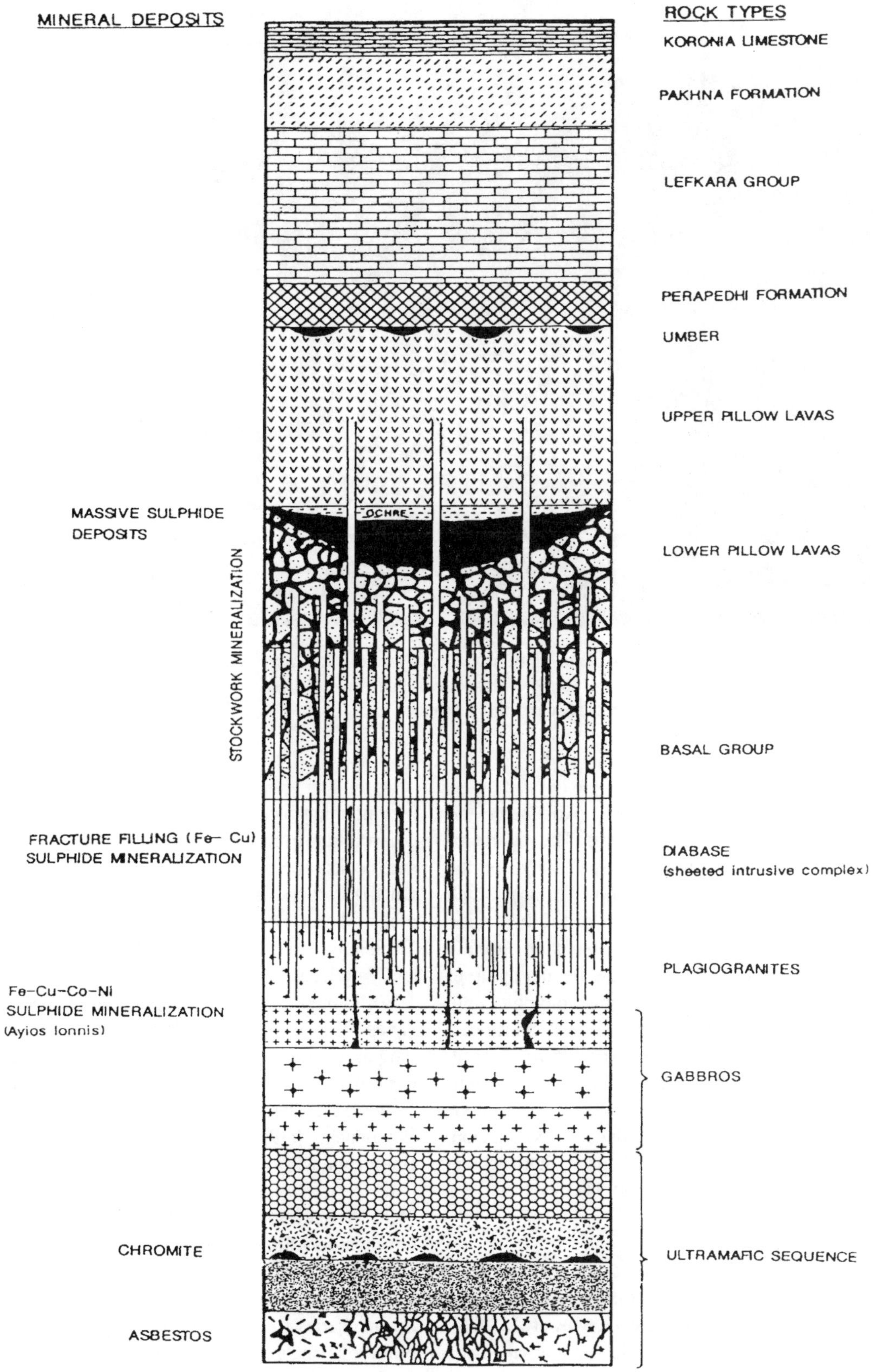

Fig. 5.3. Structure of the Troodos ophiolite massif; column after Malpas et al. (1998).

Fig. 5.4. Kathikas olistostrome near Kritou Marotto (photograph by K. Krylov). At this locality, fragments in the Kathikas olistostrome portray all the lithologies of Mamonia composite terrane. Matrix to the olistostrome is composed of both redeposited red clayey rocks after the red Mamonia mudstones and greenish gray bentonitic clays of the Kannaviou Formation.

the Kathikas Formation olistostromal deposits (Robertson and Woodcock, 1979; Swarbrick and Naylor, 1980). The matrix of the Kathikas olistostrome consists of redeposited clays of the Mavrokolymbos Formation (the Ayios Photios Group) which comprises part of the Mamonia complex, and of the bentonitic clays of the Kannaviou Formation. The abundant, usually unrounded clasts are as large as 1 m across, less frequently up to 10-15 m. They are composed of various Mamonia lithologies (*e.g.* metamorphic schists, Vlambouros limestones, Akamas sandstones, Mavrokolymbos mudstones) and Troodos ophiolitic rocks (**Fig. 5.4**). The areal distribution of the Kathikas Formation is non-uniform. These deposits are thickest (200 m) west of the Troodos Massif near Kathikas village. Near the village of Kannaviou, the lower portion of the Formation has yielded a Maastrichtian radiolarian assemblage (Bragin and Bragina, 1993).

Allochthon – The Mamonia complex

The allochthon is represented by the thrust package of the Mamonia composite terrane (**Figs. 5.5** and **5.6**) (Krylov et al., 1993). The lowermost nappe is referred to as the Ayia Varvara nappe and is composed of metamorphic rocks of the Ayia Varvara Formation with radiometric ages of 89 and 90 Ma (Spray and Rodick, 1981). The overlying slice consists of polymictic serpentinite melange incorporating blocks of ultramafics, including some that differ in composition from the Troodos ultramafics (Bazylev et al., 1993), and blocks of rocks of the Mamonia terrane (**Figs. 5.7** and **5.8**). The Dhiarizos nappe is made up of volcanic, siliceous, and carbonate rocks of the Dhiarizos Group ranging in age from the Late Triassic to mid-Cretaceous (Albian-Cenomanian?) (Robertson and Woodcock, 1979; Swarbrick, 1980; Swarbrick and Robertson, 1980; Lapierre, 1975). The uppermost nappe is composed of the Ayios Photios terrigenous/carbonate turbidites and radiolarites ranging in age from the Late Triassic to mid-Cretaceous (Albian-Cenomanian) with a very large stratigraphic break at the boundary between the Vlambouros and Episkopi Formations (Robertson and Woodcock, 1979; Swarbrick, 1980; Swarbrick, Robertson, 1980; Bragin and Krylov, 1991; Krylov, Kaleda, and Bragin, 1995).

The Neoautochthon - Sedimentary deposits of Upper Cretaceous - Pleistocene age

The neoautochthonous deposits comprise a lithologically diverse series of sedimentary formations. A coherent succession of these formations can be seen in the N-S trending basin between the Mamonia and Limassol Forest uplifts. This series includes the Lower Lefkara, Middle Lefkara, and Upper Lefkara Formations of carbonate composition (Maastrichtian-Eocene), the clays and

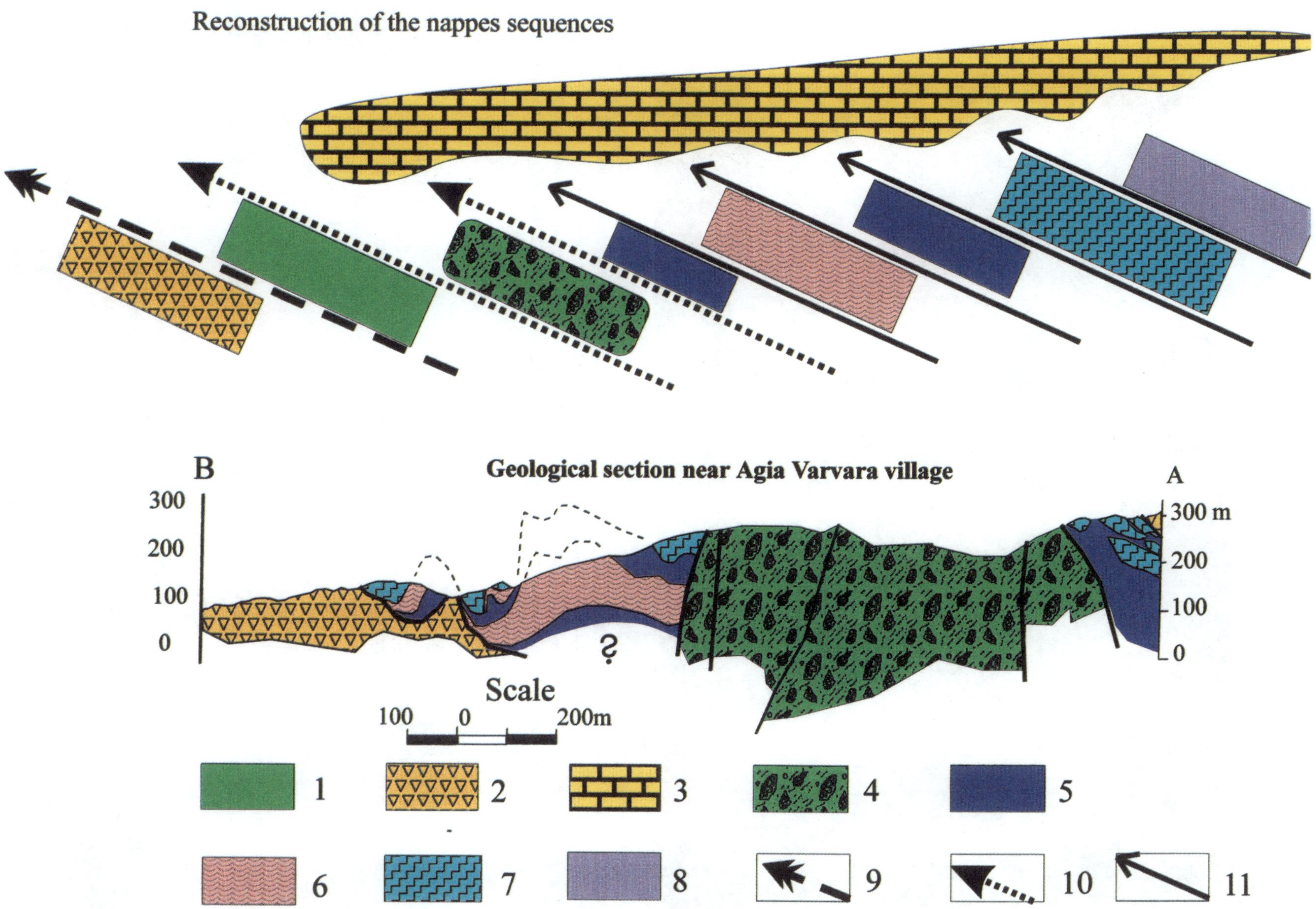

Fig. 5.5. Reconstruction of the nappe sequences in SW Cyprus. 1 - Relative subautochthon-Troodos ophiolite assemblage; 2 - Terrigenous melanges; 3 - Neoautochthonous complex; 4 - Monomictic serpentinite melange; 5 - Nappe composed of polymictic serpentinite melange; 6 - Nappe composed of Ayia Varvara metamorphics; 7 - Nappe composed of volcanogenic, chert, and carbonate deposits of the Dhiarizos Group; 8 - Nappe composed of terrigenous, chert, and carbonate deposits of the Ayios Photios Group; 9 - Latest tectonic movements (presumably related to formation of terrigenous melanges); 10 - Thrusting of the Mamonia terrane as a single thrust package onto the relative autochthon (Troodos terrane), whose upper portion is in melange; 11 - Formation of the fold-and-thrust inner structure of the Mamonia terrane.

Fig. 5.6. Relations between the Mamonia allochthon (A, B) and the Troodos autochthon (C) near Ayia Varvara village (photograph by K. Krylov). The greater part of the photograph is occupied by the nappes of the Mamonia composite terrane. The lower thin nappe is composed of a polymictic serpentinite melange (B), structurally overlain by a nappe composed of the Dhiarizos Group rocks (C). Center and bottom right, a monomictic serpentinite melange after Troodos rocks (C). Troodos extrusives and sedimentary rocks of its cover make up blocks suspended in serpentinite matrix. Contact between the Mamonia thrust package and the subautochthon on the photograph dips gently SSE. The next Figure 5.7 is eastward continuation of Fig. 5.6.

carbonates of the Kilani Formation (Oligocene), the carbonate Arsos Formation (Lower Miocene), the Pakhna Formation with carbonate deposits in the lower portion and terrigenous/carbonate in the upper (upper part of the Lower Miocene to the Tortonian stage of the Upper Miocene), the evaporitic Kalavasos Formation (Messinian stage of the Upper Miocene), and the carbonate/terrigenous Pissouri Formation (Pliocene). The total thickness of deposits is as great as 1,400 m. In general, from bottom to top the deposits change their character from deep-water lower-bathyal foraminiferal-radiolarian oozes (Maastrichtian, Paleocene, Eocene) to lagoonal and shelf sediments of Messinian and Pliocene age.

On the Mamonia uplift, an identical series of formations is exposed. These deposits, however, formed in a shallower environment; they show several hiatuses, and their total thickness is no greater than 700 m. At the base of the neoautochthonous deposits a major angular unconformity is documented, and the age of the basal autochthonous horizons varies along the strike of the structure (**Fig. 5.9**).

Terrigenous melanges

We attribute the terrigenous melanges of south-western Cyprus to a separate tectonic category. We excluded these deposits from the composite Moni melanges (Lapierre, 1975) and from the Kathikas terrigenous melange (Robertson and Woodcock, 1979) in that, in contrast to the melanges described previously, these are purely tectonic in origin and have different age (see below). Note that the study areas display deposits referred to as Kathikas melange (Robertson and Woodcock, 1979). As applied to this sequence, this term is strongly misleading because the above works advocate a clearly depositional genesis for these rocks, which is in keeping with our observations. These deposits may have originated through submarine slumping and can thus be classed with olistostromes, whereas the term ‘melange’ is more relevant in reference to chaotically structured geological bodies in which rock complexes are mixed together tectonically (Sokolov et al., 1996). With respect to Cyprus, these terms were used somewhat formally:

Fig. 5.7. A large block of highly tectonized rocks of the Dhiarizos Group (A), overlain and underlain by serpentinite (B) (photograph by K. Krylov). The contact between the serpentinite and Dhiarizos volcanogenic, chert, and carbonate deposits is tectonic and dips gently to the south. Boxed locality (2 km NE of Ayia Varvara village) is shown at a larger scale in Fig. 5.8.

bodies of tectonic origin were not distinguished from sedimentary bodies, and the former were assumed to have originated through tectonization of the latter. Our studies show that this area, with its complex fold-and-thrust structure, displays both types of deposits, intrinsically contrasting in origin. We will try to demonstrate this in further discussion. To avoid confusion, the term 'melange' should not be used in reference to sedimentary deposits, and its application should be restricted to tectonics-related entities only.

Neotectonic movements complicate the features described above, causing young faults and forming terrigenous melanges partly documented earlier (Moni, etc.). Besides, these movements reactivate the old faults and often change their sense. Thus, the Cenozoic normal fault that bounds the Polis graben in the south currently functions as a thrust fault (Makris et al., 1998).

Relationships between the subautochthonous and allochthonous complexes

These relationships were documented in several localities of south-western Cyprus.

The Ayia Varvara locality

This area has a fold-and-thrust structure modified by a system of transtensional strike-slip faults trending northwest-southeast and roughly east-west (Figs. 5.1 and 2.1). The main folds in which the thrust units are involved trend roughly E-W with dip angles on limbs ranging from 20° to 90°. Their southern limbs are often overturned, emphasizing the overall southerly convergence of the structure, although in places the relationships are reversed. The overall structure of this area can be interpreted as a major E-W striking antiform, in whose core are exposed rocks that are subautochthonous or relatively autochthonous to this area. This major antiform is modified by smaller synforms and antiforms in many places.

The lowermost structural position is occupied by the subautochthonous rocks, represented by boninitic extrusives and associated hydrothermal-sedimentary ferruginous rocks of the Perapedhi Formation and the Kannaviou Formation volcaniclastic bentonitic clays. This assemblage clearly represents disintegrated fragments of the Troodos ophiolite sequence; in some places it is mappable as a monomictic serpentinite melange. By the term

Fig. 5.8 A zoom in on the preceding photograph. Intense tectonization of basalt and radiolarite to the extent of disruption and cataclasite (photograph by K. Krylov).

'monomictic' we mean a tectonically reworked rock assemblage that has a serpentinite matrix and consists exclusively of rocks that originally represented a single ophiolite assemblage. This monomictic melange, unlike polymictic serpentinite melanges (or 'color melanges'), contains blocks with only those lithologies composing the original unreworked sequence. The lithologies of the blocks from this melange could be shown to be identical to the existing intact sequence of the Troodos ophiolite, studied by previous workers and by ourselves in the area further northward — *i.e.*, within the Troodos Massif proper (Fig. 5.3).

Subautochthonous deposits are exposed on both sides of the Ezousa River. They have a complex inner structure in which individual blocks and tectonic slices are discernible, and even individual thrust sheets can be observed. In the tectonic stratigraphy of the subautochthon proper, the lowest structural position is occupied by Troodos-type boninitic lavas and related Perapedhi and Kannaviou deposits. They are strongly tectonized and broken into numerous blocks by high-angle faults trending mostly E-W and NW-SE. The contacts between individual blocks are essentially vertical and are marked by zones of crush breccia tens of centimeters to tens of meters in width. Along certain larger fractures, serpentinite stripes are seen. Individual blocks are often rotated with respect to one another, indicating relative displacements and rotation. On the west side of the Ezousa River, the blocks composed dominantly of lavas are usually laying right-side-up. Further up the section a serpentinite melange complex occurs in which strongly tectonized serpentinites (virtually, serpentinite schists) enclose massive ultramafites that predominate. These ultramafites are composed of strongly depleted harzburgites which resemble their counterparts in the Troodos Massif (Bazylev et al., 1993). The contacts between these two 'subdivisions' within the subautochthon have different character in the northern and southern limbs of the structure.

The southern contact of the subautochthon with the overlying Mamonia nappe pile runs along a high-angle E-W trending fault. This fault has steep to subvertical dip angles on the west side of the Ezousa River, decreasing in a westerly direction along strike to acquire, 1.5-2 km further west, southerly dips of 40°-50°. Along the southern contact, the subautochthon shares a common boundary with several lithotectonic complexes of the autochthon. The fault itself is likely to be a transtensional strike-slip, relatively young in age. In places, however, primary relationships are preserved intact; thus, on the watershed between the Ezousa and Xeropothamos Rivers, the autochthon is structurally overlain by a flat lying

Fig. 5.9. Petra tou Romiou locality; neoautochthonous deposits (the Middle Lefkara Formation) overlying the Petra tou Romiou terrigenous melange (photograph by K. Krylov). Dark-colored blocks in the photograph are Triassic alkaline volcanites of the Dhiarizos Group. Reddish brown matrix formed after the Ayios Photios Group rocks. The abrupt younging of basal horizons of the neoautochthonous complex has been documented.

polymictic serpentinite melange (Fig. 5.6) enclosing minor blocks with Dhiarizos and Ayia Varvara lithologies.

By contrast, the northern contact is marked by a strip of strongly foliated serpentinites and is clearly visible on both sides of the Ezousa River; it is less steep, dipping northward at 30° to 50°. The subautochthonous rock assemblage (monomictic serpentinite melange) occurs in contact with a polymictic serpentinite melange. The northern contact between these two 'varieties' (in fact, thrust slices) dips consistently to the north. The dip angles range from 15° to 60°, and the ultramafic slice is greatly reduced in thickness. The blocks show a preferred orientation with long axes striking NW-SE, in accordance with the overall structural grain. These blocks range in size from tens of meters to 1-1.5 km. Block surfaces frequently show slickensides and tectonic striations. Near their contact with the matrix, the ultramafites are strongly foliated, becoming more intensely serpentinized, and grading into serpentinite schists, with all the primary minerals disappearing (Bazylev et al., 1993). At the Ayia Varvara locality the subautochthon as a whole represents a major asymmetric antiform with a gentle northern and a more steep and faulted southern limb. The subautochthon proper is split into individual slices. The serpentinized ultramafites (the ultramafite slice) occurring structurally above the lava sequence and its overlying sedimentary rocks. This relationship confirms that in those areas where it is in contact with the Mamonia terrane nappes the subautochthon is tectonically stratified. Based on the above data, we propose that the subautochthon should be defined as a subautochthonous monomictic serpentinite melange composed of Troodos lithologies.

The relationships between the various slices within the Mamonia nappe pile are also clearly seen near Ayia Varvara village. Thus, on the west side of the Ezousa River, 400 m away from the channel and 1.5 km upstream from Ayia Varvara, a large antiform feature occurs whose core is made up of metamorphic rocks of the Ayia Varvara sequence. Along most of its periphery, it is rimmed by a polymictic serpentinite melange (with serpentinite matrix) containing blocks consisting of various fragments of the dismembered ophiolite suite. This suite consists of lherzolites, pyroxenites, harzburgites, basalts, and their overlying sedimentary rocks (limestone and radiolarite) derived from the Dhiarizos Group, whose matrix markedly predominates over the clasts. This polymictic melange overlies the metamorphic rocks at an angle of 20° to 50°, which is especially evident on the eastern periclinal closure of the antiform structure (Fig. 5.1).

A likely continuation of this structural feature is another narrow antiform structure, about 200 m long, exposed closer to the channel. It has a steep periclinal closure in the west, where the contact of the amphibolites and serpentinites plunges beneath the serpentinites at an angle of 70°. Its northern contact is shallower; here the amphibolites are directly overlain by nearly horizontal micritic limestones of the Dhiarizos Group. The southern contact

Fig. 5.10. The Petra tou Romiou terrigenous melange (photograph by K. Krylov). Photograph shows clearly a mixture of two types of matrix: reddish brown after the Ayios Photios Group rocks (Mamonia terrane) and greenish gray after the Kannaviou Formation clays (Troodos terrane). The sequence is strongly tectonized and contains numerous shear zones. The clastics consist of diverse lithologies of the Mamonia terrane and include serpentinized ultramafites.

of the amphibolite is also structural and dips southward beneath the Dhiarizos basalts.

Similar relationships between the polymictic melange and the metamorphic rocks are observed on the east side of the Ezousa River as well. There lies the Ayia Varvara block, the largest metamorphic block in Cyprus. At its eastern periclinal closure it dips underneath a polymictic serpentinite melange at angles of 20° to 40° which become steeper (>50°) along the northern contact (Fig. 5.1).

The fabric of the metamorphic rocks is rather intricate and has been thoroughly studied (Spray and Roddick, 1981). The rocks display several generations of folds – F_1 and F_2 – with related planar mineral orientations. The orientation of principal fabrics inside the Ayia Varvara metamorphic block is observed to be dramatically discordant to the overall structural grain, reflecting earlier stages of tectonic movements which predated formation of the entire nappe pile. Note that typically the metamorphic rocks do not make up large exposures, but constitute blocks incorporated into the polymictic serpentinite melange, in which they occupy the lower levels. These relationships suggest that the polymictic serpentinite melange occurs higher in the tectonic stratigraphy than the Ayia Varvara metamorphic sequence.

An even higher structural position is occupied by the nappe composed of the undifferentiated sequence of volcanics, carbonate rocks, and cherts of the Dhiarizos Group. Relations between the Dhiarizos nappe and the polymictic serpentinite melange are frequently complicated because the Dhiarizos rocks are themselves incorporated into the melange. It is often hard to decide whether one is dealing with an outlier of a nappe or a melange-hosted block directly unrelated to the nappe. For this reason, the lower boundary of the Dhiarizos nappe is somewhat arbitrary (Fig. 5.8 and **Fig. 5.10**)

The detailed mapping of the Dhiarizos nappe showed that it cannot be represented as a single complex. Our studies reinforce the previous conclusion that the Dhiarizos Group displays at least two or three distinctive types of stratigraphy. One type comprises alkali basalts associated with reefal piles or reef-derived clastic sediments – the Fasoula Sequence of Malpas et al. (1998). The other type of stratigraphy of the Dhiarizos Group was documented and paleontologically dated by us in the vicinity of Ayia Varvara village. It is represented by a fixed succession of basalts overlain by thinly laminated pelagic sediments 5-6 m thick. These pass into pure cherts-red radiolarites via a zone of alternating limestone and gray chert (6-10 m thick). The limestones are dated as Triassic, whereas the red radiolarites in this particular block yielded no faunal ages, but elsewhere they yielded Late Jurassic-Early Cretaceous radiolarians. Consequently, we are dealing here with coeval but lithologically diversified deposits, as well as deposits of

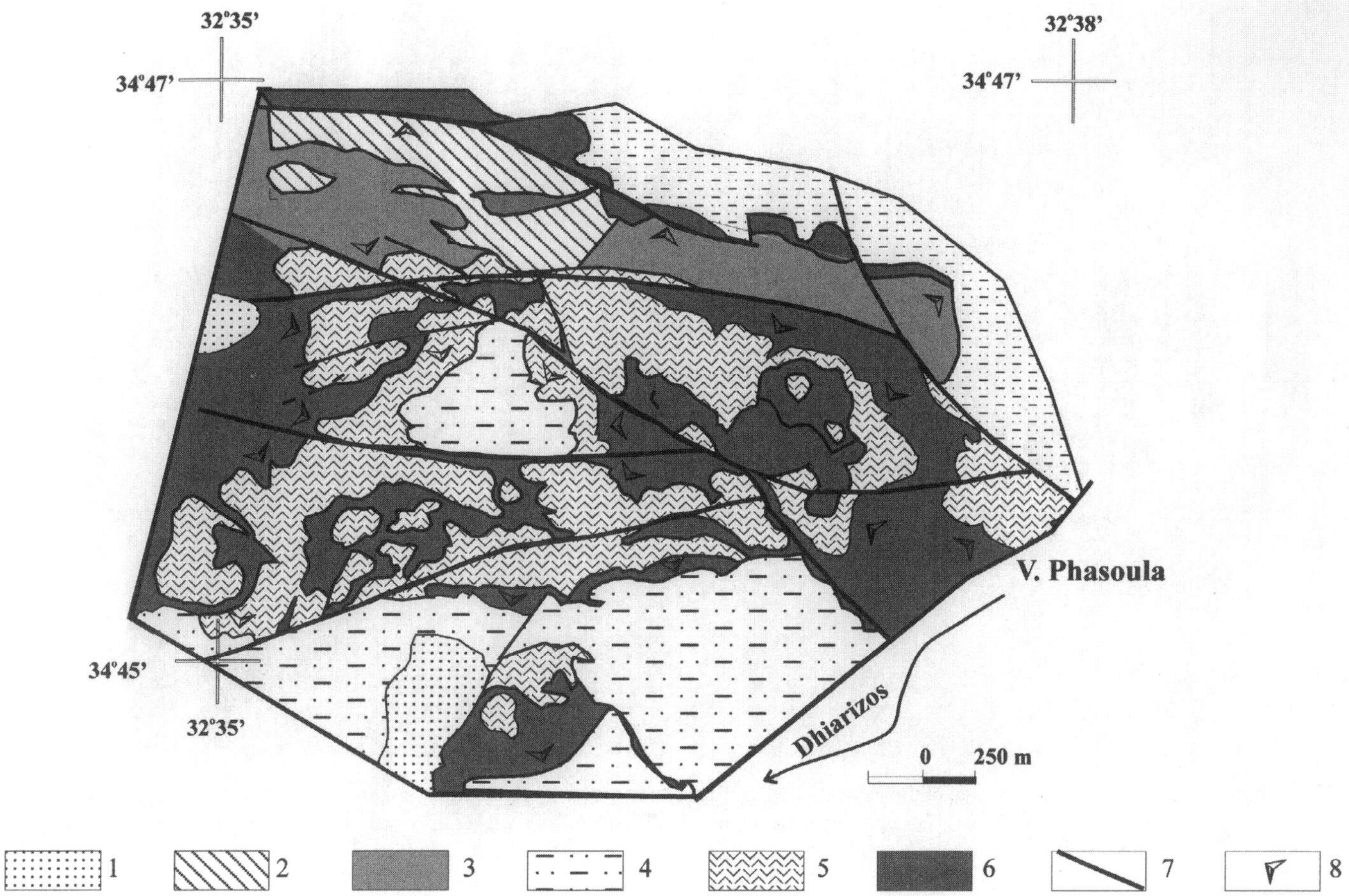

Fig. 5.11. Tectonic sketch of the interfluve between the Dhiarizos and Xeropothamos Rivers. Modified from (Sharaskin et al., 1995). Symbols: 1 - Neoautochthon-Late Cretaceous to Cenozoic deposits, undifferentiated; 2-4 - Allochthon: 2 - Polymictic serpentinite melange; 3 - Nappe composed of volcanogenic, siliceous, and carbonate rocks of the Dhiarizos Group; 4 - Nappe composed of terrigenous, siliceous, and carbonate rocks of the Ayios Photios Group; 5-6 - Subautochthonous complex: 5 - Monomictic serpentinite melange, 6 - Sedimentary deposits of the Troodos ophiolite assemblage; 7 - Faults; 8 - Dips in the serpentinite melange.

various ages, all grouped within the Dhiarizos Group. This Group is characterized by intense tectonization which has produced sequences resembling autoclastic melanges, such as the one on the east side of the Ezousa River. Here, in a road cut leading to a pit, a spectacular example is found of a Triassic limestone tectonically incorporated into Late Jurassic-Early Cretaceous cherts. Similar relationships are observed throughout the area (Figs. 5.7 and 5.8). The uppermost structural position is occupied by a nappe composed of terrigenous, carbonate, and chert deposits of the Ayios Photios Formation.

The Dhiarizos – Xeropotamos interfluve

Geological mapping of the interfluve area between the Dhiarizos and Xeropothamos Rivers was carried out by A. Ya. Sharaskin, T. N. Kheraskova, and Yu. V. Karyakin from the Geological Institute (**Fig. 5.11**) and was described by Knipper and Sharaskin (1995). Krylov and Silantyev made only a few isolated traverses.

These investigations revealed a similar structural setup in which the relative autochthon to the deformed Mamonia thrust package is expressed as an antiform, exposed in the north and undulating in a N-S direction. Its core is composed of strongly altered boninitic pillow lavas similar to the Troodos boninites and their overlying Kannaviou green clays. The subautochthon comprises terrigenous mixtites of sedimentary origin hosting basaltic, serpentinitic, and limestone clasts derived from the Dhiarizos Group, and sandstones from the Vlambouros Formation. Their composition, appearance, and definitely sedimentary origin suggests a resemblance with the Kathikas mixtites. As in the vicinity of Ayia Varvara, the subautochthon in the Dhiarizos-Xeropothamos interfluve is overlain by a polymictic serpentinite melange incorporating blocks of the Ayia Varvara metamorphics and Dhiarizos rocks. Only in some localities along the northern boundary does the subautochthon occur in tectonic contact with the rocks of the Ayios Photios Group, which comprise the upper nappe. The structural assemblage of this locality lacks an independent

Fig. 5.12. Perspective view on the right side of the Mavrokolymbos River (photograph by K. Krylov). Polymictic serpentinite melange (A) overlying rocks of the relative subautochthon (B). The nappe that is structurally higher still (C) and makes up a synform, is composed of deposits of the Ayios Photios Group. Arrow indicates the area shown in Fig. 5.13.

nappe composed of the Ayia Varvara metamorphic rocks. The remaining members of the thrust package are present, and their order is the same, suggesting the same causative mechanism for the structural piles in both the Dhiarizos-Xeropothamos interfluve and near Ayia Varvara village.

The Mavrokolymbos River area

In this area the subautochthonous complex, largely in melange, is exposed below the dam in the lower part of the Mavrokolymbos valley. It is chiefly represented by boninitic lavas and their overlying sediments of the Perapedhi and Kannaviou Formations. The subautochthon is strongly tectonized and consists of blocks showing rotation relative to one another. Along a low-angle north-dipping thrust, this assemblage is overridden by a serpentinite melange containing blocks of the Dhiarizos Formation (directly near the dam; Swarbrick, 1993). In this locality, the serpentinites dominate, accounting for more than 50% of the melange by volume. On the west side of the Mavrokolymbos River, the contact between the subautochthon and serpentinite melange runs along a thrust (or a system of thrusts) (**Figs. 5.12, 5.13** and **5.14**), whose fault plane is gently folded with dips ranging from 40° toward the north to 10° to the south. This locality is one of the most difficult to map because one type of melange overlaps the other. For a guide horizon we used the radiolarite blocks tectonically incorporated into the shear zone separating the two types of melange (Fig. 5.14).

On the west side of the Mavrokolymbos River, about 1 km downstream from the dam, subautochthonous rocks occur which share a gently dipping tectonic contact with the terrigenous, cherty, and calcareous deposits of the Ayios Photios Formation. The Ayios Photios rocks make up a gentle syncline that opens up the slope on the west side of the Mavrokolymbos valley. The core of the synform is composed of rocks of the Mavrokolymbos Formation and the limbs of rocks of the Vlambouros Formation. Therefore, in the lower part of the Mavrokolymbos valley, a tectonic window exposes strongly dislocated rocks of the Troodos Massif which are both right-side-up and overturned. The structural assemblage of this locality should be considered as a thrust package. In this area, the thrust package involves only two nappes: the lower is a polymictic serpentinite melange with blocks of Dhiarizos rocks (Swarbrick, 1993), and the upper is composed of rocks of the Ayios Photios Group.

The Akamas Peninsula and the road to Polis

In the Akamas Peninsula, moving uphill northward from Lara Bay, one observes exposures of a monomictic serpentinite melange (showing a low degree of tectonic reworking), in which the blocks are composed of Troodos-type boninitic lavas separated by isolated serpentinite strips. The serpentinite contains small blocks of

Fig. 5.13. Relations between the subautochthon and serpentinite melange on the right side of the Mavrokolymbos River (photograph by K. Krylov). Subautochthon (A) is represented by the Kannaviou Formation clays and the Perapedhi Formation umbers. Near the contact, the rocks are strongly tectonized and host tectonic inclusions of radiolarite blocks (B) (supposedly of the Dhiarizos Group) and serpentinites (C).

strongly depleted harzburgite (Bazylev et al., 1993). Some 2.5 km north of Lara Bay, a polymictic serpentinite melange with small blocks of metamorphics and Dhiarizos rocks (volcanics/cherts and volcanic/chert/carbonate deposits) overlies these boninites and serpentinites, along a contact which dips gently to the NE. In this area, although blocks made up of these rocks are present, it is no longer possible to isolate a separate thrust sheet composed of the Ayia Varvara metamorphics. This is because the blocks are up to tens of meters across, and besides, they are surrounded on all sides by serpentinite matrix. Blocks composed of Dhiarizos rocks can be mapped. The thickness of the thrust sheet composed of polymictic serpentinite melange is insignificant, usually not exceeding 70-100 m.

Structurally upsection, larger blocks occur which consist of volcanics, cherts, and carbonate rocks of the Dhiarizos Group (resembling those exposed near Ayia Varvara village). These have a structural thickness of up to 100 m. They can be conveniently interpreted as fragments of the Dhiarizos thrust sheet. These rocks are overlain by the calcareous turbidites and terrigenous rocks of the Ayios Photios Group along a gentle tectonic contact with a thin (3-4 m) zone of crush breccia. The plane of the tectonic contact is low-angle (10° to 50°), and is gently folded with widely ranging dip directions. The Ayios Photios rocks (the upper thrust sheet) make up virtually all the major elevated areas. Similar relationships are spectacularly exposed in road cuts on the way to Polis, where strongly tectonized and altered Troodos-type volcanites (boninites) are overthrust by a serpentinite melange containing amphibolite blocks and blocks of Dhiarizos volcanites, cherts, and limestones. The thrust plane is gentle, up to 45°, and is strongly folded with dip angles ranging from 0° to 45°. Higher in the structure this slice is overlapped along the same folded contact by cherts and terrigenous rocks of the Ayios Photios Group. There is no conformity between the surfaces of the upper and lower thrusts.

The relationship of the neoautochthon to the subautochthonous and allochthonous complexes

The relationship of the neoautochthonous sedimentary series to the subautochthon is the same throughout the southern slope of the Troodos Massif. In the N-S trending basin between the Mamonia and Limassol Forest uplifts, the basal neoautochthonous strata (Maastrichtian deposits of the Lower Lefkara Formation) rest

Fig. 5.14. One type of serpentinite melange overlying the other; right side of the Mavrokolymbos River (photograph by K. Krylov). Shear zone is marked by radiolarite blocks that are incorporated into it tectonically. The upper serpentinite melange is polymictic, with Dhiarizos Group rocks; the lower is monomictic and contains Troodos ophiolitic rocks. The fault plane is gently folded.

with apparent conformity on clays of the Kannaviou Formation (*viz.* the sections near the villages of Perapedhi, Kouka, Trimiklini). Unfortunately, here the rocks of the Kannaviou and Lower Lefkara Formations are obscured due to overgrowth and orchards, so the contact itself cannot be clearly observed. However, the top of the Kannaviou Formation and the base of the Lower Lefkara Formation yield virtually identical, abundant radiolarian assemblages of the *Amphipyndax tylotus* Zone, a fact that precludes any significant stratigraphic break or unconformity between the subautochthon and neoautochthon.

Further west, on the southern margins of the Troodos Massif where Mamonia rocks are present, the subautochthon and neoautochthon show a different kind of relationship. In the Ayios Nikolaos section, the Kannaviou Formation is overlain by the Lower Lefkara Formation, but the basal strata of the latter belong to the Upper Paleocene *Morozovella velascoensis* Zone – *i.e.*, the Maastrichtian and nearly all the Paleocene are missing. Besides, a minor angular unconformity must exist between the Kannaviou and Lower Lefkara Formations. Still further west, near Pano Panayia village, the subautochthon and neoautochthon are also separated by an unconformity. There, the Lower Lefkara Formation commences with Maastrichtian strata, but limestones of this age erosionally overlie the thin (2.5-30 cm) reddish brown umbers of the Perapedhi Formation, preserved in paleo-topographic lows (filled with pillow lavas). Therefore, the Kannaviou Formation has been destroyed by erosion.

The allochthonous rocks are overlapped by the neoautochthon with a regional unconformity. Near Phiti village, limestones of the *Globotruncana aegyptiaca* Zone (the upper part of the Lower Maastrichtian, the Lower Lefkara Formation) rest on the Kathikas mixtites. Similar relationships are observed near Kathikas village, where pink limestones of the *Globotruncana aegyptiaca* Zone erosionally overlie the Kathikas mixtites. The Maastrichtian limestones are in turn erosionally overlain by white limestones of Middle Eocene age. The Maastrichtian and Eocene rocks, whose total thickness is only a few meters, form a recumbent slump fold. Following heavy erosion, they were overlain by limestones of the Lower Miocene *Globigerinatella insueta-Globigerinoides trilobus* Zone. Near Simou village, the Kathikas limestone gives way to the limestones of the *Gansserina gansseri* Zone (the lower

Fig. 5.15. Ophiolitoclastic rocks of Early Jurassic age from the outcrop area of polymictic melange: direct evidence for the existence of 'pre-Troodos' ophiolites (photograph by K. Krylov). These rocks were encountered in Akamas Peninsula, where they are overlain by radiolarites of the Early Jurassic age. Lithologically similar rocks were encountered in the Louthra tis Aphroditis area, yet their poor state of preservation precludes reliable dating.

part of the Upper Maastrichtian), and near Drimou village, to the limestones of the *Acarinina uncinata* Zone of the Danian Stage. Unfortunately, near Phiti, Simou, and Drimou, the Lower Lefkara limestone outcrops are separated from the underlying Kathikas deposit outcrops by a narrow area of overgrowth, so that a thin stratigraphic interval may have been overlooked.

In the south-west of the Mamonia uplift, the break between the allochthonous and neoautochthonous rocks is considerable. In a riverside bluff near Petra tou Romiou, a sequence of limestone and chert of the Lower Eocene *Acarinina pentacamerata* Zone (the Middle Lefkara Formation), occurs above Triassic rocks. In the section along the Mavrokolymbos River (Mylos tou Mavrokolymbos, near Kili village), above the sequence of the typical Kathikas mixtites, limestones of the Middle Eocene *Morozovella lehneri* Zone (the Upper Lefkara Formation) overlie a 4.5 m thick gravelstone member .

Therefore, the Kathikas mixtites (submarine slump deposits) are confined geographically to within the Mamonia complex — *i.e.*, they are intimately related to the allochthon. Since they have yielded Maastrichtian radiolarians, these rocks must laterally replace the basal strata of the Maastrichtian portion of the Lower Lefkara Formation, which is attributed to the neoautochthon. Near Ayios Dymitrianos village (a large slump basin), the section begins with a hard tuffaceous sandstone, followed by Kannaviou clays and bentonitic mudstones. These are in turn overlain by the Kathikas red and brown mixtites which are associated here with the subautochthonous rocks. These intricate relations of the submarine slump deposits with the subautochthon, allochthon, and neoautochthon likely result from a complex history of structural, depositional, and geodynamic events in south-western Cyprus.

Tectonic interpretation

In south-western Cyprus, the Troodos ophiolites, together with their overlying pelagic Perapedhi and Kannaviou sediments and the olistostromal Kathikas Formation, are exposed in the cores of southward plunging synclines. In the area of the Akamas Peninsula and Ayia Varvara village (Fig. 5.1), the entire rock assemblage is intensely deformed, fractured into a series of slices, and frequently converted into melange near the contact with the allochthon.

The lower nappe in the Mamonia thrust package is composed of metamorphic rocks of the Ayia Varvara Sequence. This nappe was defined as an independent structural unit, based on geological mapping as well as a study of the metamorphic overprint on these rocks. According to S. Silantyev's data (1993), the P-T conditions of metamorphism correspond to 6-8 kbars and 600-800°C. These rocks could not have coexisted with the rest of the lithotectonic complexes making up the other thrust sheets. This is because none of those rocks display metamorphic grades that high (except for the rare finds of amphibole lherzolites, which were described above and by Bazylev et al., 1993). Thus this whole structural unit was formed under similar conditions as evidenced by the fact that all the metamorphic rocks sampled by S. Silantyev (1993) from various outcrops showed identical metamorphic grades. Besides, geochemical studies of the Ayia Varvara rocks discovered that metamorphism is chiefly confined to volcanic rocks of the MORB and West Philippine Basin (WPB) affinities and, to a lesser degree, to sedimentary rocks (cherts and carbonates). This led S. Silantyev (1993) to conclude that these rocks may have belonged to the Dhiarizos Group. Apparently, having been metamorphosed at the same depth and at similar temperatures, this sequence was emplaced into the upper crust as a single slab (possibly together with the amphibole lherzolites) and separately from, yet virtually simultaneously with, the rest of the nappes. This is evidenced by the radiometric age of the rocks (Spray and Rodick, 1981), which we attribute to the time of cooling and the closure of the Ar system.

Detailed geological mapping delineated two serpentinite melanges in this area, of which the polymictic melange belongs to the thrust package. Recognition of that type of melange is critical because of the greater lithologic diversity of the blocks within it. These include metamorphic rocks, various volcanites, cherts, and limestones from the Dhiarizos Group, as well as occasional blocks of Troodos-type boninites, and even sandstones resembling the Akamas sandstones are found (which we attribute to the Ayios Photios Group).

In addition, recent data (Bazylev et al., 1993) for the mapped area of the Mamonia complex show that serpentinites enclose 'outcrops' of very peculiar ultramafic rocks, never reported in Cyprus. These are spinel lherzolites and amphibole lherzolites of high metamorphic grade. Geochemical and mineralogical studies show these rocks to strongly differ from the typical Troodos rocks, which are commonly represented by very depleted lherzolites (Bazylev et al., 1993). The spinel lherzolites and high-grade amphibole lherzolites, judging from their mineralogic and geochemical peculiarities, have not undergone any significant degree of melting. Mineralogically, the spinel lherzolites and high-grade amphibole lherzolites are identical. This suggests they are refractory residues of enriched within-plate or MORB-like melts. Enriched lherzolites of this type have not been reported from the Troodos Massif. Therefore, the Mamonia complex, along with Troodos-type peridotites, includes mantle restites that were not described for the Cyprean ophiolite assemblage (Bazylev et al., 1993). These ultramafites are strongly serpentinized and were converted to melange. They might represent a fragment of the oceanic lithosphere of the Neotethys, and therefore are not genetically linked to the Troodos Massif. In other words, they can be considered as refractory residues of melts which produced the Dhiarizos volcanites (MORB and WPB).

The Mamonia complex also contains mixtites consisting of ultramafic clasts which are overlain by Early Jurassic radiolarites. This relationship explicitly suggests the existence of pre-Troodos ultramafites. These are spectacularly manifested near Louthra tis Aphroditis and in the central part of the Akamas Peninsula (**Fig. 5.15**).

Unfortunately, until now no detailed mapping showing the precise location of all the ultramafites types has been done there. Hence, in a number of cases (*e.g.*, near the Mavrokolymbos River), we could not confidently draw the boundary between the different types of melange proposed in this work. The Law of Contraries suggests that since one of these types has only been encountered within the Mamonia Complex, and is lacking in the Troodos Massif, it should be classed with the second, polymictic melange, type. This polymictic serpentinite melange clearly implies that, rocks of different ages, and of various groups and nappes, were tectonically mixed together. All these observations compel us to propose the existence in Cyprus of an independent nappe composed of a polymictic serpentinite melange. One should also take into account the data on the Mamonia amphibole lherzolites (Bazylev et al., 1993), which are similar to the Ayia Varvara metamorphics in terms of their formative metamorphic conditions. It seems likely that in the future, the polymictic melange will itself be divided into two separate complexes.

The serpentinite melanges are developed mainly along the boundary between the autochthon and allochthon. The largest melange terranes occur near the Akamas Peninsula and Ayia Varvara village. Detailed mapping in the Ayia Varvara area resulted in the identification of two melange types, each forming an independent thrust sheet, and each having a distinctive lithology for its blocks and structural position.

The structurally lower monomictic melange occurs in the upper portion of the allochthon (Krylov et al., 1993). It has a serpentinite matrix enclosing blocks of boninites, Perapedhi hydrothermal-sedimentary rocks, Kannaviou green clays, and ultramafites. In the vicinity of Ayia Varvara village, the monomictic serpentinite melange makes up a large asymmetric syncline with a gentle northern limb and a more steep, faulted southern limb. The disrupted blocks can be reassembled into a coherent section identical to the Troodos ophiolite (Gass, 1980; Thy and Moores, 1988; Sobolev et al., 1993; Bazylev et al., 1993; Knipper and Sharaskin, 1994). Compositionally, the ultramafites correspond to the strongly depleted Troodos harzburgites (Bazylev et al., 1993; Gass, 1980; Thy and Moores, 1988; Sobolev et al., 1993). The blocks show preferential orientations with their long axes parallel to the overall trend of the structure. The contacts between individual blocks are near vertical and are marked by zones of intense crush brecciation several meters to several tens of meters wide. This type of melange, resulting from disintegration of a once continuous ophiolite assemblage, can be viewed as a monomictic serpentinite melange (Ruzhentsev, 1976). It contains no exotic blocks, unlike the polymictic serpentinite melange occupying a higher structural position (Krylov et al., 1993, 1995).

In the Mamonia thrust complex another type of melange is recognized, which has a polymictic composition and is confined to the base of the allochthon (Fig. 5.5). It contains blocks of Ayia Varvara amphibolites, Dhiarizos volcanites, limestones, and cherts, and Ayios Photios sedimentary rocks. The melange sheet is estimated to be 70-100 m thick. The ultramafites are strongly serpentinized and converted to melange, represented by spinel lherzolites and high-grade amphibole lherzolites. Peridotites similar to the Troodos harzburgites occur as well. The spinel lherzolites differ from the Troodos ultramafites both geochemically and mineralogically. They are interpreted as mantle restites of enriched intraplate melts or MORB-like melts (Bazylev et al., 1993). In terms of their metamorphic parameters, the amphibole lherzolites differ from the Troodos ultramafites, and resemble the Ayia Varvara amphibolites.

In conclusion, we are compelled to recognize in Cyprus a separate nappe composed of a polymictic serpentinite melange.

Serpentinite melanges are developed mainly in the southwestern part of Cyprus (Krylov et al., 1993, 1995; Malpas et al., 1992; Knipper and Sharaskin, 1995; Sharaskin et al., 1995). It was not until recently that serpentinite melanges began to be identified as independent structural units (Krylov et al., 1995, 1993; Knipper and Sharaskin, 1995), whereas previously they had been regarded as serpentinite protrusions (Turner, 1973; Lapierre, 1975; Swarbrick, 1980, 1993; Malpas et al., 1992). This area displays a complex fold-and-thrust assemblage modified by a system of transtensional strike-slip faults trending NW-SE and roughly E-W (Figs. 5.1, 5.5, and 5.11).

In the thrust package of southwestern Cyprus, as mentioned above, two sheets of serpentinite melange are recognized. The lower structural position is occupied by a monomictic serpentinite melange formed through disintegration of the Troodos ophiolites. This melange was formed through thrusting of the whole Mamonia thrust package over the Troodos ophiolite terrane (of supra-subduction zone affinity), but the monomictic melange *per se* should not be viewed as part of the Mamonia terrane. The higher structural position is occupied by a polymictic serpentinite melange related to formation of the thrust package of the Mamonia composite terrane during accretion.

The diverse serpentinite melanges have been formed at different times and out of rock complexes originating from a variety of geodynamic settings. Thus, the polymictic melange related to the accretionary event is restricted to the late Albian-Paleocene time interval, whereas the monomictic melange is related to the collisional stage (Krylov et al., 1993, 1995) and was formed in Maastrichtian-Early Paleogene time. In those localities where the polymictic melange rests directly on the monomictic melange (the Mavrokolymbos River area), one may get the incorrect impression that it is a single geological body. Detailed large-scale mapping resolved an orderly internal structure, but smaller scale mapping leaves the two melange slices as a single feature. This is the reason why on the published maps, the rock complexes originating from contrasting geodynamic settings are united, spuriously, into a single structural unit (Swarbrick, 1993).

Therefore, the Cyprean structured melanges originated during accretion of the Mamonia terrane onto the supra-subduction zone of the Troodos Massif, and ultimately, through tectonic juxtaposition of these terranes. The monomictic melange originated through disintegration of the Troodos ophiolite during collision of the Troodos microplate with the Mamonia composite terrane (Krylov et al., 1995) and the polymictic melange, through accretion of the oceanic complexes of Meso-Tethys and its southern surroundings.

Palinspastic reconstructions

In restoring the geological history, it is vital to perform palinspastic reconstructions – *i.e.*, to successively remove the results of all the movements, from younger to older, and then to restore the Mesozoic rock complexes to their original relations.

In performing the palinspastic reconstructions, the effect of the opening of the back-arc basin should first of all be undone. This basin opened in Cenozoic time. Reliable datings for this event are scanty. Magnetic lineations in that area have not been established with any degree of confidence, and correlation with the magnetostratigraphic scale would hardly be feasible. Therefore the seismostratigraphic and gravimetric data of Makris and Wang (1994) and Krasheninnikov and Hall (1994) are of utmost importance. By analyzing these data, and based upon the morphology, one can confidently assume a spreading process and date its onset to the end of the Paleogene or to Neogene time. Opening of the Cilicia-Adana basin resulted in southward drift of Cyprus with respect to the relative reference frame – *i.e.*, to the Turkey block. Evidence exists for the presence of the oceanic crust in the Antalya basin (Robertson et al., 1998) (Fig. 5.1). In later reconstructions (Robertson et al., 1998), the inception of subduction beneath Cyprus is attributed to the Miocene, and opening of the Cilicia-Adana basin can thus be interpreted as being due to back-arc spreading that began in Miocene time. By compensating for these movements, we arrive at the conclusion that prior to their onset, Cyprus was part of the continental margin of the Turkey block. This inference is supported by the similarity of the facies characteristics of the rocks of the Kyrenia Range in Cyprus to the coeval rocks of Antalya Bay (Robertson et al., 1998). At present the crust in the Cilicia-Adana back-arc basin is either continental (in the west) or, in places, quasi-continental or else attenuated continental (in the east). No detailed mapping across the area of this basin was ever done. Our analysis is based on the presence of sedimentary bodies of considerable thickness and on the interpretation of the gravity field.

Palinspastic reconstructions for this stage should take into account partial consumption of the base of the African continental margin in the subduction zone dipping beneath Cyprus, in compliance with the kinematic scenario proposed for the African plate. This reconstruction must compensate for the northerly drift

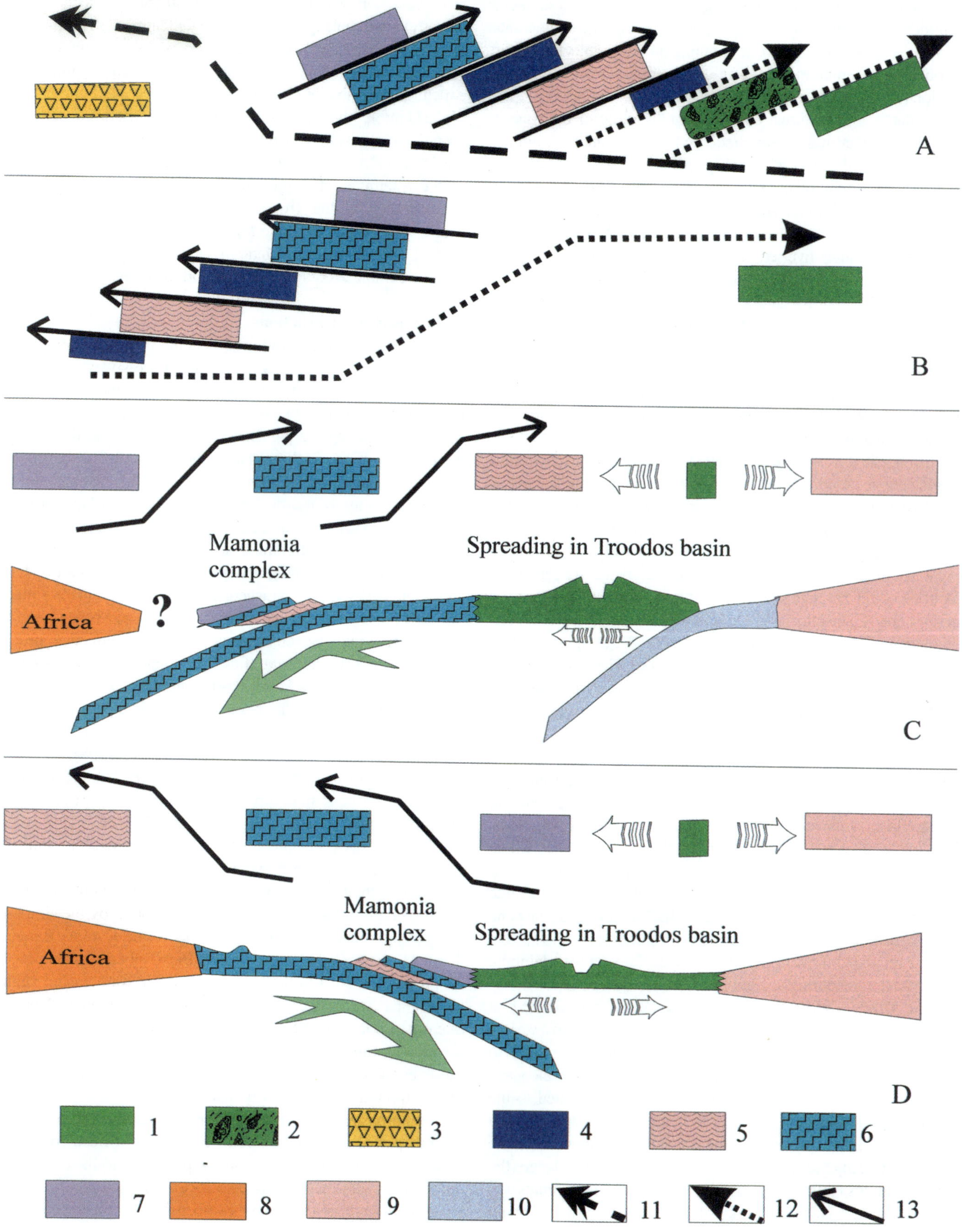
A
B
Mamonia
complex
Spreading in Troodos basin
Africa
?
C
Mamonia
complex
Spreading in Troodos basin
Africa
D
1
2
3
4
5
6
7
8
9
10
11
12
13

of the African continent and for the closure of the southern branch of the Tethys (Robertson et al., 1998; Robertson, 1998a, b; Savostin et al., 1986). Note that in the latest work of A. L. Knipper and A. Ya. Sharaskin (1995), these motions were not taken into account. Realistic assessments of the amount of this drift suggest that a few hundreds of kilometers of the African continental margin must have been consumed beneath the Cyprus arc.

Embarking on reconstructions for the Mesozoic, it is vital to restore the order of events and motions in this region. The most important of these were:

1) Juxtaposition of the Troodos and Mamonia terranes.

2) Rotation of the Troodos microplate with concurrent formation of olistostromes (the beginning of collision of the Troodos and Mamonia terranes).

3) The Troodos sea-floor spreading and magmatism in an 'oceanic-type' basin.

4) Formation of the independent Mamonia thrust package as a structural accretionary ensemble above a subduction zone, and metamorphism of the Ayia Varvara complexes and lherzolites.

5) Inception of a subduction zone (preceding that of the present-day), above which the Troodos basin with its peculiar magmatism was eventually to form. Beginning of accretion of various portions of the Meso-Tethyan crust, currently incorporated in various lithotectonic complexes of the Mamonia composite terrane.

6) Sea-floor spreading in the southern portion of the Meso-Tethys and opening of an oceanic-type basin, whose fragments are documented in the lithotectonic complexes of the Mamonia composite terrane.

As indicated in the above list, the motions were of a repeated character, and straightforward attempts to unfold the nappes would be of little value. The unfolding of the Mamonia thrust package should be done in two stages: first, the more recent motions are to be undone (the juxtaposition of Troodos and Mamonia) and then the earlier ones, which gave rise to the Mamonia fold-and-thrust package. The record of palinspastic reconstructions shows that, for the earlier stages, structural criteria such as convergence are not reliable because they may have been upset by subsequent motions. In establishing the relative motions of the Mamonia and Troodos terranes, it should be kept in mind that the spreading axis in the Troodos basin trended roughly E-W, based on paleomagnetic evidence (Clube and Robertson, 1986). This caused the basin to grow (expand) in a N-S direction, and ophiolites of similar age and geodynamic affinity are reported from elsewhere in the eastern Mediterranean as well (Turkey, Syria, and further on into Oman). Rock complexes resembling those comprising the Mamonia composite terrane exist throughout the length of the ophiolitic strip, where they occur mostly south of 'Troodos-type' ophiolites. North of this strip, the Kyrenia island-arc deposits have existed since Maastrichtian time, indirectly defining the polarity of subduction, which must have plunged to the north. Precise paleomagnetic measurements to evaluate block rotation are available only for the rocks of the Troodos terrane (Clube and Robertson, 1986; MacLeod, 1990; Varga and Moores, 1985), but not for the Mamonia terrane.

Conceivably, mutual position of the Mamonia and Troodos terranes roughly preceded the current one. It thus follows that even as the Mamonia terrane was being overthrust onto the Troodos terrane or, conversely, as the Kizil Dagh and Baer Bassit ophiolites in Syria were being thrust onto sedimentary sequences, these complexes were converging in a roughly N-S direction, with Mamonia rocks originally having occurred south of Troodos rocks. This agrees well with the vector of motion of the African plate (Savostin et al., 1986). The rotation of the Troodos microplate itself occurs at the moment of collision with the Mamonia terrane, and oblique collision should be assumed, based on real analogies (Harbert et al., 1990). Unfolding the current Cyprean structural pile restores the following features along a profile from south to north for pre-Maastrichtian time: Africa-African continental margin – southern branch of the Neo-Tethys - subduction zone – accretionary prism (Mamonia composite terrane) – Troodos supra–subduction zone basin – southern continental margin of the Turkey block (Antalya and Alania microcontinents) incorporating the Kyrenia terrane.

The formation of the Mamonia thrust package and spreading in the Troodos basin were closely synchronous; in any case the age determinations for these events show an overlap.

Removing the effect of spreading in the Troodos basin leads to the following picture: the African continental margin was separated

Fig. 5.16. Palinspastic reconstructions. A - Removing the latest movements; B - Removing the effect of amalgamation of the Troodos and Mamonia terranes; the latter by the instant of amalgamation already had had the fold-and-thrust structure. C and D - two options for unfolding the thrust package of the Mamonia terrane for Cenomanian-Turonian age interval: C - option in which the upper nappe, Ayios Photios, is shifted southward onto the African continental margin. In this case, to obtain the current succession of nappes, a south-dipping subduction zone should be pictured in the Mamonia terrane following Dilek et al. (1990, 1992), and another subduction zone should account for the Troodos supra-subduction zone ophiolite; D - option in which the upper nappe of the Mamonia terrane is shifted northward. In this case, simultaneous formation of the Troodos ophiolite massif and thrust pile of the Mamonia terrane would require a single, north-dipping subduction zone, after MacLeod et al. (1992). We favor the latter option because it satisfies the Occam's razor principle and does not require a second, south-dipping subduction zone. 1- Crust generated in a supra-subduction zone setting-Troodos basin; 2 - Kathikas olistostrome; 3 - Terrigenous melanges; 4 - Nappe composed of polymictic serpentinite melange; 5 - Nappe composed of the Ayia Varvara Group metamorphics; 6 - Oceanic crust and volcanic rises in the Neotethys basin and also in the nappe composed of the Dhiarizos Group rocks; 7 - Deposits making up the Ayios Photios Group and also encountered in the Ayios Photios nappe; 8 - African continental margin; 9 - Microcontinental blocks of the Taurian plate; 10 - Hypothetical oceanic crust subducted beneath Troodos; 11-13 Arrows indicating tectonic movements: 11 - Relatively young movements of a southerly convergence that gave rise to terrigenous melanges (?); 12 - Thrusting of the Mamonia thrust package onto the Troodos terrane; 13 - Formation of the nappes of the Mamonia composite terrane.

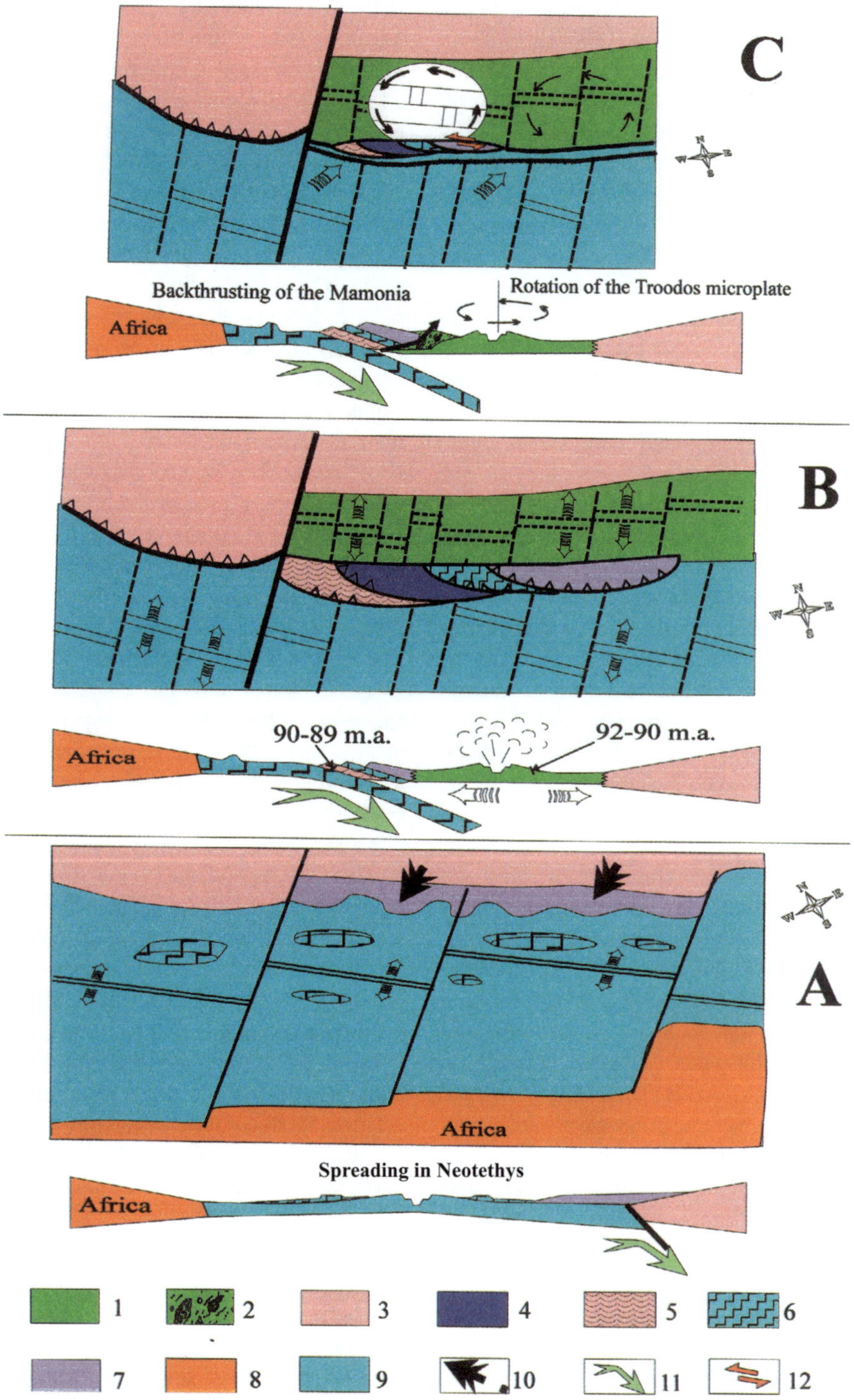

Fig. 5.17. Palinspastic reconstructions. Series of profiles: A - Profile and reconstruction for the Late Triassic-Early Cretaceous stage. B - Profile and reconstruction for the Cenomanian-Campanian stage. C - Profile and reconstruction for the early Maastrichtian stage. 1- Crust generated in a supra-subduction zone setting - Troodos basin; 2 - Kathikas olistostrome; 3 - Microcontinental blocks of the Taurian plate; 4 - Nappe composed of polymictic serpentinite melange; 5 - Nappe composed of the Ayia Varvara Group metamorphics; 6 - Oceanic crust and volcanic rises in the Neo-Tethys basin and also in the nappe composed of the Dhiarizos Group rocks; 7 - Deposits making up the Ayios Photios Group and also encountered in the Ayios Photios nappe; 8 - African continental margin; 9 - Oceanic crust of the Neo-Tethys; 10 - Arrows indicating terrigenous supply; 11 - Sense of underthrusting of block; 12 - Strike-slip motions at the boundary between the Mamonia and Troodos terranes. Arrow in oceanic domain shows inferred motion of the Neotethys oceanic crust.

from the Turkish continental block by the southern part of the Meso-Tethys, whose fragments survive currently as the Dhiarizos and Ayios Photios Groups, composing nappes in the Mamonia composite terrane. The mutual spatial relationships of the rocks of these groups are critical to the paleogeography of the southern branch of the Neo-Tethys.

Establishing a scenario for the formation of the inner structure of the Mamonia complex terrane is much more painful. As shown above, this structural pile is composed of a number of nappes. Of these, the Ayia Varvara metamorphic complex, Dhiarizos Group, and Ayios Photios Group are the most critical, bearing the majority of the data required for palinspastic reconstructions. The Ayia Varvara metamorphics occupy the lowermost structural position, while the Ayios Photios Group makes up the thrust sheet at the highest level. As noted above, convergence in the Mamonia thrust package cannot be used in the reconstructions because it was distorted by subsequent movements. Thus the correlation of these complexes to their lithologic counterparts in neighboring regions becomes of utmost importance. Of such counterparts, the Cretaceous sequences drilled on Eratosthenes Seamount and the sequences exposed in Antalya Bay are the most suitable for the purposes of comparison.

The Ayia Varvara rocks are geochemically closest to the Dhiarizos rocks – their likely metamorphic sources (protoliths – see Silantyev, this volume). The Dhiarizos Group is represented by two types of sections: one consisting of carbonate and volcanic rocks with some alkali basalts, interpreted as fragments of volcanic rises or intraplate piles, and the other more basinal, represented by MORB-like volcanites overlain by fine pelagic carbonate sediments passing upward into deep-water cherts.

Facies analysis reveals that the Ayios Photios rocks most likely correspond to the base of a large continental block, whereas the pelagic deep-sea cherts encountered in the Dhiarizos Group were deposited in the inner parts of this basin far from the continental margin.

In unfolding the thrust package, it is critical to decide to which of the two continental margins the Ayios Photios deposits belong. Based on structural evidence, they can be shifted equally well to the south (African continental margin) or to the north (Turkish continental margin). The former option would imply that the Ayios Photios Group was part of the base of the continental margin of the African plate, with the Dhiarizos rocks occurring farther north and the Ayia Varvara rocks still farther north (**Fig. 5.16**). This option can be verified by 'reassembling the thrust package', for which purpose one should assume a southerly plunge for the subduction zone beneath the margin of the African continent. This reconstruction coincides with those currently existing for the Troodos ophiolites (Dilek and Moores, 1990). All the misfits, flaws, and negative consequences ensuing from this version have already been discussed by MacLeod et al. (1992). The recent deep-sea drilling data (ODP Leg 160) from Eratosthenes Seamount (African continental margin) clearly suggest that the Late Cretaceous (pre-Maastrichtian) sections have nothing in common with those in Cyprus, in either the Ayios Photios or the Dhiarizos Groups (Robertson et al., 1998). The sedimentary pile drilled in the trench separating Eratosthenes Seamount from Cyprus (Hole 967) records deposition of shallow marine Cretaceous strata at the base of the sequence, followed by strata that reflect progressive deepening of the trench, whereas in Cyprus the Late Cretaceous sequence commences with deep-water (below CCD) sediments, followed by continual shoaling throughout the Cenozoic (Robertson et al., 1998; Krasheninnikov and Kaleda, 1994). At the same time, the Cretaceous sequence on Eratosthenes Seamount correlates clearly to the sequences of western Syria and Lebanon, which developed at the northern margin of the Afro-Arabian continental plate. That Eratosthenes Seamount belongs to the northern margin of the African craton is currently challenged by no one, whereas the essential dissimilarity from the Cyprus sequences precludes their attribution to this margin. Recently Robertson (1998a, b) arrived at the same conclusion.

Alternatively, the Mamonia thrust package could be unfolded from north to south in the following order: Ayios Photios Group – Dhiarizos Group – Ayia Varvara Group; *i.e.*, in this reconstruction the Ayios Photios rocks are assumed to have formed near the continental slope of the Turkish microcontinent (Fig. 5.16 and **Fig. 5.17**). The thrust package is interpreted to have formed in an accretionary prism above a northward-dipping subduction zone. The advantages of this model were analyzed by McLeod et al. (1992) and Robertson (1998), and this model was adopted in our early works (Kaleda and Krylov, 1992; Krylov et al., 1993, 1995). It accommodates data from facies correlation of coeval deposits and makes the second south-dipping subduction zone unnecessary because one zone is enough to assemble the whole thrust package and to generate the Troodos boninitic melts, and later the calcalkaline suites existing in the Kyrenia terrane and in Turkey (Baroz, 1980; Aktas and Robertson, 1990; Robertson, 1998 b). Since this model satisfies the principle Occam's razor (to wit, one should not increase, beyond what is necessary, the number of entities required to explain anything), we opt for it.

The onset of subduction and initiation of a subduction zone is constrained (within error) by the age of formation of the accretionary structure *i.e.* the pre-Mamonia thrust package. Its upper age limit is constrained by the age of the sequence hosting rock fragments derived from the nappes in the Mamonia terrane. These fragments were first found among the clasts of the Kathikas olistostrome (Krylov et al., Chapter 4, this volume). This sequence is dated with confidence to Late Campanian-Early Maastrichtian. Therefore, by Late Campanian, the Mamonia thrust package had been formed, and all the rocks contained in it (including the Ayia Varvara metamorphites) had been exposed to denudation. The Mamonia thrust package was expressed morphologically as a submarine ridge with submarine slumps shedding from its flanks (the Kathikas Formation). The timing of emplacement of the Ayia Varvara metamorphic rocks into the shallow crustal horizons is established as well from the $^{39}Ar/^{40}Ar$ determinations at 90 Ma (Spray and Rodick, 1981). This date can be taken to constrain the upper age limit of formation of the structural pile because it reflects closure of the isotopic system as the rocks were emplaced into the

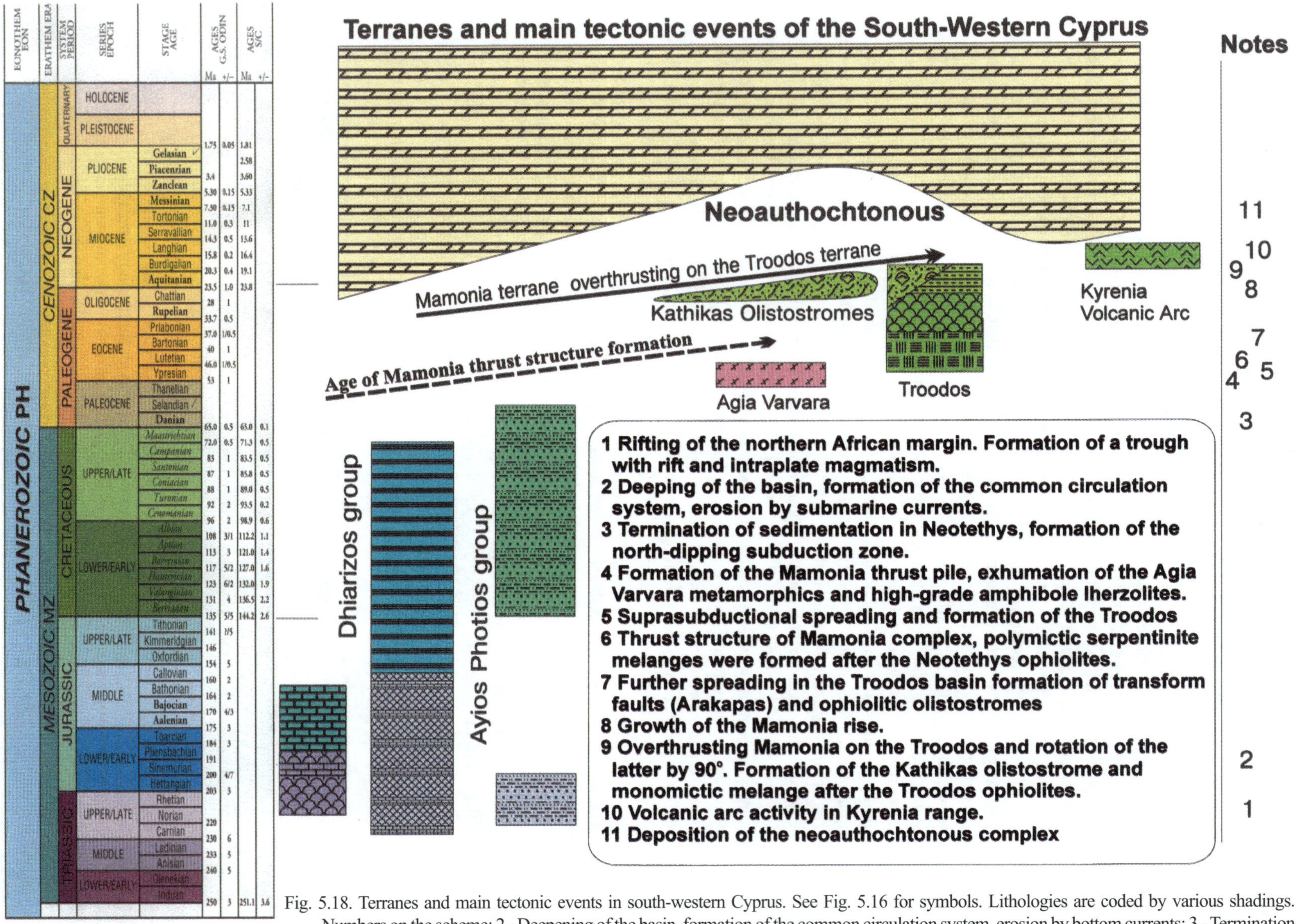

Fig. 5.18. Terranes and main tectonic events in south-western Cyprus. See Fig. 5.16 for symbols. Lithologies are coded by various shadings. Numbers on the scheme: 2 - Deepening of the basin, formation of the common circulation system, erosion by bottom currents; 3 - Termination of sedimentation in Neo-Tethys, formation of the north-dipping subduction zone; 4 - Formation of the Mamonia thrust pile, exhumation of the Ayia Varvara metamorphics and high-grade amphibole lherzolites; 5 - Supra-subductional spreading and formation of the Troodos; 6 - Thrust structure of the Mamonia Complex, polymictic serpentinite melanges forming after the Neo-Tethys ophiolites; 7 - Further spreading in the Troodos basin, formation of transform faults (Arakapas) and ophiolitic olistostromes; 8 - Growth of the Mamonia Rise; 9 - Overthrusting of Mamonia on the Troodos and rotation of the latter by 90°. Formation of the Kathikas olistostrome and monomictic melange after the Troodos ophiolites; 10 - Volcanic arc activity in Kyrenia Range; 11 - Deposition of the neoautochthonous complex.

upper crustal horizons (**Fig. 5.18**).

The lower age limit for the thrust package is estimated roughly from the youngest dates for various rocks comprising nappes in the Mamonia composite terrane. For the Ayios Photios Group, these dates fall within Albian-Cenomanian time. Similar data were obtained from the rocks of the Dhiarizos Group; they also fall within Albian-Cenomanian time.

Therefore, the time interval is established within which the Mamonia thrust package was formed. Reliable data for the formation of the Troodos ophiolite are much more abundant. The early stage of its formation based on both stratigraphic and radiometric data is attributed to the Turonian (Mukasa and Ludden, 1987), which also falls within the above interval (Blome and Irwin; 1985). A vital conclusion follows that the accretionary-type thrust package of the Mamonia composite terrane and the Troodos spreading basin formed simultaneously, most likely above the same single north-dipping subduction zone. In the compressional environment existing in the frontal part of this zone an accretionary prism formed, which was to eventually evolve into an oceanic rise, while in the rear a spreading basin with supra-subduction zone magmatism was located. Earlier, Dilek et al. (1990, 1992) discussed alternative models, in which the subduction zone dipped south, which were later analyzed in detail and criticized by McLeod et al. (1992). The characteristics of the rock complexes comprising the Mamonia composite terrane suggest that the individual thrust sheets formed almost simultaneously and in a stepwise manner. Accretion of fragments derived over a long period of time from various parts of the southern Meso-Tethys basin proceeded at various crustal levels and in a stepwise manner, resulting in nappe complexes showing internal lithofacies uniformity that is revealed through the geological mapping. This generalized scheme of palinspastic reconstructions and the scenario of events enables us to construct a series of palinspastic profiles for the beginning of the Cenozoic and for the Maastrichtian, pre-Maastrichtian, Cenomanian-Turonian, Jurassic, and Late Triassic (Fig. 5.17).

Principal geological events

The above reconstructions and the data on ages and structural relations suggest a scenario for the crustal evolution of the Cyprean segment. The main stages of this process are as follows.

Late Triassic. This time interval most likely corresponds to the formation of the lherzolites and their subsequent high-temperature amphibolization in an environment of spreading and breakup of the African continental margin. Later on, melts separated from these amphibolized ultramafites, giving rise to intraplate mafic rocks and, as a result, depleted ultramafites (including harzburgites) were formed alongside the enriched lherzolites (Bazylev, this volume, Chapter 2). On the surface, this led to the formation of volcanic piles with carbonate cover known from the Dhiarizos Group, while the intervening troughs were being filled with turbidite fans. The volcanism had both intra-plate and oceanic spreading affinities (Portnyagin, this volume, Chapter 2; Malpas et al., 1993). Reference by some workers to the presence of island-arc volcanic series at that time, appear to be poorly justified by stratigraphic and geochemical data; for instance these are simply not mentioned by Sharaskin et al. (1995). During the same period, a morphologically well-defined continental slope was developing, while the whole basin experienced gradual subsidence and deepening. The base of the continental slope was filled by turbidite fans derived from three sources:

1) calcareous oolitic material supplied from carbonate rises and shoals,

2) well-rounded subarkosic material, from inner areas of the split-off continental block, and,

3) newly added intra-basinal material precipitated from contour currents which diluted and redeposited the materials in (1) and (2) above. This is well documented in the upper portion of the Vlambouros Formation (Bragin, Krylov, and Bragina, this volume, Chapter 2). Depressed parts of the basin in the pre-Rhaetian time sagged below the CCD, and the Rhaetian deposits were already represented only by red radiolarites (Bragin and Krylov, 1998 paleontological collection).

The beginning of Jurassic time is marked by a deep-water basin forming in the southern part of the Meso-Tethys, with vigorous bottom currents that obstructed deposition. This situation must not have been ubiquitous because in Syria (the Bassit Massif) deep-water Jurassic sediments are found, whereas in Cyprus almost no sediments of that type are reported. The magmatism had an oceanic character. Melts generated at that time were similar to normal oceanic basalts. Simultaneously, the first deformations were registered in the basin: ophiolitic olistostromes were formed. Previously, these deposits were either not reported from Cyprus, or else were documented in the Louthra tis Aphrodithis Sequence (Swarbrick and Robertson, 1980). Figure 5.15 shows photographs of these rocks, in which the clasts consist of ultramafites and a matrix, of pyroxene fragments, serpentine, and clayey mass. These rocks are overlain by ophicalcite superposed by red radiolarites. The latter yielded (from Krylov and Bragin's 1988 paleontological samples) an Early Jurassic radiolarian assemblage (Bragin, Krylov, this volume, Chapter 2). These rocks definitely point to the existence of 'pre-Troodos' ophiolites.

By the close of the Middle Jurassic, the impact of bottom currents waned. The basin retained its oceanic identity, including good aeration. Deep-water siliceous sedimentation below the CCD is recorded throughout the basin. Both siliceous turbidites and normal background radiolarites are encountered. Hydrothermal activity is known to have existed in the basin.

In Late Jurassic time and especially by the beginning of the Cretaceous period, differential movements on contiguous blocks were reactivated. These movements exposed some blocks to denudation, and carbonate turbidites (calciturbidites) began to be supplied to the basin. These were buried below the CCD, judging from corroded foraminiferal tests found in these deposits. The calciturbidites are interbedded with normal background sediments of the oceanic southern Neo-Tethys basin.

In Neocomian time, very mature quartzose turbidites and arkoses appeared that make up the channel facies in turbidite fans. In Cyprus these are referred to as the Akamas sandstones.

Subsequently, sedimentation became predominately terrigenous. The whole basin was involved in a regressive cycle, and cherts were no longer to be found in the Ayios Photios Group. The inner parts of the basin, however, kept evolving in a pelagic environment, and chert deposition persisted until Albian-Cenomanian time.

To summarize, three stages can be discerned:

(1) the Callovian-Early Oxfordian, dominated by normal background siliceous sedimentation (resembling the paleo-Pacific pelagic basins, but with greater sedimentation rates), and with widespread coarse-clastic turbidites;

(2) the Middle-Late Oxfordian, during which background pelitic sedimentation increased, chert deposition dropped dramatically, and coarse clastic turbidites ceased to form;

(3) the Kimmeridgian-Tithonian, in which all three types of deposition – siliceous, background terrigenous, and turbidite – intensified.

Conceivably, this part of the Tethys in Late Jurassic time represented a relatively deep-water, actively developing basin with rugged topography akin to a backarc basin or a minor (young) ocean. This basin probably lacked environments suitable for pelagic (abyssal), oceanic *sensu stricto* sedimentation, although it retained unobstructed links with the world ocean. The overall character of sedimentation can be defined as hemipelagic for most of this period, and only at the first stage did it have an eupelagic character. The new data obtained in 1998 from field studies in the Dhiarizos Formation suggest the existence of an eupelagic environment over considerable tracts of the southern Tethys basin from the Rhaetian Stage of the Triassic through to at least the end of the Middle Jurassic (N. Bragin and K. Krylov, 1998 field collections).

The Albian-Cenomanian interval shows a major rearrangement in the whole Meso-Tethyan – the Austrian orogenic phase (Knipper, 1985). It may have coincided with a global change in plate motions: at 95 Ma, the drift direction of the African plate changed (Savostin et al., 1986). It is to this event that the complete rearrangement of the Cyprus basin, and initiation of the subduction zone, seems to be related. In any case, the development of all the complexes constituting the Mamonia composite terrane ceased at this boundary. A significant depositional break recorded in the ODP Hole 967 (Robertson et al., 1998) is attributed to the same Late Aptian-Early Cenomanian time. During that interval, the passive continental margin of the Taurian platform was transformed into an active one.

In Cenomanian–Turonian time spreading began over the newly formed subduction zone, to eventually give rise to island-arc magmatism with the opening of the Troodos basin along a roughly E-W trending spreading axis. The contribution from the subduction component was manifested in many ways: it is detectable in the Troodos ultramafites, lavas, and even their overlying sediments (Bazylev and Portnyagin, this volume, Chapter 2; Kaleda and Krylov, 1992). These motions resulted in a dynamic couple, in which the frontal part experienced compression and the rear part, extension. Well-known counterparts exist along the eastern rim of the Asian continent, and the most suitable analogy is the Izu-Bonin island-arc system with an accretionary prism in front of the arc and spreading-related boninitic magmatism in the forearc basin (Taylor and Nesbit, 1992).

In front of the Cyprean subduction zone, accretion initiated with sequential underthrusting under the continental rise of the most 'oceanic' fragments. This accretion proceeded at different crustal levels and had a stepwise character, resulting in inner lithologic homogeneity of each particular slice being scraped off. At the deeper levels, deep portions of the oceanic lithosphere suffered accretion, and peculiar ophiolitic assemblages specific to the Mamonia complex alone were formed. At that stage of the geological evolution of the Mamonia peridotites, the rocks were tectonized and recrystallized under the upper greenschist to amphibolite facies conditions (450°-600°C), and the lherzolites became tectonically juxtaposed with the harzburgites (Bazylev, this volume, Chapter 2). These movements took place synchronously with the spreading in the Troodos basin.

In Campanian time, the frontal part acquired the form of a oceanic rise with a complex thrust-related structure; a new entity that can now be referred to as the Mamonia composite terrane came into being. This may have resulted from a single process that encompassed a large area because movements of this age are also known in other localities with lherzolitic mantle restites (Silantyev, 1993). Emplacement of the metamorphic rocks into the upper crustal horizons must have been rather rapid and possibly simultaneous with emplacement of enriched lherzolites, as evidenced by the lack of retrograde recrystallization in both rock groups (Silantyev, this volume, Chapter 2). It is these movements that initiated formation of the first serpentinite melanges, which were to eventually turn into the polymictic serpentinite melanges. Ultramafic diapirism may have been responsible for the tectonic disintegration of the inverted thermal aureole in the subduction zone and transportation of its fragments as tectonic blocks to the surface, where they were incorporated together with unmetamorphosed continental-margin deposits and Troodos ophiolites (Silantyev, this volume, Chapter 2).

In Late Campanian-Early Maastrichtian time, one more rearrangement in the plate motions took place. By that period, active spreading in the Troodos basin had ceased. The basin as such retained its deep-water character, as evidenced by the presence of radiolarites (occasional radiolarians or representative assemblages thereof in the Perapedhi Formation, and extremely rich radiolarian assemblages in the Kannaviou Formation) and absence of carbonate deposits. The basin began to accumulate terrigenous and submarine slump material derived from the Mamonia rise. The topography was so rugged that huge slumps formed, individual olistoplates reaching giant dimensions (up to 1.5 km). During that stage the Troodos plate experienced a counterclockwise rotation through 90°, which is well documented by paleomagnetic measurements. Concurrent with this rotation, the whole thrust package of the Mamonia composite terrane was thrust onto the southern portion of the Troodos basin. Supposedly at that moment in time the convergence of the nappes making up the Mamonia complex changed, the causative mechanism most closely resembling back-thrusting of an accretionary prism onto a forearc basin by analogy with well-studied modern examples (Willet et al., 1993). Oceanic rocks flooring the Troodos basin suffered tectonic disintegration and serpentinization. This in turn gave rise to the structurally lowermost and compositionally monomictic serpentinite melange after the Troodos rocks. Further movements brought about the tectonic layering of the oceanic lithosphere, its individual portions separating in the form of thrust sheets. As the thrust sheets disintegrated, they experienced mixing, a process largely facilitated by the plasticity of serpentinite melange. This resulted in the coexistence of ultramafic blocks clearly derived from contrasting geodynamic settings (Fig. 5.14). Subsequently, no more catastrophic events like that took place, and the newly formed structures were overlapped by the neoautochthonous cover.

The Late Maastrichtian-Cenozoic. The lithofacies pattern and structural characteristics of the neoautochthon provide clues to the last stages of the geological history of south-western Cyprus over this time interval.

The basal strata of the neoautochthon (the Upper Maastrichtian portion of the Lower Lefkara Formation) are represented by either calcareous sediments with radiolarians deposited above the CCD but below the planktonic foraminifera lysocline, or by pelagic foraminiferal oozes. This whole rock assemblage provides evidence that the deep-water (near the CCD) depositional environment continued to exist.

The Late Cretaceous tectonic events shaped the structural grain and submarine paleotopography of that strip of south-western Cyprus where the neoautochthonous rocks were developed. Structurally, the central part of south-western Cyprus is occupied by a N-S trending basin (between the Dhiarizos and Germasogea Rivers). To the west it is bounded by an uplift where the Mamonia complex is exposed and to the east, by the Limassol Forest uplift, where the ophiolites were developed.

The character of the sedimentary cover (neoautochthon) in the basin differs from that of the uplifts. In the basin one finds a continuous sequence from the Campanian Stage (Upper Cretaceous) to the Pliocene. In Cretaceous and Paleogene time, the basin was the site of formation of pelagic, rather deep-water sediments with radiolarians and planktonic foraminifers. In Early and Middle Miocene time, the sediments were still pelagic, but less deep-water (only calcareous plankton was present, and radiolarians and diatoms had disappeared). In Late Miocene and Pliocene time, the deposits became markedly shallow-water.

On the Mamonia and Limassol Forest uplifts, the Upper Cretaceous and Cenozoic sequences are incomplete, showing a number of hiatuses and unconformities. The Maastrichtian, Paleogene, and Neogene sediments are more shallow-water, not infrequently containing abundant turbidites and scattered pebbles.

As noted above, the Upper Maastrichtian sediments of the basin (near Perapedhi village) were formed in a deep-water environment below the planktonic foraminifera lysocline. The foraminifers are abundant in the overlying Danian limestones, but their tests are completely dissolved (replaced by silica). In the north of the Mamonia uplift (near Pano Panayia village), basal strata of the Lower Lefkara Formation are composed of limestones with Maastrichtian radiolarians and planktonic foraminifers affected very strongly by selective dissolution. In the western and central Mamonia uplift (the villages of Simou, Phiti, Kathikas), the Maastrichtian is represented by chalk-like limestones with excellently preserved planktonic foraminifers. Therefore, while radiolarian-bearing calcareous oozes in the basin were deposited in a lower bathyal environment, the coeval foraminiferal oozes of the Mamonia uplift are indicative of an upper bathyal milieu. The difference in water depth can be estimated on the order of 1,000-1,500 m.

The thicknesses of the deposits and their sedimentation rates in the basin and on the Mamonia uplift were different. In the continuous Upper Cretaceous and Cenozoic sections (along the Kouris, Kryos, Hapothami, Evdimou Rivers), the thickness is as great as 1,400 m. In the composite section on the Mamonia uplift (including a number of breaks) the thickness does not exceed 700 m.

The rates of sedimentation for the continuous sections of the basin were determined to a high degree of accuracy. Based on this parameter, two stages can be discerned. In the Late Campanian-Eocene interval, the rates were very low (from 1 to 13 mm/1,000 yr), increasing gradually upwards. Rates that low are characteristic of sedimentation in oceanic basins. In the Oligocene to Neogene interval, the rate of deposition of calcareous and terri-genous/calcareous sediments became much greater – between 20 and 40 mm/1,000 yr. Precise estimates for the Mamonia uplift present difficulties. It is obvious, however, that rates of sedimentation there were much greater. This follows from comparison of particular members of the sections. Thus, the thickness of Campanian-Maastrichtian sediments in the basin (near Perapedhi village) is 12 m, and on the Mamonia uplift (near Pano Panayia village) it was 92 m – *i.e.*, the rate of sedimentation on the uplift was almost eight times greater. The low sedimentation rates in the basin were likely due to partial dissolution of carbonate material at bathyal depths in this basin.

One peculiarity of the neoautochthon is the trend of bottom-to-top variation in the character of the sediment facies – from deep-water to shallow-water the same being true of both the basinal sections and the Mamonia uplift. In the basin, this transition is recorded by a range of sediments from abyssal radiolarian oozes and lower bathyal radiolarian-foraminiferal oozes of Campanian, Maastrichtian, and Paleogene age to shallow-water terrigenous-carbonate and evaporitic deposits of Upper Miocene and Pliocene age. On the Mamonia uplift, this range is only a little narrower – -from deep-water pelagic foraminiferal oozes of Maastrichtian age to similar terrigenous-carbonate and evaporitic deposits of Upper Miocene and Pliocene age. The uplifting (by about 3,500-4,000 m) should be ascribed to the continuing horizontal drift of the crustal blocks: it was driven either by the pushing Troodos Massif or, which seems more likely, by subduction of the African plate.

The Upper Cretaceous and Cenozoic deposits of the neoautochthon formed in a single Mediterranean basin (Neo-Tethys) under the impact of paleohydrological, paleogeographic, eustatic, and tectonic factors of sub-global, regional (the eastern

Mediterranean), and local scale. Hence the similarities and contrasts in the character of sedimentary formations and geological histories of different regions of the Near East and north-eastern Africa (Egypt).

The neoautochthonous sedimentary cover of south-western Cyprus clearly displays the same formations as in the Near East and Egypt:

1) limestones, clayey limestones, and marls of the Maastrichtian, Paleocene, and the greater part of the Eocene;

2) a sequence of cherts and limestones of the Lower Eocene, in places including the lowermost portion of the Middle Eocene;

3) chalk-like massive and organic (nummulitic) limestones of the Middle and Upper Eocene;

4) shallow-water limestones, marls, and sand-clay rocks of the Oligocene;

5) chalk-like and platy limestones of the Lower Miocene;

6) terrigenous-carbonate rocks of the Middle and Upper Miocene (Tortonian);

7) evaporitic formation of the Upper Miocene (Messinian);

8) carbonate-terrigenous sequence of the Pliocene.

However, the deep-water environment in the western Cyprus basin gave rise to specific lithofacies (such as radiolarian and radiolarian-foraminiferal oozes) of the Late Cretaceous, Paleocene, and Eocene time that are totally absent from the Near East and Egyptian sections.

During the Cenozoic evolutional stage, horizontal shortening of the previously generated crust continued (primarily, remnants of the Troodos microplate), as evidenced by the Late Eocene melanges discovered on the Kyrenia Range (Robertson, 1998a).

The tectonic evolution of south-western Cyprus was influenced greatly by the sub-global Late Cretaceous and Cenozoic eustatic changes, due to which the late stages of its geologic history are shared by the entire eastern Mediterranean. Maastrichtian, Paleocene, Early and Middle Eocene sediments are developed throughout southern Cyprus and belong to a transgressive stage. The limited spread of the Upper Eocene deposits implies the beginning of a regression. This is distinctly manifest in the Oligocene (sediments of this age being confined to the basin). The Early Miocene is indicated by a new transgression that culminated in the terminal Early Miocene to Middle Miocene. The Late Miocene (Tortonian and Messinian Stages) is related to a major regression. The Pliocene transgression was of a limited extent.

The terrigenous material in the Paleogene and Neogene rocks likely results from disintegration of the lithologically diverse Mamonia rocks. The Troodos ophiolite itself must have become an erosional surface as late as Quaternary time, deposits of this age being replete with ophiolitic clastics.

Tectonic movements supposedly related to underthrusting of the African plate beneath Cyprus gave rise to the Petra tou Romiou terrigenous/volcaniclastic tectonic melange, overlapped by a Lower Eocene chert and limestone sequence. Similar movements along major transpressional strike-slip faults trending WNW-ESE produced the Neo-Moni terrigenous tectonic melange (Eocene-Miocene).

The current subduction, plunging in a north-easterly direction underneath Cyprus, gave rise to an accretionary prism at the base of the continental slope of southern Cyprus (Robertson, 1998). At present, tectonic underplating is probably taking place beneath Cyprus, ensuring the rapid growth of the Troodos Massif (Poole and Robertson, 1998). The same movements are responsible for subsidence and faulting of Eratosthenes Seamount and are to result, ultimately, in a new collisional assemblage.

REFERENCES CITED

Aktas, G., and Robertson, A. H. F., 1990. Tectonic evolution of the Tethys suture zone in S. E. Turkey: Evidence from the petrology and geochemistry of Late Cretaceous and Middle Eocene Extrusives. In: J. Malpas, E. M. Moores, A. Panayiotou, and C. Xenophontos (eds.), Ophiolites; Oceanic Crustal Analogues. Proc. Symp. Troodos 1987. Geological Survey Department, Ministry of Agriculture and Natural Resources, Nicosia, p. 311-329.

Bamler, R., 1999. The SRTM Mission: A world-wide 30 m resolution DEM from SAR interferometry in 11 days. In: D. Fritsch and R. Spiller (eds.), Photogrammetric Week 99, Wichmann Verlag, Heidelberg, p. 145-154.

Baroz, F. 1980. Volcanism and continent-island arc collision in the Pentadaktylos range, Cyprus. In: A. Panayiotou (ed.), Ophiolites, Proceedings International Ophiolite Symposium Cyprus, 1979, Ministry of Agriculture and Natural Resources Geological Survey Department, Nicosia, p. 73-75.

Bazylev, B. A., Magakyan, R., Silantyev, S. A., Ignatenko, K. I., Romashova, T., and Xenophontos, C., 1993. Petrology of peridotites from the Mamonia Complex, Southwestern Cyprus. Petrology, 1(4), p. 348-378.

Bear, L. M., 1960. The geology and mineral resources of the Akaki-Lythrodondha area. Cyprus Geol. Survey Department Memoir, No 3, 122 p.

Bernarz, U., Gotte, P., and Schminke, H. -U., 1991. The petrography of altered submarine lavas and major element mobility in Drill-holes CY-1A, Cyprus. In: Cyprus Crustal Study Project Init. Rep., Holes CY-1 and CY-1A. Geological Survey Canada, Paper 90-20, p. 95-116.

Berry, P. A. M., Hilton, R. D. & Johnson, C. P. D. 2002; ACE: first full release of the new 30" GDEM incorporating satellite altimeter derived heights. European Geophysical Assembly, Nice France, April 2002.

Blome, C. D., and Irwin, W. R., 1985. Equivalent radiolarian ages from ophiolite terrains of Cyprus and Oman. Geology, 13(6), p. 401-404.

Bragin, N. Yu., and L. G. Bragina, 1993. Radiolarian biostratigraphy of Upper Cretaceous deposits in the southwestern Cyprus. In: Abstr. L. P. Zonenshain Mem. Conf. on Plate tectonics, Moscow, November 17-20, 1993, p. 44-45.

Bragin, N. Yu., and Krylov, K. A., 1991. Radiolarian biostratigraphy of the Ayios Photios Group (Mamonia complex, Cyprus). Abstr. of INTERRAD 6th, Firenze, September 30 - October 3, 1991, p. 18.

Bragin, N. Yu., and Krylov, K. A., 1995. New data on lithology and stratigraphy of Upper Triassic sediments of Vlambouros formation (south-western Cyprus). Abstr. IAS-16th Regional meeting of Sedimentology, 5 Congr. France de Sedimentologie, ASF Book of Abstr., Publ. ASF, Paris, No 22, p. 89.

Bragin, N. Yu., and Krylov, K. A., 1996. The stratigraphy and sedimentology of Upper Triassic deposits of southwestern Cyprus (Vlambouros Formation). Stratigr. Geol. Korrelyatsiya, 4(2), p. 28-37.

Clube, T. M. M., and Robertson, A. H. F., 1986. The palaeorotation of the Troodos microplate, Cyprus, in the Late Mesozoic Early Cenozoic and plate tectonic framework of the Eastern Mediterranean. Surv. Geophys., No 8, p. 375-437.

Constantinou, G., and Govet, G. J. S., 1972. Genesis of sulfide deposits, ochre and umber of Cyprus. Trans. Inst. Mining and metal., Sect. B, v. 81, p. 834-846.

Dilek, Y., and Moores, E. M., 1990. Regional tectonic of the Eastern Mediterranean ophiolites. In: J. Malpas, E. M. Moores, A. Panayiotou, and C. Xenophontos (eds.), Ophiolites; Oceanic Crustal Analogues. Proc. Symp. Troodos 1987. Geological Survey Department, Ministry of Agriculture and Natural Resources, Nicosia, p. 295-309.

Dilek, Y., Thy, P., Moores, E. M., and Ramsden, T. W., 1990. Tectonic evolution of the Troodos ophiolite within the Tethyan framework. Tectonics, 9(4), p. 811-823.

Dilek, Y., Thy, P., Moores, E. M., and Ramsden, T. W., 1992. Tectonic evolution of the Troodos ophiolite within the Tethyan framework; comment and reply. Tectonics, 11(4), p. 910-923.

Gass, I. G., 1968. Is the Troodos massif of Cyprus a fragment of Mesozoic ocean floor? Nature, 220(5162), p. 39-42.

Gass, I. G., 1980. The Troodos massif: Its role in the unraveling of the ophiolite problem and its significance in understanding of constructive plate margin processes. In: A. Panayiotou (ed.), Ophiolites, Proceedings International Ophiolite Symposium Cyprus, 1979, Ministry of Agriculture and Natural Resources Geological Survey Department, Nicosia, p. 182-192.

Hall, J. K., 1997. Topography and bathymetry of the Dead Sea depression, Tectonophysics, 266(1-4): p. 177-185.

Hall, J. K., 1998. Digital terrain model (DTM) of the island of Cyprus. In: T. Weissbrod (ed.), Geol. Surv. Isr. Current Research, Vol. 11, Jerusalem p. 47-52.

Harbert, W., Frei, L., Jarrard, R., et al., 1990. Paleomagnetic and plate-tectonic constraints on the evolution of the Alaskan - Eastern Siberian Arctic. In: A. Grantz, G. L. Johnson, and J. F. Sweeney (eds.), The Arctic Ocean Region, The Geology of North America, DNAG Volume L, Geol. Soc. Amer., p. 567-592.

Kaleda, K. G., and Krylov, K. A., 1992. The Cyprus umbers: Composition, geodynamic affinity, and origin. Dokl. Ros. Akad. Nauk, 326(5), p. 865-870. (In Russian).

Kazmin, V. G., 1991. On the tectonic environment of formation of the Troodos and Mamonia complexes, Cyprus. Geotektonika, No 6, p. 104-116. (In Russian).

Kazmin, V. G., Ricou, L. E., and Sborshchikov, I. M., 1987. The structure and development of the passive margin of the Eastern Tethys. In: History of the Tethys Ocean. Moscow: Nauka, p. 39-57. (In Russian).

Kazmin, V. G., and Sborshchikov, I. M., 1989. Evolution of the Tethys Ocean in Late Paleozoic-Early Mesozoic time and the problem of reconstructions of the Pangea. Izv. Akad. Nauk SSSR, Ser. Geol., No 5, p. 55-67. (In Russian).

Knipper, A. L., 1985. Movements of the Tethys oceanic crust during the Austrian phase; their character, results, and possible causes. Geotectonics, 19(2), p. 85-94.

Knipper, A. L., and Sharaskin, A. Ya., 1994. Tectonic evolution of the western part of the Peri-Arabian ophiolite arc. In: V. A. Krasheninnikov and J. K. Hall (eds.), Geological Structure of the Northeastern Mediterranean (Cruise 5 of the Research Vessel 'Akademik Nikolaj Strakhov'), Historical Productions-Hall, Jerusalem, p. 295-306.

Knipper, A. L., and Sharaskin, A. Ya., 1995. Correlating tectonic events in the Mesozoic history of the northeastern Mediterranean. Geotektonika, No 1, p. 12-22.

Krasheninnikov, V. A., and Kaleda, K. G., 1994. Stratigraphy and lithology of the Upper Cretaceous and Cenozoic deposits of the key Perapedhi section (Southern Cyprus). In: V. A. Krasheninnikov and J. K. Hall (eds.), Geological Structure of the Northeastern Mediterranean (Cruise 5 of the Research Vessel 'Akademik Nikolaj Strakhov'), Historical Productions-Hall, Jerusalem, p. 195-218.

Krasheninnikov, V. A., and Hall J. K., (eds.). 1994. Geological structure of the Northeastern Mediterranean (Cruise 5 of the Research vessel Akademik Nikolaj Strakhov). Historical Productions-Hall, Jerusalem, Historical Productions-Hall Ltd., 396 p.

Krylov, K. A., Kaleda, K. G., and Bragin, N. Yu., 1993. Structure of South-Western Cyprus. In: L. P. Zonenshain Memorial Conference on plate tectonics, Moscow, November 17-20, 1993. Abstr., p. 91-92.

Krylov, K. A., Kaleda, K. G., and Bragin, N. Yu., 1995. Tectonic evolution of south-western Cyprus in Mesozoic. In: 5th L. P. Zonenshain Memorial Conference on plate tectonics, Moscow, November 22-25, 1995, Abstr., p. 77.

Krylov, K. A., Sokolov, S. D., Grigoriev, V. N., Batanova, V. G., and Peive, A. A., 1993. Ophiolite (serpentinite) melanges in the structure of the Koryak Highland.1993. In: L. P. Zonenchain Memorial Conference on plate tectonics, Moscow, November 17-20, 1993. Abstr., p. 92.

Lapierre, H., 1975. Les formations sédimentaires et éruptives des nappes de Mamonia et leur relations avec le Massif du Troodos. Mém. Soc. Géol. Fr., 123 p.

MacLeod, C. J., 1990. Role of the Southern Troodos Transform fault in the rotation of the Cyprus microplate: Evidence from the Eastern Limassol Forest Complex. In: J. Malpas, E. M. Moores, A. Panayiotou, and C. Xenophontos (eds.), Ophiolites; Oceanic Crustal Analogues. Proc. Symp. Troodos 1987. Geological Survey Department, Ministry of Agriculture and Natural Resources, Nicosia, p. 75-85.

MacLeod, C. J., Robertson, A. H. F., Allerton, S., Browning, P., Gass, I. G., Taylor, R. N., Vine, F. J., and Xenophontos, C., 1992. Comment on 'Tectonic evolution of the Troodos ophiolite within the Tethyan framework' by Y. Dilek, P. Thy, E. M. Moores and T. W. Ramsden. Tectonics, 11(4), p. 910-915.

Makris, J., Stacker, J., and Rosenkranz, C., 1998. Microseismic studies and tectonic implication of Cyprus. In: I. Panayides and C. Xenophontos (eds.), Third International Conference on the Geology of the Eastern Mediterranean, Nicosia, Cyprus, September 23-26, 1998, Abstr., p. 37.

Makris, J., and Wang, J., 1994. Bouguer gravity anomalies of the Eastern Mediterranean Sea. In: V. A. Krasheninnikov and J. K. Hall (eds.), Geological Structure of the Northeastern Mediterranean (Cruise 5 of the Research Vessel 'Akademik Nikolaj Strakhov'), Historical Productions-Hall, Jerusalem, p. 75-85.

Malpas, J., Calon, T., and Squires, G., 1993. The development of a late Cretaceous microplate suture zone in SW Cyprus. Magmatic processes and plate tectonics. Geol. Soc. Spec. Publ. No 76, p. 177-195.

Malpas, J., Xenophontos, C., and Williams, D., 1992. The Ayia Varvara Formation of SW Cyprus: a product of complex collisional tectonics. Tectonophysics, 212(3/4), p. 193-213.

Malpas, J., Xenophontos, C., and Robertson, A. H. F., 1998. Mamonia complex and its relation to the Troodos ophiolite. In: I. Panayides and C. Xenophontos (eds.), Third International Conference on the Geology of the Eastern Mediterranean. A field trip guide. Excursion to the Southwest, 26-27 September, 1998, Cyprus, Nicosia, p. 1-13.

Miyashiro, A., 1973. The Troodos ophiolite complex was probably formed in an island arc. Earth Planetary Science Letters, 19(2), p. 218-224.

Miyazaki, Y., Kaku, M., and Tsu, H., 1996. DEMs (digital terrain models) from ASTER VNIR stereo data. International Geological Congress, Abstracts, Congrés Géologique Internationale, Résumés, vol.30, Vol.1, p. 463.

Mukasa, S. B., and Ludden, J. N., 1987. Uranium-lead isotopic ages of plagiogranites from the Troodos ophiolite, Cyprus, and their tectonic significance. Geology, 15(9), p. 825-828.

Pearce, J. A., 1975. Basalt geochemistry used to investigate past tectonic environment on Cyprus. Tectonophysics, 25(1-2), p.41-67.

Pearce, J. A., Lippard, S. J., and Roberts, S., 1984. Characteristics and tectonic significances of supra - subduction zone ophiolites. Marginal basin geology, Blackwell, Oxford, p. 77-94.

Poole, A., and Robertson, A. H. F., 1998. Pleistocene fanglomerate deposition related to uplift of the Troodos ophiolite, Cyprus. In: A. H. F. Robertson, K. -C. Emeis, C. Richter, and A. Camerlenghi (eds.), Proc. of the Ocean Drilling Program, Scientific Results, v. 160, College Station, TX, p. 545-566.

Robinson, P. T., Melson, W. G., O'Hearn, T., and Schmincke, H. -U., 1983. Volcanic glass composition of the Troodos ophiolite, Cyprus. Geology, 11(7), p. 400-404.

Robertson, A. H. F., 1977. The Moni melange, Cyprus: An olistostrome formed at a destructive plate margin. J. Geol. Soc. London, 133(5), p. 447-466.

Robertson, A. H. F., 1990. Tectono-sedimentary evolution of the Eastern Mediterranean Neatethys: Summaries, questions and answers. Intern. Earth Sci. Cong. on Aegean Reg., 1-6 October. 1990, Izmir, Turkey, IESCA Publ., v. 2, p. 236-278.

Robertson, A. H. F., 1998a. The Mesozoic - Tertiary tectonic evolution of the easternmost Mediterranean area. In: I. Panayides and C. Xenophontos (eds.). Third International Conference on the Geology of the Eastern Mediterranean, Nicosia, Cyprus, September 23-26, 1998, Abstr., p. 61.

Robertson, A. H. F., 1998b. Mesozoic-Tertiary tectonic evolution of the Easternmost Mediterranean area: integration of marine and land evidence. In: A. H. F. Robertson, K. -C. Emeis, C. Richter, A. Camerlenghi (eds.), Proc. of the Ocean Drilling Program, Scientific Results, vol. 160: College Station, TX, p. 723-782.

Robertson, A. H. F., Emeis, K. -C., Richter, C., Camerlenghi, A., (eds.), 1998. Proceedings of the Ocean Drilling Program, Scientific Results, vol. 160: College Station, TX (Ocean Drilling Program), 821 p.

Robertson, A. H. F., and Woodcock, N. H., 1979. Mamonia Complex, southwest Cyprus: Evolution and emplacement of a Mesozoic continental margin. Geol. Soc. Am. Bull., 90(7), p. 651 - 655.

Robertson, A. H. F., and Woodcock, N. H., 1981. Goden Zone, Antalya Complex: Volcanism and sedimentation along Mesozoic continental margin, SW Turkey. Geol. Rundsch., Bd. 70, No 3, p. 1177-1214.

Robinson, P. T., and Malpas, J., 1990. The Troodos ophiolite, new perspectives. In: J. Malpas, E. M. Moores, A. Panayiotou, and C. Xenophontos (eds.), Ophiolites; Oceanic Crustal Analogues. Proc. Symp. Troodos 1987. Geological Survey Department, Ministry of Agriculture and Natural Resources, Nicosia, p. 65-74.

Robinson, P. T., and Malpas, J., 1998. The Troodos ophiolite of Cyprus: New perspective on its origin and emplacement. In: Third International Conf. on the Geology of the Eastern Mediterranean, 23-26 September, 1998, Nicosia, Cyprus. A field trip guide, p. 13 - 26.

Ruzhentsev, S. V., 1976. Marginal ophiolitic allochthons. Moscow: Nauka, 171 p. (In Russian).

Savostin, L. A., Sibuet, J. C., Zonenshain, L. P., Le Pichon, X., and Roulet, M. J., 1986. Kinematic evolution of the Tethys belt from the Atlantic ocean to the Pamirs since the Triassic. Tectonophysics, 123(1-4), p. 1-35.

Sharaskin, A. Ya., Karyakin, Yu. V., Knipper, A. L., and Kheraskova, T. N., 1995. The structural peculiarities, composition, and tectonic implications of Mesozoic deposits of the Dhiarizos Group, southwestern Cyprus. Geotektonika, No 5, p. 83-91.

Silantyev, S. A., 1993. Metamorphic complexes of the Eastern Mediterranean: Metamorphic conditions, protoliths, and original geodynamic settings. Petrologiya, 1(4), p. 449-472.

Sobolev, A. V., Portnyagin, M. V., Dmitriev, L. V., Tsameryan, O. P., Danyushevski, L. V., Kononkova, N. N., Shimizu, N., and Robinson, P., 1993. The petrology of ultramafic lavas and associated rocks of the Troodos Massif, Cyprus. Petrologiya, 1(4), p. 379-412.

Sokolov, S. D., Grigoriev, V. N., Peive, A. A., Batanova, V. G., Krylov, K. A., Luchitskaya, M. V., and Aleksyutin, A. A., 1996. Elements of structural and lithologic ordering in serpentinite melanges. Geotektonika, No. 1, p. 1-16.

Spray, J. G., and Roddick, J. C., 1981. Evidence for Upper Cretaceous transform fault metamorphism in west Cyprus. Earth Planet. Sci. Lett., 55(2), p. 273-291.

Swarbrick, R. E., 1980. The Mamonia complex of S. W. Cyprus; A Mesozoic continental margin and its relationship to the Troodos complex. In: A. Panayiotou (ed.), Ophiolites, Proceedings International Ophiolite Symposium Cyprus, 1979, Ministry of Agriculture and Natural Resources Geological Survey Department, Nicosia, p. 86-92.

Swarbrick, R. E., 1993. Sinistral strike-slip and transpressional tectonics in an ancient oceanic setting: the Mamonia Complex, southwest Cyprus. J. Geol. Soc., London, 150(2), p. 381-392.

Swarbrick, R. E., and Naylor, M. A., 1980. The Kathikas Melange, southwest Cyprus: Late Cretaceous submarine debris flows. Sedimentology, 27(1), p. 63-78.

Swarbrick, R. E., and Robertson, A. H. F., 1980. Revised stratigraphy of the Mesozoic rocks of southern Cyprus. Geol. Mag., 117(5), p. 547-563.

Taylor, R. N., and Nesbitt, R. W., 1992. A geochemical transect of the Izu-Bonin arc-trench system. Ofioliti, 17(1), p. 57-71.

Thy, P., and Moores, E. M., 1988. Crustal accretion and tectonic setting of Troodos ophiolite, Cyprus. Tectonophysics, 147(3-4), p. 221-245.

Turner, W. M., 1973. The Cyprean Gravity nappe and the autochthonous basement of Cyprus. In: K. A. De Jong and R. Scholten (eds.), Gravity and Tectonics, John Wiley & Sons, New York, p. 287-301.

Varga, R. J., and Moores, E. M., 1985. Spreading structure of the Troodos ophiolite, Cyprus. Geology, 13(12), p. 846-850.

Willett, S., Beaumont, C., and Fullsack, P., 1993. Mechanical model for the tectonics of doubly vergent compressional orogens. Geology, 21(4), p. 371-374.

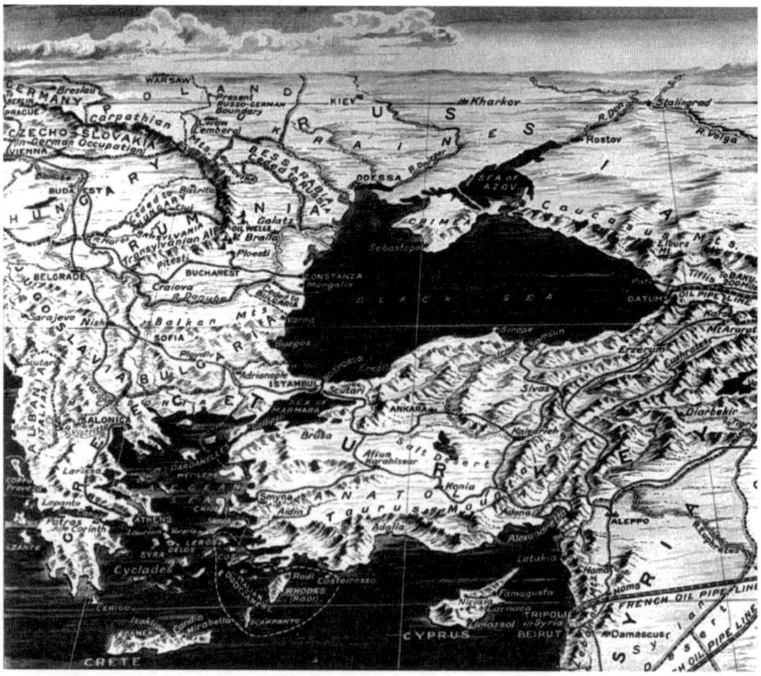

The Levant: Battleground of the Empire Today. World War II map from the Illustrated London News, 16 November 1940, p. 621. Map image (No. 81811) courtesy of the University of San Diego History Department website at: http://history.sandiego.edu/gen/WW2Index/picindexmaps.html.

PART II

Syria
Introduction

Valery A. Krasheninnikov
Geological Institute of the Russian Academy of Sciences, Pyzhevskiy pereulok 7, 109017 Moscow, Russia

The Syrian Arab Republic is located in the northwestern part of the Arabian Peninsula, and is one of the countries of the Near East. It is bounded to the north by Turkey, to the east and southeast by Iraq, to the south by Jordan, to the southwest by Israel, and to the west by Lebanon and the Mediterranean Sea. The total area of Syria is about 185,000 km^2. In the discussion below it is worthwhile noting that the translation of Arabic geographical terms into French or English can differ considerably between maps.

MAIN FEATURES OF THE LANDSCAPE OF SYRIA

Syria is characterized by strong contrasts of landscape. The highest point is Jebel El-Sheikh, or Mt. Hermon (2814 m), which crowns the mountain massif at the border with Lebanon and Israel. The lowest altitude is in the lower reach of the Yarmouk River, which flows to the Sea of Galilee (Kinneret) in Israel, located about 220 m below the Mediterranean sea level. Thus the range of topographic relief in Syria reaches a very significant value of about 3,000 m.

The territory of Syria can be subdivided into several provinces with vastly differing landscapes. These areas can be defined as follows:

1) The mountain areas of Jebel An-Nusseiriyeh and Kurd-Dagh,
2) The mountain areas of the Anti-Lebanon and Palmyrides,
3) The Aleppo Plateau,
4) The plateau of the Syrian Desert,
5) The Mesopotamian (or Al-Furat or Jezireh) plain,
6) The volcanic mountains and plateau of Jebel Ad-Drouz and Hauran.

The mountain areas of Jebel An-Nusseiriyeh and Kurd-Dagh are located in the northwestern part of Syria. The Jebel An-Nusseiriyeh range (or the Coastal Range) extends north-south parallel to the Mediterranean coast. It is asymmetrical, with the western slope of the range gradually dropping in the direction of the sea, while its eastern slope is cut by a steep escarpment, up to 1,400 m in height, which separates this range from the Al-Ghab depression. Therefore the axis of the range (the water divide) is displaced to the east. The Jebel An-Nusseiriyeh range is characterized by a smooth watershed with karstic sinkholes as well as other forms of karst relief, created on the surface of Jurassic and Cretaceous limestones and dolomites. The slopes of Jebel An-Nusseiriyeh are cut by narrow cliff-bounded canyons. To the south of Jebel An-Nusseiriyeh, as a result of intensive Pliocene-Quaternary volcanism, vast plateau areas are overlain by the Shin and Tel Kallakh volcanics.

To the north of the Jebel An-Nusseiriyeh range lies the Baer-Bassit Massif, with the Nahr El-Kabir River valley in between. Near the Baer-Bassit Massif are the highly dissected mountains of Jebel Aqraa. The main part of Jebel Aqraa, whose maximum height is 1,739 m, lies in the territory of Turkey. Similarly the Kurd Dagh ridge also extends into Turkey, and within Syria it rises slightly above 1,000 m.

The mountain ranges of the Anti-Lebanon and Palmyrides extend in a NE-SW direction. The northwestern part of this area is occupied by large mountain massifs having an oval form. These include – the Hermon (Jebel El-Sheikh), Anti-Lebanon, Jebel Nebek, Jebel Ash-Shoumariyeh, Jebel El-Balas, Jebel Shaar, and Jebel Bouaida. They correspond to large box-like and dome-like anticlinal folds and are well-expressed in the landscape. Usually these ranges exhibit smooth topography on their watersheds, while their slopes are cut by steep ravines. The highest points of the Anti-Lebanon are Mount Hermon (Jebel El-Sheikh) (2,814 m), Talaat-Mussa (2,616 m), and Jebel Nabi-Baruch (2,508 m). The peaks of the other ranges reach 1,000-1,800 m. The southeastern part of the Palmyrides is characterized by narrow and cliff-lined ridges, called by the general name Jebel Ash-Sharqiyeh. These elongated ridges are produced by linear anticlinal folds. The southeastern slopes of these ridges are usually very steep and are frequently cut by large faults. The highest elevations of this part of the Palmyrides are at Mount Abou-El-Ata (1,515 m) and Jebel Zubeideh (1,401 m). In the northeast the Palmyrides terminate as the low Jebel El-Bichri Massif which has the form of a flattened dome with an oval outline. The highest part of this uplifted area, called Jebel Karbi, reaches only 867 m.

Between the ridges of the Anti-Lebanon and the Palmyrides as well as along the periphery of these mountains are located a series of synclinal depressions: Ad-Daww, Jairoud, Homs, Tourkmaniyeh, Sabkhet-Moukh, Damascus, and others. Usually they are expressed in the landscape as flat-bottomed depressions with internal drainage. Their interiors are at elevations of 500-800-m.

The Aleppo Plateau in the north of Syria occupies a vast, comparatively flat area between the Palmyrides in the south and Jebel An-Nusseiriyeh and Kurd Dagh in the west, and the Al-Furat Valley in the east. It is expressed as an almost flat surface with elevations of 600-800 m. Along its western rim the plateau is bounded by the low mountains of Jebel Az-Zawiyeh (Jebel Az-

Zaouiye), with elevations of about 800 m and relict cones of ancient volcanoes. The central part of the Aleppo Plateau is occupied by the table-like mountains of Jebel Hass and Jebel Shbeit. Their flat summits are covered by basaltic flows. Some depressions, with interior drainage, are located near these mesa-like elevations. The largest of these depressions is occupied by the salty lake of Jabboul. In the north the relatively flat landscape of the Aleppo Plateau has some higher elevations – Jebel Smaan, Parsa-Dagh (855 m), El-Bab (505 m), and Mennbidj (516 m), and some others. These elevations are composed of hard rocks (limestones) with an uneven weathered surface. Near its eastern margin the Aleppo Plateau is gradually replaced by the valley of the Al-Furat River.

The plateau of the Syrian Desert encompasses the southern part of Syria. It is located south of the Palmyrides. This desert represents a flattened surface with absolute elevations of 500-800 m, stretching over hundreds of kilometers due to the practically horizontal bedding of the hard Paleogene limestones and flints. Sometimes this surface appears almost armored because of a dense several meters thick carbonate-siliceous crust which overlies softer limestone and marls. This crust has a Quaternary age and originated under very hot desertic conditions. The surface of the desert is covered by an ubiquitous blanket of sharp pebbles (mainly flints with desert varnish). Consequently the Syrian Desert is a classic example of the typical stony desert (Hamad) of the Arabian Peninsula. The uniform nature of the desert is broken by structural escarpments rising some tens of meters (Tarag-El-Aalab, Roumanna, Jebel Tanf), sabkhas with flat clayey surfaces (Khabra-Lokhta, Khabra-Oum-Moueil, Khabra-Mredifah, and many others), depressions with steep slopes and internal drainage (Ej Jouaif), and wadis which can be traced for tens to a few hundreds of kilometers (Wadi Miah, Wadi Swab (Souab), Wadi Ouahar and others). In the west the Syrian Desert is bounded by the volcanic massif of Jebel Ad-Drouz (Jebel Arab) and in the east it gradually merges with the valley of the Al-Furat River.

The volcanic mountains and plateaus of Jebel Ad-Drouz, Hauran, and Safa occupy the southwestern corner of Syria. Their origin is connected with the intensive young Neogene-Quaternary volcanism. This area is a single mountain massif with numerous volcanic cones. The central part of Jebel Ad-Drouz has elevations reaching 1,698 m at Tell Qualad, 1,546 m at Tell el Massie, 1,360 m at Salkhad, and 1,220 m at Chaqqa. On the sides of this massif these elevations are gradually replaced by volcanic plateaus, with heights of 600-900 m. The relief of this volcanic region is extremely varied. In the areas of ancient lavas the surface is comparatively smooth. Where young, especially Holocene, lava flows exist, the relief is very dissected, and fresh forms of typical volcanic relief can be observed such as wavy surfaces of lava flows, small ridges of distorted lavas, traces of blasted gas vesicles, lava tubes, and channels. The surface of this plateau is marked in many places by numerous and well-preserved volcanic cones. Near its western margin (the Golan Heights) the volcanic plateau is bounded by an escarpment which drops down to the Jordan Valley and the Sea of Galilee. Near its eastern margin Jebel Ad-Drouz also has an escarpment but it is not as pronounced as on the west, and the landscape has a gradual transition to the Syrian Desert.

The Mesopotamian Plain, also called Jezireh, occupies a vast area in the east and northeast of Syria. It is characterized by a very flat surface with elevations of 200-300 m. In the northern part of this flat plain there are some low and slightly dissected mountains and hills: Jebel Abd El-Aziz, Jebel Beida, Jebel Jeribe, Jebel Tchembe, Toual Aba (Tuwal El-Aba), and Qaratchouk Dagh (Qaratchok Darh). The most significant with respect to size and height is Jebel Abd El-Aziz, with a maximum elevation of 920 m (El-Rhara). Some small elevations result from extrusions of young volcanic rocks. In the southeast of the Mesopotamian Plain, near the border with Iraq, there are some closed flat-bottomed depressions with internal drainage (Sabkha el Barrhouite, Roda, Bouara, Sabkha-at-Tauil) where all elevations are less than 200 m (184 m near Bir-Mrechasen and 167 m near Sabkha Al-Boghars). The Mesopotamian Plain is dissected by the wide flat valley of the Euphrates (Al-Furat) River and its two tributaries (Nahr El-Khabour and Nahr El-Balich Rivers). These rivers have numerous terraces with abundant stone implements, arrowheads, etc. of the earliest human civilizations.

From our knowledge of the geological structure of Syria, it is quite evident that the variety of relief can be explained by the existence of various tectonic zones which have different histories of geological development. Northern Syria lies adjacent to the mountain chains of the Alpine-Himalayan belt with its complicated nappes. Southern Syria is bounded by the northern slope of the Arabian platform. Central Syria has the system of intra-platform folded chains of the Palmyrides. Eastern Syria is occupied by the vast depression of the Mesopotamian Basin where the world's largest oil and gas provinces are located. Western Syria is more complicated. Near the Mediterranean coast there are the folds of the Anti-Lebanon and the dome-like uplift of Jebel An-Nusseiriyeh. Here one can also see the northern extension of the transform fault subdividing the African and Arabian Plates, which is a continuation of the gigantic system of rift valleys of the Red and Dead Seas.

Thus it is no great surprise that this rich landscape attracted several generations of geologists who have left their imprint on the study of the geology of Syria. Their professional analysis of the entire geological framework allow an understanding of the geological structure of this country which is important not only for the solution of regional Near Eastern problems, but also for analysis of the general global geological questions. Therefore we have to feel a great respect for those who preceded us.

HISTORY OF GEOLOGICAL INVESTIGATIONS IN SYRIA

The history of geological investigations in Syria can be arbitrarily subdivided into four periods.

The First Period (mid 19th to beginning 20th century)

During this period the territory of Syria was studied by some well known European scientists – L. Lartet, M. Blanckenhorn, L. Kober, O. Fraas, C. Diener, F. Noetling, B. Doss, G. Tyrrell, G.

Prof. Max Ludwig Paul Blanckenhorn (1861–1947), Germany. A pioneer in the study of the geology of the Levant. He produced numerous publications devoted to the general geology, stratigraphy, and paleontology of the Cretaceous, Paleogene and Neogene sediments of Syria.
A biography and autobiography of Prof. Blanckenhorn may be found at the end of this introduction.

Zumoffen, and E. Passemard.

The earliest data relevant to the geology of Syria were given in the publications of Lartet (1866, 1869, 1884). In a general survey of the geology of Palestine Lartet mentioned the wide distribution of volcanic rocks in southwestern Syria and the presence of siliceous and bituminous rocks.

Fraas (1877), Diener (1887), and Noetling (1886, 1887) can be treated as pioneers in the investigation of the stratigraphy of the Jurassic and Cretaceous deposits of the Anti-Lebanon.

Numerous papers devoted to various problems of the geology of Syria were published by Blanckenhorn between 1889 and 1940. He paid special attention to the stratigraphy of Cretaceous, Eocene, Oligocene, Miocene, and Pliocene-Quaternary sediments of the Anti-Lebanon, Jebel An-Nusseiriyeh, and the Nahr El-Kabir valley (1889, 1890a, b, c, 1897, 1925, 1927, 1934, 1935). For the first time he described the presence of Danian sediments (1916), and the general characteristics of the Neogene of the Levant countries (Blanckenhorn and Oppenheim, 1927). Blanckenhorn also described the exposures of serpentinites and gabbroids in the Baer-Bassit Massif and the longitudinal faults of western Syria. In his opinion, the mountain structures of Syria and Lebanon as well as the outflows of basalts were connected with vertical motions along these faults (1891, 1914). In later papers Blanckenhorn analyzed the geology of Syria within the framework of the general geological situation of the whole Near East (1912, 1914, 1931, 1940).

Mirroring the new tectonic ideas of the Alpine geologists,

Prof. Leopold Kober (1883–1970), Austria. He considered the ophiolite complex of the Baer-Bassit Massif of Syria as an allochthonous nappe, displaced from the folded belt of Taurus. Later investigations confirmed his idea (though global tectonics gave another explanation to this process).

Kober (1915) presented a little-acknowledged prescient and iconoclastic scenario for the origin of these features. He interpreted the mountain structures of Western Syria and Lebanon as folded edifices resulting from tectonic movements in the area of Taurus. The boundary between the Arabian platform and the Alpine folded belt was drawn in the lower reaches of the Nahr El-Aasi (Oront) River. Kober explained the distribution of the 'green-colored rocks' of the ophiolite complex south-east of this line by the development of nappes – *i.e.* displacement in the direction of the platform of terranes from the Taurus geosyncline. Kober's mobilistic ideas were of great importance to subsequent tectonic reconstructions.

The publications of Doss (1886) and Tyrrell (1930) dealt with the petrography and morphology of the basaltic cover of Hauran and Jebel Ad-Drouz. The monograph of Zumoffen (1926) touched on some general problems of the geology of Syria and Lebanon (with emphasis on describing the stratigraphy of the Jurassic and Cretaceous deposits). Passemard (1926) studied the stratigraphic sequence of the Quaternary terraces in the Al-Furat Valley.

The Second Period (beginning of 20th century to about 1950)

The second period of geological investigations in Syria runs from the beginning of the 20th century approximately up to 1950, *i.e.* the first years of an independent Syria. It was characterized by intensive geological studies carried out by scientists of various countries with the main contribution by French geologists. The driving force of these investigations was apparently connected with the great interest in the petroleum basins of the Near and Middle East.

In 1924 a group headed by L. Dubertret of mostly French geologists and paleontologists (H. Vautrin, A. Keller, E. David, E. Basse, C. Dechaseaux, J. Pfender, L. Doncieux, G. Delpey, J. Roger, F. Roman, R. Dubertret, R. Wetzel, J. Haller, A. Bourgoin, G. Cottreau, H. Cizancourt, W. Fish, and E. Vaumas) initiated a geological mapping of various parts of the Levant countries. Simultaneously they were engaged in more academic investigations of general stratigraphy, lithology, tectonics, petrology of volcanic rocks, geomorphology, and petroleum geology. These studies included not only the territory of Syria but also Lebanon and the southwestern part of modern Turkey (Hatay, Kurd-Dagh, and Antakya basin).

A main achievement of this activity was the creation of a set of geological maps at various scales. The first geological map of Syria and Lebanon at a scale of 1:1,000,000 was published in 1933 (Dubertret, 1933a). Later it was reissued, with corrections and additional remarks (Dubertret, 1943). The explanatory note to this map contains a description of the main geological structures and the sequence of lithological units. The note was primarily concerned with the western part of Syria (Anti-Lebanon, Jebel An-Nusseiriyeh and Kurd-Dagh). The coverage of the eastern part of this country (the Palmyrides, Syrian Desert, Jebel Ad-Drouz and Mesopotamia) reflects the reduced interest of the French geologists, and the geological data obtained, along selected traverses, are less impressive. Thereafter the following maps were published: the geological map of the Middle East at a scale of 1: 2,000,000 (Dubertret, 1942); the lithologic map of Syria and Lebanon (Dubertret, 1944) at 1:1,000,000 scale; geological maps at a scale of 1:50,000 of the Rachaya-N (Dubertret, 1945), Zabdani (Dubertret, 1949a), Saida (Dubertret, 1949b), Rayak (Dubertret, 1950a), Baalbek (Dubertret, 1950b), Beyrouth (Dubertret, 1951a), Tripoli (Dubertret, 1951b), Zahle (Dubertret, 1953b), Tartous (Dubertret, 1958), and Rachaya-Sud (Hermon) (Dubertret, 1960) sheets. Geological maps at a scale of 1:200,000 with explanatory notes were compiled for the Antakya (Dubertret, 1953a), Tripoli (Dubertret and Wetzel, 1949), and Beyrouth (Dubertret, 1954) sheets, with a summary sheet for all of Lebanon (Dubertret, 1955a). Almost all these maps include some areas of Syria (Anti-Lebanon, Jebel An-Nusseiriyeh, Baer-Bassit, Kurd-Dagh) or are devoted to adjacent areas with similar geology. The map of the Hermon Massif, which is characterized by a complicated geological structure, is of special importance. Dubertret's (1942) geological map of the Middle East attempts to summarize the geological data of this vast realm.

In the explanatory notes to the above geological maps and in numerous publications, Dubertret and his collaborators expressed their views on a wide variety of subjects: *e.g.* the problems of stratigraphy; lithological composition of sediments; petrography of the basaltic rocks in various parts of Syria and the igneous-metamorphic ophiolite complex of Baer-Bassit; tectonic structure of Syria; peculiarities of its geological development; perspectives concerning the petroleum resources and the distribution of other kinds of raw materials and minerals.

Very serious attention was given to questions of stratigraphy as a basis for general geological reconstructions. This direction of investigation was accompanied by paleontological work with an analysis of the fauna of ammonites, sea urchins, pelecypods, gastropods, and foraminifera (nummulitids).

As a result, the Jurassic, Cretaceous, Paleogene, and Neogene deposits of Anti-Lebanon, Jebel An-Nusseiriyeh, the Palmyrides, Jebel Bishri and the Nahr El-Kabir Valley were subdivided into series, subseries, and sometimes stages substantiated by the study of various groups of fossils. The following stratigraphic subdivisions have been identified: Lower (?), Middle and Upper Jurassic, Aptian, Albian, Cenomanian-Turonian, Lower Senonian, Campanian-Maastrichtian, Lower, Middle and Upper Eocene, Oligocene, Lower Miocene (Burdigalian), Middle Miocene (Vindobonian), Upper Miocene (Pontian), and Pliocene (Dubertret, 1933c, d, 1934b, 1936, 1937a, b, e, 1938, 1940b, 1950c, 1951c; Dubertret, Keller and Vautrin, 1932a, b; Dubertret and Vautrin, 1937a, b, c; Dubertret and Roger, 1938; Dubertret and Cotreau, 1938; Dubertret R., Doncieux, Vautrin and Dubertret L., 1938; Jacquet, 1933; Keller, 1933; Keller and Vautrin, 1937; Pfender, 1937; Roger and Dubertret, 1938; Roger, 1940; Roman and Doncieux, 1927; Roman, 1940; Vautrin, 1933a, b, c, 1934a; Cizancourt, 1934). As a rule, it was determined that the Mesozoic–Cenozoic boundary is located within a monotonous sequence of marls, calcareous clays, and limestones.

Studies on the stratigraphy of marine and continental facies of the Pliocene and Quaternary sediments were carried out in various parts of Syria. The investigations of Dubertret (1937b) and Wetzel and Haller (1945-1948) created the basis for the stratigraphy of marine Pliocene and Quaternary sediments on the Levantine coast of the Mediterranean Sea. Neogene and Quaternary deposits of the coastal area were also studied by Bourcart (1940). In the vicinity of Al-Latheqiyeh he identified Miocene, Pliocene, and Quaternary sediments. Vaumas (1953a,b, 1954) described marine abrasional terraces on the Mediterranean coast located near Al-Latheqiyeh, Tartous, and Jebleh at elevations of 15-20 m, 34-45 m, 60-80 m, and 90-120 m. He noticed considerable deformation of the surface of these terraces. Young continental deposits have been studied in the Damascus basin and in the Al-Ghab (El-Rhab) graben (Dubertret and Vautrin, 1938; Vaumas, 1957). In the Damascus basin two formations of continental deposits were singled out. The lower subdivision was referred to the Miocene (the Pontian Stage) and the upper one was treated as Pliocene-Quaternary. According to Vaumas, the Al-Ghab graben originated from neotectonic motions and was filled by sediments of the Upper Pliocene and Quaternary; according to Dubertret, the onset of formation of this graben occurs a bit earlier (the Pontian Stage of the Miocene).

Prof. Louis Dubertret (1904–1979), France. His investigations opened a new epoch in our knowledge of the Syrian geology. They addressed a wide circle of geological problems, including geological mapping, general geology, stratigraphy, paleontology, lithology, tectonics, magmatic activity, and mineral resources.

Perves (1946–1948) started investigation of Late Quaternary formations – alluvial sediments with flinty tools of terraces of the Al-Furat and Nahr El-Khabour Rivers and proluvial cones with flinty tools exposed near Dmeir (the southwestern Palmyrides).

Some papers by Dubertret (1929, 1933e, 1940a, 1952) were devoted to the problem of the basaltic volcanism. He distinguished three main periods of magmatic activity:

1) basaltic volcanism of the Late Jurassic–Early Cretaceous;

2) formation of mafic and ultramafic rocks of the ophiolite complex of the Maastrichtian;

3) basaltic volcanism of Neogene-Quaternary time.

The last period has been the most significant with respect to its volume. It started in the Miocene but the peak of this process coincides with the Pliocene-Quaternary interval. Dubertret discovered basaltic cover of different ages in Hauran and Jebel Ad-Drouz; he estimated their thickness and described their morphology. He believed that although basaltic flows are located not far away from the deep longitudinal faults of the Jordan Valley, outpourings of basaltic lavas have no direct relation to these faults.

The general tectonic setting of Syria was considered by Dubertret in a series of publications (1930, 1932a,b, 1933b, 1947a,b). His views evolved with time. At first, Dubertret shared Kober's ideas. The area of Amanus was treated by him as a frontal zone of the Taurus belt; an assemblage of igneous basic and 'green-colored' metamorphic rocks in Baer-Bassit and Kurd-Dagh was explained as a result of shifting of nappes, and the depressions of the Lower Oront River and Qara Su River were referred to a marginal fore-deep. Dubertret considered the Anti-Lebanon and Jebel An-Nusseiriyeh mountain ranges to be dome-like uplifts connected with sublongitudinal faults, whereas the Palmyrides fold system was treated as an *en echelon* structure, subordinate to the main longitudinal structures. Dubertret suggested the possibility of significant vertical displacements along the large longitudinal faults, which then resulted in formation of the Anti-Lebanon and Lebanon folds. Later on, Dubertret believed that vertical motions of separate blocks of the basement formed these structures. As far as the anticlinal folds of Jebel Abd El-Aziz and Jebel Sinjar were concerned, Dubertret supposed that they were a continuation of the folded zone of Northern Iraq.

The problem of the origin of an assemblage of 'green-colored' ultramafic and mafic magmatic and metamorphic rocks, developed in the north-west of Syria (Baer-Bassit, Kurd-Dagh) and the adjacent areas of Turkey (Hatay), occupied a significant place in Dubertret's investigations (1936, 1937c,d, 1939, 1947b, 1951c, 1952, 1953a, 1955b). He refuted Kober's hypothesis of the displacement of the 'green-colored' rocks from the flank of the folded (geosynclinal) Taurus belt, as well as the idea about their intrusive origin. Dubertret proposed his own hypothesis of 'underwater formation' of the ophiolitic assemblage. According to this hypothesis, the ultramafic and basic rocks of Baer-Bassit and Kurd-Dagh originated in the Maastrichtian as a result of underwater outpourings. The intrusive appearance of these rocks is explained by assuming that a considerable part of the melted magmatic mass was intruded into the basal horizons of a pile of effusive rocks and was crystallized under nearly hypabyssal conditions. These magmatic outflows took place in the marginal part of the Arabian platform. Consequently the belt of 'green-colored rocks' of Baer-Bassit and Kurd-Dagh does not indicate a transition to the folded Alpine belt. The southern boundary of this belt was located much to the north.

Dubertret suggested that in the mountains of Amanus there was an angular unconformity between the Ordovician and Devonian that was connected with the Caledonian folding. He concluded that within Syria the folded basement of the Arabian platform is heterogeneous *e.g.* the northwestern marginal part of the platform has a Caledonian age and the main southeastern part of the platform is characterized by a more ancient Precambrian age. According to Dubertret, the intrusive and metamorphic assemblage of Baer-Bassit and Kurd-Dagh was connected with the younger Caledonian zone of the Arabian platform.

Referring to the petroleum resources of Syria, Dubertret (1934a, 1935) wrote that the most important area, where accumulations of oil and gas can be expected, is the northeastern part (Mesopotamia). He indicated that the most favorable structures for drilling were Qaratchouk, Jbissa, Tchembe and Ghouna. According to Dubertret, the occurrence of bitumens (Kfarieh) in the coastal belt of the Mediterranean Sea was related to the secondary process of migration of crude oil into a succession of Cretaceous limestones.

During this second period of investigation into its geological history, scientists of other countries also carried out similar studies in Syria. They left a noticeable imprint in the field of petrology, stratigraphy, and tectonics.

Some publications were devoted to the stratigraphy of the Jurassic deposits of Syria. Arkell (1956) restudied a collection of ammonites, which had been gathered by French geologists. He made more precise age determinations for some rock units of the Hermon Massif. Haas (1955) published a monograph with a detailed description of a rich collection of ammonites taken from the Jurassic deposits of the Mejdal Shams section.

The stratigraphy of the Upper Cretaceous of the Palmyrides was the subject of Chenevart's (1950a,b) investigation which created a facies map combined with isolines. Ammonites of the Cretaceous deposits of Syria were studied by Basse (1937, 1940).

Following in the footsteps of Kober, Krenkel (1924, 1941) treated the tectonic structures of Western Syria and Lebanon as folded edifices associated with horizontal tectonic movements in the area of Taurus. He formulated the idea that the Syrian Arc is a frontal part of the folded Taurus system. This arc originated as a result of compression from the margin of the African craton.

The complicated problem of the position of the contact between the Arabian platform and the Alpine folded belt was discussed by Stchepinsky (1947), who drew this boundary along a line from Al-Latheqiyeh (Syria) to Kilis (Turkey).

The petrographic composition of the metamorphic rocks of the Baer-Bassit Massif (amphibolites, amphibolitic shists, quarzites, and marmors) was studied by Chenevoy (1952, 1959). Based on the comparison of this assemblage of rocks with those of Turkey, he wrongly concluded they were of Pre-Cambrian age.

Consequently, between 1920 and 1950 many investigations of the geological structure of Syria were carried out. These studies progressed in almost all the geological disciplines and defined those remaining geological problems of this country that needed to be solved. While in many cases modern geological science treats the structure and the geological history of Syria in a different way, we must acknowledge with great respect and gratitude the huge contribution made by our predecessors during this period.

The Third Period (from about 1950 until about 1990)

The third period in the geological study of Syria lasted from a few years after Syria became independent in 1944 until some time around 1990. With independence, two Syrian geological organizations were created, the General Establishment of Geology and Mineral Resources, and the Syrian Petroleum Company. The first was responsible for general studies of Syrian surface geology, *e.g.* geological mapping, geological research, prospecting for mineral resources (iron ores, phosphorites, oil shales, building stone, cement material, etc.), and for geophysical and geochemical surveys. The second organization carried out various subsurface geological and geophysical studies related to petroleum geology, and mainly confined to the eastern and northeastern regions of Syria. Although initially consisting of a limited number of geologists, in the 50 plus years of their existence, these two organizations have grown into large organizations with the capability of working in different branches of scientific and applied geology.

Later on, geological departments were formed in the Universities of Damascus, Aleppo, and Al-Latheqiyeh. Due to dry climatic conditions and vast desert areas, the problem of water resources is of extremely important significance to Syria. Therefore two additional organizations have been created in Damascus and Homs which are responsible for hydrogeological studies. Consequently, Syria now possesses a network of capable geological organizations with experienced and highly qualified personnel.

Over the years, the General Establishment of Geology and Mineral Resources and the Syrian Petroleum Company have collaborated on the study of some areas of Syrian geology with specialists from other countries like France, the former West and East Germany, USSR, Czechoslovakia, Yugoslavia, etc., on the basis of contracts, licenses for exploratory drilling, and inter-governmental agreements about scientific-technical cooperation.

In 1958–1961 a group of Soviet geologists, headed by V. P. Ponikarov of the USSR Ministry of Geology, carried out the geological mapping at a scale of 1:200,000 of all of Syria. This expedition included about 70 geologists and was subdivided into 14 geological field parties. A senior geologist, well acquainted with the methods of geological mapping, was in charge of each field party. The party chiefs were V. G. Kazmin, V. N. Shatsky, V. V. Kozlov, A. V. Razlalyaev, E. D. Soulidi-Kondratyev, V. A. Faradzhev, I. A. Mikhailov, A. K. Oufland, A. B. Galaktionov, A. A. Krasnov, A. A. Maksimov, V. Ya. Medvedev, B. Ya. Ponomarev, and V. I. Petrov.

During this period the geological mapping at a scale of 1: 200,000 was carried out. The mapping produced 21 sheets; some regions with especially complicated tectonic structure were mapped at a scale of 1:50,000. These were: the Hermon Massif in Anti-Lebanon, the Baer-Bassit Massif; and Kurd-Dagh and the Aafrine graben in northwestern Syria. Before mapping began the entire area of Syria was topographically mapped by Soviet aerial photogrammetrists at a scale of 1:25,000. These topographic maps allowed the geological data to be reliably located within the geographical setting, something which was not always possible with the earlier French maps.

Very serious attention was given to substantiating the stratigraphy of the Mesozoic and Cenozoic deposits. Rich collections of macrofauna were taking to the USSR where paleontological determinations were made by the following experienced scientists: G. Ya. Krimgolts (Jurassic ammonites and belemnites), V. P. Makridin (Jurassic brachiopods), A. Yungherman (Jurassic pelecypods), A. A. Atabekian and V. P. Rengarten (Cretaceous ammonites), N. N. Bobkova, N.A. Cheltsova, and M. R. Jalilov (Cretaceous pelecypods and gastropods), N. A. Poslavskaya and A. Kallis (Jurassic and Cretaceous sea urchins), L. V. Mironova (Paleogene molluscs), R. L. Merklin, G. F. Lunghersgauzen, and I. A. Bertels-Uspenskaya (Neogene molluscs), N. A. Naidina (Neogene ostracods), and G. I. Nemkov (Paleogene nummulitids). After completion of the age determination these fossil collections were returned to Syria, where they have since been lost.

Flinty tools of Late Quaternary age were identified and studied by K. M. Mirzaev, A. V. Vinogradov, and M. A. Itina.

Vast areas of Syria are covered by Paleogene deposits. They were subdivided by means of the subglobal planktonic foraminifera zonal scale (V. A. Krasheninnikov). The detailed schemes for the Neogene of the Mediterranean and the Miocene of the Mesopotamian Basin (V. A. Krasheninnikov) were worked out with the help of planktonic and benthic foraminifera. The boundaries of the Paleogene and Miocene mapping units were regularly checked by reference to planktonic foraminifera assemblages that allowed correlation of these boundaries with chrono-stratigraphic subglobal levels.

At the same time Soviet geophysicists carried out aeromagnetic mapping at the scale of 1:200,000 for all of Syria and gravimetric mapping at the same scale for most of the country. The Lower Cretaceous iron ores of Kurd-Dagh (Radjou) were prospected with the aim of determining the economic significance of this occurrence. An investigation of the phosphate-bearing layers of the Upper Cretaceous and Lower Eocene of the Palmyrides and the Rutbah uplift resulted in the discovery of some deposits of phosphorites with enormous commercial potential.

The USSR geologists carried out all these investigations together with their Syrian colleagues, who included Soutki Al-Roch, S. Atfeh, N. Malki, M. Midani, I. Katmeh, and others. Dr. S. Atfeh played a very significant role in the investigation of the Upper Cretaceous phosphorite deposits of the Palmyrides. Quarries in this area now produce a great amount of high quality phosphatic ore.

These studies resulted in the publication in the USSR (in English) of 21 sheets of the geological map of Syria at the scale of 1:200,000 (1966) and 3 sheets at the scale of 1:50,000 (1968) with appropriate explanatory notes. The geological maps of Syria with explanatory notes at the scales of 1:500,000 (1967) and 1:1,000,000 (1966) and the tectonic map of Syria at the scale of 1:1,000,000 (1966) were compiled from these 24 more detailed sheets.

Monographs in Russian, devoted to the general geology of Syria (Ponikarov et al., 1969), the Paleogene zonal stratigraphy of Syria (Krasheninnikov, 1965a,b) and the Miocene stratigraphy of the Mediterranean by means of foraminifera (Krasheninnikov, 1971), were published in the USSR along with numerous other articles. Various aspects of the geology of Syria were studied in the doctoral dissertations of E. D. Soulidi-Kondratyev, V. G. Kazmin, A. V. Razvalyaev, V. V. Kozlov, A. K. Oufland, A. B. Galaktionov, V. I. Petrov and V. I. Pokryshkin.

Prof. Victor Pavlovitch Ponikarov (1917–1996), Russia. Under his leadership the geological mapping at the scale of 1:200 000 of the entire territory of Syria has been carried out. Based on these data, the geological maps of Syria at the scales of 1:500,000 and 1: 1,000,000 have been compiled. He is the author of the tectonic map of Syria at the scale of 1:1,000,000 (1966).

In the 1950s and 60s the German 'Concordia' concern received 49 concessional parcels of land in a small part of north-eastern Syria on which it carried out geological mapping at the scale of 1: 50,000, geophysical studies, and drilling which discovered several oil fields. Concordia's geologists also published on their work. Summarizing the work of earlier geologists, R. Wolfart (1966, 1967) published two monographs dealing with general features of the geology, hydrogeology, and the mineral resources of Syria. Two very important articles were published by H. Weber (1963, 1964) concerning the subsurface geology of the northeastern Mesopotamia based upon drilling data and seismic studies. Stoppel (1970), Weber (1968), Wolfart (1965), and Plumhoff (1970) have published on paleontological studies devoted to various groups of fauna from the Paleozoic deposits of Jebel Abd El-Aziz.

Some interesting geological structures and deposits are located in Syria which are important for the solution of some principal questions about the regional geology of the eastern Mediterranean as well as certain general geological problems. Therefore, in the period from 1983 to 1990 a geological group of the USSR Academy of Sciences headed by the author, together with geologists of the

General Establishment of Geology and Mineral Resources and the Syrian Petroleum Company, have been working on some scientific topics in various regions of Syria.

The following results should be mentioned:

1) The tectonic structure of the Baer-Bassit ophiolitic Massif has been studied (Knipper et al., 1988);

2) A new tectonic map of Syria at the scale of 1:500,000 has been completed, with accompanying description (Kopp and Leonov, 2000);

3) A monograph about the stratigraphy, lithology and geochronology (K/Ar and Rb/Sr methods by glauconite) of the Paleogene deposits of Syria was published (Krasheninnikov, Golovin, Mouraviov et al., 1996);

4) A stratigraphic correlation of pelagic Paleogene sediments with planktonic foraminifera and shallow-water limestones with nummulitids has been done (Krasheninnikov and Ashourov, 1995);

5) The problem of correlation of Miocene deposits within a system consisting of an oceanic basin (North Atlantic), a mediterranean type basin (eastern Mediterranean), and a semi-closed basin (the Mesopotamian Basin) has been addressed in the monograph by V. A. Krasheninnikov et al. (1999);

6) The stratigraphy, lithology and technical properties of oil shale deposits on the northern bank of the Yarmouk River have been studied by Puura et al. (1984);

7) The stratigraphy of the Jurassic deposits of Anti-Lebanon, Jebel An-Nusseiriyeh, and the Palmyrides has been discussed, accompanied by a description of the numerous species of foraminifera, in the monograph of K. Kuznetsova, A. Grigelis et al. (1996);

8) The Lower Cretaceous deposits of Anti-Lebanon have been subdivided based on palynological assemblages (Kotova, 1990). Their lithological peculiarities, as well as those of the weathering crust with bauxite and kaolinite at the Triassic/Jurassic boundary, were described by Tsekhovskiy et al. (1993, 1994);

9) The petrology of the cover basalts of Jebel Ad-Drouz, Hauran, and Shin Plateaus as well as of basalts in some areas of the Palmyrides, Aleppo Plateau, Jebel An-Nusseiriyeh and the Al-Furat Valley has been studied. The isotopic age of basalts in the interval of the Early Miocene to Quaternary was determined by means of the K/Ar method. Special attention was paid to the mineralogy of kimberlite-like rocks from explosion pipes (diatrems) with inclusions of mantle rocks (Sharkov et al., 1987, 1989a, b, 1994, 1996, 1998; Sharkov, 2000; Snyder et al., 1993; Sobolev et al., 1990);

10) The stratigraphy of Triassic deposits of the Mesopotamian Basin, based on palynological analysis, has been elaborated by Yaroshenko and Bash Imam (1982, 1989, 1994, 1995). Sediments of this age were lithologically and paleogeographically characterized by Bebeshev et al. (1988, 1989) and Bebeshev (1995);

11) A wide variety of geological and geophysical investigations in the eastern Mediterranean Sea, between Cyprus and the coast of Syria, were carried out during the Cruise 5 of the Research Vessel 'Akademic Nikolay Strakhov' and summarized in a book edited by Krasheninnikov and Hall (1994);

12) The stratigraphy and lithology of the continental Pliocene-Quaternary deposits of the Damascus depression (Gouta Basin), the Palmyrides (Ad-Daww depression), and Al-Ghab, as well as marine sediments of this age in the area of Al-Latheqiyeh was studied by Devyatkin et al. (1996, 1997) and by Devyatkin and Dodonov (2000).

The Syrian geologists, R. Helou, H. Nseir, A. Shweki, I. Jarmakani, K. Al-Khatib, D. Shalash, J. Adjamiyan, M. Baalbaki, Z. Jallad, K. Jabbur, A. Al-Hafez, Z. Al-Mounajed, Sh. Al-Jassem, H. Zeinab, I. Bash Imam, M. Roukiye, T. Zaza, A. Yussef, A. Moualla, and others actively participated in the above investigations and are co-authors of many of these publications.

The Near East and Middle East are considered to be the cradle of human civilization. Consequently geologists and archeologists of many countries such as France, Japan, England, USA and the Netherlands have studied the Paleolithic and Neolithic locations in the terraces of the Al-Furat, Nahr El-Khabour and Nahr El-Balikh Rivers in Mesopotamia and in the northwestern area of Syria (the Nahr El-Kebir and Oront Rivers), as well as flint tools from the alluvial-proluvial sediments of the Al-Ghab, Damascus and Ad-Daww depressions. Numerous monographs and articles were devoted to the Paleolithic and Neolithic of Syria. Among them are the investigations of Akazava (1974, 1979), Besançon (1981a,b), Bordes (1955, 1960), Clark (1967, 1968), Clason and Buitenhuis (1978), Contenson (1966, 1976, 1981), Copeland (1979, 1981), Copeland and Hours (1978), Endo (1973, 1978), Fujimoto (1979), Hanihara et al. (1978, 1979, 1983), Heinzelin (1965, 1968), Hooijer (1961, 1966, 1968, 1975), Hours (1975, 1979, 1981), Kaizer (1965, 1973, 1976), Kobori (1973) and van Liere (1960, 1963, 1966).

More recently in some areas of Syria the General Establishment of Geology and Mineral Resources has carried out additional geological mapping at the scale of 1:50,000. These sheets by Al Baca (1983), Khatib (1980), Ruske, (1978a,b), Shalash and Al-Hafez (1983), and Yousef (1979, 1980a,b) cover the north of the Anti-Lebanon, many regions of Jebel An-Nusseiriyeh, and the Al-Ghab depression. All these maps are accompanied by explanatory notes.

The Syrian Petroleum Company has carried out drillings, and geological and geophysical investigations which have helped to decipher some of the main features of the geology of northeastern Mesopotamia and which have discovered many oil fields such as Qaratchouk, Souedie, Roumeylan, El-Bardeh, El-Bouab, Ghouna, Gbeibeh, Jbissa, Tishrin, Al Hole, Oudeh, Laylak, Derik, Khourbet, etc.

The Fourth Period (~1990ff)

The fourth stage of the geological investigations in Syria relates to the last 10-15 years. Its principal accomplishment has been the acquisition of new geological and geophysical data on the subsurface rock sequence which was previously unknown. The area of drilling was expanded and now embraces the whole territory of Jezireh, the southern slope of the Aleppo Plateau, the Palmyrides, and the Euphrates depression. Oil companies from the

USA, France, Canada and England are working together with the Syrian Petroleum Company. A large number of publications have appeared, among them Chaimov, Barazangi et al. (1990, 1992, 1993), Al-Saad, Sawaf et al. (1992), Best, Barazangi et al. (1990, 1993), Sawaf, Al-Saad et al. (1993), Brew, Litak et al. (1997a, b), Alsdorf, Barazangi et al. (1995), Seber, Barazangi et al.(1993), Caron et al. (2000), Litak et al. (1997, 1998), Salel and Seguret (1994), Brew (2001), Brew et al. (1997a,b), McBride et al. (1990), and Searle (1994).

The main contribution to these studies was made by scientists from Cornell University in the USA, the Syrian Petroleum Company, and Damascus University, who have been working in the frame of the 'Cornell Syria Project'. This project investigates the regional tectonics of Syria and the northern part of the Arabian platform using drilling data and geophysical (seismic) materials. New data on the subsurface Paleozoic and Lower Mesozoic deposits have revealed the structural peculiarities of different areas of Syria and allowed the formulation of the principal features of the tectonic development of this country. Lithostratigraphic units (formations) were used to subdivide the Paleozoic and Lower Mesozoic sedimentary cover. Unfortunately, this scientific approach was not accompanied by detailed biostratigraphic studies, and many events of the Paleozoic-Lower Mesozoic geological history still await clarification.

Thus, the territory of Syria has incurred a high level of geological investigation. However there is a disproportionate balance between the amount of published surface and subsurface information. Data on the surface geology of Syria have appeared in numerous articles and monographs, while data on the subsurface geology of Syria are more modest and restricted, likely reflecting the understandable policy of commercial oil companies.

In the sections below the reader must keep in mind that the above-mentioned lack of detail about the subsurface has restricted the amount of insight that might be culled from the synthesis of the surface data presented in this monograph.

We have tried to use all the available data but special attention has been given to those original materials which were obtained by geologists of the USSR (Russia) during the joint investigations with geologists of the General Establishment of Geology and Mineral Resources and the Syrian Petroleum Company.

In concluding this introduction it is necessary to add that very complete bibliographies of articles and monographs dealing with the geology of Syria have been published by Wolfart (1966, 1967) and Arad, Ehrlich and Nathan (1985). Additional bibliographic sources are described in Appendix A of Volume II. In addition, the nineteen geological map sheets at a scale of 1:200,000 have been scanned into digital form and mosaicked and are discussed in Appendices D and E.

REFERENCES CITED

(See References of Chapter 6)

MAX BLANCKENHORN (16.4.1861-13.1.1947)

Article by Prof. Moshe A. Avnimelech (Israel J. Earth-Sci., 12(1), 1963, p. 1-11).
Hebrew University of Jerusalem, Jerusalem, Israel

The Autobiographical Notes of Max Blanckenhorn, the noted German geologist, were written by him in December 1934, thirteen years before his death on January 13th, 1947.

During his long scientific career he succeeded in collecting rich geological-paleontological material from Palestine, Egypt, Lebanon and Syria. His collections were acquired in 1929 by the Hebrew University thanks to the initiative and efforts of the known botanist and Zionist leader, the late Prof. Otto Warburg. These collections constitute the nucleus of the Hebrew University's Geological Museum and serve as a monument to his memory. A few years later, in 1935, this museum was greatly enlarged by an important collection of the late Prof. Paul Oppenheim of Berlin.

In his time Blanckenhorn was one of the leading authorities on all the problems of Middle East geology. The bibliographical list of his publications on the regions of the Middle East is the evidence of his tremendous scientific productivity.

Blankenhorn symbolizes a period in the history of geological research in the Middle East. Apart from the older, rather non-scientific observations at the end of the 18th century, this history may be divided into four periods.

The first period, dating approximately from the beginning to the middle of the 19th century is best characterized by the adventurous personalities of Seetzen and Russegger who were real pioneers in the field of Middle East geography and geology. The virgin soil of the almost unknown countries presented many exciting problems which have been discussed by such distinguished scientists as Ehrenberg, Agassiz, Haeckel, Botta, Blanche and many others.

The second period lasted for about 30 years up to 1880. This was a period of several expeditions organized by government and scientific bodies, or sponsored by rich individuals who were eager to acquire fame by promoting scientific exploration. The first of these expeditions was organized by the U. S. Navy under the command of W. F. Lynch and its purpose was to solve the mysteries of the Dead Sea and of the Jordan River. The last expedition of that period was that of Duc de Luynes. Smaller expeditions but no less important were those of Hull and O. Fraas. All of them gave an excellent opportunity to study the collected material. The studies were carried out by the best scientists of that period, such as T. A. Conrad, Coquand, Pictet, Gaudry, Lartet, Holland, and many others.

Blanckenhorn belongs to the third period which ended after the World War I, approximately in 1920. He entered the scene of scientific explorations in the Middle East when their foundation had already been laid down. At that time the Middle East countries were in the limelight of political interest, and they aroused a great scientific curiosity. Several of the European countries had started planned scientific explorations which were carried out by "biblical" organizations such as the Palestine Exploration Fund and Deutsche Palaestina Verein. The topographic survey of the region served as the basis for more detailed geological work. After the summing-up of the previous investigations (*cf.* by Huddlestone) came the time for a broader outlook and for the revision of the previous work. With this task are connected the names of K. Zittel, C. Diener and Fr. Noetling. A few years later Blanckenhorn started his researches in the Middle East countries which lasted for almost 50 years. During those years many important scientists have worked in the same field, among them J. Boehm, Zumoffen, H. Douvillé, Fourtau, Felix, Hume, Kober, Range and others. In this great ensemble Blanckenhorn was one of the leading figures, especially as his activities embraced almost all the problems pertaining to the geology and paleontology of the region.

World War I basically changed the political landscape of the Middle East. Although new independent or semi-independent states were created, they none the less remained under the European influence. This was also felt in the progress of Geological Research. In some of the countries of the Middle East nuclei of State Geological Surveys were laid down (Blake in Palestine, Dubertret in Lebanon-Syria); in Palestine – the local Hebrew University geologist (Picard, Avnimelech) and individuals (Shalem, Loewengart) have greatly contributed to the development of geological knowledge. Blanckenhorn was still active during the fourth period, summarizing the results of his numerous journeys and arranging his paleontological material. Until the Nazi regime he was in close contact with his younger colleagues in the countries of the Middle East.

Blanckenhorn's autobiography, written especially for the Hebrew University, remained for more than ten years in the archives of the Geological Department on Mount Scopus, and only recently was it returned to us. It presents an interesting document which illustrates the history of the development of geological research in Palestine and other countries of the Middle East. Readers belonging to the new generations of young geologists will certainly benefit from it by comparing the former conditions of scientific work in these regions with those of present day.

MAX BLANCKENHORN AUTOBIOGRAPHICAL NOTES Marburg, December 12th, 1934

I was born on April 16th, 1861, at Siegen, Westphalia, to railway construction master Carl Blanckenhorn and his wife Sophie, née Budach.

From April 1870, I lived with my parents at Cassel where I studied, until autumn 1879, at the Lyceum Friedericianum high school. (In my matriculation certificate all marks read "good", with only the exception of French).

I studied both philosophy and natural sciences, and especially botany, geology, paleontology, geography, meteorology, prehistory, etc., at the universities of Goettingen, Bonn, Strasbourg, Berlin and Vienna. For my doctorate thesis I chose the geological mapping of the Trias formations along the northern rim of the Eiffel Hills, in the region of Commern, Zuelpich and the Roer valley.

The paper, including a coloured map, geological profiles, and a plate of fossils was published in 1885 in the Abh. Preuss. Geol. Landesanst. Although the oral examination (which was preceded by a "magister" examination in the subjects of philosophy, mathematics, history and Latin) took place in 1884, the graduation ceremony was held only on July 5th, 1885, at Bonn; I shall therefore celebrate the golden jubilee of my doctorate in 1935.

Since 1867 I made many excursions to Western and Southern Germany as well as to Austria and Switzerland, at first together with my parents and later alone or with university professors.

During 1886-1887 I was employed as an assistant at the Geological-Paleontological Institute at Bonn under Prof. Schlueter, and later, in 1890, at the Mineralogical-Geological Institute at Erlangen under Prof. Oebbeke.

In the years 1887-1888 I prepared myself for a geological and geographical study trip to Northern Syria for which purpose I also took a course in the Turkish language at the Oriental Seminary in Berlin.

Inspired by the work of Ed. Suess "The Face of the Earth", I undertook my first exploration journey to Asiatic Turkey. The caravan was formed in Beirut; thence we travelled along the coast to Lattakieh, then inland over the Nusairi mountains to Djisr esh-Shughur, Aleppo, and Biredjik on the Euphrates, continuing by way of Ain Tab, Sindjirli, Islahiye, the Kurd Dagh and Killis into the valley of Afrin; then via Antiochia, Seleucia, Pieria, Djebel Akra, again to Djisr esh-Shughur and thence to Qalaat et Medik, Hama, Salamiye, Horns, Qalaat et Husn in the southern part of the Nusairi mountains, to return to Beirut through the northern Lebanon mountains and Baalbek.

This first journey to the Middle-East bore many scientific results which were published in a number of papers between 1888 and 1897. The most important of these was a topographical and the first ever issued geological map of Northern Syria. This map, in the scale of 1:500,000, comprised the area from Tripolis and Homs in the south to Antiochia, Ain Tab and Biredjik in the north; it appeared in 1891 in "Grundzuege der Geologic and physikal. Geographic von Nordsyrien", accompanied by a travel diary. Other publications dealt with the Cretaceous, the Eocene, the Miocene, the marine Pliocene, the fresh water Pliocene, and their fauna in Northern Syria, as well as with the basaltic rocks of the region.

In 1891 1 took up a position as lecturer in geology, paleontology and geography at Erlangen University where I lectured until 1897 on invertebrate paleontology, geology of Northern Bavaria, geology of Africa, general geography and, as from 1894, also on the geography of Palestine.

In 1894 I undertook my second journey to the Middle East, this time to Palestine. This journey was supported by the German Association for the Exploration of Palestine (Deutscher Verein zur Erforschug Palaestinas) with its seat in Leipzig. Geological mapping was carried out of the Dead Sea and the Judean region, with special emphasis on the city of Jerusalem and its surroundings.

At the request of the "Deutscher Palaestina Verrein" I set up a number of meteorological stations at different localities.

In 1897 I joined the "Preuss. Geolog. Landesanstalt" as a voluntary worker, and completed the geological mapping of the Wilhelmshoehe-Cassel region which was already begun by Beyschlag.

In October I moved to Egypt, at first without my family (I married in 1894), where on Zittel's recommendation, I was engaged by the Geological Survey of Egypt for two years as surveyor and paleontologist responsible for collection of fossils. My family followed me from Erlangen a year after. In the two years of my activity in Egypt I had the opportunity, on many caravan expeditions, to carry out research and prepare maps of the Nile-valley, of Wadi Tumilat and the Isthmus region; I was the only non-British member of the team.

In 1899 1 returned with my family to Germany, this time to Berlin where I could best work on my own rich collections as well as on the geological and paleontological collections of G. Schweinfurth which were kept at the Berlin Museum for Natural History. I lived first in the northern Pankow suburb and later, until 1912, in the suburb of Halensee.

In October 1899 I undertook, together with Professor Oebbeke from Munich, a geological study tour of Transylvania and Rumania.

In 1902 I was again in Egypt, together with my colleague Stromer von Reichenbach under the auspices of the Bavarian Academy of Sciences, to collect vertebrate fossils in the Nile valley, in the Fayum and in Wadi Natrun.

In 1903 I took part in the International Geological Congress in Vienna, in excursions to Galicia and, under the guidance of Penck, to the Alps.

In 1904 a Jewish organization, and especially Professor O. Warburg, lent its support to another journey to Palestine, with the aim of investigating soils and minerals and primarily of the phosphates I had discovered already in 1894 in Trans-jordan and on the Jericho terrace. I was accompanied by the botanist A. Aaronsohn of Zikhron Yaaqov. This time we travelled through Samaria and Galilee.

I was again in Palestine in 1905. Together with A. Aaronsohn I travelled from Haifa along the Hedjaz railway which was then under construction, reaching Batn el Ghul. From Ma'an I made a detour to Petra and thence continued to Aneze, a station on the Hedjaz railway. From Haifa, too, I undertook a caravan journey into the Jordan valley but because of strong rains, at Jericho we had to turn back to Jerusalem which we reached in heavy snowfall. I also visited Beirut whence I made excursion in the vicinity, and especially to the prehistoric sites, together with Prof. Zumoffen and Prof. Day. At the turn of 1905-1906 I went on from Beirut to Egypt in order to study – between January and March 1906 – the Diluvium and the prehistoric sites along the whole Nile valley, up to Shellal at the First Cataract. For quite a time I stayed at Thebes (Luxor) and at Helwan.

In 1907, I visited in Belgium the geologist and prehistorian Prof. Rutot who showed me and to my companion, Dr. Hahne, his prehistoric finds stored in the Brussels Museum, and also took us to the sites of his excavations.

My eighth tour of the Middle East – my fourth one of Palestine – took up the entire first half of 1908. This was the great "Sultan Voyage" which I undertook on orders of Abdul Hamid, then Sultan of Turkey. Beginning on January 6th, it ended on July with a two month's stay in Constantinople.

Here, too, I was everywhere accompanied by the botanist A. Aaronsohn as well as by the zoologist Aharoni from Jerusalem who had the task of collecting living animals in the Jordan valley for the Yildiz Kiosk Zoological Garden and preparing stuffed animals for the Zoological Museum in Yildiz Park. This journey has been described in detail in my book "Naturwissenschaftliche Studien am Toten Meer and im Jordantal, Bericht uber eine im Jahre 1908 unternommene Forschungreise in Palaestina" which appeared in 1912. My new geological map is included in that book so that I need not add further details about that journey.

In 1910 I participated at the International Geological Congress in Stockholm where I had opportunity to visit different parts of Sweden. The rock samples and petrifacts collected have all been handed over to the Geological Museum of the Hebrew University in Jerusalem.

In 1912 I moved with my family to Marburg.

World War I brought me, in 1917, as military geologist to Macedonia (Radulevo, Drama, mouth of the Struma river) and, in 1918, to Strasbourg in Alsace.

In 1930 I made, together with my wife, my ninth and last journey to the Middle East, the fifth to Palestine. From Jerusalem I set out several times by car, together with the English geologist Blake, to places such as Ramla, Kubeiba, Nablus, Anebta, Wadi Fari'a, to Transjordan and, together with the French geologist Dubertret, from Haifa to Beirut, Damascus, Horns and Tripolis.

Throughout the years 1901 to 1934 1 was each summer busy with the geological mapping (scale 1:25,000) of the district of Cassel (Prussian province Hessen-Nassau) for the "Preusiche Geologische Spezialkarte".

I prepared and wrote explanations to the following fifteen sheets: Ostheim vor der Rhön (Sandheim v.d. Rhön), Hünfeld, Grossenlüder, Neuhoff, Gudensberg, Homberg an der Efze, Schwarzenborn (Knüllgebirge), Neukirchen, Niederaula, Borken, Ziegenhain, Schrecksbach, Neustadt, Kirchhain, Amöneburg. Several of the other sheets have been mapped but not yet printed (Wetter, Frankenberg, Gilserberg, Frankenau). In this part of Northern Hessen (Rhön, Vogelsberg, Knüllgebirge, Burgwald and Ketterwald) the following geological formations have been distinguished: Zechstein, Trias, Lias, Oligocene, Miocene, Pliocene, Pleistocene and basalts.

HONOURS CONFERRED

The Prussian Minister of Culture, v. Studt, appointed Blanckenhorn Professor in 1906 "on account of his services to science".

The Prussian Academy of Sciences in its Leibnitz meeting on June 28th, 1923, bestowed upon him the (silver) Leibnitz Medal "to express gratitude and appreciation to a scientist whose life was dedicated primarily to the exploration of foreign continents". His works dealing with the geology of Syria, Palestine, the Lower Nile valley and its surroundings for which he collected rich documentary material on his frequent tours undertaken inter alia on behalf of the Geological survey of Egypt, are of great and permanent value. No less deserving are his studies on the Trias regions of Central and Western Germany where he began his scientific career in his youth and to which he has now returned as a voluntary member of the Prussian Geological Survey (*cf.* minutes of the session of the Prussian Academy of Science, 1923, p. CXXIX). (This honour was proposed by the paleontologist Prof. Pompeckij and, in secret session, received the consent of the Egyptologist Prof. Erman, the historian Eduard Meyer, the meteorologist Hellmann, the theologian Stoll and the geographer Penck).

The Natural History Society and the Geographical Society in Cassel, as well as the "Institut Egyptien" in Cairo made Blanckenhorn an honorary member.

LIST OF PUBLICATIONS ON THE MIDDLE EAST (LEVANT)* 1889-1940

1889. Beitrag zur Kenntnis der Binnenconchylien - Fauna von Mittel- und Nord-Syrien, *Nachrbl. dtsch. malakol. Ges.*, 21, 76-90.

1889a. Pteropodenreste aus der oberen Kreide Nord-Syriens und aus dem hessischen Oligocaen, *Z. dtsch. geol. Ges.*, 41, 593-602, ill.

1890. Das Eocaen in Syrien mit besonderer Beruecksichtigung Nord-Syriens, *Z. dtsch. geol. Ges.*, 42, 318-359, ill.

1890a. Beitraege zur Geologie Syriens: Die Entwicklung des Kreidesystems im Mittel- und Nord-Syrien, nebst einem Anhang ueber den jurassischen Glandarienkalk, Cassel, 1-136, ill.

1890b. Das marine Miocaen in Syrien, *Denkschr. Akad. Wiss. Wien*, 57, 591-620, ill.

1891. Grundzuege der Geologie und physikalischen Geographie von Nord-Syrien, Berlin, 1-102, ill.

1891a. Karte von Nord-Syrien: 1/500,000, Berlin, Friedlaender & Sohn.

1891b. Syrien in seiner geologischen Vergangenheit, *Ber. Ver. Naturk. Kassel* (36-37).

1891c. Das marine Pliozaen in Syrien, *S. B. phys.med.Soc. Erlangen*, 24, 1-53, ill.

1892. Syrien in seiner geologischen Vergangenheit. Eine geologisch-geschichtliche Studie, zum Teil nach Resultaten eigener Forschung, *Z. dtsch. Palaest. Ver.*, 15, 40-62.

1893. Die Strukturlinien Syriens und des Roten Meeres. Eine geotektonische Studie, Richthofen-Festschr., Berlin, 115-180, map: 1/2,400,000.

1895. Bericht ueber meine Reise nach Palaestina im Fruehjahr 1894, *Mitt. dtsch. Palaestina Ver.*, 1-6.

1895a. Notizen von meinen geologischen Streifzuegen in Palaestina, *Mitt. dtsch. Palaestina Ver.*, 35-40, ill.

1896. Entstehung und Geschichte des Toten Meeres. Ein Beitrag zur Geologie Palaestinas, *Z. dtsch. Palast. Ver.*, 19, 1-59, ill. & 3 maps.

1897. Nuetzbare Mineralien am Toten Meer, *Z. prakt. Geol.*, p. 360.

1897a. Zur Kenntnis der Suesswasserablagerungen und Mollusken Syriens, *Palaeontographica*, 1(44), 71-144, ill.

1898. *Das Tote Meer und der Untergang von Sodom und Gomorrha*, Berlin, Reimer, 1-44, ill & map

1898a. Noch einmal Sodom und Gomorrha, *Z. dtsch. Palast Ver.*, 21, 65-83.

1899. Neues zur Geologie und Palaeontologie Aegyptens, *Z. prakt. Geol.*, 392-396.

1900. Zur Frage des Untergangs von Sodom und Gomorrha, *Mitt. geogr. Ges. Wien*, (5-6), 194-197.

1900a. Neues zur Geologie und Palaeontologie Aegyptens, I-II, *Z. dtsch. geol. Ges.*, 52, 21-47; 403-479.

1901. Neues zur Geologie und Palaeontologie Aegyptens, III, IV, *Z. dtsch. geol. Ges.*, 53, 52-132 307-502, ill.

1901a. Geologie Aegyptens, ein Fuehrer durch die geologische Vergangenheit Aegyptens von der Steinkohlenperiode bis zur Jetztzeit, Berlin, 1-341, ill.

1901b. Nachtraege zur Kenntnis des Palaeogens in Aegypten, *Zbl. Miner. etc.*, 265-275, ill.

1902. Die Mineralschaetze Palaestinas, *Mitt. dtsch. Palastina Ver.*, 65-70.

1902a. Carte Géologique Internationale de l'Europe: 1/1,500,000. Feuille 36 (AVI)z.t.43 (AVII), Dietrich Reimer, Berlin.

1902b. Die Geschichte des Nilstroms in der Tertiaer- und Quartaer-Periode sowie des palaeolitischen Menschen in Aegypten, *Z. Ges.Erdk, Berlin*, 694-722; 753-762, ill.

1902c. Neue geologisch-stratigraphische Beobachtungen in Aegypten, *Sitzb. math. phys. KL., kgl. bayr. Akad. Wiss.*, 32, 353-433.

1903. Die Vola-Arten des aegyptischen und syrischen Neogens, *N. Jb. Min. Geol. Palaeont.*, 17, 163-186, ill.

1903a. Ueber das Vorkommen von Phosphaten, Asphaltkalk, Asphalt und Petroleum in Palaestina und Aegypten, *Z. prakt. Geol.*, 11(8), 294-298, ill.

1904. Abriss der Geologie Syriens, Altneuland, (Mschr.f.d. wirtschaftl. Erschliessung Paleastinas) Berlin, 289-301.

1905. Geologie der naeheren Umgebung von Jerusalem, *Z. dtsch. Palast. Ver.*, 28, 75-120, ill.

1905a. Ueber die Geologie der naeheren Umgebung von Jerusalem, *Mber. dtsch. geol.Ger.*, 35-43.

1905b. Ueber die letzten Erdbeben in Palaestina und die Erforschung etwaiger kuenftiger, *Z. dtsch. Palaest. Ver.*, 28, 206-221.

1905c. Ueber die Steinzeit und die Feuersteinartenfakte in Syrien-Palaestina, *Z. Ethn.*, 37, 447-471, ill.

1907. Nachtrag zum Kapitel: Erdbeben in Palaestina, *Z. dtsch. Palaest. Ver.*, 30, 106-111.

1907a. Der Haupt-Bundsandstein ist keine echte Wuestenbildung, *Mber. dtsch. geol. Ges.*, 59, (12), 297-311.

1907b. Die Hedschazbahn. Auf Grund eigener Reisestudien, *Z. Ges. Erdk. Berlin*, 218-245; 288-320, ill.

1909a. Carte géologique internationale de l'Europe: 1/1,500,000. Feuilles 44 (D VII) z. T., 45 (CVII) z.T. und 46 (D VII), Dietrich Reimer, Berlin.

1909a. Eine seismologische Station in Palaestina? *Z. dtsch. Palast. Ver.*, 32.

1910. Das Klima der Quartaerperiode in Syrien-Palaestina und Aegypten, In: Postglaziale Klimaaenderungen, *Congr. Geol. International, 11*^{eme} *session, Stockholm* (1910): fasc. I: 140-143 (425-428).

1910a. Neues zur Geologie Palaestinas und des aegyptischen Niltals. *Z. dtsch. geol. Ges.*, 62, 405-461, ill.

1911. Carte géologique internationale de l'Europe: 1/1,500,000. Feuilles 41 (FVI) z. T., 47 (EVII) und 48 (FVII). Dietrich Reimer, Berlin.

1912. Die Hedschazbahn, *Geogr.Z.Berlin*, 18, 15-29, ill.

1912a. Kurzer Abriss der Geologie Palaestinas. Begleitworte zu einer neuen geologischen Karte von Palaestina, *Z. dtsch. Palaest. Ver.*, 35, 113-139, ill. Geologische Karte: 1/700,000.

1912b. Naturwissenschaftliche Studien am Toten Meer und im Jordantal. Bericht ueber eine im Jahre 1908 (im Auftrag S. Maj. d. Sultans Abdul Hamid II) unternommenen Forschungsreise in Palaestina, Friedlaender & Sohn, Berlin, 1-478, ill. /Bericht des Zoologen der Expedition Aharoni als Appendix.

1913/15 (Editor of:) Beitraege zur Palaeontologie und Geologie von Palaestina und Syrien, (Stuttgart, Schweizbart) containing: Felix, J., 1913: Korallen der Kreideformation; Dienemann, W., 1915: Aelteres Palaeozoikum: Fuchs, E., 1915: Petrographie Palaestinas.

1914. Syrien, Arabien und Mesopotamien, *Handb. region. Geol. Heidelberg. 5*(4), 1-159, ill.

1915. Das Danien in Palaestina mit der Leitforen *Pecten obrutus* CONR (-P. *farafrensis* ZtTTMeyer-Eymeri NEWTON), *Z. dtsch. geol. Ges. Abt. B.*, 67, 187-191.

1918. Der Boden Palaestinas, seine Entstehung, Beschaffenheit, Bearbeitung und

Ertragfaehigkeit, *Pro-Palaestina*, (Schriften d. dtsch. Komm. z. Foerder. jued. Palaestinasiedlung), Berlin, V, 1-32.
1921. Aegypten, *Handb. region.Geol. Heidelberg*, 7(9), 1-244.
1921/22. Die Steinzeit Palaestina-Syriens und Nord-Afri as, *Das Land der Bibel, Leipzig*, 3(5), 1-48; (6), 1-26, 4, (1922) (1), 1-45.
1925. Die eeigelfauna der Kreide Palaestinas, *Palaeontographica*, 67, 83-113, ill.
1926. Der sogenannte Syrische Bogen und die erythraeische Geosynklinale, *Mber. dtsch. geol. Ges.*, (8-10), 206-226, ill.
1926a. and OPPENHEIM, P., Nachtraege zur Kenntnis des Neogens in Syrien, *Palaeontographica*, 68.
1927. Die fossilen Gastropoden und Scapholoden der Kreide von Syrien-Palaestina, *Palaeontographica, Stuttgart*, 69, 111-186, ill.
1927a. Die Erdbeben im Juli 1927 in Palaestina, *Z. dtsch. Palast. Ver.*, 50, 288-296.
1927b. and OPPENHEIM, P., Neue Beitraege zur Kenntnis des Neogens in Syrien und Palaestina, *Geol. Palaeont. Abh.*, Neue Folge, 15(4), 321-358, ill.
1928. Neuere vorgeschichtliche steinzeitliche Forschungen und Literatur betreffend Palaestina-Syrien, *Z. dtsch. Palast. Ver.*, 51, 143-149.
1928a. Kali im Toten Meer, *Palaestina Z. Wien*, 11, 496-499.
1928b. Neue Erdbeben in Jerusalem, *Z. dtsch. Palast. Ver.*, 51, 123-125.
1928c. Die geologische Forschung und Literatur der letzten Jahre, *Z. dtsch. Palast. Ver.*, 51, 137-142.
1929. Besprechung vona. Sieberg: Das Korinther Erdbeben vom 22. April 1928 in seinen Beziehungen zu den Erdbeben und dein geologischen Aufbau des oestlichen Mittelmeergebietes, *Z. dtsch. Palast. Ver.*, 52.
1929a. Der marine Ursprung des Toten Meeres und seiner Salze, *Z. dtsch. geol. Ges.*, 81(3-4), 81-93.
1930. Das Schwefelvorkommen suedlich Gaza, *Z. prakt. Geologie, Halle/Saale*, 38, 183-184.
1931. Geologie Palaestinas nach heutiger Auffassung, *Z. dtsch. Palast. Ver.*, 54, 1-50.
1934. Die Bivalven der Kreideformation von Syrien-Palaestina nebst ergaenzendem Nachtrag ueber Brachiopoden, Gastropoden und Pteropoden und Ueberblick ueber die gesamte Molluskenfauna, *Palaeontographica*, 81 A, (4-6), 161-302, ill.
1934a. Tauteiche, Palaestina (Mschr. f. den Aufbau Palaestinas), 17, p.337 ff.
1935. Ueber die Faunen der Syrischen Kreide, des Eocaens und Oligocaens, *Zbl. Miner. geol. Palaeont.* Abt. B, 148-161.
1939. Bericht ueber die Fortschritte der geologischen Erforschung Palaestinas von 1931-1938, *Z. dtsch. Palast. Ver.*, 62, p. 22-52.
1940. Syrien und Palaestina, *Geol. Jber. Berlin*, 2, 351-361.

---

*Excerpted from Avnimelech, M. A., 1965. "Bibliography of Levant geology, including Cyprus, Hatay, Israel, Jordania, Lebanon, Sinai and Syria". Israel Program of Scientific Translations, Jerusalem. This is a bibliography of more than 4,500 titles, listed both alphabetically by author and chronologically, with a classified subject index. The references date from "ca. 1250" through 1963. Peripheral subjects such as botany, zoology, and archeology are included. The chronological index shows that publications increased from five papers a year in the period 1849 to 1865 to more than 120 a year in the period 1949 to 1963.
Abbreviations are according to *World list of Scientific Publications*, 1900-1950, 3rd. ed., Butterworths Scientific Publications, London, 1952.

SUBJECT INDEX

ASPHALT:	1903a
BRACHIOPODS:	1934
CRETACEOUS:	1890a: (Entwicklung); 1915: (Danian); 1925: (Seeigelfauna); 1927: (Gastropoden u.Scaphopoden); 1934: (Bivalven etc.); 1935: (Faunen)
DANIAN:	1915
DEAD SEA:	1896: (Entstehung u. Geschichte); 1897: (Mineralien); 1898: (Sodom u. Gomorrha); 1ß98a: (Sodom u. Gomorrha); 1900: (Sodom u. Gomorrha): 1912b: (Naturwiss. Studien); 1928a: (Kali); 1929a; (Mariner Ursprung)
EARTHQUAKES:	1905b; 1907; 1909: (Seismol. Stat.); 1927a; 1928b; 1929: (Besprechung v. Sieberg: Korinth. Erdbeben)
ECHINOIDEA:	1925
ECONOMIC GEOLOGY:	1897: (Miner. Tot. Meer); 1902: (Mineral. Palaestina); 1903a: (Phosph. Asphalt u. Petrol.); 1928a: (Kali, Tot. Meer); 1930: (Schwefel, Gaza)
EOCENE:	1890: (Syrien); 1935: (Faunen)
FRESHWATER SEDIMENTS and MOLLUSCS:	1897a: (Syrien)
GASTROPODA:	1927; 1934.
GLANDARIENKALK:	see Jurassic, 1890a.
JERUSALEM:	1905: (Naehere Umgebung); 1905a: (Naehere Umgebung)
JORDAN VALLEY:	1912b: (Naturwiss. Studien)
JURASSIC, SYRIA:	1890a: (Beitraege etc.; Jurass. Glandarienkalk)
LEVANT in general:	1893: (Strukturlinien); 1899: (Geol. u. Palaeont. Aegypt.); 1900, 1901: (Geol. u. Palaeont. Aegypt. I, II, III, IV); 1901a: (Geol. Aegypt.); 1902a: (Carte-intern. 36); 1902b: (Tert. u. Quart.); 1902e: (Geol.-strat. Beobachtungen); 1907b: (Haupt-Buntsandstein); 1909: (Carte-intern. 44); 1911: (Carte-intern.41); 1912: (Hedschazbahn); 1914: (Syrien, Arabien, Mesopotam.); 1921: (Aegypten); 1925: (Syrischer Bogen); 1928c: (Geol. Forschung); 1929: (Besprechung: Geol. Aufbau)
MIOCENE, marine:	1890b: (Syrien)
MOLLUSCA, freshwater, recent:	1889: (Binnenconchylien, Syrien)
NEOGENE:	see also: MIOCENE; PLIOCENE: 1903: (Vola-Arten); 1926a: (with P. Oppenheim-Syrien); 1927b: (with P. Oppenheim-Syrien u. Palaestina):
NUBIAN SANDSTONE	see Levant: 1907b
OLIGOCENE:	1935: (Faunen, Syrien)
PALEOCENE:	1901b:(Aegypten)
PALEONTOLOGY:	
BIVALVA:	See PELECYPODA
BRACHIOPODA:	1934: (Bivalv., Brach. Kreideform.)
ECHINOIDEA:	1925: (Seeigelfauna, Kreide)
GASTROPODA:	1927: (Kreide, Syr.-Pal.); 1934: (Bivalv., Gastrop., Kreide)
LAMELLIBRANCHIA:	See PELECYPODA
PELECYPODA:	1934: (Bivalv., Kreideform.); 1903: (Vola-Arten, Neogen); 1915: (Pesten obrutus, Danian)
PTEROPODA:	1889a; 1934a a.SCAPHOPODA: 1927
PALESTINE:	in general: 1895 (Ber. ueber Reise); 1895: (Notizen u. geol. Streifzuege); 1910a: (Neues z.Geol.); 1912a: (Kurzer Abriss); 1931: (Geol.nach heutig. Auffassung); 1930: (Fortschritte); 1940: (Syrien u. Palaestina)
PEDOLOGY:	1915:(Palaestina)
PETROLEUM:	1903a
PHOSPHATE:	1903 a
PLEISTOCENE:	1910: (Klima)
PLIOCENE:	1891c: (Marin,Syrien)
PREHISTORY:	1905c: (Syrien-Palaestina); 1921-2: (Palaest.-Syrien); 1928: (Neu. Forschungen)
PTEROPODA:	1889a: (Ob. Kreide, Nord-Syr.); 1934a: (Bivalven-Kreide)
RED SEA:	1893:(Strukturlinien)
SCAPHOPODA:	1927
SYRIA:	in general: 1891: (Grundzuege); 1891a: (Karte); 1891b: (Marin. Pliocene); 1892: (Geol. Vergangenheit); 1904; (Abriss d. Geologie); 1940:(Syr. u. Palaest.) 1893 (Strukturlinien);
STRUCTURE:	1926: (Syrisch.Bogen).

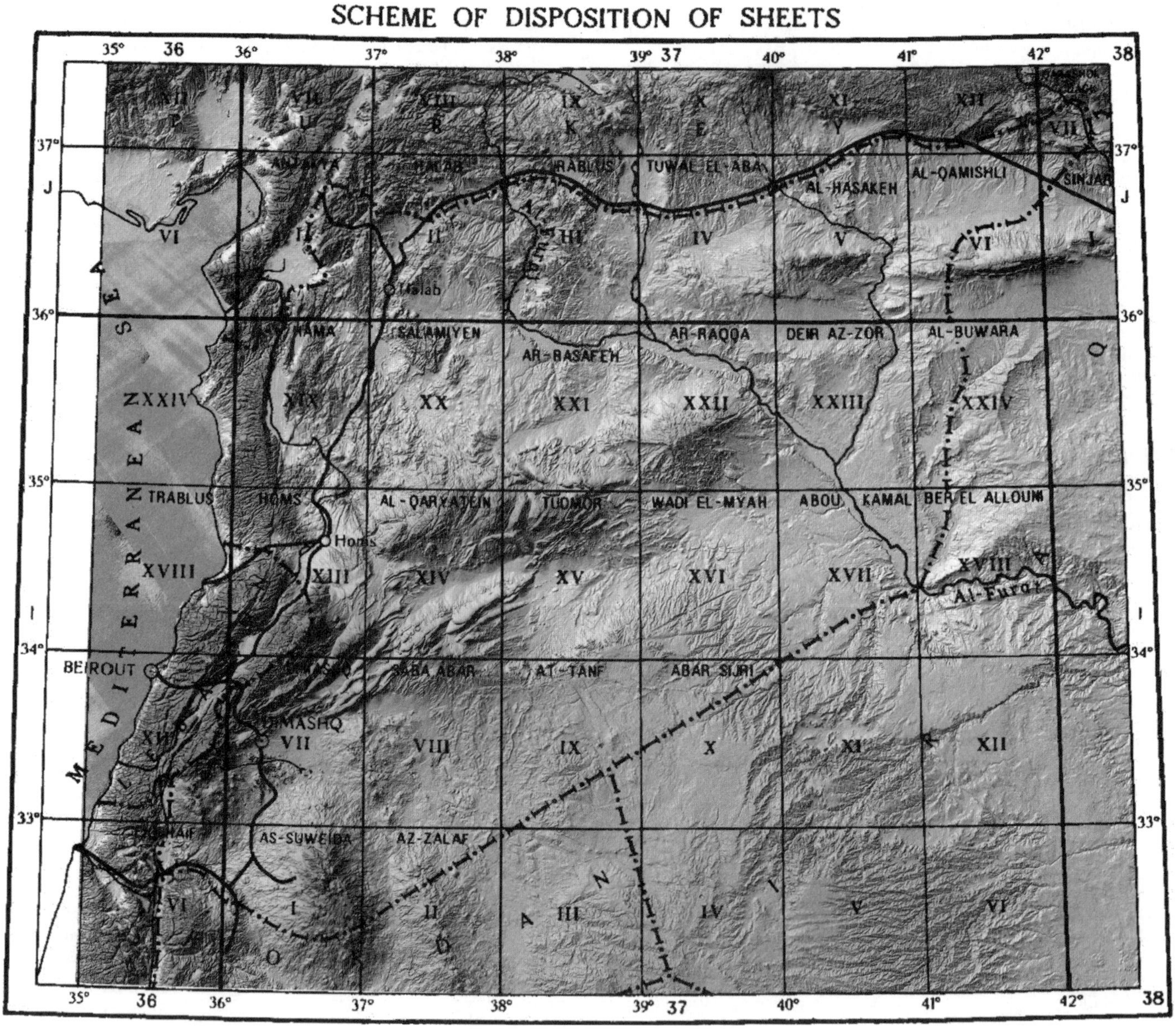

Scheme of Disposition of the 21 Geological Map Sheets at 1:200,000 scale prepared by the USSR Ministry of Geology in 1958-1961. The shaded relief in the background is derived from the 90 m gridded Shuttle Radar Topographic Mission (SRTM) coverage obtained by the Space Shuttle Endeavour (STS-99) in February 2000. This is the same imagery shown in color on the back cover of this volume. In the following chapters this topographic data set, together with lower definition ACE satellite altimeter derived topography is used to generally represent the morphological background in some of the location maps. The geological map sheets are discussed in Appendix D of Volume II, and are available as scanned digital files on a CDROM upon request. References for these map sheets are given on page 415 of this volume.

Chapter 6

Stratigraphy and Lithology of the Sedimentary Cover

Valery A. Krasheninnikov
Geological Institute of the Russian Academy of Sciences, Pyzhevskiy pereulok 7, 109017 Moscow, Russia

The sedimentary cover of the Arabian platform in Syria consists of the whole assemblage of deposits of the Phanerozoic, beginning from the Cambrian up to the Quaternary. However the Paleozoic deposits are not exposed in Syria, and those of the Triassic occupy limited areas. In the region of Jebel Abd El-Aziz, limestones and sandstones of the Lower Carboniferous are present as exotic blocks. Their mechanism of emplacement into sediments of Upper Cretaceous age will be discussed below. For the first time Triassic deposits were found as separate blocks with Upper Triassic fauna in the Baer-Bassit Massif. They are a part of the volcanogenic-sedimentary complex forming the Tamima Nappe, which occupies the lower structural position within the allochthon of the ophiolite association of Baer-Bassit. Triassic deposits of the Baer-Bassit Massif were shifted from more northern areas of the Tethys belt. *In situ* deposits of this age are developed in the Anti-Lebanon, the Palmyrides and Jebel An-Nusseiriyeh, but they are restricted in their distribution.

The larger part of Syria is occupied by exposures of Jurassic, Cretaceous, Paleogene, Neogene and Quaternary rocks. They will be considered in accordance with the main structural units of Syria: the Anti-Lebanon, Jebel An-Nusseiriyeh, the Baer-Bassit Massif, Kurd Dagh, the Palmyrides, Rutbah, the Aleppo Plateau, and the Mesopotamian Basin.

From the above, it is evident that the stratigraphy and lithology of the Paleozoic and Triassic can only be outlined in general terms. In contrast, the lithologic and stratigraphic patterns and the history of the geological development of the Jurassic, Cretaceous and Cenozoic rock can be deciphered in great detail and at high resolution.

A: Paleozoic

Paleozoic deposits have been penetrated by some tens of bore-holes, but the most valuable information was given by a series of five deep bore-holes: Marcada-101 (5007 m); Swab-1 (5009 m); Jeribe-101 (5007 m); Tanf-1 (3367 m) and Gbeibe-101 (5010 m). They were drilled in 1975-1976 by the geological organizations of Rumania and Yugoslavia under the supervision of the United Nations (**Fig. 6.1**). These bore-holes are located in a sublongitudinal direction and give a general picture of the relationship between the basement and the sedimentary cover (**Fig. 6.2**).

Paleontological determinations have been carried out by specialists of the Soviet Union, England and Syria (Yankauskas and Talli, 1978; Husri and Austin, 1985).

Paleozoic deposits have also been penetrated by the Jbissa-207, Roumelan-6, Khanasser-1, Afendi-1, Qamichliye-1, Abba-1, Ratka-1, Ahmar-101, Ahmar-102, Shedade-1, Athar-1, El-Bouab-1, El-Bardeh-1, Doubayyat-1, Doubayyat-2, Soukhneh-1, Meskene-1, Syrom-3, Aleppo-1, Arak-1, Roumelan-7, Qariatein-1, Ramtha-1, Ressafe-1, Sawanet-2, Sayad-1 and other bore-holes. Unfortunately, reliable paleontological determinations are only given for some of these bore-holes (Yankauskas and Talli, 1978; Sudbury, 1957; Wolfart, 1967; Weber, 1963). In other cases, these determinations are of a general character or we have no data at all because results are the property of the oil companies.

Cambrian

In Syria, Cambrian deposits have only been penetrated by two bore-holes: Khanasser-1 and Swab- 1. In the first hole a 1015 m thickness of Cambrian sediments was drilled, and in the second, 969 m (**Fig. 6.3**).

With respect to the stratigraphic succession, the most complete section of Cambrian is found in the Khanasser-1 bore-hole, in the southern margin of the Aleppo Plateau. In this bore-hole three formations have been singled out.

The lower subdivision is the Zabuk Formation (3848-3780 m interval), which is represented by hard, dense red quartz sandstones with massive texture and rare interbeds of lilac colored shales, salt, and volcanic tuffs of basic composition. The penetrated thickness is 68 m.

Upwards in the section within the interval 3780-3620 m the Burj Formation was defined (160 m thick), composed of massive black fine-grained dolomites and gray, massive limestones. In the upper part of this formation, poorly preserved fragments of trilobites and brachiopods were found.

The succession of the Cambrian deposits is topped by the Sosink Formation (3620-2870 m interval), made up of thick

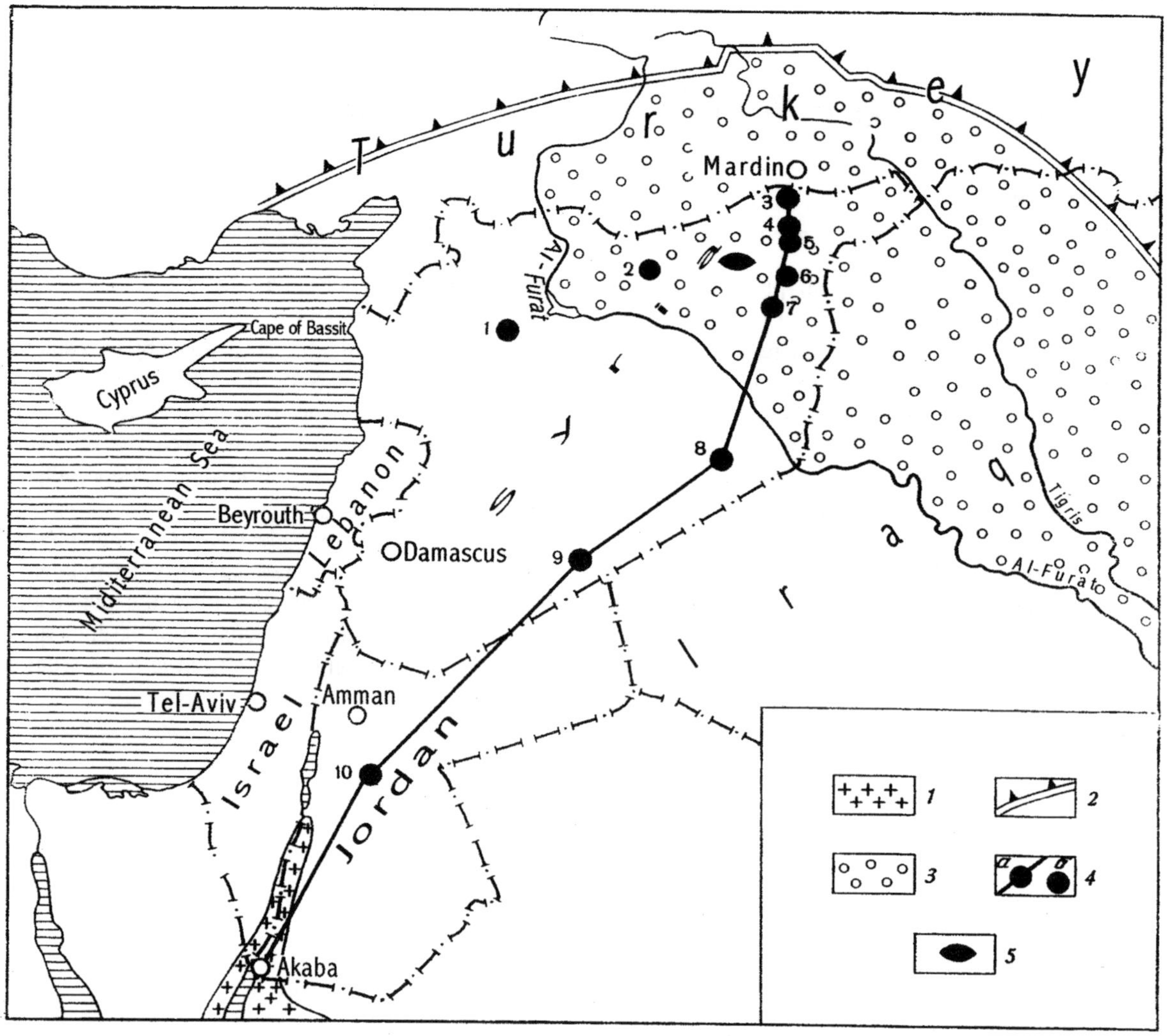

Fig. 6.1. Location of main deep bore-holes which penetrated Paleozoic deposits of Syria (after Yankauscas and Talli, 1978).
Legend: 1 – the Arabian-Nubian shield; 2 – the boundary between the Arabian plate and the Alpine folded belt; 3 – the Mesopotamian marginal fordeep; 4 – orientation of the profile (a), location of bore-holes and their numbers (b) on this figure: 1 – Khanasser, 2 – Abba, 3 – Qamichliye, 4 – Afendi, 5 – El-Bouab, 6 – Tishrin, 7 – Markada, 8 – Swab, 9 – Tanf, 10 – Safra; 5 – the Jebel Abd El-Aziz exposure.

members of red quartz sandstones, with gray and green siltstones intercalated, as well as subordinate interlayers of hard phyllite.

In the Swab-1 bore-hole, the Cambrian deposits are represented by the Sosink Formation (5009-4030 m interval) consisting of massive red quartz sandstones with subordinate intercalations of argillites and siltstones. The thickness of the Sosink Formation in the Khanasser-1 bore-hole is 750 m, but in the Swab-1 bore-hole it increases to 1000 m.

Strictly speaking, there is no paleontological proof of the Cambrian age of these deposits in Syria. The assignment is based on three factors:

1) the position of the sequence in question at the base of the Phanerozoic,

2) these deposits are overlain by the Lower Ordovician which is dated by fauna, and

3) there is a comparatively reliable correlation with Cambrian sections of Jordan and Turkey where sediments contain faunal remnants which have been reliably dated. The names of these formations in Syria have been taken from the stratigraphical terminology of these countries.

In Jordan, the above-mentioned formations are exposed on the surface and are penetrated by the Safra-1 bore-hole (Bender, 1968, 1974, 1975). The Cambrian succession in Jordan also has a three-fold structure: the lower and upper members are red sandstones with intercalations of argillites, and are separated by a member of dark, hard limestones and dolomites (the Burj Formation) with trilobites and brachiopods dated at the Lower to Middle Cambrian transition.

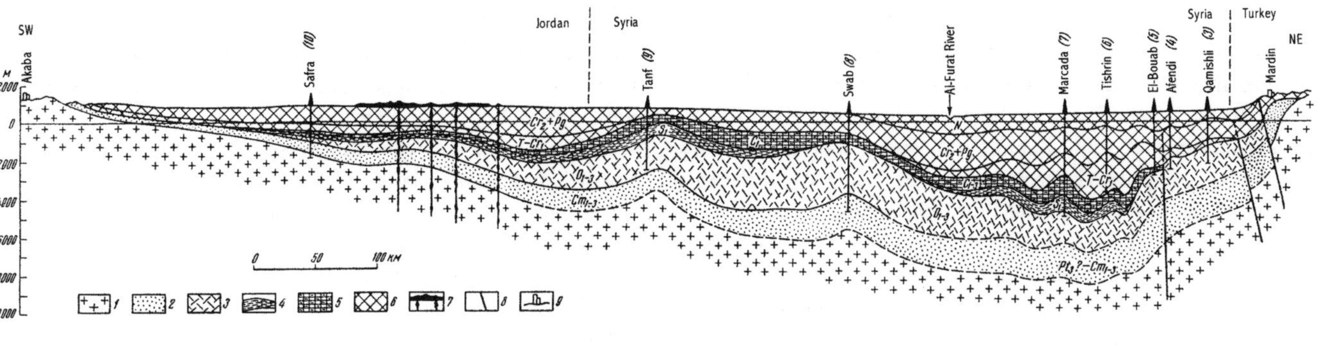

Fig. 6.2. Geological profile along the line Mardin (Turkey) – Akaba (Jordan) via the territory of Syria and Jordan (after Yankauscas and Talli, 1978, including data of Bender, 1975 and Weber, 1963).
Legend: 1 – crystalline basement; 2 – Cambrian (and possible partly Late Pre-Cambrian) deposits; 3 – Ordovician deposits; 4 – Lower Silurian deposits; 5 – Carboniferous deposits; 6 – undifferentiated Permian, Mesozoic and Cenozoic deposits; 7 – Neogene and Quaternary basalts with feeding channels; 8 – faults; 9 – geographical points.

In Turkey, a very thick section of Cambrian deposits is located to the west of Mardin (Rigassi, 1971; Dean, 1975, 1982; Dean, Monod and Perincek, 1981). It begins with coarse-grained sandstones and conglomerates of the Sadan Formation (300 m thick) which pass upwards to sandstones of the Zabuk Formation (500 m thick). This sequence of terrigenous rocks is overlain by calcareous rocks of the Koruk Formation, which is an analogue of the Burj Formation. Higher up in the sequence are sandstones and argillites of the Sosink Formation, with rare intercalations of limestones. These limestones contain trilobites indicative of the Middle Cambrian (Dean and Krummenacher, 1961).

Based on these data, the lower terrigenous Zabuk Formation, discovered in the Khanasser-1 bore-hole in Syria was referred to the upper part of the Lower Cambrian. Calcareous rocks of the Burj Formation have been considered as Middle Cambrian; however it is possible that the Lower/Middle Cambrian boundary is in fact located within this formation. The upper terrigenous Sosink Formation is considered to be undifferentiated Middle-Upper Cambrian in age.

It is necessary to note that according to their lithological patterns, the Cambrian deposits of Syria are similar to rocks of this age exposed in the southern part of Israel (the area of Elat and Timna). The Lower Cambrian starts here with red conglomerates, coarse-grained subarkosic sandstones and quartzites (the Amudei Shelomo Formation), which are overlain by a unit of alternating sandstones, dolomitic limestones, and shales with trilobites of Lower Cambrian age (the Timna Formation). The Middle Cambrian corresponds to the Shehoret Formation consisting of alternating white subarkosic sandstones, siltstones, and reddish shales. The next overlying formation is the Netafim Formation, composed of quartzites and which are referred to the Upper Cambrian. The age of these two formations is determined conventionally because organic remains are absent.

Cambrian deposits in Syria were penetrated by only two bore-holes. Therefore, there are no data on the regional distribution of the Cambrian sediments. In the same way, we have no data on the relationship of the Cambrian deposits to the Precambrian basement of the platform, or about the presence and character of Late Precambrian (Vendian) sedimentary rocks.

Based on seismic data, Cambrian sediments in Syria exist almost everywhere, and are characterized by great thickness. According to the opinion of some oil company specialists, the surface of the crystalline basement of the platform on seismic time sections does not form a clear reflecting surface. The reflecting horizon within the lower part of the sedimentary cover, which had once been considered to be the basement surface, in fact corresponds to the carbonate unit of the upper part of the Lower Cambrian–Middle Cambrian (the Burj Formation). Consequently the true depth of the crystalline basement is still unknown, or can only be estimated in a preliminary fashion. Thus the thickness of the sedimentary cover is estimated to extend down to the reference seismic horizon 'E' which is considered to be the top of the Burj Formation, corresponding to the top of the Middle Cambrian. Evidently, horizon 'E' coincides with the top of the Burj Formation of Jordan, the Koruk Formation of Turkey, as well as the Tabuk Formation of Saudi Arabia. The top of the Timna Formation of Israel may occupy more or less the same stratigraphic position. All these listed formations are predominantly composed of calcareous rocks (hard limestones and dolomites) which have thicknesses ranging up to 150-200 m. They reflect the quiet tectonic environment of shallow-water calcareous sedimentation, which existed on the Arabian Plate during Middle Cambrian time.

Analyses of seismic time-sections as well as gravity and deep drilling data in Syria, Jordan, Israel, Turkey, and Saudi Arabia permit us to suggest that below the calcareous units of the Bur, Koruk, and Tabuk Formations a volcanic sedimentary formation is present, which varies greatly in thickness from 400 to 2000 m. Evidently this formation, of the Late Precambrian or Lower Cambrian, fills grabens in the crystalline basement of Syria. Here volcanic sedimentary formations were not reached by drilling, but their presence is marked on the seismic time-sections which emphasize a block-structure for the basement.

Variations in the thicknesses and the stratigraphic volume of Precambrian rocks are demonstrated by the sections in southern Turkey. To the west of Mardin, below sediments of the Middle–Lower Cambrian Koruk, Zabuk and Sadan Formations, and below a poorly expressed unconformity, the volcanic sedimentary Tebelsmi Formation of the Late Precambrian age is exposed. Its thickness exceeds 1000 m. In the Aksakele bore-hole, located on the prolongation of the Jrablus uplift of the crystalline basement near the border with Syria, one can observe another sequence of sediments. This bore-hole reached the unconformity at the shallow depth of 3442 m. The thickness of pre-Middle Cambrian rocks does not exceed 300 m, and volcanic sedimentary Precambrian formations are absent.

In Jordan considerable variations in thickness of the pre-Middle Cambrian deposits are also reported. In the Safra-1 bore-hole, which reached the basement, the thickness of the succession is around 500 m. In the Suweibth bore-hole, which did not reach the basement, it is 931 m thick.

In Israel, the Ramon bore-hole penetrated 2014 m of the volcanic sedimentary Zenifim Formation of the Precambrian but did not reach the basement.

It is possible to suggest that terrigenous sedimentary and terrigenic-volcanogenic units of Late Riphean (Vendian) age fill grabens within the crystalline basement, and form a special structural stage in the geological development of Syria. These deposits are analogues to terrigenous, volcanogenous, carbonate (stromatolitic limestones and dolomites) and evaporitic formations of the Upper Riphean which are often termed 'Infracambrian' in many regions of the Arabian-Nubian and Aden-Somalian shields. Rocks of these formations are slightly deformed, weakly metamorphosed, with rather insignificant diagenetic changes. Formation of the Vendian rocks took place during the orogenic stage of development of the so-called 'Baikal structures' which are a part of the basement of the Arabian Platform. Later uplifts and erosional episodes resulted in destruction of the Vendian deposits on vast areas of the platform. They were preserved only in the deepest tectonic depressions (Dolginov, Kazmin and Ponikarov, 1969).

Ordovician

Ordovician deposits are widely distributed in Syria and are represented by a thick succession of up to 2500 m of terrigenous rocks, with rare intercalations of calcareous and evaporitic (anhydrite) rocks. The complete sections of the Ordovician were penetrated in the Swab-1 bore-hole (1900 m thickness) and the Tanf-1 bore-hole (2500 m thickness), located on the northern slope of the Rutbah uplift, as well as in the Khanasser-1 bore-hole (1600 m thickness) south of the Aleppo uplift. In addition they were also penetrated by the Markada-101, Abba-1, Afendi-1, and Qamichliye-1 bore-holes in northeastern Syria (Weber, 1963, 1964; Yankauskas and Talli, 1978).

According to its lithological characteristics the Ordivician is subdivided into three rock units (Lababidi and Hamdan, 1985). The lower unit is predominantly sandstone (the Khanasser Formation), the middle unit is predominantly mudstones and shales (the Swab Formation), and the upper unit is alternating sandy and clayey rocks (the Afendi Formation).

The Khanasser Formation was identified in the Khanasser-1, Swab-1, and Tanf-1 bore-holes. It is composed of white quartz sandstones with rare interbeds of black, strongly phyllitized shales and phyllites. Thicknesses range from 381 m in the Rutbah uplift (Swab-1 bore-hole) to 625 m on the Aleppo uplift (Khanasser-1 bore-hole). In the Tanf-1 bore-hole the 390 m thick section of the formation is incomplete.

The most ancient, basal layers of this formation are in the Khanasser-1 bore-hole (at 2790 m depth). Here a thin interlayer of shales contains numerous Tremadocian acritarchs (the Earliest Ordovician). Species present include *Vulcanisphaera africana* Dff., *V. nebulosa* Dff., *Acanthodiacrodium ubui* Mart., *Cymatiosphaera gassiense* Dff., *C. boulouradi* Dff., *Cymatiogalea margaritata* Dff., *Prisocgalea simplex* Dff., *P. cortinula* Dff., *P.* cf. *multarea* Dff., *Poikylofusa squama* (Dff.), and some others.

In the Tanf-1 bore-hole (at depth 3222 m) rare acritarchs of the Arenig Stage have been recovered.

At depths of 3100 and 3200 m in this bore-hole, acritarchs and chitinozoa are much richer and more diverse. The assemblage includes *Frankea longiuscula* Brm., *Dicrodiacrodium normale* Brm., *Priscogalea striatula* Vavrd., *Veryhachium trisculum* Dff., *V. reductum* Dff., *V. trispinosum* (Eis.), V. *lairdi* (De *.), V. horridum* P. et Dff., *Conochitina simplex* Eis., *C.* ex gr. *brevis* T. et J., *Lagenochitina prussica* Eis., *L.* cf. *estonica* Eis., *Rhabdochitina gotlandica* Brm., and *Desmochitina tyare* Crm. They indicate a Llanvirnian age for these sediments. Acritarchs of the Llanvirn have also been found in the Swab-1 bore-hole.

Consequently in the succession of sandstones of the Khanasser Formation all three stages of the Lower Ordovician are present. Evidently the Tremadoc is not developed in all localities, because in Swab-1 it was not identified. These data, as well as the significant change in the character of the sediments at the boundary between red-colored Cambrian and gray-colored Ordovician, suggest a stratigraphic unconformity and erosion at this level.

The Swab Formation was discovered in the Swab-1, Khanasser-1, Tanf-1, Markada-101, Abba-1, Afendi-1, and Qamichliye-1 bore-holes. The first three of these bore-holes penetrated the complete thickness of this formation, a maximum of 932 m in the Swab-1 bore-hole, and considerably less in the Tanf-1 (428 m) and Khanasser-1 (465m) bore-holes. Bore-holes in northeastern Syria mainly terminated in sediments of the Swab Formation. The maximum thickness of the Swab Formation (528 m) was found in the Qamichliye-1 bore-hole.

The Swab Formation is composed of monotonous thin-bedded strongly phyllitized black shales. Rare layers of dolomite and anhydrites are observed in the lower half of this unit. In its upper part rare intercalations of siltstones and fine-grained quartz sandstones of gray and light-gray color occur. In the northeast of the country (Afendi-1 and Qamichliye-1 bore-holes) the sediments are more silty and the transition to the Afendi Formation is very gradual. The contact of the Swab Formation with the Lower Ordovician Khanasser Formation is very distinct, but completely conformable.

At a depth of 2160 m in the Khanasser-1 bore-hole remnants of graptolites were found. Such forms as *Hallograptus* cf. *mucronatus* (Hall.) and *Glyptograptus* cf. *teretiusculus* (His.), place the age of the lower half of these shales in the Llandeilo Stage of the Middle Ordovician (det. E. Pashkevichus, Estonia). At a depth of 3420 m in the Swab-1 bore-hole a *Hallograptus* graptolite was also discovered in the overlying layers of the Swab Formation. In the middle part of the shale sequence of the Abba-1 bore-hole a graptolite identified as *Diplograptus spinulosus* of the Llandeilo Stage was described (Sudbury, 1957). The upper layers of the Swab Formation are black shales, usually containing rare acritarchs and chitinozoans of the Middle Ordovician. Therefore, the entire Swab Formation was referred to the Llandeilo, probably to its lower half.

In the Qamichliye-1 bore-hole near the Turkish border, the Swab Formation is characterized by dense gray shales, sometimes transitional to phyllites, and with intercalations of sandstones. These sediments contain a shelly fauna predominantly consisting of brachiopods and orthoceratids which include: *Craniops* aff. *trentonensis* Hall., *Leptobolus* sp., *Ctenodonta* aff. *ala* (Barr.), *Tomaculum problematicum* Groom., *Hyolites* cf. *cinctus* Barr., *Lingula* cf. *jejuna* Barr., *L. attenuata* Murch., *Orthis (Plectorthis?)* aff. *redux* Barr., *Dalmanitina* cf. *socialis* Barr., *D.* aff. *deshayesi* (Barr.), *Phacites* aff. *gotlandicus* Wahl., *Sowerbyella* cf. *sericea* Sow., and *Orthoceras* sp. (det. Nicolaus and Fluegel, Germany; Weber, 1963, 1964).

The Afendi Formation was recognized in all the aforementioned Tanf-1, Swab-1, Khanasser-1, Markada-101, Abba-1, Afendi-1, Qamichliye-1 bore-holes. Sediments of this formation conformably overlie the Swab Formation and show a gradual lithological transition. The Afendi Formation consists of white and light-gray sandstones and siltstones which intercalate with black, strongly phyllitized shales. The matrix of the siltstones is calcareous or clayey. The rocks are massive, sometimes thin-bedded. Sandstones are usually fine-grained and look like quartzites. Shales are thin-bedded, horizontally or wavy-bedded, and phyllitized. Siltstones predominate in the lower and middle parts of this formation, while sandstones occupy the bulk of the upper part of this unit. In some places thick members of sandstones and siltstones devoid of shaley intercalations are very common.

The thickness of the Afendi Formation is highly variable. It is 540 m at the Khanasser-1 bore-hole on the Aleppo uplift, a maximal thickness of 932 m on the Rutbah uplift at the Swab-1 bore-hole, and 1137 m in the Tanf-1 bore-hole. In northeastern Syria the thicknesses of this formation are minimal, ranging from 300 m at the Qamichliye-1 bore-hole, to 471 m at the Afendi-1 bore-hole.

Sediments of the Afendi Formation contain numerous and various acritarchs and chitinozoans. In the lower part of this formation, at the 2973-2770 m interval of the Afendi-1 bore-hole, the assemblage includes: *Peteinosphaeridium nudum* (Eis.), *P. trifurcatum* (Eis.), *P. micranthum* (Eis.), *Baltisphaeridium longispinosum* Eis., *B. latiradiatum* Eis., *Multiplicisphaeridium irregulare* S., and *Aremoricanium* sp., indicating the Llandeilo Stage of the Middle Ordovician.

A similar assemblage of acritarchs was also identified in the overlying layers of this bore-hole. In the uppermost part of the Afendi Formation the assemblage is impoverished, and the age of the rocks is broadly put as Middle to Upper Ordovician.

In the Swab-1, Khanasser-1, Tanf-1, Abba-1 bore-holes, the Llandeilo acritarch assemblage is typical for the main part of the Afendi Formation. In Abba-1 bore-hole, such species as *Colpocoryphe arago* (Roualt), *Pseudobasilicus* cf. *nobilis* (Barr.), and *Diplograptus spinulosus* Sudbury confirm the Llandeilo age.

The uppermost layers of the Afendi Formation in the Tanf-1 and Khanasser-1 bore-holes contain a different acritarch and chitinozoan assemblage, characterized by appearance of numerous *Multiplicisphaeridium lobeznum* (Crm.), *M.* ex gr. *denticulatum* (Crm.), *Conochitina brebicollis* T. et J., and *C. communis* Tau., which are typical of the Caradoc (Upper Ordovician).

Consequently, the Afendi Formation belongs to the upper part of the Llandeilo (Middle Ordovician) and Caradoc (Upper Ordovician), although the latter was only identified in the Tanf-1 and Khanasser-1 bore-holes. In other bore-holes, sediments of this age are absent. The Ashgill Stage, which tops the Ordovician, also has not been identified. Absence of this stage and episodic distribution of the Caradoc can be explained by deep erosion and unconformity below the younger deposits.

Silurian

Silurian deposits of Syria are characterized by a restricted distribution. They were discovered in deep bore-holes in the southern (Tanf-1) and central (Markada-101) regions of Syria. According to Weber (1963, 1964), in northeastern Syria, Silurian sediments have been penetrated in the Khanasser-1 and El-Bouab-1 bore-holes, but he did not mention paleontological evidence for this age. Evidently, a considerable part of the Silurian was removed by subsequent erosion. One can suggest that the most complete sections of Silurian are preserved within the Euphrates and Sinjar paleodepressions.

The monotonous composition of Silurian deposits, shales with subordinate intercalations of siltstones and sandstones, permits us to consider them as a single unit, called the Tanf Formation (Lababidi and Hamdan, 1985). Silurian deposits rest on underlying Ordovician rocks with a hiatus which is emphasized by a large change in the microfossil assemblage, and the total absence of the Ashgill, and in places the Caradoc deposits.

The Tanf Formation in its stratotype (the Tanf-1 bore-hole) is represented by monotonous, dark gray and black, thin-bedded shales. Interbeds of siltstones and sandstones are subordinate. This formation was penetrated between 975-1406 m in the Tanf-1 bore-hole, for a thickness of 431 m. Due to its monotonous, clayey composition, the Tanf Formation is very distinct in the bore-hole logs. Through the entire sequence, these shales are characterized by extremely rich assemblages of acritarchs

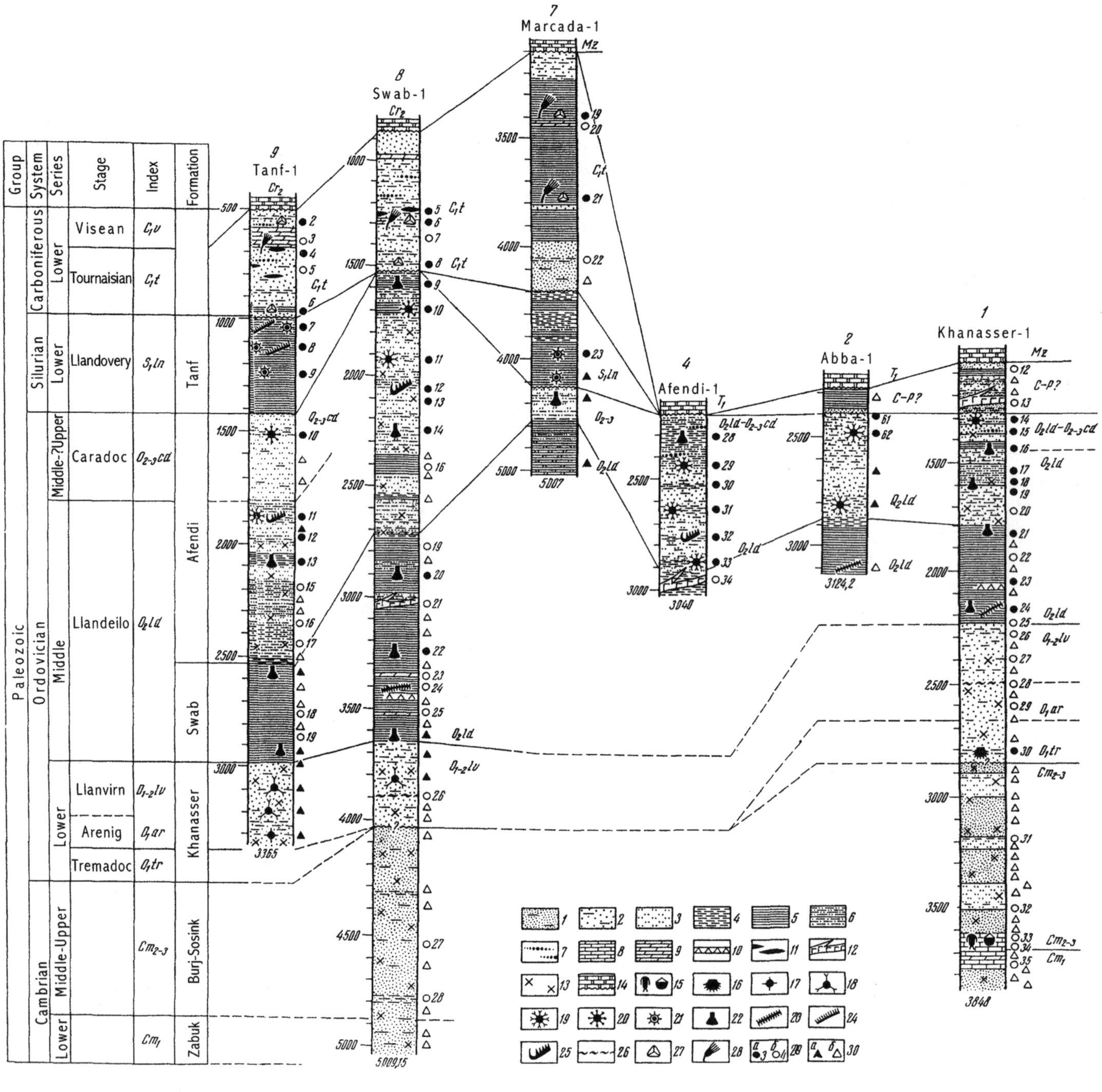

Fig. 6.3. Stratigraphic subdivision and correlation of the Paleozoic deposits of Syria (after Yankauscas and Talli, 1978).

Legend: 1 – red-colored sandstone (with intercalations of mottled sandstone, siltstone and shale); 2 – light-gray quartz sandstone (with rare interbeds of black shale); 3 – siltstone; 4 – shales (with intercalations of siltstone and sandstone which constitute 5-15 %); 5 – shales (with intercalations of siltstone and sandstone which constitute 0-5 %); 6 – alternation of shales with siltstones and sandstones; 7 – rare interbeds of grits; 8 – limestone; 9 – dolomite; 10 – interbeds of gypsum and anhydrite; 11 – interbeds and lenses of coal; 12 – dykes of dolerites; 13 – quartzite and siltstone; 14 – undifferentiated Mesozoic deposits; 15 – remnants of trilobites and brachiopods; 16-22 – assemblages of acritarchs and chitinozoans: 16 – Tremadok, 17 – Arenig, 18 – Llanvirn, 19 – Llandeilo, 20 – Middle-Upper Ordovician, 21 – Llandovery, 22 – rare specimens of Ordovician chitinozoans; 23 – graptolites of Llandeilo; 24 – graptolites of Llandovery; 25 – assemblages of scolecodonts; 26 – beds with traces of mud-eaters; 27 – spores of Tournaisian terrestrial plants; 28 – crinoids; 29 – location of samples taken and numbers of cores (a – a core with microfossils, b – a core barren of microfossils); 30 – samples of cuttings taken from intervals drilled without coring (a – with microfossils, b – barren of microfossils).

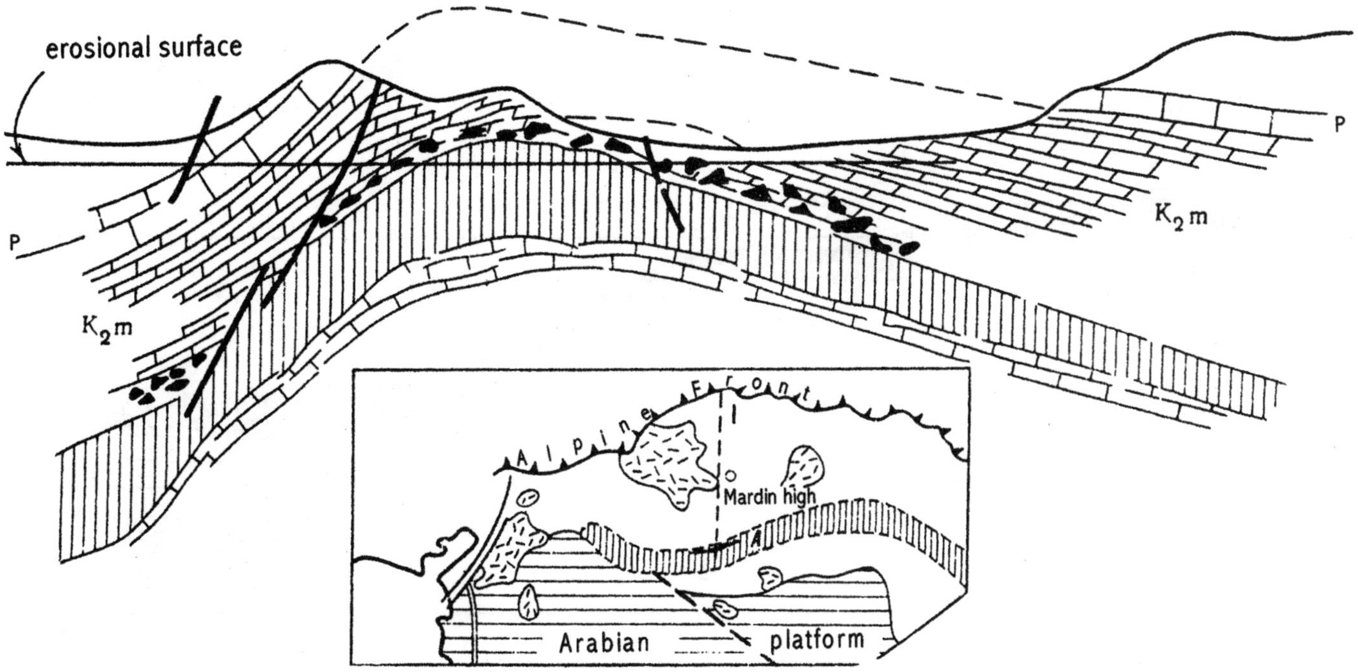

Fig. 6.4. Schematic structure of the central part of the Jebel Abd El-Aziz anticline (after Leonov, Makarem and Zaza, 1986). Vertical hatching shows tentatively a member which corresponds to the main olistostrome including large blocks and plates of exotic rocks. It is overlain by marls and limestones with debris of exotic rocks (black ovals and triangles) replaced up section by pure marls of Maastrichtian (K2m) and limestones of Paleogene (P). At the base of this figure the location of the Jebel Abd El-Aziz anticline is shown (vertical hatching – the Jebel Abd El-Aziz – Sinjar rampart, A – the Jebel Abd El-Aziz anticline, I – the line of profiles shown on Fig. 6.5).

and chitinozoans, indicating the Llandovery Stage (Lower Silurian). At a depth of 1259 m, the following microfossil assemblage was found: *Multiplicisphaeridium denticulatum* (St. et Will.)., *M. ramusculosum* (Eis.), *M. oligofurcatum* (Eis.), *M. corallinum* (Crm.), *Pterospermopsis onondagaensis* Dff., and *Domasia elongata* Downie. Higher up, at a depth of 1127 m, they are accompanied by *Veryhachium scabratum* Crm. and *Baltisphaeridium piriferum* (Downie).

Within the same interval, the graptolites *Monograptus priodon* Bronn., *Streptograptus* sp., *Oktavites* sp. were found. They confirm the Llandovery age of these sediments (det. I. Pashkevichus, Estonia).

At a depth of 994 m, the shales contain graptolites of the uppermost part of the Llandovery: *Monograptus priodon* (Bronn.), *Oktavites spiralis* (Geinitz), *Streptograptus* sp., and *Monoclimacis linnarssoni* (Tullberg). They are associated with acritarchs which were already found in the underlying layers, but for the first time in association with *Helios* cf. *aranaides* Crm., *Membranilarnacia* sp., *Baltisphaeridium* (?) *gothlandicum* (Eis.), *B.*(?) *erraticum* (Eis.), *Plectechitina carminae* Crm., and *P. resende* Crm.

In the Markada-101 bore-hole, the Lower Silurian deposits were penetrated in the 4635-4200 m interval. Here the Tanf Formation is composed of typical black shales which in the upper part of the formation alternate with gray siltstones. At levels of 4470 m and 4580 m, very rich assemblages of microfossils were recovered. They are identical in composition to those found in the Tanf-1 bore-hole and indicate the lower half of the Llandovery.

In bore-holes such as Swab-1, Afendi-1, Abba-1 and Khanasser-1, Silurian deposits are clearly missing, and the Ordovician strata are directly overlain by the Late Paleozoic or the Mesozoic.

Devonian

In Syria paleontologically-proven deposits of the Devonian System have not been encountered. However a thin unit of light-gray sandstones penetrated in the Dola-1 and Doubayyat-1 bore-holes in the northeastern Palmyrides was tentatively referred to the Devonian. Obviously, this time in Syria was marked by a continental regime when erosional processes predominated, and formation of terrigenous continental sediments took place within separate, small depressions. Intensive processes of erosion resulted in the destruction of a significant part of the underlying Silurian deposits.

The whole succession of terrigenous (sandy, clayey) deposits of the lower Paleozoic (Cambrian, Ordovician, Silurian, and Devonian) constitutes a definite sedimentary cycle. The thickness of sediments laid down in this time interval varies from 1500 m to 5000 m. The onset of this cycle is characterized by the Cambrian transgression, when deposition of evaporites and marine, shallow-water platformal limestones and dolomites took place. The

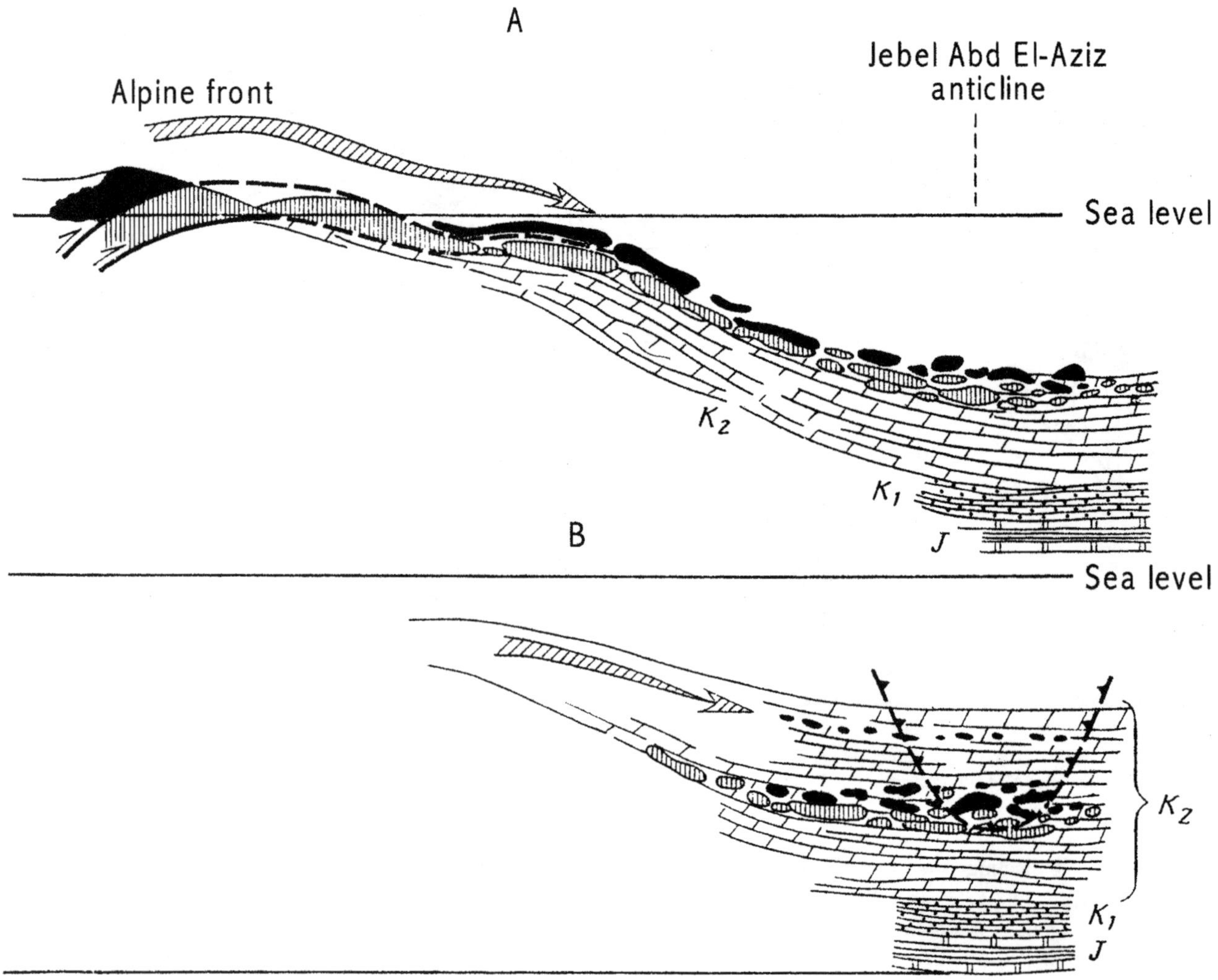

Fig. 6.5. Suggested scheme of formation of the Jebel Abd El-Aziz olistostrome (after Leonov, Makarem and Zaza, 1986). The arrow shows direction of input of exotic material (from the north). Succession of deposits below the olistostrome is hypothetical: K2 – Shiranish and Soukhne Formations, K1 – Ghouna Formation, J – Kamchouka Formation.

A. The Campanian – Early Maastrichtian interval. The time of displacement of tectonic nappes of the frontal zone of the Alpine belt to the south and formation of the olistostrome.

B. The Late Maastrichtian. Continuation of sedimentation in the area of the Jebel Abd El-Aziz anticline after origination of a sequence with olistostromes. An incised line indicates that part of the section, which is exposed in the core of the Jebel Abd El-Aziz anticline.

maximum of the transgression corresponds to the Ordovician and Lower Silurian. During this period of time deposition of shales with graptolites, acritarchs and chitinozoans took place. This cycle was terminated by the Devonian regression with intensive processes of denudation and formation in some areas of thin terrigenous continental deposits.

Carboniferous

Carboniferous deposits were discovered in deep bore-holes in many regions of Syria *e.g.* on the slopes of the Rutbah uplift (Tanf-1 and Swab-1 bore-holes), in the northeastern part of the Palmyra trough (Doubayyat-1 and Dola-1 bore-holes), in the south of the Aleppo uplift (Athar-1 and Khanasser-1), in the Euphrates depression (Ahmar-101, Ahmar -102, Ward N-102, Ratka-101, and Madabe-1), on the northern slope of the Rawda uplift (Markada-101), and in the Sinjar depression (Jbissa-107, El-Bardeh-1, El-Bouab-1, and Toueman-1 bore-holes). These data point to a wide distribution of Carboniferous deposits all over Syria. Sediments of this age are only absent in some of the bore-holes such as Abba-1, Qamichliye-1.

Drilling data and seismic investigations show that usually the Carboniferous deposits transgressively overlie the eroded surface of the Ordovician and Silurian rocks. In rare cases they are underlain by continental sandstones barren of organic remains and conventionally referred to the Devonian. The Carboniferous of Syria is represented by various terrigenous and marginal marine calcareous rocks with rich and varied organic remains, including brachiopods, bryozoans, ostracods, conodonts, foraminifera, crinoids, and spores. Unfortunately, they were poorly documented paleontologically in the bore-hole records, although allochthonous blocks found in Jebel Abd El-Aziz contain well determined fossils of this age. Evidently it was one reason why this very thick succession, corresponding to a considerable interval of geological time, was treated as a single unit, called the Markada Formation (Lababidi and Hamdan, 1985). The stratotype of the formation is located on the northwestern plunge of the Rawda uplift (Markada-101 bore-hole). Here, this formation is subdivided into three members.

The lower member (625 m thick) is composed of terrigenous rocks – gray sandstones and siltstones with intercalations of black micaceous shales. At the base of this member sandstones predominate. They rest on the eroded surface of the Lower Silurian.

The middle member (350 m thick) includes calcareous and terrigenous rocks represented by alternating limestones with foraminifera (Endothyridae), and siltstones and shales with conodonts and spores.

The upper member (165 m thick) is again represented by terrigenous sediments - intercalations of sandstones, siltstones and shales.

The total thickness of the Markada Formation in it stratotype is 1105 m.

Only four cores were taken within this huge interval. In the lower member (Core 22), organic remains have not been detected. In the lower part of the middle member (Core 21), only spores and crinoid ossicles were identified. In the top of the middle member (Cores 20, 19) foraminfera and conodonts were discovered, together with spores and crinoids. The assemblage of conodonts is rather diverse: *Gnathodus pseudosemiglaber*, *G. cuneiformis*, *G. texanus*, *Apatognathus geminus*, *A.* sp, *Spathognathodus scitulus*, *Lonchodina furnishi*, *L.* sp., and *Neoprioniodus scitulus*, *N.* sp. These conodonts place the age of the sediments in the Latest Tournaisian-Early Visean (Husri and Austin, 1985). They also indicate shallow water conditions of sedimentation. The upper terrigenous member was drilled without coring. Evidently, it corresponds to the Visean. In any case, the Markada Formation in its stratotype section is located within the Lower Carboniferous.

The three members of the formation cannot be traced in some areas of Syria. For example, the lower member is missing on the Aleppo uplift (Athar-1 bore-hole), which is explained by the unconformity between the Carboniferous and the Lower Paleozoic. In the Shedade-1 bore-hole, the lower member was not reached by drilling. The middle member is 680 m thick, and in that bore-hole the upper member is 220 m thick. In the southern part of the Euphrates depression (Ratka-101, Ahmar-102 bore-holes), only the lower member is present, on account of intensive pre-Permian erosion. In the Sinjar trough, the thickness of the Carboniferous reaches 1500 m or more. In this depression, and adjacent areas located to the south, the upper member was well preserved. In other regions, sections of the Markada Formation consist of only the two lower members. Lastly, in some bore-holes, the Carboniferous was not subdivided into separate members at all.

At a later time, the Markada Formation in its stratotype was renamed the Markada Group and subdivided into five formations (Al Youssef and Ayed, 1992). They are as follows (from base to top):

1) The Sayad Formation – alternation of siltstone, silty shale and shale, 358 m thick;

2) The Athar Formation, 230 m thick, where sandstones and siltstones predominate;

3) The Halul Formation, 125 m thick, which is composed mainly of limestones and dolomitic limestones;

4) The Sawanet Formation, 740 m thick, consisting of alternated sandstones, siltstones and shales;

5) The Najeeb Formation, 412 m thick, which is made up of interbedded sandstones and sandy shales. The latter was identified in the Sawanet-2bore-hole; in the stratotype itself in the Markada-101 bore-hole, this formation was not identified (on account of a gap in coring or a hiatus with the overlying Middle Triassic deposits).

The palynological analyses indicated the presence of three assemblages of miospores (Ravn et al., 1994). The oldest assemblage from sandy-clayey rocks of the Sayad Formation yielded *Retispora lepidophyta,* typical for the Late Famenian (Late Devonian). The middle assemblage from the Halul Formation confirmed the Tournaisian-Lower Visean age of the sediments, given earlier by conodonts. The youngest miospore assemblage from the Sawanet Formation contains *Aratrisporite,s* which is characteristic of the Visean deposits.

In the Tanf-1 bore-hole, a unit of soft, light-gray sandstones with intercalations of lilac colored and gray shales, light-pinkish dolomites, and lenses of coal with woody remains overlies the eroded surface of the Lower Silurian. Here the thickness of the Carboniferous reaches 500 m. At the very base of this unit (Core 6, at 953 m depth) spores of the Early Tournaisian have been identified. They include: *Calamosphaera microrugosa* Sch.W., B., *Wallatisporites wallatus* Hacq. (=*Hymenozonotriletes pusillites* Kedo), *Dictyotriletes submarginatus* Plauf., and *D. trivialis* Kedo (det. T. Bivsheva and G. K. Vaitekunene, USSR, Estonia). Similar assemblages of spores were also found in overlying layers (Cores 4 and 2) (Yankauskas and Talli, 1978). Terrigenous rocks of the Lower Carboniferous (Tournaisian) are unconformably overlain by limestones of the Upper Cretaceous.

A similar but not identical picture is observed in the Swab-1 bore-hole, where the thickness of the Carboniferous is 650 m. The lower part of this succession is composed of alternating light-gray siltstones, dark-gray shales, and white and gray sandstones with beds of coarse crystalline dolomites (Cores 5, 6, and 8). These sediments are characterized by having the same Tournaisian assemblage of spores as in the Tanf-1 bore-hole (Yankauskas and Talli, 1978). Core 6 contains a rather diverse assemblage of conodonts, including: *Gnathodus bilineatus*, *G. girtyi girtyi*, *G. cuneiformis*, *G. texanus*, *G. pseudosemiglaber*, *G. homopunctatus*, *G. symmutatus symmutatus*, *Euprioniodina microdentata*, *E.* sp., *Hindeodella ibergensis*, *H.* sp., *Ligonodina roundyi*, *L. furnishi*, *L.* sp., *Neoprioniodus* sp., and *Ozarkodina plana* (Husri and Austin, 1985). According to these authors, the boundary of the Tournaisian-Visean is located within Core 6, which is characterized by a similar fauna of conodonts. In this case, a 400 m thick section of sandy clay rock overlying Core 6, and penetrated without coring, should be referred to the Visean (Lower Carboniferous). The Carboniferous deposits are unconformably overlain by Upper Cretaceous limestones.

In the Doubayyat-1 bore-hole to the east of Palmyra, the thickness of the Carboniferous deposits is 310 m. According to an unpublished report on this bore-hole, these deposits are subdivided into three members:

The lower member is 52 m thick, composed of black shales with interbeds of sandstones and sandy clays. This member is conditionally referred to the Lower Carboniferous.

The middle member consists of gray shales with intercalations of limestones, white and gray sandstones, and some thin interbeds of coal. Limestones contain foraminifera (fusulinids), brachiopods (productids), bryozoans, and ostracods (*Bairdia auricula* Knight). The thickness is 135 m. The sediments belong to the Middle Carboniferous.

The upper member is made up of gray sandstones and shales with subordinate intercalations of limestones, containing the ostracod *Pentagonocyclopae* aff *dispar*. Thickness is 123 m. It belongs to the Upper Carboniferous.

In the Dola-1 bore-hole to the northwest of Palmyra, undifferentiated Carboniferous sediments consist of alternating sandstones, shales, and limestones with foraminifera (*Triticites*, *Fusulinella*, *Palaeotextularia*, *Hemigordius*), brachiopods (*Linoproductus*), and bryozoans (*Fenestella*, *Rhombopora*). The thickness exceeds 300 m.

Data on the Doubayyat-1 and Dola-1 bore-holes indicate that in the Palmyra depression there are Middle and Upper Carboniferous deposits, in contrast to the Rutbah uplift. Such a situation is possible; nevertheless, this conclusion is preliminary, because paleontological substantiation of the age of sediments in these two bore-holes cannot be considered definitive.

In northeastern Syria, Carboniferous deposits were penetrated by the Toueman-Nord, El-Bardeh-1, El-Bouab-1, Jbissa-207, and Roumelan-6 bore-holes. They consist of a succession of black shales and sandstones with subordinate intercalations of black and dark-gray limestones and sometimes dolomites. The thickness varies from 648 m in the El-Bouab-1 bore-hole, to 462 m in the El-Bardeh-1 bore-hole.

Only the age of the deposits of the Jbissa-207 borehole are reliably substantiated by paleontology. Here comparatively rare conodonts have been found in Core 1 (3869 m depth), Core 3 (4147 m depth), and Core 4 (4178 m depth). They consist of *Gnathodus cunciformis*, *G. pseudosemiglaber*, and *Lonchodina furnishi*. These species determine the age of the sediments as transitional from the Tournaisian to Visean of the Lower Carboniferous (Husri and Austin, 1985).

In the Toueman-Nord bore-hole the Lower Carboniferous deposits include beds of Namurian age (Weber, 1964). They are unconformably overlain by Triassic beds.

In the El-Bardeh-1 bore-hole sediments of the Tournaisian, Visean, Namurian, Westphalian and (?)Stephanian have been identified. They are unconformably overlain by Triassic rocks.

In the El-Bouab-1 bore-hole, deposits of the Tournaisian, Visean, Namurian, and conventional Westphalian are present (Weber, 1964; Wolfart, 1967) and are unconformably overlain by the Triassic.

Thus, in northeastern Syria, the Carboniferous sections are more complete as compared with those of the Rutbah or Rawda uplifts. Here, all subseries of the Carboniferous are represented, as well as the Lower Carboniferous, including Tournaisian, Visean, and Namurian. These ages are determined by assemblages of spores; unfortunately, lists of spore species in the cited references are not given, and it is not possible to judge the reliability of the Namurian, Middle, and Late Carboniferous ages as given.

As was mentioned above, rootless exotic blocks of Carboniferous rocks are known to exist in the core of the anticlinal uplift of Jebel Abd El-Aziz in northeastern Syria. These outcrops are located 1.8 km to the northeast of the highest point of this uplift (920 m) along a wadi which cuts the northern slope of Jebel Abd El-Aziz, midway between Sheikh Abd El-Aziz in the west and Beer El-Khazne villages in the east. The block of Carboniferous deposits is 40 m wide and is composed of limestones, shales, and sandstones some 30 m in thickness (Ponomarev and Bivshev, 1966). The shales are gray and dark-gray, and somewhat sandy. The sandstones are fine-grained, yellowish or gray with ferruginous calcareous matrix. Limestones are brownish and gray, thin-layered, fine-grained, with abundant organic remains. These intercalated rocks are covered by a 4-5 m thick unit of ocherous-yellow, fine- and medium-grained sandstones with the same ferruginous calcareous matrix.

Outcrops of Carboniferous rocks in the core of the Jebel Abd El-Aziz anticline were the subject of numerous geological studies. The first study was by Dubertret, Keller, and Vautrin (1932a, b). Later the Carboniferous exposures were mentioned in the publications of Dubertret (1933b), Tasman (1949), and Weber (1963). Special attention was paid to these Carboniferous outcrops by geologists of the USSR (Ponomarev and Bivshev, 1966; Ponikarov et al., 1967, 1969) during the geological mapping at the scale of 1:200,000. Paleontological investigation of the sediments of this age in Jebel Abd El-Aziz was very thorough as compared to the Carboniferous deposits penetrated by bore-holes, and their age was determined with considerable precision. Similar important studies were also carried out by German scientists (Wolfart, 1965, 1967; Weber, 1968; Stoppel, 1970; Plumhoff, 1970), but without the aid of the detailed geological mapping. However the views of different scientists concerning the age of these rocks are very similar.

Dubertret et al. (1932a, b) determined the following faunal

and floral species: *Spirifer tornacensis* Kon., *Orthotetes crenistria* Phil., *Ambocoelia urei* Flem., *Lepidodendron* aff. *veltheimianum* Sternb., and *Archaeosigillaria* aff. *vanuxemi* Goepp., indicating the Tournaisian Stage of the Lower Carboniferous.

According to more recent determinations, the assemblage of fauna includes *Chonetes (Rugosochonetes) laguessianus* Kon., *Spirifer* aff. *tornacensis* Kon., *Buxtonia scabricula* Mart., *Camarotoechia* sp., and *Syringothyris* sp., which confirms this Tournaisian age (det. N. Litvinovitch, in Ponomarev and Bivshev, 1966; Ponikarov et al., 1967).

A long list of species of corals, brachiopods and bryozoa was given by Wolfart (1965). The most common are *Cladoconus* cf. *crassus* (M'Coy), *Saffordotaxis* cf. *incrassata* (Ulrich), *Leptagonia analoga* (Phillips), *Marginicinctus syriensis* Wolfahrt, *Prospira* aff. *typa* Maxwell, *Syringothyris cuspidata* (Martin), *Composita* aff. *subpyriformis* (von Semenov et Moeller), *Dielasma* cf. *insigne* de Koninck, *Zaphrentites* sp., *Schuchertella* sp., *Chonetes* cf. *hemisphaericus* von Semenov., *Plicatifera* sp., *Echinoconchus* ex gr. *punctata* (Sowerby), *Camarotoechia acutirugata* (de Koninck), and *Athyris lamellosa* (l'Eveille). Besides these, the following fossils have also been found: *Fenestella* sp., *Buxtonia* sp., *Martiniopsis* sp., *Punctospirifer* sp., *Dictoclostus* sp., *Linoproductus* sp., *Plicatella* sp., tubes of annelids (*Serpula*, *Serpulospira*), stems of crinoids, and shells of gastropods (*Straparollus* sp., *Platyceras* sp.). According to Wolfart, these limestones may be referred to the upper part of the Tournaisian-Lower Visean (Lower Carboniferous).

The ostracod fauna is poorly preserved: *Geisina rotunda* Plumh., *Hypotetragona* sp., *Sanniolus* sp., *Savagellites lacunatus* Plumh., *Cytherella* sp., *Cavellina* sp., *Paracavellina syrica* Plumh., *Healdia* sp., *Silenites marginiferus* (Geis), *Bairdiocypris* sp., *Pseudobythocypris* sp., *Graphiadactyllis* sp., and *Cryptobairdia* cf. *compacta* (Geis). This fauna was studied by Plumhoff (1970), who concluded that these deposits belonged to the lower and upper parts of the Lower Carboniferous (Mississippian).

The assemblage of conodonts is of diverse taxonomic composition (Stoppel, 1970). It includes *Angulodus walrathi* (Hibbard), *Diplododella* cf. *plana* (Thomas), *Hindeodella fragilis* Haas, *H. gladiola* E. R. Branson, *Ligonodina princeps* (Hinde), L. cf. *typa* (Gunnell), *Lonchodina* cf. *tenuis* Huddle, *Neoprioniodus alternatus* (Ulrich et Bassler), *N. lanceolatus* Haas, *N. varians* (Branson et Mehl), *Ozarkodina* sp., *Polygnathus communis communis* Branson et Mehl, *P.* cf. *inornatus* E. R. Branson, *P. longiposticus* Branson et Mehl, *P. symmetricus* E. R. Branson, *Pseudopolygnathus* cf. *dentilineatus* E. R. Branson, *P.* cf. *nodomarginatus* (E. R. Branson), *P.* cf. *primus* Branson et Mehl, *Spathognathodus abnormis* (Branson et Mehl), *S.* cf. *aciedentatus* (E.R. Branson), and *S. crassidentatus* (Branson et Mehl). A peculiarity of this assemblage is the complete absence of representatives of genera *Elictognathus*, *Gnathodus*, and *Siphonodella*. D. Stoppel determined the age of the limestones of Jebel Abd El-Aziz as the Lower Carboniferous (Mississippian).

Consequently, there is agreement regarding the Lower Carboniferous (Upper Tournaisian-Lower Visean) age of these limestones between all geologists and paleontologists who studied the organic remains from the exotic blocks of Jebel Abd El-Aziz.

A short distance to the east from the main exposures of the Lower Carboniferous rocks, along a wadi located a few km from the village of Rhara, Wolfart (1965) discovered (among marls and limestones of the Campanian and Maastrichtian) small blocks of yellowish and brownish fine-grained clayey sandstones with brachiopods (casts) and bryozoans: *Choristites* sp., *Productus* sp., *Echinoconchus* sp., *Stenocisma* sp., *Fenestella* sp., and *Meekella* sp. The presence of *Choristites* indicates a Middle-Upper Carboniferous age.

As mentioned above, the views of different scientists on the age (Lower Carboniferous) of the main exotic blocks are similar. However, opinions are quite varied concerning the origin of these blocks.

Most geologists believe that the surface exposures of Carboniferous rocks can be explained by their uplift along deep subvertical faults. According to the geological mapping such a sublatitudinal lineament stretches along the northern flank of the Jebel Abd El-Aziz anticline (Dubertret et al., 1932a, b; Ponomarev and Bivshev, 1966; Ponikarov et al., 1967; Weber, 1963; Wolfart, 1965). This fault is also confirmed by a sharp gradient in the gravity field. Subsequent sliding events resulted in these huge Carboniferous blocks being included in a unit of Campanian-Maastrichtian marls and limestones. Tasman (1949) assumed that during the Late Cretaceous, processes of submarine sliding were very common.

A quite different point of view was expressed by Yu. Leonov, Makarem, and Zaza (1986). These researchers considered Jebel Abd El-Aziz to be an asymmetrical anticlinal fold with a steep (up to 50°) northern flank, flat subhorizontal axial dome, and gently dipping southern flank (**Fig. 6.4**). According to their view, the lineament in the north corresponds to a discontinuous fault which does not involve significant vertical upward displacement (to the surface) of deeply buried layers. Outcrops of Carboniferous rocks in the center of this fold (south of the lineament) do not indicate large vertical displacement because they are observed, not in their normal stratigraphic sequence, but as inclusions (blocks and tabular bodies) within the Upper Cretaceous succession.

In the outcropping section which contains the Carboniferous blocks, Yu. Leonov et al. (1986) distinguished two units. The upper unit consists of three members, subdivided by hiatuses (Maastrichtian; Paleocene-Eocene; Miocene) preserving features of normal platform sedimentary formations. It is represented by monotonous, light-gray Maastrictian marls and clayey limestones up to 300 m thick. These deposits have horizons and lenses with angular debris of exotic material which sometimes resembles true breccias with clasts ranging from very small to 10 cm in diameter. Contemporaneous input of coarse and angular material into a marine basin together with pelagic sedimentation can be connected with processes of submarine sliding and disintegration of this material by turbidity currents.

The lower unit is also composed of limestones and clayey limestones of the Campanian-Maastrictian, but they contain layers of chaotic structure characterized by a mixture of blocks and debris of rocks of different ages (Carboniferous; Neocomian; Cenomanian), as well as of different compositions and sizes. The fragments of rocks appear as tabular bodies, and sometimes equidimensional blocks from tens of meters down to small clasts. The largest tabular units are composed of dolomitic limestones of the Cenomanian, as thick as 50 m, and massive cryptocrystalline limestones barren of fauna and of unknown age, which are up to 30 m in thickness. The chaotic structure of this unit is shown on **Fig. 6.5**. This figure demonstrates an alternation of normal Campanian-Maastrichtian marls (layer 1), sedimentary breccia with chaotic structure (layer 2, thickness 15 m), sandy limestones with blocks of limestones up to 6 m (layer 3), quartz sandstones and marls (layer 4, 2 m thick), and Campanian-Maastrictian marls with separate debris of pelitomorphic limestones (layer 5, thickness 40 m).

An analysis of morphological and stratigraphical patterns of deposits exposed in the core of the Jebel Abd El-Aziz anticline led Yu. Leonov et al. (1986) to conclude that the lower unit is a huge olistostrome body, confined within the pile of marls and clayey limestones of Campanian-Maastrictian age. This age (Campanian-Early Maastrictian) determines the time of origin of the olistostrome.

The above age of the olistostrome pile of Jebel Abd El-Aziz is in accordance with the time of widespread development of processes of nappe and olistostrome formation in the southern chains of the Alpine belt on its contact with the Arabian platform. An analogous event is well observed in the Baer-Bassit ophiolite Massif in northwestern Syria. The olistostrome of Jebel Abd El-Aziz evidently originated during this epoch of nappe-formation as a result of horizontal shifting and sliding of ancient rocks in the southern direction from the Taurus zone towards the marine basin of the Arabian platform with Late Cretaceous pelagic deposition (Fig. 6.5). A special feature of the olistostrome pile of Jebel Abd El-Aziz is connected with the fact that it is located not within the limits of the frontal Alpine nappe, but within the platform and a significant distance from the nearest Alpine front.

This explanation of the structure and genesis of the rock units with exotic blocks of Carboniferous age suggested by Yu. Leonov et al. (1986) is convincing and corresponds to the general processes of the geological development of the Peri-Arabian arc in Late Cretaceous time. According to this explanation, Carboniferous deposits cropping out in the core of the Jebel Abd El-Aziz anticline were transported from the north and do not reflect indigenous deposition within Syria. Nevertheless we cannot forget that limestones and sandstones of Tournaisian and Lower Visean age at Jebel Abd El-Aziz are widely distributed all over Syria and were penetrated by many bore-holes. Evidently, the final answer will be found when:

1) the entire sequence of the Carboniferous in the vicinity of Jebel Abd El-Aziz will be drilled, and

2) when the lithological and stratigraphic character in the sections of southern Turkey become known. We assume that comparative analysis of all such data will allow a definitive conclusion regarding the distribution and origin of Carboniferous rocks in this region.

Permian

The Permian deposits of Syria are predominantly represented by continental red-colored sediments, and intercalations of rocks with marine fauna only appear in the northeastern part of the country. These deposits were lumped together under the name Amanus Formation. Unfortunately, paleontological data of the Amanus Formation are very minimal. Faunistic and palynological data exist only for northeastern Syria, resulting in differing opinions regarding the age of this formation.

The Amanus Formation was penetrated by many boreholes: Roumelan-6, Markada-101, Tell-Zrab-1, Azar-1, Sfaye-1, Amala-1, Ressafeh-1, Soukhne-1, Qaryatein-1, Doubayyat-1, Meskene-1, Syrom-3, Aleppo-1, Khanasser-1, Abba-1, Aoda-107, Aoda-116, Shadadi-1, El-Bouab-1, Qamichliye-1, etc. It is made up of mottled (brown, red, gray) soft sandstones, siltstones, and clays of sub-continental origin. The facies consists of non-sorted clayey sandstones, and the presence of some intercalations of brown coal, as well as conglomerates. In the southern part of the Aleppo Plateau (Azar-1, Habari-1 bore-holes), fine-grained red quartz sandstones with cross-bedding, and interlayers of silty mudstones predominate, with gray and dark-gray patchy coloring due to hydrated Fe oxides. In the western part of the Palmyrides (Qaryatein-1 bore-hole) red clayey limestones and calcareous mudstones with thin interlayers of mottled quartz sandstones are developed. Intervals of pure gray quartz sandstones with thin intercalations of dark gray mudstones have been penetrated by many bore-holes. On the northwestern flank of the Rawda uplift (Markada-101 bore-hole) the Amanus Formation is characterized by alternating red and green quartz and graywacke sandstones and gray mudstones. In northeastern Syria (Aoda-107, Aoda-116, Roumelan-6; El-Bouab-1; Qamichliye-1; Shadadi-1) this formation consists of gray sandstones intercalated with silty mudstones. Upsection they become calcareous and some limestone interbeds appear.

In central Syria the thickness of the Amanus Formation is highly variable. Usually the thickness is around 250-350 m. However in the Palmyrides the thicknesses are greater; thus, in the Doubayyat-1 bore-hole, the thickness reaches 635 m, and in the Soukhne-1 bore-hole, 656 m. Here the formation is subdivided into two members. The lower one is up to 500 m thick, and is composed of monotonous brown and reddish mudstones with rare and thin intercalations of limestones. The upper member is up to 200 m thick and is represented by mottled sandy clayey rocks which are common for this interval.

Sediments of the Amanus Formation usually rest transgressively on Carboniferous Markada Formation rocks, frequently separated by an erosional surface. In places, they overlie the Silurian Tanf Formation (Maskane-1 and Syrom-3 bore-holes), and even the Upper Ordovician Afendi Formation (Aleppo-1, Khanasser-1, and Abba-1 bore-holes). Evidently in these bore-holes, various parts of the Amanus Formation can be underlain by rocks of different ages.

In the extreme northeast of Syria the Amanus Formation was penetrated by many bore-holes (Shadadi-1, El-Bouab-1,

Roumelan-6, Aoda-107, Aoda-116, Qamichliye-1, etc.). In El-Bouab-1, the total thickness of this formation is 653 m (Ponomarev and Bivshev, 1966). Here it is subdivided into two units. The lower one (interval 3415-3500 m, thickness 85 m) is composed of dark-gray sandy shales, with basal sandstones. No organic remains have been discovered, and the sediments are tentatively considered as Lower Permian.

The upper unit (interval 2847-3415 m) is very thick (568 m). It belongs to the Upper Permian, and consists of three members, from base to top:

1) Dark-gray shales with thick (50-80m) intercalations of gray fine-grained sandstones. Clay interbeds are characterized by *Spirifer alatus* Goldfuss, *Terebratuloida* cf. *depressa* Waagen, and *Heliodopsis* cf. *abreviata* Waagen. Thickness is 345 m.

2) Dark-gray shales with beds of fine-grained sandstones and limestones, containing a rather rich fauna: *Productus* cf. *burlingtonesis* Hall, *P.* cf. *semireticulatus* Martin, *Lingula credneri* Geinitz, *Edmondia* cf. *sulcata* Phill., *Mytilus* cf. *squamosus* Sow., *Bekewelina* cf. *sedgwickiana* King, *Pleurophorus* aff. *jonnstonei* Dickins, *Nuculana subacuta* Waagen, *Albrisma* aff *pleuromyoides* Waagen, *Streparollus permianus* King., and *Bellerophon* cf. *permianus* Netschajew. Thickness is 155 m.

3) Gray and dark-gray limestones with beds of dark-gray shales containing *Chonetes* cf. *kirkbyi* Trechm. Thickness is 68 m.

In the Qamichliye-1 bore-hole, the Amanus Formation is composed of gray fine-grained sandstones with intercalations of siltstones. The thickness is 300 m (interval 1850-2150 m). These sediments were referred to an undifferentiated Permian age.

In Shadadi-1 bore-hole, sandy-clayey sediments of this formation are characterized by gray, massive limestones with abundant remnants of brachiopods, corals, and crinoids (Bebeshev et al., 1988). Among the brachiopods, *Derbyia* cf. *grandis* Waagen, and *Pseudoderbyia* cf. *netshaejewi* Licharev, *Ps.* sp. were determined by V. Ovcharenko.

The uppermost part of the Amanus Formation was penetrated by some bore-holes: Aoda-107 (interval 2728-2748 m), Aoda-116 (interval 2880-2929 m), and Roumelan-6 (interval 3460-3690 m). It is composed of alternating gray sandstones, siltstones and shales. In sediments of these three bore-holes, identical and comparatively rich assemblages of miospores have been identified (Yaroshenko and Bash Imam, 1994). In all samples studied, spores (59.5-92%) predominate over pollen (8-40.5%). Spores include trilete (21.5-66.5%) and monolete (24.5-43.5%) forms. Among pollen, bisaccate non-striate (0.5-32.0%), striate (1.5-15.5%), and monosaccate (1.5-15%) forms are present. Representatives of genera having wide stratigraphic ranges are common:

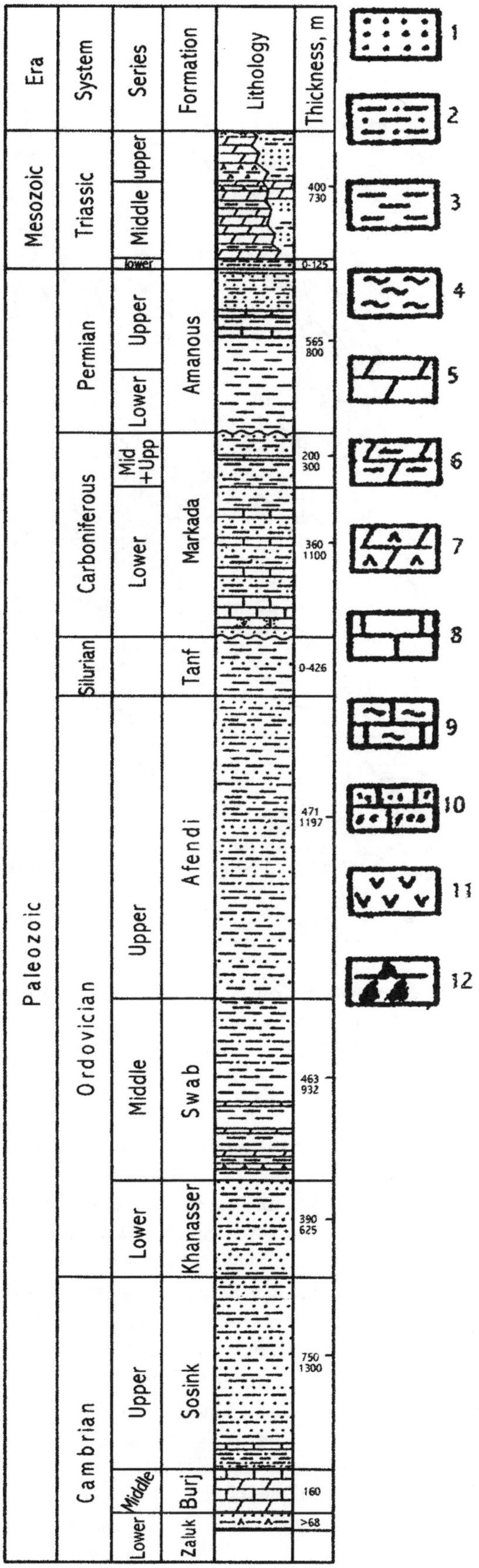

Fig. 6.6. Schematic composite section of the Paleozoic deposits of Syria and overlying Triassic sediments.
Legend: 1 – sandstone; 2 – siltstone; 3 – shale; 4 – marl; 5 – dolomite; 6 – clayey dolomite; 7 – gypsiferous dolomite; 8 – limestone; 9 – clayey limestone; 10 – sandy limestone; 11 – gypsum and anhydrite; 12 – silicified rocks

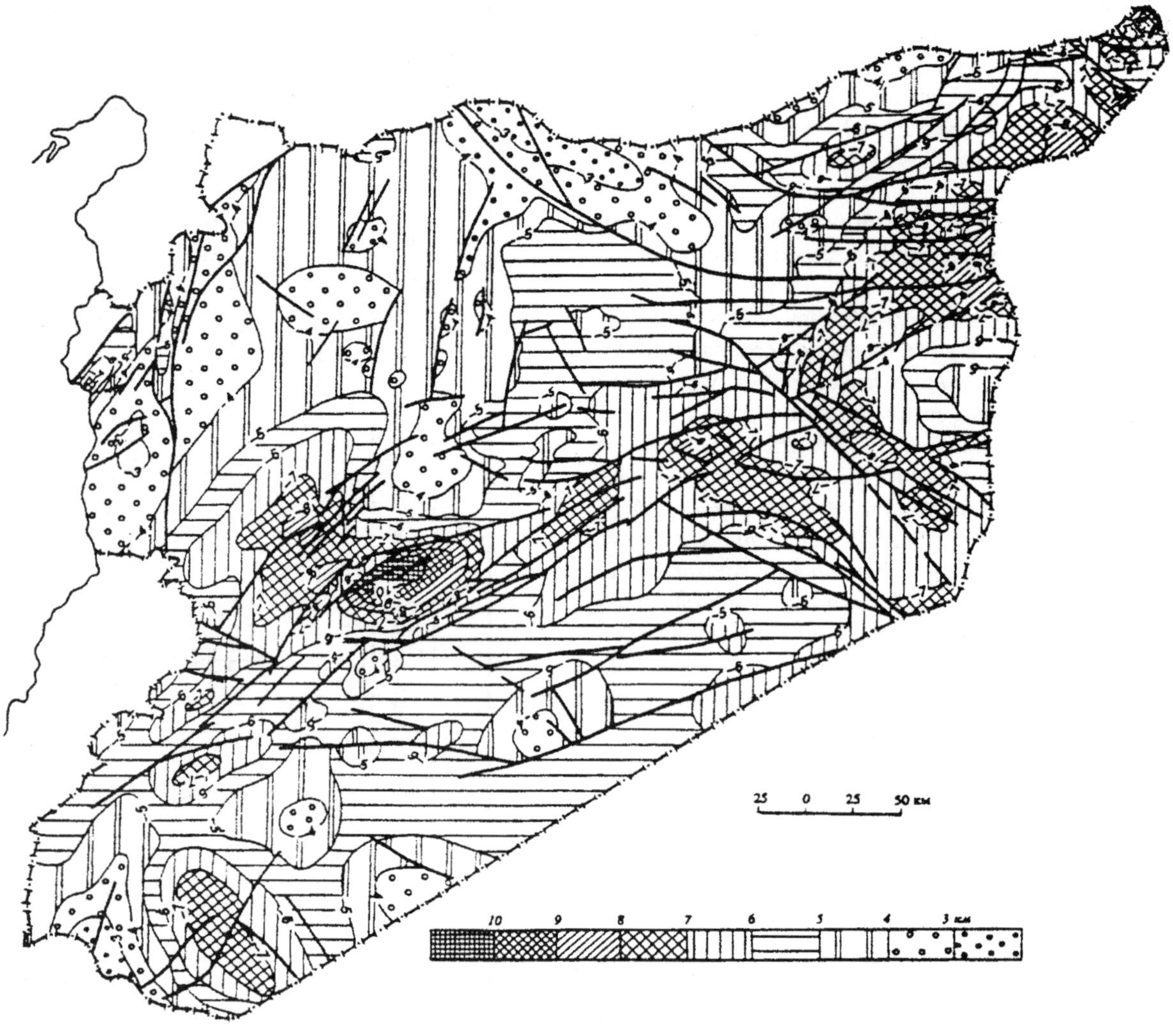

Fig. 6.7. Structural map of the base of the Paleozoic sedimentary cover of the Arabian platform, Syria (compiled by M. Outri, A. Youssef and T. Sawaf, Syrian Petroleum Company, after Kopp and Leonov, 2000).

Legend: Isolines with figures – contour lines of the platform cover base; figures – depth below the sea level (in km); thick lines – faults. Scale – depth, km.

Leiotriletes, *Punctatisporites*, *Calamospora*, *Verrucosisporites*, *Punctatosporites*, *Alisporites*, and *Vitreisporites*. They are accompanied by taxa typical for Permian and Permo-Carboniferous deposits: *Triquitrites iraqiensis* Singh, *Stellisporites inflatus* Alpern, *Schopfites lateralis* Singh, *Polymorphisporites ornatus* Alpern, *Knoxisporites* sp., *Laevigatosporites vulgaris* Alpern & Doubinger, *Torispora laevigata* Bharadwaj, *T. securis* (Balme), *Thymospora cicatricosa* (Balme & Hennelli), *Striatoabieites multistriatus* (Balme & Hennelli), *Hamiapollenites insolitus* (Bharadwaj & Salujha), *Potonieisporites novicus* Bharadwaj, *Punctatosporites granifer* Potonie & Kremp, *Iraqispora labrata* Singh, *Pinuspollenites thoracatus* Balme, *Pretricolpipollenites bharadwaji* Balme, etc. All these forms have been identified in the Permian deposits of China (Kaiser, 1976), Pakistan (Balme, 1970), Iraq (Singh, 1964), Turkey (Agrali and Akyol, 1967), and in some western European countries.

A special close resemblance is observed with palyno-assemblages of the Permian sediments of the Hazro section located near Diyarbakir in southern Turkey (Agrali and Akyol, 1967). This resemblance is expressed both in the taxonomic composition and in the quantitative ratio of the main groups of miospores. Taxa common for the palyno-assemblages of northeastern Syria and the Hazro section are represented by *Cancavissimisporites granulus, Concavisporites trilobatus, C. crassus, C. granulatus, Iraqispora labrata, Kyrtomisporis* sp., and *Tuberculatosporites modicus*, as well as forms of *Knoxisporites, Alisporites, Hamiapollenites,* and *Lueckisporites*. The similarity of these palyno-assemblages is especially emphasized by presence of considerable amounts of monolete spores – *Punctatosporites, Thymospora,* and *Torispora.* The last two genera include forms with various ornamentation,

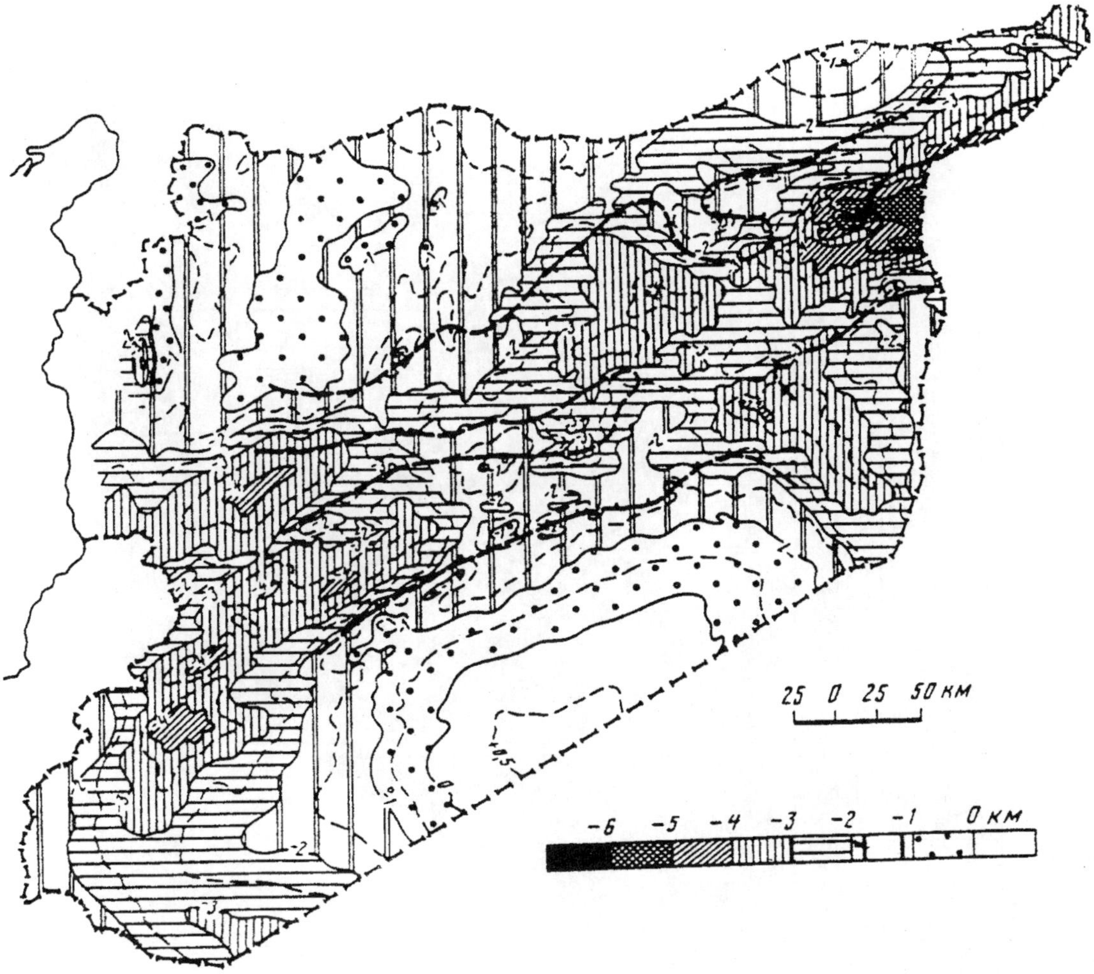

Fig. 6.8. Structural map of the top of the Paleozoic sedimentary cover of theArabian platform, Syria (compiled by M. Outri, A. Youssef, T. Zaza and T. Sawaf, Syrian Petroleum Company, after Kopp and Leonov, 2000).
Legend: Isolines with figures – contour lines for the top of the Paleozoic – the base of the Mesozoic; figures – depth of this surface below the sea level (in km); a thick dotted line with bergstrichs or hachures - contour of a Triassic trough (graben?) with Triassic deposits over 0.5 km thick; a line without bergstrichs or hachures – the axial zones of troughs where the thickness of Triassic is on average more than 1.5 km. Scale – depth, km.

from smooth to highly sculptured.

As a result, Yaroshenko and Bash Imam (1994) concluded that the Amanus Formation in the Aoda-107, Aoda-116, and Roumelan-6 bore-holes was of Permian (evidently Early Permian) age. Some authors have suggested (Bebeshev et al., 1988) that the absence of typical Early Triassic taxa excludes this age. The palynoflora of northeastern Syria contains a mixed composition of taxa known from various paleobiogeographic provinces, such as Euro-American, Cath-Asian, and Gondwanian. Such composition indicates a transitional (ecotone) character for this palynoflora, and close connection of these provinces in the Early Permian. Close resemblance of palyno-assemblages of the Amanus Formation of Syria and southern Turkey (the Hazro section) indicates that these areas were located in the southern Tethys realm, between the Eurasian and African continents (Guevenc, 1985).

In the Roumelan-6 (interval 3578-3629 m) bore-hole, in the same interval where the miospore assemblage was encountered, conodonts were looked for, but none were found (Hussri and Austin, 1985). In an intercalation of black limestones (core 20, interval 3618-3629 m), only brachiopod debris was recovered. These findings at least indicate a marine environment for the genesis of these sediments.

Permian deposits in the Aoda-107, Aoda-116, and Roumelan-6 bore-holes are unconformably overlain by Middle Triassic sediments of the Kurrachine Dolomite Formation (Yaroshenko and Bash Imam, 1994).

Various opinions exist concerning the age of the Amanus Formation. Weber (1963, 1964) suggested a Carboniferous age for this subdivision. Yankauskas and Talli (1978) treated this formation as Permian, but also suggested that a part of the Late Carboniferous may be present at its base. Wolfart (1967) and Ponikarov et al. (1969) considered the Amanus Formation as Permian or Upper

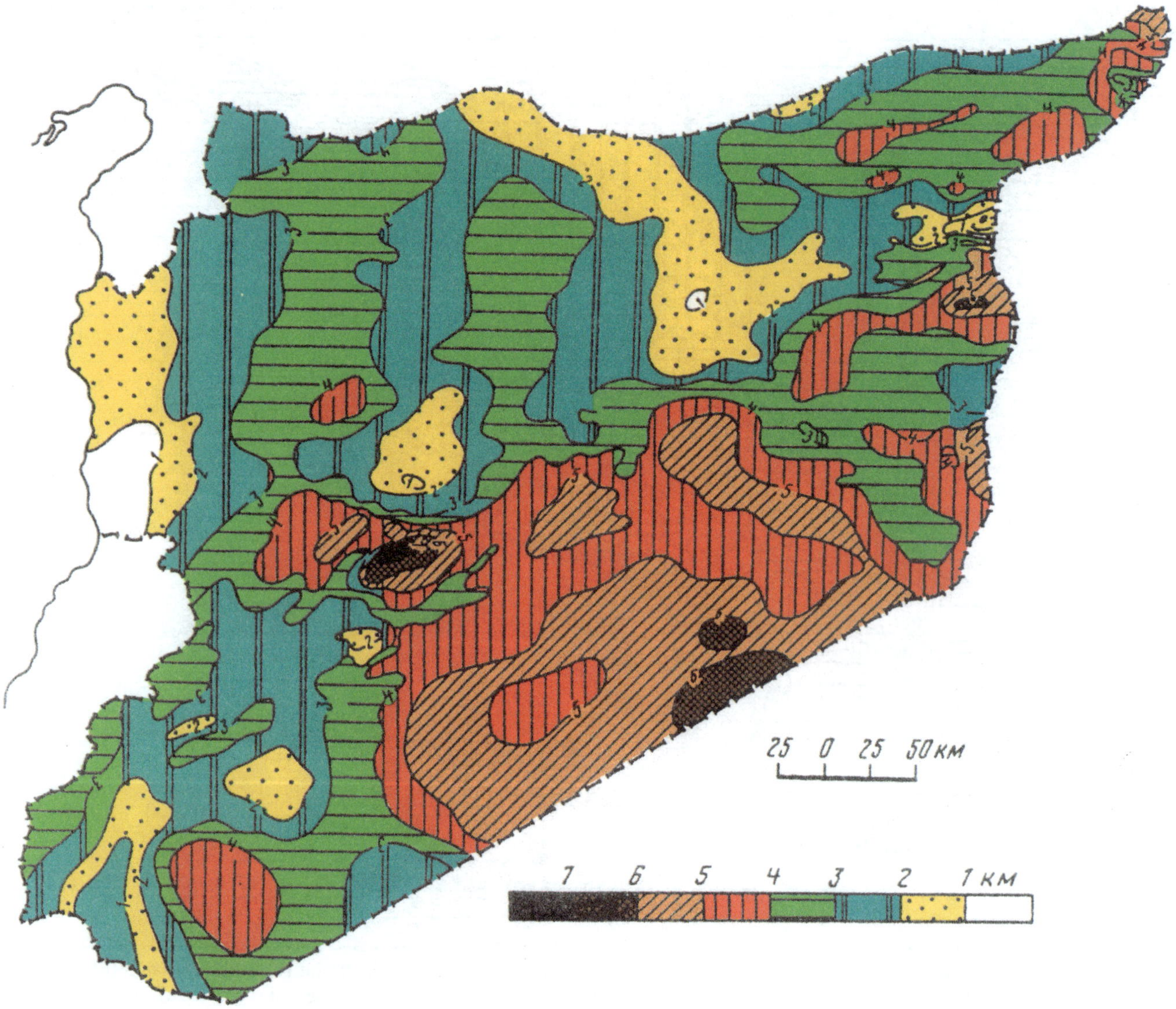

Fig. 6.9. Schematic map of the total thickness (isolines with figures, in km) of the Paleozoic succession of the Arabian platform, Syria (compiled by S. P. Sigachev, after Kopp and Leonov, 2000).
Legend: Scale, thickness, km.

Permian. Bebeshev et al. (1988) considered the formation to be Lower Triassic.

New paleontological data clarifies the age of the Amanus Formation, although not all problems were solved. The widely distributed upper clayey-sandy unit with intercalations of limestones can be referred to the Late Permian. Macrofauna (brachiopods, pelecypods, gastropods) found in the El-Bouab-1 and Shadadi-1 boreholes is very informative. The lower, predominantly clayey unit in northeastern Syria contains rich palyno-assemblages of the Permian (evidently Lower Permian), discovered in the Aoda-107, Aoda-116, and Roumelan-6 bore-holes. The Lower Triassic age for the Amanus Formation of northeastern Syria is based on mistaken determinations of miospores (Bebeshev et al., 1988) and is not supported by this new data. In central Syria (Palmyrides), where reliable published paleontological data are absent, a somewhat different uncertainty exists regarding the age of the Amanus Formation. Information on the presence of Late Carboniferous-Early Permian miospores in the Doubayyat-1 and Arak-1 boreholes is of the most general kind. It is possible that in the Palmyrides, the Amanus Formation includes not only the Permian deposits, but also the upper part of the Carboniferous. In the southwestern Palmyrides, the Qaryatein-1 bore-hole penetrated a sequence of terrigenous rocks as thick as 1200 m below the Kurrachine Formation. These deposits have been singled out as an independent unit, named the Qaryatein Formation. Its relation with the Amanus Formation is still unknown. Evidently, the Qaryatein Formation contains Permian sediments but its stratigraphic scope can be very wide (Upper Carboniferous-Lower Triassic).

The distribution of Permian sediments in Syria suggests that in the south of the country the Permian is represented by red and mottled sandy-clayey rocks of continental origin. In the northeastern part of Syria, which occupies the marginal part of the Arabian platform, continental sediments were replaced by marine terrigenous sediments, but with intercalations of limestones characterized sometimes by rich macrofauna. Intercalations of marine rocks are especially typical for the Upper Permian.

The names of the Paleozoic formations used in this chapter (Zabuk, Burj, Sosink, Khanasser, Swab, Afendi, Tanf, Markada and Amanus) were adopted by petroleum companies working in Syria and mentioned in the publications of many authors (Litak et al., 1997; Brew et al., 1997a, b; Leonov et al., 1989 and others). However, sometimes the term Abba Formation is used instead of the name Tanf Formation (Silurian), and the Markada Formation (Carboniferous) is replaced by the Doubayyat Formation (Caron et al., 2000).

CONCLUSIONS

In conclusion, published material allows formulation of only the general pattern of the composite section of the Paleozoic (**Fig. 6.6**) and derivation of the general features of the development of Syria in the Paleozoic. Accordingly the maximal thicknesses of the Paleozoic formations of Syria are as follows: Zabuk – more than 68 m, Burj – 160 m, Sosink – 1300 m, Khanasser – 625 m, Swab – 932 m, Afendi – 1200 m, Tanf – 426 m, Markada – 1300 m, and Amanus – 800 m. These figures are close to those given in the recent publication of Litak et al. (1997) which are: Burj – 150 m, Sosink – more than 750 m, Khanasser – 1500 m, Swab – 1000 m, Afendi – 1000 m, Tanf – 750 m, Markada – 1200 m, and Amanus – 750 m.

Geologists of the Syrian Petroleum Company (M. Outri, T. Zaza, A. Youssef, and T. Sawaf) compiled structural maps of the base of the sedimentary mantle of the Arabian platform in Syria (**Fig. 6.7**), and of the top of the Paleozoic sequence (**Fig. 6.8**). These maps were presented in the publication of Kopp and Leonov (2000). Comparison of these maps allowed S. P. Sigachev to create a schematic map of the total thickness of the Paleozoic deposits of Syria (**Fig. 6.9** after Kopp and Leonov, 2000). Data on the spatial distribution of the Paleozoic deposits of Syria led Leonov et al. (1989) and Kopp and Leonov (2000) to the conclusion that:

1) the Paleozoic deposits increases thicken from north to south *e.g.* from the Aleppo uplift (3-3.5 km) to the Rutbah uplift (5-7 km). A Paleozoic depression exhibiting the highest rate of sedimentation occupied the present location of the recent Rutbah uplift;

2) the distribution of the Paleozoic deposits is characterized by a mosaic pattern with an absence of sharp contrasts;

3) the Paleozoic deposits do not display linear structures typical of the Mesozoic and Cenozoic. The Palmyrides (with the Paleozoic complex from 3 km to 5-6 km thick) and the Euphrates and Drouze basins do not resemble independent structural depressions, and the Rutbah and Rawda uplifts were not separated from each other;

4) the views concerning the ancient (Early Paleozoic) origin of the main structural elements of Syria (Ponikarov et al., 1966, 1967, 1969) should be reconsidered; these events have taken place in the Late Paleozoic and especially in Mesozoic time.

Consequently, during the Paleozoic Era, Syria was an arena of intensive terrigenous platform sedimentation and calcareous deposits were definitely subordinate. A very thick succession of sandy-clayey rocks up to 5000 m thick was formed during the Cambrian and Ordovician, as well as in the Early Silurian.

This predominantly marine regime was replaced by a stage of continental development in the Late Silurian (Wenlock, Ludlow) and in the Devonian. Deposits of this age are either absent or found in limited areas and conditionally identified as Devonian.

The next cycle of sedimentation belongs to the Carboniferous and Permian. It commences with a widespread transgression of the Early Carboniferous, when a thick succession of sandy-clayey rocks with subordinate intercalations of limestones accumulated. In the Middle and Late Carboniferous, and in the Permian, marine sedimentation was localized in the marginal part of the platform (Mesopotamia), and in the most depressed structures of central Syria (the Palmyrides). During this time in the southern part of Syria continental red terrigenous sediments with rare intercalations of marine limestones were formed. In northern Syria, gray sandy-clayey sediments and marine intercalated rocks are common (the upper part of Carboniferous; Upper Permian).

Paleozoic deposits have recently been penetrated by many new bore-holes connected with oil and gas exploration in northeastern and central Syria (Khourbet block, Sinjar trough, Euphrates depression and Palmyran trough). Unfortunately, only limited stratigraphic and lithological data has been available due to the competitive and often secretive policies of the commercial oil companies. Undoubtedly, publication of new data will change our ideas regarding the stratigraphy, lithology and paleogeography of the Paleozoic of Syria, as well as our views on the geological development of this region at that time.

The western quarter of Palmyra (2nd Century CE). Background is a hill composed of Upper Cretaceous limestones with ruins of a Turkish fortress on the summit. From a Polish book (in Russian) on "Palmyra", published by "Arcady" Publishing House, Warsaw, 1968. Text by Cazimezh Mikhalovsky and photographs by Anjzei Dzevanovsky.

B: Triassic

The Triassic is poorly exposed in Syria. Outcrops were first discovered in the allochthonous Baer-Bassit Massif (Dubertret, 1955b).

Later, outcrops were found in the autochthonous Arabian Plate. These comparatively small but very important exposures have been identified in the Anti-Lebanon, the Palmyrides and Jebel An-Nusseiriyeh, where deposits of this age are exposed in cores of anticlinal folds and dome-like structures.

Most of the information on Triassic strata in Syria is derived from numerous bore-holes in the Mesopotamian basin, Aleppo Plateau, and the Palmyrides.

I. OUTCROPS OF TRIASSIC ROCKS

The Baer-Bassit Massif

The area of Baer-Bassit in northwestern Syria encompasses a succession of sediments, including radiolarites, red silicified mudstones and silicified pelitomorphic limestones, as well as calcarenites. It is very difficult to estimate the true thickness or normal sequence of these deposits, due to their confinement within deformed and complex folded tectonic sheets, broken by numerous faults. Within isolated outcrops the thickness of the sediments may vary from a few meters to some tens of meters.

Typical exposures of these sediments are found to the northeast of the Ras El-Bassit cape, on the Mediterranean coast (Shatsky et al., 1966). With some degree of uncertainty isolated outcrops can be combined into one composite section (from base to top):

1) Red argillites and cherts with rare, thin (10-15 cm) intercalations of silicified mudstones; the thickness is about 100 m.

2) Frequent alternation of gray and greenish-gray bedded flints and red silicified mudstones; the thickness is 12 m.

3) Red mudstones and cherts with massive texture, containing lenses of manganese ores; the thickness is 4-6 m.

4) Silicified limestones and radiolarites, red and brown on the surface, thin-bedded and fine-grained; the thickness is 5-6 m.

5) Cherts and radiolarites, red and red-brown, thin-bedded, with intercalations of gray cherts; the thickness is 30-40 m.

6) Gray, dark-gray, and yellowish platy limestones with shells of pelecypods on the surfaces: *Halobia* cf. *cassiana* Mojs. and *H. comata* Bittn. (det. L. D. Kiparisova); the thickness is 30 m.

These deposits belong to the Upper Triassic (possibly Carnian). The total thickness of the sequence is 180-195 m.

Lithologically similar deposits have been also found in other areas of the Baer-Bassit ophiolite Massif - near the villages of Aakchak, Baksa, Qasmin, etc. Layers of gray thin-bedded limestones in these sections usually contain shells of *Halobia* cf. *cassiana* Mojs., *H. comata* Bittn., and *Schafhautlia* sp. of the Late Triassic.

Dubertret (1955b) reported on assemblages of pelecypods from the area of Baer-Bassit in a sequence of alternating red cherts, radiolarites, silicified mudstones and gray pelitomorphic limestones exposed on the right bank of the Nahr El-Qandil River. He identified *Halobia paraceltica* Mojs., *H. norica* Mojs., *H. pliasa* Mojs., *Daonella imperialis* Kittl., indicating an Upper Triassic (Carnian-Norian) age.

Sedimentary rocks of the Upper Triassic, together with those of the Jurassic and Cretaceous (to Cenomanian-Turonian inclusive), constitute a part of the Tamima Nappe, which occupies a lower structural position in the Baer-Bassit ophiolite allochthon. These deep-water sediments originated at the base of the continental slope within the transition zone between the thinned continental crust and oceanic lithosphere of the southern Neo-Tethys. In the Maastrichtian, together with rocks of the upper structural complex (peridotites, gabbro, amphibolites) they were displaced towards the south and emplaced on the autochthon of the Arabian Platform (Knipper et al., 1988; Knipper and Sharaskin, 1994). Consequently, the Upper Triassic deposits of the Baer-Bassit Massif have no genetic links with the Arabian Platform and do not reflect depositional conditions on its northern margin.

Anti-Lebanon

In the southwest of the Anti-Lebanon (the Hermon Massif) Triassic rocks were discovered near the village of Rime (Mouty and Zaninetti, 1998). The section includes two members. The lower (A1) is composed of clayey marls with intercalations of sandstones, lignites, and micritic limestones with ostracods, crinoid stems, algal oncoliths, and small shells of pelecypods. The upper member (A2) is characterized by a rhythmic structure, and consists of three subunits, each starting with marls and terminated by banks of limestones. The total thickness is about 200 m. Intercalations of limestones from the lower and upper members contain benthic foraminifera belonging to the genus *Aulotortus* that allowed Mouty and Zaninetti (1998) to single out the *Aulotortus sinuosus* Zone, and to refer these deposits to the Upper Triassic (Norian and Rhaetian). The recent discovery that this unit of rocks is of Triassic age required establishment of an independent Rime Formation. Earlier it constituted the lower part of the overlying Arne Formation as defined by Dubertret (1960). Previous studies referred the lower part (A1) of the Rime Formation to the Lower Jurassic, and the upper part (A2) to the Bajocian of the Middle Jurassic (Vautrin, 1934a; Dubertret, 1966, 1975, Razvalyaev, 1966a).

Palmyrides

At Jebel Zubeide (Jebel Zbeidiye), to the northeast of Ad-Dmeir in the southwestern part of the Palmyrides, a formation of dolomites and dolomitic limestones as thick as 100 m is exposed at the base of the section. The surface of these calcareous rocks is deeply eroded. The karst hollows contain bauxites with inclusions of Fe-ores, kaolinite clays, and allite (of hydrothermal origin). The thickness is very variable, from 1 to 25 m. It has been named the Zbeidiye Formation. The latter is covered by sandstones and limestones of the Mikhk Formation, with Early Bathonian foraminifera. The origin of the bauxite-bearing unit is related to the regional regression of the Near East. This took place in the interval of the Early Jurassic-early Middle Jurassic, before the Bathonian, with possible onset of regression as early as the Late Triassic (Norian or Rhaetian). Therefore, according to their stratigraphic position, the section of dolomites and dolomitic limestones can be referred to the Triassic, although organic remains in these rocks have not been discovered (Tsekhovsky, Bogatyrev and Jabbour, 1993).

In the central part of the Palmyrides, exposures of Triassic rocks have been found in the core of the anticlinal structure of Jebel Hayyan (to the west of Palmyra). The Triassic here is represented by an intensively displaced and faulted member of gypsum with thin clay intercalations. The latter contain an assemblage of miospores characterized by taxa of the group *Circumpolles*: *Camerosporites secatus, Duplicisporites granulatus, Paracirculina scurrilis, P. tenebrosa.* Less abundant are *Patinasporites densus, Samaropollenites speciosus, Enzonalasporites* sp., *Vallasporites* sp., *Triadispora* sp. together with rare specimens of the genus *Lunatisporites* (Yaroshenko and Bach Imam, 1995). These authors found a similar miospore assemblage in the Abba-1 bore-hole in northern Syria (the Tuwal El-Aba uplift), and in the Habari-1 bore-hole in the southern periphery of the Aleppo Plateau. The assemblage permits reliable dating of the gypsiferous clays of Jebel Hayyan as Upper Triassic (possibly Carnian).

Mouty (1997a) suggested that this unit be named the Hayyan gypsum Formation.

Jebel An-Nusseiriyeh

Triassic deposits were recognized in some parts of this area: in the south (at the villages of Machta Helou, Nasirah, Jweikhate, Ash Shams and in Wadi Al-Ouyoun), in the center (the villages of Dalyeh and Ksebyeh) and in the north (the villages of Slenfeh and Joubeh). They are represented by 200 to 300 m of dolomites and dolomitic limestones with subordinate intercalations of limestones and thin interbeds of marls. The latter contain benthic foraminifera of Late Triassic (Norian-Rhetian) age: *Gsollbergella spiriloculiformis* (Oroveszne-Scheffer), *Trochammina alpina* Kristan-Tollmann, *Ophtalmidium* sp., *Planiinvoluta carinata* Leischner, and *Galeanella* aff. *laticarinata* Al Shaibani. For these deposits Mouty (1997b) introduced the name Jweikhate Formation.

II. SUBSURFACE TRIASSIC ROCKS

Triassic subsurface deposits have been penetrated by numerous bore-holes. Drilling data indicate a very wide distribution of Triassic sediments over all of Syria. Large petroleum and gas deposits are connected with the Triassic deposits of Syria. Therefore, many bore-holes were drilled aimed at penetratng Triassic sediments, especially in northeastern Syria – in the Sinjar Trough, between the Qamichliye uplift in the north and the Rawda uplift in the south (**Fig. 6.10**). Here a series of oil and gas fields extends along the border with Iraq – Qaratchok Darh, Laylak (Lelac), Souedie (Soueweidiyah), Khourbet (Khirba), Oudeh (Aoda), Bardeh, Sheikh Mansour, Al Hol, Jeribe, Tishrin (Tishreen), Ghouna, Gbeibe, Jbissa (Jebissa), Markada, Roumelan (Rumaylan), etc., and every field has a series of bore-holes. During the last few years numerous bore-holes have also been drilled in the Euphrates graben, where new oil discoveries have been made: Ratka, Al Ahmar, Al Ward, Jida (Jido), Tanek (Tanak), Omar, Tayyem (Thayyem), Al Nishan, Derro, Al Kharrata, Oufour, etc. A series of bore-holes is also located on the southeastern slope of the Aleppo Plateau – at the oil fields of Habari (Habbari), Wahab, Sfayeh (Sfaiyeh), Nobay, etc. Some bore-holes which penetrated Triassic deposits are also situated in the Palmyrides and in the north of the Aleppo Plateau. These are connected with gas occurrences (Cheriffe, Soukhne, Al Shaier, Najeeb, Bishri).

The presence of oil and gas within the Triassic deposits of Syria has aroused significant interest. A number of publications have appeared devoted to biostratigraphy, regional stratigraphy, lithology, and paleogeography of the Triassic (Khoja, 1972; Bach Imam, Khoja and Sigal, 1978, 1980; Bach Imam and Sigal, 1985; Husri and Austin, 1985; Yaroshenko, Bach Imam and Nassri, 1982; Yaroshenko and Bach Imam, 1989, 1995; Bebeshev, Jalilov, Portnyagina, Iodin, Mualla, Zaza and Yussef, 1988; Bebeshev, Jalilov, Iodin, Suriyya, Rustom and Shaban, 1989; Bebeshev, 1995).

The data of these authors contribute significantly to the investigation of the Triassic sequences. Nevertheless, published materials are only a small part of the abundant information which could be derived from the numerous bore-holes in various regions of Syria which were drilled into the Triassic. The policy and interests of the petroleum companies of the different countries carrying out research and exploitation of oil and gas occurrences, is evidently to limit disclosure of this information. For this reason, a number of problems related to the stratigraphic subdivision and correlation of the Triassic deposits unfortunately remain unsolved.

The Permian/Triassic Boundary

The position of the Paleozoic-Mesozoic, or Permian/Triassic (P/T) boundary is one of these unsolved problems. Three locations have been suggested for the placement of the P/T boundary, corresponding to:

1) The base of the Amanus Formation;
2) The base of the Amanus Shale Formation;
3) The base of the Kurachine (Kurra Chine) Dolomite Formation.

The first suggestion was presented in 1988 by some geologists of the USSR (Bebeshev, Jalilov, Portnyagina, and Yudin) and Syria (Mualla, Zaza, and Youssef). It is based on analysis of the lithological composition of rocks and bore-hole logs, with biostratigraphic data considered to be of less significance. These authors placed the Dolaa Formation at the top of the Upper Permian. The Dolaa Formation is composed of organic and organo-detritic limestones. In the bore-holes of El-Bardeh and El-Bouab, these limestones are indeed Late Permian in age (Ponikarov et al.,

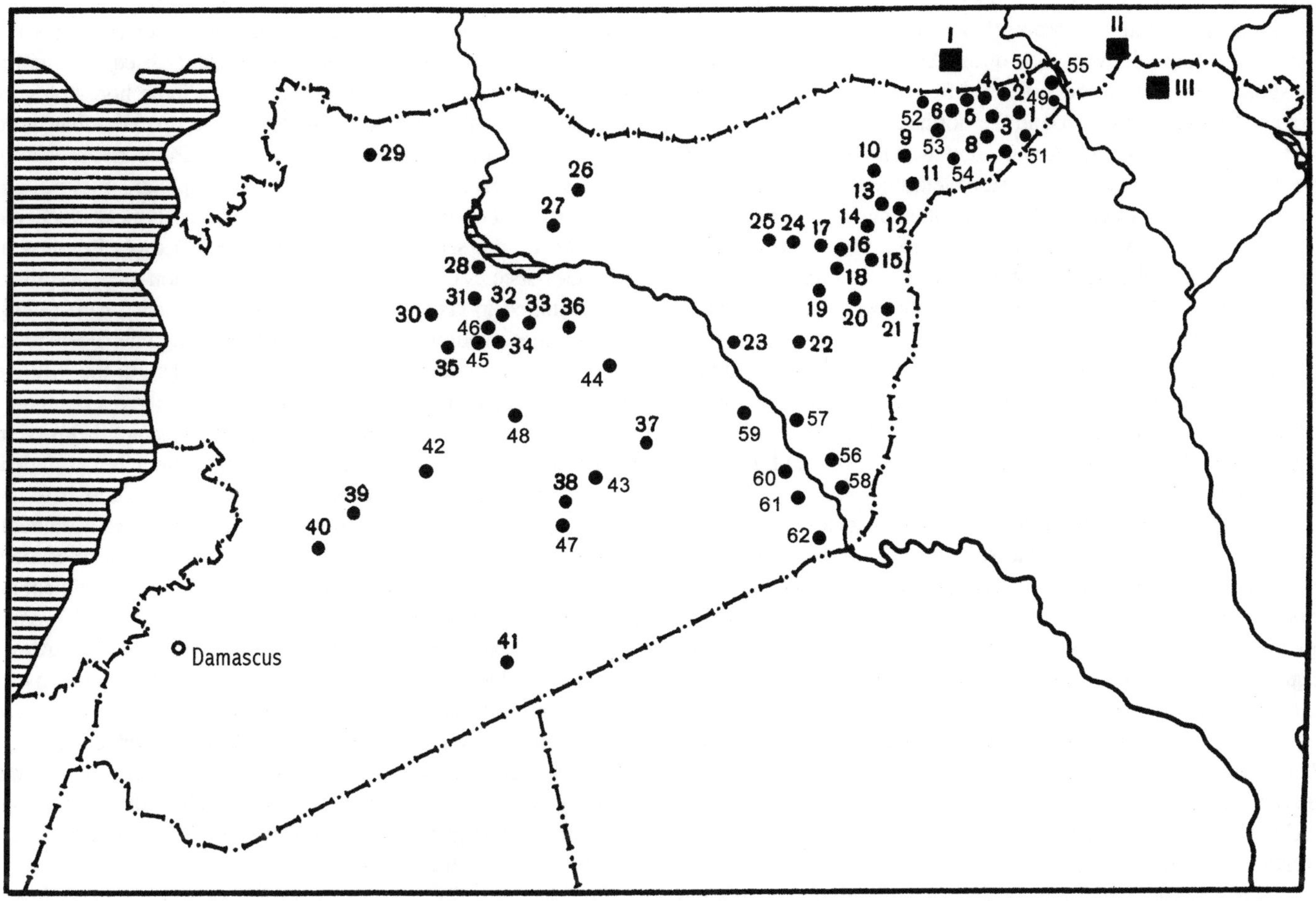

Fig. 6.10. Locations of bore-holes which penetrated Triassic deposits on the territory of Syria and distribution of outcrops on the adjacent areas of Turkey and Iraq.
Bore-holes: 1 – Roumelan-6; 2 - Alayan-101; 3 – Khadadi-1; 4 – Badran-1; 5 – Oudeh-107; 6 – Shamo-1; 7 – Naour-1; 8 – Khourber-3; 9 – Badry-101; 10 – Sheih Mansour-1; 11 – Nora-1; 12 – Al Hol-1; 13 – Abu-Khaidar-1; 14 – Tishrin-5; 15 – Gbeibe-101; 16 – Jbissa-1; 17 – Salhieh-1; 18 – Rim-1; 19 – Maaza-1; 20 – Markada-101; Bakhri-1; 22 – Tell Zrab-1; 23 – Derro-2; 24 – Syrom-1; 25 – Syrom-4; 26 – Raqqa-1; 27 – Kenaan-1; 28 – Meskene-1; 29 – Aleppo-1; 30 – Azar-1; 31 – Jib Ganem-1; 32 – Sfayeh-1; 33 – Ammala-1; 34 – Ziddan-1; 35 – Habari-1; 36 – Ressafe-1; 37 – Didi-1; 38 – Soukhne-1; 39 – Cheriffe-1; 40 – Qaruateine-1; 41 – Tanf-1; 42 – Al Shaier-1; 43 – Najeeb-1; 44 – Bishri-1; 45 – Wahab-1; 46 – Nobay-1; 47 – Doubayyat-1; 48 – Dolaa-1; 49 – Qaratchok Darh-1; 50 – Laylak-1; 51 – Souedia-1; 52 – El-Bardeh-1; 53 – El-Bouab-1; 54 – Ghouna-1; 55 – Hamzah-1; 56 – Tanak; 57 – Omar; 58 – Jido; 59 – Tayyem; 60 – Al Ward N; 61 – Al Ahmar; 62 – Ratka. Exposures: I – Khazro and II – Kharbol (Turkey), III – Northern Iraq.

1969). In the Khadadi-1 bore-hole, they can also be referred to the Late Permian because they contain an impoverished assemblage of Permian brachiopods (Bebeshev et al., 1988). However, in the Dolaa and Doubayyat bore-holes, the limestones contain Middle and Upper Carboniferous brachiopods and fusulinids (Ponikarov et al, 1969), and in the Markada-101 bore-hole, they yield very informative assemblages of Lower Carboniferous conodonts (Husri and Austin, 1985). Thus, the Dolaa Formation is an artificial complex of rocks of various ages (from Lower Carboniferous to Upper Permian). This lithostrati-graphic unit cannot be used for deciphering stratigraphical problems. Bebeshev et al. (1988) assigned limestones penetrated by the Raqqa-1, Azar-1, Sfayeh-1, Qaryatein-1, Tell Zrab-1, and Soukhne-1 bore-holes to the Dolaa Formation, but gave no paleontological data dealing with the age of these rock. The Tell Zrab-1 bore-hole is located not far from the Markada-101 bore-hole, and the Soukhne-1 site is situated near the Doubayyat bore-hole. Therefore, one can surmise that these bore-holes penetrated different Carboniferous units. It is also necessary to emphasize that the Upper Permian limestones in northeastern Syria do not form a single unit, but constitute separate members and layers within the sandy-clayey Amanus Formation.

Bebeshev et al. (1988) assigned the Amanus and Amanus Shale Formations to the Lower Triassic. The Triassic age of the former, according to these authors, is based on the transgressive character of the sediments of this Formation, which overlies the eroded surface of the Dolaa Limestone Formation, as well as on correlation with similar rocks in Turkey and Iraq. In the preceding subchapter (Paleozoic) we have already mentioned that the red-colored, sandy clay of the Amanus Formation indeed transgressively overlies deposits of different ages, but that it is of proven Permian age because in the northeast of Syria, it includes intercalations with marine Permian faunas. It is also possible that in the Palmyrides, this formation incorporates continental sediments of the Upper Carboniferous. Therefore, the opinion of Bebeshev et al. (1988) that the Amanus Formation is of Lower Triassic age is not supported by the geological data. Also it is unreasonable that the boundary of the Amanus Formation with 'the underlying Dolaa Formation' represents the P/T boundary.

In accordance with the stratigraphic scheme of the Syrian Petroleum Company, the Amanus Formation was elevated to the status of the Amanus Group, and subdivided into three independent formations, the Heil, Amanus Sand, and Amanus Shale Formations.

The stratotype of the Heil Formation is located in a bore-hole in the Heil oilfield in the southern Palmyrides. Here this formation is represented by calcareous mudstones and siltstones with subordinate interbeds of quartz sandstones, dolomites, and limestones, up to 500-600 m in thickness. These sediments were deposited under alluvial and lagoonal environments and assigned to the Lower Permian.

The Amanus Sand Formation has its stratotype in the eastern Palmyrides in the Ressafe-1 bore-hole, where it is composed of alternating fine- and coarse-grained, poorly sorted sandstones and shales of continental origin, with a thickness of 125-165 m. Its age is estimated as Upper Permian.

The Triassic begins with the Amanus Shale Formation (**Fig. 6.11**), which is found in the central areas of Syria (the Palmyrides and the southern slope of the Aleppo uplift: Dolaa-1, Azar-1, Habari-1, Qaryatein-1, etc.). It consists predominantly of mudstones and calcareous mudstones with subordinate interbeds of sandstones, siltstones, and limestones, with thickness varying from 150 to 380 m. In sediments of this formation, coalified floral remnants, foraminifera, and ostracods were found, indicating a Lower Triassic (Scythian) age. In the oil-bearing areas of northeastern Syria, this formation has not been encountered. Here, the Kurachine Dolomite (Middle Triassic) unconformably overlies Paleozoic deposits. Thus the entire Lower Triassic section appears to be absent. Unfortunately, no reliable paleontological data has been released by the Syrian Petroleum Company to confirm the Lower Triassic age of the Amanus Shale Formation, or the character of its contact with the underlying sediments of the Amanus (s.l.) Formation (Permian). For the time being we have to accept that the bottom of the Amanus Shale Formation represents the Paleozoic/Mesozoic boundary, but this supposition needs to be substantiated paleontologically.

Bebeshev et al. (1988) suggested that the Amanus Shale Formation is also to be found in northern and northeastern Syria, being gradually transitional with the underlying and overlying sediments. It was assigned to the upper part of the Lower Triassic. In the Oudeh-107, Oudeh-116, and Roumeylan-6 bore-holes, assemblages of Lower Triassic spores and pollen were found (det. L. A. Portnyagina). Paleozoic miospores which accompanied these taxa were considered to be redeposited. Subsequent studies of spore and pollen assemblages from these three bore-holes by palynologists of the former USSR and Syria revealed that these conclusions were mistaken (Yaroshenko and Bach Imam, 1994). Diagnostic forms of the Early Triassic are completely missing, and a very rich association of miospores containing typical Paleozoic taxa allowed the sediments to be considered Lower Permian in age. Index species of miospores in this publication were described and illustrated in six plates. Yaroshenko and Bach Imam (1994) consider these sediments to be part of the Amanus Formation, which was unconformably truncated by rocks of the Middle Triassic Kurachine Dolomite Formation.

Nevertheless, the presence of Lower Triassic deposits in the central and northern regions of Syria cannot be excluded. Thus, in the Wahab-1 bore-hole, in the southeastern corner of the Aleppo uplift, sediments of the Moulussa (Mulussa) Formation are characterized by the presence of an assemblage of miospores *Densoisporites neiburgii–Lunatisporites pellucidus,* indicating Upper Scythian (Lower Triassic)–Lower Anisian (Middle Triassic) ages (Yaroshenko and Bach Imam, 1989, 1995). In the Abba-1 bore-hole, located in the Tuwal-Aba uplift, the basal layers of this formation contains the pelecypod *Pseudomonotis clarai* Emmr. indicating an Early Triassic age. Middle Triassic *Daonella moussoni* Mer. and *Daonella indica* Bittn. occur upwards in the section (Ponikarov et al., 1969). However, these data have been taken from an unpublished report on the Abba-1 bore-hole and must be treated with caution.

As far as the third point of view is concerned, we now know with confidence that the Kurachine Dolomite Formation has a Middle Triassic age and that the lower boundary of this formation cannot be treated as the Paleozoic/Mesozoic boundary.

Data presented below indicate that the information about the transitional layers between the Permian and Triassic in Syria is not very reliable. Carboniferous and Middle Triassic deposits, well dated by different groups of flora and fauna, are separated in the north of Syria by a thick unit of sandy-clayey sediments with subordinate limestone intercalations. This unit corresponds to an extended interval of geological time extending over 43 m.y of the Permian and 6 Ma of the Early Triassic. As we know, this ~50 Ma interval under consideration was characterized by various large-scale geological and climatic transformations of regional and subglobal scale, as well as by the evolutionary transformation of the organic world. The very poor paleontological characteristics of the Amanus (s.l.) Formation and the Amanus Shale Formation, in combination with the paucity of cores from the bore-holes (as a rule, only every 200-300 m), do not permit detailed stratigraphic subdivision and tracing of the sequence of depositional or biotic events of this time.

From the above we can conclude that Lower Triassic deposits are present (the Amanus Shale Formation and its analogues) within the tectonic depressions of central and northern Syria. However their relationship with the Permian sediments is poorly known.

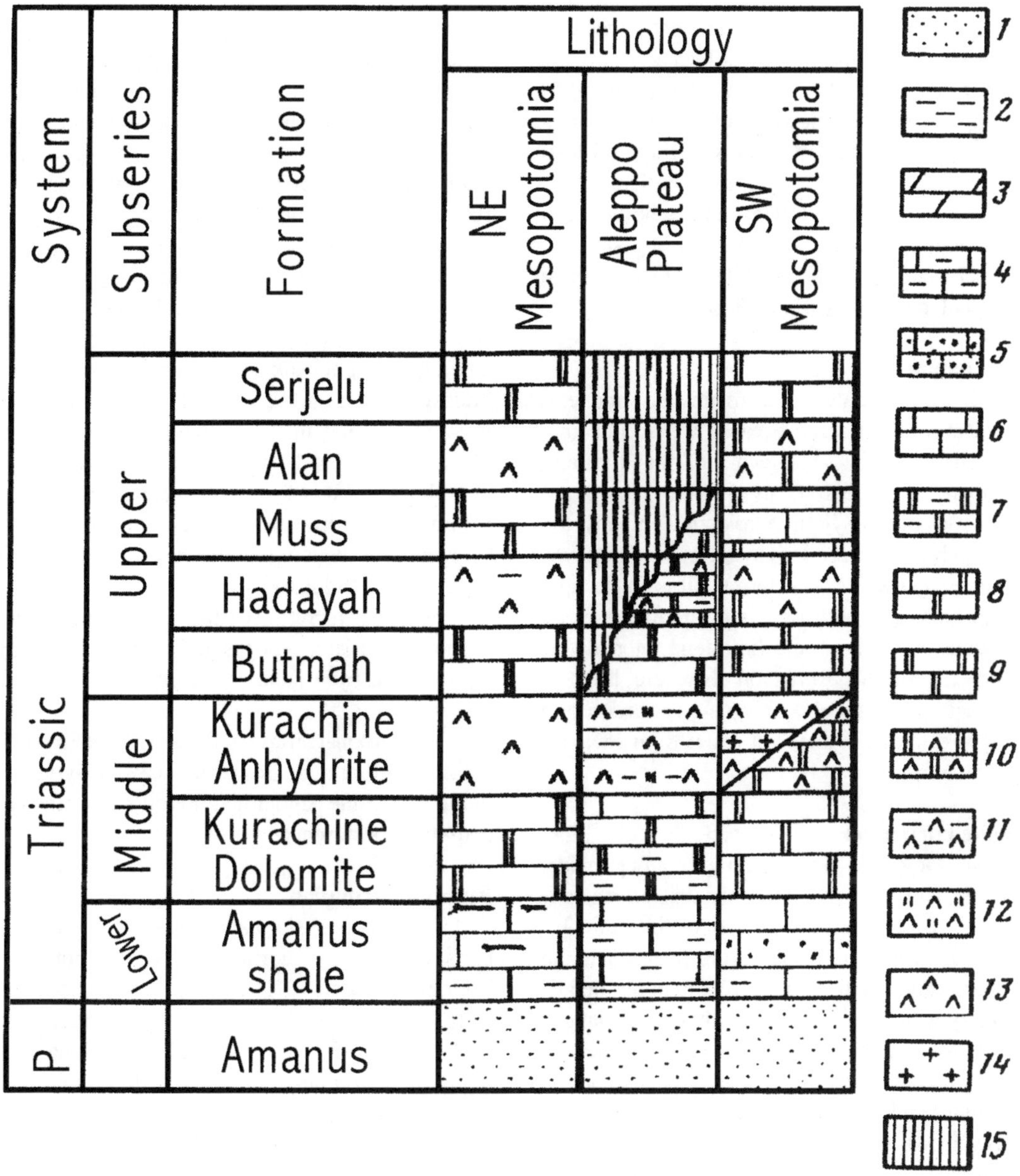

Fig. 6.11. Schematic subdivision of the Triassic deposits of Syria (according to Bebeshev et al., 1988, with modification).
Legend: 1 – sandstone; 2 – mudstone; 3 – marl; 4 – clayey limestone; 5 – sandy limestone; 6 – limestone; 7 – clayey dolomite; 8 – calcareous dolomite; 9 – dolomite; 10 – anhydritic dolomite; 11 – clayey anhydrite; 12 – dolomitic anhydrite; 13 – anhydrite; 14 – halite; 15 – absence of Triassic deposits (non-deposition or subsequent erosion).

Over most of Syria, the Middle Triassic Kurachine Dolomite Formation unconformably overlies rocks of various ages, and the Paleozoic-Triassic boundary coincides with this pronounced unconformity.

Lithostratigraphic Subdivisions of the Triassic of Syria

The lithostratigraphic scale for the Triassic used by the Syrian Petroleum Company includes eight formations, the Amanus Shale, Kurachine (Kurra Chine) Dolomite, Kurachine (Kurra Chine) Anhydrite, Butmah, Hadayah (Adaiyah), Muss (Mus), Alan, and Serjelu (Sargelu) Formations. The first was introduced in the stratigraphic nomenclature by geologists of this company. The Kurachine (s.l.) Formation was suggested by Wetzel (in Dunnington et al., 1959) in the area of Amadia to the east of Karatchok Darh. Later on it was subdivided into two lithostratigraphic units, the

Kurachine Dolomite and Kurachine Anhydrite. The Butmah, Hadayah, Muss, and Alan Formations were introduced by Dunnington et al. (1959). The type locality is located in the area of Mosul in northern Iraq. Wetzel (in Dunnington et al., 1959) established the Serjelu Formation, whose stratotype region is in the vicinity of Suleimaniya, about 100 km to the southeast of Mosul. Brief descriptions of these formations were given by Dunnington et al. (1959), Bach Imam and Sigal (1985), and Bebeshev et al. (1988, 1989). Their general features are schematically shown in Fig. 6.11. Regional peculiarities of these Middle and Upper Triassic formations are represented in **Figs. 6.12, 6.13,** and **6.14**).

Amanus Shale Formation

The Amanus Shale Formation, as mentioned above, is conditionally placed into the Triassic, because we do not know its paleontological characteristics. Sediments of this formation are definitely only found in the central part of Syria. According to the stratigraphic schemes of the Syrian Petroleum Company, they are absent in nearly all bore-hole sections located to the northwest of the Al-Furat (Euphrates) River. The formation is made up of alternating gray sandstones and mudstones. Upward in the section, the mudstone component increases and limestone intercalations appear. At the very top of this formation there are layers of dolomitized limestones. It is difficult to estimate its thickness on account of the scarcity of cores; evidently it varies from some tens to some hundreds of meters.

Kurachine Dolomite Formation

The Kurachine Dolomite Formation is more widely distributed in Syria than the overlying Triassic formations, indicating the strongly transgressive character of this unit. It is composed of dolomites of various types – from fine-grained and pelitomorphic to coarse-grained; from massively to thinly-bedded; from calcareous to clayey or gypsiferous. Usually this formation is characterized by two-fold structure: in bore-holes of northeastern Syria (Mesopotamia) the lower unit is represented by horizontally-bedded clayey dolomites and pure gray dolomites; in its upper half anhydritic dolomites appear (Oudeh-103, Oudeh-107, Roumelan-6, Jbissa-20, Tishrin-5, Badran-1, Khourbet-3, Gbeibe-101, Salhieh-1, Markada-101, Tell Zrab-1, Derro-2, Raqqa-1, etc.). The thickness ranges from 200 to 350 m.

To the west of the Al-Furat River, the formation preserves its main lithological properties, although partial changes in facies take place. In the southeast of the Aleppo Plateau (bore-holes Azar-1, Sfayeh-1, Ammala-1, Habari-1, Ressafe-1) the formation consists of alternations of dolomites and calcareous dolomites, replaced upsection by a member of pure dolomite. In the Palmyrides (Soukhne-1 and Qaryatein-1 bore-holes) the clayey dolomites predominate in the lower part of this formation, replaced by pure dolomites upsection. The thickness ranges from 100 to 550 m.

Kurachine Anhydrite Formation

The Kurachine Anhydrite Formation occupies practically the same area as the underlying Kurachine Dolomite Formation. The contact between these two formations is transitional. The Kurachine Anhydrite Formation is very monotonous and is composed of various anhydrites – gray and dark-gray, dense and massive, with thin intercalations of dolomitic mudstones, and anhydritic dolomites. Such a lithologic composition of the formation is typical for the sections in northeastern Syria (bore-holes Roumelan-6, Badran-1, Oudeh-107, Naour-1, Khourbet-3, Tishrin-5, Gbeibe-101, Jbissa-208, Salhieh-1, Markada-101, Bakhri-1, Tell Zrab-1, Oudeh-116, etc.). The thickness ranges widely, from 50 to 250 m.

The formation is composed of clayey and dolomitic anhydrites, and was sampled on the right bank of the Khabour River (Syrom-4 bore-hole), to the south of the Tuwal El-Aba uplift (Raqqa-1 bore-hole), in the right bank of the Al-Furat River near the artificial lake (Meskene-1 bore-hole), in the north of the Aleppo Plateau (Aleppo-1 bore-hole) and in the southwestern slope of this uplift (Djib-Ganem-1, Ziddan-1, Ammala-1 bore-holes). In the Palmyrides (Soukhne-1, Cheriffe-1, Qaryatein-1 bore-holes) interbeds of dolomitic anhydrites and clayey anhydrites are observed within the anhydrite section, the thickness increases to 500 m, and halite layers appear (Soukhne-1 bore-hole).

In southern Syria (northeastern slope of the Rutbah uplift), evaporitic (sulfate) deposits of the Kurachine Anhydrite Formation are replaced by a facies of terrigenous sediments of continental origin.

The carbonate rocks of the Kurachine Dolomite Formation are the main oil- and gas-reservoir rocks of Syria, and the anhydrites of the Kurachine Anhydrite Formation serve as the cap rock for the petroleum occurrences.

Butmah Formation

The Butmah Formation has been identified in almost all parts of Syria. In northeastern Syria (bore-holes Roumelan-6, Badran-1, Oudeh-107, Naour-1, Khourbet-3, Tishrin-5, Gbeibe-101, Jbissa-208, Salhieh-1, Rim-1, Markada-101, Bakhri-1, Tell Zrab-1, Derro-2, etc.) rocks of this formation are represented by gray horizontally- or wavy-bedded dolomites, usually more clayey in the lower part of the formation. Sometimes this unit starts with a member of dolomitic mudstones, and in its upper part thin intercalations of anhydrites are common. In the southeastern part of the Aleppo uplift (bore-holes Ziddan-1, Ammala-1, Sfayeh-1) the Butmah Formation is characterized by alternating mudstones and clayey dolomites. In the eastern part of the Palmyrides (Soukhne-1 bore-hole) intercalations of anhydrites appear among the dolomites (up to 30% of the total thickness) as well as fine terrigenous (clayey and silty) material. In the more eastern region (bore-hole Didy-1) the ratio of sulfatic rocks reaches 50%. In the western margin of the Palmyrides (bore-holes Cheriffe-1, Qaryatein-1) dolomites and clayey dolomites predominate. On the Rutbah uplift, near the border with Iraq, a thin member of continental sandstones and mudstones about 20 m in thickness apparently corresponds to the Butmah Formation. The thickness of this formation is comparatively minor, ranging from some tens to 100 m (bore-holes Alayan-101, Shamo-1, Raqqa-1). The maximal thickness was in the Cheriffe-1 bore-hole, reaching almost 200 m.

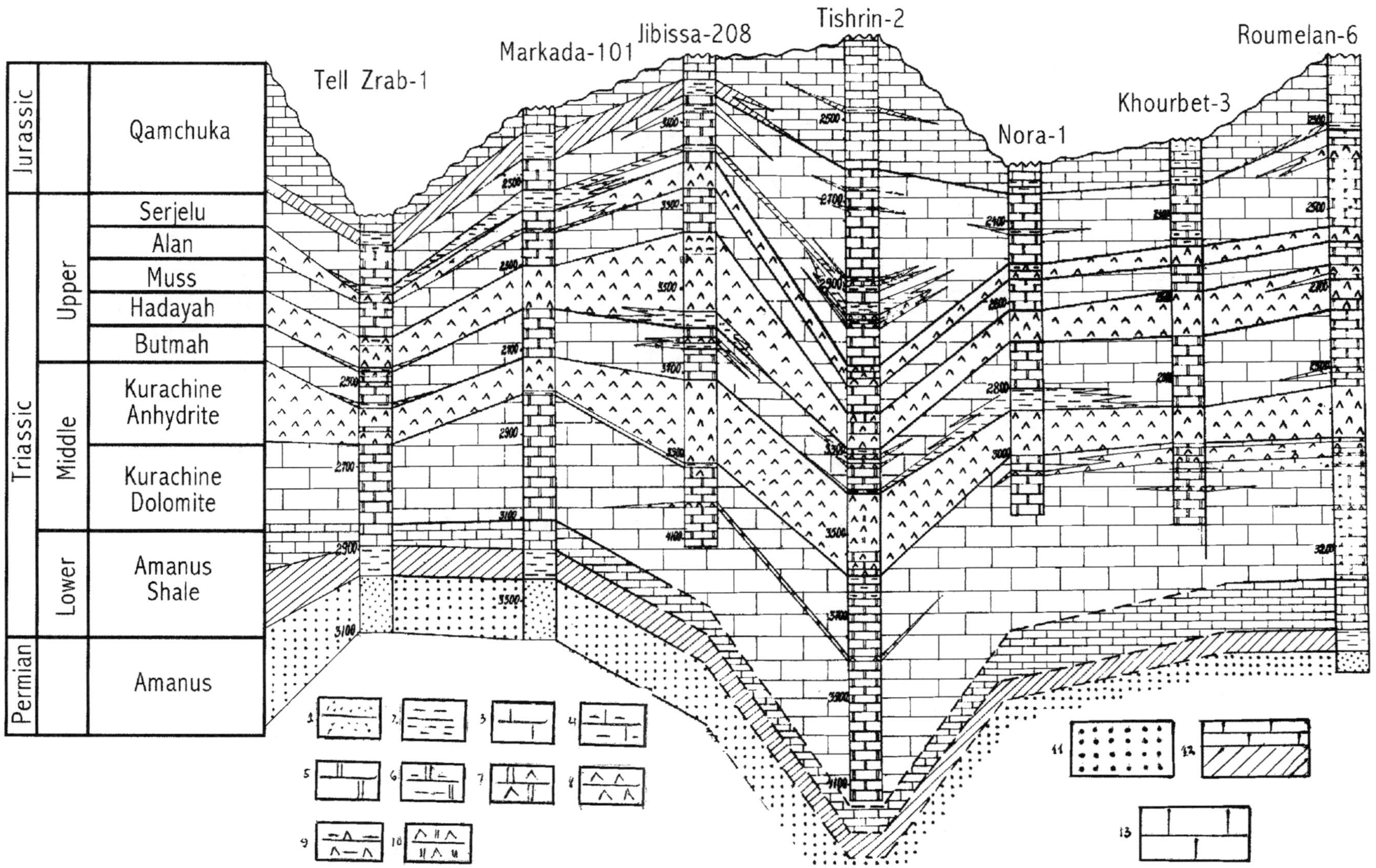

Fig. 6.12. Scheme of subdivision and correlation of Triassic deposits of Northeastern Mesopotamia (Syria) (after Bebeshev et al., 1988, with modification). Legend. In holes: 1 – sandstone and clayey sandstone; 2 – mudstone; 3 – limestone; 4 – clayey limestone; 5 – dolomite; 6 – clayey dolomite; 7 – anhydritic dolomite; 8 – anhydrite; 9 – clayey anhydrite; 10 – dolomitic anhydrite. Suggested lithology of rocks between holes: 11 – sandy-clayey rocks of the Amanus Formation; 12 - mudstones and clayey limestones of the Amanus Shale and Qamshouka Formations; 13 – dolomites of different types of the Kurachine Dolomite, Butmah, Muss and Serjelu Formations. Position of Holes is shown on Fig. 6.10.

Hadayah Formation

The Hadayah Formation was penetrated by many bore-holes; nevertheless it is not found in all areas of Syria. For example, it is absent in the north (bore-hole Aleppo-1) and southeast (bore-holes Meskene-1, Azar-1) of the Aleppo uplift, and in some structures in Mesopotamia (bore-hole Syrom-4). Apparently, sediments of the Hadayah Formation in these regions have been removed either prior to the Jurassic or in the pre- Cretaceous erosional event. Deposits of the Hadayah Formation are gradually transitional from the underlying Butmah Formation. In general, the Hadayah Formation is represented by sulfatic rocks (anhydrites) but along the strike prominent facies changes can be observed. In the northeast of Syria (Al Hol-1, Nora-1 bore-holes) and in the east of the Palmyrides (Soukhne-1 bore-hole) pure anhydrites with thin intercalations of dolomitic mudstones are observed. In the southeastern part of the Aleppo uplift, (Sfayeh-1, Ziddan-1 bore-holes) and to the south of the Tuwal El-Aba uplift (Raqqa-1 bore-hole), this formation is made up of dolomitic anhydrites and dolomites. Lastly, some bore-holes on the slope of the Aleppo uplift and in Mesopotamia drilled a section of mudstones with thin interbeds of anhydrites and dolomites. The thickness of this formation ranges from some tens of meters to 100 m.

Carbonate rocks of the Butmah Formation are also reservoirs of oil and gas and the sulfatic rocks of the Hadayah Formation are cap-rocks for the petroleum occurrences.

Muss Formation

The Muss Formation is characterized by monotonous light-gray and gray fine-grained dolomites with subordinate interbeds of anhydritic dolomites. The Muss Formation was cut by bore-holes in many areas of Syria (Roumelan-6, Badran-1, Oudeh-107, Naour-1, Khourbet-3, Tishrin-5, Gbeibe-101, Jbissa-208, Salhieh-1, Rim-1, Markada-101, Bakhri-1, Tell Zrab-1, Raqqa-1, Ammala-1, Habari-1, Ressafe-1, Soukhne-1, Cheriffe-1, etc.). The monotonous composition of this formation may be traced along the strike as well as upsection. Due to this feature, the Muss Formation serves as a good marker for stratigraphic correlation. The thickness ranges from 40 to 100 m. In the southwest of the Palmyrides (Qaryatein-1 bore-hole) Mg-carbonates (dolomites) are replaced by calcareous dolomites and dolomitic algal limestones. On the Aleppo uplift, deposits of the Muss Formation were identified only in the Sfayeh-1 and Ammala-1 bore-holes. In other regions of this structure, the formation was not found; evidently rocks of this age were truncated by pre-Cretaceous erosion.

Alan Formation

The Alan Formation is well developed in the northeastern part of Syria (bore-holes Roumelan-6, Badran-1, Naour-1, Alayan-101, Tishrin-5, Gbeibe-101, Salhieh-1, Markada-101, Tell Zrab-1, Raqqa-1). It is represented by a thin member (40-50 m) of anhydrites. In some bore-holes, this formation has a two-fold structure. The lower member consists of pure medium- and fine-grained anhydrites, dense and massively bedded. The upper member is composed of alternating clay, anhydrite, dolomitic anhydrite and mudstone lithologies. The maximum thickness reaches 100 m. Towards anticlinal structures of the second order (paleo-uplifts) pure anhydrites are replaced by a facies of anhydritic dolomites, dolomites, and dolomitic mudstones (Oudeh-107, Shamo-1, Syrom-1, Gbeibe-101, Markada-101 bore-holes). The same facies replacement from west to east can be observed in the southern areas of the Palmyrides (bore-holes Didy-1, Ressafe-1, Soukhne-1, Marife-1). In some bore-holes, the Alan Formation was not found (Bakhri-1, Azar-1, Ammala-1, Habari-1, and Qaryatein-1).

Serjelu Formation

The Serjelu Formation is distributed only in northeastern Syria (Mesopotamia), but even here it is rather sporadically developed. This formation was only found in some bore-holes (Roumelan-6, Naour-1, Khourbet-3, Tishrin-5, Gbeibe-101, Jbissa-208, Salhieh-1, Markada-101, Tell Zrab-1). In other bore-holes it was not observed (Badran-1, Oudeh-107, Rim-1, Bakhri-1). The Serjelu Formation is characterized by a varied lithological composition of alternating dolomites, anhydrites, and dolomitic mudstones. The thickness is very variable but sometimes it reaches 600 m. In the southeast of the Aleppo Plateau, this formation is absent (Sfayeh-1, Ammala-1, Habari-1 bore-holes). It is practically unknown in the Palmyrides – here sediments of the Serjelu Formation were definitely found only in the Cheriffe bore-hole, and conditionally in the Ressafe-1 bore-hole.

Deposits of the Serjelu Formation (together with older Triassic rocks) are overlain by calcareous, terrigenous, and marine sediments of the Jurassic(?) Qamchuka (Qamchuqa) Formation or terrigenous deposits of the Lower Cretaceous Ghouna Formation with a considerable stratigraphic hiatus.

BIOSTRATIGRAPHY OF TRIASSIC DEPOSITS

The Triassic deposits of Syria are very poor in organic remains, due to peculiarities in the lithological composition (dolomites, dolomitic limestones, mudstones, and anhydrites). This makes subdivision, correlation, and age determination of these deposits very difficult. In sediments of the Butmah Formation, rare foraminifera have been encountered, and rare ostracods in those of the Butmah, Hadayah, Muss, and Alan Formations (Bach Imam et al., 1978, 1980; Bach Imam and Sigal, 1985). Sometimes pelecypods are present. Findings of conodonts are extremely rare (Husri and Austin, 1985). The main tool for stratigraphic subdivision of the Triassic deposits of Syria are spore and pollen assemblages.

As has already been pointed out, the Triassic subdivisions used in Syria were defined in Iraq. The Kurachine s.l. Formation was referred to the Late Triassic, and the Butmah, Hadayah, Muss, Alan, and Serjelu were treated as Jurassic. The formations mentioned were singled out in Syria based on lithological features and their age was determined by means of correlation with the Mesozoic deposits of Iraq (Ponikarov et al., 1969). Investigation of miospore assemblages proved that these datings were erroneous

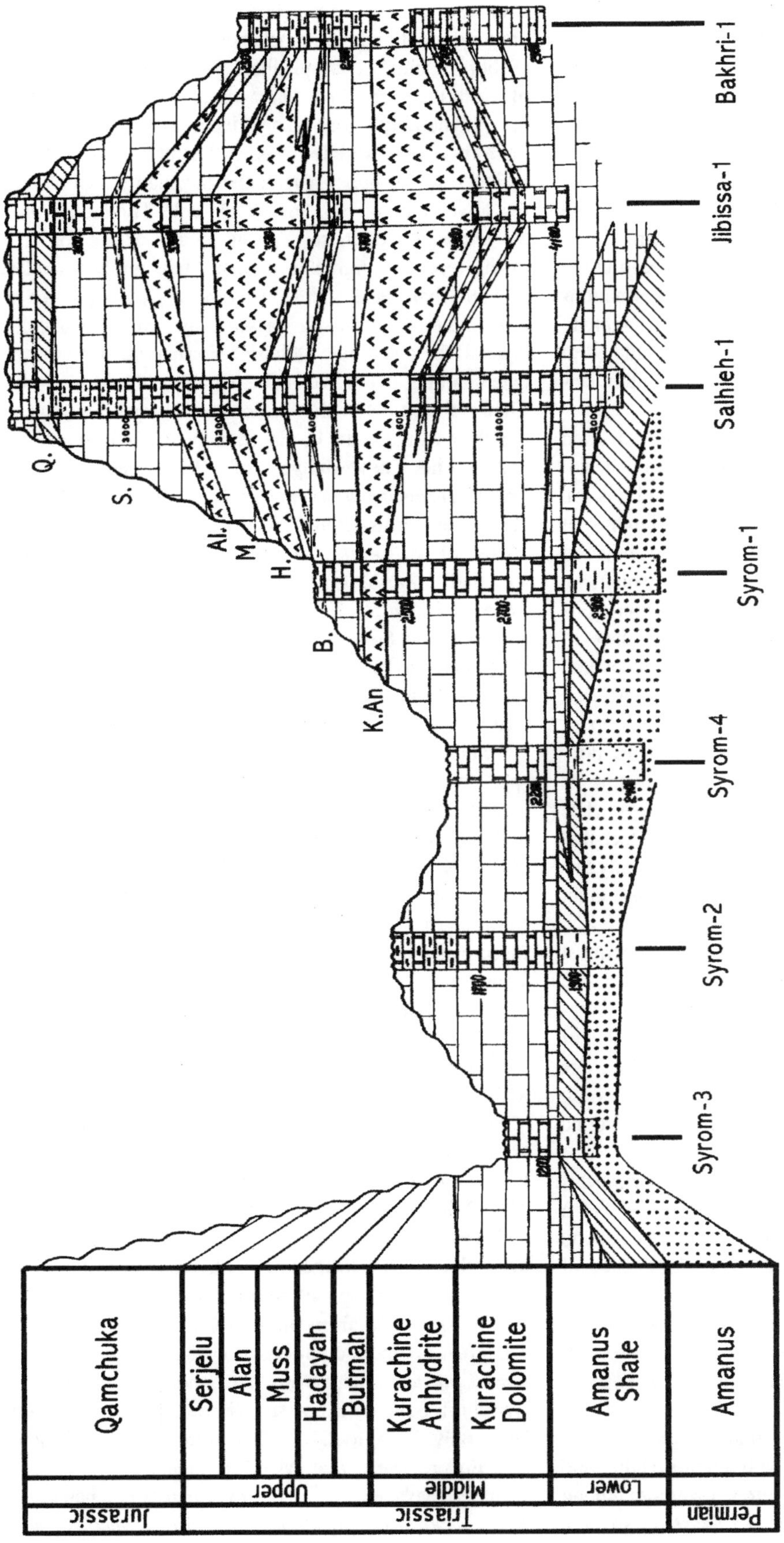

Fig. 6.13. Scheme of subdivision and correlation of Triassic deposits of southwestern Mesopotamia (Syria) (after Bebeshev et al., 1988, with modification).
Location of bore-holes is shown on Fig. 6.10;
Legend is given on Fig. 6.12.

(Khoja, 1972; Bach Imam et al., 1978, 1980; Bach Imam and Sigal, 1985). These authors completely revised the stratigraphy of the Lower Mesozoic of Syria. At the same time, the synchroneity of the Syrian and Iraqi formations with similar names has become uncertain. It is very difficult to check this correlation because very poor paleontological data is available for these formations in Iraq.

Unfortunately, differences in ages for formations in Syria and Iraq (the Zagros basin) bearing the same names continues up to the present. Thus, in the monograph of Al Sharhan and Nairn (1997, p. 695, fig. 12.32) the Kurachine s.l. Formation was placed in the Upper Triassic, the Butmah, Hadayah, Muss, and Alan Formations were assigned to the Lower Jurassic, the Serjelu Formation to the Middle Jurassic, and the Qamchuqa Formation considered to represent the greater part of the Lower Cretaceous (Hauterivian-Albian).

Investigations by palynologists of the former U.S.S.R. and Syria (Yaroshenko, Bach Imam and Nassri, 1982; Yaroshenko and Bach Imam, 1989, 1994, 1995; Yaroshenko and Nassri, 1980) were extremely important for defining the stratigraphy of the Triassic deposits of Syria. There are some difficulties inherent in palynological analysis of the Triassic sediments, which are represented by carbonate and sulfate lithologies. Palynomorphs are difficult to extract from these sediments, and their state of preservation is poor. Palynomorphs have only been recovered from thin intercalations of mudstones and clays. The evaporitic environment of sedimentation is responsible for their poor preservation and thus rare occurrence of good specimens. Poor preservation of miospores also precludes a precise diagnosis; therefore some taxa were determined only to a generic level. The level of stratigraphic precision is demonstrated by the fact that miospore assemblages may place the age of the sediments only to a stage, and sometimes even to a subseries. A succession of only eight miospore assemblages was traced in the Triassic deposits of Syria. This succession corresponds to the interval from the uppermost part of the Lower Triassic up to the top of the Upper Triassic (Rhaetian). These data are mostly derived from a series of bore-holes in the northeast of Syria. However, tracing the succession of miospores in any one bore-hole is not possible due to the rare cores and the sporadic presence of the clay intercalations. The succession of palynological assemblages and their age determinations can be outlined as follows:

1) The assemblage of *Densoisporites nejburgii–Lunatisporites pellucidus* was found in the Mulussa Formation in the 2010-2040 m interval of the Wahab-1 bore-hole, located southeast of the Aleppo Plateau. This assemblage is characterized by predominance of bisaccate pollen, *e.g.*, *Alisporites grouvogeli* Klaus, *Alisporites* sp., *Platysaccus queenslandi*, single specimens of *Vitreisporites* sp., and monosaccate pollen belonging to *Accinctisporites* sp. and *Monosaccites* sp. The presence of pollen of genus *Triadispora* is noted. Very rare specimens of *Stellapollenites thiergartii* (Maedler) Clement-Westerhof et al. are found. Typical costate forms such as *Lunatisporites pellucidus* (Goubin) Helbi and *Lunatisporites* sp. were identified. Spores are uncommon; among them *Punctatisporites* sp., *Todisporites* sp., *Cyclotriletes* sp., and *Aratrisporites*, sp. were recorded. The most important form for drawing stratigraphic conclusions is *Densoisporites nejburgii* (Schulz) Balme. Phytoplankton is present as single specimens of *Tasmanites* sp., and *Micrhystridium* sp. The composition of the palynomorph assemblage includes taxa known from both the Lower and Middle Triassic. Therefore, these deposits were assigned to the upper part of the Scythian (Lower Triassic), and the lower part of the Anisian (Middle Triassic). Unfortunately, the attribution of these deposits to any of the formations used in the recent stratigraphic nomenclature remains unclear. In the Wahab-1 bore-hole, sediments with the *Densoisporites nejburgii–Lunatisporites pellucidus* assemblage were recovered 200 m beneath the Kurachine Dolomite Formation, and were previously dated as Permo-Triassic, or belonging to the Mulussa Formation. Their position below the Kurachine Dolomite Formation suggests that they should better be placed in the Amanus Shale Formation.

2) The assemblage of *Rewanispora vermiculata–Voltziaceasporites heteromorphus* is not characterized by a great variety of forms. Pollen recovered predominately include the forms of *Alisporites* spp., *Triadispora* spp., *Angustisulcites* sp., and *Lunatisporites* spp. They are accompanied by rare *Microcachrydites* cf. *doubingeri* Klaus, *Striatobietites aytugii* Visscher, and *Stellapollenites* cf. *thiergartii*. Many specimens of the pollen species *Voltziaceaesporites heteromorphus* Klaus were found. The taxonomic variability of spores is relatively poor. A characteristic of this association is the presence of *Rewanispora vermiculata* Ant. et Taug.-Lantz, an increase in *Aratrisporites* spp., and the presence of single specimens of *Perotriletes minor* (Maedler) Ant. et Taug.-Lantz. This assemblage was encountered in the 1791-1795 m interval of the Wahab-1 bore-hole in deposits of the Kurachine Dolomite Formation (Yaroshenko and Bach Imam, 1995, p. 54, fig. 2). These authors place deposits with the *Rewanispora vermiculata–Voltziaceasporites heteromorphus* assemblage in the lower or middle part of the Anisian (Middle Triassic).

The Middle Triassic age of the Kurachine Dolomite Formation was confirmed by conodont data. In the Markada-101 bore-hole some specimens of *Neogondolella mombergensis* (Core 18, 3117-3125 m) were identified (Husri and Austin, 1985).

3) The assemblage of *Stellapollenites thiergartii–Heliosaccus dimorphus* was identified in sediments of the Kurachine (s.l.) Formation, penetrated by the Abba-1 bore-hole (in the 2119-2121; 2166-2181; 2193-2214 m intervals) on the Tuwal El-Aba uplift, and in the Afendi-1 bore-hole (2160-2166; 2178-2184; 2184-2190 m intervals) which is located further to the east in Mesopotamia.

This assemblage of miospores differs from the preceding assemblages, and from all subsequent ones, by large numbers and great taxonomic variety of costate palynospecies, including *Lunatisporites noviaulensis* var. *mollis* Scheuring, *L. acutus* (Leschik) Scheuring, *L.* sp., *Striatites* spp., etc. Some forms have only been found in this assemblage – *Dyupetalum* sp., *Heliosaccus dimorphus* Maedler; others continue upwards into younger levels, *e.g.*, *Uvaesportes gadensis* Praehauser-Enzenberg, *Keuperisporites baculatus* Schultz, *Microreticulatisporites*

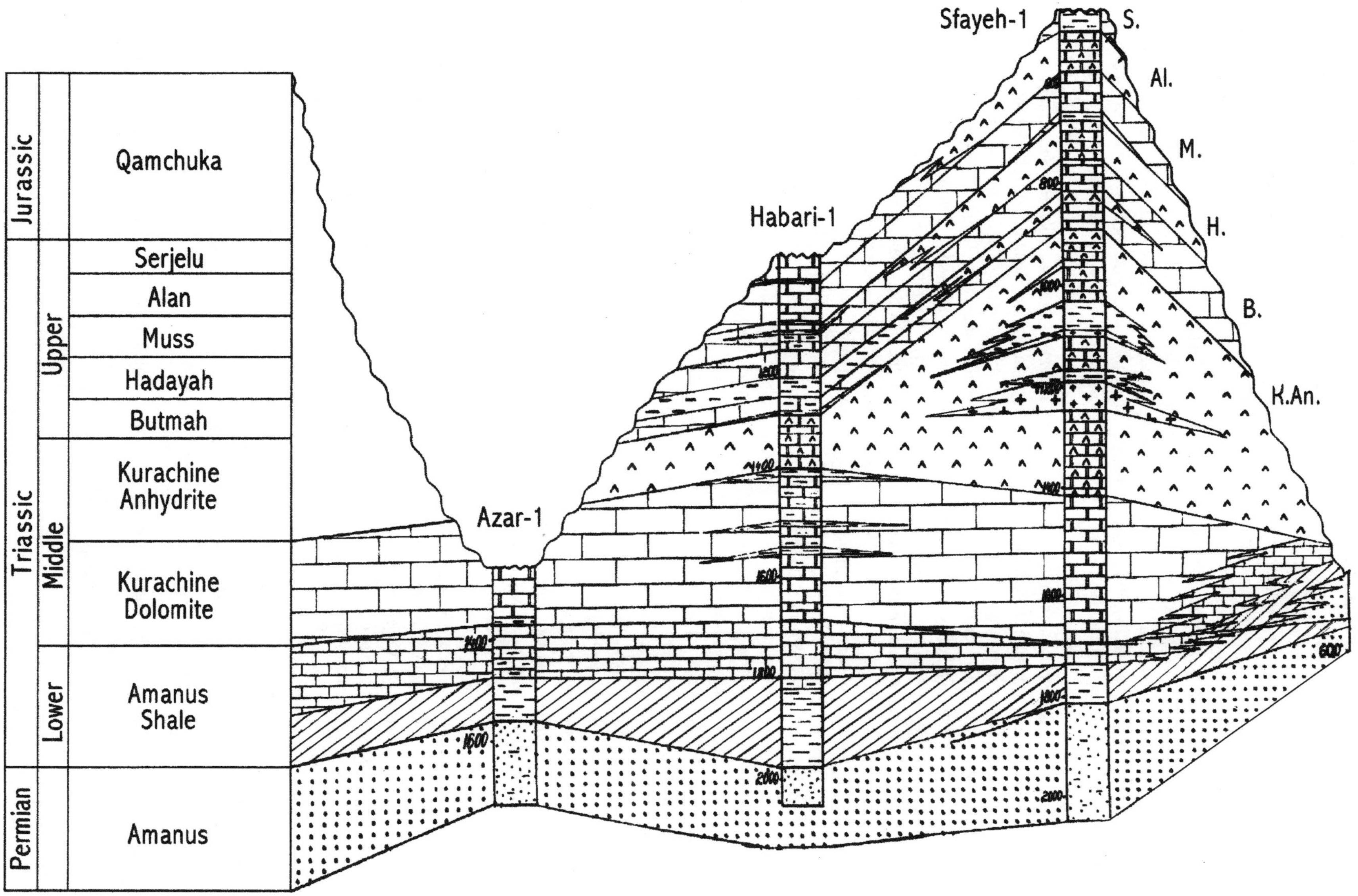

Fig. 6.14. Scheme of subdivision and correlation of Triassic deposits on the southern slope of the Aleppo Plateau (after Bebeshev et al., 1988).
Position of Holes is shown on Fig. 6.10;
Legend is given on Fig. 6.12.

galli Adloff et Doubinger, *Triplexisporites playfordii* (de Jersey et Hamilton) Foster, *Staurosaccites quadrifidus* Dolby, *Doubingerispora filamentosa* Scheuring, *Kuglerina meieri* Scheuring, *Podocarpidites keuperianus* (Maedler) Schuurman, and *Rimaesporites aquilonalis* Goubin. The index species of this assemblage, *Stellapollenites thiergartii*, is abundant. Some taxa, *e.g. Aratrisporites* sp., *Densoisporites nejburgii*, *Microcachryidites* cf. *doubingeri*, *Platysaccus queenslandi*, *Voltziaceae-sporites heteromorpha* continue upward from underlying strata.

Phytoplankton are rare in samples from the Afendi-1 bore-hole but are much more common in sediments from the Abba-1 bore-hole: *Leiosphaeridia* sp., *Micrhystridium* sp., *Cymatio-sphaera* sp.

Analysis of the stratigraphic distribution of this miospore assemblage indicates that the bulk of these forms are characteristic of the Middle Triassic of the western European stratotype regions and adjacent areas. Some taxa are most typical for the Anisian, while others predominate in the Ladinian. Therefore, sediments with the assemblage of *Stellapollenites thiergartii–Heliosaccus dimorphus* are dated to the Late Anisian-Early Ladinian interval of the Middle Triassic.

4) The assemblage of *Camerosporites secatus–Kuglerina meieri* was identified in the upper part(?) of the Kurachine (s.l.) Formation from the Abba-1 bore-hole (the 1893-1896; 1912-1914; 1914-1915 m intervals). Its taxonomic composition has some similarity with the preceding assemblage, with some forms , *e.g. Lunatisporites acutus* (Leschik) Scheuring, *L. noviaulensis mollis* Scheuring, *Triadispora* spp., *Podocarpidites keuperianus* (Maedler) Schuurman, and *Protodiploxypinus gracilis* Scheuring, continuing upwards from the underlying strata. Nevertheless, the ranges of some previously occurring taxa, *e.g. Densoisporites nejburgii, Retitriletes* sp., etc. do not continue into this assemblage.

The main distinguishing characteristic of the assemblage is the appearance of pollen of the *Circumpolles* group, represented by initial occurrence of the earliest forms, *Camerosporites secatus* (Leschik) Scheuring, *Paracirculina scurrilis* Scheuring, and *Duplicisporites granulatus* (Leschik) Scheuring.

The composition of this assemblage includes some taxa which are very important for correlation with the Ladinian of the Alpine region (Austria, Switzerland): *Uvaesporites gadensis*, *Aratrisporites* spp., *Triadispora* spp., *Lunatisporites acutus*, *Kuglerina meieri*, *Doubingerispora filamentosa*, and species previously mentioned belonging to *Lunatisporites*, *Proto-diploxypinus*, and *Triadispora*.

Phytoplankton including *Cymatiosphaera* sp., *Micrhystridium* sp, and some others, are somewhat more abundant than in the underlying palynoassemblage.

The assemblage with *Camerosporites secatus–Kuglerina meieri* is assugned to the Ladinian (Middle Triassic), and probably to its upper part.

5) The assemblage of *Patinasporites densus–Samaropollenites speciosus* was identified in the Butmah(?) Formation in the Abba-1 bore-hole (1873-1875 m interval) and in a similar stratigraphic level in the Habari-1 bore-hole (1312-1337 m interval) south of the Aleppo Plateau. There is a noticeable change of the miospore composition. The quantity of some palynospecies, *e.g. Triadispora* spp., *Lunatisporites* spp., *Kuglerina meieri, Doubingerispora filamentosa,* and *Staurosaccites quadrifidus* diminishes, and some taxa of the preceding assemblage are absent.

The most essential feature of the *Patinasporites densus–Samaropollenites speciosus* assemblage is the significant development of species of the *Circumpolles* group, the earliest species of which appeared in moderate quantities in the Ladinian assemblage (in the Abba-1 bore-hole, sediments with both these assemblages are subdivided by an 18 m interval drilled without coring). The most abundant species of the *Circumpolles* group belong to *Paracirculina scurrilis* and *Camerosporites secatus*, which are accompanied by less common *Duplicisporites granulatus* and *Paracirculina tenebrosa* Scheuring.

A very important feature of this assemblage is the appearance of *Patinasporites densus* (Leschik) Scheuring, *Samaropollenites speciosus* Goubin, *Minutosaccus crenulatus* Dolby, *Enzonalasporites* sp., *Vallasporites* sp., *Paracirculina tenebrosa,* and *Praecirculina granifer* (Leschik) Klaus. These taxa, and especially the first two, have great significance for correlation with Triassic deposits of other regions.

As mentioned above, the miospore assemblage with *Patinasporites densus–Samaropollenites speciosus* has been found in the Triassic exposures of Jebel Hayyan in the Palmyrides. This assemblage belongs to the Carnian Stage (Upper Triassic).

6) The assemblage of *Corollina meyeriana–Kyrtomisporis corrugatus* was identified in the Serjelu Formation in the Khourbet-3 (2425-2429 m interval) and Roumelan-6 (2504-2524 m interval) bore-holes, in the extreme northeast of Syria. Here the miospore assemblages change in comparison with those from the underlying sediments. Costate *Lunatisporites* spp. disappear, as well as *Douberingerispora filamentosa, Kuglerina meieri, Camerosporites secatus,* and *Duplicisporites granulatus*. The abundance of *Samaropollenites speciosus* increases, and taxa such as *Patinasporites densus*, *Enzonalasporites* sp., *Vallasporites* sp., *Paracirculina tenebrosa*, *Triadispora* spp., *Protodiploxypinus gracilis*, *Minutosaccus crenulatus* Dolby, and *Rimaesporites aquilonalis* continue upwards from the underlying deposits. There is a significant genetic link with the preceding miospore assemblage.

The main characteristic of this assemblage is the appearance of *Corollina meyeriana* Klaus, *Corollina* sp., and rare specimens of *Granuloperculatipollis* sp. and *Classopollis* sp. Consequently the earliest elements of the *Circumpollis* group are here replaced by younger representatives. Together with this change, various spores of genus *Kyrtomisporis* (*K. laevigatus* Maedler, *K. speciosus* Maedler, *K. corrugatus* Cameron) become significantly more abundant. *Elongatosaccites triassicus* Cameron is present as rare specimens.

The assemblage with *Corollina meyeriana–Kyrtomisporis corrugatus* was assigned to the lower part of the Norian (Upper Triassic).

7) The assemblage of *Corollina meyeriana–Vesicaspora schemeli* was identified in the Serjelu Formation in the Maasa-1 bore-hole (2865-2873 m interval) in the central part of

Mesopotamia. Its main feature is a great abundance of *Corollina* and *Classopollis* specimens, which may constitute more than half of the miospore assemblage, as well as the presence of numerous *Vesicaspora schemeli* Klaus in association with *Tsugaepollenites? pseudomassulae* Maedler and the dinoflagellate *Rhaetogonyaulax rhaetica* (Sarjeant) Loeblich et Loeblich.

Numerous *Kyrtomisporis corrugatus*, *K. laevigatus*, *K. speciosus*, and less common *Enzonalasporites* sp., *Patinasporites densus*, and *Elongatosaccites triassicus* continue upwards from the preceding assemblage. There is a noticeable increase in *Ovalipollis* spp., and distribution of *Rimaesporites aquilonalis* reaches maximal abundance. In some samples this taxon constitutes half of the miospore assemblage. Representatives of the genera *Deltoidospora, Dictyophyllidites, Concavisporites,* and *Todisporites* with triangular and rounded smooth spores have been recovered in greater quantities here. Taxa of the genus *Duplexisporites* are very rare.

As in the underlying assemblage, costate forms, representatives of the genus *Triadispora*, and earlier elements of the *Circumpolles* group, all become extinct.

The assemblage with *Corollina meyeriana–Vesicaspora schemeli* is placed in the Upper Norian (Upper Triassic).

8) The assemblage of *Corollina meyeriana–Converru-cosisporites cameroni* is connected with sediments of the Serjelu Formation penetrated by the Gbeibe-101 (3810-4050 m interval) and Jeribe-101 (4780-4790 m interval) bore-holes, which are located in northeastern Syria near the boundary with Iraq. Here pollen of *Corollina* dominate and *Classopollis* species are very common. Among spores, the most characteristic are triangular smooth taxa of the genera *Dictyophyllidites* and *Concavisporites*, as well slightly ornamented *Converru-cosisporites cameroni* (de Jersey) Playford et Dettman. Other taxa represented are *Inaperturopollenites* sp., *Ovalipollis* sp., *Vitreisporites* sp., *Cycadopites* sp., *Kyrtomisporis speciosus*, *Trachysporites asper* Nilsson, and *Perinopollenites* cf. *elatoides* Couper. Many taxa typical of the preceding assemblage disappear, and bisaccate pollen are completely absent.

The taxonomic composition of this assemblage is impoverished. It is characterized by taxa known from both Upper Triassic and Lower Jurassic sediments, but guide species for both the Rhaetian and Liassic are absent. Therefore, the age of the *Corollina meyeriana–Converrucosisporites cameroni* assemblage is considered transitional between the Rhaetian (Terminal Triassic)-Liassic (Lower Jurassic) (Yaroshenko and Bach Imam, 1995). Taking into consideration the geological events at the Triassic-Jurassic boundary typical for the entire Near East – a sharp change in sediment character, a hiatus and unconformity, it is most probable that the upper layers of the Serjelu Formation with this assemblage of miospores should be assigned to the Rhaetian (Terminal Triassic).

It is necessary to emphasize that the publications of Yaroshenko and Bach Imam (1989, 1995) include paleontological plates illustrating guide taxa of palynomorphs that reliably substantiate their stratigraphic conclusions.

Nevertheless, there are still many uncertainties regarding the presence of Lower Triassic deposits, their lithologic features and paleontological characteristics. However, the Middle and Upper Triassic stratigraphy is more definitive, as palynological studies proved that the Middle and Upper Triassic section is quite complete. The Kurachine Dolomite and Kurachine Anhydrite Formations belong to the Middle Triassic (Anisian and Ladinian) while the Butmah, Hadayah, Muss, Alan and Serjelu Formations have Late Triassic (Carnian, Norian, Rhaetian) ages.

Unfortunately, the scope of the paleontological investigations is clearly insufficient for complete characterization of the Triassic of Syria. First of all, lithological and stratigraphical studies to some extent were carried out in isolation from one another. Figs. 6.12, 6.13 and 6.14 show the schematic correlation of the Middle and Upper Triassic, based on numerous bore-holes in various areas of Syria. Possible facies changes of Triassic sediments may be traced in the northeast of Syria (Fig. 6.12), in the basin of the Nahr El-Khabour River (Fig. 6.13), and in the southeastern slope of the Aleppo Plateau (Fig. 6.14) (Bebeshev et al., 1988).

Nevertheless, the correlation of sections is based only on lithological data (cores and cuttings) and log interpretation. These correlations were not corroborated by paleontological data, and therefore the figures present only a generalized picture of the lateral variations in the Triassic stratigraphy of this region. Because of rare coring, and the possibility of facies changes in the sediments along the strike, mistakes in determination of formational names cannot be excluded. Thus in the Kurachine (s.l.) Formation, in the Afendi-1 bore-hole (2200-2240 m interval) the assemblage of miospores with the *Rewanispora vermiculata-Voltziaceaesporites heteromorphus* assemblage (Anisian) was determined, while in the Tishrin-2 bore-hole the assemblage of the Norian was found, and in Gbeibe-101 bore-hole (interval 4715-4980 m) an association of the Upper Triassic (Rhaetian)-Lower Jurassic was identified. Yaroshenko and Bach Imam (1982, p. 76) state that "it is possible that the same named formation may be of different ages in various areas and bore-holes, and synchronous deposits may belong to different formations".

LITHOLOGY OF TRIASSIC DEPOSITS

The Triassic of Syria is characterized by three types of rocks: terrigenous (sandstone, siltstone, and shale), carbonate (limestone and dolomite), and evaporite (anhydrites) (Bebeshev et al., 1989). Terrigenous rocks encompass the Amanus Shale Formation, conditionally assigned to the Lower Triassic in the central areas of Syria. Carbonate and sulfate rocks strongly predominate in the Middle and Upper Triassic, with clays present as rare thin interbeds.

Terrigenous rocks

Sandstones are subordinately developed in the Amanus Shale Formation. There, gray, quartz sandstone beds are monomictic, poorly sorted and rounded, fine- to medium-grained, with calcareous or clayey-calcareous cement, not exceeding a few cm in thickness.

Clays and mudstones are most characteristic for the Amanus Shale Formation, where they appear as subunits a few tens of meters in thickness. Dark gray mudstones with ferruginous stains contain sandy and calcareous admixtures.

The carbonate and sulfatic rocks are found in the Kurachine Dolomite, Kurachine Anhydrite, Butmah, Hadaya, Muss, Alan and Serjelu Formations. They may contain black dolomitic mudstones as thin interbeds a few cm thick. Calcareous content of these rocks reaches 40%. Clay minerals are represented by hydromica and kaolinite.

In the lower parts of the Butmah, Hadaya, and Serjelu Formations clays form thicker beds, associated with grayish, platy thin-bedded dolomitic mudstones or marlstones.

Clay minerals consist of kaolinite, hydromica and mixed-layered minerals of hydromica-smectite composition. In the lower part of the Amanus Shale Formation, hydromica and mixed layered minerals of hydromica-smectite composition predominate. In the upper part of this formation hydromica dominates. Clays in the Middle and Upper Triassic consist of hydromica and kaolinite.

Carbonate rocks

Among the carbonate rocks, precipitated chemogenic, organogenic, and clayey varieties can be distinguished.

Chemogenic limestones, according to Bebeshev et al. (1989) were formed by direct precipitation of $CaCO_3$ from the water mass, and include pelitomorphic, clotted fabrics, and micro-, middle-, and coarse-grained varieties.

Pelitomorphic limestones consist of minute crystals of calcite of isometric form, and some larger crystals or groups of crystals, connected with the process of post-sedimentary recrystallization. These rocks are paragenetically combined with dolomites and sulfates, emphasizing their formation by chemical precipitation. Pelitomorphic limestones are found throughout the Triassic, beginning from the Amanus Shale Formation.

Micro-clotted limestones consist of small clots of rounded form of pelitomorphic limestone cemented by fine-grained calcite. Evidently, the origin of these rocks is also associated with chemical precipitation of calcite with subsequent appearance of clots as a result of coagulation.

Fine-grained limestones are composed of microcrystals of calcite of isometric form with an insignificant admixture of quartz and clay. Very often they are dolomitized. Such rocks are typical for the upper part of the Amanus Shale Formation, although they are present as thin intercalations in the Middle and Upper Triassic.

Medium-grained limestones are composed of calcite crystals of isometric or irregular form with an admixture of rare grains of quartz and glauconite, and with thin (2-3 mm) lenses of clay. These limestones are also very often dolomitized, and are found throughout the entire Triassic section, sometimes forming thick units.

Organogenic limestones are represented by algal stromatolite limestones, as a result of cyanophycean algal deposition. Wavy bedding reflected in alternating light and darker layers is a characteristic of these limestones, which are usually dolomitized. Stromatolite limestones usually underlie evaporitic members of the Middle and Upper Triassic.

Clayey limestones with pelitomorphic structure contain 5-30% of clay matrix. They form thin intercalations (0.5-2 cm) within the Middle and Upper Triassic sequence.

Dolomites predominate in the Middle and Upper Triassic. According to structural and textural features, limestones and dolomites do not display principal differences except in chemical composition. Evidently the dolomitization took place immediately after sedimentation during the first stage of diagenesis, and they are of secondary nature. The ratio of Mg/Ca in the carbonate rocks, and the clay content, permit differentiation between Mg-dolomites, pure dolomites, Ca-rich dolomites, limestones and Mg-limestones, clay limestones, dolomitic marls, calcareous clays, and clayey dolomites. The pure middle- and coarse-grained dolomites of the Kurachine Dolomite and Butmah Formations make very good reservoir rock for the accumulation of petroleum. Dolomitic marls are common in the Serjelu Formation.

Evaporitic (sulfate) rocks

Sulfate rocks in the Triassic of Syria are represented by anhydrites. Among them are primary and secondary varieties.

Primary anhydrites are usually gray in color, fine- and medium-grained with crystal orientation parallel to bedding, with horizontal or wavy bedding, sometimes with intercalations of clays. They form very thick units, and almost completely comprise the Kurachine Anhydrite, Hadayah, and Alan Formations. These same units, but of lesser thickness, are also found within the Kurachine Dolomite, Butmah, and Serjelu Formations. These sediments were formed in shallow-water marine basins located at a distance from the coast, under conditions of limited inflow of water of normal marine salinity, including a contribution of water from terrestrial runoff.

The primary anhydrites are also represented by varieties with thin intercalations of anhydritic mudstones and mudstones with imprints of large crystals of gypsum and anhydrite. Crystals have orientation parallel to bedding, and are gray, dark-gray, and greenish in color. Such anhydrites are of insignificant thickness (up to 1 m) and underlie stromatolitic, dolomitized limestones. Their origin is near the coastline, under shallow water, lagoonal conditions, evidently with periodic desiccation.

Secondary anhydrites were formed by post-sedimentational processes of migration and redistribution of sulfates. During the transition from sedimentary gypsum to anhydrite, water is released and migrates along the section, taking with it some part of the sulfate and precipitating it along the way in pores, fissures and microfissures of the calcareous rocks. The entire suite of anhydritic dolomites and dolomitic anhydrites can be observed, in which fine-grained and medium-grained anhydrite occupies some 5-25%, 30-35%, and up to 50-80% of the rock volume. These rocks usually preserve small but highly corroded particles of fine-grained carbonate. Secondary dolomites have no fixed stratigraphic position and can be found throughout the entire Middle and Upper Triassic sequence. Usually they are located immediately near thick layers of primary anhydrite, or are connected with layers of dolomite which are found within anhydritic successions.

CLIMATIC CONDITIONS OF THE TRIASSIC IN SYRIA

The distribution and taxonomic composition of spores and pollen in the sediments of the Middle and Upper Triassic of Syria permits conclusions to be drawn about the climatic conditions during this interval (Yaroshenko and Bach Imam, 1995).

Bisaccate pollen, especially costate *Lunatisporites* spp., predominate in the Anisian and Ladinian. In the Late Anisian and the Early Ladinian, monosaccate *Stellapollenites thiergartii* is common, and in the Late Ladinian, *Triadispora* spp. Representatives of these groups of pollen can be classified as xerophytic groups of miospores, and are considered as indicators of arid or semiarid climates. Hydrophytic elements (spores of pteridophytes) are less adapted to such conditions, and their content in assemblages is significantly lower than that of the pollen. At the end of the Ladinian and in the Carnian, together with the bisaccate, costate species of *Triadispora*, non-saccate pollen of the *Circumpolles* group become more widely distributed, and quantities of spores and acritarchs are strongly diminished. Arid climatic conditions persisted at that time. In the Early Norian, transformation of the palynoflora is observed, as early elements of the *Circumpolles* group are replaced by younger elements, with predominance of species of *Corollina* and *Classopollis*. Such predominance of *Classopollis*, which is produced by xerophytic cheirolepidian plants together with large quantities of bisaccate non-costate and monosaccate pollen evidently point to further aridization.

In the Late Norian, dinoflagellates appeared. *Botryococcus* became numerous, and the quantity of pteridophytic spores increased. These features characterize a general humidization of the climate in the Late Triassic.

Consequently xerophytic elements dominate in the palynoassemblages: in the Middle Triassic, 75-98%; in the Carnian, 95-96%; in the Norian, 58-95%; and in the Rhaetian, 42-87%. Hydrophytic elements are strongly subordinate. Their role increases in the Rhaetian. These data indicate arid to sub-arid climates throughout the whole Middle and Late Triassic of Syria.

The xerophytic character of the palynofloras of the Middle and Late Triassic of Syria, and their similarity with coeval spore and pollen assemblages of the adjacent areas, lead to the conclusion that the region was a part of the Mediterranean belt of the arid Equatorial phytogeographic realm.

Against this general background of arid climate, alternation of arid phases with comparatively more humid ones may be observed, reflecting climatic fluctuations and local transgressive-regressive events. Such periodicity is traced by changes in the ratios of xerophytic/hydrophytic elements. The Triassic section in the Maaza-1 bore-hole (lower reaches of the Nahr El-Khabour River) demonstrates such fluctuations (Yaroshenko and Bach Imam, 1995). In the interval 2873-2866 m (7 m of section) the palynoflora of the *Corollina meyeriana–Vesicaspora schemeli* assemblage (Upper Norian) undergoes a 4-fold change, with two alternating phases of aridization and two of humidization. The arid phases are characterized by strong increase of *Classipollis* pollen, while the humid phases are indicated by predominance of spores and phytoplankton (algae, *Botryococcus*, and the dinoflagellate *Rhaetogonyaulax rhaetica*). These two groups predominate in alternating phases.

FACIES AND PALEOGEOGRAPHY OF THE TRIASSIC OF SYRIA

Scanty data dealing with the sediments of the Early Triassic do not allow conclusions to be drawn concerning the facies and paleogeographic conditions at that time. If sediments of the Amanus Shale Formation are indeed of Early Triassic age, then deposition of clayey calcareous sediments of marine and continental origin took place in the southern areas of Syria (the Palmyrides). In the Tuwal El-Aba uplift in the north of Syria, marine clayey and calcareous sediments with *Pseudomonotis clarai* were definitely developed. Apparently the Early Triassic of Syria was marked by sedimentation within limited areas.

Better evidence exists for describing the facies and paleogeographic conditions for the Middle and Late Triassic. Our degree of knowledge on the Triassic differs for various parts of Syria – published data on the subsurface geology are mainly relevant for the northeast and north, while information on the Triassic of southwestern Syria is very limited (Bebeshev et al., 1989). Nevertheless, two different facies belts can be distinguished.

The first includes sediments deposited in a semi-isolated lagoonal, coastal, or shallow marine basin. Marginal marine shallows, and isolated embayments and lagoons, some periodically desiccated, characterize this region. Two types of sediments were deposited in this facies belt, with similar main properties: semi-isolated lagoons and coastal shallow-water zones are represented by Mg-bearing carbonate sediments such as fine-grained dolomites, calcareous dolomites, and clayey dolomites, as well as by algal stromatolites. They were formed under conditions of arid climate in shallow embayments with limited input of marine waters.

Within periodically isolated lagoons, as well as in sabkhas, dark gray and black mudstones with imprints of gypsum crystals, and dark-gray clotted anhydrites with irregular textures were deposited. These sediments precipitated in stagnant basins, partly or completely isolated from the open sea, in lagoons and sabkhas with periodic input of marine water, and under arid, hot climate conditions. Evidently vast lowlands covered with extremely shallow water and flat islands existed along the coasts of the continent. The main evaporite sediments, especially the sulfates, precipitated on these lowland flats.

In general the facies of semi-isolated lagoons and coastal marine basins consist of sediments of clayey and sulfate composition, with clotted or irregular texture of sulfates. Thin, horizontally bedded mudstones are characterized by the absence

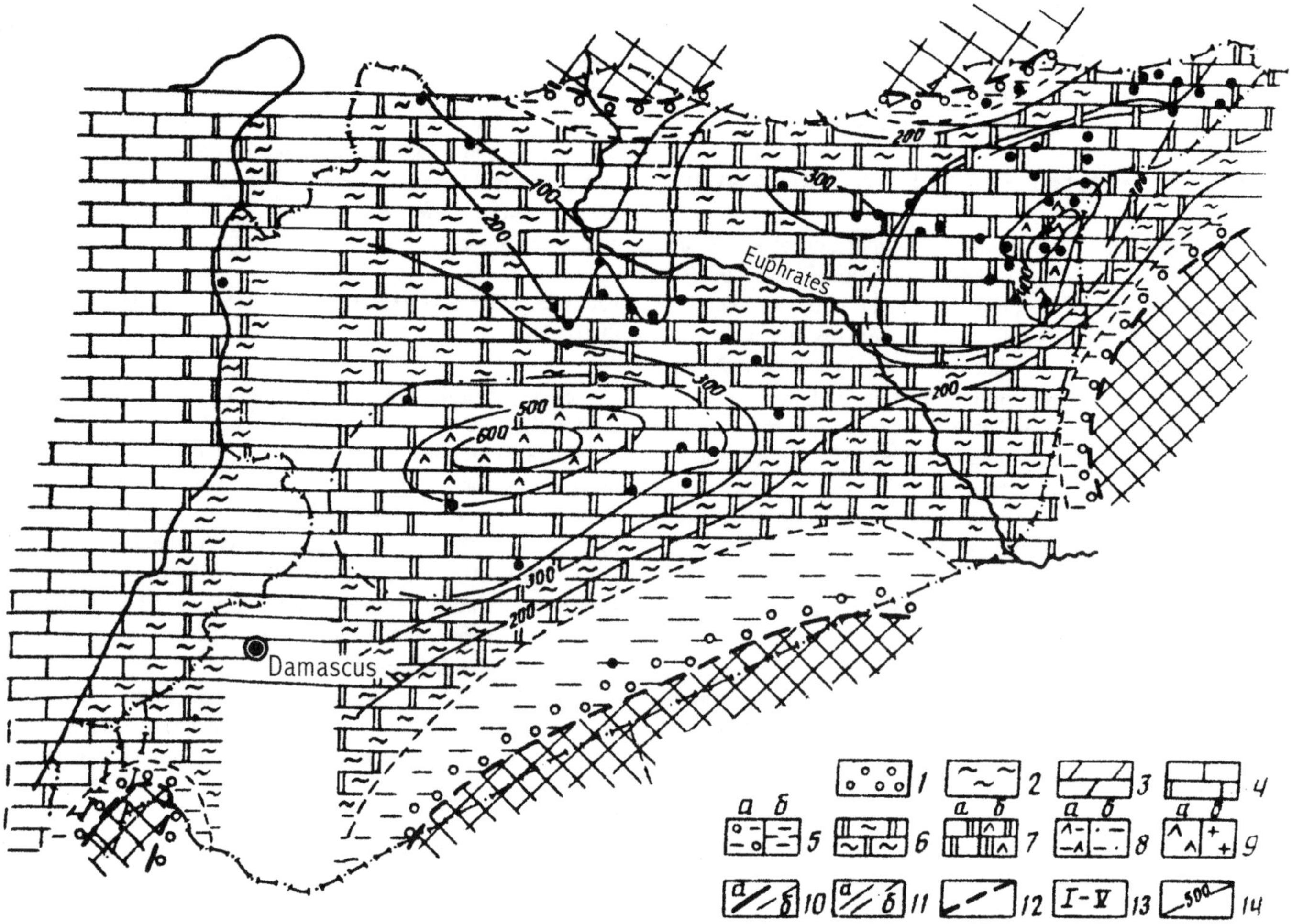

Fig. 6.15. Schematic paleogeographic map of the early Middle Triassic of Syria (the Kurachne Dolomite Formation) (after Bebeshev et al., 1989). Legend: 1-9 - paleo-landscapes: 1 – coastal shallow-water area with silty-sandy sedimentation; 2 – open shallow-water marine basin with calcareous-silty-clayey sedimentation; 3 – remote (central) part of a shallow-water marine basin with clayey-calcareous sedimentation; 4 – open marine basin with calcareous sedimentation; 5 – coastal shallow-water area with sandy-clayey-calcareous sedimentation (a – the outer zone, b – the inner zone); 6 – open shallow-water area with clayey-magnesium sedimentation; 7 – remote (central) area of a marine basin with magnesium-calcareous and magnesium-sulfate sedimentation (a – magnesium-calcareous, b – magnesium-sulfate); 8 – coastal shallow-water area with sulfate-clayey-silty sedimentation (a – the outer zone, b – the inner zone); 9 – remote from the coastal line, comparatively deep-water zone of a marine basin with sulfate (a) and halite (b) sedimentation; 10 and 11 – boundaries between paleo-landscapes: 10 – between coastal shallow-water area and shallow-water zone of an open marine basin (a – confirmed, b – suggested), 11 – between open shallow-water zone and remote from coastal line areas of a marine basin (a – confirmed, b – suggested); 12 – boundaries of paleo-uplifts; 13 – paleo-uplifts (I – Jordan, II – Rutbah, III – Rawda, IV – Qamychliye-Syrom, V – Aleppo); 14 – isolines of thickness, m.

of fauna and dinoflagellate phytoplankton.

Sediments deposited in regions more remote from the coastline are chemogenic with an insignificant content of terrigenous matter. Here, the marine environment was dominated by quiet hydrodynamic conditions. Sediments of this facies type occupy vast areas of central as well as northeast Syria. Within this facies, three rather similar rock assemblages can be distinguished.

1) Gray and dark gray calcareous mudstones and clayey limestones with rare pelecypods, ostracods, and dinoflagellates. These rocks are characterized by thin, dispersed materials and an absence of detrital and clastic particles. These sediments were formed in remote parts of the marine basin, under quiet hydrodynamic and normal hydrochemical conditions. They are of limited distribution, being observed only in the northern and northeastern parts of Syria (the Serjelu Formation; partly the Butmah Formation).

2) Dolomites, calcareous dolomites, and sometimes anhydritic dolomites with various structures and textures, ranging from pelitomorphic to coarse-grained, from massive to thin-bedded, mottled and clotted. The admixture of clay matrix is

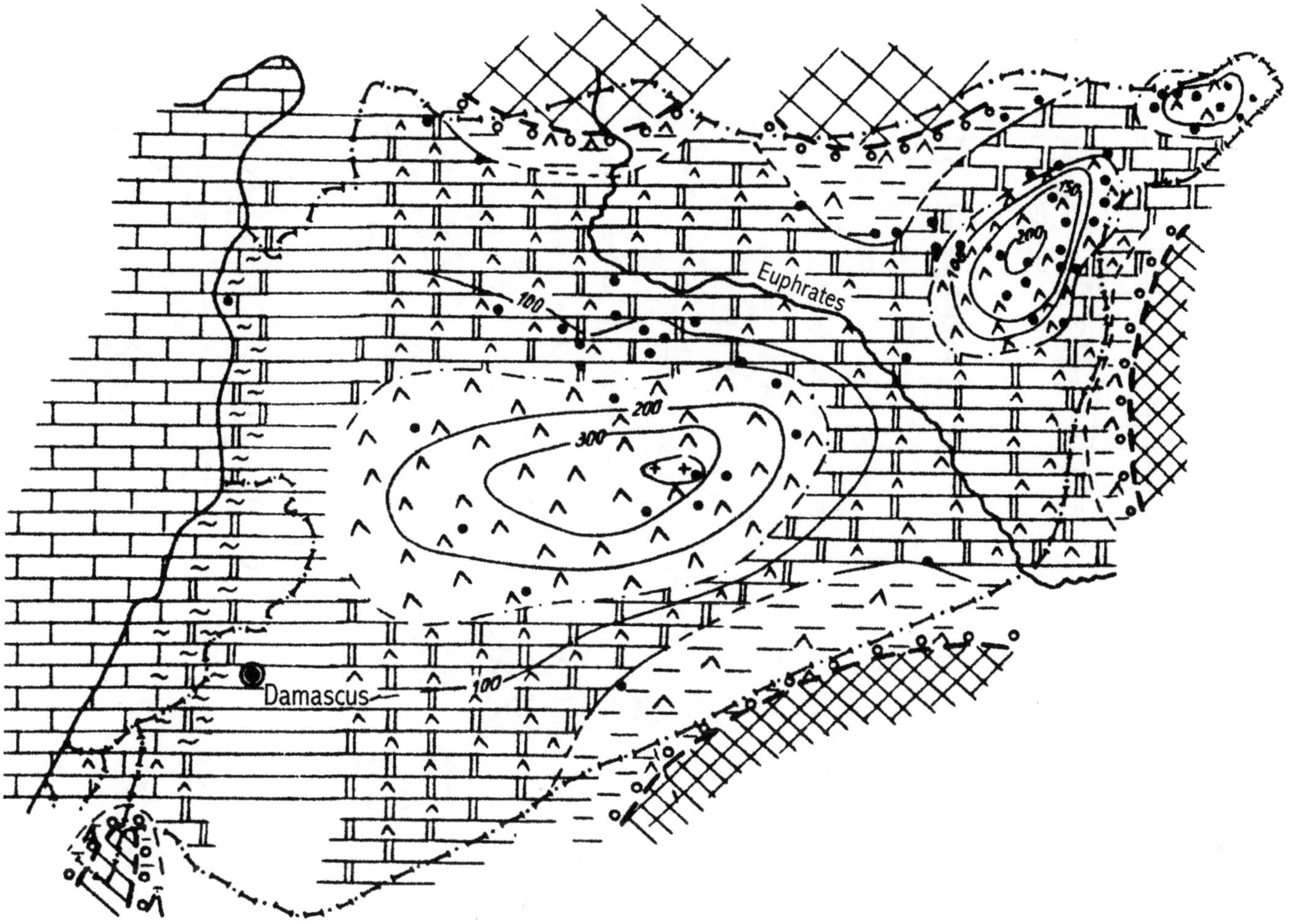

Fig. 6.16. Schematic paleogeographic map of the upper Middle Triassic (the Kurachine Anhydrite Formation) (after Bebeshev et al., 1989). Legend: see Fig 6.15.

not significant. Coarse clastic material is absent. Remnants of fauna and phytoplankton are exceedingly rare. Precipitation of these sediments proceeded in the marine basin, remote from the coastline, which may have been partly isolated and with periodic input of normal marine water under conditions of arid climate.

3) Gray, fine- and medium-grained banded anhydrites and halite. For these sulfatic Mg-rich and halitic (chloride) sediments, a complete absence of terrigeneous material and organic remains is typical. Their formation took place in the central parts of an isolated marine basin, remote from the coastline, within periods of intensive evaporation, and without input of marine waters (or with strongly limited input). Sediments of this assemblage often form concentric (in plane view) structures with halite and sulfate sediments in the center and Mg-bearing sediments along the periphery. Unfortunately, in this case, secondary redistribution of plastic halite-sulfate materials cannot be ruled out. Sediments of such a rock assemblage, compared to coastal evaporites, differ by their large thicknesses and absence of clays. They are most characteristic of the Kurachine Anhydrite, and the Hadayah and Alan Formations.

To summarize, the Middle and Late Triassic of most of Syria was covered by a shallow marine basin with sedimentation of evaporitic sediments – Mg-bearing carbonates, sulfates, and clayey sulfates. Their facies types are very variable, from coastal sediments of lagoons and sabkhas, to sediments of remote marine basins, and to shallow marine basins with limited or completely isolated connections to the open sea. Apparently, eustatic changes of sea level resulted in cyclical alternation of predominantly Mg-calcareous sediments (dolomites) and sulfatic rocks (anhydrites). Local transgressions and regressions were responsible for complex reorganization of the rock facies.

Schematic maps illustrate the paleogeographic and facies conditions for the time of formation of the Middle Triassic Kurachine Dolomite (**Fig. 6.15**) and Kurachine Anhydrite (**Fig. 6.16**), as well as for the Upper Triassic Butmah (**Fig. 6.17**) and Hadayah (**Fig. 6.18**) Formations.

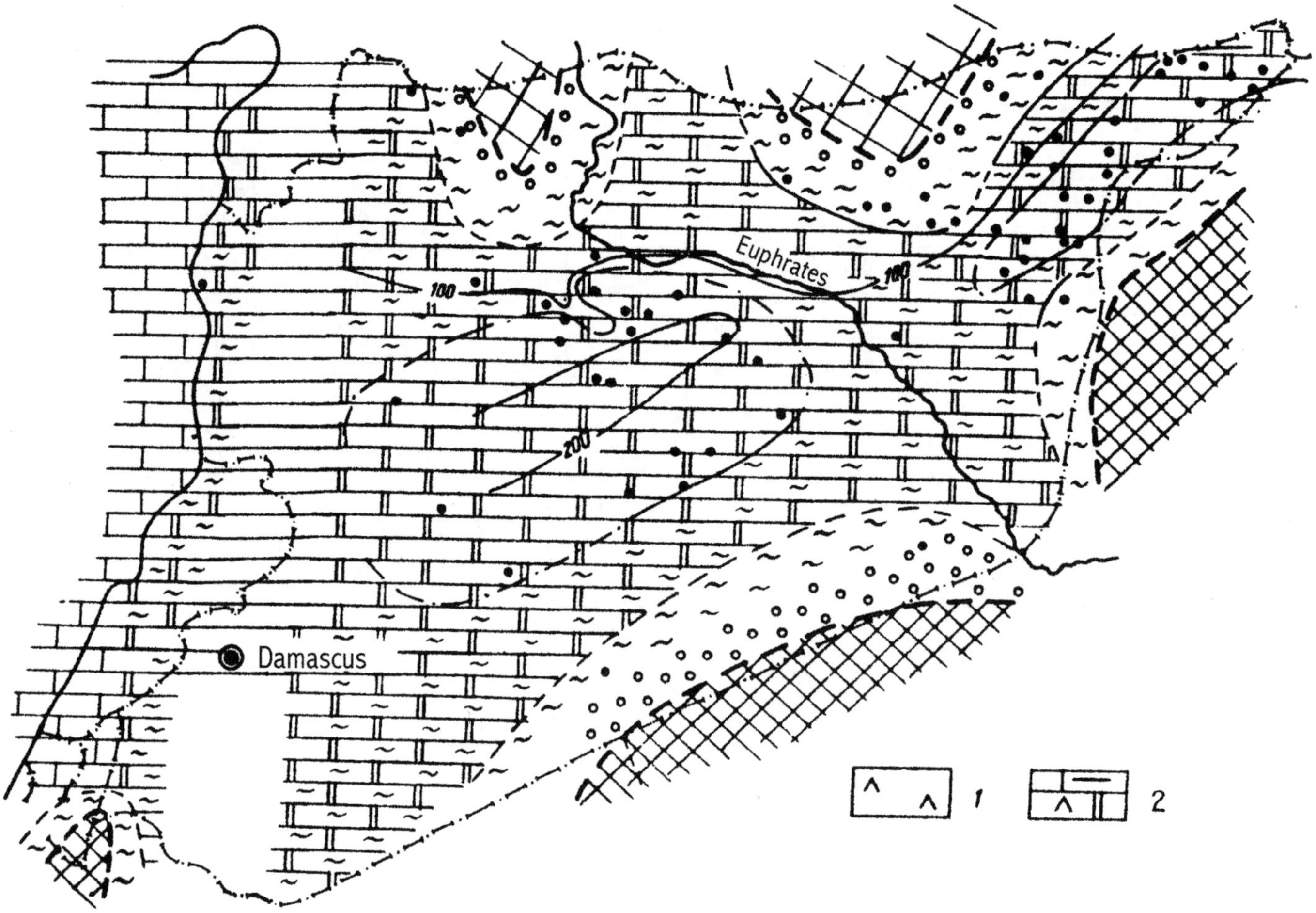

Fig. 6.17. Schematic paleogeographic map of the early Late Triassic (the Butmah Formation) (after Bebeshev et al., 1989).
Legend: 1, 2 – paleo-landscapes: 1 – coastal shallow-water area with sediments of lagoons and sabkhas; 2 – open shallow-water area with sulfate-clayey-magnesium sediments. Other symbols are given on Fig. 6.15.

STAGES OF GEOLOGICAL DEVELOPMENT OF THE STAGES IN THE TRIASSIC BASIN OF SYRIA

The history of geological development of the Triassic basin of Syria was described in the publications of Bebeshev et al. (1989) and Bebeshev (1995). Four stages of development were outlined. Unfortunately, the events of the first stage deserve critical reconsideration, because these geologists erroneously assigned the whole complex of sediments of the Amanus (s.l.) Formation to the Lower Triassic. In fact, only the Amanus Shale Formation should be included in the Lower Triassic. To repeat, conditional placement of this formation in the Lower Triassic is based on insignificant paleontological substantiation. In contrast, the interpretation of geological events of the Middle and Late Triassic is much more reliable.

The first stage of development embraces the Early Triassic. Data concerning the events and environments of the Lower Triassic - the time of deposition of the Amanus Shale Formation as defined by *Claraia (Pseudomonotis) clarai* in northern Syria, are very scanty. Apparently at this time, a shallow water marine basin of limited size was present in Syria. It had a free connection with the open sea. Within this basin, precipitation of marine terrigenous (clayey) and carbonate sediments took place. In the southern areas of Syria, they were replaced by sediments of continental origin.

The second stage of development corresponds to the Middle Triassic. The paleogeography strongly changed at the Lower/Middle Triassic boundary. In the Anisian, when the Kurachine Dolomite Formation was deposited, the sea expanded – in many areas of Syria, rocks of the Kurachine Dolomite Formation cover older deposits both transgressively and unconformably. This was also the time of formation of the Syrian platform evaporitic basin, which was encircled, from the south by the Jordan and Rutbah uplifts, on the east by the Rawda (or Iraqi) uplift, and from the north by the Mardin and Qamichliye-Syrom uplifts (the main parts of which are

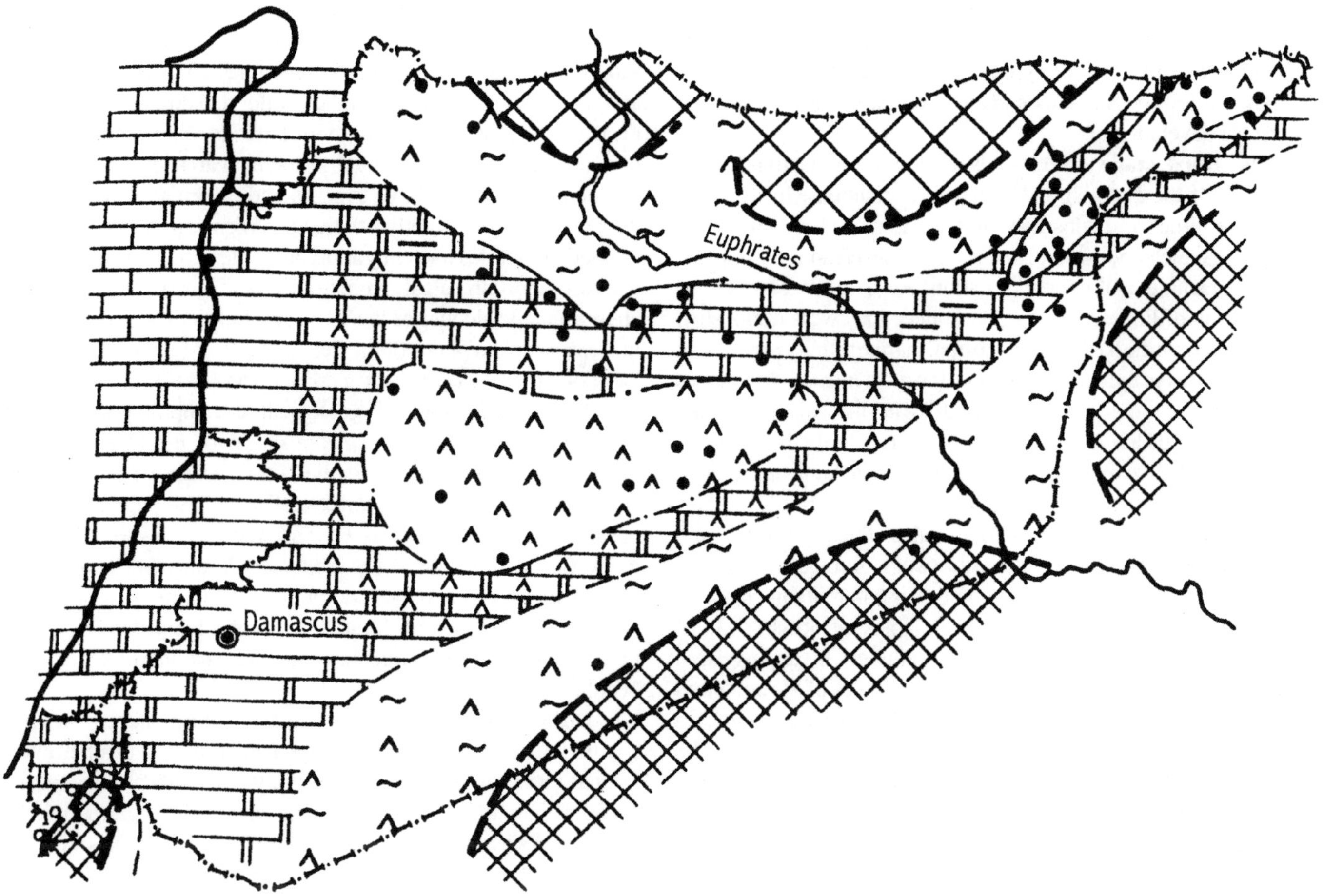

Fig. 6.18. Schematic paleogeographic map of the middle Late Triassic (the Hadayah Formation) (after Bebeshev et al., 1989). Legend: see Fig. 6.15 and Fig. 6.17.

in Turkey). To the north-west, the ridge of Jebel An-Nusseiriyeh may have formed a longitudinal paleo-uplift. This evaporitic basin had a limited connection with the Neotethyan Ocean, only to the west, through the Palmyra-Embayment.

In narrow belts or shallow coastal zones, along the ancient uplifts, terrigenous material from adjacent terrestrial areas accumulated. On the northern slope of the Rutbah uplift, the Swab-1 and Tanf-1 bore-holes penetrated a thin unit of cross-bedded sandstones with intercalations of siltstones and mudstones. In the north (the Afendi-1 bore-hole) and in the northeast (the Oudeh-101 bore-hole) deposits of the Kurachine Dolomite Formation also contain coarse terrigenous material which was brought to the basin from the Qamichliye-Syrom uplift and from adjacent minor local highs.

Towards the central part of the basin, the coastal marine sediments grade into open, shallow-water environmental sediments. At first, they consisted of calcareous dolomites, clayey dolomites, and clays, which gradually were replaced by clayey dolomites, massive monotonous pure dolomites, and sulfatic dolomitic rocks. During the Anisian, the single evaporitic basin was split into the two independent sub-basins of Palmyra in central Syria and Nahr El-Khabour River in Mesopotamia. They were characterized by maximal rates of subsidence and the greatest thickness of sediments (up to 600 m) of the Kurachine Dolomite Formation (Cherife-1, Salhieh-1, and Jbissa-203 bore-holes).

The end of the second stage (Ladinian), at the time of accumulation of the Kurachine Anhydrite Formation, was marked by almost complete isolation of the Syrian evaporite basin, and its very limited connection with the Neotethys. Further paleogeographic diversifications took place with the development of a third sub-basin of intensive subsidence alongside the Palmyrian and Mesopotamian basins, namely, the Qaratchok Darh sub-basin in the extreme northeast of Syria. Under conditions of shallow water environment and strong aridization of the climate, thick units (up to 300 m), of pure monotonous gypsum and anhydrites, lacking clastic admixtures, were deposited. Towards the more open marine environment their thickness decreases and anhydritic dolomites with lenses of gypsum and anhydrite appear. Along the ancient uplifts, the periodically flooded shallow water coastal regions contain a constant alternation of sandy and silty clayey sediments with thin beds and lenses of gypsum and anhydrite.

The Late Triassic encompasses the next three stages of geological development.

The third stage corresponds to the Early Carnian, with deposition of the Butmah and Hadayah Formations. In its initial phase, the time of deposition of the Butmah Formation, the paleogeographic conditions of sedimentation were similar to those of the Kurachine Dolomite Formation. As a result of renewed transgression of limited scope, freer connections were established with the open marine basin with its normal marine salinity regimes. In the same way as in the preceding stage, the most intensive processes of sedimentation took place in central Syria (the Palmyrides, the southern slope of the Aleppo uplift) as well as in the northeast of the country, where two depressions of the Mesopotamian area were evidently united in a northeast striking zone.

In this zone of Mesopotamia, rather thick (200 m) and predominantly Mg-bearing deposits were formed. They are composed of dark gray, massive monotonous dolomites intercalated with light gray, clotted varieties (bore-holes of Jbissa-203, El Bardeh-1, Roumelan-6, etc.). In the northwestern and western direction, from this zone of subsidence (bore-holes of Syrom-1 and Syrom-3) as well as to the southeast (the Bakhri-1 bore-hole), an admixture of silty and sandy material appears in the dolomites. The first admixture was derived from the Qamichliye-Syrom paleo-uplift, while the source of the latter was the Rawda uplift.

In the Palmyrides depression (bore-holes of Soukhne-1, Dolaa-1, Arak-1 etc.), and on the southern margin of the Aleppo uplift (the Wahab-1 bore-hole), formation of similar dolomites took place in a process of intensive subsidence. In the Palmyrides an insignificant input of terrigenous material was derived from the Rutbah uplift, located to the south; and in the area south of the Aleppo uplift a similar input was derived from the north (the Mardin uplift in Turkey).

In the rest of Syria, clayey dolomites and dolomitic mudstones were deposited under shallow-water open basin conditions. Their characteristic color is dark gray to black, caused by large amounts of bituminous organic matter. In the central, most intensively subsiding depressions, at the top of the Butmah Formation, pure Mg-bearing and Mg-sulfate sediments indicate further evaporitization of the Triassic basin.

The end of the third stage, the time of deposition of the Hadayah Formation, is characterized by partial isolation of the basin from the open marine environment. Conditions of stagnation were again restored. Apparently the paleo-relief at this time was reduced, with peneplanization of the uplifts, because sediments of the Hadayah Formation contain insignificant terrigenous input.

In the most intensively subsided central areas of the Mesopotamian basin, most distant from the basin margins (Jbissa-203, Roumelan-6, Nora-1, Al Hol-1 bore-holes), and in the Palmyrides (Soukhne-1, Cheriffe-1 bore-holes), monotonous massive sulfates (anhydrites) devoid of terrigenous material were deposited. In the open, shallow water zone, monotonous massive dolomites and anhydritic dolomites were formed.

Along the ancient uplifts, extended zones of lagoonal and sabkha environments existed. These zones constituted vast plains and sabkhas, which were periodically flooded during high tides and desiccated at low tides. Evaporation of marine water under conditions of a hot, arid climate, resulted in the appearance of concentrated solutions from which large gypsum and dolomite crystals grew in the pores of these slightly lithified sediments. In this area, coarse-grained anhydrites and anhydritic dolomites, mudstones with large anhydrite crystals, algal stromatolites, and dolomitized limestone are present. Analogues of the Triassic sabkhas of Syria can be seen in Recent sabkha plains of the Persian Gulf, extending for over 900 km along the coast (Senders and Friedman, 1970).

The fourth stage of geological development corresponds to Late Carnian time, the time of origin of the Muss and Alan Formations. In its initial period (the Muss Formation) the connection with the open sea was intermittent. The smooth paleorelief of the basin and its surroundings, as well as the slow process of subsidence, led to the precipitation of a considerably thinner (40-100 m thick) sequence of monotonous dolomites, sometimes replaced by anhydritic dolomites. At the southernmost edge of the basin, close to the Rutbah uplift, terrigenous material was intermixed with the evaporites.

During the final interval of the fourth stage (the Alan Formation), the connection with the Neo-Tethys ocean became so limited that anhydrites predominated. Its insignificant thickness of 40-50 m is witness to the slow subsidence, although the relief of the basin floor apparently became more dissected. This is shown by replacement of anhydrites by anhydritic dolomites, pure dolomites, and dolomitic mudstones.

The fifth stage of development refers to Norian-Rhaetian time, when sediments of the Serjelu Formation were deposited. The infra-Jurassic and/or pre-Cretaceous phases of denudation destroyed a considerable part of the Serjelu Formation sequence, making it very difficult to assess its nature. The terminal Triassic of Syria is characterized by gradual, though very unstable, restoration of the links with the open sea. Therefore the Serjelu Formation consists of sediments that reflect paleo-environmental conditions changing from evaporitic to open marine: anhydrite, dolomite, dolomitic mudstone, calcareous dolomite, and limestone. The limited distribution of these deposits in Mesopatamia and in the Palmyrides can be explained by the regressive character of the terminal Triassic of Syria, and not only by the subsequent processes of denudation.

Following this, during the entire Middle-Late Triassic, the Syrian basin represented a huge embayment with alternating degrees of isolation, from lagoon to shallow open marine. Narrow, shallow straits were responsible for periodic connection of the Syrian basin and the Neo-Tethys ocean, permitting the influx of waters of normal salinity. Intensive evaporation, close to the paleo-highs surrounding the basin, as well as in uplifted areas within the basin, led to the high sulfate and magnesium content of denser waters that subsequently descended into the deeper parts of the basin where they precipitated as thick evaporites.

OIL AND GAS POTENTIAL OF THE TRIASSIC DEPOSITS OF SYRIA

The high hydrocarbon content of the Middle and Upper Triassic deposits of Syria could potentially produce huge quantities of oil and gas. Favorable maturation and preservation conditions are provided by lithologic and structural features *i.e.* the development of reservoirs and cap-rocks. The Kurachine Dolomite, Butmah, and Muss Formations that encompass dolomites and clayey dolomites form reservoirs for hydrocarbons. The evaporites of the Kurachine Anhydrite, and the Hadaya and Alan Formations, ranging from anhydrite to halite, provide the necessary cap rocks.

A high degree of fracturing and porosity necessary for reservoir rocks is typical in dolomites, which attain 5-8%. The production of up to hundreds of thousands of cubic meters of gas per day is controlled by the distribution of the porosity within the reservoir rocks.

Presently a number of oil, gas, gas-condensate, and gas and oil condensate occurrences have been discovered in Syria. The highest hydrocarbon concentration occurs in northeastern and central Syria, with modern development of the predominantly gas-condensate, gas, and oil accumulations. Here in Triassic time, vast paleo-depressions existed, which subsided both constantly and rapidly. That process is responsible for the accumulation of organic matter and its transformation into hydrocarbons. The source rocks are mainly the Upper Paleozoic bituminous and calcareous clays with a comparatively high content (up to 3.7 %) of organic matter of lipid type.

The main gas and oil occurrences in Triassic deposits are located in the north-east of Syria, in Mesopatamia *e.g.* the oil fields of Qaratchok Darh, Hamzah, Al Hawi, Souedie, Oudeh, Khourbet, Bardeh, Matlout, Roumelan, Tishrin, Ghouna, Gbeibe, Jbissa and others. Heavy oil occurs in the Triassic of central Syria - on the southeastern slope of the Aleppo uplift, where the oilfields of Sfayeh, Wahab, Habari, Nobay, Ammala, Najeeb, and others are located. Gas occurrences are typical for the Palmyrides *e.g.* the gas--fields of Cheriffe, Soukhne, Bishri, Doubayyat, Al Shaier, etc. The Euphrates depression between Deir Ez-Zor and Abu Kamal represents a new field, with oil discovered in the Triassic at Tanak, Omar, Tayyem, Jido, and other places.

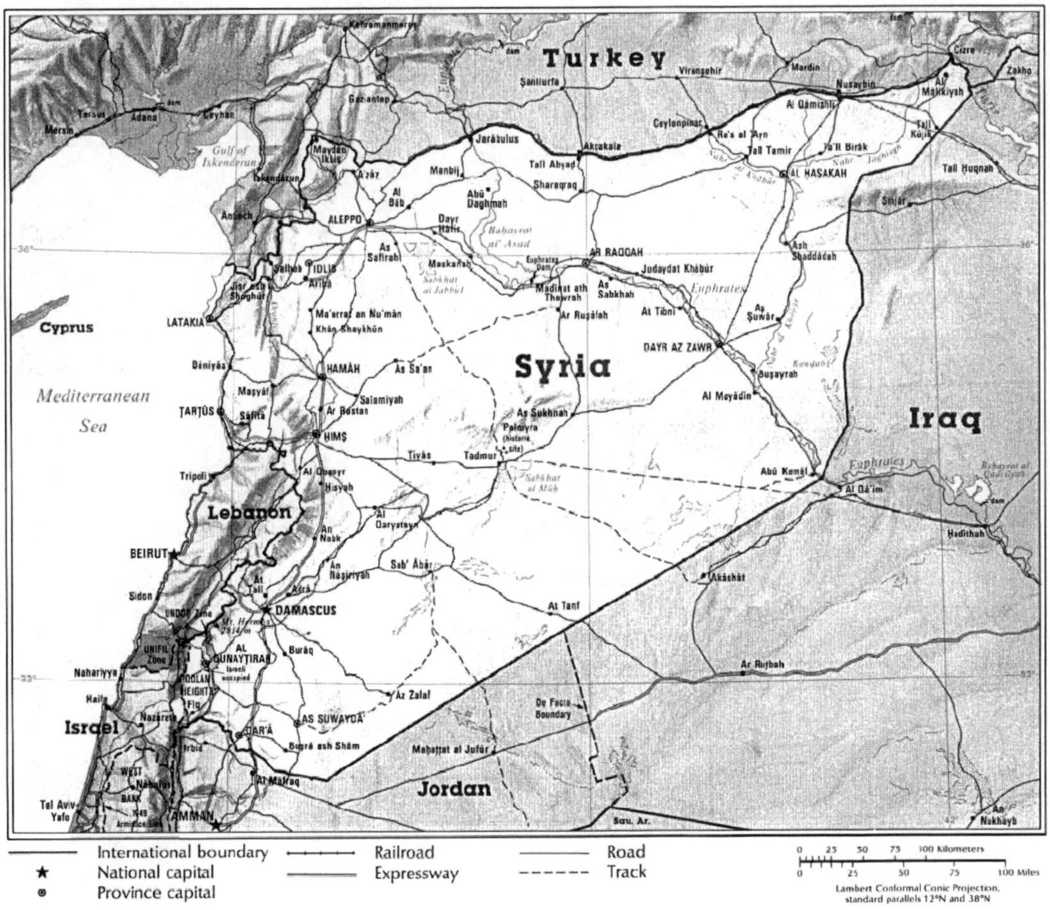

Image from www.lib.utexas.edu/Libs/PCL/Map_collection/middle_east_and_asia/syria map.pdf

Radar Imagery - Syrian Desert: Tadmor (Palmyra). European Space Agency (ESA). Image from: earth.esa.int/ers/ers_action/ syria.jpg Orbit 34687 Frame 2912 8 December 2001 5:45 GMT. Image shows the Northern edge of the Syrian desert. Upper center: halfway between the Mediterranean and the Euphrates, on an age-old caravan route, 150 miles northeast of Damascus, 1300 feet a.s.l., lies the oasis of Palmyra - the City of Palms -, the pearl of the Syrian Desert, in Semitic "Tadmor", founded by Solomon, King of Israel. Here rise the magnificent remains of the city, a settlement dating back to early second millennium BCE. Still standing, tall Corinthian columns (building the one km long Grand Colonnade), ancient walls and towers, monumental arches, and the well known Bel Temple appear as a surprise at the end of an extremely long and even boring journey. Alongside it, to the right: new Tadmor. North and south: ranges of naked and barren mountains. Upper left: the highly fertile belt of Taifah Gharbiyah, a densely cultivated area. Bottom centre: segment of a long road crossing the desert.

C: Jurassic

INTRODUCTION

Jurassic deposits are well exposed in the axial parts of anticlines and raised tectonic blocks of the western regions of Syria *e.g.* the Hermon Massif, the the Anti-Lebanon, Jebel An-Nusseiriyeh, Kurd-Dagh and Jebel Aqraa. They are also exposed as narrow strips in the cores of such Palmyrides folds as Jebel Hayyan, Jebel Abtar, Jebel Mazar, Jebel Satteh, etc. In the rest of Syria, Jurassic deposits are hidden by younger sediments. Drilling on the Aleppo Plateau, in the Syrian Desert, and in some areas of the Mesopotamian Basin has revealed various Jurassic sequences.

In the past excellent Jurassic sections in the Anti-Lebanon and Jebel An-Nusseiriyeh have attracted the attention of outstanding geologists. Lartet (1866), O. Fraas (1877), F. Noetling (1887), C. Diener (1886) and G. Zumoffen (1899, 1926) devoted their publications to the lithology, stratigraphy and paleontology of sediments of this age.

The next phase of investigation of the Jurassic rocks in Syria and Lebanon is associated with the names of L. Dubertret (1933a, 1934b, 1937a, 1942, 1943, 1949a, 1950a, b, c, 1954, 1955a, 1960, 1963, 1966), Dubertret and Vautrin (1937b), and H. Vautrin (1934a, b). H. Vautrin carried out a preliminary survey of the Zebdani district. The geological mapping of the Zebdani and Rayak sheets, under the supervision of L. Dubertret, resulted in geological maps at the scale of 1:50,000 and explanatory notes. These contain the stratigraphic subdivision of the Jurassic sections of the Hermon Massif, based on ammonites, brachiopods, and pelecypods. This research was supplemented by a thorough study of the ammonite fauna of the Hermon (Haas, 1955; Arkell, 1956), establishing a framework for the Jurassic biostratigraphy of the Anti-Lebanon.

Within the framework of the geological mapping of Syria at a scale of 1:200,000, carried out in 1958-1961 by a group of USSR geologists led by V. P. Ponikarov, the Jurassic deposits were studied in detail. The lithological and stratigraphic characteristics of the Jurassic appear on the map sheets of Dimashq and Beirouth (Razvalyaev, 1966a) Trablus and Homs (Kozlov, 1966a) Al-Latheqiyeh and Hama (Shatsky et al., 1966), Halab and Antakya (Protasevich and Maksimov, 1966), Al-Qaryatein (Faradzhev, 1966) and Tudmor (Soulidi-Kondratiev, 1966a). Geological mapping was also conducted at a scale of 1:50,000 for the Zebdani sheet (Kurbanov and Zarjanov, 1968), the area of the Baer-Bassit Ophiolite Massif (Kazmin and Kulakov, 1968) and Kurd-Dagh (Protasevich and Kurbatov, 1968), where the Jurassic deposits are imbricated in complicated tectonic structures. Geological mapping was accompanied by the study of rich collections of macrofauna; the cephalopods by G. Ya. Krimgolts, the brachiopods by V. P. Makridin, and the pelecypods by A. Yungerman. These investigations allowed the subdivision of the Jurassic system into correlatable series and stages, the establishment of facies types, and the construction of schematic isopach and lithofacies maps for the Middle and Late Jurassic rocks (Ponikarov et al., 1967, 1969). It should be emphasized that abundant ammonites found at Majdal Shams in the Mt. Hermon area allowed to G. Ya. Krymgolts to recognize the main zones of the Oxfordian Stage (in Ponikarov et al., 1967).

Until 1970 the stratigraphic subdivision of the Jurassic deposits was based on various groups of macrofauna. From the beginning of the geological mapping at scale 1:50,000, initiated by the General Establishment of Geology and Mineral Resources of Syria, in Jebel An-Nusseiriyeh (also known as Jebel As-Sahilyeh or Coastal Mountains), a new scientific approach was manifested by the application of micropaleontological methods to Jurassic stratigraphy. M. Mouty (1976) proposed the first zonal scheme for the Jurassic deposits of Syria. It included four biozones:

Haurania amiji-deserta (Middle Jurassic)
Pfenderina salernitana (Middle Jurassic)
Kurnubia palastiniensis (Upper Jurassic)
Pseudocyclammina jaccardi (Upper Jurassic).

Lower Jurassic rocks were also identified by the presence of *Lituosepta recoarensis* and *Orbitopsella* sp.

The study of foraminifera and ostracods allowed to subdivision of the monotonous carbonate sequences of the Lower-Middle Jurassic and the Upper Jurassic on the geological maps. It also permitted one to discriminate between the Bathonian, Callovian, and Oxfordian Stages (Adjamian and Al-Bakka, 1984; Adjamian and Jamal, 1983; Yousef, 1979, 1980a, b; Ruske, 1978a). Later, M. Mouty published his data in several articles (Mouty, 1997a, b, 2000; Fourcade and Mouty, 1995; Mouty and Zaninetti, 1998). The latter paper contains a more detailed biozonation of the Jurassic deposits of Syria, by the addition of several biozones:

Pseudocyclammina liasica (Early Jurassic);
Timidonella sarda (Middle Jurassic);
Haurania deserta and Amijiella amiji (Middle Jurassic);
Paleopfenderina salernitana–P. trochoidea (Middle Jurassic);
Kilianina blancheti (Middle Jurassic);
Steinekella steinekei–S. crusei (Middle Jurassic);
Levantinella egyptiensis (Late Jurassic);
Alveosepta jaccardi (Late Jurassic).

The same succession of biozones, with the exception of the *Levantinella egyptiensis* biozone, was identified in Jebel An-Nusseiriyeh (Mouty, 1997b).

During 1986-1990, a group of micropaleontologists and geologists from Russia, Lithuania, and Syria carried out investigations in the Jurassic deposits of the Anti-Lebanon, Jebel An-Nusseiriyeh, Kurd-Dagh, and the Palmyrides. They studied foraminifera and ostracods in the stratigraphic sequence, identifying facies types and estimating their paleobiogeographic significance. The results of this campaign appeared in a series of publications devoted to various aspects of the Jurassic stratigraphy of Syria (Kuznetsova, Grigelis, Adjamian, Jarmakani and Hallaq, 1991,

1996; Kuznetsova and Dobrova, 1995, 1997, 2000; Gorbachik and Kuznetsova, 1994; Dobrova, 1995). As a result, the existence of Lower Jurassic deposits was confirmed, and the Middle and Upper Jurassic were subdivided into stages and foraminiferal zones. Local layers with ostracods were correlated with foraminiferal zones.

The zonation suggested by Kutznetsova includes three categories of biostratigraphic units: a zone defined by an assemblage of species; an ecozone determined by fossils affected by local environmental conditions; beds defined by a particular fossil and bounded by erosional surfaces or by barren sediments. In the Anti-Lebanon, the complete successon of these units consists of:

Beds with *Trochammina nana* (Early Jurassic);

The *Praelamarckina humilis* Zone (Aalenian-Bajocian);

The *Lenticulina volubilis–Epistomina coronata* Zone (Bajocian);

The *Timidonella sarda* Ecozone (Bajocian);

The *Lenticulina polymorpha–Globuligerina bathoniana* Zone (Bathonian);

The *Kurnubia palastiniensis-Globuligerina calloviensis* Zone (Callovian);

The *Lenticulina brueckmanni–Globuligerina oxfordiana* Zone (Oxfordian);

The *Alveosepta jaccardi–Lenticulina quenstedti* Zone (Oxfordian);

The *Alveosepta personata* Zone (Kimmeridgian);

The *Anchispirocyclina lusitanica* Zone (Tithonian).

The biozonation of the Jurassic deposits of Jebel An-Nusseiriyeh is significantly different from that of the Anti-Lebanon. It encompasses the following subdivisions:

Beds with *Involutina cf. liassica* (Early Jurassic);

Beds with *Lenticulina centralis* (Bajocian);

Beds with *Haurania deserta–Protopeneroplis striata* (Bathonian);

The *Kilialina blancheti–Meyendorffina bathonica* Ecozone *(Bathonian);*

The *Kurnubia palastiniensis–Paleopfenderina salernitana* Zone (Callovian);

The *Lenticulina brueckmanni–Globuligerina oxfordiana* Zone (Oxfordian);

The *Alveosepta jaccardi-Lenticulina quenstedti* Zone (Oxfordian);

The *Alveosepta personata Zone* (Kimmeridgian);

The *Bramkampella arabica* Zone (Tithonian).

Consequently M. Mouty and K. Kuznetsova applied different methods and approaches to the definition of zonal scales for the Middle and Upper Jurassic. The correlation of Jurassic sections described in the respective publications of Mouty and Kuznetsova, is further complicated by the fact that the latter author did not mention lithostratigraphic units, *e.g.* formations or members, which are very important for local and regional stratigraphy.

In our opinion, the biozonation proposed in the works of Mouty and his collaborators has proved itself to be more applicable to the zonal subdivision within Syria and for correlation with adjacent countries, *e.g.* Israel (Perelis-Grossowicz, Bassoulet, Hirsch and Peri, 2000).

In 1987, L. Picard and F. Hirsch summarized stratigraphic and paleontological data relevant to the Jurassic deposits of the Anti-Lebanon with special emphasis on the Jurassic of the Hermon Massif. These authors also considered the problems of correlation of the Jurassic of this area with that of other regions of Syria, such as Palmyrides, Jebel An Nusseiriyeh, Northern Anti-Lebanon as well as with the adjacent countries of Lebanon, Israel, Jordan, and Egypt.

At the present time the regional Jurassic stratigraphy of the western part of Syria, as well as the Palmyrides, is very detailed and the age of each rock-unit is substantiated by assemblages of macro- and microfauna. We can not say the same about the stratigraphy of the Jurassic subsurface deposits. Data have only been published in a generalized way without a thorough subdivision of the Jurassic succession and strict age determinations of the formations. This precludes an understanding of the geological history during Jurassic time of a large part of Syria.

In general, all these investigations showed that the Jurassic of Syria is represented by a thick sequence of over 2000 m of mainly limestones, dolomites, and dolomitic limestones with rare intercalations of terrigenous-calcareous sediments. Volcanic rocks are rare and localized in the lower part of the Jurassic sequence in westernmost Syria.

All three of the Jurassic series were encountered in Syria. However their spatial distribution differs widely. Sediments of the Lower Jurassic occupy limited areas. The Middle Jurassic is widespread, and is characterized by Bajocian and Bathonian transgressions, followed by regressive motions towards the end of the Bathonian and in the Callovian time. The onset of the Late Jurassic is characterized by a new Oxfordian transgression and sediments of this age were found in many areas of Syria. The Late Kimmeridgian and Tithonian are marked by regression, when sedimentation became restricted to the western parts of Syria.

Lower Jurassic (Liassic)

It should be emphasized that Lower Jurassic exposures are of limited distribution in Syria. Though present in various areas, such as the Anti-Lebanon, Jebel An-Nusseiriyeh, the Palmyrides, and Kurd Dagh, it is only exposed in a few outcrops in each of these regions.

In the Anti-Lebanon (the Hermon Massif) Lower Jurassic deposits were identified near the village of Arneh where the stratotype of the **Arneh Formation** is composed of black, gray and yellow, thin-bedded and platy marls, clays, limestones and dolomitic limestones up to 170 m thick. Limestones predominate in the upper part of this formation; its lower half contains a member of underwater basaltic lavas (spilites) about 50 m thick. Intercalations of limestones above the spilites include *Pseudocyclammina liasica* Hottinger, which is characteristic for the Upper Liassic (Pliensbachian-Toarcian) (Mouty and Zaninetti, 1998; Mouty, 2000). Clayey interbeds yielded rare agglutinated

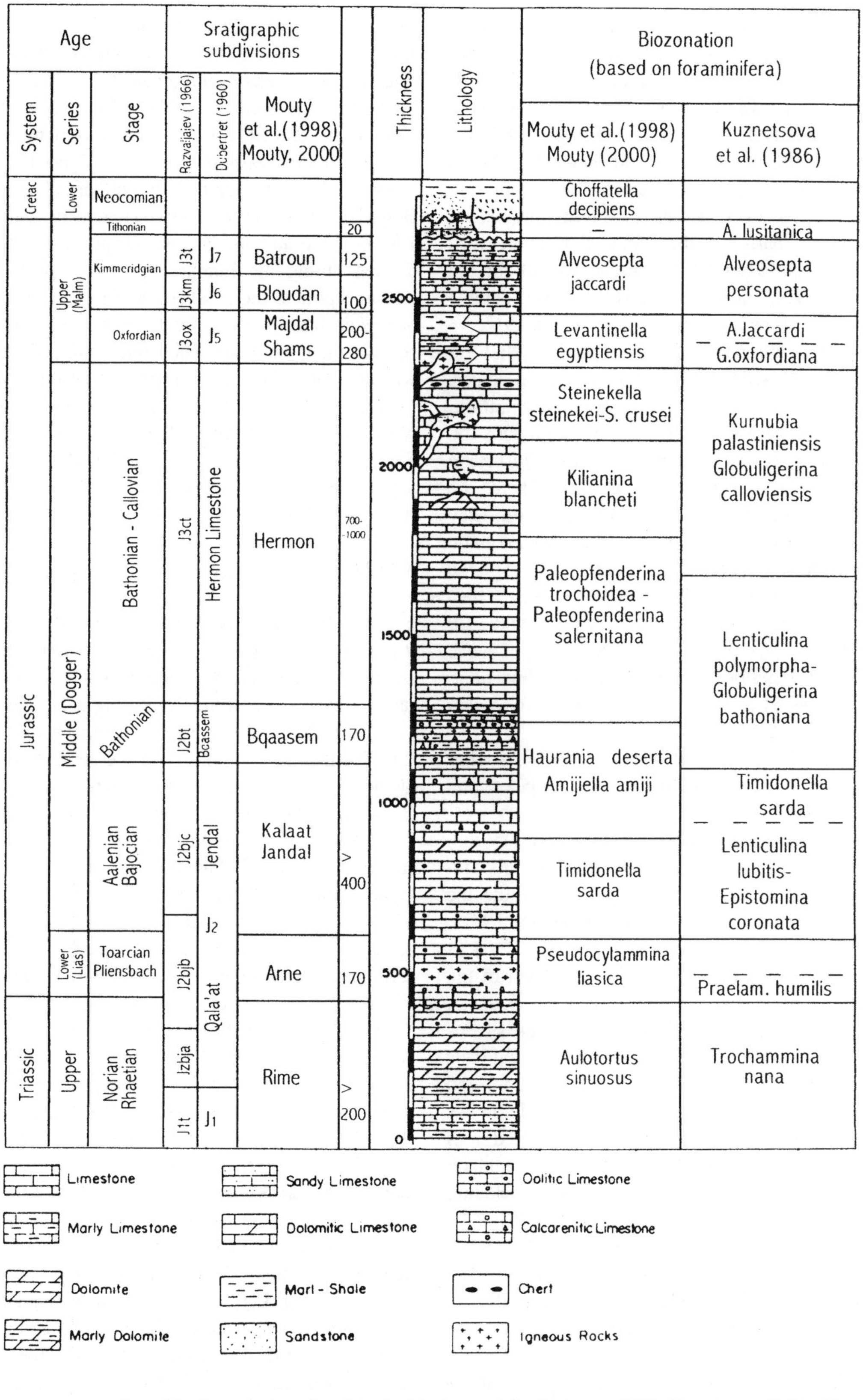

Fig. 6.19. Schematic composite section of the Jurassic deposits of the Anti-Lebanon (after Dubertret, 1960; Razvalyaev, 1966a; Picard and Hirsch, 1987; Kuznetsova et al., 1996; Mouty and Zaninetti, 1998; and Mouty, 2000).

foraminifera *Trochammina nana* (Brady), undetermined remains of small pelecypods and gastropods, and a rich assemblage of the ostracods *Progonocythere stilla* Sylvester-Bradley, *Aphelocythere kuhni* Triebel et Klinger, *Procytheridea sermoisensis* Apost., *P. vermiculata* Apost., *P.* aff. *magnicoitensis* Apost., *Limnocythere alata* Dobrova, *L. lobata* Dobrova, and *L. improcera* Dobrova (Kuznetsova et al., 1996; Kuznetsova and Dobrova, 2000). As mentioned above, deposits of the Arneh Formation with *Pseudocyclammina liasica* are underlain by black thin-bedded limestones and clayey marls of the Rime Formation with foraminifera *Aulotortus sinuosus* (Weynschenk) typical for the Upper Triassic (Norian-Rhaetian). Consequently, a considerable part of the Lower Jurassic is absent in this section (**Fig. 6.19**).

A similar picture was observed in Jebel An-Nusseiriyeh. Here the Lower Jurassic is composed by light-yellow marls, limestones, and clayey limestones of the 40-200 m thick **Treize Formation,** with foraminifera *Pseudocyclammina liasica* Hottinger and *Involutina liassica* (Jones) as well as the brachiopod *Liospiriferina undulata* (Seguenza). These deposits were assigned to the Pliensbachian-Toarcian (Mouty, 1997b, 2000) (**Fig. 6.20**). They rest on an about 300 m thick sequence of dolomites and dolomitic limestones (the **Jweikhate Formation**) with the Upper Triassic foraminifera *Gsollbergella spiriloculiformis* (Oroveczne-Scheffer) and *Galeanella* aff. *laticarinata* Al Shaibani. A more than 200 m thick Early Jurassic succession of alternating gray-yellowish dolomites, dolomitized limestones, and clays, with *Involutina* cf. *liassica* (Jones), was also found in the vicinity of Jdaida (Kuznetsova and Dobrova, 2000).

In the Kurd Dagh, the lower part of the **Dodo Formation** consists of Early Jurassic micritic limestones with the pelecypod *Lithiotis* sp. (Mouty, 2000).

In the Palmyrides, the Lower Jurassic is represented by thin layer (20-25 m) of limestones, dolomitic limestones and marls with foraminifera *Mayncina termieri* Hottinger which constitute the lower part (member 'A') of the **Satih Formation** (Mouty, 1997a, 2000) (**Fig. 6.21**).

This unit stretches along the whole southern branch of the Palmyrides. In the southwest it is exposed in on Jebel Zounnar, Jebel Rmah, and Jebel Naqnaquiyeh. In the central part of this chain it was also found on Jebel Abtar and Jebel Hayyan, and in its north-eastern part, in Jebel Mazar, Jebel Satih (Jebel As-Satteh), and Jebel Taniet As Safra. Unfortunately the limestones and dolomites are barren of fossils. Their age attribution is therefore derived from their correlation with similar Early Jurassic deposits found on Jebel An-Nusseiriyeh. An assemblage of pelecypods, including *Pholadomya* aff. *lyrata* Sow., *P. murchisoni* Sow., *Homomya gibbosa* Sow. *Pleuromya uniformis* Sow., *Lopha marshii* Sow., *Modiolus bipartitus* Sow., *Spondylopecten* sp., and *Panopea* sp., apparently already belongs to the Middle Jurassic (Mouty, 1997a). It was found in the uppermost layers of this unit, on Jebel Satih and Jebel Taniet As Safra,.

In Jebel Zubeideh (Jebel Al-Zbeidi) and Jebel Naqnaqiyeh, the **Zubeideh Formation** is an unfossiliferous sequence of dolomitic limestones, clays, and siltstones with kaolinite and bauxite, considered to be Early Jurassic-Bajocian (pre-Bathonian) age. These strata are the result of continental deluvium, proluvium, and small lakes and bogs, filling karst hollows within Triassic dolomites. The Triassic karren or solution grooves in the landscape were filled by ferruginous palaeosol. This developed under humid tropical climatic conditions, and is characterized by clay, kaolinite and bauxite. Alunite is also present, and is considered to be the product of low-temperature sulphatic hydrothermal activity (Tsekhovskiy, Bogatyrev and Jabbur, 1993: Tsekhovskiy, Muraviov and Jabbur, 1994). These authors suggest that the Zubeideh Formation was formed during the Liassic regressive epoch, that reduced the Rutbah uplift in the southernmost part of Syria to a continental peneplain which was encircled on its north by the shallow water zone of the Palmyrides. This zone consisted of bogs and lagoons with alternating marine and continental sedimentation. Late Liassic marine sediments only developed in the western part of Syria. In other parts of the Arabian platform, formations like the Zubeideh Formation, with its kaolinite and bauxite, are widely distributed and given different names, *e.g.* the Mishhor Formation in Israel (Picard and Hirsch, 1987).

Middle Jurassic (Dogger)

The Middle Jurassic transgression inundated most of Syria. It is represented by a sequence about 1000 m thick of limestones and dolomites with subordinate intercalations of marls and clayey-sandy deposits. According to drilling data (Weber, 1964), Middle Jurassic rocks are only absent in the Deir Ez-Zor (Iraqi) uplift (the Tueman bore-hole) and in the Jebel Abd El-Aziz area (the Qamichliye bore-hole). Reliable data allow the identification of Bajocian, Bathonian and Callovian strata in many areas of Syria. The lower part of the Middle Jurassic is usually considered as an undifferentiated Aalenian-Bajocian interval. The Middle-Upper Jurassic boundary is well established, defined with high precision by several fossil assemblages. In contrast, due to poor paleontological data, the lower boundary of the Middle Jurassic is still tentative and subject to two different interpretations.

For Kuznetsova et al. (1996) and Kuznetsova and Dobrova (2000), the base of the Middle Jurassic should be drawn in the Arneh section of the Anti-Lebanon, 70 m below the unit of volcanic-sedimentary rocks containing spilites and tuff-breccia, which is considered a part of the undifferentiated Aalenian-Bajocian *Praelamarckina humilis* Zone, or Beds.

Mouty (2000) attributes his *Pseudocyclammina liasica* biozone, which occurs above the spilites in the upper part of the Arneh Formation, to the Lower Jurassic. He puts the boundary with the Middle Jurassic at the base of the Kalaat Jandal Formation, well defined by *Timidonella sarda,* and we share this point of view.

In the Anti-Lebanon, the Aalenian-Bajocian and Bathonian-Callovian intervals represent the huge dome of the Hermon Massif, extending from Jebel Ash-Sheikh, to Jebel Sheqif El-Khorn, Jebel Mannsour, and Jebel Mazar. They are subdivided into three formations.

The Aalenian-Bajocian **Kalaat Jandal Formation** is made of

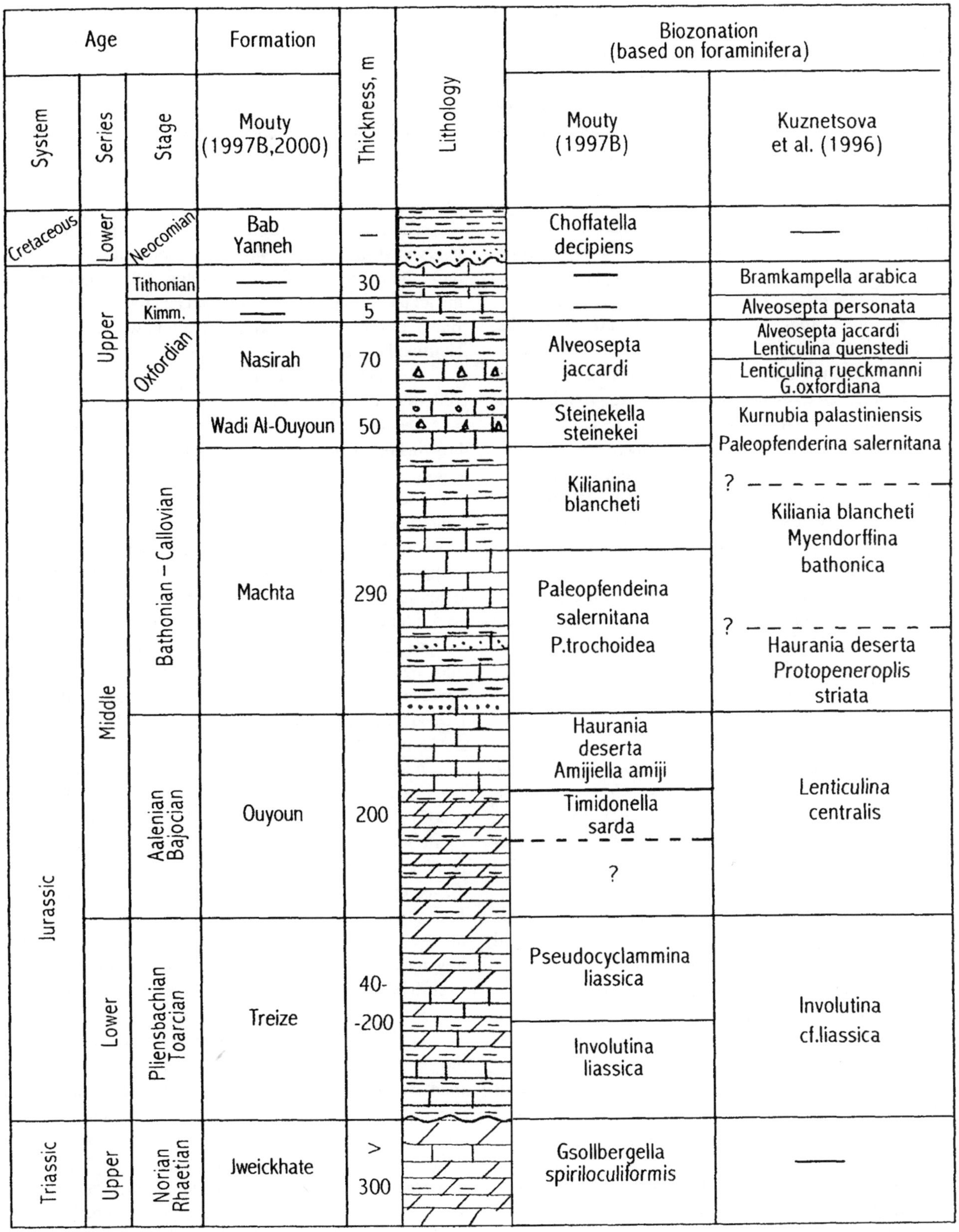

Fig. 6.20. Schematic composite section of the Jurassic deposits of Jebel An-Nusseiriyeh (after Shatsky et al., 1966; Kozlov, 1966a; Kuznetsova et al., 1996; and Mouty, 1997b, 2000).
Legend is as shown on Fig. 6.19.

light-gray, medium- and thick-bedded, micritic, bioclastic and oolitic limestones and dolomites with interbeds of black pelitomorphic limestones and marls; about 400 m thick. These deposits are subdivided, according to Mouty (2000), into the *Timidonella sarda* biozone and the *Haurania deserta-Amijiella amiji* biozone. Evidently the *Lenticulina volubitis-Epistomina coronata* Zone and the *Timidonella sarda* Ecozone of Kuznetsova's scale correspond to the above biozones of Mouty. However, Kuznetsova treats these units only within the bounds of the Bajocian Stage taking into account the foraminiferal assemblage of *Lenticulina volubitis* Dain, *L. varians* (Bornem.), *L. protrata* (Bornem.), *Citharina clathrata* (Terq.), *C. colliezi* (Terq.), *C. pauperata* (Terq.), *C. biangulata* (Terq.), *Lamarckella media* Kapt., *Epistomina coronata* Terq., *Eoguttulina polygona* (Terq.), and *Timidonella sarda* Bassoullet, Chabr. et Fourcade.

The next **Bqassem Formation** is separated from the Kalaat Jandal Formation by an erosional surface and hard ground (of the littoral zone). It is composed of limestone with a lesser amount of marls and calcareous clays, 150-170 m thick. Limestones are mostly medium-bedded, oolitic, oolitic cross-bedded, organo-clastic, or pelitomorphic. An overall peculiar yellow, yellowish-gray or rusty-brown (ochrous) color characterizes this unit, contrasting strongly with the underlying Kalaat Jandal limestones with black and dark-gray coloration. Due to this difference the boundary between the Kalaat Jandal and Bqassem formations is easily traced in the field.

Deposits of the Bqassem Formation contain an abundant fauna of brachiopods, pelecypods, gastropods, foraminifera, ostracods and rare corals, indicating the shallow-water marine environment of sedimentation with relatively high wave energy.

The following fossils are known from the publications of H. Vautrin (1934a) and L. Dubertret's (1960): *Pseudocidaris drogriaca* Lam., *Pholadomya bellona* d'Orb., *Ph. inornata* Sow., *Ceromya concentrica* Sow., *Heligmus polytipus* Des., *H. rollandi* Douv., *Rhynchonella hopkinsi* M'Coy, *R. obsoleta* Sow., *Terebratula (Ptyctothyris) superstes* Douv., and *T. quillyensis* Baile, which together indicate a Bathonian age. W. Arkell (1956) even restricted the age of these fossils to the Late Bathonian.

Investigations by V. Makridin and A. Yungerman (in Razvalyaev, 1966a) broadened this list of brachiopods and pelecypods to *Lobothyris ventricosa* (Ziet.), *Cererithyris rotzoana* Schaur., *Kallirhynchia lanta* Buckm., *K. orbis* Buckm., *Burmirhynchia hopkinsi* (Dav.), *Rhynchonelloidella* aff. *smithi* (Dav.), *Opis linulata* Sow., *Anisocardia tenera* (Sow.), *Camptonectes lens* (Sow.), *Lima duplicata* Sow., *Pholadomya murchisoni* Sow., *Ph. fidicula* Sow., *Ctenostreon* aff. *proboscideus* (Sow.), *Trigonia* aff. *elongata* Sow., and *Modiola* aff. *gibbosa* (Sow.) which together were referred to the Late Bajocian-Early Bathonian.

Using foraminifera, Mouty (2000) placed the Bqassem Formation into the upper part of the *Haurania deserta-Amijiella amiji* biozone and the lower part of the *Paleopfenderina trochoidea-Paleopfenderina salernitana* biozone (Late Bajocian-Bathonian). One can suggest that this interval coincides with the *Lenticulina polymorpha-Globuligerina bathoniana* Zone of Kuznetsova et al.'s (1996) scheme. The zone in question is characterized by *Lenticulina polymorpha* (Terq.), *Citharina proxima* (Terq.), *C. macilenta* (Terq.), *Lamarckella antiqua* Kapt., *Recurvoides bartouxi* Said et Barak., *Ammobaculites suprajurassicus* (Schwag.), and *Globuligerina bathoniana* (Pazdro), and was assigned to the Bathonian Stage. In reality a species *Amijiella amiji* (Henson), considered by Kuznetsova et al. (1996) to appear at the base of the Bathonian, ranges from the Late Sinemurian to Late Bathonian (Perelis-Grossowicz et al., 2000).

Thus, the age attribution of the Bqassem Formation is interpreted in different ways by various authors. An uncertainty remains whether this formation starts within the Upper Bajocian, but all authors agree to a Bathonian age for the main part of it. The Bqassem Formation is everywhere followed conformably by the Hermon Formation.

The **Hermon Formation** is composed of a very thick, 700-1000 m, succession of thick- to coarse-bedded limestones. The pelitomorphic, aphanitic, oolitic, detrital limestones are brownish-gray or gray in fresh fracture, whereas they are typically bluish-gray on a weathered surface. Fragments and unbroken shells of brachiopods, gastropods, pelecypods and sometimes solitary and colonial corals are common. They greatly differ from the underlying limestones of the Bqassem Formation, in color, composition, and thickness. There are minor interbeds of dolomites, dolomitic limestones, and marls. According to G. Krymgolts and V. Makridin (in Razvalyaev, 1966a), the ammonites and brachiopods include *Erymnoceras* sp., *Praecyclothyris badensis* (Opp.), and *Septaliphoria moravica* Uhlig, which are indicative of the Callovian. Earlier, a similar opinion was expressed by H. Vautrin (1934a). L. Dubertret (1954) assigned the Hermon Formation to the Upper Bathonian-Callovian.

Assemblages of foraminifera differ in the lower and upper parts of the formation. Mouty and Zaninetti (1998) and Mouty (2000) identified a lower unit with *Paleopfenderina salernitana* (Sartoni et Crescenti), *P. trochoidea* (Smout et Sugden), *Kilianina blancheti* Pfender, *Endersonella palastiniensis* (Henson), and an upper unit rich in *Steinekella steinekei* Redmond, *S. crusei* Redmond, *Praekurnubia crusei* Redmond, and *Redmondoides lugeoni* (Septfontaine). This microfaunal difference between the lower and upper halfs of the succession probably convinced Mouty to attribute the age of the Hermon Formation to an undifferentiated Late Bathonian-Early Callovian interval.

The investigation of Kuznetsova et al. (1996) showed that the Hermon Formation is characterized by *Paleopfenderina salernitana* (Sartoni et Crescenti), *Redmondoides lugeoni* (Septfontaine), *Kurnubia palastiniensis* Henson, *Lituotuba nodus* Kosyr., *Ammobaculites suprajurassicus* (Schwag.), *Nautiloculina oolithica* Mohler, *Lenticulina polonica* (Wisn.), *Citharina entypomatus* Loebl. et Tapp., *Spirillina polyggrata* Guembel, and *Globuligerina calloviensis* K. Kuznetsova. These deposits were defined as the *Kurnubia palastiniensis-Globuligerina calloviensis* Zone and referred to the Callovian Stage.

Near the Drouze village of Majdal Shams, the top of the formation, more precisely its upper 17 m (Razvalyaev, 1966a),

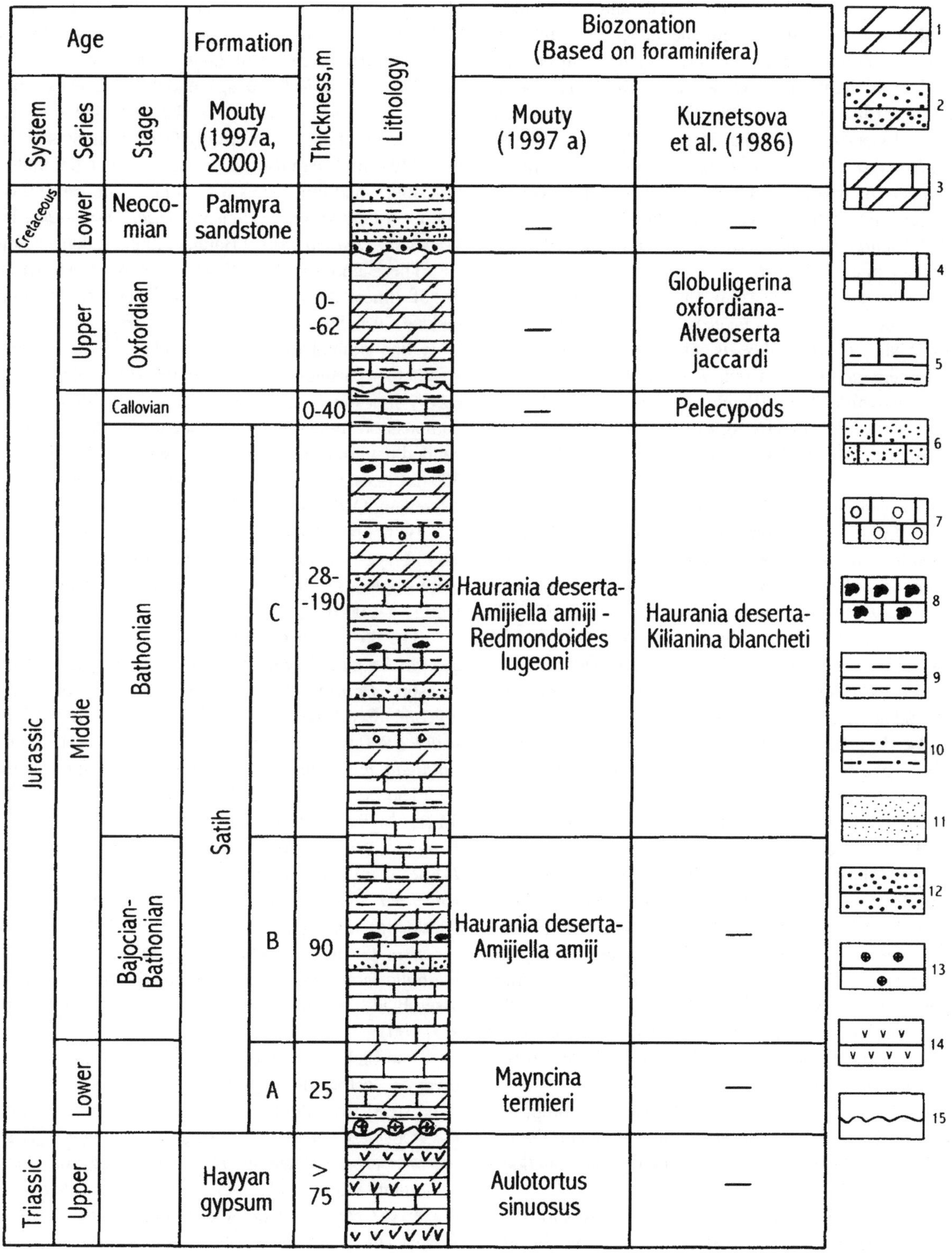

Fig. 6.21. Schematic composite section of the Jurassic deposits of the Palmyrides (after Razvalyaev, 1966a; Faradzhev, 1966; Soulidi-Kondratiev, 1966a; Kuznetsova et al., 1996; and Mouty, 1997a, 2000).

Legend: 1 – dolomite; 2 – sandy dolomite; 3 – calcareous dolomite and dolomitic limestone; 4 – limestone; 5 – clayey limestone; 6 – sandy limestone; 7 – oolitic limestone; 8 – silicified limestone; 9 – marl and calcareous clay; 10 – siltstone; 11 – sandstone; 12 – conglomerate and grit; 13 – kaolinite and bauxite; 14 – gypsum; 15 – erosional boundary.

consists of limestones with marly interlayers and identifiable fossils. This thin member differs from the underlying massive limestones by its pelitomorphic, clayey, usually thin-bedded character, and it is light-gray and white or yellowish on a weathered surface. Limestones yielded a rich fauna of ammonites, belemnites, brachiopods, and pelecypods. H. Vautrin (1934a) collected *Stepheoceras coronatum* Brug., *Reineckia* cf. *decorata* Waag., *Hecticoceras rossiensis* Teyss., *H. trilineatum* Waag., *Perisphinctes* cf. *steinmani* Par. et Bonn., *Belemnites hastatus* Blainv., and *Rhynchonella moravica* (Uhlig), and assigned these limestones to the Upper Callovian.

G. Krymgolts and V. Makridin (in Razvalyaev, 1966a) identified *Hecticoceras (Sublunuloceras) lunuloides* Kil., *H. (S.) dynastes* Waag., *H. (Putealiceras)* cf. *subrossiensis* Teyss., *Peltoceratoides regularis* Ger. et Cont., *Ivanoviella* ex gr. *arcuata* (Roll.), *Hibolites semihastatus* Blainv., *H. latesulcatus* d'Orb., *H. hastatus* Blainv., *Ptyctothyris subcanaliculata* (Opp.), and *Lobothyris* ex gr. *breviplicata (*Kitch.) and also concluded it was of Late Callovian age.

In a thin horizon (30 cm thick) of condensed platy limestone with phosphatic grains, rich in ammonites, Cariou (in Hirsch et al., 1998) has identified mostly flattened molds of Hecticoceratinae: *Pseudobrightia* sp., *Hecticoceras (Putealiceras) intermedium* (Spath), *H. (Kheraites) ferrugineus* (Spath), *H. (Brightia)* aff. *salvadorii* (P.et B.), *H. (Putealiceras) douvillei* (Jean.) and *Pachyerymnoceras levantinense* Lewy. *Pseudobrightia s.st.* precisely defines the base of the *Lamberti* Zone in Europe. The association found at the Hermon is of the same age, corresponding partly to the *Poculum* Subzone of the Submediterranean province. This strongly suggests the absence of the *Lamberti* Subzone that represents the uppermost part of the *Lamberti* Zone. The sharp contact with the overlying Majdal Shams Formation is characterizes by a sedimentary hiatus.

The 'Yellow Marls' of Wadi Al-Kharn (Dubertret, 1950c), exposed along the road Damascus Beyrouth near the village of Rowda, were studied in more detail by Mouty and Zaninetti (1998), Mouty (2000), and Almeras and Mouty (2001). These marls yield such Callovian brachiopods as *Somalirhrynchia africana* Uhlig and *Bihenithyris weiri* Muir-Wood and were correlated with the Majdal Shams Formation further south in the Anti-Lebanon. Since the Majdal Shams Formation is exclusively Oxfordian, such a correlation seems erroneous to us.

All the data indicate that the Hermon Formation encompasses the upper part of the Bathonian and the entire Callovian, the top of which consists of hard limestone with marly intercallations, yielding the Late Callovian ammonites, mentioned above.

In Jebel An-Nusseiriyeh the Middle Jurassic deposits form the top of the range. They are spread along its eastern slope, forming the steep escarpment of the Lebanon-Syrian fault, that bounds the Al-Ghab depression. On the western slope of Jebel An-Nusseiriyeh the Middle Jurassic deposits can be seen only in some deep valleys.

Aalenian-Bajocian (or Bajocian) dolomites, dolomitic limestones and aphanitic and detrital limestones with rare interbeds of clays, about 200 m thick, represent the Jdaida section. Organic remains are very poor. The impoverished assemblage of foraminifera consists of *Lenticulina centralis* (Terq.), *L. veta* Hoffman, *Pseudobolivina* sp., and *Globuligerina* sp. Beds with *Lenticulina centralis* confer a Bajocian age on these deposits (Kuznetsova et al., 1996). Mouty (1997b) assigns the up to 110 m thick **Ouyoun Formation,** consisting of thick dolomites, dolomitic limestones, and limestones with *Timidonella sarda* Bass., Chabr.et Fourc., *Amijiella amiji* (Henson), *Haurania deserta* Henson, and *Verneuilina minuta* Said et Barakat, to the Aalenian-Bajocian.

The Bathonian deposits of Jebel An-Nusseiriyeh are made up of limestones and dolomites. They are medium- to thin-bedded, usually light-gray, more rarely dark-gray or brown-gray, fine-grained to aphanitic, with peculiar banded, table-like weathering partings. These alternate with subordinate detrital and clayey limestones, marls, and clays. The difference in lithologic composition between the adjacent layers causes the unusual banded pattern of the whole succession, whose thickness is about 300 m (Shatsky et al., 1966). They are well exposed in Jdaida, Wadi Jahannam, Qadmous, Bab Janneh, and some other areas.

Bathonian shallow-water limestones and marls contain rich assemblages of foraminifera and ostracods. Two subdivisions were singled out by Kuznetsova et al.(1996) and Kuznetsova and Dobrova (2000):

The 85-140 m thick Lower Bathonian beds with *Haurania deserta-Protopeneroplis striata*, contain common *Haurania deserta* Henson, *Protopeneroplis striata* Weynschenk, *Redmondoides lugeoni* (Septfontaine), *R. primitivus* (Redmond), *Riyadhella arabica* Redmond, *Paleopfenderina trochoidea* (Smout et Sugden), *Amijiella amiji* (Henson) and *Dhrumella evoluta* Redmond.

The 90-150 m thick Upper Bathonian *Kilianina blancheti-Meyendorffina bathonica* Ecozone, includes *Kilianina blancheti* Pfender, *Meyendorffina bathonica* Aurouze et Bizon, *Paracoskinolina occinatica* Peyb., *Praekurnubia crusei* Redmond, *Kurnubia bramkampi* Redmond, *Berthelinella paradoxa* (Berthelin), *Limagnella dufaurei* Pell. et Peyb., *Mesoendothyra croatica* Gusic, *Pseudomarssonella bipartita* Redmond, and *Globuligerina bathoniana* (Pazdro). It should be noted that this assemblage may range into the Callovian (Perelis-Grossowicz et al., 2000).

The Bathonian deposits yielded numerous and taxonomically diverse ostracods. Among them were *Progonocythere honigsteini* Rosenfeld et Gerry, *Fastigatocythere bakeri* (Basha), *Glyptocythere huniensis* Basha, *G. oblonga* (Basha), *Procitheridea crassa* Peterson, *P. exempla* Peterson, *Ektyphocythere schulamitae* Rosenfeld et Gerry, *Praeschuleridea hornei* Rosenfeld et Gerry, *Cytherella index* Oertli, *Gliptogatocythere malzi* Basha, *Pleurocythere favosa* Tried., *Terquemula gublerae* (Bizon), *Fabanella sarda* Malz, and *Virgulacythere sherifensis* Oertli.

In its type locality, the lower 35 m of the up to 50 m thick **Machta Formation** consists of alternating nodular limestones and marls. Mouty (1997b) has identified *Amijiella amiji* (Henson), *Haurania deserta* Henson, *Paleopfenderina salernitana* (Sartoni et Crescenti), *P. trochoidea (*Smout et Sugden), *Nautiloculina* *circularis* (Said et Barakat), *Redmondoides lugeoni* (Septfontaine), and *Praekurnubia crusei* Redmond of Late Bathonian age.

In the 15-25 m thick upper part of the Machta Formation, Mouty (1997b) also found *Kilianina blancheti* Pfender, *Trocholina palastiniensis* Henson, *Praekurnubia crusei* Redmond, *Redmondoides lugeoni* (Septfontaine), and *Nautiloculina circularis* (Said et Barakat), indicating a Callovian age.

As a rule, the Callovian deposits at Jebel An-Nusseiriyeh show a gradual transition to the underlying Bathonian rocks. However in Bab Janneh, Braj, and Qadmous villages, an erosional unconformity separates Bathonian from Callovian strata. The Callovian of this area is composed of cryptocrystalline, oolitic, aphanitic, and bioclastic limestones with debris of brachiopods, pelecypods, echinoids, crinoids, algae, and solitary and colonial corals. Sometimes the limestones contain nodules and lenses of flints. Dolomites are subordinate. As is the case in the Qadmous area, intercalations of clayey limestones, marls, and clays are very rare. The absence of terrigenous admixture significantly differentiates the Callovian succession from the Bathonian.

On account of its wide distribution, the calcareous rocks of the Callovian are marked by abundant karst phenomena. One can see large and small holes, deep open fissures, and odd-shaped remnants. The thickness of the Callovian deposits does not exceed 60-80 m, and is sometimes reduced to 10-15 m. This feature strongly distinguishes the Callovian sequence of Jebel An-Nusseiriyeh from that of the Anti-Lebanon. At Shuekha and Bab Janneh, an insignifant amount of Callovian sediments was spared from Infra-Cretaceous erosional processes.

A high taxonomic diversity characterizes the foraminifera of the Callovian of this area. Its peculiarity is the appearance of tropical Pfenderinidae, including the genera *Praekurnubia, Paleopfenderina, Pfenderella, Sanderella*, and *Steinekella*. These define the *Kurnubia palastiniensis-Paleopfenderina salernitana* Ecozone of Kuznetsova et al. (1996). It is characterized by *Kurnubia palastiniensis* Henson, *K. bramcampi* Redmond, *Paleopfenderina salernitana* (Sartoni et Crescenti), *P. gracilis* (Redmond), *Ammodiscoides magharaensis* Said et Barakat, *Steinekella cruisei* Redmond, *Redmondoides lugeoni* (Septfontaine), *Reinholdella dreheri* (Bart.), *R. crebra* Pazdro, *Ammobaculites fontinensis* Terq., *Astacolus pellucidus* Said et Barakat, *Palaeogaudryina magharaensis* Said et Barakat, *Dorothia terquemi* Said et Barakat, *Verneuilinoides minuta* Said et Barakat, *Orthomorphina terquemi* Said et Barakat, *Rectoglandulina vulgata* (Born.), and *Globuligerina calloviensis* K. Kuzn.

Ostracods are also abundant in the Callovian of Jebel An-Nusseiriyeh. In the beds with *Ektyphocythere zoharensis*, Kuznetsova and Dobrova (2000) identified, among others, *E. zoharensis* Rosenfeld and Gerry, *Progonocythere* aff. *parastilla* Whatley, *Schulerideu triangularis* Swartz et Swain, and *Afrocythere faveolata* Bate.

At Wadi Al Ouyoun and near the village of Dalyeh, in the central part of Jebel An-Nusseiriyeh, solid 25-50 m thick biomicrites with rare intercalations of marls of the **Wadi Al Ouyoun Formation** overlie the Machta Formation and yield Callovian foraminifera (Mouty, 1997b). He identified foraminifera of the *Steinekella steinekei* biozone including *S. crusei* Redmond, *Kurnubia palastiniensis* Henson, *Trocholina palastiniensis* Henson, *Redmondoides lugeoni* (Septfontaine), *Nautiloculina circularis* (Said et Barakat), and *Syriana khouryi* Fourcade et Mouty. A diverse macrofauna encompasses *Gymnocidaris* aff. *pustulosa* (Agassiz), *Cidaropsis minor* (Agassiz), *Rhabocidaris* sp., *Kutchirynchia indica* (Orb.), *Tradigithyris thadiquiensis* Nazer, *Striithyris somaliensis* Muir-Wood, *Bihenithyris* sp., as well as pelecypods, gastropods, stromatoporoids, corals and the algae *Salpyngoporella annulata* Carozzi. This unit is absent near Machta, in the south of Jebel An-Nusseiriyeh, and near Slenfeh, in the north.

In the southwestern part of Kurd-Dagh, a continuous 1-2 km wide and 11 km long band of Jurassic deposits extends from the village of Kuemut, near the Syrian-Turkish border, to the northeast in the direction of Kerri and Banik Mazraa villages. Further to the northeast, isolated small Jurassic exposures occur near Rajou, Alamdar, and Samalek villages.

A lower unit consists predominantly of gray and dark-gray, aphanitic, fine-grained and medium-grained, medium-bedded (0.3-0.8 m) dolomites with rare intercalations of gray, microgranular and coarse-bedded limestones.

An upper unit, frequently forming steep slopes, is composed of rather thick to indistinctly bedded light-gray, almost white cryptocrystalline or bioclastic limestones with fragments of algae, corals, gastropods and pelecypods. They also contain occasional siliceous nodules. The top of the limestone has an irregular, intensively eroded surface with karst hollows. Due to numerous faults, it is difficult to estimate the true thickness, which is not less than a hundred meters. The lack of identifiable macrofauna led Protasevich and Maksimov (1966) to consider this Kurd-Dagh sequence as undifferentiated Middle-Upper Jurassic in age.

Micropaleontological studies have clarified the Jurassic stratigraphy of Kurd-Dagh. The oldest Middle Jurassic deposits are obviously Aalenian-Bajocian. These consist of micritic and organo-detritic limestones of the middle part of the **Dodo Formation** with *Timidonella sarda* Bassoulet (Mouty, 2000). Limestones and dolomites with *Haurania deserta* Henson, *Paleopfenderina trochoidea* (Smout et Sugden), *Dhrumella evoluta* Redmond, *Redmondoides primitivus* (Redmond), *R. inflatus* (Redmond), *Protopeneroplis striata* Weynschenk, and *Kurnubia* ex gr. *palastiniensis* Henson of the Rajou and Samalek sections were attributed to the Bathonian. A 5 m thick bed of pale-gray limestones with *Redmondoides lugeoni* (Septfontaine) and *Kurnubia* cf. *palastiniensis* Henson at the top of the Samalek section belongs to the Callovian (Kuznetsova et al., 1996). The Bathonian-Callovian deposits of Kurd-Dagh are directly overlain by Lower Cretaceous sediments, indicating a long hiatus.

In the southwestern part of the Palmyrides, the Middle Jurassic consists of the middle 'B' and upper 'C' members of the **Satih Formation** (Mouty, 1997a). They are exposed along the margin of the Palmyrides chain, on the southeastern slope of Jebel Ad-Dmeir and Jebel Zubeideh, and then along a northeastern direction to Jebel Rmah, Jebel An-Naqnaqiyeh, and Jebel Om Journ. They consist of alternating dolomites and limestones with subordinate sandstones and marls. The dolomites vary in texture and color. They are cryptocrystalline to coarse-grained, thin-platy to thick-bedded, sparkling, locally ferruginous, silicified or strongly bituminous. Their color is gray, dark-gray, brownish, yellowish-green to yellowish-gray. The diversity of limestones ranges from cryptocrystalline to oolitic, organic stromatoporoid, from medium- to thick-bedded and gray to dark-gray in color, sometimes clayey. Some thin interbeds of quartz sandstones and clays also occur. The total exposed thickness in Jebel Zubeideh is 113 m. A Middle Jurassic age was assumed for these unfossiliferous rocks, which are overlain by the Lower Cretaceous quartz sandstones. They can be correlated with the lithologically similar succession of the Anti-Lebanon (Razvalyaev, 1966a).

Later investigations of Kuznetsova et al. (1996) and Mouty (1997a, 2000) revealed an assemblage of foraminifera in the clayey, oolitic and poorly consolidated limestones, mainly of the upper member 'C' of the Satih Formation. It includes *Haurania deserta* Henson, *Amijiella amiji* (Henson), *Kilianina blancheti* Pfender, *Redmondoides lugeoni* (Septfontaine), *R. primitivus* (Redmond), *R. plicatus* (Redmond), *R. medius* (Redmond), *R. reflexus* (Redmond), *Paulina paula* (Pazdro), *Ammobaculites agglutinans* d'Orb., and *Nautiloculina oolithica* Mohler. This assemblage has a Bathonian age. Brachiopods *Dahanirhynchia subversabilis* (Weir.), *Kallirhynchia concinna* Sow., and *Burminrhynchia* aff. *tumida* Buckm., found in the same layers also identify these sediments as Bathonian (Kuznetsova et al., 1996). Mouty (1997a) added the brachiopod *Burmirhynchia termierae* Rousselle, the pelecypods *Sphaeriola madridi* d'Arch., *Plagiostoma* aff. *helica* (d'Orb), the gastropod *Nerinella scalaris* d'Orb. and the echinoids *Acrosalenia* aff. *angularis* (Agassiz), and *Trochotiara* sp. The middle member 'B' with *Haurania deserta* Henson and *Amijiella amiji* (Henson) is treated as Bajocian-Bathonian.

In the adjacent ridges of the Southern Palmyrides, Jebel Rmah, Jebel An-Naqnaqiyeh and Jebel Om Journ, unfossiliferous dolomites predominate. They consist of a 34 m thick gray, hard, well-bedded, aphanitic to medium-grained lower unit and a 18-20 m thick, pink and in places yellowish or nearly white, thick-bedded, hard, massive micrograined, slightly porous upper unit, sometimes silicified with dark varnish. They were conditionally assigned to the Bathonian. Above a strong erosional unconformity these dolomites are overlain by sandy-clayey Lower Cretaceous sediments, indicating a Callovian-Upper Jurassic hiatus.

In the northeastern part of the Palmyrides, small exposures of Middle Jurassic occur in the narrow cores of the anticlinal folds of Jebel Hayyan, Jebel El-Qayed, Jebel El-Mazar, Jebel Qalet El-Hirri, Jebel As-Satteh, and Jebel Thaniyyet As-Safra. Bathonian and Callovian have been encountered there.

Most common are the Bathonian deposits in Jebel Hayyan, Jebel El-Qayed, Jebel El-Mazar, and Jebel Qalet El-Hirri. They are composed of essentially unfossiliferous gray and yellow-gray, fine-banded dolomites and calcareous dolomites passing upward in the section into gray and brown, pelitomorphic and fine-crystalline, thick- and coarse-bedded, silicified dolomitic limestones and calcareous dolomites. One horizon of clayey and sandy dolomites contains numerous pelecypods: *Sphaeriola madridi* Arch., *Sph. obovata* Laube, *Sph.* aff. *crassicostata* d'Orb., *Astarte* sp., and *Panopaea* sp., which indicate the Bathonian according to A. Yungerman (in Soulidi-Kondratiev, 1966a). These deposits abruptly change from a thickness of 190 m in Jebel El-Mazar to 28 m in Jebel Hayyan. Such thickness variations are due to the truncation of the Jurassic rocks by the subsequent Lower Cretaceous transgression. Callovian and Upper Jurassic sediments are entirely missing in this region.

Callovian deposits in the northeastern Palmyrides are very limited. Only in Jebel As-Satteh and Jebel Thaniyyet As-Safra, are there gray and light-gray micrograined limestones, marls and clayey limestones about 10 m thick, yielding an assemblage of pelecypods: *Anisocardia laubei* Roll., *Ceromya calloviensis* Kazansk., *Pholadomya murchisoni* Sow., *Ph. subexaltata* Kazansk., *Ph.* ex gr. *hemicardia* Roem., *Heligmus* ex gr. *polyptypus* Desl., *Lima* ex gr. *acutirostris* Eichw., and *Idma antirostris* Richmond., indicating a Callovian age (det. Z. V. Koshelkina, in Soulidi-Kondratiev, 1966a).

In Jebel An-Naqnaqiyeh, Jebel El-Abtar, Jebel Hayyan, Jebel El-Mazar, Jebel El-Qayed, and Jebel Qalet El-Hirri, the up to 75 m thick alternating gypsum, gypsiferous clays and dolomites, lying below the Bajocian-Bathonian deposits and barren of fauna, were tentatively assigned to the Bajocian (Faradzhev, 1966; Soulidi-Kondratiev, 1966a). More recently, in the section at Jebel Hayyan, palynological investigations by O. Yaroshenko and I. Bash Imam (1995) revealed a complex of miospores indicating a Late Triassic (and probably Carnian) age.

Upper Jurassic (Malm)

The Upper Jurassic deposits of Syria consist of Oxfordian, Kimmeridgian, and Tithonian strata. Their very distinct lithology and characteristic paleontological constituents delineate the palaeogeography of the Upper Jurassic in Syria. Their spatial distribution of facies is highly variable.

Oxfordian

Oxfordian deposits have a wider distribution than the Kimmeridgian and Tithonian ones. They are well-exposed in the Anti-Lebanon, Jebel An-Nusseiriyeh, and the Palmyrides. In Kurd-Dagh Oxfordian sediments are absent.

The most complete sections of the Oxfordian were encountered in the Anti-Lebanon where they follow the strike of the southeastern flank. The southern deep-water type of deposits, bordering the Hermon Massif (also known as Jebel Ash-Sheikh) differs significantly from the northern shallow-water type at Jebel Sheqif El-Khorn and Jebel Mannsour.

The classical Oxfordian section is located in the southwestern part of Jebel Ash-Sheikh, near the village of Majdal Shams, where this stage is composed of marls with thin intercalations of organo-clastic limestones. Dark-gray and black marls, which form thin platy debris in the process of weathering, predominate. Higher up in the section, the number of limestone interbeds increases. This sequence is known as the **Majdal Shams Formation** (Dubertret, 1960). It contains extremely rich and diverse faunas which have been studied by many scientists. In chronological order these include: Noetling, 1887; Vautrin, 1934a; Haas, 1955; Arkell, 1956; Dubertret, 1960; G. Krymgolts, V. Makridin and A. Yungerman (in Razvalyaev, 1966a and Ponikarov et al., 1967); A. Parnes (in Picard and Hirsch, 1987); Moshkovitz and Ehrlich, 1976b; Rosenfeld et al., 1987; Conway, 1990; and Cariou (in Hirsch et al., 1998).

Based on these paleontological studies, the Oxfordian can be subdivided into substages and several ammonite zones.

The present description is mainly based on the stratigraphic investigations carried out in 1958–1961 during the geological mapping by USSR geologists (Razvalyaev, 1966a; Ponikarov et al., 1967).

The Oxfordian section was measured in the vicinity of the Drouze village of Majdal Shams, where it reaches a thickness of 279 m.

The Lower Oxfordian consists of two ammonite zones, as based on the determinations of G. Krymgolts:

The *Creniceras renggeri* Zone

1) At the base of this zone, a 3 m thick bed of dark-gray marls contains the following large ammonites: *Hecticoceras (Putealiceras) schumacheri* Noetl., *H. (P.) caelatum* Lor., *H. (P.) solare* Haas, *H. (P.) separandum* Haas, *H. (P.) punctatum* Stahl., and *Euaspidoceras lyra* Spath.

2) A thin (0.1 m) layer of gray organo-clastic limestone.

3) An 8 m thick member of dark-gray marls with numerous ammonites: *Phylloceras* cf. *plicatum* Neum., *Calliphylloceras schems* Noetl., *Sowerbiceras helios* Noetl., *Hecticoceras (Lunuloceras)* cf. *lunula* Rein., *H. (Sublunuloceras) guttei* Noetl., *H. (S.) socium* Haas, *Taramelliceras hermonis* Noetl., *T.* cf. *richei* Lor., *Creniceras renggeri* Opp., *Mirosphinctes syriacus* Noetl., *M. regularis* Noetl., *Euaspidoceras subcostatum* Spath, *E. lyra* Spath, *E. perisphinctoides* Sintzov, *E. subbabeanum* Sintzov, and *E. hermonis* Haas.

4) A thin (0.1-0.2 m) layer of gray organo-clastic limestones.

5) Dark-gray and black, platy marls, 32 m thick, with small ammonites, belemnites and pelecypods: *Creniceras renggeri* Opp., *Lissoceras erato* d'Orb., *Hecticoceras (Sublunuloceras) socium* Haas, *Hecticoceras (Brightia) syriacum* Haas, *H. (B.) socini* Noetl., *H. (B.) kautzschi* Noetl., *Ochetoceras (Campilites) evolutum* Haas, *O. (C.) delmontanum* Opp., *Taramelliceras (Taramelliceras) richei* Lor., *T. (T.) regenularis* Noetl., *T. (Proscaphites) hermonis* Noetl., *T. (P.) globosum* Lor., *Properisphinctes delicatulus* Haas, *P. vicinus* Haas, *P. bernensis* Lor., *P.* cf. *filocostatus* Haas, *P. radiocostatus* Haas, *P. hermonis* Haas, *P. latilimguatus* Noetl., *Euaspidoceras subcostatum* Spath, *E. perisphinctoides* Sintzov, *E. hermonis* Haas, *E.* cf. *douvillei* Collot., *E. lyra* Spath, *E. subbabeanum* Sintzov, *Phylloceras riazi* Lor., *Ph.* cf. *plicatum* Neum., *Sowerbyceras helios* Noetl., *Scaphitoides scaphitoides* Coq., *Calliphylloceras schems* Noetl., *Microsphinctes regularis* Noetl., *M. syriacus* Noetl., *Hibolites gillieroni* Mayer, and *H.* cf. *bzoviensis* Leuschn.

The *Aspidoceras perarmatum* Zone

6) Thin intercalation of gray, dark-gray and light-gray, thin-bedded limestones and medium-bedded marls containing plant imprints and fragments of carbonaceous matter. The thickness is 57 m. The fauna is not as abundant as in the underlying beds. The following ammonites have been identified here: *Aspidoceras perarmatum* Sow., *Collyrites bicordata* Leske, *Euaspidoceras waageni* Spath, and *Peltoceratoides indicus* Spath. The same assemblage of ammonites was recorded earlier by H. Vautrin (1934a).

The total thickness of the Lower Oxfordian is 119 m.

The Upper Oxfordian is also subdivided into two ammonite zones, according to G. Krymgolts. Brachiopods and pelecypods are very common in these layers.

The *Gregoriceras transversarium* Zone

7) Thin-bedded, organo-clastic limestones with interbeds of marls. The most common fossils are *Collyrites bicordata* (Leske), *Hibolites astartinus* Thurm., *Lobothyris bisuffarcinata* (Schloth.), *L. subsella* (Leym.), *L.* ex gr. *zieteni* (Lor.), *Postephyris* cf. *haasi* (Roll.), *Septaliphoria (Septaliphoria)* cf. *jordanica* (Noetl.), and *Chlamys (Aequipecten) fibrosa* (Sow.). The thickness is 21 m.

8) Black marls with rare and thin (0.1 m) intercalations of brown pelitomorphic limestones. Marls contain carbonaceous plant debris and shells of small ammonites and pelecypods: *Perisphinctes (Dichotomosphinctes)* cf. *falcuta* Rouch., *P. (Arisphinctes)* cf. *laevipickeringius* Arkell, *Nucula inconstans* Roeder, and *N.* aff. *menkei* Roeder. The thickness is 70 m.

The *Peltoceras bimammatum* Zone

9) A series of intercalations of dark-gray and black marls, gray pelitomorphic and clayey limestones with *Perisphinctes (Progeronia)* cf. *aladiensis* Schoffat., *Lobothyris zieteni* Lor., *Pholadomya potei* Agass., *Chlamys (Aequipecten) laurae* (Etallon), *Ch. (A.) subarmatus* Muenst., *Pleuromya* cf. *tellina* Agass., *Goniomya* ex gr. *dubois* Agass., *Astarte* sp., *Parallelodon* sp., and *Plicatula* cf. *kobyi* Lor. The thickness is 42 m.

10) Gray and dark-gray, detrital and pelitomorphic limestones and clayey limestones with *Lobothyris subsella* (Leym.), *L. bisuffarcinata* (Schloth.), *Septaliphoria (Septali-phoria) jordanica* (Noetl.), *Somalirhynchia africana* Weir., *S. somalica* (Dacque), *Camptonectes lens* (Sow.), *Nucleata* cf. *nucleata* (Quenst.), *Posterithyris* sp., and *Isocardia* sp. The thickness is 27 m.

The total thickness of the Upper Oxfordian reaches 160 m.

To the northeast, along the strike of the Anti-Lebanon, the Oxfordian deposits display pronounced facies changes as the clayey-marly sediments with ammonites and belemnites are replaced by more shallow-water limestones with brachiopods, pelecypods, corals, algae and crinoids. They also decrease in thickness.

Below Jebel Ash-Sheikh near Hadar village, northeast of Majdal Shams, the Lower Oxfordian and basal part of the Upper

Oxfordian consists predominantly of gray, thin-bedded and laminated clays and marls. Intercalations of cryptocrystalline, clayey limestones with abundant fragments of pelecypods and debris of ammonites and bioclastic limestones with algae and corals remain constant elements in this sequence. The upper part of the Hadar section consists of various limestones including reefal edifices. The thickness of the Oxfordian Hadar section is 87 m. This section is very rich in foraminifera which Kuznetsova et al. (1996) have divided into two separate zones:

1) The *Lenticulina brueckmanni-Globuligerina oxfordiana* Zone (Lower Oxfordian) with *Ammobaculites coprolithiformis* Schwag., *Pseudocyclammina mainci* Hottinger, *Lenticulina quenstedti* (Guemb.), *L. russiensis* Mjatl., *Astacolus vacillantes* Esp. et Sigal, *Planularia beierana* (Guemb.), *Citharina entypomatus* Loebl. et Tapp., *C. paralella* Biel. et Poz., *Marginulinopsis suprajurassicus* (Schwag.), *Conorboides paravalendisensis* Reiss, *Spirillina kuebleri* Mjatl., *Epistomina nemunensis* Grig., *E. porcellanea* Brueckm., and *Gloguligerina oxfordiana* (Grig.);

2) The *Alveosepta jaccardi-Lenticulina quenstedti* Zone (Upper Oxfordian) with *Reophax sterkii* Haeusl., *Ammobaculites suprajurassicus* (Schwag.), *Palaeogaudryina varsoviensis* (Biel. et Poz.), *Trocholina transversarii* Paalz., *Lenticulina quenstedti* (Guemb.), *Astacolus vacillantes* Esp. et Sigal, *Citharina entypomatus* Loebl. et Tapp., *C. lepida* (Schwag.), and *C. paralella* (Biel. et Poz.) Farther to the northeast in Jebel Sheikh Mannsour, Jebel Sheqif El Khorn, Jebel Ain Ansour, Jebel Sheqif, and Jebel Ayoun Berdi, the marly sediments are increasingly replaced by limestones.

In Wadi El Khorn (Wadi El Kharn) along the Damascus-Beyrouth highway, a 105 m thick lower unit consists of dense cryptocrystalline or oolitic and bioclastic limestones with fragments of pelecypods, brachiopods, crinoids and foraminifera. Kuznetsova et al. (1996) identified *Reophax sterkii* Haeusl., *Alveosepta jaccardi* (Schroedt.), *Marssonella jurassica* Mitjan., and *Epistomina* cf. *volgensis* Mjatl. The 60 m thick upper unit consists of alternations of greenish-yellow marls, dense cryptocrystalline or bioclastic limestones and clayey limestones with abundant foraminifera such as *Alveosepta jaccardi* (Schroedt.), *Reophax horridus* (Schwag.), *Haplophragmoides rotundatus* Bart., *Paulina furssenkoi* Grig., *Ammobaculites coprolithoides* Schwag., *A. hagni* Bhalla et Abbas, *A. sequanus* Mohler, *Nautiloculina oolithica* Mohler, *Haplophragmium monstratus* (Dain), *Spirillina kuebleri* Mjatl., *Turrispirillina amoena* Dain, *Marssonella metaeformis* Esp. et Sigal, *Verneuilinoides minuta* Said et Barakat, *Palaeogaudryina varsoviensis* (Biel. et Poz.), *Lenticulina hebetata* (Schwag.), *L. nostra* Grig., *L. quenstedti* (Guemb.), *Astacolus vacillantes* Esp. et Sigal, *Planularia cordiformis* (Terq.), and *Citharina sokolovi* Mjatl. Kuznetsova et al.(1996) put the lower and upper units into the Lower and Upper Oxfordian respectively.

From the limestones in the uppermost layers V. Makridin identified the following brachiopods: *Somalirhynchia africana* Weir., *Terebratula* ex gr. *rauraciensis* Roll., *T. subinsignis* Etall., and *Eobothyris bisuffarcinata* (Shloth.) which confirm the Late Oxfordian age of these sediments (Razvalyaev, 1966a).

In the same Oxfordian succession L. Dubertret (1949a) found *Rhynchonella moravica* Uhlig, *Terebratula bisuffarcinata* Zieten, *T. pyroidea* Kitchin, *Alectryonia hastellata* (Schloth.), and *A. pulligera* Sow.

The Upper Oxfordian section near Rowda is characterized by some peculiar features. Its 60 m thick lower part consists of alternating clayey medium-bedded limestones, greenish-gray marls, and clays. In places limestones are loaded with shells and debris of brachiopods, pelecypods, corals, and algae. The clayey rocks contain numerous foraminifera such as *Reophax sterkii* Haeusl., *Alveosepta jaccardi* (Schroedt.), *Haplophragmium subaequale* (Mjatl.), *H. lutzei* Hanzl., *Pseudobolivina aegyptiaca* (Said et Barakat), *Marssonella dumortieri* (Schwag.), *Lenticulina hebetata* (Schwag.), *L. russiensis* (Mjatl.), *Astacolus vacillantes* Esp. et Sigal, *Planularia tricarinella* (Reuss), *Marginulinopsis suprajurassicus* (Schwag.), *Citharina sokolovae* Mjatl., *C. flabelloides* (Terq.), *Paalzowella feifeli* (Paalz.), and *Epistomina praereticulata* Mjatl. Continuing upwards there are 25 m of dense, cryptocrystalline and slightly clayey limestones with *Mesoendo-thyra croatica* Gusic, *Quinqueloculina tersa* Danitch, and *Protopeneroplis striata* Weynsch. The Upper Oxfordian terminates with a 120 m thick member of dense, cryptocrystalline, unfossiliferous limestones, dolomitized limestones and dolomites (Kuznetsova et al., 1996).

In the area of Zebdani, Bloudan, and Serrhaya villages, in the northeastern part of Jebel Cheqif, the Oxfordian is completely represented by carbonatic deposits. These 160 m thick oolitic, fine-detrital organogenic, coral and dense cryptocrystalline limestones form topographic escarpments. The best sections are located in Wadi Fawuar and Wadi Mannsour. Foraminifera include *Globuligerina oxfordiana* (Grig.), *Trocholina* cf. *transversaria* Paalz., and *Citharina lepida* (Schwag.) which are the most characteristic species (Kuznetsova et al., 1996). V. Makridin identified the brachiopods *Lobothyris valfinensis* (Lor.), *L. bisuffarcinata* (Schloth.), *Septaliphoria (Septaliphoria) jordanica* (Noetl.), *Moisseevia* aff. *rollieri* (Lor.), *Somalirhynchia africana* Weir., *Terebratula subinsignis* Etall., *T.* ex gr. *rauraciensis* Roll., *T.* aff. *dobrogiaca* Sim., *T. hossingensis* Roll., and *T. subformosa* Suess (Razvalyaev, 1966a).

Evidently, H. Vautrin (1934a) was the first to notice the difference in lithology and thickness of the Oxfordian deposits between the Hermon Massif (Jebel Ash-Sheikh) and the Jebel Sheqif region. He explained this geological situation as a result of the transgresive nature of the Oxfordian, leading to the absence of the lower marly unit in some sections at Jebel Sheqif. On the other hand, L. Dubertret's (1949a) observations during his mapping activity led him to assume that the sediments changed facies. A thorough study of the macrofauna, ammonites and brachiopods in the process of our geological mapping proved that Dubertret's assumption was correct (Razvalyaev, 1966a). Later, the same conclusion was reached on the basis of micropaleontologic studies (Kuznetsova et al., 1996).

In Jebel An-Nusseiriyeh the Oxfordian rests conformably on the Callovian. Its distribution is restricted to the southern part of this range. It is exposed in the vicinity of Marmarita, Kafroun Dash Shout, and Ash Shmeiseh villages (Kozlov, 1966a). North of this region, Oxfordian outcrops are located near Ain Khlaken, Sindianeh, and Bekraka villages, and along Wadi Shkeir (Kuznetsova et al., 1996).

A 10 to 70 m thick lower unit, represented by alternating yellow and yellowish-gray lumpy clays and various limestones (gray, dense, cryptocrystalline, soft clayey, thick-bedded and with fragments of pelecypod and brachiopod shells), yielded the foraminifera of the *Lenticulina brueckmanni–Globuligerina oxfordiana* Zone with *Alveosepta jaccardi* (Schroedt.), *Kurnubia palastiniensis* Henson, *K. wellingsi* Henson, *Lenticulina brueckmanni* (Mjatl.), *Pseudocyclammina* sp., *Everticyclammina* sp., *Nautiloculina oolithica* Mohler, *Verneuilinoides* cf. *osowiensis* Biel., *Marssonella dumortieri* (Schwag.), and *Globuligerina oxfordiana* (Grig.) The ~20 m thick upper unit yielded foraminifera of the *Lenticulina quenstedti-Alveosepta jaccardi* Zone (Kuznetsova et al., 1996). In the Jebel An-Nusseiriyeh area this thin upper unit is absent in some sections at Al Bustan, Ain Khlaken, and Wadi Shkeir. Their absence, together with the Kimmeridgian and the Tithonian, is due to pre-Cretaceous erosion. The Early Cretaceous terrigenous sediments overlie Oxfordian or even Callovian rocks with sharp erosional unconformity.

Mouty (1997b) refers the Oxfordian deposits of Jebel An-Nusseiriyeh to the 15-30 m thick **Nasirah Formation**, consisting of two members. The lower member is made up of biomicritic limestones with subordinate interlayers of marls. In the upper member marls predominate. The stratotype of this formation is located near the village of Nasirah, in the vicinity of the ruins of the Krak des Chevaliers fortress. Abundant foraminifera such as *Alveosepta jaccardi* (Schroedt), *A. powersi* (Redmond), *A. praelusitanica* (Maync), *Kurnubia palastiniensis* Henson, *K. willengsi* (Henson), *Levantinella egyptiensis* Fourcade et al., *Everticyclammina virguliana* (Koechlin), and *Nautiloculina circularis* (Said et Barakat) belong to the *Alveosepta jaccardi* biozone (Late Oxfordian-Kimmeridgian). Lower Aptian ferruginous clays and sands overlie the Nasirah Formation.

Small outcrops of the Oxfordian are known in the ridges of Jebel As-Satteh and Jebel Thaniyyet As Safra in the northeastern part of the <u>Palmyrides</u>, where ~12 m thick clayey and dolomitic limestones grade upwards into massive dense unfossiliferous dolomites, forming a cliff about 50 m high (Soulidi-Kondratiev, 1966a). At the base of this unit, in a 3 m thick layer of marls and clayey limestones Kuznetsova et al.(1996) found the Oxfordian foraminifera *Everticyclammina virguliana* (Koechlin), *Ammo-baculites coprolithiformis* Schwag., *Palaeogaudryina magharaensis* Said et Barakat, *Verneuilinoides minuta* Said et Barakat, *Lenticulina subalata* (Reuss), *Planularia beierana* (Guemb.), *Globulina oolithica* Terq., *Patellina douvillei* Said et Barakat, *Conorboides* sp., and *Globuligerina oxfordiana* (Grig.). Another small outcrop of Oxfordian deposits was found in the southwestern margin of the Palmyrides where dolomites and oolitic, bioclastic and stromatoporoid limestones contain the foraminifera *Globuligerina oxfordiana* (Grig.), *Pseudocyclam-mina maynci* Hottinger, *Alveosepta jaccardi* (Schroedt.), *Everticyclammina virguliana* (Koechlin), and *Planularia beierana* (Guemb.).

Kimmeridgian

In Syria, the Kimmeridgian has a limited distribution and is composed exclusively of limestones and lesser amounts of marls. They are encountered only in the Anti-Lebanon, in Mt. Hermon-Jebel Ash-Sheikh, Jebel Mannsour, and Jebel Sheqif as well as in the southern part of Jebel An-Nusseiriyeh. Everywhere the Kimmeridgian rests conformably on the Oxfordian.

In the <u>Anti-Lebanon</u>, two units are attributed to the Kimmeridgian. They are equivalent respectively to the subdivisions J6 and J7 of Dubertret (1960), the subdivisions J3km and J3t of Razvalyaev (1966a), the Bloudane Formation of Mouty and Zaninetti (1998) and the Batroun Formation (Dubertret, 1960; Mouty and Zaninetti, 1998), and are lumped together in the Nahal Sa'ar Formation of Saltzman et al. (1968) and Hirsch et al. (1998).

The **Bloudane Formation** is built up of gray and light-gray, massive, very dense organogenic limestones with abundant colonial and solitary corals, brachiopods, pelecypods, gastropods, stromatopores, and numerous debris of sea urchins. Among the organogenic limestones are found bands of cryptocrystalline laminated limestones (0.8-1.2 m) and very thin impersistent seams of marls. The massive reefal limestones are expressed as cliffs and escarpments with steep slopes, forming an excellent marker on the landscape. F. Noetling (1887) gave the name 'the Zone with *Cidaris glandaria* Goldf.' to the reefal limestones of the Kimmeridgian exposed near the village of Majdal Shams; sea urchins are associated here with the brachiopod *Terebratula bisuffarcinata* Zieten. H. Vautrin (1934a) called the reefal limestones as 'limestones with *Balanocidaris glandifera* Muenst'. The cliff reaches a height of 38 m at Jebel Ash-Sheikh and 30 m at Jebel Mannsour and Jebel Sheqif El-Khorn (Razvalyaev, 1966a).

Various groups of fossils have been identified in the Kimmeridgian of the Anti-Lebanon.

In the lower member of the Wadi Mannsour section (the Jebel Sheqif El-Khorn area) V. Makridin identified *Somalirhynchia africana* Weir., *Lobothyris subsella* (Leym.), and *Septaliphoria* ex gr. *inconstans* (Sow.) (Razvalyaev, 1966a). L. From limestones of this outcrop Dubertret (1955a) collected *Pholadomya moravica* Uhlig, *Ph. acuminata* Zieten, and *Ph. cardissoides* Agass.

According to Mouty and Zaninetti (1998), in the area of Rawda and Bloudane, the Bloudane Formation is characterized by *Alveosepta jaccardi* (Shroedt), *A. powersi* (Redmond), *Levantinella egyptiensis* Fourcade, Arafa et Sigal, *Parurgonina caelinensis* Cuvillier et al., and *Kurnubia palastiniensis* Henson which belong to the Late Oxfordian-Kimmeridgian.

At the top of the Kimmeridgian Mouty and Zaninetti (1998) singled out the **Batroun Formation** ('Calcaire de Batroun', according to Dubertret, 1960). These mainly organo-clastic and oolitic yellow and rusty-yellow colored limestones, which are sometimes clearly cross-bedded, reach a thickness of 34 m at Jebel Ash-Sheikh, 15-20 m in Jebel Mannsour and Jebel Sheqif El-Khorn (Razvalyaev, 1966a).

Mouty and Zaninetti (1998) report that *Alveosepta jaccardi* (Shroedt), *A. powersi* (Redmond), and *Kurnubia palastiniensis* Henson are accompanied by *Everticyclammina virguliana* (Koechlin) and *Alveosepta praelusitanica* (Maync). From these oolitic limestones near Majdal Shams, Vautrin (1934a) collected *Pyguropsis noetlingi* Lor., *Terebratula subsella* Leym., *Ceromya excentrica* Agass., *Natica dido* Krumb., and *Mytilus subpectinatus* d'Orb. South of Qalaat Nimrod, near the village of Ein Qinya at the southwestern end of Mt. Hermon, organo-clastic limestones with *Pinna loustani* Lor., *Thracia incerta* (Desh.), *Inoperta perplicata* (Etallon) and *Pholadomya paucicosta* Roem. have been found (det. A. Yungerman).

Interbeds of marls and clayey limestones in the Wadi El-Khorn and Wadi Fawuar sections yielded an assemblage of foraminifera consisting of *Alveosepta personata* (Tobler), *Reophax hounstoutensis* Lloyd, *Triplasia jurassica* Mjatl., *Marssonella doneziana* Dain, *M. hechti* Dieni et Mass., *Citharina lepida* (Schwag.), *Verneuilinoides* cf. *kirillae* Dain, *Mesoendothyra croatica* Gusic, *Quinqueloculina jurassica* Biel. et Styk., *Torinosuella peneropliformis* (Yabe et Hanzawa), *Haplophragmoides nonioninoides* (Reuss), *Triplasia elegans* (Mjatl.), and *Gaudryina vadazi* Cushm. et Glaz. Layers with this microfauna are treated by Kuznetsova et al. (1996) as the *Alveosepta personata* Zone of Kimmeridgian age.

In his monograph W. Arkell (1956) summarized data of some previous authors, and pointed out that the reefal limestones with *Balanocidaris glandifera* Muenst. also contain *Phylloceras sabina* Humb., *Nautilus turcicus* Krumb., *Aspidoceras* sp. and that overlying oolitic limestones included *Terebratula subsella* Leym. and *Mytilus subpectinatus* d'Orb. Everyone agrees that the deposits are Kimmeridgian in age.

In Jebel An-Nusseiriyeh the Kimmeridgian has only been found in the very south of this range, near the village of Marmarita in the Nicola Quarry. Here it is represented by thin (5 m) brownish clays and pale-gray microcrystalline limestones with *Alveosepta personata* (Tobler), *Proteonina difflugiformis* (Brady), *Ammobaculites pellucidus* Said et Barakat, *A. magharaensis* Said et Barakat, *Lenticulina subtilis* (Wisn.), *Nautiloculina oolithica* Mohler, and *Pseudocyclammina* sp. which make up the *Alveosepta personata* Zone (Kuznetsova et al., 1996).

Tithonian

Tithonian exposures occupy very limited areas in Syria. They have been encountered only in the central part of the Anti-Lebanon and at Jebel An-Nusseiriyeh.

In the central part of the Anti-Lebanon the Tithonian sequence is made up of gray pelitomorphic limestones, consisting of a 6 m thick band of greenish-gray marls. The total thickness is 20 m (Razvalyaev, 1966a).

Its most complete section can be observed along Wadi El-Khorn and Wadi Fawuar, as well as near Rowda. In the Wadi Fawuar section, the Tithonian rests on an erosional surface, filling pockets within the underlying Kimmeridgian deposits.

Macrofauna in the Tithonian deposits of the Anti-Lebanon is scarce. Rare *Ceromya* ex gr. *excentrica* Agassiz are present in limestones near Bloudan (Razvalyaev, 1966a; Kurbanov and Zarjanov, 1968).

Tintinnids occur in microcrystalline limestones of the Wadi El-Khorn section: *Calpionella alpina* Lorenz, *C. elliptica* Cadisch., *Crassicolaria intermedia* Durand-Delga, and *Tintinnopsella carpatica* (Murg. et Filip.).

Foraminifera are abundant in the marls and clayey limestones of the sections at Wadi El-Khorn, Wadi Fawuar, and Rowda. They include: *Anchispirocyclina lusitanica* (Egger), *Alveosepta powersi* (Redmond), *Gaudryina vadazi* Cushm. et Glaz., *G. neocomica* Chalilov, *Pseudocyclammina parvula* Hottinger, *P. lituus* (Yokoyama), *Melathrokerion eospiralis* Gorb., *Stomatostoecha compressa* Gorb., *Verneuilina angularis* Gorb., *Trochammina neocomiana* Mjatl., *Ophthalmidium inflatum* (Anton.), *Globospirillina caucasica* (Hoffm.), *G. neocomiana* (Moull.), *Reophax folkestonensis* Chapm., *Ammobaculites aequale* (Roem.), *Recurvoides limbatus* (Chapm.), *Palaeotextularia crimica* Gorb., *Nautiloculina oolithica* Mohler, *Choffatella decipiens* Schlumb., *Dorothia praehauteriviana* Dieni et Massari, and *Belorussiella taurica* Gorb. This microfauna is considered to belong to the *Anchispirocyclina lusitanica* Zone (Kuznetsova et al., 1996).

In Jebel An-Nusseiriyeh Tithonian deposits were only found in the vicinity of Qadmous. They consist of ~30 m of alternating yellowish-gray clay, sandy clay, dense cryptocrystalline thick-bedded limestones and gray clayey limestone. They rest with sharp erosional unconformity on Callovian deposits and are overlain by Lower Cretaceous strata. Soft clays are rich in foraminifera such as *Pseudocyclammina sulaiyana* Redmond, *P. cylindrica* Redmond, *Stomatostoecha* aff. *enisalensis* Gorb., *Melathrokerion eospiralis* Gorb., *Charentia* aff. *evoluta* Gorb., *Choffatella decipiens* Schlumb., *Bramkampella arabica* Redmond, and *Feurtillia frequens* Maync. According to Kuznetsova et al. (1996), the microfauna already includes Neocomian foraminifera and the section should be considered as Tithonian-Berriasian.

Upper Jurassic rocks are evidently present in the Jebel Aqraa and the Qara Douran mountains on the border with Turkey. In this area a sequence of monotonous pelitomorphic and microgranular, very hard and dense limestones about 450 m thick is found. On the Turkish side L. Dubertret (1936, 1955b) reported *Mytilus (Pachymytilus) crassissimus* Boehm. The whole succession of these monotonous limestones was assigned to the undifferentiated Middle-Upper Jurassic (Shatsky et al., 1966; Kazmin and Kulakov, 1968).

CONCLUDING REMARKS

The stratigraphically most complete sections of the Jurassic are located in the Anti-Lebanon, where their thickness is nearly 2000 m. The Jurassic deposits of Jebel An-Nusseiriyeh have much in common with those of the Anti-Lebanon, but some facies changes are obvious; the thickness decreases, and the Kimmeridgian and Tithonian are essentially missing. Schematic composite sections of the Jurassic deposits of these two areas (**Figs. 6.19** and **Fig. 6.20**) show these similarities and differences. The Bathonian and

Callovian are developed mainly in the Palmyrides (**Fig. 6.21**) and Kurd-Dagh, and the Upper Jurassic, if present at all, forms only small exposures.

The geological history of the Jurassic marine basin in Syria can be outlined as three stages of evolution:

1) During the Early Jurassic and early Middle Jurassic, the regressive phase of the Triassic/Jurassic transitional interval is reflected in the dominance of a terrigenous-calcareous regime, within a palaeogeographically limited area of Syria;

2) In general the late Middle Jurassic (Bajocian, Bathonian, Callovian) is characterized by widespread transgression and the development of dominant calcareous facies. In the early Upper Jurassic (Oxfordian) this transgressive trend culminates in the development of the deeper Majdal Shams marls, in the south-westernmost part of Syria;

3) The Kimmeridgian and Tithonian are marked by pronounced regression and the development of shallow-water calcareous sediments, including organo-detrital and reefal limestones.

Relief map of Lebanon from:
www.lib.utexas.edu/Libs/PCL/Map_collection/middle_east_and_asia/lebanon_rel_2002.jpg.

The Hermon Massif, composed of Jurassic limestones and dolerites. A large slump produced the pediment east of the mountain in the direction of Damascus. Oblique computer generated image from the DTM of Israel and adjacent areas (Hall, 1997) and the 10 m resolution merged LANDSAT TM and SPOT (STM) panchromatic imagery. Reproduced from Cleave (1999), The Holy Land Satellite Atlas, Volume II. Satellite imagery ©1999 ROHR Productions Ltd. and C.N.E.S.

Cleave, R. L. W., 1999. The Holy Land Satellite Atlas, Volume I: Terrain recognition, 152 p., Volume II: The regions, 248 p., Rohr Productions Ltd., Nicosia, Cyprus.

D: Cretaceous

INTRODUCTION

Cretaceous rocks are widely exposed in the mountainous regions of Syria, such as in the Anti-Lebanon, Jebel An-Nusseiriyeh, Kurd-Dagh, and the Palmyrides. Small exposures of Cretaceous deposits have also been discovered in the Jebel Abd El-Aziz region and on the left bank of the Al-Furat River to the southeast of Jrablus (**Fig. 6.22**). According to drilling data, Cretaceous rocks are also present almost everywhere in Syria, but hidden under the mantle of Cenozoic sediment. Their absence has been noted only in the southeastern part of the Syrian Desert, in the Khabra Oum Moueil, bore-hole N-46, near the border with Iraq, where the Paleocene rocks lie directly on the Jurassic limestones and dolomites.

The Cretaceous in Syria is represented by a wide spectrum of sedimentary rocks including soft and hard limestones, marls, clays, and dolomites with phosphorites, and flints and oil shales. Terrigenous rocks (sandstones, conglomerates, siltstones and clays) as well as effusive ones (basalts, basic tuffs and tuffites) are typical for the Lower Cretaceous. Abundant fossils (macro- and microfauna, palynological assemblages) serve as a good basis for stratigraphic subdivision of the Cretaceous system.

Studies of Syrian Cretaceous deposits date back to the investigations of O. Fraas (1877, 1878), C. Diener (1886), F. Noetling (1886), H. Douville (1910), and M. Blanckenhorn (1889, 1890a, 1914, 1925, 1927, 1934, 1935). However, in the study of the sediments of this period the main role belongs to L. Dubertret and his colleagues H. Vautrin, A. Keller, C. Arambourg, and others. Numerous publications have been devoted to the lithology and stratigraphy of the Syrian Cretaceous (Dubertret, 1932b, 1940a,b; Dubertret and Vautrin, 1937a,b; Keller, 1933; Keller and Vautrin, 1937; Vautrin, 1933; Arambourg et al., 1959, etc.). In the explanatory notes to individual sheets of the geological map of Syria (and Lebanon), these authors outlined the principal features of the stratigraphy of the Cretaceous deposits in Syria (and Lebanon). The ages of the stratigraphic subdivisions were substantiated by fossils, which include numerous species of ammonites, echinides, pelecypods, and gastropods.

Even today these paleontologic findings retain their significance and below we shall list some of these fossils. L. Dubertret's studies mainly encompassed the western part of Syria (the Anti-Lebanon, Jebel An-Nusseiriyeh, and Kurd-Dagh); less attention was given to the Cretaceous of the Palmyrides. The epicontinental and relatively shallow-water deposits of the Lower Cretaceous, the Cenomanian, and the Turonian received the most detailed paleontologic attention (based on macrofauna). The Senonian part of the Upper Cretaceous is marked by the development of pelagic, deeper-water facies with macrofossils poorly represented. The stratigraphic boundaries between Santonian, Campanian, and Maastrichtian Stages therefore appear rather vague, and so does the boundary with the Cenozoic.

In the process of geological mapping work conducted by Soviet geologists in 1958-1961, great attention was paid to the stratigraphy of the Cretaceous of Syria. Huge collections of fossils had previously been amassed in the course of the study of the numerous sections and outcrops of the Syrian Cretaceous. These fossils were sent to the USSR for specific analysis and determination. The following specialists took part in this paleontologic work - A. A. Atabekyan and V. P. Rengarten (ammonites); N. N. Bobkova, N. A. Cheltsova, and M. R. Dzhalilov (pelecypods and gastropods); N. A. Poslavskaya and A. E. Kalis (sea urchins); and V. A. Krasheninnikov (foraminifera). After determination of the fauna the entire collection of fossils was returned to the General Establishment of Geology and Mineral Resources of Syria.

The explanatory notes of the Geological Map of Syria (at a scale of 1:200,000) contain abundant data on lithology, stratigraphy, and paleontology of the Cretaceous deposits. These include the following sheets: Dimashq and Beirouth (Razvalyaev, 1966a); Saba Abar (Razvalyaev, 1966b); Trablus and Homs (Kozlov, 1966a), Al-Qariatein (Faradzhev, 1966), Tudmor (Soulidi-Kondratiev, 1966a), Al-Latheqiyeh and Hama (Shatsky, Kazmin and Kulakov, 1966), Salamiyeh (Mikhailov, 1966), Ar-Rasafeh (Oufland, 1966a), Halab and Antakya (Protasevich and Maksimov, 1966), Jrablus (Oufland, 1966b), Ar-Raqqa (Medvedev, 1966a), and Al-Hasakeh (Ponomarev and Bivshev, 1966).

Monographs entitled 'The Geology of Syria' (Ponikarov, Kazmin et al., 1967) and 'Geology and Mineral Resources of Foreign Countries. Syria' (Ponikarov, Kazmin et al., 1969) as well as an article entitled 'Cretaceous Deposits of Syria' (Kozlov, Ponikarov et al., 1965) summarize our knowledge of the Cretaceous deposits of Syria.

From 1984-1990, K. I. Kuznetsova et al. (1996) carried out additional stratigraphic studies in order to determine the Jurassic/Cretaceous boundary. I. S. Kotova (1990) was involved in a palynological study of Lower Cretaceous sediments and V. A. Krasheninnikov worked on the zonal subdivision of the Upper Cretaceous sediments by means of planktonic foraminifera.

The Cretaceous of Syria is subdivided into Neocomian (Beriassian-Barremian) [K_1n], Aptian [K_1ap], Albian [K_1al], Cenomanian [K_2cm], Turonian [K_2t], Coniacian [K_2cn], Santonian [K_2st], Campanian [K_2cp], and the Maastrichtian [K_2m]. Planktonic foraminifera in intercalations of pelagic sediment in the Albian and in the Upper Cretaceous (especially in Senonian chalk) serve to make further zonal subdivisions. According to lithologic peculiarities, the following mapping units were shown on various sheets of the geological map of Syria on a scale of 1:200,000: Neocomian (sometimes undifferentiated Lower Cretaceous), Cenomanian-Turonian, Coniacian-Santonian-Campanian, and finally Maastrichtian.

Our description of the Cretaceous stratigraphy is mainly based on surface geology. Unfortunately, the geological literature contains very scanty data concerning subsurface Cretaceous sediments in

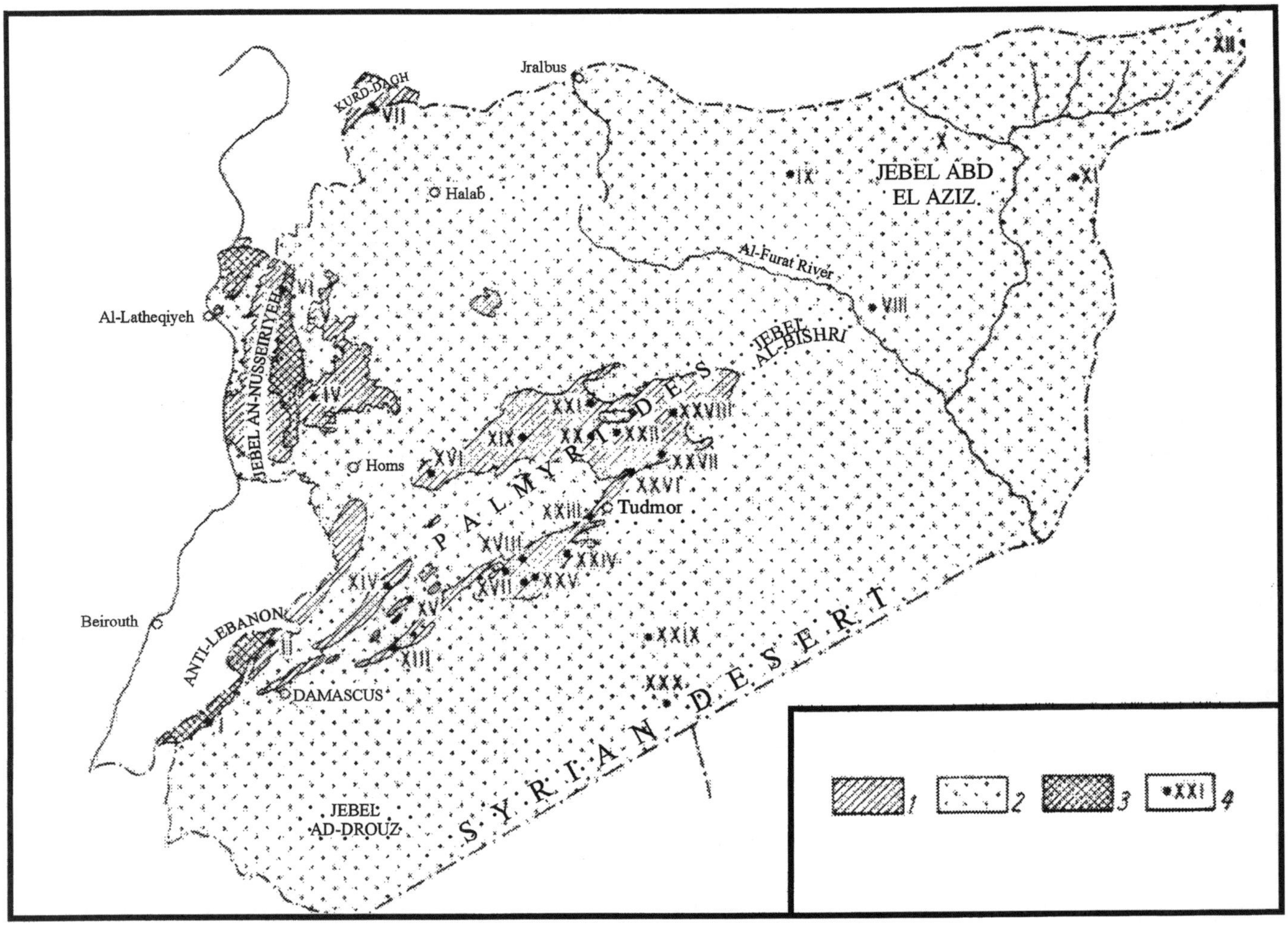

Fig. 6.22. Schematic distribution map of the Cretaceous deposits in Syria and locations of reference sections.
Legend: 1 - Outcrops of Cretaceous rocks; 2 - Areas where Cretaceous deposits are overlain by Cenozoic sediments; 3 - Outcrops of pre-Cretaceous (mainly Jurassic rocks); 4 - Location of sections: I - Beit Jann and Jebel Ash Sheikh; II - Hureira, Deir Qanoun, Bloudan and Sheqif El-Khorn; III - Jebel An-Nusseiriyeh (the southern Kafroun area); IV - Jebel An-Nusseiriyeh (the eastern Missiaf and Hama areas); V - Jebel Az-Zawiyeh; VI - Jebel An-Nusseiriyeh (the northern Qerdaha, Slenfeh and Haffe areas); VII - Meidanki (Kurd-Dagh); VIII - Derro Hole (near Deir Az-Zor); IX - Aba Hole (Jebel Tuwal El-Aba); X - Jebel Abd El-Aziz; XI - Gouna Hole (southeast of Al-Hassetche); XII - Qaratchok Hole (northeastern Syria); XIII - Jebel Ad-Dmeir; XIV - Jebel Sharqi and An-Nabek areas; XV - Jebel Ghorab and Jebel Maksar An Nimr (southwestern Palmyrides); XVI - Jebel Ash-Shoumariyeh (northern Palmyrides); XVII - Jebel Rmah, Jebel El-Bardeh and Jebel El-Bassiri (central Palmyrides); XVIII - Jebel An-Naqnaqiyeh (central Palmyrides); XIX - Jebel Ash-Shaer; XX - Jebel El-Mrah; XXI - Wadi Amsha Raddeh, Jebel El-Khashshabiyeyeh and Beer Barbar (northern Palmyrides); XXII - Jebel Abou Rajmein; XXIII - Jebel Hayyan; XXIV - Jebel El-Abtar; XXV - Jebel Sawwanet el Mhasseh; XXVI - Jebel El Harraziyeh; XXVII - Jebel As-Satih; XXVIII - Jebel El-Minshar (northeastern Palmyrides); XXIX - Hole 41 (Syrian Desert); XXX - At-Tanf Hole.

Syria. This fact creates difficulties in understanding the regional distribution of the facies types, thickness and completeness of the succession. Nevertheless, in this paper we have also attempted to use materials dealing with subsurface Cretaceous deposits.

Reference sections of the Lower and Upper Cretaceous in Syria are illustrated by **Figs. 6.23** and **6.24**). Additional sections give additional details for selected intervals of the Cretaceous sequences, and some intervals of the reference sections will be shown in even greater details.

Lower Cretaceous [K_1]

Lower Cretaceous rocks (usually with Jurassic deposits) are exposed in the axial parts of the anticlinal folds in the mountainous regions of Syria: the Anti-Lebanon, Jebel An-Nusseiriyeh, Kurd-Dagh and the Palmyrides. Small outcrops of Lower Cretaceous deposits are also located in the northeast in the Jebel Abd El-Aziz Range.

The boundary between the Lower Cretaceous sediments and

underlying Jurassic rocks is strongly disconformable. Neocomian rocks usually rest on Oxfordian shallow-water limestones, though they may overlie Callovian to Tithonian rocks (Kuznetsova et al., 1996). Consequently, the duration of the break in sedimentation and intensity of the erosional processes differ in various areas of Syria.

Paleontological data permit us to subdivide the Lower Cretaceous into three units: Neocomian, Aptian and Albian. Neocomian sediments consist of terrigenous rocks; Aptian deposits are alternating terrigenous and carbonate sediments; and among Albian rocks limestones and dolomites predominate.

The thickness of the Lower Cretaceous deposits is very irregular, varying from a few dozen meters to 300-400 m.

Neocomian [K_1n]

Paleontologically dated Neocomian deposits have been recognized in a limited number of areas (Ponikarov et al., 1967).

To the south of **Jebel An-Nusseiriyeh** (south of the village of Ash-Shmeiseh), Upper Jurassic limestones are overlain without visible traces of unconformity by light-gray and white pelitomorphic limestones containing small fragments of pelecypod and gastropod shells. The thickness of limestones does not exceed 5 m. These limestones contain the Valanginian foraminifera: *Haplophragmoides* aff. *inconstans gracilis* (Bart. et Brand), *H. uruhiensis* Shokh., *Ammobaculites eocretaceus* Bart. et Brand, *Amm. subaequalis* Mjatl., *Verneuilina neocomiensis* Mjatl., *Lenticulina muensteri* (Roem.), and *Trocholina* sp. This horizon has only been traced for little more than 100 m. It is unconformably overlain by Aptian clays. In the neighboring outcrops, clays of this age cut the Valanginian limestones and rest directly on Jurassic rocks (Kozlov, 1966a).

In the northeast of Syria (**Jebel Abd El-Aziz**), Neocomian deposits consist of three units. The lower unit is made up of a boulder-pebble conglomerate and quartz sandstones with lenses of clayey sandstones and marls with the foraminifera - *Cyclammina* aff. *greisi* Henson, *Pseudocyclammina* sp., *Choffatella* sp., and *Orbitolina* sp., and the pelecypod *Lima* ex gr. *tunetana* Perv. Its thickness is 62 m. The middle unit is characterized by an alternation of marls, limy and clayey dolomites, oolitic and microgranular limestones and unconsolidated quartz sandstones with *Pseudocyclammina* aff. *grandis* Romanova and *Choffatella* sp. Its thickness is about 13 m. The upper unit is composed of fine- to coarse-grained quartz sandstones with calcareous or ferruginous-calcareous cement. The thickness here is ~50 m. All varieties of rocks are marked by intense, bright and variegated colors of yellow, gray, white, pink and red. The total thickness of the Lower Cretaceous deposits reaches up to 125 m. The two lower units are assigned to the Valanginian-Hauterivian, and the upper unit has been provisionally put in the Barremian (Ponomarev and Bivshev, 1966).

More recently additional Neocomian deposits were found in the **Anti-Lebanon** (Rowda, Wadi Fawuar) and the **Jebel An-Nusseiriyeh Range** (Al Bustan, Ain Khlaken, Sindianeh, Teir Jamleh, Wadi Shkeir, and Qadmous) (Kuznetsova et al., 1996).

Transitional Tithonian-Berriasian layers were encountered in the Qadmous section. They are composed of a unit of interlayered clayey nodular limestones, silicified hard pink limestones and unconsolidated sandy clays of a yellowish-green color with a thickness of 29 m. This unit overlies Callovian strata with sharp erosional unconformity; it is covered disconformably by green clays with fragments of charred wood (Neocomian) and then by clays of the Aptian Bab Janneh Formation. The layers contain a very rich foraminiferal assemblage which includes both Late Jurassic and Early Cretaceous forms: *Stomatostoecha compressa* Gorb., *Phenacophragma* sp., *Charentia evoluta* Gorb., *Melath-rokerion eospirialis* Gorb., *Pseudocyclammina sulaiyana* Redmond, *Feurtillia frequens* Maync, and *Bramkampella arabica* Redmond.

In the Al Bustan section, Oxfordian organogenic limestones are overlain with a sharp erosional unconformity by a unit of interbedded yellow, reddish and green clays, marls and limestones with rich brachiopods including *Loriolithyris valdensis* and foraminifera - *Kutsevella albustanensis* Kuznets., *K. infravol-gensis* (Mjatl.), *Choffatella decipiens* Schlumb., *Pseudo-cyclammina* sp., *Melathrokerion spirialis* Gorb., *Charentia evoluta* Gorb., and *Bullopora rostrata* (Quenst.); the thickness here does not exceed 15 m. The age of these strata has been defined as Berriasian-Hauterivian.

In the Teir Jamleh section (near Missyaf), Oxfordian limestones are unconformably overlain by a unit of marls and loose calcareous clays with lenses of microcrystalline, well-cemented limestones containing remnants of pelecypods, brachiopods, and corals; their thickness is over 20 m. The foraminiferal assemblage consists of mixed Oxfordian *Kurnubia palastiniensis* Hens., *Alveosepta jaccardi* (Schroedt), Neocomian *Choffatella decipiens* Schlumb., *Feurtillia frequens* Maync, and *Stomatostoecha compressa* Gorb. Evidently the Jurassic fossils were redeposited in overlying layers of Neocomian age.

In general, the duration of the pre-Cretaceous sedimentary break and its erosional interval was shorter in the Anti-Lebanon than in Jebel An-Nusseiriyeh and Kurd-Dagh. The Anti-Lebanon is characterized by more complete Jurassic and Cretaceous (Berriasian-Barremian) sections.

Aptian [K_1ap]

Aptian sediments are widely distributed throughout Syria. They are transgressive on underlying rocks of different ages and are mainly represented by shallow-water littoral marine deposits.

In the classical sections of the **Anti-Lebanon**, the Aptian sequence consists of three formations: the variegated terrigenous 'Gres de base' at its base, followed by the predominantly calcareous 'Falaise de Blanche' and the recurrence of variegated unnamed terrigenous rocks at the top (Razvalyaev, 1966a; Kurbanov and Zarjanov, 1968).

1) The '<u>Gres de base</u>' (Dubertret, 1937, 1949a, 1955a) is composed of ferruginous quartz sandstones and clays and, at times, of effusive rocks. This formation is well-exposed in Jebel Ash-Sheikh (near the village of Jabbate Az-Zet) and in Jebel Sheqif

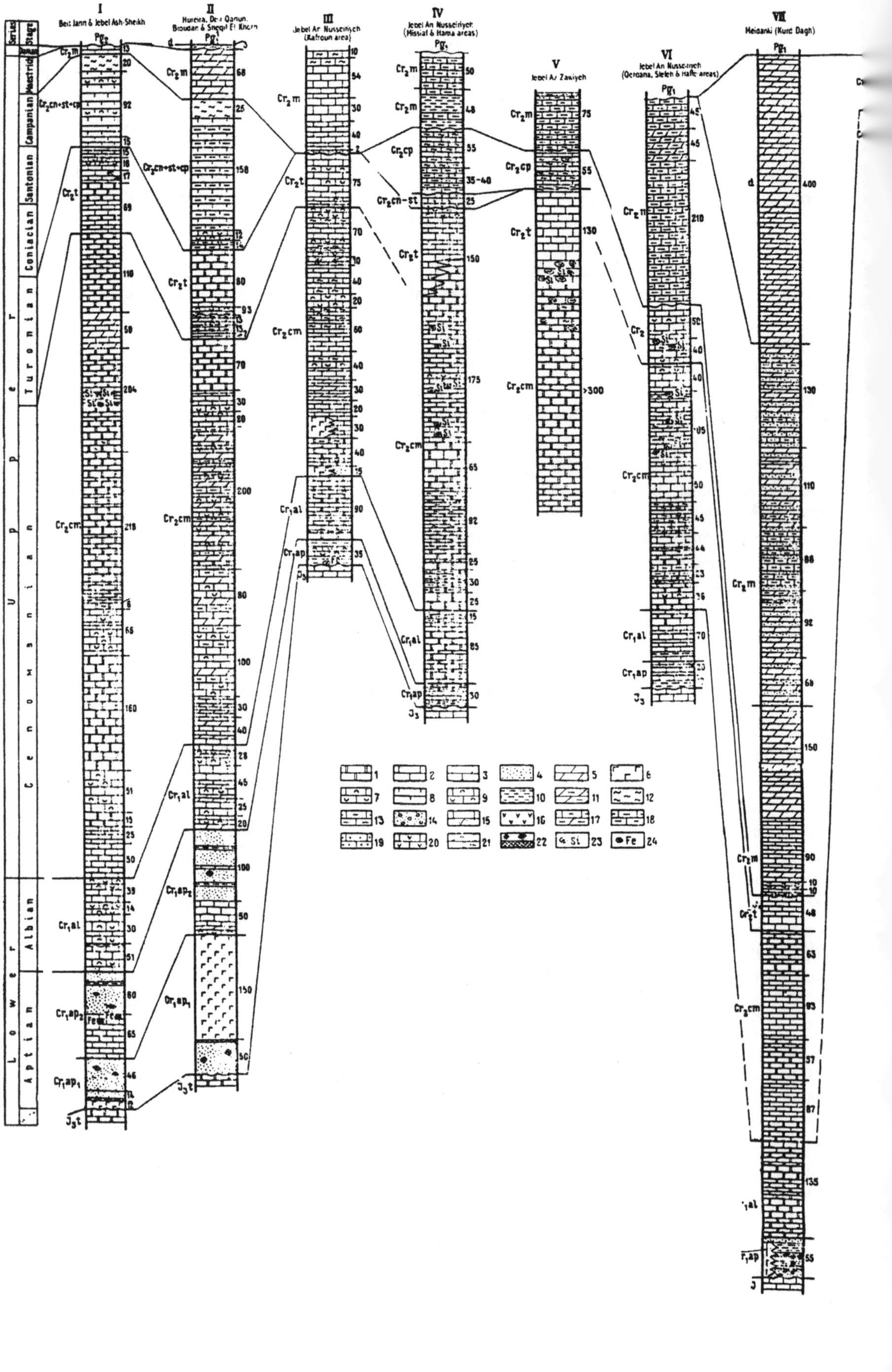
I
Beit Jann & Jebel Ash-Sheikh
II
III
Jebel Ar Nusseiriyeh
(Kafroun area)
IV
Jebel An Nusseiriyeh
(Missiaf & Hama areas)
V
Jebel Az Zawiyeh
VI
VII
Meidanki (Kurd Dagh)
Series
Stage
Maestrichtian
Campanian
Santonian
Coniacian
Turonian
Cenomanian
Albian
Aptian
Upper
Lower
Cr₂m
Cr₂cn+st+cp
Cr₂t
Cr₂cm
Cr₁al
Cr₁ap₂
Cr₁ap₁
Cr₂cp
Cr₂cn-st
Cr₁ap
J₃t
>300

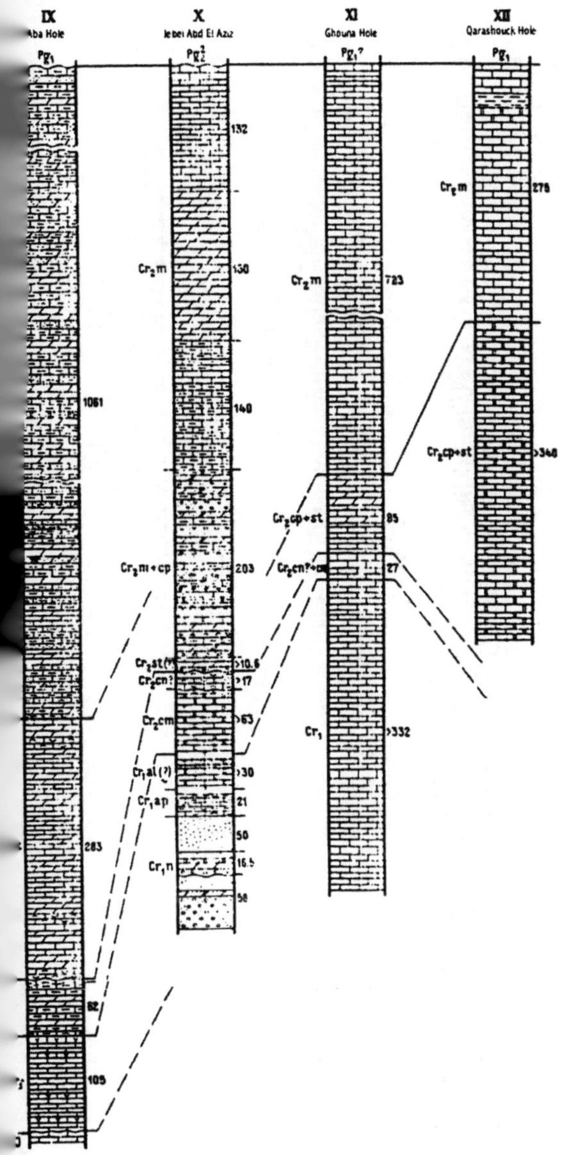

Fig. 6.23. Correlation of the Cretaceous deposits reference sections: the Anti-Lebanon, Jebel An-Nusseiriyeh, Kurd-Dagh and Mesopotamia (modified from Ponikarov et al., 1967).

Lithologic symbols: 1 - dolomite; 2 - limestone; 3 - calcareous dolomite; 4 - sand and sandstone; 5 - marl; 6 - basalt; 7 - organogenic limestone; 8 - chalk-like limestone; 9 - organic chalk-like limestone; 10 - clay; 11 - clayey marl; 12 - flint; 13 - clayey chalk limestone; 14 - gritty sandstone and conglomerate; 15 - sandy limestone and marl; 16 - gypsum; 17 - silicified limestone and marl; 18 - clayey limestone; 19 - thin phosphate streaks and lenses in limestone and marl beds; 20 - gypsiferous limestone; 21 - silicified dolomite; 22 - ferrugination and brown iron ore nodules; 23 - flint nodules and concretions; 24 - brown iron ore concretions.

Section names: I - Beit Jann and Jebel Ash-Sheikh; II - Hureira, Deir Qanun, Bloudan and Sheqif El-Khorn; III - Jebel An-Nusseiriyeh (the Kafroun area); IV - Jebel An-Nusseiriyeh (the Missiaf and Hama areas); V - Jebel Az-Zawiyeh; VI - Jebel An-Nusseiriyeh (the Qerdaha, Slenfeh and Haffe areas); VII - Meidanki (Kurd-Dagh); VIII - Derro Hole; IX - Aba Hole; X - Jebel Abd El-Aziz; XI - Ghouna Hole; XII - Qaratchok Hole.

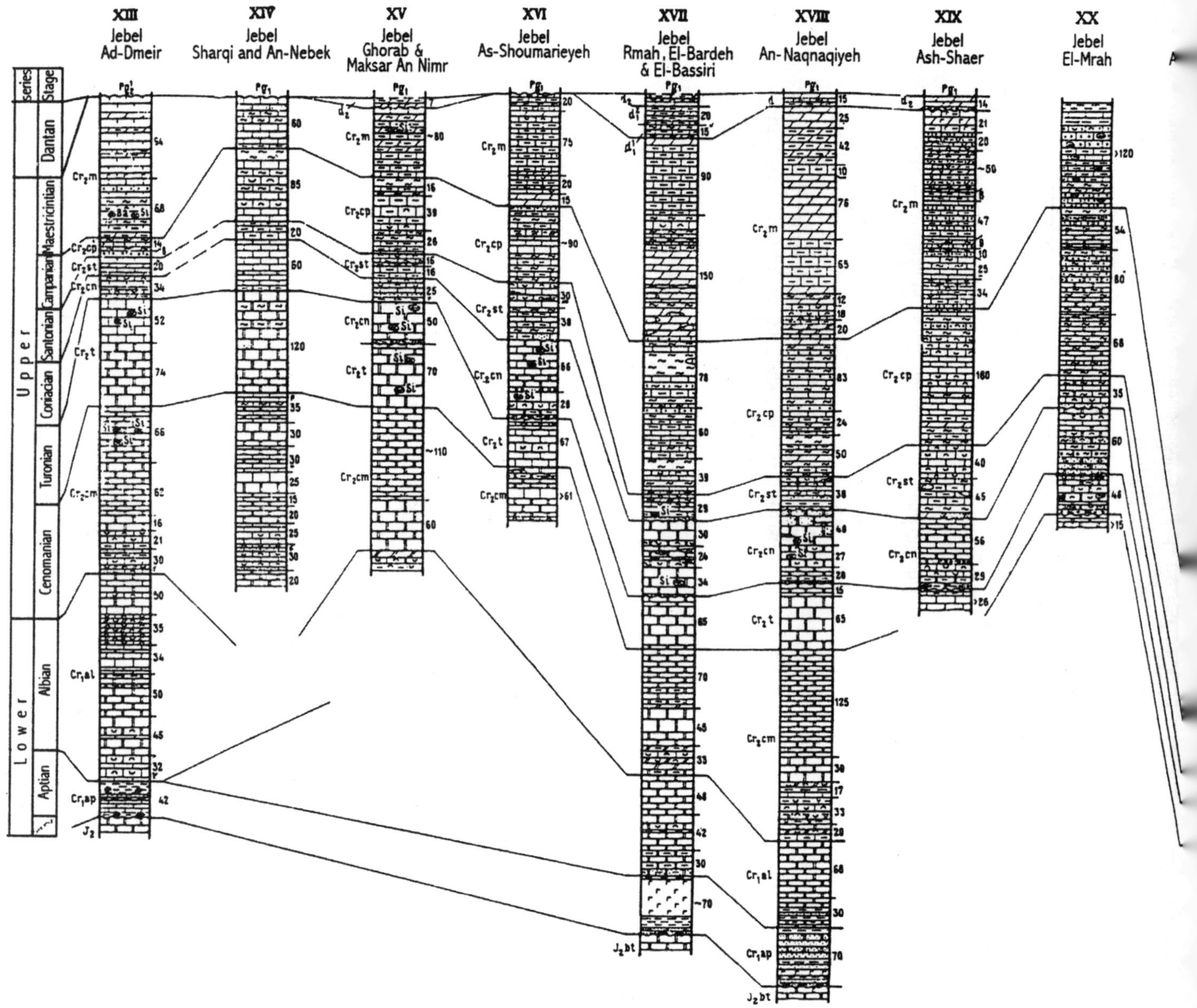
XIII
Jebel
Ad-Dmeir
XIV
Jebel
Sharqi and An-Nebek
XV
Jebel
Ghorab &
Maksar An Nimr
XVI
Jebel
As-Shoumarieyeh
XVII
Jebel
Rmah, El-Bardeh
& El-Bassiri
XVIII
Jebel
An-Naqnaqiyeh
XIX
Jebel
Ash-Shaer
XX
Jebel
El-Mrah
series
Stage
Upper
Lower
Dantan
Maestrichtian
Campanian
Santonian
Coniacian
Turonian
Cenomanian
Albian
Aptian

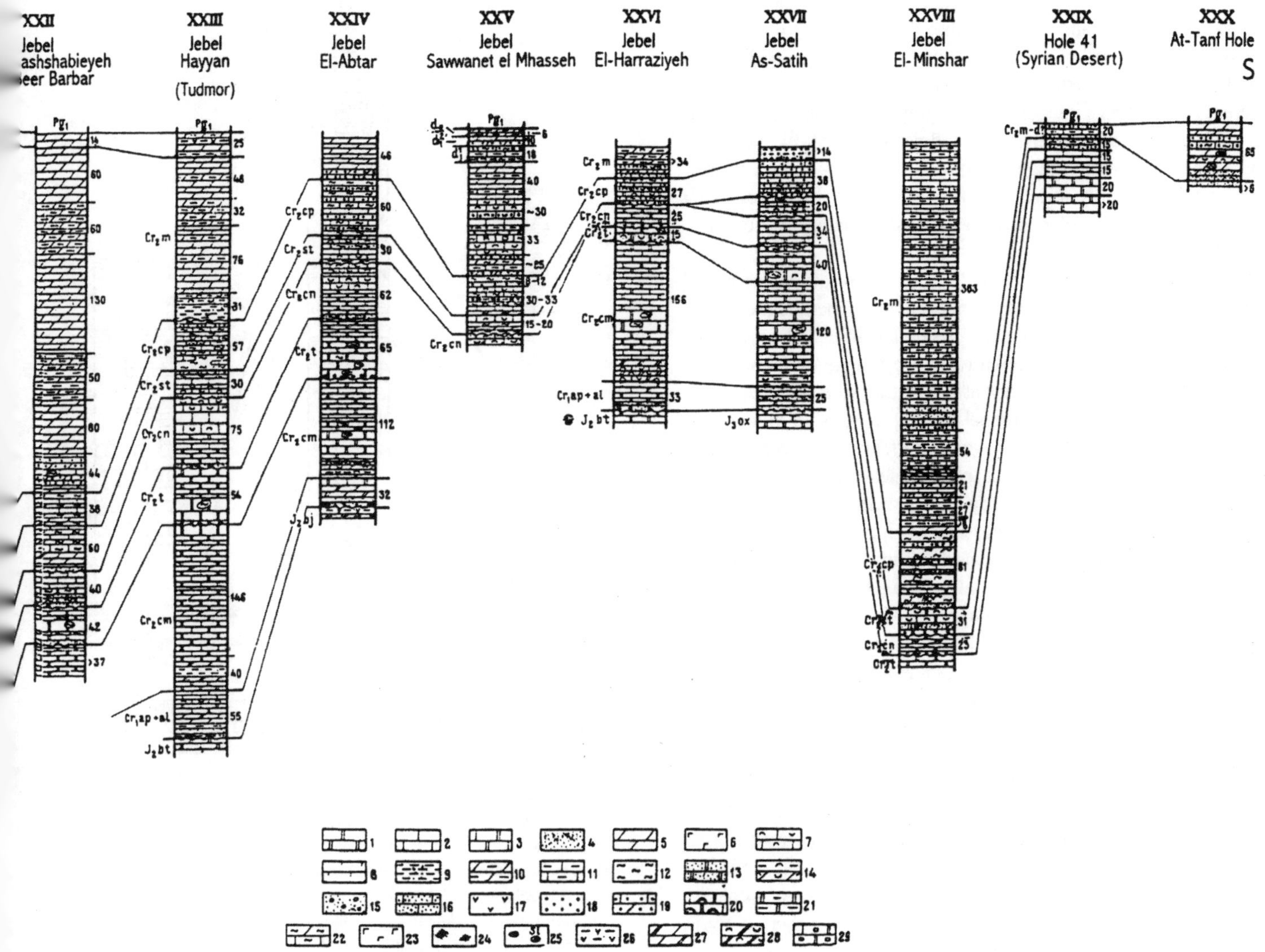

Fig. 6.24. Correlation of the Cretaceous deposits reference sections of the Palmyrides and Syrian Desert (modified from Ponikarov et al., 1967).

Lithologic symbols: 1 - dolomite; 2 - limestone; 3 - calcareous dolomite; 4 - sand and sandstone; 5 - marl; 6 - basalt; 7 - organogenic limestone; 8 - chalk-like limestone; 9 - clay; 10 - clayey marl; 11 - clayey limestone; 12 - flint; 13 - sandy dolomite; 14 - marl and clay with organic remains; 15 - gritty sandstone and conglomerate; 16 - sandy limestone; 17 - gypsum; 18 - phosphate rock; 19 - thin phosphate streaks and lenses in limestone and marls beds; 20 - algal dolomitic limestone; 21 - clayey dolomite; 22 - silicified limestone and marls; 23 - glauconitization; 24 - ferrugination and brown ore nodules; 25 - flint nodules and concretions; 26 - gypsiferous clay; 27 - magnesium-bearing marl; 28 - organic magnesium marl; 29 - breccia-like calcareous dolomite.

Section names: XIII - Jebel Ad-Dmeir; XIV - Jebel Sharqi and An-Nebek; XV - Jebel Ghorab and Jebel Maksar An Nimr; XVI - Jebel As-Shoumarieyeh; XVII - Jebel Rmah, Jebel El-Bardeh and Jebel El-Bassiri; XVIII - Jebel An-Naqnaqiyeh; XIX - Jebel Ash-Shaer; XX - Jebel El-Mrah; XXI - Wadi Amsha Raddeh, Jebel El-Khashshabieyeh and Beer Barbar; XXII - Jebel Abou Rajmein; XXIII - Jebel Hayyan; XXIV - Jebel El-Abtar; XXV - Jebel Sawwanet el Mhasseh; XXVI - Jebel El-Harraziyeh; XXVII - Jebel As-Satih; XXVIII - Jebel El-Minshar; XXIX - Hole 41 (the Syrian Desert); XXX - At-Tanf Hole.

(near the villages of Zebdani, Bloudan, and Serghaya, at the heads of Wadi Mannsour and Wadi El-Khorn). It is mainly represented by brown, gray and reddish, medium- to coarse-grained quartz sandstones with a ferruginous clayey matrix. Among sandstones there are subordinate intercalations of gray and yellow clays as well as thick beds of basalts, anamesites, dolerites, and tuffs. Very thin intercalations consist of lignites (up to 0.2 m). Beds of pisolitic iron ores are considerably thicker (up to 2-2.5 m). Pyrite segregations and fine grains of amber (copal) are disseminated in the clays, lignites, and iron ores.

The lithology, thicknesses and the ratio of sedimentary/effusive rocks in 'Gres de Base' vary laterally. The thickness here varies from 70 m up to 200 m. In some localities volcanic rocks are absent. In Jebel Ash-Sheikh (near the village of Jabbata Az-Zet), compact black basalts rest at the bottom of a 12-meter-thick pile covered by sandstone. In Jebel Sheqif (along Wadi Mannsour), this section starts with sandstones, while effusive rocks (basalts and tuff-breccia) crown a 150-meter thick sequence (**Fig. 6.25**). In the vicinity of the village of Serghaya, lava sheets occupy the entire interval between the Upper Jurassic and Upper Aptian limestone. These lava flows are unstable themselves, and laterally the dolerites may be replaced first by anamesites and then by basalts.

The terrigenous rocks contain no faunal remains, a fact which creates uncertainty in the determination of age. Their upper age limit is evidenced by the Upper Aptian limestones which conformably cover the unit. Therefore, the age of this unit in the geological literature has usually been assumed to be pre-Upper Aptian. The matter of their lower limit is rather complicated as can be realized from the next paragraph.

The 'Gres de base' Formation was singled out for the first time in Lebanon and was attributed to the Neocomian and Lower Aptian based upon paleontological data (Dubertret, 1937, 1949a,b, 1955a). L. Dubertret (1955a) suggested that the 'Gres de base' of Lebanon did not correspond to a lithologically similar member in Syria (which he called 'Gres de Liban'). L. Dubertret dated this 'Gres de Liban' as Lower Aptian and explained the change in the age of the basal Cretaceous unit in Syria as a result of trans-gressive motion of the Aptian sea in an easterly direction. L. Dubertret's point of view was adopted by many geologists, including V. Ponikarov et al. (1967). Nevertheless, Ponikarov and his colleagues admitted, "that the basal Cretaceous beds in the Anti-Lebanon may have a wide age range since they do not contain any identifiable organic remains" (Ponikarov et al., 1967, p. 41).

Palynological investigation of the Lower Cretaceous deposits in Syria suggests the presence of Neocomian strata within the 'Gres de base' (Kotova, 1990). In the section along Wadi El-Khorn (Anti-Lebanon and Jebel Sheqif), a basal layer of the 'Gres de base' includes rich assemblages of pollen and spores. Among the gymnosperm pollen there are numerous *Inaperturopollenites* sp., *Classopolis* sp., and *Callialasporites dampieri*, together with rare *Ephedripites* sp. Spores include *Cicatricosisporites* spp., *Contignisporites* sp., *Klukisporites* sp., *Appendicisporites jansonii*, *Patellasporites distaverrucosus*, *Leptolepidites* sp., and *Uvaesporites* aff. *pseudocingulatus*. There are neither typical

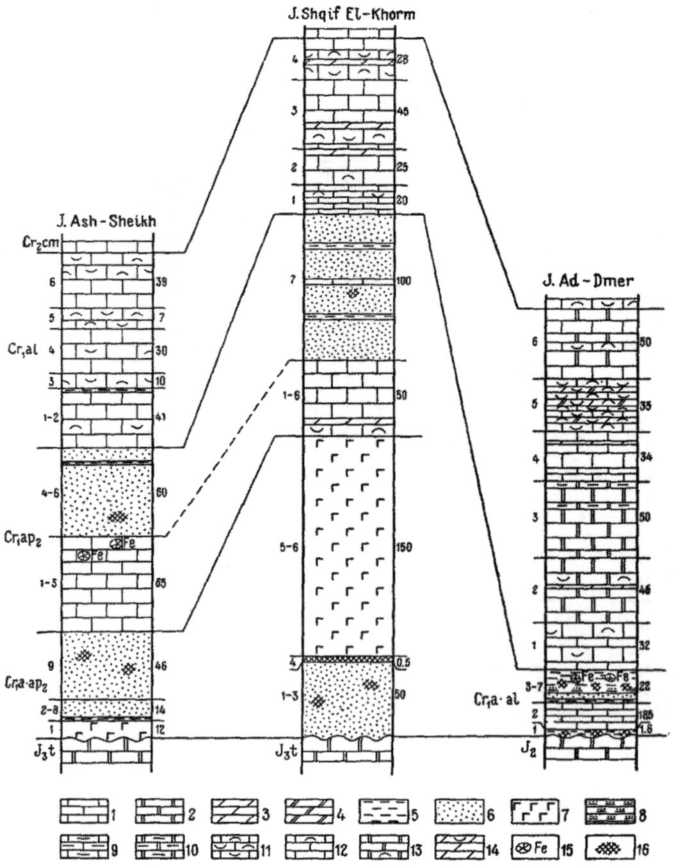

Fig. 6.25. Lower Cretaceous deposits of the Anti-Lebanon and southwestern margin of the Palmyrides (after Razvalyaev, 1966a).
Legend: 1 - limestone; 2 - dolomite; 3 - marl; 4 - dolomitic marl; 5 - clay; 6 - sandstone; 7 - basalt; 8 - sandy clay; 9 - clayey limestone; 10 - clayey dolomite; 11 - shelly limestone; 12 - limestone with fauna; 13 - dolomite with fauna; 14 - marl with fauna; 15 - brown iron-stone concretions; 16 - bands and lenses of iron-stone.

Barremian species, nor any pollen of the early angiosperm. The composition of such palynological assemblages indicates a pre-Barremian age for these sediments.

In deposits which overlie the 'Gres de base' Formation, *Classopolis* pollen predominate among gymnosperm. They are encountered in conjunction with *Eucommiidites* sp., *Inaperturopollenites* sp., *I. limbatus*, and *Ephedripites* sp. The spores are marked by the appearance of *Rugulatisporites* sp., *Cicatricosisporites nuni*, *C. avnimelechi*, *Appendicisporites potomacensis*, and *Murospora* sp. This assemblage is associated with another important feature, *i.e.* the first appearance of the angiosperm pollen *Retimonocolpites*, known from the Barremian Stage of England and the Moroccan Basin. Consequently, the middle layers of the 'Gres de base' in Syria can be attributed to the Barremian Stage.

It is therefore quite possible that the terrigenous sediments of the Lower Cretaceous basal layers of Lebanon and Syria (the 'Gres de base') are synchronous and correspond to a part of the

Neocomian and the Lower Aptian.

2) The 'Falaise de Blanche' of the Anti-Lebanon Range is composed of white and gray, massive, pelitomorphic and organogenic detrital limestone. It is known under both the names of 'Falaise de Blanche' or 'Muraille de Blanche' (Dubertret, 1947, 1949a, 1954, 1955a). Thin interbeds of marls, clays, and sandy limestones are subordinate. The limestones of the 'Falaise de Blanche' are hard and light in color, forming a distinct escarpment or cliff providing the mountain-slopes with an easily discernable marker against the background of brown and mottled layers characterizing the underlying and overlying sandstones and clays. These rocks are well-exposed in many sections of Jebel Ash-Sheikh (Hermon Massif), Jebel Sheqif, and Jebel Mannsour along the Damascus to Beyrouth highway. The thickness of the 'Falaise de Blanche' varies from 20 to 65 m, mostly about 40-50 m. These limestones encompass numerous mollusc and sea urchin shells, including *Heteraster oblongus* Brongn. *syriacus* Vautrin et Keller and have been dated as Upper Aptian (Dubertret, 1954).

The Aptian age of the 'Falaise de Blanche' limestones has been confirmed by a study of pollen and spores (Kotova, 1990). Pollen of gymnosperm plants (*Classopolis*) are the most common. They are also associated with angiosperm pollen: *Clavati-pollenites* sp., *Stellatopollis* sp., *Retimonocolpites peroreticulatus*, *Afropollis zonatus*, and *A. operculatus*, which are typical for Aptian sediments in many areas of Europe and Gondwanaland. Among the spores are very diverse representatives of *Cicatri-cosisporites (C. salardii*, etc.), species of the genus *Appendici-sporites* (*A.* aff. *ordtmani, A. matesovae), Costatoperfosporites* sp., *Camarozonosporites insignis,* and *Foraminisporis assymetricus*.

3) The upper, as yet unnamed formation represents the uppermost Aptian-lowermost Albian layers. It is made up of quartz sandstones - white, loosely cemented by carbonate matter, containing up to 95% quartz grains and sometimes devoid of argillaceous and ferruginous groundmass. Thin interbeds are composed of clays and limestones. The thicknesses here range from 50 to 100 m, increasing somewhat in a southerly direction. This formation differs from its lower terrigenous counterpart in the white color of the sandstones, lower content of ferruginous rocks, and higher number of clayey and limy intercalations (Fig. 6.25).

Due to poor fossiliferous content the age of the upper terrigenous formation is as yet unclear. According to L. Dubertret's and H. Vautrin's scheme (1937), this unit was attributed to the Upper Aptian. Traditionally, the sharp lithologic boundary between this unit of terrigenous rocks and the overlying limestones with *Knemiceras* is treated as the Aptian/Albian boundary. This level is easily recognizable during geological mapping. However, in one marly interbed which occurs near the village of Beit Jann (the Hermon Massif) among quartz sandstones at the top of the upper terrigenous member, some pelecypods have been found - *Amphidonta* cf. *launesi* Lartet and *Trigonia* cf. *pseudocaudata* Thomas et Peron (Razvalyaev, 1966a, determined by N. Cheltsova). These fossils testify to the Albian age of the sediments.

A palynological analysis has given similar results (Kotova, 1990). In some samples taken from the lower half of the upper terrigenous member, *Classopolis* predominates among the pollen of gymnosperm plants. The spores are represented by numerous *Cicatricosisporites nuni, C. avnimelechi*, and *Cyathidites*; such taxons as *Clavatisporites* sp., *Balmeisporites* sp., *Appendici-sporitea* aff. *crimensis*, and *Trilites* sp. are encountered here for the first time. The composition of the angiosperm pollen changes - *Afropollis operculatus* and *A. zonatus* disappear and *A. jardinus* and *Retitricolpites* sp. appear, all of which indicate an Early Albian age of the sediments.

In the overlying deposits of the third formation, the quantity and diversity of the angiosperm pollen increase. Especially numerous are *Retimonocolpites peroreticulatus* and *Afropollis jardinus*, which are also accompanied by *Foveotricolpites* sp., *Retitricolpites* aff. *geranioides, Striatopollis* sp., *Clavatipollenites minutus*, and *Striatricolpites reticulatus*. Forms typical for the Upper Albian are absent, as are any representatives of the genera *Elaterocolpites* and *Elaterosporites*, and this makes it possible to confine the age of these deposits to the Middle Albian interval.

In palynological assemblages from the upper layers of the upper terrigenous formation, the ratio of gymnosperm/angiosperm relative to spores of ferns shifts strongly, indicating an increase in gymnosperm pollens (*Classopolis*). This means that the humid tropical climate of the Aptian was supplanted by an arid climate at the beginning of the Albian.

Summing up, the upper terrigenous formation may already belong (either partly or completely) to the Albian Stage. It is interesting to note that in Lebanon sandstones overlying the 'Falaise de Blanche' have also been attributed to the Albian Stage by G. Zumoffen (1926). Considering the Aptian/Albian boundary in Syria to be the boundary between the unnamed terrigenous rocks and the overlying formation of limestones and dolomites, V. Ponikarov et al. (1967) as well as A. Razvalyaev (1966a) did not exclude the possibility "that the boundary between the Aptian and Albian Stages may be drawn lower" (Razvalyaev, 1966a, p. 25).

This three-formational Early Cretacerous succession can be recognized in some sections of the southern part of the Jebel An-Nusseiriyeh Range and in the Palmyrides, but this stratigraphic subdivision has not been traced into the northern part of Jebel An-Nusseiriyeh, the Kurd-Dagh ridge, or into the Jebel Abd El-Aziz ridge.

South of **Jebel An-Nusseiriyeh**, near the village of Kafroun (section III in Fig. 6.23) and to the northeast of Safita, Upper Jurassic and Valanginian limestones are disconformably overlain by a member about 8 m thick of greenish-gray calcareous clays with thin limestone interbeds, limonite concretions, and pebbles of underlying limestones (Kozlov, 1966a) (**Figs. 6.26** and **6.27**). These clays are very rich in Aptian fossil remains (Dubertret, 1937): *Heteraster oblongus* Brong. var. *syriaca* Vautrin et Keller, *Exogyra boussingaulti* d'Orb, *Ex. minos* Coq., *Ostrea* cf. *heintzi* Th. et P., *Nucula glaustriticea* Whitefield, *Trigonia lewesi* Bl., *Astarte (Platopis?) triangularis* Whitefield, *As. formosa* Fitton, *Crassatella cranulicosta* Whitefield, *Cr.* aff. *baudeti* Coq., *Cardium (Protocardium) biseriatum* Noetling, *Corbiculopsis* sp., *Venus (?) globulus* Whitefield, *Cytherea lybanotica* Fraas, *C.*

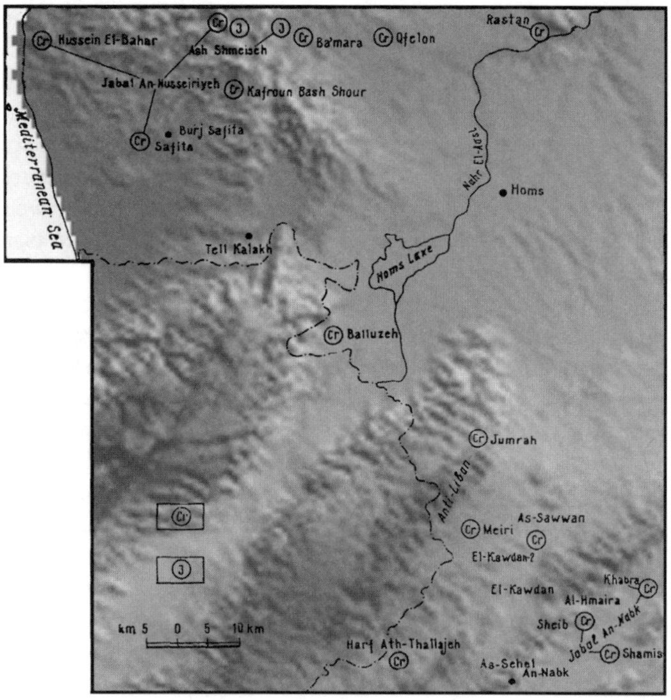

Fig. 6.26. Schematic location map showing position of Jurassic and Cretaceous deposits sections in southern Jebel An-Nusseiriyeh, northeastern Anti-Lebanon and the northwestern Palmyrides (Jebel An-Nabek).

subbrongniartina d'Orb., *Turritella sub-scetzeni* Bl., *Hauconia* aff. *lujani* de Verneuil, *Calumbellina verneuilli* Cossmann, *Aporrhais pleurotomoides* Bl., and *Phasianella abeihensis* Bl. Apparently, these clays correspond to the uppermost layers of the 'Gres de base' of the Anti-Lebanon.

The middle part of this section is composed of light-gray and yellowish organogenic limestones with abundant fossils which are similar to the Aptian fauna of the Falaise de Blanche' of Lebanon: *Orbitolina lenticularis* Blum., *Choffatella decipiens* Schlumb., *Heteraster oblongus* Brongn. var. *syriaca* Vautrin et Keller, *Cardita rawsoni* Whitefield, *Cardium syriacum* Conr., *Venus reynesi* Coq., *Cytherea syriaca* Conr., *Cercomya* cf. *deserti* Douville, *Liopistha (Psilomya) ligeriensis* d'Orb., *Coronatica ornata* Fraas (Dubertret, 1937a). The thickness of these limestones reaches up to 15 m. The limestones can be correlated to the middle calcareous unit of the Anti-Lebanon ridges.

The Kafroun section terminates in green, compact, lumpy calcareous clays with dolomitic limestone interbeds, up to 12 m in thickness. This section is equivalent to the upper terrigenous unnamed unit of the Anti-Lebanon sections and can be viewed as transitional from the Aptian to the Albian Stage.

Similar Aptian deposits can be observed in the vicinity of the villages of Bamara and Ofelon, located to the northeast of Kafroun.

In the northern part of Jebel An-Nusseiriyeh, the character of the Aptian deposits is different and variable (**Fig. 6.28).** On the western slope near the village of Slenfeh, the deposits consist of alternating green clays and lumpy breccia-like limestones up to 37 m thick, containing *Trigonia undulatocostata* Bl., *Nerinea coaundiana* d'Orb., and *N. forojuliensis* Pirona (Dubertret, 1937a). On the eastern slope of the same ridge near the town of Missyaf, the Aptian limestones are almost entirely replaced by dolomites which alternate with green clays and sandstones; their thickness does not exceed 30 m.

To the east of the village of Qerdaha, the Aptian Stage starts with a thin conglomerate bed containing pebbles of underlying Jurassic limestones. Near the village of Qadmous at the base of the Aptian Stage, there occur beds and lenses of brown ironstones and quartz sandstones with a ferruginous matrix. Red clays laterally replace the ironstones. The main pile of the Aptian unit in these sections is composed of green clays intercalated with lenses and beds of fine-grained aphanitic limestones, dolomites and sometimes oolitic limestones. In the vicinity of the village of Bab Janneh, dolomites and limestones alternate with small-grained polymictic calcareous sandstones. The thickness of the Aptian sediments in this area varies from 25 to 40 m. No palynological studies were carried out here. Though organic remains in the Aptian green clays and dolomites are very rare, near Qadmous, *Biauris praelongum* Scharpe and *Exogyra* (?) *falco* Coq. have been found, indicating an Aptian age (Shatsky et al., 1966).

Very different Aptian deposits occur in **the Kurd-Dagh block**. Here, the Aptian succession comprises highly ferruginous variegated terrigenous rocks with thin beds and lenses of limonite without intercalation of calcareous rocks. Fossil remains are essentially lacking in these ferruginous deposits. Only L. Dubertret (1953a) has reported findings of *Orbitolina* sp. here. This formation rests on the erosional surface of the Upper Jurassic limestones and is conformably covered by Albian rocks. According to its stratigraphic position, it has been dated as Aptian, but we should not exclude the possibility that this formation might also comprise a portion of the Neocomian.

The northwestern part of Kurd-Dagh is broken by a series of faults of various magnitudes, creating a complicated block structure. Aptian sediments are exposed over small distances along these faults near the villages of Kerri, Alamdar, Maydenli, and especially Rajou. The ferruginous Aptian unit varies strongly in its lithology and thickness (10-60 m) along the strike. The composite Aptian section (Protasevich and Maksimov, 1966; Ponikarov et al., 1967) of Kurd-Dagh consists from base to top of:

1) Greenish-brown basic tuff and tuffite with polymictic pebbles and boulders (up to 0.3 m in size) encompassing sedimentary, metamorphic and intrusive rocks, which are apparently derived from the Giaour-Dagh and Anti-Taurus ridges. The thickness is 20 m;

2) Yellow-brown and gray, sandy clay with flint and markasite nodules. The thickness is 0.5-13 m;

3) Hematite-goethite-hydrogoethite iron ores having an oolitic-conglomerate texture with streaks and lenses of ferruginous clays and sandstones. In a lateral direction, these terrigenous rocks may almost completely replace the iron ores. The thickness is 19 m;

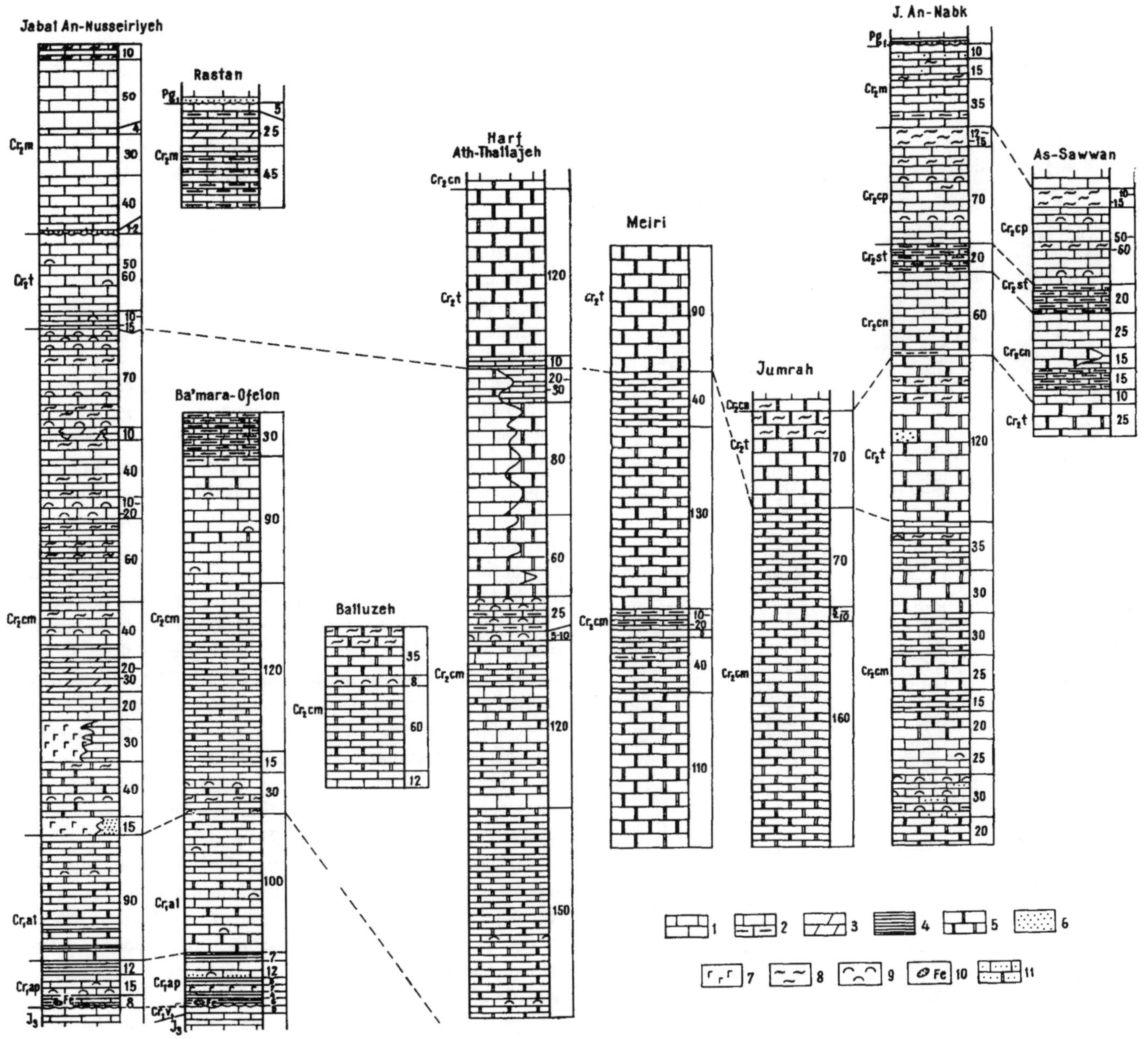

Fig. 6.27. Cretaceous deposits sections of southern Jebel An-Nusseiriyeh, northeastern Anti-Lebanon and the northwestern Palmyrides (after Kozlov, 1966a).
Legend: 1 - limestone; 2 - clayey limestone; 3 - marl; 4 - clay; 5 - dolomite; 6 - sandstone; 7 - basalt; 8 - flint; 9 - shelly limestone; 10 - ferruginous concretions; 11 - sandy limestone and dolomite.

4) Alternations of ferruginous clays with iron ore fragments, compact silty clays with limonite concretions and conglomerates of ferruginous and clayey pebbles. The thickness is 2-15 m;

5) Ferruginous dolomitic quartz sandstones intercalated with ferruginous detrital limestones. The thickness is up to 11 m, but in places this layer pinches out completely;

6) Variegated ferruginized clay with thin iron ore interbeds. The thickness varies from 1-2 to 6 m;

7) Alternation of ferruginous sandstones and clayey dolomite with quartz and glauconite grains. The thickness is 2-6 m.

The ferruginous ore sediments occur in lenses. These are composed of hematite and hydrogoethite ores, in which the latter prevails. Two types of ores occur on the base of their structure: cemented (conglomeratic) and concretional. The former is detritic (semi-rounded pebbles of clay and iron oxides) immersed in an argillo-ferruginous cement. The amount of detrital material in these ores ranges from 10% to 40-50%. The ratio of metallic and non-metallic parts of the cement varies considerably - from 20-30% up to 80%. By contrast, the concretional ores consist of an accumulation of ferruginous concretions (5-20 cm) in an arenaceous or argillaceous matrix. The concretions are irregularly distributed throughout the ore horizons, without any traces of gravity grading. The ores of concretional composition have a rather low iron content varying from 20% to 37%. In the cementational ores, the metal content is more consistent and usually exceeds 35%.

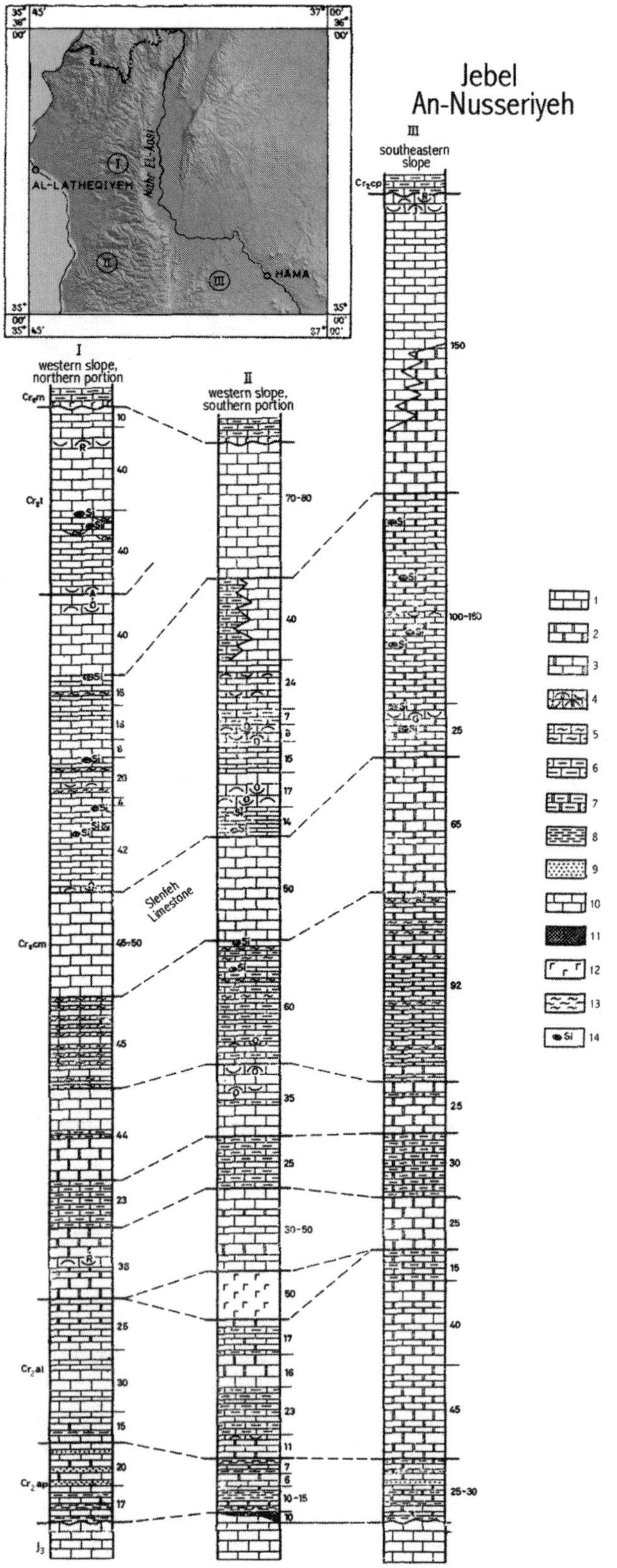

In the southwestern part of the **Palmyrides** (Jebel Ad-Dmeir), the Aptian subdivision into three formations can be clearly observed; though over most of these vast mountains, it remains very indistinct or may disappear entirely.

In Jebel Ad-Dmeir (Fig. 6.24, Section XIII) Aptian deposits lie on a strongly denuded surface of Middle Jurassic limestones and dolomites (Razvalyaev, 1966a). At the base of this section there is a 1.6 m thick bed of violet and greenish-yellow clay repleted with pisolites and limonitic oolites as well as pelecypod shells, identified as *Ceratostreon callimorphe* Coq. by N. Cheltsova. These clays probably correspond to the 'Gres de base' of the Anti-Lebanon. The overlying deposits are composed of 18 m of yellow and yellow-gray aphanitic, medium- to thick-bedded limestones with poorly preserved pelecypod shells. This interval is best correlated with the 'Falaise de Blanche' in the Anti-Lebanon. The ~23 m thick upper part of the section consists of mottled terrigenous reddish-brown and violet clays with rounded dolomite pebbles and fine pisolites of limonite; clays saturated with ferric hydroxides; gray sandy fine-spotted lumpy clays; violet quartz sandstones with ferruginous-carbonate and dolomitic-ferruginous cement. This upper sequence is evidently equivalent to the third, unnamed terrigenous formation of the Anti-Lebanon. The total thickness of the Aptian deposits in the Jebel Ad-Dmeir section is nearly 43 m.

Along the entire Palmyrides chain, the Aptian consists of terrigenous deposits which disconformably overlie various horizons of Jurassic rocks (**Fig. 6.29**).

In Jebel Rmah (to the southeast of Al-Qaryatein), the Aptian succession is represented by variegated clays and quartz sandstones, reaching up to 84 m in thickness (Faradzhev, 1966). In the lower portion of this section a thin layer (4.5 m) of yellow dolomite occurs. The upper portion consists of diabase and hydrothermally altered mandelstones (up to 40 m).

In the Central Palmyrides (to the east of Wadi Bardeh), the Aptian strata are well-exposed in Jebel An-Naqnaqiyeh (near the

Fig. 6.28. Composite sections of Cretaceous deposits of the northern half of Jebel An-Nusseriyeh (after Shatsky et al., 1966).

Legend: Sections: I – the Jebel An-Nusseiriyeh range, western slope, northern portion (the Haffeh, Slenfeh and Qerdaha area); II – the Jebel An-Nusseiriyeh range, western slope, southern portion (the Qadmous, Banyas and Tartous area); III – the Jebel An-Nusseiriyeh range, southeastern slope (the Massiaf and Hama areas).

Lithologic symbols: 1 - limestone; 2 - dolomite; 3 - dolomitized limestone; 4 - shell limestone: O - oyster limestone, R - rudists limestone, G - gastropods limestone; 5 - silicified limestone; 6 - clayey limestone and marl; 7 - dolomitic marl; 8 - clay; 9 - sandstone; 10 - glauconite limestone; 11 - brown ironstone; 12 - basalt; 13 - flint; 14 - flint in concretions.

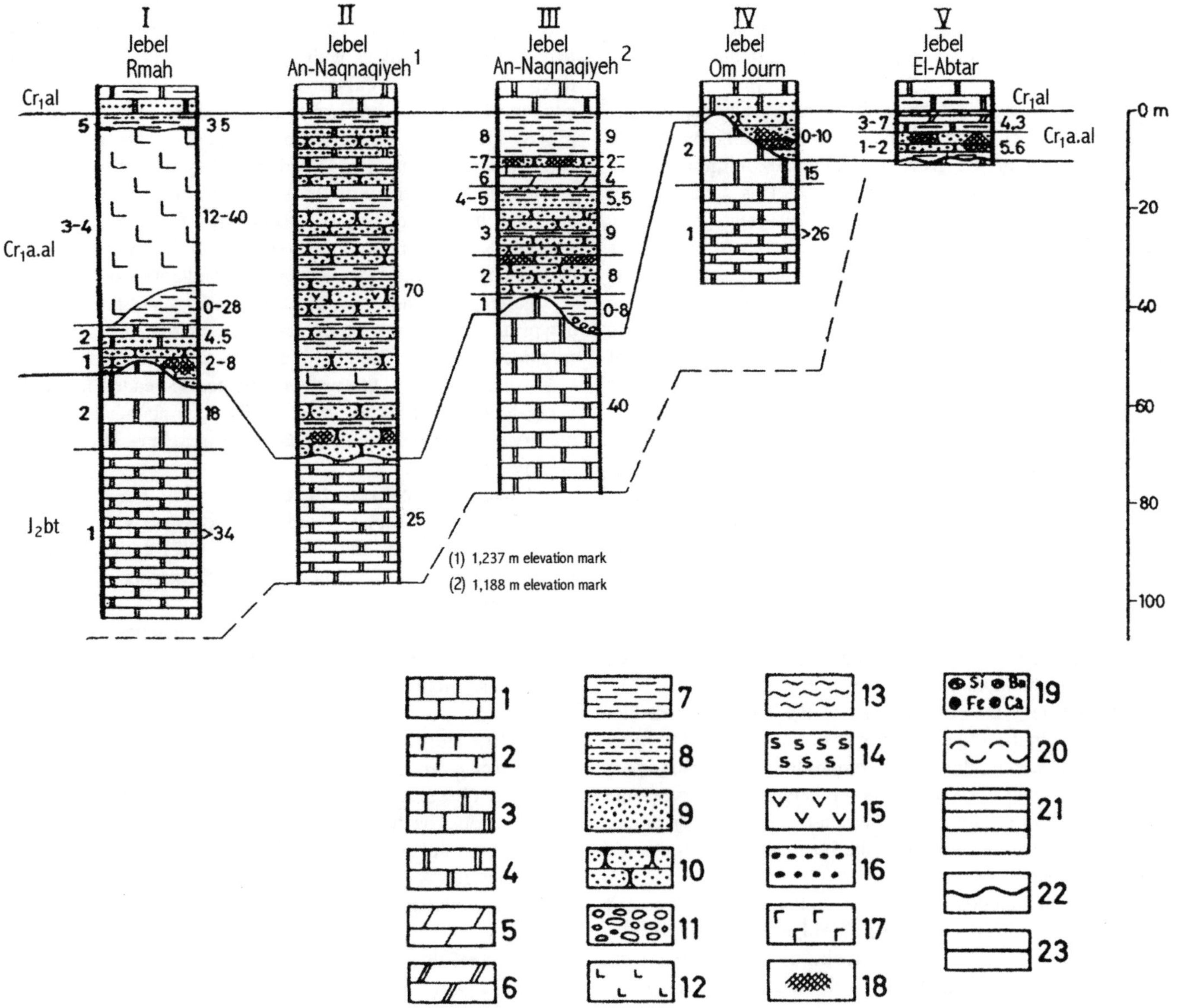

Fig. 6.29. Neocomian and Aptian deposits sections of the Central Palmyrides, the southern branch (modified from Faradzhev, 1966,).

Legend: Sections: I - Jebel Rmah; II - Jebel An-Naqnaqiyeh (near 1237 m mark); III - Jebel An-Naqnaqiyeh (near 1188 m elevation mark); IV - Jebel Om Journ; V - Jebel El-Abtar.

Lithologic symbols (for Figs. 6.29, 6.30, 6.35; 6.40; 6.41 and 6.44): 1 - limestone; 2 - chalky limestone; 3 - dolomitized limestone; 4 - dolomite; 5 - marl; 6 - dolomitic marl; 7 - clay and clayey varieties of rocks; 8 - siltstone; 9 - sand and sandy varieties of rocks; 10 - sandstone; 11 - conglomerate; 12 - basalt, diabase; 13 - silicite; 14 - opoka; 15 - gypsum; 16 - phosphate; 17 - glauconite; 18 - limonite and ferruginous rocks; 19 - flint, barite, iron and calcareous concretions; 20 - microfossils; 21 - bedding: thin, medium, thick, coarse and massive; 22 - unconformities; 23 - stratigraphic units boundaries. Number of strata shown to left, thickness in meters shown to right.

geodetic elevation markers at 1,188 m and 1,237 m), where they rest directly on Bathonian rocks. The basal layer (up to 8 m thick) is composed of brown calcareous clays with pebbles and boulders. These sediments fill pockets in the eroded surface of the Jurassic limestones. The basal layer is then followed by a sequence of alternating mottled clays, sandy clays and sandstones, which are brown, red-brown, pink, white, yellow and green in color. The latter are sometimes cross-bedded, containing plant remains and lumps and veins of iron hydroxides. In the upper portion of the section, there is a layer (4 m thick) of yellow and greenish clayey limestones and marls, but their correlation to the 'Falaise de Blanche' of the Anti-Lebanon is doubtful. The total thickness of Aptian deposits in the Jebel An-Naqnaqiyeh section is about 46 m.

To the northeast, Aptian layers preserve their sandy-clayey

character, but become gradually thinner, measuring 10 m in thickness at Jebel El-Abtar. In some localities in the eastern part of Jebel Om Journ, they are completely absent (Faradzhev, 1966; Ponikarov et al., 1967).

To the southwest of Tudmor (Palmyra), Aptian deposits are developed in Jebel Hayyan. They disconformably overlie the erosional surface of the Bathonian dolomites, forming eroded pockets up to 0.5 m wide on their surface. This section begins with a 1-1.5 m thick bed of green and yellow-green gypsiferous, highly arenaceous clays with 1-5 cm concretions of brown iron ore and lenses of fine-grained brown or pink quartz sand. They are followed by white, red, brown and pink, fine- and medium-grained quartz sandstones with cross lamination and gypsum cement, green and yellow clays and highly arenaceous white and gray marls. Near the top, thin- and medium-bedded, dark-gray and gray dolomitic marls and clayey dolomites appear. The total thickness is about 22 m (Soulidi-Kondratiev, 1966a).

In the eastern-most section of the Palmyrides, to the north of Tudmor (Jebel El-Mazar) and to the west of As-Sukhneh (Jebel As-Satih and Jebel Thaniyyet As-Safra), the Aptian deposits become more arenaceous: pink, violet and yellow, medium- to coarse-grained quartz sandstones, dolomitic or ferruginized sandstones with interlayers of green clays, yellow dolomites and gray 2.5 m thick gypsum. The thicknesses decrease from about 25 to sometimes 4-5 m. It is evident that here the sections are derived from the marginal littoral environment of the Aptian sea.

Unfortunately, the Aptian rocks of the Palmyrides do not contain any identifiable organic remains. The age of these rocks is derived from their lithologic composition and stratigraphic position. The terrigenous sediments are intercalated between the carbonate units of the Jurassic at their base and the Albian at top, though in the Eastern Palmyrides the boundary with the Albian Stage still remains somewhat unclear.

In the extreme northeast of Syria, Aptian rocks are exposed in the center of the **Jebel Abd El-Aziz ridge**, where they overlie Hauterivian-Barremian variegated rocks and are overlain by the Albian limestones. The Aptian of this area is represented by green calcareous clays with thin streaks of white limestones and poorly preserved fossils: *Tylostoma rochatianum* d'Orb., *Trochus* sp., *Akera* sp., *Biornis* aff. *praelongum* Sharpe, *Ceratostreon* aff. *tysiphone* Coq., *Roloboceras* sp., *Pholadomya* sp., and *Dosinia* sp. (Ponomarev and Bivshev, 1966).

Albian [K_1al]

Albian sediments have been encountered almost everywhere in Syria, with the exception of the Rutbah dome in the Syrian Desert (according to drilling data). Albian strata conformably overlie the Aptian ones. Southeast of Tudmor, as the result of local erosional events, Albian carbonates rest transgressively on Bathonian (Middle Jurassic) strata.

As compared to the widespread Neocomian and Aptian-earliest Albian terrigenous formations, the main Albian sequence is now represented only by carbonate deposits of various types. In the Anti-Lebanon and the southern part of Jebel An-Nusseiriyeh, limestones predominate. In the northern part of Jebel An-Nusseiriyeh, Kurd-Dagh, and the Palmyrides, the most common rock-types are dolomites with lesser amounts of limestones and marls. In northeastern Syria (Jebel Abd El-Aziz), the Albian consists of dolomites, limestones and clays.

In the **Anti-Lebanon**, Albian deposits remain similar in composition and character of bedding. They consist of brownish, greenish and yellowish limestones with some marls. The limestones are often built up of coquinas, sometimes made of pelecypods, gastropods, or a mixture of both. Clayey limestones are less common (Razvalyaev, 1966a; Kurbanov and Zarjanov, 1968).

Near the village of Beit Jann (Jebel Ash-Sheikh Range) a typical section of Albian organogenic detritic limestones was measured (Fig. 6.23, Section I). From base to top it is made up of:

1) Slightly clayey, greenish-yellow, coarse-bedded limestones alternating with brownish varieties. They are abundant in poorly preserved pelecypods, gastropods, and sea urchins. The thickness is 40 m;

2) Green calcareous clay, 1 m thick;

3) Clayey limestones with the pelecypods *Neithea* cf. *quinquecostata* Sow. and *Protocardia hillana* Sow. (det. N. Bobkova) and the sea urchin *Heteraster* aff. *delgadoi* Lor. (det. A. Kalis). The thickness is 10 m;

4) Gray pelitomorphic massive-bedded limestones with ochrous thick-bedded pelitomorphic limestone intercalation, containing unidentifiable remains of molluscs and sea urchins. The thickness is 30 m;

5) Detrital gray limestones with numerous shells of pelecypods - *Neithea* cf. *quinquecostata* Sow., *Trigonia (Buchotrigonia) fraasi* Cox, and *Ceratostreon flabellatum* Goldf., and the sea urchin *Heteraster* sp. The thickness is 14 m;

6) Brownish ochrous pelitomorphic limestones with the ammonites *Knemiceras syriacum* Conr., *K. compressum* Basse, and *K.* aff. *dubertreti* Basse (det. V. Rengarten). The thickness is 39 m.

The total thickness of the Albian deposits in the Beit Jann section is 134 m.

Albian deposits are also well-exposed on the northwestern slope of Jebel Shaqif El-Khorn (along Wadi Mannsour) (Fig. 6.23, Section II). They are characterized by a similar lithology, differing only slightly from the Beit Jann section in the intercalation of marls, thin-platy limestones at the base of the section, and lumpy limestones in its upper part. The thickness is almost the same at 118 m. Among pelecypods *Exogyra flabellata* Goldf., *Trigonoarca* sp., and *Trigonia (Buchotrigonia) fraasi* Cox are usually present (det. N. Bobkova).

In the westernmost part of the Anti-Lebanon (the Chir Mannsour Range, near the village of Batroun), organogenic limestones are replaced by massive pelitomorphic limestones and clayey limestones. Only at the top of this section can a member of yellowish lumpy organogenic limestones with abundant fossils be traced. In these layers, L. Dubertret (1949a) identified *Heteraster delgadoi* Loriol and *Ostrea flabellata* Goldf. The thickness of the Albian deposits in this range reaches up to 170 m.

Albian deposits are also well-developed in the southern parts of **Jebel An-Nusseiriyeh** (Kozlov, 1966a) (Figs. 6.26 and 6.27). Near the village of Kafroun (Kafroun Bash Shour), they are characterized by well-bedded (up to 0.3-0.5 m) light-gray, compact, fine-grained dolomites. At the bottom of the section dolomites are intercalated with 0.1-0.2 m thick green clays and at the top, layers of light-gray fine-grained dolomitic limestones are present. The thickness is about 90 m. Only fragments of unidentifiable pelecypods have been found in this section, so that its Albian age has been derived from its lithology and stratigraphic position. Similar deposits have been encountered on the southeastern slope of Jebel An-Nusseiriyeh as well as in the vicinity of the villages of Bamara and Ofelon, where they consist of gray and yellow dolomites with interbeds of lumpy limestones. Their thickness is practically the same at 100 m.

In the northern-most areas of Jebel An-Nusseiriyeh, the Albian deposits preserve their main features and are represented by limestones, clayey limestones, and dolomites, but the ratio of these three rock-types varies (Shatsky et al., 1966) (Fig. 6.28). Near the villages of Bab Janneh and Qadmous, aphanitic and clayey limestones predominate and intercalations of dolomites are comparatively rare; the thickness is about 70 m. On the eastern slope of this range (to the west of Hama), limestones are almost entirely replaced by dolomites and magnesian marls; their thickness reaches up to 100 m. In the surroundings of Slenfeh, aphanitic and lumpy light-gray limestones prevail and dolomites are typical for the upper part of the section; the thickness is 70 m. The Albian sediments of Jebel An-Nusseiriyeh are almost unfossiliferous. Therefore, L. Dubertret's (1937a) paleontological findings in the Slenfeh section of *Knemiceras syriacus* Conr., *Trigonia ethra* Coq., *Turbo moralli* Fraas, and *Nerinea cochlalaformis* Conr. are quite important.

In the northwestern part of **Kurd-Dagh** (a section along the railroad, in a narrow V-shaped valley near the village of Baraband) dolomites approaching 120 m in thickness belong to the Albian. These dolomites vary in lithology: dark-gray, yellowish-gray and gray, massive, fine-grained, coarse- to medium-bedded, sometimes aphanitic, slightly bituminous and brecciated. In the upper half of the formation, thin-bedded microgranular dolomites appear, containing fine cavities filled with fibrous chalcedony and iron oxides. No organic remains occur in these dolomites and the Albian/Cenomanian boundary is arbitrary (Protasevich and Maksimov, 1966).

In the central regions of Syria, the **Palmyrides** are very important for understanding the general stratigraphy of the Albian interval. Here the mostly dolomitic section is intercalated with organogenic limestones containing abundant fauna.

The best Albian section is located in the eastern part of Jebel Ad-Dmeir, where it has a maximum thickness of 270 m (Razvalyaev, 1966a). The following members can be traced from the base of the section upwards (Fig. 6.25):

1) Alternation of brownish-gray, massive pelitomorphic organogenic limestones and greenish-gray lumpy clayey limestones with the pelecypods *Ceratostreon flabellatum* Goldf., *Amphydonta (Aetostreon) arduennensis* Coq., *Protocardia hillana* Sow., and *Cytherea plana* Sow. (det. N. Bobkova and N. Cheltsova). The thickness is 32 m;

2) Greenish-gray and gray dolomites, clayey and calcareous, compact, fine-grained, massive, sometimes lumpy, alternating with bright yellow, fine- to medium-grained varieties and subordinate thin layers of yellow magnesian marl. These sediments are characterized by numerous pelecypod shells such as *Cardita blackei* Blanck., *Neithea shawi* Perv., and *Venus reynesi* Coq. The thickness is 45 m;

3) Brownish-gray massive, clayey and calcareous, unfossiliferous dolomites. The thickness is 50 m;

4) White, light-gray and yellowish, massive, medium-grained limestones with interbeds of thin platy varieties at the bottom of the member. The thickness is 34 m;

5) Alternation of green and greenish-gray, foliated magnesian marls and light-gray, yellowish and greenish clayey organogenic limestones. These rocks contain abundant pelecypods, gastropods, sea urchins, and ammonites. Especially numerous are pelecypods: *Nucula perobliqua* Conr., *Liopistha (Psilomya) gigantea* Sow., *L. (P.) alta* Roemer, *Cardita blackei* Blanck., *C.* cf. *forgemoli* Coq., *C. forgemoli* Coq. var. *delettrei* Coq., *Neithea* cf. *subatavus* Blanck., *N. dutrugei* Coq., *Pholadomya* cf. *costazari* Almonte, *Ph. pedernalis* Rom., *Ph.* cf. *vignesi* Lartet, *Cardium (Trachycardium)* aff. *productum* Sow., *Amphidonta luynesi* Lartet, *A. conica* Sow., *A.* cf. *mermeti* Lartet, *Ceratostreon flabellatum* Goldf., *C.* aff. *boussigaulti* d'Orb., *Lopha syphax* Coq., *Anisocardia* cf. *aquilina* Coq., *Venus* cf. *mauretanicus* Coq., *V.* cf. *perovalis* Conr., and *Exogyra* spp. Due to its abundance of *Exogyra* shells, this member has been named 'the Exogyra layers'. This unit is clearly visible in the field and serves as a good marker horizon. Among sea urchins, *Hemiaster syriacus* Conr., *Caenholectypus subpentagonalis* Blanck., and *Heterodeadema libycum* Cott. were identified by A. Kalis. The ammonites *Knemiceras syriacum* Conr. and *Knemiceras* sp., discovered by V. Rengarten, are reliable for assuming an Albian age. The thickness is 36 m;

6) Gray and brownish-gray dolomitic, medium- to massive-bedded limestones intercalated with clayey calcareous dolomites and clayey organogenic limestones. The thickness is 50 m.

In Jebel Ghorab and Jebel Shaq Semri, to the northeast of Jebel Ad-Dmeir, only the uppermost part of the Albian is exposed, consisting of dolomites and some limestones, not exceeding 10-20 m (Razvalyaev, 1966 b).

In the Central Palmyrides (between Al-Qaryatein and Tudmor), the Albian is almost exclusively composed of dolomites. A maximum thickness of 120 m for these Albian rocks has been observed in Jebel Rmah, where eight members of dolomites have been recognized, distinguishable by their color, structure, texture and clayey admixture (Faradzhev, 1966) (**Fig. 6.30**).

To the east of Wadi El-Bardeh in the Jebel An-Naqnaqiyeh Range, the Albian begins with alternating yellow dolomites, green calcareous clays, and white loose quartz sandstones. The main part of the succession is made up of very monotonous, gray and yellow dolomites; the slightly smaller thickness varies here from 80 to 100 m. In Jebel Om Journ, the dolomites and clayey dolomites thin to

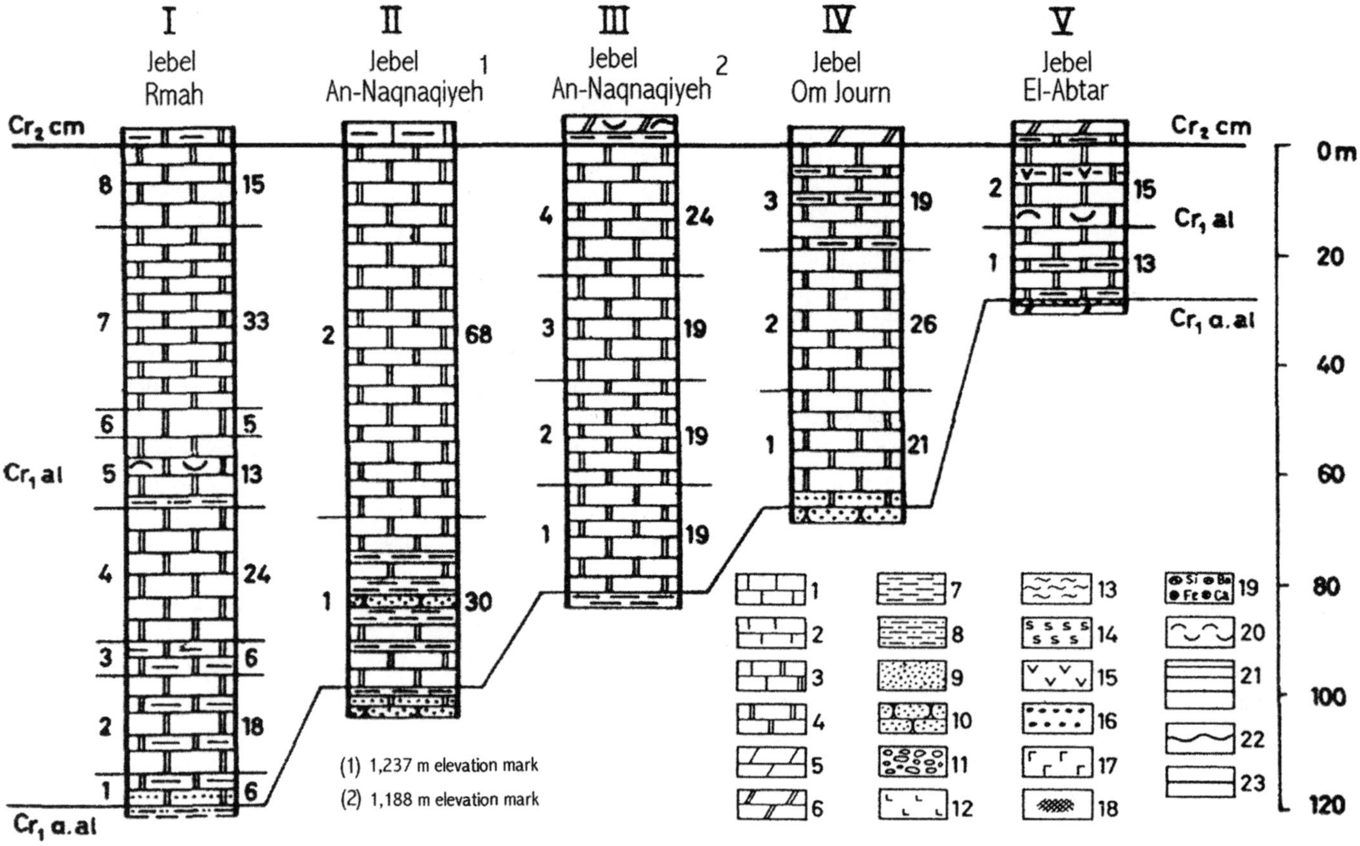

Fig. 6.30. Albian deposits sections of the Central Palmyrides, southern branch (after Faradzhev, 1966).

Legend: Sections: I - Jebel Rmah; II - Jebel An-Naqnaqiyeh (1237 m elevation mark); III - Jebel An-Naqnaqiyeh (1188 m elevation mark); IV - Jebel Om Journ; V - Jebel El-Abtar.

Lithologic symbols: the same as on Fig. 6.29.

66 m.

A minimum thickness of 28 m has been observed at Jebel El-Abtar, where the Albian consists of two members. The lower member is represented by white, yellow, cream and pink dolomites, thin- to medium-bedded, fine- to micrograined, fine-porous, generally calcareous (13%) and sometimes clayey admixtures are present, but never exceed 11%. The upper member is characterized by harder dolomites, medium- to thick-bedded, with intercalations of completely dolomitized crinoidal limestone and detrital limestone containing gastropod and pelecypod casts (*Trigonoarca* cf. *trigona* Seguenza). In places, the dolomites are sandy and incorporate lenses and ball concretions of gypsum (up to 20 cm in diameter).

The eastern Palmyrides are characterized by unfossiliferous Albian deposits which are difficult to differentiate from Aptian rocks (Soulidi-Kondratiev, 1966a). In Jebel Hayyan, about 42 m of alternating white, yellow and gray, clayey, thin- and medium-bedded dolomites, dolomitic marls and dolomitized limestones with pelecypod and gastropod casts are conventionally attributed to the Albian. To the northeast of Tudmor (Palmyra) in the sections of Jebel El-Mazar and Jebel As-Satih, unfossiliferous, medium- to coarse-grained, thin- and medium-bedded yellow dolomites reach a thickness of about 20 m, with intercalations of pink and yellow quartz dolomitic sandstones, greenish clays and white and gray gypsum.

In the extreme northeast of Syria (**Jebel Abd El-Aziz**), yellow and yellow-brown compact dolomites and dolomitized limestones with thin interbeds of dolomitic clay and sandy limestones were assigned to the Albian without paleontological substantiation.

Upper Cretaceous [K_2]

As a rule the Upper Cretaceous deposits in Syria rest conformably on Albian limestones and dolomites. They are well-exposed in the axial parts and on slopes of the anticlinal structures of the Anti-Lebanon, the Palmyrides, and Kurd-Dagh, as well as on the western and eastern slopes of the Jebel An-Nusseiriyeh Uplift. Upper Cretaceous deposits are predominantly carbonatic and terrigenous and effusive rocks play an insignificant role.

Lithologically, the Upper Cretaceous can be divided into two main parts.

The lower part includes the Cenomanian, Turonian, and part of the Coniacian. It is essentially composed of carbonate rocks - dolomites and hard limestones with subordinate intercalations of marls, calcareous clays and practically no chalk. In the old geological literature, this subdivision was frequently designated as the Middle Cretaceous (including the Albian).

The upper part (Senonian) is predominantly formed of soft chalky limestones and marls representing part of the the Coniacian, and the Santonian, Campanian, and Maastrichtian. Three lithological subdivisions can be made: lower soft chalky limestones of Coniacian-Early Campanian age, a persistent middle series of black and brown flints of Late Campanian age and an upper unit of chalks of Maastrichtian age. Within the Upper Campanian-Lower Maastrichtian, huge accumulations of phosphorites are present and oil shales are typical for the Maastrichtian of some areas of Syria.

The Upper Cretaceous contains very rich assemblages of macrofauna (ammonites, pelecypods, gastropods, sea urchins) and microfossils (foraminifera, calcareous nannoplankton). Macro-faunas are especially important for biostratigraphy of the Cenomanian-Coniacian interval, and retain some significance for the Santonian-Maastrichtian interval. Microfaunas (abundant planktonic and benthic foraminifera) provide reliable subdivision for the latter.

The best continuous and most complete (Cenomanian-Maastrichtian) sections are located in the Anti-Lebanon (Razvalyaev, 1966a; Kurbanov and Zarjanov, 1968) and in the Palmyrides (Razvalyaev, 1966a; Faradzhev, 1966; Soulidi-Kondratiev, 1966a; Oufland, 1966a). In Jebel An-Nusseiriyeh, excellent sections can also be observed, but they are only complete in the southern part of this uplift (Kozlov, 1966a) whereas to the north of Jebel An-Nusseiriyeh, sections are continuous in the Cenomanian-Coniacian interval, with some gaps in the Senonian (Shatsky et al., 1966).

Upper Cretaceous sections are incomplete, due to poor exposure or as a result of breaks in the sedimentation in the Kurd-Dagh (Protasevich and Maksimov, 1966; Protasevich and Kurbatov, 1968), southwestern slope of the Palmyrides (Razvalyaev, 1966b), northern slope of Jebel El-Balas (Mikhailov, 1966), the Jrablus Uplift (Oufland, 1966b), and the center of Jebel Abd-El-Aziz (Ponomarev and Bivshev, 1966).

Important data on the Upper Cretaceous stratigraphy of Jebel An-Nusseiriyeh are also given in the explanatory notes to the Geological Map of Syria at a scale of 1:50,000 by geologists of the General Establishment of Geology and Mineral Resources of Syria (Yousef, 1979, 1980a,b; Ruske, 1978a,b; Adjamian and Jamal, 1983; Adjamian and Al-Bakka, 1984).

All of the above data, along with the classical studies of L. Dubertret and our observations on planktonic and benthic foraminifera, make it possible to provide a very detailed picture of the Upper Cretaceous stratigraphy of Syria.

Cenomanian [K_{2c}cm]

As already pointed out, the Albian-Cenomanian boundary is rather vague. The Cenomanian deposits are connected with Albian ones by a gradual transition. The Cenomanian is represented by hard limestones and dolomites with subordinate intercalations of soft marls and calcareous clays. Limestones predominate in the western regions of Syria, whereas dolomites are widely developed in the rest of Syria. Therefore, three facies types can be singled out:

1) A preponderance of limestones;
2) A preponderance of dolomites; and
3) Transitional mixed facies.

Among the limestones, organogenic varieties are very common and contain various groups of fossils. The guide fossils of the Cenomanian in Syria include ammonites – *Mantellicoceras mantelli* d'Orb., *Neolobites fourtani* Perv., *Calycoceras boulei* Collign., and *Entrephoceras sublaevigatum* d'Orb., sea urchins – *Hemiaster syriacus* Conr., *H. ibilensis* Loriol, and *Diplopodia hermonensis* Loriol, and molluscs (including rudists) – *Amphidonta africana* Coq., *A. mermeti* Lartet, *A. columba* Desh., *A. suborbiculata* Lam., *A. (Aetostreon) conica* Sow., *A. (A.) haliotidea* d'Orb., *Ceratostreon flabellatum* Goldf., *Eoradiolites syriacus* Conr., and *E. lyratus* Conr. Planktonic foraminifera have a subordinate significance because of the rarity of washable soft clayey limestones and marls.

In the **Anti-Lebanon**, Cenomanian deposits are exposed as narrow belts and spots in the most eroded and uplifted areas of the Jebel Ash-Sheikh, Jebel Hassiyya, Jebel Qasyoun, and Jebel Suekai anticlines (Razvalyaev, 1966a; Kurbanov and Zarjanov, 1968).

Near the village of Beit Jann on the southeastern slope of Jebel Ash-Sheikh, the southwestern-most section of the Cenomanian is exposed as a succession of rather monotonous dolomites and some limestones up to 900 m thick (**Fig. 6.31**). The limestones mainly occur in the lower part of the sequence, which (from base to top - numbers refer to original text of the explanatory notes) consists of:

1) Organogenic, thick-bedded brownish limestones with *Ceratostreon flabellatum* Goldf. and *Neithea* cf. *quinquecostata* Sow. (det. N. Bobkova). This marker-band of shelly limestones was adopted in the field as the Albian/Cenomanian boundary. The thickness is 50 m.

2) Pelitomorphic limestones, gray, thick-bedded with yellow, platy interbeds, yielding *Cardita* ex gr. *forgemoli* Coq., *Liopistha (Psilomya) pervinquieri* Greco, and *Pholadomya* ex gr. *vignesi* Lartet. The thickness is 25 m.

3) Calcareous, thick-bedded, pelitomorphic, light-gray dolomites containing *Ceratostreon flabellatum* Goldf. The thickness is 15 m.

4) Pelitomorphic, gray, massive-bedded limestones with poorly preserved sea urchins and pelecypods. The thickness is 51 m.

5) Monotonous gray, coarse-bedded, fine-grained dolomites without fossil remains. The thickness is 160 m.

6) Gray and yellowish, massive-bedded, pelitomorphic limestones with numerous shells of *Pholadomya* ex gr. *vignesi*

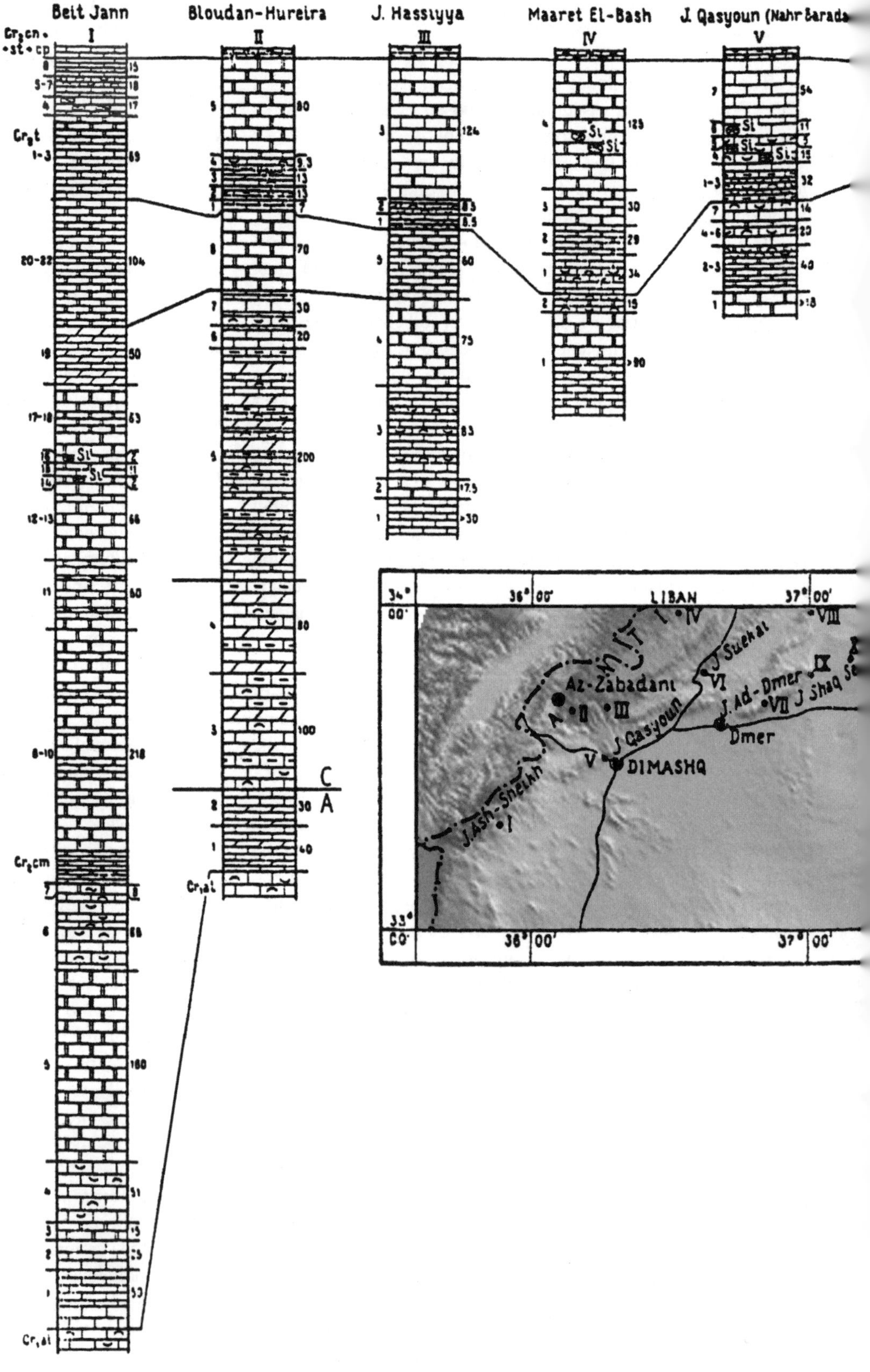
Beit Jann
I
Bloudan-Hureira
II
J. Hassiyya
III
Maaret El-Bash
IV
J. Qasyoun (Nahr Barada
V
C
A
LIBAN
Az-Zabadani
J. Qasyoun
DIMASHQ
J. Ash-Sheikh
J. Suekat
J. Ad-Dmer
Dmer
J. Shaq
34° 00'
36° 00'
37° 00'
33° 00'

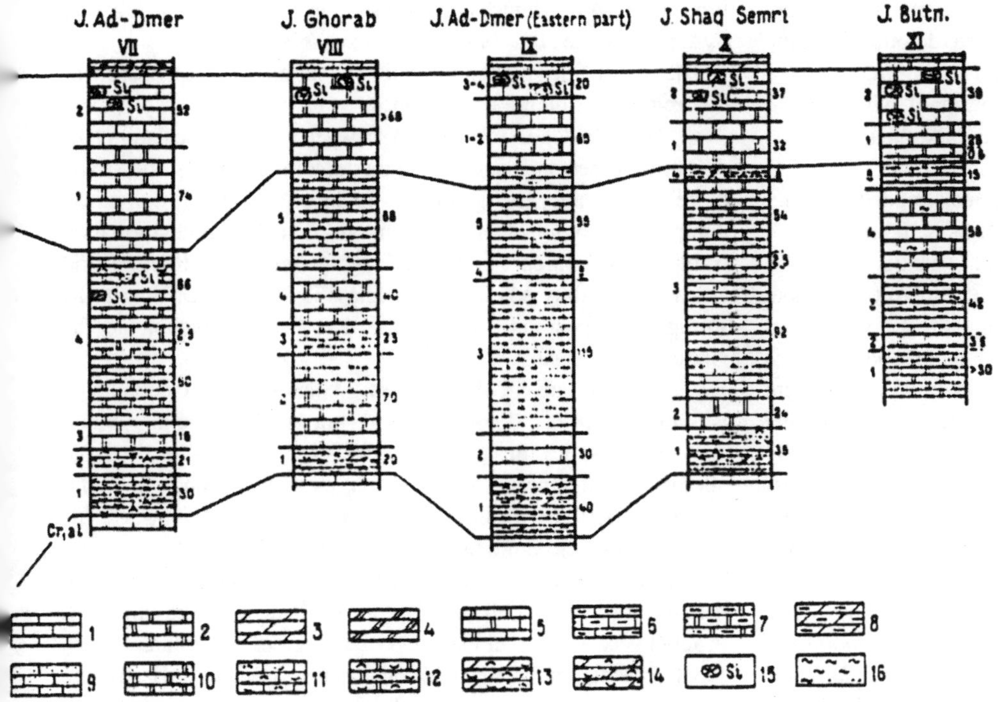

Fig. 6.31. Cenomanian and Turonian deposits sections: the Anti-Lebanon and the southwestern Palmyrides (after Razvalyaev, 1966a).
Legend: 1 - limestone; 2 - dolomite; 3 - marl; 4 - dolomitic marl; 5 - dolomitic limestone; 6 - clayey limestone; 7 - clayey dolomite; 8 - clayey marl; 9 - sandy limestone; 10 - sandy dolomite; 11 - shelly limestone; 12 - dolomite with fauna; 13 - marl with fauna; 14 - dolomitic marl with fauna; 15 - flinty concretions and nodules; 16 - bands and lenses of flints.

Lartet, *Liopistha (Psilomya)* cf. *pervinquieri* Greco, and *Ceratostreon flabellatum* Goldf. The thickness is 66 m.

7-18) A very thick formation of alternating pelitomorphic, thick, gray, bedded and coarse-bedded fine-grained dolomites. Some layers contain siliceous nodules. The thickness is 378 m.

19) Alternating light-gray marls and limestones with *Cardita* cf. *forgemoli* Coq., *Trigonoarca trigeri* Coq., and *Hemiaster* aff. *saulegi* d'Orb. The thickness is 50 m.

20-22) Alternation of gray, pelitomorphic thick-bedded and fine-grained coarse-bedded dolomites. The thickness is 104 m.

The Cenomanian deposits thin and change facies in a northeasterly direction along the trend of the Anti-Lebanon, On the slope of the Cheqif-el-Khorn Range (in the vicinity of the village of Bloudan), the Albian/Cenomanian boundary is defined by the transition from organogenic Albian limestones into a formation of rhythmic alternating Cenomanian limestones and marls. At Bloudan, the Cenomanian deposits can be lithologically subdivided into two formations which vary greatly in thickness (Fig. 6.31). The lower formation (subunits 1-7) is represented by a regular alternation of various limestones and marls, with a total thickness of 500 m, conposed of:

1) Rhythmic alternation of yellowish-gray, lumpy marls and gray, platy, pelitomorphic limestones with *Cardita forgemoli* Coq. var. *delettrei* Coq., *Protocardia* cf. *hillana* Sow., and *Pholadomya* ex gr. *vignesi* Lartet (det. N. Bobkova). The thickness is 40 m.

2) Rhythmic alternation of thin-platy, microbedded, pelitomorphic and coprogenic limestones and yellowish, lumpy marls with *Ceratostreon flabellatum* Goldf. and *Pholadomya vignesi* Lartet. The thickness is 30 m.

3) Rhythmic alternation of pelitomorphic, fine-grained, coprogenic and coprogenic-organogenic, thin-platy limestones and yellowish thick-layered marls. The thickness is 100 m.

4) Rhythmic alternation of gray, clayey, organogenic, pelitomorphic dolomitic limestones and white massive, coarse-bedded marls. The thickness is 80 m.

5) Alternation of organogenic, pelitomorphic, argillaceous and dolomitic limestones, calcareous dolomites and white chalk-like compact, massive-bedded marls. The thickness is 200 m.

Layers 3-5 in the limestones yielded numerous shells of *Ceratostreon flabellatum* Goldf., *Pholadomya vignesi* Lartet, and *Protocardia* ex gr. *hillana* Sow., and in the soft marls rich assemblages of planktonic foraminifera such as *Rotalipora appenninica* (Renz), *Praeglobotruncana stephani* (Gandolfi), *P. delrioensis* (Plummer), *Hedbergella planispira* (Tappan), *H. delrioensis* (Carsey), *H. simplex* (Morrow), *Schackoina cenomana* (Schacko), *Heterohelix moremani* (Cushman), and *Globigerinelloides ultramicra* (Subb.). In layers 3 and 4, these species are associated with *Rotalipora brotzeni* (Sigal), which assigns these sediments to the *Rotalipora brotzeni* Zone; *Rotalipora reicheli* Mornod in layer 5 marks the base of the *Rotalipora reicheli* Zone. Thus, layers 3-5 are definitely of Early-Middle Cenomanian age.

6) Light-gray, pelitomorphic, organogenic, thick-layered limestones intercalated with 1.5-2.0 m thick beds of clayey limestones, topped by a 3 m thick bed with oyster shells. The thickness is 20 m.

7) Alternation of clayey, thin-platy to thick-bedded, pelitomorphic limestones with greenish and reddish tints, with a 1-m-thick bed of oyster limestone at the base. The thickness is 30 m.

The 70 m thick upper formation (subunit 8) is composed of dolomites and dolomitic limestones:

8) Thick-bedded, massive dolomites with intercalations of dolomitic limestones and calcareous dolomites. Along the strike, these limestones are gradually replaced by pure dolomites, completely devoid of organic remains. This dolomite formation is clearly identified in the field by its massive structure and characteristic gray color. The thickness is 70 m. According to its stratigraphic position, this formation is best attributable to the Upper Cenomanian.

In other areas of the Anti-Lebanon, the Cenomanian is composed of dolomites and limestones (pelitomorphic and organogenic) in a very inconsistent ratio. The rhythmic alternations of limestones and marls, as well as bands of shelly (oyster) limestones at the base, are at times absent, rendering its discrimination from the underlying Albian difficult. This situation is observed in the southeastern slope of Jebel Shu'ayb Mannsour, near the border with Lebanon. Due to the character of the exposures in Jebel Hassiyya, Jebel Qasyoun (the gorge of the Barada River), Jebel Suekai, and Jebel Maaloula (near Maaret El-Bash), the Cenomanian is represented only by its upper part. The thickness of these incomplete sequences varies from 92 m (at Jebel Qasyoun) to 250 m (at Jebel Suekai). Here the calcareous dolomites are sometimes characterized by a 'vermicular' structure, pointing to bio-turbation by mud-eating organisms.

Rich faunas have been collected in the above-mentioned localities of the Anti-Lebanon. Among them are the pelecypods: *Exogyra africana* Lam., *Amphidonta columba* Lam., *Cardita forgemoli* Coq., *Pholadomya vignesi* Lartet, *Liostrea delettrei* Coq., *Ceratostreon flabellatum* Goldf., *Dosinia delettrei* Coq., and *Lopha syphax* Coq. (det. N. Bobkova); sea urchins: *Hemiaster syriacus* Conr., *Heterodiadema libycum* Cott., *Diplopodia hermonensis* Lor., *Coptodiscus larteti* Cott., *Hemiaster luynesi* Cott., and *Caenholectypus subpentagonalis* (Blanck.) (det. N. Poslavskaya). The most common ammonite is *Neolobites fourtani* Perv. (det. A. Atabekyan). It should also be noted that for the southwestern region of the Anti-Lebanon, long lists of Cenomanian fossils were given by L. Dubertret (1949a, 1954, 1955a).

In the extreme northeast of the Anti-Lebanon, the Albian/Cenomanian contact is not exposed and the Cenomanian is only represented by its upper part. Near the village of Harf Ath-Thallajeh this upper part consists of 475 m of dolomites. Directly south of Jebel Meiri on the slope of the valley, the sections thin to 345 m and even to 240 m, as observed near the village of Jumrah (Figs. 6.26 and 6.27). Throughout this region the Cenomanian is composed of gray and yellowish-gray, fine-grained, compact, massive, 0.3-0.5 m thick beds of sometimes pelitomorphic or medium-grained dolomites with a sugary structure. They are intercalated by sedimentary breccias, whose large 2-10 cm angular fragments are cemented by dolomitic matrix showing conchoidal fracture. In places, these secondary dolomites still contain interbeds of dolomitic limestones with poorly preserved fragments of oysters and sea urchins (Kozlov, 1966a).

Approximately in the middle of these sections at Jebel Meiri, and in the Al-Koroum and Martabya Valleys, a 15 to 25 m thick member is found consisting of variable light-gray, fine-grained, compact and massive, clayey, shelly limestones, with numerous fragments of fossils. This so-called 'Al-Koroum member', is well-exposed in the Al-Koroum Valley and can be followed over large distances within the Northern Anti-Lebanon. The following assemblage of fossils is particular for the Cenomanian Al-Koroum shelly limestones: *Neolobites fourtani* Perv., *N.* cf. *vibrayeanus* d'Orb. (det. A. Atabekyan), *Hemiaster syriacus* Conr., *H. laynesi* Cott., *Ceptodiscus larteti* (Cott.), *Heterodiadema libycum* (Cott.), *Cidaris beyrouthensis* Lor. (det. A. Kalis), *Tapes* ex gr. *numidicum* Coq., *Dosinia* ex. gr. *delettrei* Coq., *Dosinopsis* sp., *Gryphaea* cf. *vesiculosa* Sow., *Exogyra* ex gr. *flabellata* Goldf., *Trigonia* sp., *Pholadomya vignesi* Lartet, *Loipistha (Psilomya)* cf. *pervinquieria* Greco, *Cardita* cf. *forgemoli* var. *begnesi* Coq., *Anisocardia aquilina* Coq., *Ceratostreon flabellatum* (Goldf.), *C. olisiponensis* (Scharpe), *Amphidonta africana* (Coq.), *A. suborbiculata* (Lam.), and *A. (Aetostreon)* cf. *conica* (Sow.) (det. N. Bobkova and N. Cheltsova).

Good exposures on the southwestern slope of **Jebel An-Nusseiriyeh**, in the vicinity of the villages of Safita, Kafroun Bash Shour, Hussein El-Bahar, etc. (Figs 6.26 and 6.27) reveal a Cenomanian succession (about 370 m thick) of limestones and marls with some intercalations of dolomites. The following subunits were measured (from base to top) (Kozlov, 1966a):

1) Basalts (anamesites), strongly weathered, appear as dark-gray spheroids up to 1 m in diameter. Unaltered rock is only preserved in their centers. This flow is present almost everywhere in the southern part of Jebel An-Nusseiriyeh and, as such, is a good marker for geological mapping. Only near the village of Marmarita is this basalt replaced by sandstone, composed of slightly cemented basalt grains, with intercalations of tuffs of basic effusive rocks and fine-layered dolomites. The latter are contaminated by fragments of volcanic glass and ferric hydroxides. Ripple marks can be clearly seen on the bedding planes of the dolomite. The thickness is 15 m.

2) Dolomite, light-gray, compact, fine-grained, usually thick-bedded and massive, with rare flint concretions. The thickness is 40 m.

3) Basalt, strongly weathered with typical spherical texture, similar in composition to the lower one (subunit 1). It pinches out in a southerly direction and is replaced by light-gray, fine-grained and thick-bedded limestones, sometimes with poorly preserved pelecypod casts. The thickness is 30 m.

4) Light-gray, thick-bedded limestones with sporadic flint concretions. Near the village of Kafroun Bash Shour, L. Dubertret (1937a) evidently collected the following fossils from the beds directly overlying the basalt: *Septifer lineatus* Sow., *Arca (Macrodon) cenomanensis* d'Orb., *Cerithium (Pirenella) rudistarum* Basse., *Cardita lacunai* Hamlin, *Cardium productum* Sow., *Eoradiolitus lyratus* Conr., and *E. syriacus* Conr. The thickness is 20 m.

5) White, pelitomorphic, soft, fine-bedded and platy marls with intercalations of compact clayey limestones. This marl bed is a pronounced marker throughout the southern Jebel An-Nusseiriyeh and was very useful during the geological mapping. It is named the 'Maisra member' after the nearby locality. Intercalations of shelly limestones near the village of Beshraib contain pelecypods: *Exogyra* cf. *mermeti* Lartet, *E. flabellata* Goldf., *Neithea* ex gr. *shawi* Perv., *N. quinquecostata* Sow., *Pholadomya* sp., *Trigonia* sp., and *Protocardia* sp. (det. N. Bobkova), and the ammonite *Mantelliceras mantelli* d'Orb. (det. A. Atabekyan). The thickness is 20-30 m.

6) White and cream, pelitomorphic and fine-grained, compact, about 0.5 m thick bedded limestones with thin interbeds of marls, organogenic limestones and silicified limestones with conchoidal texture. The thickness is 40 m.

7) Light-gray, fine-grained, 0.2-0.3 m thick well-bedded limestones with intercalations of clayey limestones and very compact brown limestones with numerous flint concretions. The thickness is 60 m.

8) Gray, massive, thick-bedded limestones, forming a topographic scarp. This limestone is locally abundant in shell remains, which may form a coquina. This massive and shelly limestone is well observed over the entire Jebel An-Nusseiriyeh area and carries the local name of the 'Slenfeh limestones' (Dubertret, 1937a). To the west of the village of Safita, this member contains the sea urchin *Hemiaster saulcyanus* d'Orb. (det. A. Kalis) and various pelecypods such as *Protocardia pauli* Coq., *Trigonia* sp., *Neithea* ex gr. *dutrugei* Coq., *Gryphaea vesiculosa* Sow., *Amphidonta* aff. *suborbiculata* Lam., *A.* aff. *mermeti* Coq., *A. columba* Lam., *A. (Aetostreon) haliotidea* d'Orb., *A.* cf. *conica* Sow., and *Ceratostreon olisiponensis* Scharpe (det. N. Bobkova and N. Cheltsova). The thickness is 10 to 20 m.

9) White and light-gray, fine-grained, compact, thick-bedded limestones with ball-shaped flint concretions and thin interbeds of shelly limestones. The thickness is 40 m.

10) White, clayey, thin-platy, fine-grained limestones, becoming laterally thick-bedded. The thickness is 10 m.

11) Gray and light-gray, fine-grained and pelitomorphic, thick-bedded 0.5-1.2 m thick limestones with flint concretions, intercalated with white and greenish-gray, clayey, soft, thin-bedded limestones. The latter contain sea urchins: *Diplopodia hermonensis* Lor., *Hemiaster ibelensis* Lor., and *H.* cf. *serutiger* Forb. (det A. Kalis), and ammonites: *Calycoceras boulei* Collign. and *Eutrephoceras sublaevigatum* (d'Orb.) (det. A. Atabekyan). At the top of this subunit interbeds of shelly and rudist limestones form isolated bioherms containing: *Eoradiolites* ex gr *syriacus* Conr., *Radiolites* sp., *Bournonia* sp., *Neithea dutrugei* Coq., *N.* cf. *laevis* Drouet., *Protocardia pauli* Coq., *P.* ex gr. *hillana* Sow., *Liopistha (Psilomya) pervinquierei* Greco, *L.* (P.) ex gr. *alta* Roem., *Trigonia ethra* Coq., *Lopha* sp., and *Exogyra* sp. (det. N. Bobkova). The thickness is 70 m.

On the southeastern slope of Jebel An-Nusseiriyeh in the vicinity of the town of Missyaf and between the villages of

Ba'mara and Ofelon, the Cenomanian succession assumes a different character. Here, basaltic rocks are absent and are replaced by dolomites. The limestones are light-gray, compact, fine-grained, usually thick-bedded, sometimes clayey pelitomorphic or silicified with numerous flint concretions. The dolomites and dolomitic limestones are yellowish-gray, thick-bedded, with numerous karst hollows and caverns. At the base of the succession, a shelly limestone yields the pelecypods *Amphidonta suborbiculata* Lam., *A. africana* Coq., *A. (Aetostreon) conica* Sow., *Trigonia ethra* Coq., and *Cardita forgemoli* Coq. (det. N. Cheltsova). The uppermost layers (30 m thick) consist of white clayey chalky limestones with planktonic foraminifera – *Schakoina cenomana* (Schacko), *Rotalipora brotzeni* (Sigal), *R. appenninica* Renz, *Hedbergella infracretacea* (Glaessner), *H. planispira* (Tappan), *H. globigerinellinoides* (Subb.), *Praeglobotruncana delrioensis* (Plummer), and *Globi-gerinelloides ultramicra* (Subb.). The upper part of the Cenomanian is truncated here by a mantle of Pliocene basalts .The exposed Cenomanian is 285 m thick.

In the northern part of Jebel An-Nusseiriyeh, on its western slope, Cenomanian strata are well-exposed and represented by massive and bedded limestones and dolomites intercalated with oyster limestones, chalk-like limestones and marls. In contrast, on the eastern slope, dolomites predominate. Throughout this region, the contact with the Albian is conformable and gradational.

The northwestern slope of Jebel An-Nusseiriyeh exposes Cenomanian strata near the villages of Slenfeh, Bab Janneh, Ain Leiloun, Qerdaha, Qal'at Mahalbeh, etc. The sequence can be subdivided into three formations which are very uniform in lithology and thickness; changes along the strike are very minor.

The Cenomanian <u>Lower Formation</u> is about 150 m thick and consists of four members.

Member 1 begins with light-gray, thick-bedded limestones and dolomites containing rare rudist *Eoradiolites syriacus* Conr and is 36 m thick (Fig. 6. 28). The dolomites and limestones of this lower member are rather unfossiliferous and the lithology of the transitional layers is monotonous.

The marly member 2 is 23 m thick and consists of light-gray marls and chalk-like clayey soft limestones with flint lenses and spherical nodules. These sediments have yielded the sea urchin *Hemiaster kfourensis* Loriol and an assemblage of planktonic foraminifera including: *Rotalipora appenninica* (Renz), *R. brotzeni* (Sigal), *Praeglobotruncana delrioensis* (Plummer), *Schakoina cenomana* (Schacko), *S. multispinata* (Cushm. et Wick.), *Hedbergella planispira* (Tappan), *H. delrioensis* (Carsey), *H. simplex* (Morrow), and *Globigerinelloides ultramicra* (Subb.). This interval can be attributed to the *Rotalipora brotzeni* Zone of the Early Cenomanian age. This marly member is seen as a clear white band in the field and represents a useful marker since it is comparatively close to the Albian/Cenomanian boundary. Member 3 consists of characteristic massive and thick-bedded limestones and dolomites with a thin intercalation of marls that form a double stepped cliff. Dolomites are confined to the basal part, while limestones of the upper part yielded *Ceratostreon flabellatum* Goldf., *Amphidonta columba* Lam., and *Inoceramus* ex gr. *lynchi* Conr.; the thickness is 44 m.

Member 4 is composed of alternating thin-bedded dolomitic clayey limestones, microlaminated silicified limestones, loose detrital and oyster limestones, and marls; 45 m thick. Ammonites and pelecypods are rare but included: *Sharpeiceras laticlavium* Sharpe. *Mantelliceras* ex gr. *mantelli* Sow., and *Neithea quinquecostata* Sow.

The Cenomanian <u>Middle Formation</u> has a total thickness of about 150 m.and comprises two members. The lower member of up to 50 m thick massive, thick-bedded limestones and slightly dolomitic limestones was named by L. Dubertret (1937a) as the 'Slenfeh horizon'. This marker is shown in the 1:200 000 scale geological map of the Al-Latheqiyeh-Hama sheet . The upper 100 m thick member is composed of alternating, medium-bedded aphanitic, silicified limestones with flint nodules and chalk-like clayey and organic limestones with oyster shells, dolomite being subordinate. Ammonites, pelecypods, and sea urchins include *Acanthoceras* ex gr. *newboldi* Kossmat, *Calycoceras alaouitense* Basse, *C.* ex gr. *naviculare* Mant., *Gryphaea* cf. *vesiculosa* Sow., *Amphidonta canaliculata* d'Orb., *A. africana* Coq., *Arctostrea carinata* Lam., *Ceratostrea olisiponensis* Sharpe, *Neithea laevis* Drouet, *N. shawi* Perv., *Liopistha (Psilomya) pervinquieri* Greco, *Trigonia* cf. *ethra* Coq., *Protocardia pauli* Coq., *P. hillana* Sow., and *Carpinula cedrorum* Blanck. (Shatsky et al., 1966). Rare interbeds of soft limestones and marls include assemblages of planktonic foraminifera with *Rotalipora brotzeni* (Sigal), *R. reicheli* Mornod, *R. greenhornensis* (Morrow), *Hedbergella planispira* (Tappan), *H. delrioensis* (Carsey), *Schackoina cenomana* (Schacko), and *Sch. multispinata* (Cushm et Wick.), all of which indicate the *Rotalipora reicheli* Zone of Middle Cenomanian age.

The Cenomanian <u>Upper Formation</u> is composed of monotonous gray, massive and thick-bedded limestones, locally containing oyster beds with *Amphidonta* sp. whose thickness is 40 m. The basal layers of this unit contain the Late Cenomanian ammonites *Acanthoceras rhotomagense* Defr., *A.* cf. *breistrofferi* Collign., *Turrilites costatus* Lam., and *Eumphaloceras* cf. *meridionale* Stol. (Shatsky et al., 1966).

The total thickness of the Cenomanian deposits northwest of the Jebel An-Nusseiriyeh Range approaches 350 m.

On the southwestern slope of northern Jebel An-Nusseiriyeh, near the village of Qadmous, in the Nahr El-Qaqi Valley, and at Wadi El-Ain, the character of the Cenomanian deposits remains much the same (Fig. 6.28). This section starts with a sheet of basalt, followed by porphyrites and albitized porphyrites, which have a characteristic spheroid shape. The central part of these spheres consists of mostly unaltered anamesite, with highly weathered periphery, suggesting an underwater origin for these effusive rocks. The thickness of the basaltic member varies and is about 50 m. In the overlying strata, the Cenomanian sequence is the same as in northwestern Jebel An-Nusseiriyeh, reaching a thickness of about 400 m.

In the Missyaf-Hama area, on the eastern slope of the Jebel An-Nusseiriyeh Range the Cenomanian sequence is composed almost entirely of dolomites (Fig. 6.28). Despite this change in facies some main members of the Cenomanian deposits which are typical of the western slope can also be traced on the eastern slope (Shatsky et al., 1966). Fossils are almost completely absent from these dolomitized rocks. Rare intercalations of oyster limestones in the middle part of the succession have yielded *Amphidonta luynesi* Lart. and *Plicatula* cf. *anressensis* Coq. The thickness of these Cenomanian rocks is estimated to be 450-480 m.

In northwestern Syria (the area of the Baer-Bassit ophiolite Massif and Kurd-Dagh), Cenomanian calcareous sediments are poorly fossiliferous. They constitute a part of the thick Albian-Turonian succession and, as a result, their lower and upper boundaries are inadequately documented.

North of the **Baer-Bassit Ophiolite Massif** two formations were singled out (Kazmin and Kulakov, 1968): The Albian-Cenomanian Lower Formation makes up the greater part of the Qara Dourane and Saldirene Dagh Ranges (at the border with Turkey). A complete section of this formation can be observed on the northern scarp of Qara Dourane. This very thick (850-900 m) and extremely monotonous sequence consists of various limestones – gray and light-gray, hard, medium- and thick-bedded; dark and brownish-gray, bituminous, pelitomorphic, laminated; bituminous organogenic (comprised of up to 60-70% organogenic material), as well as hard aphanitic silicified; crinoidal and oyster coquina. Interbeds of brownish, fine-grained, hard and massive dolomites are rare. In the upper horizons of this succession, pelecypods *Amphidonta (Aetostreon) haliotidea* d'Orb. and *Liostrea* aff. *delettrei* Coq. (det. N. Cheltsova) have been found, indicating a probable Cenomanian age.

The Upper Formation is developed on the southern slopes of Jebel Aqraa, Qara Dourane, and Saldirene Dagh. The contact with the underlying formation is conformable as observed near the village of Badroussiiye. The upper formation is also represented by a somewhat monotonous sequence of massive and vaguely-bedded pelitomorphic limestones, very frequently with irregular nodular flint concretions which in places are strongly bituminized and black. This formation differs from the underlying Albian-Cenomanian deposits only in its greater massiveness and the characteristically porcelain-like appearance of the rocks. The thickness does not exceed 170-180 m. The uppermost layers of the upper formation contain hippurites, resembling those of the Turonian. This allows us to assume that the lower half of this formation still belongs to the Cenomanian.

The Albian-Cenomanian and Cenomanian-Turonian formations of Jebel Aqraa, Qara Dourane, and Saldirene Dagh are para-autochthonous with respect to the Baer-Bassit ophiolite Complex.

In the **Kurd-Dagh** region, Cenomanian deposits are exposed along the railway line running through a narrow, V-shaped valley between the villages of Berbannd and Meidanki (Protasevich and Maximov, 1966; Protasevich and Kurbatov, 1968). The contact with the Albian dolomites is conformable and indistinct. The base of the Cenomanian is placed beneath a bed of organic limestones with thin-valved oyster shells; among these have been identified fragments of *Gryphaea* ex gr. *vesiculosa* Sow. Here the following rock succession can be observed (from the bottom upwards):

1) Gray, fine-grained, thin-bedded limestones, dolomitized limestones and dolomites. The thickness is 87 m.

2) Gray and light-gray dolomites, calcareous dolomites and highly sandy dolomites. The thickness is 57 m.

3) Gray and brownish-gray, thick- and medium-bedded, silicified limestones and dolomitic limestones with some interbeds of organo-detrital limestones containing pelecypods *Amphidonta columba* Lam., *Ceratostreon bellatum* Goldf., *Gryphaea* ex gr. *vesiculosa* Sow., *Trigonia ethra* Coq., *Neithea shawi* Perv., *Plicatula* ex gr. *paucicostata* Seg., *Liopistha (Psilomya)* sp., and *Limatula* sp. (det. N. Bobkova). The thickness is 93 m.

4) Dark, massive, medium-bedded dolomites. The thickness is 63 m.

The total thickness of the Cenomanian deposits is 300 m. As can be seen from the above, the lower and upper parts of this stage are composed mainly of dolomites, and the middle part is made up of an irregular alternation of dolomites and limestones, with the latter predominating.

As compared to the Cenomanian deposits of the Anti-Lebanon and Jebel An-Nusseiriyeh, those of the **Palmyrides** differ in the broad development of facies of dolomites and a decrease in thickness. The replacement of the limestones by the dolomite facies begins in the extreme southwestern corner of the Palmyrides. In this area (the Jebel Ad-Dmeir Range), the sections demonstrate those primary features of the Cenomanian deposits which are typical of the entire Palmyrides area (Razvalyaev, 1966a). In the eastern part of Jebel Ad-Dmeir, the Cenomanian succession (from the base upwards) consists of the following members (Fig. 6.31):

1) Greenish pelitomorphic, clayey, dolomitic limestones with intercalations of yellowish-gray marls and compact organogenic limestones with abundant oysters. Due to this, the unit has been designated as 'the layers with oysters' or 'the layers with *Exogyra* (= *Amphidonta*)'. Among pelecypods, N. Bobkova identified *Amphidonta africana* Lam., *A.* cf. *mermeti* Lartet, *Dosinia delettrei* Coq., *Liostrea delettrei* Coq., and *Cardita forgemoli* Coq. Ammonites are represented by *Neolobites fourtani* Perv. (det. A. Atabekyan). The thickness is 40 m.

2) Gray, massive, medium-grained dolomites, in places with concretions of gray flints. These hard rocks form scarps in the surface relief. The thickness is 30 m.

3) A thick unit of light-gray, medium- to thin-bedded calcareous dolomites, and hard micro-grained dolomites. The thickness is 115 m.

4) Dark-gray, massive and compact, fine-grained dolomites. This bed is a good marker due to its dark coloring and massive texture since it can be clearly observed between the underlying and overlying light-gray dolomites. The thickness is 8 m.

5) Light-gray, micrograined calcareous dolomites. At their base, they contain numerous tracks of mud-eaters, which creates a peculiar 'vermiculate' structure. The thickness is 55 m.

The total thickness of the Cenomanian deposits is 248 m.

Lithological units of this section have also been identified in the western part of Jebel Ad-Dmeir and have been traced in the

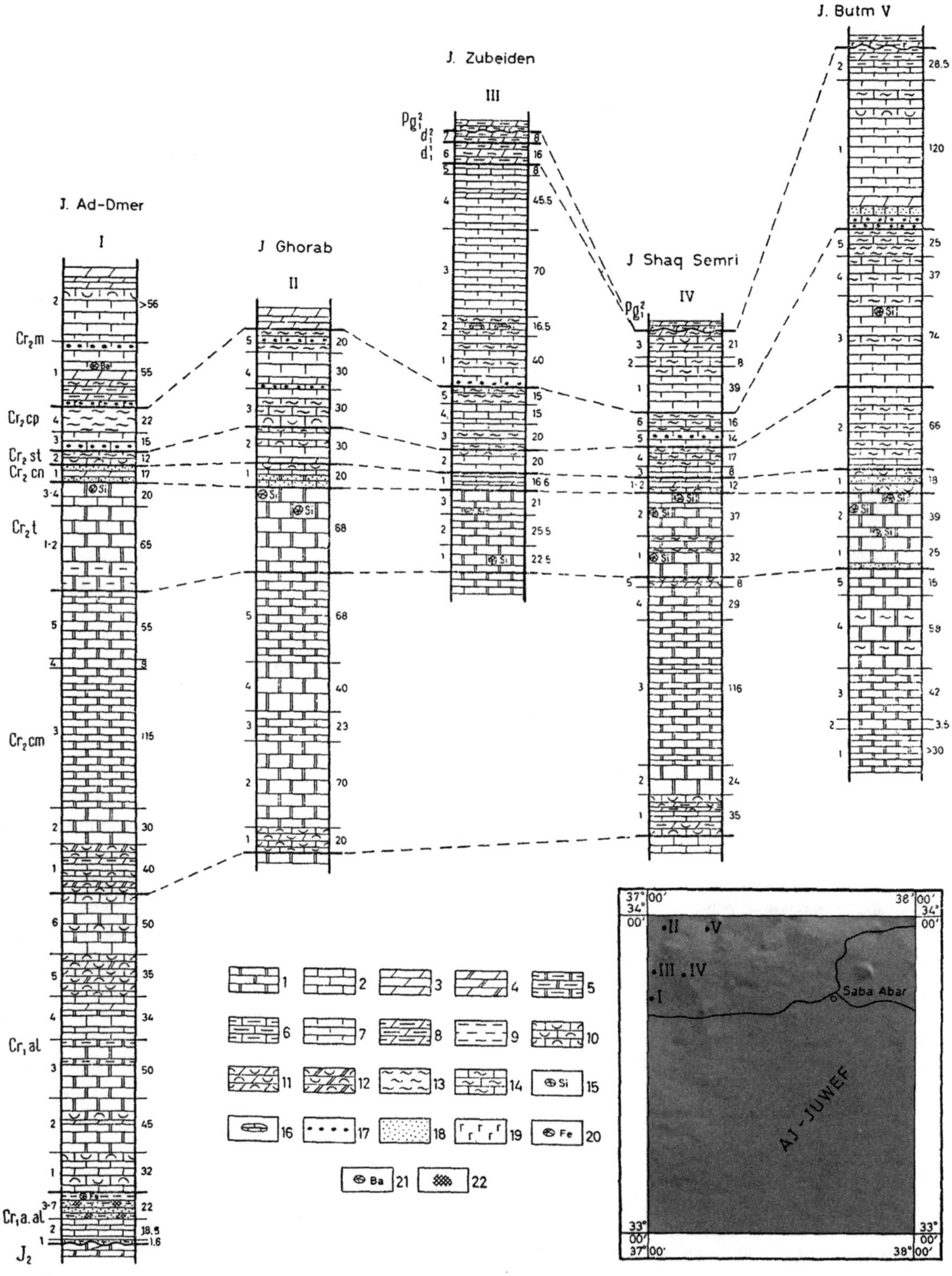

Fig. 6.32. Cretaceous deposits sections of the southwestern Palmyrides, southern branch (after Razvalyaev, 1966b).

Legend: 1 - dolomite; 2 - limestone; 3 - marl; 4 - magnesium marl; 5 - clayey dolomite; 6 - clayey limestone; 7 - chalk-like limestone; 8 - clayey marl; 9 - clay; 10 - fossiliferous limestone; 11 - fossiliferous marl; 12 - magnesium fossiliferous marl; 13 - flint; 14 - flinty limestone; 15 - flinty concretions; 16 - limestone buns; 17 - intercalations of phosphorites; 18 - sandy rocks; 19 - glauconite; 20 - concretions of bog iron ore; 21 - barite concretions; 22 - intercalations and lenses of bog iron ore.

nearby folds of Jebel Ghorab, Jebel Shaq Semri, and Jebel Butm (**Fig. 6.32**). The basal *Exogyra* layers are characterized throughout by having rich assemblages of fossils. In addition to those species of pelecypods and ammonites previously mentioned, this unit has also yielded *Amphidonta columba* Lam., *Pholadomya vignesi* Lartet, *Ceratostreon flabellatum* Goldf., *C. olisiponensis* Sharpe, *Lopha syphax* Coq., *Neithea zakarjensis* Blanck., *N.* ex gr. *dutrugei* Coq., *Venus perovalis* Conr., *Protocardia* ex gr. *hillana* Sow., *Midola* ex gr. *transjordanica* Blanck., and *Anisocardia aquinina* Coq. among pelecypods, and *Hemiaster syriacus* Corn., *Heterodiadema libycum* Cott. and *Coptodiscus larteti* (Cott.) among sea urchins (Razvalyaev, 1966a,b).

A thick sequence of dolomites overlies the layers with *Exogyra* and is devoid of any fossil remains. Only in Jebel Shaq Semri does the Cenomanian end in a bed of organogenic limestones and marls (8 m thick), that includes numerous pelecypods – *Amphidonta* cf. *luynesi* Coq., *Ceratostreon* cf. *olisiponensis* Sharpe, *Liopistha (Psilomya) ligeriensis* d'Orb., and *Lima* aff. *obliquistriata* Forbes (det. N. Cheltsova).

In the central Palmyrides, in the area between Al-Qaryatein in the west, and T-4 Station and Al-Beida in the east, Cenomanian rocks are present throughout the entire area, but outcrops are confined to the core of anticlinal folds as erosional windows (**Fig. 6.33**). Cenomanian deposits are well-exposed in two regions: in the southern branch of the Palmyrides, the so-called Ash-Sharqi Mountain Chain to the south of the Ad-Daww intermountain depression; and in the northern branch of the Palmyrides to the north of the Ad-Daww Depression (the mountains of Jebel Abou Dhour, Jebel El-Balas, and Jebel Ash-Shoumariyeh) (**Fig. 6.34**) (Faradzhev, 1966).

Complete continuous sections of the Cenomanian deposits are found in the ridges of the Southern Palmyrides at Jebel Rmah, Jebel El-Bassiri, Jebel An-Naqnaqiyeh, Jebel Om Journ, and Jebel El-Abtar (**Fig. 6.35**), where the Albian undergoes a gradual upward transition into Cenomanian sediments. The Cenomanian sequence clearly consists of three units. The lower unit corresponds to the *Exogyra* layers – limestones, marls, dolomites and clays with pelecypods and sea urchins. The middle unit is composed of dark, massive, thick-bedded dolomites which form easily seen cuestas in the field. The upper unit is made up of various, distinctly bedded dolomites with spotted 'vermicular' and porous (cavernous) 'spongy' structures. They can be correlated to similar lithological units which have been observed in the Jebel Ad-Dmeir section located to the southwest of the Palmyrides.

The Cenomanian sequence on the slope of the Jebel An-Naqnaqiyeh Range (near the 1,237 m elevation marker) represents a typical section for the Central Palmyrides. Here the Cenomanian sediments reach a maximum thickness of 225 m and (from the base upwards) consist of:

1) Gray, aphanitic, thick-bedded, very compact limestones, usually dolomitized (26%) and clayey (13%). They contain interbeds of medium-bedded and micro-grained dolomites and yellow magnesium marls. The rocks are saturated with abundant oysters. The thickness is 20 m.

2) Green clays with interbeds of organogenic limestones containing numerous oysters. The thickness is 33 m.

3) A member of alternating gray aphanitic limestones, green clays and magnesium marls (26% of insoluble residue) with remains of pelecypods, gastropods, sea urchins and bryozoa in a poor state of preservation. The thickness is 17 m.

4) Gray, massive dolomites, very hard, with a sparkling, sandy fracture and a 'spongy' weathered surface. The thickness is 30 m.

5) A thick formation of dolomites – white, yellow and gray, hard, compact, usually laminated, 'ribby' in relief, with 'vermicular' and 'spongy' varieties. The thickness is 125 m.

In the sections of Jebel Maksar An Nimr, Jebel Rous At-Twal, Jebel Butm, Jebel Ash-Sheikh, and Jebel Ghantous Cenomanian deposits are mainly represented by the middle and upper units.

The basal unit of the Cenomanian (the *Exogyra* layers) incorporates rich assemblages of pelecypods and sea urchins: *Amphidonta africana* Coq., *Ceratostreon flabellatum* Goldf., *C. olisiponensis* Sharpe, *Cardita forgemoli* var. *bequei* Coq., *Dosinia* ex gr. *delettrei* Coq., *Hemiaster syriacus* Conr., and *Heterodiadema libycum* Cott. The middle unit of hard massive dolomites is unfossiliferous. The upper unit of laminated dolomites includes only casts of poorly preserved *Protocardia, Isocardia, Anisocardia, Cardita, Trigonoarca, Corbula,* and *Meretrix* (Faradzhev, 1966).

In the northern branch of the Palmyrides (Jebel El-Balas and Jebel Abou Ad-Dhour), only the upper horizons of the Cenomanian outcrop. Here, the lithology of the rocks, which do not exceed 60 m in thickness, is somewhat special. As in the southern Palmyrides, gray, yellow and white, compact and massive dolomites and calcareous dolomites predominate, sometimes with a 'spongy' weathered surface. However all of the rocks are highly gypsiferous and contain lenses of gypsum up to 3 m thick. Interbeds of aphanitic limestones are extremely rare. Fossils collected from them are very poorly preserved; *Ceratostreon flabellatum* Goldf., *Liostrea* aff. *baylei* Coq., *Dosinia* ex gr. *delettrei* Coq., *Cardita* sp., and *Isocardia* sp.

In the northeastern part of the Palmyrides between Palmyra (Tudmor) and As-Sukhneh, Cenomanian deposits are exposed in the cores of the most eroded anticlinal folds which form the Palmyrian chain. The following ridges (from southwest to northeast) are marked by perfect sections of Cenomanian deposits: Jebel El-Abtar, Jebel Hayyan, Jebel El-Qayeh, Jebel El-Mazar, Jebel As-Safra, Jebel El-Harraziyeh, Jebel El-Waariyeh, Jebel As-Satih, Jebel El-Mrah, Jebel Abou Rajmein, Jebel As-Sayyah, and Jebel Thaniyyet As-Safra (**Fig. 6.36**).

The entire succession of Cenomanian deposits in this area of the Palmyrides can be clearly subdivided into two formations which vary in lithology and thickness.

The lower formation is composed of green and yellow-green clays and marls with thin intercalations of limestone bands. The rocks contain numerous pelecypod shells. The thickness ranges from 20 to 40 m. In general, this marker unit corresponds to those layers with *Exogyra* in the more western areas of the Palmyrides.

The upper formation is represented by light-gray, dark and yellowish dolomites and dolomitized limestones, medium- and

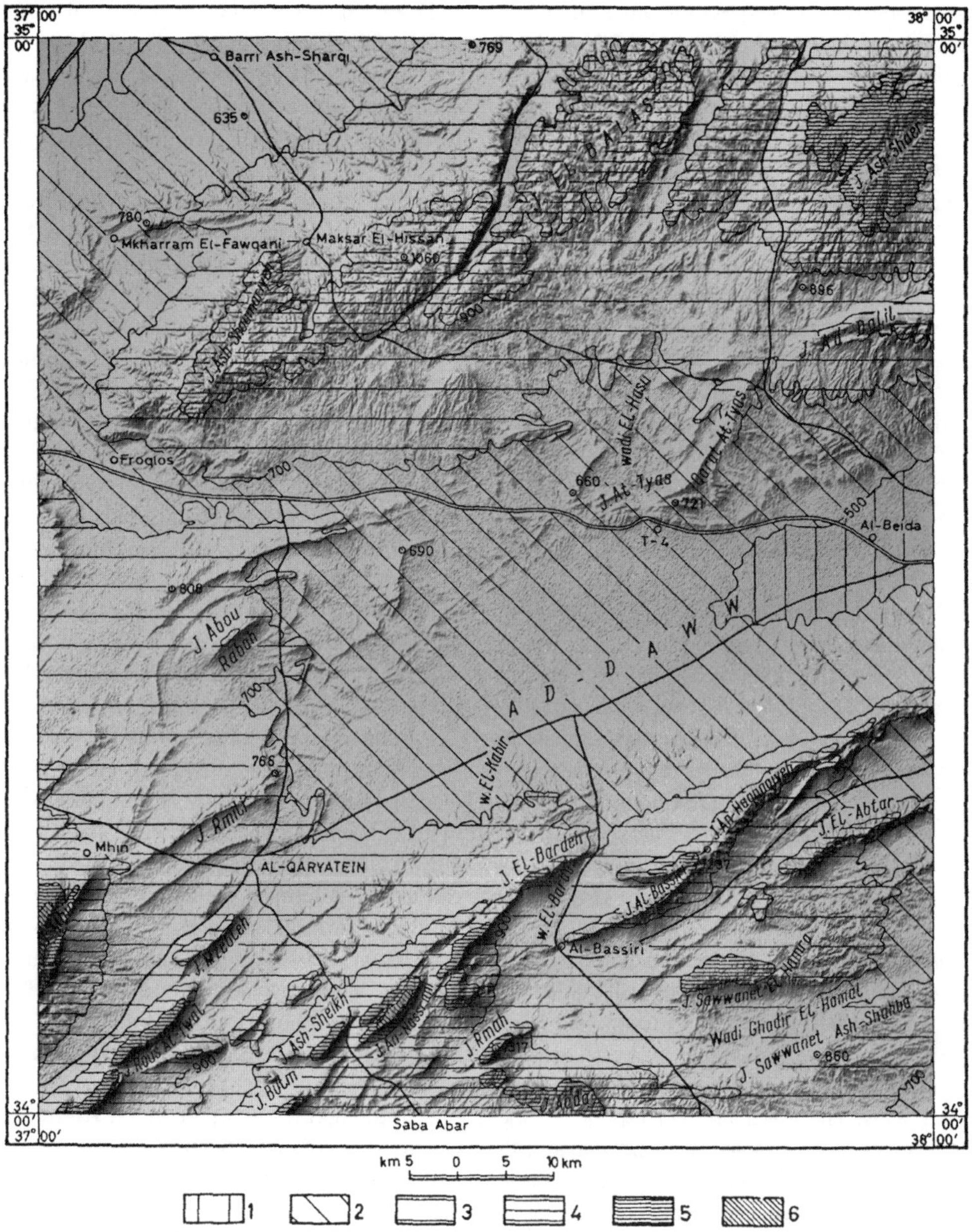

Fig. 6.33. Schematic landscape map of the Central Palmyrides. Exposures of the Cretaceous deposits are located on the elevations of southern and northern branches of the Palmyrides (divided by the Ad-Daww Depression). SRTM 0.05' topography, with the sun in the northwest at 60° elevation. Legend: Relief in meters above sea level: 1 - 0-500; 2 - 500-700; 3 - 700-900; 4 - 900-1100; 5 - 1100-1300; 6 - over 1300.

thick-bedded, massive and hard. In some sections (Jebel El-Abtar, Jebel El-Mazar, and Jebel As-Safra), these calcareous rocks bear flinty concretions. Their thickness varies from 70 to 130 m (**Fig. 6.37**).

A section of Cenomanian deposits typical for the eastern Palmyrides has been described thus (from the base upwards) for the Jebel Hayyan Range (Soulidi-Kondratiev, 1966a):

1) Brown and yellow, thin- and medium-bedded dolomites and dolomitic marls with rare clayey bands. The thickness is 15 m.

2) Green clays and marls with abundant pelecypods and sea urchins: *Amphidonta africana* Coq., *A.* cf. *suborbiculata* Lam., *A. conica* Sow., *A.* cf. *mermeti* Lartet, *A. columba* Lam., *Ceratostreon flabellatum* Goldf., *C. (Exogyra) olisiponensis* Sharpe, *Trigonoarca* sp., *Cardium* sp., *Leda* sp., *Dosinia delettrei* Coq., *Anisocardia* sp., *Pholadomya* cf. *vignesi* Lartet, *Cytherea* sp., *Cardita forgemoli* Coq. (det. N. Cheltsova and N. Bobkova), *Hemiaster syriacus* Conr., *Heterodiadema libycum* Cott., and *Coptodiscus* sp. (det. A. Kalis). The thickness is 10 m.

3) Gray and greenish, calcareous and dolomitic, thin-bedded marls. The thickness is 15 m.

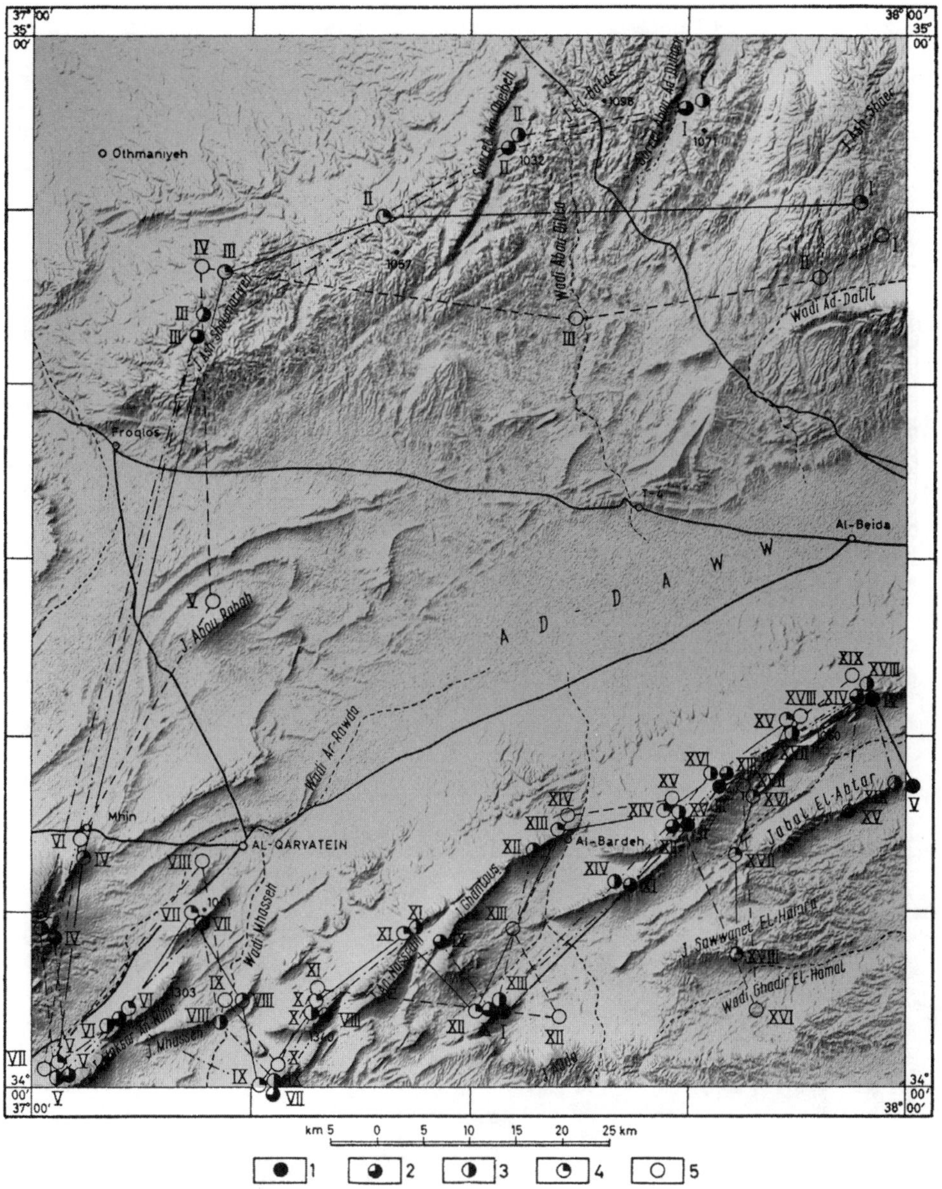

Fig. 6.34. Locality map of section of the Lower and Upper Cretaceous deposits, the Central Palmyrides (these sections are shown in Figs.6.29, 6.30, 6.35, 6.40, 6.41, and 6.44) (after Faradzhev, 1966). SRTM 0.05' topography, with the sun in the northwest at 60° elevation.
Legend: Section of: 1 - Lower Cretaceous; 2 - Cenomanian and Turonian; 3 - Coniacian; 4 - Santonian and Campanian; 5 - Maastrichtian. The Latin figures show the numbers of the sections in corresponding illustrations.

4) Monotonous gray and green-gray dolomites with well defined bedding. Interbeds of calcareous and dolomitic marls occur, as well as lenses and thin (20-30 cm) bands of highly silicified dolomites. The thickness is 130 m.

5) An alternation of clayey limestones and various dolomites (medium- and thick-bedded, calcareous, silicified) with a characteristic 'vermicular' structure. The thickness is 16 m.

The total thickness of Cenomanian rocks in the Jebel Hayyan section is 186 m. In Jebel-El-Mazar, the maximum thickness of the Cenomanian has been determined to be 265 m, with a minimum thickness of 112 m in Jebel El-Abtar.

As a rule, dolomites of the main part of the Cenomanian sequence lack fossils. Fossils have only been found in dolomitized limestones and calcareous dolomites in the Jebel Thaniyyet As-Safra and Jebel As-Safra sections (12-15 m below the contact with Turonian beds). These fossils are in a poor state of preservation and include *Dosinia* ex gr. *delettrei* Coq., *Cardita forgemoli* var. *bequei* Coq., *Venus* ex gr. *perovalis* Conr., *Trigonoarca* sp., and *Pholadomya* sp. (det. N. Cheltsova).

In the extreme northeastern part of the Palmyrides,

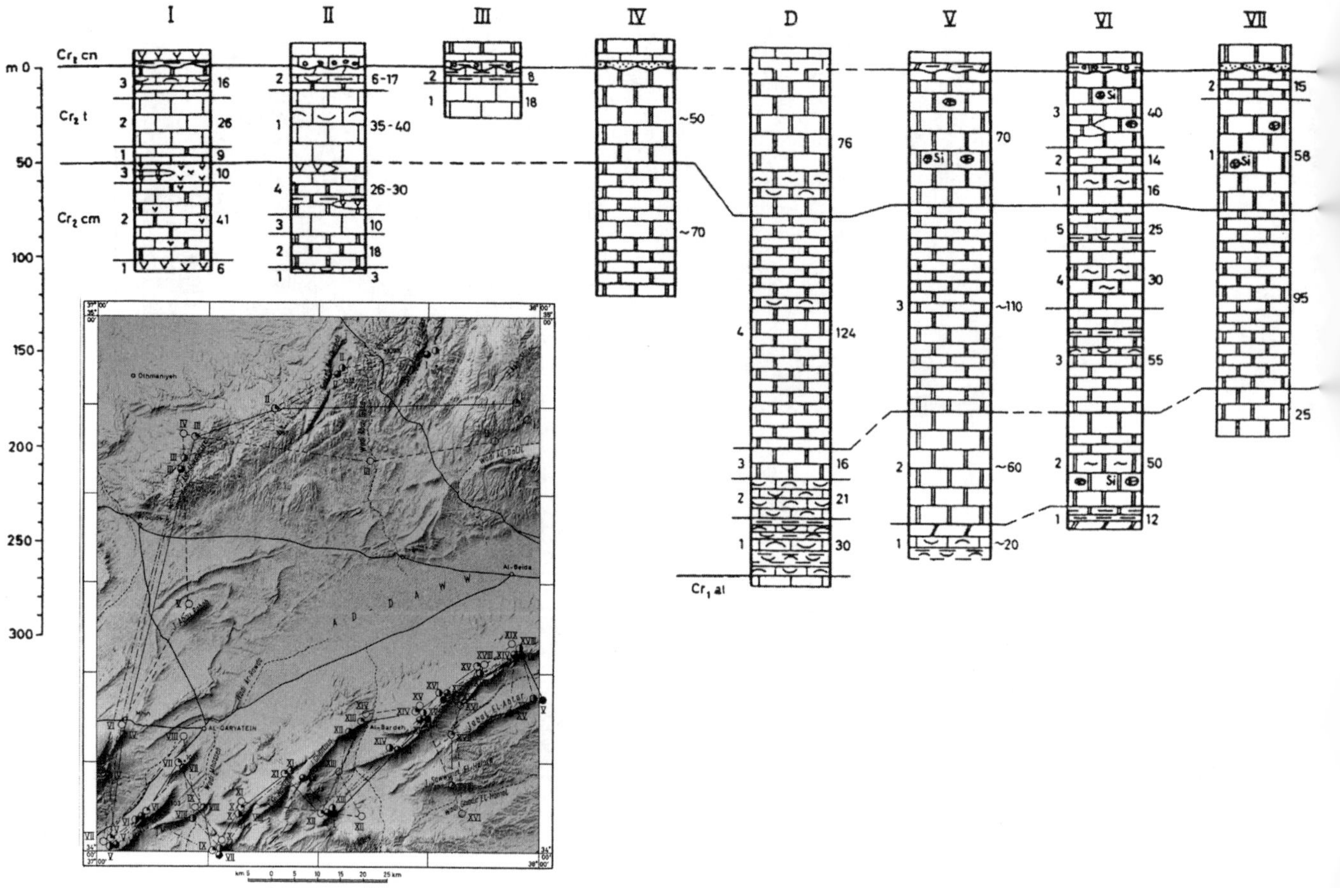
I
II
III
IV
D
V
VI
VII
m 0
50
100
150
200
250
300
Cr₂ cn
Cr₂ t
Cr₂ cm
Cr₁ al
Si
AD DAWW
Al-Qaryatein
Al-Beida
km 5 0 5 10 15 20 25 km

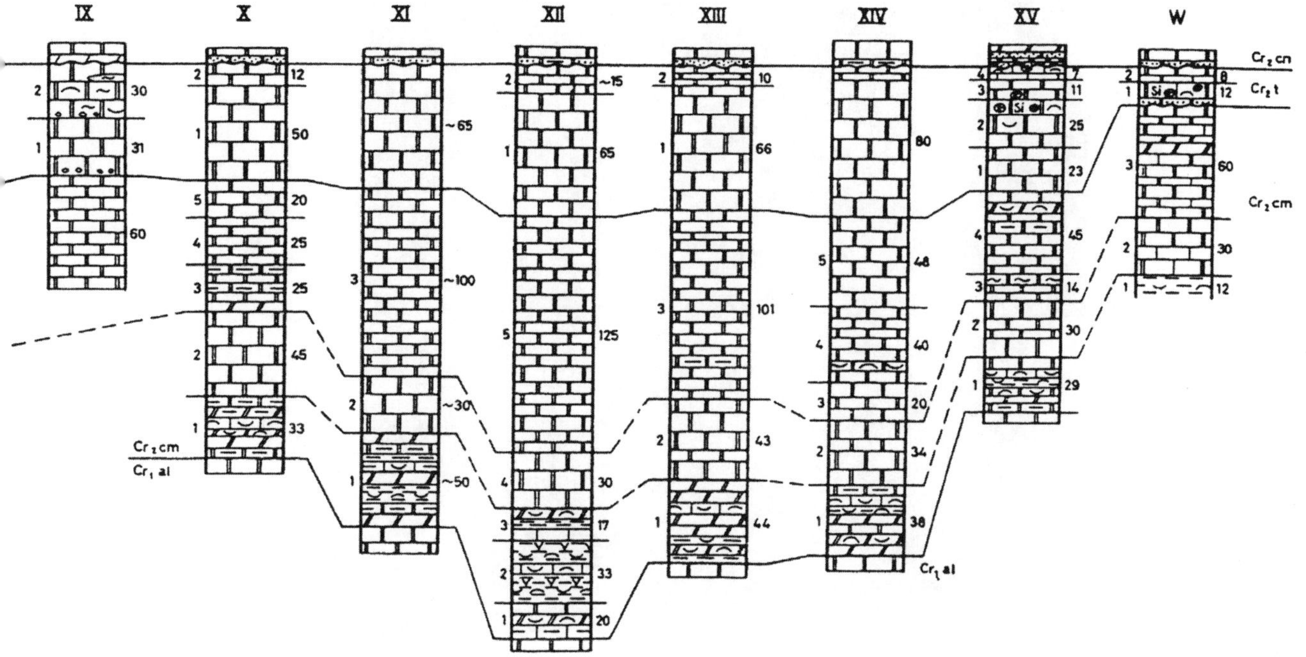

Fig. 6.35. Cenomanian and Turonian deposits sections of the Central Palmyrides (after Faradzhev, 1966).

Legend: Sections: I - Jebel Abou Dhour; II - Jebel El-Balas; III - Jebel Ash-Shoumariyeh; IV - Jebel Mhin (and Jebel An-Nabek); D - Jebel Dmeir (for comparison, see Fig. 6.31); V - Jebel Maksar An Nimr; VI - Jebel Rous At-Twal; VII - Jebel Butm; VIII - Jebel Ash-Sheikh; IX - Jebel Ghantous; X - Jebel Rmah; XI - Jebel El-Bassiri; XII - Jebel An-Naqnaqiyeh (1237 m elevation mark); XIII - Jebel An-Naqnaqiyeh (1188 m elevation mark); XIV - Jebel Om Journ; XV - Jebel El-Abtar; W - Jebel El-Waariyeh (for comparison, see Fig. 6.37).

Location of sections: see Fig. 6.34.

Lithologic symbols: see Fig. 6.29.

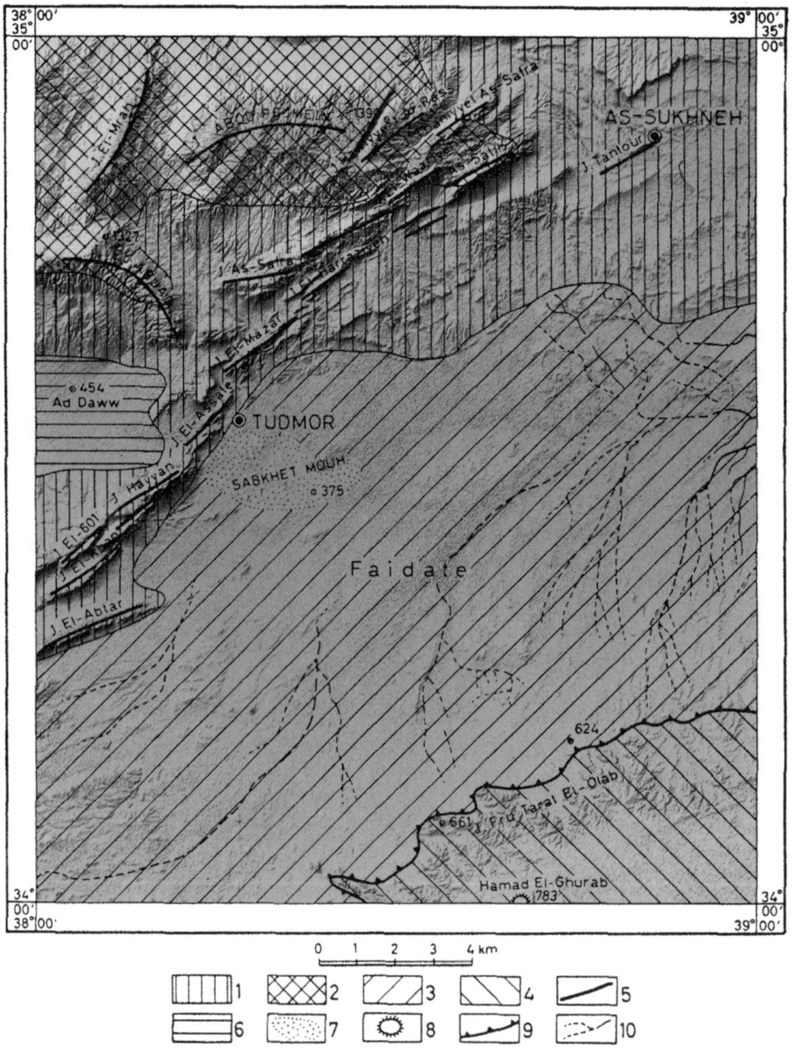

Fig. 6.36. Schematic landscape map of the northeastern Palmyrides, and position of ridges with outcrops of Cretaceous deposits.

Legend: Mountain regions (in meters above sea level): 1 - 450-900 m; 2 - 900-1300 m. Plains regions (the Syrian Desert): 3 - 370-625 m; 4 - 660-783 m. 5 - Axial lines of main mountain ridges, the northeastern Palmyrides. 6 - the intermountain Ad-Daww basin. 7 - the Sabkhet Mouh basin. 8 - Extinct volcanos. 9 - the Fru Tarat El-Olab Plateau escarpment. 10 - Main wadis.

Cenomanian rocks outcrop as a narrow band extending along the axial zone of the Jebel Mqebra Range and in the area of Beer Berber (**Fig. 6.38**). These rocks are composed of yellowish-gray and brown, thin-bedded, fine-grained and massive dolomites with bands of ferruginous and sandy limestones. Their visible thickness does not exceed 30-35 m (Oufland, 1966a). These sediments are unfossiliferous. Their age has been determined based on a lithological resemblance to Cenomanian sections in adjacent regions of the Palmyrides and to their stratigraphic position below dolomites with Turonian rudists.

Cenomanian deposits are also exposed in the central part of the **Jebel Abd El-Aziz** mountains in northeastern Syria. No contact with the underlying Lower Cretaceous rocks has been observed. The Cenomanian sequence consists of gray, massive-bedded, compact dolomites, clayey dolomites and dolomitic limestones. Some layers are characterized by a particular 'vermicular' texture. The visible thickness is 63 m (Ponomarev and Bivshev, 1966). Practically all the rocks are devoid of fossils and only near the 682 m elevation have *Gryphaea* ex gr. *vesiculosa* Sow. and *Neithea* sp. (det. N. Bobkova) been found.

Turonian [K_2t]

Turonian sediments are widely distributed in the Anti-Lebanon, Jebel An-Nusseiriyeh, Palmyrides, and Kurd-Dagh ranges. They are exposed in the cores of all the anticlines of the Anti-Lebanon and the Palmyrides as well as on the slopes of Jebel An-Nusseiriyeh and Kurd-Dagh. The Turonian is composed of limestones, dolomites, and a variety of transitional rocks. Layers and lenses of marls, sandstones and flints are subordinate. The contact between the Turonian deposits and the underlying Cenomanian rocks is almost everywhere conformable and gradual, though lithologically very distinct. Only in some areas of the Palmyrides is this boundary accompanied by a bed of quartz sandstones, indicating local erosional events. Similar erosional surfaces characterized by thin lens-like non-continuous sandstone layers have also been observed within the Turonian sediments. Turonian deposits are not uniformly fossiliferous. Limestones and marls of the lower part of the sections, especially in the Anti-Lebanon and Jebel An-Nusseiriyeh, usually contain rich assemblages of ammonites and pelecypods. At the same time, the dolomitic facies of the Turonian is poor in fossils and its age is only derivable from its stratigraphic position and lithology. The thickness of the Turonian deposits varies widely from 20 to 217 m.

In the **Anti-Lebanon** sections, Turonian deposits consist of two distinct formations. The lower formation is made up of thin- and medium-bedded limestones, chalk-like and clayey limestones, and calcareous dolomites. A permanent feature of these rocks is their softness, which is manifested in a gentle morphology and smoothed-out slopes. The contrast in lithology with the underlying rocks renders it easy to trace the Cenomanian/Turonian boundary. The upper formation is made up of massive, thick- and coarse-bedded, monotonous, hard dolomites and dolomitic limestones.

A representative section of the Turonian deposits can be observed in the Jebel Cheqif El-Khorn Range near the village of Hureire (Razvalyaev, 1966a; Kurbanov and Zarjanov, 1968) (Fig. 6.31). From bottom to top it consists of:

1) Light-gray, almost white, pelitomorphic, thin-platy limestones with thin and rare interbeds of dolomitic varieties and streaks of marls. The thickness is 7 m.

2) Yellowish and white, pelitomorphic, clayey, platy and thin-platy limestones. The thickness is 13 m.

3) Compact, clayey, platy, organogenic limestones with intercalations of marls. The thickness is 13 m.

Abundant ammonites in these beds include: *Thomasites* ex gr. *rollandi* (Thomas et Peron), *T. meslei* Perv., *Vascoceras globosum* Reyment, *V.* aff. *ellipticum* Barber, *V. higeriensis* Woods, *V.* cf. *robustum* Barber, *Leoniceras quassi* Peron, *L.* cf. *liciae* Perv., *Pachidiscus durandi* (Thomas et Peron), *Pseudoaspidoceras footeanum* Stolk., *P.* cf. *paganum* Reyment, *Mammites pervinquerei* Gross., *Hoplitoides* cf. *ingens* Roem., *Paramammites polymorphum* Perv., *Neoptychites xetriformis* Perv., *Fagesia* aff. *superotes* Cossm., *Gombeoceras gongilense* Woods, and *Fallotites* sp. (det. G. Krimgolts and A. Atabekyan). Among pelecypods *Astarte* cf. *planissima* Forbes, *Amphidonta rhadomantus* Coq., *Dosinia delettrei* Coq., *Liostrea* cf. *thevestensis* Coq., *Arctostrea carinata* Lam., and *Ceratostreon olisiponensis* Sharpe (det. N. Cheltsova) have been identified.

4) Greenish-yellow dolomitic and clayey friable sandstones rest on an uneven erosional surface of the underlying limestones. Upward they gradually become more compact and grade into yellowish dolomitic limestone. The latter contain a lenticular intercalation of oyster limestones with *Astarte subnumismalis* Thomas et Peron, *Plicatula ferryi* Coq., *P.* cf. *flattersi* Coq., *Lopha nicaisei* Coq., and *Corbula elegans* Sow. The thickness is 9.3 m.

5) Gray and brownish, compact, massive, obscure-bedded dolomites and dolomitic limestones with some thin bands of organic limestones containing undeterminable fossil debris. The thickness is 80 m.

Subunits 1-4 represent the lower formation, whereas subunit 5 relates to the upper formation. The total thickness of the Turonian deposits is 122.3 m.

This Turonian two-formation sequence can be distinctly observed in other sections of the Anti-Lebanon (Fig. 6.31).

On the eastern slope of Jebel Qasyoun (the Nahr Barada Valley near Damascus), the lower formation, producing a gentle slope, is represented by 32 m of white and light-gray, usually soft, pelitomorphic clayey limestones with thin intercalations of marls, organogenic detrital limestones and calcareous dolomites. A rich complex of ammonites has been collected from these rocks: *Vascoceras barcoicense* Choffat, *V. ellipticum* Barber, *V. turkestanense* Arkhang., *V.* cf. *mundae* Choffat, *Scaphites* ex gr. *geinitzi* d'Orb., *Austeniceras mobergi* Gross., *A.* cf. *austeni* Sharpe, *Paravascoceras crassum* Furon, *Leoniceras alaouitense* Basse, *L.* cf. *segne* Solger, *Neoptychites xetriformis* Perv., and *Paramammites* cf. *polymorphum* Perv. (det V. Rengarten and A. Atabekyan). The upper 10 m of this formation are massive-bedded gray, pelitomorphic and clayey limestones and dolomitic limestones, with decimeter-sized flint concretions and casts of large gastropods and ammonites among which *Neoptychites gourguechoni* Perv. has been determined.

The nonfossiliferous upper formation is made up of thick bedded and massive aphanitic dolomites and dolomitic limestones with flint concretions which form a prominent cliff-like steep scarp 55 m high.

On the southeastern slope of Jebel Hassiyya (to the north of the village of Halboun and along Wadi Mahdi), the lower formation is thin (17 m) and composed of greenish-gray, friable, calcareous dolomites and aphanitic and organogenic limestones. Soft intercalations here are absent. Ammonites and pelecypods include *Vascoceras nigeriensis* Woods, *V.* aff. *ellipticum* Barber, *V.* cf. *robustum* Barber, *Pseudoaspidoceras footeanum* Stol., *P.* cf. *paganum* Reyment, *Neoptychites xetriformis* Perv., *Hoplitoides* cf. *ingens* Koen., *Paramammites polymorphus* Perv., *Gombeoceras gongilense* Woods, *Mammites perviquierei* Gross. (det. A. Atabekyan), *Arctostrea carinata* Lam., *Amphidonta rhadamantus* Coq., *Ceratostreon* aff. *olisiponensis* Sharpe, *Dosinia delettrei* Coq., and *Astarte* cf. *planissima* Forbes (det. N. Cheltsova). The upper formation is 124 m thick and consists of gray thick-bedded unfossiliferous dolomites and limestones.

To the northeast, along the strike of the Anti-Lebanon, the Turonian deposits preserve their two-formation identity, but it becomes less pronounced as the limestones are replaced with dolomites at Maaret El-Bash and in the area of Jebel Suekai. This results in a general impoverishment of the fossil assemblages. In Maaret El-Bash, Turonian deposits reach a maximum thickness of 217 m. In the Jebel Suekai (Smekai) sections, the basal member

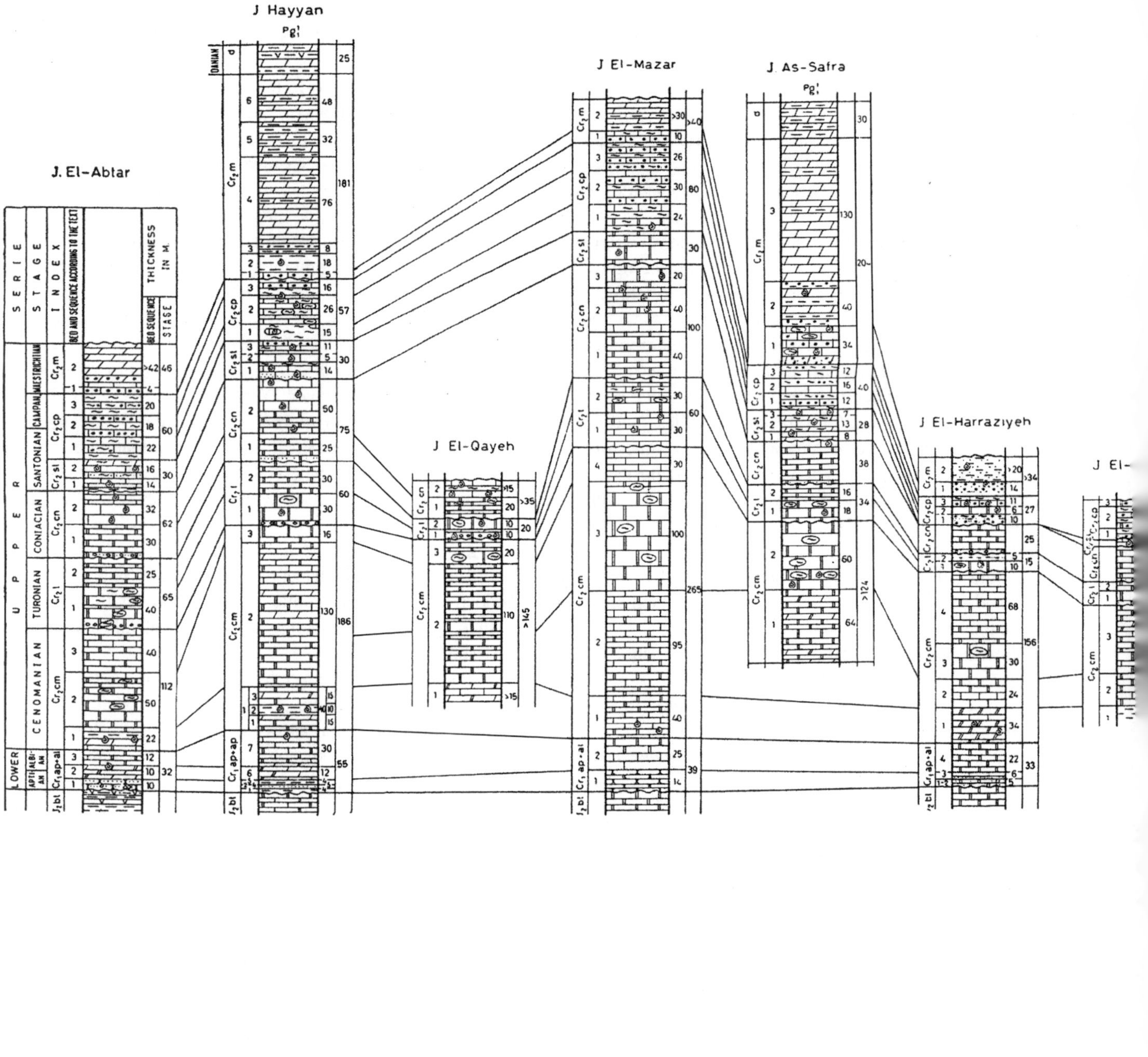
J Hayyan
J. El-Abtar
J El-Mazar
J. As-Safra
J El-Qayeh
J El-Harraziyeh
J El-
SERIE
STAGE
INDEX
BED AND SEQUENCE ACCORDING TO THE TEXT
THICKNESS IN M.
BED SEQUENCE
STAGE
UPPER
LOWER
CENOMANIAN
TURONIAN
CONIACIAN
SANTONIAN
CAMPANIAN
MAESTRICHTIAN
DANIAN

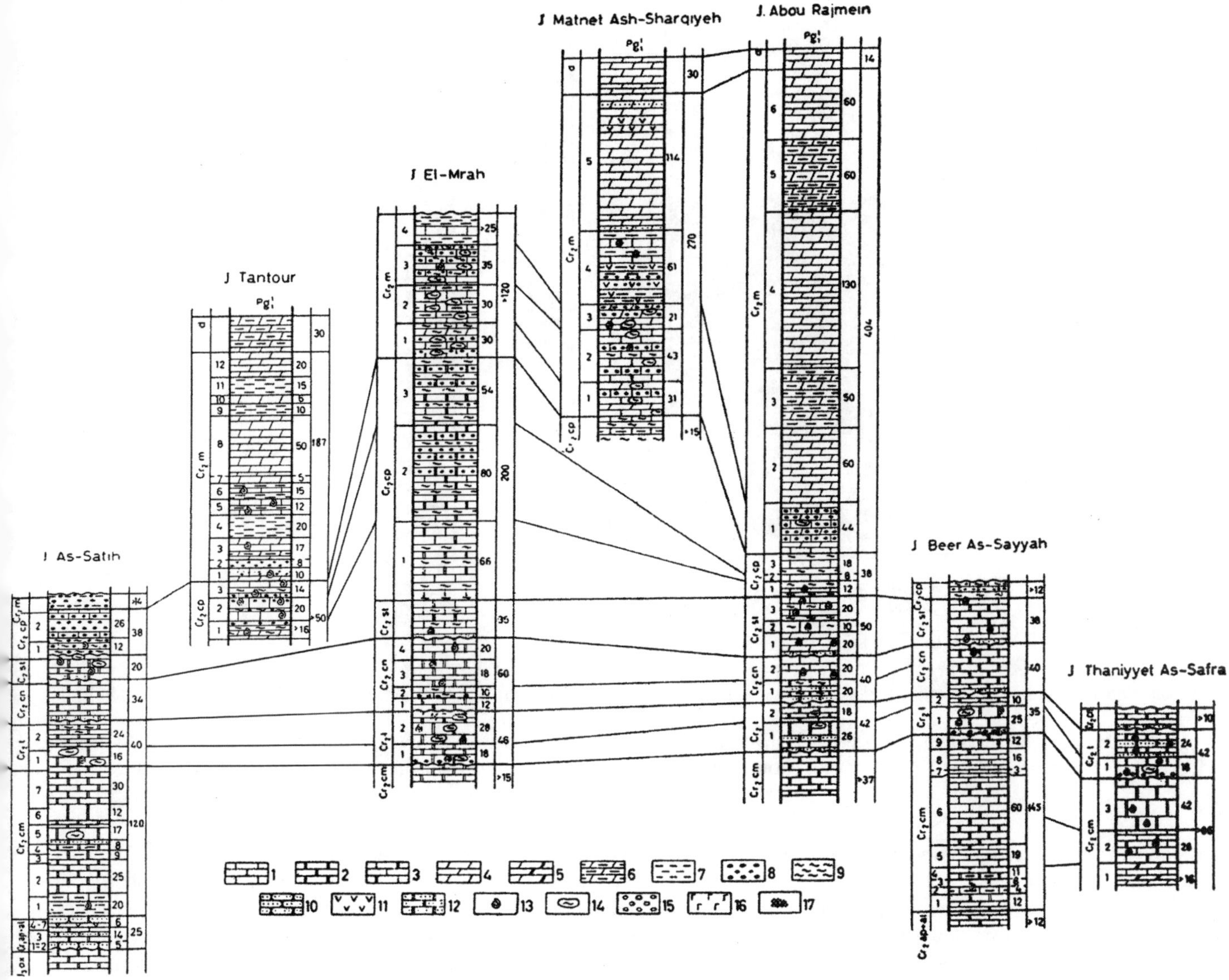

Fig. 6.37. Cretaceous deposits sections of the northeastern Palmyrides (modified from Soulidi-Kondratiev, 1966a).

Legend: 1 - limestone; 2 - dolomite; 3 - calcareous dolomite; 4 - marl; 5 - magnesian marl; 6 - marly clay; 7 - clay; 8 - phosphorite; 9 - flint; 10 - red sandstone; 11 - gypsum; 12 - sandy dolomite; 13 - fauna; 14 - flint concretions; 15 - brecciated rocks; 16 - glauconite; 17 - brown iron ore concretions.

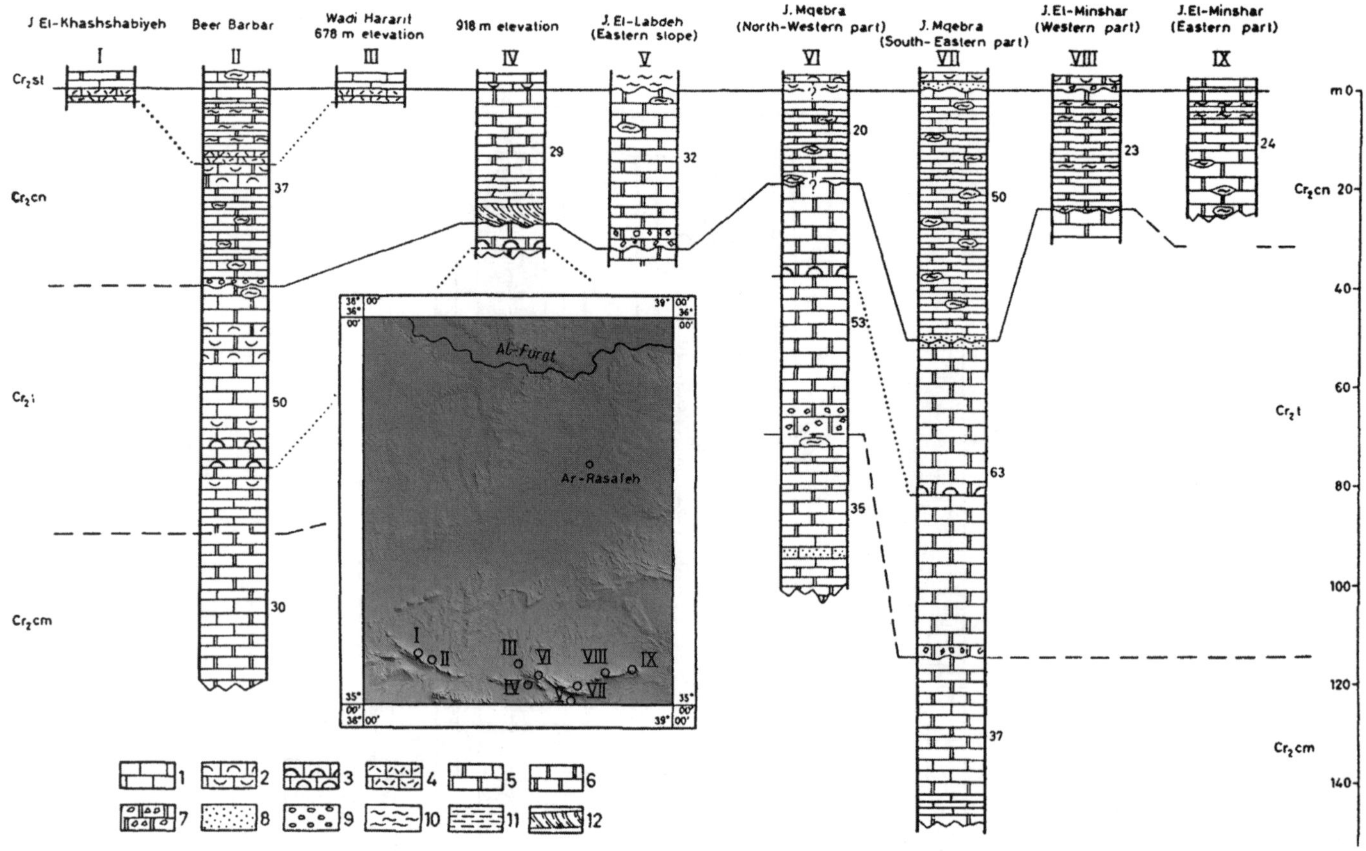

Fig. 6.38. Cenomanian, Turonian and Coniacian deposits sections of the Palmyrides, northeastern marginal portions (south of Ar-Rasafeh) (after Oufland, 1966a).

Legend: 1 - limestone; 2 - organogenic limestone; 3 - limestone rich in rudists; 4 - limestone with Serpula; 5 - dolomitic limestone (and calcareous dolomite); 7 - breccia-like dolomite; 8 - sandstone, sand and sandy ingredients; 9 - conglomerate; 10 - flint; 11 - clay; 12 - cross-bedded rocks.

contains only *Paravascoceras crassum* Furon and *Thomasites* cf. *rollandi* Thomas et Peron.

In the area of the northeastern plunging chains of the Anti-Lebanon, the Turonian sequence loses its two-formation structure. The entire succession (near the villages of Harf Ath-Thallajeh, Meiri, Jumrah, and As-Sawwan) is represented by a monotonous unit of gray and dark-gray, massive, thick-bedded, fine-grained dolomites covering the tops and slopes of the mountains (Figs. 6.26 and 6.27) (Kozlov, 1966a). As a rule, these rocks are devoid of fossils. Only at the base of the Turonian pile, in the intercalations of pelitomorphic clayey, slightly dolomitized limestones, exposed on the slopes of the Harf Ath-Thallajeh Massif, have rare ammonites been found. Among them A. Atabekyan determined *Mammites petraschecki* Gross., *Romaniceras* sp. as well as forms from the Acanthoceratidae family. The complete Turonian thickness of the Harf Ath-Thallajeh section is 120 m and visible thicknesses in other sections range from 70 to 90 m.

Summing up, the limestones in the Lower Formation of the Turonian developed in the Anti-Lebanon contain rich assemblages of ammonites, including the guide-species *Thomasites rollandi* Thomas et Peron, which undoubtedly confers an Early Turonian age to these sediments. The author discovered planktonic foraminifera in intercalations of soft marls and chalky limestones of the Jebel Qasioun section which confirm this age. They are: *Helvetoglobotruncana praehelvetica* (Trujillo), *Dicarinella imbricata* (Mornod), *D. canaliculata* (Reuss), *Hedbergella planispira* (Tappan), *H. delrioensis* (Carsey), *H. simplex* (Morrow), *Whiteinella archeocretacea* Pessagno, *Globigeri-nelloides ultramicra* (Subb.), and *Heterohelix reussi* (Cushman).

A more complicated problem emerges concerning the age of the upper formation of hard, thick-bedded dolomites which are essentially unfossiliferous. Two points of view exist regarding their age. According to N. Kurbanov and Yu. Zarjanov (1968), the entire sequence comprising these two formations belongs to the Lower Turonian. This implies that Upper Turonian sediments are absent, having been completely removed by pre-Coniacian erosional processes. According to A. Razvalyaev (1966a), the correlation with Upper Cretaceous rocks in adjacent regions of Syria and Lebanon shows that the upper formation of the thick-bedded dolomites has an Upper Turonian age. In Syria, the western-most outcrop of this formation is located in Jebel Chir Mannsour, just near the border with Lebanon. A mere 4 km northwest from this point, but within Lebanon near the village of Nabi Shite, the well-known section described by C. Diener (1886), contains a Turonian succession, subdivided into two units by Vautrin (1933a) and Keller and Vautrin (1937). The lower 100 m thick unit, composed of marls and clayey limestones, yielded a rich fauna of ammonites: *Thomasites rollandi* Thomas et Peron, *T. jordani* Perv., *Neoptychites cephalotus* Conr., *Leoniceras quaasi* Peron, and *Hoplitoides ingens* Koen., which indicate an Early Turonian age. The upper unit consists of 100 m of hard massive reef and reef-like limestones with hippurites fauna: *Hippurites grossouvrei* Douv., *Biradiolites lumbricalis* d'Orb., *Durania laevis* Douv., and *Nerinea pseudonobilis* Choffat, which are considered to be Late Turonian. In the area of Baalule village on the Rayak sheet, L. Dubertret (1950) described a similar sequence:

1) 10 m of marls with *Thomasites rollandi* Thomas et Peron and other ammonites (Early Turonian), and

2) reef limestones with *Hippurites resectus* Defr., *Eoradiolites lyratus* Conr. and *Nerinea pseudonobilis* Choffat; Late Turonian.

In the author's opinion, there are good reasons to correlate the massive unfossiliferous dolomites of Syria with the hippurite bearing limestones of Lebanon and to consider them as Late Turonian. Indeed, the massive unfossiliferous dolomites of Syria occupy a similar stratigraphic position between the Lower Turonian sediments with *Thomasites rollandi* and the Lower Coniacian layers.

Turonian deposits are well-exposed on the western slope of **Jebel An-Nusseiriyeh** where they form a wide belt that stretches parallel to the Mediterranean Sea. In this area, the Turonian succession comprises two units, in which limestones predominate.

Near the village of Kafroun Bash Shour on the southwestern slope of this uplift (Figs. 6.26 and 6.27), the lower unit is composed of white, pelitomorphic clayey limestones containing poorly preserved remains of pelecypods and gastropods and assemblages of the following Lower Turonian ammonites: *Vascoceras ellipticum* Barber, *Mammites petrascheki* Gross., *Leoniceras alauitensis* Basse, *L. segne* Solger, and *Pseudoaspidoceras footeanus* Stol. The thickness is about 20 m. These white limestones appear as a light band in the field (Kozlov, 1966a). The upper unit consists of light-gray, thick-bedded and massive, fine-grained limestones with interbeds of organo-detrital material with poorly preserved pelecypod and sea urchin shells; the thickness is about 60 m. Here the Turonian sections are incomplete because deposits of this age are either overlain unconformably with Maastrichtian layers or covered by Pliocene and Quaternary coastal plain sediments.

Further north at Jebel An-Nusseiriyeh, the Turonian sequence again appears as two formations, although in some sections they are indistinct (Shatsky et al., 1966). Though the lower contact of this section is not exposed in the vicinity of the village of Qerdaha, the sequence consists (from bottom to top) of the following units (Fig. 6.28):

1) Gray, medium- and thin-bedded, aphanitic limestones with numerous lenses and concretions of black and gray flints. These limestones contain abundant Early Turonian ammonites: *Pseudoaspidoceras paganum* Reyment, *P. footeanus* Stol., *Thomasites rollandi* Thomas et Peron, *Leoniceras alaouitensis* Basse, *L.* cf. *segne discoidalis* Perv., *Hoplitoides* cf. *ingens* Koen., *Broggiiceras olssoni* Benevid-Cacer, *Fagesia* sp., *Vascoceras* sp., and *Discovascoceras* sp., together with pelecypods *Protocardia biseriata* Conr. and *Ceratostreon* aff. *cadierensis* Coq. The thickness is 40 m

2) Massive, thick-bedded limestones with rare flint nodules. In the middle part of this member there is a layer of detrital organogenic limestones with rudists: *Eoradiolites syriacus* Conr., *E. lyratus* Conr., *Radiolites* cf. *peroni* Choffat, *Sphaerulites* aff. *depressus* Blanck., and *Hippurites resectus* Defr. The latter are associated with *Amphidonta* cf. *lateralis* Nils., *Plicatula ferryi* Coq. *P.* var. *desjardini* Coq., *Protocardia hillana* Sow., and *Pycnodonta* aff. *canaliculata* d'Orb. The thickness is 50 m.

The measured Turonian section reaches 90 m.

On the southeastern slope of Jebel An-Nusseiriyeh (the Missyaf Plateau), the Cenomanian/Turonian boundary is unclear due to its position within a thick sequence of monotonous unfossiliferous dolomites. A gradual replacement of dolomites by limestones occurs both in the upward and the northern direction. At the top of this unit, dolomites and limestones are interbedded with some organogenic limestones with the rudists *Eoradiolites syriacus* Conr. and *Radiolites* cf. *peroni* Choffat. Evidently these hard and massive rocks about 100 m thick belong to the Lower Turonian. They are unconformably covered by Campanian sediments.

To the north of the **Baer-Bassit Ophiolite Massif** in Jebel Aqraa and the Qara Douran mountains, Turonian deposits are poorly dated. They apparently constitute a part of a massive, poorly-bedded, sequence of aphanitic limestones with inter-layers of sometimes bituminous thin-bedded limestones with rare flint concretions. The observed thickness of Turonian rocks exceeds 180 m and they are unconformably overlain by Maastrichtian deposits (Shatsky et al., 1966; Kazmin and Kulakov, 1968). These hard, thick-bedded limestones correspond to the massive limestones and dolomites that are exposed on the eastern slope of Jebel An-Nusseiriyeh near the town of Hama, and they can be attributed to the Upper Cenomanian-Turonian. In the Jebel Aqraa area, the massive limestones are essentially unfossiliferous and only casts of gastropods have been found at the top of the section. In these upper layers, L. Dubertret (1936) discovered *Nerinea pseudonobilis* Choffat and hippurites resembling Turonian species. One can suggest that Turonian rocks are a part of the autochthon of the Baer-Bassit Ophiolite Massif.

On the southeastern slope of **Kurd-Dagh**, along the railroad from Aafrine to Rajou, near the villages of Gueubek, Qoda Keuy, and Meidanki, the Turonian sequence is composed of gray and brownish, very hard and massive, aphanitic, medium- and thick-bedded limestones, sometimes bituminous and dolomitized. A limestone bed, in the middle of the section, yields abundant poorly preserved rudists – *Praeradiolites* sp.and *Radiolites* sp. (det. N. Bobkova). The thickness varies from 40 to 60 m. The Turonian rocks are unconformably truncated by Maastrichtian sediments. Hard Turonian limestones are well-exposed in the

Kurd-Dagh region, forming steep scarps and benches (Protasevich and Maksimov, 1966; Protasevich and Kurbatov, 1968). Unfortunately, the massive limestones in this area do not contain any stratigraphically reliable fossils. Only L. Dubertret (1953) mentioned finding Turonian *Hippurites* sp.

The Turonian deposits of the **Palmyrides** are almost entirely built up of dolomites. In the central part of the southern branch of the Palmyrides (to the west and east of Al-Qaryatein), they rest on the top of thin-bedded, well stratified, micro-grained Cenomanian dolomites (Faradzhev, 1966). The Turonian rocks typically subdivide into two units. In Jebel An-Naqnaquiyeh, a 65 m thick lower member consists of gray, coarse-bedded and massive, medium- to fine-grained, very hard, dolomites with sparkling fracture. A 15 m thick upper member is composed of medium- and fine-bedded, aphanitic, light-gray dolomites. A similar succession was found in Jebel El-Abtar, where a lower 48 m thick member is characterized by white, light-gray and pink, thick-bedded, aphanitic and hard dolomites, in places with spotted-vermicular and spongy-texture, which contain only undetermined casts and broken shells of hippurites. An upper 18 m thick member is made up of white, pink and cream, medium-bedded and laminated, hard, fine-grained to compact dolomites and dolomitic limestones with poorly preserved casts of pelecypods and gastropods. These two Turonian units are also found in Jebel Maksar An-Nimr, Jebel Butm, Jebel Rmah, etc (Fig. 6.35). Nevertheless, this succession is not the same in other areas such as Jebel Mhin, Jebel Rous At-Twal, Jebel Ghantous, Jebel El-Bassiri and Jebel Om Journ. There the entire Turonian is represented by a monotonous 60-80 m succession of thick-bedded dolomites. Fossil remains did not provide sufficient paleontological data for an age determination. Furthermore these Turonian deposits are truncated by an erosional surface, above which follow Coniacian sediments.

In the northeastern Palmyrides, dolomitized limestones with fossils supplement the Turonian dolomites. It is possible to follow the same two rock units between Palmyra (Tudmor) and As-Sukhneh (Soulidi-Kondratiev, 1966a). The lower unit is composed of massive dolomites and dolomitized limestones containing rudists. The upper unit is represented by clearly bedded and sometimes thin-bedded dolomites (Fig. 6.37).

Just to the west of Palmyra (the Jebel Hayyan section), thick-bedded silicified Turonian dolomites rest on an erosional surface of Cenomanian rocks and include pockets (up to 1 m in size) filled with clastic material and lenses of conglomeratic and brecciated dolomites. Calcareous dolomites include numerous fragments and imprints of rudists and gastropods, such as *Plesioplocus* cf. *olisiponensis* (Sharpe) and *Plesioptygmatis* sp. (det. M. Djalilov). These rocks are followed by thin- and medium-bedded dolomites. The total thickness of Turonian sediments reaches 60 m.

The erosional surface at the Cenomanian-Turonian boundary can be followed in many other sections of the eastern Palmyrides (Jebel El-Qayeh, Jebel El-Mazar, Jebel As-Safra, Jebel El-Harraziyeh, Jebel El-Mrach, Jebel Abou Rajmein, etc.).

In Jebel El-Mazar (to the north of Palmyra), dolomites include intercalations of dolomitized organogenic limestones with abundant *Praeradiolites* sp., *Biradiolites* sp., *Eoradiolites* ex gr. *syriacus* Conr., *Lucina* sp., and *Neithea* sp. (det. N. Bobkova).

To the southwest of As-Sukhneh (Jebel El-Harraziyeh and Jebel El-Waariyeh), massive and thick-bedded organogenic dolomitized limestones are characterized by numerous rudists – *Caprinula* cf. *cedrorum* Blanck. and *Sauvagesia* aff. *schaperi* (Bayle) (det. N. Bobkova). These fossils have been found both at the base and at the top of the Turonian dolomite sequence.

The top of the Turonian in the area of the eastern Palmyrides is generally an uneven erosional surface. Where the sections are complete, the sediments reach a thickness that ranges from 15-20 m (Jebel El-Qayeh, Jebel El-Harraziyeh, and Jebel El-Waariyeh) to 60-65 m (Jebel El-Abtar, Jebel Hayyan, and Jebel El-Mazar).

In the northern branch of the Palmyrides (Jebel Abou Dhour, Jebel El-Balas,and Jebel Ash-Shoumariyeh), Turonian deposits are relatively thin (50-60 m thick) and are mainly made up of limestones, with dolomites in a subordinate role. The massive white, slightly clayey, partly dolomitized limestones contain molds of ammonites, casts of pelecypods and gastropods and remnants of echinoid armor. The ammonites are Early Turonian: *Leoniceras pavillieri* Perv., *L.* cf. *alaouitensis* Basse, *L. segne* Solger, *Vascoceras* cf. *gamai* Choffat, *Discovascoceras* cf. *tavense* Faraud, *Paravascoceras cauvini* Chud., and *Neoptychites* sp. (Faradzhev, 1966) (Fig. 6.35).

To the east of Jebel El-Balas in the vicinity of Beer Barbar, the Turonian succession consists of gray and light-gray, massive, thick- and medium-bedded, aphanitic dolomites, with a lesser amount of calcareous dolomites, dolomitic, aphanitic and detrital limestones. The latter contain remains of rudists, oysters, and sea urchins, but their poor state of preservation does not allow a precise identification. Farther to the east, the limestones are replaced by hard, sparkling, thick-bedded dolomite with rare beds of aphanitic limestones. The thickness is 50-63 m (Fig. 6.38). In this marginal part of the Palmyrides, the Turonian deposits are unconformably covered by Coniacian rocks (Oufland. 1966a).

An analysis of the dolomitic facies in the Palmyrides shows that these predominantly unfossiliferous deposits are not representative of the stratigraphy of the Turonian Stage in Syria. Rare fauna (Jebel El-Balas, Jebel Abou Dhour, Jebel El-Mazar, and Jebel El-Harraziyeh) indicate an Early Turonian or Turonian age. No data confirm the presence of any Late Turonian deposits. Taking into consideration the transgressive character of the Coniacian deposits in the Palmyrides, we cannot exclude the possibility that a portion of the Turonian rocks may have been removed by pre-Coniacian erosion.

At the same time, we should also note that at least in the southwestern Palmyrides, the Turonian stratigraphy remains comparable with that of the Anti-Lebanon. Transitional sections between the Turonian facies of the Anti-Lebanon and those of the northeastern Palmyrides exist *e.g.* in Jebel An-Nabek, where they consist of entirely unfossiliferous dolomites of significant thickness (120 m) (Fig. 6.27) (Kozlov, 1966a). In Jebel Ad-Dmeir, Jebel Ghorab, and Jebel Shaq Semri, dolomites also predominate, but they alternate with limestones containing Turonian fossils

– *Amphidonta* cf. *luynesi* Coq., *Protocardia* cf. *biseriata* Conr., and *Astarte seguenzae* Thomas et Peron (Fig. 6.32) (Razvalyaev, 1966a,b); the thickness of these rocks decreases to 60-70 m. The Turonian limestone/dolomite formation of Jebel Ghorab and Jebel Ad-Dmeir is overlain by Early Coniacian sediments with the ammonite *Reesideoceras gallicum* Basse (det. A. Atabekyan). Limestone beds with Early Turonian ammonites, outcropping in the southwestern part of Jebel Suekai (the southeastern fold of the Anti-Lebanon system) are replaced at distances of 2-3 km along the strike by thick-bedded dolomites which produce one monotonous formation of massive dolomites. Farther to the northeast, in Wadi Cheb, this succession overlies calcareous dolomites of Cenomanian age. A Turonian age can therefore be attributed to these massive dolomites in the Palmyrides.

In **Jebel Abd El-Aziz** in northeastern Syrian, the Turonian deposits have been completely eroded.

Generally speaking the subject of Turonian stratigraphy in Syria, and that of the Upper Turonian in particular, still requires thorough investigation.

Coniacian

In the context of the regional geology of Syria, Coniacian sediments are regressive. They are absent in the northwest of the country, in the Baer-Bassit Massif and Kurd-Dagh, and locally in Jebel An-Nusseiriyeh. On the other hand, Coniacian rocks are fairly widespread in the Anti-Lebanon and the Palmyrides. A very prominent feature of the Coniacian Stage in the Anti-Lebanon is the appearance of a facies of soft white chalk-like limestones with abundant planktonic foraminifera. This type of facies varies along the strike. In the Western Palmyrides, chalky limestones are replaced by limestones, marls, clays, and sandstones and in the Eastern Palmyrides by dolomites and shallow-water limestones. Where present, the Coniacian deposits range from 20 to 90 m in thickness. Faunas of ammonites, pelecypods, and foraminifera make it possible to work out a good biostratigtaphic subdivision.

In the **Anti-Lebanon**, Coniacian rocks constitute the lower part of a homogenous sequence of white and light-gray, soft chalk-like limestones.This sequence also embraces the Santonian and Campanian Stages (Razvalyaev, 1966a). The Coniacian rocks of the Anti-Lebanon conformably overlie the massive-bedded, compact aphanitic Turonian limestones. Representative sections of this type are located near the villages of Beit Jann, Hureira, and Maaret El-Bash (**Fig. 6.39**). The Coniacian age of the chalky limestones has been determined by the presence of planktonic foraminifera evidently belonging to the *Dicarinella concavata* Zone – *D. concavata* (Brotzen), *Marginotruncana marginata* (Reuss), *M. coronata* (Bolli), *M. angusticarinata* (Gandolfi), *M. pseudolinneiana* Pessagno, and *Hedbergella holmdelensis* Olsson, accompanied by taxa already found in more ancient sediments: *Hedbergella delrioensis* (Carsey), *Globigerinelloides ultramicra* (Subb.), *Schackoina multispinata* (Cushm. et Wick.), *Whiteinella inornata* (Bolli), and *Heterohelix reussi* (Cushman).

Near the village of Maaraba on the western slope of Jebel Qasyoun, the base of the Coniacian Stage is well-exposed and the entire Coniacian sediments are 83 m thick. The following units were identified:

1) 23 m of tobacco-green clays with greenish-gray foliated marls and interbeds of chalk-like clayey limestones. Planktonic foraminifera include rare specimens of *Marginotruncana schneegansi* (Sigal), together with *M. marginata* (Reuss), *M. angusticarinata* (Gandolfi), *M. renzi* (Gandolfi), *Dicarinella imbricata* (Mornod), *Whiteinella archeocretacea* Pessagno, *W. baltica* Douglas et Rankin, and *Heterohelix reussi* (Cushman). According to Bolli et al.'s (1985) zonal scheme, this assemblage belongs to the *Marginotruncana schneegansi* Zone, which represents the lower part of the Coniacian.

2) 42 m of white chalk-like soft limestones with fish remains, pelecypods – *Gryphaea* cf. *flicki* Perv. and *Lucina safedensis* Conr. (det. N. Bobkova) and foraminifera of the *Dicarinella concavata* Zone – *D. concavata* (Brotzen), *Marginotruncana coronata* (Bolli), *M. marginata* (Reuss), *M. pseudolinneiana* Pessagno, and *M. angusticarinata* (Gandolfi), in association with rare specimens of *Hedbergella holmdelensis* Olsson, *Dicarinella imbricata* (Mornod), and *Marginotruncana renzi* (Gandolfi).

3) 18 m of white marls and clayey limestones with shells of oysters, ammonites *Texanites* sp. (det. A. Atabekyan) and planktonic foraminifera of the *Dicarinella concavata* Zone.

In the northern Anti-Lebanon (Jebel Suekai), a section near the village of Ain At-Tineh displays a similar character. The section also begins with greenish-gray clay and yellow marl (13 m thick), overlain by light-gray, compact organogenic limestones and black bituminous limestones (25 m thick) with pelecypods – *Meretrix andergoni* Newton and *Crassatella africana* Woods (det. N. Bobkova), and ammonites – *Gauthiericeras* cf. *bajuvaricum* Redt. and *Peroniceras* ex gr. *moureti* Gross. (det. A. Atabekyan). The upper part of the Coniacian Stage is represented by 50 m of white chalky soft limestones with planktonic foraminifera of the *Dicarinella concavata* Zone.

The position of the Turonian/Coniacian boundary in the Anti-Lebanon is the subject of debate. At Wadi Hubeide, to the north of the village of Hureire, a bed of dolomitic sandstones underlies dolomitic limestones and dolomites that contain an assemblage of pelecypods *Astarte subnumismalis* Thomas et Peron, *Plicatula ferryi* Coq., *P.* cf. *flattersi* Coq., *Lopha nicaisei* Coq., and *Corbula elegans* Sow., which according to N. Bobkova, who identified the fauna, are of Early Coniacian age (Kurbanov and Zarjanov, 1968). Usually this type of compact limestones and dolomites is attributed to the Turonian Stage (Razvalyaev, 1966a). They are covered by chalky, soft, thick-bedded, white limestones and marls containing Coniacian ammonites – *Baculites incurvatus* Duj. and *Hoplitoplacenticeras* sp. Consequently, the clear lithologic boundary of hard limestones/dolomites and soft chalky limestones shown as the Turonian/Coniacian boundary on the 1:200,000 scale Damascus and Beirouth geological map sheets (Razvalyaev, 1966a) does not coincide with the chronostratigraphic boundary between these two stages.

In **Jebel An-Nusseiriyeh,** according to V. Shatsky et al. (1966), undifferentiated Coniacian-Santonian rocks are only

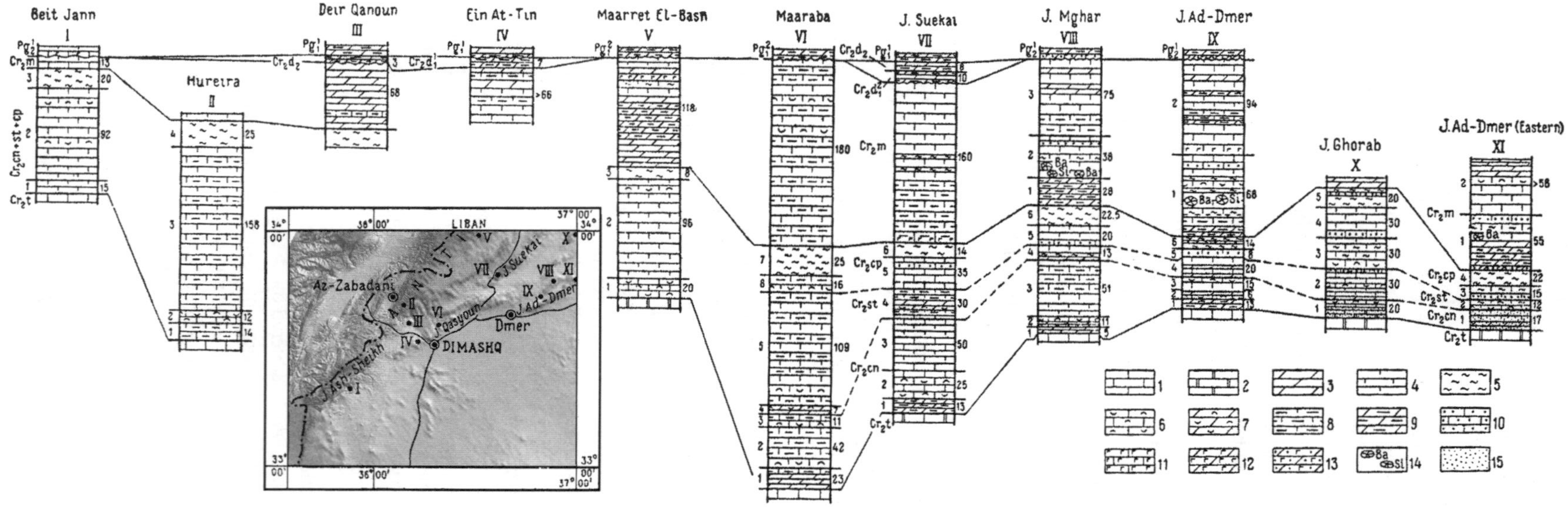

Fig. 6.39. Coniacian, Santonian, Campanian and Maastrichtian deposits sections of the Anti-Lebanon and the Palmyrides (southwestern margin) (modified from Razvalyaev, 1966a).

Legend: 1 - limestone; 2 - dolomite; 3 - marl; 4 - chalky limestone; 5 - flint; 6 - shelly limestone; 7 - marl with fauna; 8 - clayey limestone; 9 - clayey marl; 10 - limestone with phosphate; 11 - glauconitic limestone; 12 - glauconitic marl; 13 - glauconitic marl with phosphate; 14 - flinty and baryte concretions and nodules; 15 - sandstones and sandy ingredients.

encountered along the right bank of the Nahr El-Aasi River, near the town of Hama. The 25 m thick sequence starts with a layer containing boulders of reworked underlying rocks followed by mottled gray-green, dark-gray, and red gypsum-carbonate sediments. A Coniacian-Santonian age has been attributed due to its stratigraphic position between Turonian and Campanian rocks. The gypsum-bearing sediments are interpreted as the marginal facies of a regressive Coniacian sea in the realm of Jebel An-Nusseiriyeh because "in the northwestern portion of the country there was a break in sedimentation during the Coniacian time" (Ponikarov et al., 1967, p. 57). These previous views were changed by the 1:50,000 scale geological maps compiled by geologists of the General Establishment of Geology and Mineral Resources of Syria

Lower Senonian (Coniacian-Santonian) deposits have been found in some areas of the western slope of Jebel An-Nusseiriyeh in the vicinity of the towns of Haffe (the villages of Salma, Ain ett Tine, and Ain Layloun), Qerdaha (the villages of Boody and Ain El-Charkie), Banyas (the village of Zillo), and Tartous (the village of Houssain El-Baher). The Lower Senonian begins with a thin marly layer grading up into chalky or aphanitic limestones which are obscurely-bedded, light-colored and with smooth, weathered surfaces (Youssef, 1979, 1980a, b; Ruske, 1978a, b). Near the village of Ain Layloun, they include layers with phosphate ore. The thickness of the Lower Senonian ranges within wide limits, from 0 to 80 m. Deposits of this age rest conformably on Upper Turonian hard limestones and dolomites with *Hippurites resectus* Defr., *Radiolites* sp., and *Nerinea cochlaeaformis* Conr. The contact is indistinct and was placed where the bioclastic elements of the Turonian rocks disappear. This lithologic similarity serves to explain why in previous studies the deposits in question were included in the Turonian Stage. These Coniacian-Santonian limestones are unconformably overlain by Campanian or Maastrichtian sediments.

The Coniacian-Santonian age of these limestones is based on planktonic foraminifera consisting of *Marginotruncana coronata* (Bolli), *M. angusticarinata* (Gandolfi), *M. marginata* (Reuss), *Dicarinella concavata* (Brotzen), *Globotruncana linneiana* (d'Orb.), and *Rosita fornicata* (Plummer) (det. F. Ejel, Syria, and P. Saint Marc, France). These assemblages definitely indicate a Coniacian age, but layers with *Globotruncana linneiana* and *Rosita fornicata* may also be of Santonian age.

The differences in thicknesses of the Lower Senonian sediments and the transgressive character of the Campanian-Maastrichtian are evidently related to erosional processes in pre-Late Campanian-Maastrichtian times. Nevertheless, the possibility should not be excluded that some areas of Jebel An-Nusseiriyeh could have been uplifted in Early Coniacian time (with formation of gypsiferous sediments on the eastern slope of this range).

Coniacian deposits are widely distributed throughout the **Palmyrides** and are well-exposed in many sections. Their lithology and the structure of the sequences are variable along the strike of this mountain chain. In the southwestern part, pelagic facies of chalky limestones with planktonic foraminifera are developed; in the central and northeastern Palmyrides, these facies are replaced by shallow-water limestones and dolomites.

The Coniacian sections of the southwestern Palmyrides (the western margin of Jebel Ad-Dmeir and Jebel Mghar) preserve a similarity to those in the Anti-Lebanon (Fig. 6.32). Green clays and marls at the base, grading upwards into detrital organogenic limestones and, finally into soft chalky limestones (Razvalyaev, 1966a, b). Their thickness varies from 34 m at Jebel Ad-Dmeir to 67 m at Jebel Mghar. The most common pelecypods from these sediments are *Protocardia biseriata* Conr., *P. hillana* Sow., *P. moutai* Rennie, *Lucina safedensis* Conr., *Gryphaea* cf. *flicki* Perv., *Astarte seguenzae* Thomas et Peron, and *A. similis* Muenster (det. N. Bobkova). The lower layers of these sections are characterized by the presence of ammonites: *Reesideoceras gallicum* Basse, *Gauthiericeras bajuvaricum* Redt., *Peroniceras* ex gr. *moureti* Gross., and *P. czoernigi* Redt. In the upper beds *Texanites quinquenodosum* Redt. has been found. According to A. Atabekyan, the entire Coniacian stage is represented in this assemblage of ammonites.

To the northeast, the thickness of the Coniacian deposits decreases and terrigenous material appears. In Jebel Ghorab, greenish-yellow, medium-grained, sandy dolomites grade upwards into organogenic limestones with 0.1-0.2 m intercalations of quartz sandstones and shelly limestones. The latter contain pelecypods *Protocardia biseriata* Conr., *P. moutai* Rennie, and *Pycnodonta clavata* Nils. (det. N. Cheltsova), and the ammonite *Reesideoceras gallicum* Basse (det. A. Atabekyan). The thickness is 20 m. In the eastern part of Jebel Ad-Dmeir, the thickness of the Coniacian sandy limestones and marls does not exceed 17 m (Fig. 6.32).

The Coniacian succession in Jebel Zubeide is represented by a distinct alternation of thin-bedded, tobacco-green marls and light-gray limestones. At the base a band of organogenic limestone about 1.5 m thick occurs. The overall thickness is 16.5 m (Razvalyaev, 1966b) (Fig. 6.32). Marls contain planktonic foraminifera: *Marginotruncana marginata* (Reuss), *M. pseudolinneiana* Pessagno, *M. coronata* (Bolli), *Dicarinella imbricata* (Mornod), *Hedbergella holmdelensis* Olsson, *Archaeoglobigerina cretacea* (d'Orb.), *Globigerinelloides ultramicra* (Subb.), and *Heterohelix reussi* (Cushman). In the adjacent Jebel Shaq Semri, the same band of organogenic limestones contains numerous pelecypods: *Protocardia biseriata* Conr., *Astarte seguenzae* Thomas et Peron, and *A. similis* Muenster (det. N. Cheltsova). The 12 m thickness of the Coniacian rocks in this section is minimal. In Jebel Butm, the sequence of marls is replaced by organogenic limestones and dolomites which are enriched with quartz grains; the thickness is essentially the same (at 18 m).

To the north of Jebel Ad-Dmeir, Coniacian deposits are known in the axial parts as well as on the northwestern slopes of Jebel An-Nabek and Jebel As-Sawwan (Kozlov, 1966a). Here, the lithology is different. Usually the Coniacian sequences begin with a thin 1-2 m layer of green calcareous clays, primarily composed of alternating white pelitomorphic limestones and yellowish-gray, fine-grained dolomites. The sections everywhere terminate in gray, medium-bedded (about 0.5 m), compact, fine-grained lime-stones (Figs.

6.26 and 6.27). In the section at Jebel An-Nabek (the right slope of Wadi Khabra), the typical Coniacian ammonite *Subbarroisiceras mahafalense* Basse occurs in a layer located 15 m above the top of the Turonian dolomites. In a bed 10 m higher, *Peroniceras* sp. has also been encountered (det. A. Atabekyan). The total thickness of the Coniacian deposits at Jebel An-Nabek is 60 m and at Jebel As-Sawwan it is almost the same at 65 m.

In the central part of the southern branch of the **Palmyrides**, the Ash-Sharqi Mountain system shows a change in the lithology of the Coniacian deposits (Faradzhev, 1966), which here rest on an eroded Turonian surface. The upper boundary of the Coniacian is also usually eroded and uncomformably covered by Santonian beds of different lithology. From Jebel Mhin and Jebel Maksar An-Nimr in the west and as far as Jebel An-Nassrani and Jebel Ghantous in the east, the Coniacian consists of two formations (**Fig. 6.40**).

The lower formation is made up of fairly monotonous dolomites which can be distinguished from their Turonian counterparts only by a terrigenous bed at its base. This bed is represented by yellow-pink and white, highly dolomitic sandstones with gravels and pebbles of underlying dolomites; not exceeding a thickness of 2 m. The sandstones pass upwards into yellowish, pinkish and dark-gray, massive or medium-bedded, micro-grained to aphanitic dolomites, sometimes full of gray flint inclusions. In some cases the dolomites preserve an organic texture which confirms their secondary origin; ranging in thickness from 30 m at Jebel Mhin to 70 m at Jebel Mhasseh, Jebel Ghantous, and Jebel Butm.

The upper formation is built up of alternating marl, sandstone, clay, dolomite and various limestones (thick-bedded, aphanitic or fine-grained, clayey, detrital with a relict-organic texture). At the base of this formation, a bed of sandstones or dolomitic sandstones is quite common. Concretions of flint and interbeds of kerogen limestones with kerosene-H_2S smell are frequently present. A special feature is the rather numerous oyster-beds containing poorly preserved pelecypod shells and molds or imprints of ammonites. Nevertheless, the Coniacian ammonite *Reesideoceras gallicum* Basse and the pelecypods *Protocardia coquandi* Seguenza, *P.* ex gr. *biseriata* Conr. *P. moutai* Rennie, *Liostrea* cf. *boucheroni* Coq., and *Pycnodonta clavata* Nils., as well as the foraminifera *Dicarinella concavata* (Brotzen), *Marginotruncana coronata* (Bolli), *M. marginata* (Reuss), *M. pseudolinneiana* Pessagno, *Archaeoglobigerina cretacea* (d'Orb.), and *Heterohelix reussi* (Cushman) have been identified in sediments of this formation. The thickness and relationship of the sandy, clayey and calcareous rocks change from place to place: from 12-15 m at Jebel Mhasseh, Jebel Ash-Sheikh, and Jebel An-Nassrani, to 44-50 m at Jebel Mzebleh and Jebel Mhin. The boundary between the lower and upper formations is usually very distinct, but without any indication of a break or erosion. The incompleteness of paleontological characteristics prevents our concluding whether these formations are chronostratigraphic units or whether they replace each other laterally.

To the east of Jebel Ghantous, all Coniacian sections of Jebel El-Bardeh, Jebel Rmah, Jebel Bassiri, Jebel An-Naqnaqiyeh, Jebel Om Journ, and Jebel El-Abtar are characterized by undifferentiated dolomites and limestones of varying lithology. The dolomites here are sandy, compact, coarse-bedded, medium-grained, with thin lenses and small concretions of brown and gray chert and with an H_2S smell. The hard white, gray, yellow and greenish organic limestones are dolomitized, and medium- to massive-bedded. In some sections, a sporadic layer of sandstones, gravels, and magnesial marls is developed. At the top of the Coniacian deposits, *Serpula*-biostromes are very useful for local correlation. The fairly constant thickness ranges from 70 to 90 m. All fossils are poorly preserved and species-wise undeterminable. The following pelecypods could be recognized along the southern slope of Jebel El-Abtar: *Plicatula* ex gr. *nummidica* Coq., *Liostrea* aff. *bourguignati* Coq., *Cytherea* cf. *plana* Sow., *Ostraea* sp., *Exogyra* sp., *Amphidonta* sp., *Gryphaea* sp., *Protocardia* sp., *Venus* sp., *Crassatella* sp., *Lopha* cf. *deshayesi* Coq., and *L.* cf. *dichotoma* Bayle.

In the northeastern part of the Palmyrides, the lithology of the Coniacian deposits is changeable (Soulidi-Kondtatiev, 1966a). The coastal facies is developed as a coquina, coarse-detrital limestones, and shallow-water carbonate sediments with oysters and colonies of *Serpula*, with strong dolomitization.

In the vicinity of Tudmor (Palmyra), the Coniacian succession is again built up of two units (Fig. 6.37). An essentially dolomitic lower unit encompasses a band of quartz sandstones and gravelites at its base. The upper unit is represented by coarse-detrital and shelly, thick-bedded, white and light-gray limestones. Coniacian deposits of this type are observed in the eastern margin of Jebel El-Abtar, and also in Jebel Hayyan and Jebel El-Mazar. They overlie Turonian dolomites with some hiatus, but without visible angular unconformity. A maximum thickness of about 100 m has been observed at Jebel El-Mazar, with the thickness decreasing to 62 m at Jebel El-Abtar.

Farther to the northeast, at Jebel As-Safra, Jebel El-Harraziyeh, Jebel El-Waariyeh, Jebel El-Mrah, and Jebel Beer As-Sayyah (Fig. 6.36), the limestones of the upper unit are replaced by medium- and thin-bedded dolomites so that the entire Coniacian is composed of monotonous, unfossiliferous dolomites (Fig. 6.37). The two-formation lithology is preserved in only one section at Jebel Abou Rajmein, with the dolomites and sandy dolomites of the lower 20 m thick unit being covered by the 20 m thick limestones of the upper unit. Here pelecypods are present: *Lopha* ex gr. *deshayesi* Coq., *L.* cf. *semiplana* Sow., *Plicatula batnensis* Coq., *P.* cf. *flattersi* Coq., *Liostrea delettrei* Coq., *Trigonoarca* sp., *Venus* sp., *Lucina* sp., *Protocardia* cf. *biseriata* Conr., and *Roudairea undata* Conr. (det. N. Cheltsova and N. Bobkova). The thickness of the Coniacian deposits in this part of the Palmyrides decreases to 25-40 m.

South of the Byzantine ruins of Ar-Resafeh, along the northeastern margin of the Palmyrides, Coniacian rocks are poorly exposed. Small outcrops are found in Jebel El-Minshar, Jebel Mqebra, and Jebel El-Labdeh as well as in the area of Beer Barbar (Oufland, 1966a). A bed of coarse-grained, sometimes cross-bedded quartz sandstones overlies the eroded underlying Turonian carbonates. In some places they grade into small-pebble conglomerates. This terrigenous basal layer is covered with a sequence of hard, fine-grained or aphanitic dolomites and dolomitic limestones. Limestones with numerous tubes of *Serpula* and porous algal dolomitic limestones are typical for the upper portion of the sequence. Where the sections are complete, the thickness of the Coniacian deposits varies from 20 to 50 m (Fig. 6.38).

The Coniacian deposits in the southern branch of the Palmyrides are essentially unfossiliferous. Their age has been determined by means of correlation with the transitional sediments of the southwestern termination of the Palmyrides (the area of Jebel Ad-Dmeir), where macrofauna are common and pelagic chalky limestones with planktonic foraminifera appear.

The Coniacian rocks of the northern branch of the Palmyrides, in Jebel Abou Dhour, Jebel El-Balas, and Jebel Ash-Shoumariyeh resemble those of the southern branch, between Jebel An-Nimr in the west and Jebel Ghantous in the east (Faradzhev, 1966; Mikhailov, 1966). These rocks are also subdivided into two formations (Fig. 6.40).

The base of the lower formation consists of a 5-15 m thick band of white or gray, and in some places pink and green, non-equigranular gypsum. It is covered by a 25-29 m thick member of white and light-gray, medium- to coarse-bedded, micrograined or aphanitic limestones with interbeds of ferruginous dolomites containing casts of *Exogyra*, *Neithea,* and *Plicatula*. The upper layers are composed of white and pinkish, massive and thick-bedded, very hard aphanitic limestones, forming an escarpment at Jebel Abou Dhour and Jebel El-Balas. As a rule sediments of the lower formation do not contain any fossils suitable for age determination.

The upper formation is considerably different in its lithology and paleontology. As in the southern branch of the Palmyrides, this upper formation consists of two members.

The lower member is represented by light-gray, thin- to thick-bedded limestones, very often with an H_2S smell. The composition and outlook of these limestones are quite variable: aphanitic or distinctly grained; detrital or oolitic; very compact or porous and even cavernous; strongly silicified or with lenses and concretions of flint and highly dolomitic (40% admixture).

The upper member is characterized by medium-bedded micrograined limestones with ammonites, pelecypods, gastropods and tubes of *Serpula*. Among the ammonites *Barroisiceras onilahyensis* Basse, *Reesideoceras* sp., and *Scaphites* sp. have been determined by A. Atabekyan. Pelecypods such as *Plicatula flattersi* Coq., *Protocardia coquandi* Seg., *P. hillana* Sow., *P.* cf. *biseriata* Conr., *Pycnodonta clavata* Nils., *Amphidonta* cf. *auricularis* Gein., *Cytherea plana* Sow. var. *judaica* Picard, *Crassatella matercula* Mayer-Eymar, and *Liostrea* cf. *delettrei* Coq. have been identified (det. N. Bobkova and N. Cheltsova).

The thickness of the upper formation ranges from 60 to 90 m.

To the southwest of Jebel El-Balas, in Jebel Ash-Shoumariyeh, the Coniacian decreases in thickness to 85 m and the limestones, especially in the lower formation, are gradually replaced by dolomites. In the southwestern corner of this range, the morphology of the Coniacian deposits resembles that of Jebel Mhin (the southern branch of the Palmyrides), where dolomites predominate.

In northeastern Syria, **Jebel Abd El-Aziz**, 15-22 m thick, white, laminated, organo-detrital and aphanitic limestones with remains of *Pycnodonta* cf. *clavata* Nils. and *P.* cf. *vesicularis* Lam. (det. N. Cheltsova) have been tentatively referred to the Coniacian (Ponomarev and Bivshev, 1966). Near the elevation marker for 682 m, these limestones rest unconformably on Cenomanian dolomitic limestones and Aptian clays. In turn, they are truncated by Santonian rocks with pelecypods.

Santonian [K_2st]

Santonian deposits are well-exposed in the mountains of the Anti-Lebanon and the Palmyrides. In Jebel An-Nusseiriyeh, these deposits are characterized by a spotty distribution, either due to non-deposition, or subsequent erosional processes. In the region of the Baer-Bassit Massif and the Kurd-Dagh, sediments of this age are missing.

Pelagic facies of white soft chalky limestones with planktonic foraminifera occupy larger areas than were found in the Coniacian. These sediments are not only common in the Anti-Lebanon, but also in the western part of the Palmyrides. In the northeastern Palmyrides, they are replaced by more shallow-water detritic limestone and the ratio of dolomites to limestones strongly decreases. Another peculiar feature is the development of bituminous limestones and marls. The thickness of these Santonian deposits is variable, ranging from 0 to 100 m.

In the northwestern ridges of the **Anti-Lebanon** (Jebel Ash-Sheikh and Jebel Hassiyya) the Santonian deposits represent an interval within the monotonous Senonian sequence of white, chalky, massive-bedded, soft limestones (Razvalyaev, 1966a). Sections near the villages of Beit Jann, Hureira, and Maaret El-Bash display this facies-type very well (Fig. 6.39). Abundant planktonic foraminifera occur: *Dicarinella asymetrica* (Sigal), *D. concavata* (Brotzen), *Marginotruncana pseudolinneiana* Pessagno, *M. marginata* (Reuss), *M. coronata* (Bolli), *Globotruncana* aff. *linneiana* (d'Orb.), *Rosita fornicata* (Plummer), *Archaeoglobigerina cretacea* (d'Orb.), *A. bosquensis* Pessagno, *Pseudoguembelina costulata* (Cushman), *Hedbergella holmdelensis* Olsson, and *Heterohelix reussi* (Cushman). The lower and upper boundaries of the Santonian strata are not discernable in the field and the contacts with the Coniacian and Campanian deposits should be considered as conformable. This creates difficulties in exactly assessing the thicknesses of the Santonian strata (30-40 m).

In the southeastern ridges of the Anti-Lebanon (Jebel Qasyoun and Jebel Suekai), white chalky limestones alternate with soft gray clayey limestones and marls with the same rich assemblages of planktonic foraminifera. Rare intercalations of organogenic detrital limestones are present among them. A maximum thickness of 116 m for these Santonian deposits has been measured near the village of Maaraba. To the northeast at Jebel Suekai, this figure decreases to 30 m. Macrofauna (pelecypods) are very rare in pelagic sediments – *Leda perdita* Conr., *Cytherea* cf. *andersoni* Bull.-Newt., and *Gryphaea flicki* Perv. (det. N. Bobkova).

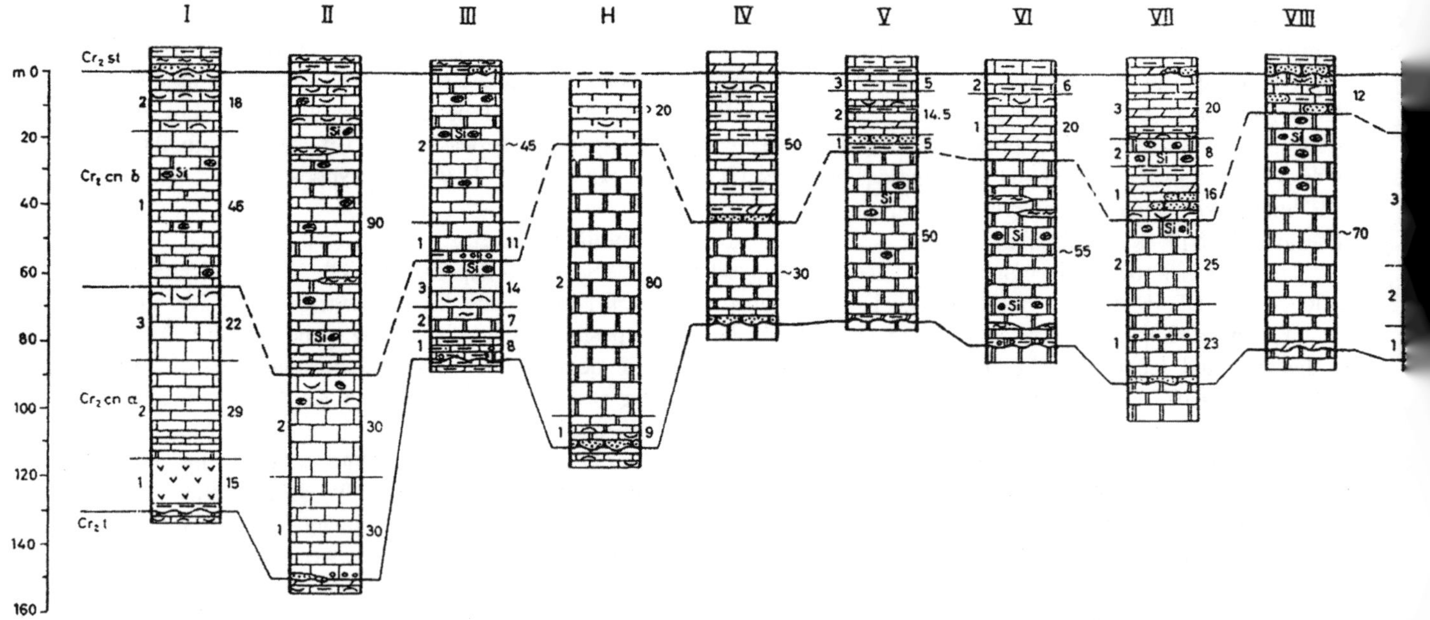
I
II
III
H
IV
V
VI
VII
VIII
m 0
20
40
60
80
100
120
140
160
Cr₂ st
Cr₂ cn б
Cr₂ cn a
Cr₂ t

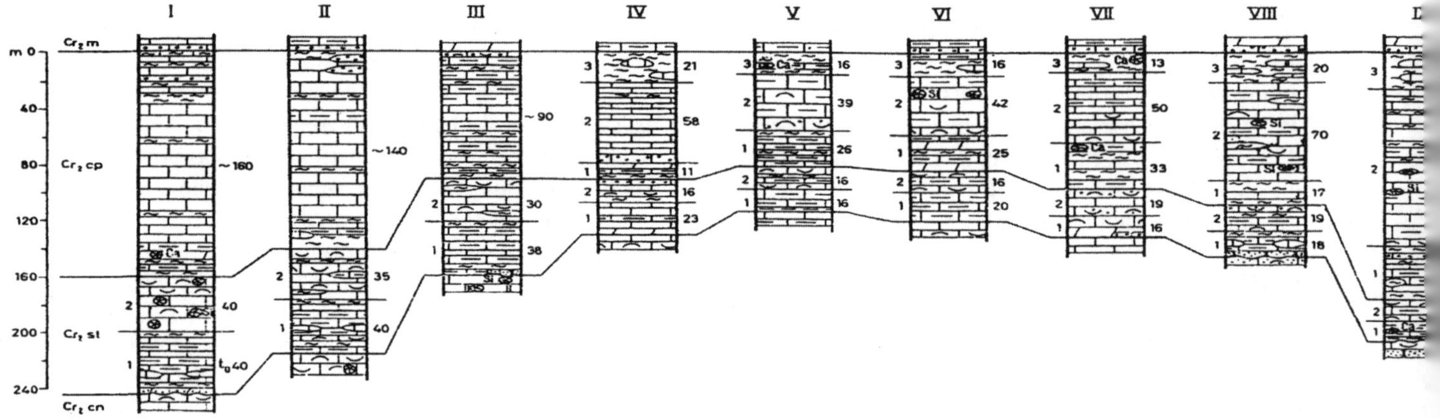
I
II
III
IV
V
VI
VII
VIII
m 0
40
80
120
160
200
240
Cr₂ m
Cr₂ cp
Cr₂ st
Cr₂ cn

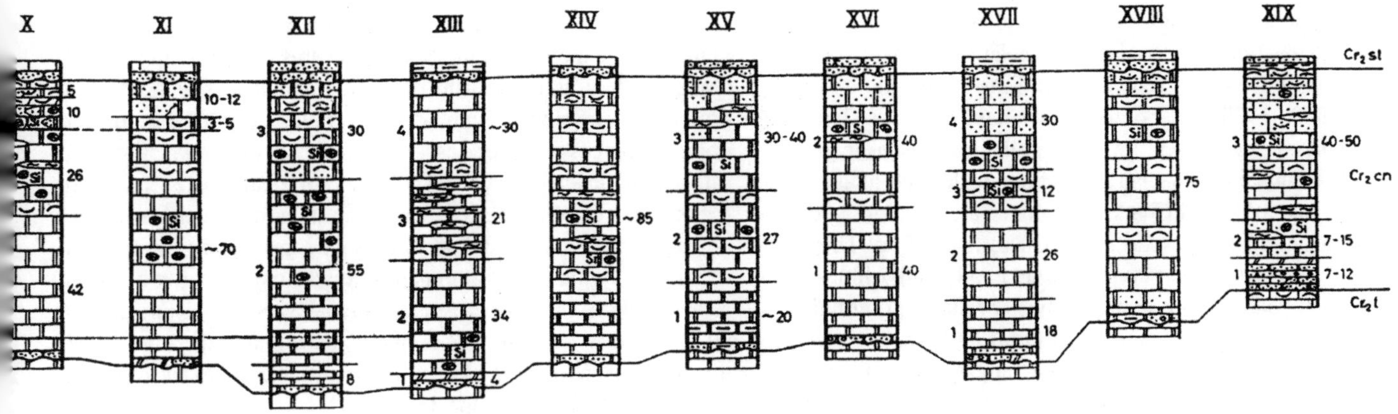

Fig. 6.40. Coniacian deposits sections of the Central Palmyrides (after Faradzhev, 1966).

Legend: Sections: I - Jebel Abou Dhour; II - Jebel El-Balas; III - Jebel Shoumariyeh; H - Hureire (for comparison, the Anti-Lebanon, see Fig. 6.39); IV - Jebel Mhin (and Jebel An-Nabek); V - Jebel Maksar An-Nimr; VI - Jebel Rous At-Twal; VII - Jebel Mzebleh; VIII - Jebel Mhasseh; IX - Jebel Butm; X - Jebel Ash-Sheikh; XI - Jebel An-Nassrani and Jebel Ghantous; XII - Jebel El-Bardeh; XIII - Jebel Rmah; XIV - Jebel Bassiri; XV - Jebel An-Naqnaqiyeh, 1237 m elevation mark; XVI - Jebel An-Naqnaqiyeh, 1188 m elevation mark; XVII - Jebel Om Journ (western portion); XVIII - Jebel Om Journ (eastern portion); XIX - Jebel El-Abtar.

Location of sections: see Fig. 6.34.

Lithologic symbols: see Fig. 6.29.

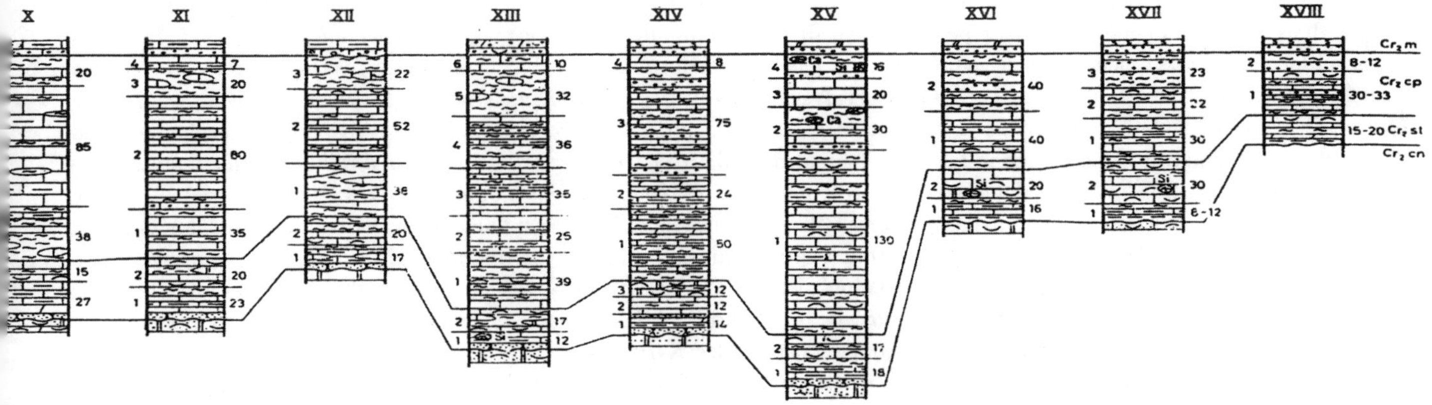

Fig. 6.41. Santonian and Campanian deposits sections of the Central Palmyrides (after Faradzhev, 1966).

Legend: Sections: I - Jebel Ash-Shaer; II - Jebel El-Balas; III - Jebel Ash-Shoumariyeh; IV - Jebel Mhin; V - Jebel Maksar An-Nimr; VI - Jebel Rous At-Twal; VII - Jebel Mzebleh; VIII - Jebel Sawwanet El-Mhasseh; IX - Jebel Butm; X - Jebel Ash-Sheikh; XI - Jebel An-Nassrani; XII - Jebel Rmah; XIII - Jebel El-Bardeh; XIV - Jebel An-Naqnaqiyeh, 1237 m elevation mark; XV - Jebel Om Journ; XVI - Jebel An-Naqnaqiyeh, south-eastern slope; XVII - Jebel El-Abtar; XVIII - Jebel Sawwanet El-Hamra.

Locations of sections: see Fig. 6.34.

Lithologic symbols: see Fig. 6.29.

As already mentioned, the Coniacian and Santonian deposits in the region of **Jebel An-Nusseiriyeh** were not identified during the 1:200,000 scale geological mapping (Shatsky et al., 1966; Ponikarov et al., 1967). Subsequently, geologists of the General Establishment of Geology and Mineral Resources of Syria singled out the Lower Senonian (Coniacian-Santonian) rocks in the vicinity of Qerdaha, Banyas, and Tartous (Ruske, 1978a,b; Yousef, 1979, 1980a,b) and their data have been used in the compilation of the newer 1:50,000 scale geological mapping. The undifferentiated Coniacian and Santonian deposits attain 45 m near the village of Ain al Charkiye, south of the town of Qerdaha. Eastward of Qerdaha they decrease to less than 15 m. This unit is composed of hard microcrystalline and soft chalky limestone, poorly or not bedded, with thin interlayers of marls. Foraminifera, such as *Globotruncana linneiana* (d'Orb.) and *Rosita fornicata* (Plummer) are indicators for a Santonian age. Maastrichtian strata unconformably overlie the Lower Senonian unit. Nevertheless, these data only provide preliminary and incomplete information concerning the Santonian rocks of Jebel An-Nusseiriyeh.

In the **Palmyrides** the widespread distribution of Santonian deposits exhibits a large variety of lithologies and facies over a vast territory.

On the southwestern margin of the southern branch of the Palmyrides, the Santonian is characterized by white and light-gray, soft, thick-bedded chalk-like limestones, with abundant planktonic foraminifera and fish scales. These soft rocks form a slight depression which is clearly seen in the landscape. Intercalations of compact, dense, organo-detrital or shelly (with oysters) limestones are also present. Very often limestones contain nodules of brown and black flints. Santonian deposits of this type are known from sections in Jebel Ad-Dmeir, Jebel Mghar, Jebel Ghorab, Jebel Zubeideh, Jebel Shaq Semri, and Jebel Butm (Razvalyaev, 1966a,b). The contact with the underlying Coniacian rocks is conformable. A maximum thickness of 66 m has been estimated at Jebel Butm where interbeds of silicified limestones are common (Fig. 6.32). A minimum thickness of 12-13 m has been observed at Jebel Mghar and in the eastern part of Jebel Ad-Dmeir (Fig. 6.39). The soft chalky limestones of Jebel Ghorab and Jebel Zubeideh include rich assemblages of foraminifera: *Dicarinella asymetrica* (Sigal), *Marginotruncana marginata* (Reuss), *M. coronata* (Bolli), *M. angusticarinata* (Gandolfi), *M. pseudolinneiana* Pessagno, *Globotruncana linneiana* (d'Orb.), *Rosita fornicata* (Plummer), *Archaeoglobigerina cretacea* (d'Orb.), and *Heterohelix reussi* (Cushman).

At Jebel An-Nabek and Jebel As-Sawwan, a high bitumen content accounts for the black and brown coloration of the thin-bedded clayey pelitomorphic limestones that yield fish teeth and scales (Kozlov, 1966a). Their thickness does not exceed 20 m (Fig. 6.27).

In the central part of the southern branch of the Palmyrides, the Santonian consists of two lithologically different units (Faradzhev, 1966). The lower unit is composed of soft clayey limestones and marls with abundant fish scales. Bands of organic limestones with flint and breadloaf-shaped lenses of hard limestones are subordinate. In the southwestern-most area, at Jebel Mhin, Jebel Maksar An-Nimr, Jebel Rous At-Twal, and Jebel Mzebleh, the Coniacian-Santonian contact is conformable. Beginning at Jebel Sawwanet El-Mhasseh in the west and as far as Jebel El-Abtar and Jebel Sawwanet El-Hamra in the east, a terrigenous layer appears at the base of the lower unit. Sometimes this layer is composed of orange-yellow and greenish, non-equigranular, easily weathered dolomitic sandstones with flinty pebbles and coarse quartz grains; whereas in other cases it is built up of brown, lustrous, salt-bearing calcareous clay. The underlying carbonate Coniacian rocks reveal definite traces of ancient weathering which created an uneven, hammock-shaped surface. As a consequence, in the area of the Palmyrides, pre-Santonian time was marked by a break in sedimentation, but this erosional interval was very short and in all areas the Santonian rests on the Coniacian and does not cover the more ancient rocks. These easily weathered rocks result in depressions in the landscape. (**Fig. 6.41**).

The upper member is composed of hard gray, coarse-bedded, shelly (oysters) limestones and grained dolomites containing flint concretions of a rounded or irregular shape and of varying sizes. Sometimes numerous bands of flint intercalate these calcareous rocks, giving the Santonian unit a lithologic aspect similar to that of the Campanian.

The total thickness of the Santonian rocks varies in the regions of Jebel Ash-Sheikh, Jebel Rmah, Jebel An-Naqnaqiyeh, and Jebel Om Journ between 30 and 45m. At Jebel Sawwanet El-Hamra, it drops to 15-20 m.

Organic remains are common in both units. The chalky limestones of the lower unit include rich assemblages of foraminifera: *Globotruncana fundiconulosa* Subb., *G. linneiana* (d'Orb.), *Rosita fornicata* (Plummer), *Marginotruncana coronata* (Bolli), *M. marginata* (Reuss), *M. angusticarinata* (Gandolfi), *M. pseudolinneiana* Pessagno, *Archaeoglobigerina cretacea* (d'Orb.), *Globigerinelloides ultramicra* (Subb.), *Heterohelix reussi* (Cushman), and *Pseudoguembelina costulata* (Cushman). Some pelecypods have been found in the hard organogenic limestones of this unit: *Corbula* aff. *distefanoi* Greco, *Protocardia hillana* Sow., *Cytherea plana* Sow. var. *judaica* Picard, *Crassatella* cf. *rothi* Fraas, *Pseudocucculea lens* Solger, and *Roudaireia undata* Conr. (det. N. Bobkova).

Planktonic foraminifera cannot always be extracted from the compact rocks of the upper unit. Nevertheless *Globotruncana linneiana* (d'Orb.), *Rosita fornicata* (Plummer), *Marginotruncana marginata* (Reuss), *M. coronata* (Bolli), *Dicarinella concavata* (Brotzen), and *Archaeoglobigerina cretacea* (d'Orb.) have been identified. Pelecypods are very common here: *Cardium ovulum* Conr., *Protocardia* cf. *picardi* Blanck., *P. coquande* Seg., *P. silicea* Blanck., *Crassatella matercula* May.-Eym., *C. haughtoni* Rennie, *C.* cf. *rothi* Fraas, *Cytherea plana* Sow. var. *judaica* Picard, *Roudaireia drui* Mun.-Chalm., *R.* cf. *undata* Conr., *Pseudocucculea lens* Solger, *Lima (Plagiostima) pseudohoernesi* Riedel, *Agelasina* cf. *plenodonta* Riedel, *Lopha semiplana* Sow., *L.* cf. *sollieri* Coq., and *Liostrea heinzi* Thomas et Peron.

In the area of the northeastern Palmyrides between T-4 Station and As-Sukhneh, the Santonian deposits undergo facies changes (Soulidi-Kondratiev, 1966a). To the west and north of Palmyra, they preserve a two-formational character, as in Jebel El-Khan, Jebel El-Qal, and Jebel Hayyan.

The lower member is represented by gray quartz sandstones and loose sandy limestones at the base, followed by thin- and medium-bedded aphanitic limestones which alternate with thin bands of gray flint and breadloaf-shaped silicified limestones. Their disconformable contact to the underlying eroded surface of the Coniacian deposits is clearly expressed in a number of places by the appearance of quartz sandstones and gravelites.

The upper member is composed of thick-bedded organogenic medium-bedded limestones with intercalations of breadloaf-shaped silicified limestones, bands of gray flints and very thin (1-5 cm) interbeds of nodular-granular phosphorites (Fig. 6.37). Thicknesses here do not exceed 30 m. Poorly preserved pelecypod shells are numerous: *Protocardia pauli* Coq., *Crassatella matercula* May.-Eym., *Agelasina* sp., *Lopha dichotoma* Bayle, *Ostraea* cf. *bella* Conr., and *Roudaireia undata* Conr. (det. N. Cheltsova).

The facies of the Santonian deposits change abruptly in Jebel El-Mazar, northeast of Tudmor (Palmyra). They are represented by a single dolomitic unit, which overlies an eroded surface of the Coniacian sequence. Quartz-dolomitic sandstones occur at their base. In Jebel As-Safra, Jebel Waariyeh, Jebel As-Satih, and Jebel Beer As-Sayah, the Santonian consists of gray and light-gray, massive, medium- to thin-bedded dolomites and strongly dolomitized limestones with remnants of pelecypods (*Protocardia* sp., *Crassatella* sp.); the thickness varies from 6 to 40 m (Fig. 6.37). Santonian sediments in Jebel El-Harraziyeh are completely missing and the basal beds of the Campanian Stage directly overlie eroded Coniacian rocks.

In Jebel Hawiyyet Ar-Ras, Jebel Thaniyyet As-Safra, Jebel As-Satih, and Jebel Mhashsha, unfossiliferous Coniacian and Santonian dolomites, as well as dolomitized limestones, cannot be differentiated.

In Jebel El-Mrah and Jebel Abou Rajmein, north of Tudmor, the thickness of the Santonian increases to 35-50 m. Organogenic and silicified limestones intercalate the dolomites. They contain *Lopha* cf. *dichostoma* Bayle, *Protocardia pauli* Coq., *Crassatella matercula* May.-Eym., *Aphrodina angustosinuosa* Riedel, and *Cardita* sp. (det. N. Cheltsova). Here the Santonian deposits also rest on an eroded surface of Coniacian rocks (Fig. 6.37).

The pelagic facies of chalky limestones is completely missing in the northeastern Palmyrides. It is replaced by shallow-water organogenic limestones and secondary dolomites, with casts of oysters.

Along the northeastern margin of the Palmyrides, adjacent to the Ar-Rassafeh Plain, Santonian deposits are developed in Jebel El-Khashshabiyeh, Jebel Mqebra, Jebel El-Minshar, Jebel El-Labdeh, and in the vicinity of Beer Barbar (Oufland, 1966a). Their appearance changes from east to west.

The Santonian deposits at Jebel El-Minshar consist of an alternation of fine-grained dolomites with irregular zones of silicification, sandy dolomitic limestones with quartz grains and rounded phosphate concretions and organogenic limestones. The latter contain many cavities as the result of the leaching out of pelecypod shells. All rock types contain numerous small nodules of light-gray flint. Near the eastern margin of this ridge, the Coniacian-Santonian contact is conformable; however, on the western side, the Santonian is transgressive over a visible erosional surface of Coniacian dolomites. The basal bed of the Santonian sequence starts with a breccia of flint debris (3-4 cm in size) and sandy, dolomitic, organo-detrital limestones. The thickness of the Santonian in Jebel Minshar ranges from 18 to 31 m (**Fig. 6.42**). Primary organogenic sediments can be classified as of shallow-water origin. Due to the processes of secondary dolomitization and recrystallization, abundant organic remains are generally present in the form of poorly defined detritus and casts. Therefore, only a limited number of species were identified: *Lopha* cf. *dichotoma* Bayle, *Pycnodonta flicki* Perv., *Liostrea* cf. *boucheroni* Coq., *Astarte similis* Muenst., and *Crassatella* cf. *desvauxi* Coq. (det. N. Cheltsova).

The Santonian deposits of the Jebel Mqebra range to the west of Jebel El-Minshar also rest unconformably on Coniacian dolomites. The entire Santonian succession is represented here by limestones – aphanitic, microgranular, organo-detrital or sandy. Intercalations of dolomites are completely absent. On the eastern slope of Jebel Mqebra the Santonian deposits are missing, possibly eroded, and Campanian flints directly overlie Coniacian dolomites.

Further west, in Jebel El-Labdeh, Wadi Hararit, Bir Qdeim, Beer Barbar, and Jebel El-Khashshabiyeh, Santonian deposits are composed of various limestones – hard aphanitic, clayey, organic detrital, coquina with silicified thick-walled oysters and phosphate grains. Interbeds of dolomites are absent. In this westerly direction, the thickness of the Santonian deposits increases from 8 m in the northeastern slope of Jebel El-Labdeh, to 60 m in Jebel El-Khashshabiyeh. Macrofauna in shallow-water limestones, not affected by dolomitization, have been encountered at different levels. These include pelecypods: *Protocardia* cf. *biseriata* Conr., *P. pauli* Coq., *Crassatella* cf. *desvauxi* Coq., *Roudaireia* sp., *Liostrea* cf. *boucheroni* Coq., *Lopha dichotoma* Bayle, *Astarte similis* Muenst., and *Pycnodonta flicki* Perv., and the gastropod *Dicroloma* sp. (det. N. Cheltsova and M. Dzhalilov). Pelagic chalky limestones with planktonic foraminifera are missing in this region.

To the west of Jebel El-Labdeh, the Coniacian-Santonian contact is conformable. At Wadi Hararit, Beer Barbar, and Jebel El-Khashshabiyeh, limestones full of *Serpula* tubes characterize the end of the Coniacian sequence, clearly marking the boundary.

At Jebel Ash-Shaer, Jebel El-Balas, Jebel Ash-Shoumariyeh, and Jebel Shafa Othunein, in the northern branch of the Palmyrides, a gradual transition from Coniacian to Santonian occurs. Only in one section at Jebel Ash-Shaer is the eroded surface of the Coniacian limestones covered by a bed of sandy limestones, marking the base of the Santonian (Faradzhev, 1966; Mikhailov, 1966). In this area, two members characterize the Santonian deposits, as in the southern branch of the Palmyrides (Fig. 6.41).

The lower member is composed of alternations of soft pink,

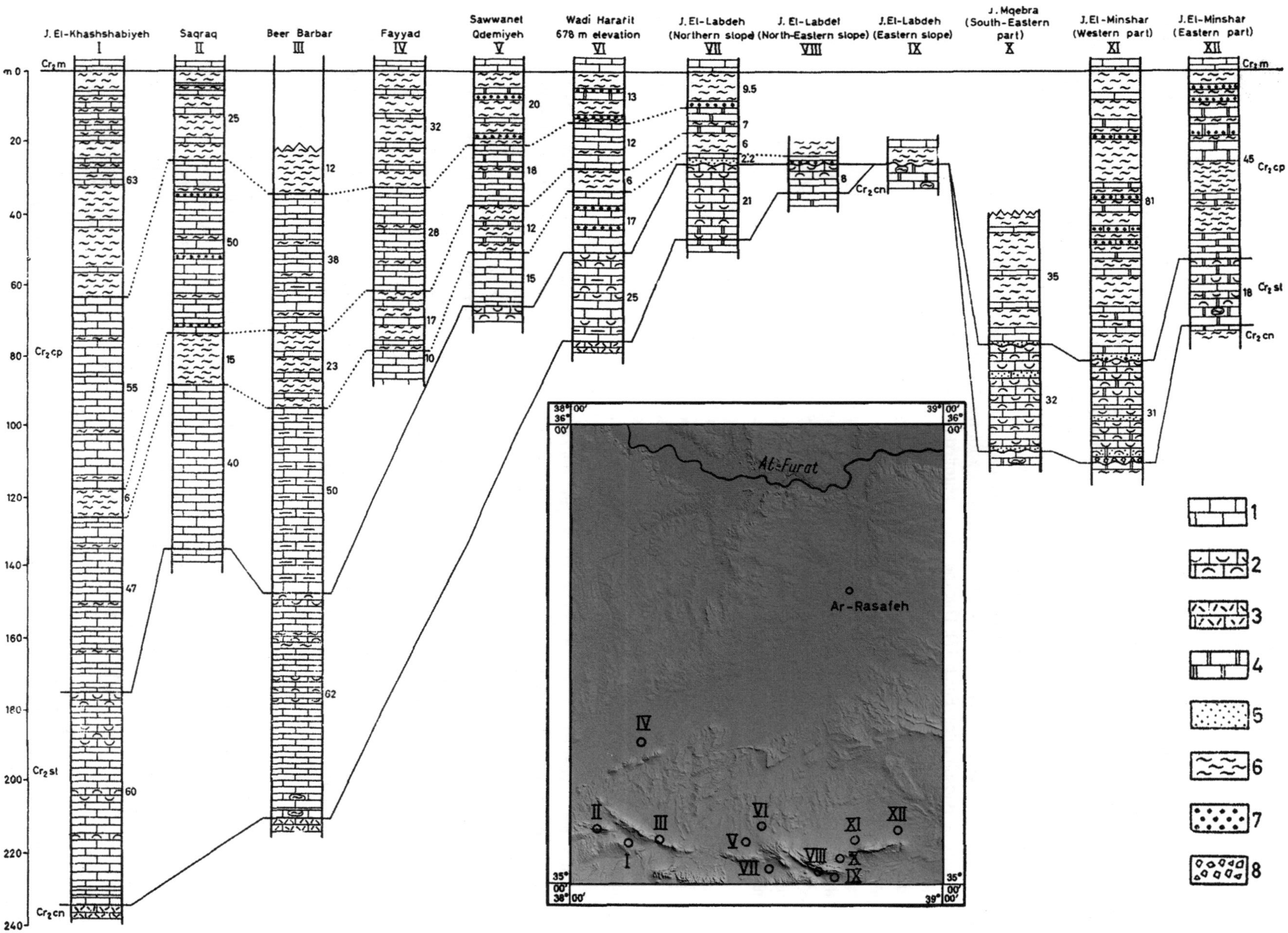

Fig. 6.42. Santonian and Campanian deposits sections of the Palmyrides (northeastern marginal portion, south of Ar-Resafeh) (after Oufland, 1966a). Legend: 1 - limestone; 2 - organogenic limestone; 3 - limestone with Serpula; 4 - dolomite; 5 - sand, sandstone and sandy varieties of rocks; 6 - flint; 7 - phosphate and phosphate-bearing rocks; 8 - breccia.

red, yellow and white clayey limestones with fishbones and scales and of gray, micro-grained, very dense and firm limestones which, at times, form horizons of breadloaf-shaped bodies. Thin (5-10 cm) intercalations of flint are common. The thickness of this member is about 40 m.

The upper member consists of light-gray, medium- to thick-bedded and massive limestones, frequently with an odor of H_2S. The limestones possess various textures – aphanitic, micro-granular, and detrital. Siliceous concretions and lenses are also very common here. The thickness ranges from 30 to 50 m. The following pelecypods have been identified in this member: *Cardium ovulum* Conr., *Parmicorbula incurvata* Koenen, *Roudaireia* cf. *undata* Conr., *R. drui* Mun.-Chalm., *Cytherea plana* Sow. var. *judaica* Picard, *Protocardia* cf. *silicea* Blanck., *Crassatella* sp., and *Leda* sp. (det. N. Bobkova and N. Cheltsova).

Campanian [K_2cp]

Campanian deposits are well-exposed in many parts of the Anti-Lebanon, the Palmyrides, and Jebel Az-Zawiyeh (north of Hama). They occur sporadically in Jebel An-Nusseiriyeh, but are entirely absent in Kurd-Dagh. In general, the Santonian-Campanian contact appears to be conformable. In the Anti-Lebanon, this boundary is not represented by any lithological change within the monotonous Senonian soft chalky limestones. The transgressive character of the Campanian is only found in the eastern part of the Palmyrides and Jebel Az-Zawiyeh.

The Campanian lithology encompasses three facies-types:

1) The widely distributed pelagic soft white chalky limestones and marls;

2) Shallow marine cherts and siliceous carbonates and

3) Phosphorites.

Their degree of combination varies from location to location. The maximum thickness of the Campanian sediments reaches 200 m.

The biostratigraphy of the Campanian pelagic sediments is provided by the planktonic foraminifera. These consist of the *Globotruncanita elevata*, *Globotruncana ventricosa* and *Globotruncanita calcarata* Zones. Ammonites are present, and benthic foraminifera are abundant. The more shallow-water detrital limestones contain numerous temporally long-ranging pelecypods.

In the **Anti-Lebanon**, the Campanian is usually about 50-60 m thick. It is subdivided into two members (Razvalyaev, 1966a).

The Lower member: In the sections at Beit Jann, Hureira, Deir Qanoun, the Barada River Valley, and Maaret El-Bash white and light-gray, soft, massive-bedded, chalk-like limestones predominate. In the sections at Maaraba, Jebel Mannsour, and Jebel Suekai, chalky limestones alternate with fairly hard organo-detrital limestones. These contain large oyster shells of *Gryphaea vesicularis* Lam., and sometimes thin interbeds of phosphorites (Fig. 6.39). Two planktonic foraminiferal zones subdivide the lower member:

1) The *Globotruncanita elevata* Zone with the index-species, *G. stuartiformis* (Dalbies), *Globotruncana linneiana* (d'Orb.), *G. bulloides* Vogler, *G. arca* (Cushman), *G. orientalis* El Naggar, *G. lapparenti* Brotzen, *Rosita fornicata* (Plummer), *Marginotruncana marginata* (Reuss), *Archaeoglobigerina cretacea* (d'Orb.), *Globigerinelloides ultramicra* (Subb.), *Pseudoguembelina costulata* (Cushman), *Hedbergella holmdelensis* Olsson, *Heterohelix globulosa* (Ehrenb.), and *H. striata* (Ehrenb.), and

2) The *Globotruncana ventricosa* Zone, which is characterized by the appearance of the index-species, *G. rosetta* (Carsey), *G. rugosa* (Marie), *Rosita plummerae* (Gandolfi), and *Globotruncanita subspinosa* (Pessagno), in conjunction with *Globotruncana linneiana* (d'Orb.), *G. arca* (Cushman), *G. fornicata* (Plummer), *Hedbergella holmdelensis* Olsson, *Heterohelix globulosa* (Ehrenb.), *H. striata* (Ehrenb.), and *Pseudoguembelina costulata* (Cushman), already encountered in underlying layers.

The Upper Member is made up of brownish and dark-gray flints alternating with compact clayey limestones. It can be traced easily in the field, as it forms a clearly visible and mappable scarp. Interbeds of calcareous rocks contain planktonic foraminifera of the Upper Campanian *Globotruncanita calcarata* Zone – *G. calcarata* (Cushman), *Globotruncana linneiana* (d'Orb.), *G. arca* (Cushman), *G. orientalis* El Naggar, *G. rosetta* (Carsey), *G. ventricosa* (White), *Rosita fornicata* (Plummer), *R. patelliformis* (Gandolfi), *Pseudotextularia elegans* (Rzehak), *Heterohelix striata* (Ehrenb.), *H. globulosa* (Ehrenb.), and *Pseudoguembelina costulata* (Cushman).

The total thickness of the Campanian Stage is usually about 50-60 m.

In the region of the **Jebel An-Nusseiriyeh**, Campanian deposits are generally missing (Kozlov, 1966a; Shatsky et al., 1966). In the area of Haffeh, in the northwestern part of these mountains, near the villages of Kushba and Ain Layloun, Santonian rocks are overlain by chalky and more compact pelitomorphic limestones with *Globotruncanita elevata* (Brotzen), *Globotruncana arca* (Cushman), *Marginotruncana marginata* (Reuss), and *Rosita fornicata* (Plummer) (Youssef, 1979). These limestones are transgressively covered by Maastrichtian sediments. Microfauna indicates that the Lower Campanian *Globotruncanita elevata* Zone and the younger sediments of Campanian age are evidently absent.

East of Jebel An-Nusseiriyeh, at **Jebel Az-Zawiyeh** (Shatsky et al., 1966) and along the right bank of the Nahr El-Aasi River, Campanian rocks are quite well-developed. Usually they rest unconformably on Cenomanian-Turonian hard limestones. However, in the Nahr El-Aasi Valley (5 km below the town of Hama), the Campanian deposits cover Coniacian-Santonian gypsum-bearing rocks. The two Campanian members are identifiable as in the Campanian Stage of the Anti-Lebanon.

The Lower Member consists of white, cream and pinkish, soft, chalky limestones and more compact clayey limestones with numerous fish scales; the thickness is 35-40 m.

The Upper Member is made up of the same limestone, but well bedded (3-25 cm thick), with lenses and irregularly-shaped concretions of dark-brown and black flints; the thickness is up to 55 m.

Macrofauna is very rare in the chalky limestone. The lower

sequence yields the ammonite *Menabites breistrofensis* Collign. and the pelecypods *Pycnodonta* ex gr. *canaliculata* d'Orb., *P. clavata* Nils., *Liostrea* sp., and *Amphidonta* sp. On the other hand, abundant planktonic foraminifera belong to the *Globotruncanita elevata, Globotruncana ventricosa* and *Globotruncanita calcarata* Zones. Thus, at Jebel Az-Zawiyeh, the Campanian Stage is represented by a complete stratigraphic sequence.

In the **Palmyrides,** Campanian deposits are very well-exposed. Three main lithologies, calcareous, siliceous, and phosphatic, occur here:

1) The calcareous rocks are very diverse – from soft chalky limestones and marls up to hard microgranular and organo-detritic limestones.

2) Siliceous rocks are extremely abundant as beds, lenses or concretions. Chert concretions (chalcedony flint) are rounded, irregular, disc- or bun-shaped. Layered and lens-like flint is present in two varieties. One is finely crystallized, with uniform banded and brecciated structure, which is found in many outcrops. The other consists of metasomatic silifications of organic relicts within limestones – the silica replacing entire shells or their fragments. The Campanian outcrops are quite well marked by debris of silicified rocks, with a dark-brown desert varnish.

3) Phosphatic rocks form huge commercial occurrences in the central Palmyrides.

The lithology and thickness of the Campanian deposits in the Palmyrides are controlled by the tectono-environmental conditions of local anticlines and troughs.

In the southwestern margin of the southern branch of the Palmyrides the Campanian deposits resemble those of the Anti-Lebanon and can again be subdivided into two members (Razvalyaev, 1966a,b).

The Lower Member is composed of white and light-gray, chalky, soft limestones and marls, in which interbeds of compact fine-grained, fine-detrital, phosphate-bearing and silicified limestones also appear. These deposits contain rare pelecypods – *Leda perdita* Conr., *Cytherea* cf. *andersoni* Bull.-Newt., and *Protocardia* ex gr. *biseriata* Conr. (det. N. Bobkova), and planktonic foraminifera of the *Globotruncanita elevata* and *Globotruncana ventricosa* Zones. To the west, at Jebel Mghar and Jebel Ad-Dmeir, chalky limestones are common (Fig. 6.39). To the east, at Jebel Ghorab, Jebel Zubeideh and Jebel Butm, intercalations of hard microgranular and silicified limestones increase (Figs. 6.32 and 6.39).

The Upper Member forms a 15-20 m thick marker which includes medium- to thick-bedded brownish and black chert with thin layers (20-30 cm) or bun-shaped inclusions of hard dense silicified, organo-detritic or phosphatic limestone.

The total thickness of these deposits reflects the Campanian paleo-landscape. On uplifted blocks like Jebel Ad-Dmeir and Jebel Shaq Semri, it is reduced to 22-30 m, whereas thicknesses reach a maximum of 50-136 m in the structural troughs, such as Jebel Ghorab, Jebel Zubeideh, and Jebel Butm.

Further north of Jebel Ad-Dmeir, in the area of Jebel Sharqi An-Nabek and Jebel As-Sawwan, the Campanian still preserves its double-member subdivision (Kozlov, 1966a). The lower member is calcareous, but the chalky limestones are practically absent and the succession consists of gray and black, thin-bedded, platy, compact limestones. Their dark color is associated with organic matter, which is especially intensive in the Sheib Valley. Thin intercalations of brown and black flints, as well as shelly limestones with pelecypods – *Amphidonta lateralis* Nils., *A. arietina* d'Orb., *Meretrix* sp., *Liostrea* sp., and *Gryphaea vesicularis* Lam. are also present. In the upper member, limestones are replaced by black and gray, compact, layered cherts with thin interbeds of fine-grained silicified limestones. Disharmonic contortions characterize these beds. The total thickness of the deposits reaches up to 70-90 m (Fig. 6.27).

In the central part of the southern branch of the Palmyrides, between Al-Qaryatein and Al-Beida, the Campanian is characterized by a single unit of irregularly alternating limestones of various types, cherts and phosphatic deposits. Medium-bedded, compact, hard, micrograined limestones predominate; soft, white chalky varieties with abundant fish remains and other organic debris are comparatively rare. The cherts are layered, sometimes brecciated, with a relict-organic texture or in the form of lenses and nodules. In the uppermost part of this unit, hard, brown to dark-black chert, with interbeds of compact limestones and phosphatic rocks form a marker-horizon in the landscape. The Campanian in this area varies from 13 to 30 m in thickness (Faradzhev, 1966).

The Campanian deposits of this type form splendid exposures in many ridges of the Palmyrides – Jebel Mhin, Jebel Maksar An-Nimr, Jebel Rous At-Twal, Jebel Mzebleh, Jebel Rmah, etc. On some ridges the Campanian acquires a distinctive three-member subdivision: the chert beds being mainly confined to the lower and upper parts of these sections with various limestones in the middle *e.g.* at Jebel Sawwanet El-Mhasseh, Jebel Ash-Sheikh, Jebel An-Nassrani, Jebel El-Bardeh and others (Fig. 6.41).

A distinctive feature of the uppermost siliceous part of the Campanian is the huge accumulation of phosphatic matter. Commercial occurrences of phosphorites have been discovered in the areas of Jebel El-Abtar, Jebel Sawwanet El-Hamra, Jebel An-Naqnaqiyeh, and Jebel Om Journ. To the west of these ridges, phosphate beds become thinner and poorer, finally pinching out. The compact bedded phosphate rock consists of various types of phosphatic matter: microconcretions, oolites, oolite-like bodies (grains), and organic remains. The phosphatic beds are generally 1 m thick, but sometimes reach up to 3-12 m. They extend laterally for several kilometers. The best-known occurrences are the Sharqiyeh and Khneifiss. In each case the total extension of the phosphate formation measures several dozens of kilometers.

The total reserves of the Sharqiyeh deposit have been estimated to 266 million tons, with an average bed thickness of about 9 m, an average P_2O_5 content of 24.2%, extreme P_2O_5 values of 19-33% and an overburden not exceeding 22 m. However, only about 140 million tons are of commercial value, with an average ore bed thickness of about 11 m, an average P_2O_5 content of 24.5%, and an average depth of the overburden of 3 m.

The total phosphate reserves of the Khneifiss deposit, located at the western end of the Jebel El-Abtar range, amount to 24.6 million tons, with an average P_2O_5 content of 27.8% and extreme P_2O_5 values of 19.84-34.82%. The commercial reserves constitute 6.7 million tons, with an average P_2O_5 content of 28.58%.

In reality, the actual reserves of these two occurrences, as well as some other deposits, are greater because their boundaries have not yet been fully outlined. As far as is known at present, additional estimations based on bore-hole research have not shown any decrease in the ore thickness or any lessening in its quality.

The Campanian sediments reach a maximum thickness of 196 m in Jebel Om Journ. Similar figures (142-177 m) have been measured in the near-by ridges at Jebel An-Naqnaqiyeh, Jebel El-Bardeh, and Jebel An-Nassrani. In Jebel Rmah, Jebel Mzebleh, Jebel Rous At-Twal, etc. the thickness usually ranges around 80 to 90 m. A minimum thickness of 40-45 m has been observed in Jebel Sawwanet El-Hamra, near the northern margin of the Rutbah Uplift.

The Campanian sediments were first dated by ammonites – *Pseudoschloenbachia* aff. *umbulazi* Baily, *Baculites* cf. *vagina cazadoriana* Paulcke, *B. anceps* Lam., and *Clyptoxoceras* sp. (det. A. Atabekyan), and pelecypods – *Cyprimera aurilitoralis* Cox, *Roudaireia undata Corn., Pycnodonta vesicularis* Lam. var. *hippopodium* Nils., *Cytherea plana* Sow. var. *judaica* Picard, *Leda perdita* Conr., *Exogyra spinosa* Math., *Crassatella matercula* May.-Eym., *Inoceramus regularis* d'Orb., *Nucula belzonii* Greco, and *Meretrix* sp. (det. N. Bobkova). Fish teeth include *Scapanorhynhus rapax* (Quaas) (det. L. Glikman).

Micropaleontological investigation of the Campanian deposits has augmented the information received from the macrofauna.

Although only extractable from soft chalky limestones, planktonic foraminifera indicated the presence of the following three zones.

1) The *Globotruncanita elevata* Zone with the index-species *Globotruncana arca* (Cushman), *G. linneiana* (d'Orb.), *G. bulloides* Vogler, *Rosita fornicata* (Plummer), *R. patelliformis* (Gandolfi), *Heterohelix globulosa* (Ehrenb.), and *Pseudoguembelina costulata* (Cushman);

2) The *Globotruncana ventricosa* Zone with the index-species, *G. rosetta* (Carsey), *G. rugosa* (Marie), *G. arca* (Cushman), *Globotruncanita subspinosa* (Pessagno), *Rosita plummerae* (Gandolfi), *Pseudotextularia elegans* (Rzehak), and *Heterohelix striata* (Ehrenb.).

3) The *Globotruncanita calcarata* Zone corresponds to the upper member, composed of chert and hard limestone with very poor microfauna. Usually the Late Campanian age of this member is based on thin soft chalk which underlies the upper member and contains planktonic foraminifera of the zone in question.

Benthic foraminifera in the Campanian deposits of the Central Palmyrides are also numerous and diverse: *Marssonella oxycona* (Reuss), *Arenobulimina obesa* (Reuss), *Ataxophragmium compactum* Brotzen, *Anomalina clementiana* (d'Orb.), *A. stelligera* (Marie), *A. umbilicatula* Mjatl., *A. bartensteini* (Hofker), *A. pertusa* (Marsson), *A. monterelensis* Marie, *Cibicides temirensis* Vass., *C. excavatus* Brotzen, *Gyroidinoides umbilicatus* (d'Orb.), *Eponides biconvexus* Marie, *Parrella cordieriana* (d'Orb.), *Globorotalites michelinianus* (d'Orb.), *Bolivinoides decoratus* (Jones), *Bolivinita planata* Cushman, *Neoflabellina cordata* (Reuss), *N. rugosa* (d'Orb.), and many others.

In the northeastern part of the southern branch of the Palmyrides (between Palmyra and As-Sukhneh), Campanian deposits are exposed on the slopes of Jebel El-Mrah, Jebel Abou Rajmein, and Jebel Tlol El-Arida, as well as in the central part of Jebel Tantour. They extend as a narrow belt along the northern slopes of Jebel Hayyan, Jebel El-Mazar, Jebel As-Safra, Jebel El-Harraziye, Jebel El-Waariyeh, and Jebel As-Satih. The southern flanks of these anticlines are frequently cut by faults. (Fig. 6.36).

As a rule, Campanian layers conformably overlie Santonian deposits. A break in sedimentation (in pre-Campanian time) was noted in Jebel El-Waariyeh and Jebel As-Satih, where Campanian deposits overlie an irregular surface of Santonian rocks with an erosional hiatus, as well as in Jebel El-Harraziyeh, where such a disconformity exists between the Campanian and Coniacian deposits (Soulidi-Kondratiev, 1966a).

The presence of persistent flint bands and phosphorites, associated with hard organogenic limestone distinguishes the Campanian deposits from the underlying formations. The appearance of dolomites differentiates the Campanian rocks of this area from coeval sequences in the southwestern part of the Palmyrides.

As in some sections of the central Palmyrides, deposits of this age in the northeastern Palmyrides can be divided into three members –predominantly siliceous lower and upper units and a predominantly calcareous middle unit. This Campanian three-member structure can be observed very well in the Jebel Hayyan section, southwest of Palmyra (Figs. 6.36 and 6.37).

The lower member (15 m thick) consists of gray and dark-gray bedded flint (2 to 3 m thick) with intercalations of hard aphanitic and organogenic silicified limestones with *Lopha* aff. *nicaisei* Coq., *Cytherea* ex gr. *andersoni* Bull.-Newt., and *Astarte* sp. (det. N. Cheltsova).

The middle member (26 m thick) is composed of silicified, arenaceous or organogenic limestones with subordinate bands and nodules of black and brown flint. Limestones have yielded some rare ammonites – *Baculites* cf. *bouilei* Collign. and *Pseudoschloenbachia* sp. (det. A. Atabekyan).

The upper member is essentially flinty (16 m thick), represented by black and gray flints (silicites) with interbeds of silicified organogenic limestones containing shells of *Baculites anceps* Lam. and inclusions of granular phosphorites.

The total thickness of the Campanian Stage in this section approaches 57 m.

To the northeast of Palmyra, the general features of the Campanian are well preserved, but the number of phosphorite bands notably increases and their thicknesses become quite variable. In Jebel El-Mazar, the up to 80 m thick middle and upper members contain numerous phosphorite layers. All three members of the Jebel As-Safra section are phosphate-bearing. Twelve bands of granular silicified phosphorites reach approximately 1 m thickness each, though the total thickness decreases to 40 m. The Campanian deposits of Jebel El-Harraziyeh and Jebel As-Satih are the thinnest, being 27 m and 38 m respectively; the number of phosphorite bands drops to 5 and 7, but their thickness increases to 2-5 m. A maximum number of 20 phosphorite beds has been found in Jebel El-Waariyeh, where some bands are 1 to 1.5 m thick. The Campanian deposits of Jebel Abou Rajmein essentially lack phosphoritic intercalations.

The thicknesses of the Campanian deposits in the northeastern Palmyrides are less significant (27-80 m) as compared to those in the central and southwestern parts. By contrast, a very peculiar section has been observed in Jebel El-Mrah, where its thickness reaches up to 200 m and dolomites appear. Notwithstanding its greater_thickness, this section is also characterized by a three-member structure: a lower member of black and brown flint beds with intercalation of dark-gray dolomites (66 m thick); a middle member of yellowish and light-gray dolomites with phosphorites (80 m); and an upper member of intercalations of dark-gray flint, light-gray phosphatized and silicified limestones and thin bands of nodular phosphorites (54 m).

In the northeastern part of the Palmyrides, fauna is very rare. Most common are pelecypods, although their state of preservation is rather poor: *Astarte similis* Muenster, *Agelasina* sp., *Aphrodina angustosinuosa* Riedel, *Amphidonta (Aetostreon)* aff. *langloisi* Coq., *Lopha* cf. *nicaisei* Coq., *L. villei* Coq., *L. renoui* Coq., *L. tissoti* Thomas et Peron, *Fragum subperobliquum* Riedel, *Liostrea* cf. *incurva* Nils., *L. stoliczkai* Basse, and *Lima* sp. (det. N. Bobkova).

Campanian deposits stretch as a continuous narrow strip along the northeastern margin of the southern branch of the Palmyrides. They are exposed from Jebel El-Minshar in the east to Jebel Khashshabiyeh in the west, as well as in the Fayyad Ridge, an isolated elevation near the northern slope of the Palmyrides (Fig. 6.42). In the area of Jebel Mqebra and Jebel Labdeh, Campanian and Santonian sediments are separated by an erosional surface. To the east and west of this uplifted area, Campanian deposits rest conformably on Santonian dolomites and limestones with *Serpula* (Oufland, 1966a). Both sides of the uplifted area exhibit different lithologies. To the east of Jebel Mqebra, the Campanian formation comprises frequent alternations of pinkish fine-grained dolomites, gray limestones with abundant fish remains, granular phosphates and organogenic phosphate rocks as well as numerous bands and lenses of brown and black chert. The latter are especially abundant at the top of the sequence. The thickness in Jebel El-Minshar varies from 45 to 81 m.

West of Jebel Mqebra and Jebel El-Labdeh, the character of the Campanian changes:

1) The number of phosphate beds diminishes;

2) The dolomites are replaced by limestones;

3) The quantity of organic remains increases to include shells of Baculites sp.;

4) The thickness increases up to 171 m in Jebel El-Khashshabiyeh; and

5) In almost all localities the Campanian succession consists of four members. Among these members, the first and third members are carbonate rocks, while the second and fourth members are made up mainly of flint with separate layers of limestones, dolomites and phosphates. The carbonate members are white, platy, fine-grained and hard limestones, bun-shaped aphanitic dolomites and rarer white and yellowish soft limestones with fish remains, while streaks of gray and brown flints are subordinate.

Macrofossils in the Campanian deposits of this area are rather rare. Present only in the western part, they are unfortunately poorly preserved: *Lucina* ex gr. *dachelensis* Conr., *Aphrodina* cf. *angustosinuosa* Riedel, *Crassatella* sp., and *Protocardia* sp.

The Campanian deposits in the northern branch of the Palmyrides are characterized by similar limestones, silicites and phosphorites, though dolomites are lacking here. Their relationship varies from section to section (Faradzhev, 1966). In Jebel Ash-Shaer, limestones predominate and thin layers of chert appear only at the base. These cherts almost disappear from the section forming a distinct member only at the top (Figs 6. 33 and 6.41). On the western and southern flanks of Jebel El-Balas, a band of chert about 10 m thick occurs at the base of the section. In Jebel Ash-Shoumariyeh, numerous very thin, 15-20 cm thick chert bands are distributed throughout the section.

Phosphatic concentrations here are confined to the top of the sections. The thickest phosphate beds have been found on the southern slope of Jebel El-Balas, where their thickness ranges from 0.7 to 1.2 m.

The total thickness of the Campanian deposits in the Jebel El-Balas mountains varies from 90 to 160 m.

East of Salamiyeh, the northern rim of Jebel El-Balas and Jebel El-Shaer is fringed by foothills with small outcrops of Campanian. Drill-holes also penetrated sediments of this age in the table mountains of Jebel Abisan (Mikhailov, 1966). Various limestones with flints and rare intercalations of phosphorites conformably overlying the Santonian were attributed to the Campanian. The limestones are extremely diverse: microgranular, detrital, slightly phosphatized with remnants of fish bones, soft chalk-like, pelitomorphic clayey with fish scales, hard aphanitic, or bun-shaped silicified. Gray, brown and black flints are present in the form of layers and lenses. Their greatest accumulation is at the bottom and top of the Campanian, reaching a thickness of up to 10-15 m, with individual flint beds up to 1 m thick. In the middle, rare flints appear as thin lenses and concretions. Granular organogenic phosphorites are developed as rare intercalations up to 15-20 cm thick.

Due to the smooth relief, it is difficult to estimate the thickness of the Campanian. Tentative figures are about 140-160 m.

The Campanian age of these sediments has been confirmed by the ammonites *Hoplitoplacenticeras plasticum* var. *laevis* Paulcke and *Baculites* cf. *vagina* Forb. var. *cazadorianus* Paulcke, as well as by planktonic foraminifera of the *Globotruncana ventricosa* Zone, assemblages of which include *G. ventricosa* (White), *G. rosetta* (Carsey), *G. arca* (Cushman), *G. hilli* Pessagno, *Rosita plummerae* (Gandolfi), *R. fornicata* (Plummer), *Heterohelix globulosa* (Ehrenb.), and *H. striata* (Ehrenb.).

In the **Mesopotamian Basin,** two localities with Campanian deposits have been encountered. Their exposures are confined to the Jrablus - Tuwal El-Aba - Jebel Abd El-Aziz - Sinjar Rampart (or Swell) chain of structures, which stretches along the border with Turkey.

Near **Jrablus**, in the core of the Tellik anticline, a member not exceeding 25-30 m in thickness of black and brown, thin bedded and highly brecciated flints with rare interlayers of firm and massive dolomites (up to 1 m thick) is exposed (Oufland, 1966b). These rocks have yielded remnants of pelecypods and gastropods in a poor state of preservation among which N. Cheltsova recorded the Santonian-Maastrichtian species *Aphrodina* cf. *angustosinuosa* Riedel. Flints and dolomites are unconformably overlain by Miocene detrital limestones, but in the adjacent area of Jub Mahly typical chalky white Maastrichtian marls and dolomites are also developed (**Fig. 6.43**). Bearing this huge gap in mind and, most importantly, its lithology, this member should be considered as Campanian.

In the **Jebel Abd El-Aziz** Anticline, undivided rocks of the Santonian, Campanian and Lower Maastrichtian rest on an eroded surface ranging from Neocomian to Coniacian in age. This about 200 m thick succession starts with basal conglomerates which are replaced upwards by limestones and marls (Ponomarev and Bivshev. 1966). The upper beds of this formation contain Senonian pelecypods: *Trigonoarca camerunensis* Riedel, *T. curvatodonta* Riedel, *Arca (Eonavicula) bogaerti* Dart., and *Neithea quinquecostata* Sow. (det. N. Cheltsova), and the planktonic foraminifera *Globotruncana rosetta* (Carsey), *G. arca* (Cushman), and *Rosita fornicata* (Plummer), for which a Campanian age can be suggested, although not yet proven.

Maastrichtian [K_2m]

Maastrichtian deposits are widespread throughout Syria. They are well-exposed in the Anti-Lebanon, Jebel An-Nusseiriyeh, Jebel Az-Zawiyeh, the Baer-Bassit Massif, Kurd-Dagh, the Palmyrides, the Jrablus Rampart, and in Jebel Abd El-Aziz.

Sediments of this age have been penetrated by numerous bore-holes on the Aleppo Plateau in the northern part of Syria, the Mesopotamian Basin, the Yarmouk River Valley, and the Syrian Desert to the south of the Palmyrides.

In the Anti-Lebanon and in the Palmyrides, Maastrichtian chalks and limestones lie on Campanian hard limestones and flints, either conformably or with some slight traces of erosion.

The Maastrichtian deposits of the northwestern part of Syria, in Jebel An-Nusseiriyeh, Jebel Az-Zawiyeh, Kurd-Dagh, the Baer-Bassit Massif, as well as those in eastern Syria in Jebel Abd El-Aziz, are distinctly transgressive and rest on the eroded surface of older rocks of various age.

The main Maastrichtian lithology consists of a wide distribution of white and light-gray soft chalky limestones, clayey limestones and marls. A very important lithologic feature of the lower part of the Maastrichtian sediments is the existence in the southern areas of Syria of rich bands of phosphorites and black oil shales throughout the section. Intercalations of hard microgranular and detrital limestones and also of silicified limestones and flints are subordinate. To a great extent these peculiarities distinguish Maastrichtian deposits from those of the Campanian.

In the Anti-Lebanon the Maastrichtian consists of a single unit of monotonous chalky limestones and marls, but nearly everywhere else in Syria it can be subdivided into two formations. The lower formation is composed of various rocks, within which chalky limestones predominate. They alternate with compact silicified limestones, phosphate beds, lenses, and streaks of flint. This appears to be a transition between the Campanian facies and the typical Maastrichtian facies of the upper formation, which consists of uniform white, light-gray and yellowish soft chalky limestones and marls.

The thickness of these Maastrichtian deposits varies broadly from several dozens of meters up to 750 m. Drilling data in the Mesopotamian Basin revealed a thickness of up to 1200 m of Maastrichtian sediments (El-Bouab and Aba bore-holes).

Planktonic foraminifera are the most reliable and important tool for unraveling the biostratigraphy of the Maastrichtian pelagic sediments. Among the various zonal schemes for the Maastrichtian we will follow that of Sigal (1977). It includes three zones: *Globotruncana falsostuarti, Gansserina gansseri,* and *Abathomphalus mayaroensis*. In some sections, according to Caron (1985), the abundance of microfauna in the *Globotruncana falsostuarti* Zone permits its subdivision into two, the *Globotruncanella havanensis* Zone and the *Globotruncana aegyptiaca* Zone. Benthic foraminifera are also numerous, and pelecypods are sometimes present.

Ammonites and belemnites are absent in the Maastrichtian of Syria. Therefore, based on foraminiferal biostratigraphy, the Campanian/Maastrichtian boundary is put at the contact of the lower *Globotruncanita calcarata* Zone with the upper *Globotruncana falsostuarti* Zone.

The regional stratigraphy of the Maastrichtian in Syria can be outlined as follows:

In the **Anti-Lebanon**, the Maastrichtian is represented by a uniform formation of white and light-gray chalk-like limestones and marls – soft, thick-bedded, with rare intercalations of organo-detrital clayey limestones (Razvalyaev, 1966a; Kurbanov and Zarjanov, 1968). The thickness is quite variable, reflecting the structural position of the section and the transgressive character of the Paleocene deposits. From a minimum thickness of 13 m in the Beit Jann section on the slope of the Hermon Massif, the thickness increases up to 66-68 m in the sections at Deir Qanoun, Ein et Tine, Doummar, and Ain El-Khadra along the Nahr Barada River Valley. Near the village of Hureire on the southeastern slope of Jebel El-Khorn, the thickness reaches up to 128 m. The most complete sections of Maastrichtian deposits have been encountered on the slopes of Jebel Suekai and near the village of Maaraba of Jebel Qasioun, where values of 160 m and 180 m respectively, were measured (Fig. 6.39). According to planktonic foraminifera, all four Maastrichtian zones are present here:

The *Globotruncanella havanensis* Zone, where *G. havanensis* (Woorw.), *Globotruncana falsostuarti* Sigal, *Globotruncanita*

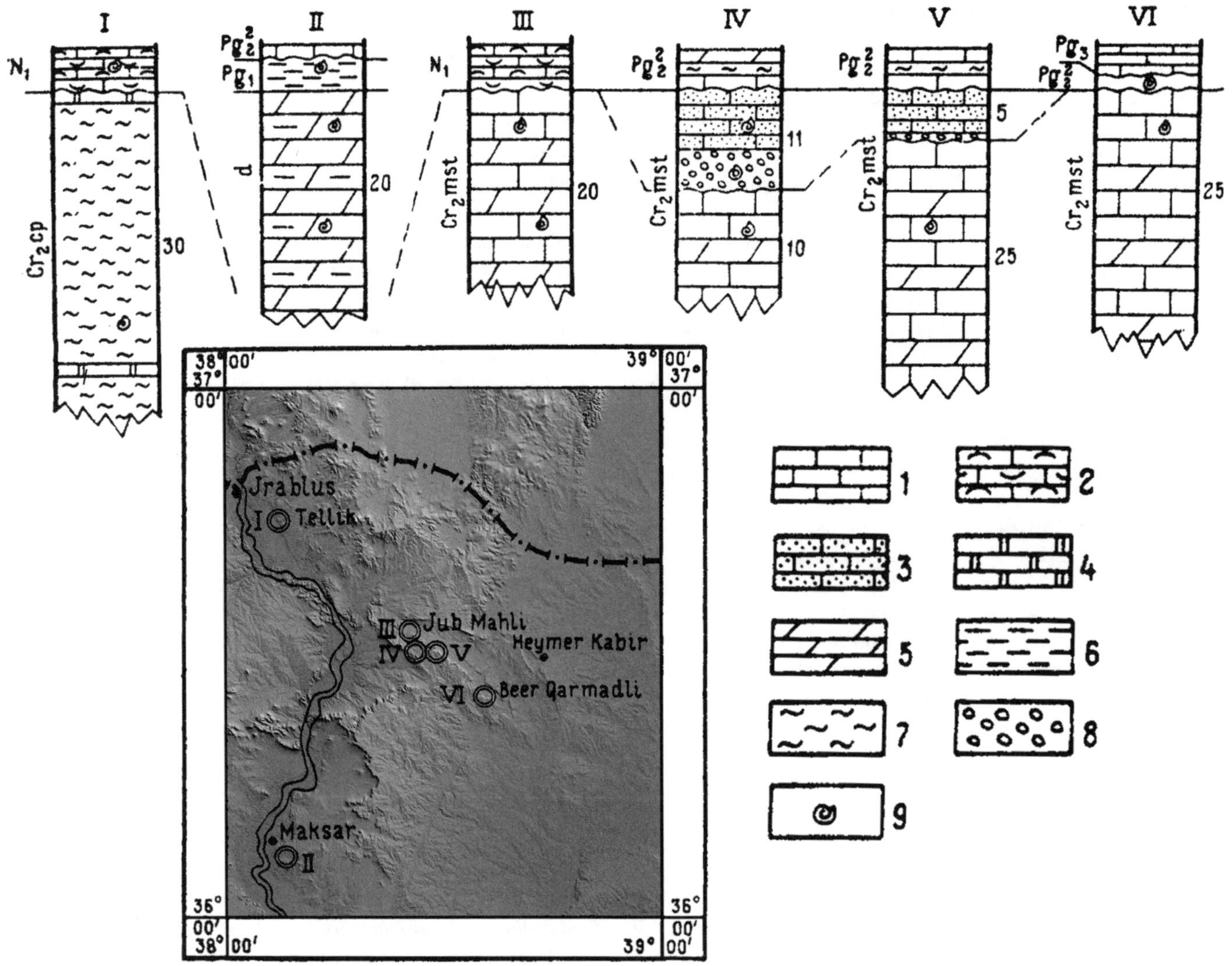

Fig. 6.43. Campanian and Maastrichtian deposits sections of the Jrablus rampart (after Oufland, 1966b).

Legend: 1 - limestone; 2 - organogenic limestone; 3 - clayey limestone; 4 - sandy limestone; 5 - dolomite; 6 - clay and clayey varieties of rocks; 7 - flint; 8 - conglomerate; 9 - fauna.

stuarti (Lapparent), and *Rugotruncana subcircumnodifer* (Gandolfi) appear, accompanied by *Globotruncana arca* (Cushman), *G. bulloides* Vogler, *G. linneiana* (d'Orb.), *Globotruncanita stuartiformis* (Dalbiez), *Rugoglobigerina rugosa* (Marie), *Pseudotextularia elegans* (Rzehak), *Pseudoguembelina costulata* (Cushman), *Heterohelix striata* (Ehrenb.), *H. globulosa* (Ehrenb.), and *Hedbergella holmdelensis* Olsson;

The *Globotruncana aegyptiaca* Zone, which is characterized by the appearance of *G. aegyptiaca* Nakkady, *Globotruncanella petaloides* (Gandolfi), and *Pseudoguembelina excolata* (Cushman), and some species of *Rugoglobigerina* – *R. hexa-camerata* Bronn., *R. macrocephala* Bronn. in association with *Globotruncana falsostuarti* Sigal, *G. arca* (Cushman), *Globotruncanita stuarti* (Lapparent), *G. subspinosa* (Pessagno), *Globotruncanella havanensis* (Woorw.), and *Rugotruncana subcircumnodifer* (Gandolfi), as well as some others;

The *Gansserina gansseri* Zone, where *G. gansseri* (Bolli), *Rosita contusa* (Cushman), *Globotruncanita conica* (White), *Rugoglobigerina reicheli* Bronn., *Abathomphalus intermedius* (Bolli), and *Guembelitria cretacea* Cushman belong to new elements of microfauna. A very rich assemblage also includes numerous species found upwards from the underlying sediments – *Globotruncana aegyptiaca* Nakkady, *G. falsostuarti* Sigal, *G. arca* (Cushman), *Globotruncanita stuartiformis* (Dalbiez), *G. stuarti* (Lapparent), *Globotruncanella havanensis* (Woorw.), *G. pschade* (Keller), *G. petaloidea* (Gandolfi), *Rugoglobigerina rugosa* (Plummer), *R. scotti* (Bronn.), *R. macrocephala* Bronn., *Pseudoguembelina costulata* (Cushman), *R. excolata* (Cushman),

Pseudotextularia elegans (Rzehak), and *Heterohelix striata* (Ehrenb.);

The *Abathomphalus mayaroensis* Zone is marked by the same taxonomically diverse assemblages of foraminifera. Among these, the most important are: *A. mayaroensis* (Bolli), *Plummerita hantkeninoides* (Bronn.), *Ventilabrella glabrata* (Cushman), *V. eggeri* Cushman, *Planoglobulina acervulinoides* (Egger), *P. brazoensis* Martin, *Racemiguembelina fructicosa* (Egger), and *R. powelli* Smith et Pessagno, accompanied by various species of *Rugoglobigerina (rugosa, hexacamerata, macrocephala)*, *Rosita (contusa)*, *Globotruncanita (stuarti, conica)*, *Globotruncana (aegyptiaca, falsostuarti)*, *Globotruncanella (havanensis, petaloidea)*, *Pseudoguembelina (costulata, excolata)*, *Pseudotextularia (elegans)*, and *Guembelitria (cretacea)*. The latter were already developed in older layers of the Maastrichtian.

Benthic foraminifera in the Maastrichtian deposits in the Anti-Lebanon are also very diverse: *Anomalina taylorensis* Carsey, *A. complanata* Reuss, *Cibicides voltzianus* (d'Orb.), *C. actulagayensis* Vass., *C. bembix* (Marsson), *Grammostomum incrassatum* (Reuss), *G. decurrens* (Ehrenb.), *Bolivinoides draco* (Marsson), *B. decoratus* (Jones), *Eponides frankei* Brotzen, *Neoflabellina reticulata* (Reuss), *Gyroidina umbilicata* (d'Orb.), *Stensioina caucasica* (Subb.), *S. exculpta* (Reuss), and *Globorotalites michelinianus* (d'Orb.), as well as many others.

On the southwestern slope of **Jebel An-Nusseiriyeh** in the area of the town of Safita, Maastrichtian deposits rest on an uneven surface of Cenomanian and Turonian limestones, with denudational pockets at the base (Kozlov, 1966a). The succession starts with a 1-2 m thick basal layer of green glauconite calcareous sandstones with small nodules of phosphates. They are replaced by a monotonous unit of chalky limestones – white and light-gray, soft, massive or obscurely bedded, friable, with thin interbeds of compact fine-grained varieties. The succession is topped by a member, 10 m thick, of intercalation of chalky limestones and brown hard silicified limestones with a conchoidal structure (Fig. 6.27). The total thickness is 135 m. According to planktonic foraminifera, these deposits belong to the *Globotruncana falsostuarti* and *Gansserina gansseri* Zones. The preservation of foraminifera in the upper layers is rather poor and the possible presence of the *Abathomphalus mayaroensis* Zone should not be excluded. Nevertheless, it is evident that a part of the Maastrichtian is missing here, due to regressional non-deposition during Latest Maastrichtian or Cenozoic erosion. Paleogene deposits have not been found in the southwestern part of Jebel An-Nusseiriyeh, but to the north, in the Nahr Marqiyeh section near Banyas on the Mediterranean coast, Maastrichtian and Upper Paleocene rocks occur, separated by a large hiatus.

To the north of Safita, along the western slope of Jebel An-Nusseiriyeh facing the Mediterranean Sea, the Maastrichtian preserves its lithological and stratigraphical peculiarities. It is made up of a rather uniform unit of light-gray clayey chalk-like limestones and bluish-gray marls with ball-shaped barite concretions (Shatsky et al., 1966). Paleogene or Neogene sediments transgressively overlie these Maastrichtian rocks. In general, the length of the erosional hiatus increases toward the axial part of the uplift. Due to this phenomenon, the Maastrichtian deposits vary both in thickness and stratigraphic volume from location to location.

Along the upper course of the Nahr As-Saneh River, near Qerdaha, the 315 m thick Maastrichtian sequence is complete:

The *Globotruncana falsostuarti* Zone can be readily identified by the index-species, *G. stuarti* (Lapparent), *G. arca* (Cushman), *G. aegyptiaca* Nakkady, *Globotruncanita stuarti-formis* (Dalbiez), and *Globotruncanella havanensis* (Woorw.).

The *Gansserina gansseri* Zone can be readily identified by the index-species, *Abathomphalus intermedius* (Bolli), *Globotruncanita conica* (White), *Rosita contusa* (Cushman), and *Globotruncanella petaloidea* (Gandolfi).

The *Abathomphalus mayaroensis* Zone can be readily identified by the index-species, *Ventilabrella glabrata* (Cushman), *Planoglobulina acervulinoides* (Egger), *Racemiguembelina fructicosa* (Egger), *Rosita contusa* (Cushman), and *Globotruncanita conica* (White).

These three zones of planktonic foraminifera were also recorded in the sections near Brianis, Sakhabe, and Jebleh. Here, however, the uppermost part of the *Abathomphalus mayaroensis* Zone is missing and Paleogene deposits overlie the Maastrichtian deposits with an erosional gap.

In other sections in the Qerdaha and Haffeh areas, the thickness of the Maastrichtian unit diminishes to 100-250 m (the villages such as Boody, Maallaqa, Mzeraa etc.). Here only the *Globotruncana falsostuarti* and *Gansserina gansseri* Zones have been recorded.

On the southeastern slope of Jebel An-Nusseiriyeh, Maastrichtian rocks outcrop near the town of Rastan, in the deep valley of the Nahr El-Aasi River. The lower contact is not exposed and the overlying Paleogene deposits rest on the Maastrichtian rocks with a hiatus. The Maastrichtian is composed of a series of alternating greenish-gray, platy, micrograined, clayey limestones and white, soft, massive chalk-like limestones with numerous fish scales. Its visible thickness is about 75 m. Planktonic foraminifera attest to the *Globotruncana falsostuarti* and *Gansserina gansseri* Zones. Benthic foraminifera are also numerous: *Cibicides voltzianus* (d'Orb.), *C. excavatus* Brotzen, *Anomalina welleri* Plummer, *A. taylorensis* Carsey, *Stensioina stellaris* (Vass.), *S. pommerana* Brotzen, *S. caucasica* (Subb.), *Bolivinoides draco* (Marsson), *Eponides frankei* Brotzen, *Grammostomum plaita* (Carsey), *G. incrassatum* (Reuss), *Bolivinita planata* Cushman, and *Pullenia quaternaria* (Reuss).

To the north of Rastan, along the eastern slope of Jebel An-Nusseiriyeh, the Maastrichtian unit displays some very interesting and complicated changes. It is divided here into two formations.

In the Hama district the lower formation is composed of greenish-gray marls and calcareous clays with rare interbeds of silicified limestones and a bed of poorly cemented glauconitic sandstones at its base. The thickness is about 50-55 m. Maastrichtian deposits rest conformably on Campanian rocks. Marls and clays contain planktonic foraminifera – *Globotruncana aegyptiaca* Nakkady, *G. rosetta* (Carsey), *G. arca* (Cushman), and

Globotruncanella havanensis (Woorw.) of Early Maastrichtian age.

The upper formation starts with a 0.2-1.0 m thick bed of glauconitic limestones which can be traced throughout the area. This formation consists of dominating white, gray and light-gray chalk-like soft limestones with interbeds of clayey limestones (5-15% of clayey admixture).

To the south and east of Hama, the ~50 m thick upper formation conformably overlies the lower one. North of Hama, the upper formation is missing due to pre-Middle Eocene erosion, but it reappears near the villages of Maaret An-Noman and Ariha, where its thickness increases up to 75 m. In the Jebel Az-Zawiyeh range, the upper formation disconformably overlies Cenomanian-Turonian limestones. Planktonic foraminifera of the *Gansserina gansseri* and *Abathomphalus mayaroensis* Zones attribute a Late Maastrichtian age to these sediments.

Evidently, the calcareous sediments of the Jebel An-Nusseiriyeh Uplift were shallower than the pelagic chalky Maastrichtian limestones of the eastern slope of the Anti-Lebanon. In any case, the Maastrichtian rocks of Jebel An-Nusseiriyeh are characterized by extremely diverse benthic foraminifera: *Anomalina taylorensis* Carsey, *A. complanata* Reuss, *A. pertusa* Marsson, *A. monterelensis* Marie, *Cibicides voltzianus* (d'Orb.), *C. bembix* (Marsson), *C. excavatus* Brotzen, *C. pinguis* Henn., *C. aktulagayensis* Vass., *Stensioina caucasica* (Subb.), *S. pommerana* Brotzen, *S. whitei* (Moroz.), *S. exculpta* (Reuss), *Grammostomum incrassatum* (Reuss), *G. decurrens* (Ehrenb.), *G. plaita* (Carsey), *Bolivinoides draco* (Marsson), *B. decoratus* (Jones), *B. peterssoni* Brotzen, *Pseudouvigerina plummerae* Cushman, *Neoflabellina cordata* (Reuss), *N. reticulata* (Reuss), *Globorotalites michelinianus* (d'Orb.), *Gyroidinoides turgidus* (Hagenow), *Eponides frankei* Brotzen, *Allomorphina obliqua* (Reuss), etc.

Quite unusual Maastrichtian deposits have also been observed in the northeastern corner of Jebel An-Nusseiriyeh, near the village of Tibat and along the steep eastern escarpment of this uplift. Here, the sequence of Maastrichtian clayey limestones and marls is capped by a member of very hard fine-grained, fine-detrital and algal limestones with numerous larger foraminifera: *Siderolites calcitrapoides* Lam., *S. rhomboidalis* Osimo, *Pseudosiderolites vidali* (Douv.), *Orbitoides media* d'Arch., *O. apiculata* Schlumb., *O. tissoti* Schlumb., *O. erdagi* Meric, *O. gensasicus* (Leym.), *Lepidorbitoides mammillata* (Schlumb.), *L. minima* (Leym.), *Omphalocyclus macroporus* (Leym.), *Sulcoperculina globosa* Cizanc., *S. dickersoni vermunti* (Thiadens), *Nemkovites nummulitispira* (Osimo), *N. quadrataformata* (Ashurov), *Daviesina primitiva* Hofker, *Lockhartia roestae* (Leym.), *Orbitolina* sp., and *Goupillaudina* sp. (det. A. Ashourov). This microfauna defines the age of these limestones as Late Maastrichtian. On account of some faults in the area, it is difficult to estimate the thickness of these limestones, but it reaches up to several tens of meters. Nummulitic limestones of the Middle Eocene overlie these Maastrichtian rocks.

The development of such shallow-water algal sediments suggests that local positive tectonic motions of the Jebel An-Nusseiriyeh block controlled the Latest Maastrichtian deposition. These motions caused a hiatus of varying duration between the Maastrichtian and Paleogene formations, both from nondeposition and subsequent erosion.

Two formations have been identified in the extreme northwestern part of Syria (the **Baer-Bassit Massif**). A very complicated structural relationship exists between these units.

On the southern slope of Jebel Aqraa and the coastal side of the Qara Douran mountains the lower formation rests with a break in sedimentation, but without angular unconformity, directly on a Cenomanian-Turonian sequence. The lower formation is incomplete, since it is truncated by a thrust surface. The most continuous section is located to the west of the village of Kassab. It begins with a 5 m thick layer of brown and reddish, quartz-glauconite sandstones, fine-grained, with phosphatic organic remains. Higher up, there follow: dark-gray, massive, fine-grained limestones, 25 m thick; bluish-gray, medium-bedded, clayey limestones, 25 m thick; greenish-gray and brown marls and mudstones with *Globotruncana arca* (Cushman), *G. rosetta* (Carsey), and *Rugoglobigerina ordinaria* (Subb.), 35 m thick. The total thickness approaches 90 m (Shatsky et al., 1966; Kazmin and Kulakov, 1968).

Rocks of this formation constitute the uppermost part of the autochthon on which the ophiolite formation was emplaced. Therefore, determining the age of these limestones and marls is very important. Traditionally they were assigned to the Early Maastrichtian. Unfortunately, the assemblage of planktonic foraminifera in these deposits is very poor and only indicates ages within the interval of the Campanian-Early Maastrichtian.

The upper formation rests unconformably on the so-called green-colored rocks of the allochthonous Baer-Bassit ophiolites and constitutes the basal part of the neoautochthon. The contact can be observed in the village of Qara Jalie, on the southeastern slope of the Baer-Bassit Massif, and in the village of Qafer Fawkani on the northwestern slope of Kyzyl-Jahr, which is to the north of the Nahr El-Qandil River. South of this river, near the villages of Beit Nasser, Makhous, and Mashkita, Maastrichtian rocks are well-exposed but the contact between the sedimentary neoautochthon and the allochthonous ophiolites is poorly exposed.

The base of the upper formation consists of a horizon of sandstones and conglomerates. The well-rounded pebbles (up to 10-15 cm in size) incorporate gabbro, basic effusives, serpentinites and radiolarites, which were derived from the magmatic, metamorphic and sedimentary substrate. These rocks alternate with sandy limestones containing debris of oysters, gastropods, and corals. In the outskirts of the village of Guiaour Qrane, the thickness of this basal member reaches up to 20-30 m.

Near the villages of Qara Godja and Troundje, the conglomerates of the upper formation are replaced by coarse- and medium-grained calcareous polymictic sandstones with shells of *Modiolus* sp., *Amphidonta* sp., *Pycnodonta* sp., and *Ceratostreon* sp. The thickness here drops to 10-15 m.

In the vicinity of the villages of Bisnada, Kesmine, Markhous, Aakrak, and Faqi Hassane, the basal layer is formed of red, green and greenish-gray soft clays with thin interbeds of fine-grained sandstones. Clays fill pockets within the ophiolitic substrate; their thickness varies from a few meters to 12-15 m.

Thus, the upper formation has a basal horizon of terrigenous rocks changing from clays to coarse conglomerates.

In all sections of the Baer-Bassit area, the terrigeneous member is followed by a member of yellow and gray-yellow, rather hard, distinctly bedded limestones with subsidiary beds of softer clayey limestones. This calcareous member is very well-exposed near Damata on the right bank of the Nahr El-Qandil River. The section begins on the eastern slope of Jebel Rhouz at the margin of the small village of Qafer Fawkani and extends along the path to the top of this hill (Krasheninnikov, 1994). The clayey limestones are characterized by abundant, though poorly preserved, planktonic foraminifera of the *Abathomphalus mayaroensis* Zone (Upper Maastrichtian) – *A. mayaroensis* (Bolli), *Globotruncana aegyptiaca* Nakkady, *G. falsostuarti* Sigal, *Globotruncanita conica* (White), *G. stuarti* (Lapparent), *Rosita contusa* (Cushman), *Globotruncanella havanensis* (Woorw.), *G. petaloidea* (Gandolfi), *Rugoglobigerina rugosa* (Plummer), *R. hexacamerata* Bronn., *R. macrocephala* Bronn., *Hedbergella holmdelensis* Olsson, *Globigerinelloides subcarinata* (Bronn.), *Pseudoguembelina costulata* (Cushman), *P. excolata* (Cushman), *Pseudotextularia elegans* (Rzehak), *Heterohelix striata* (Ehrenb.), and *Racemiguembelina fructicosa* (Egger). Rare specimens of *Gansserina gansseri* (Bolli) have also been found in the lower layers of this formation. The thickness of the upper formation in the Jebel Rhouz section is about 60 m. It is conformably covered by yellow clayey limestones of the Danian Stage (Paleogene).

Consequently, the emplacement of the ophiolite series can be determined as being pre-Late Maastrichtian. A precise age for the lower limit of this event in the Baer-Bassit Massif can unfortunately not be given. It is only defined as Campanian-Early Maastrichtian.

On the southeastern slope of **Kurd-Dagh** the Maastrichtian sediments are thick, reaching about 770 m and resting unconformably on Turonian limestones. Upwards they pass gradually into a marly unit of the Danian Stage (Krasheninnikov, 1965; Protasevich and Maksimov, 1966; Protasevich and Kurbanov, 1968). Lithologically, the Maastrichtian sequence is subdivided into two formations.

The Lower Formation consists of two members. The lower member displays a 0.5 m thick basal layer of detritic, sandy, glauconitic bituminous limestones with pebble fragments (about 2 cm in size) which are derived from the underlying rocks. They pass laterally into conglomerates with abundant glauconite. Above, brownish, bituminous, fine-grained, medium-bedded limestones with phosphatized organic matter follow. The upper member encompasses thick black and gray, aphanitic, bituminous marls, which are usually irregularly plated and thinly laminated, with dispersed organic matter and glauconite as well as interbeds of argillaceous limestones. The lowermost layer of this marly member is a yellow marl with numerous imprints of thin shells of *Inoceramus* ex gr. *balticus* Boehm. The thickness of the lower formation, near the village of Meidanki, reaches up to 260 m, and in the vicinity of Korkan Fawqani this thickness is reduced to 170 m. The lack of reliable paleontological data precludes precise dating. It has traditionally been assigned to the Lower Maastrichtian, although Campanian deposits in the basal part of this sequence should not be excluded.

The Upper Formation, on the outskirts of the villages of Meidanki and Khalil Kolko, is extremely thick (510 m) and is represented by rhythmic alternations of marls, clayey limestones, and calcareous sandstones. The light-gray and greenish marls and clayey limestones are hard, aphanitic, and sometimes with imprints of worm tracks and shells of *Inoceramus* cf. *regularis* d'Orb. and *I.* ex gr. *balticus* Boehm. The sandstones are brownish, very hard, generally medium-grained, sometimes dolomitized or silicified with lenticular concretions of dark flints.

In the lower and upper parts of the succession, various types of alternating calcareous and sandy rocks are observed. In the lower part of the section, beds of marls and limestones (from 1.5-2.5 to 8 m thick) are intercalated by 15-35 cm thick sandy interbeds. In the middle part, limestones and marls form 0.5-1.5 m thick bands, separated by thin (2-8 cm) sandstone interbeds. Finally, repeated rhythmic flysch-like alternations of thin sandstone and limestone beds (0.5 to 8 cm thick) and marls (3-20 cm thick) represent the upper part of this succession.

Among the planktonic foraminifera are *Abathomphalus mayaroensis* (Bolli), *Globotruncanita conica* (White), *Globotruncana arca* (Cushman), and *Pseudotextularia elegans* (Rzehak), all of which indicate a Late Maastrichtian age (Protasevich and Maksimov, 1966). However, the presence of *Globotruncana rosetta* (Carsey), *G. ventricosa* (White), *G. lapparenti* Brotzen, *Heterohelix striata* (Ehrenb.), and *H. globulosa* (Ehrenb.) may indicate that the upper formation also includes Early Maastrichtian beds in its lowermost part.

The rhythmic alternations of pelagic marls and terrigenous material suggest a terrigenous input in the form of turbidites and sharp changes in environmental conditions. Such succession essentially differentiates the thick Maastrichtian deposits of Kurd-Dagh from all the other areas in Syria.

In the **Palmyrides** Maastrichtian deposits are very well-exposed and have a quite diverse lithology. The soft chalky limestones and marls undoubtedly are the predominating rock type, but the development of bands with phosphorites and barite concretions as well as intercalations of hard detrital or silicified limestones highlight the Maastrichtian of the various regions of the Palmyrides.

In the southwestern Palmyrides, the Maastrichtian can be subdivided into two formations (Razvalyaev, 1966a,b).

The lithology of the Lower Formation is manifold: chalk-like, white and pale-gray soft limestones are most common, but they also alternate with thin lenticular interbeds of gray flints and various limestones: with bun-shaped barite concretions bearing compact organo-detrital limestones with *Inoceramus regularis* d'Orb. and *I.* ex gr. *balticus* Boehm, as well as strongly phosphatized rocks. Sometimes a member of foliated marls is located at the base of this formation as in Jebel Ad-Dmeir and Jebel Mghar.

The Upper Formation is composed of monotonous white, light-gray and yellowish, soft, compact, thick-bedded chalky limestones with subordinate layers of yellowish-green marls.

Such sections of the entire Maastrichtian Stage have been observed in Jebel Mghar, Jebel Ad-Dmeir, Jebel Zubeideh, Jebel Shaq Semri, and Jebel Butm (Figs. 6.32 and 6.39). The thickness varies from 68 m at Jebel Shaq Semri to 180 m at Jebel Zubeideh. Detailed sampling at Jebel Mghar showed that the *Globotruncana falsostuarti*, *Gansserina gansseri*, and *Abathomphalus mayaroensis* Zones are fully represented.

Further north, Maastrichtian sequences lose their two-fold subdivision. At Jebel An-Nebek (Fig. 6.27) 60 m thick white and light-gray, compact, chalk-like limestones with rare interbeds of brownish bituminous limestones and gray silicified limestones represent the Maastrichtian (Kozlov, 1966a).

In the central Palmyrides, between Al-Qaryatein and Al-Beida, the Maastrichtian deposits conformably overlie the Campanian. Indications of erosion have been noticed at Jebel Rmah, where the basal member of the Maastrichtian contains breccia consisting of fragments from its substratum. Geographically the lithology and thickness of the Maastrichtian deposits vary significantly, though here it also subdivides into lower and upper formations. This structure is especially evident to the east of Wadi Bardeh (Faradzhev, 1966).

The Lower Formation is exposed in the sections at Jebel Bardeh, Jebel Om Journ, Jebel An-Naqnaqiyeh, Jebel Al-Bassiri, Jebel Aada, and Wadi Ghadir El-Hamal and is represented by a broad complex of various rocks (**Fig. 6.44**). The following types can be distinguished among them:

1) Gray and yellowish, soft, clayey limestones and marls;

2) Fine-grained and aphanitic, very hard, dolomitic and silicified limestones;

3) Phosphate-bearing members, containing up to 11 phosphatic beds (0.1-1.1 m thick) and separate oolitic phosphate layers;

4) Opoka-like rock which consists of very fine clayey-siliceous material (up to 87-94% of unsoluble residue), light in weight, with conchoidal fracture;

5) Organogenic limestones with shells, often broken, of bivalves and ammonites: *Lopha villei* Coq., *L. tissoti* Thomas et Peron, *L. sollieri* Coq., *L.* aff. *pomeli* Coq., *L.* cf. *morgani* Douv., *Picnodonta vesicularis* Lam., *P. flicki* Perv., *P.* cf. *subspathulata* Sow., *Amphidonta decussata* Goldf., *Lucina* cf. *safedensis* Conr., *Inoceramus regularis* d'Orb., *I. balticus* Boehm, *I.* ex gr. *siccensis* Perv., *Neancyloceras* sp., and *Diplomoceras* sp.

Organogenic limestones with oyster shells are well-developed in Jebel Om Journ, Jebel An-Naqnaqiyeh, and Wadi Ghadir El-Hamal where these compact rocks form a small escarpment. Opoka beds are typical in Jebel An-Naqnaqiyeh and Wadi Ghadir El-Hamal. Phosphatic deposits are well-developed in Jebel Abtar, Jebel An-Naqnaqiyeh, and Jebel Om Journ. A member of dark-gray, non-plastic, dolomitic and pyritized clays, with abundant fragments of phosphatized fish bones is common for some sections on the outskirts of Jebel Aada.

The Upper Formation is represented by more monotonous, white, yellowish-gray and light-yellow clayey limestones and marls, sometimes dolomitic, with barite nodules.

A maximum thickness of 312 m has been observed for the Maastrichtian deposits in the Wadi El-Bardeh section. This continuous and well-exposed section starts just to the north of the ancient Roman dam, where all three planktonic foraminifera zones have been identified.

The *Globotruncana falsostuarti* Zone has the index-species, *G. aegyptiaca* Nakkady, *G. arca* (Cushman), *Globotruncanita stuarti* (Lapparent), *Globotrun-canella havanensis* (Woorw.), *G. petaloidea* (Gandolfi), *Rugoglobigerina macrocephala* Bronn., *R. hexacamerata* Bronn., *Pseudoguembelina costulata* (Cushman), and *P. excolata* (Cushman).

The *Gansserina gansseri* Zone with the zonal marker, *Abathomphalus intermedius* (Bolli), *Rosita contusa* (Cushman), *Globotruncanella havanensis* (Woorw.), *G. pschade* (Keller), *Globotruncana aegyptiaca* Nakkady, *G. falsostuarti* Sigal, *Globotruncanita conica* (White), *Rugoglobigerina rugosa* (Plummer), *R. scotti* (Bronn.), *R. reicheli* Bronn., *Rugotruncana circumnodifer* (Gandolfi).

The *Abathomphalus mayaroensis* Zone, with the index-species, *Rosita contusa* (Cushman), *Globotruncanita conica* (White), *Globotruncana aegyptiaca* Nakkady, *Racemiguembelina fructicosa* (Egger), *Planoglobulina acervulinoides* (Egger), *Ventilabrella glabrata* (Cushman), *Gueblerina cuvillieri* Kikoine, *Rugoglobigerina hexacamerata* Bronn., *Guembelitria cretacea* Cushman, *Globotruncanella petaloidea* (Gandolfi), and *Heterohelix striata* (Ehrenb.).

Benthic foraminifera in the sections of Jebel El-Bardeh, Jebel Om Journ, Jebel An-Naqnaqiyeh, and Jebel Bassiri are very diverse and numerous: *Grammostomum incrassatum* (Reuss), *G. plaita* (Carsey), *G. decurrens* (Ehrenb.), *Bolivinoides draco* (Marsson), *B. decoratus* (Jones), *Bulimina quadrata* Plummer, *Anomalina taylorensis* Carsey, *A. complanata* Reuss, *A. welleri* Plummer, *A. pertusa* Marsson, *A. lorneiana* d'Orb., *A. praeacuta* Vass., *Cibicides voltzianus* (d'Orb.), *C. bembix* (Marsson), *C. excavatus* Brotzen, *Neoflabellina reticulata* (Reuss), *Globorotalites michelinianus* (d'Orb.), *Stensioina caucasica* (Subb.), *S. exculpta* (Reuss), *S. pommerana* Brotzen, *Gyroidina depressa* Alth, *Eponides frankei* Brotzen, *Allomorphina cretacea* Reuss, *A. trigona* Reuss, *Gyroidinoides globosus* (Hag.), and *G. turgidus* (Hag.), as well as many others.

To the west of Wadi El-Bardeh, the Maastrichtian succession becomes more monotonous and is represented at Jebel As-Sheikh, Jebel An-Nassrani, and Jebel Mhin by chalky, clayey limestones and marls with rare intercalations of aphanitic limestones and thin chert beds. The phosphate layers at the base disappear or their thickness becomes insignificant at Jebel Mhasseh, Jebel Butm, and Jebel As-Sheikh. Oyster beds also pinch out. The thickness diminishes to 60-80 m at Jebel Mhin and Jebel Maksar An-Nimr and even to a minimal thickness of 35 m, as observed in Jebel Abou Rabah. It should be noted that in the Jebel Ad-Dmeir area on the southwestern margin of the Palmyrides, the two-formation nature of the Maastrichtian again reappears.

In the northeastern part of the Palmyrides, between Palmyra and As-Sukhneh, Maastrichtian deposits are well-exposed along the flanks of almost all the anticlinal folds and sometimes in their cores, as happens near the village of Arak.

A typical section is exposed on the northern slope of Jebel Hayyan (Soulidi-Kondratiev, 1966a). At its base the lower 31 m thick unit consists of alternating green and gray-green, calcareous clays, fine-pebble conglomerates, organogenic silicified limestones and bands of granular phosphorites.

The 150 m thick upper unit is composed of light-gray and yellow marls and gray-green clays. Chalky and clayey limestone are essentially absent here.

To the east and north, the Maastrichtian sequence resembles that of Jebel Hayyan, although the thickness and lithology of both units vary here. For instance this is expressed by a sharp increase in the thickness of the lower member – up to 74 m at Jebel As-Safra, 87 m at Jebel Tantour and 120 m at Jebel El-Mrah (Fig. 6.37). Adding to this diversity is the development of numerous thick layers of phosphorites at Jebel Abou Rajmein, Jebel Matnet Ash-Sharqiyeh, and Jebel El-Mrah, and interbeds of hard silicified limestones with macrofauna at Jebel Matnet Ash-Sharqiyeh, Jebel El-Mrah, and Jebel As-Safra., In contrast, marls strongly predominate at Jebel Abou-Rajmein, where they reach a thickness of up to 360 m.

An interesting feature of the lower unit in the northeastern Palmyrides is the existence of relatively rich macrofauna. Various ammonites have been found – *Baculites inornatus* Meek, *B.* ex gr. *vertebralis* Lam., *B. asper larteti* Blanck., *Menuites* aff. *menu* Forb., and *Hoplitoplacenticeras* sp.; gastropods – *Helicaulax* aff. *costata* Sow.; and various pelecypods – *Cardita* aff. *lybica* Zitt., *Amphidonta (Aetostreon) langloisi* Coq., *Liostrea* cf. *incurva* Nils., *L.* cf. *wegmanniana* d'Orb., *Lopha* aff. *barrandei* Coq., *L. nicaisei* Coq., *L. renoui* Coq., *L.* aff. *aucapitainei* Coq., *L. villei* Coq., *L. forgemolli* Coq., *L.* cf. *semiplana* Sow., *Ostrea bella* Conr., *Lucina* cf. *dachelensis* Wann., *L.* aff. *egitoensis* Rennie, *L.* cf. *safedensis* Conr., *Cytherea plana* Sow. var. *judaica* Picard, *C. andersoni* Bull.-Newt., *Venus* sp., *Inoceramus balticus* Boehm., *I. regularis* d'Orb., *I.* cf. *sagensis* Meek, *Crassatella* aff. *apoloniensis* Cox., *Aphrodina angustosinuosa* Riedel, *Astarte similis* Muenst., and *Corbula olivae* Whitef. (det. A. Atabekyan, M. Djalilov and N. Bobkova). Planktonic foraminifera are poor in the phosphorite-bearing layers and preclude the determination of any precise zonal attribution.

The marls of the upper unit in the northeastern Palmyrides contain planktonic foraminifera of the *Gansserina gansseri* and *Abathomphalus mayaroensis* Zones.

The Jebel Al-Bishri uplift forms the northeastern closure of the folds of the Palmyrides. Here Maastrichtian deposits are exposed only in its western-most part, to the north of the ruins of El-Hier. In the direction of the Al-Furat Valley, before it plunges beneath Paleogene rocks (Medvedev, 1966a), the Maastrichtian is represented by gray marls and clayey limestones ($CaCO_3$ content ranges from 50% to 95%). Their visible thickness does not exceed 200 m. Planktonic foraminifera of the *Gansserina gansseri* Zone and the *Abathomphalus mayaroensis* Zone are also present here.

Along the northern margin of the northeastern Palmyrides, south of Ar-Ressafeh, Maastrichtian deposits are rather peculiar, being characterized at Jebel El-Minshar, Jebel Mqebra, and Jebel El-Khashshabiyeh by the increase in very thick clayey rocks and the appearance of a sandy component (Oufland, 1966a). They conformably overlie the uppermost thick flinty member of the Campanian and in general consist of marls, clays and limestones with intercalations of phosphate and opoka-like rocks, flints and dolomites (**Fig. 6.45**). This succession is subdivided into 5 members.

Such a five-member subdivision is especially evident in Jebel El-Minshar. The first and third members are composed of light-gray, thin-laminated marls and clayey limestones with bands of medium-bedded, compact limestones, phosphate and opoka. The second and fourth members consist of tobacco-green, ferruginous, calcareous clays with rare interbeds of clayey limestones. The fifth member, which is the thickest, is made up of various types of limestones – medium- to thick-bedded, soft clayey or hard aphanitic, with intercalations of marls and calcareous clays. In Jebel El-Khashshabiyeh, bands of sandstones, sandy clays, and limestones appear. Along the strike from east to west the thickness of these members changes and their individual lithological characters fade away, as the boundaries between them become indistinct. The total thickness remains rather constant between 450-480 m.

In the area of the northern branch of the Palmyrides, Maastrichtian deposits are exposed along the southern slopes of Jebel Ash-Shoumariyeh, Jebel El-Balas, and Jebel Ash-Shaer. In all these sections, they rest conformably on Campanian flints (Faradzhev, 1966). In the west, at Jebel Ash-Shoumariyeh, a unit of rather monotonous marls and clayey limestones with interbeds of calcareous clays and glauconite rocks is about 120 m thick. To the east it increases up to 240 m and its lithology diversifies (Fig. 6. 44). The clay content increases and phosphate beds, glauconite-bearing horizons and opoka beds all appear. On the southern side of Jebel Ash-Shaer, opoka is present in the lower half, whereas clayey marls with conchoidal structure dominate in the upper part.

On the northern slope of Jebel El-Balas, the Maastrichtian deposits are widespread in a belt extending from Wadi Az-Zaroub eastward to Wadi Hsayyeh. A very complete section is observed on the left side of Wadi Hsayyeh. Its lithology closely resembles that of the Maastrichtian deposits on the southern slopes of Jebel El-Balas and Jebel Ash-Shaer, but the thickness increases here, reaching up to 412 m. Light-gray, greenish-yellow and reddish marls, clayey marls, and calcareous clays constitute the main part of the section. They intercalate with thin-bedded, hard, white and ash-gray, slightly calcareous opoka and clayey pelitomorphic limestones. Phosphorites, flints and glauconitic rocks are present as thin layers. (Mikhailov, 1966).

To the north of Jebel El-Balas, the Maastrichtian deposits are overlain by Paleogene sediments. In Jebel Abisan Hills they have been penetrated by bore-holes. Here, the Maastrichtian is represented by tobacco-green clayey limestones and brown and cream-colored marls with lenses of black flints; the thickness decreases to 150-200 m.

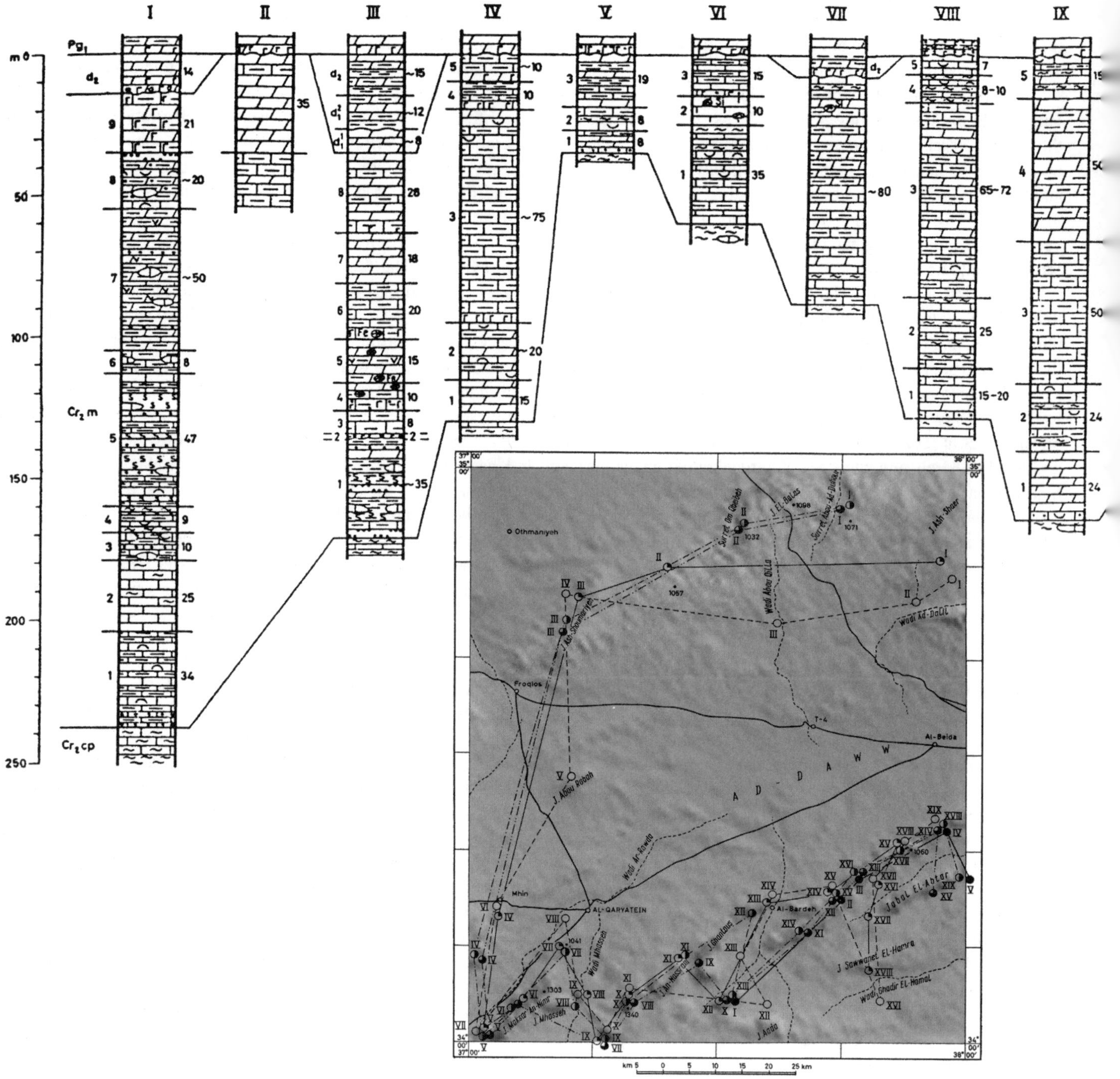

Fig. 6.44. Correlation of Maastrichtian deposits of the Central Palmyrides (modified from Faradzhev, 1966).

Legend: Sections: I - Jebel Ash-Shaer; II - Wadi Ad-Dalil valley; III - Wadi Abou Qilla valley; IV - Jebel Ash-Shoumariyeh; V - Jebel Abou Rabah; VI - Jebel Mhin; VII - Jebel Maksar An-Nimr; VIII - Ghabriyat; IX - Wadi Mhasseh valley; X - Jebel Butm; XI - Jebel Ash-Sheikh; XII - Wadi Aada valley; XIII - Jebel Al-Bassiri; XIV - Jebel El-Bardeh; XV - Jebel An-Naqnaqiyeh, 1237 m elevation mark; XVI - Wadi Ghadir El-Hamal valley; XVII - Jebel An-Naqnaqiyeh, 1188 m elevation mark; XVIII - Jebel Om Journ, 1071 m elevation mark; XIX - Wadi Om Journ valley.

Location of sections: see Fig. 6.34.

Lithologic symbols: see Fig. 6.29.

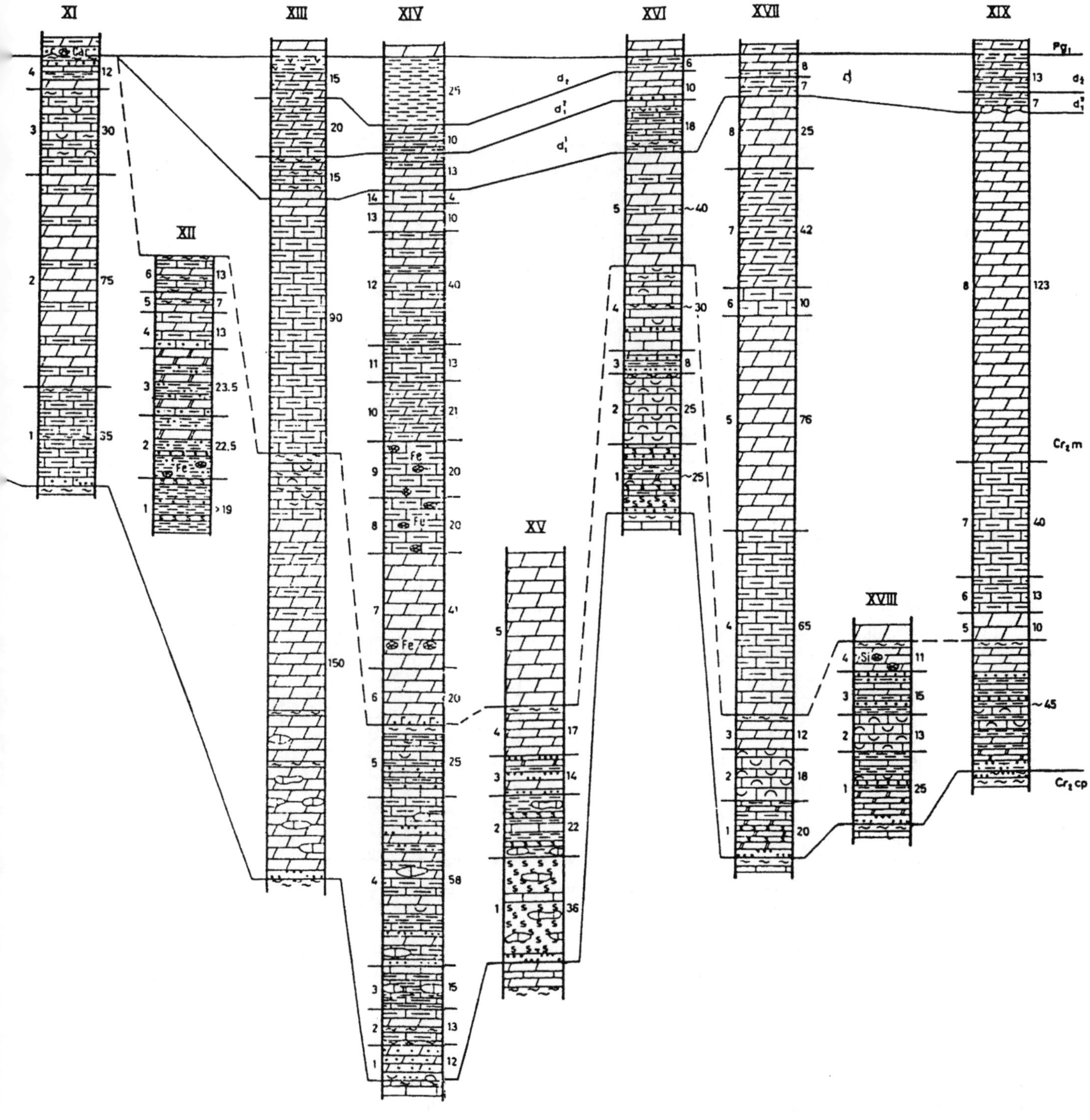
XI
XII
XIII
XIV
XV
XVI
XVII
XVIII
XIX
Pg_1
d_2
d_1^2
d_1^1
Cr_2m
Cr_2cp
Car
Fe
Si

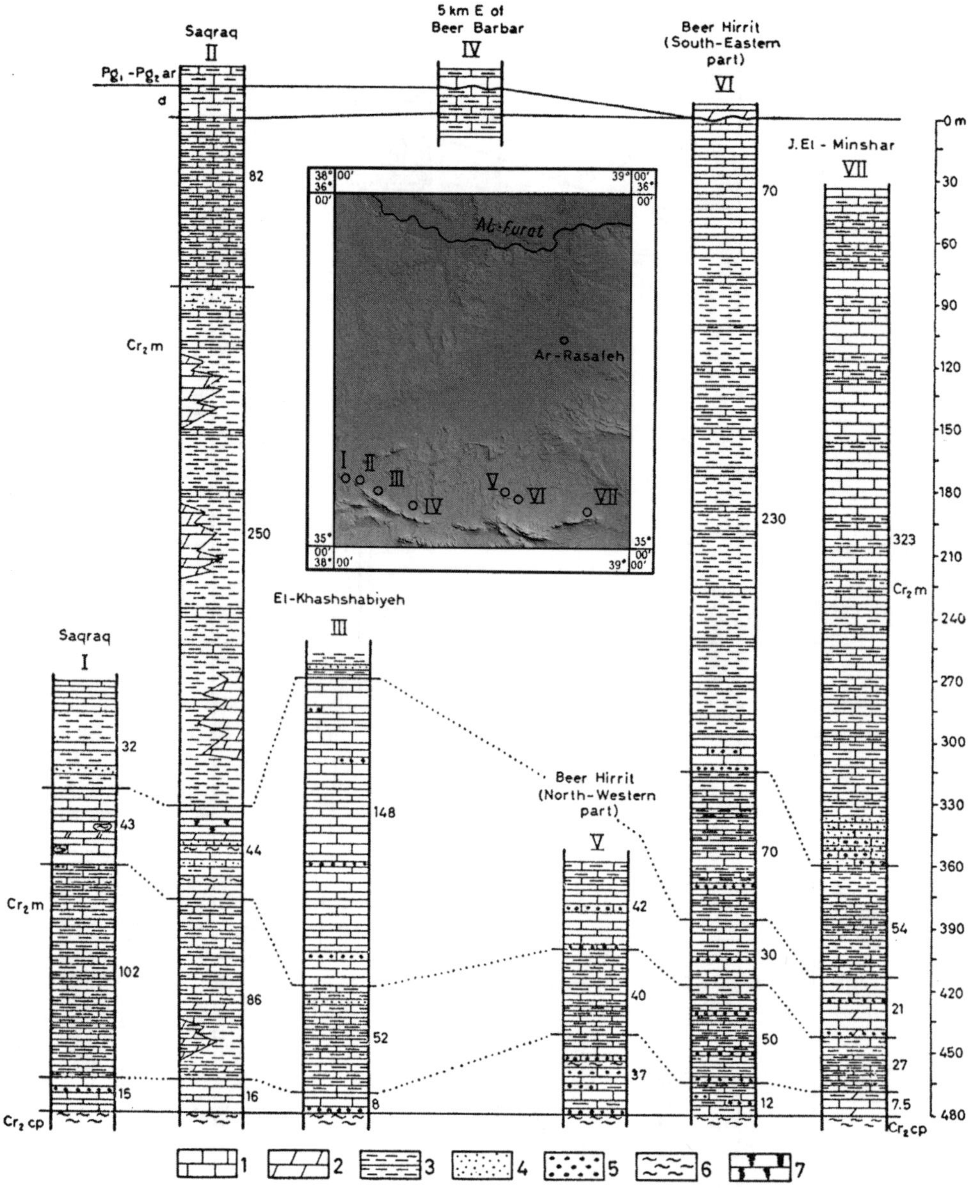

Fig. 6.45. Maastrichtian deposits sections of the Palmyrides (northeastern marginal portion, south of Ar-Resafeh) (modified from Oufland, 1966a). Legend: 1 - limestone; 2 - marl; 3 - clay and clayey varieties of rocks; 4 - sand, sandstone and sandy varieties of rocks; 5 - phosphate and phosphate-bearing rocks; 6 - flint; 7 - opoka-like rock

In northern Syria, Maastrichtian deposits are exposed in two areas – the Jrablus Rampart and the Jebel Abd El-Aziz mountains.

Within the limits of the **Jrablus Rampart**, the Maastrichtian sequence is built up of monotonous, white and yellowish, soft, thick-bedded, chalk-like limestones and marls, near the villages of Jub Mahli, Kabareb, Safari, and Beer Qarmadli. In some sections near Jub Mahli, the Maastrichtian terminates with a member of sandy glauconitic limestones and small-pebble conglomerates (Oufland, 1966b). The lower contact is not exposed; the Maastrichtian is very unconformably overlain by various Paleogene and Miocene horizons. Its visible thickness does not exceed 30 m (Fig. 6.43).

Planktonic foraminifera of the *Gansserina gansseri* Zone and *Abathomphalus mayaroensis* Zone define these sediments as Late Maastrichtian.

Benthic foraminifera are very diverse: *Gaudryina retusa* Cushman, *G. pyramidata* Cushman, *Arenobulimina presli* (Reuss), *Neoflabellina reticulata* (Reuss), *Buliminella carseyae* Plummer, *Grammostomum plaita* (Carsey), *G. tegulata* (Reuss), *G. decurrens*

(Ehrenb.), *G. incrassatum* (Reuss), *Bolivinoides draco* (Marsson), *B. decoratus* (Jones), *B. delicatula* Cushman, *Anomalina taylorensis* Carsey, *A. complanata* Reuss, *A. monterelensis* (Marie), *Cibicides bembix* (Marsson), *C. pinguis* (Jenn.), *C. actulagayensis* Vass., *C. voltzianus* (d'Orb.), C. *excavatus* Brotzen, *Stensioina stellaria* Vass., *S. exculpta* (Reuss), *S. pommerana* Brotzen, *S. whitei* Moroz., *Gyroidina umbilicata* (d'Orb.), *Eponides frankei* Brotzen, *Globorotalites michelinianus* (d'Orb.), *Gyromorphina allomorphinoides* (Reuss), and *Bulimina quadrata* Plummer, as well as many others.

In the sandy limestone layers near the village of Jub Mahli, we found larger benthic foraminifera, including *Loftusia persica* Brady, *L. farsensis* Manutscher, *L. morgani* Douville, *Siderolites calcitrapoides* Lam., *Omphalocyclus macroporus* (Leymerie), and *Orbitoides media* (d'Arch.) (det. A. Ashourov). This shallow water facies recalls one found in the northeastern margin of Jebel An-Nusseiriyeh. In the Latest Maastrichtian, the area of the Jrablus Rampart was also tectonic uplifted with the development of shallow-water deposits and subsequent erosion.

The Maastrichtian deposits of **Jebel Abd El-Aziz** are made up of a ~450 m thick series of calcareous rocks. Three members are distinguished.

The lower member is composed of calcareous clays with rare bands of dolomitic limestones and marls; all of them are slightly sandy. The middle member consists of white unconsolidated marls with rare intercalations of clayey limestones; the content of sandy material diminishes. The upper member includes white, compact limestones and marls without sandy admixture. Planktonic and benthic foraminifera confirm the Maastrichtian age (Ponomarev and Bivshev, 1966).

Very peculiar Maastrichtian deposits are developed in the southwestern part of Syria on the northern slope of the **Jordan Uplift** in the vicinity of the town of Deraa and along the right bank of the Yarmouk River. The Maastrichtian deposits, only known from bore-holes, are represented by dark-gray, black and brown, calcareous and clayey sediments with a high content of organic matter.

In those instances when the organic matter of these Maastrichtian rocks constitutes 14-24%, the C_{org}/H ratio (by mass) is equal to 6.0-8.0 and the caloric value accounts for 5.65-7.58 kJ/kg, these rock can be classified as oil shales. And when the content of its organic matter does not exceed 6-8% and the C_{org}/H ratio (by mass) is 5.4-6.2, the rock belongs to kerogen limestones (Puura et al., 1984). These rock varieties alternate very often within an oil shale formation.

Two bore-holes reached Campanian flints and the maximum thickness of this organic rich Maastrichtian formation is 280 m (Hole D-8). These oil shales contain poorly preserved planktonic foraminifera; benthic foraminifera are very rare or entirely absent (*e.g.* buliminids, some agglutinated species). Nevertheless, planktonic foraminifera in limestones with a low content of organic matter make it possible to recognize the *Globotruncana falsostuarti, Gansserina gansseri* and *Abathomphalus mayaroensis* Zones. The Maastrichtian, here, is transgressively overlain by Paleocene and Eocene deposits.

The Maastrichtian formation of black oil shales is also developed in Jordan and Israel (Abed and Amireh, 1983; Almogi-Labin, Bein, and Sass, 1990, 1993).

The Campanian and Maastrichtian deposits on the northern slopes of the Rutbah anteclise (southwestern Iraq and southeastern Syria) and on the Jordan Uplift (southwestern Syria and the northern part of Jordan) are characterized by remarkable features such as a wide distribution of siliceous rocks, phosphorites, and oil shales. A belt of these deposits stretches from Syria to Jordan and then to Israel. According to some authors (Kemper and Zimmerle, 1983; Almogi-Labin, Bein, and Sass, 1990, 1993; Reiss, 1988), the origin of the huge accumulations of phosphorites, oil shales, and flints is associated with the existence of a subtropical upwelling system along the southern Tethys margin. This system was responsible for the high productivity of the water mass, the bottom water conditions (oxygen depletion and anoxic environment), the preservation of organic and phosphatic matter, certain geochemical peculiarities of the sediments, and the character of the benthic and planktonic foraminifera assemblages. Taking into consideration the lithological, paleontological and paleogeographical data on the Campanian and Maastrichtian of Syria, we completely support the idea of the existence of the upwelling system along the North Arabian shelf.

Some questions regarding the Maastrichtian stratigraphy in Syria still remain unclear. The Campanian-Maastrichtian boundary can be distinctly seen in the field, and it is shown on geological maps as the contact between a unit of thick black flints alternating with hard limestones and the overlying formation of soft white chalky limestones. However, in some sections the soft limestones lying above the black flints contain microfauna of the *Globotruncanita calcarata* Zone (Late Campanian). Evidently, the flints and hard limestones member covering the Campanian sequence can be heterochronous along the strike. In many areas of Syria, the Maastrichtian deposits are subdivided into litho-stratigraphic units (members and formations), but their relation to chronostratigraphic units (planktonic foraminifera zones) is poorly understood. Therefore, for the time being, these members and formations should be only considered as additional regional units which are useful for local subdivision and correlation. These and other ambiguous problems of the Maastrichtian stratigraphy can be solved with the help of zonal biostratigraphy, based on planktonic foraminifera. However at our present state in the investigation these zonal scales have only been applied to some selected reference sections of the Maastrichtian. A great number of the excellent Maastrichtian sections of Syria still await additional study.

Shaded relief of the southwestern quadrant of Syria. This image clearly shows the complex topography associated with the Palmyrides, the Syrian Desert, the volcanism around Jebel Druze, and the continuation and northward splaying of the Dead Sea Transform. The topography is from the 90 m (3") Shuttle Radar Topographic Mission (SRTM) datasets for Africa and Eurasia, with interpolation of areas of no data using the 900 m (30") SRTM dataset. The white areas along parts of the Mediterranean coast are shadow zones for the radar interferometry. The image is generated by version 5.10 of Global Mapper®, on the UTM projection, using daylight shader, 48x vertical exaggeration, dim illumination of the sun at altitude of 45°, and azimuth N315°E.

E: Paleogene

INTRODUCTION

Paleogene deposits are extremely widespread in Syria (**Fig. 6.46**). The northern (Aleppo Plateau) and southern (Syrian Desert) areas of this country are occupied by an unbroken cover of sediments of this age. The Paleogene deposits are also very well-exposed on the flanks of anticlinal folds and uplifts of the Anti-Lebanon, the Palmyrides, Jebel An-Nusseiriyeh and as a part of the neoautochthon of the Baer-Bassit Massif and Kurd-Dagh. In the Mesopotamian Basin and in the western marginal part of the Aleppo Plateau the Paleogene deposits are overlain by Neogene-Quaternary sedimentary rocks; their outcrops are observed only in the Jebel Abd El-Aziz and Jrablus Uplifts. In the southwestern corner of Syria the Paleogene is hidden under the thick basaltic Neogene-Quaternary flows of Jebel Drouz and the Hauran Plateaus and is only exposed in the deep gorge of the Nahr Yarmouk River.

The first information about the Paleogene deposits of Syria appeared at the end of the 19th and beginning of the 20th Century and is connected with the investigations of M. Blanckenhorn (1890b, 1912, 1914, 1916, 1931, 1935). Systematic study of the stratigraphy of the Syrian Paleogene started in the 1930s, when L. Dubertret and his colleagues carried out geological investigations in various areas of the country (L. Dubertret, 1933d, 1937e, 1938, 1942, 1945, 1949a, 1950a, b, 1953a, 1960, 1963, 1966; R. Dubertret, Doncieux, Vautrin, and L. Dubertret, 1938; L. Dubertret and Vautrin, 1937c; Doncieux, 1937, 1938; Doncieux and Roman, 1927; Doncieux, L. Dubertret, and Vautrin, 1936; L. Dubertret, Keller, and Vautrin, 1932a; David, 1933a, b; Cizancourt, 1934). These studies primarily took place in the western part of Syria (the Anti-Lebanon, Jebel An-Nusseiriyeh, the Aafrine graben) but very important information was also obtained for some Paleogene sections of the Palmyrides and the Syrian Desert.

As a result, the French geologists determined the main lithological units of the Paleogene in Syria, and their approximate age. The lower unit includes marls and clays of the undifferentiated Senonian (Maastrichtian) and the lower part of the Paleogene. The second unit consists of alternating limestones, marls and flints and was referred to the Lower Eocene (the Ypresian Stage). The third unit encompasses chalky and organogenic (nummulitic) limestones of the Middle and Upper Eocene. The fourth, uppermost unit is represented by hard limestones and (in some areas of Syria) calcareous sandstones with mollusks, sea urchins and the larger foraminifera (*Lepidocyclina*) of Oligocene age. The most widespread facies of the Paleogene belongs to pelagic sediments with planktonic foraminifera. In the 1930s to 1950s there were no zonal schemes based on this group of fossils. Therefore, important boundaries such as those between the Maastrichtian and Paleogene, Paleocene and Eocene, Middle and Upper Eocene, Eocene and Oligocene (in Mesopotamia) could not be identified.

When the 1:200,000 scale geological mapping in Syria was carried out in 1958-1961, the detailed zonal schemes had already been created for the Paleogene of the Crimea-Caucasian region of the USSR (Subbotina, 1947, 1953), and for the Caribbean basin (Bolli, 1957a, b, c). Application of the zonal stratigraphy method, based on planktonic foraminifera, allowed us to subdivide numerous sections of the Syrian Paleogene, to correlate these sections, to identify the peculiarities of the Paleogene sediments in various tectonic zones of Syria, to draw lithofacies maps, and to decipher the geological history of Syria during Paleogene time (Krasheninnikov, 1964, 1965a, b, 1969a, 1978; Krasheninnikov et al., 1964; Krasheninnikov and Nemkov, 1975; Ponikarov et al., 1966, 1967, 1969). Later, in 1983-1990, investigations were devoted to a) correlating the pelagic chalky limestones and marls with shallow-water nummulitic limestones (Krasheninnikov and Ashourov, 1995) and b) determining the distribution of glauconite horizons in the Syrian Paleogene and the isotopic age (using K/Ar and Rb/Sr methods) of some levels within the Paleocene and Eocene (Krasheninnikov et al., 1996; Golovin et al., 1986; Nikolaeva et al., 1985; Vinogradov et al., 1988, 1990a, b).

The following pages contain the description of:

1) the paleontologic basis of the Paleogene stratigraphy in Syria,

2) the general features of the lithostratigraphy,

3) the special patterns of the Paleogene regional stratigraphy of various tectonic areas in Syria (the Anti-Lebanon, Jebel An-Nusseiriyeh, the Baer-Bassit Massif, Kurd-Dagh, the Palmyrides, the Aleppo Plateau, the Rutbah Uplift, and parts of the Mesopotamian Basin and the Jordan Uplift); and

4) stages in the geological development of Syria during the Paleogene.

THE PALEONTOLOGIC BASIS OF THE SYRIAN PALEOGENE STRATIGRAPHY

Planktonic foraminifera are the main stratigraphic tool because they are extremely abundant in the comparatively deep-water pelagic sediments (soft chalky limestones, marls, and calcareous clays) which are the predominant type of deposit. They are characterized by mixed assemblages – the species of the tropical belt prevail but the species of the subtropical/temperate belt are also common. Therefore, both the tropical Caribbean and the subtropical/temperate Crimea-Caucasian zonal schemes can be applied to the subdivision of the Paleogene deposits of Syria.

In this respect the Syrian Paleogene is very important because it provides a direct correlation between the Paleogene sediments of two paleoclimatic belts – the tropical and subtropical/temperate. We chose the tropical (Caribbean) zonal scheme whose main features were outlined in the publications of Bolli (1957a, b, c),

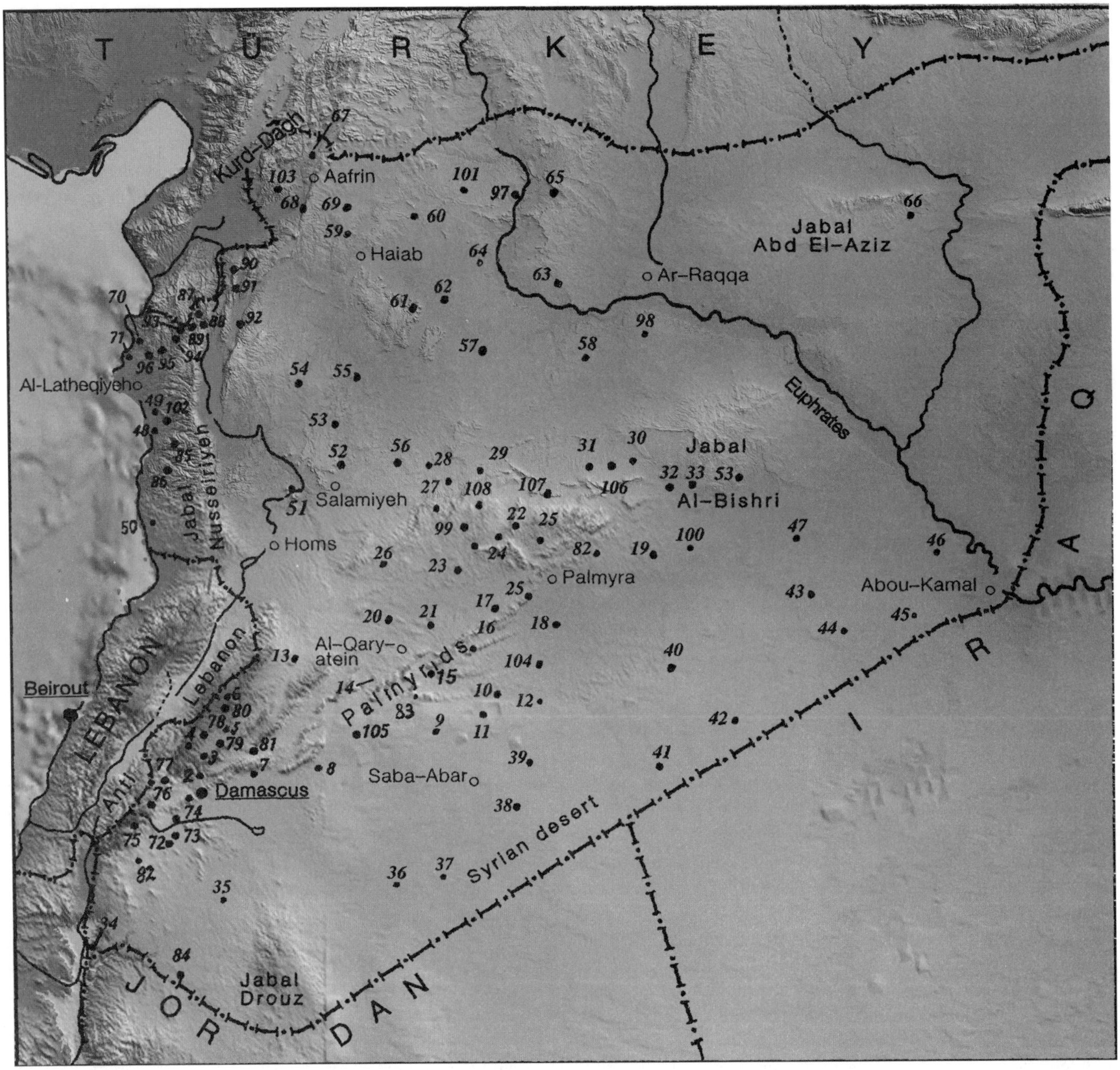

Fig. 6.46. Location of stratigraphic sections of Syrian Paleogene deposits and bore-holes penetrating sediments of this age.
1. Damascus (Doummar), 2. Maaraba, 3. Mnin, 4. Halboun, 5. Jebel Maaloula, 6. Rass El-Marra, 7. Lebel Dmeir, 8. Jebel Zubeideh, 9. Jebel Ash-Sheikh, 10. Jebel Kohle, 11. Jebel Aade, 12. Wadi Ghadir El-Hamal, 13. Hmeir (Wadi Khabra), 14. Al-Qaryatein (Rarbiyate), 15. Al-Qaryatein (Wadi Lebtah), 16. Wadi El-Bardeh, 17. Wadi Erzeq, 18. Jebel Hayyan, 19. Arak (Wadi Harrar), 20. Abou Rahba, 21. Ain Qomqom, 22. Jebel Tyas, 23. Wadi Jhar, 24. Jebel Heite (Wadi Al Jizel), 25. Jebel El-Abuad (Rass El-Rechem), 26. Jebel Habli, 27. Jebel Balas, 28. Tell Snime, 29. Bir Hasseye, 30. Beer Qdem bore-hole, 31. Jebel Dilaa, 32. Wadi As-Souq, 33. Jebel Al-Bishri, 34. The Nahr El-Yarmouk River, 35. Jebab, 36. Jebel Sis bore-hole, 37. The Joeif Depression, 38. Saba Abar bore-hole, 39. Jebel Tarag El-Olab, 40. Beer Alayyaniyeh, 41. Khabra Oum Moueil bore hole, 42. Tell Doueiye, 43. Bir Oum Rabba bore-hole, 44. Wadi El-Miyah, 45. Wadi As-Suwab and Wadi El-Waar, 46. Wadi As-Suwab bore-hole, 47. Bir Nedjib bore-hole, 48. Safarqiyeh, 49. Nqourou, 50. Nahr Marqiyeh River, 51. Jebel Abou Dardeh, 52. Jebel Phanate bore-hole, 53. Wadi Khanazir bore-hole, 54. M'arret An-No'man, 55. Meniss bore-hole, 56. Juba bore-hole, 57. Jebel Shbit, 58. Beer An-Naam bore-hole, 59. Aanadane bore-hole, 60. El-Bab, 61. Jebel El-Hass, 62. Sabkhet Aj-Jabboul, 63. Maksar, 64. Sheikh Aroud, 65. Serrin Esh-Shamaliyeh and Milha, 66. Jebel Abd El-Aziz, 67. Meidanki, 68. Khalil-Kolko, 69. Tcholaqane, 70. Damata, 71. Salib Turukman bore-hole, 72. Wadi Maesra, 73. Oz-Aantar, 74. Rass El-Aabid, 75. Beit Jann, 76. Kafar Qouq, 77. Deir Qanoun, 78. Rankous (Nabi Wees), 79. Hafir At-Tahta, 80. The Bissa gorgr, 81. Jebel Qarn, 82. Jebel Antar, 83. Jebel Bassiri, 84. Dar'a bore-hole, 85. Brianis, 86. Sakhabe, 87. zainiye, 88. the Abou Gazi Hill, 89. Bdama, 90. Jebel Dweleh, 91. Jebel El-Wastani, 92. Ariha, 93. Shambar Faouqani, 94. Ain-Gazal, 95. Kealadjouq, 96. Bachoura, 97. Tishrin-300 bore-hole, 98. Ar-Raqqa (RB-524 and RB-212) bore-holes, 99. Wadi Terfaoui, 100. As-Soukhneh, 101. Jrablus, 102. Qerdaha, 103. Arab-Oshari, 104. Jebel El-Naqnaqiyeh, 105. Jebel Labdeh, 106. Bir Berber, 107. Wadi Ech Chefe, 108. Wadi Erqeime.

Blow (1969), and Bolli, Saunders and Perch-Nielsen (1985). In cases of impoverished microfauna or its poor preservation we used combined zonal subdivisions.

We consider the succession of stages (the Danian, Selandian, Thanetian, Ypresian, Lutetian, Bartonian, Priabonian, Rupelian, and Chattian) and the stratigraphic scopes of the Paleocene, Eocene and Oligocene subseries in accordance with the recommendations of the 28th International Geological Congress held in 1989 in Washington, DC, USA.

The problem of the taxonomy of the Paleogene planktonic foraminifera is very complicated. We follow the classification of Loeblich and Tappan (1988). After thorough morphological investigations Olsson et al. (1999) developed a more advanced and reliable classification, but unfortunately it refers just to the Paleocene planktonic foraminifera. Studies on the Eocene planktonic foraminifera are in progress, but are of course unavailable for the present study.

Below we give a short description of the zonal scheme of the Paleogene deposits of Syria based on planktonic foraminifera.

The Lower Paleocene consists of the Danian and Selandian Stages.

The Danian Stage includes four foraminiferal zones but only three of them were detected in Syria.

The lowest, ***Parvularugoglobigerina eugubina* Zone (P1a)** has not been found in Syria, though we studied a number of sections with transitional layers between the Maastrichtian and Paleocene. Usually sediments of these ages are separated by an obscure unconformity and the Danian deposits start the transgressive series. There is an erosional gap in the sections of variable duration. Very often it is insignificant and the Danian Stage is represented by its next *Turborotalia pseudobulloides* Zone. In other sections the hiatus is more pronounced and some zones are missing. The unconformity can be reliably revealed only by microfauna, because above and below this boundary the character of the sediments is essentially the same. As a rule, a transgressive unit contains at its base a glauconite horizon but soft clayey limestones, marls and calcareous clays are sometimes poorly exposed.

Evidently, the process of sedimentation at the Cretaceous/Paleogene boundary was continuous in some of the structural depressions of the Palmyrides. In the Wadi El-Bardeh section we collected samples every few centimeters. There is no lithological change in the sediments at this boundary, and no traces of erosion, mud-eater activity, or glauconite accumulation. Sediments with rich planktonic foraminifera of the *Abathomphalus mayaroensis* Zone (the Upper Maastrichtian) and those of the *Turborotalia pseudobulloides* Zone (the Danian Stage, Paleocene) are separated by a layer of clayey limestones, 2.5 m thick containing only benthic foraminifera of Paleocene age such as *Clavulina aspera, Gaudryina retusa, Ammodiscus incertus, Marssonella indentata, Spiroplectammina spectabilis, Stensioina caucasica, Anomalina acuta, A. danica, Cibicides favorabilis, Nuttallides truempyi, Karreria fallax, Alabamina wilcoxensis, Loxostomum applinae, Gyroidina globosa, G. octocamerata, Flabellina rugosa, Pullenia coryelli*, and *Bulimina paleocenica*. We believe that this layer, according to its stratigraphic position is a shallow-water analogue of the *Parvularugoglobigerina eugubina* Zone interval. It corresponds to the hiatus which is observed in many other sections of the Syrian Paleogene. We don't exclude the possibility that sediments of the *Parvularugoglobigerina eugubina* Zone can be found in some regions of Syria (the Palmyrides, the Aafrine and Al-Latheqiyeh areas) with well-expressed indications of continuing subsidence.

In many sections located in various areas of Syria, all the other zones of the Danian Stage were identified

The ***Turborotalia pseudobulloides* Zone (P1b)** is represented by an assemblage of foraminifera consisting of *T. pseudobulloides, Subbotina triloculinoides, Globoconusa daubjergensis, Eoglobigerina taurica, E. eobulloides, E. tetragona, E. pentagona, E. fringa, Chiloguembelina midwayensis, Ch. morsei,* and *Guembelitria irregularis*. Deposits of this zone were found in the Palmyrides at Wadi El-Bardeh, Jebel Hayyan, Jebel El-Abuad, Jebel Antar, Jebel El-Bassiri, Wadi Ghadir El-Hamal, Jebel Zubeideh, Jebel Kohle, Arak, and Bir Berber; in the Anti-Lebanon at Doummar, Jebel Maaloula; and in the Al-Latheqiyeh and Aafrine depressions at Meidanki, Khalil-Kolko, and Damata.

The ***Turborotalia trinidadensis* Zone (P1c)** contains numerous specimens of *T. trinidadensis, T. varianta, T. pseudobulloides, T. quadrata, Planorotalites compressa,* and *Subbotina triloculinoides* together with *Chiloguembelina midwayensis* and *Guembelitria irregularis*. Sediments of this zone were recognized in sections of the Palmyrides at Wadi El-Bardeh, Jebel Hayyan, Jebel El-Abiad, Jebel Antar, Wadi Ghaadir El-Hamal, Jebel El-Bassiri, Wadi Terfaoui, Wadi Erzeq, Jebel Zubeideh, Jebel Kohle, Jebel El-Abtar, Arak, As-Sukhneh, Jebel Al-Bishri, Al-Kom, Wadi Aafou, Wadi Ech Chefe, Mrhaizel; in the Anti-Lebanon at Jebel Maaloula; in the Al-Latheqiyeh and Aafrine depressions at Meidanki, Khalil-Kolko, and Damata; and in the Mesopotamian basin at Maksar.

In the ***Acarinina uncinata* Zone (P2)** some new species of the genus *Acarinina* appeared such as *A. uncinata, A. spiralis, A. praecursoria, A. inconstans, A. indolensis,* which are accompanied by *Turborotalia pseudobulloides, T. varianta, T. quadrata, Subbotina triloculinoides, S. trivialis, Planorotalites compressa, Eoglobigerina edita*, and *Chiloguembelina wilcoxensis*. This zone was traced in sections of the Palmyrides at Wadi El-Bardeh, Jebel Hayyan, Jebel Al-Abuad, Jebel Antar, Jebel Al-Bassiri, Wadi Ghadir El-Hamal, Wadi Erzeq, Bir Hasseiye, Wadi Jhar, Wadi Terfaoui, Beer Qdem, Jebel Labdeh, As-Sukhneh, Al-Kom, Jebel Bishri, and Wadi Ech Chefe; in the Anti-Lebanon at Jebel Maaloula; in the Al-Latheqiyeh and Aafrine depressions at Damata, Khalil-Kolko and Meidanki; and in the Mesopotamian basin at Maksar.

The Selandian Stage corresponds to the ***Morozovella angulata* Zone s.l. (P3)** which is subdivided into two zones or subzones – the ***Morozovella angulata* Zone s. str. (P3a)** and the ***Igorina pusilla* Zone (P3b)**. According to the Russian stratigraphic nomenclature the latter is called the *Morozovella conicotruncata* Zone. The foraminiferal assemblage of this zone (P3) consists of *Morozovella angulata, Planorotalites ehrenbergi, P. compressa, Turborotalia pseudobulloides, T. varianta, T. quadrata, Subbotina* *triloculinoides,* and *S. trivialis*. The upper part (P3b) of this zone distinguishes the appearance of *Morozovella conicotruncata, M. cubanensis, Igorina pusilla* and *I. tadjikistanensis djanensis*. This zone was observed in the Palmyrides at Jebel Ash-Sheikh, Wadi El-Bardeh, Jebel Hayyan, Wadi Ghadir El-Hamal, Jebel El-Abuad, Wadi Jhar, Jebel Al-Bishri, Bir Hasseye, and Bir Oum Rabba (Hole 24); in the Anti-Lebanon at Jebel Maaloula: in the Aafrine graben at Meidanki; in the Al-Latheqiyeh area at Damata; and in the Mesopotamian basin at Maksar.

The Upper Paleocene includes the Thanetian Stage which is subdivided into two zonal units.

The ***Planorotalites pseudomenardii* Zone (P4)** is characterized by *P. pseudomenardii, Igorina laevigata, I. convexa, Morozovella velascoensis, M. occlusa, Turborotalia imitata, Acarinina subsphaerica, A. intermedia, Muricoglobigerina mckannai, Subbotina velascoensis, S. nana, S. bacuana, S. pileata,* and *S. quadritriloculinoides*.

The ***Morozovella velascoensis* Zone (P5)** is defined by numerous specimens of *M. velascoensis, M. acuta, M. apanthesma, M. aequa, Acarinina intermedia, A. subsphaerica, A. acarinata, A. esnaensis, A. trichotrocha, A. hispidicidaris, Muricoglobigerina mckannai, M. chascanona, M. aquiensis, M. soldadoensis, Subbotina velascoensis, S. nana, S. compressaformis, S. bacuana,* and *S. pileata.*

Deposits of these two zones are well-exposed in many sections of the Anti-Lebanon at Rass El-Aabid, Doummar, Maaraba, Jebel Maaloula, and Rass El-Marra; in the Palmyrides at Wadi Ghadir El-Hamal, Jebel Zubeideh, Jebel Ash-Sheikh, Abou Rahba, Wadi Jhar, Jebel Al-Abuad, Wadi Khabra, Al-Qaryatein, Wadi Erzeq, Jebel Hayyan, Wadi El-Bardeh, Jebel Antar, Bir Hasseiye, and Jebel Al-Bishri; on the Jebel An-Nusseiriyeh Uplift at Safarqiyeh, Nqourou, Nahr Marqiyeh, Sakhabe, and Brianis; and in the Al-Latheqiyeh and Aafrine depressions at Meidanki, Damata and Khalil-Kolko.

The Lower Eocene (Ypresian Stage) is subdivided into four foraminiferal zones – *Morozovella subbotinae* and *Morozovella formosa* Zones (the Lower Ypresian), *Morozovella aragonensis* and *Acarinina pentacamerata* Zones (the Upper Ypresian).

The ***Morozovella subbotinae* Zone (P6)** includes the planktonic foraminifera assemblage consisting of *Morozovella subbotinae, M. formosa gracilis, M. aequa, M. apanthesma, Acarinina pseudotopilensis, A. acarinata, A. wilcoxensis, A. trichotrocha, A. intermedia, Muricoglobigerina soldadoensis, M. mckannai, Pseudohastigerina wilcoxensis, Subbotina eocaenica, S. compressaformis*, and *S. nana*. Very rare specimens of *Morozovella acuta* and *M. velascoensis* can be found in the basal layers of this zone.

The ***Morozovella formosa* Zone (P7)** is characterized by *M. formosa formosa, M. lensiformis, M. marginodentata, Acarinina triplex, A. camerata,* and *A. broedermanni* which are accompanied by species already developed in the underlying sediments – *Morozovella formosa gracilis, M. subbotinae, Acarinina pseudotopilensis, A. wilcoxensis, Pseudohastigerina wilcoxensis, Subbotina compressaformis* and *S. eocaenica.*

The zonal subdivisions of the Lower Ypresian were recognized in many sections of the Anti-Lebanon at Maaraba, Doummar, Jebel Maaloula, Wadi Maesra, Kafar-Qouq, Rass El-Aabid, and Beit Jann; in the Palmyrides at Jebel Dmeir, Jebel Ash-Sheikh, Wadi El-Bardeh, Jebel Hayyan, Jebel Al-Abuad, Wadi Jhar, Arak, Jebel Al-Bishri, Bir Hasseye, Jebel Zubeideh, Abou Rahba, Ain Qomqom, Jebel Tyas, Al-Qaryatein (Rarbiyate), Al-Qaryatein (Wadi Lebtah), Wadi Erzeq, Jebel Balas, and Wadi As-Souq; on the Jebel An-Nusseiriyeh Uplift at Nahr Marqiyeh, Nqourou, Safarqiyeh, Sakhabe, and Brianis; and in the Aafrine graben at Meidanki.

The ***Morozovella aragonensis* Zone (P8)** is determined by numerous specimens of *M. aragonensis, Acarinina interposita, A. pseudotopilensis, A. triplex, A. broedermanni, Subbotina pseudo-eocaena , S. inaequispira, S. eocaenica* which associate with *Morozovella quetra, Turborotalia planoconica, Muricoglobigerina gravelli, M. soldadoensis*, and *Pseudohastigerina wilcoxensis*. Such species as *Morozovella caucasica* and *Acarinina pentacamerata* are present in every sample but as rare specimens.

In **the *Acarinina pentacamerata* Zone (P9)** the assemblage is made up of *A. pentacamerata, A. interposita, A. pseudotopilensis, A. aspensis, A. triplex, Subbotina pseudo-eocaena* which are accompanied by less abundant *Morozovella caucasica, M. aragonensis, Turborotalia planoconica, Subbotina eocaena, S. eocaenica, S. inaequispira, Pseudohastigerina micra, Acarinina broedermanni, Muricoglobigerina senni*, and *Guembelitrioides higginsi*.

The stratigraphic distribution of *Astrorotalia palmerae,* which is limited by this zone, is very irregular. In the section of Maksar on the eastern bank of the Al-Furat River specimens of this species are abundant. They appear at the base of this zone and disappear at its very top. It means that the stratigraphic range of the *Acarinina pentacamerata* Zone (Krasheninnikov, 1965b) exactly corresponds to that of the *Globorotalia palmerae* Zone of Bolli (1957a). Taking into consideration the priority of the definition of this stratigraphic subdivision, it would be reasonable to return to the original name of this zone (with the appropriate change in generic name) *i.e.* the *Astrorotalia palmerae* Zone. Downstream on the Al-Furat River near Ar-Raqqa, deposits of the *Acarinina pentacamerata* Zone, with occasional specimens of *Astrorotalia palmerae,* were penetrated by some bore-holes. Sporadic specimens of this species were also encountered in some sections of the Anti-Lebanon (Doummar, Maaloula). Due to the needs of the geological mapping, over 1500 samples with rich planktonic foraminifera of the *Acarinina pentacamerata* Zone were investigated but *Astrorotalia palmerae* was lacking. For this reason the name *Acarinina pentacamerata* Zone was suggested for the uppermost Lower Eocene sediments.

The deposits of the Upper Ypresian were accumulated during the peak of the Paleogene transgression. Therefore the *Morozovella aragonensis* Zone and the *Acarinina pentacamerata* Zone were identified in many sections of the Anti-Lebanon at Doummar, Maaraba, Mnin, Halboun, Rankous, Jebel Maaloula, Rass El-Marra, Wadi Maesra, Kafar Qouq, Rass El-Aabid, Beit Jann, Hafir At-Tahta, and the Bissan gorge; in the Palmyrides at Jebel Dmeir, Jebel Ash-Sheikh, Jebel Kohle, Wadi El-Bardeh, Jebel Hayyan, Wadi Ghadir El-Hamal, Jebel El-Abuad, Jebel Heite, Wadi Jhar,

Arak, Jebel Al-Bishri, Bir Hasseye, Jebel Zubeideh, Jebel Aade, Abou Rahba, Ain Qomqom, Jebel Tyas, Hmair (Wadi Khabra), Al-Qaryatein (Rarbiyate), Al-Qaryatein (Wadi Lebtah), Wadi Erzeq, Jebel Balas, Tell Snime, Jebel Dilaa, Wadi As-Souq; Jebel An-Nusseiriyeh: Safarqiyeh, Nqourou, Nahr Marqiyeh, Brianis, Sakhabe, and Qerdaha; on the Baer-Bassit Massif at Damata; in the Aafrine graben at Meidanki; on the southern margin of the Aleppo Plateau at Abou Darde, Jebel Phanate, Jebel Shbit, and Sabkhet Aj-Jabboul; on the Jordan Uplift at Nahr El-Yarmouk and Jebel Sis; on the Rutbah Uplift at Jebel Fru Tarag El-Olab, Beer Alayyaniyeh, Tell-Doueiye, and Wadi El-Myah; and in the Mesopotamian basin at Maksar.

The Middle Eocene consists of five planktonic foraminifera zones. The *Hantkenina aragonensis, Globigerapsis kugleri* and *Morozovella lehneri* Zones constitute the Lutetian Stage while the *Orbulinoides beckmanni* and *Truncorotaloides rohri* Zones belong to the Bartonian Stage.

The *Hantkenina aragonensis* Zone (P10) is characterized by plentiful specimens of *Acarinina bullbrooki, Subbotina boweri,* and *S. pseudotriloculinoides* together with rarer *Globigerapsis index, Morozovella spinulosa, Truncorotaloides rohri,* and *Turborotalia pseudomayeri*. Specimens of *Hantkenina aragonensis* are sporadic and in some sections are completely absent. One cannot exclude the possibility that for some taxonomical reasons this species should be called *Hantkenina nuttalli* (Bolli et al., 1985). The assemblage also yields *Acarinina aspensis, A. pentacamerata, A. triplex, A. broedermanni, Subbotina pseudoeocaena, S. eocaena, S. inaequispira, Pseudohastigerina micra, Muricoglobigerina senni, Guembelitrioides higginsi, Morozovella aragonensis*, and *M. caucasica*.

The *Globigerapsis kugleri* Zone (P11) differs from the P10 Zone by the appearance of *Globigerapsis kugleri, Truncorotaloides topilensis, Clavigerinella akersi, C. jarvisi* and more numerous specimens of *Globigerapsis index, Truncorotaloides rohri*, and *Morozovella spinulosa*. Associated species are similar to those from the underlying sediments.

In some sections the index-species of these two zones were not detected and we defined the undifferentiated interval of the P10 and P11 Zones as the *Acarinina bullbrooki* Zone. This zonal unit was suggested by Subbotina (1953) and is very useful for the stratigraphic subdivision of the Paleogene deposits of the subtropical paleoclimatic belt.

The *Morozovella lehneri* Zone (P12) is marked by a common presence of *Morozovella lehneri* and the appearance of *Acarinina rotundimarginata, Globigerinatheca barri, Subbotina frontosa, Turborotalia possagnoensis, Hantkenina liebusi*, and *H. lehneri*. Rare and small-sized specimens of *Turborotalia centralis* appear in the upper part of this zone. The accompanying planktonic foraminifera are very diverse and their assemblage consists of *Acarinina bullbrooki, A. aspensis, Subbotina boweri, S. pseudoeocaena, S. eocaena, Pseudohastigerina micra, Globigerapsis index, G. kugleri, Morozovella spinulosa, Muricoglobigerina senni, Truncorotaloides topilensis*, and *T. rohri*.

The Lutetian *Hantkenina aragonensis, Globigerapsis kugleri* and *Morozovella lehneri* Zones were recognized in a number of sections composed by chalky limestones in the Anti-Lebanon at Doummar, Maaraba, Mnine, Halboun, Rankous, Jebel Maaloula, Rass El-Marra, Wadi Maesra, Rass El-Aabid, and Hafir At-Tahta; in the Palmyrides at Jebel Dmeir, Jebel Kohle, Wadi El-Bardeh, Jebel Hayyan, Jebel El-Abuad (Rass El-Rechem, Jebel Heite (Wadi Jizel), Arak, Jebel Al-Bishri (Wadi Souq), Bir Hasseye, Jebel Zubeideh, Jebel Aade, Abou Rahba, Ain Qomqom, Jebel Tyas, Hmair (Wadi Khabra), Al-Qariatein (Wadi Lebtah), Wadi Erzeq, Jebel Khabli, Tell Snime, and Jebel Dilaa; on the Jordan Uplift at Nahr El-Yarmouk and Jebab; on the Rutbah Uplift at Jebel Sis, in the Joeif depression at Saba Abar, Jebel Fru Tarag El-Olab, Beer Aalayyaniyeh, Tell Doueye, and Wadi El-Miyah; on Jebel An-Nusseiriyeh at Nqourou; on the southern margin of the Aleppo Plateau at Jebel Abou Dardeh, Wadi Khanazir, Meniss, Jebel Shbit, Beer An-Naam, Jebel El-Hass, and Sabkhet Aj-Jabboul; in the Mesopotamian basin at Maksar; in the Aafrine graben at Meidanki; and on the Baer-Bassit Massif at Damata and Salib Turukman.

The Orbulinoides beckmanni Zone (P13) has extremely abundant planktonic foraminifera. It is recognized by the presence of numerous big specimens of *Turborotalia centralis* associated with *Hantkenina alabamensis, H. dumblei, Dentoglobigerina pseudovenezuelana, Globigerinita martini, Globorotaloides suteri, Globigerina praebulloides,* and *Chiloguembelina cubensis*. The index-species *Orbulinoides beckmanni* is common but is usually represented by a limited number of specimens. This microfauna is accompanied by *Globigerapsis index, G. kugleri, Globigerinatheca barri, Subbotina frontosa, S. pseudoeocaena, S. eocaena, Acarinina bullbrooki, A. rotundimarginata, Morozovella spinulosa, Truncorotaloides topilensis, T. rohri,* and *Pseudohastigerina micra*, and rare specimens of *Morozovella lehneri* which already existed in the older *Morozovella lehneri* Zone.

The *Truncorotaloides rohri* Zone (P14) is characterized by numerous specimens of *T. rohri* in combination with *Hantkenina longispina, Acarinina rugosoaculeata, Globigerina incretacea, G. azerbaidjanica,* and *Globigerinita echinata*. Such species as *Globigerina praebulloides, Subbotina pseudoeocaena, Globigerapsis index, Dentoglobigerina pseudovenezuelana, Turborotalia centralis, T. pomeroli, Pseudohastigerina micra,* and *Chiloguembelina cubensis* continue to exist in this zone. The last specimens of *Truncorotaloides topilensis, Acarinina bullbrooki, Morozovella lehneri, M. spinulosa* and *Subbotina frontosa* disappear at the top of this zone.

The Bartonian *Orbulinoides beckmanni* and *Trunco-rotaloides rohri* Zones were traced in many continuous sections of the Anti-Lebanon at Doummar, Maaraba, Mnine, Halboun, Rankous (Nabi Wees), Jebel Maaloula, Rass El-Marra, Wadi Maesra, Oz-Aantar, and Rass El-Aabid; in the Palmyrides at Jebel Dmeir, Jebel Kohle, Wadi El-Bardeh, Jebel Hayyan, Jebel Al-Abuad, Jebel Heite, Arak, Jebel Al-Bishri (Wadi Souq), Bir Hasseye, Jebel Kohle, Jebel Aade, Abou Rahba, Ain Qomqom, Jebel Tyas, Heite, Hmair (Wadi Khabra), Al-Qariatein (Wadi Lebtah), and Wadi Erzeq; on the Jordan Uplift at the Naht El-Yarmouk River and Jebel Sis; on the northern slope of the Rutbah Uplift at the Joeif depression, Saba

Abar, Jebel Fru Tarag El-Olab, Beer Alayyaniyeh, Tell Doueiye, Wadi El-Miyah, and Wadi El-Waar; on the southern slope of the Aleppo Plateau at M'arret An-No'man, Meniss, and Jebel Shbit; in the Mesopotamian basin at Cheikh Aroud, Serrin Esh-Shamaliyeh, and Milha; and in the Aafrine graben at Meidanki. Nevertheless, the tectonic movements of the late Middle Eocene resulted in the absence of *Truncorotaloides rohri* Zone sediments in some sections at Arak, Jebel Zubeideh, Bir Hassiye, Jebel Dilaa, and Juba. Sometimes both the *Orbulinoides beckmanni* and *Truncorotaloides rohri* Zones are absent as at Jebel Khabli and Tell Snime.

The Upper Eocene (Priabonian Stage) includes three zones: the *Globigerapsis semiinvoluta, Turborotalia cocoaensis* and *Turborotalia centralis-Globigerina gortanii* Zones.

The *Globigerapsis semiinvoluta* Zone (P15) is represented by microfauna which includes *Globigerapsis semiinvoluta, G. index, G. tropicalis, Turborotalia cerroazulensis, T. centralis, Dentoglobigerina corpulenta, D. galavisi, D. pseudovenezuelana, D. tripartita, Globigerina praebulloides, G. ouachitaensis, Hantkenina suprasuturalis, H. alabamensis,* and *Pseudohastigerina micra,* and less numerous *Globigerinatheca barri, G, lindiensis, Globigerinita martini, Globorotaloides suteri* and *Chiloguembelina cubensis.*

The *Turborotalia cocoaensis* Zone (P16) is defined by the appearance of *T. cocoaensis, T. cunialensis, Cribrohantkenina inflata, Hantkenina primitiva* and *Globigerina gortanii.* Also numerous are *Globigerapsis tropicalis, G. index, Dentoglobigerina corpulenta, D. galavisi, D. tripartita, Turborotalia cerroazulensis, T. centralis, Pseudohastigerina micra*, and *Globigerina praebulloides* which were already encountered in the underlying sediments of the older zone. Specimens of *Globigerapsis semiinvoluta* are completely missing.

The *Turborotalia centralis - Globigerina gortanii* Zone (P17) contains an assemblage of planktonic foraminifera transitional from the Eocene to the Oligocene microfauna. The bulk of this assemblage consists of *Turborotalia centralis* (small-sized specimens), *T. nana, Globigerina gortanii, G. praebulloides, G. ouachitaensis, Dentoglobigerina corpulenta, G. galavisi, D. tripartita, G. pseudovenezuelana, Globorotaloides suteri, Pseudohastigerina micra*, and *Chiloguembelina cubensis*. Typical Eocene taxons, such as representatives of *Globigerinatheca, Globigerapsis, Cribrohantkenina, Hantkenina* and also *Turborotalia cerroazulensis* and *T. cocoaensis* disappear. In some sections only singular specimens of *Turborotalia cunialensis, Globigerapsis index* and *Hantkenina* aff. *alabamensis* were identified. The above foraminifera associate with rather rare specimens of *Globigerina officinalis, G. ampliapertura, G. prasaepis, Turborotalia permicra* and *T. gemma* which are very common for the Oligocene sediments.

The *Globigerapsis semiinvoluta* and *Turborotalia cocoaensis* Zones were identified in some areas of Syria with pelagic sedimentation, such as the Anti-Lebanon at Doummar, Maaraba, Mnine, Rankous, Jebel Maaloula, Rass El-Marra, Wadi Maesra, Jebel Oz-Aantar, and Rass El-Aabid; in the Palmyrides at Jebel Dmeir, Jebel Kohle, Wadi El-Bardeh, Jebel El-Abuad, Jebel Heite, Arak, Jebel Al-Bishri (Wadi Souq), Bir Hasseye, Ain Qomqom, Jebel Tyas, Hmair (Wadi Khabra), and Al-Qariatein (Wadi Lebtah); on the Aleppo Plateau at Aanadane and El-Bab; in the Mesopotamian basin at Serrin Esh-Shamaliyeh and Milha, and in some bore-holes near Raqqa.

The *Turborotalia centralis-Globigerina gortanii* Zone was found in a limited number of sections in the Anti-Lebanon at Maaraba, Mnine, Rankous, Jebel Maaloula, Rass El-Marra, Jebel Oz-Aantar, Rass El-Aabid, and Doummar; in the Palmyrides at Wadi El-Bardeh, Jebel El-Abuad, Jebel El-Bishri, and Bir Hasseye; and in the Mesopotamian basin at Serrin Esh-Shamaliyeh, Qarah-Qozaq, and in some bore-holes near Raqqa. In the Anti-Lebanon and the Palmyrides determination of this zone is rather complicated because pelagic sediments of the Upper Eocene are replaced by shallow-water hard limestones of the Oligocene and the transitional layers (P17 Zone) contain very poor planktonic foraminifera.

The area of the Syrian Desert, the Anti-Lebanon, and the Palmyrides are occupied by shallow-water Oligocene deposits. The pelagic facies of the Oligocene with planktonic foraminifera is common for the Aleppo Plateau and the Mesopotamian Basin. Sediments of this age crop out on the Aleppo Plateau and in the Al-Furat River valley between the towns of Jrablous and Maskene where they are also penetrated by some bore-holes drilled for hydrogeological purposes. The entire sequence of the Oligocene foraminiferal zones can be observed in this region.

The ***Globigerina tapuriensis* Zone (P18)** is characterized by *Globigerina tapuriensis, G. ampliapertura, G. prasaepis, G. ouachitaensis, G. angustiumbilicata, G. officinalis, G. gortanii, Dentoglobigerina pseudovenezuelana, D. galavisi, D. tripartita, Catapsydrax unicava, C. martini, Turborotalia gemma, T. nana, Pseudohastigerina micra, P. barbadoensis, Cassigerinella chipolensis, Globorotaloides suteri*, and *Chiloguembelina cubensis*.

The ***Globigerina sellii* Zone (P19)** contains a similar assemblage of the planktonic foraminifera together with first findings of *G. sellii.* In some sections this species is absent and these two zones are considered as a single undifferentiated subdivision.

The ***Globigerina ampliapertura* Zone (P20)** includes the majority of species which were encountered in the underlying sediments of the two above-mentioned zones. Its main feature is defined by numerous specimens of *G. ampliapertura* and the absence of representatives of the genus *Pseudohastigerina.*

The above three zones, together with rather similar assemblages of planktonic foraminifera, belong to the Rupelian Stage, Lower Oligocene.

The ***Turborotalia opima* Zone (P21)** is marked by the first appearance of *Turborotalia opima, Globigerina angulisuturalis, G. euapertura*, and *G. senilis* which are accompanied by *Globigerina angustiumbilicata, G. prasaepis, G. officinalis, G. ouachitaensis, Catapsydrax unicava, C. martini, Globorotaloides suteri, Cassigerinella chipolensis*, and *Chiloguembelina cubensis*.

The ***Globigerina ciperoensis* Zone (P22)** is defined by abundant specimens of *G. ciperoensis* and *G. angulisuturalis* together with a strong decrease of the number of specimens of

Turborotalia opima.

These two zones are referred to the Chattian Stage, Upper Oligocene.

The *Turborotalia kugleri* Zone, which terminates the Oligocene in complete sections, and contains assemblages of planktonic foraminifera transitional to those of the Lower Miocene, was not discovered in the Al-Furat area on account of the unconformity with the Lower Miocene.

The stratigraphic subdivision of the Paleogene shallow-water deposits of Syria is based on the fauna of nummulitids. Data on their distribution appear in publications of Cizancourt (1934), Dubertret et al. (1938), Abdoulrahim and Nemkov (1972), and Krasheninnikov and Nemkov (1975).

Krasheninnikov and Ashourov (1995) carried out a special study dealing with the correlation of the pelagic and shallow-water facies of the Syrian Paleogene. In the section at Safarqiyeh in the area of Jebel An-Nusseiriyeh, the first small rare *Nummulites* were found, together with species of *Assilina, Operculina, Discocyclina* and *Ranikothalia*, in the upper layers of the *Igorina pusilla* Zone (P3b) which belongs to the Selandian Stage, Upper Paleocene. In sediments of the next *Planorotalites pseudomenardii* Zone (P4) and *Morozovella velascoensis* Zone (P5) of the Upper Paleocene, representatives of *Nummulites* are more common and were identified in the sections of Safarqiyeh, Wadi Gamiq, Brianis, Ein El-Shikak, Damata, and Tcholaqane, in the areas of Jebel An-Nusseiriyeh, the Baer-Bassit Massif, and Aafrine.

In the Early Eocene, pelagic sediments predominate and nummulitic limestones were found only in some exposures located in Jebel An-Nusseiriyeh and the Baer-Bassit Massif. The *Morozovella subbotinae* Zone was found at Kesladjouk, the *Morozovella formosa* Zone at Nahr Marqiyeh, and the *Morozovella aragonensis* Zone at Nahr Marqiyeh, Wadi Gamiq, Safarqiyeh, and Damata. In the interval represented by the *Acarinina pentacamerata* Zone, shallow-water deposits also appeared in the Anti-Lebanon, and limestones with *Nummulites* were met in sections at Kafar Qouq, Beit Jann, Rass El-Aabid, Jebel Kashin, Doummar, Deir Qanoun, Halboun, and the Bissan gorge as well as in northwestern Syria at Nahr Marqiyeh, Wadi Gamiq, Shambar Foukani, and Damata.

In the Middle Eocene, especially in its upper part, areas with shallow-water limestones were increasing in the Anti-Lebanon, Jebel An-Nusseiriyeh, the Palmyrides, and the ridges in the extreme northwestern part of Syria. Rich assemblages of *Nummulites* were defined in sediments of all the planktonic foraminifera zones: the *Hantkenina nuttalli* Zone (Kafar Qouq, Deir Qanoun, Beit Jann and Sakhabe); the *Globigerapsis kugleri* Zone (Wadi Maesra, Rass El-Aabid, Sadyr Mkin, Halboun, and Kafar Qouq); the *Morozovella lehneri* Zone (Wadi Maesra, Rass El-Aabid, Rankous, and Safarqiyeh); the *Orbulinoides beckmanni* Zone (Mnin, Rankous, Jebel Oz-Aantar, Deir Qanoun, Sadyr Mkin, Halboun, and Jebel Hayyan); and the *Truncorotaloides rohri* Zone (Mnin, Wadi Maesra, Rankous, Jebel Oz-Aantar, Rass El-Aabid, Deir Qanoun, Sadyr Mkin, and Halboun).

The Upper Eocene assemblages of *Nummulites* are connected with outcrops of the Anti-Lebanon. They were found in sediments of the *Globigerapsis semiinvoluta* Zone (Mnin, Wadi Maesra, Rankous, and Rass El-Marra) and the *Turborotalia cocoaensis* Zone (Jebel Oz-Aantar, Doummar, Mnin, Maaloula, Rankous, Rass El-Marra, and Rass El-Aabid).

Massive recrystallized Oligocene limestones of the Anti-Lebanon and the Palmyrides contain only debris of *Nummulites*. Rich Lower Oligocene assemblages of *Nummulites* were discovered in a limited number of sections of the Anti-Lebanon at Rass El-Marra, Hafir At-Tahta, Rankous, and Wadi Khabra, and in the Aafrine graben at Meidanki. The last rare specimens of *Nummulites germanicus* and *N. vascus* before they became extinct are found in sediments of the Upper Oligocene *Globigerina ciperoensis* Zone (Wadi Khalla and Wadi Al Jizel to the north of Palmyra). The Oligocene deposits of the Palmyrides and the Syrian Desert are characterized by enormous quantities of representatives of Lepidocyclinidae.

The paleontological investigation of the Syrian *Nummulites* and their ranges (within zones of planktonic foraminifera) has not yet been completed and we present lists of *Nummulites* species relevant only to the stages and subseries of the Paleogene.

The Thanetian Stage (P4 and P5 Zones) is characterized by *Nummulites fraasi fraasi, N. fraasi densispira, N. deserti, N. solitarius, N. gamardensis, N. pernotus*, and *N. bigurdensis* which are accompanied by species of *Assilina, Operculina, Discocyclina, Miscellanea* and *Ranikothalia.*

The Lower Ypresian substage (P6 and P7 Zones) contains *Nummulites planulatus, N. exilis, N. globulus, N. pernotus, N. rotularius, N. partschi, N. bolcensis, N. spileccensis, N. ficheuri, N. silvanus, N. praelucasi*, and *N. haymanensis*.

The Upper Ypresian substage (P8 and P9 Zones) yields a richer assemblage: *Nummulites* ex gr. *manfredi, N. planulatus, N. leupoldi, N. spileccensis, N. praelucasi, N. aquitanicus, N. praelaevigatus, N. burdigalensis burdigalensis, N. burdigalensis pergranulatus, N. burdigalensis cantabricus, N. nitidus, N. formosus, N. rotularius, N. globulus, N. pavloveci, N. kapellosi, N.* aff. *pustulosus, N. irregularis, N. praediscorbinus, N.* aff *polygyratus, N.* aff. *subramondi, N. nemkovi, N. soerenbergensis, N. kaufmani, N. archiaci, N. pratti, N. partschi*, and *N.* aff. *subdistans*.

The Lutetian Stage in its lower part (P10 and P11 Zones) includes the assemblage of larger foraminifera which is represented by *Nummulites laevigatus, N. uranensis, N. alponensis, N. praediscorbinus, N. formosus, N. gallensis, N. rotularius, N. caupennensis*, and *N.* aff. *lehneri*. In the upper part of this stage (P12 Zone) appear species such as *N. gizehensis, N. millecaput, N. beneharnensis*, and *N. kugleri*. Possibly, limestones with *N. jaquoti, N. bullatus, N. stephani, N. carteri*, and *N.* aff. *carpenteri* belong also to the P12 Zone of planktonic foraminifera.

The Bartonian Stage (P13 and P14 Zones) is characterized by *Nummulites perforatus, N. gizehensis, N. lyelli, N. dufrenoyi, N. millecaput, N.* aff. *brongniarti, N. variolarius, N. cyrenaicus, N. discorbinus, N. praepuschi, N. acutus pachispira, N. aturicus, N. incrassatus, N. ptukhiani, N. striatus, N. biarritzensis, N. chavannesi hajastanica, N.* aff. *vicarii, N. migiurtinus, N.*

alponensis, N. carpenteri, N. beaumonti and *N. bullatus.*

The assemblages of nummulitids are different in the lower and middle parts of the Priabonian Stage.

In its lower part (the *Globigerapsis semiinvoluta* Zone, P15) large *Nummulites lyelli, N. maximus, N. dufrenoyi*, and *N. perforatus* are still common accompanied by *N.* aff. *fabianii, N. variolarius, N. cyrenaicus, N. incrassatus, N. orbignyi, N. chavannesi chavannesi, N. chavannesi hajastanica, N. striatus, N. stellatus, N. garnieri*, and *N.* aff. *praestwichianus*.

In the middle part of this stage (the *Turborotalia cocoaensis* Zone, P16) the first four above-mentioned species of *Nummulites* with a large shell disappear and their association consists of *N. fabianii, N. incrassatus, N. striatus, N. stellatus, N. chavannesi, N. garnieri, N. bouillei, N. orbignyi, N. retiatus, N. budensis, N. concinnus, N.* aff. *prestwichianus, N. boulangeri*, and *N. vascus initialis*. They are accompanied by various species of *Grzybowskia, Operculina, Spiroclypeus, Pellatispira, Discocyclina, Asterocyclina* and *Actinocyclina*.

Limestones of the upper part of the Priabonian Stage (the *Turborotalia centralis–Globigerina gortanii* Zone, P17) in our material do not contain representatives of the larger foraminifera.

The Lower Oligocene deposits include only five species of *Nummulites* such as *N. fichtelli, N. retiatus, N. vascus, N. orbignyi*, and *N. stellatus* though in some sections of the Anti-Lebanon they are quantitatively numerous and the rock can be treated as the nummulitic limestone.

Representatives of the macrofauna (pelecypods, gastropods and sea urchins) are rare in the predominantly pelagic Paleocene and Eocene sediments but they are very important for determining the age of the Oligocene deposits. These groups of fossils are especially common for the Oligocene deposits of the Central Palmyrides at Jebel Rmah, Jebel Aade, Jebel Tyas, Jebel El-Bardeh, and Jebel Ad-Dalil (Faradzhev, 1966), and the Eastern Palmyrides at Jebel El-Abuad, Jebel Ad-Dawwara, and Jebel El-Abtar (Soulidi-Kondratiev, 1966a).

The list of species includes *Pecten arcuatus* Brocchi var. *angulocostata* Sacco, *Pecten quinquepartitus* Blanck., *P. livoniani* Blanck., *P. pseudolivoniani* Eames et Cox,, *Ostrea khamirensis* Cox, *O. plicata* Sol., *O. flabelliformis* Schauroth., *O. (Alectrionia) martinsi* Arch., *Corbula conglobata* Koenen, *Cardium saribulachicum* Gabrielian, *C. acutum* Blanck., *C. rouyanum* d'Orb., *Strombus auriculatus* Grat., *Conus* aff. *ineditus* Mich., *Ampullina vulcani* Brong., *Pitar villanovae* Desh., *Divaricella ermenonvilensis* d'Orb., *Arca (Barbatia) cylindracea* Desh., *Panope heberti* Bosq., *Cordiopsis incrassata* Sow., *Spondylus bifrons* Muenst. var. *cisalpinus* Brongn., *Chlamys biarritzensis* Arch., *Ch. subdiscors* Arch., *Gryphaea* cf. *gigantica* Solander, *Turritella planispira* Nyst., *T. sulcifera* Desh., *T. imbricataria* Lam., *Diastoma costellatum* Lam., *Cassidea (Phalium) deshayesi* Bellardi, *Pleurotomaria nicensis* Bayan, *Meretrix longior* Boussac, *Corbula valdensis* Hobert et Renevier, *Arca globulosa* Desh., *Chama granulosa* Arch., *Ch. vicentina* Fuchs, *Miltha (Pseudomiltha) gigantea* Desh., *Crassatella vapincana* Boussac, *Tellina* ex gr. *striatissima* Desh., *Phacoides subrhomboidalis* Newton, *Ph. (Lucina) eaglesomei* Newton, *Surcula transversaria* Lam., *Miocardiopsis ovalina* Koenen, *Ficus helvetica* Mayer var. *daralagaesica* Aslanjan, *Schizaster vicinalis* Agass., *Sch. rimosus* Desor, *Echinolampas blainvilleii* Agass., *Eupatagus ghivanensis* Gauth, *Detremaster nux* Mun.-Chalm., and *Echinanthus* sp. (det. of L. P. Kakhanova, L. V. Mironova, N. A. Cheltsova, and O. I. Schmidt).

ISOTOPIC AGES OF SELECTED BIOSTRATIGRAPHIC LEVELS

Sediments of almost all the planktonic foraminifera zones of the Syrian Paleocene and Eocene contain one or several glauconite horizons that allow isotopic age determinations of some of these subglobal (tropical, subtropical, and temperate paleoclimatic belts) subdivisions. The majority of the glauconite horizons were found in the deposits of the Palmyrides, on the eastern slope of the Anti-Lebanon, the southern slope of the Aleppo Plateau, and the western slope of Jebel An-Nusseiriyeh (**Fig. 6.47**). The process of glauconite accumulation was most active during the time interval from the uppermost Lower Eocene (the *Acarinina pentacamerata* Zone) to the lowermost Upper Eocene (the *Globigerapsis semiinvoluta* Zone). In the Oligocene deposits glauconite horizons are completely absent.

The K/Ar and Rb/Sr methods were used for the age determination. It is known that the isotopic system of glauconite is not stable and can be easily distorted during processes of epigenesis and hypergenesis. Therefore our investigation was directed at searching for a closed, undistorted isotopic system of glauconite (Krasheninnikov et al., 1987, 1996; Nikolaeva et al., 1985; Vinogradov et al., 1990a, b; Golovin et al., 1986). In our opinion, such a geochemical system can be treated as closed and undistorted if it satisfies four criteria:

1) Coincidence of K/Ar ratios for monomineral fractions of glauconite with varying potassium content (the K-Ar isochrone),

2) Good approximation to a straight line for the experimental points on the Rb-Sr isochronic plot. Small dispersion of the points about the line is an indication of a closed Rb-Sr isotopic system in a monomineral fraction (the Rb-Sr isochrone),

3) Coincidence of the estimated K-Ar and Rb-Sr isochronic ages, and

4) Conformity of the estimated initial ratio of strontium isotopes ($^{87}Sr/^{86}Sr$) in glauconite (a sedimentary rock) to the isotopic composition of the marine water strontium at the corresponding time. It means that the strontium isotopic system was closed exactly at the time of sedimentation.

Converging radiological ages have been obtained for the three biostratigraphic (chronostratigraphic) levels:

1) The base of the *Planorotalites pseudomenardii* Zone, P4 (the Upper Paleocene) – 55.0±1.0 Ma.

2) The base of the *Acarinina pentacamerata* Zone, P9 (the uppermost Lower Eocene) – 46.0±1.0 Ma.

3) The base of the *Globigerapsis semiinvoluta* Zone, P15 (the bottom of the Upper Eocene) – 37±1.0 Ma.

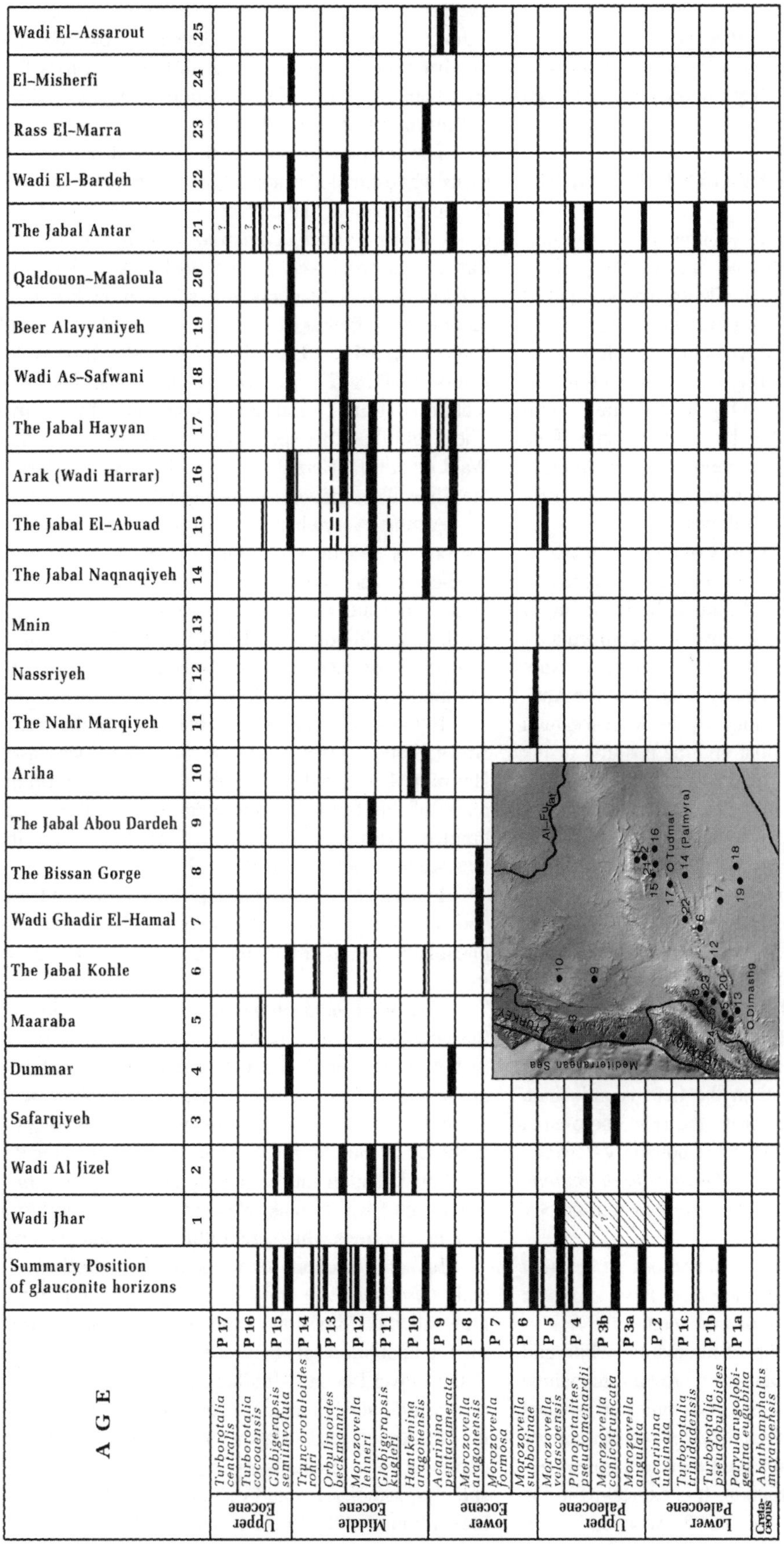

Fig. 6.47. Position of the glauconite horizons in the Paleogene sections of Syria.

Diverging K-Ar and Rb-Sr datings by glauconite, obtained from various samples of the Syrian Paleocene and Eocene indicate only different events which took place in the process of transformation of sediments to rocks. They can not be used for the chronostratigraphy.

LITHOSTRATIGRAPHY OF THE SYRIAN PALEOGENE

The Paleogene deposits of the Anti-Lebanon, the Palmyrides, Jebel An- Nusseiriyeh, the Aleppo Plateau, the Aafrine graben, the Baer-Bassit Massif, and the Syrian Desert are composed of calcareous rocks (various kinds of limestones and marls). Terrigenous (sandy-clayey) rocks and volcanic (basalts) rocks occupy very small areas. The former group is present in the Eocene of Jebel Bishri and in the Upper Oligocene of the Eastern Palmyrides, the latter was encountered in the Lower Eocene of the Aafrine graben. Against the lithologically monotonous background of these calcareous deposits five lithostratigraphic subdivisions (formations) can be defined. They are as follows.

The first formation is composed mostly of marls, calcareous clays and clayey limestones. It embraces the entire Paleocene and the main part of the Lower Eocene (up to the top of the *Morozovella aragonensis* Zone). Deposits of this rock unit rest unconformably on the erosional surface of the Maastrichtian limestones and marls with a glauconite horizon at the base. In many sections of the Anti-Lebanon and the Palmyrides the lithology of the Maastrichtian and Paleocene sediments is rather similar and the position of the Cretaceous/Paleogene boundary in the field is indistinct.

The formation is subdivided into two parts (members). The lower one consists of gray and greenish-gray thin-bedded, sometimes foliated marls and calcareous clays with subordinate interbeds of limestones. The upper part is made up of light-gray clayey limestones and marls with massive or ovoid texture; intercalations of calcareous clays are rare. Upwards in the section the amount of clayey material is drastically reduced from 50-70% in the lower member to 10-40% in the upper member. The lower member belongs to the Paleocene and the upper one to the Lower Eocene. These units are connected by a gradual transition and the lithologic boundary between them in the field is indistinct. Certainly, it can be easily determined from the rich microfauna common for these calcareous sediments as the boundary between the *Morozovella velascoensis* Zone and the *Morozovella subbotinae* Zone,. Soft rocks of this formation create a smooth topography with valleys and wadis.

A special feature of the first formation in sections on the right bank of the Nahr El-Yarmouk River is the presence of kerogene limestones with a rather high content of organic matter.

The second formation is built up of alternating white and light-gray limestones and marls with black and brown flints which form layers, lenses, and concretions. This rock unit corresponds to the uppermost Lower Eocene (the *Acarinina pentacamerata* Zone). The presence of hard erosion-resistant flints and limestones provides better exposures of this formation. It is expressed in the topography as slopes of anticlines and uplifts with additional questas or small ridges. The flints look like black stripes on the light-gray background of the calcareous rocks. The thickness of the interbeds of flints and siliceous limestones varies from 1-3 cm to 30-40 cm. Very often the formation terminates with several layers of flints up to 0.5-1.0 m thick.

The total thickness and the ratio of siliceous to calcareous rocks was estimated in some typical sections of the Palmyrides and the Anti-Lebanon. In the Wadi El-Bardeh section the total thickness of the *Acarinina pentacamerata* Zone is 170 m, and there are 88 flint layers whose composite thickness is 5.11 m, constituting 3.0% of the total thickness of this formation. For the Jebel Hayyan section these figures are 120 m. 49 flint layers totalling 4.14 m, comprising 3.45%. The Jebel El-Abuad section is 100 m, 41 flint layers totalling 2.9 m, comprising 2.9%. The Wadi El-Harrar section is 40 m, 18 flint layers totalling 1.4 m, comprising 3.5%. These and the other sections studied at Jebel Kohle, Jebel Antar, Wadi Ghadil El-Hamal, etc. are all different in these respects but the siliceous rocks are usually around 3% of the total.

Almost everywhere at the base of the second formation there are one or several glauconite-phosphorate marls with traces of mud-eaters and underwater erosion, but the foraminiferal assemblages indicate continuous sedimentation at this stratigraphic level. One additional feature of the basal layers of the *Acarinina pentacamerata* Zone is the appearance of syngenetic breccia containing numerous nummulitids.

In the area of structural highs, such as the central part of the Anti-Lebanon, Jebel An-Nusseiriyeh, the Baer-Bassit Massif, the Hermon Massif, and the Syrian Desert, pelagic soft limestones and marls of this formation are replaced by shallow-water hard fine-grained, detritic, or nummulitic limestones. Here the flints form lenses and irregular shaped nodules.

In the area of the Rutbah Uplift in the Syrian Desert calcareous and siliceous rocks of the second formation contain layers of phosphatic ore of commercial grade which are only exploited in Iraq.

The **third formation** consists of pelagic white and light-gray, thick-bedded, soft chalky limestones and marls. It corresponds to the entire Middle Eocene, from the base of the *Hantkenina aragonensis* Zone up to the top of the *Truncorotaloides rohri* Zone. As a rule, a gradual transition is observed between deposits of the Lower and Middle Eocene. In sections of the Palmyrides and the Anti-Lebanon this formation appears as a very mono-tonous sequence of white chalk. Nevertheless, its subdivisions have their own peculiarities which are useful during the geological mapping.

In some sections deposits of the *Hantkenina aragonensis* and *Globigerapsis kugleri* Zones contain intercalations of gray silicified limestones and lenses of black flints. Here only micropaleontological analysis allows one to determine the position of the Lower Eocene/Middle Eocene boundary.

Marls and limestones of the *Morozovella lehneri* Zone in the Arak (Wadi Harrar) and Jebel El-Abuad sections are enriched with remnants of siliceous organisms such as sponge spicules, radiolarians and diatoms. As a result of dissolution and redistribution of the siliceous matter, these calcareous rocks contain small isolated flinty nodules.

Very often the sediments of the *Morozovella lehneri* and *Orbulinoides beckmanni* Zones are characterized by numerous ball-like barite concretions.

The *Morozovella lehneri* Zone in the area of the Rutbah Uplift is represented by nummulitic limestones.

The *Orbulinoides beckmanni* Zone is composed of classical white soft chalk but in almost all the sections studied this zone also contains intercalations of hard light-gray silicified limestones of 2-10 cm in thickness, and chalk. The total thickness of this member varies from 5 to 10 m. In the relief it produces a rather steep slope. In the Jebel Hayyan section this zone is made up of detrital, massive, or porous limestones.

The Jebel Antar section, located in the Central Palmyrides, demonstrates very unusual, individual features. The entire Middle Eocene is composed here of greenish-gray carbonate-phosphate, carbonate-phosphate-glauconite sands, gravelites and conglomerates; the thickness is reduced to 45 m. Zonal subdivision of these rocks by means of planktonic foraminifera is impossible.

On the structural highs the pelagic chalk facies is replaced by shallow-water limestones – nummulitic, nummulitic-alveolinid, algal, organogenic-detrital and massive fine-grained. This facies of the third formation is widely developed in the Hermon Massif, the Central Anti-Lebanon, Jebel An-Nusseiriyeh, the Baer-Bassit Massif, Jebel Az-Zawiyeh, the Rutbah Uplift in the Syrian Desert, and the ridges of northwestern Syria such as Jebel Wastani and Jebel Douela.

The **fourth formation** of the Syrian Paleogene consists of a variety of different limestones – soft pelitic, massive biogenic, and fine-grained – which may or may not alternate. The formation belongs to the Upper Eocene.

The tectonic position of the area determines whether pelagic or shallow-water limestones predominate. On the eastern slope of the Anti-Lebanon and in the Palmyrides one can observe a combination of pelagic chalky limestones and shallow-water organogenic limestones. In the central part of the Anti-Lebanon and in the Rutbah Uplift, massive biogenic shallow-water limestones, containing numerous nummulites and discocyclines strongly predominate. The Aleppo Plateau and the western margin of the Mesopotamian Basin along the Al-Furat River valley are covered by outcrops of white, soft, pelitic chalky-like limestones, although with an increasing amount of neritic foraminifera. On the northeastern flank of the Palmyrides and especially on their northeastern plunge at Jebel Al-Bishri calcareous rocks alternate with sandy-clayey sediments and the source of the sandy quartz material remains unknown.

The Upper Eocene deposits of the fourth formation display a very complicated contact with the underlying sediments. In many sections of the Palmyrides and the Anti-Lebanon the transition of the Middle Eocene to the Upper Eocene is gradual but paleo-geographic changes at this level resulted in the appearance of a thick glauconite horizon at the base of the *Globigerapsis semiinvoluta* Zone. In addition, the lower part of this foraminiferal zone is occupied by a 5-10 m member of calcareous rocks with a block structure in the sections at Doummar, Maaraba, Wadi El-Bardeh, Jebel El-Abuad, Arak, etc. Separate blocks up to 2-5 m thick and 15-20 m long have slid over limestone blocks of the same layer which are themselves displaced. The relation between these blocks, the break-up in the continuity of the layer, the wavy nature of the layers, and the signs of motion of the limestone blocks indicate that the sliding events took place while the sediments were weakly consolidated (syn-sedimentary structures).

In some sections of the Palmyrides (Wadi Harrar, Tell Snime, Bir Hasseye, Bir Qdeim, Jebel Dilaa, and Jebel Zubeideh) the Upper Eocene rocks unconformably overlie sediments of various Middle Eocene stratigraphic zones.

The deposits of the fourth formation were not found in the exposures of Jebel An-Nusseiriyeh and some of the ridges of northwestern Syria.

Where hard limestones are found in the Upper Eocene (Mnin, Jebel El-Abuad, Wadi Harrar, etc.), they form a steep escarpment together with massive Oligocene limestones (the fifth formation). When soft chalky limestones and marls dominate as at Doummar, Maaraba, Maaloula, Wadi El-Bardeh, etc., they resemble the underlying Middle Eocene rocks (the third formation). These lithological similarities of the fourth formation created serious difficulties during the geological mapping but they were easily solved with the help of foraminiferal zonal stratigraphy.

All the lithological and paleogeographical patterns of the Upper Eocene fourth formation are transitional between the different geological settings of the Middle Eocene and Oligocene. They are 'predecessors' of those significant changes which took place at the Eocene/Oligocene boundary.

The **fifth formation** includes Oligocene deposits. In the Anti-Lebanon, the Palmyrides, and the Rutbah Uplift this formation is composed of massive hard limestones which create impressive steep escarpments. Among the limestones different varieties can be distinguished: coarse-layered and unlayered, fine-grained and aphanitic, algal, organogenic-detrital, and foraminiferal (nummulitic in the lower part of this formation, and lepidocyclinid in upper layers). In the Anti-Lebanon the entire Oligocene succession consists of massive limestones. In the Palmyrides they are confined to the lower part of this formation; the upper half is composed of alternating hard limestones, detrital limestones with abundant mud-eater traces, sandy marls, sands, and calcareous clays. In the Rutbah Uplift, sections of the fifth formation are characterized by the alternation of hard and foraminiferal (lepido-cyclinid) limestones with greenish clays.

The paleogeographic conditions at the Eocene/Oligocene boundary changed drastically but in all the sequences of the Anti-Lebanon, the Palmyrides, and the Syrian Desert the fourth (Upper Eocene) and the fifth (Oligocene) formations are joined by a gradual transition. This gradual transition is confirmed by a member of well-bedded limestones which is usually present at the base of the fifth formation in many of the above-mentioned sections.

On the northern margin of the Aleppo Plateau and in the western Mesopotamian Basin (the outcrops along the Al-Furat

River) the shallow-water deposits of the fifth formation are replaced by soft pelagic chalky limestones and marls. Here the transition from the white chalky Eocene limestones to the yellowish chalky Oligocene limestones is gradational, although in the field this lithological boundary is difficult to see.

Deposits of the fifth formation were not identified in the area of Jebel An-Nusseiriyeh and the Baer-Bassit Massif.

These five formations are well-traced throughout Syria and also in some other eastern Mediterranean countries – in southwestern Cyprus, southern Turkey, north-western Iraq (Jebel Sinjar), Israel, Jordan, and Egypt (the Nile valley, the Eastern and Western Deserts). Their peculiarities reflect subglobal and regional (Mediterranean) geological events such as the regression at the Cretaceous/Paleogene boundary; the Lower-Middle Eocene transgression; the wide distribution of pelagic facies of soft white chalk and shallow-water facies of nummulitic limestones of the Middle Eocene; the prevalence of siliceous deposits (flints) in the uppermost Lower Eocene; the Oligocene regression, and a large stratigraphic and structural unconformity at the Paleogene/Neogene boundary. Therefore the boundaries of these five lithostratigraphic units (formations) coincide rather closely with the chronostratigraphic boundaries.

These five formations were used during our geological mapping in Syria. The following four mapping units were shown on sheets of the 1:200,000 scale Geological Map of Syria:

1) the interval incorporating the Paleocene and a part of the Lower Eocene including the *Morozovella aragonensis* Zone (the first formation);

2) the *Acarinina pentacamerata* Zone of the uppermost Lower Eocene (the second formation);

3) the Middle Eocene-Upper Eocene (the third and fourth formations), and

4) the Oligocene (the fifth formation).

The territory of Syria is subdivided into several tectonic (structural) units. The geological development of every structural unit during Paleogene time was characterized by some individual features. These local factors were influenced by events of subglobal and regional (Mediterranean) rank. The combination of all these factors resulted in facies replacements so that the lithological boundaries ceased to coincide with the chrono-stratigraphic ones. The local displacement of the formational boundaries was controlled and corrected by the paleontological method. From the pelagic Paleogene sediments many thousands of samples with planktonic foraminifera were analyzed in order to correct the stratigraphic position of the boundaries of the mapping units. Within the shallow-water Paleogene deposits these boundaries were checked by means of the larger foraminifera (nummulitids). Therefore, boundaries of the Paleogene mapping units occupy fixed stratigraphic positions and these units can be treated as regional subdivisions with monotonous lithology and chronostratigraphic boundaries. In the Russian stratigraphic nomenclature they are termed a **suite**.

Below we shall outline the main features of the Paleogene deposits in various structural regions of Syria.

REGIONAL STRATIGRAPHY OF THE SYRIAN PALEOGENE

These regions include the Palmyrides, the Anti-Lebanon, Jebel An-Nusseiriyeh, the Baer-Bassit Massif, the Aafrine Graben, the Aleppo Plateau, the Rutbah Uplift, the Drouze Depression, the Jordan Uplift, and the Mesopotamian Basin.

The Palmyrides

This area is characterized by classical and very complete successions of Paleogene sediments (**Figs. 6.48, 6.49, 6.50, and 6.51**).

The Danian Stage is made up of soft light-gray, gray and greenish, thin-laminated to foliated (leaf-like) marls and calcareous clays. Light-gray clayey limestones were encountered as separate bands. They differ from the underlying Maastrichtian rocks by a greater clay content but sometimes it is difficult to observe the boundary between the Upper Cretaceous and Paleogene in the field on account of similar lithology (Wadi El-Bardeh, Jebel El-Bassiri, Wadi Ghadir El-Hamal). The boundary becomes more visible when the transgressive Danian Stage contains a glauconite layer at its base. It is made up of marl with plenty of glauconite grains, sharks teeth and the debris of Maastrichtian rocks. The underlying rocks were pierced by the tracks of digging organisms and these cavities were filled by glauconite. The basal glauconite layer is from 5-10 cm to 50-70 cm thick and is located either at the bottom of the *Turborotalia pseudobulloides* Zone (Jebel Antar, Jebel El-Abuad), at the base of the *Turborotalia trinidadensis* Zone (Beer Qdem), or at the bottom of the *Acarinina uncinata* Zone (Wadi Jhar, Bir Hasseiye). Consequently, the duration of the stratigraphic gap between the Cretaceous and Paleogene was different in various areas of the Palmyrides. It the most complete sections at Wadi El-Bardeh, Jebel El-Bassiri, Jebel El-Abuad, and Wadi Ghadir El-Hamal the Danian Stage is 30-50 m thick. Where it is represented by its upper zones (or zone) the thickness drops to 8-25 m as at Wadi Jhar, Bir Hasseiye, and Beer Qdem.

The Selandian Stage sediments are monotonous gray and dark-gray thin-bedded soft marls with occasional layers of clayey limestones (Wadi El-Bardeh, Abou Rahba, Balas, Jebel El-Abuad, Bir Hasseiye, Wadi Ghadir El-Hamal). In the sections of Wadi Jhar, Hnezir, Wadi Erzeq, Jebel Hayyan, As-Soukhneh, and Al-Qom foliated dark and gray-green calcareous clays prevail. In the central part of the Palmyrides, the Selandian Stage is 30-50 m thick (Wadi El-Bardeh, Jabal Hayyan, and Wadi Jhar) but in the south (Wadi Ghadir El-Hamal) and in the north (Bir Hasseiye) the thickness reduces to 8-10 m. Usually the Danian and Selandian Stages are connected by a gradual transition. In the Ash-Sheikh and Abou Rahba sections the Selandian Stage (the *Morozovella angulata* Zone s.l.) rests directly on Maastrichtian rocks. In some sections such as Jebel Zubeideh, Jebel Dmeir, and Wadi Khabra the sediments of this stage are absent.

The Thanetian Stage is represented by gray and dark-gray, soft, thin-laminated marls but in some areas they are replaced by light-gray clayey limestones. Calcareous clays are extremely rare.

Dark-gray and green marls of the *Planorotalites pseudomenardii* Zone in the Jebel Zubeideh, Wadi Erzeq, Jebel Hayyan, Jebel El-Abuad, and Jebel Al-Bishri sections are slightly dolomitized. They contain abundant calcareous and agglutinated foraminifera but are very poor in planktonic forms. In the more calcareous sediments of the next *Morozovella velascoensis* Zone the planktonic foraminifera are very diverse. In the center of the Palmyrides depression the thickness of the Thanetian Stage reaches 80-150 m. In the north of the Palmyrides at Bir Hasseiye, and Wadi As-Souq and in the south at Jebel Zubeideh, Wadi Ghadir El-Hamal, and Jebel Ash-Sheikh the thickness is reduced to 22-45 m. In most sections the Selandian and Thanetian Stages are characterized by continuous sedimentation. The deposits of the Thanetian Stage are only transgressive in a limited number of areas where they lie unconformably on the Danian rocks (Jebel Zubeideh) or even on the Maastrichtian ones (Wadi Khabra, Al-Qaryatein, and Jebel Dilaa).

The main part of the Ypresian Stage (the *Morozovella subbotinae*, the *Morozovella formosa,* and the *Morozovella aragonensis* Zones) is represented by gray and greenish-gray marls and soft clayey limestones which are pelitic, well-bedded (sometimes with concentric parting), and containing fish scales and barite concretions. Here the process of sedimentation at the Paleocene/Eocene boundary was continuous and sections of the Palmyrides are excellent for studying events which occurred near this level. In the center of the Palmyrides at Abou Rahba, Wadi Jhar, and Al-Qaryatein the thickness of this Lower Eocene unit is 130-250 m. In some sections along the northern and southern margins of this depression, as well as in the anticlinal folds within the Palmyrides, the thickness drops to 18-50 m as at Bir Hasseiye, Wadi As-Souq, and Jebel El-Abuad. Some zones of the Lower Eocene are missing in the sections at Wadi Ghadir El-Hamal, Wadi Beer Qdem, and Jebel Dilaa. These gaps reflect local erosional processes and can not be traced laterally.

The upper part of the Ypresian Stage (the *Acarinina pentacamerata* Zone) is composed of alternating white, gray and yellowish marls, soft pelitic limestones, hard fine-grained limestones and numerous layers, lenses, and nodules of black and brown flints. The thickness of the flinty beds varies greatly from 1-3 cm to 30-45 cm. The nodules form beaded bands.

At the base of this zone in many areas of the Palmyrides (Jebel Dmeir, Jebel Aade, Jebel Hayyan, Wadi Harrar, Jebel El-Abuad, Jebel Antar) there is a layer (or several layers) of sandy marls with abundant glauconite and phosphorite nodules. The sediments were permeated by burrows of mud-eaters filled with glauconite. These erosional processes at the boundary between the *Morozovella aragonensis* Zone and the *Acarinina pentacamerata* Zone were at most very minor, and not accompanied by destruction and absence of any stratigraphic unit.

In some sections of the Northern Palmyrides (Bir Jhar, Bir Hasseiye, Hmair) layers of black and brown flints are relatively rare and intercalations of very hard, gray silicified limestones are more common. In the Hmair section (Wadi Khabra) the distribution of siliceous rocks is irregular, therefore permitting the deposits of the *Acarinina pentacamerata* Zone to be subdivided into three members:

1) The lower one is built up of dense, well-bedded limestones with very small and rare chert concretions and tiny flint bands, whose thickness is 120 m;

2) The 40 m thick middle member consists of alternating light-gray limestones and thick beds (15-20 cm) of brown flints

3) The upper member is represented by 90 m of soft chalky limestones with barite concretions and again rare flint lenses and nodules. This section illustrates very well the lateral changes in the sediments and the necessity of applying the micropaleontological method.

The thickness of the Upper Ypresian (the *Acarinina pentacamerata* Zone) formation of limestones and flints in the Palmyrides is highly variable but is usually between 100 and 250 m. Only in some sections, such as Wadi Harrar, Jebel Zubeideh, and Wadi Qdem, does it decrease to 40-80 m.

The Lutetian Stage yields a typical sequence of white and light-gray, soft, thick-bedded and unbedded chalk with block parting. Within this monotonous pile are thin members of light-gray marls, hard fine-grained and soft, dark, clayey limestones. In the sections at Wadi Harrar, Jebel Al-Bishri, Rass El-Rechem, Jebel El-Abuad, Bir Hasseiye, and others the lower part of this unit (the *Hantkenina aragonensis* and the *Globigerapsis kugleri* Zones) contains thin interbeds of massive, gray silicified limestones. They are completely absent among the white chalky and clayey limestones of the next *Morozovella kugleri* Zone. The Lutetian deposits of the Jebel Hayyan section are very unusual in that they consist of green sandy glauconite limestones and glauconite sandstones with admixture of small gravel; the thickness is very reduced at 16 m. The deposits of this age conformably overlie the Lower Eocene limestones, though in many sections the basal layers of the Lutetian Stage were marked by a glauconite horizon with phosphorate nodules, chert pebbles and shark teeth (*e.g.* Jebel Dmeir, Jebel Aade, Ain-Bardeh, Rass El-Rechem, Wadi Harrar, Jebel Tyas, Bir Hasseiye). A similar glauconite layer can be traced at the base of the *Morozovella lehneri* Zone at Jebel Dmeir, Jebel Kohle, Jebel El-Aabid, Wadi Harrar, Jebel Aade, and Jebel Al-Bishri.

Maximum thicknesses (260-370 m) of the Lutetian Stage were observed in the central part of the Palmyrides depression at Wadi El-Bardeh, Wadi Lebtah, Ain Qomqom, Jebel Tyas, and Abou Rahba. On the southern flank of the Palmyrides, in the sections of Jebel Zubeideh, Jebel Kohle, Jebel Aade, Jebel Dmeir, and Wadi Erzeq the thickness drops to 36-100 m. A similar situation occurs in the area of Al-Kom and Bir Hasseiye on the northern flank of the Palmyrides where the thickness does not exceed 50-65 m. In this region the reduced thickness is related to the absence of the *Morozovella lehneri* Zone sediments, due to the transgressive character of the overlying Middle Eocene layers.

The Bartonian Stage deposits have many things in common with those of the Lutetian Stage. White, soft chalky limestones predominate. Among these rocks, in the lower part of this stage (the *Orbulinoides beckmanni* Zone), there is a member of alternating

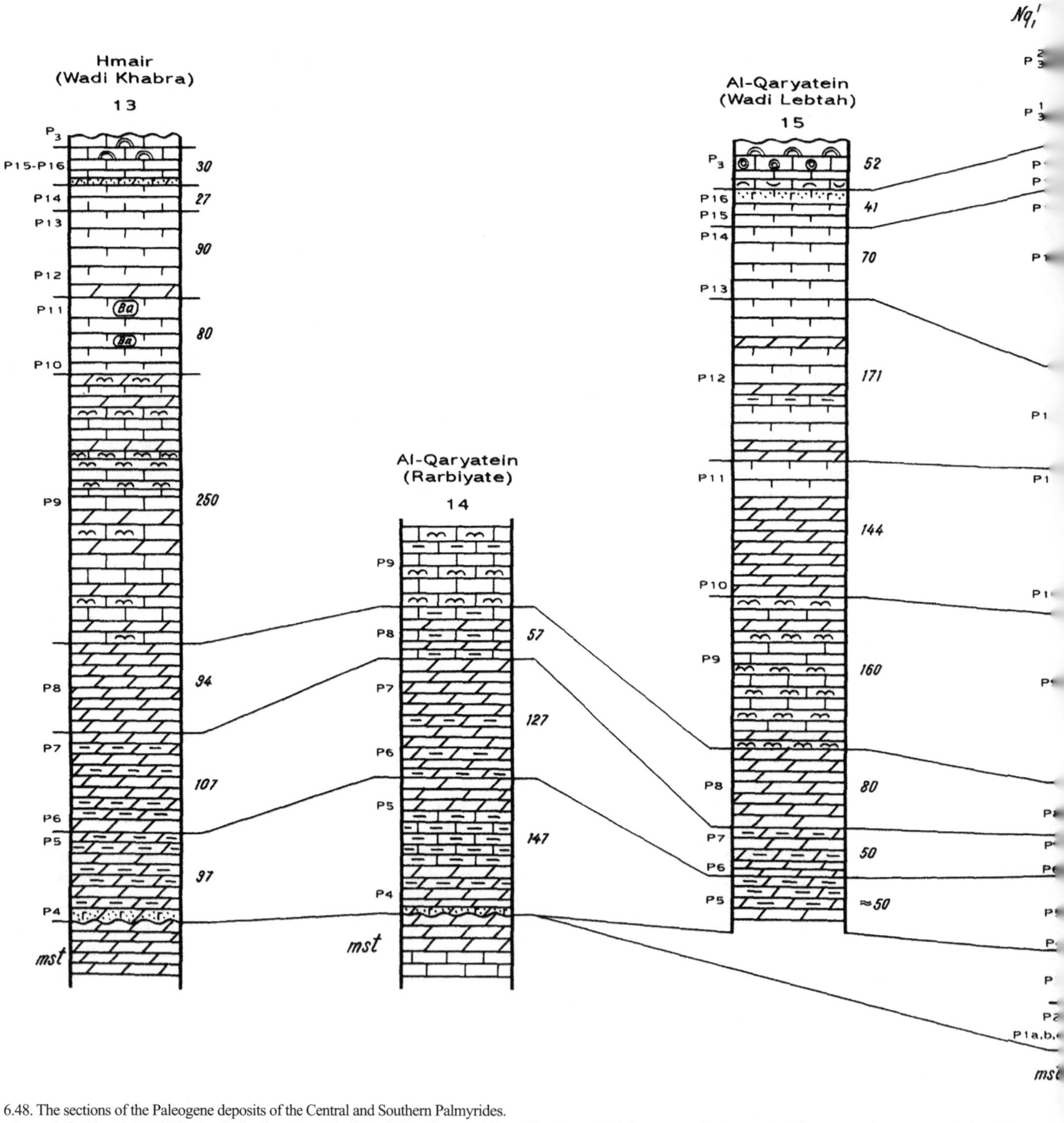

Fig. 6.48. The sections of the Paleogene deposits of the Central and Southern Palmyrides.

Legend (for Figs. 6.48 – 6.50): 1 – chalky limestones; 2 – foraminiferal (nummulitic and lepidocyclinid) limestones; 3 – fine-grained limestones; 4 – organogenic-detrital limestones; 5 – clayey limestones; 6 – algal and coralline limestones; 7 – dolomites; 8 – marls; 9 – clayey marls; 10 – flints; 11 – silicified rocks; 12 – clays; 13 – sands; 14 – sandstones; 15 – sandy rocks; 16 – glauconite; 17 – phosphorate; 18 – barite nodules.

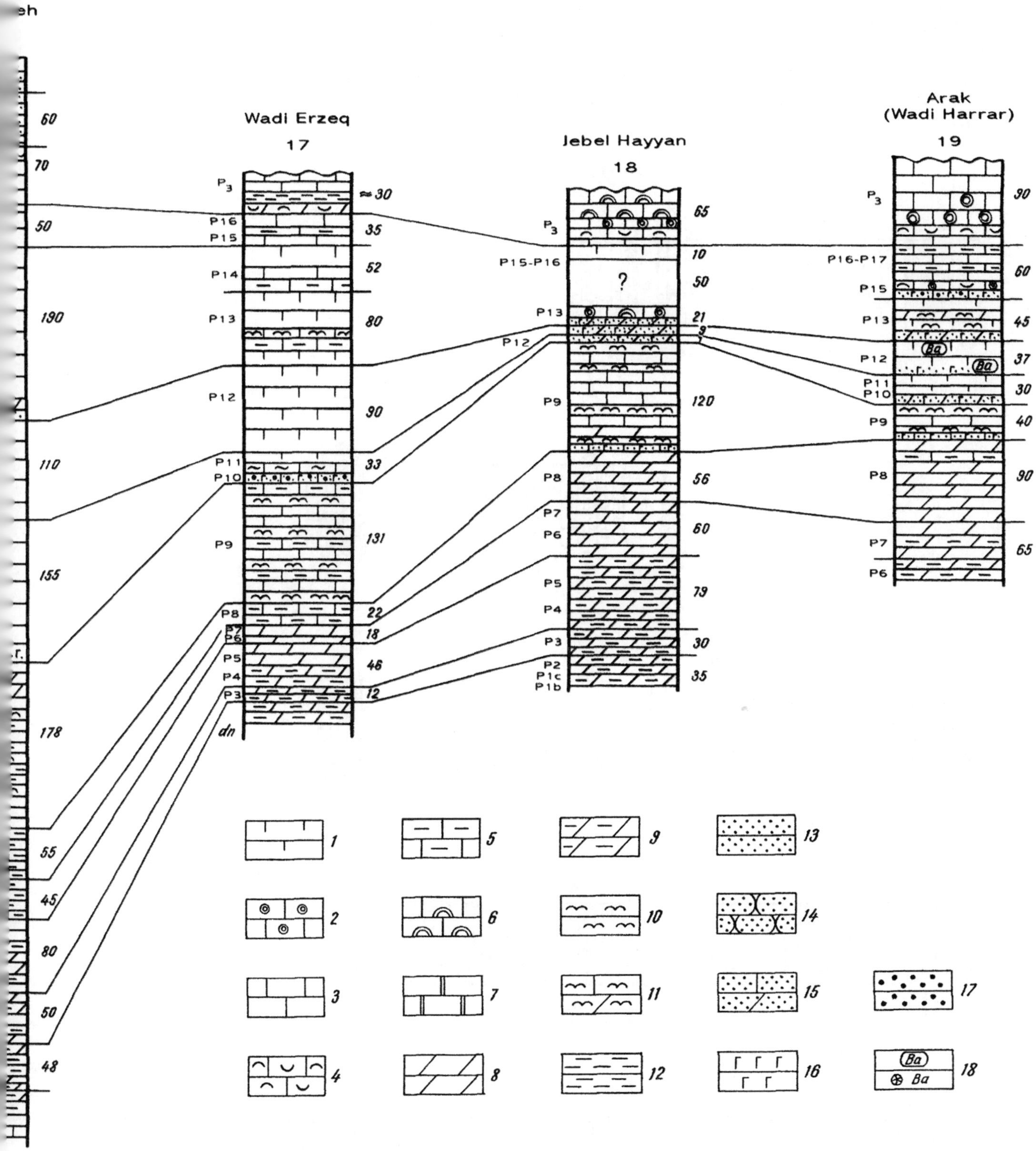

Wadi Erzeq
17
Jebel Hayyan
18
Arak
(Wadi Harrar)
19
P3
P16
P15
P14
P13
P12
P11
P10
P9
P8
P7
P6
P5
P4
P3
dn
P15-P16
P2
P1c
P1b
P16-P17
?
60
70
50
190
110
155
178
55
45
80
50
48
≈30
35
52
80
90
33
131
22
18
46
12
65
10
50
21
120
56
60
79
30
35
90
60
45
37
30
40
90
65
Ba

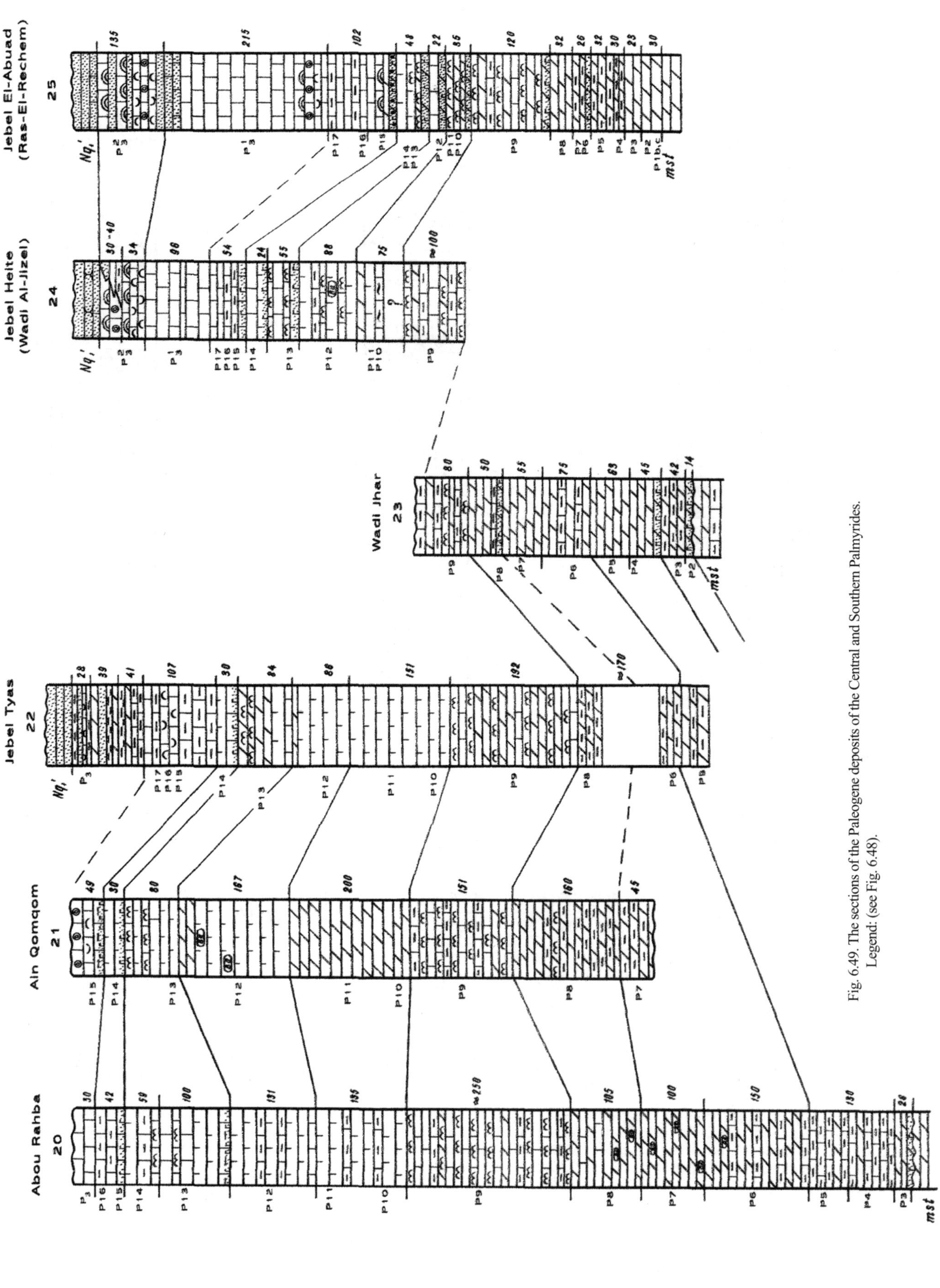

Fig. 6.49. The sections of the Paleogene deposits of the Central and Southern Palmyrides.
Legend: (see Fig. 6.48).

chalk and hard silicified limestones. The upper part of this stage (the *Truncorotaloides rohri* Zone) consists exclusively of thick-bedded soft chalk. Along the northern slope of Jebel Al-Bishri, chalky limestones are replaced by yellowish and greenish sandy marls and quartz sandstones with calcareous matrix, and then on the opposite slope return again to chalks. In many sections such as Jebel Hayyan, Wadi Harrar, Jebel El-Abuad, Wadi El-Bardeh, Jebel Aade, Jebel Al-Bishri, Bir Hasseiye, Abou Rahba and others, a glauconite-phosphorite bed is traced at the bottom of the Bartonian Stage.

In the complete sections of the Central Palmyrides (Wadi El-Bardeh, Jebel Tyas, Abou Rahba, Al-Qaryatein, Ain Qomqom, and Jebel El-Abuad) the thickness of the Bartonian Stage sediments varies from 90 to 190 m. They are much thinner (to 20-60 m) in sections of the southern and especially northern flanks of the Palmyrides where some foraminiferal zones are missing.

The Priabonian Stage (Upper Eocene) in the Palmyrides has a very complicated and changeable structure.

In the Central Palmyrides its lithological composition varies laterally. White and light-gray, soft chalky limestones with abundant planktonic foraminifera predominate in the sections at Wadi El-Bardeh, Al-Qaryatein, and Jebel Hayyan. In the Jebel El-Abuad section hard and massive recrystallized limestones and organogenic (algal, detrital, and nummulitic) limestones dominate; soft chalky limestones are present as rare interbeds. Near Arak village in Wadi Harrar the entire succession is represented by coarse-bedded detrital and algal limestones with a thin layer of soft limestones at its base. The same combination of various types of limestones was discovered in other sections in the Central Palmyrides at Ain Qomqom, Jebel Tyas, Jebel Heite, Wadi Lebtah, and Wadi Khabra.

The Upper Eocene was found in all the sections of the Southern Palmyrides *e.g.* Jebel Dmeir, Jebel Zubeideh, Jebel Kohle, and Jebel Aade as well as in the Northern Palmyrides at Jebel Khabli, Bir Hasseiye, and Jebel Dilaa. Aphanitic and fine-grained hard limestones are common here and soft chalky and clayey limestones are subordinate.

Complex facies substitutions were observed in the eastern part of the Palmyrides. On the northern slope of the Jebel Al-Bishri Uplift, deposits of this age are represented by a member of alternating clayey limestones, sandstones, and quartz sands. On the southern slope of this uplift the entire sequence is composed of chalky and clayey limestones with abundant planktonic foraminifera.

A very pronounced (2-3 m thick) layer of glauconite-phosphorite sediments is present at the base of the Upper Eocene and the underlying beds of the Middle Eocene are pierced by traces of mud-eaters with dissiminated glauconite. Nevertheless, the micropaleontological analysis shows that in many sections the Middle Eocene sediments continue into the Upper Eocene. In contrast, in some sections of the Central (Wadi Harrar), Northern (Jebel Dilaa, Bir Hasseiye, Tell Snime, Jebel Khabli) and Southern (Jebel Zubeideh) Palmyrides this contact is unconformable and one or two of the Middle Eocene foraminiferal zones are missing.

The thickness of the Upper Eocene varies from 40 to 110 m.

The Oligocene in the Palmyrides is subdivided into two lithological units (members).

The lower member is characterized by massive and hard, marble-like limestones. They include a number of varieties – aphanitic, massive, fine-grained, coarse- and unclearly-bedded with a block parting and vugged surface, clotted-algal, reef-like algal-coralline, detrital, foraminiferal (nummulitic at the base of this member, but mainly lepidocyclinid), and coquina with numerous shells of mollusks and sea urchins. These rocks form a steep escarpment and are well-exposed in many sections such as at Jebel Dmeir, Adra, Jebel Zubeideh, Jebel Kohle, Jebel Aade, Ain Qomqom, Jebel Tyas, Wadi Al-Jizel, Jebel El-Abuad, Hmair (Wadi Khabra), Al-Qaryatein (Wadi Lebtah), Wadi El-Bardeh, Jebel Hayyan, Jebel Antar, Arak, Bir Hasseiye, Jebel Dilaa, and many other.

Deposits of the upper member are distributed to the north of Palmyra at Jebel Abuad, Jebel Heite, and Jebel Tyas. This member consists of various limestones: detrital, foraminiferal (with abundant lepidocyclinids and operculinas), and clayey fine-grained, alternating with yellowish sandy limestones, greenish calcareous and non-calcareous clays, soft quartz calcareous sands and hard sandstones containing numerous shells of mollusks and sea urchins and algal balls. All these rocks are permeated by tubular tracks of burrowing organisms. At the top there is a 4-6 m thick unit of massive reef-like algal-coraline limestones with lenses of detrital and foraminiferal limestones and quartz sandstones. These lithologies replace each other both laterally and upwards in the section. Clayey intercalations within the upper member include assemblages of smaller benthic foraminifera and rare planktonic species of the Chattian Stage.

We suggest that in the area of Palmyra the lower member of massive limestones and the upper member of alternating limestones, sandstones and clays belong, correspondingly, to the Lower and Upper Oligocene.

In the western part of the Palmyrides, between Dmeir and Al-Qaryatein, the upper member was not observed, although layers of greenish clays and marls are present among massive limestones in the area of Adra. The Oligocene sequences terminate here as hard aphanitic non-fossiliferous limestones containing interbeds of sandy limestones and sandstones. Numerous vertical burrows of irregular form filled with siliceous concretions produce an unusual structure in these rocks. Evidently, these burrows were left by roots of plants growing in the coastal zone of the marine basin. The unfossiliferous limestones are covered by Lower Miocene quartz sands. It seems likely that in the western part of the Palmyrides the upper member of limestones, clays, and sands is replaced by monotonous massive limestones.

The dependence of the Oligocene facies upon the underwater paleorelief is demonstrated in some sections located on the anticlinal fold of Jebel Heite near Palmyra. The axial part of this structure is occupied by massive and algal solid biohermal limestones. Downdip on the southern limb of this fold they are replaced by various bedded limestones – detrital, foraminiferal,

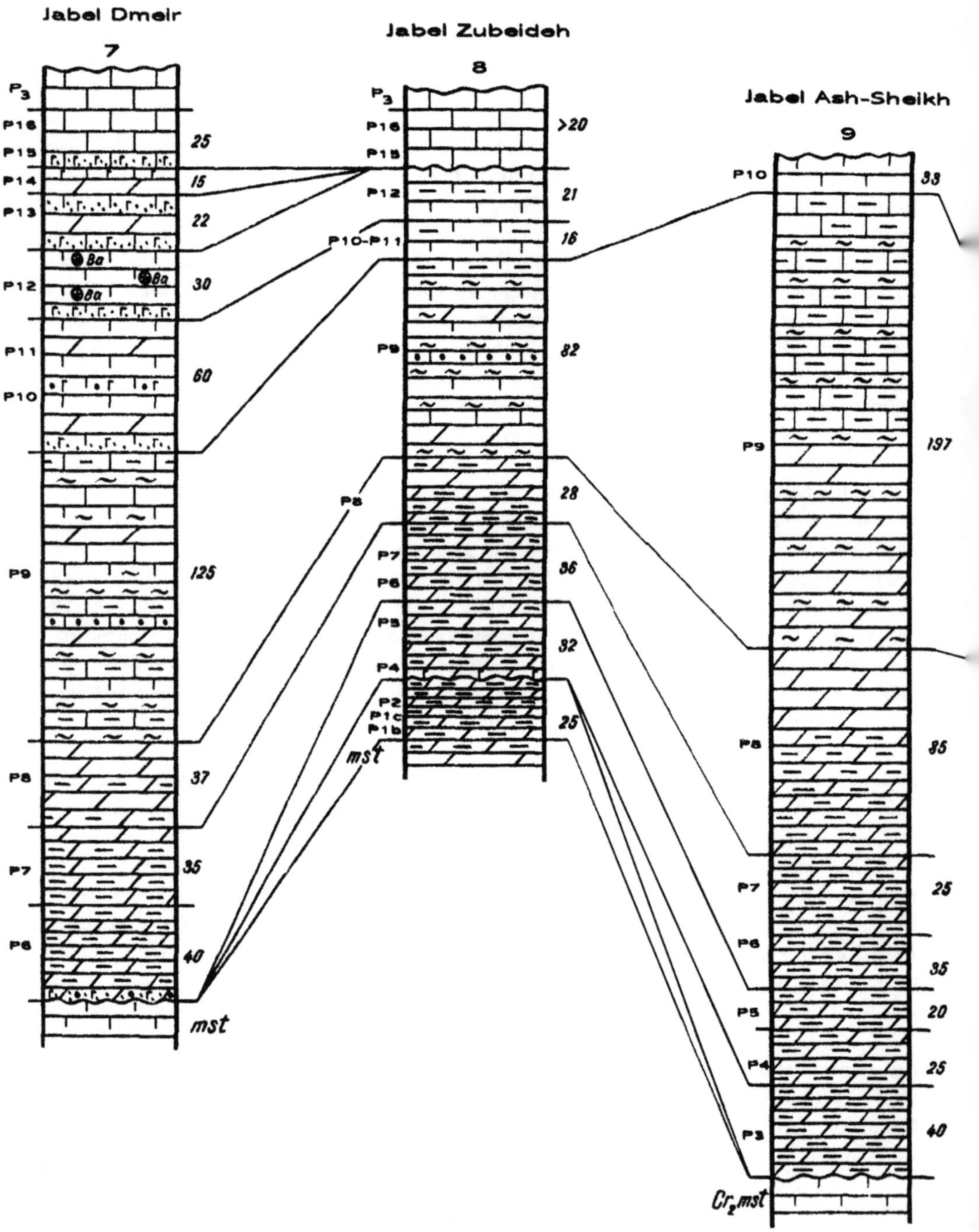
Jabel Dmeir
7
P3
P16
25
P15
P14
15
P13
22
Ba
P12
30
P11
60
P10
P9
125
P8
37
P7
35
P6
40
mst
Jabel Zubeideh
8
P3
P16
>20
P15
P12
21
P10-P11
16
P9
82
P8
28
P7
P6
36
P5
32
P4
P2
P1c
P1b
25
mst
Jabel Ash-Sheikh
9
P10
33
P9
197
P8
85
P7
25
P6
35
P5
20
P4
25
P3
40
Cr2mst

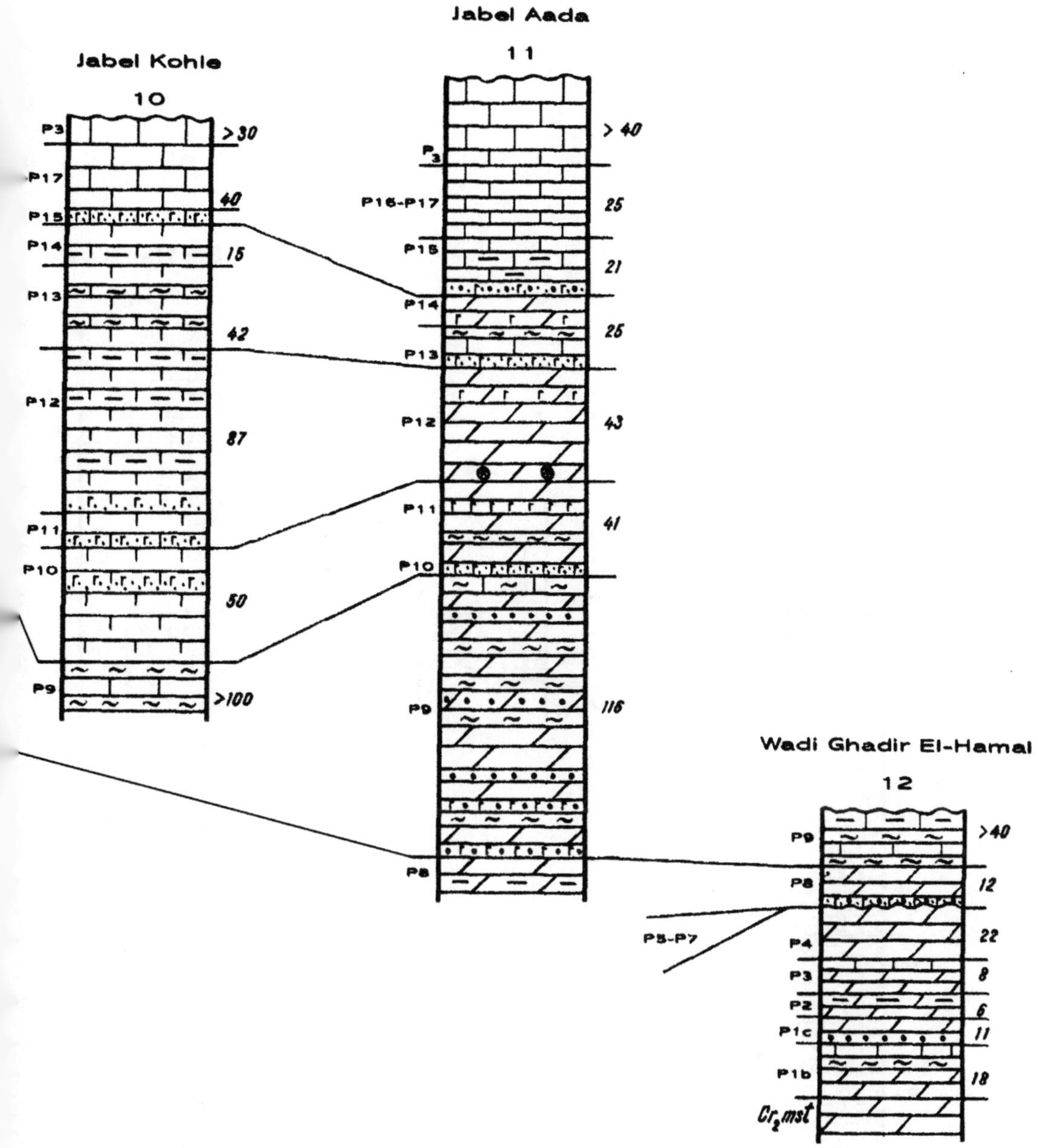

Fig. 6.50. The sections of the Paleogene deposits of the Central and Southern Palmyrides.
Legend: (see Fig. 6.48).

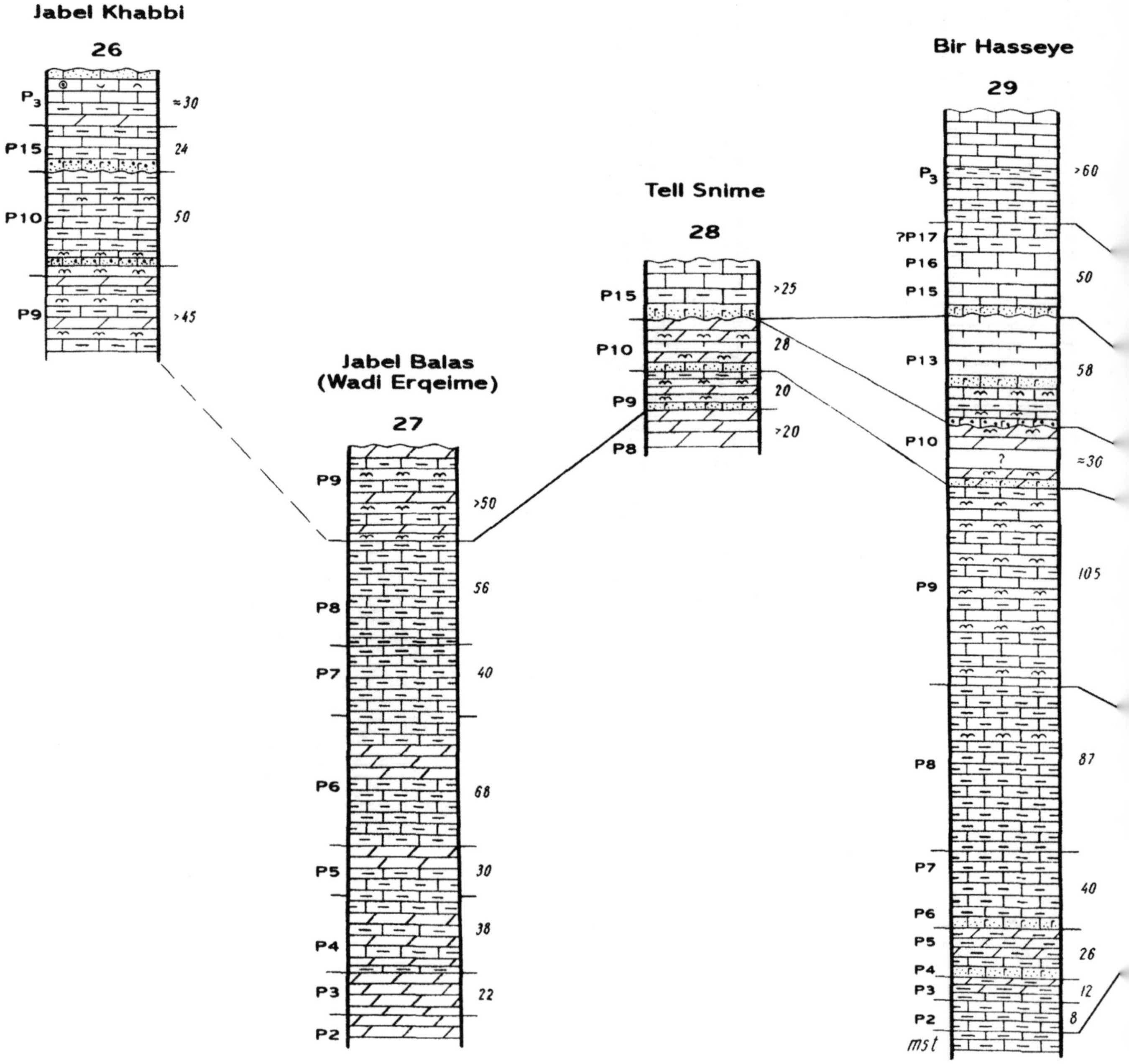

Fig. 6.51. The sections of the Paleogene deposits of the Northern Palmyrides. Legend: (see Fig. 6.48).

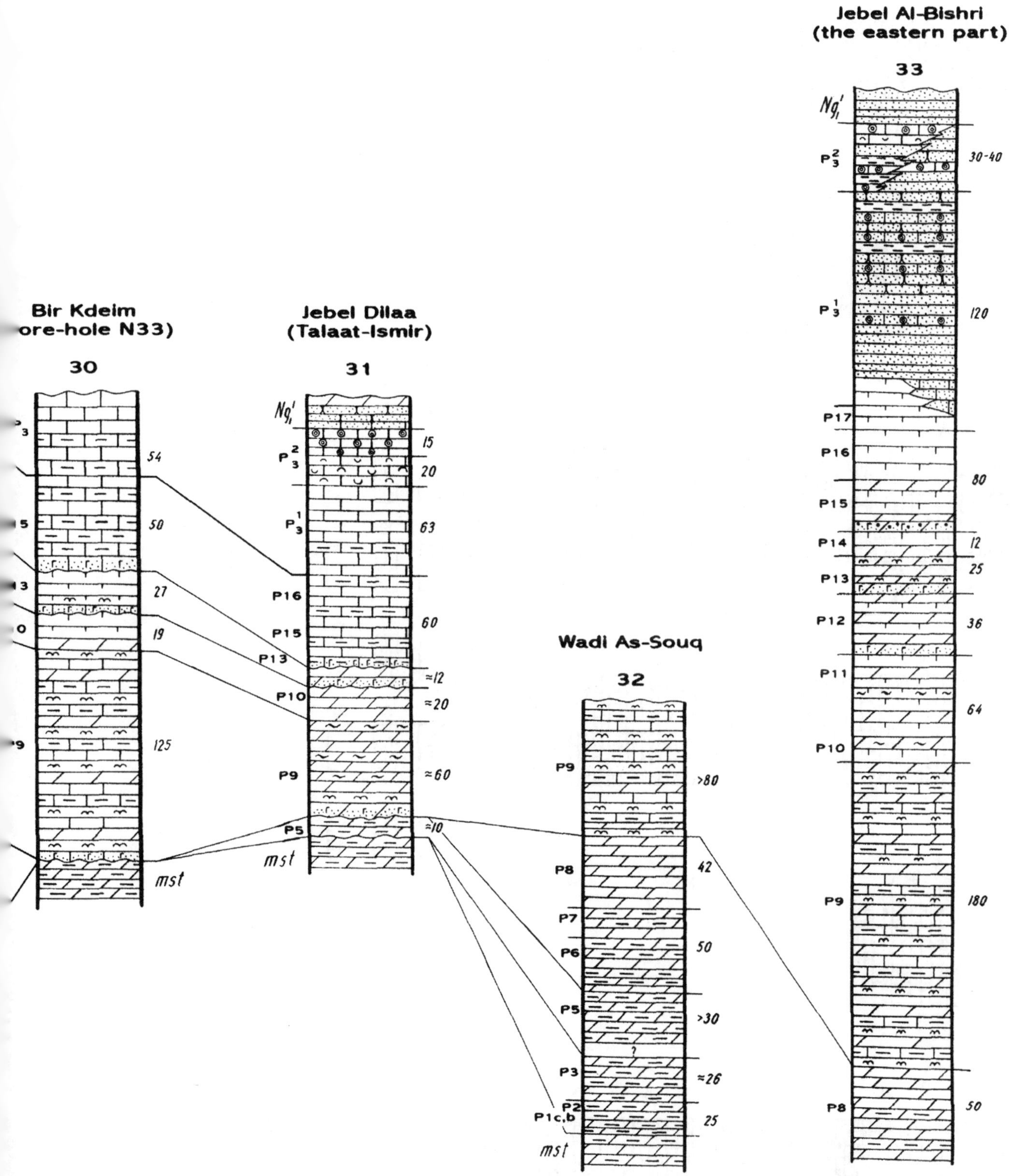
Bir Kdeim
ore-hole N33)
30
54
50
27
19
125
mst
Jebel Dilaa
(Talaat-Ismir)
31
15
20
63
P16
60
P15
P13
≈12
P10
≈20
P9
≈60
P5
≈10
mst
Wadi As-Souq
32
P9
>80
P8
42
P7
P6
50
P5
>30
P3
≈26
P2
P1c,b
25
mst
Jebel Al-Bishri
(the eastern part)
33
30-40
120
P17
P16
80
P15
P14
12
P13
25
P12
36
P11
64
P10
P9
180
P8
50

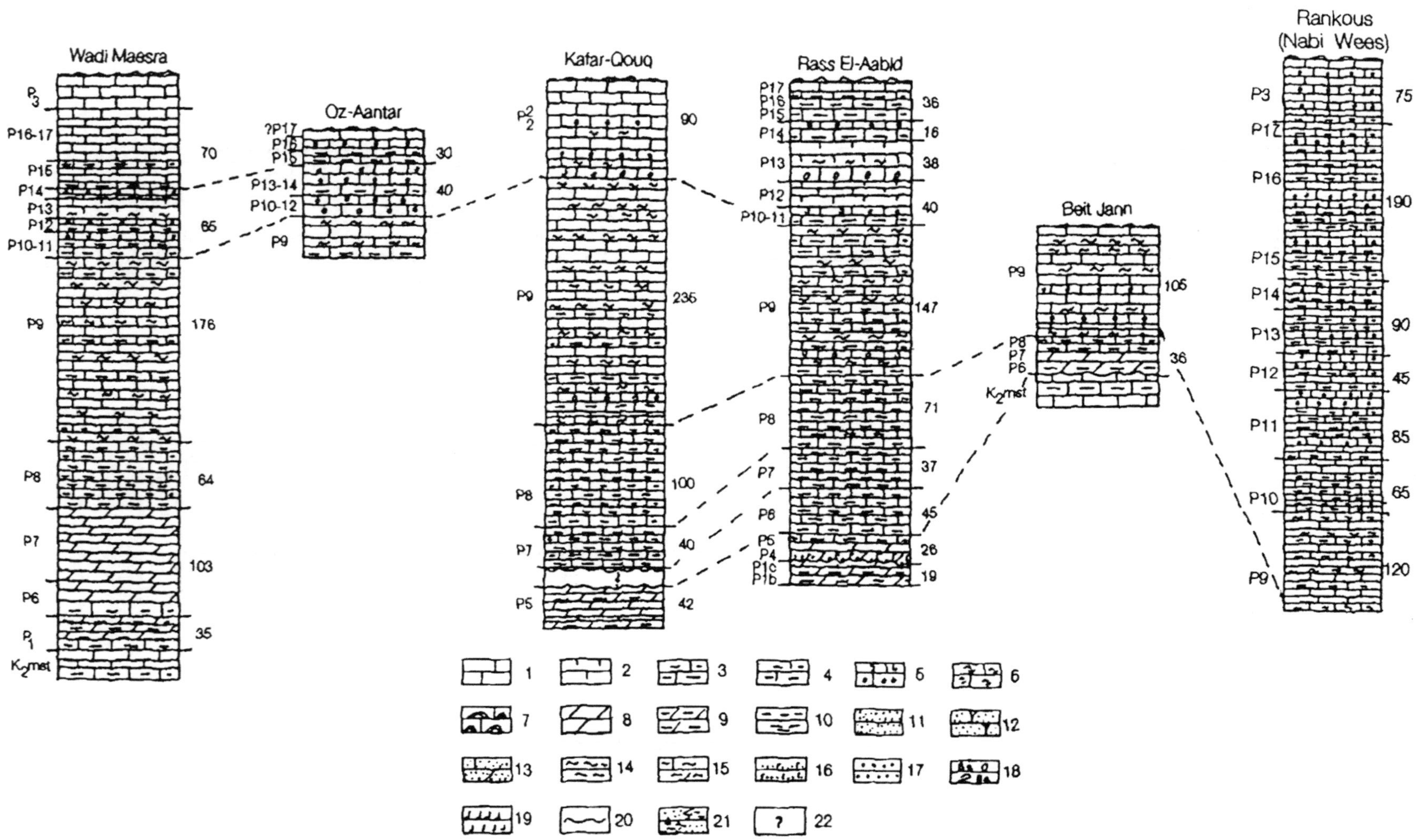

Fig. 6.52. The sections of the Paleogene deposits of the Anti-Lebanon.

Legend: 1 – fine-grained limestones; 2 – chalky limestones; 3 – clayey limestones; 4 –chalky clayey limestones; 5 - foraminiferal (nummulitic, alveolinid, lepidocyclinid) limestones; 6 – organogenic-detrital limestones; 7 – algal and corallime limestones; 8 – marls; 9 – clayey marls and calcareous clays; 10 – weakly calcareous and noncalcareous clays; 11 – sands; 12 – sandstones; 13 – rocks with sand; 14 – cherts; 15 – silicified rocks; 16 – glauconites; 17 – phosphorites; 18 – barite concretions; 19 – effusive rocks; 20 – eroded surfaces; 21 – facies replacements; 22 – absence of exposures.

laminated algal and oyster beds. Further downdip terrigenous and mixed deposits appear. Here the limestones interfinger with greenish, gray and brown marls and calcareous clays, light-blue and red non-carbonate clays with ferruginous concretions and fine-grained greenish sandstones. Such a distribution of sediments on the crest and flank of the structure may also point to growth of this fold during the process of sedimentation.

The Anti-Lebanon

As is the case in the Palmyrides, the Paleogene of the Anti-Lebanon is represented by a complete succession beginning from the Danian Stage and continuing up to the Oligocene (**Figs. 6.52, 6.53,** and **6.54**).

The full sequence of the Danian Stage was identified in the Jebel Maaloula section near the village of Qaldoun, where it unconformably covers Maastrichtian limestones. The stage starts with gray-greenish and yellowish-gray, soft, thin-laminated to foliated calcareous clays of the *Turborotalia pseudobulloides* Zone with a 30 cm thick glauconite bed at its base. Gray marls occur among these clays in the *Turborotalia trinidadensis* and *Acarinina uncinata* Zones. The thickness of the Danian Stage does not exceed 18 m. In other sections the Danian Stage is incomplete. Near the village of Doummar the Danian Stage is composed of a 7 m thick member of white and light-gray marls of the *Turborotalia pseudobulloides* Zone; there are no sediments of the two overlying Danian Stage zones. In contrast, in the vicinity of Deir Qanoun village the Lower Danian sediments are absent and the Maastrichtian rocks are unconformable overlain by yellow-greenish thin-laminated calcareous clays of the *Acarinina uncinata* Zone. In the sections at Maaraba, Beit Jann, Houreira, and Rass El-Marra the Danian Stage is definitely missing and Cretaceous and Paleocene layers are separated by an obvious gap. Near the villages of Mnin, Hafir At-Tahta, and Nabi Wees soft clays and marls of this transitional interval are very poorly exposed and the character of the Cretaceous/Paleogene boundary is unclear.

Deposits of the Selandian Stage with rich planktonic foraminifera of the *Morozovella angulata* Zone s.str. and the *Igorina pusilla* Zone are very well-represented in Jebel Maaloula as a member of foliated soft clayey marls and greenish and yellowish calcareous clays 32 m in thickness. Sediments of similar lithology belong to the Selandian Stage to the west of Damascus (near Deir Qanoun village) but here their thickness decreases to 8 m. Deposits of this stage are apparently absent in the Rass El-Marra, Maaraba, Doummar, and Beit Jann sections

Sediments of the Thanetian Stage were recovered in all sections of the Anti-Lebanon. The only exception is the Beit Jann section. In the Jebel Maaloula, near the village of Qaldoun, this stage is made up of a 120 m thick pile of soft light-gray limestones and clayey, slightly dolomitized limestones with abundant benthic foraminifera. Upward in the section, they pass into greenish and dark-gray calcareous clays and marls with interbeds of limestones containing plentiful planktonic foraminifera. The Thanetian Stage preserves its lithological features along the whole eastern slope of the Anti-Lebanon but the thickness diminishes to 40-80 m. Closer to the axial part of the Anti-Lebanon it is reduced to 15-25 m near the villages of Deir Qanoun and Rass El-Marra.

The Lower Ypresian deposits (the *Morozovella subbotinae* and *Morozovella formosa* Zones) in Jebel Maaloula are formed by soft gray and greenish-gray marls with subordinate intercalations of lighter clayey limestones. To the southwest of this area, near Maaraba, Doummar, and Rass El-Aabid, light-gray laminated marls with small barite nodules are more common. The largest thickness of sediments was measured in Jebel Maaloula – 176 m. In other sections it ranges from 100 to 125 m but in the axial part of the Anti-Lebanon and in the Hermon Massif at the villages of Deir Qanoun, Houreira, and Beit Jann it drops to 40-60 m.

The Upper Ypresian deposits, as in the Palmyrides, are subdivided into two rock units. The lower one (the *Morozovella aragonensis* Zone) consists of light-gray and yellowish, soft pelitic limestones, clayey limestones and marls with fish scales and rare and small barite nodules. In sections along the eastern slope of the Anti-Lebanon at Doummar, Maaraba, Jebel Maaloula, Wadi Maesra, Kafar Qouq, and Rass El-Aabud, its thickness ranges from 64 to 100 m and the deposits conformably overlie the Lower Ypresian rocks. Near the central part of the Anti-Lebanon (Beit Jann, Rass El-Marra) the thickness decreases to 8-20 m and limestones of the *Morozovella aragonensis* Zone with a glauconite horizon at their base rest on the washed out surface of the underlying rocks (the Bissan gorge, Rass El-Marra).

In the sections along the eastern slope of the Anti-Lebanon (Doummar, Maaraba, Jebel Maaloula, Rass El-Marra, Mnin, Halboun, Rankous, Wadi Maesra, Oz-Aantar, Kafer Qouq, and Beit Jann) the upper unit of the Upper Ypresian (the *Acarinina pentacamerata* Zone) consists of alternating white and light-gray soft clayey limestones, marls, hard fine-grained limestones and layers and lenses of black and brown cherts (flints). Interbeds of chalky limestones are comparatively rare. The beds of calcareous rocks are 1 to 5m thick; the bands of siliceous rocks vary from 2-3 cm to 30-40 cm in thickness. In some sections (Doummar, Maaraba, Beit Jann, and Rass El-Aabid) at the base of this unit there is a 5 to 15 m thick pile of intraformational syngenetic conglomerates and nummulitic limestones. The upper part of this formation is occupied by an up to 20-30 m thick pile of frequent alternation of limestones with flint layers up to 60 cm thick.

To the west towards the axial part of the Anti-Lebanon, the soft pelagic limestones gradually disappear as the hard fine-grained, aphanitic and detrital limestones increase in importance (*e.g.* Halboun, Kafar Qouk, and Rass El-Marra). At the core of the Anti-Lebanon the entire succession is represented by massive hard limestones and the layering of the siliceous rocks is replaced by lenses and nodules (Beit Jann, the Bissan gorge).

During the interval of the *Acarinina pentacamerata* Zone the rates of sedimentation were very significant. A maximum thickness of 315 m was recorded in the Jebel Maaloula section. In the Maaraba, Halboun, and Kafar Qouk sections the thickness is 230-250 m and only near the axial part of the Anti-Lebanon at the village of Houreira does it decrease to 50 m.

The Lutetian Stage on the eastern slope of the Anti-Lebanon

Rates of sedimentation during Lutetian time continued to be as high as those in the late Early Eocene. The thickness of the Lutetian Stage in the pelagic sediment facies reaches 460 m at Jebel Maaloula. In the section of Halboun the thickness of the shallow-water limestone facies is similar at 361 m. In other sections such as Maaraba, Mnin, Rankous, Rass El-Marra, and Doummar, the thickness ranges between 150-200 m. The thickness is reduced to 40-80 m only in those sections located not far away from the axial part of the Anti-Lebanon, at Houreira, Beit Jann, and Kafar Qouq.

The deposits of the Bartonian Stage differ from those of the Lutetian Stage by their wider variety of facies types that reflect the tectonic movements of the late Middle Eocene which created diverse paleogeographic settings. Three facies types can be singled out in the Anti-Lebanon.

The first facies type consists of monotonous chalky limestones – white and light-gray, soft, unclearly-bedded or thick-bedded, with rare intercalations of marls and hard fine-grained limestones and rich assemblages of planktonic foraminifera. As in the Palmyrides, the chalk of the *Orbulinoides beckmanni* Zone contains thin bands (usually 2-5 cm, sometimes up to 10 cm thick) of gray, aphanitic, hard silicified limestones. Exposures of such sediments are located in the vicinity of Maaraba, Maaloula, Doummar, and Hafir At-Tahta villages.

The second facies type is characterized by the alternation of chalky limestones, which are full of planktonic foraminifera, with various shallow-water limestones – algal, nummulitic, detrital and hard fine-grained. Exposures with such rocks sequences are found in the vicinity of Rass El-Marra, Mnin, Halboun, Rankous, Rass El-Aabid, and Wadi Maesra.

The third facies type is common to the axial part of the Anti-Lebanon and includes shallow-water algal, nummulitic, and massive fine-grained limestones. It can be observed near the villages of Houreira and Ain Fije as well as in the Bissan gorge.

In general, going from east to the west, *viz* from the marginal folds of the Anti-Lebanon to the crest of this uplift, pelagic soft chalk is replaced by shallow-water biogenic limestones. Such substitutions can sometimes be traced within a single SW-NE trending anticlinal structure. For example, the eastern flanks of the Jebel Qasyoun, Jebel Suekai, and Jebel El-Hassiya folds are covered by pelagic deposits, while the Bartonian deposits of the western flanks consist of alternating chalk and nummulitic limestones. Near the border with Lebanon shallow-water limestones strongly predominate.

Complicated facies changes are responsible for the variations in the thickness of the Bartonian deposits. The maximum thicknesses (130-140 m) were recorded in the Halboun and Jebel Maaloula sections. Thicknesses of 50 to 80 m were identified in the sections at Maaraba, Mnin, Rass El-Marra, Rass El-Aabid, and Wadi Maesra. The minimum thickness of 20 m was measured at Jebel Aantar.

In the Late Eocene the same three facies types continued to exist in the Anti-Lebanon but their spatial distribution changed as the area occupied by pelagic sediments was reduced and the area of shallow-water limestones increased.

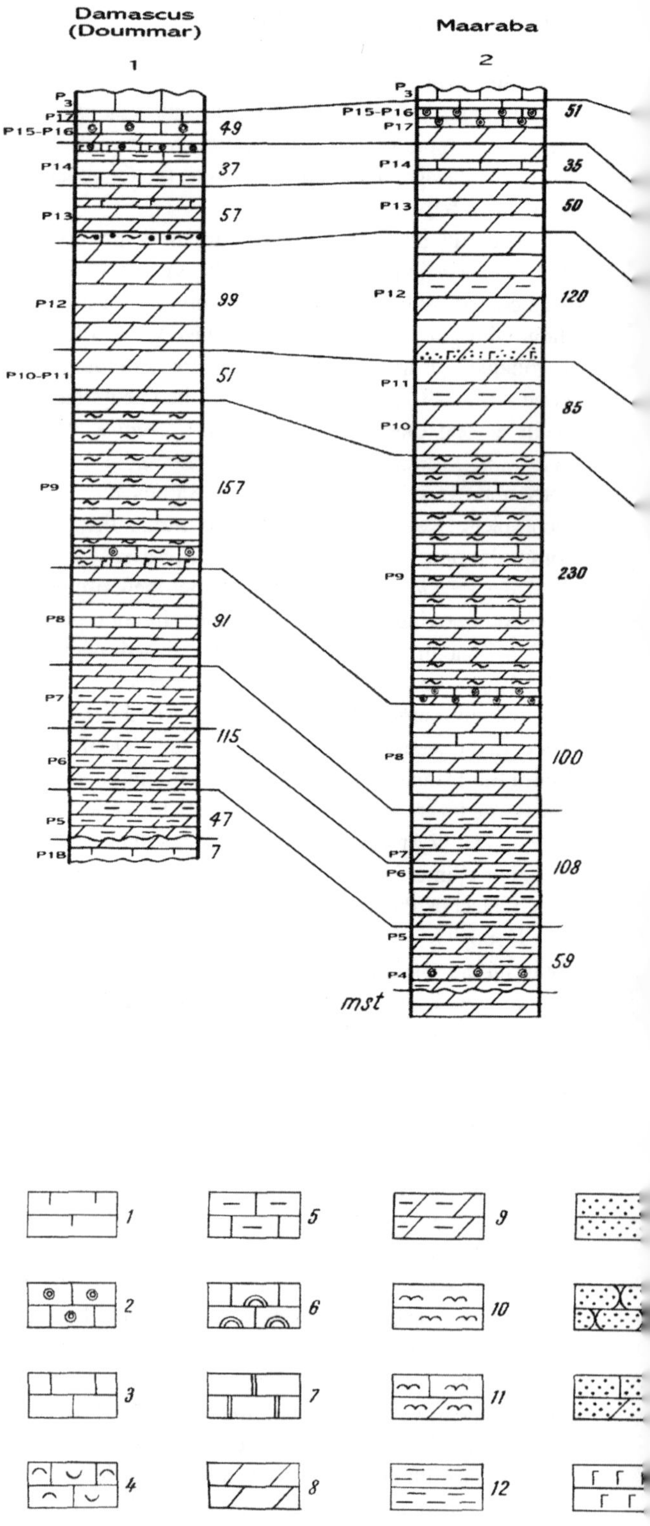

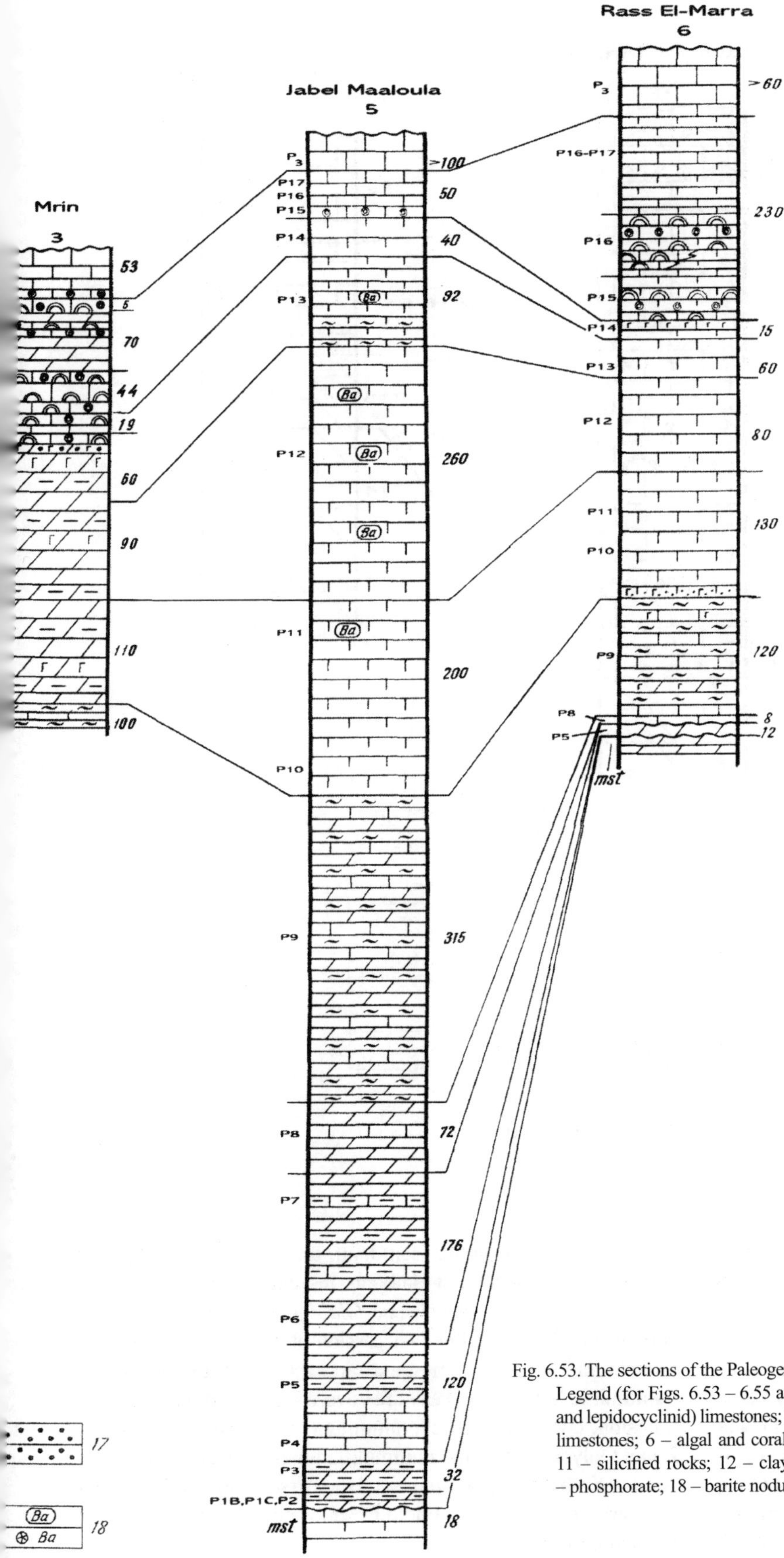

Fig. 6.53. The sections of the Paleogene deposits of the Anti-Lebanon.

Legend (for Figs. 6.53 – 6.55 and Figs 6.57 –6.65): 1 – chalky limestones; 2 – foraminiferal (nummulitic and lepidocyclinid) limestones; 3 – fine-grained limestones; 4 – organogenic-detrital limestones; 5 – clayey limestones; 6 – algal and coralline limestones; 7 – dolomites; 8 – marls; 9 – clayey marls; 10 – flints; 11 – silicified rocks; 12 – clays; 13 – sands; 14 – sandstones; 15 – sandy rocks; 16 – glauconite; 17 – phosphorate; 18 – barite nodules.

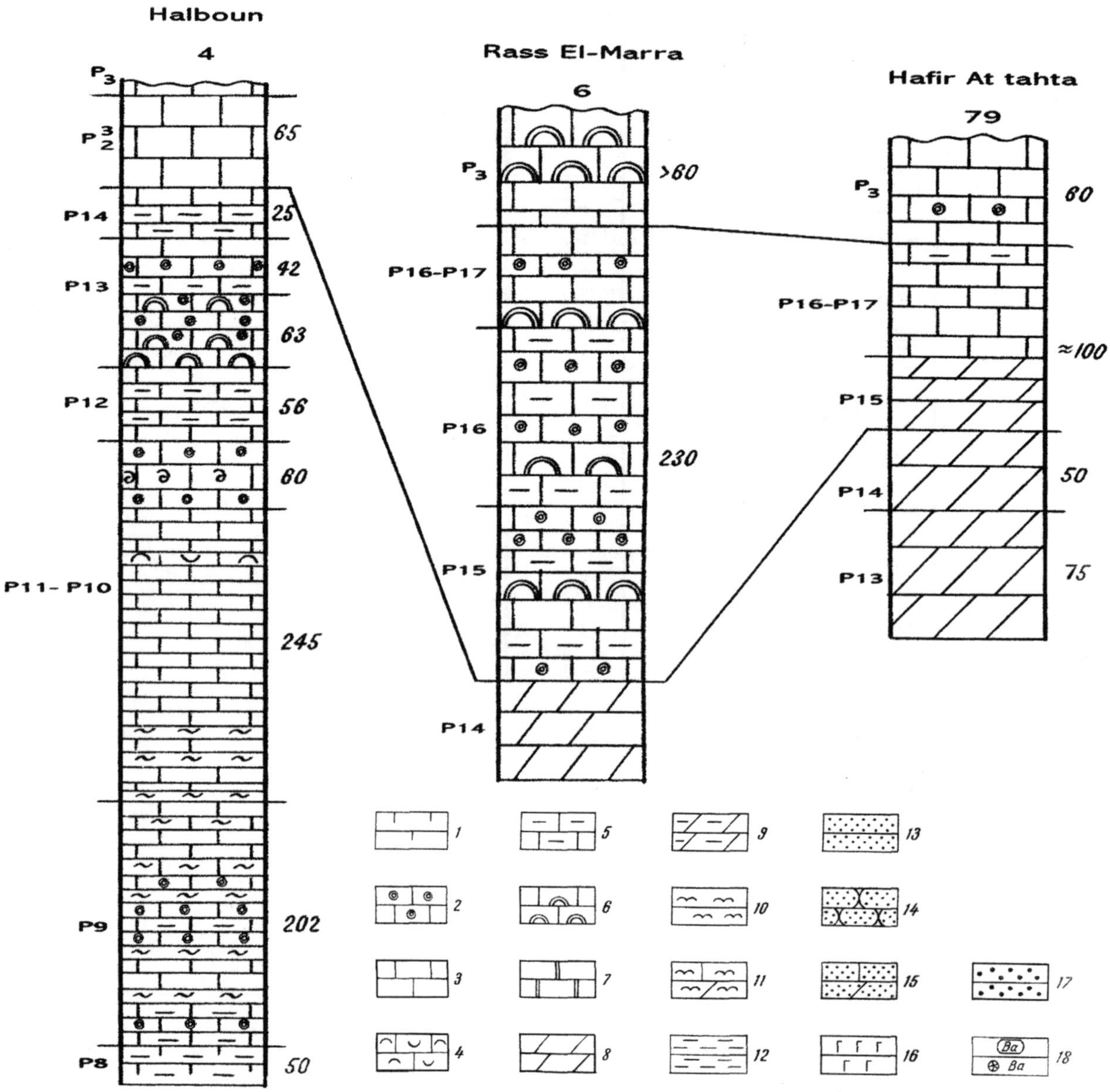

Fig. 6.54. The sections of the Paleogene deposits of the Anti-Lebanon.
Legend: (see Fig. 6.53).

includes a monotonous sequence of white and light-gray, soft, thick-bedded chalky limestones with spherical or ovoidal barite concretions and interbeds of harder fine-grained limestones. The typical sections with such lithology are located near the villages of Maaraba, Maaloula, Rass El-Marra, and Rankous. Near the village of Doummar these rocks are impregnated by bituminous matter and acquire a dark color. To the west pelagic sediments grade into shallow-water deposits. To begin, layers of nummulitic and hard fine-grained limestones appear within chalky layers at Kafar Qouq, Deir Qanoun, Wadi Maesra, Rass El-Aabid, Sadyr Mkin, and Rankous. Further west, the pelagic sediments partly or completely disappear as at Halboun, Beit Jann, and the Bissan gorge. For example, in the section at Halboun, the interval corresponding to the *Hantkenina aragonensis* and *Globigerapsis kugleri* Zones, is composed of a 245 m thick succession of hard, thin-bedded, fine-grained and aphanitic limestones and a 60 m thick sequence of nummulitic limestones. They are covered by a 56 m unit of clayey limestones with planktonic foraminifera of the *Morozovella lehneri* Zone. Only aphanitic and nummulitic limestones occur in the Oz-Aantar, Kafar Qouq, Beit Jann, and the Bissan gorge sections. In all sections the Lower and Middle Eocene deposits are joined by the gradual transition.

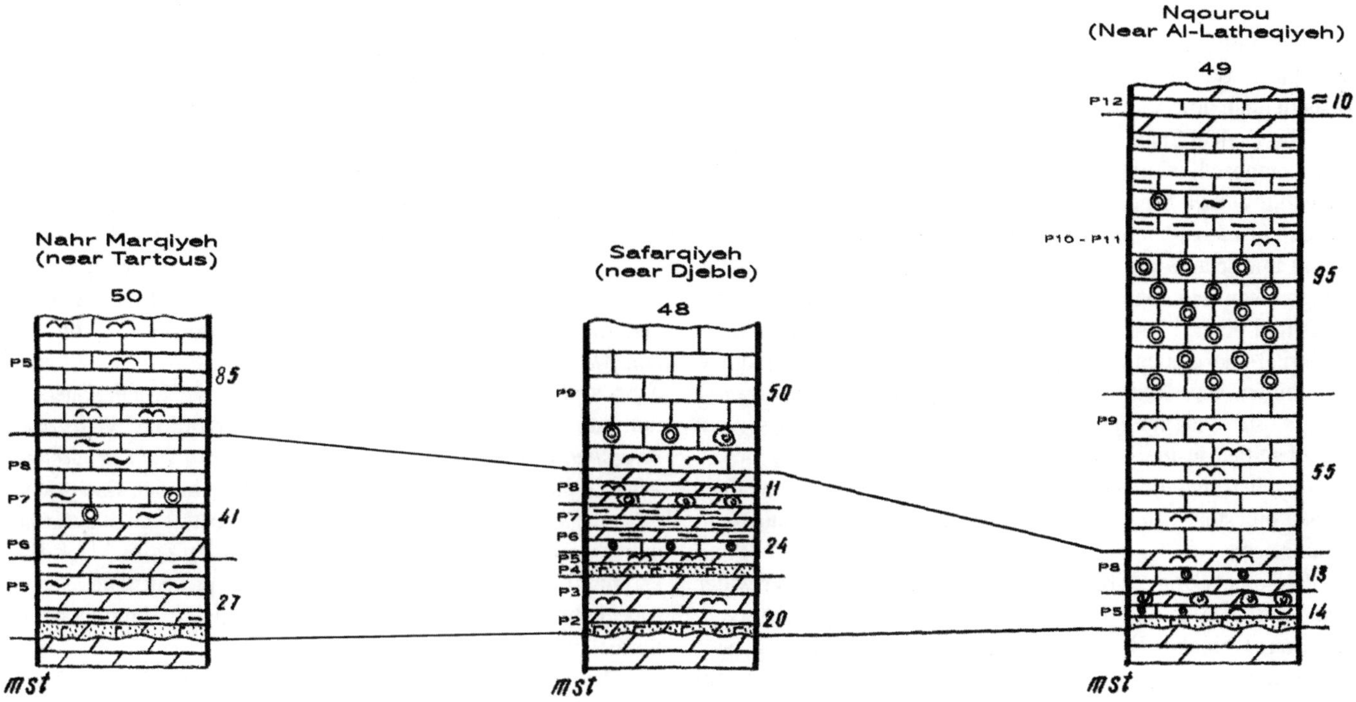

Fig. 6. 55. The sections of the Paleogene deposits on the western slope of Jebel An-Nusseiriyeh.
Legend: (see Fig. 6.53).

The first type – white and light-gray, soft chalky and clayey limestones with abundant planktonic foraminifera dominated in the areas of Doummar, Maaraba, Hafir At-Tahta and Maaloula. At the base of this formation there is a member with syngenetic conglomerates and breccia with traces of submarine slumping, algal crusts, and lenses of nummulitic limestones. This member is especially well-exposed in the Doummar and Maaloula. sections

The second type is represented by alternation of rocks with different paleoecological features – chalky limestones with planktonic foraminifera, and various shallow-water limestones such as nummulitic, algal-coralline, detrital, oyster coquina and fine-grained and aphanitic limestones. The most indicative sections are located at Mnin, Rass El-Marra, Rankous, Wadi Maesra, Jabal Oz-Aantar, and Rass El-Aabid.

The third Upper Eocene facies type consists of strongly dominating massive and very hard, coarse-bedded, fine-grained and aphanitic limestones with thin lenses of nummulitic and detritic limestones.

Evidently, various local factors influenced the rate of accumulation of biogenic calcareous matter, because the thicknesses of the Upper Eocene deposits in the Anti-Lebanon are highly variable. The maximum thicknesses in the sections of Rass El-Marra and Rankous reach 190-230 m. Usually they are between 50 and 120 m as occur in the sections at Maaraba, Mnin, Halboun, Doummar, Maaloula, and Wadi Maesra. The minimal thicknesses of around 30-40 m were observed in the sections at Jebel Az-Aantar and Rass El-Aabid.

The Oligocene deposits of the Anti-Lebanon are very monotonous. They belong to the fifth formation made up of massive, very hard, marble-like overcrystallized limestones – coarse- and thick-bedded, aphanitic, with a block parting and vuggy surface. These limestones form impressive steep escarpments along the trend of the Anti-Lebanon folds. At the base of this formation there is a member of clearly-bedded hard aphanitic limestones. In its lower part are lenses of nummulitic limestones which are present near the villages of Hafir At-Tahta, Rass El-Marra, Rankous, and Maaloula, but mainly the limestones are barren of fossils. They were destroyed by secondary processes of dissolution and recrystallization.

The maximum thickness of the Oligocene limestones reaches 170 m. Visible thicknesses are usually much less due to the erosional processes at the Oligocene/Neogene boundary, where a considerable part of the Oligocene deposits was washed out. Near the village of Doummar, not far from the cement factory, one can see that the Miocene continental deposits completely displace the Oligocene limestones and transgressively overlie the Upper Eocene chalk. Thus, the thickness of the Oligocene deposits drops here to zero.

Jebel An-Nusseiriyeh

The Paleogene sequences of this uplifted area are incomplete, with some pronounced hiatuses and with reduced thickness (**Fig.**

6.55). The most significant gap was discovered in the vicinity of the village of Tibat, not far from the town of Jisr Ash-Shaghour. Here the Middle Eocene nummulitic limestones rest directly on Maastrichtian limestones with numerous orbitoids.

The Danian Stage was found near the village of Safarqiyeh. It is composed of gray-greenish soft marls with disseminated glauconite at their base. The marls contain planktonic foraminifera of the *Acarinina uncinata* Zone and their thickness is only 3-4 m.

In this section the Selandian Stage follows conformably the Danian layers and consists of a 15 m thick unit of black and dark-green thin-bedded calcareous clays and marls with some bands of detrital and clotted algal limestones. According to planktonic foraminifera, these sediments belong to the *Morozovella angulata* Zone s. str. and the *Igorina pusilla* Zone. These zones were also identified in the section at Sakhabe, located to the south of Safarqiyeh. Here sediments are more calcareous – light-green, thin-bedded limestones and marls with rich assemblages of planktonic foraminifera.

Sediments of the Thanetian Stage are distributed over wider areas and seen in the sections at Nahr Marqiyeh, Safarqiyeh, Nqourou, Qerdaha, Brianis, Ain El-Sheqaq, Wadi Gamiq, and Sakhabe. They consist of gray, light-gray and greenish marls and clayey limestones with numerous beds of organogenic-detrital, nummulitic-discocyclinid, and clotted algal limestones. There is usually a glauconite bed at the bottom of the sections. Some layers contain small nodules of flint. Only in the sections at Safarqiyeh and Sakhabe do the Thanetian sediments rest conformable on the Selandian clays and marls. In other sections they transgressively overlie the Maastrichtian limestones. In the section at Ain El-Shiqaq the Thanetian Stage begins with sediments of the *Planorotalites pseudomenardii* Zone, and in sections at Nahr Marqiyeh, Brianis, and Nqourou it begins with sediments containing microfauna of the *Morozovella velascoensis* Zone. Their thicknesses are insignificant. Near Qerdaha and Brianis they are 3-5 m increasing up to 14-27 m in the Nqourou and Nahr Marqiyeh sections.

The Lower Eocene deposits are well-represented on the western slope of Jebel An-Nusseiriyeh. As in the Anti-Lebanon and the Palmyrides, the lithological composition of their lower and upper parts is different.

The lower part (the *Morozovella subbotinae, Morozovella formosa*, and *Morozovella aragonensis* Zones) is made up predominantly of soft marls and clayey limestones but displays some lateral changes.

To the south of the town of Banyas, in the area of the Nahr Marqiyeh River, white and light-gray, soft, well-bedded marls and clayey limestones of the *Morozovella subboinae* and *Morozovella formosa* Zones with a basal glauconite horizon conformably overlie the Upper Paleocene rocks. There are some interbeds of nummulitic limestones with *Nummulites planulatus* within the *Morozovella formosa* Zone rocks. The thickness is about 20 m. This unit is followed by a 21 m thick member of more compact fine-grained and clayey limestones with interbeds of nummulitic limestones and rare flinty nodules. According to the microfauna, it belongs to the *Morozovella aragonensis* Zone.

In the area of Safarqiyeh, Qerdaha, and Brianis (Wadi Yarte) to the southeast of Al-Latheqiyeh, the transition from Paleocene rocks to those of the Lower Eocene is also gradual. In the sections mentioned the *Morozovella subbotinae* and *Morozovella formosa* Zones are formed of gray and light-gray, soft marls and clayey limestones with subordinate intercalations of detrital, nummulitic-alveolinid, and clotted algal limestones. Another form of contact is observed in the Nqourou section where marls of the *Morozovella velascoensis* Zone (Upper Paleocene) are unconformably covered by the *Morozovella aragonensis* Zone sediments (Lower Eocene), and the Lower Ypresian deposits (the *Morozovella subbotinae* and *Morozovella formosa* Zones) are missing. The same type of contact is observed in Wadi Gamiq, where the Upper Paleocene nummulitic limestones are overlain by marls of the *Morozovella aragonensis* Zone.The deposits of the *Morozovella aragonensis* Zone in all the sections are characterized by harder, fine-grained, gray and yellowish limestones with subordinate beds of soft marls and clayey limestones. Among them, thin lenses of black flints and interbeds of nummulitic limestones are present. The latter are especially common in the section at Safarqiyeh. The thickness of these three zones varies from 35 to 50 m.

The upper part of the Lower Eocene (the *Acarinina pentacamerata* Zone) is represented by the second formation, made up of hard, thick-bedded limestones – aphanitic, fine-grained, detrital, detrital-algal, with rare intercalations of nummulitic and clayey varieties. These calcareous rocks contain numerous lenses up to 50-70 cm long and 10-20 cm thick of black and brown cherts. Sometimes the cherts form nodules with very irregular outlines. At the base of this formation are lenses of intraformational (syngenetic) conglomerates. Visible thicknesses in the sections at Nahr Marqiyeh, Safarqiyeh, Brianis, and Sakhabe are up to 50-80 m. These massive rocks cap the elevated landscape of the An-Nusseiriyeh western slope. Certainly, the lithology of the calcareous-siliceous formation of Jebel An-Nusseiriyeh is rather similar to that of the Palmyrides and the Anti-Lebanon but the thick flint layers are absent and here the siliceous matter forms lenses, nodules or bands of nodules.

The Middle Eocene deposits are exposed both on the western and eastern slopes of Jebel An-Nusseiriyeh. Their distribution is spotty because they were strongly eroded and only preserved as outliers. The shallow-water limestones predominate.

In the outcrops near Nqourou the Lutetian deposits conformably overlie the Lower Eocene rocks. They begin with massive, thick-bedded, biogenic-detrital and nummulitic limestones with *Nummulites laevigatus* which are covered by a member of alternating hard, fine-grained limestones with rare nodules of flint and soft, white chalky limestones with planktonic foraminifera of the *Globigerapsis kugleri* Zone. The section terminates with a 10 m thick layer of chalk and marls, containing the microfauna of the *Morozovella lehneri* Zone. The total thickness of the Lutetian Stage is over 100 m.

To the south, near Safarqiyeh, the Lutetian Stage consists mainly of soft chalky limestones with the planktonic foraminifera of the *Acarinina bullbrooki* Zone (P10-P11). They alternate with massive, hard, fine-grained limestones. This unit is covered by chalky limestones with the microfauna of the *Morozovella lehneri* Zone (P12). A peculiar lithological feature of this zone is the presence of gray silicified limestones. Poor and spotty outcrops hidden by agricultural terraces prevent an exact determination of the thickness, but evidently it is around 70-80 m. An obscure unconformity separates the underlying Lutetian Stage chalky limestones from the overlying very similar Lower Miocene (Aquitanian) chalks.

In the northern regions of Jebel An-Nusseiriyeh the Lutetian deposits outcrop as separate promontories. They are located along the valleys of the Nahr El-Kabir and Nahr El-Abyad Rivers near the villages of Ain-Gazal, Bdama, Zainiye, and Bekfala as well as on the steep northeastern slope of this uplift near Jisr Ash-Shaghour and Jourin. In all the exposures they are represented by massive, hard, thick-bedded, aphanitic or nummulitic-alveolinid limestones with *Nummulites laevigatus, N. gallensis*, and *N. uroniensis*. Sometimes the limestones are algal, with irregular bedding. In the process of weathering they break down into huge blocks of irregular form and with vuggy surface. The visible thickness varies from 15 to 70 m.

The distribution of the Bartonian Stage deposits is very limited. Exposures of nummulitic and algal limestones with abundant larger foraminifera of the *Nummulites brongniarti* Zone were found near the village of Zainiye and at the Abou Gasi hill, where the Al-Latheqiyeh–Aleppo highway descends to the Al-Ghab graben. It may be possible that careful study of the fauna of the nummulitic limestones of Jebel An-Nusseiriyeh will reveal additional exposures of Bartonian deposits, but it is evident that the distribution area of sediments of this age is considerably smaller than that of the Lutetian Stage.

Sediments of the Upper Eocene and Oligocene were not discovered within the limits of Jebel An-Nusseiriyeh. Also according to Walley (1997) they were not identified in the adjoining areas of Lebanon.

The main features of the Paleogene stratigraphy of Jebel An-Nusseiriyeh can be summarized as follows. The Cretaceous and Paleogene are separated by an unconformity and a gap of variable duration. The Paleocene and Lower Eocene sequence (up to the *Morozovella aragonensis* Zone) is characterized by development of a mixed complex of pelagic and shallow-water sediments with reduced thickness and some gaps. The sediments of the uppermost Lower Eocene (the *Acarinina pentacamerata* Zone) and the Lutetian Stage are the most widely distributed but are mainly represented by shallow-water limestones. The areas occupied by Bartonian nummulitic limestones are limited and no deposits of the Upper Eocene or Oligocene were discovered.

The Baer-Bassit Massif

Within this area the Paleogene deposits constitute a part of the autochthonous structural complex. A section in the vicinity of the village of Damata on the right bank of the Nahr El-Qandil River reveals the main patterns of the Paleogene stratigraphy of this area (**Fig. 6.56**).

The section begins on the eastern slope of Jebel Rhouz Hill, at the edge of the small village of Qafer Fawkani and climbs along a path through the forest to the top of this hill. There are no traces of unconformity at the boundary between the Upper Maastrichtian (the *Abathomphalus mayaroensis* Zone) and the Danian Stage (Lower Paleocene).

The Danian Stage consists of two rock units with different lithologies. The lower one is 130 m thick and is made up of alternating yellow and yellow-gray soft marls and clayey limestones and more compact distinctly-bedded limestones. They contain planktonic foraminifera of the *Turborotalia pseudobulloides* and *Turborotalia trinidadensis* Zones. The upper unit is 70 m thick and consists of the alternation of these yellow marls with dark laminated clays; it belongs to the *Acarinina uncinata* Zone.

The Selandian Stage is very thick (380 m) and is represented by a monotonous pile of black and dark-gray, thin-bedded calcareous mudstones with rare interbeds of softer marls. The *Morozovella angulata* and the *Igorina pusilla* Zones can be recognized according to their planktonic foraminifera. A band of Selandian black rocks is clearly seen against a background of the underlying and overlying lighter calcareous rocks. With weathering the mudstones produce fine platy debris and the gentle slopes are covered by a clayey crust.

The Thanetian Stage is composed of gray compact limestones with subordinate layers of soft light-gray marls and rare thin (5-10 cm thick) interbeds of organogenic limestones. This member is 190 m thick and is referred to the *Planorotalites pseudomenardii* Zone. Hard limestones create a steep scarp in the relief. The overlying deposits belong to the *Morozovella velascoensis* Zone which is made up of alternating hard gray marls with conchoidal structure and massive yellow-gray limestones; the thickness is 125 m.

The Lower Eocene rests conformably on the Paleocene deposits. Its succession is incomplete due to faulting within this stratigraphic subdivision. The Ypresian Stage begins with a 110 m thick member of alternating marls and limestones, containing planktonic foraminifera of the *Morozovella subbotinae* Zone. Sediments of the overlying *Morozovella formosa* Zone and perhaps a part of the next *Morozovella aragonensis* Zone were not identified here due to the faulting. The upper part of the *Morozovella aragonensis* Zone is represented by yellow, light-yellow and light-gray massive limestones, which are well-bedded, fine-grained or aphanitic and include thin and rare interbeds of black flints; the thickness is 70 m.

The Lower Eocene terminates with a member of massive and solid, gray and light-gray, coarse-bedded, fine-grained and aphanitic limestones with numerous layers (10-25 cm thick), lenses and nodules of black and brown flints and gray silicified limestones. The basal part of this unit includes syngenetic breccias, slumps with distorted beds and thin layers (5-15 cm thick) of organogenic limestones formed by debris of algae, bryozoa, mollusks and shells of nummulitids and alveolinids. Sparse planktonic foraminifera, separated from these massive limestones assign this member to the *Acarinina pentacamerata* Zone. These hard rocks also crop out on the sea coast and form very steep cliffs.

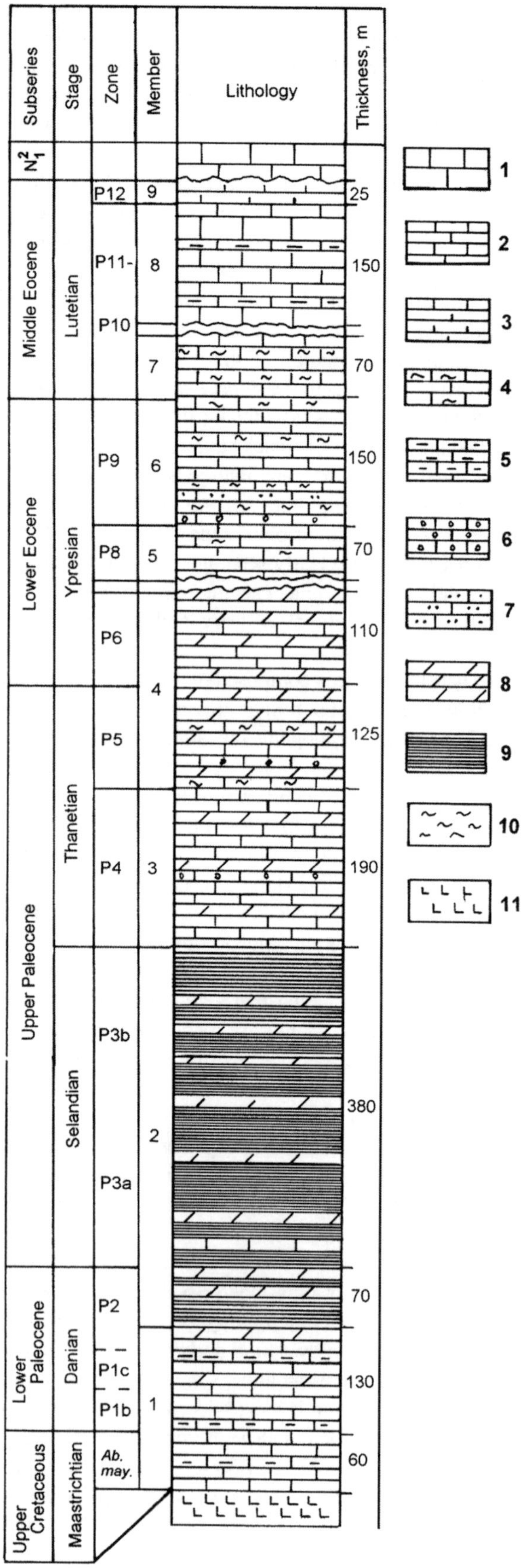

The continuation of the Jebel Rhouz Hill section is observed in the coastal scarps on the Mediterranean Sea as well as in exposures near the villages of Slayeb Tourkmanie, Borj Islam, and Chabatliye. They display the succession of the Middle Eocene deposits.

The top of Jebel Rhouz Hill is covered by a 70 m thick member of white and gray, hard and solid, aphanitic and fine-grained limestones with intercalations of silicified or leached limestones. Rare interbeds of clayey limestones contain poor microfauna of the *Acarinina bullbrooki* Zone (P10-P11), of the Middle Eocene.

The continuation of the Jebel Rhouz Hill section is observed in the coastal scarps along the Mediterranean Sea as well as in exposures near the villages of Slayeb Tourkmanie, Borj Islam, and Chabatliye. They display the succession of the Middle Eocene deposits.

On the Mediterranean coast, to the west of Slayeb Tourkmanie, the Lower Lutetian (the *Acarinina bullbrooki* Zone) is composed of the same white, gray and yellowish limestones which are clearly-bedded, very hard, aphanitic or fine-grained, with occasional interbeds of clayey limestones; the thickness is about 150 m. These rocks are overlain by a 25 m thick member of soft, white and gray chalky limestones with subsidiary layers of aphanitic and fine-grained limestones. The deposits are leached but the chalk yielded poorly preserved planktonic foraminifera of the *Morozovella lehneri* Zone, Upper Lutetian. These chalky limestones, in outcrops at the western margin of the village of Chabatliye and in a quarry of the Borj Islam cement factory, are not so severely leached and contain abundant and well-preserved planktonic foraminifera of the *Morozovella lehneri* Zone. Near these two villages the visible thickness of the Upper Lutetian chalk varies from 10 to 30 m. At Borj Islam village, the Middle Eocene deposits are unconformably covered by organogenic and aphanitic Middle Miocene limestones. Thus, sediments of the upper Middle Eocene, Upper Eocene, and Oligocene are absent in this area.

It is worthwhile noting that to the east of the Al-Latheqiyeh–Antakya highway, the Paleogene sequence is terminated by chalky limestones with planktonic foraminifera of the *Orbuli-noides beckmanni* Zone, the Bartonian Stage, but younger Paleogene deposits are also absent (Krasheninnikov, 1965b).

The Aafrine Graben

The Paleogene rocks are beautifully exposed in the Aafrine Graben. The entire continuous and very thick (1743 m) succession of the Paleogene, from the Danian Stage to the Oligocene, can be observed in the section at Meidanki (**Fig. 6.57**). The continuous and thick (942 m) Paleocene sequence is present in the section near the village of Khalil-Kolko (**Fig. 6.58**).

The Danian Stage begins with a member of alternating hard

Fig. 6.56. The section of the Paleogene deposits near the village of Damata, the Baer-Bassit Massif area.

Legend: 1 – thick-bedded limestones; 2 – thin-bedded limestones; 3 – chalky limestones; 4 – silicified limestones; 5 – clayey limestones; 6 – organogenic limestones; 7 – detrital limestones; 8 – marls; 9 – dark and black argillites; 10 – flints; 11 – ophiolite rocks.

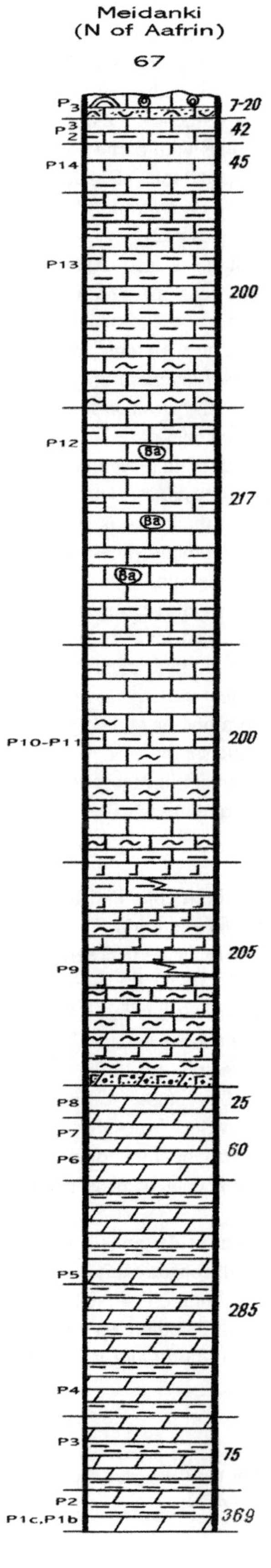

black limestones, marls and dark mudstones with the microfauna of the *Turborotalia pseudobulloides* Zone. The next *Turborotalia trinidadensis* Zone consists of alternating black and dark-gray thin-bedded marls, calcareous mudstones and soft, foliated, gray-greenish calcareous clays; interbeds of hard limestones disappear. The *Acarinina uncinata* Zone is made up of a monotonous pile of black calcareous mudstones with subordinate bands of greenish marls and calcareous clays. The thickness reaches 370 m. In the Khalil-Kolko section the thickness of the Danian Stage is even greater at 420 m.

In the Meidanki section the lithologic composition of the Upper Paleocene (the Selandian and Thanetian Stages) is uniform throughout the whole interval. It consists of black and dark-black, rather hard mudstones with interbeds of greenish calcareous mudstones. Rocks of this 360 m thick unit include very poor planktonic microfauna. It indicates that all foraminiferal zones are present in this sequence but that the boundaries of these zonal subdivisions, and the boundary between the Selandian and Thanetian Stages can only be approximated. To the southwest of Aafrine, in the sections at Khalil-Kolko and Tcholaqane hard black mudstones are replaced by light-green and yellowish marls and foliated clays. The assemblages of planktonic foraminifera are here much richer. They enable us to define the *Morozovella angulata* Zone s. str. and the *Igorina pusilla* Zone in the Selandian Stage and the *Planorotalites pseudomenardii* and the *Morozovella velascoensis* Zones in the Thanetian Stage, as well as the boundary between these stages. The thickness of the Selandian Stage is 150 m and that of the Thanetian Stage 250 m.

The Paleocene is overlain in the Meidanki section by Lower Eocene gray thin-layered calcareous mudstones and marls. The rates of sedimentation ceased to be so rapid and the thickness of this unit does not exceed 85 m. The sediments contain planktonic foraminifera of the *Morozovella subbotinae, Morozovella formosa*, and *Morozovella aragonensis* Zones. In the section at Khalil Kolko deposits of this age are absent and the *Morozovella velascoensis* Zone marlstones (Paleocene) are unconformably covered by the *Acarinina pentacamerata* Zone limestones (upper Lower Eocene).

The upper unit of the Lower Eocene at Meidanki corresponds to a formation observed throughout Syria. In this exposure it consists of gray hard fine-grained and aphanitic silicified limestones with numerous lenses and concretions (10-50 cm) of black and brown flints. At the base of this formation there is a layer of glauconite limestones and lenses of intraformational conglomerates. A special feature of this section is the presence of several layers of dense black basalts and weathered volcanic tuffs. Rare interbeds of clayey limestones contain the microfauna of the *Acarinina pentacamerata* Zone. The thickness of the calcareous-siliceous formation is 205 m.

Fig. 6. 57. The section of the Paleogene deposits at the village of Meidanki, the Aafrine Graben.
Legend: (see Fig. 6.53)

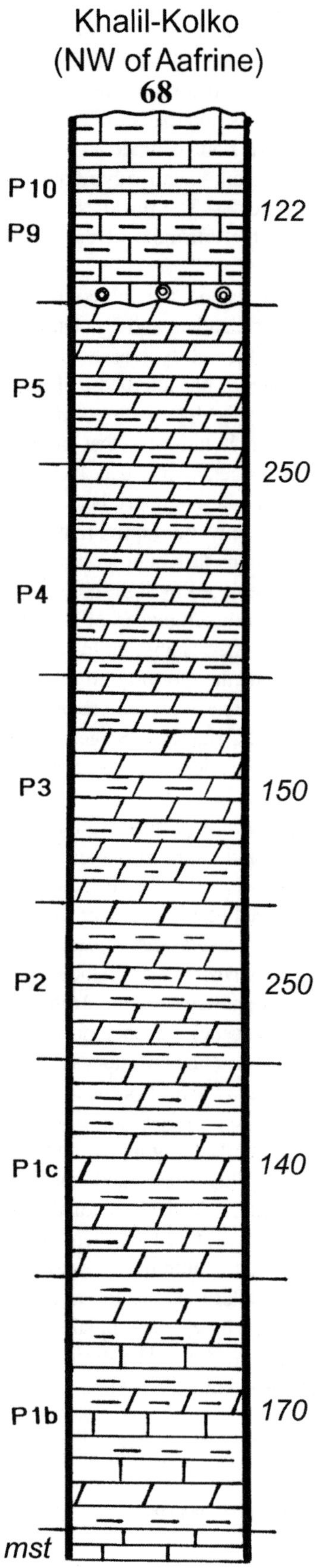

The Lutetian Stage is perfectly exposed in the Meidanki section where it is represented by a thick unit (417 m) of hard limestones and clayey limestones of light-gray and yellowish color. They are medium-bedded, with small lenses and nodules of flints in the lower part of the sequence; the upper part consists of coarse-bedded brown limestones with barite concretions. The planktonic foraminifera from separate soft intercalations of marls assign the deposits to the *Acarinina bullbrooki* Zone (P10-P11) and the *Morozovella lehneri* Zone (P12).

The Bartonian Stage is also characterized by a significant thickness of sediments (245 m). Its lower part (the *Orbulinoides beckmanni* Zone) is composed of white, thick-bedded soft chalky limestones with abundant planktonic foraminifera. At the base of this unit there are several layers of gray silicified limestones. The upper part of this stage (the *Truncorotaloides rohri* Zone) consists of pure soft thick-bedded chalk.

The Upper Eocene of this section is formed of hard, thick-bedded fine-grained limestones with subordinate interbeds of clayey limestones; the thickness is 42 m. A poor microfauna prevents any zonal subdivision. These hard rocks create a single distinct questa in the landscape.

The Meidanki Paleogene section is crowned by about 20 m of Oligocene rocks. The basal layer of the Oligocene is composed of hard, massive, fine-grained, algal or detrital limestones with *Nummulites intermedius* and *N. incrassatus*. They are followed by aphanitic, algal-coralline or *Lepidocyclina*-bearing limestones. Sometimes the rocks are sandy.

The Oligocene is unconformably covered by sediments of the Lower Miocene (the Aquitanian Stage).

The Aleppo Plateau

The entire area of the Aleppo Plateau is covered by Paleogene deposits (**Figs. 6.59, 6.60** and **6.61**). They rest almost horizontally, with a small inclination to the north. Therefore, the older Paleogene layers (the Lower and Middle Eocene) are developed along the southern margin of this plateau (Figs. 6.60 and 6.61). The northern areas of the Aleppo Plateau are occupied by sediments of the upper Middle Eocene and, mainly, by those of the Upper Eocene and Oligocene (Fig. 6.59). Continuous sections here are rare due to the smooth relief of the plateau. Exposures of the Paleogene deposits appear as outliers involving limited intervals of the Paleogene succession. The drilling data help us to understand the general geological setting.

Along the southern margin of the Aleppo Plateau, the Paleogene deposits unconformably overlie Maastrichtian or Campanian rocks. The duration of the hiatus varies from place to place. At Jebel Abou Dardeh the Upper Cretaceous limestones are covered by sediments of the *Planorotalites pseudomenardii* Zone,

Fig. 6.58. The section of the Paleogene deposits at the village of Khalil Kolko, the Aafrine Graben.
Legend: (see Fig. 6. 53)

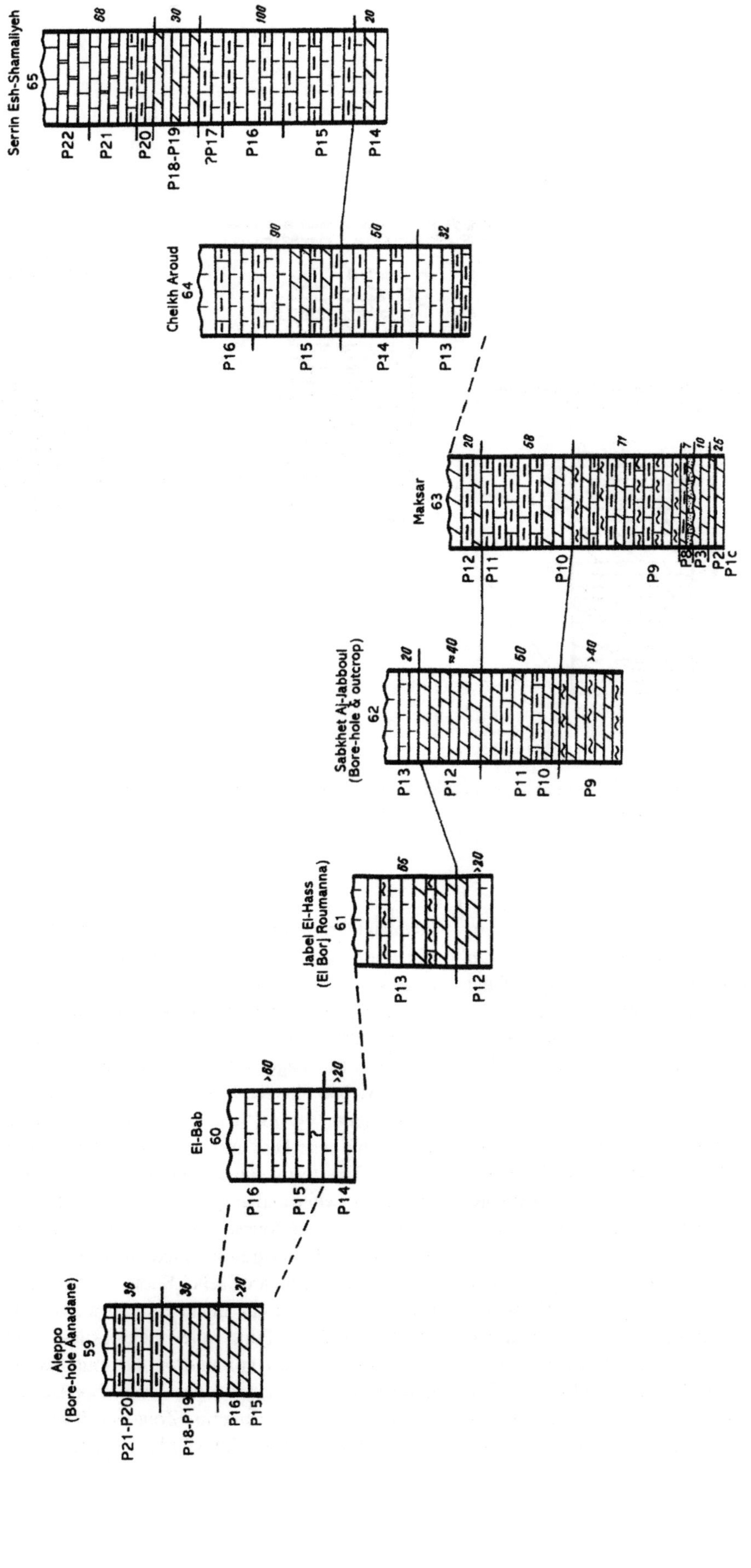

Fig. 6. 59. The section of the Paleogene deposits of the Aleppo Plateau.
Legend: (see Fig. 6.53)

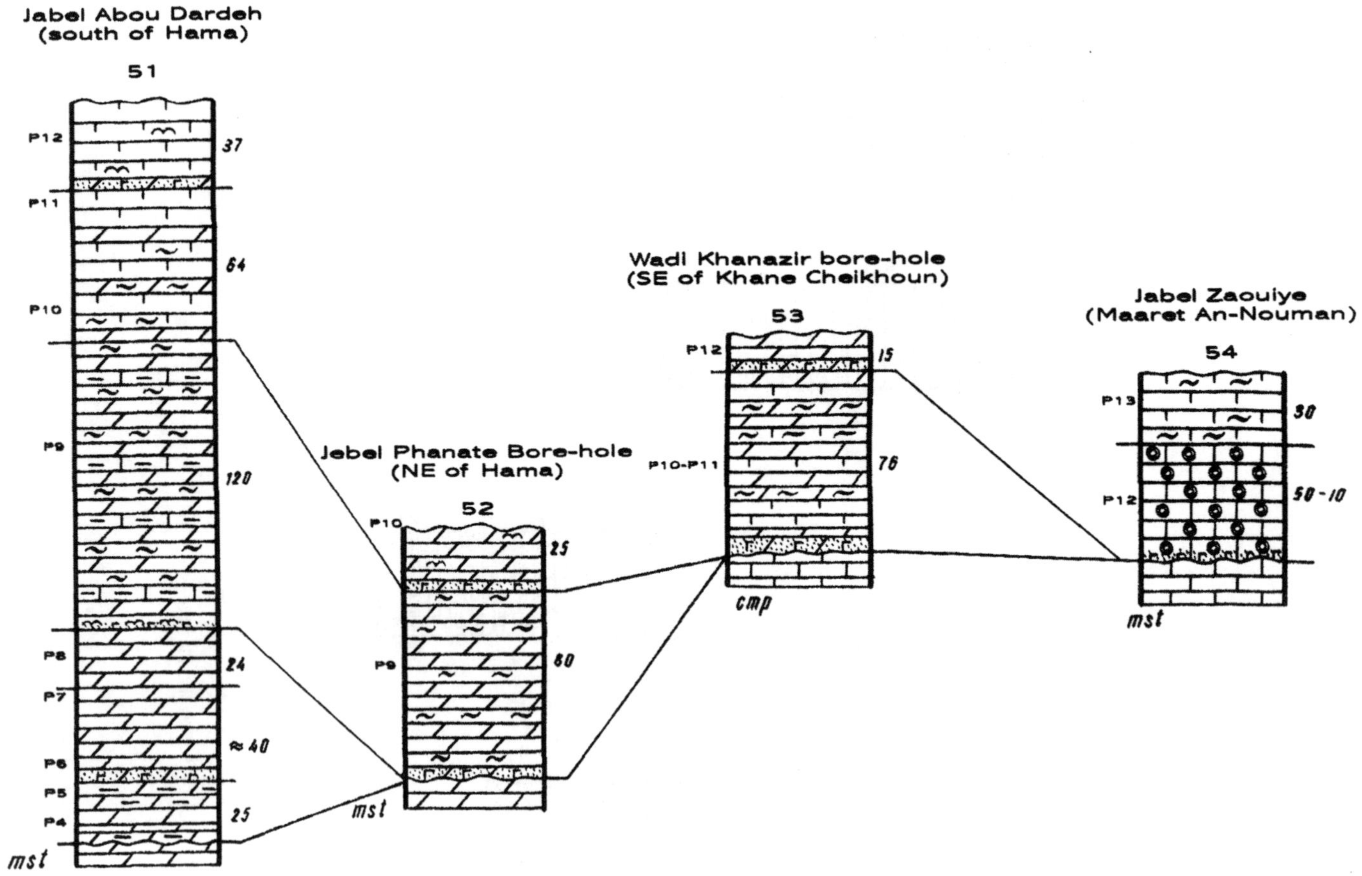

Fig. 6.60. The section of the Paleogene deposits of the southwestern part of the Aleppo Plateau. Legend: (see Fig. 6.53)

Upper Paleocene; at Jebel Phanate, Jebel Shbit, and Bir An-Naam they are overlain by sediments of the *Acarinina pentacamerata* Zone, upper Lower Eocene; in Wadi Khanazir, southeast of the town of Khane Cheikhoun, they are overlain by sediments of the *Acarinina bullbrooki* Zone sediments (P10-P11), Middle Eocene, and in the Jebel Zaouiye High, near the village of Maaret An-Nouman, they are overlain by sediments of the next *Morozovella lehneri* Zone, Middle Eocene.

Paleocene deposits were encountered south of Hama at Jebel Abou Dardeh, which looks like a high outlier on the flat plain. They consist of a 25 m thick unit of greenish-gray calcareous clays and marls with planktonic foraminifera of the *Planorotalites pseudomenardii* and the *Morozovella velascoensis* Zones.

The sediments of the Lower Eocene are more widespread. The lower part of the Lower Eocene is only well-exposed in the south-western part of the Aleppo Plateau, in the vicinity of Hama and Salamiyeh, in the deep valleys of the right-hand tributories of the Nahr El-Aasi (Orontes) River, and at the base of flat table-like hills. It is represented by a monotonous pile of white, gray and greenish, soft and foliated marls, calcareous clays and clayey limestones with abundant foraminifera of the *Morozovella subbotinae*, *Morozovella Formosa*, and *Morozovella aragonensis* Zones. The thickness ranges from 50 to 70 m.

The peak of the transgression coincides with the termination of the Early Eocene (the *Acarinina pentacamerata* Zone). The deposits of this age were traced along the entire southern margin of the Aleppo Plateau, from Jebel Abou Dardeh and Jebel Phanate in the southeast, to Sabkhet Aj-Jabboul and Jebel Shbit in the center, and near Bir An-Naam on the southeastern margin of this plateau. This formation, as in other areas of Syria, is composed of alternating white and gray soft clayey limestones and marls, harder fine-grained limestones, and numerous layers and lenses of brown and black flints and gray silicified limestones. Its thickness is highly variable. In the area of Jebel Shbit, where the Upper Cretaceous rocks are exposed, the thickness varies from 10 to 30 m. In the direction to the Al-Furat valley the thickness increases to 60-80 m. The same tendency is found on the south-western margin of the plateau, where in the section at Jebel Abou Dardeh the thickness of the *Acarinina pentacamerata* Zone is 120 m.

The structure of the Middle Eocene on the Aleppo Plateau

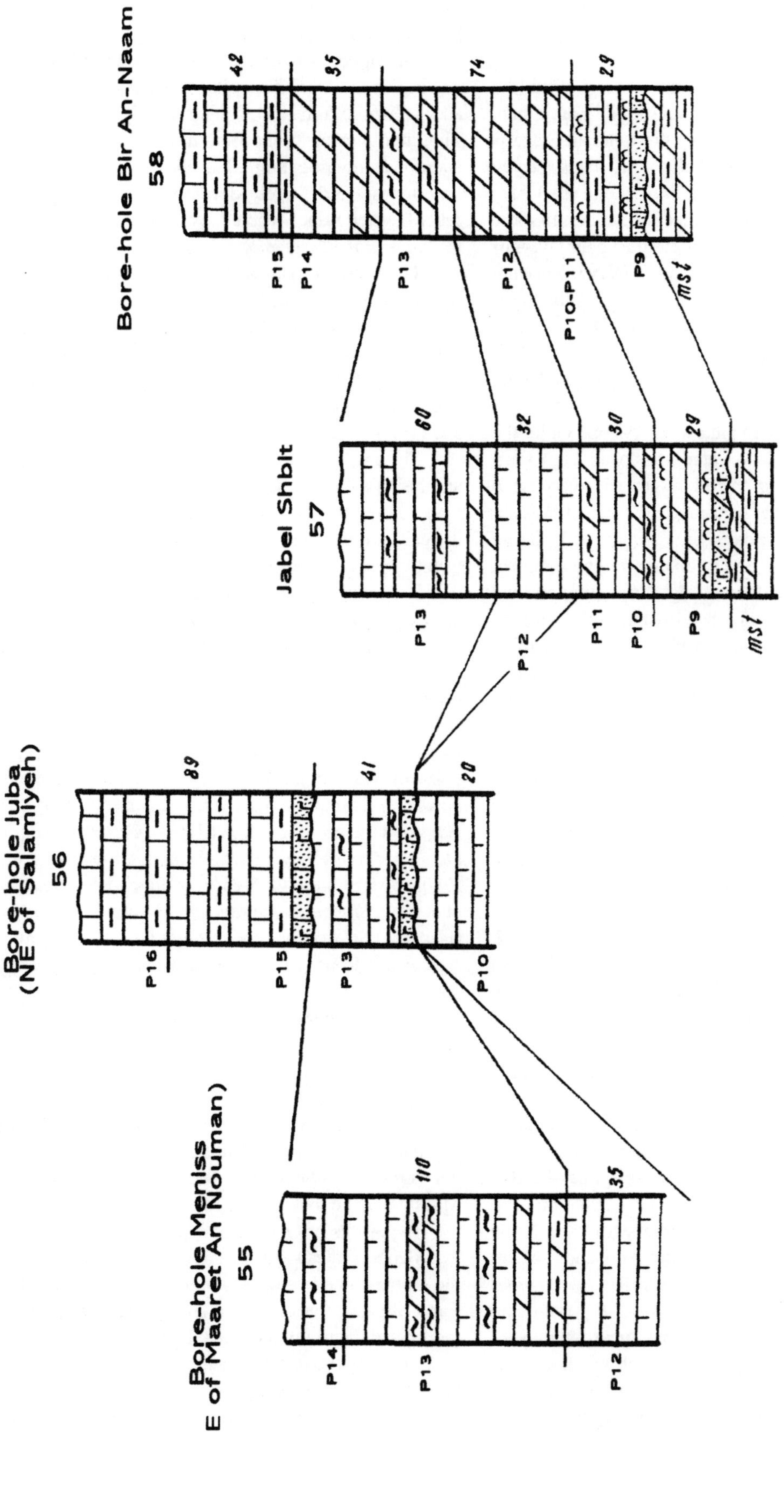

Fig. 6. 61. The section of the Paleogene deposits of the southern and southeastern parts of the Aleppo Plateau.
Legend: (see Fig. 6.53)

is rather complicated due to the interrelation between the pelagic facies of chalky limestones and the shallow-water facies of nummulitic limestones.

In the southwestern and southeastern part of this plateau (the villages and towns of Khan Sheikhoun, M'arret An-No'man, Hama, Salamiyeh, Abou Dardeh, and Biar Esriye) the Lutetian Stage is made up of light-gray and white, soft chalky limestones, clayey limestones and marls. In the *Acarinina bullbrooki* Zone (P10-P11) these rocks intercalate with gray, hard silicified limestones. In some sections of the Salamiyeh and Biar Esriye areas the upper layers of this zone and the deposits of the next *Morozovella lehneri* Zone are missing due to local erosional processes. The maximum thickness of the Lutetian deposits in the complete Jebel Abou Dardeh, Jebel Shbit and Sabkhet Al-Jabboul sections is about 90-130 m. It is reduced to 50-60 m closer to the center of the uplift where Upper Cretaceous rocks are exposed.

In the western and northwestern parts of the Aleppo Uplift the pelagic sediments grade into shallow-water limestones. They form the Jebel Az-Zawiyeh Massif and narrow ridges with very steep slopes such as Jebel Dweleh, Jebel Mrasras, and Jebel El-Wastani between the towns of Jist Ash-Shaghour, Salqein, and Harim. Lithologically these limestones are very diverse *i.e.* nummulitic, nummulitic-alveolinid, biogenic-detrital, algal, fine-grained, and aphanitic. Usually the rocks are thick-bedded but with indistinct bedding. The algal limestones create reef-like bodies. Very often the limestones are over-crystallized and converted into marble-like rocks in which all organic remains were dissolved, leaving empty cavities. When weathered, they break down into huge irregular blocks with uneven, fluted, and even whimsical surfaces. These limestones, with a basal glauconite bed, rest unconformably on different units of the Upper Cretaceous. A glauconite bed exposed near the town of Ariha is very thick. It varies laterally from 4 to 6 m. In the Jebel Az-Zawiyeh Uplift the thickness of the Lutetian limestones ranges from 50 to 100 m. Their lower part contains *Nummulites laevigatus, N. gallensis,* and *N. uroniensis*. The overlying layers include the first specimens of *N. gizehensis* and, evidently, belong to the Upper Lutetian. It is difficult to estimate the thickness of the Lutetian rocks in the ridges of Jebel Dweleh, Jebel Mrasras, and Jebel El-Wastani due to numerous faults and the unclear position of the boundary with the Bartonian Stage. In separate blocks, the thickness of the Lutetian limestones exceeds 100-150 m.

The Bartonian deposits are widespread along the southern margin of the Aleppo Plateau. The pelagic sediments predominate and consist of white, light-gray and yellowish, soft, thick-bedded chalky limestones with a block parting. The most complete sections are common in the eastern and southeastern parts of the plateau (the area of Cheikh Aaroud and Bir An-Naam) where both the *Orbulinoides beckmanni* Zone and the *Truncorotaloides rohri* Zone were identified. Here the thickness of the Bartonian Stage reaches 110 m. In more western regions (near Salamiyeh) sediments of the *Truncorotaloides rohri* Zone can be absent and the thickness of the *Orbulinoides beckmanni* Zone varies from 40 to 60 m. Chalky limestones of this zone very frequently contain intercalations of gray silicified limestones (at Meniss, Juba, Jebel Shbit, Bir An-Naam, and Jebel El-Hass).

The upper part of the shallow-water Middle Eocene rocks making up the Jebel Zawiyeh Massif consists of hard fine-grained and aphanitic limestones with poorly preserved nummulitids. They may belong to the Bartonian Stage but this question needs additional investigation. The Bartonian nummulitic-alveolinid limestones are definitely developed in the Jebel El-Wastani and Jebel Dweleh ridges where they contain the larger foraminifera of the *Nummulites brongniarti* Zone.

The northern part of the Aleppo Plateau is characterized by wide distributed Upper Eocene rocks. To the north of the latitudinal line through Aleppo, El-Bab, and Sheikh-Aroud they form an almost unbroken cover. The sediments of this age are represented by chalky limestones – white, light-gray and yellowish, soft, coarse-laminated, with intercalations of medium-bedded varieties and clayey limestones. Their thickness ranges from 60 to 100 m.

The chalky limestones contain rich assemblages of planktonic and benthic foraminifera. However the benthic assemblages have many more specimens which are taxonomically diverse. Evidently the calcareous oozes of the Aleppo Plateau are of more shallow-water origin than those of the Palmyrides and the Anti-Lebanon.

Planktonic foraminifera enabled us to identify the *Globigerapsis semiinvoluta* and the *Turborotalia cocoaensis* Zones.

The list of benthic foraminifera is very long: *Bulimina sculptilis* Cushman, *B. acsuatica* Moroz., *Bolivina antegressa* Subb., *Bifarina millepunctata* (Tutkovs.), *Uvigerina hispida* Schwager, *Anomalina affinis* (Hantken), *A. granosa* (Hantken), *Planulina costata* (Hantken), *Cibicides perlucidus* Nuttall, *C. pigmeus* (Hantken), *Frondicularia budensis* (Hantken), *Vaginulina mexicana* Nuttall, *Saracenaria arcuata* (d'Orb.), *Marginulina fragaria* Guembel, *M. behmi* (Reuss), *Cristellaria limbosa* (Reuss), *Dentalina consobrina* d'Orb., *Siphonodosaria longiscata* (d'Orb.), *S. volgensis* Samoilova, *S. exilis* (Neugeb.), *S. annulifera* (Cushman et Berm.), *Siphonina subreticulata* Mjatliuk, *Clavulinoides czaboi* (Hantken), *Clavulina cyclostomata* Gall. et Morr., *Karreriella siphonella* (Reuss), *Bolivinoides reticulatus* (Hantken), and many others.

Near the southwestern margin of the Aleppo Plateau (the area of Esriye and Tanahe) chalky limestones alternate with hard fine-grained and detrital limestones. The thickness of the Upper Eocene decreases here to 60-90 m.

Usually the Middle and Upper Eocene are connected by a gradual transition (at El-Bab, Cheikh-Aroud, and Bir An-Naam) but sometimes the Upper Eocene rests unconformably on the Middle Eocene rocks and contains a basal glauconite layer.

Exposures of Oligocene deposits occupy the northern part of the Aleppo Plateau. Their lithological features resemble those of the Upper Eocene and they were strongly eroded before the late Early Miocene transgression. Therefore it is difficult to estimate the real spatial distribution of the Oligocene sediments.

To the northwest of Aleppo the Oligocene was penetrated by a bore-hole near the village of Aanadane. The Oligocene starts with a 35 m thick member of gray and greenish-gray marls containing rich assemblages of planktonic foraminifera of the *Globigerina tapuriensis* and *Globigerina sellii* Zones. They conformably overlie similar marls of the Upper Eocene. The upper 36 m thick member of the Oligocene is composed of alternating marls with yellowish harder clayey limestones characterized by the microfauna of the *Globigerina ampliapertura* and the *Turborotalia opima* Zones.

Very nice sections of the Oligocene are located in the eastern part of the Aleppo Plateau, along the Al-Furat valley, but it is already in the marginal part of the Mesopotamian Basin which is described below.

The Mesopotamian Basin

The Paleogene deposits of this basin are covered by a thick pile of Neogene sediments. They only crop out on the western and eastern banks of the Al-Furat River valley between the towns of Jrablous and Ar-Raqqa. A small anticline near the village of Maksar shows the succession of Paleocene, Lower Eocene and lower Middle Eocene deposits. Cliffs of the Al-Furat valley yield the sequence of the upper Middle Eocene, Upper Eocene, and Oligocene (Fig. 6.59). Data relevant to the natural outcrops were supplemented by the results of drilling some bore-holes (Tishrin-300 near Qara-Qozaq, and RB-405, RB-212 and RB-524 near Raqqa) to support the hydroengineering infrastructure for the Euphrates Dam.

At Maksar the contact of the Upper Cretaceous with the Paleogene is not exposed. The section starts with a 35 m thick member of soft, foliated, greenish-gray marls with planktonic foraminifera of the Danian Stage (the *Turborotalia trinidadensis* Zone and the *Acarinina uncinata* Zone) and the Selandian Stage (the *Morozovella angulata* Zone s. l.).

These deposits are unconformably overlain by a 7 m thick unit of glauconite marls of the *Morozovella aragonensis* Zone, Lower Eocene. Consequently, the stratigraphic gap represents a considerable time interval *i.e.* from the *Planorotalites pseudomenardii* Zone, Paleocene to the *Morozovella formosa* Zone, Lower Eocene. The Lower Eocene terminates with a 71 m thick formation made up of alternating greenish marls and yellowish limestones with plentiful layers and concretions of black flints. This formation includes rich assemblages of planktonic foraminifera of the *Acarinina pentacamerata* Zone with numerous specimens of *Astrorotalia palmerae*.

The Lutetian Stage is exposed on the northern slope of the Maksar fold and comprises 80 m of white and yellowish, soft, thick-bedded chalky limestones with interbeds of gray marls. The complete sequence of zonal subdivisions can be observed here – the *Hantkenina aragonensis*, the *Globigerapsis kugleri* and the *Morozovella lehneri* Zones. To the north, along the Al-Furat valley on the Jrablus rampart (near the village of Kibaret), the chalky Lutetian limestones overlie the Maastrichtian chalk with an obscure unconformity.

The Bartonian Stage is characterized by similar white thick-bedded chalky limestones with subordinate interbeds of darker clayey limestones. Near the villages of Cheikh-Aroud and Maskene their thickness ranges from 60 to 80 m. The sediments contain planktonic foraminifera of the *Orbulinoides beckmanni* and *Truncorotaloides rohri* Zones.

The Upper Eocene is well-exposed in the cliffs of the western and eastern banks of the Al-Furat River near the villages of Serrin Esh-Shamaliyeh, Milha, Qara-Qosaq, and on the Jebel Esh-Sheikh Sem Hill (Fig. 6.59). In all these areas it is made up of white, light-gray and yellowish, soft, thick-bedded chalky limestones and marls with planktonic foraminifera of the *Globigerapsis semiinvoluta* Zone and the *Turborotalia cocoaensis* Zone. The thickness varies from 70 to 100 m.

The distribution area of the Oligocene sediments coincides with that of the Upper Eocene. The Oligocene sequence consists of monotonous yellowish and light-gray, soft chalky limestones with more dense clayey limestones. According to drilling data, their thickness reaches 170 m. The entire set of the zonal subdivisions – from the *Globigerina tapuriensis* Zone (P18) to the *Globigerina ciperoensis* Zone (P22) was identified in the most complete sections. There is a gradual transition from the Eocene to the Oligocene deposits. It is not easy to find this boundary in the field on account of the similarity of the rocks. The Oligocene chalk differs from the white Eocene chalk by its yellow or yellowish color.

The pelagic calcareous Oligocene oozes of the Al-Furat valley contain extremely rich assemblages of planktonic and benthic foraminifera. Such facies is unique for the Oligocene of the eastern Mediterranean. Therefore, the Oligocene sections of the Al-Furat valley are of great importance to the Paleogene stratigraphy of the Near East.

Near the village of Bir El-Khasne on the Jebel Abd El-Aziz Uplift the Upper Cretaceous deposits are unconformably covered by massive reef-like algal and layered aphanitic and fine-grained limestones. Their normal succession is broken by some faults. The basal beds of this sequence contain *Nummulites fraasi* and have an Upper Paleocene age. The next member, consisting of limestones with remnants of nummulits, discocyclinids, and alveolinids, belongs to the Eocene.

The Jordan Uplift and the Drouz Depression

The oldest layers of the Paleogene are exposed in the core of the Jordan Uplift. They consist of sediments of the *Morozovella aragonensis* Zone of the Lower Eocene. They are followed by the upper Lower Eocene and Middle Eocene succession of rocks. The oldest exposed rocks of the Drouz Depression were assigned to the *Orbulinoides beckmanni* Zone of the Middle Eocene. More ancient deposits of the Paleogene were reached in these two regions by drilling (**Fig. 6.62**).

In the deep canyon of the Nahr El-Yarmouk River the Lower Eocene *Morozovella aragonensis* Zone is composed of gray, yellowish and brown, pelitic and soft limestones enriched with organic matter (kerogen) which alternate with combustible shales and marls. Their visible thickness is about 30 m. Drilling near the town of Dar'a showed that these sediments unconformably overlie the Maastrichtian black shales with their high content of organic matter.

This unit is followed by a 190 m thick formation of alternating

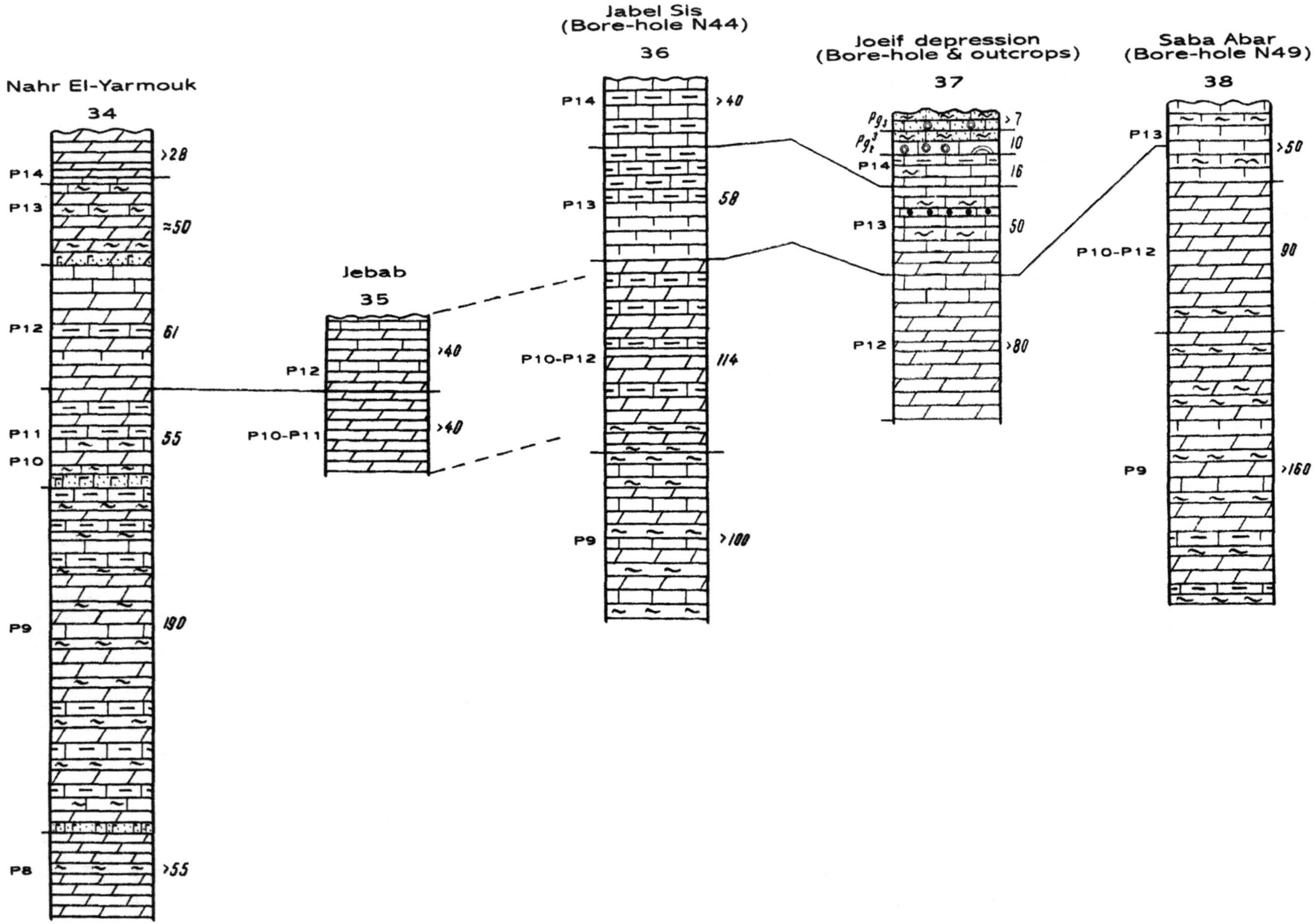

Fig. 6. 62. The section of the Paleogene deposits of the Jordan Uplift and the Drouz Depression.
Legend: (see Fig. 6.53)

white and gray clayey limestones, brown kerogen limestones with dark weathering crusts, hard aphanitic silicified limestones and layers (2-10 cm thck) of black cherts. The sediments contain extremely rich assemblages of planktonic foraminifera of the *Acarinina pentacamerata* Zone together with rare benthic foraminifera. They are very well-exposed in the Nahr El-Yarmouk valley itself and also in the valleys of the right-hand tributaries of this river (*e.g.* Wadi Zeizoun, Wadi Hrer, and Wadi Khit).

In the Drouz Depression lithologically similar deposits of the *Acarinina pentacamerata* Zone were penetrated by Bore-hole 44 at Jebel Sis (or Jebel Seiss) near the eastern margin of the Safa basaltic cover, and by Bore-hole 49 to the southeast of Saba-Abar. The thickness of this formation of alternating limestones, marls and flints is over 160 m.

On the western slope of the Jordan Uplift the Middle Eocene is exposed in the Nahr El-Yarmouk valley and also in the cliffs above the eastern shore of Lake Tiberias (the Sea of Galilee) at Wadi Samaq, Wadi Ar-Reqqad, Wadi Massaoud and Wadi Al-Kursi.

The Lutetian Stage starts with a 55 m member of light-gray marls and soft clayey limestones with rare intercalations of silicified limestones. There is a glauconite layer at the base. These sediments yielded the microfauna of the *Hantkenina aragonensis* and the *Globigerapsis kugleri* Zones. They are followed by white marls and clayey limestones of the *Morozovella lehneri* Zone whose thickness is 60 m. White chalky-like marls of the Lutetian Stage are also exposed on the northern plunge of the Jordan Uplift near the town of Jebab where the thickness is about 80 m.

The Bartonian Stage consists of two members. The lower 40-50 m thick member is built up of white soft marls with rather numerous thin interbeds of solid gray silicified limestones and a basal glauconite bed. This member is characterized by planktonic foraminifera of the *Orbulinoides beckmanni* Zone. The upper member, consisting of white marls with microfauna of the *Truncorotaloides rohri* Zone, showed a visible thickness of 28 m in Wadi Massaoud.

According to drilling data, at Jebel Sis, near Saba Abar, and in the closed Joeif Depression the Lutetian Stage is composed of soft white marls and clayey limestones up to 114 m thick. The deposits of the Bartonian Stage are present in the Joeif Depression as natural outcrops. They can be subdivided, as on the Jordan Uplift, into two members. The lower member consists of alternating white and yellowish clayey limestones, hard fine-grained limestones and chalky marls with foraminifera of the *Orbulinoides beckmanni* Zone. This unit also includes layers of gray silicified limestones and black flints. One layer of these black flints is 35 cm thick and forms the flat bottom of the depression. The thickness of the lower member is 50 m. The upper member includes only white soft limestones and marls with microfauna of the *Truncorotaloides rohri* Zone whose thickness does not exceed 20 m.

The Upper Eocene in the Joeif Depression is represented by a thin (10 m) unit of nummulitic, algal and solid fine-grained limestones.

The section is terminated by a member of hard limestones with lepidocyclinids intercalating with solid sandy limestones, which belongs to the Oligocene.

The Rutbah Uplift

The Paleogene deposits of the southeastern part of Syria (the Syrian Desert) are characterized by shallow-water sediments of reduced thickness. They are almost horizontal. Therefore, the Paleogene sections can be mainly observed in the steep escarpment of Jebel Fru Tarag El-Olab, and in the valleys of Wadi El-Miyah, Wadi El-Waar, and Wadi As-Suwab, as well as on slopes of small heights, such as Roumanna and Tell Doueiye (**Figs. 6.63** and **6.64**). Usually, the small hills on the flat surface of the Syrian Desert display only limited intervals of the Paleogene succession.

In outcrops, the oldest Paleogene deposits belong to the *Morozovella aragonensis* Zone of the Lower Eocene. At Bir Alayyaniyeh this consists of a 20 m thick unit of greenish thin-layered marls and white, gray and yellowish soft limestones. Sediments of this age (up to 50 m thick) were also discovered in some bore-holes drilled along the road from Palmyra to Abou-Kamal and near Abar Sejiri.

These bore-holes also penetrated greenish and gray, laminated and soft marls and calcareous clays of the *Morozovella velascoensis* Zone of the Paleocene and the *Morozovella subbotinae* and the *Morozovella formosa* Zones of the Lower Eocene. Thus, in the Rutbah Uplift the sediments of the stratigraphic interval from the Paleocene to the *Morozovella aragonensis* Zone of the Lower Eocene inclusive (in our opinion the first formation of the Syrian Paleogene) preserve their lithological features. This formation rests unconformably on various Upper Cretaceous subdivisions and in the area of Khabra Oum Mouell (Bore-hole 46) even on dolomites and limestones with rich Jurassic foraminifera (Fig. 6.63).

The deposits of the *Acarinina pentacamerata* Zone of the terminal Lower Eocene are exposed over vast areas of the Rutbah Uplift. They consist of alternating white and gray hard fine-grained limestones, soft clayey limestones, marls and numerous layers and lenses of black and brown flints; at the base there is a bed of glauconite limestones. Near Bir Alayyaniyeh the thickness of this formation is 87 m. To the east and southeast of Bir Aaleyyaniyeh the lithological composition of this unit is different. In the area of Tell Ar-Roumanna, Tell Doueyeh, and Wadi Miyah the *Acarinina pentacamerata* Zone is composed of gray and yellowish, fine-grained and aphanitic limestones with numerous lenses (up to 50 m thick) of black flints and plenty of phosphorite layers. Occurrences of phosphorites have commercial value. Interbeds of soft marls and clayey limestones are rarely seen. The thickness of the formation decreases to 40-50 m. The shallow-water character of the sediments is emphasized by the composition of the microfauna because planktonic foraminifera are rare and benthic species prevail.

The whole surface of the Syrian Desert is covered by shards of Lower Eocene flints burnished with a glossy 'desert sunburn' or 'desert varnish'. This abundant flinty debris in the Syrian Desert is described by the local name 'khamad' which means 'a rocky desert'.

The lithological composition of the Middle Eocene deposits varies along the northern slope of the Rutbah Uplift.

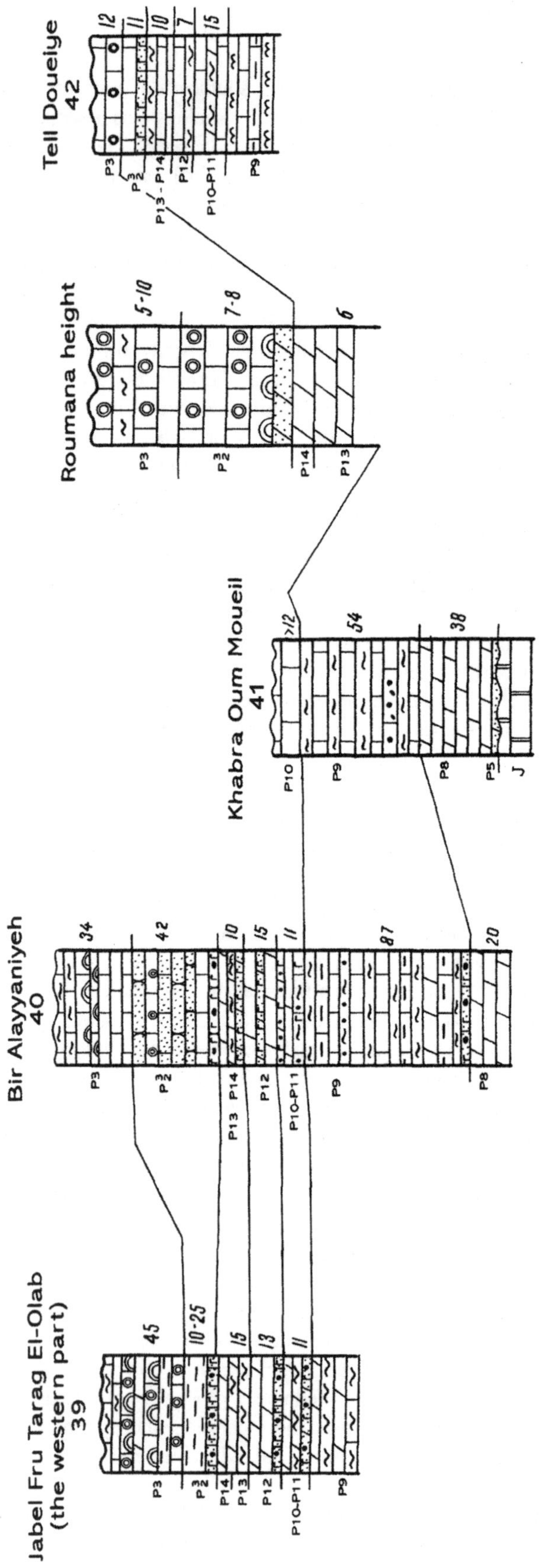

In the western part of Jebel Fru Tarag El-Olab and near Bir Aaleyyaniyeh the *Acarinina bullbrooki* Zone (P10-P11) consists of soft white marls with interbeds of hard silicified limestones and gluconitic limestones. The *Morozovella lehneri* Zone incorporates soft marls and clayey limestones. The thickness of the Lutetian Stage is about 25 m.

To the south, at Jebel At-Tanf, Tell Ar-Roumanna, and Tell Doueiye, these soft rocks grade into a member of light-colored, hard, aphanitic and fine-grained limestones with very thin bands of soft clayey limestones whose thickness decreases to 15 m.

To the east in outcrops along Wadi El-Miyah, Wadi El-Waar, and Wadi As-Suwab the *Acarinina bullbrooki* Zone deposits are replaced by white and light-gray, strongly leached limestones with poor microfauna and a member of nummulitic limestones corresponding to the *Morozovella lehneri* Zone (Fig. 6.64). The thickness of the Lutetian Stage here is 30 to 40 m.

Similar lithological replacements are common for the deposits of the Bartonian Stage. In the west, at Jebel Fru Tarad El-Olab and Bir Alayyaniyeh the *Orbulinoides beckmanni* Zone is represented by soft white marls with interbeds of various limestones – silicified, glauconitic and hard fine-grained. The *Truncorotaloides rohri* Zone is only made up of soft marls and clayey limestones. Soft intercalations contain rather good assemblages of planktonic foraminifera. The thickness of the Bartonian Stage is insignificant – from 10 to 15 m.

To the south, in the area of Jebel At-Tanf, hard, well-bedded, aphanitic and fine-grained limestones predominate. Interbeds of soft marls are extremely rare. The thickness is 5 to 10 m. Here it is very difficult to distinguish deposits of the *Orbulinoides beckmanni* and *Truncorotaloides rohri* Zones.

On the north-eastern slope of the Rutbah Uplift a lateral facies change in the Bartonian deposits takes place over a short distance (Fig. 6.64). The *Orbulinoides beckmanni* Zone in the outcrops along Wadi El-Miyah, Wadi El-Waar, and Wadi As-Suwab is composed of a 14 m thick unit of soft white marls and limestones with rich assemblages of planktonic foraminifera. The *Truncorotaloides rohri* Zone in the Wadi El-Miyah section includes soft clayey limestones with planktonic foraminifera and hard aphanitic limestones, whose total thickness is about 10 m. In Wadi As-Suwab these deposits of the *Truncorotaloides rohri* Zone were replaced by a 20 m thick member of hard aphanitic, fine-grained and nummulitic limestones with abundant large specimens of *Nummulites lyelli.* Many years ago Cizancourt (1934) described her visit to this remote exposure of nummulitic limestones in the Syrian Desert.

The Upper Eocene of the Rutbah Uplift incorporates shallow-water deposits of insignificant thickness which vary horizontally. Near the western margin of the Jebel Fru Tarag El-Olab escarpment,

⇦

Fig. 6.63. The section of the Paleogene deposits of the northwestern part of Rutbah Uplift.
Legend: (see Fig. 6.53)

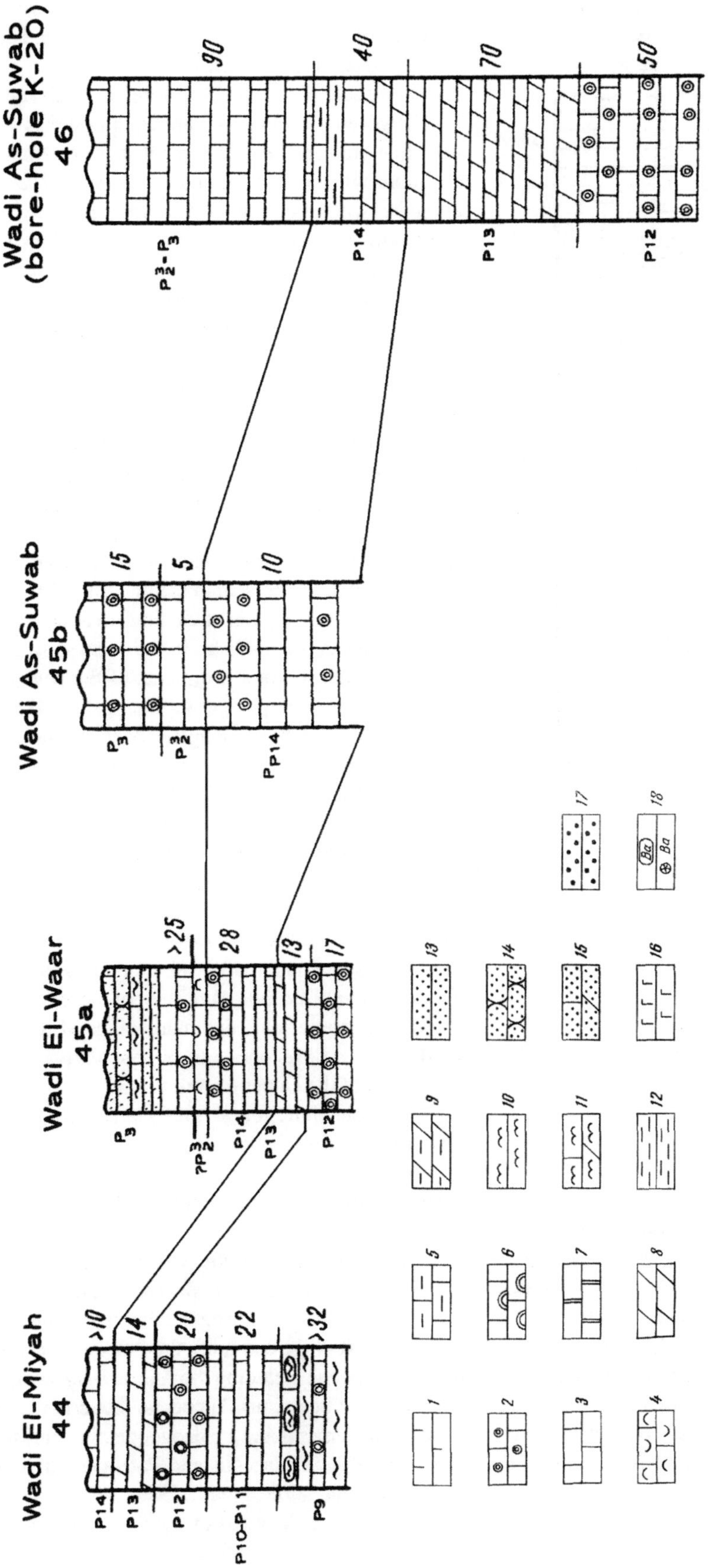

Fig. 6.64. The section of the Paleogene deposits of the northeastern part of the Rutbah Uplift.
Legend: (see Fig. 6.53)

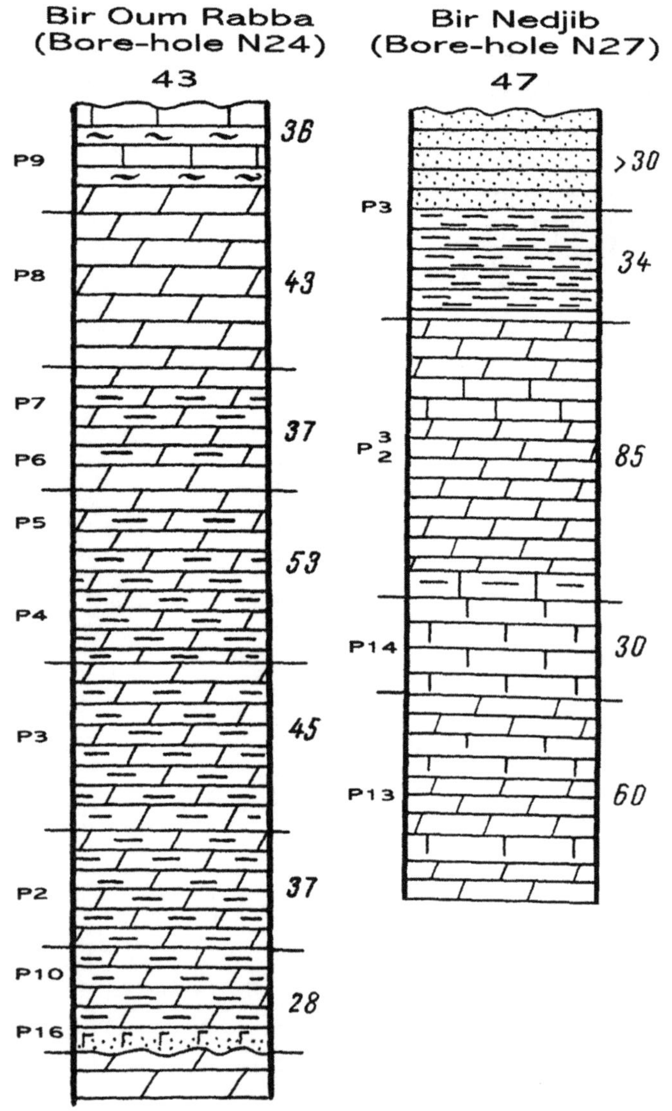

Fig. 6.65. The section of the Paleogene deposits of the southeastern part of the Syria (to the north of the Rutbah Uplift).
Legend: (see Fig. 6.53)

and at the village of Bir Alayyaniyeh, the Upper Eocene consists of different limestones – hard and massive aphanitic, fine-grained, algal, nummulitic, and sandy. They alternate with terrigenous rocks – calcareous sandstones and mudstones. At the base of this formation there is a bed of glauconite rocks up to 1 m thick. The maximum thickness of the Upper Eocene reaches 42 m. To the south and to the east terrigenous rocks disappear and the entire thin (5-10 m) Upper Eocene succession is represented by massive limestones which are frequently barren of fossils (*e.g.* at Jebel At-Tanf, Tell Doueiye, Tell Ar-Roumanna, Wadi El-Waar, and Wadi As-Suwab).

The Oligocene of the Rutbah Uplift is characterized by our fifth formation which is made up of shallow-water calcareous-terrigenous rocks. Its visible thickness varies, reaching 30-45 m in the western part of Jebel Fru Tarag El-Olab, and decreasing to 5-25 m to the south and to the east.

In the sections at Jebel Fru Tarag El-Olab and Bir Alayyaniyeh the Oligocene comprises various limestones – hard aphanitic and fine-grained, reef-like algal and foraminiferal consisting of abundant shells of lepidocyclinids and operculinids. They alternate with greenish and yellowish clays and sandy clays containing impoverished assemblages of planktonic foraminifera of the Rupelian Stage (from the *Globigerina tapuriensis* Zone to the *Globigerina ampliapertura* Zone). The lower layers of this formation include lenses of nummulitic limestones with *Nummulites intermedius* and *N. vascus*. The upper layers are composed of massive aphanitic and sandy limestones very poor in organic remains.

The Oligocene deposits of similar lithology are widespread in the north-eastern margin of the Rutbah Uplift (Wadi El-Waar, Wadi As-Suwab, and Tell Doueiye). In some sections limestones predominate, in other areas quartz sandstones and sandy limestones are very common. Intercalations of limestones with abundant lepidocyclinids are present in the majority of sections.

Drilling on the plain stretching between the Rutbah Uplift in the south and Jebel Al-Bishri in the north showed that towards the north the shallow-water sediments of the Rutbah Uplift are very quickly replaced by pelagic sediments typical to the Palmyrides. The thickness of the Paleogene deposits increases in the same direction. Three bore-holes, No. 24 at Bir Oum Rabba, No. 27 at Bir Nejib (**Fig. 6.65**), and K-20 at the entrance to Wadi As-Suwab (Fig. 6.64) allow us to see the same five formations we observed in other areas of Syria. From bottom to top they are as follows:

1) marls and calcareous clays of the Danian, Selandian, Thanetian, and Ypresian Stages with all zones of planktonic foraminifera up to the *Morozovella aragonensis* Zone;

2) limestones and marls with numerous bands of black flints of the *Acarinina pentacamerata* Zone of the Upper Ypresian;

3) chalky limestones and marls of the Middle Eocene;

4) chalky limestones and marls with intercalations of hard limestones of the Upper Eocene;

5) various limestones, green and yellow clays, sands and sandstones of the Oligocene.

Stages in the Paleogene geological development of Syria

The varied stratigraphy and lithology of the five formations listed above reflect five different stages in the geological development of Syria during the Paleogene interval.

The **first stage** embraces the Paleogene and the larger part of the Early Eocene including the *Morozovella aragonensis* Zone.

At the Cretaceous/Paleogene boundary Syria experienced tectonic movements and a paleogeographic transformation. Almost everywhere the Cretaceous and Paleogene deposits are separated by an erosional surface and a hiatus. These tectonic events were most intensive in the areas of uplift such as Jebel An-Nusseiriyeh, Rutbah, Jordan, and Jebel Adb El-Aziz. Here the duration of the hiatus was significant and the various subdivisions of the Upper Cretaceous (sometimes even Jurassic) were transgressively overlain by various Paleocene, Lower Eocene and even Middle Eocene units. Evidently, after the post-Cretaceous regression these uplifted areas became islands or submerged shallow banks. These were subject to a vigorous hydrodynamic regime which resulted in nondeposition or destruction of weakly consolidated sediments.

In the depressed areas, such as the Palmyrides, the eastern slope of the Anti-Lebanon, and the Mesopotamian Basin, the hiatus was of short duration and the Paleogene sequence starts with the Danian *Turborotalia pseudobulloides* or *Turborotalia trinidadensis* Zones. The fact that the sediments below and above this erosional surface are lithologically similar emphasizes that the sedimentartary environment in these subsiding areas was not drastically changed by these events.

We suggest that the sedimentation at the Cretaceous/Paleogene boundary was continuous in those definitely subsiding areas, such as the Aafrine Graben, the depression to the south of the Baer-Bassit Massif, and the El-Bardeh area of the Palmyrides. Here the sediments of the Danian Stage are very thick and there is no evidence of a gap. It is not impossible that future thorough investigations will discover the presence of the *Parvularugoglobigerina eugubina* Zone.

During Paleocene–Early Eocene time (up to the *Morozovella aragonensis* Zone) a very monotonous pile of pelagic marly-clayey sediments with abundant planktonic foraminifera was deposited (the open-sea facies). These monotonous oozes testify to the constant position of the shore-line and relative tectonic quiescence.

The upwelling continued to exist along the northern margin of the Jordan Uplift. The abundant phytoplankton accumulated in the sediments and this so-called planktonic rain resulted in the formation of layers with a high content of organic matter (kerogen limestones and black combustible shales).

The intercalations of shallow-water deposits (nummulitic and detrital limestones with algal fragments) were only discovered in the Jebel An-Nusseiriyeh Uplift and in the axial part of the Anti-Lebanon. Such findings indicate that even during the Paleocene-Early Eocene transgression with its predominantly pelagic sedimentation these two areas were subjected to positive tectonic motions.

By the end of the Paleocene, elevated areas around Syria were peneplained and input of terrigenous silty-clayey material ceased. In the lower part of the Early Eocene (the *Morozovella subbotinae* and *Morozovella formosa* Zones) calcareous clays were gradually replaced by marls and clayey limestones and in the *Morozovella aragonensis* Zone chalky limestones appeared.

The **second stage** of the geological development refers to the latest Early Eocene (the *Acarinina pentacamerata* Zone). Deposits of this age are encountered in all regions of Syria and the Paleogene transgression reached its peak. The character of the sediments changed radically. As before, the monotonous pelagic sediments dominated but the accumulation of calcareous oozes was accompanied by deposition of silica that resulted in the appearance of numerous layers, lenses, and concretions of black and brown flints and grey silicified limestones. Areas with shallow-water organogenic limestones in Jebel An-Nusseiriyeh, the Baer-Bassit Massif, in the axial part of the Anti-Lebanon, and especially in the Rutbah Uplift were enlarged. In the Rutbah Uplift the precipitation of phosphatic matter also took place and created commercial phosphorite deposits. This process was a result of upwelling at the flank of this uplift. The short time interval of the *Acarinina pentacamerata* Zone was characterized by high rates of sedimentation and this zone accumulated a considerable thickness. Thus, the end of the Early Eocene was marked by differential tectonic motions – rapid subsidence in depressions coexisted with enlargement of the elevated areas.

The **third stage** of the geological development corresponds to the Middle Eocene. Vast areas continued to be occupied by the marine basin. Middle Eocene deposits covered almost the same areas as those of the Lower Eocene. However, the facies of the Middle Eocene sediments are more diverse in appearance. The Palmyrides, the Mesopotamian Basin, the Aleppo Plateau, the eastern slope of the Anti-Lebanon, the Jordan Uplift, and the Drouz Depression were characterized by the development of calcareous pelagic oozes – classical white, soft, thick-bedded chalk with abundant planktonic foraminifera. The areas with shallow-water nummulitic, algal, detrital and hard fine-grained limestones increased in size on Jebel An-Nusseiriyeh, the central Anti-Lebanon, and the Rutbah Uplift. In places at Jebel An-Nusseriyeh, during the upper half of the Middle Eocene, the sea shrank and sandy-clayey sediments appeared in the Eastern Palmyrides (Jebel Al-Bishri). There are some gaps in sections along the northern slope of the Palmyrides, which lack the *Morozovella lehneri* Zone or the *Truncorotaloides rohri* Zone.

The **fourth stage** embraces all the Late Eocene. It is characterized by two features:

1) An increase in the diversity of the sediment facies types. In the Palmyrides and on the eastern slope of the Anti-Lebanon one can see both pelagic chalky limestones and shallow-water nummulitic or detrital limestones which replace each other laterally or upwards in the section. Accumulation of shallow-water sandy-clayey sediments continued in Jebel Al-Bishri. Only on the Halab Plateau and in the Mesopotamian Basin were pure chalky limestone facies typical;

2) The area of distribution of the Upper Eocene sediments was reduced – they were not found within Jebel An-Nusseiriyeh, the area of Al-Latheqiyeh, the western margin of the Aleppo Plateau (at Jebel Dweleh, Jebel Mrasras, and Jebel El-Wastani), and the Jordan Uplift. Taking into account the lithological and paleogeographical peculiarities, the Late Eocene was a forerunner of those events which were to take place in the Oligocene.

The **fifth stage** occupied the Oligocene. Its main features are a regression and development of shallow-water facies including foraminiferal (lepidocyclinid), algal-coralline, detrital, hard fine-grained and massive aphanitic limestones. In those areas of Syria, where the Middle-Upper Eocene sequence was represented by pelagic facies, it was replaced by a shallow-water sequence in the

Anti-Lebanon, the Palmyrides, and the Aafrine Graben. In those areas of Syria where the Middle-Upper Eocene succession was composed of shallow-water sediments, the Oligocene deposits are missing as at Jebel An-Nusseiryieh, the Baer-Bassit Massif, and the ridges of Jebel Dweleh and Jebel El-Wastani. The upper layers of the Oligocene in the Palmyrides and the Rutbah Uplift show an increase in terrigenous material. The pelagic sedimentation only continued in the Mesopotamian Basin and in the northern part of the Aleppo Plateau.

Throughout Syria, the Paleogene and the Neogene are separated by a **regional unconformity (**which is essentially a sixth stage).

Summing up, the geological development of Syria during the Paleogene represents a single megacycle. After a short regression and paleogeographic reorganization at the Cretaceous/Paleogene boundary, the vast Paleocene-Early Eocene transgression reached its peak in the terminal Early Eocene. Pelagic sedimentation predominated. Such conditions prevailed in the Middle Eocene although the first signals of the oncoming regression appeared in Bartonian time. In the Late Eocene the regression was underway, and this megacycle ended with the vast regression of the Oligocene.

Damascus is located at the foot of Jebel Qasioun which is composed of Upper Cretaceous limestones. (Photo V. A. Krasheninnikov, 1958).

F: Neogene

INTRODUCTION

The eastern Mediterranean Neogene deposits are developed in southern Turkey (the Adana or Seyhan depression, the Alexandretta depression, Antiochia or Lower Orontes depression), in Syria and partly in Lebanon. On the Mediterranean coast of Israel, sediments of this age are overlain by younger Quaternary sediments, and penetrated by many bore-holes. Sections of the Neogene of Syria yield very good data on the properties of sediments of this age in the Near and Middle East.

In Syria, Neogene sediments occupy vast areas (**Fig. 6.65**), and belong to three facies/paleogeographic types:

The first facies type – Mediterranean, is limited to the depressions (Al-Latheqiyeh) and grabens (Aafrine) in northwestern Syria as well as the Aleppo Plateau. During the Neogene, that region of the eastern Mediterranean had a free connection with the World Ocean (primarily the North Atlantic) and sedimentation took place under conditions of normal marine salinity, with appropriate assemblages of fauna and nannoflora. This connection was only broken at the end of the Miocene (Messinian time) and the evaporitic column of the Messinian Stage deposited under conditions of almost complete isolation except for an influx of marine waters of normal salinity from the Atlantic.

The second facies type – Mesopotamian, embraces the northeastern part of Syria (Mesopotamian depression), including the Al-Furat River valley, with its left tributaries (Nahr El-Balikh, Nahr El-Khabour) as well as the Jezireh plain up to the valley of the Tigris River (Nahr El-Dajleh). During the whole interval of the Miocene the Mesopotamian Basin had a restricted connection with the Mediterranean Sea (and the North Atlantic) as well as with the Indian Ocean. At the end of the Miocene and in the Pliocene, these links were completely severed. Therefore, the Mesopotamean Neogene is characterized by alternation of sediments of various salinity: normal marine salinity, homogeneous hypersaline, brackish, and fresh water, and at the end of the Miocene and in Pliocene, sediments of continental origin.

Lastly, the third facies type is represented by exclusively continental deposits: gray and red-colored terrigenous sequences of conglomerates, sandstones and clays developed in the intra-mountain depressions of the Palmyrides and the Anti-Lebanon.

A special, fourth type of Neogene rocks belongs to the volcanic formations (basalts). They are thickest in southwestern Syria (Jebel Drouz) but are also present in other areas of Syria (the Homs depression, Aafrine Graben, Aleppo Plateau, and Jezireh plain). The Neogene basaltic volcanism of Syria is considered in Chapter 7 of this volume.

The stratigraphy and history of paleogeographic links of the Neogene basins of the Mediterranean, Mesopotamian, and inner regions of Syria will be addressed separately.

MEDITERRANEAN BASIN

In the Mediterranean Basin marine Miocene sediments cover rather vast areas, whereas the marine Pliocene is developed as a narrow belt or strip along the Mediterranean seacoast and within the northern Al-Ghab Graben (**Fig. 6.66**).

In northwestern Syria Neogene deposits fill the narrow Al-Latheqiyeh Depression (Nahr El-Kabir valley) and the Aafrine Graben (Nahr Aafrine valley). The origin of these structures is connected with the Al-Latheqiyeh-Kilis system of northeast-trending faults (Kazmin, 1965).

In the area of the Al-Latheqiyeh Depression, this fault separates the Mesozoic limestones of the Jebel An-Nusseiriyeh horst to the south and the rocks of the Baer-Bassit ophiolitic complex in the north. The depression formed at the time of the Cretaceous/Paleogene transition, just following emplacement of the ophiolitic series (oceanic crust) onto the continental basement of the Arabian Plate autochthon. This conclusion is supported by the fact that the facies of the Upper Maastrichtian (neoautochthon) is the same (pelagic limestones) both to the north and the south of this fault. In the Paleogene, the facies patterns of the deposits are marked by a major change: to the south of this fault, a thin less than 200 m unit of shallow-water limestones with nummulitids was deposited, while to the north of the fault, a thick (>1 km) formation of clays, marls and limestones with planktonic foraminifera and radiolarians can be observed (Krasheninnikov, 1994). The amplitude of displacement can be estimated at 1-2 km. Nevertheless, the Al-Latheqiyeh Depression itself originated considerably later, in the Early Miocene, as a result of intensive subsidence.

The Aafrine Graben is characterized by a similar history of formation. It is located in a more northern part of the Al-Latheqiyeh-Kilis lineament, which separates the ophiolitic complex of Kurd-Dagh (allochthon) and the marginal part of the Arabian Plate (autochthon). A very thick clayey Paleogene sequence is developed to the northwest of this fault, whereas shallow-water limestones of insignificant thickness can be observed to the southeast of this lineament. It is necessary to add that in this area the Paleogene is poorly exposed, being covered by Miocene rocks. Here, the formation of the graben is also connected with the very beginning of the Miocene (Aquitanian).

Between these two subsiding structures, in the vicinity of Jisr Ash Shaghour and Harim (the area of the Pliocene-Quaternary Al-Ghab Graben), Neogene sediments occupy the marginal part of the vast Antiochia depression, located within Turkey.

In these areas of northwestern Syria, the Miocene is represented by a wide spectrum of litho-facies types – from pelagic chalk-like white limestones up to boulder conglomerates and reef limestones. Another type of Miocene sediment can be observed on the Aleppo Plateau, stretching from the Al-Ghab and Aafrine Grabens in the

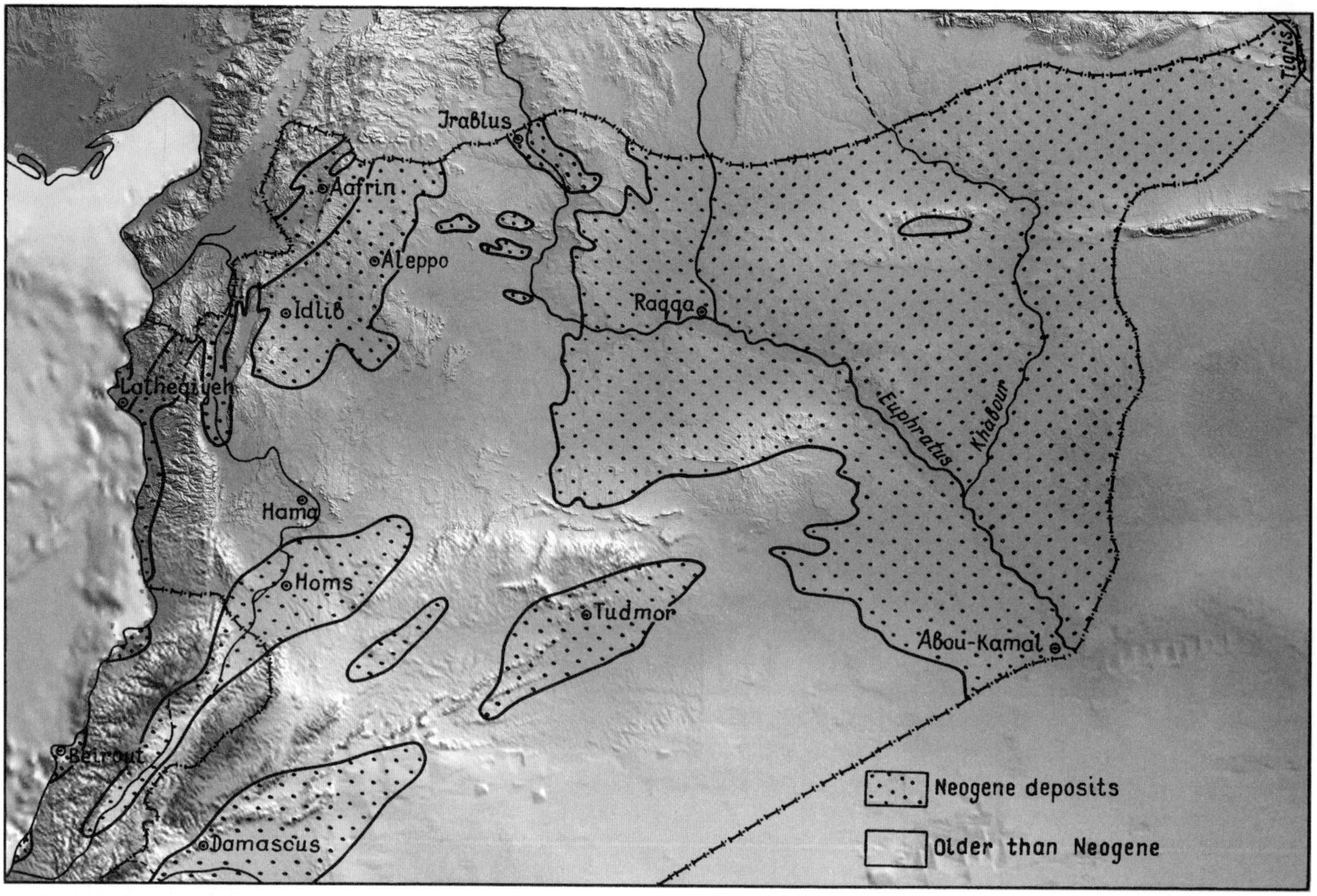

Fig. 6.66a. The distribution of the Neogene deposits in Syria.
A – the Mediterranean Basin; B – the Mesopotamian Basin; C – inner basins with continental sedimentation (Damascus, Homs, Ad-Daww and Feidet El-Mazra'a depressions).

west up to the Al-Furat (Euphrates) River in the east. Here the Miocene is composed of hard, thick-layered limestones which overlie and protect the soft marls and limestones of the Middle and Upper Eocene and Oligocene. To the west of Aleppo and El Bab they form a continuous cover (the Idlib Plateau) while to the east of these towns, and as far as the Al-Furat valley, they were preserved as outliers protecting the underlying soft Eocene rocks. In contrast, the lithology of the Pliocene sediments is monotonous: clay and marl units with subordinate interlayers of sandstones.

French geologists and paleontologists contributed significantly to the investigation of the Mediterranean Neogene of Syria (Dubertret, 1945, 1953a, b; Dubertret and Roger, 1938; Dubertret and Cottreau, 1938; David 1933a). Their stratigraphic conclusions were based mainly on molluscan faunas. They singled out:

- the marly-terriginous pile of the Lower Miocene (the Burdigalian Stage) in the Aafrine Graben and on the Baflioun Swell,
- massive limestones of the Middle Miocene (the Helvetian Stage) of the Idlib Plateau,
- clayey-marly deposits of the Middle Miocene (the Tortonian Stage) in the Nahr El-Kabir valley,
- gypsiferous formation of the Upper Miocene (the Pontian Stage) in the lower reach of this river,
- marly, clayey and sandy sediments of the Pliocene in the vicinity of Al-Latheqiyeh.

The peculiarities of the Neogene scale of that time are reflected in the ideas of these French geologists.

In 1958-1961 geologists of the Soviet Union carried out geological mapping of Syria at a scale of 1:200,000, including the sheets of Al-Latheqiyeh (Shatsky, Kazmin, and Kulakov, 1966) and Aleppo (Protasevich and Maksimov, 1966). The main sections of the Mediterranean Neogene are located on these two sheets. These two regions were covered by 1:50,000 scale geological mapping (Kazmin and Kulakov, 1968; Protasevich and Kurbatov, 1968) on account of the complicated structure of the Al-Latheqiyeh Depression and the adjacent area of the Baer-Bassit Ophiolite Massif, as well as the region of the Aafrine Graben and the nearby Kurd-Dagh Ophiolite Massif,

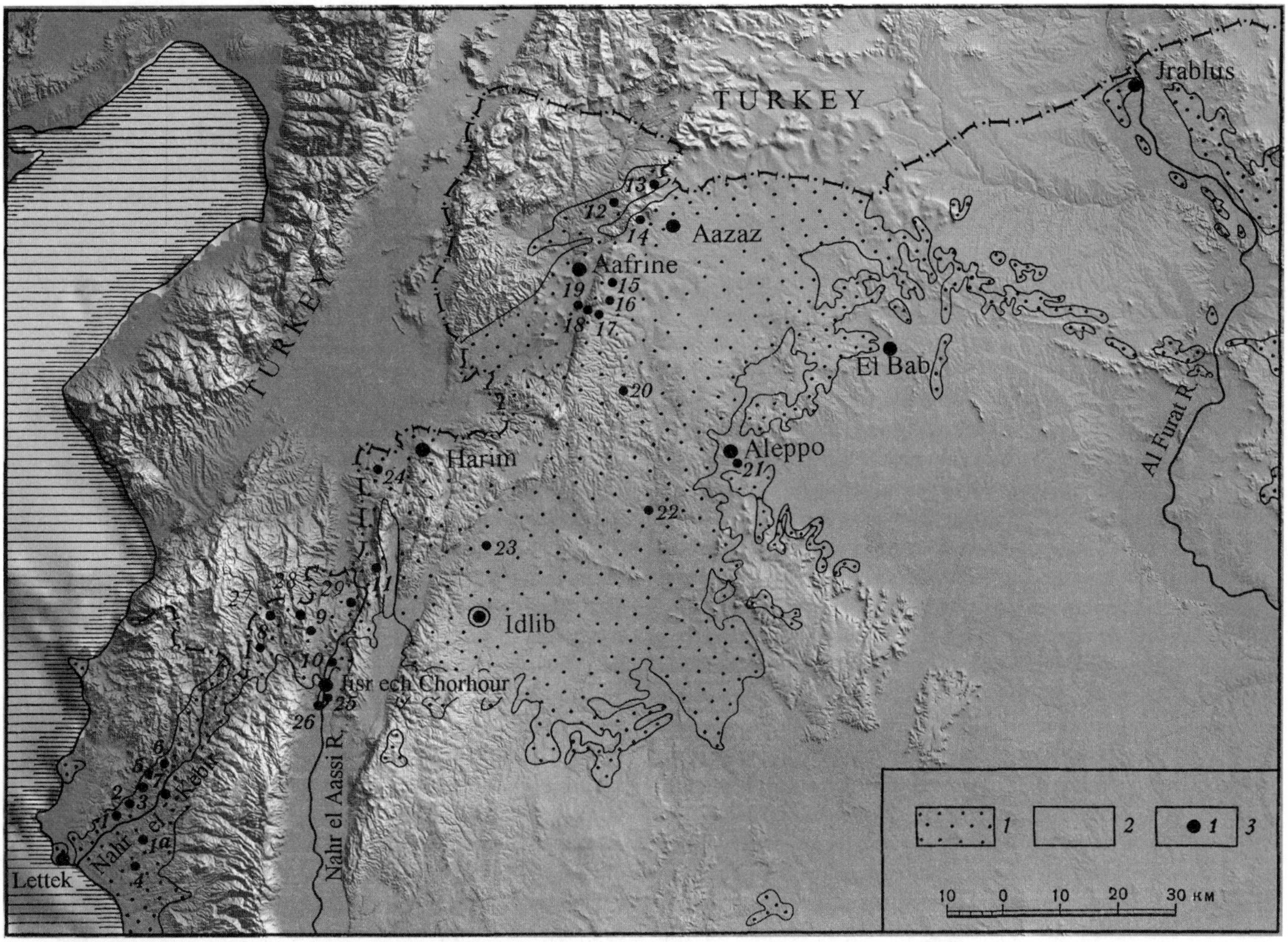

Fig. 6.66b. The distribution of the marine Miocene deposits in northwestern Syria (the Mediterranean Basin) and location of the main sections.
1 – areas occupied by marine Miocene deposits; 2 – areas where marine Miocene deposits are absent; 3 – sections (outcrops and bore-holes) of Miocene deposits shown on figures and mentioned in the text: 1 – Kesladjouk, 1a – Zoubar, 2 – Tcheqourdjaq, 3 – Qara Jalie, 4 – Nqoura, 5 and 6 – Guebelli-1 and Guebelli-2, 7 – Bachoura, 8 – Zeynieh, 9 – Enn Nahr (Beckfala), 10 – Al-Ghab (Abou Hasi hill), 11 – Kafer Hinnd Foani, 12 – Buyuk-Haloubi, 13 – Parsa-Dagh, 14 – Baflioun, 15 – Mashaale, 16 – Kourt-Farmoz, 17 – Goballi, 18 – Bteine (bore-hole), 19 – Bassouta, 20 – Aanadane (bore-hole), 21 – Aleppo, 22 – Souarhiye (bore-hole), 23 – Berne (bore-hole), 24 – Salkine, 25 – Mhannbel, 26 – Ain El-Aarouz, 27 – Kheurbet ej Jaouz, 28 – Melahnd, 29 – Hammam ech Sheikh Issa.

Based on all these materials, geological maps of Syria at scales of 1:500,000 and 1:1,000,000 were prepared with appropriate explanatory notes (Ponikarov et al., 1966, 1967). These explanatory notes contain quite detailed lithological and stratigraphical descriptions of the Mediterranean Neogene. Planktonic and benthic foraminifera provided the paleontological basis which allowed subdivision of the Neogene deposits into stages, subseries, and series (Krasheninnikov, 1966, 1968, 1971; Ponikarov et al., 1969).

In 1988-1990 the Neogene stratigraphy of northwestern Syria was investigated and the sediments of this age were subdivided based on the planktonic foraminifera zonal scale (Krasheninnikov, 1994; Krasheninnikov et al., 1999) and the Mediterranean Neogene sediments correlated with those of the Mesopotamian Basin.

Miocene

The Miocene is represented in its complete scope by all three subseries, although the spatial distribution of the Lower, Middle, and Upper subseries varies reflecting the various transgressions and regressions of this time.

Lower Miocene

The Lower Miocene succession is very complete in some sections, including both the Aquitanian and Burdigalian Stages. Due to the complicated disjunctive tectonics of northwestern Syria, with vertical displacements of tectonic blocks and consequent erosion of sediments, only some intervals of the Lower Miocene were preserved in other sections. The Aquitanian and Burdigalian

Stages are represented by very different lithologies in the Al-Latheqiyeh Depression, including adjacent areas of the northern part of the Al-Ghab Graben (**Fig. 6.67**), and in the Aafrine Graben (**Fig. 6.68**). Calcareous rocks predominate in the first and second structures and terrigenous rocks in the third.

The Aquitanian Stage

In the **Al-Latheqiyeh Depression**, an excellent section of the Aquitanian is located near the village of Zeynieh, on the left bank of the Nahr El-Abiad River. Sediments of this age unconformably overlie an uneven surface of organogenic and hard fine-grained limestones of the Middle Eocene. At the base of the section lies a layer of white, light-gray and yellowish clayey limestones with interbeds of hard fine-grained limestones and detrital limestones with operculinids, lepidocyclinids and miogypsinids. This basal layer contains terrigenous (sand and gravel) and detrital material. Upwards in the section, the number of hard limestone interbeds diminishes, gradually replaced by a member of layered, soft marls and more compact light-gray, clayey limestones. The Aquitanian is terminated by a thick column of monotonous white and light-gray soft chalky limestones, poorly-bedded, with blocky vertical jointing. The total thickness of the Aquitanian is 144m.

Aquitanian sediments appear in outcrops to the southwest and northeast of Zeynieh. To the southwest of Zeynieh near the village of Bachoura (in the Nahr El-Kabir valley) this stage comprises white and light-gray clayey limestones and marls lacking distinct bedding. Due to numerous faults, it is difficult to estimate the true thickness of the sediments, but they are clearly >60 m.

Near the village of Guebelli (northwestern slope of the depression) the Aquitanian Stage, as at Zeynieh, is represented in its entirety. It is composed here of a monotonous column of white, light-gray and light-yellow soft clayey and chalky limestones with subordinate intercalations of marls which separate beds of hard, fine-grained limestones. The thickness reaches 150 m.

To the southwest of Guebelli, near the village of Qara-Jalieh, the lithology of the Aquitanian is somewhat different. As before, soft, clayey, thick-bedded limestones and marls, light-gray and gray in color predominate, but interbeds of terrigeneous rocks – brownish and dark-gray polymictic fine- and medium-grained sandstones appear. The number of sandstone intercalations increases upwards. In the Aquitanian outcrops area, there are some faults, and estimation of the true thickness becomes difficult. The composite thickness is probably not less than 100 m. In all outcrops, Aquitanian rocks unconformably overlie Middle Eocene limestones.

To the southwest of Qara-Jalieh, the Lower Miocene is overlain by younger Miocene and Pliocene deposits. On the coastal plain, 20 km to the northeast of Al-Latheqiyeh, the Aquitanian was penetrated by bore-holes near Machkita village – soft clayey limestones of white and cream color, up to 30 m in thickness. They unconformably overlie Middle Eocene limestones.

On the Mediterranean coast of Syria to the south of the Nahr El-Kabir valley, Aquitanian deposits crop out near Nqoura (12 km to the south of Haffe). They are represented by monotonous soft clayey pelitomorphic limestones which are white, light-gray and yellowish in color. Sometimes they are replaced with white chalky limestones, poorly-bedded and with blocky vertical partings, up to 55 m in thickness. Here, the sediments of the Aquitanian disconformably overlie chalky limestones of the Middle Eocene, and only microfauna allows distinguishing between these units of similar lithology but considerably differing age. The contact of the Eocene/Aquitanian is concealed within the valley of a small creek, and is difficult to see.

Aquitanian deposits are well-exposed to the northeast of Zeynieh, unconformably overlying different horizons of the Eocene. In outcrops of the Nahr El-Abiad River (near Enn-Nahr and Bekfala villages) the Aquitanian is represented by white clayey limestones and light-gray marls. Here they are characterized by distinct layering, and poorly-bedded soft chalky limestones with blocky partings are comparatively rare. The thickness reaches up to 120 m. The same lithogies characterize the Aquitanian in outcrops along the highway Al-Latheqiyeh-Aleppo, near the descent to the Al-Ghab Graben (Abou-Hasi hill). In this area the Aquitanian consists of well-bedded soft light-gray and yellowish marls and clayey limestones with interbeds of harder fine-grained limestones which form low escarpments, and some layers of thin (4-6 cm) detrital limestones and brown medium-grained sandstones. The thickness may reach 140 m.

Aquitanian deposits occupy vast areas of the Al-Ghab Graben, between the towns of Harim and Salqine in the north and Jisr Ash-Shaghour in the south. They overlie Eocene limestones unconformably in all localities. The lithological features of the Aquitanian rocks are somewhat different in comparison with those of the Al-Latheqiyeh Depression. The Aquitanian here includes yellow and light-gray marls and clayey limestones, well-bedded, comparatively soft, with interbeds of yellow massive and hard clayey limestones. As a result of vertical tectonic movements, Aquitanian sediments are sometimes located on the surface of horst blocks where they may be strongly eroded and weathered (the Jebel Douele and Jebel Amin ridges). Therefore, their visible thicknesses are very variable, from several meters to 50 m. Typical exposures of yellow limestones and marls of the Aquitanian can be observed near the villages of Ein Ghazal, Shambar-Faouqani, Douele, Hammam ech Sheikh-Issa, and Zakhar-Ashir mountain.

Sediments of the Aquitanian Stage contain extremely rich assemblages of particularly well-preserved planktonic and benthic foraminifera. Especially abundant microfauna is present in the white, soft, chalky limestones typical of some sections of the Al-Latheqiyeh Depression.

The Aquitanian Stage consists of two planktonic foraminiferal zonal subdivisions. The lower is the *Globigerinoides primordius–Globorotalia kugleri* Zone (N4b). This zone is characterized here by rare specimens of *Globigerinoides primordius* and much more common *Globigerina venezuelana, G. praebulloides, G. juvenilis, G. woodi woodi, Globoquadrina praedehiscens, Catapsydrax dissimilis, C. unicavus, Globorotalia nana,* and *G. siakensis*. Foraminifera of this zone were identified in the basal layers of the Aquitanian Stage.

The second zonal subdivision combines the *Catapsydrax dissimilis* and *Catapsydrax stainforthi* Zones (N5-N6), which are undifferentiated in the Mediterranean region. The assemblage of foraminfera includes these index-species, accompanied by *C. unicavus, Globigerina venezuelana, G. praebulloides, G. woodi woodi, G. woodi connecta, G. juvenilis, G. angustiumbilicata, G. falconenesis, Globoquadrina praedehiscens, Globorotalia siakensis, G. nana* and rare specimens of *Globigerinoides trilobus*. The number of specimens of the latter species gradually increases upwards in the section. These two undifferentiated zones constitute the main part of the Aquitanian Stage.

Benthic foraminifera are quantitatively not as abundant as the planktonic ones, but they are extremely diverse. More than 500 species, including agglutinated and calcareous species, are present, and it is not possible to list them all. Some important species belong to the following: *Rheophax* sp., *Hormosina* sp., *Ammobaculites* sp., *Ammodiscus parialanus* Hedb., *Vulvulina pennatula* (Batsch) var. *italica* Cushman, *Textularia vautrini* Cuv. et Sz., *T. lalickeri* Cushman et Renz, *Siphotextularia* spp, *Saccammina sphaerica* Sars, *Gaudryina flintii* Cushman, *Clavulinoides jarvisi* Cushman, *C. eucarinatus* Cushman et Bermudez, *Schenckiella suteri* Cushman et Stainforth, *Dorothia brevis* Cushman et Stainforth, *Karreriella subcylindrica* (Nuttall), *K. gaudryinoides* (Forn.), *Uvigerina gallowayi* Cushman, *U. rustica* Cushman et Edwards, *Siphogenerina multicostata* Cushman et Jarvis, *Neobulimina* sp., *Angulogerina vicksburgensis* Cushman, *Bulimina alazanensis* Cushman, *Globobulimina* sp., *Robulus nuttalli* Cushman et Renz, *Buliminella grata* Parker et Bermudez, *Anomalina dorri* Cole, *A. alazanensis spissiformis* Cushman et Stainforth, *Planulina renzi* Cushman et Stainforth, *P. marialana* Hadley, *Cibicides mexicanus* Nuttall, *C. isidroensis* Cushman et Renz, *Marginulina cristellaroides* Cz., *Pullenia bulloides* (d'Orb.), *P. aff. compressiuscula* Reuss, *Melonis nicobarense* (Cushman), *Gyroidina girardana perampla* Cushman et Stainforth, *G. laevigata* d'Orb., *G. altispira* Cushman et Stainforth, *G. complanata* Cushman et Stainforth, *Ellipsonodosaria mappa* Cushman et Jarvis, *Ellipsoglandulina multicostata* (Gall. et Morr.), *Pleurostomella ecuadorana* Cushman et Stainforth, *P. brevis* Schwager, *P. alternans* Schwager, *Quinqueloculina rugosa* d'Orb., *Nodosarella robusta* Cushman, *N. subnodosa* (Guppy), *N. salmojraghii* Martinotti, *Cassidulinoides erecta* Cushman et Renz, *Sphaeroidina chilostomata* Gall. et Morr., *Sph. ciperana* Cushman et Renz, *Chilostomella cylindroides* Reuss, *Siphonina pulchra* Cushman, *Lingulina grimsdalei* Cushman et Renz, *Nodosaria lamellata* Cushman et Stainforth, *N. boffalorae* Martinotti, *N. stainforthi* Cushman et Renz, *Siphonodosaria verneuili* (d'Orb.}, *S. paucistriata* (Gall. et Morr.), *Chrysalogonium elogatum* Cushman et Jarvis, *Chr. tenuicostatum* Cushman et Bermudez, *Chr. lanceolum* Cushman et Jarvis, *Chr. longicostatum* Cushman et Jarvis, *Bolivina arta* Macf., *Parrella mexicana* Cole, *Plectofrondicularia mexicana* Cushman, *P. raricosta* (Karr.), and *Almaena alticosta* (Sigal).

The strong predominance of planktonic foraminifera and the systematic composition of the benthic ones (buliminids, lagenids, nodosariids, cassidulinids, chilostomellids, anomalinids) indicate that the Aquitanian Stage of the Al-Latheqiyeh Depression is represented by pelagic calcareous oozes deposited under comparatively deep-water conditions of the upper bathyal zone (200-700 m), with minimal input of terrigenous material.

In the **Aafrine Graben**, sediments of the Aquitanian Stage unconformably overlie Oligocene lepidocyclinid and algal-coral limestones, but along the strike, successively overlie older clayey and fine-grained Eocene limestones.

The most typical section of the Aquitanian Stage is located near the village of Buyuk-Haloubi. This section is about 340 m thick and contains several members consisting of calcareous and terrigenous rocks.

The calcareous members comprise rhythmic alternations of light-gray clayey limestones, gray-green marls, and dark gray clays, 10-40 cm in thickness. The harder limestone layers form small ribbed escarpments on the valley slopes. Intercalations of greenish sandstones and gravel lenses occur infrequently. The section contains three members of calcareous rocks, from 27 to 65 m thick.

Members of terrigeneous rocks are composed of alternating sandstones, gravels, and conglomerates with subordinate intercalations of marls and clays. Sandstones are gray and gray-green, from fine- to coarse-grained, often cross-bedded, sometimes with tuffaceous material, plant detritus, and fossilized tree trunks. Conglomerates and gravels are present as lenses which quickly wedge-out laterally, frequently cross-cutting each other. Pebbles of flint and limestone predominate, 1-3 cm in diameter. Boulder conglomerates formed by assorted and unoriented blocks of sedimentary and volcanic lithologies, 5-50 cm in diameter, form a special lithological feature. Individual boulders may be 1-1.5 m in diameter. These rocks were formed by periodic high energy streams on the steep slopes of the graben. Dark-gray and black clays, and sometimes coalified plant-rich sediments are present. The thickness of the terrigenous members stands out, from 50 to 90 m.

In the section of Buyuk-Haloubi the Aquitanian Stage begins with a basal conglomerate, followed by coarse sands and sandstones with thin 30-40 cm thick intercalations of basalt. It is terminated by a unit of light-gray hard slightly sandy limestones.

To the southwest of Buyuk-Haloubi (in the Aafrine River valley) the Aquitanian Stage generally preserves its composition – alternating limestones, marls, clays, sandstones and conglomerates, but the above-mentioned calcareous and terrigenous members of the Buyuk-Haloubi section cannot be traced along the strike. To the northeast of this village, the facies changes along the strike of the graben become more significant. Near the village of Aarab-Virane, sandstone, clay and marl alternate with greenish tuffaceous sandstones and 2-4 m thick layers of olivine basalt. Further to the northeast the terrigeneous rocks gradually wedge out and the Aquitanian is represented by gray clayey limestones and marls with subordinate clay and fine-grained sandstone intercalations.

Foraminifera are usually comparatively numerous in the calcareous clay and marl layers, whereas their content in the terrigeneous rocks is very reduced and they are poorly preserved.

The planktonic foraminiferal asssemblage consists of small

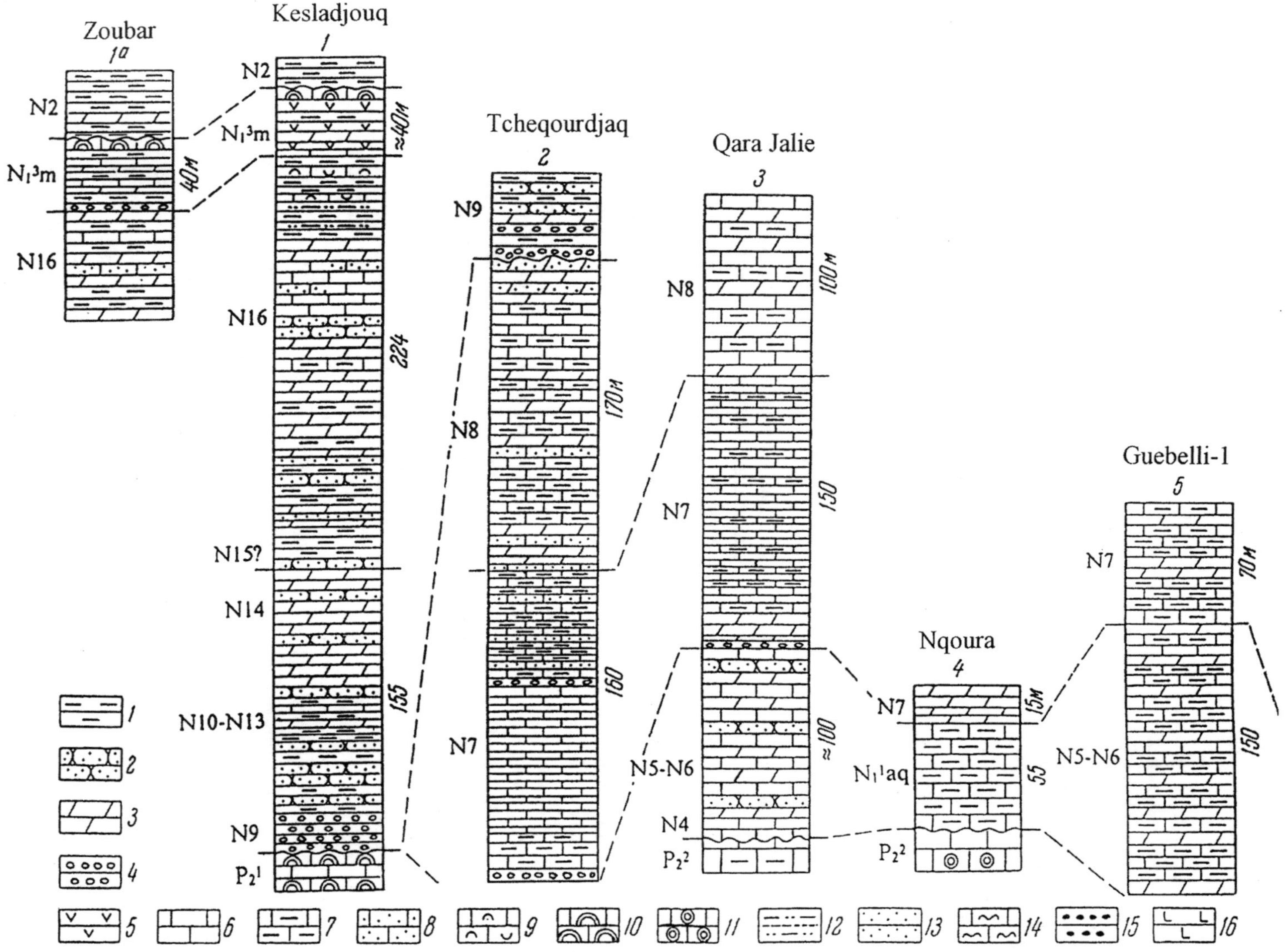

Fig. 6.67a. Sections of the Miocene deposts of the Al-Latheqiyeh Depression and the western part of the Al-Ghab Graben.

Legend (for Figs. 6.67 and 6.68):

Lithologic symbols: 1 – clays; 2 – sandstones; 3 – marls; 4 – conglomerates; 5 – gypsum; 6 – limestones; 7 – clayey limestones; 8 – sandy limestones; 9 – organogenic limestones (detrital and coquinas); 10 – algal limestones; 11 – nummulitic limestones; 12 – siltstones; 13 – sands and weakly cemented sandstones; 14 – silicified limestones; 15 – gritstones and conglomerates; 16 – effusive rocks.

Stratigraphic symbols: P1/2 – Lower Eocene; P2/2 – Middle Eocene; P3/2 – Upper Eocene; P3 – Oligocene; N1/1aq – Aquitanian Stage; N2/1 – Middle Miocene; N3/1 – Tortonian Stage, Upper Miocene; N3/1 – Messinian Stage, Upper Miocene; N2 – Pliocene. The planktonic foraminifera zonation: N4 – the *Globigerinoides primordius-Globorotalia kugleri* Zone; N5-N6 – undifferentiated interval of the *Catapsydrax dissimilis* and *Catapsydrax stainforthi* Zones; N7 – the *Globigerinatella insueta- Globigerinoides trilobus* Zone; N8 – the *Praeorbulina glomerosa* Zone; N9 – the *Orbulina suturalis-Globorotalia peripheroronda* Zone; N10-N13 – undifferentiated interval from the *Globorotalia peripheroacuta* Zone to the *Sphaeroidinellopsis subdehiscens-Globigerina druryi* Zone; N14 – the *Globigerina nepenthes-Globorotalia siakensis* Zone; N15 – the *Globorotalia continuosa* Zone; N16 – the *Globorotalia acostaensis* Zone; N8-N1/2 – undifferentiated interval of the *Praeorbulina glomerosa* Zone (Lower Miocene) - Middle Miocene.

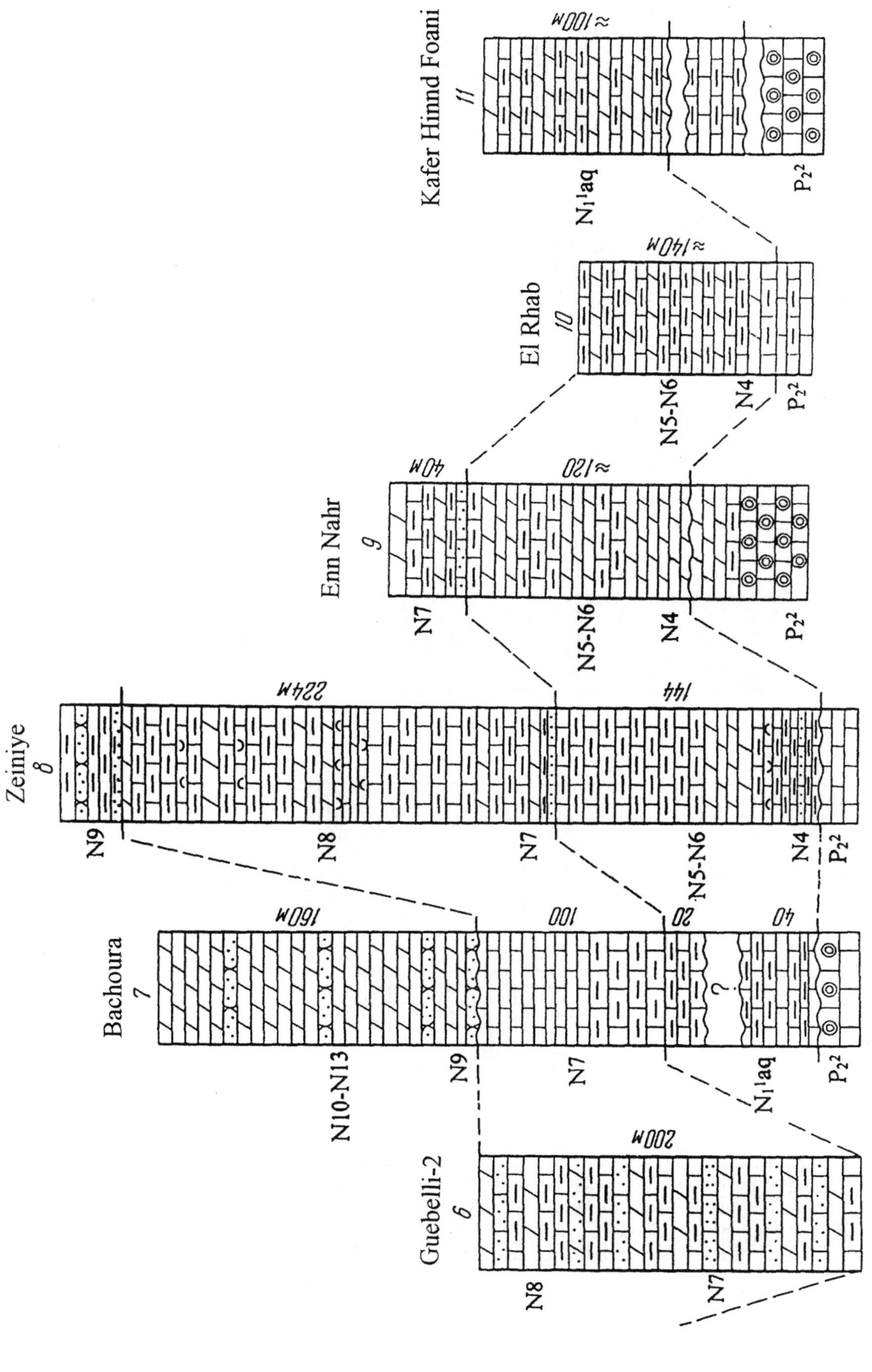

Fig. 6.67b (continuation). Sections of the Miocene deposits of the Al-Latheqiyeh Depression and the western part of the Al-Ghab Graben.

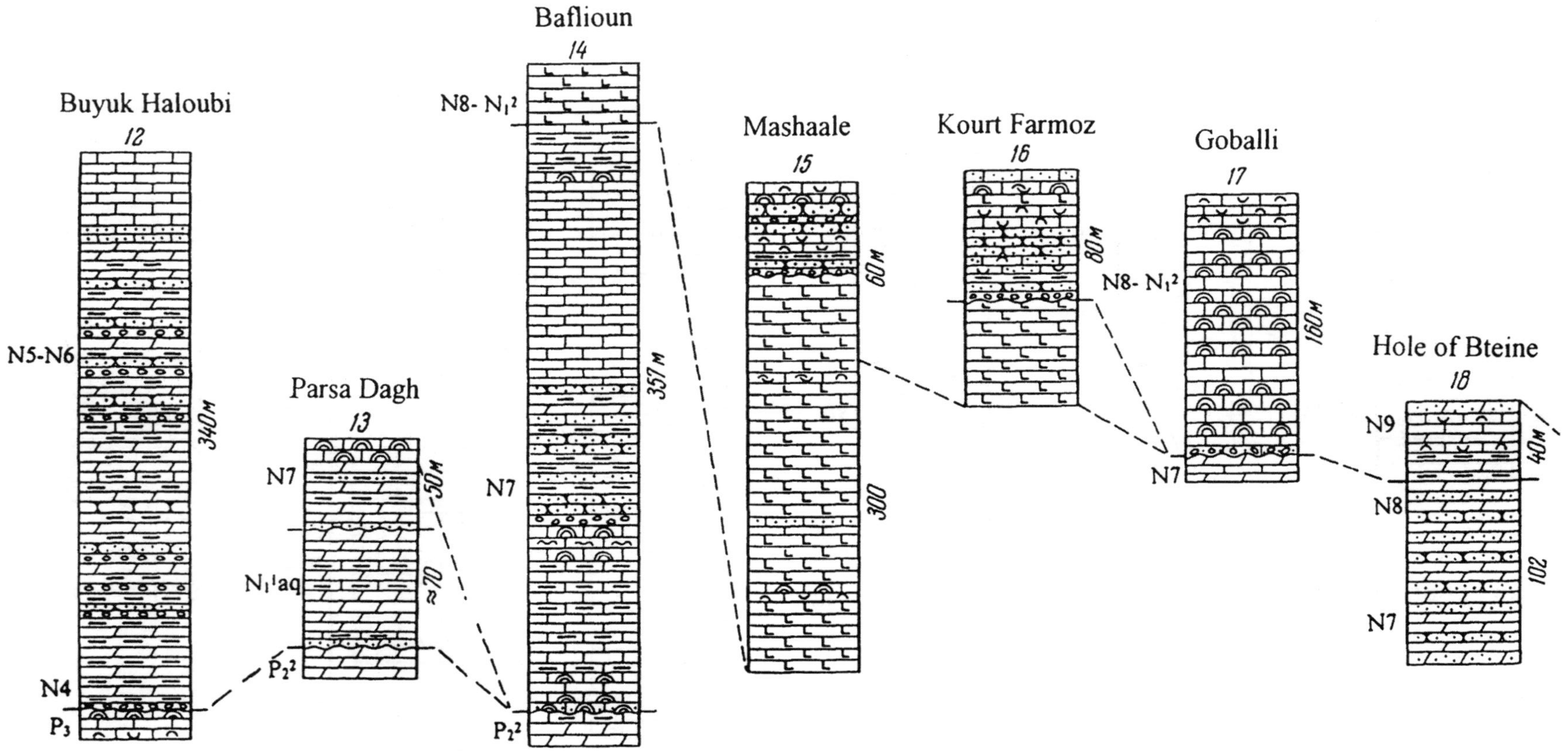

Fig. 6.68a. Sections of the Miocene deposits of the Aafrine Graben and the eastern part of the Al-Ghab Graben. Legend (see Fig. 6.67).

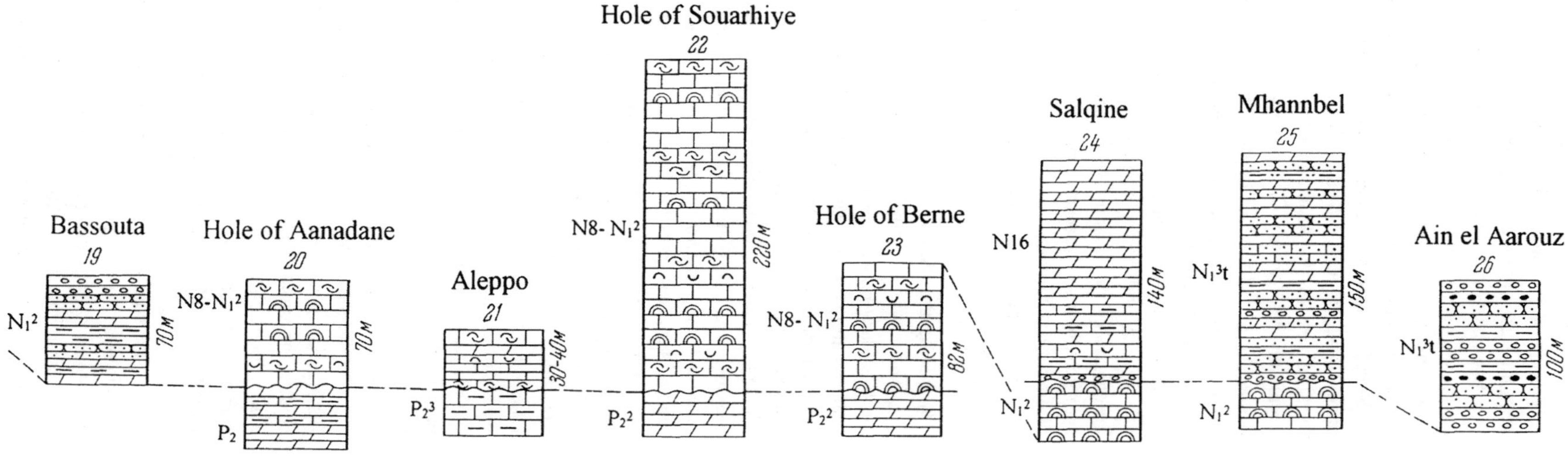

Fig. 6.68b (continuation). Sections of the Miocene deposits of the Aafrine Graben and the eastern part of the Al-Ghab Graben.
Legend (see Fig. 6.67).

specimens of *Globigerina juvenilis, G. woodi connecta, G. praebulloides, G. angustiumblicata, G. falconensis, Cassigerinella chipolensis,* and *Globorotalia nana.* Such species as *Globigerina venezuelana, Globoquadrina praedehiscens, Globigrinoides trilobus,* and *Globorotalia siakensis* are present as rare specimens. In the basal part of the section, planktonic foraminifera are very rare, and the *Globigerinoides primordius–Globorotalia kugleri* Zone was not established with certainty. In any case, the main part of the Aquitanian deposits of the Aafrine Graben belongs to the undifferentiated *Catapsydrax dissimilis–Catapsydrax stainforthi* Zones (N5-N6).

Benthic foraminifera are numerous, but their ratio to the planktonic forms changes in favor of benthic forms in comparison to the Al-Latheqiyeh assemblages. Among them, the same species occur as were found in the calcareous facies of the Al-Latheqiyeh Depression, but in association with forms that are more typical of shallow-water deposits.

The very large number of species found include the following more prominent species: *Vulvulina pennatula* (Batsch) var. *italica* Cushman, *V.* aff. *pectinata* Hantk., *Gaudryina flintii* Cushman, *Clavulinoides jarvisi* Cushman, *Dorothia brevis* Cushman et Stainforth, *Karreriella subcylindrica* (Nuttall), *Almaena hieroglyphica* (Sigal), *A. alticosta* (Sigal), *Bolivina algeriana* Glac., Magne et Mur., *B. miocenica* Macf., *B. arta* Macf., *B. alata* (Seg.), *B. imporcata* Cushman et Renz, *Rectobolivina virgata* (Cushman), *Bolivinella* spp., *Bifarina reticulosa* Cushman, *Plectrofrondicularia mexicana* Cushman, *P. alazanensis* Cushman, *P. raricosta* (Karr.), *P. vaughani* Cushman, *Cassidulina horizontalis* Cushman et Renz, *C. havanensis* Cushman et Reinhold, *Nodosarella subnodosa* (Guppy), *Trifarina bradyi* Cushman, *Siphogenerina multicostata* Cushman et Jarvis, *Angulogerina vicksburgensis* Cushman, *Bulimina alazanensis* Cushman, *Elphidium lobatum* Gall. et Hem., *E. kvesanensis* Artch., *Nonion dilatatum* Gall. et Hem., *Cibicides isidroensis* Cushman et Renz, *C. mantaensis* (Gall. et Morr.), *C. falconensis* Renz, *Anomalina alazanensis spissiformis* Cushman et Stainforth, *Gyroidina girardana perampla* Cushman et Stainforth, *G. complanata* Cushman et Stainforth, *Stomatorbina* ex gr. *concentrica* (Parker et Jones), *Siphonina pulchra* Cushman, *Cancris auriculus* (F. et M.), *Discorbis* ex gr. *mira* Cushman, *Siphonodosaria verneuili* (d'Orb.), *S. paucistriata* (Gall. et Morr.), *Chrysalogonium elongatum* (Cushman et Jarvis), *Ch. tenuicostatum* Cushman et Berm., *Ch. longicostatum* Cushman et Jarvis, *Nodosaria lamellata* Cushman et Stainforth, *N. adelinensis* Palmer et Bermudez, *Lagena trinitatensis* Nuttall, *L. pulcherrima enitens* Cushman et Stainforth, *Quinqueloculina rugosa* d'Orb., *Ellipsonodosaria mappa* Cushman et Jarvis, *Chilostomella cylindroides* Reuss, *Sphaeroidina chilostomata* Gall. et Morr., *Sph. ciperana* Cushman et Todd, *Miogypsina tani* Drooger, *M. gunteri* Cole, *Miogypsinoides complanatus* (Schlumberger), and small lepidocyclinas.

The lithology of the Aquitanian in the Aafrine Graben, the predominance of benthic foraminifera, and the appearance among them of shallow-water representatives (elphidiids, nonionids, discorbids, cibicidids, miogypsinids, etc.) all serve to indicate that deposition took place under unstable shallow-water conditions, accompanied by considerable input of coarse terrigenous material from the steep slopes of the graben walls.

Southeast of the Aafrine Graben, on the Baflioun Swell and on the Parsa-Dagh High to the north of the town of Aazaz, terrigenous sediments of the Aquitanian (beyond the graben) are replaced by soft white and yellowish, well-bedded marls and more compact yellowish clayey limestones. Near the village of Qastal they rest unconformably above Middle Eocene chalky limestones. Here the thickness of the Aquitanian exceeds 70 m. The basal layers apparently refer to the *Globigerinoides primordius–Globorotalia kugleri* Zone (N4b), while the overlying marls and limestones with assemblages of abundant planktonic foraminifera already belong to the undifferentiated *Catapsydrax dissimilis–Catapsydrax stainforthi* Zones (N5-N6). Benthic foraminifera here are numerous and diverse, and their systematic composition is closer to that of the calcareous facies of the Al-Latheqiyeh Depression. They differ significantly from the benthic assemblages of the terrigenous facies of the Aafrine Graben.

During the entire Aquitanian interval the **Aleppo Plateau** was an uplifted area affected by intensive erosional processes.

The Burdigalian Stage

There was essentially no change in the paleogeography at the boundary between Aquitanian and Burdigalian time, and the sedimentation process was transitional (sections near the villages of Zeynieh, Bachoura, Guebelli, Tcheqourjaq, etc.). Sometimes traces of erosion may be observed at the Aquitanian/Burdigalian contact (Parsa Dagh), or a layer of small pebble conglomerate may be present (Qara Jalieh). In the Early Burdigalian (the *Globigerinatella insueta–Globigerinoides trilobus* Zone, N7), the outline of the marine basin was preserved in the Al-Latheqiyeh Depression and the Aafrine Graben. Burdigalian sediments transgressively overlie Middle Eocene limestones only near the eastern slope of the Aafrine Graben (the village of Kazreha) and on the Baflioun Swell (the village of Kafer Janeh). Pronounced changes took place in the Late Burdigalian (the *Praeorbulina glomerosa* Zone, N8) when waters of the marine basin covered the Aleppo Plateau as well.

In the **Al-Latheqiyeh Depression**, the Burdigalian deposits preserved their lithologic patterns and calcareous rocks (chalky and clayey limestones, marls) strongly predominate. Sandy-clayey sediments are limited in distribution.

On the southern slope of this depression (near Zeynieh) the Burdigalian is represented by monotonous white pelitomorphic chalky limestone and light-gray and yellowish more compact clayey limestones, usually with poor, indistinct bedding and blocky partings. Intercalations of greenish marls, brownish detrital hard limestones, and fine-grained limestones also occur. The thickness reaches 224 m.

To the southwest of Zeynieh the Burdigalian crops out along the highway to Al-Latheqiyeh. Near Bachoura, the Burdigalian comprises two members. The lower one is composed of soft white and light-gray clayey limestones, relatively massive and with blocky partings, some 40 m in thickness. The upper member consists of well-bedded to hard platy fine-grained limestones and clayey limestones of light-gray and yellowish color. The thickness is up to 60 m. The section of the Burdigalian is incomplete here, as it is cut by several faults.

Along the northern slope of the Al-Latheqiyeh Depression, the Burdigalian deposits form a narrow belt. Here, two types of lithology can be distinguished.

The first type is characterized by the section near Guebelli. The Burdialian is represented here by white, light-gray, and yellowish marls, chalky and clayey limestones, which are soft and pelitomorphic, with minor quantities of detrital material. Intercalations of harder light-brown limestones with organogenic detrital material are common. Sometimes they contain sandy material, as well as small flint pebbles and large quantities of siliceous sponge spicules. In general, the lithological rock patterns are similar to those in the sections of Zeynieh and Bachoura.

The second lithofacies type is marked by a different pattern. Its typical features are observed in sections near Tcheqourjaq and Qara-Jalieh. Sandstones and conglomerates are widely distributed, as well as platy limestones. Soft clayey limestones are subordinate and white chalky limestones are absent.

In the section near Tcheqourjaq the Burdigalian has a two-fold structure. The lower unit, about 150 m thick, consists of bituminous clayey limestones and sandstones, with abundant siliceous sponge spicules. The limestones are of different shades of brown, their surfaces are covered by light-gray crusts, comparatively hard, thin layered to micro-laminated, easily split into thin plates, and sometimes sandy. Among them are thin intercalations (4-7 cm) and up to 5 m thick members of dark brown bituminous coarse-grained sandstones, as well as 1-1.5 m thick layers and lenses (to 2.5 m) of coarse conglomerates.

The upper member, about 170 m thick, consists of gray, greenish-gray, and brown clayey limestones and marls, often characterized by platy layering or vague nodular bedding. Siliceous sponge spicules are rare here. As above, lenses of polymictic sandstones, gravels, and conglomerates, with debris of molluscs, echinoids, bryozoa, and rhodophycean algae are common. Several boulders within the conglomerate horizons were 25-30 cm in diameter. Some conglomerate lenses with unsorted pebbles and irregular layering are associated with submarine sliding.

A similar structure for the Burdigalian was found in the neighboring section of Qara Jalieh, although the role of terrigenous rocks here is less significant. A bed about 0.7 m thick of small pebble conglomerates and gravels is located at the base of the sequence, although the microfauna indicates that there was no notable hiatus at the Aquitanian/Burdigalian boundary. This succession, about 250 m thick, includes three members. The lower one, about 20 m thick, is of light-gray, soft, poorly-bedded clayey limestones and marls. The middle member is of hard, thin-layered to platy fine-grained limestones and clayey limestones, dark gray and brown in color. Interbeds of brown polymictic sandstones are very common. They contain large quantities of siliceous sponge remains. The upper member (visible thickness about 100 m) consists of light-gray, soft clayey limestones and marls, with interbeds of hard yellowish limestones.

The southernmost exposure of the Burdigalian is located to the southeast of Al-Latheqiyeh near the village of Nqoura, where the coastal plain passes into the hilly piedmonts of Jebel An-Nusseiriyeh. Here, Aquitanian sediments are conformably overlain by a unit of thin-bedded, soft light-gray, gray, and yellowish marls, with intercalations of more massive clayey limestones with common Burdigalian microfaunas. The thickness in outcrops does not exceed 25 m.

As in the Aquitanian, foraminifera of the Burdigalian deposits are abundant and taxonomically diverse. Their asso-ciations are very different from calcareous and terrigenous rocks. Planktonic foraminifera predominate in the soft chalky and clayey limestones and marls. In all sections of the Al-Latheqiyeh Depression they allow subdivision of the Burdigalian into two zonal units: the *Globigerinatella insueta–Globigerinoides trilobus*, N7, and the *Praeorbulina glomerosa*, N8, Zones.

The *Globigerinatella insueta–Globigerinoides trilobus* Zone is characterized by abundant specimens of *G. trilobus,* together with *G. subquadratus, G. diminutus, G. altiaperturus, Globoquadrina dehiscens, G. altispira, G. quadraria, Globigerina bollii, G. foliata, G. falconensis, G. venezuelana, G. praebulloides, G. woodi woodi, G. angustiumbilicata, Cassigerinella chipolensis,* and *Globorotalia siakensis*. Rare specimens of *Catapsydrax dissimilis, C. unicavus,* and *Globorotaloides suteri* also occur. In the section of Zeynieh, single specimens of *Globigerinatella insueta* were found.

The *Praeorbulina glomerosa* Zone is marked by the appearance of *P. glomerosa glomerosa, P. transitoria, Globi-gerinoides bisphaericus, Globorotalia peripheroronda,* and *Globigerina druryi,* accompanied by continuing occurrence of *Globoquadrina dehiscens, G. altispira, Globigerinoides trilobus, G. diminutus, G. subquadratus, G. altiaperturus, Globigerina bollii, G. falconensis, G. foliata, G. angustiumbilicata, Cassigerinella chipolensis, Globorotalia siakensis,* and *Catapsydrax unicavus.*

Benthic foraminifera are exceedingly diverse, with about 600 species present. In the calcareous pelagic sediments, the most common species are *Textularia angularis* d'Orb., *Siphotextularia* spp., *Vulvulina pennatula* (Batsch) var. *italica* Cushman, *Gaudryina pseudocollinsi* Cushman et Stainforth, *Karreriella* aff. *gaudryinoides* (Forn.), *Quinqueloculina rugosa* d'Orb., *Sigmoilina* aff. *tenuis (Cz.), Chrysalogonium elongatum* Cushman et Jarvis, *Siphonodosaria verneuili* (d'Orb.), *S. paucistriata* (Gall. et Morr.), *Pseudoglandulina gallowayi* Cushman, *Lingulina ponceana* Gall. et Hem., *Planularia venezuelana* Hedb., *Ceratobulimina evoluta* Cushman et Jarvis, *Anomalina pseudogrosserugosa* Colom, *Cibicides falconensis* Renz, *Pullenia* aff. *bulloides* d'Orb., *Globobulimina hannai* Cushman et Ellis, *Melonis soldanii (*d'Orb.*), Elphidium ortenburgensis (*Egger*), Bulimina alazanensis* Cushman, *Uvigerina barbatula* Macf., *U. striatissima* Perconig, *U. rustica* Cushman et Edwards, *U. gallowayi* Cushman, *Valvulineria venezuelana* Hedberg, *Cibicorbis herricki* Hadl., *Pleurostomella bierigi* Palmer et Bermudez, *P. rimosa* Cushman et Bermudez, *P. brevis* Schwager, *Cassidulina tricamerata* Gall. et Hem., *Plectofrondicularia jarvisi* Cushman, *Chilostomella globata* Gall.

et Hem., *Ch. fakrounensis* Marks, *Chilostomelloides blanchardi* Deb. et Sac., *Bolivina scalprata miocenica* Macf., *B. arta* Macf., *B. ventricosa* Gall. et Hem., and *B. alata* (Seg.)

Sandy-clayey sediments and limestones with detrital material contain a rather different assemblage of benthic foraminifera. Common species include *Operculina* sp., *Elphidium onerosum* Bogd., *E.* aff *crispum* (L.), *E. kvesanensis* Artch., *Florilus communis* (d'Orb), *Asterigerina planorbis* d'Orb., *Caucasina* aff. *buliminoides* Bogd., *Loxostomum digitale* (d'Orb.), *Rotalia audouini* d'Orb., *Discorbis havanensis* Cushman et Bermudez, *Gyroidina perampla* Cushman et Stainforth, *Eponides parantillarum* Gall. et. Hem., *Almaena alticosta* (Sigal), *Siphonina tenuicarinata* Cushman, *Cibicorbis herricki* Hadl., *Bolivinella margaritacea* Cushman, *Plectofrondicularia raricosta* (Karr.), *P. trinitatensis* Cushman et Jarvis, and *P. morreyae* Cushman.

Complicated lateral substitutions of lithofacies and variability of paleocoenosis of foraminifera clearly indicate changes in sedimentary conditions in the Burdigalian of the Al-Latheqiyeh Depression. Areas of pelagic sediments with abundant planktonic foraminifera (chalky limestones and marls) are located not far from areas of terrigenous sediments (sandy clay to conglomerate lithofacies), where shallow-water benthic foraminifera predominate. Evidently turbidity currents and submarine sliding events played a significant role in the process of sedimentation.

To the northeast of Zeynie, at the junction of the Al-Latheqiyeh Depression with the Al-Ghab rift, the structure of the Burdigalian Stage is different. In sections near Bekfala and to the north of the Al-Latheqiyeh to Aleppo highway where it descends into the Al-Ghab Graben (the Abou Hasi hill and the nearby unnamed hill with a 547 m elevation marker) the *Globigerinatella insueta-Globigerinoides trilobus* Zone is composed of monotonous yellowish and light-gray marls which are soft, well-bedded, and sometimes with blocky joint partings. They contain separate interbeds of harder fine-grained limestones of similar coloration. In the section of Bekfala the most common lithology is white, poorly-bedded chalky limestones. Their thickness reaches 110-120 m. In the upper part of this unit (but below the top) there occurs a layer of hard algal limestone consisting of *Lithothamnion* rhodolites, molluscan and echinoid debris, and operculinid larger foraminfera. The thickness is 2-3 m. Above this layer the yellowish marls recur, with faunas of the *Globigerinatella insueta-Globigerinoides trilobus* Zone.

These sediments are replaced by a unit of rocks of different lithology – very hard, brown aphanitic and yellowish fine-grained limestones with subordinate interbeds of detrital algal and pure algal limestones, as well as softer yellow marly limestones. Their thickness is about 25 m. This unit of hard limestones, expressed as distinct cliffs in the general slope relief, is covered by the greenish marls, calcareous clays, clayey sandstones, and yellowish limestones with foraminifera of the *Globorotalia peripheroacuta* Zone (N10) of the Middle Miocene. Thus, the unit of hard limestones corresponds to the *Praeorbulina glomerosa* (N8) Zone of the Lower Miocene, and the *Orbulina suturalis–Globorotalia peripheroronda* (N9) Zone of the Middle Miocene.

A similar structure of the Burdigalian is observed in the northern part of **the Al-Ghab Graben**. South of the town of Salqine on the Jebel Amin ridge (Zahar Ashir mountain), the *Globigerinatella insueta-Globigerinoides trilobus* Zone is represented by yellow marls with intercalations of yellow and light-yellow hard fine-grained limestones. They are overlain by massive thick-bedded limestones with blocky joint partings and uneven vugged surface. They are about 145 m thick. Their lithological features are very diverse: white, light-gray and yellowish varieties, fine-grained or well-crystallized marble-like limestones, aphanitic limestones, small nodular algal limestones. The lower part of these massive limestones corresponds to the *Praeorbulina glomerosa* Zone of the Lower Miocene, because any traces of erosion and hiatus at the boundary with the underlying sediments of the *Globigerinatella insueta–Globi-gerinoides trilobus* Zone are missing.

Proceding up the section in the high cliff located above the left bank of the Orontes River near the village of Hammam ech Sheikh Issa, the yellow marls and clayey limestones of the *Globigerinatella insueta–Globigerinoides trilobus* Zone are replaced by massive white and light-gray limestones. The latter are very hard algal-bryozan, aphanitic or fine-grained, with echinoids and molluscs. The basal part of this unit must be referred to the *Praeorbulina glomerosa* Zone. Due to the poor exposure of the outcrop, the character of the contact between the soft marls and the hard limestones is difficult to determine.

Consequently, in the eastern part of the Al-Latheqiyeh Depression and in the north of the Al-Ghab Graben, the Burdigalian is marked by significant facies changes. The lower part of the Burdigalian (the *Globigerinatella insueta-Globi-gerinoides trilobus* Zone) preserves the lithology (soft marls and clayey limestones with planktonic foraminifera), but in the upper half of this stage (the *Praeorbulina glomerosa* Zone), pelagic marls and soft clayey limestones are rapidly replaced along the strike by shallow-water massive fine-grained and algal limestones.

In the **Aafrine Graben**, Burdigalian deposits are poorly exposed. Information on the Burdigalian sediments here is given by the Bteine bore-hole, which was drilled on the right bank of the Aafrine River across from Bassouta village. The Burdigalian Stage is comprised here of a unit of alternating gray-greenish clays and marls, and fine- and medium-grained clayey sandstones. According to drilling data, the thickness does not exceed 100 m, however the contact with the Aquitanian deposits was not reached. The planktonic foraminfera allowed us to single out the *Globigerinatella insueta-Globigerinoides trilobus* Zone with *G. trilobus, G. subquadratus, Globoquadrina dehiscens, G. altispira, Globigerina bollii,* and *Globorotalia siakensis*, and the *Praeorbulina glomerosa* Zone is proved by the index-species, *P. transitoria, Globigerinoides bisphaericus*, and *Globorotalia peripheroronda*.

Among benthic foraminfera, the most common are *Rotalia audouini* d'Orb., *Hopkinsina bononiensis* (Forn.), *Florilus communis* (d'Orb.), *Asterigerina planorbis* d'Orb., *Elphidium kvesanensis* Artch., *E. ortenburgensis* (Egger), *E. onerosum* Bogd., *E. miusserensis* Djan., *Cibicides concentricus* Cushman,

C. falconensis Renz, *Cribrononion dollfusi* (Cushman), *Reussella laevigata* Cushman, *Loxostomum digitale* (d'Orb.), *Valvulineria globularis bradyi* Cushman, *V. venezuelana* Hedberg, *Cibicorbis herricki* Hadl., *Baggina indica* (Cushman), *B. paucilocula* (Cushman), *Stomatorbina* ex gr. *concentrica* (Parker et Jones), *Gyroidina girardana perampla* Cushman et Stainforth, *Eponides parantillarum* Gall. et. Hem., *Bolivina carapitana* Hedberg, *Uvigerina striatissima* Perconig, *U. capayana* Hedberg, *Plectofrondicularia jarvisi* Cushman et Todd, *Pl. raricostata* (Karr.), *Bolivinella margaritacea* Cushman, *B. subpectinata* Cushman, and *Amphistegina lessonii* d'Orb.

As in the Aquitanian Stage, the Burdigalian deposits of the Aafrine Graben are of more shallow-water type, as compared to the coeval deposits of the Al-Latheqiyeh Depression.

Outside of the Aafrine Graben the Burdigalian deposits are characterized by complex facies replacements, similar to those in the Al-Latheqiyeh Depression. The *Globigerinatella insueta–Globigerinoides trilobus* Zone, of predominantly soft clayey limestones and marls, has been identified in all sections, whereas the microfauna of the *Praeorbulina glomerosa* Zone was not found. Evidently the massive limestones which conformably overlie sediments of the *Globigerinatella insueta–Globigerinoides trilobus* Zone correspond in part to the *Praeorbulina glomerosa* Zone.

On the eastern slope of the Aafrine Graben near the village of Kazreha (the Jebel ech Smaane hills), yellow and light-yellow marls and clayey limestones of the *Globigerinatella insueta–Globigerinoides trilobus* Zone were uplifted along a fault, and are overlain by thick hard thick-bedded fine-grained and aphanitic limestones. Their lower part probably corresponds to the *Praeorbulina glomerosa* Zone, while higher layers have a Middle Miocene age.

Within the limits of the Parsa Dagh High (near the village of Qastal) Burdigalian deposits disconformably overlie Aquitanian marls. The Burdigalian consists here of alternating greenish platy marls and calcareous clays, light-gray limestones and brownish clayey sandstones, some 42 m thick. Planktonic foraminifera of the *Globigerinatella insueta-Globigerinoides trilobus* Zone are rather numerous and include the latter nominate species, along with *G. subquadratus, Globigerina bollii, G. foliata, Globo-quadrina dehiscens, G. altispira, Cassigerinella chipolensis*, and other species.

Benthic foraminifera are very rich and diverse, and include: *Hopkinsina bononiensis* (Forn.), *Bulimina tumidula* Bogd., *Virgulinella pertusa* (Reuss), *Caucasina* ex gr. *buliminoides* Bogd., *Baggina indica* (Cushman), *B. paucilocula* (Cushman), *Valvulineria globularis bradyi* Cushman, *Gyroidina girardana perampla* Cushman et Stainforth, *Florilus communis* (d'Orb.), *Almaena alticosta* (Sigal), *Rotalia audouini* d'Orb., *Planularia venezuelana* Hedberg, *Cassidulina tricamerata* Gall. et Hem., *Cibicides concentricus* Cushman, *C. mexicanus* Nuttall, *Planulina zigzag* Gall. et Hem., *P. marialana* Hadley, *Loxostomum digitale* (d'Orb.), *Cribrononion dollfusi* (Cushman), *Pseudoglandulina comatula* (Cushman), *Lingulina ponceana* Gall. et Hem., *Globobulimina hannai* Cushman et Ellisor, *Siphonodosaria verneuili* (d'Orb.), *Chrysalogonium longicostatum* Cushman et Jarvis, *Plectofrondicularia interrupta* (Karr.), *Uvigerina capayana* Hedberg, *Bolivina alata* (Seg.), *B. hebes* Macf., *B. ventricosa* Gall. et Hem., and *Chilostomella globata* Gall. et Hem.

Sediments of the *Globigerinatella insueta-Globigerinoides trilobus* Zone bearing this microfauna are covered conformably by a unit about 7 m thick of light-gray hard shallow-water limestones with debris of *Lithothamnium* (up to 1 cm), corals, crinoids and pelecypods. Evidently they already belong to the *Praeorbulina glomerosa* Zone.

Burdigalian deposits have a very unusual structure in the vicinity of the villages of Qatma and Kafer-Janeh on the southern slope of the **Baflioun Swell**, where they transgressively overlie Middle Eocene limestones. The succession (357 m thick) may be subdivided into three members. The lower member (111 m thick) is composed of aphanitic and fine-grained limestones with intercalations of softer clayey limestones. The second member (85 m thick) is represented by alternating conglomerates and sandstones with subordinate layers of sandy clays, marls and platy limestones. The third member (161 m thick) consists of monotonous fine-detrital and aphanitic limestones – hard, light-yellow, well-bedded (beds 7-20 cm thick), and sometimes platy with beds 2-4 cm thick. The upper part of this member contains interbeds of lithothamnium limestones and very thin intercalations of light-greenish marls and calcareous clays.

The first and the second members are characterized by a rather poor assemblage of planktonic foraminifera of the *Globigerinatella insueta–Globigerinoides trilobus* Zone including the latter nominate species, *Globoquadrina dehiscens, Globi-gerina bollii, G. falconensis,* and *Globorotalia siakensis.*

Benthic foraminifera in these shallow-water deposits are much more numerous and diverse: *Virgulinella pertusa* (Reuss), *Hopkinsina bononiensis* (Forn.), *Bulimina tumidula* Bogd., *Florilus communis* (d'Orb.), *Reophax* aff. *morrisoni* Cushman et Ellisor, *Textularia ovulata* Lalick., *T. consecta* d'Orb., *T. angularis* d'Orb., *T. dollfusi* Lalick., *Quinqueloculina pseudoseminulina* Bogd., *Q. rugosa* d'Orb., *Spiroloculina alveata* Cushman et Todd., *Caucasina* ex gr. *buliminoides* Bogd., *Melonis soldanii* (d'Orb.), *Elphidium kvesanensis* Artch., *E. ortenburgensis* (Egger), *Almaena alticosta* (Sigal), *Planulina zigzag* Gall. et Hem., *Anomalina pseudogrosserugosa* Colom, *Stomatorbina* ex gr. *concentrica* (Park. et Jon.), *Gyroidina jarvisi* Cushman et Stainforth, *Discorbis havanensis* Cushman et Bermudez, *Rotalia rimosa* Reuss, *Reussella laevigata* Cushman, *Operculina* sp., *Carpenteria bulloides* Gall. et Hem., *Miogypsina irregularis* (Mich.), and *M. intermedia* Drooger.

Microfauna of the third member is sparse and poorly preserved, preventing a precise age determination. These deposits probably belong to the Lower Miocene (Burdigalian) but it is unclear whether they still correspond to the *Globigerinatella insueta–Globigerinoides trilobus* Zone or already to the *Praeorbulina glomerosa* Zone.

This indicates that in the Burdigalian the sedimentary

environment was the same as in the Aquitanian, in both the Al-Latheqiyeh Depression and the Aafrine Graben. In the Al-Latheqiyeh Depression, pelagic calcareous sedimentation (soft chalky limestones with abundant planktonic foraminifera) prevailed, with a minor local terrigenous influx. In the Aafrine Graben, more shallow-water calcareous sandy-clayey sediments with neritic benthic foraminifera (elphidiids, nonionids, miogypsinids, miliolids, etc.) accumulated.

During the first half of the Burdigalian (the *Globigerinatella insueta–Globigerinoides trilobus* Zone) the contours of the basin essentially did not differ from those of the Aquitanian. During the second half of the Burdigalian (the *Praeorbulina glomerosa* Zone) the paleogeographic setting changed substantially. The sea expanded to the east and covered the entire Aleppo Plateau. The comparatively deep-water deposits with planktonic foraminifera of the *Praeorbulina glomerosa* Zone were also replaced to the east by hard aphanitic, fine-grained and organogenic (algal and detrital) limestones of more shallow-water origin, which were transgressively overlain on soft chalks and marls of the Upper Eocene and Oligocene.

The first intercalations of shallow-water organogenic limestones already appeared in the upper part of the *Globigerinatella insueta–Globigerinoides trilobus* Zone (the hills of Abou Hasi and that with the 547 m elevation marker). Exposures near Bekfala in the south and near Salqine in the north show a gradual transition of the soft marls and clayey limestones of the *Globigerinatella insueta–Globigerinoides trilobus* Zone, to massive and hard limestones. These apparently correspond to the *Praeorbulina glomerosa* Zone (Lower Miocene) and the *Orbulina suturalis–Globorotalia peripheroronda* Zone (Middle Miocene) because in the village of Darkouch on the Orontes River, the eastern rim of Salqine, and the village of Ej Janoudiye, they are overlain by soft marls and limestones of the *Globorotalia peripheroacuta* Zone, which contains rare specimens of the nominate species. This 'wedge' of shallow-water, very often organogenic limestones increases its thickness in an eastern direction and encompasses the entire Middle Miocene, as will be discussed below.

Unfortunately, the monotonous lithology of the hard limestones, and the poor preservation of their fauna, makes it impossible to separate the Upper Burdigalian (the *Praeorbulina glomerosa* Zone) from the Middle Miocene. Therefore deposits of this age are undifferentiated, and will be considered in the part of this chapter devoted to the Middle Miocene.

Middle Miocene

The transgression which began in the Late Burdigalian (the *Praeorbulina glomerosa* Zone) also embraces the entire Middle Miocene interval. The distribution of sediments of the Middle Miocene in northwestern Syria was therefore at a maximum. They are developed not only in the narrow Al-Latheqiyeh Depression and in the Aafrine Graben, but also over vast areas of the Aleppo Plateau, up to the Al-Furat River in the east. The most widespread outcrops of Middle Miocene limestones are located in the western part of the Aleppo Plateau, known as the Idlib Plateau.

A change in paleogeographic setting took place in the Late Burdigalian (near the lower boundary of the *Praeorbulina glomerosa* Zone). For this reason, Late Burdigalian and Middle Miocene deposits are transgressive upon older rocks, while a gradual transition may be observed from the Lower to the Middle Miocene. Nevertheless, on the steep slopes of the Al-Latheqiyeh Depression, one can sometimes observe unconformable relations between the Middle Miocene and underlying Paleogene and in the vicinity of the village of Kesladjouq even Upper Cretaceous rocks.

The change in paleogeography is reflected strongly in the facies patterns of the Middle Miocene sediments. White and light gray, soft, chalky limestones and marls, very typical for the Lower Miocene, no longer occur. Three facies types can be singled out in the Middle Miocene:

1) alternating clays, marls, limestones, sandstones, grits and conglomerates;

2) massive hard thick-bedded limestones (aphanitic, fine-grained, algal and detrital);

3) volcanogenic- terrigenous rocks.

Deposits of the first type are restricted to the Al-Latheqiyeh Depression and the Aafrine Graben. The second facies type is widely distributed on the Aleppo and Idlib Plateaus. The third type occupies a limited area near the southeastern slope of the Aafrine Graben.

In the **Al-Latheqiyeh Depression** the Middle Miocene is exposed in excellent outcrops near the village of Kesladjouq in the Nahr El-Kabir valley, along the road from the village of Baksariyeh to the village of Kheurbet ej-Jaouz, as well as between the villages of Ej Janoudiye and Melahnd. In these outcrops the entire succession of the Middle Miocene is represented, from the top of the Burdigalian to the base of the Tortonian. The Middle Miocene sections near Salqine, in the vicinity of the Darkoush, Zof, Tcheqourdjaq, and Bachoura, and in the valley of the Nahr Al-Yamana River near the villages of Zetn-Jouq and Shakhrura are very important for understanding the stratigraphic relations with the under- and overlying deposits, and the microfaunal associations.

In the section of Kesladjouq, the Middle Miocene begins with a basal member consisting of coarse terrigenous rocks – conglomerates, conglo-breccias (boulders to 35 cm), grits, and sandstones with interbeds and lenses of greenish marls and clays. They are overlain by a unit of frequent intercalations of gray grits, gray-green coarse and poorly-sorted sandstones, and greenish clays and marls. Units of sandstone predominate, from 1.5 to 8 m in thickness. These deposits are replaced by a comparatively thin unit (11 m in thickness) of light-gray hard limestones, light-greenish marls, and calcareous clays. The section is capped by a 76 m thick unit of alternating light-gray marls, greenish clays and brown sandstones. Sandy beds increase in the upper part. The total thickness of the Middle Miocene reaches 160 m.

The basal member of conglomerates, grits and sandstones, reaching 25 m in thickness and formed by cross-cutting lenses, is also found near the village of Tcheqourjaq. Overlying this unit are alternating light-gray and gray-greenish marls, calcareous clays and sandstones. In other sections, the basal coarse-grained unit is more poorly expressed, with only a few meters of sandstones present, as near the village of Bachoura. Sometimes this subdivision is completely absent, as near the village of Baksariyeh, where at the base of the Middle Miocene, gray, thin-bedded calcareous clays with rare sand and siltstone intercalations can be observed.

In the section from the village of Baksariyeh to the village of Kheurbet ej Jaouz, the Middle Miocene is represented by a monotonous alternation of marls, calcareous clays, siltstones, sandstones, and limestones. Marls and clays predominate. They are gray-greenish, soft, thin-bedded, sometimes sandy. Brown fine-grained sandstones and siltstones are subordinate, in interbeds from 3 to 10 cm in thickness. Rarely they form beds of 0.5 m thickness. Intercalations of light-gray limestones are very rare. Bedding is very flat and some parts of the section is not exposed. For this reason, an estimation of the thickness is sometimes problematic. In any case it exceeds 200 m.

The section of the Middle Miocene between the villages of Ej Janoudiye and Melahnd is similar – alternating light-green soft marls, green calcareous clays, hard brownish fine-grained limestones, softer clayey limestones and very rarely, gray fine-grained sandstones. The sandstones are a minor component, and the limestone interbeds are common. Very often intercalation of hard limestones and softer clays results in a ribbed topographic expression of the slopes.

In the vicinity of the **Baer-Bassit Ophiolite Massif** to the northeast of the Al-Latheqiyeh Depression, the lithological composition of the Middle Miocene deposits undergoes a pronounced change. Coarse-grained rocks are widely distributed. In the Nahr Al-Yamana valley near the villages of Shakhrura and Zetn Jouq, the Middle Miocene comprises alternating soft unsorted conglomerates, grits, and sandstones, with subordinate beds of light-gray clayey limestones, greenish and brownish marls and clays. Bedding appears to be lenticular, with cross-cutting lenses. As a rule, the conglomerates are unsorted and where there is abundant clay, they are like puddingstones. The thickness reaches 230 m.

The southernmost outcrops of the Middle Miocene occur on the Mediterranean coast to the southeast of Al-Latheqiyeh on the slope of the Jebel An-Nusseiriyeh horst. Near the village of Nqoura, the Middle Miocene comprises a variety of limestones – clayey, leached, vuggy, detrital, and algal, with intercalations of soft, thin-bedded light-greenish marls and calcareous clays, sandstones and conglomerates. Visible thicknesses do not exceed 25 m.

Clayey limestones, marls and calcareous clays of the Middle Miocene of the Al-Latheqiyeh Depression are characterized by very rich assemblages of planktonic foraminifera. Surprisingly, even the interbeds between terrigenous sandstones, grits and conglomerates may be rich in planktonic forms. *Orbulina* shell accumulations have a similar appearance to well-sorted and rounded sand grains. The following zonal subdivisions can be singled out:

The *Orbulina suturalis–Globorotalia peripheroronda* Zone, where *Orbulina suturalis* is always abundant, accompanied by *Biorbulina bilobata, Praeorbulina glomerosa, P. transitoria, Globigerinella aequilateralis, Globigerinoides trilobus, G. bisphaericus, G. obliquus, Globoquadrina dehiscens, G. altispira, G. larmeui, Globorotalia peripheroronda, G. obesa, G. continuosa, G. siakensis, Globigerina concinna, G. druryi,* and *G. bulloides.*

The *Globorotalia peripheroacuta* Zone, marked by the index-species (or the nominate species) above the extinction of the genus *Praeorbulina*, as well as *Globigerinoides bisphaericus*. Other species of foraminifera continue upwards from the underlying deposits. This zone was identified in a very limited number of outcrops (Darkoush, Salqine, and Ej Janoudiye) because *G. peripheroacuta* is quite rare. It should be also noted that even in these listed sections the zone was defined rather conditionally, because its upper boundary cannot be determined with confidence as other species of the *Globorotalia fohsi* group are absent.

The undifferentiated *Globorotalia fohsi–G. fohsi lobata–Sphaeroidinellopsis subdehiscens–Globigerina druryi* Zones contain numerous planktonic foraminifera. Unfortunately, the nominate species (except *Globigerina druryi*) were not found, and such a taxonomically impoverished assemblage does not permit a clear distinction of these zonal subdivisions. The interval is characterized mainly by transitional species such as *Orbulina suturalis, Biorbulina bilobata, Globigerinella aequilateralis, Globigerinoides trilobus, G. obliquus, Globigerina bulloides, G. druryi, G. eamesi, G. concinna, Globoquadrina dehiscens, G. altispira, Globorotalia obesa, G. continuosa,* and *G. siakensis*. New species such as *Orbulina universa, Globigerinoides bollii, Globorotalia lenguaensis* only appear in the upper part of this interval.

The *Globigerina nepenthes–Globorotalia siakensis* Zone contains a similar assemblage of foraminifera. Among them appears *Globigerina nepenthes*. It was found in all sections, indicating the upper level of the Middle Miocene.

Benthic foraminifera are less abundant than planktonic, and their taxonomic composition varies with lithology. The most common species are: *Pseudoclavulina rudis* (Costa), *Textularia deperdita* d'Orb., *T. mayeriana* d'Orb., *T. haueri* d'Orb., *Spiroplectammina carinata* (d'Orb.), *Martinottiella communis* (d'Orb.), *Marginulina hirsuta* d'Orb., *Lenticulina imperatora* (d'Orb.), *L. calcar* (d'Orb.), *Robulus crassa* (d'Orb.), *Globulina spinosa* d'Orb., *Gyroidina soldanii* (d'Orb.), *Cancris brongnartii* (d'Orb.), *Eponides schreibersi* (d'Orb.), *E.* aff. *probatus* Krash., *Epistomina partschiana* (d'Orb.), *Siphonina reticulata* Cz., *Amphistegina hauerina* d'Orb., *Cibicides nucleatus* (Seg.), *C. dutemplei* (d'Orb.), *C. ungerianus* (d'Orb.), *C. boueanus* (d'Orb.), *Melonis soldanii* (d'Orb.), *Florilus communis* (d'Orb.), *Streblus beccarii* (L.), *Elphidium fichtellianum* (d'Orb.), *E. macellum* (F. et M.), *Uvigerina spherula* Cz., *U. semiornata* d'Orb., *Globobulimina pyrula* (d'Orb.), *Bulimina ovata* d'Orb., *Sphaeroidina bulloides* d'Orb., and *Bolivina dilatata* Reuss.

The Middle Miocene sediments of the **Aafrine Graben** resemble the coeval strata of the Al-Latheqiyeh Depression. Here alternating clayey limestones, marls, calcareous clays, sandstones, grits, conglomerates and coquina rocks are also found. They conformably overlie the Burdigalian and have two-fold structure.

The lower half, 43 m thick, is represented by a unit of greenish marls and clays with interbeds of sandy clays, comparatively soft clayey coquina-like limestones, and clayey limestones with small lithothamnion spherules (data from a bore-hole on the right bank of the Aafrine River to the south of the village of Bteine).

The section penetrated by this bore-hole is overshadowed by a steep cliff outcrop above the Aafrine River, opposite the village of Bassouta. In this part of the Middle Miocene, about 70 m thick, the content of coarse-grained shallow-water rocks increases. They contain abundant macrofauna – bivalves, gastropods, echinoids, bryozoans, corals, and serpulids. Lithothamnion spherules 5 to 10 cm in diameter are very common. This section terminates with hard conglomerates and sandstones 6 m in thickness.

Planktonic foraminifera are numerous in marls and clays of the lower member. In sandy rocks of the upper member their quantity decreases. The species composition of the planktonic foraminiferal assemblage is the same as in the coeval deposits of the Al-Latheqiyeh Depression.

Benthic foraminifera are very diverse in the Aafrine Graben. Here a number of species were identified which are not known from sediments of the Al-Latheqiyeh Depression, including the following: *Dentalina inornata* d'Orb., *D. pauperata* d'Orb., *Planuralia lanceolata* d'Orb., *Lenticulina simplex* (d'Orb.), *L. cymboides* (d'Orb.), *Marginulina rugosocostata* d'Orb., *Saracenaria arcuata* (d'Orb.), *Vaginulina badensis* d'Orb., *Globulina spinosa* d'Orb., *Eponides umbonatus* (Reuss), *E. haidingerii* (d'Orb.), *Asterigerina planorbis* d'Orb., *Anomalina badenensis* d'Orb., *A. rotula* d'Orb., *A. austriaca* d'Orb., *Cibicides variolatus* (d'Orb.), *Planorbulina mediterranensis* d'Orb., *Cymbalopora bradyi* Cushman, *Nonion granosus* (d'Orb.), *N. punctatus* (d'Orb.), *Pullenia bulloides* (d'Orb.), *Elphidium rugosum* (d'Orb.), *E. flexuosum* (d'Orb.), *Operculina* sp., *Heterostegina simplex* d'Orb., *H. costata* d'Orb., *Bulimina pupoides* d'Orb., *B. buchiana* d'Orb., *Reussella spinulosa* (Reuss), *Uvigerina asperula* Cz., *U. semiornata* d'Orb., *U. macrocarinata* Papp. et Turn., *Bolivina plicatella* Cushman, *B. scalprata miocenica* Macf., *B. arta* Macf., *B. pokornyi* Cicha et Zapl., *Sphaeroidina austriaca* d'Orb., *Borelis melo* (F. et M.), *B. haueri* (d'Orb.), *Quinqueloculina juleana* d'Orb., *Q. akneriana* d'Orb., *Q. verneuiliana* d'Orb., and *Q. nussdorfensis* d'Orb. The majority of these species are indicative of shallow-water environments.

On the southern slope of the **Baflioun Swell** and in the vicinity of the Mashaale and Kourt Farmoz villages along the eastern slope of the Aafrine Graben, the Middle Miocene is different. Its lower part consists of a unit of effusive rocks (pillow lavas, olivine basalts, olivine anamesites) with subordinate intercalations and members of shallow-water sedimentary rocks – poorly-bedded algal-coral massive limestones, sandy limestones with tuffogenous material, *Operculina* and detrital limestones, coquina with numerous bivalve and gastropod shells, sandstone, gritstone, and weakly cemented conglomerates. The total thickness is about 300 m. The upper member, 40 m thick, comprises organogenic limestones – detrital, bioherm-like coral-algal limestones, coquina, as well as sandstones and gritstones with some basalt interlayers.

Submarine flows of basalts occupied a rather limited area, and are probably connected with the Middle Miocene volcano located near the village of Qara-Tepe.

Planktonic foraminifera are practically absent in the shallow-water sediments of this area. Benthic foraminfera in the effusive member are rare, and are poorly preserved. They are much more diverse in the upper sedimentary rock member. The list is rather close to that of the Middle Miocene of the Aafrine Graben, but shallow-water species predominate. These include *Elphidium listeri* d'Orb., *E. rugosum* d'Orb., *E. obtusum* d'Orb., *E. crispum* (L.), *E. macellum* (F. et M.), *E. fichtellianum* (d'Orb.), *E. flexuosum* (d'Orb.), *Quinqueloculina akneriana* (d'Orb.), *Q. nussdorfensis* (d'Orb.), *Eponides schreibersii* (d'Orb.), *Operculina* sp., *Borelis melo* (F. et M.), *B. haueri* (d'Orb.), *B. rotellus* (d'Orb.), *Heterostegina costata* d'Orb., *H. simplex* d'Orb., *H. heterostegina* (Silv.), and *Asterigerina planorbis* d'Orb.

As mentioned above, in the area of the villages of Bekfala, Darkoush, and Salqine, rapid replacement of pelagic marls and clayey limestones of the Lower Burdigalian (the *Praeorbulina glomerosa* Zone, Lower Miocene) and lower Middle Miocene (the *Orbulina suturalis–Globorotalia peripheroronda* Zone) by shallow-water, massive limestones takes place. Along the southeastern slope of the Aafrine Graben, marly-clayey-sandy sediments of higher Middle Miocene horizons are analogously replaced by massive limestones. These facies transitions are especially clear in the excellent outcrops of the Jebel Samaan hills in the area of Kourt-Farmoz. Where this transition is complete, the Upper Burdigalian (the *Praeorbulina glomerosa* Zone) and the entire Middle Miocene comprise a single unit of medium- to thick-bedded, massive limestones.

These limestones are exposed over very large areas in the **Aleppo Plateau**, from the Aafrine and Al-Ghab Grabens in the west to the Al-Furat valley in the east. The western part of this plateau (usually called the Idlib Plateau) is almost completely covered by these limestones, extending to the south to the Jebel Zaouiye mountains in the area of the town of Khane Cheikoun) and to the east almost to the towns of El Bab and Aleppo. To the east of this line as far as the Al-Furat valley, the limestones are strongly eroded and were preserved only as outliers.

Superficially, these limestones appear as a monotonous, massive, hard, coarse- to unbedded unit with pockmarked and vuggy weathering surface, but closer examination shows it to have diverse features. Many lithological types can be distinguished:

1) Massive, hard, thick-bedded limestones of light-gray and yellowish color, fine- and medium-grained, and recrystallized;

2) Aphanitic bedded brownish limestones;

3) White, sugary, finely porous, leached limestones transitional to vuggy varieties. Evidently the primary lithology was calcareous sands, which were diagenetically transformed to calcarenites with massive or porous texture;

4) Detrital limestones of molluscan, echinoid, or algal fragments;

5) Coquina limestones, sometimes consisting only of oyster valves, usually strongly leached;

6) Limestones with small algal spheroids (0.5-0.7 cm), embedded in a fine-grained matrix;

7) Limestones of ball-like lithothamnion accumulations, with spheroids 1-3 cm in diameter;

8) Unbedded algal-coral limestone bioherms with structures of overlapping;

9) Rare occurrences of pelitomorphic, and comparatively soft limestones.

The maximum thickness of this limestone unit is found on the Idlib Plateau, where they were penetrated by numerous bore-holes near the villages of Ibbineh, Kafer-Nuraneh, Maaret El-Khane, Batabou, Kelli, Souarhiye. Here the thicknesses range between 140 and 250 m. Only in some bore-holes (Hraitane and Berne) do they diminish to between 60 and 80 m. Natural outcrops near Harim, Eriha and Idlib demonstrate thicknesses from 100 to 150 m. In the eastern part of the Aleppo Plateau, the thicknesses are reduced to 35-40 m, but these figures are a minimum as the outcrops represent a part of the section, with the amount of denudation over this broad region being unknown.

Micropaleontological characterization of the hard limestones is very difficult. It is virtually impossible to isolate foraminiferal shells from solid rocks, and in the soft leached limestones they are usually preserved only as casts, which preclude exact species determination.

In some outcrops of the Jebel Smaane mountains (the areas of Salqine and Eriha), thin section examination of the lower part of this limestone unit displays *Miogypsina*, indicating an Early Miocene age for these limestones. In outcrops near Aleppo, Idlib, Mareah, Mennbidj, Djerablous, and the Jebel Hass mountains rather compact marly limestones corresponding to higher horizons of the limestone unit contain benthic foraminifera which belong to two shallow-water paleocoenoses.

The first elphidiid paleocoenosis yields representatives of elphidiids, nonionids, rotaliids and amphisteginids, including the following species: *Elphidium crispum* (L.), *E. macellum* (F. et M.), *E. fichtellianum* (d'Orb.), *E. listeri* (d'Orb.), *E. rugosum* (d'Orb.), *E. flexuosum* (d'Orb.), *E.* aff. *antonina* (d'Orb.), *Nonion granosus* (d'Orb.), *Florilus boueanus* (d'Orb), *Streblus beccarii* (L.), *Amphistegina hauerina* d'Orb.), *Asterigerina planorbis* d'Orb., and *Reussella spinulosa* (Reuss).

The second, miliolid paleocoenosis is dominated by alveolinids, miliolids and peneropliids, such as: *Borelis melo* (F. et M.), *B. haueri* (d'Orb.), *B. rotellus* (d'Orb.), *Quinqueloculina akneriana* d'Orb., *Q. nussdorfensis* d'Orb., *Q. verneuiliana* d'Orb., *Pyrgo inornata* d'Orb., *Peneroplis* aff. *evolutus* Hens., *Dendritina rangi* d'Orb., *Meandropsina* aff. *anahensis* Hens., *Operculina* sp., and *Heterostegina costata* d'Orb.

All listed species are typical for the Middle Miocene. Nevertheless, rare findings of microfauna and their poor state of preservation preclude determination with certainty of the Lower Miocene (Upper Burdigalian) and Middle Miocene within this massive, hard limestone succession of the Aleppo Plateau. Their age was reliably determined with the help of calibration against the pelagic sediments of the Al-Latheqiyeh Depression and the Aafrine Graben, which contain rich assemblages of planktonic foraminifera.

Resolution of the question of the upper age limit of the massive limestone unit is very important, particularly whether this unit includes the entire Middle Miocene, or only part of it. Sections with Miocene deposits in the Al-Latheqiyeh Depression and the Aafrine Graben gave a definitive answer.

In sections near the villages of Ej Janoudiye, Kheurbet ej Jaouz, Zof, and Astara, Middle Miocene sediments with planktonic foraminifera of the *Globigerina nepenthes–Globorotalia siakensis* Zone (N14) are overlain by a member with shallow-water (possibly littoral) rocks. These include coarse-porous lithothaminion limestones, composed of algal spheroids with interstices filled with molluscan debris; lithothaminion limestones with coral colonies; aphanitic stromatolitic limestone composed by algal crusts; leached coquinas of bivalve and gastropod accumulations; conglo-breccias consisting of clasts of underlying lithologies, with various quantities of molluscan material; coarse-grained calcareous sandstones and sandy limestones. Their thickness near the village of Kheurbet ej Jaouz reaches 20-25 m. These rocks are remnants of a hard armored surface found on hilltops in this region, which preserved the underlying soft marly and clayey lithologies from erosion. Younger sediments are absent in all the studied sections of this region. These hard limestones are relicts of a broader distribution: for example near the village of Baksariyeh, they form a large discontinous field of erratic blocks (up to 10 m in length) lying directly on marls of the *Orbulina suturalis–Globorotalia peripheroronda* Zone. Consequently the main part of the Middle Miocene was removed by erosion, and massive limestone blocks were preserved as exotic allochthonous blocks on the surface of the underlying softer sediments.

The same organogenic and aphanitic limestones may be identified near the villages of Melahnd, Gaurgou, and Msibine. They contain a noticeable content of sandstone and grit. This unit is well-expressed in the landscape, but in the vicinity of these villages the limestones are covered by the member of marls, clays and sandstones with planktonic foraminifera of the Upper Miocene Tortonian *Globorotalia acostaensis* Zone (N16). Therefore, these shallow-water organogenic limestones and sandstones apparently belong to the *Globorotalia continuosa* Zone (N15). It is known that the assemblage of planktonic foraminifera of this zone is transitional between the Middle Miocene microfauna (the *Globigerina nepenthes–Globorotalia siakensis* Zone) and the Tortonian microfauna of the Upper Miocene (the *Globorotalia acostaensis* Zone). For this reason, various stratigraphic schemes place the *Globorotalia continuosa* Zone either at the top of the Middle Miocene or at the base of the Upper Miocene.

In summary, the geological history of northwestern Syria during the entire Middle Miocene is as follows: The Al-Latheqiyeh Depression, the Aafrine Graben, and the Aleppo Plateau were all located within a very stable marine basin. The transgression began during the *Praeorbulina glomerosa* Zone of the Lower Miocene, and represents the largest marine transgression of the Miocene. The limestones of the Aleppo Plateau correspond to the *Praeorbulina glomerosa* Zone of the Upper Burdigalian along with the entire Middle Miocene.

The boundary of the Middle and Late Miocene was marked by regression. The sea receded from the Aleppo Plateau, the Aafrine Graben, and the eastern part of the Al-Latheqiyeh Depression. This event is expressed by regressive shallow-water organogenic limestones and coarse terrigenous rocks (probably the *Globorotalia continuosa* Zone, N15). In the northwestern part of Syria the overlying sediments of the Tortonian stage of the Upper Miocene are already characterized by a rather limited distribution.

Upper Miocene

The two subdivisions of the Upper Miocene, the Tortonian and Messinian Stages, are characterized by very different deposits and organic remains, which reflect the different conditions of sedimentation and biota of the marine basins of that time.

The Tortonian Stage

Sediments of the Tortonian in the northwestern part of Syria are limited in distribution. Outcrops occur in the southwestern part of the Al-Latheqiyeh Depression and north of the Al-Ghab Graben, but these areas were not connected. The Al-Latheqiyeh Depression was directly connected to the Miocene Mediterranean Sea, while the Al-Ghab Graben was an embayment of the Antiochia Basin, located in Turkey, which was also indirectly linked to the Miocene Mediterranean.

The best section of the Tortonian is located in the Nahr El-Kabir valley of the **Al-Latheqiyeh Depression**. Sediments from the Middle to Upper Miocene are characterized by a gradual transition. The contact itself can be seen in the upper part of the slope of the Nahr El-Kabir valley near the village of Kesladjouq. Because of the southwesterly dip, the boundary descends to the level of the river and can be observed near the Jebel Safkoun Massif.

The Tortonian section is made up of alternating marls, calcareous clays, various limestones, sandstones, grits, and small-pebble conglomerates, up to 230 m in thickness. The section along the Nahr El-Kabir River where both the base and the top of the succession are present is the only one in which the true thickness is exposed.

The marls and calcareous clays are light-gray and greenish, soft, thin-bedded, often with detrital material. They form thin intercalations and thick (20-46 m) members. Sandstones are gray and brownish, ranging from fine-grained to coarse-grained and polymictic, as well as from hard and massive to weakly cemented soft sandstones, and they sometimes display cross-bedding. They pass to gritstones and small-pebble conglomerates, with pebbles (1-4 cm) of flints, limestone and effusive rocks. Occasionally, large boulders to 30 cm are present. Limestones are variable – hard, fine-grained, soft and clayey, detrital varieties with debris of molluscs, echinoids, bryozoans, lithothamnion, and leached coquina. Limestone units may reach considerable thicknesses of 6 to 30 m. They form the steep slopes of the Nahr El-Kabir valley and the Jebel Safkoun Massif as well as the flat upper surface of this massif.

Local erosional surfaces, lenticular bodies, cross-cutting relationships between lenses, cross-bedding and intensive bioturbation of sediments may be observed within this unit of shallow-water origin.

Clay and marl intercalations in this shallow-water section unexpectedly contain rich planktonic foraminiferal faunas. These faunas are evidence for their belonging to the *Globorotalia acostaensis* Zone. The assemblage contains the nominate species, along with *G. merotumida, G. menardii, G. scitula, G. lenguaensis, G. pseudopachyderma, Orbulina universa, Hastigerina siphonifera, Globigerinoides obliquus, G. extremus, G. trilobus, Globigerina bulloides, G. microstoma, G. nepenthes, G. decoraperta, G. eamesi, G. globorotaloidea, G. quinqueloba, G. bulbosa,* and *Biorbulina bilobata.*

Benthic foraminifera are numerous and diverse, reflecting the variety of environments and lithological types of sediments. Species present include *Reophax palpillosus* (Neug.) var. *compressus* Seg., *Vulvulina pennatula* (Batsch) var. *italica* Cushman, *Gaudryina tumidula* Cushman, *Bigenerina agglutinans* d'Orb., *Martinottiella communis* (d'Orb.), *Spiroplectammina carinata* (d'Orb.), *Siphotextularia concava* (Karr.), *Textularia deperdita* d'Orb., *T. gramen* d'Orb., *Spiroloculina depressa* (d'Orb.), *Robulus dilectus* (Seg.), *R. echinatus* (d'Orb.) var. *curvicostus* (Seg.), *Planularia auris* (Defr.), *Gyroidina soldanii* (d'Orb.), *Valvulineria obtusa* (d'Orb.), *V. complanata* (d'Orb.), *Discorbis semiorbis* (Karr.), *D. squamulus* (Reuss), *Eponides umbonatus* (Reuss), *E. schreibersii* (d'Orb.), *Cancris brongnartii* (d'Orb.), *C. oblongus* (Will.), *Siphonina reticulata* Cz., *Asterigerina planorbis* d'Orb., *Epistomina partschiana* (d'Orb.), *Stomatorbina* ex gr. *concentrica* (Parker et Jones), *Heronallenia lingulata* (Burr. et Holl.), *Alliatina excentrica* di Nap. All., *Lamarckina erinacea* (Karr.), *Planorbulina mediterranensis* d'Orb., *Cibicides lobatulus* (W. et Jac.), *C. ungerianus* (d'Orb.), C. *dutemplei* (d'Orb.), *C. boueanus* (d'Orb.), *Ceratobulimina haueri* (d'Orb.), *Florilus communis* (d'Orb.), *Melonis soldanii* (d'Orb.), *Nonion granosus* (d'Orb.), *N. umbostelligerum* Serova, *Porosononion subgranosus* (Egger), *Criborononion dollfusi* var. *vigneauxi* Caralp et Jul., *Elphidium hauerinum* (d'Orb.), *E. angulatum* (Egger), *E. antonina* (d'Orb.), *E. aculeatum* (d'Orb.), *E. crispum* (L.), *E. macellum* (F. et M.), *E. listeri* (d'Orb.), *E. fichtellianum* (d'Orb.), *Streblus beccarii* (L.), *Reussella spinulosa* (Reuss), *Bulimina elongata* d'Orb., *B. buchiana* d'Orb., *B. subulata* Park. et Jon., *Virgulina schreibersiana* Cz., *Uvigerina venusta* Franc., *U. acuminata* Hos., *U. semiornata* d'Orb., *Bolivina dilatata* Reuss, *B. antiqua* d'Orb., *Cassidulina laevigata* d'Orb., *C. marginata* Karr., *Ehrenbergina dinapolii* Gian., and *Borelis melo* (F. et M.).

The alternation of calcareous and terrigenous rocks in the Tortonian sequence is rather monotonous but the latter become more frequent upwards, and their grain-size increases. For example, to the south of the Nahr El-Kabir valley in outcrops along the El Haffe wadi near the pumping station, the Tortonian terminates with a member of massive coral limestones, grits, unsorted boulder conglomerates, and sandstones with rare interbeds of marls and clayey limestones. The observed thickness of this terminal regressive unit of the Tortonian is not less than 35 m.

In the **northern part of the Al-Ghab Graben** and adjacent areas Tortonian deposits are exposed near the villages of Melahnd, Gaurgou, Jdaide, Salqine, and Abreiteh. The sections here are incomplete, and only the lower part of the Tortonian is exposed. Contacts with younger sediments are absent because they are already located on Turkish territory near the villages of Gaurgou, Hatia, and others.

The lithological patterns of Tortonian deposition in this region resemble those of the southwestern part of the Al-Latheqiyeh Depression. Near the village of Melahnd, the Tortonian is composed of alternating gray marls, greenish clays, hard organogenic limestones, brown sandstones of variable induration, grits, small-pebble conglomerates, and puddingstones. Pebbles are common in all these lithologies, and are diverse in composition. They also include pebbles derived from the ophiolite complex. Some intervals of this section demonstrate rhythmic alternations (1-5 cm in thickness) of marls, clays, limestones, and sandstones. The visible thickness reaches 90 m. A similar section of Tortonian age can be found between the villages of Hatia and Gaurgou (within the neutral zone along the Syria-Turkey boundary). Here, the same rhythmic alternating pattern of brown clays, greenish calcareous clays and marls, hard sandstones, soft sands, and detrital limestones is present. Upwards in the succession the quantity of coarse terrigenous material increases. In the area of Salqine and Harim, marls and limestones predominate and terrigenous deposits are subordinate.

Assemblages of planktonic and benthic foraminifera in the northern part of the Al-Ghab Graben are less rich than in the Al-Latheqiyeh Depression. The presence of such species as *Globorotalia menardii, G. acostaensis, Orbulina universa, Hastigerina siphonifera, Globigerinoides obliquus, G. extremus, Globigerina nepenthes, G. quinqueloba,* and *G. bulloides*, place this unit in the *Globorotalia acostaensis* Zone (N16).

In more southerly areas of the Al-Ghab Graben (near Idlib and Eriha), Tortonian deposits acquire a more littoral aspect. Intercalations of clays and marls lack planktonic foraminifera, and the assemblage of benthic foraminifera is impoverished, consisting of typically shallow-water species. Some examples are *Elphidium hauerinum* (d'Orb.), *E. macellum* (F. et M.), *E. crispum* (L.), *E. fichtellianum* (d'Orb.), *E. listeri* (d'Orb.), *E. aculeatum* (d'Orb.), *E. antonina* (d'Orb.), *E. angulatum* (Egger), *Borelis melo* (F. et M.), *Nonion granosus* (d'Orb.), *Streblus beccarii* (L.), *Asterigerina planorbis* d'Orb., *Reussella spinulosa* (Reuss), *Globulina gibba* d'Orb., *Cibicides lobatulus* (W. et Jac.), *Cancris brongnartii* (d'Orb.), and *Florilus communis* (d'Orb.).

Further to the south, near the villages of Mhannbel and Ain el Aarouz, marine deposits of the Tortonian are replaced by a continental unit consisting of sandstones, grits, coarse-pebble and boulder conglomerates, about 100 m thick. Microfauna is absent even in thin sandy-marly intercalations and the whole section is unfossiliferous.

The Messinian Stage

Deposits of the Messinian Stage are very limited in area. In the extreme southwestern part of the **Al-Latheqiyeh Depression**, they are only exposed in the lower reaches of the Nahr Al-Kabir and the Nahr Al-Kich Rivers, where they approach the coastal plain by the Mediterranean. Shallow water sediments of this age terminate the Late Miocene regression, which had started in the Tortonian. In general, Messinian deposits are evaporitic, represented by gypsum, serpulid bioherm-like limestones, algal bioherm limestones, leached coquinas, clays, and marls. The Messinian deposits are laterally very variable.

In the section along the Nahr Al-Kabir River, the Messinian Stage begins with a unit about 7 m thick of serpulid limestones, aphanitic and detrital limestones with numerous casts of oysters, and *Lithophaga* borings, sometimes with casts. These are overlain by a member about 30 m thick, of coarse-grained and fine-grained gypsum with intercalations of stromatolitic limestones, and clay and marl, as well as leached coquinas consisting of small bivalve and gastropod casts. A clay and marl layer about 3-3.5 m thick is observed in the middle part of this member.

The clay and marl lithofacies is characterized by a dwarfed but taxonomically rather diverse assemblage of planktonic and benthic foraminifera. These include *Globorotalia apertura* Pezz., *G. minutissima* Bolli, *G. involuta* Pezz., *G. incompta* (Ciff.), *G. gavalae* Perc., *G. scitula* (Brady), *Globigerinita apenninica* (Pezz.), *Globigerina bulloides* d'Orb., *G. apertura* Cushman, *G. nepenthes* Todd, *G. quinqueloba* Natl., *G. riveroae* Bolli et Berm., *G. picassiana* Perc., *Orbulina* sp., *Globigerinoides trilobus* (Reuss), *G. extremus* Bolli et Berm., *Hastigerina siphonifera* (d'Orb.), *Porosononion subgranosus* (Egger), *Nonion citai* di Nap. All., *Elphidium macellum* (F. et M.), *E. aculeatum* (d'Orb.), *E. advenum* (Cushman), *Streblus beccarii* (L.), *Asterigerina planorbis* d'Orb., *Quinqueloculina consobrina* (d'Orb.), *Bolivina dilatata* Reuss, *Cymbalopora vitrea* Pezz., *Bulimina aculeata* d'Orb., *B. echinata* d'Orb., *Cibicides boueanus* (d'Orb.), *Valvulineria* ex gr. *complanata* (d'Orb.), *Uvigerina guadryinoides* Lipp., and *U. siphogenerinoides* Lipp.

Near the village of Zoubar the Messinian Stage is different. It is composed of dark-gray, greenish and light-brown marl and calcareous clays with interbeds of hard, pelitomorphic limestones. These rocks are overlain by a layer 5 m thick of very hard stromatolitic limestones which form loaf-like bodies up to 2.5 m in diameter. The total thickness is about 40 m. There is no gypsum in this section. The assemblages of planktonic and benthic foraminifera in the Zoubar section are approximately the same as those in the sequence along the Nahr Al-Kabir River, but contain more foraminferal specimens.

The original Messinian deposits were more widely distributed than at present. This is evidenced by large blocks of algal stromatolitic limestones along the Haffe wadi, and on the slopes of the Jebel Qarik hills. Apparently, the soft clays and marls were destroyed by the pre-Pliocene erosion which left the hard limestones preserved on the surface as erratic blocks. Sometimes, these blocks are included in the dark clays of the Pliocene.

In the axial part of the Al-Latheqiyeh Depression Messinian deposits are unconformably overlain by dark-gray and dark-greenish Pliocene clays, with planktonic foraminifera of the *Sphaeroidinellopsis seminulina* acme-zone. On the slopes of this depression, above the Messinian deposits, the younger Pliocene *Globorotalia margaritae evoluta* Zone is present.

Nannofossil Stratigraphy of the Miocene of Zeyneih Section

The stratigraphic subdivision according to foraminiferal zonation of the Miocene deposits of northwestern Syria is confirmed by the analysis of the nannoplankton carried out by L. Golovina (Krasheninnikov, Basov, Golovina et al., 1999).

The nannoplankton distribution was thoroughly investigated in a sequence of Miocene deposits extending from the village of Zeynieh to the north towards Baksariyeh and then to Kheurbet ej Jaouz village. As already mentioned, at Zeynieh the Miocene chalky limestones unconformably overlie Middle Eocene nummulitic limestones. In this section, a complete sequence of nannoplankton zones, from the *Triquetrorhabdulus carinatus* Zone to the *Discoaster calcaris* Zone, can be observed. Unfortunately, concentrations of nannofossils and their state of preservation in the various intervals are very variable, sometimes preventing the exact determination of zonal boundaries.

In the *Triquetrorhabdulus carinatus* Zone (NN1) typical specimens of the nominate species were not found, but some layers consistently bear *T.* aff *carinatus* and *T.* sp. The basal layers of the Lower Miocene generally are poor in nannoplankton. Upwards nannoplankton become more abundant and better preserved, emphasized by the presence of intact coccospheres. The most numerous is *Cyclicargolithus floridanus*, accompanied by *C. abisectus, Helicosphaera recta, H. euphratis, H. obliqua, H.* sp., *Sphenolithus moriformis, Sph. compactus, Sph. conicus, Discolithina multipora*, and *Pontosphaera* sp., together with rarer *Coccolithus pelagicus, Discoaster adamanteus*, and *D. deflandrei*.

This assemblage of nannoplankton corresponds to the *Globigerinoides primordius-Globorotalia kugleri* planktonic foraminiferal zone. Consequently, both foraminifera and nannoflora indicate that the transgression in northwestern Syria started very early in the Miocene.

The *Discoaster druggii* Zone (NN2) is determined by the appearance of *Helicosphaera ampliaperta*. The preservation of nannoplankton in this zone deteriorates, taxonomic variety diminishes, secondary overgrowths obscure the morphology among the discoasters, and their identification becomes difficult. The assemblage includes *Discoaster* cf. *druggii, Cyclicargolithus floridanus, Coccolithus pelagicus, Discoaster deflandrei, Helicosphaera carteri*, and *Triquetrorhabdulus* sp. These nannoplankton of the *Discoaster druggii* Zone correspond to the lower part of the undifferentiated *Catapsydrax dissimilis* and *Catapsydrax stainforthi* planktonic foraminiferal zones.

The *Sphenolithus belemnos* Zone (NN3) is characterized by more diverse and better preserved nannoplankton. One peculiarity of this interval is the presence of abundant, very small specimens of *Reticulofenestra* sp. and *Dyctiococcites* sp. The assemblage consists of numerous *Coccolithus pelagicus, Cyclicargolithus floridanus, Helicosphaera euphratis, H. carteri, H. ampliaperta, H. scissura, H.* sp., *Discoaster deflandrei, Discolithina* sp., *Sphenolithus belemnos, Sph. moriformis,* and *Triquetrorhabdulus* sp.

The *Sphenolithus belemnos* Zone corresponds mostly to the upper part of the *Catapsydrax dissimilis* and *Catapsydrax stainforthi* planktonic foraminiferal zones, and only the uppermost part corresponds to the lower part of the overlying *Globigerinatella insueta-Globigerinoides trilobus* Zone.

The *Helicosphaera ampliaperta Zone* (NN4) also yields diverse nannoplankton of good preservation. The nominate species *H. ampliaperta* is associated with *Sphenolithus heteromorphus,* which appears at the very base of this zone. Planktonic foraminifera, found in the *Helicosphaera ampliaperta* zonal interval, belong to the *Globigerinatella insueta-Globigerinoides trilobus* Zone.

The main nannoflora species of the *Sphenolithus heteromorphus* Zone (NN5) consists, as in the preceding zone, of larger forms of *Coccolithus pelagicus* together with *Cyclicargolithus floridanus, Helicosphaera euphratis, H.* sp., *Discoaster deflandrei,* and *Triquetrorhabdulus* sp. New elements are represented by *Calcidiscus macintyrei* and *C. leptoporus*. This zone corresponds to the foraminiferal *Praeorbulina glomerosa* Zone and, possibly, to the lowermost part of the *Orbulina suturalis-Globorotalia peripheroronda* Zone.

The *Discoaster exilis* Zone (NN6) contains the most diverse and well-preserved nannoflora. Quantitatively, *Coccolithus pelagicus* and *Cyclicargolithus floridanus* remain dominant and are accompanied by *Reticulofenestra pseudoumbilica, Calcidiscus macintyrei, C. leptoporus,* and *Triquetrorhabdulus* sp. The appearance of *Discoaster exilis, D. variabilis, Rhabdosphaera clavigera* and rare *Scyphosphaera* sp. and *Syracosphaera mediterranica* is significant. The *Discoaster exilis* Zone can be mostly correlated with the foraminiferal *Orbulina suturalis-Globorotalia peripheroronda* Zone, although the exact coincidence of the upper boundary of these two zones remains unclear.

The lower boundary of the undifferentiated interval of the *Discoaster kugleri* and *Catinaster coalitus* Zones can be identified by the appearance of rare specimens of *D. kugleri* together with abundant, diverse and well-preserved nannoplankton. The nominate species *C. coalitus* was not found. Numerous specimens of *Coccolithus pelagicus*, *Calcidiscus macintyrei*, *C. leptoporus*, *Reticulofenestra pseudoumbilica, Rhabdosphaera clavigera, Discoaster variabilis, D. exilis, Helicosphaera carteri,* small *Helicosphaera* sp., *Sphenolithus moriformis*, and very rare *Scyphosphaera* sp. and *Thoracosphaera heimi* continue to be present. Small placoliths belonging to genera *Reticulofenestra* and *Umbilicosphaera* are common in sediments of these zones.

In the upper part of this interval, taxonomic diversity and the quality of preservation of nannoflora deteriorate. The quantity of terrigenous influx increases. The following association of species occurs here: *Coccolithus pelagicus, Reticulofenestra pseudo-umbilica, Calcidiscus macintyrei, C. leptoporus,* and rare *Discoaster* sp. and *Thoracosphaera mediterranica.* The uppermost samples are practically barren of nannoflora.

The interval of the *Discoaster kugleri* and *Catinaster coalitus* Zones corresponds to the undifferentiated interval of the foraminferal zones from the *Globorotalia peripheroacuta* Zone to the *Sphaeroidinellopsis subdehiscens–Globigerina druryi* Zone.

The *Discoaster hamatus* Zone is characterized by a significant increase in the abundance and taxonomic diversity of nannoplankton. The lower boundary of this zone is determined by the appearance of the 5-rayed discoasters, such as *Discoaster bellus, D.* cf. *pentaradiatus*, accompanied by rare specimens of *Discoaster hamatus* and numerous *Coccolithus pelagicus, Calcidiscus macintyrei, C. leptoporus, Helicosphaera carteri, Reticulofenestra pseudoumbilica, Rhabdosphaera clavigera, Sphenolithus moriformis, Syracosphaera mediterranica, Umbilicosphaera* sp., and rarer specimens of *Discoaster brouweri, D.* sp., and *Scyphosphaera amphora.*

The sediments of the *Discoaster hamatus* Zone contain planktonic foraminifera belonging to the *Globigerina nepenthes–Globorotalia siakensis* Zone.

The *Discoaster calcaris* Zone was conditionally determined, because nannoplankton were only found in some individual horizons. Guide species of discoasters are very rare, and their determination is problematic due to secondary overgrowths. The first appearance of *Helicosphaera sellii* is used for determination of the lower boundary of this zone. The assemblage of nannoflora, typical for the preceding zone continues into the *Discoaster calcaris* Zone, with the addition of some new discoasters, such as *D. neohamatus, D.* cf. *asymmetricus*, and some varieties of *D. variabilis*, including 4-rayed forms. The uppermost part of the section possibly belongs to the upper part of the *Discoaster calcaris* Zone, due to the presence here of some specimens of *Discoaster* cf. *berggrenii.* The *Discoaster calcaris* Zone interval corresponds to the *Globorotalia acostaensis* foraminiferal zone (Tortonian, Upper Miocene).

Pliocene

Marine Pliocene deposits are limited in distribution. Their exposures are located in the extreme southwestern part of the **Al-Latheqiyeh Depression** to the east of Al-Latheqiyeh. A narrow belt of Pliocene sediments is also developed on the Mediterranean coast to the north and south of Al-Latheqiyeh. Continuous seismic profiling which took place during Cruise 5 of the R/V Akademik Nikolay Strakhov showed that a very pronounced fault coincides with the shoreline. A submerged block is located beneath the Mediterranean Sea to the west of this fault, while to the east lies the Jebel An-Nusseriyeh uplift. Apparently the sedimentary and effusive rock sequences of the Pliocene of the coastal area are strongly affected by the large displacement and the submergence of the western block along this fault.

The Pliocene sediments are transgressive. In the Nahr El- Kabir valley they overlie an erosional surface above the gypsiferous unit of the Messinian, and in the southern slope of the Al-Latheqiyeh Depression they overlap the Tortonian rocks. To the north of Al-Latheqiyeh in the area of the Baer-Bassit Massif, Pliocene sediments unconformably overlie the green rocks of the ophiolite assemblage.

A series of Pliocene outcrops extends along the Nahr El-Kabir River below the village of Arras El-Ahmar. Their thickness is probably about 500 m, but low dip angles combined with intervening Quaternary outcrops made estimation of the true thickness difficult. This is apparently close to the true value, as in the Fidio-1 bore-hole drilled some kilometers from the coast, the top of the Messinian gypsum was reached at a depth of 550 m (**Fig. 6.69**).

The Pliocene is composed of comparatively uniform dark-gray, gray-green and green clays and calcareous clays, well-bedded, usually with silty material and sometimes with fine sandy admixture. On the weathered surface, clays are light-gray in color. These rocks alternate with yellowish and gray fine-grained sandstones (0.3 to 6 m in thickness) and light-gray clayey limestones, the latter occurring as thin interbeds or members up to 7-10 m thick. Usually the limestones are soft, although sometimes they are relatively hard and form small escarpments on the valley slopes.

Pliocene deposits are very rich in macrofauna. Bivalve and gastropod shells are sometimes abundant, forming lenses of clayey coquinal limestone. Layers of clayey detrital limestones are sometimes present. Roman (1940) and Roger (1940) determined about 200 molluscan species in these sediments. The most common are *Glycimeris rogeri* Eruenal-Erentoez, *Lucina orbicularis* (Desh.), *L. borealis* Linne, *Chlamys bollensis* Mayer-Eymar, *Ch. (Grandipecten) latissima* (Brocchi), *Chama gryphoides* (Linne), *Aporrhais uttingerianus* (Risso), *Limopsis (Pectunculina) anomala* Eichwald, *Natica dillwyni plicata* Bronn., *Conus (Conospira) antideluvianum* (Borson), *Pycnodonta cochlear nawicularis* (Brocchi), *Ostrea plicata germanitata* (Di Gregori), *Turritella (Haustator) marginalis* (Brocchi), *T. (H.) tornatus* (Brocchi), *Cerithium (Vulgocerithium) vulgatum* Brug., *Xenophora deshayesi* Michelotti, *Terebra (Subula) fuscata* (Brocchi), *Dentalium (Entale) novemcostatus* Lam., and *D. michelotti* Hoernes (det. R. L. Merklin, in Shatsky et al., 1966; Kazmin and Kulakov, 1968).

The area of Al-Latheqiyeh in the Pliocene was a small embayment of the sea, and planktonic foraminfera are numerous, but impoverished in taxonomic composition. They are more diverse in the Lower Pliocene than in the Upper Pliocene, where their stratigraphic resolution is diminished (Krasheninnikov, 1994).

At the base of the Pliocene the *Sphaeroidinellopsis seminulina* Acme-zone is determined, with numerous *Sphaeroidinellopsisi seminulina* and *Sph. subdehiscens*, which are accompanied by *Globigerinoides sacculifer, G. ruber, G. conglobatus, G. obliquus, G. extremus, G. trilobus, Orbulina universa, Globigerina nepenthes, G. bulloides, G. apertura, G. decoraperta, Globorotalia scitula, Globigerinella aequilateralis,* and *Globoquadrina altispira.* This interval was only discovered in the axial part of the Al-Latheqiyeh Depression, in the vicinity of the village of Knaissate. On the southern slope of this depression it is absent, and Tortonian rocks

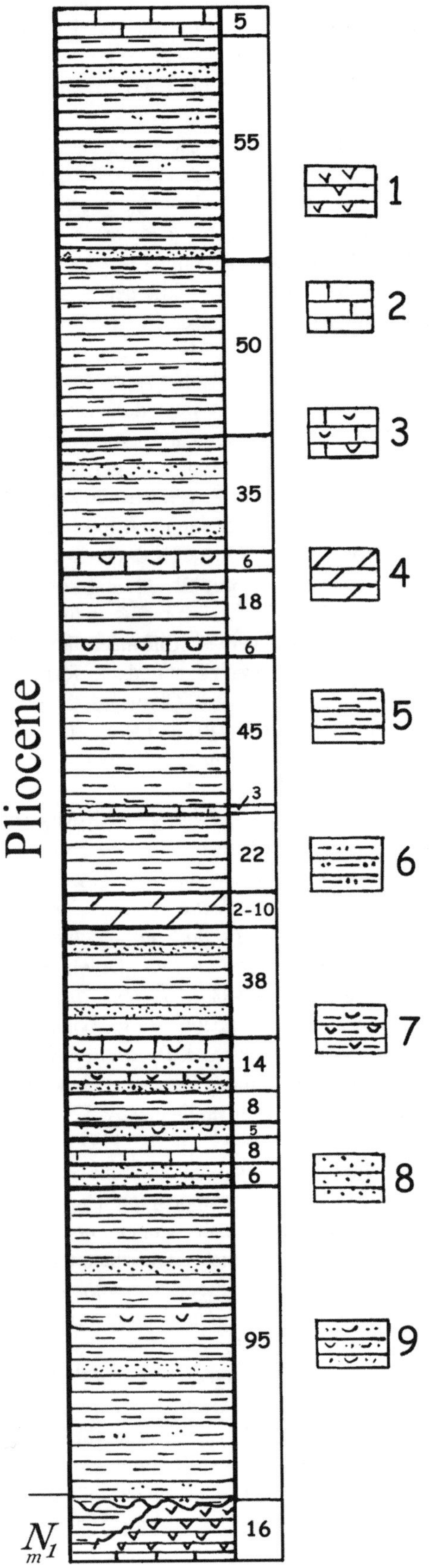

are overlain by younger Pliocene sediments.

The overlying *Globorotalia margaritae evoluta* Zone is characterized by rare specimens of *Globorotalia margaritae margaritae* and *G. margaritae evoluta*, together with numerous representatives of *Globigerinoides, Globigerina, Globoquadrina, Orbulina* and *Globigerinella*, as in the underlying sediments, but *Sphaeroidinellopsis* has nearly disappeared.

Sediments belonging to these two zonal subdivisions are referred to the Lower Pliocene (Zanclean Stage).

The Upper Pliocene (Piacenzian Stage) contains a poorer planktonic foraminiferal assemblage than the underlying deposits, and includes *Globigerinoides conglobatus, G. sacculifer, Orbulina universa, Globigerina bulloides, G. apertura, G. decoraperta, Globigerinella aequilateralis, Globorotalia scitula* and rare *G. inflata*. Representatives of the genus *Globoquadrina*, as well as *Globigerina nepenthes* and *Globorotalia margaritae*, are absent.

Benthic foraminifera in the Pliocene deposits exposed along the Nahr El-Kabir valley are very diverse: *Planulina ariminensis* (d'Orb.), *Anomalina helicina* (Costa), *Sigmoilina celata* Costa, *Uvigerina peregrina* Cushman, *Siphonina planoconvexa* (Silv.), *Bulimina costata* d'Orb., *Bolivina villalverniensis* Mart., *B. catanensis* Seg., *Cassidulina crassa* d'Orb., *Cassidulinoides bradyi* (Norm.), *Dorothia gibbosa* (d'Orb.), *Karreriella bradyi* (Cushman), *Textularia agglutinans* d'Orb., *T. aciculata* d'Orb., *T. abbreviata* (d'Orb.), *Spiroplectammina sagittula* (Defrance), and *Bigenerina nodosaria* d'Orb., among others.

To the north of Al-Latheqiyeh, small outcrops of the Pliocene are located near the cape of Ras El Bassit near the village of Qara-Tepe, where the section begins with basal conglomerates and coarse sandstones, overlying rocks of the ophiolite unit. Above them lie alternating grits, polymictic sandstones, siltstones, and calcareous clays with plant debris, and limestones with *Chlamys scabrella* Lam., *Pycnodonta cochlear naviculata* (Brocchi), *Ostrea plicatula germanitata* De Gregorie, and *Spondylus crassicostata* Lam. (Kazmin and Kulakov, 1968).

On the northern slope of the Al-Latheqiyeh Depression near the village of Jinndiriye beneath Quaternary alluvium, a blue-gray clay unit of Pliocene age up to 25 m in thickness is found. These clays retain a nannoplankton assemblage of exceptional preservation (Devyatkin et al., 1996). The lower half of this section contains nannoplankton of the *Reticulofenestra pseudoumbilica Zone* (NN 15) and just above, *Discoaster tamalis* appears, indicating the presence of the overlying *Discoaster surculus* Zone (NN 16) of the Martini scheme. According to the Central Mediterranean stratigraphic scale (Rio et al., 1991), the boundary between these two nannoplankton zones is considered as the boundary between the Lower (Zanclean) and Upper (Piacenzian) Pliocene.

⇦

Fig. 6.69. The section of the Pliocene deposits in the lower reach of the Nahr El-Kabir River (near Al-Latheqiyeh).
Legend: 1 – gypsum; 2 – limestone; 3 – coquina; 4 – marl; 5 – clay; 6 – siltstone; 7 – clay with numerous debris of molluscs; 8 – sandstones; 9 – sandstones with numerous debris of molluscs.

On the southern slope of the Al-Latheqiyeh Depression, the lower half of the Pliocene is not exposed. In sections near the villages of Mardido and Msherphy, the Pliocene is represented by its upper half. It is a unit of plastic calcareous clays with gray and bluish or greenish tints, with abundant detrital material. Among them are interbeds of yellow clayey sandstones and siltstones. Numerous shells of small bivalves, gastropods, and scaphopods sometimes form distinct interlayers. This unit is some 130-150 m in thickness. Planktonic foraminifera in these comparatively shallow-water sediments are very poor. Benthic foraminifera predominate. According to the nannoplankton, the upper part of the *Discoaster surculus* (NN 16), the *Discoaster pentaradiatus* (NN 17), and *Discoaster brouweri* (NN 18) Zones can be identified, indicating an Upper Pliocene age for these sediments (Devyatkin et al., 1996).

To the south of Al-Latheqiyeh on the Mediterranean coast, the character of the Pliocene deposits is different. At the base of the sections there occurs a member of calcareous breccias with intercalations of marls and limestones whose thickness varies from some meters to 50 m. These beds are overlain by gray-bluish clays and marls with subordinate limestone interbeds.

In the vicinity of Banias and to the south near the villages of Hussein El Bahar, Shas, etc., effusive rocks *i.e.* tuffs and tuff-breccias are widely distributed. They unconformably overlie Mesozoic, Paleogene, and Miocene rocks, filling depressions in the landscape. Tuff-breccias resembling comparatively soft black or brown rocks consist of debris of basalt, volcanic glass, and tuff materials. The thickness of the pyroclastic intercalations is 50 to 150 m. Usually they contain lenses (0.2-4 m) of gray calcarous clays, gray-yellowish sands, and detrital limestones. To the south, the thickness of intercalations of marine sediments increases to 50 to 80 m. They contain rich molluscan faunas, including *Ostrea plicatula germanitata* (Di Gregori), *Amussium cristatum* (Bronn.), *Pecten jacobeus* (L.), *P. rhegiensis* Seg., *P. medius-plioparva* Sacco, *Corbula gibba* (Oliv.) var. *curta* Loc., *Chlamys scabrella* (L.), *Ch. opercularis* L., *Ch. bollenensis* (Mayer-Eymar), *Ch. zenonis* (Cowper-Wood), *Glycimeris insubrica* (Br.), *G. inflatus* (Brocchi), *Nucula placentina* (L.), *Turritella turris* (Bast.) var. *taurolaevis* Sacco, *Strombus* cf. *coronatus* (Desh.), *Venus multilamellata* (Desh.), and *Apporhais uttingerianus* (Risso), the presence of which supports the Pliocene age of these beds (det. I. Bertels-Uspenskaya, in Kozlov et al., 1966a).

In the **northern part of the El-Ghab depression** the Pliocene along the Nahr El-Aasi River (Orontes) is represented by alternating clays, greenish tuffogenic sandstones, sandy limestones, grits and conglomerates. In the north near the Turkish border, clays and fine-grained sandstones predominate. To the south, the content of coarse terrigenous material increases. In the vicinity of the village of Mazraet Esh Shuyokh, the thickness of the Pliocene is 45 m. These shallow-water, sublittoral sediments contain impoverished assemblages of molluscs, including *Oxystele amedei* (Brong.), *Retusa decussata* (Bonelli), *Nassarius serraticostata* (Brong.), *Helix facilis* Mayer, *Melanopsis fossilis* Martini, *Vermetus intortus* Lam., *Cardium aculeatum* L., *Neritina picta* Fer., *Venus* sp., *Cardita* sp., *Spirula subtruncata triangula* Ren., *Eastonia rugosa* (Chemnitz), and *Pitar italica* (Defr.) (det. R.L. Merklin, in Protasevich and Maksimov, 1966).

The Pliocene marine embayment was only limited to the Al-Ghab depression, and did not reach the elevated areas. To the north it was open into the Antiochia depression of the Neogene Mediterranean Sea, and to the south it was closed.

Geological history of the eastern Mediterranean during the Neogene

On the basis of the spatial distribution of the Neogene deposits of northwestern Syria, correlated by foraminiferal and nannofossil stratigraphy, the transgressional/regressional history of the Neogene can be represented in the following way (**Fig. 6.70**).

The Early Miocene interval (from the *Globigerinoides primordius–Globorotalia kugleri* Zone to the *Globigerinatella insueta–Globigerinoides trilobus* Zone) is characterized by a transgression of limited extent, as sediments of this age are only found in the Al-Latheqiyeh Depression and the Aafrine Graben. During the entire interval, the shoreline of the sedimentary basin retained its original contours.

At the end of the Early Miocene (the *Praeorbulina glomerosa* Zone) and throughout the Middle Miocene, a major transgression took place. The sea advanced to cover the entire Aleppo Plateau, and sea level remained stable to the end of the Middle Miocene (the *Globigerina nepenthes–Globorotalia siakensis* Zone).

The beginning of the Late Miocene (Tortonian, the *Globorotalia acostaensis* Zone) is characterized by clear regression, with the sea receding and remaining only in the western part of the Al-Latheqiyeh Depression and in the northern part of the Al-Ghab Graben.

At the end of the Late Miocene (Messinian) the marine basin was further reduced in area and marine evaporitic sediments were only developed in the extreme southwestern part of the Al-Latheqiyeh Depression.

During the Pliocene interval the areas of the Al-Latheqiyeh Depression and the northern part of the Al-Ghab graben were marked by a local marine ingression.

Although the Neogene deposits of southwestern Cyprus and northwestern Syria were formed in different tectonic frameworks, they followed a similar pattern of geological development in this time interval.

The Early Miocene represents the beginning of the Neogene transgression, both in Cyprus and in the Al-Latheqiyeh Depression, with calcareous sediments predominating, including such types as white thick-bedded soft chalky limestones, and hard platy limestones. Terrigenous sedimentation was limited to the narrow Aafrine Graben.

The late part of the Early Miocene (*Praeorbulina glomerosa* Zone) and the Middle Miocene correspond to the transgression maximum. Sediments of this age both in Cyprus and Syria have the most widespread distribution. Calcareous rocks are predominant, but during the second half of the Middle Miocene, the content of

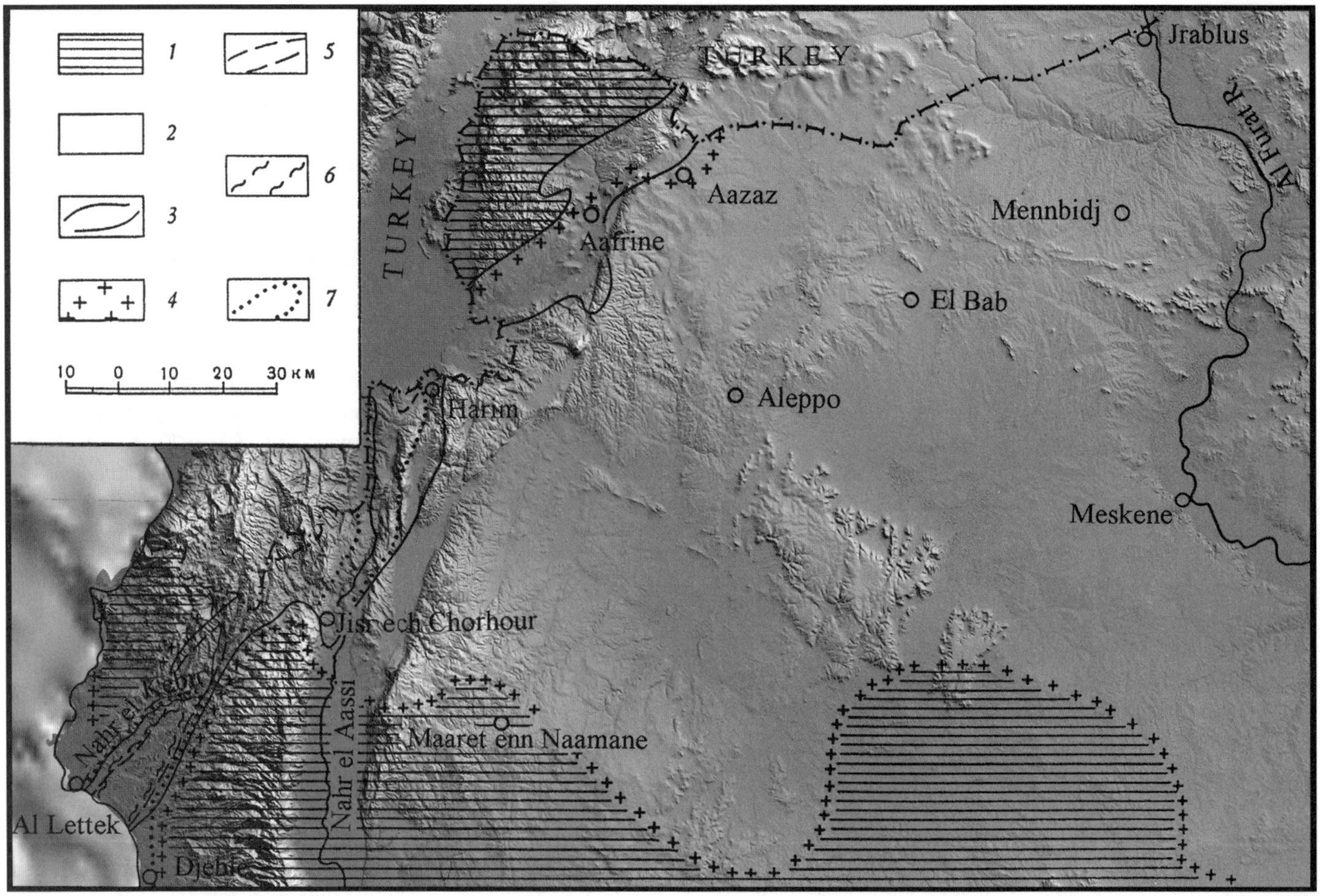

Fig. 6.70. Boundaries of distribution in north-western Syria marine deposits of: the Aquitanian-Lower Burdigalian, Upper Burdigalian-Middle Miocene, Tortonian, Messinian (Upper Miocene) and Pliocene.

Legend: 1 – exposures of the pre-Neogene rocks; 2 – areas occupied by the marine Miocene and Pliocene sediments; 3-7 – boundaries of distribution of sediments of: 3 – Aquitanian-Lower Burdigalian (N4 – N7 Zones), 4 – Upper Burdigalian-Middle Miocene (N8 – N14 Zones), 5 – Tortonian (N15 – N16 Zones, Upper Miocene), 6 – Messinian (Upper Miocene), 7 – Pliocene.

terrigenous material increases noticeably.

The onset of the Late Miocene (Tortonian) is characterized by a definite regression. Tortonian sediments in southwestern Cyprus covered limited areas, whereas in Syria marine conditions were preserved only west of the Al-Latheqiyeh Depression and in the northern part of the Al-Ghab Graben. The sediments everywhere are calcareous-terrigenous, with an increasing terrigenous component near the top of the Tortonian.

The end of the Late Miocene (Messinian) saw regression at a maximum and evaporitic sediments were laid down both in Cyprus and the southwestern part of the Al-Latheqiyeh Depression.

The Pliocene was marked by transgression over a very limited area.

MESOPOTAMIAN BASIN

During Neogene time, the Mesopotamian depression was an area of stable subsidence. For this reason, Neogene deposits are characterized by uniform distribution over this huge area, which extends over 1000 km from the northwest to the southeast. The northeastern edge of this basin is bounded by the folded Zagros terrains, and in the north by the Eastern Taurus. The southwestern limit of this basin has no natural boundary. The geological boundary of this depression may be defined as the limit of distribution of marine Miocene sediments along the right bank of the Al-Furat River. In the southeast, the Persian Gulf area was a continuation of the Miocene Mesopotamian Basin.

The tectonic structure of the Mesopotamian Basin is complicated. In general, this area corresponds to the Mesopotamian marginal fore-deep. Its northeastern part belongs to the inner folded zone of this fore-deep. The southwestern part of the depression is located within the outer undeformed zone of the Mesopotamian fore-deep. This type of tectonic structure, in combination with the existence of additional second order tectonic elements, is responsible for the considerable diversity of facies and thicknesses of the Neogene deposits of the basin.

The Mesopotamian Basin occupies areas within three countries. It includes northeastern Syria (to the east of the right bank of the Al-Furat River), northern and northwestern Iraq, and southwestern Iran.

We will be considering only the Syrian part of the Mesopotamian Basin. This part clearly represents only a small part of its total area. Nevertheless, within Syria, the complete Neogene succession of the basin can be observed. Many of the lithological and paleontological features of this succession are typical for the Neogene of the entire Mesopotamian Basin, because they were also found in Iraq and Iran.

The Neogene deposits of the Syrian part of the Mesopotamian Basin were formed in two tectonic zones: the extreme northeast of Syria is located in the inner zone of the fore-deep – to the north of the ramparts of Tuwal El-Aba - Jebel Abd El-Aziz - Jebel Jeribe (western plunge of Jebel Sinjar, which is located in northern Iraq) and Qaratchok Darh. The area to the south of this elevated belt belongs to the outer zone of the Mesopotamian fore-deep. This fore-deep conceals a number of older Mesozoic-Cenozoic rift-related basins, such as the Al-Furat depression, the Rawda Uplift, the Al-Beida High, the Derro High and others.

The publications of French geologists (Dubertret 1938, 1943, 1944; Dubertret, Keller and Vautrin, 1933 a, b; Doncieux, Dubertret and Vautrin, 1936; Cizancourt, 1934) provided the first insight into the Neogene of the Syrian part of Mesopotamia. They outlined the main succession of the sedimentary units. However the views of the French geologists reflected the stratigraphic knowledge of their time, and their approach and stratigraphic ages differed significantly from the modern methods and results.

Following the geological mapping in 1958-1961 by geologists of the USSR, the general lithological and stratigraphical characteristics of the Neogene of Mesopotamia were presented in the explanatory notes to the various maps of the region. The explanatory notes for the 1:200,000 scale geological map sheets were published by: Koslov, 1966b; Medvedev, 1966a, b; Petrov and Antonov, 1966; Oufland, 1966a, b; Ponomarev, 1966; Ponomarev and Bivshev, 1966; and Galaktionov, Demidov, and Katkova, 1966. Those for the maps at 1:1,000,000 scale were by Ponikarov et al., 1966; at 1:500,000 by Ponikarov et al., 1967; and a summary was published in the monograph about the geology of Syria by Ponikarov et al., 1969. The age of the Mesopotamian Miocene formations was based on foraminiferal investigations by Krasheninnikov (1966). Molluscan faunas were studied by I. A. Bertels-Uspenskaya under supervision of R. L. Merklin (Bertels-Uspenskaya and Oufland, 1971; Oufland and Bertels-Uspenskaya, 1971). Ostracods from the Pliocene were determined by N. N. Naidina.

During the Neogene interval, a variety of connections linked the Mesopotamian Basin with the Mediterranean Sea and the Indian Ocean. The stratigraphic succession is represented by alternations between sediments of normal salinity, brackish-water, and fresh-water deposition, hypersaline/evaporitic facies, and continental deposition. As a rule, the fossil assemblages are peculiar, impoverished, or demonstrating considerable local endemic patterns which create difficulties in determining the age of the sediments.

In 1988-1990, to resolve these problems, V. A. Krasheninnikov carried out a series of special investigations on the Aleppo Plateau in northern Syria where the junction of the Mediterranean and Mesopotamian facies occurs. These studies, in combination with observations of the Miocene biota of Mesopotamia, and analysis of the transgressive/regressive cycles of the Neogene in northwestern, northern, and northeastern Syria, allowed him to reliably determine the age of the stratigraphic subdivisions of the Mesopotamian Basin. It is now possible to correlate these subdivisions with the Mediterranean Neogene, and consequently to trace the history of the paleogeographic connections between the Atlantic and Indian Oceans via Syria.

Miocene

The Miocene deposits of that part of the Mesopotamian Basin in Syria can be subdivided into four formations. The lower Dibbane Formation comprises a variety of lithological types, including sandstones, clays, marls, limestones, and sometimes intercalations and even members of gypsum. Its age is Lower Miocene (the interval from the *Globigerinoides primordius–Globorotalia kugleri* Zone to *Globigerinatella insueta–Globi-gerinoides trilobus* Zone). It is overlain by the Jeribe Formation, which is made up of a unit of diverse limestones. It corresponds to the uppermost part of the Lower Miocene (the *Praeorbulina glomerosa* Zone) and the entire Middle Miocene. In cases where the Dibbane and Jeribe Formations are both represented by limestones, the name Euphrates Limestone Formation is used for both. Above the Jeribe Formation lies a thick unit of evaporitic deposits consisting of alternating gypsum, halite, marls, limestones, and clays. It is called the Lower Fars Formation (Tortonian Stage, Upper Miocene). The Miocene sequence is terminated by continental red-colored sediments of the Upper Fars Formation (conditionally referred to the Messinian Stage of the Upper Miocene).

These formations preserve their special features over vast areas of northeastern Syria. Their boundaries in the various structural regions of the Mesopotamian Basin are coeval, and nearly synchronous. For this reason, here the term 'formation' will correspond to the term 'suite' as used in the Russian stratigraphic terminology. All these formations were shown on the appropriate sheets of the Geological Map of Syria at 1:200,000 scale.

In Iraq the Dibbane, Jeribe, Lower Fars, and Upper Fars Formations were defined during the first half of the 1900s. For example, the Jeribe Formation was first mentioned in 1936 (Lexique Stratigraphique Internationale, 1963), and the Lower Fars Formation -- in 1918 (Lexique Stratigraphique Internationale, 1959). Application of these lithostratigraphic names in Iraq is quite understandable, as the large petroleum occurences of Kirkuk are connected with the Jeribe Formation, and the evaporitic deposits of the Lower Fars Formation are the cap rock for the petroleum reservoirs.

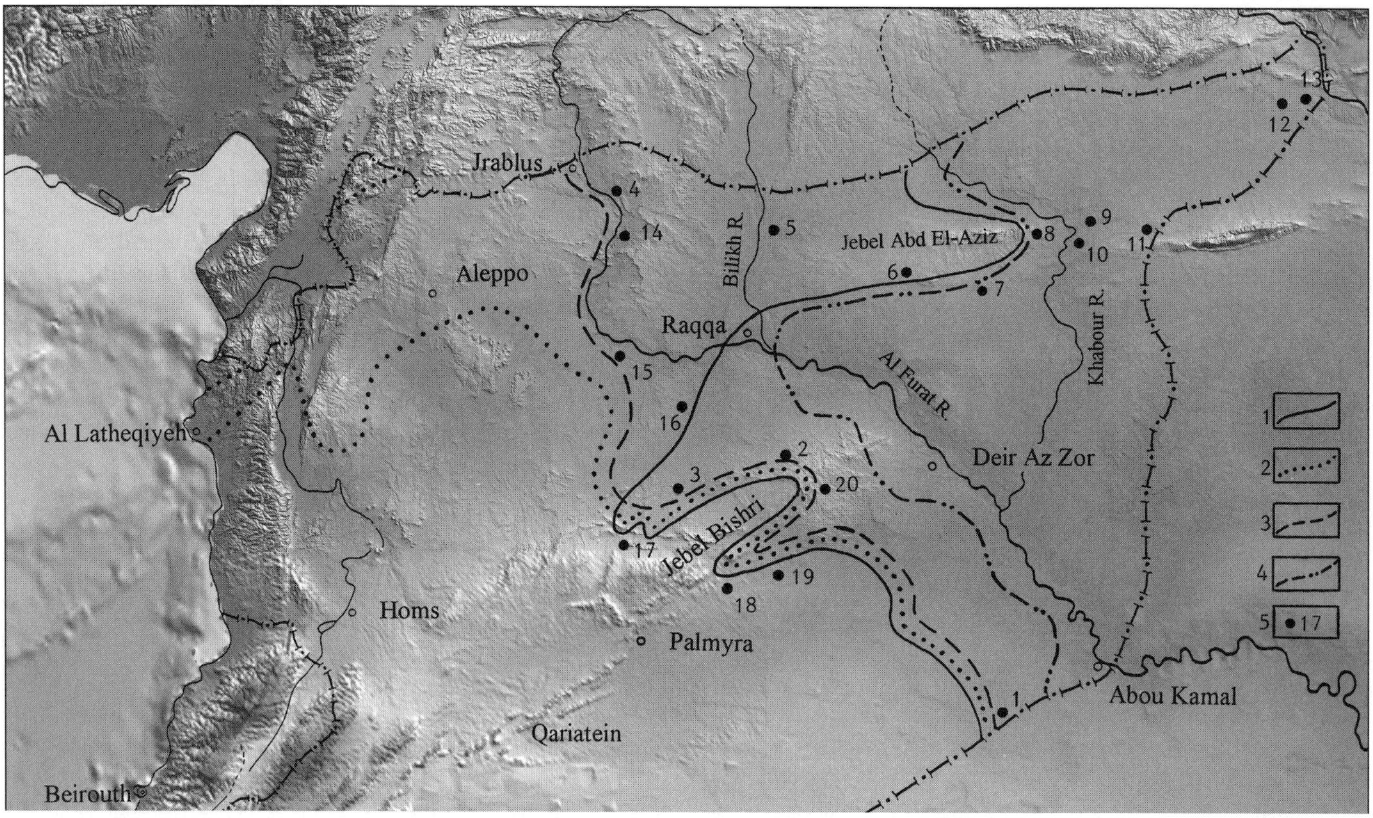

Fig. 6.71. Boundaries of distribution of marine and continental Miocene deposits in north-eastern Syria (the Mesopotamian Basin) and location of the main sections.
Note: Background from SRTM and ACE topographics – registration approximate.
Legend:
Boundaries of distribution of sediments: 1 – Lower Miocene (Aquitanian-Lower Burdigalian, N4 – N7 Zones, the Dibbane Formation); 2 – Upper Burdigalian-Middle Miocene (N8 – N14 Zones, the Jeribe Formation), the maximum of the Miocene transgression in Syria; 3 – Tortonian, Upper Miocene (the Lower Fars Formation); 4 – Messinian, Upper Miocene (the Upper Fars Formation).
Sections and outcrops (5) of the Miocene deposits of the Mesopotamian Basin shown on figures and mentioned in the text: 1 – Wadi As-Suwab; 2 – Jebel Bishri; 3 – Jebel Dilaa; 4 – Qara Qozak; 5 – Tuwal El-Aba; 6 – Jebel Beida; 7 – Jebel Abd El-Aziz; 8 – Umm Al-Khasane; 9 – Dbari; 10 – Abou Fije; 11 – Jebel Jeribe; 12 – Qaratchouk Darh; 13 – Touramish; 14 – Sirrine ech Chemaliye; 15 – Abou Houreira; 16 – Ressafe; 17 – Tourkmaniye; 18 – El-Hair; 19 – Bir Diddi; 20 – Bir Sejiri.

It is not inconceivable that in Iraq, facies changes occur within these formations from the west to the east and that their boundaries are thus not isochronous (Alsharhan and Nairn, 1997). It should be noted that the Jeribe Formation stratotype is located in northern Iraq in the village of Jaddalah on the Sinjar rampart. The western continuation of this structure in Syria is Jebel Jeribe studied by the USSR team. Because the formation received its name from this locality, it is suggested that the term Jeribe Formation as applied to both Syria and Iraq is identical. Unfortunately, the ages of the Miocene formations in Iraq as used in the geological literature are controversial, for they have not been substantiated by reliable paleontological data.

Natural outcrops of Miocene deposits of the Mesopotamian facies extend in a wide belt along the entire Al-Furat valley (mainly on the right bank), from Abou Kamal in the south in the boundary with Iraq to Jrablus in the north on the boundary with Turkey (**Fig. 6.71**). Very good Miocene sections are found on ramparts in the north and northeast of Mesopotamia at Tuwal El-Aba, Jebel Abd El-Aziz, Jebel Beida, Jebel Jeribe, Chembe, and Qaratchok-Darh. In the central part of the Mesopotamian plain at Jezireh the Miocene is overlain by continental sediments of Pliocene and Quaternary age. The Miocene deposits were penetrated by a large number of bore-holes, but publications rarely present data dealing with the biostratigraphy of these sediments, as the information is considered proprietary by the various petroleum companies. Producing horizons of petroleum in Syria are generally in the more ancient deposits (mainly Triassic) and the Miocene deposits are practically devoid of petroleum, unlike in Iraq and Iran. Therefore the sediments of this age are usually drilled without coring.

The following is a description of the lithological and paleontological characteristics of these formations (**Fig. 6.72**).

Dibbane Formation

The Dibbane Formation is exposed mainly on the right bank of the Al-Furat River. From the area of Abou Kamal in the south the strip of exposures extends to the hills of Jebel Bishri, encircling it on the east, and then stretching along the northern slope of the Palmyrides to the village of Tourkmaniye. To the north, in sections along the Al-Furat River between Jrablus and Abou Houreira, deposits of the Dibbane Formation are missing, for here Paleogene rocks are overlain directly by Jeribe Formation limestones. In the interfluve between the Al-Furat and Tigris Rivers (the Jezireh plain), there is only one outcrop of the Dibbane Formation – the eastern plunge of the Jebel Abd El-Aziz horst structure. In other areas of Jezireh in the Nahr El-Khabour region the Dibbane Formation is covered by younger sediments. The lithological features of this formation in the marginal part of Mesopotamia on the right bank of the Al-Furat River are quite different from those in the center of this basin.

In the western marginal part of the Mesopotamian depression, the Dibbane Formation is represented by coastal terrigenous and terrigenous-calcareous deposits. It is a unit of white, gray, yellowish and pinkish quartz sandstones and sands, varying from coarse-grained with flint pebbles, to fine-grained, and from hard massive sandstones to soft friable uncemented sands. Among the sandstones occur limestone and clay intercalations. Limestones are generally aphanitic, sometimes vuggy, slightly sandy, and often with leached shells of molluscs. Clays are calcareous, greenish-yellow and greenish, usually sandy, with rare shells of euryhaline foraminifera, *e.g. Elphidium* sp., *Rotalia* sp., and *Discorbis* sp., and with the algae balls (charophycean oogonia). Thicknesses vary between 50 and 85 m. In its typical appearance the formation outcrops on the northern and southern slopes of the Jebel Bishri uplift at Bir Sejiri, Wadi Rhum, Bir Qdeir, Jebel Dilaa, and El Heir, and to the west of Abou Kamal at Wadi Ouahar, Wadi Souab, and Wadi Miah.

In the central part of the Mesopotamian depression, the Dibbane Formation is characterized by quite different features. It is well-exposed on the eastern plunge of Jebel Abd El-Aziz near the village of Umm Al Khasane, where its exposed thickness reaches 115 m. The underlying contact is not exposed. The formation contains two members:

The lower member, 50 m thick, consists of greenish, yellowish, and light-gray limestones and marls with intercalations of aphanitic, foraminiferal (rock-forming *Operculina*), or algal (lithothamnion spheroids 1-3 cm across) limestones. Among these rocks is a layer of white, sugar-like gypsum, 8 m in thickness. Marls and limestones are often penetrated by burrowing organisms, with compressed skeletons of echinids, and extremely peculiar assemblages of foraminifera. The latter are entirely monospecific, consisting of planktonic foraminifera *Globigerinoides trilobus* (many hundreds of specimens/100 g of sediment), in association with very rare *Globigerina praebulloides, G. falconensis, G. foliata,* and *Globoquadrina* sp., and also rare benthic foraminifera – *Elphidium kvesanense* Archv., *E.* aff. *macellum* F. et M., *Florilus boueanus* d'Orb., *Cibicides stavropolensis* Bogd., *Cancris auriculus* F. et M., and *Textularia consecta* d'Orb.

The upper member, 65 m thick, comprises sediments with another lithological pattern. Various limestones of white, gray and brownish color (aphanitic, fine-grained vuggy, fine-detrital) predominate. They intercalate with 3 to 8 m thick layers of white, massive gypsum. Foraminifera were not found in these deposits, and casts of small bivalves and gastropods in leached limestones are indeterminable.

Drilling on the anticlinal structures of northeastern Mesopotamia at Khourbe, Tchembe, Jebissa, and Gouna showed that in this area, the Dibbane Formation is represented by marls and limestones with interbeds of anhydrite. These sediments contain rather diverse assemblages of planktonic foraminifera (Bach-Imam et al., 1976). At the base of this formation the *Globigerinoides primordius–Globorotalia kugleri* Zone can be identified, where nominate species are accompanied by *Globigerinoides immaturus, G. quadrilobatus, Globigerina woodi, G. semivera, Globorotalia siakensis,* and *Cassigerinella chipolensis* (the Aquitanian Stage, Lower Miocene). In the overlying layers of the Lower Miocene, *Globigerinoides trilobus* and *Globoquadrina praedehiscens* appear (probably the *Catapsydrax dissimilis* Zone).

The Dibbane Formation in this area is underlain by the marly Chilou Formation with rich associations of Oligocene planktonic foraminifera. These foraminifera indicate the *Globorotalia opima* Zone and *Globigerina ciperoensis* Zone of the Upper Oligocene, Chattian Stage. The latter zone contains also *Globorotalia pseudokugleri,* a species typical for the Oligocene/Miocene transition. In modern stratigraphic schemes, the *Globorotalia kugleri* Zone is placed by some workers in the top of the Oligocene, and by others, in the basal Miocene. Thus, in the submerged part of the Mesopotamian Basin, the transition from the Oligocene to the Miocene is gradual, but the environmental conditions changed significantly, as the Lower Miocene is gypsiferous. This lithological feature has been used as a marker in the logging of bore-holes for determining the Oligocene-Miocene boundary; however the appearance of the first gypsum beds does not constitute stratigraphic proof of a Miocene age for the sediments.

Deposits of the Dibbane Formation in the Jebel Abd El-Aziz section with abundant *Globigerinoides trilobus* apparently belong to a higher stratigraphic horizon in the Lower Miocene. This species appeared in sediments of the Aquitanian Stage but are only common in large numbers in the Early Burdigalian (the *Globigerinatella insueta–Globigerinoides trilobus* Zone). Very unusual, almost monospecific floods of the planktonic foraminifera *Globigerinoides trilobus* are evidence for environmental events of abnormal salinity.

In the western marginal part of the Mesopotamian Basin, sediments of the Dibbane Formation cover a smaller area than that occupied by Oligocene deposits, indicating that the Lower Miocene was regressive. Near Raqqa, bore-holes revealed that the Oligocene and Miocene are separated by a hiatus. In some bore-holes, sandstones of the Lower Miocene unconformably overlie marls with planktonic foraminifera of the *Globigerina ciperoensis* Zone which tops the Oligocene. Consequently, any hiatus present would be minimal. In other bore-holes, sandstones of the Dibbane Formation overlie marls of the *Globorotalia opima* Zone. Here, a more lengthy hiatus is clearly present. It should be noted that substantial lithological changes were found in all the bore-holes

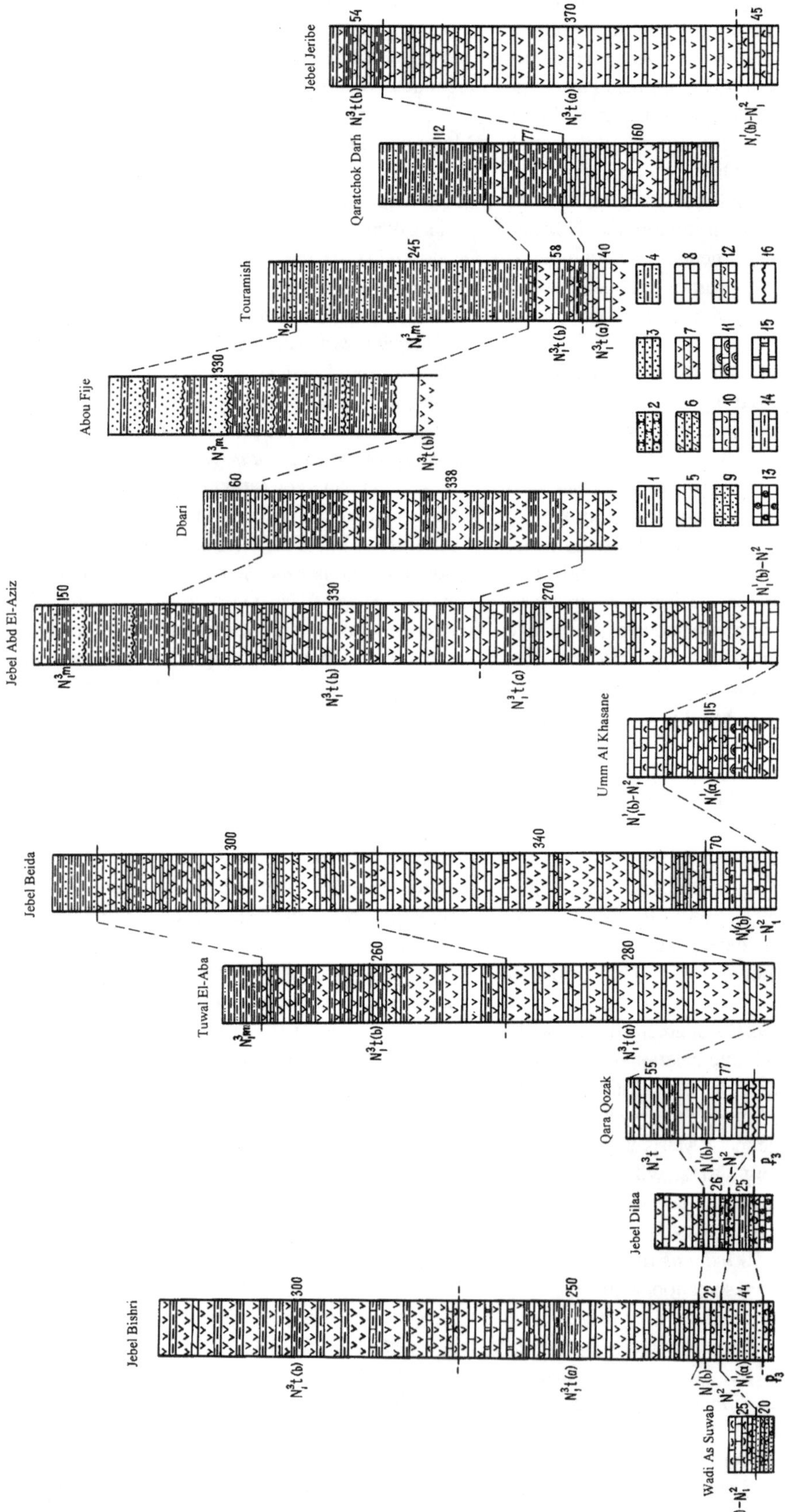

Fig. 6.72. Sections of the Miocene deposits of the Mesopotamian Basin.
Legend: 1 – clay, 2 – sandstone; 3 – sand; 4 – siltstone; 5 – marl; 6 – sandy marl; 7 – gypsum; 8 – limestone; 9 – sandy limestone; 10 – detrital limestone and coquina; 11 – algal limestone; 12 – silicified limestone; 13 – *Lepidocyclina* limestone; 14 – clayey limestone; 15 – dolomites; 16 – erosional boundary.

in the Al-Furat valley *i.e.* calcareous pelagic sediments of the Oligocene were replaced by shallow-water terrigeneous sediments of the Lower Miocene.

On the northern and southern flanks of Jebel Bishri and to the west of Abou Kamal, terrigenous rocks of the Dibbane Formation overlie an eroded surface of shallow-water limestones of the Oligocene. Therefore in this area a significant change of lithology is also indicated across that boundary. Unfortunately it is difficult to estimate the duration of this hiatus because the Oligocene is represented either by massive, recrystallized limestones, or by *Lepidocyclina* and *Operculina* limestones, whose stratigraphic position within the Oligocene is not clear.

To summarize the data regarding the Dibbane Formation, the Early Miocene age is evident, from the *Globigerinoides primordius–Globorotalia kugleri* Zone up to the *Globigerinatella insueta–Globigerinoides trilobus* Zone (inclusive), indicating the Aquitanian Stage and the lower half of the Burdigalian Stage. The character of the deposits (gypsiferous sediments) and faunal assemblages suggest that the Early Miocene Mesopotamian Basin was restricted in relation to the open ocean stenohaline regime. Biotic features and the type of sedimentation were very diverse in different regions of this basin. In the western marginal coastal area of Mesopotamia shallow-water terrigenous sediments with small brackish-water molluscan faunas, non-ornamented ostracod carapaces, and rare benthic foraminifera, as well as sandy clays with *Chara* algae were deposited under lagoonal or estuarine conditions.

In the central areas of the Mesopotamian Basin, calcareous-clayey pelagic sediments were developed. However on the Jebel Abd El-Aziz rampart they are of shallow-water origin, since there are *Operculina-* and lithothamnion-bearing limestones among the marls. The environmental conditions indicated by the fauna and sediment patterns within this interval were also changeable. It is difficult to explain a paleogeographic setting wherein layers of anhydrite and gypsum intercalate with marls containing assemblages of planktonic foraminifera. The monospecific character of the latter in the Jebel Abd El-Aziz section suggests abnormal changes in salinity. Finally, it may be shown that the Dibbane Formation is regressive when compared with the larger area of distribution of the Oligocene deposits,

Jeribe Formation

Deposits of the Jeribe Formation cover very extensive areas of the Mesopotamian depression. They accumulated during the course of the largest Miocene transgression in the entire eastern Mediterranean and Near East. Their lithological and faunistic characteristics are quite different from those of the Dibbane Formation.

Outcrops of the Jeribe Formation form an uninterrupted belt along the right bank of the Al-Furat River, beginning west of Abou Kamal at Wadi Souab and Wadi Ouahar and extending towards the southern slope of Jebel Bishri. This belt of outcrops encircles the Jebel Bishri Uplift on the east, following the northern slope of the Palmyrides as far as Tourkmaniyeh, where it turns towards the northeast and reachs the Al-Furat valley. Near the village of Abou Houreira, the outcrops of this formation shift to the left bank of the river and can be traced to the north as far as the border with Turkey. To the west of the north-south trending upper reach of the Al-Furat River, deposits of the Jeribe Formation were undoubtedly developed across the entire Aleppo Plateau in northern Syria, where the waterway between the Mesopotamian and Mediterranean basins was established. At present, outcrops of this formation are preserved as outliers on the eastern part of the Aleppo Plateau, whereas to the north and west of Aleppo, they occupy vast areas at Jebel Samaan, Jebel Douele, the Jebel Baricha uplifts, and the Idlib Plateau.

In the central part of the Mesopotamian depression (between the Al-Furat and Tigris valleys), the Jeribe Formation dips beneath younger Neogene units and only reappears in the Jebel Abd El-Aziz, Jebel Beida, Ras el Ain, and Jebel Jeribe Uplifts.

In the area of distribution of the Dibbane Formation on the right bank of the Al-Furat River from Jebel Bishri to Wadi Souab, the gradual transition to the Jeribe Formation rocks may be observed. On the Aleppo Plateau, deposits of the Jeribe Formation transgressively overlie Oligocene and Upper and Middle Eocene limestones. The same relationship is found in the Jebel Abd El-Aziz Uplift. On the eastern plunge of this structure, the Jeribe Formation conformably overlies rocks of the Dibbane Formation, whereas on the southern, gently dipping flank of this uplift the rocks of the Jeribe Formation overlie different horizons of the Paleogene.

The Jeribe Formation is characterized by two special features. First its lithological composition is rather uniform and consists almost exclusively of pure limestone. Second, these limestones are very diverse, represented by a number of types.

The composition of the Jeribe Formation on the flanks of Jebel Abd El-Aziz is very informative. A unit of limestone about 45 m thick includes several lithologies:

1) light-gray and yellowish fine-grained limestones packed with lithothamnion debris;

2) white fine-grained limestones, thick-bedded, finely porous and solid;

3) white detrital limestones, porous with casts and imprints of molluscan shells;

4) yellowish coquina limestones containing leached vugs after molluscan shells;

5) unlayered reefoidal algal limestones, very hard and aphanitic. In the process of weathering these break up into irregular blocks. All these varieties of limestones alternate upward in the section and replace each other along the strike. They do not form stable guide horizons.

A small anticlinal structure, Jebel Beida, is located some 20 km to the southwest of Jebel Abd El-Aziz. At the core of this structure Jeribe Formation limestones are exposed, revealing the same lithological properties.

Along the left bank of the Al-Furat River, outcrops of the Jeribe Formation extend from the Turkish border as far south as the village of Meskene. The typical section of the sediments of this unit can be observed in the cliffs of the left bank of Al-Furat near the village of Qara-Qozak. Here the Jeribe Formation, up to 83 m thick, can be subdivided into three members. Limestones of the lower member unconformably overlie Oligocene rocks. At the base they contain a layer of conglomerate which is laterally replaced by poorly-bedded white and yellow detrital limestones with lenses of coquinas. This unit is replaced upwards by a sequence of alternating poorly-bedded, fine detrital limestones which may sometimes be massive to vuggy, white, thin-bedded, hard and solid fine-grained limestones, platy aphanitic light-brown limestones, porous coquinas, and finally, lithothamnion-bearing limestones. In the middle member, thin interbeds of marls and greenish calcareous clays appear among hard, fine-grained and detrital limestones, as well as coquinas. White and yellowish limestones with varying quantities of lithothamnion and molluscan debris once more predominate in the upper member. These three members are of local significance and cannot be traced laterally between outcrops.

Further to the south, on the left bank of the Al-Furat River, the Jeribe Formation is exposed at the villages of Serrin-Esh-Shemaliyeh, Qassaq-Kebir, Esh-Sheikh-Annan, Joub-es-Safa, Tell Jarmiyeh, Qantara, and Mrchar. In all these exposures the formation comprises the same fine-grained, aphanitic and organogenic limestones, with lenses of leached coquinas, oyster banks, and thin beds of clayey limestones and marls. The exposed thicknesses vary from 12 to 43 m.

On the right bank of the Al-Furat River near the village of Abou Houreira, deposits of the Jeribe Formation display the above lithological features. Here they unconformably overlie Oligocene rocks. As a rule the succession starts with platy limestones with gravel beds. Upwards follow white and gray organogenic detrital limestones. They consist of foraminifera (*Operculina*, miliolids, and sometimes miogypsinids) with debris of bryozoa, molluscs, and lithothamnion algae. Usually the limestones are leached, with small vugs. They may display cross-bedding when lenses of detrital limestones with different sizes of debris cross-cut one another. The next overlying unit includes white fine-grained limestones, finely porous, unclear-bedded, and with characteristic cellular weathering surfaces. The Abou Houreira section terminates in leached coquinas typical for the Jeribe Formation. The exposed thickness does not exceed 14 m.

Along the northern slope of the Palmyrides, and on the slopes of Jebel Bishri, the deposits of the Jeribe Formation are somewhat different. All these exposures are located near the western paleo-coastal area of the Mesopotamian Basin. As in other regions of Mesopotamia, organogenic and calcareous rocks predominate, but the quantity of terrigenous material is more significant here.

For example, in the section on Jebel Taniyet-et-Tikh, the Jeribe Formation starts with basal conglomerates and calcareous sandstones. Overlying these is a member of light-gray, hard, sandy limestones and vuggy organogenic and detrital limestones passing into coquinas composed of casts of molluscs and echinoids. Among these rocks there occur layers and lenses of sands and gritstones. Their thickness is 28 m.

On Jebel Bishri, the Jeribe Formation outcrops on the northern slope at Wadi Rhoum, the village of Houaiet Abou Djir, and Bir Sejiri, the eastern slope at Shaafat-el-Beshir mountain, and southern slope at Jebel Didi and the El Hair ruins. It is made up of alternations of various limestones – hard and solid fine-grained, sandy, finely detrital, and vuggy coquinas with casts of large sea urchins. Intercalations (10-20 cm) of fine-grained and medium-grained quartz sandstones and sands are not common among them. The thickness ranges from 10 m to 20 m.

The southernmost outcrops of the Jeribe Formation in Syria are located to the west of Abou Kamal along Wadi Souab and Wadi Ouahar. In this area, a unit of hard aphanitic limestones, vuggy leached coquinas, sandy limestones, and foraminiferal limestones consisting of abundant shells of alveolinids, peneroplids and miliolids is developed. A sandy content is observed in all the rock varieties, and lenses of coarse quartz sandstones can be found as well. The thickness does not exceed 25 m.

In the northeast of Syria the Jeribe Formation is penetrated by numerous bore-holes. In all sections the formation is represented by limestones. Maximal thicknesses were observed in the bore-holes of Qaratchok-1 (135 m), Qamishliye-1 (73 m), and El Bouab-1 (62 m). Usually thicknesses are limited to 30-45 m as in bore-holes Gbeibeh-1, Ghouna-1, El Bardeh-1, Sheikh Salah-1, and others. In the Derro bore-hole, located on the left bank of the Al-Furat River somewhat to the north of the mouth of Nahr El-Khabour, the thickness of the Jeribe Formation is 48 m.

Limestones of the Jeribe Formation contain a rather diverse fauna of bivalve, gastropod, echinid and benthic foraminifera. Unfortunately as a rule the shells are leached, represented only by casts, and therefore exact determinations are very difficult. The assemblages of molluscs which have relatively satisfactory preservation were encountered in outcrops on the left and right banks of the Al-Furat River between Meskene and Raqqa. Here the list includes: *Chlamys multistriata* Poli, *Ch.* cf. *submalvinae* Blanck., *Corbula* ex gr. *carinata* Duj., *Pholadomya alpina* Math., *Venus clathrata* Duj., *Tapes* cf. *vetulus* Bast., *Miltha callipteryx* (Tourn.), *Divaricella ornata* Agass., *Diplodonta trigonula* Bronn., *Cardium* ex gr. *multicostatum* Brocc., *Monodonta* ex gr. *pygmea* Cosm. et Peyr., *Conus* ex gr. *ventricosus* Brocc., *Strombus* sp., *Fusus* ex gr. *valensiensis* Grat., *Cerithium* ex gr. *bronni* Partsch., *Turritella erronea* Cosm., *Haustator tornatus* (Br.), *H.* aff. *triplicatus* (Br.), *Patella* sp., *Psammobia ornata* Agass., *Ostrea* ex gr. *lamellosa* Brocchi, *Anadara turonica* Duj., and *Modiolus* sp.

Similar assemblages of molluscs were found in the Jeribe Formation limestones in northeastern Mesopotamia at Jebel Abd El-Aziz and Jebel Jeribe. In these sections a gradual transition may be observed from the Jeribe Formation to the evaporitic deposits of the overlying Lower Fars Formation. The first layer of gypsum is overlain here by organogenic limestones crowded with shells of *Ostrea lamellosa* Brocchi, *O. gingensis* Schloth., *O. frondosa* de Serres, *Gigantostrea crassicostata* Sow., *Chlamys* ex gr. *opercularis* L., *Pecten* cf. *fuchsi* Font., *Flabellipecten* ex gr. *larteti* Tourn., and *Anadara diluvii* Lam. According to their molluscan fauna these limestones should still be referred to the Jeribe Formation,.

Benthic foraminifera are numerous and sometimes form foraminiferal limestones. They are unfortunately always present as casts: *Borelis melo* (F. et M.), *Dendritina rangi* d'Orb., *Peneroplis evolutus* Hens., *Meandropsina iranica* Hens., *Elphidium crispum* (L.), *E. macellum* (F. et M.), *Streblus beccarii* (L.) and diverse miliolids. In the lower horizons of this formation *Miogypsina* sp. was identified. The poor state of preservation of this microfauna makes species determination difficult.

The age of the Jeribe Formation has been a matter of some controversy. In general, the molluscan and benthic foraminiferal faunas indicate the Middle Miocene, while the presence of miogypsinids near the base suggests that the lower part of the Jeribe Formation still belongs to the Lower Miocene. Here, stronger evidence for the age of the Jeribe Formation results from the correlation of the Mesopotamian and Mediterranean types of the Miocene in sections from the western margin of the Aleppo Plateau. There, shallow-water limestones of the Jeribe Formation are replaced by pelagic marls and clays with planktonic foraminifera of the *Praeorbulina glomerosa* Zone (Lower Miocene) and all the zones of the Middle Miocene (including the *Globigerina nepenthes-Globorotalia siakensis* Zone, N14).

The main lithological features of the Jeribe Formation can be summarized as follows: Shallow-water limestone deposits of this formation are strongly transgressive. The thickness of limestones usually does not exceed 100 m. Among them organogenic varieties are very common. These limestones are highly porous, with abundant minute cavities originating from the leaching (dissolution) of organic remains. Poor preservation of the stenohaline fauna makes age determination difficult, but their exceptional porosity makes them productive petroleum reservoirs. The upper producing horizons of the super-giant oil fields of Northern Iraq are mainly in the deposits of this formation.

Lower Fars Formation

Sediments of the Lower Fars Formation are found throughout the Mesopotamian Basin. The transition upwards from the underlying Jeribe Formation is gradual.Within the Mesopotamian depression the Lower Fars Formation covers a smaller area than the Jeribe Formation, indicating its regressive character. This regressive nature is obvious from the fact that the Lower Fars is completely absent in the area of the Aleppo Plateau, from the Al-Ghab Graben in the west to the Al-Furat valley in the east. Its deposits appear only in exposures along the north-south stretch of the Al-Furat River near the towns of Djerablous and Sirrine ech Chemaliye, where their lithology (shallow-water conglomerates, sandstones and clays) suggests accumulation in a coastal, marginal marine environment.

The best outcrops of the Lower Fars Formation are in the escarpments of the Al-Furat valley and along the northern boundary of the Mesopotamian depression, where they are exposed on the flanks of uplifts such as Tuwal El-Aba, Jebel Abd El-Aziz, Qaratchok Darh, Jebel Jeribe, Jebel Beida, Ras el Ain, Qamichliye, and Jebel Tchembe. Exposures in the central areas of the Djezireh plain east of the Al-Furat River and south of Jebel Abd El-Aziz are more sparse because here the Lower Fars is overlain by Late Neogene and Quaternary rocks. The lithologies of the Lower Fars Formation are complex and variable with clayey, calcareous, and evaporitic subunits formed under various salinity conditions.

In the sections from Tuwal El-Aba, Jebel Abd El-Aziz, Qaratchok Darh, Tchembe and other sites in northern Mesopotamia the Lower Fars Formation is subdivided into two members: a lower calcareous-gypsiferous member and the upper clayey-gypsiferous one.

The lower member is composed by alternating limestones, gypsum and subordinate marls and clays. The limestones are white, gray and yellowish in color, fine-grained, finely porous and vuggy, and very often leached. Layers of brown indurated aphanitic limestones and porous coquinas consisting of casts of the bivalve *Clausinella* ex gr. *persica* Cox are comparatively rare. The thickness of limestone interbeds varies from 0.5 to 4 m. The thickness of white, massive, fine-crystalline gypsum units is more significant – from 2 to 40 m. These gypsum layers are not constant, and along the strike it is possible to observe large changes in thickness. In the uppermost part of this member light-gray marls and greenish and red-brownish silty clays which pass into siltstones are present as thin intercalations.

The thickness of the lower calcareous-gypsiferous member in natural outcrops ranges from 100 m at Tchembe to 290 m at Tuwal El-Aba. The gypsum constitutes 70-90% of the total thickness of the member.

The upper clayey-gypsiferous member is represented by a monotonous, sometimes rhythmic alternation of white and light-gray gypsum in beds up to 4-10 m, and rarely to 15-30 m, with mottled clay beds commonly 0.3-7 m thick, and rarely up to 9-11 m. Two varieties of clays are present:

1) red and brown mudstones: weakly calcareous or non-calcareous, poorly layered, sometimes silty, and barren of fossils;

2) gray-green and green calcareous clays; hard, thin-bedded, with shells of molluscs, benthic foraminifera and ostracods, spines and plates of echinids. Interbeds of marls, fine-grained limestones, and leached coquinas consisting of shells of *Clausinella* ex gr. *persica* Cox, are rare. Siltstone and fine-grained sandstone appear near the top of this member. The minimum thickness of the upper member is less than 80 m (Qaratchok Darh), and the maximum can reach 300-400 m. The total thickness of gypsum exceeds that of clayey-calcareous rocks but is less than in the lower member. For example, 20 km to the south-west of Hassetche at Wadi Dbari on the southern slope of Jebel Abd El-Aziz, the total thickness of the clayey-gypsiferous member is 230 m, with gypsum layers constituting 138 m and clayey rocks 92 m.

Within some anticlinal structures the calcareous-gypsiferous and clayey-gypsiferous members are easily and locally traced. However for the entire part of Syrian Mesopotamia they do not belong to stratigraphic units with isochronous boundaries. For example, the tracing of guide horizons during our geological mapping showed that the lower 100 m of the clayey-gypsiferous member of the Tuwal El-Aba section corresponds to the upper part of the calcareous-gypsiferous member of the Jebel Abd El-Aziz section.

To the north of Jebel Abd El-Aziz in the inner zone of the

Mesopotamian fore-deep these lower and upper members cannot be identified. Here the Lower Fars Formation is made up of thin-bedded calcareous clays of red, brown and green colors with subordinate interbeds of gray-greenish marls, white fine-grained limestones and vuggy coquinas consisting of abundant *Clausinella* ex gr. *persica* Cox. Gypsum intercalations are rare, of insignificant thickness, and sometimes even absent. The thickness of the undifferentiated Lower Fars deposits in this area ranges between 350-500 m.

Nevertheless, the two-fold structure of the Lower Fars Formation in its general form can be observed in adjacent areas of Northern Iraq. The Lower Fars Formation as described in the Syrian sections, is subdivided in Iraq into the Lower Fars and Middle Fars Formations, which are treated as regional lithostratigraphic units (Dunnington et al., 1959). In this case the first unit corresponds to the calcareous-gypsiferous member and the second to the clayey-gypsiferous member.

The subdivision of the Lower Fars Formation into two members was confirmed by drilling on many of the anticlinal structures in northeastern Syria, such as Souedie, El-Bouab, El-Bardeh, Ghouna, Gbeibeh, Jbissa, Sheikh Sallah, and others. The lower member consists of alternating gypsum, anhydrite, and limestones; the upper member is composed of alternating evaporitic rocks with mottled (red, green, gray) clays and marls. A very important feature of the lower member is the appearance of halite layers. To the north-east of Hassetche the El-Bouab bore-hole penetrated only one layer of salt. To the southeast of this town they are much more numerous: 12 beds of halite in the Ghouna bore-hole and 14 beds in the Sheikh Sallah bore-hole. Some salt beds were 10-20 m thick (the Jbissa bore-hole). The main mass of salt-bearing horizons is developed in the area of maximum subsidence of the Mesopotamian depression around the Nahr El-Khabour River. According to drilling data, the total thickness of the Lower Fars Formation in northeastern Syria ranges from 345 m in the El-Bardeh-1 bore-hole to 712 m in the Gbeibeh-1 bore-hole.

Sediments of the calcareous-gypsiferous and clayey-gypsiferous members are characterized by similar assemblages of molluscs and benthic foraminifera, but green calcareous clays of the latter are mostly enriched in organic remains.

Assemblages of molluscs become poorer in comparison with the Jeribe Formation. They include *Clausinella* ex gr. *persica* Cox, *Cardium* ex gr. *gallicum* (May), *Lucina* sp., *Diplodonta* sp., *Anomya* cf. *ephippium* L., *Aloidis* cf. *gibba* Olivi, *Lithophaga lithophaga* L., *Pectunculus* cf. *glycymeris* L., *Ostrea fimbriata* Grat., *O. frondosa* de Seer, *Turritella* ex gr. *turris* Bust., and *Clementina* cf. *papiracea* Cox. Basal layers of the upper clayey-gypsiferous member very often contain numerous *Ostrea fimbriata,* which may be used as a guide horizon for local correlation.

Benthic foraminifera are represented by *Elphidium listeri* (d'Orb.), *E. antonina* (d'Orb.), *E. angulatum* (Egger), *E. crispum* (L.), *E. hauerinum* (d'Orb.), *E. macellum* (F. et M.), *Nonion granosus* (d'Orb.), *Florilus communis* (d'Orb.), *Cibicides boueanus* (d'Orb,), *Cancris brongnartii* (d'Orb.), *Reussella spinulosa* (Reuss), *Streblus beccarii* (L.), *Pyrgo inornata* d'Orb., *Quinqueloculina ungeriana* d'Orb., *Q. akneriana* d'Orb., *Triloculina austriaca* d'Orb., *Valvulineria obtusa* (d'Orb.), *Textularia mayeriana* d'Orb., *Spiroloculina canaliculata* d'Orb., *Sp. excavata* d'Orb., *Sigmoilina mediterranensis* Bogd., *Silicosigmoilina* sp. and rare specimens of *Borelis melo* (F. et M.). The species listed are accompanied by elphidiids, nonionids, miliolids, discorbids and rotaliids which are evidently endemic to the Mesopotamian Miocene, and require further study.

Molluscs and benthic foraminifera of the Lower Fars Formation are known from the Middle Miocene and the lower part (Tortonian) of the Upper Miocene of the Mediterranean, the Vienna Basin, the Fore-Carpathian, and the Trans-Carpathian. Unfortunately, they cannot be used for exact age determination within this interval. The Lower Fars Formation can be placed in the Tortonian Stage of the Upper Miocene on the basis of correlation with the Mediterranean Miocene of northwestern Syria, analysis of overall geological development of this area, and its position above the Middle Miocene Jeribe Formation.

In exposures along the southwestern margin of the Mesopotamian depression the Lower Fars Formation preserves its main characteristics but the amount of terrigenous material strongly increases as the part of the basin closest to the coast is approached.

Near Djerablous the basal layers of the Lower Fars Formation consist of conglomerates, grits and sandstones, overlain by frequent alternations of aphanitic and fine-grained limestones, coquinas, sandstones, greenish clays and marls. Gypsum layers are completely absent. The exposed thickness does not exceed 20 m. Foraminifera in clayey interbeds are rare and consist of *Elphidium listeri* (d'Orb.), *E. antonina* (d'Orb.), *E. macellum* (F. et M.), *E. crispum* (L.), and *Streblus beccarii* (L.). Some samples contain only *Chara* oogoniums indicating brackish- or fresh-water conditions along the coastal zone of the marine basin.

To the south, near the villages of El-Mrchara, Qara-Qozak and Sirrine ech Chemaliye along the north-south stretch of the Al-Furat River, the intercalations of coarse-grained rocks wedge out. Here the Lower Fars Formation is composed by alternating greenish marls and calcareous clays and red noncalcareous mud-stones. Interbeds of limestones are rare as are very thin (1-10 cm) layers of gypsum. Exposed thicknesses are 40-50 m. The pattern of sedimentation of the Lower Fars Formation here is cyclical. Each cycle starts with green clays and marls containing shells of bivalves, ostracods and foraminifera, and spines and skeletal debris of echinids: *Ostrea fimbriata* Grat., *Clausinella* ex gr. *persica* Cox, *Anomya* cf. *ephippium* L., *Cytherideis longula* Ul. et Bass., *Trachyleberis expunctata* Zalanyi, *Cypridea muelleri* (Muens.), *Elphidium hauerinum* (d'Orb.), *E. listeri* (d'Orb.), *E. antonina* (d'Orb.), *E. crispum* (L.), *Nonion granosus* (d'Orb.), *N.* aff. *punctatus* (d'Orb.), *Porosononion subgranosus* (Egger), *Discorbis semiorbis* (Karrer), *Dendritina* sp., and *Streblus beccarii* (L.). The appearance of reddish tints in the clays is accompanied by a general impoverishment of the fauna, with remnants of echinids disappearing completely. The overlying red noncalcareous mudstones yield only euryhaline foraminifera, such as *Streblus beccarii, Elphidium macellum,* and *E. crispum*. Each cycle terminates with mudstones containing *Chara* algae. Consequently, in this area terrigenous sediments appear at the very base of the Lower Fars Formation and the calcareous-gypsiferous lithofacies is absent.

On the right bank of the Al-Furat River, along the northern slope of the Palmyrides to the east of the village of Tourkmaniye, the Lower Fars Formation is composed of sandy limestones, greenish clays and marls and red mudstones with lenses of gritstones and conglomerates. Gypsum layers are absent. Outcrop thicknesses do not exceed 10-15 m. Clearly the upper part of this formation was truncated by transgressive Pliocene sediments.

On the plain between the northern slope of the Palmyrides and the Al-Furat valley, the Lower Fars deposits are overlain by Pliocene rocks. According to drilling data from Hole-153 to the northwest of Tourkmaniye, and Hole-300 to the south of Es-Safsafe, this formation is made up of alternating mottled clays, marls and limestones, but among them thick members (up to 10-20 m) of gypsum are now seen.

This type of succession is exposed in the vicinity of Ressafe. Here the hot and dry climatic conditions of Syria encouraged the use of gypsum from the Lower Fars Formation as a constructional material for the walls of the fortress, temples, and palaces of Byzantine Ressafe, or the Sergiopolis of the IV-VI centuries. Clays and marls of the Ressafe area contain good assemblages of foraminifera similar to those from other regions of Syria.

On the eastern plunge of the Palmyrides on the Jebel Bishri uplift, the Lower Fars Formation reaches considerable thickness, and is subdivided into a lower calcareous-gypsiferous and an upper clayey-gypsiferous member, as in the northeastern regions of Syria.

The lower member consists of alternating gypsum (beds of 2-7 m thick, and sometimes to 18 m), fine-grained limestones, porous coquinas formed by shells of *Clausinella* ex gr. *persica* Cox, and small gastropods (beds 0.2-0.6 m thick, and sometimes to 2 m), thin interbeds of green clays, and dense red mudstones. The quantity of clayey layers increases upward in the section. The thickness of the lower member reaches 250 m. As a rule, green calcareous clays include debris of echinids and the following benthic foraminifera: *Elphidium listeri* (d'Orb.), *E. macellum* (F. et M.), *E. hauerinum* (d'Orb.), *E. antonina* (d'Orb.), *Streblus beccarii* (L.), *Textularia mayeriana* d'Orb., *Spiroloculina canaliculata* d'Orb., *Quinqueloculina ungeriana* d'Orb., *Q. akneriana* d'Orb., and *Q. contorta* d'Orb.

The upper clayey-gypsiferous member is very well-exposed in the eastern part of Jebel Bishri along Wadi Fassayateh and Wadi Kharamiyeh, and in cliffs above the right bank of the Al-Furat River from Raqqa to Deir Ez-Zor. It is made up of alternating gypsum (beds of 3-8 m thick, and sometimes up to 15 m), greenish marls, calcareous clays, and red mudstones (beds of 3-10 m thick). Interbeds of siltstones and fine-grained sandstones appear in the upper part of this unit. The thickness of this member is about 300 m. The assemblage of foraminifera here is more diverse than in the lower calcareous-gypsiferous member. The microfauna includes, together with the above-mentioned species, *Elphidium ungeri* (Reuss), *E.* ex gr. *falunicum* Cushman, *Nonion* aff. *punctatus* (d'Orb.), *Nonion granosus* (d'Orb.), *Florilus communis* (d'Orb.), *Cibicides boueanus* (d'Orb.), *Bulimina pupoides* d'Orb., *Bolivina dilatata* Reuss, and *Reussella spinulosa* (Reuss).

The southernmost outcrops of the Lower Fars Formation are situated near Abou Kamal. Here the formation is composed of alternating green clays, red mudstones, leached coquinas, and gypsum.

In general, regressive Lower Fars deposits can be treated as Tortonian evaporitic lithofacies. It is represented by various rocks such as limestones, coquinas, mottled clays and marls, gypsum, anhydrite, and halite. All these sediments were formed under abnormal sea water salinity conditions, ranging from fresh water sediments with *Chara* algae, via sediments of more normal marine salinity with benthic foraminifera and echinids debris, to sediments consisting of hypersaline chemicals. Coarse terrigenous sediments (sandstones, gritstones, and small-pebble conglom-erates) characterize only the western coastal zone of the marine basin which produced the Lower Fars Formation. Undoubtedly this sea had restricted access to marine basins with normal salinity.

The Lower Fars Formation is widespread throughout the Mesopotamian depression. In Iraq and Iran its clays and evapo-rites form the cap rock of the petroleum occurrences of these countries, although evidently its place in the stratigraphic column varies laterally depending upon location and even country (Alsharhan and Nairn, 1997).

The Upper Fars Formation

Sediments of the Upper Fars Formation rest apparently conformably on the Lower Fars rocks. They occupy a smaller area than the Lower Fars Formation and can be considered as a regressive unit. The Upper Fars Formation deposits are confined mainly to two regions. The first coincides with the Nahr El-Khabour basin east of the Al-Furat valley and south of the Tuwal El-Aba - Jebel Sinjar Uplift. The second region is situated to the north of this uplift in the inner zone of the Mesopotamian fore-deep.

On the eastern Qaratchok Darh Uplift, and especially on the small adjacent Qanik-Touramish anticlinal structure, the Upper Fars Formation is subdivided into two members. The lower red-colored member consists of frequent alternations of red, brown, gray and greenish clays, siltstones and sandstones. At the base of the member the sandstones are fine-grained. Upward, the number of sandstone interbeds increases and they become medium-grained. Cross-bedding, ripple-marks, and desiccation cracks are very common. The upper gray-colored member is also composed of clays, siltstones, and sandstones, but here dull gray and brownish colors predominate and the sandstones are medium- to coarse-grained, poorly-sorted and are polymictic. The thickness is 230 to 250 m.

To the south of Jebel Abd El-Aziz the Upper Fars Formation preserves its lithological nature. Here also it may be subdivided into a red lower member consisting of thin-bedded clays, silt-stones and fine-grained sandstones (30-80 m thick), and a gray upper member, where coarse and poorly-sorted sandstones pre-dominate (230 to

300 m thick). Some layers of cross-bedded alluvial sandstones may reach 20 to 25 m in thickness.

On the Nahr El-Khabour basin plain the Upper Fars Formation was penetrated by a series of shallow drillings carried out for our geological mapping. The formation is composed of red clays, mudstones, and sandstones, and the gray member with coarse sandstones is present in some localities. Apparently, the formation was eroded in the course of pre-Pliocene erosion, and its thickness varies widely from 65 to 130 m.

Organic remains in the Upper Fars Formation are rare. In the areas of Tuwal El-Abba, Deir Ez-Zor, and Raqqa only one species of Foraminifera is present - *Streblus beccarii* (Lin.). Fresh water ostracodes are also not numerous, and include the species *Iliocypris bradyi* Sars., *I. gibba* (Ramd.), *Candoniella albicans* (Brady), *Darwinula stevensoni* (Brady et Rober.), *Potamocypris punctillata* Vorosh., *Cyclocypris minimus* Vorosh., *Zonocypris membranae* (Livent.), *Cyprinotus micra* Vorosh., *Limnocythere* sp., and *Eucypris* sp. (det. N. N. Naidina). *Chara* oogonia are characteristic of all the outcrops of the Upper Fars Formation.

The lithology and faunal remains of the Upper Fars Formation clearly indicate the continental, mainly lacustrine and alluvial, nature of these sediments. Thus the Mesopotamian depression was not connected with the open sea at this time.

The sparse paleontological material does not allow a direct age determination of these sediments. According to its stratigraphic position, the formation is referred to the uppermost part of the Miocene. Based on an analysis of the geological history of all of Syria, the Upper Fars Formation should be correlative with the Messinian Stage of the Mediterranean (Upper Miocene).

The Upper Fars Formation is transgressively and erosively overlain by Pliocene continental grits, sandstones, and siltstones, with subordinate clay intercalations.

Pliocene

Although Pliocene sediments occupy vast areas of the Mesopotamian depression, they do not form a continuous cover **(Fig. 6.73)**. They are absent on some highs or slopes of structures, such as the Djerablous High, the adjacent area on the left bank of the Al-Furat River to the north of Meskene, Jebel Abd El-Aziz and its continuation eastward to the Iraqi border at Jebel Jeribe, the eastern flank of the Jebel Bishri uplift, the valley of the Al-Furat River between Raqqa and Deir Ez-Zor, and also the western slope of the Rawda Uplift (Ponikarov et al., 1969).

Pliocene deposits are clearly transgressive with respect to the underlying Miocene Upper Fars Formation. As a rule, an erosional surface is found at their base, together with a basal layer of coarse terrigenous rocks, and sometimes an angular unconformity.

The Pliocene is solely represented by continental deposits of lacustrine, alluvial, and proluvial types of facies. Different combinations of these facies in various parts of Mesopotamia demonstrate a very diverse pattern of deposition. At the same time there are some features in common.

The Pliocene deposits consist of two members, which are also separated by an unconformity. The lower member includes clays, sandy marls, siltstones, and sandstones with subordinate interbeds of grits and conglomerates. Coarse terrigenous rocks consisting of conglomerates, grits and sandstones predominate in the upper member, with limestones and clays subordinate.

On the northwestern margin of the Mesopotamian depression on the left bank of the Al-Furat River between Djerablous and Meskene, only the lower member of the Pliocene is present. It directly overlies limestones of the Eocene, Oligocene and Middle Miocene. The Pliocene is represented here by alluvial and proluvial conglomerates, grits, and sandstones in the sections near Meskene and Khalijaq. Intercalations of lacustrine limestones sometimes occur within this succession near Djerablous The thicknesses are minimal (15-40 m). To the east in the area of Kherbet Shomba at an elevation of 358 m, bore-holes penetrated a unit up to 134 m thick of alternating limestones and marls, which are usually sandy, as well as dolomites, clays and sandstones. Such an increase in thickness is evidently due to a depression in the pre-Pliocene landscape of this locality (Oufland, 1966a).

In the area of Tuwal El-Aba the Pliocene succession is more complete. Here both lower and upper members can be distinguished. Clearly the Pliocene transgressively overlies older rocks. To the south of the Bteman and Beida hills, Pliocene deposits overlie the Messinian Upper Fars Formation, while to the south and southwest of Tuwal El-Aba, they rest on the upper member of the Tortonian Lower Fars Formation. To the northwest of this uplift, the Pliocene truncates the lower member of the Lower Fars Formation. Therefore, it is apparent that the Pliocene transgression truncated 300-400 m of the Lower Fars Formation.

The lower member of the Pliocene consists of alternating clays, sandy marls, limestones, sandstones, and separate interbeds of conglomerates up to 2-4 m thick. In general, this succession represents a lacustrine basin located not far from the source of the clastics. This feature is indicated by a predominance of deluvial and proluvial sediments. Thicknesses range from 20 to 40 m. Due to the morphology of the Mesopotamian plain, sections of the lower member of the Pliocene can only be observed in fairly deep wells, or in huge karstic hollows within the evaporitic rocks of the Lower Fars Formation *e.g.* the karstic hollow at Jafr El Houteh located 24 km northwest of the top of Tuwal El-Aba.

North and south of Tuwal El-Aba the upper member of the Pliocene is characterized by deposits of a different mix of lithological features and origin. To the north near the Turkish border, lacustrine algal limestones are developed, and their thickness ranges from 5 to 8 m. They form a plateau with a flat surface bounded by steep escarpments. To the south of Tuwal El-Aba, a section of alluvial cross-bedded sandstones, gritstones and conglomerates with individual lacustrine limestone beds occurs. The thickness does not exceed 20-25 m. These deposits overlie the lower member of the Pliocene with a very pronounced erosional unconformity (Ponomarev, 1966).

Pliocene deposits are poorly exposed north and south of the Jebel Abd El-Aziz Uplift. Here they were penetrated by a series of shallow bore-holes placed along three meridional profiles

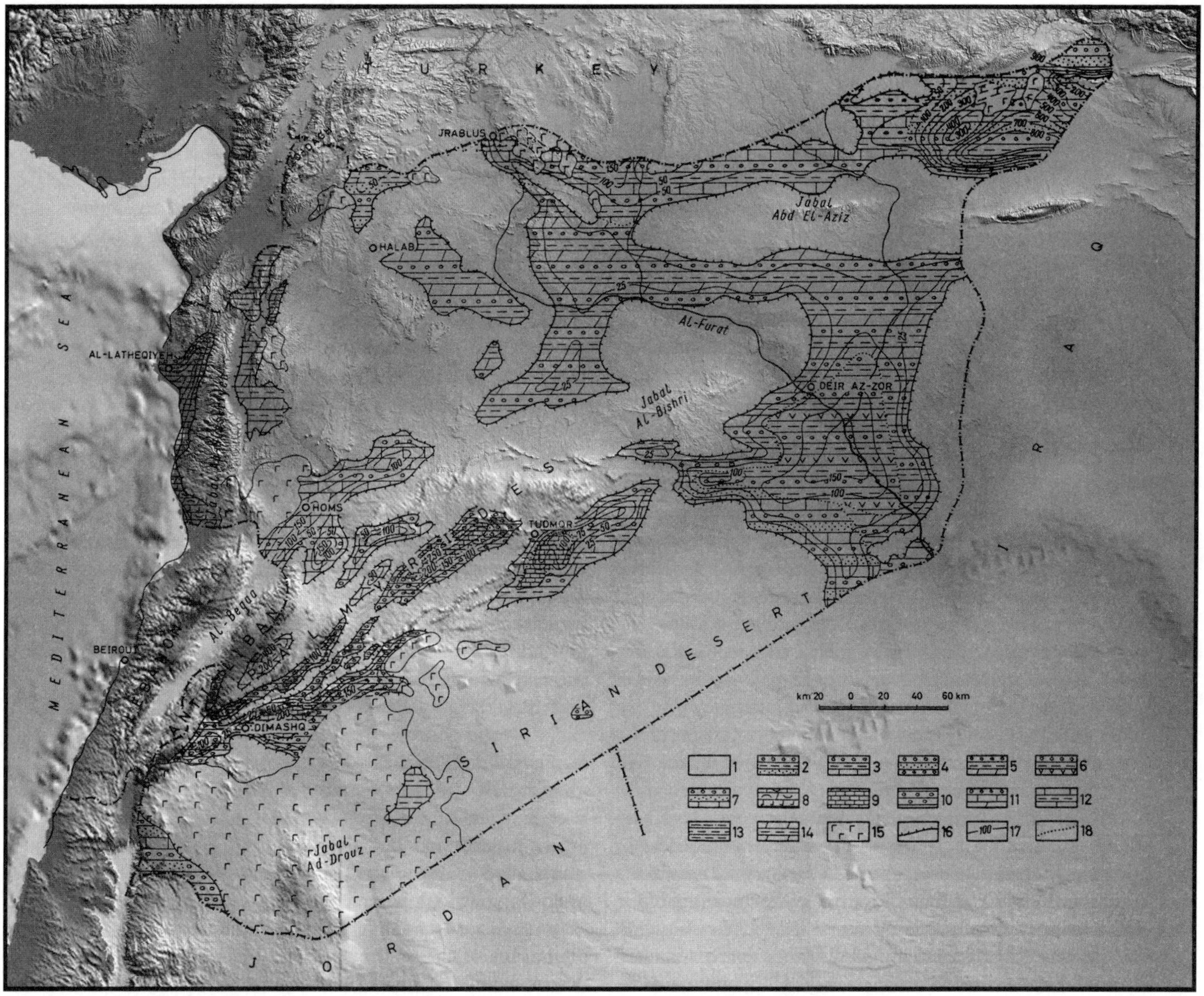

Fig. 6.73. Schematic map of the lithofacies and thicknesses of the Pliocene deposits of Syria (after Ponikarov et al., 1967, modified).
Legend: 1- exposures of the pre-Pliocene rocks; continental (proluvial, alluvial and lacustrine) deposits: 2 – conglomerate, clay and sandstone; 3 – conglomerate, clay and limestone; 4 – conglomerate and sandstone; 5 – conglomerate, clay, marl and sandstone; 6 – conglomerate, sandstone, clay and gypsum; 7 – conglomerate, sandstone and limestone; 8 – marine limestone, marl and volcanic tuff; 9 – lacustrine limestone; 10 – continental conglomerate; 11 – lacustrine limestone and marl with interbeds of conglomerates; 12 – lacustrine limestone and clay; 13 – marine clay; 14 – lacustrine marl and clay; 15 – basalts; 16 – shore line of basins of sedimentation (marine and lacustrine);17 – isopach lines; 18 – lithofacies boundaries.

(Ponomarev and Bivshev, 1966). One profile extends to the east of Jebel Abd El-Aziz from Jebel Tchembe towards the town of Qamishliye **(Fig. 6.74)**. The second profile is located to the south of Jebel Abd El-Aziz, extending to the southwest from the town of Hassetche **(Fig. 6.75)**. The third profile stretches to the south of this uplift along the right bank of the Nahr El-Khabour River.

Pliocene deposits unconformably overlie the Lower and Upper Fars Formations. They are distinctly subdivided by lithology into the lower and upper members. The lower member consists of alternating greenish-gray and brown calcareous clays, marls, fine-grained calcareous sandstones, with rare intercalations of gritstones and conglomerates. The thickness here does not exceed 35 m, but increases to the east to 50-60 m. In this region the Pliocene deposits resemble the Upper Fars Formation, and may be distinguished by the presence of a calcareous matrix, coarse terrigenous material, and by the ubiquitous presence of crusts around terrigenous clasts. The upper member is represented by gray and brownish cross-bedded sandstones, grits, and conglomerates with rare interbeds of fine-grained limestones, algal limestones, and marls. The thickness of the upper member is rather constant at 20-25 m. In this area, the contact between the lower and upper members are characterized by a gradual transition, but in some sections the upper member commences with a basal conglomerate and sandstone unit.

In the northeastern part of Syria the Pliocene fills the Ein Diwar depression (within the inner zone of the Mesopotamian fore-deep) and the Sinjar depression (within the outer zone of the Mesopotamian fore-deep). They are separated by the Qaratchok Darh-Qamishliye Uplift, and on the southern side the depression is limited by the Tuwal El-Aba - Sinjar ramparts. Pliocene sediments are absent on the Eastern Qaratchok Darh, Qamishliye, Sheikh Sallach, Tchembe, and Ghouna structures of the Qaratchok Darh - Qamishliye Uplift. The most complete Pliocene sequences are located within the Ein Diwar depression along the Soutan Dere River. In the Sinjar depression, the Pliocene was penetrated by numerous shallow (up to 150 m) structure holes, and by some deeper exploratory bore-holes (Galaktionov et al., 1966).

The Pliocene deposits overlie with slight angular unconformity the erosional surface of the Upper Miocene Upper Fars Formation. They are represented by continental terrigenous molasse: alternations of various sandstones, siltstones, grits, conglomerates and clays, with subordinate interbeds of marls and limestones. Coarse terrigenous material becomes increasingly common upwards in the section. The Pliocene is again subdivided into lower and upper members, but unlike the western areas of the Mesopotamian depression, the general lithological features and thicknesses of the Pliocene here in the Ein Diwar and Sinjar depressions are rather different.

In the Ein Diwar depression, the complete succession of the lower member can be observed on the eastern slope of the Kili Jelli uplift. It is characterized by frequent intercalations of gray, brown, and red sandstones, gritstones and calcareous clays with subordinate layers of marls and conglomerates. The thickness of the Pliocene here is considerable, up to 736 m. The thickness of the coarse sandstone layers reaches 45-50 m, but the thickness of the clay units is greater, up to 106 m.

In the Sinjar depression, to the southwest of Qaratchok Darh Uplift, the lower member thins appreciably to 150-200 m. Here clays and siltstones predominate. In the area of El-Bardeh and Tchembe the sandstones are more fine-grained, and gritstones and conglomerates form only thin interbeds.

The upper member of the Pliocene in the Ein Diwar depression is well-exposed in the Kili Jilli section on the left bank of the Soutan Dere River. Here it comprises alternating conglomerates, gritstones, and coarse-grained sandstones with thin interbeds of clays. Basaltic beds were encountered at the top of this member. The thickness is 116 m. These deposits erosionally overlie the lower member with a 5-10° angular unconformity. In the Sinjar depression, the upper member is poorly exposed. It can be traced to the east of Qamishliye and in some other localities.

The origin of the Pliocene molasse deposits is connected to the tectonic event in the Taurus-Zagros orogen. The fine-grained terrigenous rocks of the lower member originated with the onset of vertical tectonic movements. Conglomerates and coarse sandstones of the upper member reflect the period of intensive erosion of the newly-uplifted terrains.

In nearby Iraq these Pliocene molasse deposits are known by the name of the Bakhtiari Formation.

Within the western periphery of the Mesopotamian depression in the valley of the Al-Furat River, and in its central areas including the left bank of the Al-Furat, and the lower reaches of the Nahr El-Khabour River, the Pliocene is again subdivided into lower and upper members. Usually they are separated by a hiatus, and in places, an angular unconformity is observed. The Pliocene sediments are clearly transgressive, truncating the Upper Fars, the evaporitic deposits of the Lower Fars, and also older rocks including those from the Paleogene and Maastrichtian (Oufland, 1966b).

In the vicinity of Ressafe in the areas of Beer Az-Zabad, Mustahet El-Makhroum, and Iben Husein, the lower member is composed of conglomerates, gritstones, sandstones, clays and marls, with some beds of limestones. All these lithologies are alluvial, proluvial and lacustrine in origin. Especially characteristic are yellowish and light-gray porous limestones with abundant shells of *Dreissensia*, which alternate with sandy limestones. These sediments evidently accumulated in shallow fresh-water lakes. The thicknesses range between 10 and 30 m. The upper Pliocene member is only exposed in a small number of localities such as Al-Kom, Wadi Amshah Raddeh, and Iben Husein, where it is overlain by Quaternary sediments. The upper member consists of marls, clays, limestones, and travertines. Conglomerates and sandstones are very subordinate. Lacustrine sediments contain plant debris and vertebrate remains. The thicknesses are minimal, only 2-18 m.

To the southeast, in the area of Raqqa, on the right and left banks of the Al-Furat River, Pliocene deposits unconformably overlie the Upper and Lower Fars Formations (Medvedev, 1966a). The lower member in the area of Beer Qabaqeb and Beer Khdejan consists of frequent alternations of clays, limestones, siltstones and sandstones. At the base there is a bed of coarse conglomerate and sandstones, whereas the limestones at the top of this member are gypsiferous. The thicknesses range from 10 to 35 m. On the left bank of Al-Furat in the area of Tar As-Sirra and Tell Ghreta, coeval deposits were penetrated by a series of shallow structure mapping bore-holes. In this area the terrigenous component content increases. At the base of this member, a bed of gritstones 1-4 m in thickness is present.

The upper member of the Pliocene is separated from the underlying rocks by an unconformity. It was found in limited areas around Beer Goeran in the north and the region of Deir Ez-Zor in the southeast. It is composed of 30 to 35 m of poorly-sorted gritstones, polymictic cross-bedded sands, and sandstones.

In the southwest of the Mesopotamian depression, the Pliocene deposits crop out in the terraces of the Al-Furat River near Deir Ez-Zor. They were penetrated by some bore-holes in the lower reach of the Nahr El-Khabour River (Medvedev, 1966b). The Pliocene unconformably overlies Miocene rocks. Their upper surface is cut by channels which were filled later by Pliocene sediments. As in other regions of Mesopotamia, the lower and upper members can be distinguished within the Pliocene sequence. They are separated by an unconformity.

The lower member in this part of Syria is characterized by some special features. It consists of two units. The lower unit includes

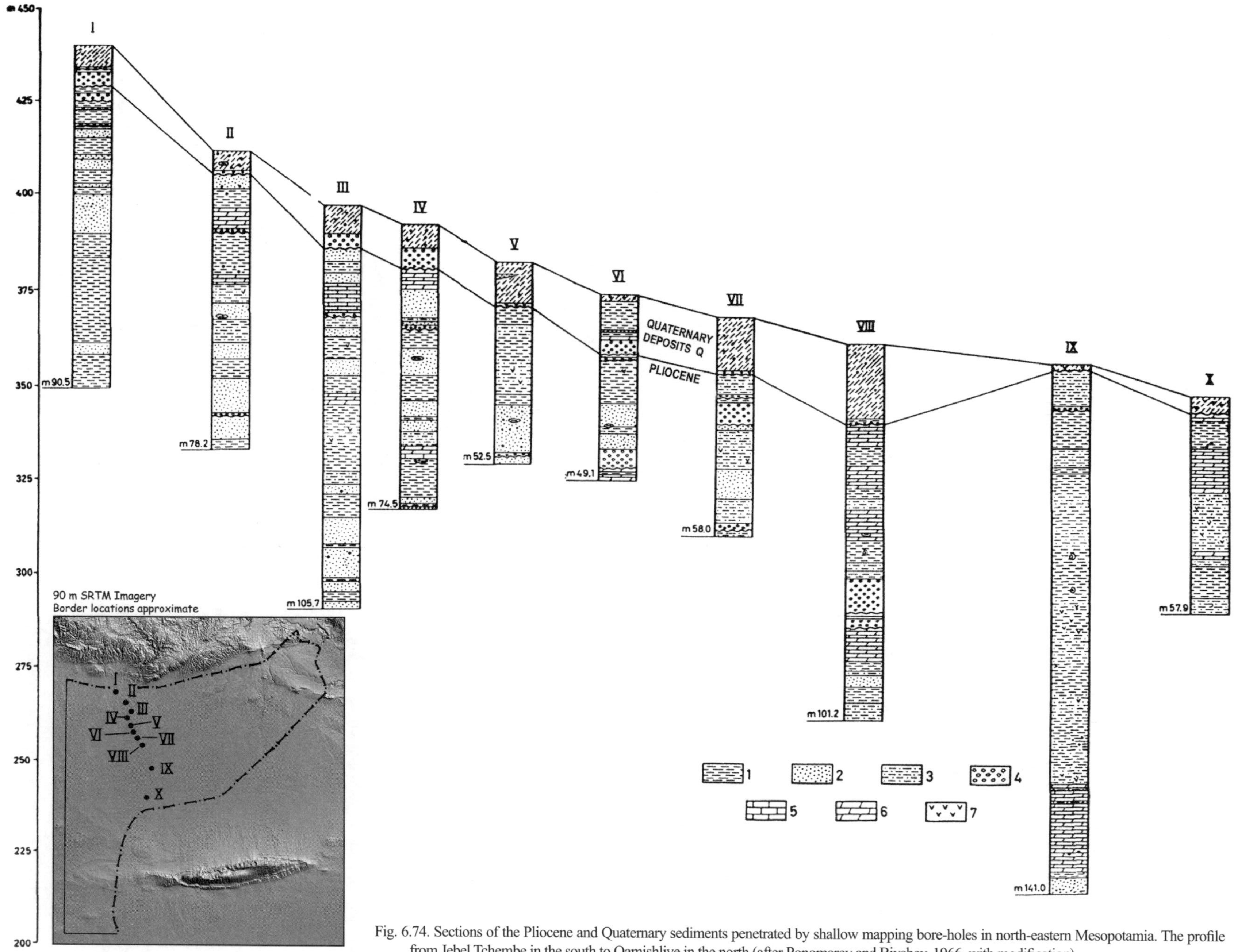

Fig. 6.74. Sections of the Pliocene and Quaternary sediments penetrated by shallow mapping bore-holes in north-eastern Mesopotamia. The profile from Jebel Tchembe in the south to Qamishliye in the north (after Ponomarev and Bivshev, 1966, with modification).
Legend: 1 – clay; 2 – sand and sandstone; 3 – siltstone; 4 – gravel and conglomerate; 5 – limestone; 6 – marl; 7 – gypsum.

the usual terrigenous rocks, grits and coarse-grained sandstones with rare intercalations of clays. The upper unit is represented by alternating terrigenous and chemogenic rocks consisting of fine-grained sandstones, clays, marls, limestones and gypsum. Beds of solid gypsum reach 1 m in thickness, but layers of calcareous-gypsiferous rock of 0.3-4.7 m thick are more common. The total thickness of the lower member in outcrops near Deir Ez-Zor is 70 m, and is 130 m in a bore-hole in the lower reaches of the Nahr El-Khabour River

The upper member of the Pliocene outcrops on the right bank of the Al-Furat near Deir Ez-Zor and Mayadin and was also penetrated by bore-holes at the mouth of the Nahr El-Khabour River. It is composed of cross-bedded sands and sandstones, grits, siltstones and clays with subordinate layers 0.1-0.5 m thick of limestones. The thickness of the upper member near Deir Ez-Zor is about 30 m and along the Nahr El-Khabour River it thins to 16 m.

The transgressive nature of the Pliocene sediments is very well-expressed in southeastern Syria near the boundary with Iraq. In this area the Pliocene deposits unconformably overlie various formations of the Miocene and onlap the Oligocene, and Upper and Middle Eocene rocks. The twofold structure of the Pliocene is very clearly observed because the lower and upper members are separated by an erosional surface (Petrov and Antonov, 1966).

Within the lower member the sediments display obvious changes in facies. A unit of gypsum with subordinate inter-calations of clays and marls can be traced in outcrops along the right bank of the Al-Furat River from Wadi Beer El-Ahmar to the town of Mayadin, as well as near the town of Abou Kamal. At the base of this member there is a layer of coarse sandstones grading laterally into small-pebble conglomerates. The thickness of the member near the village of Salihigeh is 44 m. To the northwest, at Bore-hole 34, the thickness increases up to 95 m. To the south-west *i.e.* in the direction of the Pliocene basin coastline, gypsiferous-clayey deposits are replaced by pinkish and gray-pinkish limestones which are compact and hard, sandy, and sometimes with conglomerate-like structure. At the base they contain a layer 6 m thick of calcareous-cemented conglomerates and sandstones. The total thickness is about 50 m. This member of limestones and sandy limestones is well-exposed in sections located along Wadi Ratka, Wadi As-Suwwab (Wadi Souab) and Wadi El-Waar (Wadi Ouahar) near the Iraqi border and also along Wadi El-Miyah (Wadi Miah).

The upper member of the Pliocene unconformably overlies not only the lower member, but also deposits of the Lower Fars and Upper Fars Formations. Sediments of this member are also characterized by lateral changes in facies. Conglomerates,and coarse- and middle-grained sandstones predominate in exposures along Wadi El-Miyah. On the right bank of the Al-Furat River near Mayadin soft greenish-gray sand and sandstones contain numerous interbeds of clays. On the same bank near the entrance to Wadi El-Khawt, frequent alternations of cross-bedded sand-stones, clays and gypsiferous clays can be observed. The maximum thickness reaches 30 m.

The Pliocene deposits of the Mesopotamian Basin are extremely poor in organic remains. They contain only fresh-water ostracods – *Cyprideis torosa* (Jones), *C. torosa littoralis* (Brady), *Candoniella subellipsoida* (Shar.), *C. albicans* (Brady), *C. suzini* Schn., *Loxoconcha eichwaldi* Liv., *Ilyocypris bradyi* Sars, *Candona neglecta* Sars, *C. obrutschevi* Mandel., *Cypridopsis formosa* Schn., *Herpetocypris pellucida* Schn., *Cyprinotus baturini* Schn., *C. inflatus* Schn., *Limnocythere* sp., *Eucypris pira* Mandel., *Cyclocypris* sp., and *Potamocypris* sp. (det. of N.N. Naidina). They do not permit a detailed subdivision of the continental Pliocene sediments. The lower and upper members of the Pliocene should be considered as the regional sub-divisions of the Pliocene, but not as the Lower and Upper Pliocene in the chronostratigraphic sense. The boundary between the lower and upper members of the Mesopotamian Pliocene coincides with the tectonic events then taking place in the folded system of Zagros and Taurus.

Paleogeographic links between the Mediterranean and Mesopotamian Basins

Lithological and stratigraphical data dealing with the Neogene of northwestern Syria (the Mediterranean Basin affinity) and northeastern Syria (the Mesopotamian Basin affinity) allow us to reconstruct the paleogeographic links between these two sedimentary basins in the following way (Figs. 6.70 and 6.71).

The drop in the World Ocean sea level and the regional tectonic movements (uplifts), which took place in the Terminal Oligocene, resulted in strong regression of the sea in northwestern Syria (the Mediterranean Basin) and in shrinkage of the marine Mesopotamian Basin.

The Early Miocene – the Aquitanian and Early Burdigalian (the *Globigerinatella insueta–Globigerinoides trilobus* Zone, NN7) in northwestern Syria is characterized by the marine transgression and appearance of the narrow Al-Latheqiyeh Depression and Aafrine Graben. Linkage of the Mediterranean and Mesopotamian Basins was accomplished via a very narrow strait which evidently existed to the south of Taurus (Southern Turkey). Such a restricted connection was responsible for an abnormal salinity regime in the Mesopotamian Basin where a unit of limestones and marls with intercalations of gypsum and poor microfauna (the Dibbane Formation) accumulated. Compared to the widespread distribution of the Uppermost Oligocene sediments, characterized by rich microfauna raised under normal marine salinity conditions, this formation is clearly regressive.

The interval of the Late Burdigalian and the entire Middle Miocene (up to the *Globigerina nepenthes–Globorotalia siakensis* Zone, N 14) was marked by a vast transgression. Marine waters covered the whole Aleppo Plateau. The Mediterranean Basin was openly connected with the Mesopotamian Basins, where at that time, the calcareous sediments of the Jeribe Formation were being deposited, including molluscs and foraminifera typical of normal salinity conditions. Despite this, the shallow-water over the Aleppo Plateau may have served as an ecological barrier between the two basins.

In the early Late Miocene (Tortonian) the northwestern part of Syria underwent a regression. Normal marine conditions were

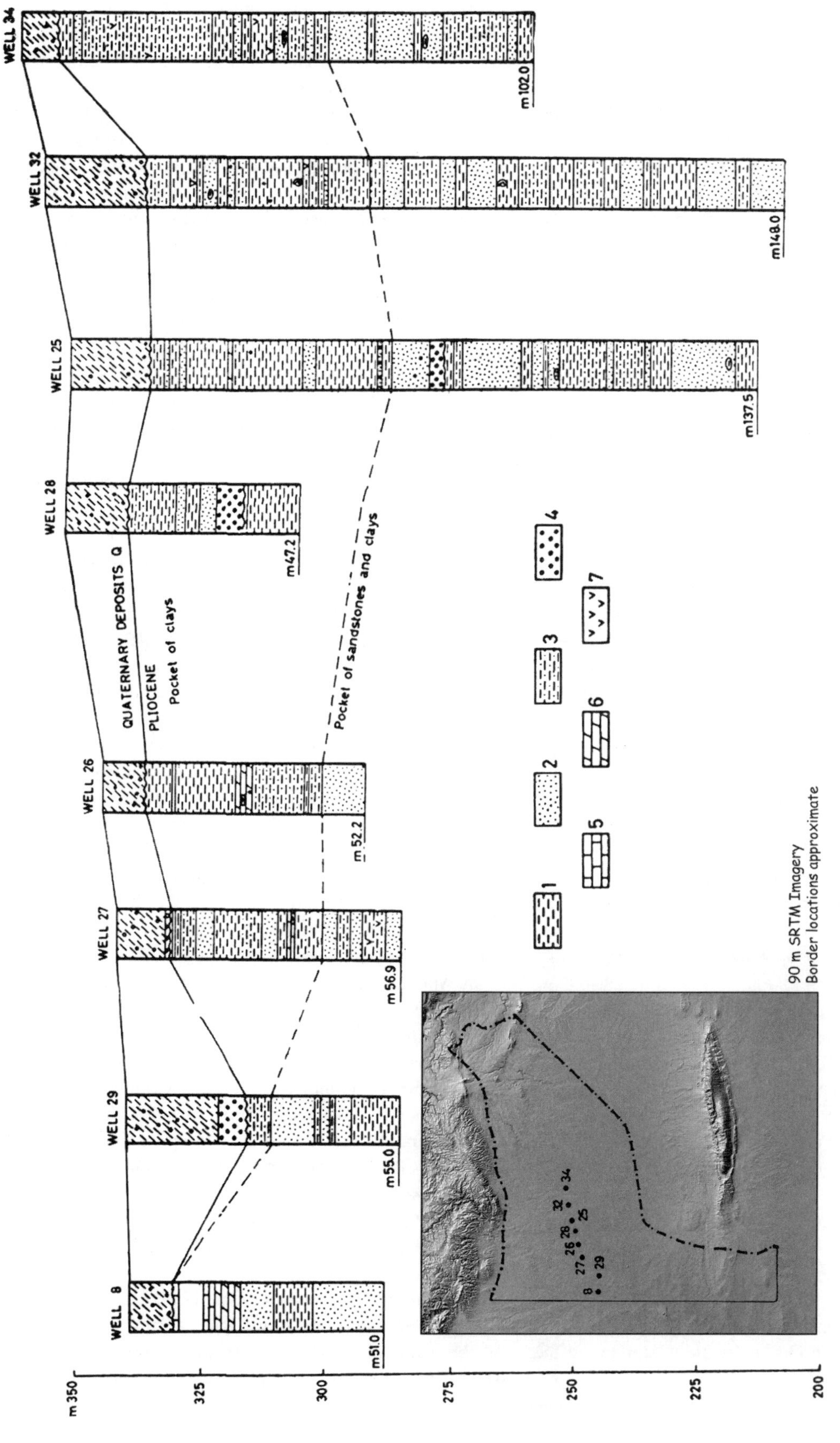

Fig. 6.75. Sections of the Pliocene and Quaternary sediments penetrated by shallow mapping bore-holes in north-eastern Mesopotamia. The profile from Hassetche to the south-west (after Ponomarev and Bivshev, 1966, with modification).

Legend: 1 – clay; 2 – sand and sandstone; 3 – siltstone; 4 – gravel and conglomerate; 5 – limestone; 6 – marl; 7 – gypsum.

preserved in only two isolated areas, the southwestern part of the Al-Latheqiyeh Depression, and in the northern part of the Al-Ghab Graben near the Turkish border. In these regions clayey-sandy sediments of the Middle Miocene (the *Globigerina nepenthes–Globorotalia siakensis* Zone, N 14) and those of the Upper Miocene (the *Globorotalia acostaensis* Zone, N 16) are separated by a regressive member of shallow-water algal and detrital limestones and coquinas., It evidently corresponds to the *Globorotalia continuosa* Zone, N 15, and this time interval corresponds to the decoupling of the Mesopotamian Basin from the Mediterranean.

Under conditions of abnormal salinity, the Lower Fars Formation was formed in the Mesopotamian Basin. It consists of a succession of clays, marls and limestones with thick intercalations of gypsum and halite salt. Benthic foraminifera are present in deposits of this formation. They are common to coeval sediments of the Mediterranean and Central Paratethys (the Vienna Basin, the Fore-Carpathian and the Trans-Carpathian depressions). However the path by which they migrated from the Mediterranean into Mesopotamia is enigmatic. We may suggest that during Tortonian time:

1) the Mesopotamian and Mediterranean Basins had a very restricted communication which resulted in the accumulation of the evaporitic sediments of the Lower Fars Formation; and that

2) this communication was ensured by a very narrow strait to the south of Taurus which allowed penetration of some tolerant groups of fossils. Traces of this strait might exist in Southern Turkey, or perhaps they were destroyed by subsequent erosional processes.

In the Terminal Miocene (Messinian) of northwestern Syria the regression reached its peak and deposits of this age occupy a limited area in the southwest of the Al-Latheqiyeh Depression where they are represented by an evaporitic succession. In the Mesopotamian Basin the continental regressive sediments of the Upper Fars Formation were deposited and all links between the Mediterranean and Mesopotamian Basins were severed.

The onset of the Pliocene in the Mediterranean was characterized by a transgression of limited extent. The continental Pliocene deposits of the Mesopotamian Basin were also transgressive.

The above analysis of the paleogeography and paleoenvironmental conditions of the Mediterranean and Mesopotamian Basins during the Neogene is based on routine paleontological and geological data (geological mapping, and stratigraphic, lithologic, and paleogeographic criteria). Some of these materials were published many years ago by Ponikarov et al. (1967, 1968, 1969) and Krasheninnikov (1966, 1968, 1969b, 1971). Unfortunately these publications have been overlooked, with the result that mistakes have been made in reconstructing the geography of these times.

In the publications of Roegl and Steininger (1983), Steininger and Roegl (1984), and Roegl (1996, 1998) the links between the Mediterranean and Mesopotamian Basins (*i.e.* the connection between the Atlantic and Indian Oceans) were interpreted in the following way.

According to these authors, the Oligocene/Miocene boundary and the Aquitanian interval are characterized by a vast transgression of tropical waters from the Indo-Pacific into the eastern Mediterranean. This event was accompanied by penetration of *Lepidocyclina* and the subsequent deposition of limestones mainly composed of these larger rock-forming foraminifera. In stark contrast, our observations revealed that the Oligocene in Syria is regressive, that pelagic sedimentation continued to exist mainly in Mesopotamia, and that *Lepidocyclina*-limestones are very typical for the Oligocene in the Palmyrides and the Syrian Desert, but extremely rare in the Lower Miocene of the Mediterranean Basin, and totally absent in the Aquitanian of Mesopotamia. The transgressive Aquitanian sediments are confined to the narrow Al-Latheqiyeh Depression and Aafrine Graben; in Mesopotamia they were precipitated under conditions of abnormal salinity.

According to Roegl and Steininger, the Early Burdigalian (the nannoplankton *Helicosphaera ampliaperta* Zone, NN4) was marked by the sharp change in the paleogeographic setting. The marine strait crossing the Near East was closed, and the Mediterranean Sea was converted into an embayment of the Atlantic Ocean, and the Mesopotamian Basin into an embayment of the Indian Ocean. The *Gomphotherium* land bridge appeared in the Near East, allowing migration of mammals between the Anatolian and Arabian-African Plates. Again in stark contrast, data dealing with the Miocene of Syria indicate that in Early Burdigalian time (the foraminiferal *Globigerinatella insueta–Globigerinoides trilobus* Zone, N7), there were not any serious changes in the paleogeographic, bionomic, and lithologic (facies) conditions in the Near East. As in Aquitanian time, the Burdigalian sediments of northwestern Syria, characterized by normal salinity fauna, were limited to the narrow Al-Latheqiyeh Depression and Aafrine Graben. Apparently, the Mediterranean Basin had a restricted connection with the Mesopotamian Basin, in which clayey-calcareous sediments with intercalation of chemogenic rocks (gypsum) continued to accumulate.

Roegl (1996, 1998) suggests that the existence of the Early Burdigalian land bridge is proven by the migration of *Gomphotherium* and *Deinotherium* to Eurasia, and of *Dicerorhinus* to the Arabian Plate. In our opinion, the large differences between the Early Burdigalian lithology and fauna of the Mediterranean and Mesopotamian Basins indicate that any connection would probably have to be a narrow strait. Either this strait was not a barrier to the migration of mammals, or it was intermittent. In other words, when the strait existed, impoverished assemblages of planktonic foraminifera penetrated into the Mesopotamian Basin; when the strait was closed, a land bridge created routes for the migration of mammals. In the Mesopotamian Basin such an unstable environment resulted in alternating layers, either with peculiar assemblages of fossils, or completely barren of organic remains, or consisting of gypsum.

According to Roegl (1996, 1998), the second Miocene transgression took place in the interval embracing the upper part of the nannoplankton *Helicosphaera ampliaperta* Zone (NN 4), and the lower part of the *Sphenolithus heteromorphus* Zone (NN 5) *i.e.* the end of the Lower Miocene to the beginning of the Middle Miocene. The Mediterranean Basin was again united with the Indo-Pacific and precipitation of evaporitic sediments in the Mesopotamian Basin ceased. In this we agree. Indeed, this time interval saw a vast transgression and, as we already know, the whole area of the Mesopotamian Basin in Syria including the Aleppo Plateau was covered by normal salinity marine waters which deposited the limestones of the lower part of the Jeribe Formation.

According to Roegl (1996, 1998) the Mesopotamian Basin lost its connection with the Mediterranean Basin in the interval embracing the upper part of the *Sphenolithus heteromorphus* Zone (NN 5) and the *Discoaster exilis* Zone (NN 6), or the Middle Miocene. The Mediterranean Sea became an extended gulf of the Atlantic Ocean. The Mesopotamian Basin became connected with the Indian Ocean, and was characterized by development of the very thick evaporitic Lower Fars Formation. In contrast, this assumption does not correspond to the geological facts. During the entire Middle Miocene, up to the top of the *Globigerina nepenthes–Globorotalia siakensis Zone (*N 14), the Mediterranean and Mesopotamian Basins were united by a wide strait whose waters covered the Aleppo Plateau. Calcareous sediments, containing fauna of normal salinity, were deposited in both basins, and the limestones of the Jeribe Formation were deposited in the Mesopotamian Basin. Evaporitic deposits (gypsum, halite) are completely lacking in Syria. They appear in the overlying sediments of the Lower Fars Formation (Upper Miocene, Tortonian), when the connection between the Mesopotamian Basin and the Mediterranean was severed, or alternatively, when it was restricted to a narrow strait to the south of the Taurus.

Consequently, this investigation into the stratigraphy of the Miocene deposits of Syria allows a different interpretion of the paleogeography and history of the links between the Mediterranean and Mesopotamian Basins than that given in the publications of Roegl and Steininger. It seems to us that their conclusions are based on an incorrect estimation of the age of the local formations as well as their spatial distribution.

Some problems related to the paleogeography and paleoenvironment of the Neogene Mesopotamian Basin remain unsolved. In Iraq, southwestern Iran, and in some other countries of the Persian (Arabian) Gulf, sections of the Neogene deposits demonstrate the same sequence of litho-stratigraphic units or formations as were observed in the northern (Syrian) area of this huge depression.

The Neogene sections start with a clayey-sandy-calcareous unit containing intercalations of gypsum. This unit bears various names – the Ghar Formation in Kuwait and Southern Iraq, the Lower Fars Formation in Qatar and the United Arab Emirates, the Jebel Cap Formation in Bahrain, the Dibbane Formation in Western Iraq, and the Gachsaran Formation in Southwestern Iran.

This unit is overlain by a pile of limestones with marine fauna and rich accumulations of petroleum – the lower layers of the Lower Fars Formation of Kuwait, the Dam Formation of Qatar, the lower part of the Mishan Formation of the United Arab Emirates, the Jeribe Formation of Iraq, and the upper layers of the Asmari Formation and the basal part of the Mishan Formation of Iran.

The next unit is represented by a succession of limestones, marls, gypsum and halite which serve as a cap rock over the petroleum occurrences – the Lower Fars Formation of Iraq, the upper part of the Lower Fars Formation of Kuwait and the United Arab Emirates, and the upper layers of the Mishan Formation of Southwestern Iran.

The Neogene terminates with fresh-water and continental terrigenous sediments which are fine-grained in the lower part of this unit (Upper Miocene) and are coarse-grained at the top (Pliocene) – the Dibdibba Formation of Kuwait, the Hofuf Formation of Qatar, the Upper Fars and Bakhtiari Formations of Southern Iraq, and the Agha Jari and Bakhtiari Formations of Southwestern Iran. In Bahrain and the United Arab Emirates there is a hiatus which occupies this interval of time.

Unfortunately, the geological literature contains very little information dealing with the paleontological characteristics of the Neogene deposits of Southern Mesopotamia. Therefore it is difficult to estimate the stratigraphic position of the boundaries of formations which have similar names *i.e.* are they coeval or 'sliding'? According to Alsharchan and Nairn (1997), lateral changes in the age of the above-mentioned formations are very significant. As examples they showed three profiles which included:

1) lateral changes of the Asmari, Gachsaran, Mishan, Agha Jari, and Bakhtiary Formations of Iran (*op. cit.* page 408, Fig. 9.8);

2) lateral changes of the Euphrates Limestone, Dibbane, Jeribe, Lower Fars, Upper Fars, and Bakhtiary Formations of Western and Eastern Iraq (*op. cit.* page 407, Fig. 9.7) and

3) changes within the Neogene deposits of the western coast of the Persian Gulf from Kuwait to the United Arab Emirates (*op. cit.* page 406, Fig. 9.6).

These correlational charts lead one to conclude that these lithologically similar units received the same names although they occupy different stratigraphic position. For example, the evaporitic Lower Fars Formation of Qatar and the United Arab Emirates has no relation to the Lower Fars Formation of Syria and Western Iraq.

It is necessary to emphasize that the distribution of Neogene deposits with various paleoecological characteristics in Kuwait, the United Arab Emirates, Qatar, Bahrain, Southern Iraq, and Southwestern Iran indicates the existence of complicated and unstable paleogeographic connections between Southern Mesopotamia and the Indian Ocean. The development of evaporitic and continental facies is evidence of periodic isolation of the Mesopotamian Basin from the Indian Ocean, mainly in Early Miocene, Late Miocene, and Pliocene times. Evidently a paleogeographical/paleoecological barrier existed in the area of the Mesozoic folded terrain of Oman and the present Straits of Hormuz. Apparently it was only in the late Early Miocene and Middle Miocene that the Southern Mesopotamia had an open connection with the Indian Ocean.

Summing up, to possess an adequate understanding of the geological history of the Mesopotamian Basin during Neogene time, it is necessary to know not only what links were possible with the Mediterranean (and thus the Atlantic Ocean), but also those that were possible with the Indian Ocean. The absence of reliable data, relevant to the detailed stratigraphy and paleogeography of the entire Persian Gulf area in the Neogene, precludes any definitive conclusions.

INNER REGIONS OF SYRIA WITH CONTINENTAL SEDIMENTATION

Continental Miocene and Pliocene deposits fill vast flat depressions and narrow intramountain basins of the inner areas of Syria. The most significant among them are represented by the Damascus and Homs depressions, intramountain basins of the Anti-Lebanon, Al-Ghab Graben, Ad-Daww depression in the Palmyrides, the flat Sabkhet Mouh and Feidet depressions to the southeast of Palmyra, and the Jabboul depression near the recent lake of the same name (Fig. 6.65). Neogene deposits of these depressions are characterized by some common features, but also have individual peculiarities.

The stratigraphic subdivision of the Neogene continental deposits and the age determination of their units is difficult because they contain very scarce organic remains of fresh-water ostracods and palynological assemblages. They have been subdivided and dated by correlation with marine sediments, and an estimation of their lithological patterns against the background of the general Neogene paleogeographic setting and geological development of Syria. This approach was supplemented by isotopic dating of basalts where they interfinger with sedimentary rocks (Sharkov et al., 1994, 1998). Unfortunately, we have to acknowledge that at present the stratigraphy of the Neogene continental deposits of Syria can not be considered as detailed.

In the **Damascus depression** and adjacent intramountain Sahl As-Sahra, El-Majar, and Sahl Sednaya **depressions of the Anti-Lebanon,** the Miocene and Pliocene deposits are well-developed (Razvalyaev, 1966a; Kurbanov and Zaryanov, 1968). It should be noted that in the Damascus depression, because of the faults along the Anti-Lebanon ridges and the large amplitude subsidence, the basal layers of the Neogene are now found in bore-holes at a depth of 800 m (Sharkov et al., 1994) (**Fig. 6.76**).

The contact of the Miocene deposits with more ancient rocks is well-pronounced along the northern periphery of the Damascus depression on the southeastern slope of the Jebel Abou El-Ata ridge. The Lower Miocene overlies Oligocene and Upper Eocene limestones with a sharp unconformity (**Fig. 6.77**). This boundary is distinctly seen as an uneven erosional surface of Paleogene limestones and a bed of Miocene basal conglomerates and coarse-grained sandstones with pebbles of Paleogene rocks. The Lower Miocene is composed of soft or weakly cemented, white and yellowish quartz sands. Very often they are with alluvial type cross-bedding. Thin interbeds (up to 0.3 m) of light-gray sandy limestones with flint nodules can be observed within a monotonous sandy unit. In some outcrops the sands are replaced upwards in the section by hard aphanitic limestones with small flint concretions. The thickness varies from 30 to 50 m.

The quartz sands are alluvial, and were accumulated under tectonically quiet conditions on a level landscape, following the retreat of the Oligocene sea. Igneous and metamorphic rocks of the basement of the Arabian Platform were the source of clastic material. That area was located in the south far away from the Anti-Lebanon. Therefore the quartz grains are rounded and well-sorted. Cross-bedding is usually inclined to the north indicating transportation of the terrigenous debris from south to north.

The pure quartz sandy unit is covered by a member of alternating sandstones and basalts whose thickness reaches 47 m. The Early Miocene (Aquitanian) age of the rocks is confirmed by isotopic K/Ar dating of basalts taken from some outcrops, which gives ages from 19.9 to 20.9 Ma (Sharkov et al., 1994).

Middle Miocene sediments are more widespread in the Damascus Depression and some depressions of the Anti-Lebanon than are those of the Lower Miocene. In the Damascus depression they replace, without visible signs of hiatus, the sandy unit of the Lower Miocene. In the Sahl As-Sahra, El-Majar, and Sahl Sednaya depressions of the Anti-Lebanon, Middle Miocene deposits overlie with strong unconformity the Oligocene and Eocene limestones. Consequently, the Middle Miocene is transgressive and the basin of sedimentation in southwestern Syria was enlarged. The Middle Miocene is represented by a thick unit of variegated conglomerates, sandstones, siltstones, clays and limestones of predominantly red and brown color.

A typical Middle Miocene section is situated near Doummar, a suburb of Damascus. It includes four members:

1) Red and brown conglomerates consisting of pebbles, boulders, and angular blocks of underlying rocks. They rest on the uneven surface of the Paleogene limestones with erosional pockets filled by sandy-clayey matter;

2) Brown and red sandstones of differing granularity, with intercalations of calcareous and sandy clays in the upper half;

3) A member of frequent intercalation of various rocks such as brown conglomerates and sandstones, red and green clays and siltstones, gray marls and limestones. A coarse cyclicity can be observed in these rocks. In the middle part of this member the alternation of the small-pebble conglomerates, sandstones, clays and marls of brown, green and gray colors is very irregular, with lens-like layers pinching out or cutting each other;

4) A ~200 m thick member of frequent intercalation of red and brown sandy clays and sandstones with subordinate interbeds of gray marls.

The total thickness of the above section is about 530 m.

The red-colored Middle Miocene unit also has a similar composition in other sections in the Anti-Lebanon. To the southwest of Doummar, near the village of El-Hame on the Jebel Aantar High, its thickness reaches 494 m. To the southeast of this village this unit becomes unconformable with the Pliocene and gradually thins to 60-100 m. In the Maaret Sednaya depression northeast of Doummar drilling data indicates that the thickness of the red-colored Middle Miocene unit ranges from 275 m to 400 m.

On the Hauran Plateau and Jebel Drouz the Middle Miocene

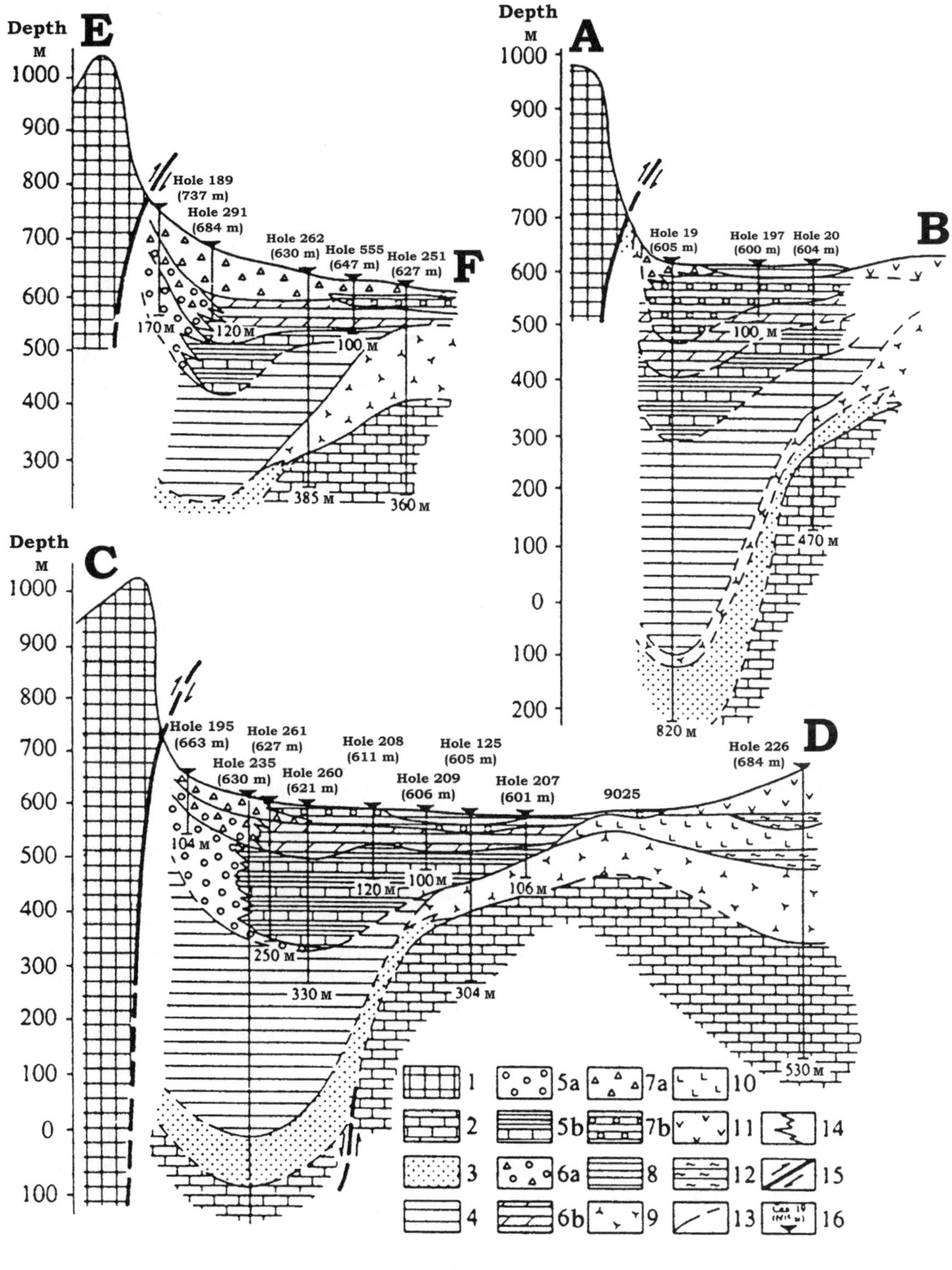

Fig. 6.76. Geological structure of the Damascus depression according to drilling data; profiles A-B, C-D and E-F (after Sharkov et al., 1994).

Legend: 1 – Cretaceous and Paleogene rocks of the Anti-Lebanon folded zone; 2 – marine limestones of the Eocene and Oligocene; 3 – continental sands of the Lower Miocene; 4 – lacustrine clays, marls and sands of the Lower Pliocene covered by sediments of the Upper Pliocene-Lower Pleistocene; 5a – red-colored alluvial and proluvial conglomerates; 5b – lacustrine clays and limestones of the Middle Pleistocene; 6a – proluvial and alluvial gritstones and gravel of the Upper Pleistocene; 6b – lacustrine clays, marls and sands of the Upper Pleistocene; 7a – proluvial gravel of the Upper Pleistocene; 7b – sands and clays of the Upper Pleistocene; 8 – lacustrine sands of the Holocene; 9 – Miocene basalts; 10 – Pliocene basalts; 11 – Quaternary basalts; 12 – alluvial and eluvial clays of the Miocene and Pliocene age; 13 – stratigraphic boundaries; 14 – facies boundaries; 15 – faults and direction of movements along these planes; 16 – bore-holes (their numbers and absolute altitudes).

sedimentary rocks are replaced by basalts. In a quarry, located 1 km to the south of the village of Mazraid Adham near the northern margin of this plateau, K/Ar analysis of a basalt sample, taken from a bed of effusive rocks among the red-colored clayey-marly unit, yielded an age of 13.6 Ma, in the upper part of the Middle Miocene (Sharkov et al., 1994). To some degree this dating confirms the Middle Miocene age of the red-colored rock formation.

The Middle Miocene includes sediments of proluvial, alluvial, and lacustrine origin. Coarse conglomerates and sandstones give evidence of a very rugged and perhaps very high landscape. The alluvial deposits are represented by clays, siltstones, and fine-grained sandstones with cross-bedding and lens-like layer structure. Limestones and marls with small gastropods and smooth ostracods have a lacustrine provenance. The abundance of lycopod and fern spores among the spore and pollen assemblages indicates swampy conditions that are confirmed by the mottled red and brown iron oxide colors of clayey-sandy rocks. The ubiquitous presence of hydrophilic plants and coniferous vegetation suggests a humid and fairly cool climate (Ponikarov et al., 1969).

The presence of Upper Miocene sediments in the Damascus and the Anti-Lebanon depressions has not been proved. There are at least three possibilities:

1) as shown above, the distribution of marine Upper Miocene deposits in northwestern Syria reflects the onset of regression in Tortonian time and its peak in the Messinian. Therefore, it is possible that there was no sedimentation at all in the Anti-Lebanon depressions during the Late Miocene;

2) deposits of this age were destroyed by pre-Pliocene erosion;

3) the upper part of the red-colored unit belongs instead to the Upper Miocene (Tortonian) and the hiatus between the Miocene and Pliocene only correspond to the Messinian interval.

The continental Pliocene deposits of southwestern Syria are subdivided into two lithological units. Sediments of the lower unit are exposed along the northern slope of the Damascus depression and were also penetrated by numerous bore-holes. They fill such intramountain depressions of the Anti-Lebanon as Sahl As-Sahra, El-Majar, Sahl Sednaya, Jeiroud, and Serghaya. In the Sahl As-Sahra depression the Pliocene rocks rest on the uneven, washed-out surface of the Miocene deposits and to the west they overlie Paleogene and Upper Cretaceous rocks. This unit is composed of conglomerates and sandstones with clay intercalations. Brown conglomerates are characterized by coarse, slightly-rounded and badly-sorted clastic material, and the presence of Cretaceous and Paleogene flint and limestone boulders. The conglomerate layers are 3-5 m thick and are separated by sandstones and sandy clays. The sandstones are yellowish-gray, medium- and rarely fine-grained, with inclusions of pebbles and small lenses of gravelites. Yellowish calcareous clays are typical for the upper part of this unit. The thickness of the Pliocene lower unit in the Sahl As-Sahra depression is 210 m, but in adjacent basins it increases up to 300 m with the same lithological patterns.

Pollen of xerophytic vegetation, consisting of grass and shrubs, prevail in palynological assemblages from the sediments of the lower Pliocene unit. They are accompanied by pollen of thermophilic Mediterranean trees and shrubs. Such vegetation types indicate that the landscape consisted of steppes and semideserts during the deposition of the coarse terrigenous rocks. Forests occupied limited areas. The mixed spore and pollen assemblage taxonomy reflects the vertical climatic zonation accompanying strongly dissected relief. The coarse clastic composition of the rocks supports this conclusion (Ponikarov et al., 1967).

The upper unit of the Pliocene crops out in the central part of the El-Majar depression. Near the village of Rass El-Ain it unconformably overlies the older Pliocene deposits and rests transgressively on Paleogene limestones. This unit consists of alternating white and light-gray limestones, greenish-gray marls, and calcareous clays. Limestones predominate, distinguished by medium- and thick-bedded, massive aphanitic, silicified, porous, varieties, and by traces of abundant mud-eaters. Its thickness reaches 260 m. The lacustrine sediments contain shells of fresh-water molluscs and ostracods. Similar deposits were penetrated by bore-holes in the Damascus depression near the villages of Harran El-Awamid, Al-Qusseir, and others. There is an insignificant admixture of terrigenous material in the calcareous rocks. Thus, in the Late Pliocene this area was a lake surrounded by low elevations where erosional processes were not intensive.

In the **Homs depression** the continental deposits of the Miocene and the lower part of the Pliocene have much in common with those of the Anti-Lebanon. However the Late Pliocene differs strongly in its intensive basaltic volcanism.

The Lower Miocene crops out on the southern slope of the Homs depression near the villages of As Schel and Flita as well as on the western slope of the Jebel El-Kawdan ridge. Sediments of this age consist of white and yellowish quartz sands which are fine- and medium-grained, friable, with well-sorted and rounded grains of quartz. Among them intercalations of light-gray hard limestones and sandy limestones with flint nodules are present. Upwards in the section the number of calcareous interbeds increases. At the base of the Lower Miocene there is a conglomerate bed which transgressively covers the erosional surface of the Oligocene and Upper Eocene limestones. The thickness of the member with white sands and limestones ranges from 10 to 25 m. Cross-bedding emphasizes the alluvial origin of the sands. The limestones with casts of thin-walled gastropods are lacustrine. Thin basalt layers yielded a K/Ar date of 17.3 Ma that confirms the Early Miocene age of this unit (Sharkov, 2000).

The Middle Miocene is made up of alternating conglomerates, sandstones, clays, and limestones with a cyclical character. The limestones have a light-gray color but due to the prevailing red and brown tints of the sandstones and clays the entire rock sequence takes on the appearance of a red-colored formation. The sediments of the Middle Miocene are transgressive. At their base they include a conglomerate bed which in different outcrops at Al-Hmaira, Flita, and Jebel El-Kawdan rests not only on the Lower Miocene sands but also on limestones of the Oligocene, Upper, and Middle Eocene. The 20 to 40 m thickness of the Middle Miocene is insignificant.

In the Homs depression the Middle Miocene sedimentary unit is overlain by basalts. Lower flows of this sequence of effusive

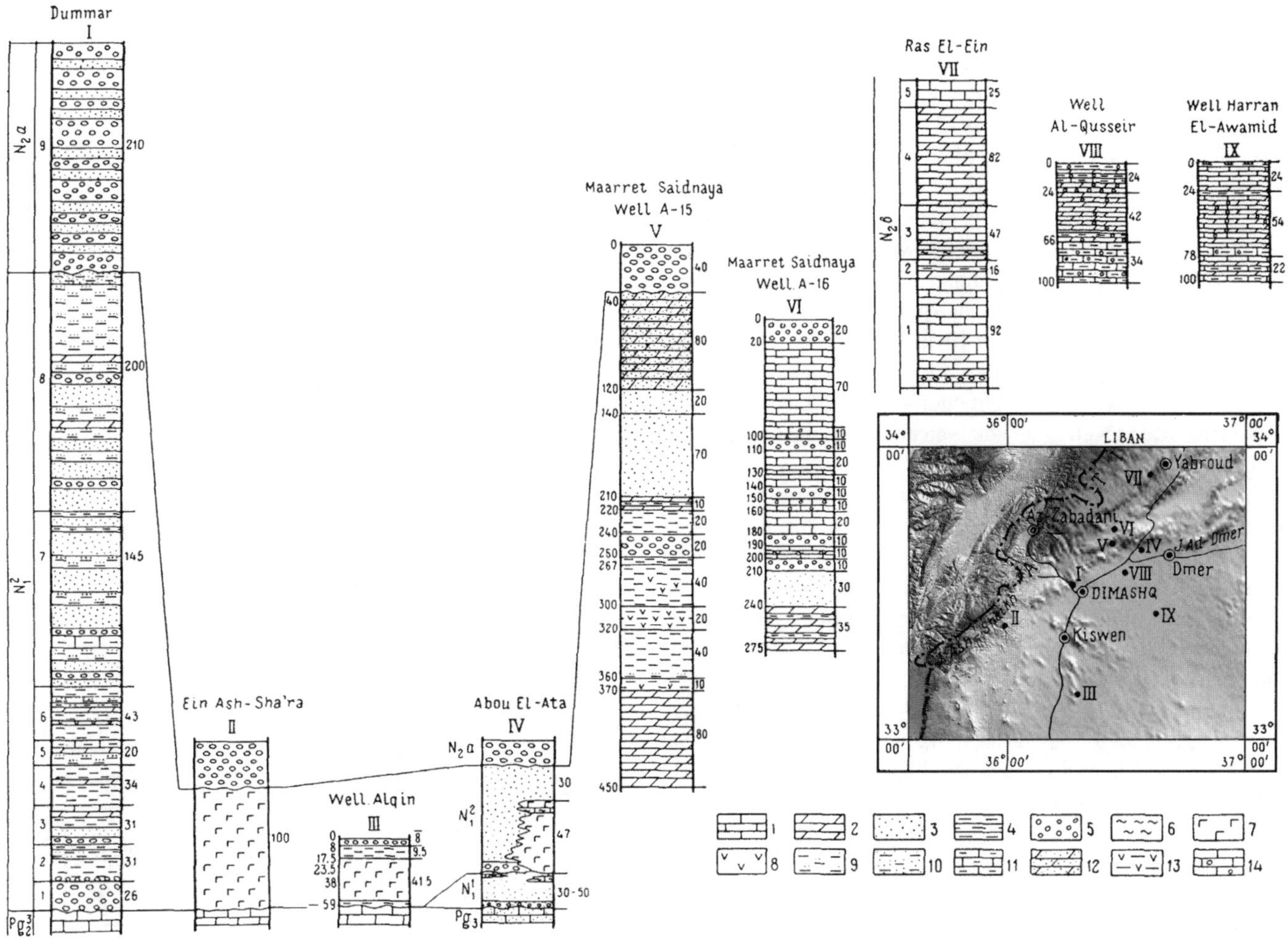

Fig. 6.77. Sections of the Neogene continental deposits of the Damascus depression and some adjacent depressions of the Anti-Lebanon (after Razvalyaev, 1966a). Legend: 1 – limestone; 2 – marl; 3 – sandstone; 4 – clay; 5 – conglomerate; 6 – flint; 7 – basalt; 8 – gypsum; 9 – sandy clay; 10 – clayey sandstone; 11 – clayey limestone; 12 – sandy marl; 13 – gypsiferous clay; 14 – limestone with scattered pebbles.

rocks gave ages of 12.8 and 9.7 Ma (Sharkov, 2000). They indicate the basalts belong in the upper part of the Middle Miocene and in the lower part of the Upper Miocene.

In the Homs depression the lower part of the Pliocene is represented by conglomerates with subordinate intercalations of sandstones, marls, and limestones. The conglomerates are badly sorted and consist of angular and subangular debris of Jurassic, Cretaceous and Paleogene rocks. They rest unconformably on rocks of various ages. The thickness varies from 30 to 100 m in exposures near the villages of Ash-Shmeiseh, Bamara, Flita, and Beer Ad-Dowleh. These sediments accumulated during a period of intense tectonic movements which resulted in the rapid growth of mountains and high rates of erosion and denudation.

In the Homs depression the upper part of the Pliocene is a sedimentary-effusive formation. It consists of lacustrine limestones and marls with fresh-water ostracods (*Ilyocypris bradyi* Sars, *Eucypris* sp., *Candona* sp., and *Candoniella suzini* Schn.) intercalated with basalts. The role of the conglomerates and sandstones is insignificant. The sedimentary-effusive formation unconformably overlies deposits of different age; near Rastan it covers Paleogene limestones. According to drilling data, the thickness of this formation near the village of Zaidal reaches 110 m. Consequently, the Pliocene of the Homs depression, like that of the Damascus depression, is terminated by lacustrine sediments which accumulated on a smooth landscape, however here it was subject to intense basaltic volcanism. It is necessary to take note of the discrepancies in the ages determined for this formation. According to its stratigraphic position, the lacustrine-effusive formation should be placed in the Late Pliocene. However the basalts from the Shin Plateau yielded isotopic ages between 4.79 to 6.3 Ma (Sharkov et al., 1994), which correspond to the Early Pliocene. In our opinion the age of this formation is most probably Late Pliocene.

The Al-Ghab depression/graben and the nearby small Masiaf depression/graben represent relatively young structures which are considered to have originated at the time of the Miocene/Pliocene boundary (Shatsky et al., 1966; Ponikarov et al., 1969; Trifonov et al., 1991; Devyatkin et al., 1997; Devyatkin and Dodonov, 2000). If so, then they were only filled up by Pliocene sediments. However, since even the deepest bore-hole (545 m in depth) did not reach the bottom of the section, the time of origin might be earlier. The origin of the grabens is connected with subsidence along the Syrian-Levant transform fault zone. The vertical displacement along the western flank of the El-Ghab graben is estimated at 2000-2500 m and on its eastern flank at 1000-1500 m. The combination of seismic data, structural profiles from bore-holes, and natural outcrops have allowed a reconstruction of the Al-Ghab Graben's general structure as well as its lithological, facies, and stratigraphic features (**Fig. 6.78** and **Fig. 6.79**).

The zones of sedimentation near the western and eastern slopes of the Al-Ghab Graben, and in the central part of this depression, differ greatly. At the foot of the steep stepped fault escarpment of the western flank of the Al-Ghab Graben a huge mass of olistostroms is found. This sedimentary pile reflects complicated processes of accumulation. It includes gravitational-proluvial deposits made up of coarse breccias of calcareous Jurassic and Cretaceous rocks. They alternate with lacustrine marls and clays and alluvial sands. Palynological studies showed that these sediments accumulated under warm and humid climate conditions typical of the Early Pliocene in the Mediterranean.

The central part of the Al-Ghab depression is occupied by lacustrine deposits. According to the geophysical data, their thickness is estimated at about 1000 m. It means that the bottom (bedrock) of the depression is located at a depth of about 700 m below sea level. The lacustrine deposits are composed of gray calcareous clays with interbeds of marls, limestones, peat, sands, and grits. Two rich assemblages of fresh-water molluscs are present in the sediments.

The lower assemblage includes *Theodoxus jordoni* (Sowerbi), *Melanopsis orientalis* Buk., *M. bitiottii* Buk., *M. binodosa* (Blank.), *Cincinna (Athropidina) sayicyi* (Bgt.), *Bithynia phialensis* (Conrad), *Anisus leucostomus* (Mill.), *Giraulus piscinarum* (Bgt.), *Planorbis antiochianus* Locard, *Planorbarius* sp., *Hippeutis euphaea* (Bgt.), *Lymnaea* sp., *Leguminaia* sp., *Rytia* sp., *Eolymnium* sp., *Corbicula* sp., *Euglesa suecica* (Clessin), and *E. casertiana casertiana* (Poli) (det. of V.S. Zykina). The sediments with such malacofauna refer to the Lower Pliocene. Sometimes clays contain numerous oogonium of *Chara* algae that confirms the lacustrine character of the sediments.

The younger assemblage of molluscs consists of *Siriomargaria apameae* (Blanck.), *Melanopsis* cf. *polyptycha* Neum., *Cincinna sayicyi* (Bgt.), *Bithynia phialensis* (Conrad), *Semisalsa longiscata* (Bgt.), *Falsipyrgula rabensis* (Blanck.), *Gyraulus piscinarum* (Bgt.), *Leguminaia* sp., *Pseudunio* sp., *Potomida delesserti* (Bgt.), *Eolymnium kincelbachi* (Schutt), *Nemrodiana* aff. *tigridis* (Bgt.), and *Dreissena bourguignati* Lokard. Some species of this assemblage are neither present in underlying sediments nor in

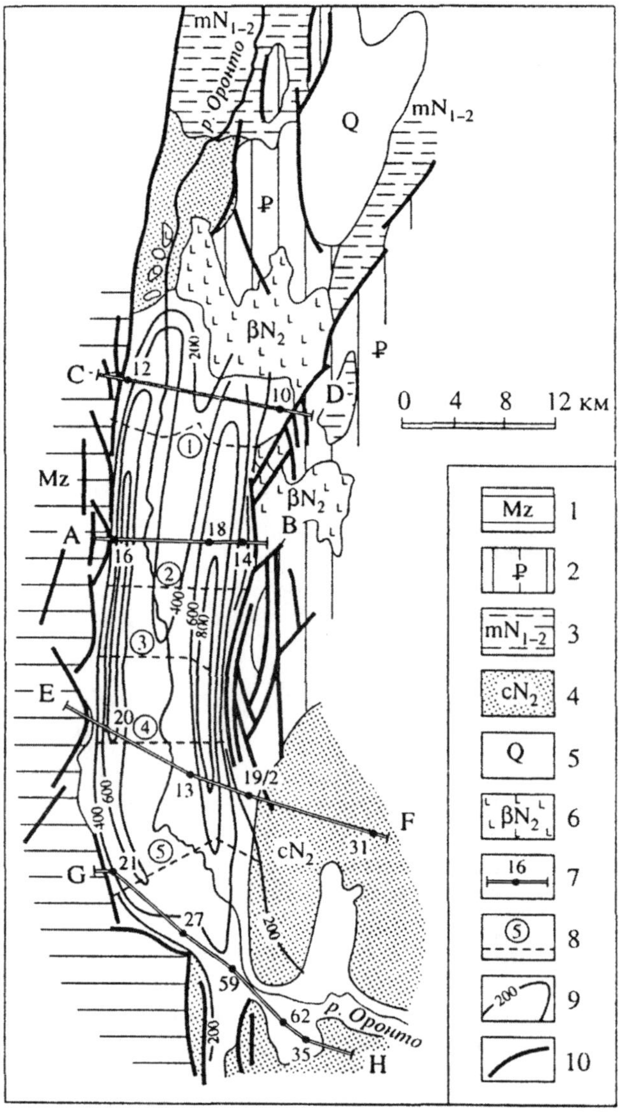

Fig. 6.78. The geological structure of the Al-Ghab rift depression (after Devyatkin et al., 1997).
Legend: 1 – Mesozoic rocks of the Jebel An-Nusseiriyeh; 2 – Paleogene and Miocene rocks of Jebel Zaouiye; 3 – marine Miocene and Pliocene deposits; 4 – continental (predominantly lacustrine) Pliocene deposits; 5 – Quaternary sediments of various origins; 6 – Pliocene and Quaternary basalts; 7 – profiles of bore-holes (A-B, C-D, E-F and G-H) and numbers of bore-holes (27); 8 – geophysical profiles and their numbers; 9 – isopach lines, in meters; 10 – faults.

the Recent fauna. Therefore the sediments with molluscs listed above, are considered to be Upper Pliocene (Schutt, 1983, 1988; Devyatkin et al., 1997). Palyno-logical assemblages of the Late Pliocene display alternations of arid and humid phases against a background of generally insignificant variations in temperature.

The deposits of the lacustrine facies also crop out along the eastern slope of the Al-Ghab Graben. They consist of gray calcareous clays, marls, sands, and small-pebble conglomerates. Near the fault escarpment on the western flank of the Jebel Az-

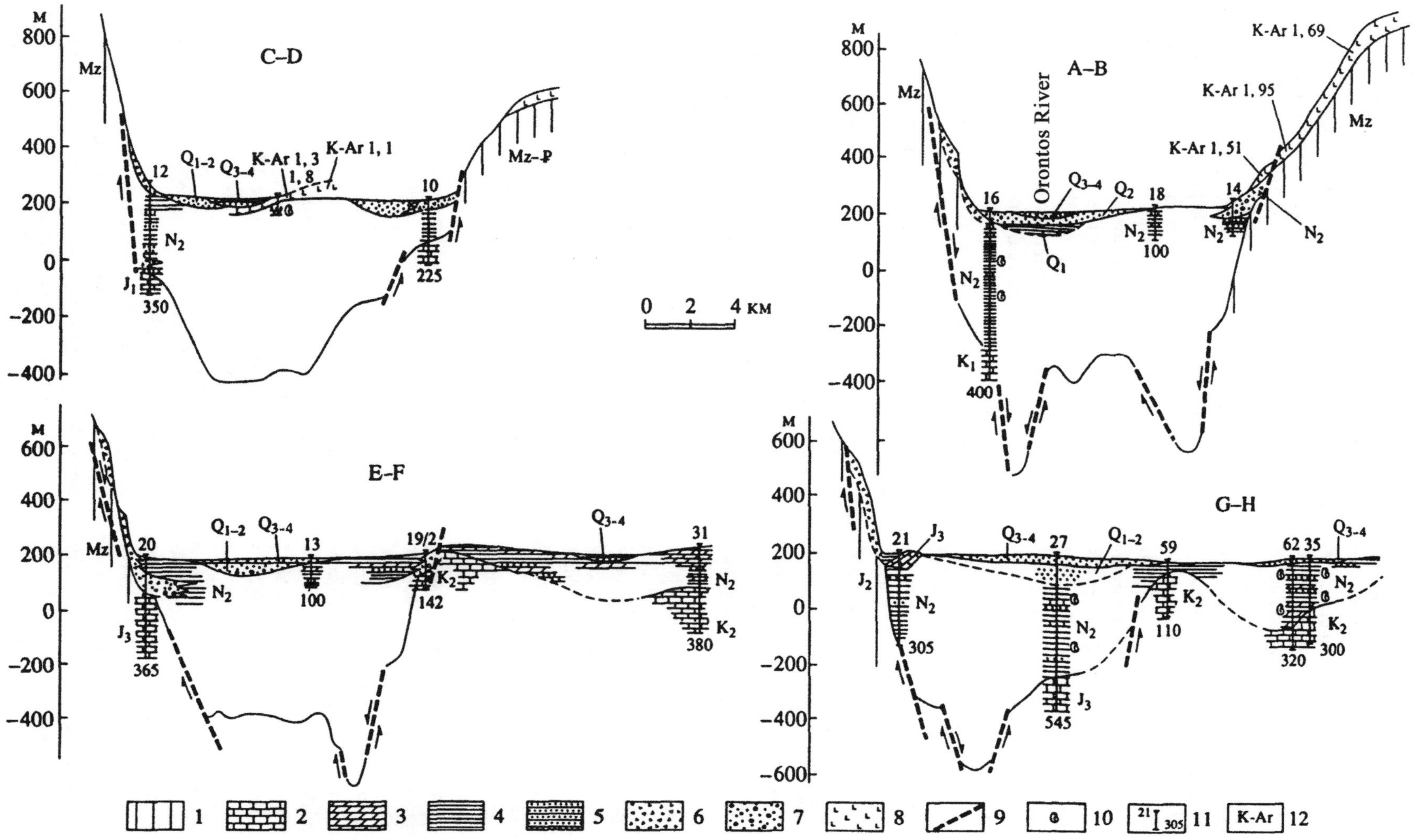

Fig. 6.79. The geological profiles (based on drilling and geophysical data) across the Al-Ghab rift depression (after Devyatkin et al., 1997).
Legend: 1 – Mesozoic and Paleogene rocks; 2-10 – Pliocene-Quaternary sediments: 2 – limestones; 3 – marls; 4 – clays; 5 – sandstones and sands; 6 – conglomerates and breccias; 7 – gravels with sand; 8 – basalts; 9 – faults; 10 – shells of molluscs; 11 – bore-holes, their numbers and depth, in meters; 12 – K/Ar datings.

Zawieh High the lacustrine deposits are replaced by a unit of brown conglomerates and gritstones with interbeds of sands and clays.

The vast flat **Sabkhet Aj-Jabboul** depression, with the Jabboul lake at its center, is located near the southeastern margin of the Aleppo Plateau. A small area of this depression is occupied by Lower Pliocene sediments which overlie the erosional surface of Middle Eocene marls. The most complete Pliocene succession was penetrated by a bore-hole on the northern coast of this lake near the village of Al-Assmeyeh. At the base there is a layer about 6 m thick of poorly-sorted conglomerates and slightly cemented coarse-grained sandstones. They are covered by white and pinkish marls and greenish and yellow clays of lacustrine origin with the fresh-water ostracods *Cyprideis torosa litoralis* (Brady), *Candoniella subellipsoida* (Sars), *Iliocypris bradyi* Sars, *Cyprideis* sp., *Eucypris* sp., and *Loxoconcha eichwaldi* Liv. The total thickness is 32 m. Upper Pliocene basalts cover the sedimentary rocks.

In the Palmyrides the continental Neogene is very well-represented in the intramountain **Ad-Daww depression**, which stretches sublatitudinally from Al-Qaryatein to Palmyra (Tudmor) (**Fig. 6.80**). At the time of the Oligocene/Miocene transition the sea left the area of the Palmyrides, orogenic movements created a dissected landscape, and intensive erosional processes resulted in formation of a Neogene continental molasse up to 600 m thick (Faradzhev, 1966; Soulidi-Kondratiev, 1966a).

The Lower Miocene is made up of white quartz sandstones (sometimes friable sands) with calcareous cement and greenish sandstones with clayey cement. At the base there is a thin bed of conglomerates. The sandstones are fine-grained, infrequently cross-bedded, with vertical root-like cavities and inclusions of massive quartzite-like sandstones of irregular form. Upwards in the section the amount of the calcareous material increases and the sandstones are replaced by gray and brownish limestones which are very hard, medium-bedded, aphanitic, usually sandy, with flint nodules of irregular shape. The thickness varies from 50 to 95 m. On the southwestern flank of the Ad-Daww depression in the Jebel Rus area the sandstones include some layers of basalts. A K/Ar dating of 24.7±1.4 Ma gives a Early Miocene age (Devyatkin and Dodonov, 2000). This formation of quartz sandstones and sandy limestones rests unconformably on the Oligocene. It is easily traced along the entire periphery of the Ad-Daww depression: on the northern slope at Jebel Tias and Jebel Ad-Dalil; on the eastern slope at Jebel Abiad and Jebel Aj-Tizel, which is cut by the valley of Wadi Az-Zakara; on the southern slope at Jebel Bardeh and Jebel Aada; and on the western slope at Jebel Rimli. To the east of Palmyra this white sand and sandy limestone formation pinches out, and in the Arak section Oligocene limestones are unconformably covered by a thin 9 m thick unit of red-colored Middle Miocene sandstones.

The Middle Miocene is composed of alternating conglomerates, sandstones and clays. A peculiar feature of these rocks is their predominating red color. Among them are interbedded layers of gypsum up to 13 m in thickness and dolomitized marlstones. The composition of this red-colored formation is laterally very changeable. It transgressively covers deposits of the Lower Miocene, Oligocene, Upper and Middle Eocene, although an angular unconformity was nowhere observed. The thickness reaches 175 m. The sediments have proluvial, alluvial and lacustrine provenance. Lacustrine clays and marls contain the fresh-water ostracods *Mediocytherideis inflata* Schneider, *Cyprideis gigantica* Vorosh., and *Eucypris* sp. In the upper part of the red-colored rocks, in the section near T-4 Pumping Station, a tooth fragment of *Gomphotherium* ex gr. *angustidens* was identified by L. I. Alexeeva. The age of the sediments can be considered as not older than the Middle Miocene (Devyatkin and Dodonov, 2000). Exposures of this formation were discovered in various regions of the Ad-Daww depression at Jebel Tyas, Jebel Abiad, Jebel Bardeh and others.

Deposits, conditionally placed in the Upper Miocene, occupy very limited areas in the Ad-Daww depression. They crop out on the flank of the Jebel Tyas ridge near the T-4 Pumping Station and consist of alternating conglomerates with poorly-rounded pebbles, brown and greenish siltstones and yellow sandstones. Interbeds of marls and clays are rare. Deposits are usually gypsiferous, with thin intercalations of gypsum. The thickness is about 235 m. This unit is separated from the underlying Middle Miocene sediments by an erosional surface and is characterized by an absence of the red rock coloration.

The Pliocene deposits of the Ad-Daww depression are poorly exposed (**Fig. 6.81**). They rest transgressively on rocks of different ages *e.g.* Miocene, Paleogene and the Senonian part of the Upper Cretaceous. The Pliocene starts with an up to 30-35 m thick member of basal conglomerates. They belong to a new orogenic sedimentary unit. The conglomerates are poorly-sorted, with large blocks of slightly-rounded rocks of local origin. They alternate with sandstones and gritstones. The amount of coarse clastic rocks diminishes upwards in the section. The basal member is covered by a unit with irregular alternation of gray and yellowish sandstones, siltstones, and sandy marls with subordinate interbeds of gritstones and conglomerates. Marls contain fauna of the lacustrine ostracods *Cyprideis torosa littoralis* (Brady), *Trachyleberis* sp., *Limnocythere* sp., and *Xestoleberis* sp. On account of poor exposures, it is difficult to estimate the total thickness of the Pliocene, but it certainly exceeds 110 m.

In the lateral direction the Pliocene deposits of the Ad-Daww depression change facies. Coarse-clastic deposits predominate in the marginal parts of this basin. The central part of the depresion is occupied by lacustrine sediments with a moderate amount of the terrigenous material. It is possible that an inner unconformity subdivides the Pliocene into two parts, but poor outcrops preclude a resolution of this question.

The flat **Subkhet Mouh** and **Feidet El-Mazra'a depressions** are situated to the south and southwest of Palmyra. They belong to young geomorphological structures and their western parts are only filled with Pliocene sediments. Small outcrops and some shallow bore-holes drilled for geological mapping purposes reveal the character of these sediments (Soulidi-Kondratiev, 1966a).

In the area of Arak and Ad-Dbayyat the lower part of the Pliocene is represented by a member with gray and yellowish sandstones and sands with separate layers of limestones and conglomerates whose thickness does not exceed 60 m. Sediments

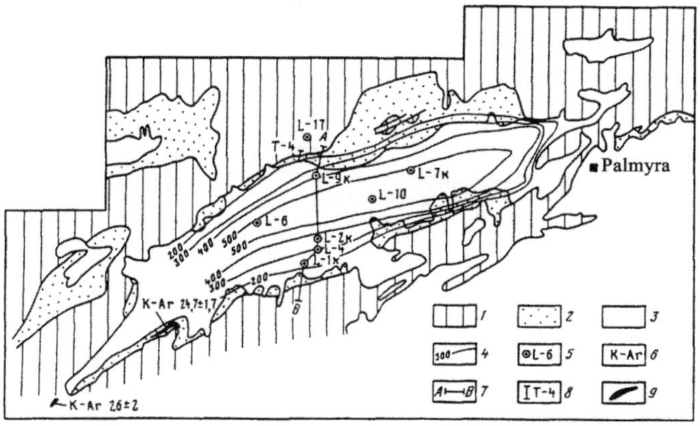

Fig. 6.80. The structure of the Ad-Daww depression in the Central Palmyrides (after Devyatkin and Dodonov, 2000).
Legend: 1 – marine Mesozoic and Paleogene deposits; 2 – continental Neogene deposits; 3 - Quaternary deposits; 4 – isopach lines of the Neogene-Quaternary deposits, in meters; 5 – bore-holes; 6 – K/Ar ages of basalts; 7 – the geological profile A-B; 8 – the reference section near T-4 Station; 9 – basalts.

of the upper part of the Pliocene occupy wider areas including parts of Ad-Dbayyat, Wadi El-Ghidrafih, Wadi Sawwanet El-Hamra, the T-3 oil pumping station, and others. They rest with 5-10° angular unconformity on the rocks of different ages ranging from Miocene to Santonian. This member is characterized by alternation of various lithologies such as sands, sandstones, sandy limestones, calcareous clays, and small-pebble conglomerates. The deposits change laterally. Conglomerates prevail in the western margin of the Feider El-Mazra'a depression, and to the east they are replaced by calcareous sandstones, pinkish limestones, and clays with the fresh-water lacustrine ostracods *Cyprideis torosa littoralis* (Brady), *Limnocythere* sp., *Candoniella albicans* Brady, *Lineocypris* sp., and *Cyprinotus* sp. The thickness varies greatly from 5 to 180 m.

In the eastern part of the Feidet El-Mazra'a depression deposits of both the Miocene and Pliocene are present. They are transitional between the Neogene continental deposits of the Palmyrides and the Neogene marine-continental deposits of the Mesopotamian depression.

On Jebel Dahek-Sharqi, the Oligocene limestones are unconformably covered by Lower Miocene quartz sands and sandstones which are white, yellow and pinkish in color, fine- and medium-grained, calcareous, well-rounded, and sorted. In the upper part of this unit intercalations of mottled clays are present (Kozlov, 1966b). The thickness is about 40 m.

This unit is overlain by marine shallow-water organogenic limestones with interbeds of sandy limestones, sandy clays, and siltstones whose thickness varies from 12 to 30 m. In the Mesopotamian Basin these deposits are known as the Jeribe Formation (the upper part of the Burdigalian Stage - the Middle Miocene).

Upwards in the section limestones are conformably replaced by a formation consisting of intercalated sandstones, siltstones, and clays with separate interbeds of limestones, which have variegated yellow, brown, white and greenish colors. At the base there is a thin conglomerate bed. The thickness of the formation ranges from 30 to 80 m. Evidently, this unit can be considered as a transition to the evaporitic Lower Fars Formation (Tortonian Stage, Upper Miocene), but intercalations of gypsum have not yet appeared.

The Miocene is crowned by a member of brown, reddish and greenish clays, siltstones and fine-grained sandstones whose thickness is about 50 m. This member is regressive and was found only near the northern margin of the Feidet El-Mazra'a depression. Lithologically it is similar to the Upper Fars Formation of the Mesopotamian Basin which is Messinian, Upper Miocene.

The Pliocene deposits of the Feidet El-Mazra'a depression are transgressive and occupy rather large areas. They include two members. The lower member is well-exposed on the Jebel Dahek-Sharqi High and along the valley of Wadi Dbayyat, and penetrated by some shallow bore-holes. It is composed of coarse conglomerates and sandstones of proluvian origin with intercalations of limestones and clays of lacustrine provenance. The thickness ranges from 45 to 105 m, but reaches 150 m in the deepest parts of the depression. The upper member fills the central part of the depression and its composition is known from drilling data. The 33 m thick member is represented by friable sands, sandstones and small-pebble conglomerates with intercalations of clays. It not only rests on the deposits of the lower member, but also transgressively overlies older rocks.

Previously we compared the Neogene deposits of the Mediterranean and Mesopotamian Basins and found a similarity in their geological development. It is a much more complicated task to carry out such a comparison of the Neogene deposits of the Mediterranean, Mesopotamian, and the inner areas of Syria where the stratigraphic subdivision is poorly substantiated by paleontological data. Nevertheless, such a comparison reveals some similar features in the geological history of these three sedimentary basins.

The Lower Miocene is transgressive but sediments of this age occupy considerably limited areas. In the Damascus, the Anti-Lebanon, and the Homs depression, as well as the Ad-Daww and Feidet El-Mazra'a depressions of the Palmyrides, the white quartz sands and sandy limestones have an irregular and spotty distribution. Evidently, they correspond to the Dibbane Formation of Mesopotamia and to the Aquitanian-Lower Burdigalian interval (Lower Miocene) of the Mediterranean.

The Late Burdigalian-Middle Miocene interval coincides with the vast transgression in the Mediterranean basin and Mesopotamia, where sediments of this age are transgressive. The inner areas of Syria were occupied by a transgressive and thick formation of red-colored and variegated proluvial and alluvial terrigenous rocks and lacustrine limestones and marls, accumulated under warm and humid climate conditions. Apparently, this assemblage of rocks can be correlated with the Jeribe Formation of Mesopotamia and the Upper Burdigalian-Middle Miocene marine sediments of the Mediterranean.

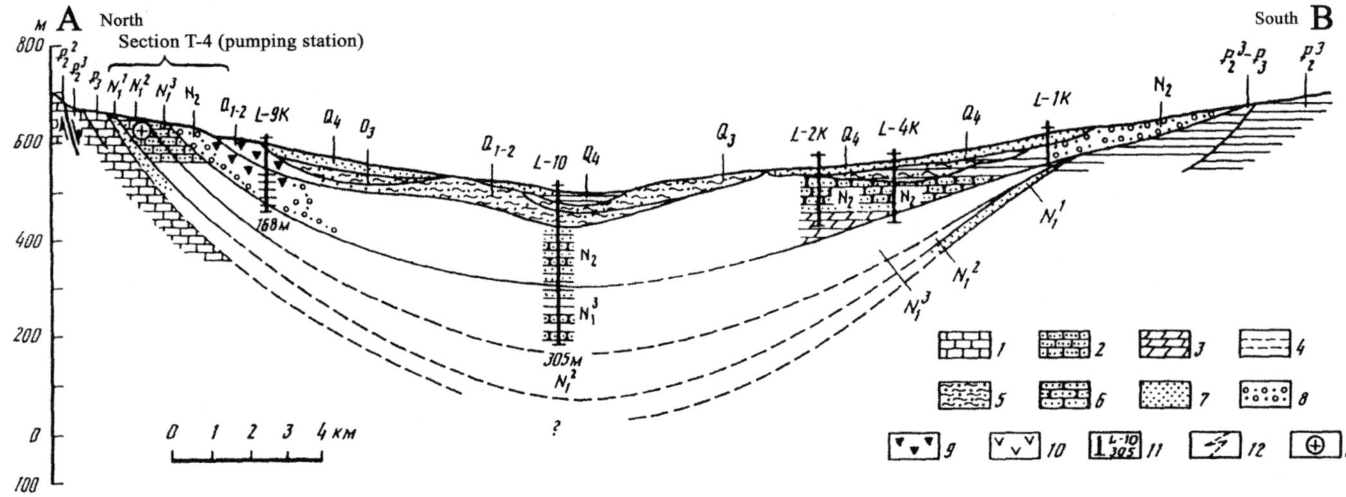

Fig. 6.81. The geological profile through the Ad-Daww depression (after Devyatkin and Dodonov, 2000).
Legend: 1 – limestones; 2 – sandy limestones; 3 – marls; 4 – clays; 5 – siltstones; 6 – sandstones; 7 – sands; 8 – conglomerates and gravel; 9 – debris; 10 – salt; 11 – boreholes, their numbers and depths; 12 – facies transition; 13 – findings of the mammalian fauna.

The Upper Miocene in the Mediterranean and Mesopotamia is regressive. The regression commenced in Tortonian time and reached its maximum extent in the Messinian. The continental sediments of the Upper Miocene in the inner areas of Syria are probably also regressive, creating difficulties in their identification. They have been conditionally identified in the Ad-Daww depression at the Jebel Tyas High, and more reliably in the eastern part of the Feidet El-Mazra'a depression near the boundary with the Mesopotamian Basin. Deposits occurring in limited areas of the inner regions of Syria can be analogues of the Lower and Upper Fars Formations of Mesopotamia. The character of the Upper Miocene deposits in the sections of the Palmyrides and the Anti-Lebanon is the most vexing problem of the stratigraphy of the continental Neogene. We reiterate that the so-called Middle Miocene of the Anti-Lebanon and the Palmyrides can include, at places, also lower layers of the Upper Miocene.

The continental Pliocene of the inner areas of Syria is also transgressive, like in the Mediterranean and Mesopotamia. As in Mesopotamia, it can be subdivided into lower and upper units. We do not exclude the possibility that they are coeval although their lithologic compositions differ diametrically. Such a difference is explained by the peculiarities in the tectonic development of the Palmyrides and the Anti-Lebanon on the one hand, and in Eastern Mesopotamia including the system of the Taurus and Zagros on the other hand. Therefore, the lower unit of the Pliocene in the Anti-Lebanon and the Palmyrides is composed of coarse terrigenous rocks and the upper unit is formed by predominantly lacustrine sediments. In northeastern Mesopotamia the lower unit consists of fine-grained terrigenous and calcareous rocks whereas the upper unit is made up of rough conglomerates and sandstones.

In summary, the stratigraphy of the continental Neogene deposits of the inner areas of Syria was not satisfactorily studied, primarily due to the scarce organic remains. Additional stratigraphic investigations are certainly required.

G: Quaternary

The Quaternary rocks of Syria are represented by marine and continental sedimentary formations as well as by basaltic magmatic rocks.

Marine sediments are exposed along the Mediterranean coast. They consist of the sand and clay of marine terraces, recent beaches, and dunes.

Continental sediments cover most of Syria. The Geomorphologic Map of Syria (**Plate 1**, **Fig. 6.82**) shows a rather complicated landscape of folded and block-type mountains, plateaus and plains formed by flat layers, volcanic mountains, erosional plateaus, and plains. Therefore the continental Quaternary deposits are very diverse with respect to their genetic types (alluvial, proluvial, eluvial, and lacustrine) and their morphology (erosional, karstic, structural, gravitational, desertic aggradation, water-aggradation, wind-borne, etc.).

In addition, considerable areas of Syria are covered by Quaternary basalts. Such areas are Jebel Drouz in southwestern Syria, the Damascus Basin, the Al-Ghab graben, the Al-Latheqiyeh depression, the Al-Furat River, and the Tigris River valleys. They are also characterized by various volcanic structural forms such as volcanic cones, necks, calderas, and lava flows.

M. Blanckenhorn (1891, 1897, 1914) was a pioneer in the investigation of the Quaternary deposits in Syria. He subdivided them into units corresponding to the Alpine glacial scale. In his opinion, pluvial epochs could be correlated with glaciations and interpluvial epochs with interglaciations. He correlated the most pronounced pluvial epoch with the Mindel glaciation. Blanckenhorn considered the whole Mindel-Riss interval as a long dry period. He also initiated the investigation of the fresh-water mollusks from the Pliocene-Quaternary sediments of the Al-Ghab depression.

L. Dubertret and his colleagues (Dubertret, 1937; Dubertret et al., 1937) studied the marine terraces of the Mediterranean coast in the vicinity of Al-Latheqiyeh. They recognized that the succession of terraces composed of hard calcareous sandstones starting from present sea-level up to 200 m elevation was a Quaternary feature.

Later on, E. Vaumas (1953a, b, 1954) investigated the marine terraces from Al-Latheqiyeh in the north to Tartous in the south. He distinguished three groups of terraces:

- low – from 10 to 20 m elevation,
- middle – from 35 to 80 m elevation and
- high – from 100 to 200 m elevation.

Vaumas (1957) subdivided the lacustrine clays of the Al-Ghab depression into the Upper Pliocene and Quaternary.

Butzer (1958) analyzed the links between climatic and geological events in Syria during the Quaternary. Three main pluvial epochs correspond to the Mindel, Riss, and Würm glaciations and are characterized by cold and humid climatic conditions, formation of alluvial terraces, and regressions of the Mediterranean Sea. They are separated by interpluvial epochs which can be associated with interglacials marked by warm and dry climate, marine transgressions, and incision of river valleys.

Great attention was given to the Quaternary deposits of Syria in 1958-1961 during the geological mapping at a scale of 1: 200,000 carried out by geologists of the USSR. Each explanatory note of the nineteen sheets in this map series contains a description of the Quaternary deposits on that sheet. Explanatory notes to the Geological Maps of Syria at scales of 1:500,000 (Ponikarov et al., 1967) and 1:1,000,000 (Ponikarov et al., 1966) as well as the Geomorphological Map of Syria, compiled by K. Mirzaev, summarize the stratigraphic, lithologic, and archaeo-logical data.

Quaternary deposits have been subdivided into four stratigraphic units:

Lower Pleistocene (Q1),
Middle Pleistocene (Q2),
Upper Pleistocene (Q3), and
Holocene or Recent (Q4).

In general, we drew the Pliocene/Quaternary boundary in Syria at the same level as in the present boundary-stratotype in Italy (found at the base of the Calabrian Stage), although in 1958, its absolute age was unknown. In retrospect this was a good choice as it agrees completely with the modern International Scheme of the Quaternary System. Geologists in our group did not know the correspondence of this level with the base of the Quaternary System adopted in the stratigraphic scheme of the Soviet Union. At that time the top of the Apsheronian Regiostage, at 0.8 Ma, was accepted as the lower boundary of the Quaternary system in the USSR.

The Quaternary sediments of Syria are extremely rich in flint tools. These findings are very important for the stratification of continental sediments of this age and for the understanding of the history of human civilization. Scientists from Japan, France, the USA, England, Canada, the Netherlands, the former USSR and elsewhere have contributed to the knowledge of this aspect of Quaternary geology. Among numerous papers, the publications of Akazawa (1973, 1974, 1979), Akazawa et al. (1993), Julig et al. (1999), Besançon (1981a, b), Contençon (1966, 1976, 1981), Copeland (1979, 1981), Endo (1973, 1978), Fujimoto (1979), Hanihara et al. (1978, 1983), Heinzelin (1965, 1968), Hours (1975, 1981), Kaizer (1965, 1976), Kobori (1973), van Liere (1960, 1963, 1966), Dodonov et al. (1993), Ponikarov et al. (1969), Clark (1966, 1968), Suzuki, Takai (1973, 1974), Coon (1957), Rust (1950) and others should be mentioned.

In 1990-1992 and 1995 scientists of the Geological Institute of the Russian Academy of Sciences and the General Establishment of Geology and Mineral Resources of Syria carried out investigations aimed at understanding the stratigraphy and correlation of the marine and continental Quaternary deposits of Syria, mainly in the western and central regions. They used a variety of methods: biostratigraphic (foraminifers, nannoplankton,

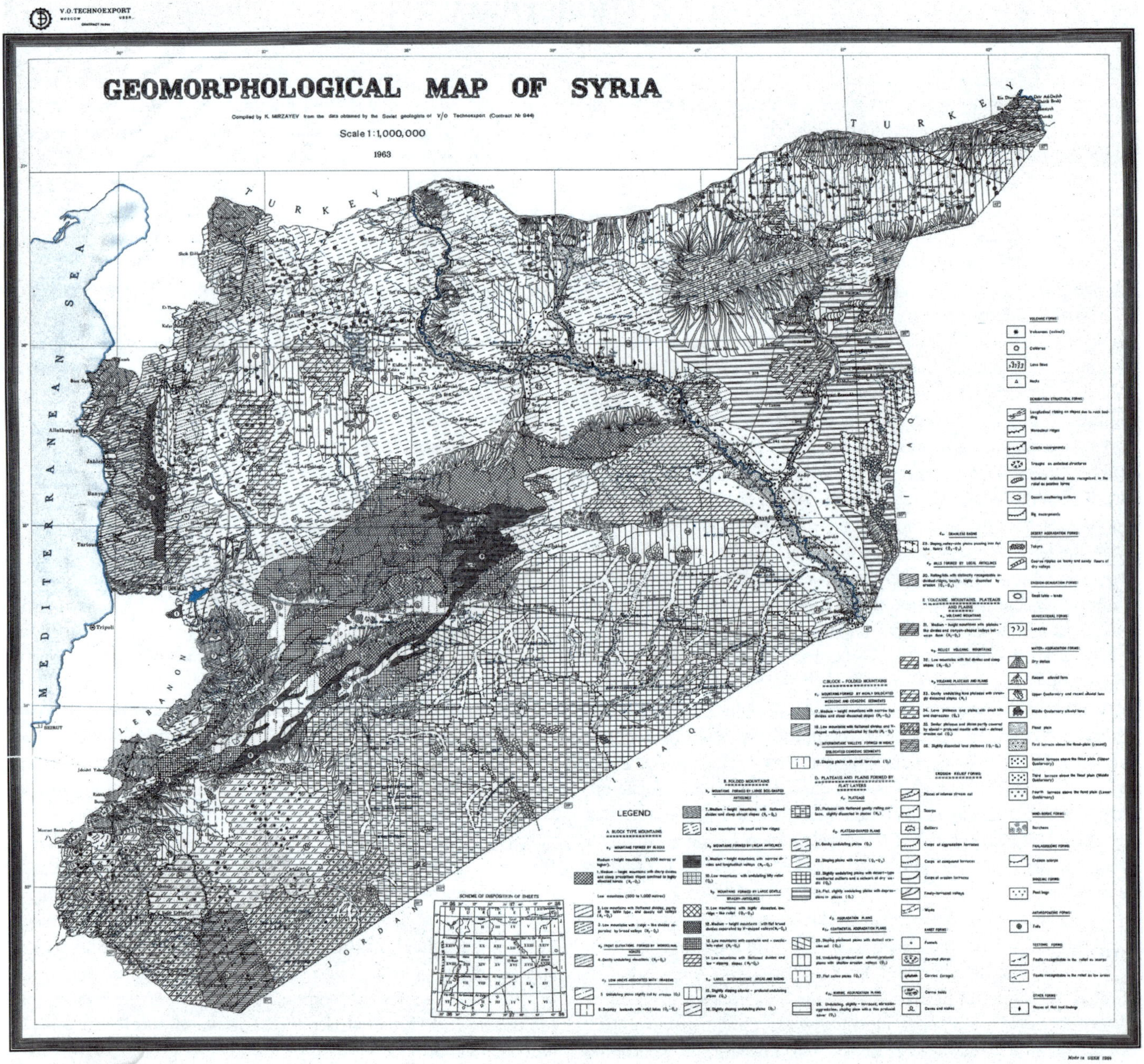

Plate 1. Fig. 6.82. **Geomorphological Map of Syria**. Compiled by K. Mirzayev from the data obtained by the Soviet geologists of V/O Technoexport (Contract No. 944). This map, originally at scale 1:1,000,000 and dated 1963, was printed in the USSR in 1964. The printed copies were sent to Syria, and several decades later could not be located. This personal copy of V. A. Krasheninnikov has been scanned and is reproduced in the plates accompanying these volumes.

The map schematically divides the morphology of Syria into five main categories: Block Type Mountains – 6 subtypes; Folded Mountains – 10 subtypes; Block-Folded Mountains – 3 subtypes; Plateaus and Plains formed by flat layers – 11 subtypes; Volcanic Mountains, Plateaus and Plains – 6 subtypes.

The map also delineates thirteen other landforms: Erosion Relief Forms – 8 subtypes; Karst Forms – 5 subtypes; Volcanic Forms – 4 subtypes; Denudation Structural Forms – 7 types; Desert Aggradation Forms – 2 types; Erosion Denudation Forms – 1 type (Small table lands); Gravitational forms – 1 type (landslides); Water-Aggradation Forms – 9 types; Wind-borne forms – 1 type (Barchans); Thalassogenic forms – 1 type (Erosion scarps); Biogenic Form – 1 type (Peat bogs); Anthropogenic Forms – 1 type (Tells); Tectonic Forms – 2 types; Other Forms – 1 type (Flint tool findings).

diatoms, palynological assemblages), geomorphology, lithology, archaeology, paleomagnetics, thermoluminescence and K-Ar isotopic dating. This study resulted in:

1) The establishment of the Pliocene/Quaternary boundary just above the Olduvai event at 1.8 Ma;

2) The subdivision of the Quaternary into the
Lower Pleistocene (1.8-0.8 Ma),
Middle Pleistocene (0.8-0.13 Ma),
Upper Pleistocene (0.13-0.01 Ma and
Holocene (0.01-0 Ma);

3) The correlation of marine, alluvial and proluvial terraces; estimation of changes in the climate and salinity of marine water; and deciphering the lithological peculiarities of the Quaternary deposits in various regions of Syria (Devyatkin and Dodonov, 2000; Golovina, Gablina, and Simakova, 1996; Devyatkin et al., 1993, 1996, 1997).

In the discussion below, all elevations will be referenced to the present Mediterranean sea level.

MARINE SEDIMENTS

Marine Quaternary sediments are only developed on the Mediterranean coast and in the mouth of the Nahr El-Kabir valley, and farther east up the valley they are replaced by alluvial terraces. According to previous studies (Ponikarov at al., 1967), the Pliocene terminates here with a member of clays and sandstone intercalations, whose sandy material content increases upwards in the section. This unit is covered by a member of hard calcareous sandstone which form a flat surface at an elevation of 200 m. This was considered as the highest (Calabrian) terrace (Shatsky et al., 1966). However biostratigraphic and paleomagnetic investigations changed these views (Devyatkin and Dodonov, 2000; Devyatkin et al., 1996).

In the sections at the villages of Msherpheh and Mardido the *Discoaster brouweri* Zone, which terminates the Pliocene, has been identified. The last specimens of *D. brouweri* some 30-40 m below the top of the clayey-sandy unit. In the same interval the first specimens of *Gephyrocapsa oceanica* were encountered and paleomagnetic studies revealed the Olduvai event of normal polarity. Consequently, these data exactly determine the position of the Pliocene/Quaternary boundary. The uppermost 10 m of the clayey-sandy unit are also characterized by the normal polarity of the Jaramillo event. Thus, the upper 40-50 m of this unit belong to the Lower Pleistocene.

Younger Quaternary sediments are observed in the marine terraces. On the Mediterranean coast and in the mouth of the Nahr El-Kabir River five terraces have been identified:

I – 2 to 4 m (Recent),
II – 10 to 20 m (Upper Pleistocene),
III - 30 to 40 m (Middle Pleistocene),
IV – 60 to 80 m (Lower-Middle Pleistocene) and
V – 90 to120 m (Lower Pleistocene) (Ponikarov et al., 1967).

In addition to these terraces, up to 7 m of marine limestones and conglomerates have been found in the Jebel An-Nusseiriyeh foothills to the east of the town of Jableh at elevations of 180-215 m. These rocks overlie soft Pliocene-Lower Quaternary clays and sands forming a kind of a flat armor for them. To the north of the Nahr El-Kabir River valley, near the village of Msherphet-es-Samouk at an elevation of 180 m, 2-7 m of marine sediments are found consisting of conglomerates, yellowish and gray organogenic detrital limestones with small rounded pebbles and sandstones. Limestones in the nearby village of Al-Hwez yielded *Loripes lacteus* (L.) and *Bittium* sp. (Shatsky et al., 1966). The Msherphet Formation found near Msherphet-es-Samouk at an elevation of 180 m consists of well-rounded gravels up to 1.5 m thick. In the geological literature this structure is considered as an abrasion surface or the uppermost terrace corresponding to the Calabrian. In any case, its age is Early Pleistocene, pre-Guenz (Besançon et al., 1978).

The fifth terrace was traced near the village of Baqsa at an elevation of 120-130 m. In this area, the flattened Paleogene limestone surface is covered by the rounded marine pebbles of the Baqsa Formation. This terrace was also assigned to the Lower Pleistocene (Ponikarov et al., 1967). Near the village of Al-Hwez and between the Nahr Snoubar River and the road to Qerdaha the sediments of the terrace are represented by a member of intercalated organogenic-detrital limestones and gritty sandstones with *Pecten jacobeus* (L.), *Cardium multicostatum* (L.), *C. echinatum* (L.), and *Echinolampas hoffmanni* Desor (Shatsky et al., 1966). Near Qerdaha conglomerates contain Abbevillean three-edged blades (Ponikarov et al., 1967).

The fourth terrace is at an elevation of 60-80 m. At the village of the same name the Khennadi Formation is made up of about 10 m of coarse-grained calcareous sandstones and gravelites with the marine mollusks *Nucula (Lamellinucula) placentina* Lmk., *Glycymeris pilosus* (L.), *Pecten jacobeus* (L.), *Lucinoma borealis* (L.), *Anodonta fragilis* (Phil.), *Venus (Ventricoloidea) multilamella* (Lmk.), *Glossus humanus* (L.), and *Thracia* sp. The assemblage of mollusks is typical of a marine basin sublittoral zone with normal salinity. A thin interlayer of clays contains nannofossils with a predominance of small *Gephyrocapsa, Gephyrocapsa caribbeanica, G. oceanica* s.l., and *Pseudo-emiliania lacunosa.* This terrace is placed at the top of the Lower Pleistocene (Ponikarov et al., 1967), but the simultaneous presence of small *Gephyrocapsa* and *Pseudoemiliania lacunosa* allows one to refer the sediments to the Middle Pleistocene (Devyatkin and Dodonov, 2000). It is worth noting that in nearby Lebanon. sediments of the fourth terrace include Abbevillean flint tools (Watzel and Haller, 1948). Therefore we cannot exclude the possibility that the fifth (Sicilian) and fourth (Milazzian) terraces constitute two stages of the Lower Quaternary Sicilian *sensu lato* terrace.

The third Tyrrhenian terrace at an elevation of 30-40 m was found in the form of a narrow strip of small remnants. Its elevation varies from 30 m near the brow to 40 m near the rear. Usually the terrace resembles a bench but in some areas, such as the mouth of the Snoubar River and near Tartous, it is replaced by a terrace with complex lithology. Near the Snoubar River the third terrace is characterized by the following succession of sediments: greenish-

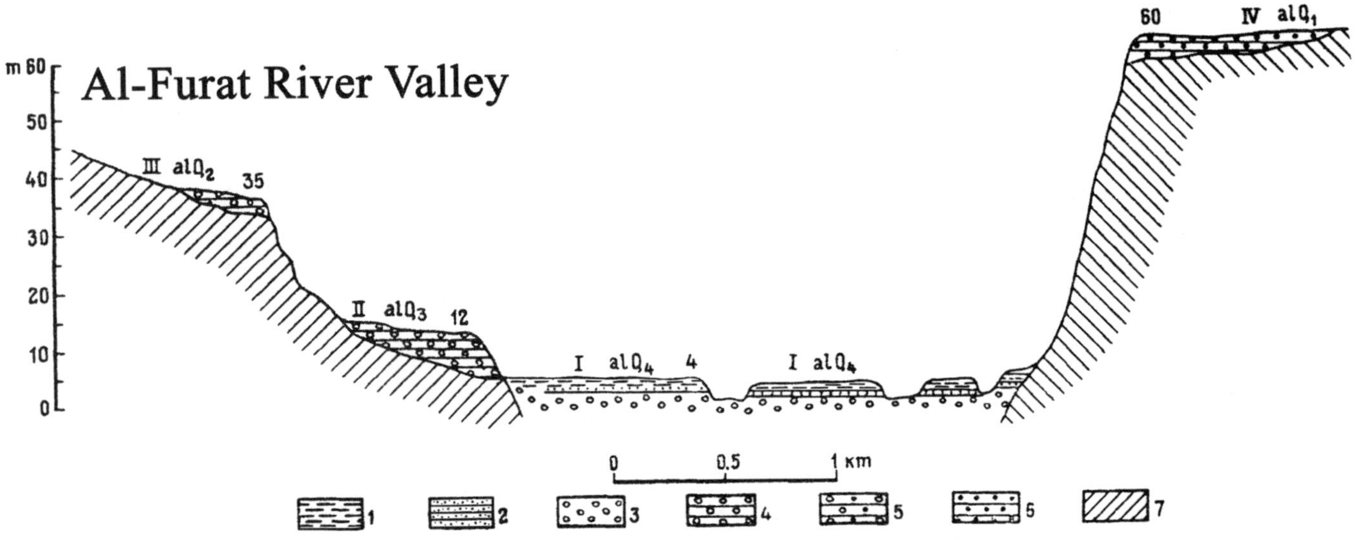

Fig. 6.83. The cross-section of the Al-Furat River valley in Djerablous (Jrablus) area (after Ponikarov et al., 1967).
Legend: 1 – alluvial loam; 2 – alluvial sandy loam; 3 – alluvial pebbles; 4 – alluvial small-pebbled, compact, bedded conglomerate; 5 – alluvial medium-pebbled, compact, bedded conglomerate; 6 – alluvial light-brown, bedded conglomerate with dominating coarse pebbles; 7 – bed-rocks.

gray well-sorted sands with rounded marine pebbles, 3-4 m thick; ancient light-brown calcareous soil, about 1 m thick; and sands and gravelites with cross-bedding, 3-4 m thick. The marine mollusks *Acanthocardia tuberculata* (Poirr.), *Glycymeris pilosus* L., and *Cerastoderma* sp. have only been found in the lower sand layers (Devyatkin and Dodonov, 2000). Conglomerates in this terrace near Tartous contain Middle Acheulean disc-shaped cores and rounded flakes of the Middle Pleistocene (Middle Paleolithic age) (Ponikarov et al., 1967).

The second terrace at an elevation on 10 to 20 m extends as a wide band between Al-Latheqiyeh and Tartous. In places it looks like a rock bench but it is mostly made up of conglomerates and compact calcareous sands with thermophilic *Strombus bubonius* Lam. In the vicinity of the Nahr El-Kabir river-mouth, the marine deposits are replaced by alluvium of the second river terrace with flint flakes and blades of the Upper Paleolithic (Upper Levallois). The thermoluminescent dating of the marine sands gave an age of 101,000±20,000 years (Devyatkin and Dodonov, 2000). All these data make it possible to correlate the second marine terrace with the Monastirian terrace of the Mediterranean (Upper Quaternary) (Ponikarov et al., 1967).

The first terrace, at an elevation of 2 to 4 m, forms a narrow band of sand beaches developed along the seashore. It is widest near the mouth of the Nahr El-Kabir River and to the south of Tartous. The terrace is composed of light-yellow bedded sand with inclusions of well-rounded small flint and limestone pebbles. Sometimes these sediments are replaced by compact calcareous sandstones. Various types of pottery are often incorporated into the Recent deposits of this terrace.

A comparison of the morphological, lithological, and archaeological features of the marine terraces helps in assigning them to their proper level.

The distinctive features of the fifth and fourth terraces are high elevations (60-200 m high), deep erosional incisions (50-60 m), dissected hills with considerable relief, small thickness of sediments (2-7 m) and their poor preservation, the presence of a carbonate crust up to 1 m thick, and Abbevillean flint tools.

The third terrace is characterized by medium elevations (30-40 m high), moderate erosional incisions (15-20 m), gentler hilly relief with moderate differences in elevation, a larger thickness of sediments (up to 10 m), a thin carbonate crust (0.3 m thick), and the presence of Middle Acheulean flint tools.

The second terrace with low 10-20 m elevations has a shallow erosional incisions (2-3 m), slightly undulating relief with distinctly expressed scarps, a small difference in relative elevation, a varying thickness of sediments, and the presence of Upper Paleolithic flint tools.

The first Recent terrace at an elevation of 2-4 m is marked by slight traces of erosion, continuous soil cover, Holocene settlements, cultivated land, and pottery fragments.

All these marine terraces belong to the abrasional/aggradational type.

CONTINENTAL SEDIMENTS

As has already been mentioned above, the main part of Syria is occupied by continental deposits of various alluvial, proluvial, lacustrine, eluvial, and mixed origins.

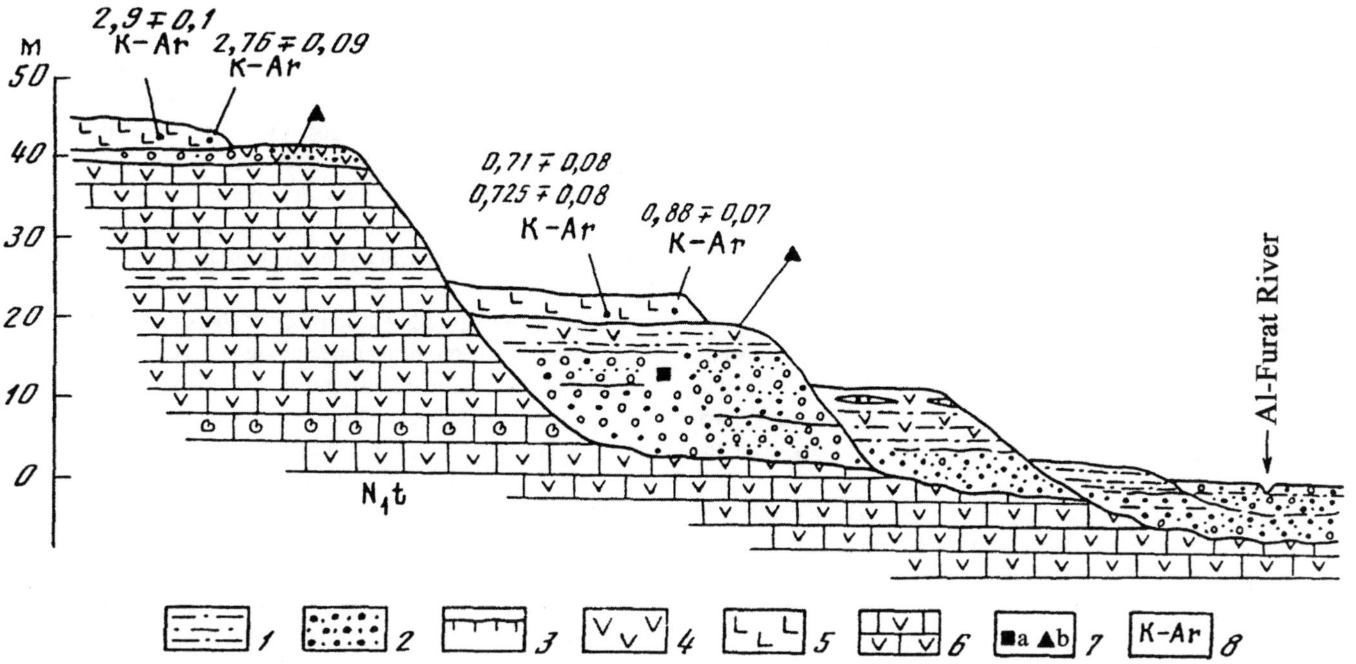

Fig. 6.84. The geologic-geomorphological cross-section of the Al-Furat River valley between Deir Ez-Zor and Raqqa (after Devyatkin and Dodonov, 2000).
Legend: 1 – silt; 2 – gravel; 3 – caliche; 4 – gypsum; 5 – basalts; 6 – the Lower Fars Formation deposits (Tortonian); 7 – archaeological findings (a – Lower Palaeolithic, b – Middle-Upper Paleolithic); 8 – K/Ar ages of basalts, Ma.

Alluvial Deposits of the Main Rivers of Syria

During the whole of the Quaternary Syria was crossed by large rivers such as the Tigris (Nahr El-Dajleh), Al-Furat, Nahr El-Kabir, Orontes (Nahr El-Aasi) and some smaller ones like the Nahr El-Khabour, Nahr El-Balikh, and Barada. Alluvial sediments of these rivers contain numerous traces of human civilization and other features that are extremely important for the correlation of the various facies types of Quaternary deposits.

In the outskirts of Jrablus near the border of Syria with Turkey, the valley of the Al-Furat River is rather narrow (**Fig. 6.83**). Downstream, near the village of Meskene, it widens to 10-12 km and at Mayadin it reaches 18-20 km. Four river terraces can be traced along the whole length of this valley. The fourth terrace is at an elevation of 60-80 and up to 120 m, the third terrace is at 25-30 m, the second terrace at 12-15 m, and the first terrace at only 3-5 m.

The fourth river terrace near Meskene lies at an elevation of 60 m, in the vicinity of Raqqa at 85 m, between Raqqa and Deir Ez-Zor at 120 m, and then between Mayadin and Abou Kamal it drops to 50 m. The deformations of the fourth terrace are caused by the movement of tectonic blocks which are crossed by the river. This terrace is composed of about 10 m of light-gray medium-grained conglomerates with well-rounded polymictic pebbles. They are distinctly bedded, with local small lenses of sandstones. The composition of the pebbles is quite variable *i.e.* flint, limestone, quartz, and igneous rocks. Clastic rocks are covered by a gypsum crust. On the right bank of the Al-Furat River, near the ancient Roman ruins of Halabiyeh, sediments of the fourth terrace include Abbevillean flint tools such as coarse three-edged hand-axes (Lower Pleistocene) (Ponikarov et al., 1967).

The fourth river terrace in the vicinity of Halabiyeh is blanketed by a sheet of basalts (**Fig. 6.84**). K-Ar dating gives the isotopic age of the basalts as 2.76±0.09 and 2.9±0.1 Ma (Devyatkin and Dodonov, 2000). However this Late Pliocene age does not match the Early Pleistocene age of the sediments containing Abbevillean tools.

The third river terrace at 25-30 m consists of alluvial medium-pebbled conglomerate, cross-bedded sandstone, and gravelite with thin intercalation of sandy clays. The thickness ranges from 12 to 30 m. The thickest (25-30 m) alluvial deposits of the terrace are confined to the area between the towns of Raqqa and Deir Ez-Zor. The third river terrace has a gentle hilly relief and can be observed along the whole length of the Al-Furat valley near Jrablus, Meskene, Raqqa, Deir Ez-Zor, and Abou Kamal. It is characterized by numerous findings of Middle Pleistocene (Middle Paleolithic, Middle Acheulean) flint tools such as hand blades, disc-shaped cores, and other rough cutting implements discovered near Raqqa, Qibli, and Al-Kasra (Ponikarov et al., 1967).

A few kilometers to the north-west of Deir Ez-Zor near Abou Jemaa village, the third river terrace is covered by basalts; three K-Ar datings determine their isotopic age as 0.71±0.08 Ma, 0.725±0.08 Ma, and 0.88±0.07 Ma (Early Pleistocene) (Devyatkin and Dodonov, 2000). These figures contradict the Middle Pleistocene age of the Middle Acheulean and even Late Acheulean tools which exhibit features of the Levallois technique (Besançon, 1983).

The second river terrace at an elevation of 12-15 m is preserved in narrow bands stretching along the sides of the Al-Furat valley. It can be observed near Jrablus, Meskene, Raqqa, and Deir Ez-Zor where it is about 15 m high. In the area of the mouth of the Nahr El-Khabour this terrace drops to 8-10 m. The material of the second terrace is light-gray loose-bedded conglomerate with well-rounded pebbles cemented by calcareous-sandy matter; sandy clays are found as rare intercalations. The thickness is not constant and varies from 5 to 12 m. Upper Paleolithic flint tools (Upper Levallois blades and flakes with sharp points and scrapers) have been found near Abou Kamal and Deir Ez-Zor in sediments of the second terrace (Upper Pleistocene).

The first river terrace or 'the upper flood plain' occupies vast areas of the Al-Furat River valley and its left tributaries. It is over 12 km wide near Deir Ez-Zor. Its sediments are composed of loam and sandy loam with interbeds of well-rounded pebbles of various composition. To the south of Meskene, these Holocene sediments contain flint tools and pottery of the Tahoun and Yamouk cultures.

Up to four similar terraces are observed in the valley of the Tigris River (Nahr El-Dajleh) in the extreme northeastern part of Syria (Galaktionov et al., 1966).

The fourth river terrace at an elevation of 100 m occupies a very limited area. Sediments of this terrace were found on the right side of the valley near Ein Diwar village. They are composed of 4-5 m of alluvial washed-out boulder-pebble conglomerate, and near Mazra Et-Bala village they contain rough Abbevillean 3-edged flint cutting tools, made from pebbles (Lower Pleistocene). The uneven washed surface of the conglomerates is covered by a 8-10 m thick basalt sheet.

The third river terrace, 40 m, is well-developed on the right bank of the Nahr El-Dajleh River. The alluvial sediments of this terrace are represented by 20-25 m thick coarse gravels and polimictic conglomerates with clayey-calcareous cement. The Middle Acheulean disc-shaped cores and flakes assign the third terrace to the Middle Pleistocene.

The second river terrace at an elevation of 15-20 m high is made up of coarse gravels, grading into conglomerates, gritstones and sands. Pebbles are usually well-rounded and composed of various sedimentary, igneous, and metamorphic rocks. Boulders of basalts up to 1 m in diameter were encountered in the lower part of this unit of alluvial sediments. The thickness reaches 12 m. Flint tools are absent here. In the area of Khanik-Touramish on the Eastern Qaratchok-Darh ridge the surface of the second terrace gradually plunges beneath the surface of proluvial fans containing Upper Levallois blades and flakes, which assigns the second terrace to the Upper Pleistocene.

The first Recent river terrace at an elevation of 4-6 m consists of greenish-gray, powder-like, well-sorted, sandy loam, in some places with small interbeds and lenses of pebbles and gritstones. Along the stream of the Nahr El- Dajleh, the loam can be replaced by fine-grained, polimictic calcareous or clayey compact sands with pottery.

The third, second, and first alluvial terraces are also observed in the valleys of the smaller rivers of Mesopotamia: the Jagh-Jagh River (Galaktionov et al., 1966), the Nahr El-Khabour River (Ponomarev and Bivshev, 1966; Medvedev, 1966b), and the Nahr El-Balikh River (Oufland, 1966b; Ponomarev, 1966).

In the western regions of Syria, the alluvial terraces were very well-studied in the valleys of the Orontes and Nahr El-Kabir Rivers.

The Quaternary alluvial terraces of the Orontes River were investigated by geological, paleontological, archaeological, and thermoluminescence methods and gave important results (Liere, 1960; Liere and Hooijer, 1961-1962; Clark, 1966; Besançon et al., 1978; Hooijer, 1962; Devyatkin and Dodonov, 2000; Dodonov et al., 1993; Heinzelin, 1966). Five terraces were identified in the middle part of the Orontes River (**Fig. 6.85**).

The highest or fifth terrace at an elevation of 80 m is developed near Shmasy village. It consist of 12-13 m of conglomerates of sub-rounded pebbles with boulders of basalts and limestones in the basal part of this unit (the Khattab Formation). The conglomerates are overlain by 4-5 m of rough angular material of proluvial origin. Early Acheulean flint tools found among pebbles of the Khattab Formation assign the terrace to a pre-Mindel Early Pleistocene age.

The fourth terrace at an elevation of 50-60 m is well-expressed near the villages of Latamne and Khattab (the Latamne Formation). In the Miramil quarry this formation consists of about 11 m of alluvial gravelites with lenses of cross-bedded sands and clayey sands. The sediments are characterized by Middle Acheulean handaxes with elongated and pointed blades (Clark, 1966; Dodonov et al., 1993), as well as by the vertebrate fauna *Stegodon* cf. *trigonocephalus*, *Elephas trogonterii*, *Equus* sp., *Dicerorhinus* cf. *hemitoechus*, *Hippopotamus amphibius*, *Orthogonoceras verticornis*, *Camelus* sp., *Antilopidarum* gen. et sp. indet, *Bison* cf. *priscus*, *Canis* cf. *auteus*, *Crocuta crocuta* (Hooijer, 1962), and *Hippopotamus* sp. (Dodonov et al., 1993). This mammalian fauna was found near the base of the Latamne Formation which dates the sediments to the Mindel glaciation; the whole succession is attributed to the Mindel-Riss interval.

Palynological investigation of the alluvium revealed a wide spectrum of pollen of the broad-leaved trees *Quercus, Carpinus, Tilia, Juglans,* and *Ulmus, Corylus* as well as the coniferous trees *Pinus* and *Cupressus*. In the herbaceous and low shrub group there are many pollen of *Compositae* associated with rare pollen of *Gramineae, Artemisia, Chenopodiaceae, Ericales, Caryophyllacea*, and *Adonis*. Palynological assemblages point to various hypsometric environments consisting of pine forests predominating in mountainous areas, oak forests with *Carpinus* and *Tilia* that were common for piedmonts, plains that were occupied by grass, steppes and forests with *Ulmu,s* and *Corylus* that were typical for valleys. The general composition of the vegetation indicates a climate that is more humid and cooler than at present (Dodonov et al., 1993).

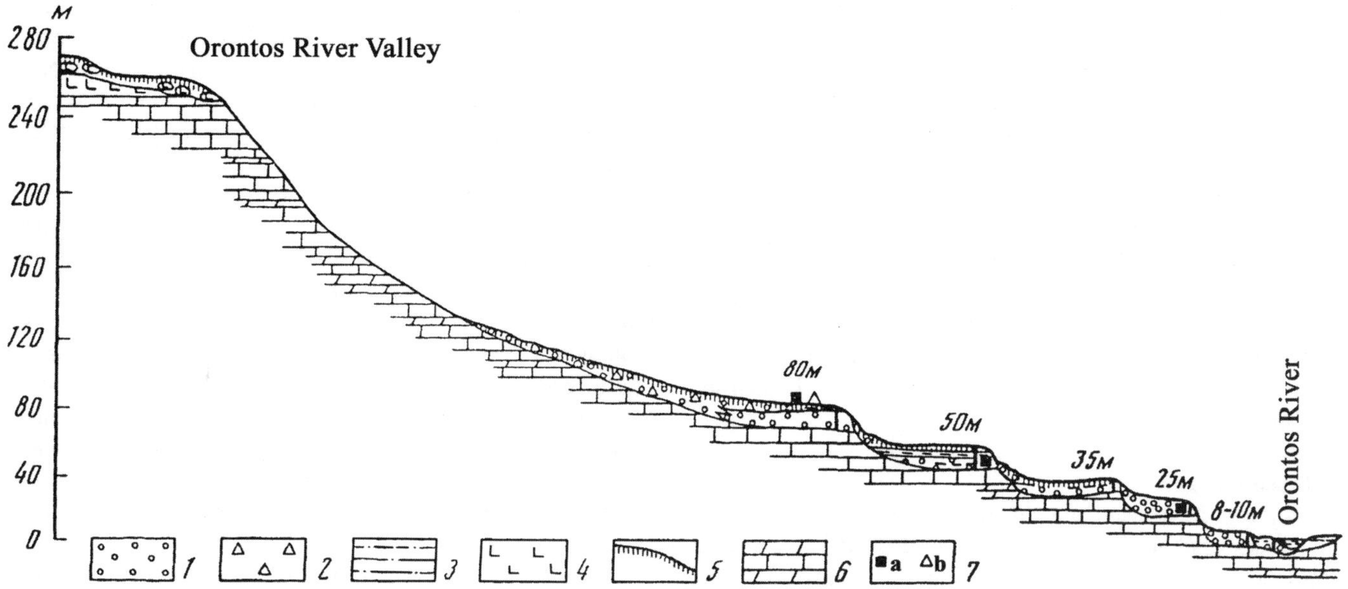

Fig. 6.85. The geologic-geomorphological cross-section of the Orontes River valley (after Devyatkin and Dodonov, 2000).
Legend: 1 – alluvial pebbles; 2 – proluvial debris; 3 – fluvial silt; 4 – basalts; 5 – caliche; 6 – limestones and marls; 7 – Paleolithic findings (a – Acheulian, b – Middle-Upper Palaeolithic).

The radio-thermoluminescence dating of the fine sand fractions from depths of 5.5 m and 8.5 m (above and below the layer with flints implements) in the Miramil quarry gave ages of 324,000±65,000 and 567,000±42,000 years. They confirm the Middle Acheulean age of the flint tools. One might suggest that the Early Acheulean technical methods were replaced by those of the Middle Acheulean about 600,000 years ago (Dodonov et al., 1993; Devyatkin and Dodonov, 2000).

The third terrace of the Orontes River at an elevation of 35 m, and the second at 25 m are composed of gravelites up to 5-7 m thick. The latter is characterized by Late Acheulean flint tools (the upper half of the Middle Pleistocene). Palynological assemblages include various pollen of broad-leaved and coniferous plants such as *Tilia, Quercus, Ulmus, Celtis, Carpinus, Corylus, Ostrya, Juglans, Fagus, Pinus, Cedrus,* and *Cupressaceae*. Subtropical forms are represented by *Oleacea, Tamarix, Pistacia, Magnolia, Palmae, Nyssa,* and *Liquidambar,* but in general the climatic conditions were colder and more humid as compared with the present (Devyatkin and Dodonov, 2000).

The first or 'lowest' terrace at an elevation of 8-10 m consists of gravelites with subordinate intercalations of sandy-clayey material incorporating mixed palynological assemblages. They reflect cool pluvial conditions when herbaceous vegetation was developed on plains, oak forests with *Tilia* and *Carpinus* in piedmonts, and vegetation with *Elaeaguus, Ilex* and *Tamarix* along river valleys. The age of this terrace is treated as Late Pleistocene because the alluvial sediments include flint implements of the Middle and Late Paleolith (Besançon et al., 1978).

The above five terraces of the paleo-Orontes reflect the processes of erosion and accumulation of this Late Pliocene-Quaternary river. A significant thickness of alluvial deposits (up to 10 m), the predominance of rough material (gravelites), and the weak differentiation of facies types indicate that the paleo-Orontes was a big river with flows capable of transporting rough pebbles. During the Quaternary the pluvial climatic conditions were replaced by a period of aridization.

The alluvial terraces of the Nahr El-Kabir River are very important for understanding the Quaternary history of Syria because to the west they are replaced by the marine terraces of the Mediterranean coast.

The highest (fifth) terrace at an elevation of 110-120 m has a sporadic distribution. It can be observed on the right bank of the river near the villages of Sitt Markho and Msherphet Es-Samouk. Its alluvial sediments are made up of gravelites with reddish clayey matrix up to 2 m thick. They contain primitive flint implements of the Early Acheulean, which is the end of the Early Pleistocene (Sanlaville, 1979; Muhesen, 1985; Devyatkin and Dodonov, 2000).

The fourth river terrace, at an elevation of 80-90 m stretches along the right bank of the Nahr El-Kabir near the village of Jinndiriye (Ponikarov et al., 1967; Devyatkin and Dodonov, 2000). Here the alluvial deposit section starts with a 5-10 m thick member of gravelites which are overlain by a 10-12 m thick member of soft sandstones and sandy clays. At its base a mandible of the big mammal *Dicerorhinus hemitoehus* was found indicating a Middle Pleistocene age for the sediments. Flint tools collected near the village of Rudoh belong to the Middle Acheulean. According to paleomagnetic measurements, the sediments of this terrace are characterized by normal polarity which can be interpreted as the lower part of the Brunes epoch. It is worth noting that Ponikarov et al. (1967) placed the sediments of the fourth terrace at the top of the Lower Pleistocene.

The lower series of terraces encompasses:

1) the third terrace at an elevation of 40 m with Middle Acheulean disc-shaped cores (Middle Pleistocene);

2) the second terrace at an elevation of 15-20 m which belongs to the Upper Pleistocene, and

3) the first terrace at an elevation of 2-4 m with pottery shards, and two-edged flint tools of the Mattock type from the Tahoun and Yamouk cultures.

Proluvial Deposits

Among the Quaternary continental deposits of Syria, the most widespread are proluvial. They surround the intermountain basins, depressions, and valleys of various size *e.g.* Ad-Daww, Damascus, Homs, Jabboul, Al-Ghab, Faydate, and many areas of the Mesopotamian depression. They form the wide piedmont fans and cones that fringe the slopes of mountain chains such as the Palmyrides, the Anti-Lebanon, Jebel An-Nusseiriyeh, Jebel Abd El-Aziz, Qaratchok, Jebel Sinjar, Jebel Al-Bishri and some other elevations. The belt of proluvial deposits also stretches along the foothills of the Taurus which is located farther north in Turkey.

The stratigraphy of the proluvial deposits is complicated since organic remains and flint implements are only rarely found. It is therefore based on correlation with the alluvial river terraces, which permit identification of Middle and Upper Pleistocene as well as Recent proluvial deposits (Ponikarov et al., 1967).

According to drilling data, there is a gradual transition between the Pliocene and Quaternary sediments in the Ad-Daww basin (Devyatkin and Dodonov, 2000). Lower Quaternary formations are not exposed, as they are confined to the deep central part of the basin. There they are represented by clay and marl (calcareous loam) with a gravel-sand and boulder-pebble admixture (Faradzhev, 1966). On the slopes of the Ad-Daww basin, deposits of this age were washed out and the sections begin with the Middle Quaternary. They rest unconformably on Pliocene beds which form a hilly landscape. Within the mountain slopes, at altitudes of 800-1000 m, the Middle Quaternary deposits overlie Cretaceous and Paleogene rocks and form flat tops on the erosional remnants and outliers of the older rocks.

Proluvial deposits are mostly composed of conglomerates which are generally well-cemented. These rocks can grade laterally into intercalated gravels, sands and sandy loams whose thickness usually does not exceed a few meters, but in the direction of the central part of the Ad-Daww basin, can reach up to 20 m.

Several generations of proluvium fans and cones can be identified. The highest levels at 30-40 m and 15-20 m above the plain are coarse solid conglomerates with sandy-gravel matrix. In several regions of the Ad-Daww basin they are covered by gray and brownish, very hard aphanitic vuggy carbonate crust. This carbonate rock yields numerous shells of continental xerophite mollusks such as *Pupilla, Orcula,* and *Helicidae* and is pierced by the vertical root-cavities of the remains of steppe vegetation. On the surface of the carbonate crust, Late Acheulean implements have been found. Therefore this generation of proluvium can be referred to the Middle Pleistocene.

The younger generation of proluvium consists of coarse brownish-gray clastic rocks *e.g.* conglomerate, brecchia, pebbles and, to a lesser degree, gravels, sandstone, and loam. The clastic material only includes rocks of local origin. Pebbles are poorly-rounded and cementation is not as strong as that in the older proluvium. The age of this generation is considered as Late Pleistocene. The Upper Pleistocene aggradation surface is located 8-10 m above the present-day erosional surface. At the center of Ad-Daww basin the surface gradually lowers and the thickness of the accumulated deposits increases from 4 to 7-10 m. The Upper Quaternary proluvium covers larger areas than the Middle Quaternary proluvium.

The Recent proluvium of the Ad-Daww basin is widespread but thin, and consists of pebbles, gravels and sands which cover the slopes and floors of the mud-laden streams.

In the central part of the Ad-Daww depression the proluvial deposits of the distal fans and cones grade laterally into sandy-clayey lacustrine sediments.

In the Damascus Basin the proluvial deposits of the Lower and Middle Pleistocene are not exposed. This is due to the large fault running along the northwestern margin of this depression (Fig. 6.76). The existence of these deposits was proven by a series of holes drilled to the east of this fault. The nearby ridges of the Palmyrides folded zone were the source of the terrigenous material. To the southeast, in the inner part of the Damascus Basin, proluvial sediments give way to lacustrine deposits.

In the northeastern part of the Damascus Basin is located a small area with Middle Pleistocene proluvial outcrops which are composed of the piedmont fans of Jebel Yabroud. Here proluvium is represented by gray and brownish-gray compact conglomerates mostly made up of angular pebbles of limestones. An absence of bedding and poor sorting of the pebbles characterizes these rocks. The thickness of the conglomerates varies from 5 to 8 m. The conglomerate-covered surface lies 30-40 m above the plain (Razvalyaev, 1966a).

In the Late Pleistocene the lake shrank considerably and the proluvial fans shifted to the central part of the Damascus Basin, covering also the peripheral margins of the depression. With respect to size, the largest is the Damascus fan or cone, located to the southwest of the city (**Fig. 6.86**). Southward it gradually passes into a vast lacustrine-takyr (or playa) plain and eastward into the Ad-Dmeir fan. The latter then extends into a belt of smaller sub-mountain cones of the southern slope of Jebel Ad-Dmeir. The morphological features of nearly all the fans are identical *i.e.* small 1-2° surface slope angles away from the mountains and shallow depth of the erosional ravines.

Along the mountain ridges the lithological composition of the proluvial deposits is monotonous *i.e.* yellow-gray loams with

Fig.6.86. The Quaternary proluvial-alluvial cone on the slope of the Hermon Massif, Anti-Lebanon, to the south-west of Damascus. The Holy Land satellite database project, the sheet of Damascus, published by J. K. Hall, (2000).

intercalations of angular predominantly limestone pebbles. In the outer portions of the fans proluvial pebble beds and loam some dozens of meters thick alternate with lacustrine deposits. Closer to the central part of the ancient lakes, proluvium is represented by loess-like loams which laterally finally grade into lacustrine deposits.

In the proluvial deposits south of the village of Ad-Dmeir, Perves (1946-1948) found flint blades and flakes of the Upper Levallois. These implements allow these sediments to be assigned to the Upper Pleistocene.

Recent proluvial deposits form numerous fans of ephemeral creeks and wadis at the foothills of the Anti-Lebanon chains. Near the higher elevations, where the creeks are full of water and energetic enough, poorly-sorted rough boulder and pebble material predominates. The central flat areas of the intermountain basins of the Anti-Lebanon and the Damascus Basin accumulate silty-sandy material.

Proluvial deposits occupy significant areas on the Faidate plain in the Syrian Desert between the Palmyrides chain in the north and the escarpment of the Jebel Fru' Tarak El-Olab Plateau in the south (Soulidi-Kondratiev, 1966a).

Lower Pleistocene proluvial conglomerates have a limited distribution and are confined to the relict hills on the Faidate plain. They are composed of coarse, poorly-sorted, medium-to-well-

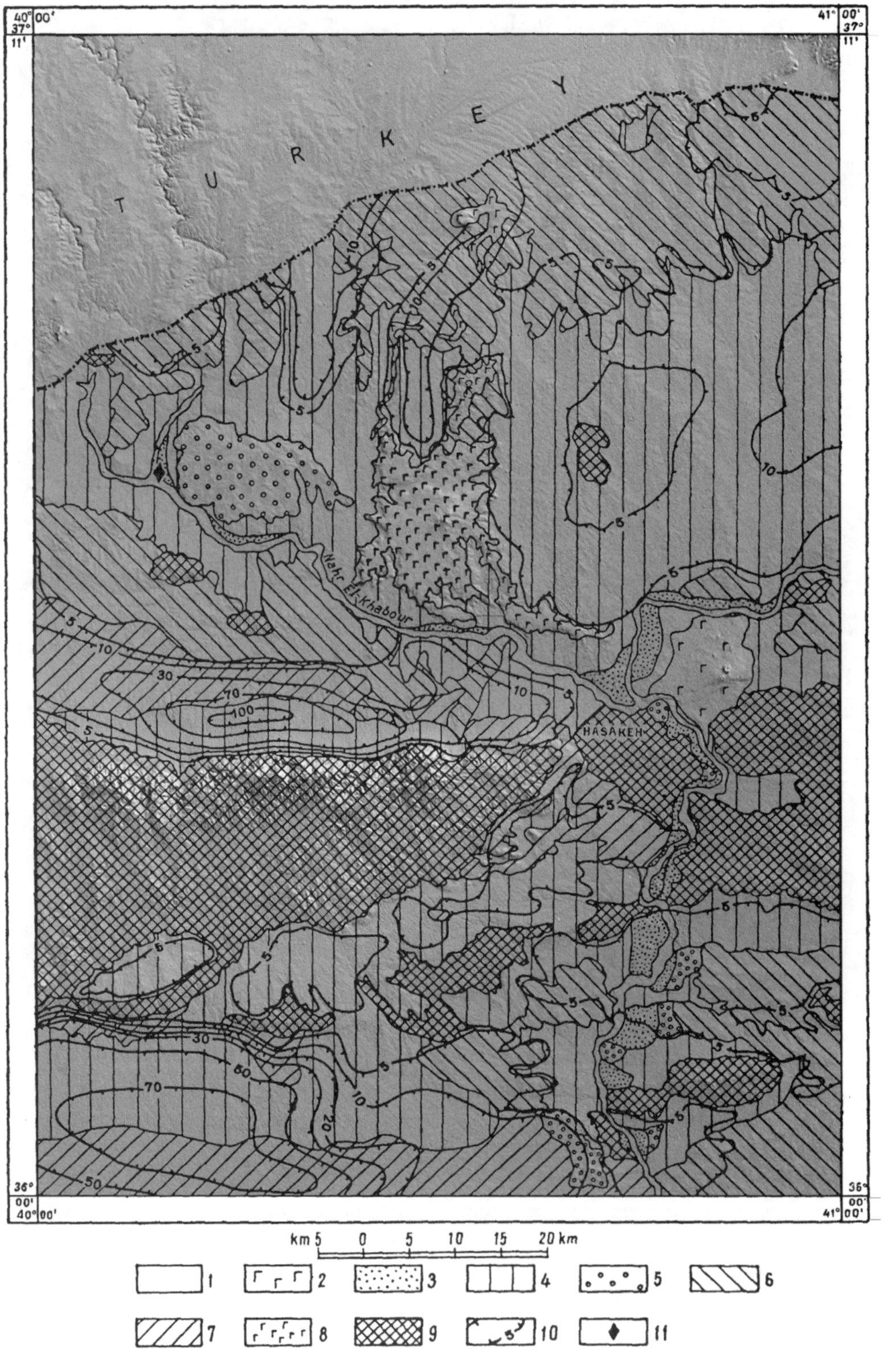

Fig. 6.87. Generalized map of the Quaternary deposits showing the distribution of the alluvial and proluvial types of sediments in the area of the eastern plunging of Jebel Abd El-Aziz (after Ponomarev and Bivshev, 1966).

Legend: **Upper Quaternary** and **Recent**: 1 – alluvial sands, loams and gravels; 2 – basalts and tuffs; 3 – alluvial pebbles, gravels and sands; 4 – proluvial loams and sandy loams with debris, gravels and pebbles. **Middle Quaternary**: 5 – alluvial pebbles and sands; 6 – proluvial conglomerates, loams with some pebbles, boulders and sand. **Lower Quaternary**: proluvial limy-clayey-gypseous rocks, loams, gypsum; 8 – basalts and tuffs. **Pre-Quaternary** rocks cropping out or concealed by a thin cover of eluvial, deluvial and proluvial deposits. 10 – isopach lines of the Quaternary deposits; 11 – places of finding stone implements.

rounded pebbles of local Paleogene limestones and flints, ranging from 5 to 10 cm in size, though some of them reach boulder size sometimes up to 1 m in diameter. The thickness varies from 8 to 10 m.

Middle Pleistocene proluvial deposits are most widely distributed on the Faidate plain. They are poorly-cemented conglomerates, which pass into loose angular debris. Sub-rounded pebbles consist only of Paleogene limestones and flints. The thickness does not exceed 5 m. Along the southeastern foothills of the Palmyrides near the village of As-Soukhneh, Middle Pleistocene conglomerates overlie relicts of Paleogene rocks rising 15-20 m above the local plain. On the Faidate plain the Middle Pleistocene proluvial fans directly cover the surface of the Lower Pleistocene deposits.

Upper Pleistocene proluvial deposits are found in fringing fans and cones to the piedmonts of the South Palmyrian ridges such as Jebel El-Abtar, Jebel Hayyan, Jebel El-Mazar, and Jebel Tantour, and the foothills of Jebel Abou Rajmein. They also cover the Faidate plain. They are represented everywhere by poorly-sorted conglomerates consisting of limestone and flint pebbles, which are cemented by a sandy-calcareous matrix. Occasionally small boulders 15-20 cm in size are found in these rocks. The Upper Pleistocene cover is a few meters thick and rarely reaches 10 m. It is characterized by its low 4-6 m elevation above the local base level.

In these proluvial conglomerates southwest of Palmyra, Perves (1946-1948) found flint tools (flakes and blades) which he assigned to the Upper Paleolithic (Upper Levallois). His finding was very important as it confirmed the Late Pleistocene age of these conglomerates. They were deposited into the eroded surface of the underlying proluvial conglomerates and the latter can thus be considered as Middle Pleistocene.

Recent proluvial deposits of Southern Syria are confined to the base of the Palmyrian ridges where they produce wide fans. A very broad proluvial cone is observed to the northeast of Palmyra fringing the folded structure of the Palmyrides. The apexes of these Recent fans are deeply incised into the Upper Pleistocene proluvial conglomerates. The upper parts of the Recent proluvial fans are composed of poorly-cemented coarse angular debris, pebbles, gravel, and sand. Downwards from the apexes the terrigenous material becomes finer, and the pebble beds are gradually replaced by sands and sandy loam with occasional angular pebbles and gravels. The maximum thickness of the Recent proluvial fans does not exceed 8-10 m. The deposits of mud-torrents and slides belong to a special type of Recent proluvium, but their distribution is limited.

Quaternary proluvial deposits are widespread out on the Mesopotamian depression. Here the proluvium deposits are correlated with the Middle and Upper Pleistocene terraces of the Al-Furat and Nahr El-Khabour Rivers.

Middle Pleistocene proluvial fans occupy the northern slope of Jebel Al-Bishri, the southern and northern slopes of Jebel Abd El-Aziz (**Fig. 6.87**), as well as the foothills of the Taurus, the slopes of Jebel Qaratchok Darh, and of Jebel Jeribe (**Fig. 6.88**). The debris cones consist of conglomerates which pass into gravels and loams in the cone-margins. Pebbles are mostly angular and vary in size from 3 to 5 cm. In the case of Jebel Al-Bishri and Jebel Abd El-Aziz the lithologic composition of the debris is of local provenance. Proluvial fans along the foothills of the Taurus are polymictic, containing various sedimentary and volcanic rocks. The thickness of the Middle Pleistocene proluvium varies from 10 to 15 m.

The Upper Pleistocene proluvial deposits are also well-developed along the slopes of Jebel Abd El-Aziz and Jebel Qaratchok Darh. The most extensive piedmont fans are common on the northern and southern slopes of Jebel Abd El-Aziz where their maximum width reaches 25 to 32 km. The surface of the proluvial cones is almost smooth, their edges flatten and all proluvial fans merge into a nearly flat and wide proluvial piedmont plain. The piedmont cones are composed of gravelly loam with thin pebble and gravel bands. In the vicinity of Qamishliye, Perves (1946-1948) found Upper Paleolithic flint tools (Upper Levallois flakes and blades). Owing to these findings, a part of these proluvial deposits can be attributed to the Upper Pleistocene.

The wide development of Middle and Upper Pleistocene proluvial deposits in Syria in the Anti-Lebanon, Palmyrides, Jebel Abd El-Aziz, Qaratchok Darh, etc., attests to the strong tectonic uplifts that took place at that time. These renovated the landscape and increased the rate of erosion.

Lacustrine Deposits

Although they are limited to small isolated localities within the large Late Cenozoic basins and depressions of Syria, lacustrine sediments are a very important part of the Quaternary deposits of Syria. They occur in the depressions of Damascus, Ad-Daww, Homs, Jabboul, Tudmore (Faidate), Yabroud and Al-Ghab. Lacustrine sediments also occur in closed localized flat-bottomed minor depressions along the left banks of the Al-Furat and Nahr El-Khabour Rivers in the El-Manassif, Sabkha El-Barrouite, Sabkha er Regasser, Bouara, and Roda areas.

In the central part of the Damascus Basin, bore-holes penetrated lacustrine clays, revealing that lake sedimentation persisted during the entire Early-Middle Pleistocene (Devyatkin and Dodonov, 2000). Near the eastern slope of this basin, the clays are covered by basalts with three K-Ar age datings of 0.36; 0.4 and 0.5 Ma (Giannerini et al., 1988). They roughly confirm the Early-Middle Pleistocene age of these sediments.

Outcrops of Lower-Middle Pleistocene lacustrine deposits were only found near the village of Takyyeh, on the slope of the Nahr Barada valley (Razvalyaev, 1966a). Here they are composed of white, yellow and grayish-white, very light, soft, friable calcareous tuff (travertine) with shells of fresh-water *Melanopsis* and remnants of stems and leaves. The thickness reaches 55 m. Laterally the lacustrine clays give way to the alluvial conglomerates of the Nahr Barada terrace.

In the Late Pleistocene the whole lacustrine depression shrank and it was divided into the smaller Buheiret El-Qtaifeh, Ard Buheiret El-Hijaneh, and Jeiroud basins. Sediments of the first two basins form a rather wide and slightly undulating lacustrine-sabkha

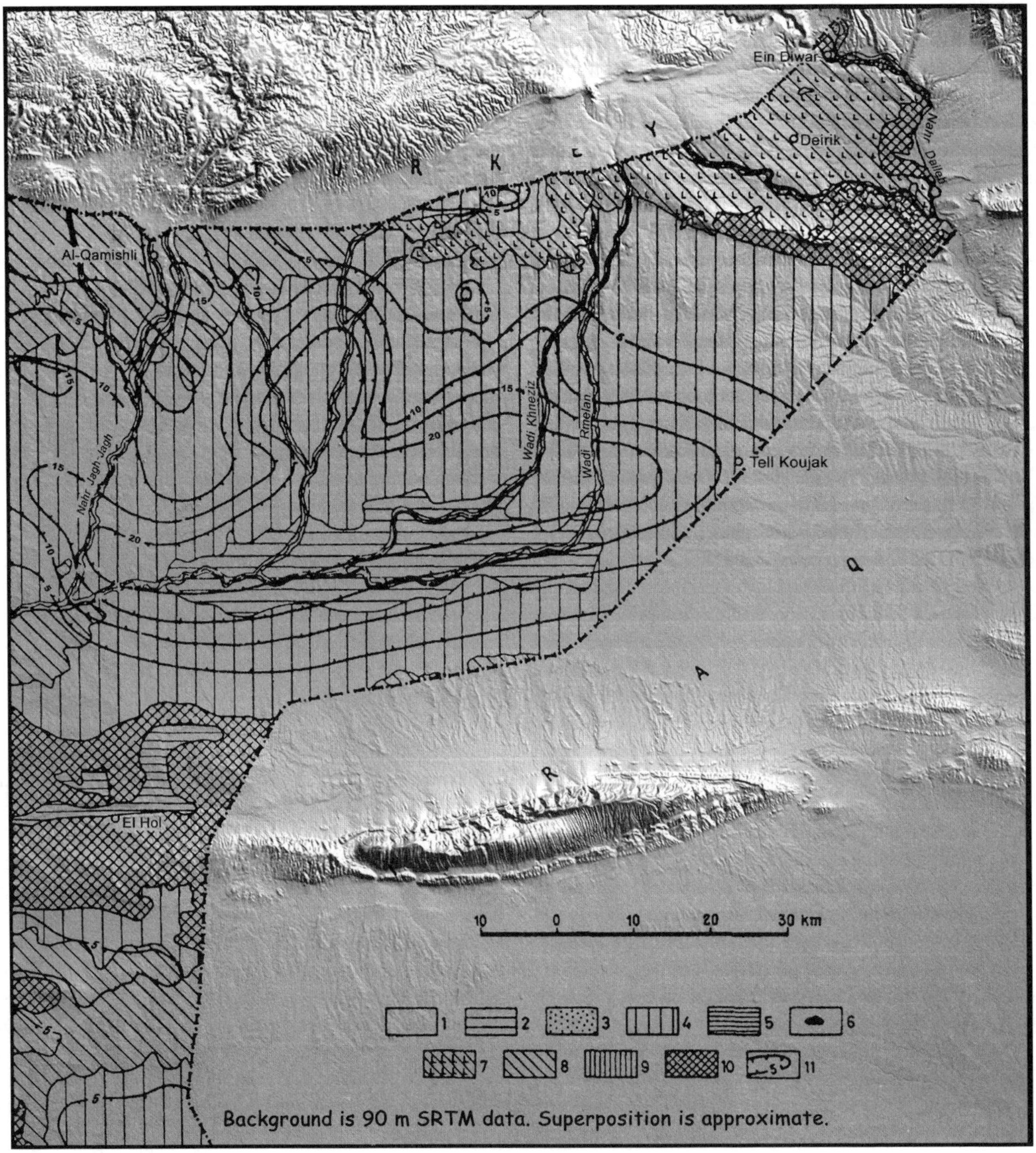

Fig. 6.88. Generalized map of the Quaternary deposits showing the distribution of the alluvial, proluvial and lacustrine types of sediments in the area of Jebel Jeribe and Qaratchok Darh (after Galaktionov et al., 1966).
Legend: **Recent**: 1 – alluvial pebble beds, gravels, sands, and loams. **Upper Quaternary**: 2 – lacustrine pebble beds, gravels, and sands; 3 – alluvial beds and sands; 4 – alluvial and proluvial fans and cone deposits; 5 – lacustrine sands and pebbles. **Middle Quaternary**: 6 – alluvial pebble beds; 7 – basalts; 8 – alluvial-proluvial fans and cone deposits. **Lower Quaternary**: 9 – alluvial pebble beds and gravels. 10 – **Neogene** rocks exposed on the surface or concealed by the thin Quaternary mantle; 11 – isopach lines of Quaternary deposits.

plain with takyrs and solonchaks (= salty soils). To the north these grade into the piedmont surface. In the central part of this plain the lacustrine deposits are represented by white compact lumpy marl intercalating with sandy loam. These sediments are characterized by numerous shells of fresh water mollusks such as *Limnaea stugualis, L. stugualis* var. *subulata, Radix ovata, Bithynia* aff. *leachi*, and *Valvata pulchella*. Near the contact with the piedmont proluvial fans, the composition of the lacustrine deposits changes.

Here sandy and sandy-loam deposits predominate. Closer to the mountains, lacustrine sediments are entirely replaced by proluvial pebble beds. As a rule the thickness of the Upper Pleistocene lacustrine sediments averages 15-20 m.

The Quaternary sediments of the Damascus Basin end with Recent lacustrine ash-gray silt and clay, with remnants of fresh water gastropods and diatoms. The marginal-lake sediments consist of sabkha and solonchaks.

The Quaternary lacustrine deposits of the Ad-Daww and Faidate (Sabkhet Mouh) Basins occupy limited areas and very little geological data is available for them.

Flat-lying lacustrine deposits were encountered in the deep erosional ravines which dissect the sides of the Khshish depression, a part of the Ad-Daww Basin. They are made up of yellow lumpy limestones, blue clays and white marls with interbeds of light-gray loose sandstones. Marly-clayey sediments, encompassing pebbles and sandy grains, belong to the littoral facies of an ancient lake. Their thickness amounts to 15 m (Faradzhev, 1966).

Southeast of Palmyra the Sabkhet Mouh Basin is filled with lacustrine deposits consisting of clearly laminated brown clays, loam and arenaceous marls with bands of sandstones and gravel. The clays are often saline. Takyrs are common in the central part of the basin (Soulidi-Kondratiev, 1966a).

The Quaternary lacustrine deposits of the Jabboul Basin to the south of the Aleppo Plateau are very peculiar. They consist of highly gypsiferous clays and massive bedded gypsum layers (Mikhailov, 1966).

Thick lacustrine deposits are developed in the central part of the Al-Ghab depression. The main part of this sequence is of Pliocene age. Within the Quaternary section, two members have been described (Devyatkin and Dodonov, 2000).

The lower member is found on the right bank of the Orontes River near Jisr Ash-Shaghour. It is composed of white bedded calcareous clays capped by a 0.5 m layer of dark-gray clays with numerous shells of the fresh-water gastropods *Melanopsis polyptycha* Neum., *M.* cf. *blanckenhorni* Shutt., *Cincinna (Athropidina) sayleyi* (Bgt.), *Bithynia* cf. *badiella* (Kuster), and *Gyraulus piscinarum* (Bgt.). These sediments are covered by basalts with K-Ar ages of 1.3±0.09 Ma and 1.1±0.2 Ma near the eastern margin of the basaltic massif. Consequently, the age of the lower member cannot be younger than Lower Pleistocene. These clays display palynological assemblages similar to those identified in the Lower Pleistocene marine sediments of the Mardido section in the Al-Latheqyieh depression. As such, an Early Pleistocene age can be assumed for the lacustrine deposits near Jisr Ash-Shaghour.

The Upper Pleistocene lacustrine clays were penetrated by bore-holes in the central part of the Al-Ghab depression to the south of Jisr Ash-Shaghour. Their age is based on isotopic ^{14}C dating of mollusks shells. The ^{14}C method cannot go beyond 47 000 years.

Eluvial Deposits

The eluvial deposits in Syria are represented by two types of weathering crusts - carbonate and gypsum.

The carbonate crusts are widely distributed all over Syria. They are typical, especially in the inner desertic areas of the country, and were already mentioned by Fraas (1867) and Blanckenhorn (1901). These crusts were formed on various rocks irrespective of their age, but mainly on the surface of the Quaternary formations

The crust itself looks like a very compact, solid, extremely hard ('concrete-like') gray or brownish rock. It incorporates local *in situ* rock-fragments cemented by cryptocrystalline calcareous matter of aphanitic character. Depending on the primary materials, carbonate crusts are conglomerate-like, laminated, and fluidal with spotted structures. The cement also yields such forms as incrustations, veinlets, and elongated particles resembling icicles. The thickness varies from 0.2 to 1.5 m but in some cases it reaches 2.5-3.0 m. The base of the crust has an uneven hummock surface and is always separated from the bedrock by a fine-earthy pillow or a plate of highly porous and loose material. Infrequently the crust rock is blanketed with thin layers of sand or sandy loam.

The carbonate crusts are most characteristic in the plain-like areas of the deserts. They play the role of a protecting armor, overlying softer deposits and creating even table-like surfaces. Subsequent processes of desquamation (=exfoliation) result in the appearance of stony deserts. Along the right bank of the Al-Furat River the carbonate crust armors the surface of the fourth alluvial terrace (Lower Pleistocene). In mountain areas crusts overlie proluvial fans and ancient erosional surfaces. The steep slopes of ridges are usually devoid of carbonate crust.

The process of carbonate crust formation is associated with the ascent of capillary water through the soil, which evaporates and precipitates its salts in the uppermost layer of the sediments.

The most ancient carbonate crusts have been found above the Lower Pleistocene terraces. They are absent within Miocene and Pliocene continental deposits as well as over Recent alluvial and proluvial fans, lacustrine sediments and takyrs. Evidently the formation of the carbonate crusts has only taken place within the interval between the end of the Early Pleistocene and the Pleistocene/Holocene boundary.

The gypsum weathering crust has a limited distribution. It was found in the southeastern corner of Syria on the left bank of the Al-Furat River downstream from the mouth of the Nahr El-Khabour River. It is composed of gypsum layers and weakly consolidated gypsiferous clays having a thickness of 0.5-1.0 m. The crust overlies Pliocene and Lower Pleistocene beds.

A gypsum crust was also identified in the Ad-Daww Basin where it is developed over the Senonian and Paleogene marls and Quaternary formations. The thickness does not exceed 1 m. The origin of the gypsum crust is again associated with the evaporation of sub-soil water under arid conditions.

The calcareous crusts have a utilitarian purpose for the indiginous population. The underlying soft sediments are excavated to produce storage areas, as well as living accomodations for the hot summer months.

GEO-ARCHAEOLOGY

The Quaternary deposits of Syria are extremely important for understanding the development of civilization. As mentioned above, numerous flint artifacts characterize the various stratigraphic units with alluvial, proluvial, and lacustrine sediments. They reflect the sequence of cultures. One can see a direct influence of the geological setting in Syria on the evolution of the stone age, both in providing a suitable habitat and the materials for a tool-based culture.

The Mezosoic and Paleogene deposits of this country include some members with siliceous rocks, but the most remarkable are two formations *i.e.* the Upper Campanian (the *Globotruncana calcarata* Zone) and the uppermost Lower Eocene (the *Acarinina pentacamerata* Zone, P 9 according to planktonic foraminifera). These subdivisions are composed of thick piles of alternating brown and black flints and limestones. Geomorphologically, they are very well-expressed along the flanks of anticlines in the Anti-Lebanon and the Palmyrides. The flint and limestone rocks of the *Acarinina pentacamerata* Zone form the horizontally bedded rocks armoring the table-like elevations of the Rutbah Uplift. The first subdivision was shown on the Geological Map of Syria (scale 1:200 000) as the Campanian/Maestrichtian boundary and deposits of the *Acarinina pentacamerata* Zone constitute an independent mapping unit.

These two siliceous-carbonate formations supplied the ancient population with an enormous quantity of ubiquitous raw material for the 'flint industry'. In the Palmyrides, following the strike of the *Acarinina pentacamerata* Zone deposits, one can observe remnants of ancient 'factories' with abundant worked-out and semi-worked-out pieces of flint.

Numerous flint implements in the Quaternary deposits of Syria allow one to trace the evolution of these tools from the trihedral Abbevillean cutters (Lower Pleistocene) to the very diverse artifacts of the Middle-Upper Pleistocene (Middle-Upper Paleolithic, Acheulian, Mousterian and Levallois assemblages).

A very significant contribution to Paleolithic history has been made based on geoarchaeological and paleo-anthropological investigations of human remains and refuse in some caves located in different parts of Syria. These studies are multidisciplinary in their approach and they address various problems in the stratigraphy, paleogeography, paleoecology, paleontology. paleoanthropology and archaeology of the Late Pleistocene. When hominids spread beyond their homeland in the Great Rift Valley of Africa, they followed the Dead Sea Valley and settled in many areas of Syria *e.g.* the Damascus Basin, the Al-Ghab Valley, the Palmyrides, the Aafrine Valley, the Al-Furat Valley, etc.

The best known caves are:

• northeast of Palmyra (Tudmor) - the Douara Cave (Akazawa, 1973, 1974, 1979, 1996; Akazawa et al., 1993; Endo, 1973, 1978; Fujimoto, 1979; Hanihara and Akazawa, 1979, 1983; Hanihara and Sakaguchi, 1978; Kobori, 1973; Sakaguchi, 1978; Suzuki and Akazawa, 1971; Julig et al., 1999) and the El-Koum Cave (Besançon et al., 1982; Besançon and Sanlaville, 1996; Le Tensorer, 1996);

• at the northern rim of the Palmyrides - the Umm El-Tiel Cave (Bourguignon, 1996);

• northwest of Palmyra - the Jerf al-Ajla Cave (Julig et al., 1999);

• northeast of Damascus - the Yabroud Cave (Farrand, 1970);

• the western slope of Jebel Samaan near the town of Aafrine - the Dederieh Cave (Akazawa et al., 1993).

All these studies answered many questions about the Middle and Upper Paleolithic cultures and their Levallois and Mousterian techniques as well as the position of the Middle/Upper Paleolithic boundary (which is now set at 35.6±3.4 ka).

And last but not least, human artifacts from the Holocene sediments of the Damascus Basin, the Nahr El-Kabir valley, Nahr Balikh valley, the Mediterranean coast, the Al-Furat valley, and other places in Syria which hosted the Neolithic culture of the eastern Mediterranean, were studied and described by Contençon (1966, 1981), Copeland (1979), Hooijer (1975), and Liere (1963).

Photograph of an ashlar (hewn stone) from a temple in Palmyra with a carving of Acanthus (thorny herb native to the Mediterranean area) with wine grape leaves. Plate 18 of Palmyra (1968), Translated from Polish to Russian. A. Dzevenovsky (photos) Mikhailovsky, K. (text). Arkada Publishing House, Warsaw, Poland. 32 p, plus 88 photos.

REFERENCES CITED (for PART II Introduction and Chapter 6)

Abdoulrahim, A. and Nemkov, G. I., 1972. Zonal subdivision of the Paleogene deposits of the Damascus region (Syria) by means of nummulites. Trans. of Educ. Institutions, Geology and Prospecting, No. 7, p. 27-31. (In Russian).

Abed, A. M. and Amireh, B. S., 1983. Petrography and geochemistry of some Jordanian oil shales from north Jordan. J. Geol. Petrol., 15(3), p. 261-274.

Adjamian, J. and Al-Bakka, F. F., 1984. Geological map of Syria. Scale 1:50,000. Explanatory Notes of Misyaf sheet (N I-37-NW-S-1-b). General Establ. of Geol. and Mineral Resources, Damascus, 42 p.

Adjamian, J. and Jamal, N. E., 1983. Geological map of Syria. Scale 1:50,000. Explanatory Notes of Qalaat Al Housen sheet (N I-37-M-3-d). General Establ. of Geol. and Mineral Resources, Damascus, 38 p.

Agrali, B. and Akyol, E., 1967. Etude palynologique des charbons de Hazro et considerations sur l'age des horizons lacustres du Permo-Carbonifere. Bull. Mineral. Res. and Explor. Inst. Turkey, No. 68, p. 1-26.

Akazawa, T., 1973. The Paleolithic assemblages from the Douara Cave Site, Syria and treir relationship to the prehistory of Western Asia. Quaternary Research, 12(3), p. 77-99.

Akazawa, T., 1974. Palaeolithic assemblages from the Douara cave site. Tokyo Univ., Univ. Mus. Bull., No. 6, 167 p.

Akazawa, T., 1979. Middle Paleolithic assemblages from Douara cave. Tokyo Univ., Univ. Mus. Bull., No. 16, p. 1-30.

Akazawa, T., 1996. A reconstruction of the Middle Paleolithic cultural ecology of the Douara Cave, Palmyra, Central Syria. In: Palmyra and the silk road. Vol. XLII. Ann. Arch. Arab. Syriennes, p. 63-73.

Akazawa, T., Dodo, Y., Muhesen, S., Abdul-Salam, A., Abe, Y., Kondo, O., and Mizoguchi, Y., 1993. The Neanderthal remains from the Dederiyeh Cave, Syria: Interim Report. Anthropol. Sci., 101(4), p. 361-387.

Al Baca, F. F., 1983. Geological map of Syria. Scale 1:50,000. Explanatory Notes of Tal Kalakh sheet (N I-37-M-3-b). General Establ. of Geol. and Mineral Resoutces, Damascus, 26 p.

Al-Maleh, K., Mouty, M., Sawaf, T., Brew, G. E., and Barazangi, M., 2000. Mesozoic depositional trends in the northern Arabian Platform in Syria; regional and tectonic implications. GeoArabia, 5(1), p. 30-31.

Almeras, Y., and Mouty, M., 2001. Les Brachiopodes du Jurassique de Syrie. Revue Paléobiol., Genève, 20(1), p. 9-17.

Almogi-Labin, A., Bein, A., and Sass, E., 1990. Agglutinated Foraminifera in organic-rich neritic carbonates (Upper Cretaceous, Israel) and their use in identifying oxygen levels in oxygen-poor environments. In: Paleoecol., Biostratigr., Paleoceanogr. and Taxonomy of Agglutinated Foraminifera, Kluver Acad. Publ., the Netherlands, p. 565-585.

Almogi-Labin, A., Bein, A., and Sass, E., 1993. Late Cretaceous upwelling system along the Southern Tethys margin (Israel): interrelationship between productivity, bottom water environments, and organic matter preservation. Paleoceanography, 8(5), p. 671-690.

Al-Saad, D., Sawaf, T., Al-Youssef, W., Khaddour, M., Barazangi, M., Chaimov, T. A., and Seber, D., 1992. Crustal structure and tectonics of the northern Levantine margin in Syria. 29th Int. Geol. Cong., Abstr., Kyoto, Japan, Vol. 29, p. 460.

Al-Saad, D. T., Sawaf, T., Al-Otri, M., Gebran, A., Barazangi, M., Best, J., Chaimov, T. A., and McBride, J., 1989. 28th Int. Geol. Cong., Abstr., Washington, DC, Vol. 28(1), p. 28.

Al-Saad, D., Sawaf, T., Gebran, A., Barazangi, M., Best, J. A., and Chaimov, T. A., 1992. Crustal structure of Central Syria: the intracontinental Palmyride mountain belt. Tectonophysics, 207(3-4), p. 345-358.

Alsdorf, D. M., Barazangi, M., Litak, R., Seber, D., Sawaf, T., and Al-Saad, D., 1995. The intraplate Euphrates depression Palmyrides mountain belt junction and relationship to Arabian plate boundary tectonics. Annali di Geofisica, 38(3-4), p. 385-397.

Al Sharhan, A. S. and Nairn, A. E. M., 1997. Sedimentary basins and petroleum geology of the Middle East. Elsevier, Amsterdam, 843 p.; Appendix, 99 p.

Al Youssef, W. and Ayed, H., 1992. Evolution of Upper Paleozoic sequences and the application of straturaphy as a tool for hydrocarbon exploration in Syria. 11th Egyption Exploration Petroleum Conf., Cairo, p. 636-645.

Arad, V., Ehrlich, A., and Nathan, Y., 1985. The geology of Syria. Bibliography. Geological Survey of Israel, Jerusalem, Report 39/85, 106 p.

Arambourg, C., Dubertret, L., Signeux, J., and Sornay, J., 1959. Contributions à la stratigraphie et à la paléontologie du Cretacé et du Nummulitique de la marge NW de la Péninsule Arabique. Notes et Mem. Moyen-Orient, Vol. 7, Paris, p. 193-252.

Arkell, W. J., 1956. Jurassic geology of the World. Oliver and Boyd, London. 806 p.

Bach Imam, I., Dagher, H., Khiami, K., and Sigal, J., 1976. Sédimentation marine ou lacune à l'Oligocène et au Miocene basal en Syrie orientale. [Marine or lacustrine sedimentation from the Oligocene to the basal Miocene in eastern Syria.] C. R. Somm. Soc. Géol. France, 18(2), p. 45-47.

Bach Imam, I., Khoja, I., and Sigal, J., 1978. Note sur l'âge principalement Triassique des formations évaporitiques et dolomitiques du Mésozoique inférieur des forages dans l'Est Syrien. Note destinée au 5eme Congr. Géol. d'Iraq, Baghdad, p. 1-15.

Bach Imam, I., Khoja, I., and Sigal, J., 1980. Note sur l'âge principalement Triassique des formations évaporitiques et dolomitiques du Mésozoique inférieur des forages dans l'Est Syrien. J. Geol. Soc. Iraq, 13(1), p. 155-160.

Bach Imam, I., and Sigal, J., 1985. Précisions nouvelles sur l'âge triassique, et non jurassique, de la majeure partie des Formations évaporitiques et dolomitiques des Forages de l'Est Syrien. Rev. Paléobiologie, 4(1), p. 35-42.

Balme, B. E., 1970. Palynology of Permian and Triassic strata in the Salt Range and Surghar Range, West Pakistan. Stratigraphic boundary problems: Permian and Triassic of West Pakistan. Kansas Univ. Press, Dept. Geol., Spec. Publ., No. 4, p. 306-453.

Barazangi, M., Chaimov, T. A., Seber, D., Al-Saad, D., Sawaf, T., Al-Youssef, W., and Khaddour, M., 1992. Structure of the intracontinental Palmyride mountain belt in Syria. 29th Int. Geol. Cong., Abstr., Kyoto, Japan, Vol. 29, p. 151.

Barazangi, M., Seber, D., Chaimov, T. A., Best, J., Litak, R., Al-Saad, D., and Sawaf, T., 1993. Tectonic evolution of the northern Arabian Plate in western Syria. In: E. Boschi, E. Mantovani, and A. Morelli (eds.), Recent evolution and seismicity of the Mediterranean region, NATO ASI Series C: Math. and Phys. Sci., Vol. 402, p. 117-140.

Basse, E., 1937. Les Céphalopodes Crétaces des massifs côtiers syriens. I. Not. et Mém. Syrie-Liban, Vol. 2, p. 165-200.

Basse, E., 1940. Les Céphalopodes Crétaces des massifs côtiers syriens. II. Not. et Mém. Syrie-Liban, Vol. 3, p. 411-472.

Bebeshev, I. I., 1995. History of development of the Arabian Platform in the Late Permian and Triassic. Lithology and Mineral deposits, No. 2, Moscow, p. 183-194. (In Russian).

Bebeshev, I. I., Dzhalilov, Yu. M., Portnyagina, L. A., Yudin, G. T., Mualla, A., Zaza, T., and Yussef, A., 1988. The Triassic deposits of Syria. Trans. of the USSR Acad. Sci., Ser. Geol., No. 11, p. 43-53. (In Russian).

Bebeshev, I. I., Dzhalilov, Yu. M., Yudin, G. T., Souriya, O., Roustom, Sh., and Shaban, F., 1989. Lithological-facies and mineralogical-petrographical characteristic of the Triassic deposits of Syria. Lithology and mineral resources, No. 5, Moscow, p. 61-73. (In Russian).

Bender, F., 1968. Geologie von Jordanien. Beitraege zur Region. Geologie der Erde, Vol. 7, Borntraeger, Berlin, 230 p.

Bender, F., 1974. Geology of Jordan. Borntraeger, Berlin, 196 p.

Bender, F., 1975. Geology of the Arabian Peninsula, Jordan. U. S. Geol. Surv. Prof. Paper, No. 560-1, 36 p.

Bertels-Uspenskaya, I. A. and Oufland, A. K., 1971. Middle Miocene stratigraphy of the north-western part of the Tigris and Euphrates Rivers interfluve. Trans. of Laboratory of geology of foreign countries, Vol. 22, p. 136-141. (In Russian).

Besançon, J., 1981a. Stratigraphie et chronologie du Quaternaire continental du proche Orient. C. N. R. S. Colloq. Intern., No. 598, p. 33-53.

Besançon, J., 1981b. Chronologie du Pleistocène au Levant, Synthese. C. N. R. S. Colloq. Intern., No. 598, p. 145-153.

Besançon, J., 1983. L'Euphrate et le Nord-Est de la Syrie: Modifications du milieu naturel au cours du Quaternaire. Ann. De géogr., Univ. Saint-Joseph, Beyrouth, p. 41-120.

Besançon, J., Copeland, L., Hours, F., and Sanlaville, P., 1978. The Paleolithic sequence in Quaternary formations of the Orontes River Valley, Nothern Syria: a Preliminary Report. Bull. London Inst. Archaeology, Vol. 15, p. 149-170.

Besançon, J., Copeland, L., Hours, F., Muchesen, S., and Sanlaville, P., 1982. Prospection géographique et préhistorique dans le bassin d'El-Komm (Syrie). Cahiers de l'Euphrate, Vol. 3, p. 9-26.

Besançon, J., and Sanlaville, F., 1991. Une oasis dans la Steppe Aride Syrienne. La cuvette d'El-Komm au Quaternaire. Contributions réunies par Jacques Cauvin. Editions Recherche sur les Civilisations. Cahiers de l'Euphrate, Paris, Vol. 5-6, p. 11-32.

Best, J. A., Barazangi, M., Al-Saad, D., Sawaf, T., and Gebran, A., 1990. Bouguer gravity trends and crustal structure of the Palmyre mountain belt and surrounding northern Arabian platform in Syria. Geology, 18(12), p. 1235-1239.

Best, J. A., Barazangi, M., Al-Saad, D., Sawaf, T., and Gebran, A., 1993. Continental margin evolution of the northern Arabian platform in Syria. Am. Assoc. Petrol. Geol. Bull., 77(2), p. 173-193.

Best, J., Barazangi, M., Chaimov, T. A., Al-Saad, D., and Sawaf, T., 1989. Crustal structure of the northern Arabian Platform in Syria as inferred from gravity observations. Geol. Soc. Am., Ann. Mtg. Abstr., 21(6), p. A81.

Best, J., Barazangi, M., McBride, J., Chaimov, T. A., Al-Saad, T., Al-Otri, M., Jajour, A., and Gebran, A., 1988. Anatomy of an intracratonic mountain belt from seismic profiling; Palmyride fold and thrust belt, Syria. Geol. Soc. Am., Ann. Mtg. Abstr., 20(7), p. 184.

Beydoun, Z. R., 1977. The Levantine countries: the geology of Syria and Lebanon (Maritime regions). In: The Ocean Basins and Margins, Vol. 4A, The Eastern Mediterranean, Nairn, A. E. M., Kanes, W. H. and Stehli, F. G. (Eds.), Plenum Publ. Corp., p. 319-353.

Beydoun, Z. R., 1988. The Middle East: regional geology and petroleum resources. Scientific Press Ltd, U. K., 292 p.

Blanckenhorn, M., 1889. Pteropodenreste aus der Oberen Kreide Nord Syriens und aus dem Hessischen Oligocaen. Deutsch. Geol. Ges. Z., Vol. 41, p. 593-602.

Blanckenhorn, M., 1890a. Die Entwicklung des Kreidesystems in Mittel und Nord Syrien, nebst einen Anhang ueber den Jurassischen Glandarienkalk.-Cassel, 135 p.

Blanckenhorn, M., 1890b. Das Eocen in Syrien, mit besonderer Beruecksichtigung Nord Syriens. Ein Beitrag zur Geologie Syriens. Deutsch. Geol. Ges. Z., Vol. 42, p. 318-359.

Blanckenhorn, M., 1890c. Das marine Miocaen in Syrien. Denksch. K. K. Acad. Wiss. Nath. Naturw. Kl., Wien, Vol. 57, p. 591-620.

Blanckenhorn, M., 1891. Grundzuege der Geologie und physicalischen Geographie von Nordsyrien. Berlin, Friedlaender, 102 p, 2 maps.

Blanckenhorn, M., 1897. Zur Kenntniss der Suesswasserablagerungen und Mollusken Syriens. Palaeontographica, No. 44, p. 71-144.

Blanckenhorn, M., 1901. Neues zur Geologie und Palaeontologie Aegyptens. III, Das Miozaen. Z. D. geol. Ges., Vol. 53.

Blanckenhorn, M., 1912. Kurzer Abriss der Geologie Palaestinas. Z. deutsch. Palaestina-Verein, Vol. 35, p. 113 139, 1 geol. Karte, Leipzig.

Blanckenhorn, M., 1914. Syrien, Arabien, Mesopotamien. Heidelberg, Winter's Universitaetsbuchhanglung, Handbuch der Regionalen Geologie, 5(4), 159 p.

Blanckenhorn, M., 1916. Das Danien in Palaestina mit der Leitform Pecten obrutus. Deutsch. Geol. Ges. Z., Vol. 67, p. 187-191.

Blanckenhorn, M., 1925. Die Seeigelfauna der Kreide Palaestinas. Palaeontographica, Vol. 67, p. 83 113, Stuttgart.

Blanckenhorn, M., 1927. Die fossilen Gastropoden und Scaphopoden der Kreide von Syrien Palaestina. Palaeontographica, Vol. 69, p. 111-186.

Blanckenhorn, M., 1931. Geologie Palaestinas nach heutiger Auffassung. Z. deutsch. Palaestina-Verein., Vol. 54, 50 p., Leipzig.

Blanckenhorn, M., 1934. Die Bivalven der Kreideformation von Syrien-Palaestina nebst einem Ergaenzendem Nachtrag ueber Brachiopoden, Gastropoden und Pteropoden und einem Ueberblick ueber die gesampe Molluskenfauna. Palaeontographica, 81-A(4-6), p. 161-302.

Blanckenhorn, M., 1935. Ueber die Faunen der Syrischen Kreide, des Eocaens and Oligocaens. Zentralb. F. Mineralogie, Abt. 8, No. 5, p. 148-161.

Blanckenhorn, M., 1940. Syrien und Palaestina. Geol. Jahresber., Bornitraeger Verlag., Berlin, Vol. 2, p. 351-361.

Blanckenhorn, M., and Oppenheim, P., 1927. Neue Beitraege zur Kenntnis des Neogens in Syrien und Palaestina. Geol. Palaeontol. Abh., N. F., 15(4), p. 321-357.

Blow, W. H., 1969. Late Middle Eocene to Recent planktonic foraminiferal biostratigraphy. Proc. 1th Intern. Conf. on planktonic microfossils, Vol. 1, Leiden, Brill, p. 209-421.

Bolli, H. M., 1957a. The genera Globigerina and Globorotalia in the Paleocene-Lower Eocene Lizard Springs Formation of Trinidad, B. W. I. U. S. Nat. Mus. Bull., Vol. 215, p. 61-81.

Bolli, H. M., 1957b. Planktonic foraminifera from the Eocene Navet and San Fernando Formations of Trinidad, B. W. I. U. S. Nat. Mus. Bull., Vol. 215, p. 155-172.

Bolli, H. M., 1957c. Planktonic foraminifera from the Oligocene-Miocene Cipero and Lengua Formations of Trinidad, B. W. I. U. S. Nat. Mus. Bull., Vol. 215, p. 97-123.

Bolli, H. M., Saunders, J. and Perch-Nielsen, K., (Eds.), 1985. Plankton stratigraphy. Cambridge, Cambridge Univ. Press, 1032 p.

Bordès, F., 1955. Le Paléolitique inférieur et moyen de Jabroud (Syrie) et la question du pré-Aurignacien. L'Antropologie, 59(5-6), p. 486-507.

Bordés, F., 1960. Le pré-Aurignacien de Jabroud (Syrie), et son incidence sur la chronologie du Quaternaire au Moyen Orient. Isr. Res. Counc., Bull., 9G(2-3), p. 91-103.

Bourguignon, L., 1996. Un Moustérien tardif sur le site d'Umm el Tiel (Bassin d'El Komm, Syrie). Exemples des niveaux II Base' et III2A': A tale about human diversity. Cultural change and human evolution: The Crisis at 40 Ka BP. In: E. A. M. V. Carbonell (Ed.). The last Neandertals. The first anatomically modern humans, Spain, Cataluna, Universitat Roviea I Virgil, p. 317-336.

Bourcart, J., 1940 (1941). Recherches stratigraphiques sur le Pliocéne et le Quaternaire du Levant. Bull. Soc. Géol. France, sér. 5, 10(7-9), p. 207-230

Brew, G. E., 1999. A tectonic model for Syria; hydrocarbon implications. Am. Assoc. Petrol. Geol. Bull., 83(8), p. 1302.

Brew, G. E., 2001. Tectonic evolution of Syria interpreted from integrated geophysical and geological analysis. Ph.D. Thesis, Chapters 1 5. Cornell University, Ithaca, NY (Unpublished).

Brew, G. E., Barazangi, M., and Sawaf, T., 1999. Intraplate tectonics of Syria; regional model and hydrocarbon implications. AAPG, Ann. Mtg. Expand. Abstr., p. A15.

Brew, G. E., Barazangi, M., Sawaf, T., and Al-Maleh, K., 2000. A new tectonic map of Syria and surrounding regions. GeoArabia, 5(1), p. 62.

Brew, G. E., Litak, R., Barazangi, M., Sawaf, T., and Zaza, T., 1998. Tectonic and geological evolution of NE Syria; implications for regional tectonics and hydrocarbons. AAPG, Ann. Mtg. Expand. Abstr. on CD.

Brew, G. E., Litak, R., Barazangi, M., Sawaf, T., Al-Imam, A., and Al-Youssef, W., 1997. Petroleum prvinces and structural evolution of eastern Syria. AAPG-SEPM, Ann. Mtg. Abstr., Vol. 6, p. 14.

Brew, G. E., Litak, R. K., Seber, D., Barazangi, M., Al-Imam, A., and Sawaf, T., 1997a. Basement depth and sedimentary velocity structure in the northern Arabian platform. Eastern Syria. Geophys. J. Int., 128(3), p. 617-631.

Brew, G. E., Litak, R. K., Seber, D., Barazangi, M., Sawaf, T., and Al-Imam, A., 1997b. Summary of the geological evolution of Syria through geophysical interpretation: implications for hydrocarbon exploration. The Leading Edge, 16(10), p. 1473-1482.

Brew, G. E., Lupa, J., Barazangi, M., Sawaf, T., Al-Imam, A., and Zaza, T., 2001. Structure and tectonic development of the Ghab Basin and the Dead Sea fault system, Syria. J. Geol. Soc. London, 158(4), p. 665-674.

Butzer, K. W., 1958. Quaternary stratigraphy and climate in the Near East. Bonn, 157 p.

Caron, C., Jamal, M., Zeinab, H., and Cerda, F., 2000. Basin development and tectonic history of the Euphrates graben (eastern Syria): a stratigraphic and seismic approach. In: S. Crasquin-Soleau and E. Barrier (Eds.), Peri-Tethys Memoir, Vol. 5, Mém. Mus. Nam. Hist. Nat., Vol. 182, p. 169-201.

Caron, M., 1985. Cretaceous planktic foraminifera. In: Plankton stratigraphy, H. Bolli, J. Saunders, and K. Perch-Nielsen (Eds.), Cambridge Univ. Press, Cambridge, p. 17-86.

Chaimov, A., Barazangi, M., Al-Saad, D., Sawaf, T., and Gebran, A., 1990. Crustal shortening in the Palmyride fold belt of Syria and implications for movement along the Dead Sea fault system. Tectonics, 9(6), p. 1369-1386.

Chaimov, T. A., Barazangi, M., Al-Saad, D., Sawaf, T., and Gebran, A., 1992. Mesozoic and Cenozoic deformation inferred from seismic stratigraphy in the southwestern intracontinental Palmyride fold-thrust belt, Syria. Geol. Soc. Am. Bull., 104(6), p. 704-715.

Chaimov, T. A., Barazangi, M., Al-Saad, D., Sawaf, T., and Khaddour, M., 1993. Seismic fabric and 3-D structure of the southwestern intracontinental Palmyride fold belt. Am. Assoc. Petr. Geol. Bull., 77(12), p. 2032-2047.

Chaimov, T. A., Barazangi, M., Best, J., Al-Saad, D., Sawaf, T., Youssef, A., and Gebran, A., 1990. Mesozoic and Cenozoic deformation inferred from seismic stratigraphy in the southwestern Palmyride fold belt of Syria. SEG Abstr., Vol. 60, p. 191-194.

Chenevart, C., 1950a. The isopach maps in oil geology, with reference to the Cretaceous of Syria, Southwestern Asia. Bull. Assoc. Suisse Géol. Ing. Pétrole, 17(52), p. 36-50.

Chenevart, C., 1950b. Problèmes de géologie du pétrole, à propos de l'usage des cartes à isopaques: quelques faits observés dans le Crétacé de Palmyrène, Syrie. Mém. Soc. Vaudoise de sci. Naturelles, 9(4), p. 243-272.

Chenevoy, M., 1952. Sur la présence d'une serie métamorphique au nord de Lattaquie, Syrie. C. R. Acad. Sci. Paris, 234(21), p. 2087-2088.

Chenevoy, M., 1959. Le substratum métamorphique des roches vertes dans le Baer et le Bassit, Syrie septentrionale. Notes et Mémoires sur le Moyen Orient, Vol. 7, p. 1-17.

Cizancourt, H., 1935. Matériaux pour la stratigraphie du Nummulitique dans le Désert de Syrie. Bull. Soc. Géol. France, ser. 5, Vol. 4, p. 737-750.

Clark, J. D., 1966. Acheulean occupation sites in the Middle East and Africa: A study in cultural variability. Am. Anthropol., 68(2), Pt. 2, p. 202-229.

Clark, J. D., 1967. The Middle Acheulean occupation site at Latamne, Northern Syria (1st paper). Quaternaria, Vol. 9, p. 1-68.

Clark, J. D., 1968. The Middle Asheulean occupation site at Latamne, Northern Syria. General results. Definition and interpretation (2nd paper). Quaternaria, Vol. 10, p. 1-71.

Clason, A. T., and Buitenhuis, H., 1978. A preliminary report on the faunal remains of Nahr el Homr, Hadidi and Ta'as in the Tabqa Dam region in Syria. J. Archaeol. Sci., 5(1), p. 75-83.

Contenson, H., 1966. Découvertes récentes dans le domaine du Néolithique en Syrie. L'Anthropologie, Vol. 70, p. 388-391, Paris.

Contenson, H., 1976. Précisions sur la stratigraphie de Tell Aswad, Syrie. Bull. Soc. Préhist. France, C. R., 73(7), p. 198-199.

Contenson, H., 1981. Le Néolithique de Damascène. C. N. R. S. Colloq. Intern., No. 598, p. 467-470.

Conway, B. H., 1990. Paleozoic-Mesozoic palinology in Israel. II: Palynostratigraphy of the Jurassic succession in the subsurface of Israel. Israel Geol. Surv., Bull, 82, 139 p.

Coon, C. S., 1957. The seven caves. Archaeological explorations in the Middle East. New York, A. F. Knopf, 338 p.

Copeland, L., 1979. Observations on the prehistory of the Balikh valley, Syria, from the 7th to the 4th millenium B. C. Paleorient, No. 5, p. 251-276.

Copeland, L., 1981. Chronology and distribution of the Middle Paleolithic as known in 1980 in Lebanon and Syria. C. N. R. S. Colloq. Intern., No. 598, p. 239-263.

Copeland, L. and Hours, F., 1978. La séquence Acheuléenne du Nahr el Kebir, région septentrionale du littoral Syrien. Paléorient, Vol. 4, p. 5-31.

Copeland, L. and Hours, F., 1981. La fin de l'Acheuléen et l'avènement du Paléolithique moyen en Syrie. C. N. R. S. Colloq. Intern., Vol. 598, p. 225-238.

David, E., 1933a. Note sur l'Oligocène et de Burdigalien de la Syrie septentrionale. C. R. Acad. Sci. France, Paris, 196(15), p. 1134-1136.

David, E., 1933b. Foraminifères Sénoniens et Eocènes de la Syrie septentrionale. Not. et Mém., Vol. 1, Paris, p. 53-60.

David, E., 1933c. Les grands Foraminifères Miocènes de la zone désertique Syrienne. Not. et Mém., Vol. 1, p. 121-139.

Dean, W. T., 1975. Cambrian and Ordovician correlation and trilobite distribution in Turkey. Fossils and Strata, Vol. 5, p. 353-373.

Dean, W. T., 1982. Middle Cambrian trilobites from the Sosink Formation, Derik-Mardin District, south-eastern Turkey. Bull. Brit. Mus. Nat. History, Geol., Vol. 36, p. 1-41.

Dean, W. T. and Krummenacher, R., 1961. Cambrian trilobites from the Amanos Mountains, Turkey. Paleontology, 4(1), p. 71-81.

Dean, W. T., Monod, O., and Perincek, D., 1981. Correlation of Cambrian and Ordovician rocks in southeast Turkey. Turkiye Cumhuriyet Petrol Isleri Genel Mueduegeluele, Vol. 25, p. 269-291.

Devyatkin, E. V. and Dodonov, A. E. 2000. Stratigraphy of the Neogene and Quaternary deposits. In: Outline of geology of Syria, trans. of the Geol. Inst. of the Russian Acad. Sci., Vol. 526, p. 129-175. (In Russian).

Devyatkin, E. V., Dodonov, A. E., Gablina, S. S., Golovina, V. G., Kurenkova, V. G., Simakova, A. N., Trubikhin, V. M., Yasamanov, N. A., Khatib, K., and Nseir, Kh., 1996. Marine Upper Pliocene-Lower Pleistocene of Western Syria: Stratigraphy and paleogeography. Stratigraphy. Geological correlation, 4(1), p. 71-81. (In Russian).

Devyatkin, E. V., Dodonov, A. E., Sharkov, E. V., Zykin, V. S., Simakova, V. G., Khatib, K., and Nseir, Kh., 1997. The rift depression El-Rhab (Syria): structure, stratigraphy, history of development. Stratigraphy. Geological correlation, 5(4), p. 55-67. (In Russian).

Diner, C., 1886. Libanon, grundlinien physischen Geographie und Geologie von Mittel Syrien. Wien, A. Hoelder, 412 p., map on scale 1:500,000.

Diner, C., 1887. Ein Beitrag zut Kenntnis der Syrischen Kreidebildungen. Deutsch. Geol. Ges. Z., Vol. 39, p. 314-342.

Dodonov, A. E., Deviatkin, E. V., Ranov, A. E., Khatib, K., and Nseir, H., 1993. The Latamne Formation in the Orontes River Valley. Le Paleolithique de la vallee moyenne de l'Oronte (Syrie), BAR Intern. Ser., No. 587, p. 189-194.

Dobrova, M. R., 1995. Various types of Ostracoda assemblages in Syrian Jurassic basins and their connection with the environment. Questions of Micropaleontology, No. 31, p. 39-46, Moscow. (In Russian).

Dolginov, E. A., Kazmin, V. G., and Ponikarov, V. P., 1969. Structure of the Upper Precambrian sections and Precambrian history of development of the Arabian Platform and its marginal folded zones. Bull. Moscow Soc. of Testers of Nature, Ser. Geol., XLIV(6), p. 61-76. (In Russian).

Doncieux, L., 1937. Les foraminifères Eocènes de la Syrie septentrionale. Not. et Mém. Syrie-Liban, Vol. 2, p. 207-226.

Doncieux, L., 1938. Au sujet de l'Oligocène et du Burdigalien en Syrie. C. R. Acad. Sci. France, Paris, 207(26), p. 1422-1424.

Doncieux, L., 1940. Le Sénonien dans les régions d'Antioche et de Lattaquie (Levant). C. R. Acad. Sci. France, Paris, 210(22), p. 737-739.

Doncieux, L., Dubertret, L., and Dubertret, R., 1938. Sur le Nummulitique du Kosseir au sud d'Antioche (Syrie). C. R. Acad. Sci. France, Paris, Vol. 206, p. 1224-1226.

Doncieux, L., Dubertret, L., Dubertret, R., and Vautrin, H., 1938. Sur le Nummulitique du versant oriental de l'Anti-Liban (region de Damas, Syrie). C. R. Acad. Sci. France, Paris, Vol. 207, p. 1230-1232.

Doncieux, L., Dubertret, L., and Vautrin, H., 1936. L'Oligocène et le Burdigalien du désert de la Syrie.-C. R. Acad. Sci. France, Paris, Vol. 202, p. 2165-2167.

Doncieux, L. and Roman, F., 1927. Observations sur l'Eocene de Syrie et de Palestine. C. R. Soc. Géol. France, Vol. 13, p. 171-173.

Doss, B.1886. Ueber die basaltischen Laven und Tuffe der Provinz Hauran und vom Diret et Tullul in Syrien. Tschermak's Miner. Petrogr. Mitt., N. F., Vol. 7, p. 461-535, Wien.

Douvillé, H., 1910. Etudes sur les rudistes. Rudistes de Sicile, d'Algérie, du Liban et de la Perse. Mém. Soc. geol. France, Paleont., Vol. 41, p. 5-83, Paris.

Dubertret, L., 1929. Etude des régions volcanique du Hauran, du Djebel Druze et Diret et Touloul, Syrie. Rev. Géogr. Phys. Géol. Dyn., 2(4), p. 275-321, 1 map 1:500.000.

Dubertret, L., 1930. Note préliminaire sur la structure géologique des Etats du Levant sous Mandat français. C. R. Somm. Soc. Géol. France, 30(6), p. 43-45.

Dubertret, L., 1932a. Les formes structurales de la Syrie et de la Palestine; leur origine. C. R. Acad. Sci. France, Vol. 195, p. 66-68.

Dubertret, L., 1932b. L' évolution structurale des Etats du Levant sous Mandat Français. C. R. Acad. Sci. France, Vol. 194, p. 1964-1966.

Dubertret, L., 1933a. La carte géologique au millionieme de la Syrie et du Liban. Rev. Géogr. Phys. et Géol. Dyn., 6(4), p. 269-318, 1 geol. Map.

Dubertret, L., 1933b. La tectonique de la Syrie Septentrionale à la fin du Crétacé et au début du Tertiaire. Not. et Mém. Syrie-Liban, Vol. 1, p. 13-28.

Dubertret, L., 1933c. Le Miocène en Syrie et au Liban. Introduction. Not. et Mém. Syrie-Liban, Vol. 1, p. 63-73.

Dubertret, L., 1933d. Le Djebel Bishri. Not. et Mém. Syrie-Liban, Vol. 1, p. 75-99.

Dubertret, L., 1933 e. Les grandes nappes basaltiques syriennes: age et relations avec la tectonique. C. R. Soc. Géol. France, Paris, Vol. 11, p. 178-180.

Dubertret, L., 1934a. Les gisements d'hydrocarbures minérauux de la Syrie et du Liban. C. R. Acad. Sci. France, Paris, 198(4), p. 379-381.

Dubertret, L., 1934b. Présentation d'une carte géologique au millionieme de la Syrie et du Liban, suivie d'une étude de l'hydrologie et de l'hydrographie. Rev. Géogr. Phys. Géol. Dyn., 6(4), 160 p., maps.

Dubertret, L., 1935. Premieres recherches sur les hydrocarbures minéraux dans les états du Levant sous Mandat Français. Ann. Office nat. combustibles liquides, p. 33-40.

Dubertret, L., 1936. Stratigraphie des régions recouvertes par les roches vertes du nord ouest de la Syrie. C. R. Acad. Sci. France, Paris, 203(22), p. 1173-1175.

Dubertret, L., 1937a. Le Cretace. 1. Le Massif Alaouite ou Djebel Ansariyeh. Not. et Mém., Vol. 2, p. 9-42

Dubertret, L., 1937b. Données diverses sur le Pliocène et le Quaternaire marins de la Syrie et du Liban. Not. et Mém., Vol. 2, p. 111-121.

Dubertret, L., 1937c. Sur la constitution et la genèse des roches vertes Syriennes. C. R. Acad. Sci. France, Paris, 204(22), p. 1663-1665.

Dubertret, L., 1937d. Sur les lambeaux de breche tectonique a la surface des roches vertes syriennes. C. R. Acad. Sci. France, Paris, 204(4), p. 283-285.

Dubertret, L., 1937 e. L'Eocène. 1. L'Eocène du nord ouest de la Syrie. 2. L'Eocène de la Syrie méridionale et de la Palestine. Not. et Mém., Vol. 2, p. 75-92.

Dubertret, L., 1937 f. La coupe type du Pliocène et du Quaternaire dans la région de Lattaquie: Extension et morphologie du Néogène et du Quaternaire pres de Lattaquie. Notes et Mémoires, Vol. 2, p. 94-97.

Dubertret, L., 1938. Au sujet de l'Oligocène et du Burdigalien en Syrie. C. R. Acad. Sci. France, Paris, 207(26), p. 1422-1424.

Dubertret, L., 1939. Sur la genèse et l'âge des roches vertes syriennes. C. R. Acad. Sci. France, Paris, 209(21), p. 763-765.

Dubertret, L., 1940a. Sur l'âge du volcanisme en Syrie au Liban. C. R. Soc. Géol. France, Vol. 6, p. 55-57.

Dubertret, L., 1940b. Le Sénonian dans les régions d'Antioche et de Lattaquie (Levant). C. R. Acad. Sci. France, Paris, 210(22), p. 737-739.

Dubertret, L., 1942. Carte géologique du Moyen Orient au 2 millionieme. 67 p. Beyrouth.

Dubertret, L., 1943. Carte géologique de la Syrie et du Liban au millionieme. 2nd Ed. Beirouth, 1941-1943, Délégation Génerale au Levant de la France Combattante, Sec. Géologique, 67 p., 1 map.

Dubertret, L., 1944. Carte lithologique de la Syrie et du Liban (1:1,000,000). Beirouth.

Dubertret, L., 1945. Carte géologique au 50,000e, feuille de Rachaya-N. Beyrouth.

Dubertret, L., 1947a. Problèmes de la géologie du Levant. Bull. Soc. Géol. France, ser. 5, Vol. 17, p. 3-31.

Dubertret, L., 1947b. Sut la limite nord du plateau Syrien. C. R. Soc. Géol. France, No. 9, p. 107-109.

Dubertret, L., 1949a. Carte géologique au 50,000e, Feuille de Zabdani. Damas, Min. des Travaux Publics, 62 p., 1 map.

Dubertret, L., 1949b. Carte géologique au 50,000e, Feuille de Saida. Beyrouth, 46 p., 1 map.

Dubertret, L., 1950a. Carte géologique au 50,000e, Feuille de Rayak. Damas, Min. des Travaux Publics, 48 p., 1 map.

Dubertret, L., 1950b. Carte géologique au 50,000e, Feuille de Baalbek. Beyrouth, 52 p., 1 map.

Dubertret, L., 1950c. Sur le Jurassique du Liban et de l'Anti-Liban. C. R. Soc. Géol. France, No. 1, 12, 13 p.

Dubertret, L., 1951a. Carte géologique au 50,000e, Feuille de Beyrouth. Beyrouth, 66 p., 1 map.

Dubertret, L., 1951b. Carte géologique au 50,000e. Feuille de Tripoli. Beyrouth, 62 p., 1 map.

Dubertret, L., 1951c. Aperçu sur la géologie du Kurd Dagh (Syrie). C. R. Soc. Géol. France, No. 5-6, 70 71 p.

Dubertret, L., 1952. Basaltes et roches vertes du Liban, de Syrie et du Hatay (ancien Sandjak d'Alexandrette, Turquie). 19 th Int. Geol. Congr., Algeria, C R., Sec. 15, Vol. 17, p. 29-36.

Dubertret, L., 1953a. Carte géologique au 200,000, Feuille d'Antioche. Damas, Min. des Travaux Publ., 67 p. 1 map.

Dubertret, L., 1953b. Carte géologique au 50,000e, Feuille de Zahle. Beyrouth, 64 p., 1 map.

Dubertret, L., 1954. Carte géologique au 200,000, Feuille de Beyrouth. 108 p., Paris.

Dubertret, L., 1955a. Carte géologique du Liban au 200,000. Beyrouth, 74 p., 1 map.

Dubertret, L., 1955b. Géologie des roches vertes du nord-ouest de la Syrie et du Hatay (Turquie). Not. et Mém. sut le Moyen Orient, Vol. 6, Paris, 179 p.

Dubertret, L., 1958. Feuille de Tartous, 1:50,000, Notice explicative. Beyrouth, Min. des Travaux Publ., 12 p., 1 map.

Dubertret, L., 1960. Carte géologique au 1:50,000e, Feuille Hermon (Rachaya Sud). Beyrouth, Min. des Travaux Publ., 46 p. 1 map.

Dubertret, L., 1963. Liban, Syrie: chaine des grands massifs côtiers et confins à l'Est. Lexique stratigraphique internat., Vol. 3, Asie, 10c (Liban, Syrie, Jordanie), 155 p., 1 geol. Karte 1:1,000,000, Paris.

Dubertret, L., 1966. Liban, Syrie et bordure des pays voisins. 1e partie: Tableau stratigraphique avec carte géologique au millionieme. Not. et Mem. sur le Moyen Orient, Vol. 8, Paris, p. 251-358, 1 map.

Dubertret, L., 1975. Introduction a la carte géologique au 1:50,000 du Liban. Not. et Mém. sur le Moyen Orient, Vol. 13, Paris, p. 345-402.

Dubertret, L., and Cottreau, J., 1938. Les Echinides miocènes des régions d'Antioche et d'Alep. Leur signification stratigraphique. C. R. Soc. Géol. France, p. 57-58.

Dubertret, L., Keller, A., and Vautrin, H., 1932a. Contribution à l'étude de la région désertique syrienne. C. R. Acad. Sci. France, Paris, Vol. 194, p. 1000-1002.

Dubertret, L., Keller, A., and Vautrin, H., 1932b. Contribution à la géologie de la Djezireh (territoires syriens de la rive gauche de l'Euphrate). C. R. Acad. Sci. France, Parus, Vol. 194, p. 1254-1256.

Dubertret, L., and Roger, J., 1938. Les pectinides néogènes des régions d'Antioche et d'Alep (Syrie) C. R. Soc. Géol. France, Paris, sér. 5, No. 8, p. 73-74.

Dubertret, L., and Vautrin, H., 1937a. Contribution à l' étude géologique de la côte Libano Syrienne: Revision de la stratigraphie du Crétacé du Liban. Not. et Mém. Syrie-Liban, Vol. 2, p. 43-73.

Dubertret, L. and Vautrin, H., 1937b. Sur la présence du Jurassique marin dans la region plissée palmyrénienne. C. R. Soc. Géol. France, No. 10-11, p. 135-136.

Dubertret, L. and Vautrin, H., 1937c. La coupe du sommet du Crétacé et la base de l'Eocène au Djebel Tenf, dans la désert de Syrie. C. R. Soc. Géol. France, Paris, No. 11, p. 160-162.

Dubertret, L., and Vautrin, H., 1938. Sur l'existence du Pontien lacustre en Syrie et sur sa signification tectonique. C. R. Acad. Sci. France, Paris, 206(1), p. 69-71.

Dubertret, L., and Wetzel, R., 1949. Carte géologique au 200,000, Feuille de Tripoli. Damas, Min. de Trav. Publ., 62 p., 1 map.

Dubertret, R., Doncieux, L,. Vautrin, H., and Dubertret, L., 1938. Sur le Nummulitique du versant oriental de l'Anti Liban (région de Damas, Syrie). C. R. Acad. Sci. France, Paris, 207(24), p. 1230-1232.

Dubertret, L., Vautrin, H. and Keller, A., 1937a. La stratigraphie du Pliocène et du Quaternaire marins de la côte Syrienne. Not. et Mém. Syrie et Liban, Vol. 2.

Dubertret, L., Vautrin, H. and Keller, A., 1937b. Stratigraphie du Pliocène et du Quaternaire de la région de Lattaquie. Not. et Mém. Syrie et Liban, Vol. 2, p. 98-110.

Dunnington, H. V. et al., 1959. Lexique stratigraphique international, Vol. III, Fasc. 10 a, Asie, Iraq. CNRS, Paris.

Endo, K., 1973. Sedimentological analysis of the deposits in the Douara cave site. Tokyo Univ., Univ. Mus. Bull., No. 5, p. 89-111.

Endo, K., 1978. Stratigraphy and paleoenvironments of the deposits in and around Douara cave site. Tokyo Univ., Univ. Mus. Bull., No. 14, p. 53-81.

Faradzhev, V. A., 1966. The Geological map of Syria. Scale 1:200,000. Sheet I-37-XIV (Al Qaryatein). Explanatory Notes. Technoexport, USSR, 146 p.

Farrand, W. R., 1970. Geology, climate and chronology of Yabroud Rock shelter I. Monographien zur Urgeschichte, Reihe A, Vol. 2, Koeln-Boehlau, p. 212-233.

Fedosova, S. P., Sharkov, E. V., Sindeev, A. S., and Hanna, S., 1987. Megacrystals of hornblende from the Tell-Danun volcano (Syria) and their origin. Rep. of the USSR Acad. Sci., Petrography, 297(2), p. 445-448. (In Russian).

Fluegel, H., 1963. Graptoliten aus dem Mittleren Ordovicium von Nord Syrien. Neues. Jahrb. Geol. Palaeontol., Abh., 118(1), p. 21-26.

Fourcade, E. and Mouty, M., 1995. Découverte d'un nouveau Grand Foraminifère du Jurassique de la marge sud téthysienne: *Syriana khouryi* n. gen. n. sp. Rev. de Micropaléontologie, 38(3), p. 217-227.

Fraas, O., 1867. Aus dem Orient: Geologische Beobachtungen am Nil, auf der Sinai Halbinsel und Syrien.-Stuttgart.

Fraas, O., 1877. Juraschichten am Hernon. Neues Jahrbuch fuer Mineralogie, Geologie. und Palaeontologie, Vol. 1, p. 17-30.

Fraas, O., 1878. Aus dem Orient. 2. Geologische Beobachtungen am Libanon. Jahrb.

Verein. Natur., Wuerttemberg, p. 257-392.

Fujimoto, T., 1979. Upper Paleolithic and Epi-Paleolithic assemblages in the Palmyra Basin: Site 50 and Site 74. Tokyo Univ., Univ. Mus. Bull., No. 16.

Galaktionov, A. B., Demidov, V. A., and Katkova, N. S., 1966. The Geological map of Syria. Scale 1:200,000. Sheets J-37-VI, XII (Al-Qamishli, Sinjar), and J-38-I, VII (Qarashouk-Dagh). Explanatory Notes. Technoexport, USSR, 84 p.

Giannerini, G., Campredon, R., Feraud, D., and Abou Zakhem, B., 1988. Déformations intraplaques et volcanisme associé: Exemple de la bordure NW de plaque Arabique au Cénozoique. Bull. Soc. Géol. France, No. 6, p. 938-947.

Golovin, D. I., Bouyakaite, M. I., Vinogradov, V. I., and Krasheninnikov, V. A., 1986. K-Ar and Rb-Sr dating of glauconite from sediments of the Acarinina pentacamerata Zone (Lower Eocene) of the Eastern Mediterranean. Trans. of the USSR Acad. Sci., Ser. Geol., No. 7, p. 75-81. (In Russian).

Golovina, L. A., Gablina, S. S., and Simakova, A. N., 1996. Paleogeographic interpretation of a paleofloristic assemblage of the Pliocene-Pleistocene of Western Syria. Stratigraphy. Geological Correlation, 4(2), p. 102-105. (In Russian).

Gorbachik, T. N. and Kuznetsova, K. I., 1994. Comparison of Tithonian foraminifera of the Crimea and Syria. Stratigraphy. Geological Correlation, 2(2), p. 51-63. (In Russian).

Grossowicz, L., Bassoulet, J.-P., Hirsch, F., and Peri, M. 2000. Jurassic large Foraminifera from Israel. GSI Current Research, Vol. 12, p. 132-144.

Guevenc., 1985. Permian of Turkey. Sixth Colloq. On Geology of the Aegean Region. Piri Reis Internat. Contribution, Publ. No. 2, p. 263-282.

Guyot, C., and Zeinab, H., 2000. Euphrates Graben, East Syria; basin evolution and petroleum habitat. GeoArabia, 5(1), p. 100-101.

Haas, O., 1955. Revision of the Jurassic Ammonite fauna of Mount Hermon, Syria. Bull. Am. Museum of Natural History, Vol. 108, 210 p.

Hanihara, K. and T. Akazawa, (Eds.), 1979. Paleolithic site of Douara cave and paleogeography of Palmyra Basin in Syria. Part II: Prehistoric occurrences and chronology in Palmyra Basin. Tokyo Univ., Univ. Mus. Bull., No. 16, 286 p.

Hanihara, K. and T. Akazawa, (Eds.), 1983. Paleolithic site of Douara cave and paleogeography of Palmyra Basin in Syria. Part III: Animal bones and further analysis of archaeological material. Tokyo Univ., Univ. Mus. Bull., No. 21, 167 p.

Hanihara, K. and Y. Sakaguchi, (Eds.), 1978. Paleolithic site of the Douara cave and paleogeography of Palmyra Basin in Syria. Part I. Tokyo Univ., Univ. Mus. Bull., No. 14, 128 p.

Haq, B. U., Hardenbol J., and Vail, P. R., 1988. Mesozoic and Cenozoic chronostratigraphy and cycles of sea-level changes. Spec. Publication of Society of Economic Paleontologists and Mineralogists, Vol. 42, p. 72-108.

Heinzelin, J. de., 1966. Observations sur les terrasses du moyen Euphrate. Bull. Soc. Géol. France, 7(1), p. 37-44.

Heinzelin, J. de., 1966. Geological observations near Latamne. Annales Archeologiques Arabes Syriennes, XVI(2), Damas, p. 115-120.

Heinzelin, J. de., 1968. Lithology of the Acheulean living floor of Latamne, Northern Syria. Quaternaria, Vol. 10, p. 48-52.

Hirsch, F., Bassoullet, J. P., Cariou, E., Conway, B. H., Feldman, H. R., Grossowicz, L., Honigstein, A., Owen, E. F., and Rosenfeld, A., 1998. The Jurassic of the Southern Levant: biostratigraphy, paleogeography and cyclic events. Mém. Mus. Natl. Hist. Nat., Vol. 179, p. 213-235.

Hooijer, D. A., 1961(1962?). Middle Pleistocene mammals from Latamne, Orontes valley, Syria. Annales Archeologiques Arabes Syriennes, Vol. 11, p. 117 132.

Hooijer, D. A., 1966. The *Dicerorhinus hemitoechus* (Falconer) at Jabroud. Annales Archeologiques Arabes Syriennes, 16(2), p. 155-156.

Hooijer, D. A., 1968. A Cretaceous dinosaur from the Syrian Arab Republic. Proc. Koninkl. Neder. Acad. Wet., 71(2), p. 150-152.

Hooijer, D. A., 1975. The identity of the bronze age Elephant of Ras Shamra, Ugarit. Neth. J. Zool., 25(3), p. 371-372.

Hours, F., 1975. The Lower Paleolithic of Lebanon and Syria. Problems in prehistory: North Africa and Levant. In: Wendorf, F. & Marks, A. (Eds.). Southern Methodist Univ. Press., Dallas, p. 249-271.

Hours, F., 1979. Le fin de l'Acheuléen en Syrie du nord, note préliminaire. Paléorient, No. 5, p. 9-16.

Hours, F., 1981. Le Paleolithique inférieur de la Syrie et du Liban le point de la question en 1980. C. N. R. S. Colloq. Intern., No. 598, p. 165-183.

Husri, S. and R. L. Austin, 1985. Carboniferous and Triassic conodonts from Syrian boreholes. J. Micropaleontol, 4,(2), p. 93-99.

Jaquet, F., 1933. Une faune du Miocène moyen dans la vallée du Nahr El Kebir (nord de Lattaquie, Syrie). C. R. Soc. Géol. France, No. 3, p. 67-68.

Julig, P. J., D. G. F. Long, H. B. Schroeder, W. J. Rink, D. Richter, and H. P. Schwarcz, 1999. Geoarchaeology and new research at the Jerf al-Ajla Cave, Syria. Geoarchaeology, 14(8), p. 821-848.

Kaiser, K., 1965. Quartaer Meeresstraende und Terrasses der Kuestenflusse an der Syrisch Libanesischen Mittelmeerkueste. INQUA, Rep. 6th Congr., Warsaw, Vol. 1, p. 185-200.

Kaiser, K., 1973. Die Quartaerbildungen im Damascus Becken und seinen Rahmenbereichen. Z. Geomorph., 17(3), p. 271-303.

Kaizer, H., 1976. Die Permische Microflora der Cathaysia-Schichten von Nordwest Schansi, China. Palaeontographica, Band 159, p. 83-157.

Kazmin, V. G., 1965. Peculiarities of junction of the ancient platform and the Alpine geosy nclinal region in the north-eastern marginal part of the Mediterranean Sea realm. Bull. Moscow Soc. of Testers of the Nature, Ser. Geol., 40(6), p. 43-56. (In Russian).

Kazmin, V. G. and V. V. Kulakov, 1968. The Geological map of Syria. Scale 1: 50,000. Sheets I-37-XXIV-4d, I-37-XIX-3c and I-36-XXIV-4b (the Bassit Massif area). Explanatory Notes. Technoexport, USSR, 124 p.

Keller, A., 1933. Sut quelques Rudistes du Djebel Ansaryeh et de l'Amanus (Syrie Septentrionale). Not. et Mém. Syrie-Liban, Vol. 1 p. 45-52.

Keller, A., and Vautrin, H., 1937. Nouvelle contribution a l'étude des Echinides de la Syrie et du Liban. 1. Crétacé. Not. et Mém., Vol. 2, p. 137-164.

Kemper, E., and Zimmerle, W., 1983. Facies patterns of a Cretaceous/Tertiary subtropical upwelling system (Great Syrian Desert) and an Aptian/Albian boreal upwelling system (NW Germany). In: Coastal upwelling, its sedimentary record; Part B, Sedimentary records of ancient coastal upwelling, Thiede, J. and Suess, E. (Eds.), NATO conference series IV, Marine science, 10 B, Plenum Press, New York, p. 501-533.

Khair, K., Khawlie, M., Barazangi, M., Chaimov, T. A., and Seber, D., 1992. Bouguer gravity and crustal structure of the Dead Sea fault zone and adjacent mountain belts in Lebanon. 29th Int. Geol. Cong., Abstr., Kyoto, Japan, Vol. 29, p. 443.

Khair, K., Khawlie, M., Haddad, F., Barazangi, M., Seber, D., and Chaimov, T. A., 1993. Bouguer gravity and crustal structure of the Dead Sea transform fault and adjacent mountain belt, Syria. Geology, 21(8), p. 739-742.

Khatib, K., 1980. Geological map of Syria. Scale 1:50,000. Explanatory Notes of of Halba-Hamedyieh sheet (N I-37-M-3-a and No. I-36-R-4-b). General Establ. of Geol. and Mineral Resources, Damascus, 18 p.

Khoja, I., 1972. Palynological investigation on Upper Triassic (Kurachine Dolomite) deposits of the northeastern part of Syria. Jahrb. Geol. Bundesanst (Oesterreich), Sonderband No. 19, p. 32-59.

Khoja, I., 1985. The use of palynology in Permo-Carbon stratigraphy of the Syrian oil wells. Izdar, E. and Nakoman, E. (Eds.). Sixth Colloquium on Geology of the Aegean Region, Piri Reis International Contribution, second edition, publ. No. 2, p. 375-384.

Knipper, A. L., Saveliev, A. A., and Rukie, M., 1988. Ophiolite association of the north-western Syria. Geotectonika, No. 1, p. 92-104. (In Russian).

Knipper, A. L. and A. Ya. Sharaskin, 1994. Tectonic evolution of the western part of the Peri-Arabian ophiolite arc. In: V. A. Krasheninnikov and J. K. Hall (Eds.), Geological structure of the North-Eastern Mediterranean (Cruise 5 of the Research Vessel "Akademik Nikolay Strakhov"), Historical Productions-Hall, Jerusalem, p. 295-305.

Kober, L., 1915. Geologische Forschungen in Vorderasien. 1. Zur Tectonik des Libanons. Denkschr. Akad. Wiss Wien, math.-naturwiss Kl., Vol. 91, p. 379-427, Wien.

Kobori, I., 1973. Historical background (Palaeolithic site at Douara cave: lacustrine evidence from the interior basins; alluvial and colluvial phenomena; evidence from cave sediments and associated faunas). Tokyo Univ., Univ. Mus. Bull., No. 5, p. 29-43.

Kopp, M. L. and Leonov, Yu. G. 2000. Tectonics. In: Outline of geology of Syria, Trans. of the Geol. Inst of the Russian Acad. Sci., Vol. 526, p. 7-104. (In Russian).

Kotova, I. Z., 1990. Palynological assemblages of the Lower Cretaceous of Syria. In: Paleofloristics and stratigraphy of the Phanerozoic, No. 2, Geolog. Inst., Moscow, p. 85-101. (In Russian).

Kozlov, V. V. (assisted by A. V. Artyemov, and A. F Kalis), 1966a. The Geological map of Syria. Scale 1:200,000. Sheets I-36-XVIII, I-37-XIII (Trablus, Homs). Explanatory Notes. Technoexport, USSR, 68 p.

Kozlov, V. V., 1966b. The Geological map of Syria. Scale 1:200,000. Sheets I-37-XVI (Wadi El-Miyah), I-37-X (Abar Sijri). Explanatory Notes. Technoexport, USSR, 48 p.

Kozlov, V. V., V. P. Ponikarov, A. V. Razvalyaev, E. D. Soulidi-Kondratiev, and V. A. Faradzhev, 1965. Cretaceous deposits of Syria. Bull. Moskow Soc. of Testers of Nature, Sec. of Geology, XL(3), p. 57-68.

Krasheninnikov, V. A., 1964. Significance of Foraminifera of the open tropical basins of the Danian and Paleogene time for elaboration of the International stratigraphic scale. Question of Micropaleontology, No. 8, p. 190-213. (In Russian).

Krasheninnikov, V. A., 1965a. Zonal stratigraphy of the Paleogene deposits. XXII Intern. Geol. Congr., Reports of the USSR geol., Probl. 16 Zh, p. 37-61. (In Russian).

Krasheninnikov, V. A., 1965b. Zonal Paleogene stratigraphy of the Eastern Mediterranean. Trans. of Geol. Inst. of the USSR Acad. Sci., Vol. 133, 75 p. (In Russian).

Krasheninnikov, V. A., 1966. Foraminifera and some questions of biostratigraphy of marine Miocene deposits of the Eastern Mediterranean. Questions of Micropaleontology, No. 10, p. 398-419. (In Russian).

Krasheninnikov, V. A., 1968. Correlation of the Miocene deposits of the Eastern Mediterranean to the stratotypical sections of the Miocene stages. Giornale di Geologia, Vol. XXXV, fasc. III, p. 167-178.

Krasheninnikov, V. A., 1969a. Geographical and stratigraphical distribution of planktonic foraminifera in Paleogene deposits of tropical and subtropical areas. Trans. of Geol. Inst. of the USSR Acad. Sci., Vol. 202, 187 p. (In Russian).

Krasheninnikov, V. A., 1969b. Stage scale of the Miocene of open marine basins of the tropical and subtropical belts. Questions of Micropaleontology, No. 11, p. 132-156. (In Russian).

Krasheninnikov, V. A., 1971. Stratigraphy of the Miocene deposits of the Mediterranean on the base of foraminifera. Trans. of the Geol. Inst. of the USSR Acad. Sci., Vol. 220, 236 p. (In Russian).

Krasheninnikov, V. A., 1978. The main stratigraphic features of Paleogene deposits of the Eastern Mediterranean. V Iraqi Geol. Congr., Baghdad, p. 60-61.

Krasheninnikov, V. A., 1994. Stratigraphy of the Maastrichtian and Cenozoic deposits of the coastal part of north-western Syria (neoautochthon of the Bassit ophiolite Massif). In: V. A. Krasheninnikov and J. K. Hall (Eds.), Geological structure of the Northeastern Mediterranean (Cruise 5 of the Research Vessel "Akademik Nikolay Strakhov"), Historical Productions-Hall, Jerusalem, p. 265-276.

Krasheninnikov, V. A., 1995. Stratigraphy and paleogeographic links of the Mediterranean and Mesopotamian Neogene Basins (Syria). II Intern. Symposium on the geology of the Eastern Mediterranean region, Jerusalem, p.12.

Krasheninnikov, V. A. and Ashourov, A. A., 1995. Relation of pelagic and shallow-water (nummulitic) facies in the Paleogene of Syria and distribution of chert-phosphate-oil shale rocks. II Intern. Symposium on the geology of the Eastern Mediterranean region, Jerusalem, p.12.

Krasheninnikov, V. A., Basov, I. A., Golovina, L. A., Dmitrenko, O. B., Kazarina, G. Kh., Kruglikova, S. B., Mukhina, V. V., and Ushakova, M. G., 1999. Northeastern Atlantic and Eastern Mediterranean Miocene (biostratigraphy, paleoecology, paleogeography). Scientific World, Moscow, 185 p., 16 phototables. (In Russian).

Krasheninnikov, V. A., and J. K. Hall, (Eds.), 1994. Geological structure of the North-Eastern Mediterranean (Cruise 5 of the Research Vessel "Akademik Nikolay Strakhov"), Historical Productions-Hall, Jerusalem, 396 p.

Krasheninnikov, V. A., D. I. Golovin, V. I. Mouravyov, R. Helou, H. Nseir, and A. Shweki, 1996. The Paleogene of Syria – Stratigraphy, Lithology, Geochronology. Geol. Jahrbuch, Reihe B (Regionale Geologie Ausland), Heft 86, Hannover, 136 p.

Krasheninnikov, V. A., V. I. Mouravyov, D. I. Golovin, R. Helou, and A. Shweki, 1987. Stratigraphy, lithology and glauconite horizons of the Paleogene deposits of the Palmyrides, Syria. J. Problems of Micropaleontology, 29, p. 92-113, Moscow. (In Russian).

Krasheninnikov, V. A. and Nemkov, G. I., 1975. Relation of faunas of planktonic foraminifera and nummulites in Syrian Paleogene sediments. Questions of Micropaleontology, Vol. 18, p. 179-211. (In Russian).

Krasheninnikov, V. A., Razvalyaev, A. V., and Ponikarov, V. P., 1964. Geological development of Syria in the Paleogene period. Bull. Moscow Soc. of Testers of Nature, Ser. Geol, 39(5), p. 75-95. (In Russian).

Krasnov, A. A., Kazmin, V. G., and Kulakov, V. V., 1966. The Geological map of Syria. Scale 1:200,000. Sheets I-36-VI (Fiq-Haifa), I-37-I (As-Suweida), I-37-II (Az-Zalaf). Explanatory Notes. Technoexport, USSR, 75 p.

Krenkel, E., 1924. Der Syrische Bogen. Central Bl. Mineral., No. 9, p. 274 281; No. 10, p. 301-313.

Krenkel, E., 1941. Syrabien (Arabien und Syrien). Deutsch, Geol. Ges. Z., 93(7-8), p. 357-376.

Kurbanov, N. and Yu. Zarjanov, 1968. The Geological map of Syria. Scale 1:50,000. Sheets I-37-VII-3 a (Zebdani) and I-37-VII-3 c (Rayak). Explanatory Notes. Technoexport, USSR, 120 p.

Kuznetsova, K. I. and M. R. Dobrova, 1995. Endemic and cosmopotitic assemblages of foraminifera and ostracoda in the Middle Jurassic basins of Syria. Stratigraphy. Geological Correlation, 3(2), p. 38-52. (In Russian).

Kuznetsova, K. I. and M. R. Dobrova, 1997. The Middle Jurassic basins of the Eastern Mediterranean and their microbiota (foraminifera, ostracoda). Stratigraphy. Geological Correlation, 5(1), p. 32-41. (In Russian).

Kuznetsova, K. I. and M. R. Dobrova, 2000. Stratigraphy of the Jurassic sediments. In: Outline of geology of Syria, trans. of the Geol. Inst. of the Russian Acad. Sci., Vol. 526, p. 105-128. (In Russian).

Kuznetsova, K. I., A. A. Grigelis, J. Adjamian, and L. Hallaq, 1991. Zonal subdivision of Jurassic deposits of Syria by means of foraminifera. Trans. of the USSR Acad. Sci., Ser. Geol., No. 10, p. 125-127. (In Russian).

Kuznetsova, K. I., A. A., Grigelis, J., Adjamian, E., Jarmakani, and L. Hallaq, 1996. Zonal stratigraphy and foraminifera of the Tethys Jurassic (Eastern Mediterranean). Gorgon and Breach Publishers, The Netherlands, 256 p.

Lababidi, M. M., and A. N. Hamdan, 1985. Preliminary lithostratigraphic correlation study in OAPEC member countries. Rep. of OAPEC, p. 1-87.

Lartet, L., 1866. Sur les gîtes bitumineux de la Judée et de la socle Syrie et sur le mode d'arrivée de l'asphalte au milieu des eaux de la mer Morte. Bull. Soc. Géol. France, sér 2, Vol. 24, p. 12-31.

Lartet, L., 1884. La Syrie d'aujourd'hui, voyage dans la Phénicie, le Liban, la Judée. Paris, Machette, 675 p.

Lartet, L., 1869. Essai sur la geologie de la Palestine. Thèse Faculté des Sciences de Paris, France, Vol. 316, Victor Masson et Fils, 229 p.

Leonov, Yu. G., Makarem, Kh., and Zaza, T., 1986. Olistostrome origin for rocks in the core of the Abd-el-Aziz Anticline, Syria. Geotectonics, No. 2, p. 85-91. (In Russian). [n English in Geotectonics, 20(2), p. 142-146].

Leonov, Yu. G., Sigachev, S. P., Otri, M., Yusef, A., Zaza, T., and Sawaf, T., 1989. New data on the Paleozoic complex of the platform cover of Syria. Geotectonics, 23(6), p. 538-542.

Le Tensorer, J. M., 1996. Les Cultures Paléolitiques de la Steppe Syrienne: L'exemple d'El-Komm. In: Palmyra and the Silk Road, Vol. XLII, Les Annales Archéologiques Arabes Syriennes, p. 43-61.

Lexique stratigraphique international (Congrès géologique international. Commission de stratigraphie), 1959. Vol. III, Asie, Fasc. 10 a, Iraq. Paris.

Lexique stratigraphique international (Congrès géologique international. Commission de stratigraphie), 1963. Vol. III, Asie, Fasc. 10 c1, Liban, Syrie, Jordanie. Paris.

Liere, W. J. van., 1960. Un gisement Paléolithique dans un niveau Pleistocène de l'Oronte a Latamne (Syrie). Ann. Archaeol. Syr., Vol. 10, p. 165-175.

Liere, W. J. van., 1963. Capital and citadels of bronze-iron age, Syria, in their relationship to land and water. Ann. Archaeol. Syr., Vol. 13, p. 109-122.

Liere, W. J. van., 1966. The Pleistocene and stone age of the Orontes River. Ann. Archaeol. Syr., Vol. 16, p. 7-29.

Liere, W. J. van and Hooijer, D. A., 1961 1962. A Paleo-Orontes level with Archidiscodon meridionalis (Arch. Nesti) at Hama. Ann. Archeol. Arab. Syr., Damas, Vol. XI-XII, p. 167-172.

Litak, R. K., Barazangi, M., Beauchamp, W., Seber, D., Brew, G., Sawaf, T., and Al-Youssef, W., 1997. Mesozoic-Cenozoic evolution of the intraplate Euphrates fault system, Syria: implication for regional tectonics. J. Geol. Soc., London, 154(4), p. 653-666.

Litak, R. K., Barazangi, M., Brew, G., Sawaf, T., Al-Imam, A., and Al-Youssef, W., 1998. Structure and evolution of the petroliferous Euphrates Graben system, southeast Syria. Am. Assoc. Petrol. Geol. Bull., 82(6) p. 1173-1190.

Loeblich, A. R., and Tappan, H., 1988. Foraminiferal genera and their classification, Vol. 1, 970 p., Vol. 2, 847 p. New York, Van Nustrand and Reinhold.

McBride, J. H., Barazangi, M., Best, J., Al-Saad, D., Sawaf, T., Al-Otri, M., and Gebran, A., 1990. Seismic reflection structure of intracratonic Palmyride fold-thrust belt and surrounding Arabian Platform, Syria. Am. Assoc. Petrol. Geol. Bull., 74(3) p. 238-259.

Medvedev, V. Ya., 1966a. The Geological map of Syria. Scale 1:200,000. Sheet I-37-XXII (Ar-Raqqa). Explanatory Notes. Technoexport, USSR, 56 p.

Medvedev, V. Ya., 1966b. The Geological map of Syria. Scale 1:200,000. Sheets I-37-XXIII, I-37-XXIV (Deir Az-Zor, Al-Buwara). Explanatory Notes. Technoexport, USSR, 42 p.

Mikhailov I. A., 1966. The Geological map of Syria. Scale 1:200,000. Sheet I-37-XX (Salamiyeh). Explanatory Notes. Texnoexport, USSR, 64 p.

Moshkovitz, S. and A. Ehrlich, 1976a. Schizosphaerella punctulata Deflandre et Dangeart and Crepidolithus crassus (Deflandre, Noel), Upper Liassic calcareous nannofossils from Israel and Northern Sinai. Israel J. of Earth Sciences, 25(2), p. 51-57.

Moshkovitz, S. and A. Erlich, 1976b. Distribution of Middle and Upper Jurassic calcareous nannofossils in the Northern Negev, Israel, and Gebel Maghara, Northern Sinai. Isr. Geol. Surv., Bull., Vol. 69, 47 p.

Mouty, M., 1976. Présence du Lias dans le Massif Alaouite, Syrie. Compt. Rend. Somm. de la Soc. Géol. France, No. 3, p. 104-105.

Mouty, M., 1997a. Le Jurassique de la Chaine des Palmyrides (Syrie Centrale). Bull. Soc. Géol. France, 168(2), p. 181-186.

Mouty, M., 1997b. Le Jurassique de la Chaine Côtière (Jibal As-Sahilyeh) de Syrie: essai de biozonation par grands foraminifères. Compt. Rend. Acad. Sci., Paris, 325(3), p. 207-213.

Mouty, M. 2000. The Jurassic in Syria: an overview. Lithostratigraphic and biostratigraphic correlations with adjacent areas. In: S. Crasquin-Soleau and E. Barrier (Eds.), Peri-Tethys Mém. 5, Mém. Mus. natn. Hist. nat., Vol. 182, p. 159-168.

Mouty, M. and Zaninetti, L., 1998. Le Jurassique du Mont Hermon (Anti-Liban). Decouverte de Trias et de Lias. Arch. Sci. Genève, 51(3), p. 295-304.

Muhesen, S. A., 1985. L'Acheuléen Recent, évolution de la Syrie. BAR Intern Ser., Vol. 248, p. 1-189.

Nikolaeva, I. V., Krasheninikov, V. A., Golovin, D. I., and Ivanovskaya, T. A., 1985. Possibilities of radiological dating of the Paleogene deposits of the Eastern Mediterranean based on glauconite. Lithology and mineral resources, No. 5, p. 76-88. (In Russian).

Noetling, F., 1886. Entwurf einer Gliederung der Kreideformation in Syrien und Palaestina. Z. Deutsch. Geol. Ges., Vol. 38, p. 824-875, Berlin.

Noetling, F., 1887. Der Jura am Hermon. Schweizerbartcshe Verlagsbuchhandling, Stuttgart, 46 p.

Novikov, V. M., Sharkov, E. V., Chernyshev, I. V., Devyatkin, E. V., Dodonov, A. E., Ivanenko, V. V., Karpenko, M. I., Hanna, S., and Akhtum H., 1993. Geochronology of crusts of wethearing of platobasalts of Syria and evolution of the climate of this area during last 20 mln years. Stratigraphy. Geological Correlation, 1(6), p. 66-74.

Olsson, D. R., Ch. Hemleben, W. H. Berggren, and B. T. Huber, (Editors), 1999. Atlas of Paleocene Planktonic Foraminifera. Smithsonian Contributions to Paleobiology, 85, 252 p., 37 figs., 71 plates.

Oufland, A. K., 1966a. The Geological map of Syria. Scale 1:200,000. Sheet I-37-XXI (Ar-Rasafeh). Explanatory Notes. Technoexport, USSR, 83 p.

Oufland, A. K., 1966b. The Geological map of Syria. Scale 1:200,000. Sheet J-37-III (Jrablus). Explanatory Notes. Technoexport, USSR, 56 p.

Oufland, A. K. and Bertels-Uspenskaya, I. A., 1971. Conditions of precipitation and age of sediments of the Lower Fars Formation in North-Eastern Syria. Trans. of Laboratory of geology of foreign countries, Vol. 22, p. 142-149.

Passemard, E., 1926. Les terrasses alluviales de l'Euphrate et les industries qu'elles contiennent. C. R. Acad. Sci. France, Paris, 183(5), p. 365-368.

Perves, M., 1946/1948. La préhistoire de la Syrie et du Liban. Syrie, Vol. 25, p. 109-129, Paris.

Petrov, V. I. and Antonov, B. S., 1966. The Geological map of Syria. Scale 1:200,000. Sheets I-37-XVII and I-37-XVIII (Abou Kamal and Beer El-Allouni). Explanatory Notes. Technoexport, USSR, 44 p.

Pfender, J., 1937. Quelques Hydrozoaires de la Syrie septentrionale. Not. et Mém., Vol. 2, p. 125-136.

Picard, L. and Hirsch, F., 1987. The Jurassic stratigraphy in Israel and the adjacent countries. Israel Acad. Sci. and Hum., Jerusalem, 106 p.

Plumhoff, F., 1970. Die Fauna des Karbons vom Djebel Abd-El-Aziz (Nordost Syrien). 2. Ostracoda. Neues Jahrbuch fuer Geol. und Palaeontol., Abh., Band 135, Heft 2, p. 190-212.

Ponikarov,V. P., Kazmin, V. G., Mikhailov, I. A., Razvaliayev, A. V., Krasheninnikov, V. A., Kozlov, V. V., Soulidi-Kondratiyev, E. D., Mikhailov, K. Ya., Kulakov, V. V., Faradzhev, V. A., and Mirzayev, K. M., 1967. The Geology of Syria. Explanatory Notes on the Geological map of Syria. Scale 1:500,000. Part I. Srtatigraphy, igneous rocks and tectonics. Technoexport, USSR, 230 p.

Ponikarov, V. P., Kazmin, V. G., Mikhailov,I. A., Razvaliayev, A. V., Krasheninnikov, V. A., Kozlov, V. V., Soulidi-Kondtatiyev, E. D., and Faradzhev, V. A., 1966. The Geological map of Syria. Scale 1:1,000,000. Explanatory Notes. Technoexport, USSR, 111 p.

Ponikarov, V. P., Kazmin, V. G., Kozlov, V. V., Krasheninnikov, V. A., Mikhailov, I. A., Razvalyaev, A. V., Soulidi-Kondratiyev, E. D., Oufland, A. K., and Faradzhev, V. A., 1969. Geology and mineral resources of foreign countries. Syria. Scientific Laborat. of geology of foreign countries, Vol. 18. Publ. Off. Nedra, Leningrad, 216 p. (In Russian).

Ponomarev, B. Ya., 1966. The Geological map of Syria. Scale 1:200,000. Sheet J-37-IV (Tuwal El-Aba). Explanatory Notes. Technoexport, USSR, 47 p.

Ponomarev, B. Ya. and Bivshev, A. S., 1966. The Geological map of Syria. Scale 1: 200,000. Sheets J-37-V, XI (Al-Hasakeh). Explanatory Notes. Technoexport, USSR, 67 p.

Protasevich, L. N. and Kurbatov, V. P., 1968. The Geological map of Syria. Scale 1: 50,000. Sheets J-37-I-4a (Radjou), J-37-I-4b (Aafrin), J-37-I-4c (Meidanki), J-37-I-4d (Bulbul). Explanatory Notes. Technoexport, USSR, 83 p.

Protasevich, L. N., and Maksimov, A. A., 1966. The Geological map of Syria. Scale 1:200,000. Sheets J-37-I, II (Halab, Antakya). Explanatory Notes. Technoexport, USSR, 74 p.

Puura, V., Martins, A., Baalbaki, M. Kh., and Al-Khatib, K., 1984. Occurence of oil shales in the south of Syrian Arab Republic (SAR). Oil shale, Tallinn, Vol. I, No. 4, p. 333-340. (In Russian).

Ravn, R. L., McPhilemi, B., Rutherford, M., Talli, S., and Bahre, G., 1994. Late Devonian-Early Carboniferous palynostratigraphy and its applications in north-eastern Syria. In: Simmons, M. D., (ed). Micropaleontology and hydrocarbon exploration in the Middle East, Chapman Hall, London, p. 5-14.

Razvalyaev, A. V., 1966a. The Geological map of Syria. Scale 1:200,000. Sheets I-37-VII (Dimashq) and I-36-XII (Beirouth). Explanatory Notes. Technoexport, USSR, 124 p.

Razvalyaev, A. V., 1966b. The Geological map of Syria. Scale 1:200,000. Sheet I-37-VIII (Saba Abar). Explanatory Notes. Technoexport, USSR, 64 p.

Reiss, Z., 1988. Assemblages from a Senonian high-productivity sea. Rev. Paleobiologie, Vol. 2, p. 323-332.

Rio, D., Sprovieri, R. and Thunnel R., 1991. Pliocene Lower Pleistocene chronostratigraphy: A re-evaluation of Mediterranean type sections. Bull. Geol. Soc. Am., 103(8), p. 1049-1058.

Rigassi. F., 1971. Petroleum geology of Turkey. In: Campbell, A. S. (Ed.), Geology and History of Turkey. Petroleum Exploration Society of Libya, Tripoli, p. 453-482.

Roegl, F., 1996. Migration pathways between Africa and Eurasia Oligocene-Miocene paleogeography. Europal., Vol. 10, p. 23-26.

Roegl, F., 1998. Paleogeographic considerations for Mediterranean and Paratethys seaways (Oligocene to Miocene). Ann. Naturhist. Mus., Vol. 99A, p. 279-310.

Roegl, F. and Steininger, F., 1983. Vom Zerfall der Tethys zu Mediterran und Paratethys. Die Neogene Palaeogeographie und Palinspastik des Zirkum-Mediterranean Raumen. Ann. Naturhist. Mus. Wien, Vol. 85, ser. A, p. 135-163.

Roger, J., 1940. Pectinides, miocènes, pliocènes, et quaternaires de Syrie. Notes et Mém. sur le Moyen Orient (Syrie-Liban), Vol. 3, Paris, p. 325-348.

Roger, J. and Dubertret, L., 1938. Les Pectinides Neogènes des régions d'Antioche et d'Alep (Syrie). Leur signification stratigraphique. C. R. Soc. Géol. France, No. 5, p. 73-74.

Roman, F., 1940. Listes raisonnées des faunes du Pliocène et du Miocène de Syrie et du Liban. Notes et Mém. sur le Moyen Orient (Syrie-Liban), Vol. 3, Paris, p. 353-400.

Roman, F. and Doncieux, L., 1927. Observation sur l'Eocène de Syrie et de Palestine. C. R. Soc. Géol. France, No. 14, p. 171-173.

Rosenfeld, A., Oertly, H., Honigstein, A. and Gerry, E., 1987. Oxfordian Ostracods from the Kidod Formation of the Majdal Shams area, Mount Hermon, Golan Heights. Bull. Centres de Recherches Explorat. Product. Elf Aquitaine, Vol. 11, p. 233-248.

Ruske, R., 1978a. Geological map of Syria. Scale 1:50,000. Sheet Qerdaha (N I-37-S-1-c). Explanatory Notes. General Establ. of Geol. and Mineral Resources, Damascus, 36 p.

Ruske, R., 1978b. Geological map of Syria. Scale 1:50,000. Sheet Jableh (N I-36-X-2-d). Explanatory Notes. General Establ. of Geol. and Mineral Resources, Damascus, 26 p.

Rust, A., 1950. Die Hoehlenfunde von Jabrud (Syrien). Neumuenster, Karl Wachholtz Verlag.

Sakaguch, Y., 1978. Palmyra pluvial lake. In: Hanihara K. and Sakaguchi Y. (Eds.). Palaeolithic site of the Doura Cave paleogeography of the Palmyra Basin in Syria; Part I. Stratigraphy and palaeogeography in the Late Quaternary, Bulletin 14, University of Tokyo.

Saint-Marc, P., 1987. Biocénoses et thanatocénoses à foraminifères dans le Miocène inférieur marin du NW Syrien. Interprétation paléogéographique. Rev. de Micropaléontologie, 30(1), p. 24-31.

Salel, J. F., and Seguret, M., 1994. Late Cretaceous to Paleogene thin-skinned tectonics of the Palmyrides Belt (Syria). Tectonophysics, 234(4), p. 265-290.

Saltzman, U., Bartov, Y., Gotlib, B., Magharitz, M., and Mimran, Y., 1968. Composite columnar section of sequences exposed at the SE flank of Mt. Hermon. In: U. Saltzman (Ed.), The geology of the Southern Hermon Slope, Internal Report, TAHAL., Tel Aviv, p. 1-46.

Sanlaville, P. (Ed.), 1979. Quaternaire et préhistoire du Nahr el Kebir septentrional: les débuts de l'occupation humaine dans la Syrie du Nord et au Levant Coll. Maison Orient, C. N. R. S., No. 9, Lyon, 162 p.

Sawaf, T., Al-Saad, D., Gebran, A. Barazangi, M., Best, J. A., and Chaimov, T. A., 1993. Stratigraphy and structure of eastern Syria across the Euphrates depression. Tectonophysics, 220(1-4), p. 267-281.

Sawaf, T., Al-Saad, D., Khaddour, M., Barazangi, M., and Chaimov, T. A., 1992. Structure and stratigraphic development of eastern Syria; a transect under the auspices of the Global Geoscience Transect (GGT) Project. 29th Int. Geol. Cong., Abstr., Kyoto, Japan, Vol. 29, p. 463.

Sawaf, T., Brew, G. E., Litak, R., and Barazangi, M., 2001. Geologic evolution of the intraplate Palmyride Basin and Euphrates fault system, Syria. In: P. A. Ziegler, W. Cavazza, A. H. F. Robertson, and S. Crasquin-Soleau (eds.), Peri-Tethys Memoir 6; Peri-Tethyan rift/wrench basins and passive margins. Mém, Mus. Nat. d'Hist. Natur., 186, p. 441-467.

Schutt, H., 1983. Die Molluskenfauna der Susswasser in Einzugsgebiet des Orontes unter Berucksichtigung benachbarter Fluss systeme. Arch. Moll., Bd. 113, No. 1/6, p. 17-91.

Schutt, H., 1988. Erganzungen zur Kenntniss der Molluskenfauna Oberpliozaener Susswasserkonglomerate Syriens. Arch. Moll., Bd. 118, No. 4/6, p. 129-143.

Searle, M. P., 1994. Structure of the interplate Eastern Palmyride fold belt, Syria. Geol. Soc. Am. Bull., 106(10), p. 1332-1350.

Seber, D., Barazangi, M., Al-Saad, D., Sawaf, T., and Gebran, A., 1991. Crustal shortening in the Palmyride fold belt, Syria, and implications for movement of the Dead Sea fault system. Geol. Soc. Am., Ann. Mtg. Abstr., 23(5), p. 316.

Seber, D., Barazangi, M., Chaimov, T. A., Al-Saad, D., Sawaf, T., and Khaddour, M., 1992. Geometry and velocity structure of the Palmyride fold-thrust belt and surrounding Arabian Platform in Syria. Proc. Abstr., 1st Intl. Symp. Eastern Mediterranean geology, Adana Turkey, Geosound, Vol. 20, p. 103-110.

Seber, D., Barazangi, M., Chaimov, T. A., Al-Saad, D., Sawaf, T., and Khaddour, M., 1993. Upper crustal velocity structure and basement morphology beneath the intracontinental thrust-fold belt and North Arabian platform in Syria. Geophys. J. Int., 113(3), p. 752-766.

Senders, J. and Fridman, J., 1970. Origin and distribution of dolomites. In: Calcareous rocks and their origin. Mir Publ. House Moscow, p. 249-309. (In Russian).

Shalash, J. and Al-Hafez, M. A., 1983. Geological map of Syria. Scale 1:50,000. Sheet Asal Al Ward (N I-37-SW-G-3-d). Explanatory Notes. General Establ. of Geol. and Mineral Resources, Damascus, 31 p.

Sharkov, E. V. 2000. Mesozoic and Cenozoic basaltic magmatism. In: Outline of geology of Syria, Trans. of the Geol. Inst. of the Russian Acad. Sci., Vol. 526, p. 177-200.

Sharkov, E. V., Chernyshev, I. V., Devyatkin, E. V., Dodonov, A. E., Ivanenko, V. V., Karpenko, M. I., Leonov, Yu. G., Novikov, V. M., Hanna, S., and Khatib, K., 1994 Geochronology of the Late Cenozoic basalts of Western Syria. Petrology, 2(4), p. 439-448. (In Russian).

Sharkov, E. V., Chernyshev, I. V., Devyatkin, E. V., Dodonov, A. E., Ivanenko, V. V., Karpenko, M. I., Lebedev, V. A., Novikov, V. M., Hanna, S., Khatib, K., 1998. New data on the geochronology of the Late Cenozoic basalts of the north-eastern periphery of the Red Sea rift zone (Northern Syria). Rep. of the Russian Acad. Sci., 358(1), p. 96-98. (In Russian). (In English, Dokl. Earth Sciences, 358(1), p. 19-22).

Sharkov, E. V. and Hanna, S., 1987. Evolution of a matter of the upper mantle in the area of the intraplate basaltic magmatism (on the example of Western Syria). Rep. of the USSR Acad. Sci., 297(3), p. 684-686. (In Russian).

Sharkov, E. E., Las'ko, E. E., and Hanna, S., 1992. Inclusions of abyssal rocks in explosive body of the Nabi-Matta volcano, North-western Syria. Geochemistry, No. 9, Moscow, p. 1241-1261. (In Russian).

Sharkov, E. V., Las'ko, E. E., Fedosova, S. P., and Hanna, S., 1989a. Depth-derived inclusions of the Quaternary volcano of Tell Danun (Southern Syria) in connection with the problem of the intraplates basaltic magmatism. Geochemistry, No. 11, Moscow, p. 1609-1622. (In Russian).

Sharkov, E. V., Las'ko, E. E., Fedosova, S. P., Hanna, S., Ali, S., Muravitskaya, G. N., and Ivanov, V. P., 1989b. Xenoliths of pegmatoid clinopyroxene-hornblende rocks with the Ba-zeolite from diatrems of explosion of North-Western Syria. Rep. of the All-Union Mineralog. Soc., No. 1, p. 50-56. (In Russian).

Sharkov, E. V., Sindeev, A. S., Ali, S., Hanna, S., and Sharaf, M., 1987. Petrlogy of the Homs basaltic area (Western Syria). Trans. USSR Acad. Sci., ser. Geol., No. 7, p. 61-71. (In Russian).

Sharkov, E. V., Snyder, G. A., Taylor, L. A., Laz'ko, E. E., Jerde, E., and Hanna, S., 1996. Geochemical peculiarities of the asthenosphere beneath the Arabian Plate: evidence from mantle xenoliths of the Quaternary Tell-Danun volcano. Geochemistry, No. 9, Moscow, p. 819-835. (In Russian).

Shatsky, V. N., Kazmin, V. G., and Kulakov, V. V., 1966. The Geological map of Syria. Scale 1:200,000. Sheets I-37-XIX and I-36-XXIV (Al-Latheqiyeh and Hama). Explanatory Notes. Technoexport, USSR, 107 p.

Sigal, J., 1977. Essai de zonation du Crétacé méditerranéen à l'aide des foraminifères planctoniques. Géologie Méditerranéenne, Vol. 4, p. 99-108.

Singh, H. P., 1964. A miospore assemblage from the Permian of Iraq. Paleontology, 7(2), p. 240-265.

Snyder, G. A., Taylor, L. A., Jerde, E. A., Sharkov, E. V., Las'ko E. E., and Hanna, S., 1993. Petrogenesis of garnet pyroxenite and spinel peridotite xenoliths of the Tell-Danun alkali basalt volcano, Harrat As Shamah, Syria. Intern. Geol. Rev., 35(12), p. 1104-1120.

Sobolev, N. V., Abou Assale, G., Kepezhinskas, V. V. et al., 1990. Lamprophirs of the Cretaceous diatrems of the Syrian Rift. Rep. of the USSR Acad. Sci., Vol. 314, p. 435-438. (In Russian).

Soulidi-Kondratiev, E. D., 1966a. The Geological map of Syria. Scale 1:200,000. Sheet I-37-XV (Tudmor). Explanatory Notes. Technoexport, USSR, 104 p.

Soulidi-Kondratiev, E. D., 1966b. The Geological map of Syria. Scale 1:200,000. Sheet I-37-IX (At-Tanf). Explanatory Notes. Technoexport, USSR, 40 p.

Suzuki, C. and Akazawa, T., 1971. Manufacturing technique of the stone artifacts from Palmyra, Syria. J. Anthrop. Soc. Nippon, Vol. 79, p. 105-127.

Suzuki, H. and Takai, F. (Eds.), 1973. The Paleolithic site at Douara cave in Syria. Part I. Bull. Univ. Museum, University of Tokyo, No. 5, 150 p.

Suzuki, H. and Takai, F. (Eds.), 1974. The Paleolithic site at Douara cave in Syria. Part II. Bull. Univ. Museum, University of Tokyo, No. 6.

Steininger, F. and Roegl, F., 1984. Paleogeography and palinspastic reconstruction of the Neogene of the Mediterranean and Paratethys. In: Dixon, J. and Robertson, A. (Eds.), The geological evolution of the Eastern Mediterranean, Oxford, Blackwell Sci. Publ., p.659-668.

Stoppel, D., 1970. Die Fauna des Karbons vom Djebel Abd El-Aziz (Nordost-Syrien). 3. Conodonta. N. Jb. Geol. Palaeont. Abh., Vol. 135, Heft 2, p. 213-225.

Stchepinsky, V., 1947. Sur la limite septentrionale du Plateau Syrien. Bull. Soc. Géol. France, Vol. 17, p. 33-38.

Subbotina, N. N., 1947. Foraminifera of the Danian and Paleogene sediments of the Northern Caucasus. In: Microfauna of oil deposits of the Caucasus, Emba and Central Asia, Leningrad, Gostoptekhizdat Publ. House, p. 39-160. (In

Russian).
Subbotina, N. N., 1953. Globigerinides, hantkeninides and globorotaliides. Proc. of the All-Union Oil Scientific and Research Prospecting Institute, a New series, Vol. 76, Leningrad, Gostoptekhizdat Publ. House, 295 p. (In Russian).
Sudburry, M., 1957. *Diplograptus spinulosus* sp. nov. from the Ordovician of Syria. Geol. Mag., Vol. 34, p. 503-506.
Tasman, C. E., 1949. Stratigraphy of south-eastern Turkey. Bull. Am. Assoc. Petrol. Geol., 33(1), p. 22-31.
Trifonov, V. G.,Trubikhin, V. M., Adjamian, J., Djallad, Z., El Khair, Y., and Ayed K., 1991. Levant fault zone in northeast Syria. Geotectonics, 25(2), p. 145-154.
Tsekhovskiy, Yu. G., Bogatyrev, B. A., and Jabbour, R., 1993. Triassic-Jurassic bauxites of Syria. Lithology and Mineral Resources, No. 4, p. 34-50. (In Russian).
Tsekhovskiy, Yu. G., Muravyov, V. I., and Jabbour, R., 1994. Hydrothermal-sedimentary lithogenesis on the Arabian Plate (in the Jurassic-Cretaceous continental landscapes). Trans. of Educ. Institutions, Geology and Prospecting, No. 2, p. 33-41.
Tyrrell, G. W., 1930. Pleistocene and Recent basalte of Transjordania and Palestine. Geol. Mag., Vol. LXVII.
Vaumas, E. de., 1953a. Sur les terrasses d'abrasion marine de la région de Lattaquie. C. R. Acad. Sci. France, Paris, Vol. 237, p. 1266-1268.
Vaumas, E. de., 1953b. Sur les terrasses d'abrasion marine de la région de Djeble et de Tartous (Syrie). C. R. Acad. Sci. France, Paris, Vol. 237, p. 1343-1344.
Vaumas, E. de., 1954. Les terrasses d'abrasion marine de la côte Syrienne. Rev. géogr. Alpine, Vol. 42, p. 633-664.
Vaumas, E. de., 1957. Sur l'évolution structurale et morphologique de la dépression du Rhab et du Bas Oronte. C. R. Acad. Sci. France, Paris, Vol. 244, p. 2946-2948.
Vautrin, H., 1933a. Sur quelques formes nouvelles de Rudistes recueillies en Syrie septentrionale. Not. et Mém. Syrie-Liban, Paris, Vol. 1, p. 29-43.
Vautrin, H., 1933b. Les échinidés burdigaliens de la zone désertique syrienne. Not. et Mém. Syrie-Liban, Paris, Vol. 1, p. 101-115.
Vautrin, H., 1933c. Le Miocène de la région côtière d'Alexandrette (comparaison avec les bassin Miocènes environnants de Syrie) . Not. et Mém. Syrie-Liban, Paris, Vol. 1, p. 141-153.
Vautrin, H., 1934a. Contribution à l'étude de la série jurassique dans la chaine de l'Anti-Liban et plus particulierement dans l'Hermon (Syrie). C. R. Acad. Sci. France, Paris, 198(16), p. 1438-1440.
Vautrin, H., 1934b. Sur l'orogenèse du Massif de l'Hermon (Syria). C. R. Acad. Sci. France, Paris, 199(1), p. 82-84.
Vinogradov, V. I., Golovin, D. I., and Krasheninnikov, V. A., 1988. Isotopic dating of glauconite from the Globigerapsis semiinvoluta Zone sediments, Upper Eocene of the Eastern Mediterranean. Rep. of the USSR Acad. Sci., Ser. Geol., 301(5), p. 1171-1174. (In Russian).
Vinogradov, V. I., Golovin, D. I., and Krasheninnikov, V. A., 1990a. Isotopic dating by means of glauconite of the Planorotalites pseudomenardii Zone sediments of the Eastern Mediterranean. Lithology and mineral resources, No. 1, p. 115-119 (in Russian).
Vinogradov, V. A., Krasheninnikov, V. A., and Golovin, D. I., 1990b. Three reliable datings of glauconite from the Paleogene deposits of the Eastern Mediterranean. Lithology and mineral resources, No. 3, p. 44-55. (In Russian).
Walley, C. D., 1997. The stratigraphy and geological history of Lebanon: an Outline. Version 1.5. The Geological Survey of Lebanon, Unpubl. Rep., 30 p.
Weber, H., 1963. Ergebnisse erdoelgeologischer Aufschlussarbeiten der DEA in Nordost Syrien. Teil 1. Schichtfolge, Fazies und Tectonic in der Haute Djesireh. Erdoel und Kohle, Erdgas, Petrochemie, 16(6), Hamburg, p. 669-682.
Weber, H., 1964. Ergebnisse erdoelgeologischer Aufschlussarbeiten der DEA in Nordost Syrien. Teil 2. Geophysikalische Untersuchungen und Tiefbohrungen in der Haute Djesireh. Erdoel und Kohle, Erdgas, Petrochemie, 17(4), Hamburg, p. 249-261.
Weber, H., 1968. Der Palaeozoische Unterbau der Mardin-schwelle in Nordost Syrien. [The Paleozoic foundation of the Mardin swell in northeastern Syria]. Z. Deutsch. Geol. Gesell., 117(2-3), p. 590.
Wetzel, R. and Haller, J., 1945-1948. Le quaternaire côtier de la region de Tripoli (Liban). Not. et Mem. Syrie-Liban, Paris, Vol. 4, p. 1-48.
Wolfart, R., 1965. Die Fauna des Karbons vom Jebel Abd-EL-Aziz (Nordost Syrien). 1. Anthozoa, Bryozoa, Brachiopoda. Geol. Jahrb., p. 277-326.
Wolfart, R., 1966. Zur Geologie und Hydrogeologie von Syrien. Beih. Geol. Jahrb., Heft 68, Hannover, 129 p.
Wolfart, R., 1967. Geologie von Syrien und dem Libanon. [Syria and Lebanon, contributions to the regional geology of the earth]. Beitr. Zur Regionalen Geologie der Erde, Vol. 6, Gebrueder Borntraeger, Berlin, 306 p.
Yankauscas, T. V. and Tally, S., 1978. The Paleozoic of Syrian Arab Republic. Trans. of the USSR Acad. Sci., Ser. Geol., No. 11, p. 92-97. (In Russian).
Yaroshenko, O. P. and Bach Imam, I., 1989. Palynoflora and stratigraphy of deposits of the Middle and Upper Triassic of Syria. In: Paleofloristics and stratigraphy of the Phanerozoic, No. 1, Geol. Inst. of the USSR Acad. Sci., Moscow, p. 32-65. (In Russian).
Yaroshenko, O. P. and Bach Imam, I., 1994. Palynological characteristics of the Lower Permian deposits of Northeastern Syria. Stratigraphy. Geological Correlation, 2(1), p. 62-73. (In Rusian).
Yaroshenko, O. P. and Bach Imam, I., 1995. Temporal variability of composition of palynological assemblages, originated from the Middle and Late Triassic of Syria, and their dependence on climate and facies conditions. Stratigraphy. Geological Correlation, 3(4), p. 44-61. (In Russian).
Yaroshenko, O. P., Bach Imam, I., and Nassri, S., 1982. Palynological assemblages of the Triassic deposits of Syria and their stratigraphic significance. Trans. USSR Acad. Sci., Ser. Geol., No. 9, p. 72-77. (In Russian).
Yaroshenko, O. P. and Nassri, S., 1980. Triassic palynology of Syria. In: Internat. Palynol. Conf., Abstracts, Cambridge, p. 441.
Yousef, Sh., 1979. Geological map of Syria. Scale 1:50,000. Sheet Haffeh (N I-37-D-3-a). Explanatory Notes. General Establ. of Geol. and Mineral Resources, Damascus, 35 p.
Yousef, Sh., 1980a. Geological map of Syria. Scale 1:50,000. Qadmous-Banyas sheets (N I-37-S-1-a and No. I-36-X-2-b). Explanatory Notes. General Establ. of Geol. and Mineral Resources, Damascus, 32 p.
Yousef, Sh., 1980b. Geological map of Syria. Scale 1:50,000. Sheets Safita and Tartous (N I-37-M-3-c and No. I-36-R-4-d). Explanatory Notes. General Establ. of Geol. and Mineral. Resources, Damascus, 34 p.
Youssef, A., Al-Saad, D., Sawaf, T., Khaddour, M., Barazangi, M., Best, J., Chaimov, T. A., and Seber, D., 1992. Mesozoic evolution of the Palmyrides, central Syria. 29th Int. Geol. Cong., Abstr., Kyoto, Japan, Vol. 29, p. 300.
Zumoffen, G., 1899. La Météorologie de la Palestine et de la Syrie. Bull. Soc. géogr. France, Vol. 20, p. 344-364, p. 482-488.
Zumoffen, G., 1926. Géologie du Liban, avec carte géologique au 1/200,000. Barrère, 165 p.

GEOLOGICAL MAPS OF SYRIA - REFERENCES (for Part II Introduction and Chapter 6)

Adjamian, J. and Al-Bakka, F. F., 1984. Geological map of Syria. Scale 1:50,000. Explanatory Notes of Misyaf sheet (N I-37-NW-S-1-b). General Establ. of Geol. and Mineral Resources, Damascus, 42 p.
Adjamian, J. and Jamal, N. E., 1983. Geological map of Syria. Scale 1:50,000. Explanatory Notes of Qalaat Al Housen sheet (N I-37-M-3-d). General Establ. of Geol. and Mineral Resources, Damascus, 38 p.
Al Baca, F. F., 1983. Geological map of Syria. Scale 1:50,000. Explanatory Notes of Tal Kalakh sheet (N I-37-M-3-b). General Establ. of Geol. and Mineral Resoutces, Damascus, 26 p.
Dubertret, L., 1929. Etude des régions volcanique du Hauran, du Djebel Druze et Diret et Touloul, Syrie. Rev. géogr. phys. Géol. dyn., 2(4), p. 275-321, 1 map 1:500.000.
Dubertret, L., 1942. Carte géologique du Moyen Orient au 2 millioniеme. 67 p. Beyrouth.
Dubertret, L., 1943. Carte géologique de la Syrie et du Liban au millionieme. 2nd Ed. Beirouth, 1941-1943, Délégation Génerale au Levant de la France Combattante, Sec. Géologique, 67 p., 1 map.
Dubertret, L., 1944. Carte lithologique de la Syrie et du Liban (1:1,000,000). Beirouth.
Dubertret, L., 1945. Carte géologique au 50,000e, feuille de Rachaya-N. Beyrouth.

Dubertret, L., 1949a. Carte géologique au 50,000e, Feuille de Zabdani. Damas, Min. des Travaux Publics, 62 p., 1 map.

Dubertret, L., 1949b. Carte géologique au 50,000e, Feuille de Saida. Beyrouth, 46 p., 1 map.

Dubertret, L., 1950a. Carte géologique au 50,000e, Feuille de Rayak. Damas, Min. des Travaux Publics, 48 p., 1 map.

Dubertret, L., 1950b. Carte géologique au 50,000e, Feuille de Baalbek. Beyrouth, 52 p., 1 map.

Dubertret, L., 1951a. Carte géologique au 50,000e, Feuille de Beyrouth. Beyrouth, 66 p., 1 map.

Dubertret, L., 1951b. Carte géologique au 50,000e. Feuille de Tripoli. Beyrouth, 62 p., 1 map.

Dubertret, L., 1953a. Carte géologique au 200,000, Feuille d'Antioche. Damas, Min. des Travaux Publ., 67 p. 1 map.

Dubertret, L., 1953b. Carte géologique au 50,000e, Feuille de Zahle. Beyrouth, 64 p., 1 map.

Dubertret, L., 1954a. Carte géologique au 200,000, Feuille de Beyrouth. 108 p., Paris.

Dubertret, L., 1955a. Carte géologique du Liban au 200,000. Beyrouth, 74 p., 1 map.

Dubertret, L., 1958. Feuille de Tartous, 1:50,000, Notice explicative. Beyrouth, Min. des Travaux Publ., 12 p., 1 map.

Dubertret, L., 1960. Carte géologique au 1:50,000e, Feuille Hermon (Rachaya Sud). Beyrouth, Min. des Travaux Publ., 46 p. 1 map.

Dubertret, L., 1963. Liban, Syrie: chaine des grands massifs côtiers et confins à l'Est. Lexique stratigraphique internat., Vol. 3, Asie, 10c (Liban, Syrie, Jordanie), 155 p., 1 geol. Karte 1:1,000,000, Paris.

Dubertret, L., 1966. Liban, Syrie et bordure des pays voisins. 1e partie: Tableau stratigraphique avec carte géologique au millionieme. Not. et Mem. sur le Moyen Orient, Vol. 8, Paris, p. 251-358, 1 map.

Dubertret, L., 1975. Introduction a la carte géologique au 1:50,000 du Liban. Not. et Mém. sur le Moyen Orient, Vol. 13, Paris, p. 345-402.

Dubertret, L., and Wetzel, R., 1949. Carte géologique au 200,000, Feuille de Tripoli. Damas, Min. de Trav. Publ., 62 p., 1 map.

Faradzhev, V. A., 1966. The Geological map of Syria. Scale 1:200,000. Sheet I-37-XIV (Al Qaryatein). Explanatory Notes. Technoexport, USSR, 146 p.

Galaktionov, A. B., Demidov, V. A., and Katkova, N. S., 1966. The Geological map of Syria. Scale 1:200,000. Sheets J-37-VI, XII (Al-Qamishli, Sinjar), and J-38-I, VII (Qarashouk-Dagh). Explanatory Notes. Technoexport, USSR, 84 p.

Kazmin, V. G. and Kulakov, V. V., 1968. The Geological map of Syria. Scale 1: 50,000. Sheets I-37-XXIV-4d, I-37-XIX-3c and I-36-XXIV-4b (the Bassit Massif area). Explanatory Notes. Technoexport, USSR, 124 p.

Khatib, K., 1980. Geological map of Syria. Scale 1:50,000. Explanatory Notes of of Halba-Hamedyieh sheet (N I-37-M-3-a and No. I-36-R-4-b). General Establ. of Geol. and Mineral Resources, Damascus, 18 p.

Kozlov, V. V. (assisted by Artyemov, A. V., and Kalis, A. F.), 1966a. The Geological map of Syria. Scale 1:200,000. Sheets I-36-XVIII, I-37-XIII (Trablus, Homs). Explanatory Notes. Technoexport, USSR, 68 p.

Kozlov, V. V., 1966b. The Geological map of Syria. Scale 1:200,000. Sheets I-37-XVI (Wadi El-Miyah), I-37-X (Abar Sijri). Explanatory Notes. Technoexport, USSR, 48 p.

Krasnov, A. A., Kazmin, V. G., and Kulakov, V. V., 1966. The Geological map of Syria. Scale 1:200,000. Sheets I-36-VI (Fiq-Haifa), I-37-I (As-Suweida), I-37-II (Az-Zalaf). Explanatory Notes. Technoexport, USSR, 75 p.

Kurbanov, N., and Zarjanov, Yu., 1968. The Geological map of Syria. Scale 1:50,000. Sheets I-37-VII-3 a (Zebdani) and I-37-VII-3 c (Rayak). Explanatory Notes. Technoexport, USSR, 120 p.

Medvedev, V. Ya., 1966a. The Geological map of Syria. Scale 1:200,000. Sheet I-37-XXII (Ar-Raqqa). Explanatory Notes. Technoexport, USSR, 56 p.

Medvedev, V. Ya., 1966b. The Geological map of Syria. Scale 1:200,000. Sheets I-37-XXIII, I-37-XXIV (Deir Az-Zor, Al-Buwara). Explanatory Notes. Technoexport, USSR, 42 p.

Mikhailov I. A., 1966. The Geological map of Syria. Scale 1:200,000. Sheet I-37-XX (Salamiyeh). Explanatory Notes. Texnoexport, USSR, 64 p.

Oufland, A. K., 1966a. The Geological map of Syria. Scale 1:200,000. Sheet I-37-XXI (Ar-Rasafeh). Explanatory Notes. Technoexport, USSR, 83 p.

Oufland, A. K., 1966b. The Geological map of Syria. Scale 1:200,000. Sheet J-37-III (Jrablus). Explanatory Notes. Technoexport, USSR, 56 p.

Petrov, V. I., and Antonov, B. S., 1966. The Geological map of Syria. Scale 1:200,000. Sheets I-37-XVII and I-37-XVIII (Abou Kamal and Beer El-Allouni). Explanatory Notes. Technoexport, USSR, 44 p.

Ponikarov, V. P., Kazmin, V. G., Mikhailov, I. A., Razvaliayev, A. V., Krasheninnikov, V. A., Kozlov, V. V., Soulidi-Kondratiyev, E. D., Mikhailov, K. Ya., Kulakov, V. V., Faradzhev, V. A., and Mirzayev, K. M., 1967. The Geology of Syria. Explanatory Notes on the Geological map of Syria. Scale 1:500,000. Part I. Srtatigraphy, igneous rocks and tectonics. Technoexport, USSR, 230 p.

Ponikarov, V. P., Kazmin, V. G., Mikhailov, I. A., Razvaliayev, A. V., Krasheninnikov, V. A., Kozlov, V. V., Soulidi-Kondtatiyev, E. D., and Faradzhev, V. A., 1966. The Geological map of Syria. Scale 1:1,000,000. Explanatory Notes. Technoexport, USSR, 111 p.

Ponikarov, V. P., Kazmin, V. G., Kozlov, V. V., Krasheninnikov, V. A., Mikhailov, I. A., Razvalyaev, A. V., Soulidi-Kondratiyev, E. D., Oufland, A. K., and Faradzhev, V. A., 1969. Geology and mineral resources of foreign countries. Syria. Scientific Laborat. of geology of foreign countries, Vol. 18. Publ. Off. Nedra, Leningrad, 216 p. (In Russian).

Ponomarev, B. Ya., 1966. The Geological map of Syria. Scale 1:200,000. Sheet J-37-IV (Tuwal El-Aba). Explanatory Notes. Technoexport, USSR, 47 p.

Ponomarev, B. Ya., and Bivshev, A. S., 1966. The Geological map of Syria. Scale 1: 200,000. Sheets J-37-V, XI (Al-Hasakeh). Explanatory Notes. Technoexport, USSR, 67 p.

Protasevich, L. N., and Kurbatov, V. P., 1968. The Geological map of Syria. Scale 1: 50,000. Sheets J-37-I-4a (Radjou), J-37-I-4b (Aafrin), J-37-I-4c (Meidanki), J-37-I-4d (Bulbul). Explanatory Notes. Technoexport, USSR, 83 p.

Protasevich, L. N., and Maksimov, A. A., 1966. The Geological map of Syria. Scale 1:200,000. Sheets J-37-I, II (Halab, Antakya). Explanatory Notes. Technoexport, USSR, 74 p.

Razvalyaev, A. V., 1966a. The Geological map of Syria. Scale 1:200,000. Sheets I-37-VII (Dimashq) and I-36-XII (Beirouth). Explanatory Notes. Technoexport, USSR, 124 p.

Razvalyaev, A. V., 1966b. The Geological map of Syria. Scale 1:200,000. Sheet I-37-VIII (Saba Abar). Explanatory Notes. Technoexport, USSR, 64 p.

Ruske, R., 1978a. Geological map of Syria. Scale 1:50,000. Sheet Qerdaha (N I-37-S-1-c). Explanatory Notes. General Establ. of Geol. and Mineral Resources, Damascus, 36 p.

Ruske, R., 1978b. Geological map of Syria. Scale 1:50,000. Sheet Jableh (N I-36-X-2-d). Explanatory Notes. General Establ. of Geol. and Mineral Resources, Damascus, 26 p.

Shalash, J., and Al-Hafez, M. A., 1983. Geological map of Syria. Scale 1:50,000. Sheet Asal Al Ward (N I-37-SW-G-3-d). Explanatory Notes. General Establ. of Geol. and Mineral Resources, Damascus, 31 p.

Shatsky, V. N., Kazmin, V. G., and Kulakov, V. V., 1966. The Geological map of Syria. Scale 1:200,000. Sheets I-37-XIX and I-36-XXIV (Al-Latheqiyeh and Hama). Explanatory Notes. Technoexport, USSR, 107 p.

Soulidi-Kondratiev, E. D., 1966a. The Geological map of Syria. Scale 1:200,000. Sheet I-37-XV (Tudmor). Explanatory Notes. Technoexport, USSR, 104 p.

Soulidi-Kondratiev, E. D., 1966b. The Geological map of Syria. Scale 1:200,000. Sheet I-37-IX (At-Tanf). Explanatory Notes. Technoexport, USSR, 40 p.

Yousef, Sh., 1979. Geological map of Syria. Scale 1:50,000. Sheet Haffeh (N I-37-D-3-a). Explanatory Notes. General Establ. of Geol. and Mineral Resources, Damascus, 35 p.

Yousef, Sh., 1980a. Geological map of Syria. Scale 1:50,000. Qadmous-Banyas sheets (N I-37-S-1-a and No. I-36-X-2-b). Explanatory Notes. General Establ. of Geol. and Mineral Resources, Damascus, 32 p.

Yousef, Sh., 1980b. Geological map of Syria. Scale 1:50,000. Sheets Safita and Tartous (N I-37-M-3-c and No. I-36-R-4-d). Explanatory Notes. General Establ. of Geol. and Mineral. Resources, Damascus, 34 p.

Zumoffen, G., 1926. Géologie du Liban, avec carte géologique au 1/200,000. Barrère, 165 p.

Chapter 7

Volcanism

Alexander V. Razvalyaev

Vladimir G. Kazmin
P. P. Shirshov Institute of Oceanology of the Russian Academy of Sciences, Nakhimovsky Prospekt 23, 117 850 Moscow, Russia
Anatoly B. Galaktionov
State Research and Production Enterprise "Aerogeologia", 35 Lenin Prospekt, Moscow 117071, Russia

INTRODUCTION

One of the typical features of the geological structures of northeastern Africa and Arabia is the wide-scale development of basaltic fields of mostly Neogene-Quaternary age. These are the volcanic fields of Saudi Arabia (Harat-Reihat, Qisb, Jebel Abiad), Libya (Tibesti, Tripolitania, Haruj), Sudan (Darfur, Bayuda), and others. In Syria the basaltic fields and isolated volcanoes can be traced from the south to the north up to the boundary with the Taurus and Zagros (**Fig. 7.1**), but the main Neogene-Quaternary volcanism is located in the southwest of the country in the uplifts of Jebel Drouze and the Hauran Plateau (Fig. 7.1, and **Fig. 7.2**). The scale of this volcanism in Syria and its longevity, especially in the southwestern part of the country, are unique among the above-mentioned basaltic fields in northeastern Africa and Arabia.

In Syria and the adjacent countries (Lebanon, Israel, and Jordan) several stages of volcanism can be defined: Lower Jurassic; Lower Cretaceous; Upper Cretaceous; Eocene; and Neogene-Quaternary. The products of the Lower Jurassic, Lower and Upper Cretaceous and also Eocene volcanic activity are developed in limited areas, located along the western margin of Arabia (the West Arabian rift system: Lebanon, the Anti-Lebanon, and Jebel An-Nusseiriyeh).

The large basaltic fields (covers) in Syria, and primarily at Jebel Drouze and Hauran, are of Neogene-Quaternary age. For a long time they have attracted the attention of researchers. The first information on volcanism in this area appeared at the end of the nineteenth century (Lartet, 1877; Doss, 1886, and others). Later, the volcanic rocks were studied by Blanckenhorn (1914). The majority of the investigations at that time addressed the different questions of morphology and composition of the covers and only dealt with small areas of the Jebel Drouze uplift and the Hauran Plateau. In addition, some attempts were made to connect the occurrences of the Neogene-Quaternary volcanism in those regions with the N-S trending faults of the Dead Sea (Blanckenhorn, 1914).

Important contributions to the study of Middle East volcanism (in particular Jebel Drouze and the Hauran Plateau) were the investigations made by Tyrrell (1930), Dubertret (1929, 1933, 1940, 1954), and Vaumas (1956). Tyrrell discovered the alkaline variety among the Hauran Plateau basalts. The studies of Dubertret based on his 1:50,000 and 1:200,000 scale geological mapping shed light on many of the questions concerning the volcanism of Syria and Lebanon. He justifiably noted that the volcanism is related to a long interval of geological time, beginning from the Middle Jurassic; the earliest occurrences of the last stage of volcanism belong to the Miocene, and the latest, to the Holocene. Dubertret, and later Bentor (1957), noted the contemporaneity of the subsidence of the grabens related to the West Arabian (Levantine) fault system (rift) with periods of volcanic activity. Taking into consideration the above, they showed that there was no direct relationship between the volcanic extrusions and the above-noted faults. In their opinion, the numerous northwest trending faults on the Jebel Drouze uplift and the Hauran Plateau were the numerous sources of the lava flows, and this northwestern system is oblique to the longitudinal system of the Dead Sea faults.

A new stage in the investigation of Syrian volcanism began with the complex 1:200,000 (and in part, 1:50,000) scale mapping done by geologists of the USSR under the supervision of V. P. Ponikarov from 1958 to 1961. These investigations were conducted all over Syria and were the basis for the geological maps at 1:500,000 and 1:1,000,000 scale. The results were published in the explanatory notes to the different map sheets and also in a few articles and monographs dealing with various geological problems of the northern part of the Arabian platform and volcanism (Soulidi-Kondratiev and Kozlov, 1963; Kozlov, 1963; Kurbanov and Zaryanov, 1963; Ponikarov and Razvalyaev, 1967; Ponikarov, Soulidi-Kondratiev, Kozlov and Kazmin, 1964; Razvalyaev, 1964, 1971, 1991; Kazmin, Shatsky, and Mikhailov, 1964). Different aspects of Syrian geology were treated in the monograph entitled 'Geology and Mineral Resources of the Foreign Countries. Syria' by Ponikarov et al. (1969).

Later many facets of the geology of Syria, including volcanism, were studied by Soviet and Syrian geologists in the context of joint investigations by the USSR Academy of Sciences and the General Establishment of Geology and Mineral Resources of Syria. Among the works related to volcanism, the articles devoted to the age of the basalts and their petrology were especially important (Sharkov et al., 1987a, b, 1989, 1992, 1994, 1996; Sharkov, 2000; Novikov et al., 1993; Sharkov and Hanna, 1986).

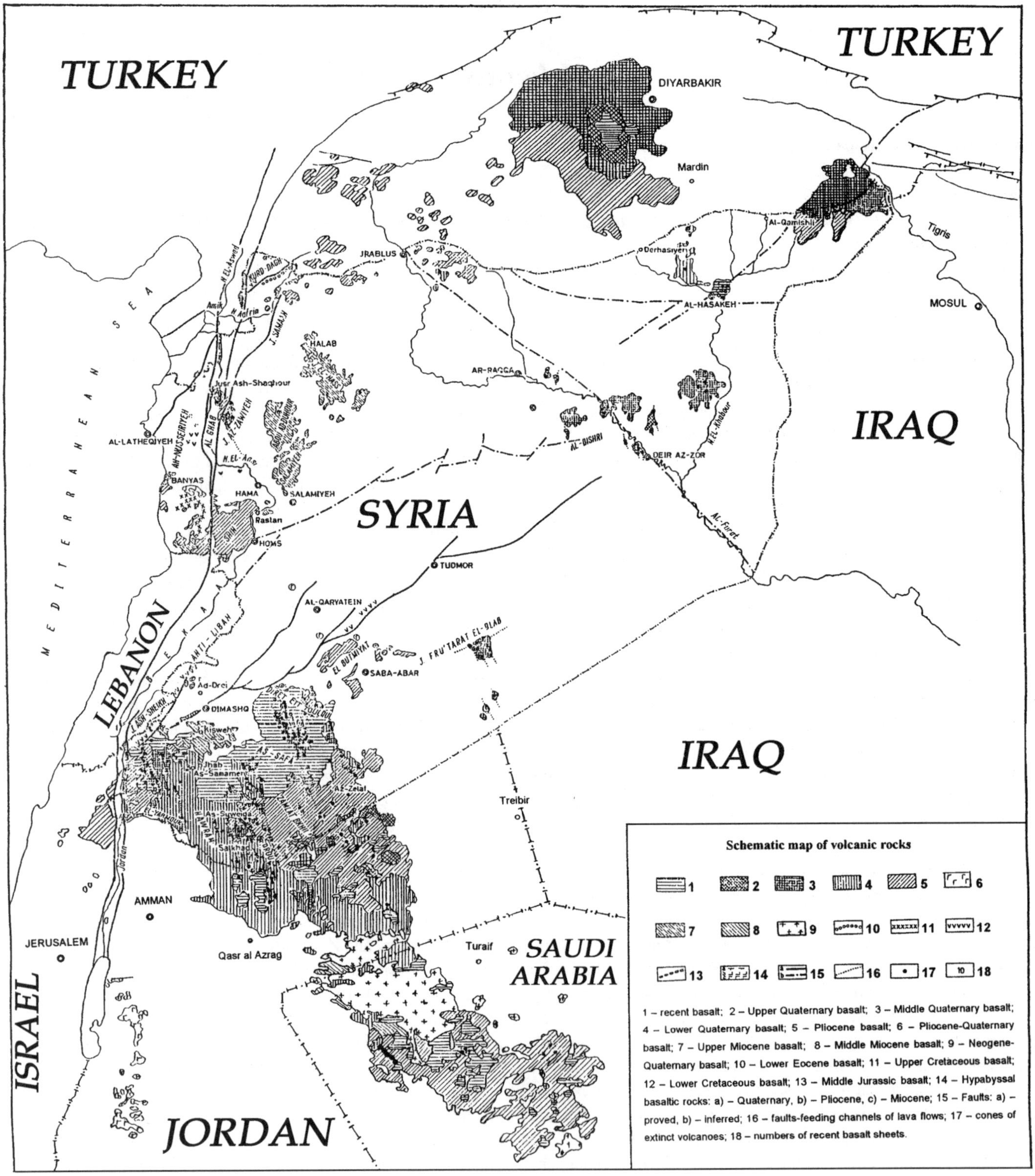

Fig. 7.1. Distribution of the volcanic rocks on the territory of Syria and neighboring countries.

Legend. Basalts: 1 – Recent; 2 – Upper Quaternary; 3 – Middle Quaternary; 4 – Lower Quaternary; 5 – Pliocene; 6 – undifferentiated Pliocene-Quaternary; 7 – Upper Miocene; 8 – Middle Miocene; 9 – undifferentiated Neogene-Quaternary; 10 – Lower Eocene; 11 – Upper Cretaceous; 12 – Lower Cretaceous; 13 – Lower Jurassic; 14 – Hypabyssal basaltic rocks: a) Quaternary. b) Pliocene, c) Miocene; 15 – faults: a) proved, b) inferred; 16 – faults-feeding channels of lava flows; 17 – cones of extinct volcanoes; 18 – numbers of Recent basalt sheets.

Jurassic, Cretaceous, and Paleogene Volcanism

Jurassic Volcanism

The Lower Jurassic basalts are well-developed in the Jebel Hermon Massif of southwestern Syria. They form a stratified body 40-60 m thick within the limestones of the Pliensbachian-Toarcian age (the Arne Formation). The rocks of the bottom and middle parts of this body are characterized by massive texture; the basalts of the upper part have an amygdaloidal texture. Basalts display a high grade of secondary changes: among the primary minerals only albite (0.2-0.6 mm laths), ilmenite (up to 1 mm) and chloritized glass could be seen; the colored minerals were replaced by a mixture of calcite, serpentine, and chalcedony. In the peripheral parts of the body, the pores are filled by zeolites, and in the internal part by quartz and chlorite rimmed with chalcedony.

Dubertret described these rocks as hypabyssal bodies of the Upper Jurassic basalts of Lebanon. But they differ from those basalts (without reference to their stratigraphic position) by the very high degree of secondary changes and stratification of their bodies, with a dominance of the amygdaloidal texture in the upper part. For these reasons these basalts should be attributed to effusive rocks, formed in sub-marine conditions. Their Early Jurassic age has been paleontologically proved (see Chapter 15 of this monograph).

Late Jurassic pyroclastic formations have been found in northwestern Syria in the axial part of Jebel An-Nusseiriyeh, near the village of Jourbet El-Burghal east of Qerdaha. Two ancient volcanic necks are described there, which breached the Oxfordian and, perhaps Lower Kimmerigian sediments of the Upper Jurassic. The necks are reflected in the relief by depressions up to 150 m deep of uneven but nearly elliptical form. The longer axes of these ellipses reach 1 km. The third volcanic edifice in this group is a relict of a neck with remnants of a volcanic cone buried under Aptian sediments. The necks and also the cones are built up of volcanic breccia and tuff of basaltic composition. Black or brownish-black rocks, sometimes with a greenish hue, are composed of angular pieces of basalt 0.01-30 cm in size, diabase, and Jurassic limestone and dolomite cemented by carbonate matter and volcanic ash. The rocks forming a volcanic cone are poorly stratified and dip towards the periphery of the cone. The thickness of the pyroclastic rocks in the central part of the cone is 40 m. On their surface, near the contact with the Aptian sediments, the volcanic rocks are covered by a reddish weathering crust. The age of these rocks is post-Oxfordian and pre-Aptian.

The explosive character of the volcanics of the Jebel An-Nusseiriyeh eruptions is similar to the Bhannes basaltic tuff in Lebanon (Dubertret, 1955). The age of this horizon is Kimmeridgian-Tithonian. Its thickness varies from 20 to 50 m, but in the Beqaa depression it reaches 150 m.

In the upper part of the Mulussa Formation, the Aba-1 bore-hole, drilled in the crest of Jebel Tuwal El-Aba, penetrated two beds of effusive rocks (dolerites) about 2 m thick (at the depth of 1714-2000 m). The age of this member was treated as Upper Triassic-Lower Jurassic. We can not exclude the idea that these effusives correspond to the Lower Jurassic basalts of the Arne Formation (the Anti-Lebanon, the Hermon Massif).

Lower Cretaceous Volcanism

The Lower Cretaceous basalts are much more widespread than those of the Jurassic. They are exposed in the mountains of the Anti-Lebanon, Jebel An-Nusseiriyeh, and on the southeastern side of the Kurd-Dagh range. Some small exposures of Lower Cretaceous volcanics have also been found in the Palmyrides (Jebel Rmah).

In the Anti-Lebanon volcanogenic formations of the Lower Cretaceous are neither uniform in their thickness nor in the extent of their exposure. In some sections they are absent or manifested by thin lenses, but in others (near the villages of Mazra'a and Serhaya) the lava flows comprise the entire section from the Upper Jurassic limestone up to the very well-known Blanche Wall limestone in Lebanon and the Anti-Lebanon (**Fig. 7.3**). The internal structure of the flows is also not homogeneous, neither in its composition nor in structure. A transformation from dolerite to microdolerite and then to basalt is recorded.

There is no record of a single stratigraphic position for the Lower Cretaceous basalts. In the Jebel Hermon area, where their thickness is 12 m, they lie at the base of the sandstone member underlying the Blanche Wall limestone, but in Jebel Shaqif their 150 m thick unit was found much higher. Another difference between the Early Cretaceous volcanism of northern and southern Lebanon is observed on the Tripoli sheet where basalts overlie the Blanche Wall limestones (Dubertret 1949, 1955). Consequently, the main part of basaltic covers belongs to the Neocimian but the Aptian basalts are also present.

The Lower Cretaceous deposits of the Palmyrides folded zone, which unconformably overlie the Middle and Upper Jurassic rocks, consist of Albian carbonates and brown, yellowish and white pre-Albian sandstones, sometimes ferruginous, in places cross-bedded, intercalating with marl layers (Faradzhev, 1966). In some exposures (*e.g.* Jebel Rmah) these deposits include 12-40 m thick covers and sills of diabase, sometimes with amygdaloidal structure. The diabase is built up of labradorite olivine, and ilmenite. Apatite is usually present. The primary minerals are hydrothermally altered and replaced by zeolites, chlorite, iddingsite, and seladonite. The volcanic glass is replaced by chlorite. The amygdaloids are filled by calcite.

Thin covers of basaltic lavas and lava breccias are known in deposits of probably Aptian age in the south of Jebel An-Nusseiriyeh, in the southern part of the Massiaf Plateau near the villages of Ba'mara and Ofelon (Kozlov, 1966a), and also in the Kurd Dagh mountains in northwesternmost Syria (Protasevich and Maksimov, 1966).

Upper Cretaceous Volcanism

Two horizons of Upper Cretaceous basalts have been found in the southwestern part of Jebel An-Nusseiriyeh. They lie at the base and in the lower part of the Cenomanian sediments (Shatsky et al., 1966; Kozlov, 1966a). The lower cover (or series of covers), 15-50 m thick, is in between faunistically dated Albian and Cenomanian

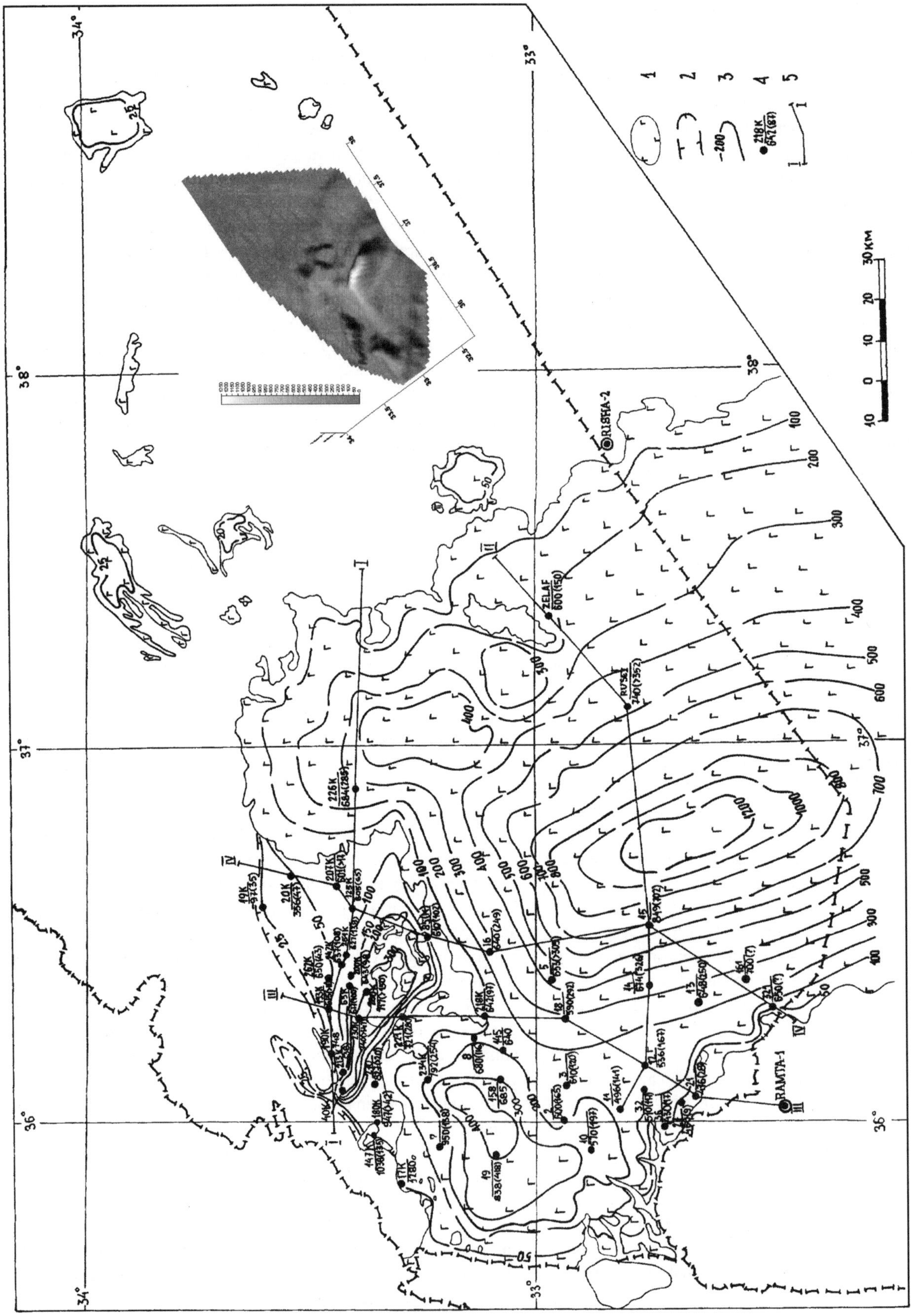

Fig. 7.2. Thickness map of the volcanic rocks of the Hauran Plateau and Jebel Drouze highland.
Legend: 1 - the region of effusive rock exposures; 2 – boundaries of the effusive rock distribution in the Damascus depression (drilling data); 3 – contour lines of the total thickness of the effusives; 4 – bore-holes that penetrated the effusive rocks: numerator = bore-hole number, denominator = absolute elevation of the top of the effusive rocks and (in parentheses) = bore-hole depth; 5 – the lines of geological profiles (cross-sections): I-I, II-II, III-III, IV-IV.

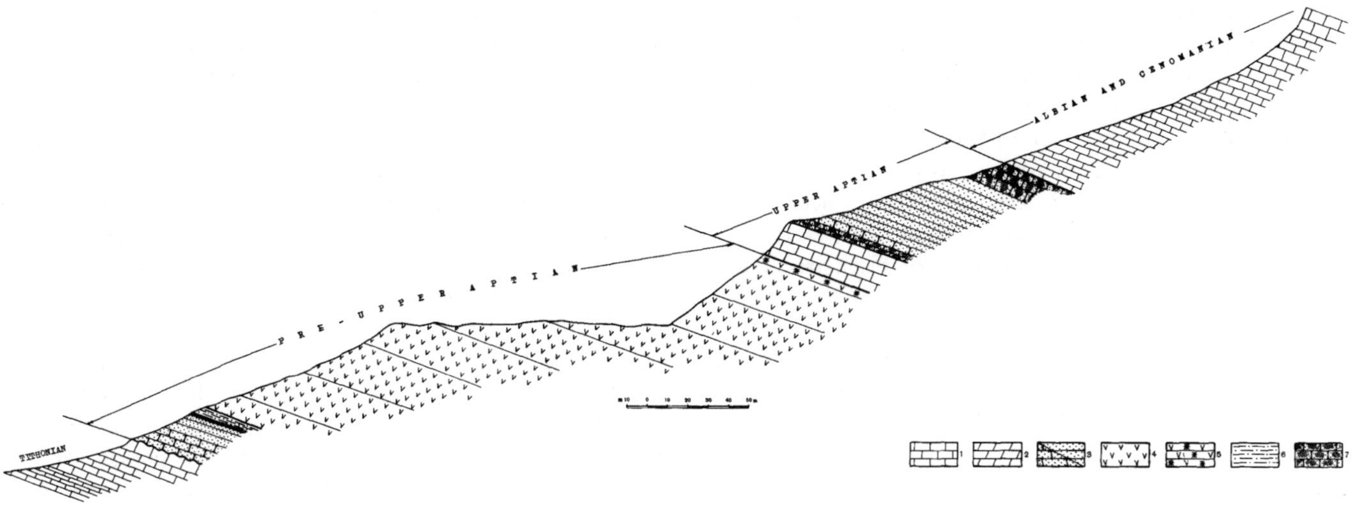

Fig. 7.3. Cross-section of the Upper Jurassic and Early Cretaceous sediments of the Jebel Hermon Massif (the Anti-Lebanon).
Legend: 1 – limestones; 2 – marls; 3 – sands and sandstones; 4 – effusive rocks; 5 – volcanogenic ferruginous rocks; 6 – sandy-clayey rocks; 7 – beds and lenses of ferruginous rocks.

sediments. The upper cover (about 30 m thick) lies within the Cenomanian carbonate rocks. The lower cover is laterally replaced by sandstones containing basalt fragments (mostly glass) with some addition of tuffogenic material.

The basalts are characterized by pillow lavas, attesting to underwater eruption. This conclusion is confirmed by the high degree of secondary changes, wherein the primary minerals have only been preserved in the central parts of the pillows. The peripheral parts of these pillows were converted to typical hyaloclastites.

The rocks have a porphyric structure with large crystals consisting of augite, olivine, and plagioclase. The main mass is predominately volcanic glass; unfortunately, there are no data on its chemical composition. According to the mineralogical description of the rock, it probably belongs to alkaline and sub-alkaline olivine basalts.

Eocene Volcanism

Basalts and tuffs of Early Eocene age are only known in the northwest of Syria, not far away from the Kurd Dagh mountains, north of the town of Aafrine. In the section southwest of the Meidanki settlement, a 205 m thick formation of intercalating white-gray marls and black and brown chert beds (3-6 m thick) conformably overlies the clayey limestones and marls of the Lower Eocene. At the base of this formation there is a bed of dark, brown-green, loosely cemented tuffs consisting of large, sharply angular fragments of volcanic glass in carbonate cement. Through the entire sequence, sedimentary rocks alternate with basaltic tuffs and black hard melanocratic basalts with amygdaloidal structure. The amount of basalts and tuffs constitutes not less than a half of the total thickness of this sedimentary-volcanic sequence. Planktonic foraminfera from intercalating limestones assign this formation to the latest Early Eocene (the *Acarinina pentacamerata* Zone)

The distribution of Eocene volcanics is very limited. The amount of tuff material decreases towards the northeast and to the southwest of Meidanki, and effusive rocks of this age haven't been found in any other area. The eruption centers are probably related to the Al-Latheqiyeh-Kilis fault.

In southwestern Syria (Hauran Plateau) the 218-K hydrogeological bore-hole near Jebab penetrated two members (in the intervals of 277-305 m and 310-375 m) of effusive rocks (basalts, dolerites and anamesites) separated by greenish-gray marls of Paleocene age. To the north of Jebab, the 230bore-hole also cut (in the interval of 450-487 m) Paleocene deposits with basalts. Unfortunately, we are not familiar with paleontological data which might confirm the age of the sediments.

Neogene-Quaternary Volcanism

A primary feature of the geological structure of the Arabian platform is the highly intensive Neogene-Quaternary volcanism. The basalt outflows in this part of the Near East occurred during Miocene, Pliocene, and Quaternary times. Another feature is the spotty lateral distribution of the youngest volcanism. The area of maximum volcanism is within or close to the bounds of the West Arabian rift system (or fault zone). The Neogene-Quaternary volcanism is not evident outside this system. Even within the West Arabian rift system the youngest volcanism is localized to two areas (Fig. 7.1). The first and largest includes a huge region (6000 km^2) of southwestern Syria - Jebel Drouze and the Hauran Plateau. The second volcanic region includes the southern margins of Jebel An-Nusseiriyeh in northwestern Syria, the area near the Lebanese border between the cities of Tripoli and Homs, the nearshore area of the Mediterranean Sea ,and the Aleppo Plateau.

Besides these regions, dispersed covers of mostly Pliocene-Quaternary basalts have been recorded between the Al-Furat (Euphrates) River and its Nahr El-Khabour left tributary near Hassetche, and in northeasternmost Syria to the east of Qamichliye. Finally, quite a large area of Pliocene-Quaternary volcanism was recorded over the border in Turkey near Mardin and Diyarbakir.

In southwestern Syria huge masses of black basalt form mountains and broad plateaus (Jebel Drouze and the Hauran Plateau) topped by numerous rows of volcanic cones. The Holocene basalt covers and flows are practically untouched by erosion and on their surface they demonstrate beautifully preserved lava features, which give an illusion of a frozen stone sea.

Miocene Volcanism

Following a period of quiescence, volcanic activity in the Near East was rejuvenated in the Miocene.

<u>Lower Miocene Basalts</u>

The Lower Miocene basalts are developed in the southwest and northwest of Syria.

In the southwest they occupy the northernmost margin of the Hauran-Drouze Plateau, south and east of Damascus. In the neighborhood of Kissouye village south of Damascus, the Lower Miocene basalts build separate outliers of a strongly dissected plateau, about 200 km^2 in size. The basalts reach 400 m in thickness. The consist of a succession of 15 to 20 flows, and include basalts of both normal and moderate alkalinity. Every flow begins with solid, massive basalts and ends with porous varieties. Among the rocks the small-grained, well-crystallized types of basalts are usually dominant; the micro-grained and glassy varieties were more rarely encountered. Thin lenses and layers of volcanic tuffs, and the weathering crusts that separate the flows, can be often seen among the basalts.

The normally alkaline basalts are usually represented by dolerites and less often by micro-dolerites. Melanocratic and leucocratic types are rare. Among the moderately alkaline basalts, analcime basalts and analcime basanites are evident, and teschenite is seldom encountered.

Basalts of the same age are also known from the southwestern part of the Ad-Daww depression (Sharkov et al., 1998).

The Miocene basalts descend in a northerly direction into the Damascus depression, where they are clearly related to the Miocene and younger sedimentary rocks (Razvalyaev, 1966a; Sharkov et al., 1994). In the eastern part of the depression, in the Abu al-Ata quarry 30 km from Damascus, the Oligocene limestones are unconformably overlain by a 25 m thick unit of white and red fine-grained quartz sands of lacustrine-alluvial genesis, which contain large (1.0-1.5 m) fragments of basalts at the top. On the sand there is a basalt flow 10-12 m thick overlain by gray sandstones which grade upward to conglomerates. This section and the data from neighboring Bore-holes 19 and 20 show that the basalts fill a large part of the Damascus depression. They are positioned exactly above the unit of quartz sands and below the sandy-conglomerate member, which probably belongs to the Lower Pliocene. The Early Miocene age of the quartz sands was determined by correlation with the continental sediments of the Palmyrides folded zone, where a similar lithologic unit rests between faunistically dated Oligocene and Middle-Miocene sediments (Ponikarov et al., 1967). Later the Early Miocene age of the basalts was also confirmed by isotopic dating of the basalt blocks from the top of these sands, *i.e.* 20-21 Ma (Sharkov et al., 1994). In the Ad-Daww depression the K/Ar age of the basalts, within the unit of quartz sands, was 24.7-26.0 Ma (Sharkov et al., 1998).

On the Hauran-Drouze Plateau, some basaltic flows gave Middle Miocene isotopic dates (14-11 Ma). Thus, in southwestern Syria the oldest basaltic covers overlying Paleogene deposits are of Early-Middle Miocene age. To our regret, at present there are still insufficient data for subdividing the basaltic covers into some regional mapping units. Because of this, the Miocene basalts of south-western Syria are usually considered as a single complex.

Hydrogeological bore-holes have shown that the Miocene basalts are developed far beyond the boundaries of the Damascus depression. They have been penetrated by a series of bore-holes south of the depression up to the western slope of Jebel Drouze. The thickness of the effusive rocks (according to drilling data) reaches 150-200 m. The maximum thickness of the Miocene basalts was recorded in the vicinity of the Kissouye village, where the visible part of their succession reaches almost 500 m.

We may assume that the Lower Miocene basalts are also present within the bounds of the Hauran Plateau and Jebel Drouze itself (**Figs. 7.4** and **7.5**) This evidence originates from Bore-hole 15, drilled near the western margin of Jebel Drouze. Here a 200 m thick unit of basalts overlying Middle Eocene marls and covered by Pliocene basalts is assigned to the Miocene. Presumably the Lower Miocene basalts were penetrated by a bore-hole near the town of Zelaf in the eastern part of Jebel Drouze as well as by Bore-holes 16, 20-K, 125-K, 190-K, 230, and 253-K. Indirect evidence for the presence of Miocene basalts within the bounds of Jebel Drouze is provided by a conglomeritic member exposed in a deep wadi east of the Tell Fezraneh village. These conglomerates consist of pebbles of strongly fractured and changed basalts which are visibly different from the fresh basalts of Pliocene age overlying the conglomerates. There is no direct data on the presence of Miocene basalts in the axial part of Jebel Drouze, but this mass of basalts about 400 m thick probably belongs to the Miocene.

The location of the sources of Miocene lava flows is not known. No volcanic cones have been found on the surface of the Miocene basaltic covers. Nevertheless, the localization of basalts near large faults of the West Arabian fault (rift) system, the Palmyra aulacogen, and the Al-Latheqiyeh-Kilis suture may indicate that these structures are the likely sources of the outflow.

<u>Middle Miocene basalts</u>

In most of northwestern Syria the Middle Miocene sediments are represented by carbonate rocks. In the Al-Latheqiyeh and Aafrine depressions they are related to the Lower Miocene by a gradual transition. On the Aleppo Plateau and some small block structures they transgressively overlie the Lower Miocene, Oligocene, and Eocene sediments with a stratigraphic hiatus.

Volcanic rocks have only been found in the northwesternmost part of this area - the valley of the Nahr Aafrine River, the mountains of Jebel Smaane, and in the vicinity of Aleppo (Protasevich and Maksimov, 1966).

The greatest thickness of volcanics was recorded south of the Qatma settlement, near the main Aleppo-Aafrine highway. Here, there is a sequence of a few basalt covers separated by layers of organogenic-detrital limestones overlying an eroded surface of Burdigalian (Lower Miocene) rocks. These limestones contain the fauna of gastropods and pelecypods of Middle Miocene age. The thickness of a single cover of basalt (or a packet of thin covers) reaches 80 m. The total thickness of the basaltic unit is about 200 m. Higher in the section there is a member of sands and conglomerates containing a single thin (0.8 m) basalt flow. The total thickness of this part of the section, which also includes Middle Miocene fauna, is 50 m.

In certain covers belonging to the effusive-sedimentary series, pillow structures were recorded, which attest to the underwater character of the outflow. In the upper part of each basaltic cover, there is a transition to effusive rocks with amygdaloidal structure, with a growing amount of volcanic glass. A red weathering crust is developed on the surface of the pillow lava cover that tops the 250 m thick lower part of the sequence. It probably indicates a dry period of continental weathering.

The basalts comprise plagioclase (labrador), augite, olivine, ilmenite, and apatite. An important part of the rock mass is glass. The porphyric phenocrysts (1-2 mm) are composed of olivine and plagioclase. Among the secondary minerals described are iddingsite, bowlingite, serpentine, talc, limonite, chlorite, calcite, zeolites, and chalcedony. A strong carbonatization of the rock is also recorded.

Towards the east and south of the Nahr Aafrine valley the quantity of volcanic rocks in the Middle Miocene section decreases. In the succession along the Aleppo-Damascus railway line (south of Aleppo) only two thin basalt covers were recorded, and further south the volcanics are completely absent (Mikhailov, 1966). Replacement of basalts and associated clastic rocks by limestones has also been noted in the western scarp of Jebel Smaane (Protasevich and Maksimov, 1966). Thus we can conclude that the outflow centers were located in the Kurd Dagh area and in the proximity of the Syrian-Levant Fault.

Upper Miocene basalts

Upper Miocene basalts are developed east of Jebel An-Nusseiriyeh and in the Jebel Az-Zaouiye mountains, where they make up the relicts of the high plateau which reaches elevations of 400 m and higher. The basalts armor the older surface of the peneplaned Paleogene carbonate rocks. In some exposures basalts overlie Middle Miocene limestones. The K/Ar age of those basalts is 7.8-12.0 Ma (Sharkov et al., 1998). The largest relicts of the basalt cover (Jebel El-Hass, Jebel Shbith, and the Abou Ad-Dahour and Selemiye plateaus) are dissected by valleys 100-150 m deep. In places, smaller exposures of basalts have been recorded in between the huge relicts, which attests to a much larger distribution of these rocks in the past. The largest of these relicts reach 800 km^2 and more. Small exposures of the Upper Miocene basalts are also known on the southern margin of the Jebel An-Nusseiriyeh mountains.

The thickness of the Upper Miocene basalts is not very significant and ranges between 10 and 50 m. They probably consist of a few covers, but there is no exact documentation. The location of the centers of eruptions is also unclear. These eruptions were probably of the fracture type and were controlled by the NW-SE-trending faults. At least, faults in this direction are clearly visible, especially on aerial photographs, crossing the sediments underlying the basalts. It is also possible that there were eruptions of the single-center type. Thus, some gently sloping low hills (up to 20 m high), built of brownish scoria, have been discovered on the surface of the basalt plateau near the village of El-Fhali.

Basalts are usually of massive, sometimes fluidal texture. Their composition is very monotonous. Small variations have only been found on the crystallization level - the main mass exhibits a transformation of micro-dolerite to hyalobasalts. The rock-forming minerals are plagioclase (labrador, bitovnite), olivine, augite, a large amount of ilmenite and magnetite; sometimes apatite and rutile are found. The age of these basalts is indirectly suggested on the basis of several observations. On the one hand, they overlie sediments of up to Middle Miocene age which are observed in the Aleppo area and in Lebanon (Dubertret, 1937), and on the other hand, Pliocene lacustrine sediments have been found in deep valleys which cut the Miocene basaltic plateau within the bounds of the Homs depression (Kozlov, 1966).

To the southwest of Syria, the Upper Miocene basalts display a special relation to the large faults of the West Arabian (Levantine) system. According to Bentor (1957), the Upper Miocene outflows are well-developed in the Jordan River rift valley, though less so than those of the Middle Miocene. In some exposures lava flows, agglomerates and tuffs intercalate with lacustrine sediments. Among the products of volcanic activity, there is much more pyroclastic material. These data, together with the rapid changes in thicknesses of the tuffs and lavas allow us to assume that the basalt flows in this area came from a single center. The eruption centers, however, haven't been found. The total thickness of the Upper Miocene basalts in the Jordan River valley is around 150 m.

Pliocene Volcanism

The Cenozoic volcanism of Syria and adjacent countries reached its peak in the Pliocene (Sharkov et al., 1988). At that time huge masses of lava flowed out, forming large basalt plateaus. The Pliocene basalts are especially well-developed in the southwest of Syria. They also build up the large Shin Plateau between the cities of Homs and Tripoli. Some comparatively thin covers of Pliocene basalts are present along the border between Syria and Turkey. They have been recorded as relicts in many places of the southern part of Jebel An-Nusseiriyeh, and were also penetrated by a few

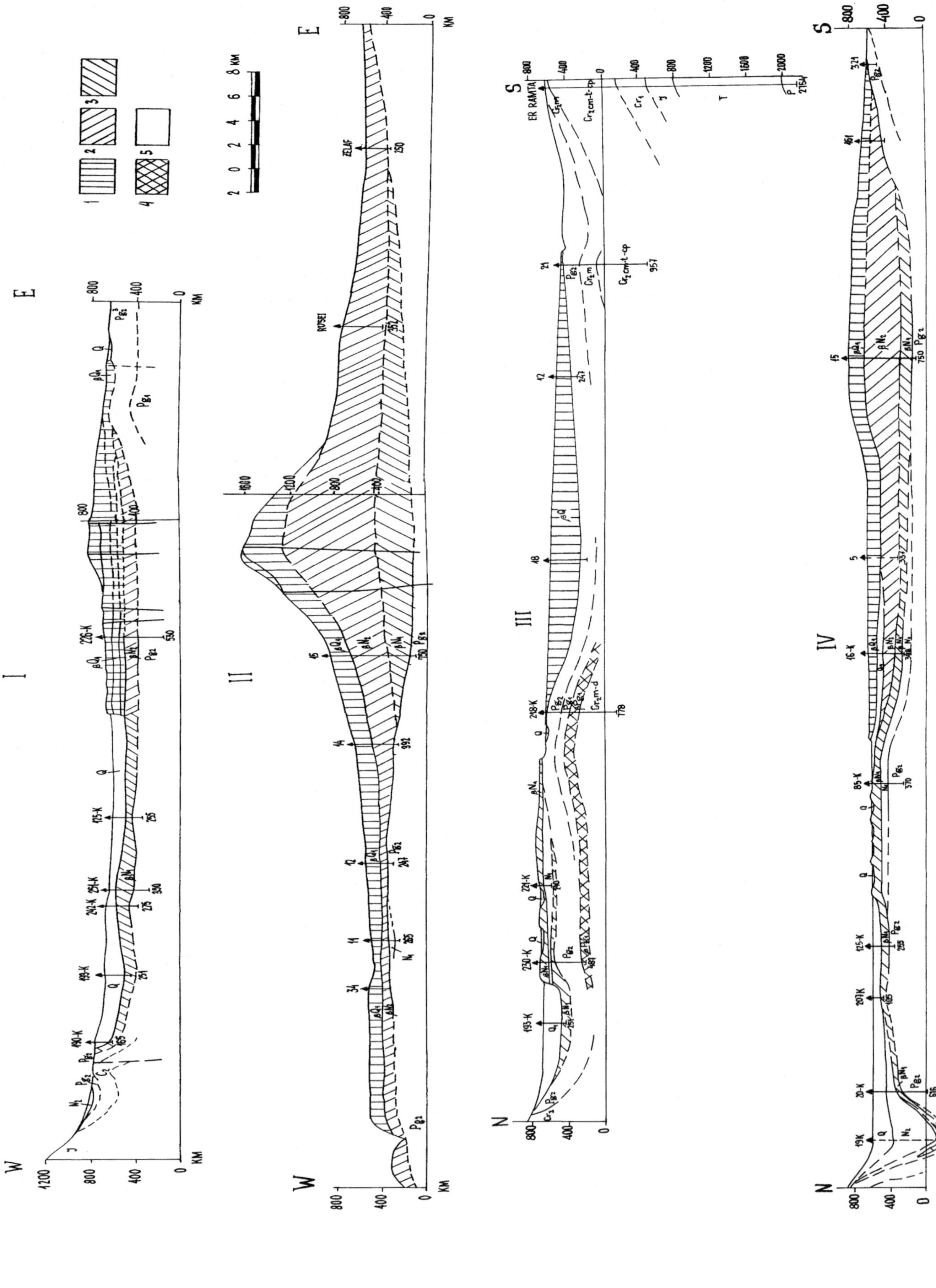
I
II
III
IV
W
E
N
S
ER RAMTA
ZELAF
ROSEI
KM

Fig. 7.6. The Upper Pliocene basalts overlying Mesozoic rocks. Southern margins of Jebel Hermon (the Anti-Lebanon).

bore-holes within lacustrine sediments in the Homs depression, where they constitute a large part of the fill. Their existence has been confirmed by aero-magnetic survey.

In southwestern Syria the Pliocene basalts built a large plateau to the east of the axial part of Jebel Drouze, as well as another on the western part of the Hauran Plateau adjacent to the Jordan and Yarmouk valleys. They are also developed in the areas with intensive manifestation of Quaternary volcanism in Diret-ett-Touloul and Hauran, where their presence at depth has been confirmed by drilling (Figs. 7.4 and 7.5).

The Pliocene basalts of southwestern Syria have some common features. All of them form covers comprised of a series of lava flows separated by sedimentary rocks, including layers of red clays. The surface of these covers is intensely eroded and overlain by piles of basaltic debris and blocks, and loam. In contrast to the Quatenary covers, no ropy or pillow lavas could be found among them. Apart from the common features, the Pliocene basalts often occupy a different structural (tectonic) position and possess some specific additional characteristics. They were influenced by climatic factors different from those of the Miocene.

In the Hauran area, along the border with Israel, the Pliocene basalts form a slightly hummocky (hilly) and sometimes flattened plateau. The surface of the plateau is covered by loam and basaltic debris. Just near the Israeli border, the plateau is gently inclined westward, towards the Jordan River. It is dissected by deep valleys or wadis, on whose thalwegs the Paleogene sediments are often exposed. In the north, at the foothills of Jebel Hermon, separate relicts of this plateau rise to an elevation of 1,500 m (**Fig. 7.6**).

Fig. 7.4. Sub-latitudinal geological cross-sections (I-I and II-II) across the Hauran and Jebel Drouze volcanic plateaus.
Legend (for Figs. 7.4 and 7.5): 1 – Quaternary basalts; 2 – Pliocene basalts; 3 – Miocene basalts; 4 – Paleogene basalts; 5 - sedimentary rocks (with age indexes). Position of cross-sections is shown on Fig. 7.2.

Fig. 7.5. Sub-longitudinal geological cross-sections (III-III and IV-IV) across the Hauran and Jebel Drouz volcanic plateaus.
Legend (see Fig. 7.4).

Fig. 7.7. Volcanic neck Salkhad of the Pliocene age. The Jebel Drouze highland, view from the west.

On the surface of the Hauran Plateau strongly eroded volcanoes were preserved, which erupted during the final stage of volcanic activity. In the eastern part of the plateau, south of Lake Traneh, there is a row of low hills (20-100 m) trending northwest, consisting of 8-10 almost completely destroyed volcanoes. In the foothills of the northern margin of the plateau, the outliers of the lava plateau can be seen with remnants of the volcanic cones, but usually they are almost completely eroded and the volcanic necks (lava tubes) are exposed (**Fig. 7.7**). A similar neck can be observed near the Majdal Shams village on the southern margin of Jebel Hermon (**Fig. 7.8**).

The Pliocene basalt covers of the Hauran Plateau are comprised of numerous intercalating lava flows built up of normal or alkaline basalts. The thickness of these flows ranges between 4 and 20 m. About 18 of such flows were counted in the upper reach of Wadi Raqqa. All flows have very similar textural and morphological features. The lower part of each is composed of porous, sometimes vesicular basalt; its central part includes well-crystallized massive types, and the porous and vesicular basalts are again predominant at the top of the flow. At the base of the flows, layers of basaltic breccias can be observed, where fragments of effusive rocks are cemented by a red-colored sandy-clayey matrix. These layers, up to 2.5-3.0 m thick, represent a special type of weathering crust which developed under conditions of a hot arid climate during periods between the lava outflows.

The Pliocene basalts of the Hauran Plateau overlie the Jurassic, Cretaceous, and Paleogene deposits and Miocene basalts with sharp angular unconformity. On the eastern nearshore of Lake Tiberias, in the deep valleys of the Yarmouk River and Wadi Raqqad and Saffa, the Pliocene is subdivided into two parts. The lower part is represented by sedimentary rocks. The upper part also starts with sedimentary rocks but already at the base of this unit separate layers of basalts appear. Upwards in this section the basalts quickly replace sedimentary rocks and form a 200 m thick sequence. Farther to the north the sedimentary rocks completely disappear from the section, and the whole unit is only comprised of basalts. It should be noted that the terms Lower Pliocene and Upper Pliocene can be arbitrarily used here. It would be better to refer these two units to the lower and upper parts of the Pliocene (Krasnov et al., 1966).

East of the Hauran Plateau, the Pliocene basalts occupy the large desert region of Harra (Kraa), whose northern part is in Syria and southern in Jordan. The Pliocene basalts are overlain by covers of Quaternary basalts in a sizable part of this region. The presence

Fig. 7.8. Neck near the Majdal Shams village, the southern foothills of Jebel Hermon (the Anti-Lebanon).

of these basalts has been confirmed by numerous drillings. They are exposed on the surface of Jebel Drouze and to its east within the bounds of the Syrian desert, up to the contact with exposed Paleogene sediments. The Pliocene basalts stretch as separate outliers to the area of Saba-Abar and further on to Jebel Boutliyate and Diret-et-Tlul (Diret-att-Touloul) near the Palmyrides mountain ranges. The entire area of exposed Pliocene basalts covers about 4,000 km^2.

The Harra mountain plateau (600-700 m elevations) looks like a typical rocky (stony) desert covered with the debris of basaltic rocks. Along the western border of this plateau, the mountain massif of Jebel Drouze rises up to elevations of 1,800 m. This massif is elongated in the northwestern direction, and separates the mountainous Harra and Hauran plateaus. It is a typical volcanic megastructure. Local centers of the Pliocene volcanism and its accumulated products make up this structure. Jebel Drouze is topped by northwestern-trending rows of volcanoes which produce its present relief. The activity of most of these ceased in the Pliocene, but some of them were active up to the Quaternary. The volcanic cones are usually not very high and are strongly eroded.

The thickness of the Pliocene basalts increases from the peripheral parts of the Harra Plateau to a maximum within Jebel Drouze. In the periphery of the plateau their thickness is several tens of meters. East of Jebel Diret-att-Touloul, according to drilling data, it is about 100 m. In the Zelaf bore-hole their thickness increases up to 200 m. On the western slope of Jebel Drouze the thickness of the basalts reaches up to 300 m (Bore-hole 15) and in the east it is considered to be about 350 m (Bore-hole Ruseyi). There is no direct data on the thickness of the Pliocene basalt in the central part of Jebel Drouze. However, according to the elevation and hypsometric pattern of the base of these basalts in the nearby bore-holes (about 450 m), their thickness will be a maximum there and should be approximately 700-800 m.

Another large area of Pliocene basalts is in northwestern Syria. The basalts are well-exposed there on the southern end of Jebel An-Nusseiriyeh where they build up the basaltic Shin and Tell Kalakh plateaus. To the north, in the eastern part of the Al-Ghab graben, on the Jebel Az-Zawiyeh (Jebel Az-Zaouiye) mountain massif, and in the northern end of the Al-Ghab graben near the town of Jisr Ash-Shaghour. there are small exposures of Pliocene basalts. In addition there are also well-developed areas with Pliocene basalts near the Syrian-Turkish border, which can be traced from Jebel Samaan (Jebel Smaane) in the west up to the right bank of the Al-Furat River near the town of Jrablus. In summary, the main

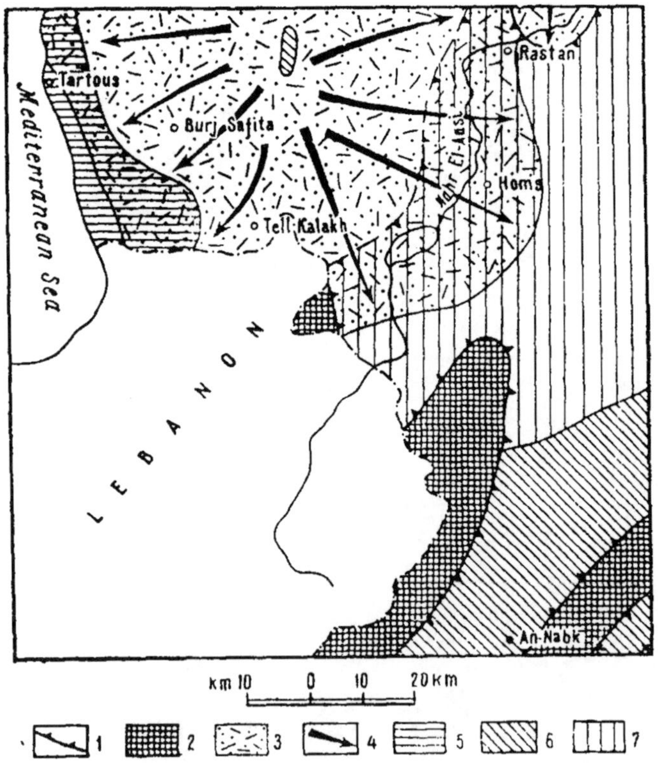

Fig. 7.9. Paleogeographic scheme of the Neogene epoch in the area of the basaltic Plateau Shin (the Homs Basin). After Kozlov (1966a).
Legend: 1 – slopes of elevations; 2 – stable elevations, denuded; 3 – covers of basic effusives; 4 - trend of volcanic flows; 5 – Lower Pliocene marine basin; 6 – Miocene – Lower Pliocene depressions; 7 – Upper Pliocene lacustrine basin.

mass of Pliocene volcanics is located near the Levantine-Syrian fault zone, mainly to its east, but also partly to the west, near the Mediterranean coast.

Both Lower and Upper Pliocene volcanics are known in the southern margins of Jebel An-Nusseiriyeh. In respect of their areal distribution the Upper Pliocene volcanics predominate. On the coast in the vicinity of the towns of Tartous and Banyas, Early Pliocene marine sediments are overlain by a pile of tuff breccia composed of basalt fragments cemented by tuff material. These rocks are 25-30 m thick and in places include lenses of marine limestones containing Early Pliocene molluscan fauna. The tuffs and tuff breccias are overlain in some of the coastal sections by a 20-25 m thick unit of basaltic lava. These basalts have been assigned to the upper part of the Pliocene because the Lower Quaternary marine terrace (at 60 m elevation above sea level) is situated 150 m below the surface overlain by the basaltic cover.

In an eastern direction the Upper Pliocene basalts are traced on the Tell Kalakh and Shin plateaus, to the west and to the east of the Levantine-Syrian fault respectively. Adjacent to the fault they are up to 300 m thick; to the west of it (seaward) and eastwards (to the Homs depression) their thickness decreases to 10-20 m. The volcanic sequence is composed of numerous covers clearly separated by glassy crusts or red weathering crusts. Some basalt covers were drilled within the Homs depression, where they intercalate with the Upper Pliocene lacustrine deposits. The centers of eruption of the Miocene and Pliocene basalts were situated on the southern margin of Jebel An-Nusseiriyeh, in the Levantine-Syrian fault zone. From this area the lava flowed to the west, south and east (**Fig. 7.9**) (Kozlov, 1966a). On the highest part of Jebel An-Nusseiriyeh the volcanics were eroded, but some small hypabyssal intrusions of dolerites, possibly volcanic necks, were preserved. It is interesting to note that the presumed centers of eruptions there coincide with a big (up to 200 γ or nT) magnetic field maximum (**Fig. 7.10**). This could indicate the presence of large basic intrusions at depth.

Absolute age determinations of basalt samples from the Shin Plateau vary from 10 to 4.5 Ma, *i.e.* within the bounds of the Late Miocene-Early Pliocene (Sharkov et al., 1998). However, in the field these basalts are known to alternate with Late Pliocene lacustrine sediments, so these isotopic findings are not in agreement with the geological setting.

Two fields of basaltic lava, respectively 54 and 45 km^2 in area,. exist at the northern end and on the eastern side of the Al-Ghab graben. In both cases the outflows are related to highly eroded volcanoes which are still visible on aerial photographs and in the field. North of the Al-Ghab graben, eight volcanoes encircle the southern plunge of the box-shaped anticline of the Jebel El-Wastani range, which separates the graben into two branches. The lava flows probably overlie the Middle Miocene limestones and deepen in a southerly direction under the Quaternary sediments filling the graben. A volcanic edifice up to 100 m high and up to 1 km in diameter is composed of intercalated lava flows, tuff and scoria. The flow thickness quickly decreases from tens of meters near the eruption center to 1-2 meters at the periphery of the lava field.

The lava field of the Jebel Az-Zawiyeh range includes 18 small volcanoes situated on its crest. This range is the eastern flank of the Al-Ghab graben. It has a steep western slope and a gentle eastern one The graben already existed during the eruptions: the lava flowed down its plunging flanks and filled the more ancient valleys within. These features and the comparatively good preservation of the volcanoes sharply differentiate the basalts of Jebel Az-Zawiyeh and Al-Ghab from the Upper Miocene plateau-basalts developed further east. Those basalts are older than some of the Quaternary sediments filling the Al-Ghab graben. According to K/Ar determinations (1.95-1.1 Ma), these basalts belong to the Uppermost Pliocene-Lower Pleistocene (Sharkov et al., 1998).

A few exposures of basalts, less than 40 m thick, are known close to the Syrian-Turkish border. They represent outliers of much larger former basaltic plateaus that were preserved on the flat summits of the mountains at elevations of 600-650 m. According to their morphology and composition, these rocks cannot be differentiated from the Upper Miocene basalts, but in certain cases (*e.g.* at Jebel Smaane) they overlie the lacustrine sediments of the lower part of the Pliocene (Protasevich and Maksimov, 1966).

A presumably Pliocene age was attributed to the outliers of

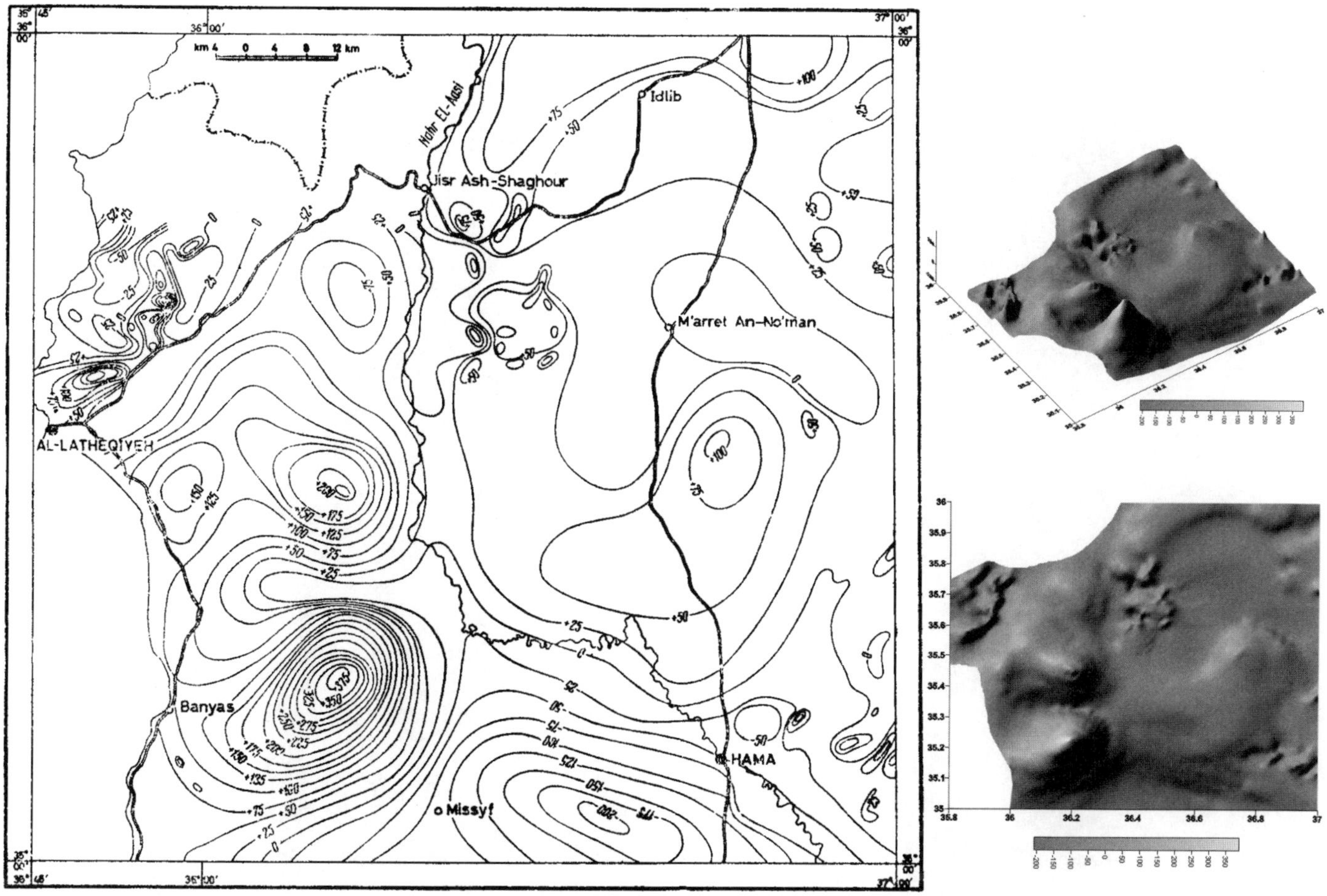

Fig. 7.10. Magnetic map of northwestern Syria in isolines T. (The isolines are drawn every 25 gammas). After V. P. Katkov (1959, unpublished materials).

a basaltic sequence on the two highest summits of the Kurd-Dagh range (at elevations of 1,269 m and 1,201 m). This sequence contain five covers totaling 160 m, comprised of massive basalts with porphyric structure, and also of porous types composed of plagio-basalts. These volcanics are older than the Quaternary basalts located a lower elevation in the Kara Su River valley, which do not look like Upper Miocene plateau-basalts. In Dubertret's (1953) opinion, these basalts are of the Pliocene age.

According to the results of petrographic investigations, the Pliocene volcanics are alkaline basalts and basanites, as indicated by the large content of analcime and nepheline in the rocks. A few chemical analyses attest to the low content of SiO_2 (42.84%) and high content of alkalines (5.57%) in which Na_2O dominates, and also to a high content of TiO_2 (2.9%). On the SiO_2–$\Sigma alkaline_s$ diagram, the analyzed rocks fit the field of alkaline basalts. It is possible that in addition to the alkaline basalts, sub-alkaline kinds also exist, but this assumption is based only on petrographic descriptions.

In the Mesopotamian Basin in northeastern Syria, Pliocene basalts are sparse. They have been observed in two areas: on the left bank of the Euphrates River in the area of Jrablus near the Turkish border; and in the northeasternmost part of Syria, to the east of the town of Qamichliye, and in the area of Qaratchok Darh (Fig. 7.1).

In the Jrablus area the Pliocene basalts form isolated covers, which are outliers of a large basalt plateau in southern Turkey, west of Diyarbakir and Mardin (Oufland, 1966b). The largest of these was mapped near the village of Kharaeb Nas (more than 120 km^2) and Bir Kani (90 km^2). Other covers are not larger than 1-8 km^2. The basalts overlie rocks of different ages: Oligocene and Miocene marine sediments; and Pliocene continental sediments near the village of Khalijaq. Thus the basalts belong to the upper part of the Pliocene. They are overlain by Pleistocene proluvial sediments. The thickness of these basalts depends on the degree of erosion and varies from 50 m (Kharabeh, Beer Rash, Koufi) to 15-20 m (Khalijaq).

In the most complete sections the basalt sequence consists of three flows (5, 15, and 25 m thick), separated by red oxidized scoria crust (15-20 cm). The lavas are composed of augite-olivine-

Fig. 7.11. The Upper Pliocene basalts, exposed on the northern limb of the Alian anticlinal fold. Effusive flows flooded low valleys, but the arch of the fold was not covered by basalts.

Fig. 7.16. The Middle Quaternary basalts which cover the flat plateau and the northern flank of the Western Qaratchok Darh structure.

Fig. 7.12. The Upper Pliocene basalts in the Tigris River valley, covered by conglomerates of the Bakhtiary Formation.

plagioclase basalts, changing into microdolerites in the central part of the flows. Olivine ankaramites have been encountered near the village of Kharaeb Nas.

East of Qamichliye the Pliocene effusive rocks occupy an area of 1,500 km^2. The Kastopan volcano, with the elevation of 693 m, is located in the center of this area. There are two basaltic lava tongues which flow down to the Leilak, Alian, and Deric anticlinal structures. These rocks are represented by normal basalts, dolerites and anamesites with flows from 3 to 20 m in thickness. They rest on Upper Pliocene conglomerates and are overlain by Middle Quaternary conglomerates or effusives (Galaktionov et al., 1966).

In the area of the Alian anticlinal fold, the northern flank of this structure is covered by Upper Pliocene basaltic flows, but on its arch effusive rocks are absent (**Fig. 7. 11**).

On the right bank of the Tigris River effusive rocks are intercalated within conglomerates of the Bakhtiary Formation (**Figs. 7.12** and **7.13**). Therefore, their age is Late Pliocene. Some outcrops here clearly display the ball-shaped structure of basalts (**Fig. 7.14**).

In the valley of the Sutan-Dere River a very distinct cover of Upper Pliocene basalts can also be observed (**Fig. 7.15**).

Quaternary Volcanism

The intensive volcanic activity of the Quaternary period created large outpourings of basalts in different parts of the northern margin of the African-Arabian plate. Quaternary basalts form isolated plateaus in northeasternmost Syria, near the Turkish border, in the Al-Furat River valley and adjacent areas, and on the mountains of Jebel An-Nusseiriyeh. However, they are even more widely distributed in the southwest of Syria (Hauran Plateau, Jebel Diret-att-Touloul, Jebel Drouze) and in the neighboring areas of Jordan (Fig. 7.1). The Quaternary volcanism products have been subdivided (with some reservations) into the Lower, Middle and Upper Quaternary (Pleistocene), and Recent basalts.

Lower Quaternary basalts

Covers of Lower Quaternary lavas, the so-called 'ancient wavy lavas', are widely distributed in the south of the Hauran Plateau. The covers are slightly dissected by erosion. Usually the surfaces of the 'ancient wavy lavas' form agricultural fields on which isolated exposures of the covers protrude every 100-500 m. In most cases these exposures belong to the arched part of the lava swells, to the exploded gas bubbles, and to the relicts of marginal bars and warped swells of single flows up to 8-10 m high. Unlike the younger effusive rocks, the Lower Quaternary covers don't have on their surfaces those features which are typical to the fresh

Fig. 7.13. The Upper Pliocene basalts in the Tigris River valley overlying conglomerates of the Bakhtiary Formation.

surfaces of lava flows, *e.g.* glassy scoria crust and structures of the ropy (pahoehoe) lavas. The absence of these features, due to erosion, attests to a comparatively ancient age for these rocks.

In places, strongly eroded volcanoes formed from intercalated scorias, lavas, and loose pyroclastic material have been preserved on the surface of the covers. As a result of continuous erosion, the volcanoes have acquired cut-off (truncated) cones with a flattened summit and moderate slopes (15-20°). They are 50-80 m high. In most cases the craters of the volcanoes are not preserved.

The Lower Quaternary basalt covers consist of a series of basalt flows separated by weathered crusts composed of red clay. Each flow begins with massive basalts, usually dolerites 6-8 m thick, gradually grading upwards to vesicular types 1-2.5 m thick.

According to their petrographic composition, the Lower Quaternary basalts are mostly composed of dolerites, and to a lesser degree, of anamesites, ankaramites and hyalobasalts. Petrochemically, these rocks belong to the normal basalt group.

The Lower Quaternary age of the basalts was determined by correlating them with the sedimentary and volcanogenic sequences in valley of the Yarmouk River. There the basalts overlie Paleogene limestones and Pliocene rocks composed of basalts and sandy-clayey deposits, and are in turn overlain by Middle Quaternary basalts, which occupy the 60 m high terrace of the Yarmouk River.

According to data from bore-holes drilled in the northern part of the Hauran Plateau the thickness of the Lower Quaternary basalts is 100 m. In the southern half of the Hauran Plateau, and on the eastern slope of Jebel Drouze (Bore-holes 12, 14, 15), it increases to 200 m. The maximum thickness of the Lower Quaternary basalts was recorded in the central Hauran Plateau (Fig. 7.2) where it reaches 350 m (Bore-hole 18). There the basalts overlie Middle Eocene marls. Their thickness gradually decreases to the north and south, and in Bore-hole 28 in the north and Bore-hole 21 in the south, they are only tens of meters. They decrease even further to only a few meters towards the periphery of the covers. Lava flows probably filled the ancient undulating relief. The thickness of the effusive rocks is greatest where they fill depressions.

In the Mesopotamian Basin Lower Quaternary effusives have been found in two areas – to the southeast of Jrablus on the left bank of the Al-Furat River, and in the upper Nahr El-Khabour valley near the town of Hassetche (Fig. 7.1).

On the left bank of the Al-Furat River volcanic rocks have been found near the villages of Kharabeh and Sirrine Ech-Chemaliye. In the area of Kharabeh the effusives are related to a central-type eruption, with relicts there of three volcanoes. The largest of them has an open crater 35 m in diameter and up to 15 m deep. Effusives are represented by porous tuffs overlain by basalts

Fig. 7.14. The ball-shaped structure of the Upper Pliocene basalts, the right bank of the Tigris River.

with amygdaloidal structure. They rest on Pliocene basalts. In the area of Sirrin Ech-Chemaliye the 10-11 m thick cover of effusive rocks consisting of two flows overlies Pliocene conglomerates. The rocks are represented by ankaramite, basanite, and augite-olivine basalts (Oufland, 1966b).

In the area of Hassetche near the village of Derbasiyeh the Lower Quaternary effusives form the hard lava which creates the Ard esh Sheikh and Kapes Darh table-mountains (Ponomarev and Bivshev, 1966).

On Ard esh Sheikh table-mountain by the left bank of the Nahr El-Khabour River, dolerites, anamesites and basalts form a cover 280 km^2 in area and about 10-15 m thick. They are related to a fracture-type outflow. The center part of this cover is occupied by the mountain Jebel Qleib – which is a relict volcanic cone. The effusive sequence includes two covers. At its base the lower cover is composed of massive basalts which are replaced upwards by porous basalts. In the western part of this plateau, the effusive rocks are only made up of massive anamesites and dolerites.

The structure of the Kapes Darh table-mountain is very peculiar. It protrudes 80 m above the surrounding plain and is 2 km in diameter; its summit is flattened and its slopes are steep. The plateau is comprised of hard ankaramites up to 15 m thick, which overlie 22 m of tuffs and tuff-sandstones with lapilli. In this area there was probably an ancient volcano which emitted the tuffs and afterwards erupted lava flows.

To the north-east of Kapes Darh effusive rocks have produced another cover (named the Fakka cover) whose central part is preserved as an isolated neck. The flow fills a valley of an ancient river. It is 0.5-5 m thick and consists of plagioclase-olivine basalts, anamesites, plagioclase-picrites and agglomerate lavas. The basalts adjoin to the Lower Quaternary terrace of the Nahr El-Khabour River and are overlain by Middle Quaternary proluvial sands.

Middle Quaternary basalts

Middle Quaternary basalts form the covers in southwestern Syria on the Hauran Plateau, in northeastern Syria north of the town of Deir Ez-Zor, and also in the northeasternmost part of the country. They have been mostly studied on the Hauran Plateau.

Fig. 7.15. Covers of effusive rocks in the valley of the Soutan Dere River. The lower cover is composed by the Upper Pliocene basalts; the upper cover consists of the Upper Pleistocene basalts.

Basalt covers of this age are less eroded than those of the Lower Quaternary. Scoria crust, ropy lavas and other signs of the lava flow are preserved as relicts on the flow surface. Usually the marginal bars are very well preserved. The covers are formed by the intercalation of separate lava flows whose structure is similar to that of the Lower Quaternary basalts.

On the surface of the Middle Quaternary basaltic covers there are a few rows of well-preserved northwest-trending volcanoes. In many of them the craters are in good shape. The volcanoes are composed of intercalated scoria, lava, and tuff. Volcanic bombs and lapilli are often encountered on the slopes of these volcanic cones.

The Middle Quaternary effusives consist of normal basalts (dolerites, anamesites, basalts), melanocratic basalts (dolerite of ankaramite composition, ankaramites, picrites), and alkaline basalts (analcime basanites and analcime basalts, as well as nepheline basalts).

The Middle Quaternary age of the basalts was determined by their relationship to the river terraces and peneplains, dated by the findings of Upper Paleolithic flint tools. Thus, in the valleys of the Yarmouk River and Wadi Raqqad, the basalts cover a 60 m high terrace which contains Lower Paleolithic flint tools. In the Syrian desert (the Tarag El-Aalab escarpment) Upper Paleolithic flint tools were found on the surface of the basalts. In northeastern Syria the basalts overlie the Middle Quaternary terraces of the Tigris and Al-Furat Rivers.

According to drilling data, the Middle Quaternary basalts in the center of the Hauran Plateau are 100 m thick. To the south their thickness decreases and in Wadi Raqqad it drops to 10 m. In the rest of Syria, the thickness of the Middle Quaternary basalts is thought to be about a few tens of meters.

In the Mesopotamian Basin the Middle Quaternary effusives are developed on the right bank of the Al-Furat River south of the town of Raqqa (Ar-Raqqa) and at the foothills of Jebel Bishri (the Jebel Qleb Hamma Plateau and Jebel As-Samman Plateau). The lava covers consist of ankaramites, augite-olivine picrites, and tuffs. A few small volcanic cones can be observed in this area (Medvedev, 1966a).

Southeast of Raqqa, not far from Deir Ez-Zor and between the Al-Furat and Nahr El-Khabour Rivers, the Middle Quaternary effusive rocks occupy much larger areas (Medvedev, 1966b). They form the Feidat El-Mieza and Himmat El-Jezira Plateaus. The former is situated on the right bank of the Nahr El-Khabour

Fig. 7.17. Occurrence of the Recent basalts on the Upper Quatenary lacustrine limestones in the Bahret Hijane area (the Damascus depression).

River; it is about 40 km across, and the thickness of the effusive cover is 5-10 m. The surface of the plateau is covered by vesicular lavas about 1 m thick. Two series of extinct volcanoes have a sub-longitudinal trend. The effusive rocks of the Feidat El-Mieza Plateau are petrographically variegated: basalts, ankaramites, picrites, basanites, volcanic scoria, and agglomerate lavas.

The Himmat El-Jezira Plateau is situated on the left bank of the Al-Furat River. The plateau's absolutely flat surface is formed by the 10 m thick covers of effusive rocks. Dominant among them are massive or slightly porous ankaramites. One strongly eroded volcano with a crater about 250 m in diameter rises up from the plateau. Its cone is built up of thin (1-3 m) flows of lava. Pyroclastics are absent.

In northeasternmost Syria, the Middle Quaternary effusive rocks are well-developed on the Lelak Dagh and Qaratchok Dagh highlands (Galaktionov et al., 1966). The volcanic rocks form the flat surface of the Qaratchok Dagh, which rises above the Jezira plain to an elevation of 769 m. The basaltic cover varies in thickness from 10 to 30 m, and occupies an area of about 800 km^2. It is elongated, stretching some 72 km in a sub-latitudinal direction. The effusive rocks overlie dislocated and eroded Upper Pliocene sandstones and conglomerates. this cover usually comprises three flows, measuring 10, 13 and 23 m in thickness in the Soutan Dere basin. Each of these flows begins with massive basalts and ends with porous, vesicular and amygdaloidal rocks. Among them basalts, dolerites, and doleritic basalts have been identified. Middle Quaternary basalts also overlie the whole northern flank of the Western Qaratchok Darh structure (**Fig. 7.16**).

Upper Quaternary basalts

Basalts of this age have a limited distribution in Syria and the neighboring countries. In the Mesopotamian Basin the Upper Quaternary volcanic rocks are well-developed in the area of Raqqa, on the left and right banks of the Al-Furat valley (Medvedev, 1966a). They overlie gravel deposits of the second and third terraces of the river. These covers don't continue below the Upper Quaternary terrace. There are three central-type volcanoes of this area: Jebel Mankhar Gharbi, Jebel Mankhar Sharqi, and Hauel Rumman. These volcanoes have complex somma-vesicular type mechanisms, which generate 2 to 3 stages of outflows.

The Mankhar Gharbi volcano has the following dimensions. The diameter of the somma is 2.6 km and it rises up to 75 m above

Fig. 7.18. Occurrence of the Recent basalt on the surface of the Upper Pleistocene (Q_3) terrace southeast of the Mazraat Beit Jinn village (the Anti-Lebanon).

the plain. Within the somma there is a crater 1.6 km in diameter with steep slopes some 15-20 m high. The atrium in the crater is no less than 70 m wide and about 1.25 km in diameter. The other complex volcanoes have similar dimensions.

The somma is filled by pyroclastic products – volcanic sands, tuffs and lapilli, among which bombs 10-30 cm in diameter are found. The volcanic fragments are poorly cemented, the rock is not compacted and can be classified as pyroclastic agglomerate and, in places, as tuff. The somma of the crater was probably formed during pulses of volcanic activity, with brief but strong explosions. The surface of the covers is eroded and hummocky. Sometimes the scoria crust, ropy lavas, and some of the other morphologic elements of the flow surfaces can be observed. The volcanoes lie along a NNE line corresponding to the sources of the outflows, which are probably related to faults with the same trend.

According to petrographic studies, the lavas from the volcanoes of this group belong to the nepheline basalts and ankaramites.

Volcanoes with similar structures are situated in the area of the town of Deir Ez-Zor, down the Al-Furat River from Raqqa (Medvedev, 1966b). Their dimensions are smaller. These volcanoes form two groups (Deir Ez-Zor and Jebel Az-Zurab). The first is situated on the right bank of the Al-Furat River and was formed by separate lava flows. It consists of a few small strongly eroded volcanic cones made of pyroclastic material with thin intercalations of lavas. The effusive rocks of the Deir Ez-Zor group of volcanoes are exclusively composed of picrites and their scoria.

The second group of Jebel Az-Zurab volcanoes is situated on the left bank of the Al-Furat River, approximately 18-20 km north of Deir Ez-Zor. They consist of three small volcanic cones, about 1 km in diameter with a small lava flow. According to their composition, the lavas, tuff-lavas and tuff-agglomerates belong to the ultra-alkaline effusives.

Generally, the Upper Quaternary basalts are a few tens of meters thick and are everywhere represented by melanocratic (dolerites, doleritic basalts and basalts) and alkaline (nepheline basalts) types.

Recent (Holocene) basalts

The Recent basalts are almost exclusively developed in the southwest of Syria, where they form a large volcanic region about

Fig. 7.19. The Hadar Recent basalt flow (βQ_4) plunging from the Jebel Hermon Massif flank (the Anti-Lebanon) into the valley.

6,000 km^2 in size (Hauran Plateau, Jebel Drouze, Jebel Diret att-Touloul, Safa, and others). They are characterized by all the possible forms of volcanic relief (Fig. 7.1). The Recent basalt covers possess beautifully expressed morphologic and structural features of the lava surfaces. Their freshness appears in sharp contrast to the more ancient, underlying covers. Typically they have excellently preserved volcanic cones. This preservation of the Holocene lava surfaces gives the impression that the outflows occurred quite recently. And in fact, south of Damascus on the northern edge of the Kraa flow (Dubertret, 1929), isotopic age determinations of the buried (baked) bones of antelopes and gazelles showed that these basalts were laid down 4,000 years ago (Burdon, 1959). Younger volcanic activity also cannot be excluded, because according to the Holy Scriptures, there are indications that volcanic eruptions took place in the Near East in King David's time (Tyrrell, 1930). In historical time young post-Neolithic basalts have also been observed in the basaltic fields of Saudi Arabia, where the sources of their outflows are controlled by northwest trending fault zones (Camp et al., 1991).

In addition, there is direct geological evidence for basalts of Holocene age. In the Bahret Hijane depression south-east of Damascus they overlie the Upper Quaternary lacustrine sediments (**Fig. 7.17**), and near the village of Beit Jinn on the southeastern slope of Jebel ech-Sheikh, basalts overlie the Upper Quaternary river terrace (**Fig. 7.18**). The very young age of such basalts is confirmed by the way in which they follow the relief, filling up the river valleys (**Fig. 7.12**).

The multi-stage type of lava outflows is very typical of the Holocene volcanism. The lava sources are related to the system of numerous sub-parallel and mainly NW-trending tectonic faults and fractures, which are manifested in the landscape by chains of volcanic cones. Geological observations in the field showed that the volcanoes, genetically related to these structures, are of different ages. These observations highlight the longevity of these fractures. During the early stages of volcanic activity, the faults served as conduits for the fracture-type lava outflows. Later, these fractures closed, provoking the generation of a chain of volcanic cones along the trace of the fracture. The complexity of the volcanic chain structures is determined by the longevity of these fractures, and differences in their development. Changes in the intensity of the

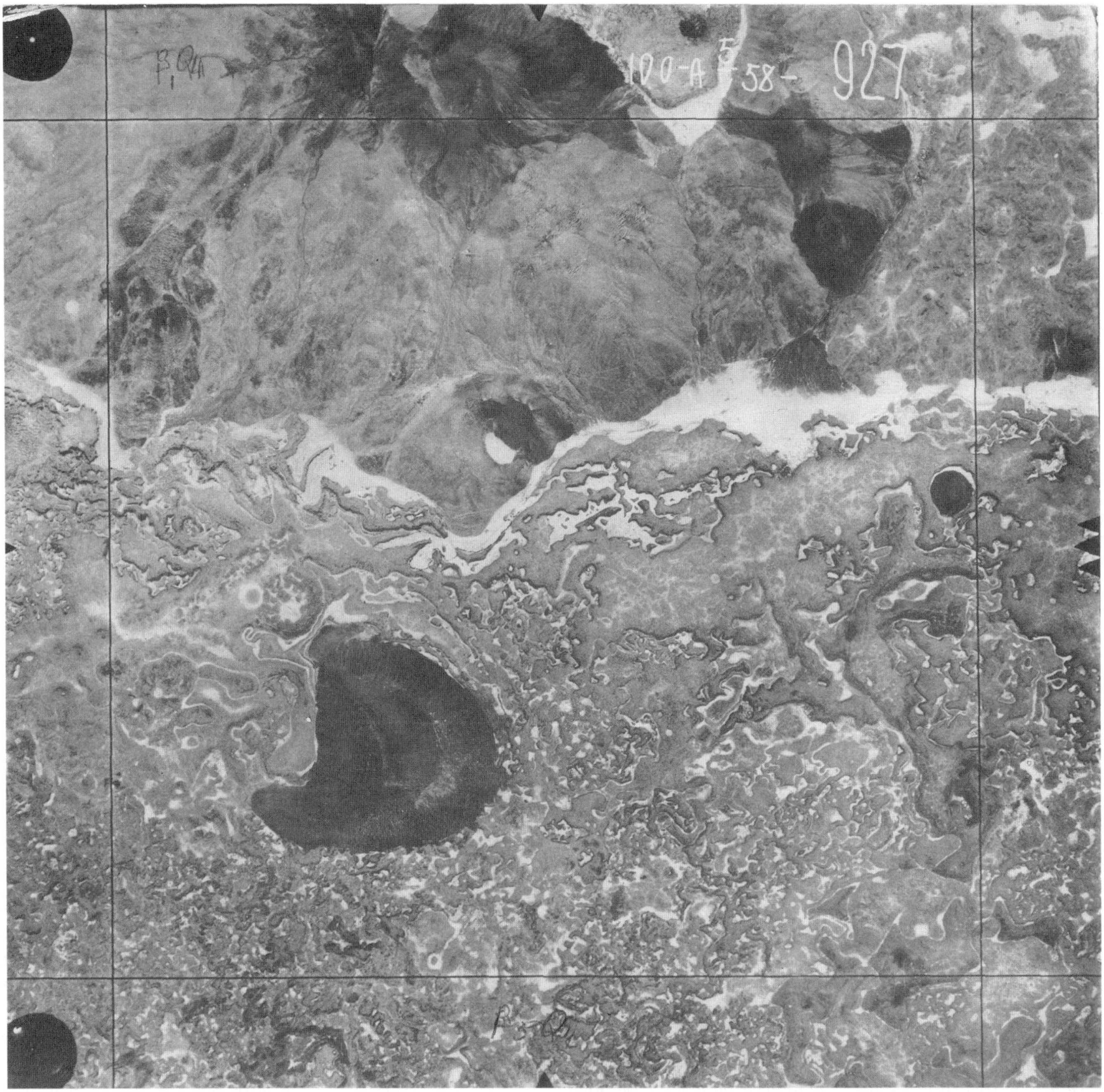

Fig. 7.20. The Safa cover (β_5Q_4) overlying the Recent basalts of the first generation (β_1Q_4). In the lower part of the photograph a volcanic cone can be seen, half-drowned by the S.afa cover. Aerial photograph 1:25,000 scale, Jebel Drouze.

volcanism can be seen along the fractures (faults): the volcanic cones and their related covers are sometimes evident, sometimes they disappear completely. In the zones where fractures developed, intercalation of flows of different ages has been recorded, as well as partial or complete submergence of more ancient volcanic cones under the younger lava flows (**Figs. 7.20** and **7.21**).

The multi-stage cyclicity of the Holocene outflows caused the formation of seven generations of lava covers of comparatively different ages: β_1Q_4, β_2Q_4, β_3Q_4, β_4Q_4, β_5Q_4, β_6Q_4, β_7Q_4 (**Fig. 7.22**).

It is remarkable that in Jordan, on the southern continuation of Jebel Drouze and the Hauran Plateau, six different ages have been differentiated among the Holocene covers (Bender, 1975), and that in near the Red Sea coast Saudi Arabia, seven covers of

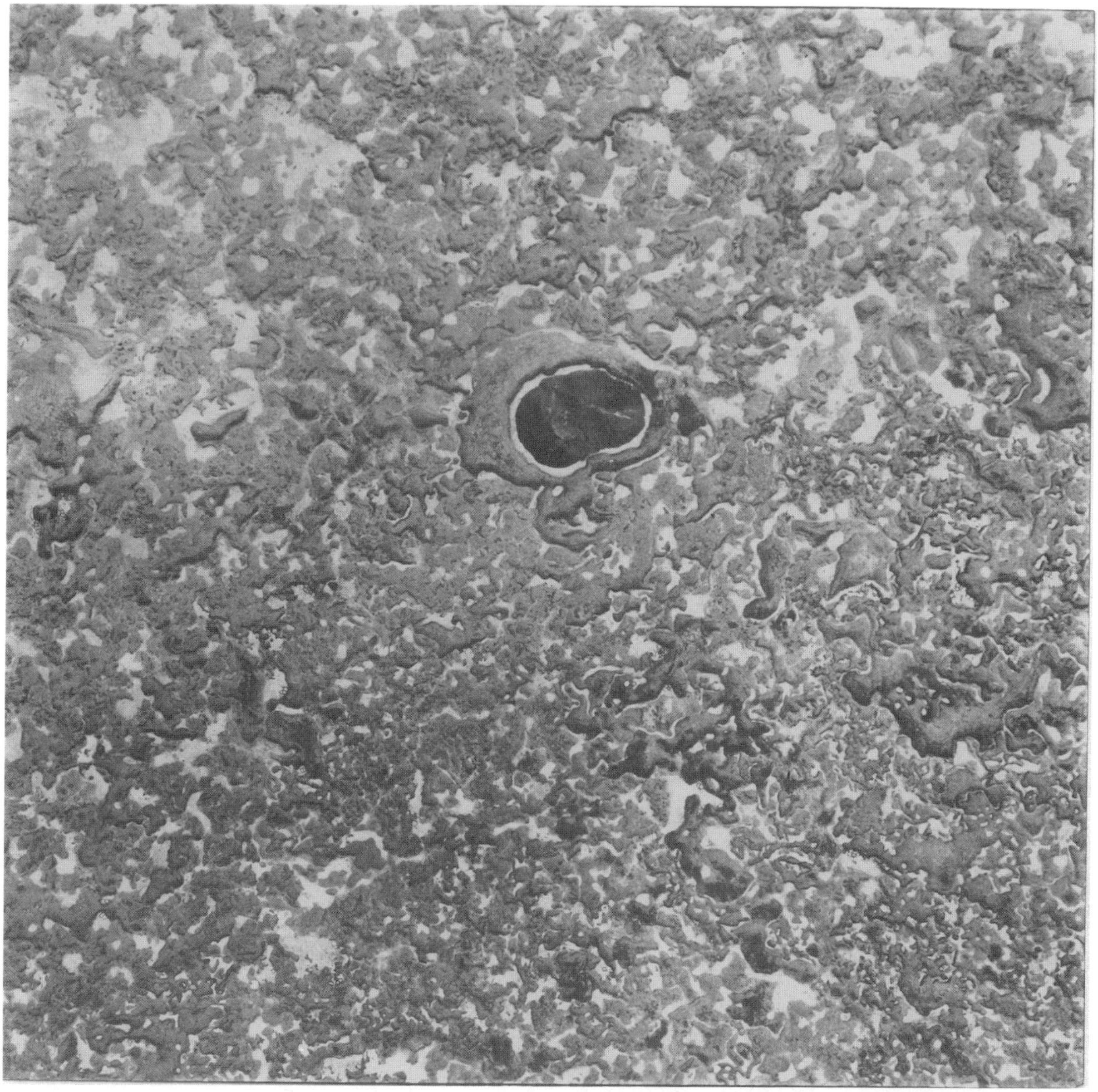

Fig. 7.21. The drowning of a Recent volcano by younger Recent lavas (β_5Q_4). Aerial photograph 1:25,000 scale, Jebel Drouze.

different age were also identified much later (Camp et al., 1991).

For each generation of Holocene basalts the same type of preservation and the same picture of the lava surface is typical. In geological mapping this allows the details of the cover structures to be deciphered by using aerial photographs. The degree of preservation of the cover surfaces increases from the early lava outflows to the later ones. Thus, on the surfaces of the early generation the scoria crust, ropy lavas and other structural elements of the surface are only preserved in places, but on the younger covers they are ubiquitous (**Figs. 7.23; 7.24; 7.25;** and **7.26**).

The typical columnar (pillar-like) structure of basaltic rocks can be observed both among the Holocene and Pleistocene covers (**Fig. 7.27**).

Because of the different degrees of preservation of the cover surfaces, they have different characteristics in the aerial photographs (**Figs. 7.28; 7.29; 7.30; 7.31;** and **7.32**).

Consequently, the differentiation between the various lava covers and their correlation are based on a number of factors, the most important of which are the spatial relation of the covers

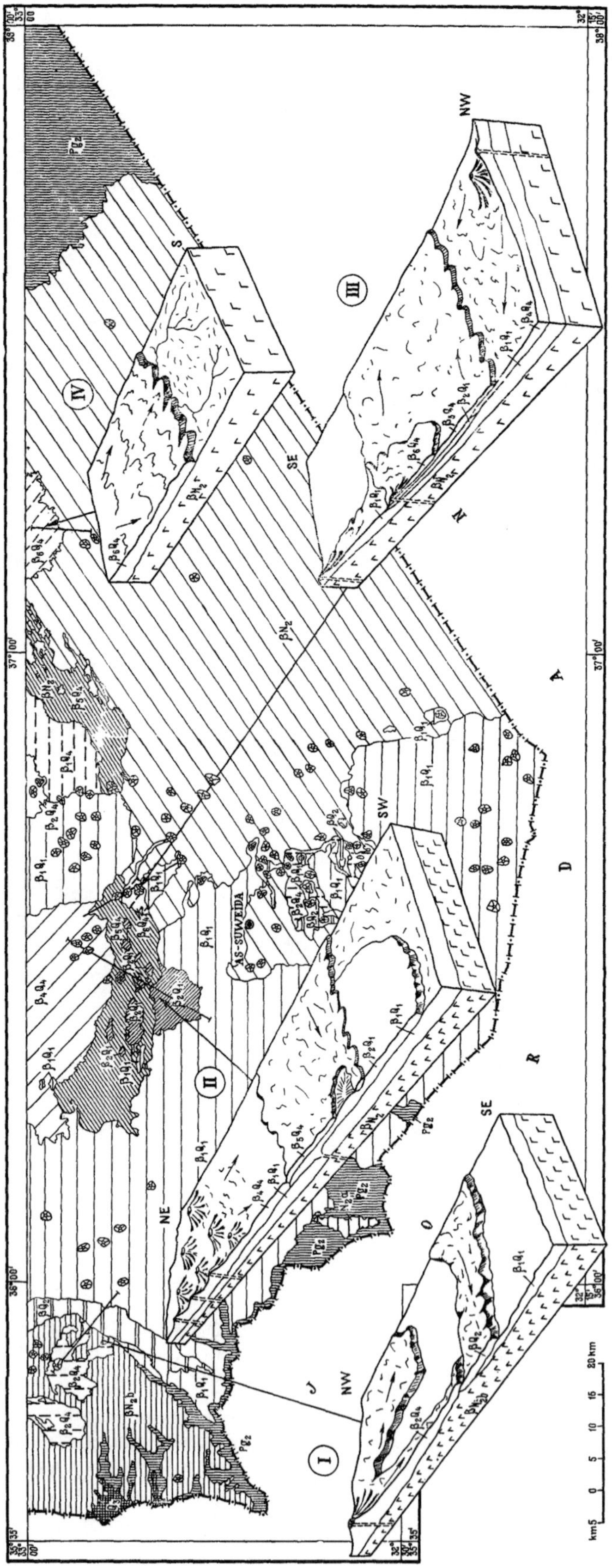

Fig. 7.22. The diagram demonstrating the relationship of the Quatenary lava sheets of different generations. The southern part of Jebel Drouze. After Krasnov et al. (1966).

Fig. 7.23. The lava flow prints on the surface of the Hadar cover (β_4Q_4), the Anti-Lebanon.

Fig. 7.24. Detail of the surface structure of one of the tumuli (a dome-like swelling) on the surface of the Tell As-Safa cover (β_6Q_4), Jebel Drouze. The ropy lavas are visible.

with each other and with the sedimentary rocks, the extent of their erosion, and the morphology of the flows. The use of aerial photographs helps in differentiating between the covers of various ages and in determining their stages of flow during the Holocene.

The Holocene covers are formed by lavas with wavy or blocky patterns. Typical are warped swells hundreds of meters long, developed in the marginal parts of the covers. The impression of waviness is given by a succession of bulges, depressions, warped swells, and exploded gas bubbles. Some of the bulges are 10 m high. The edges of the depressions separating these bulges sometimes look like vertical walls a few meters high. The arches of the bulges are often collapsed. Fractures are often seen along the axial parts of the warped swells.

Rows of beautifully preserved volcanic cones rise up above

Fig. 7.25. The bubbles' orientation in the lava flow plane. The Hadar cover (β_4Q_4), the Anti-Lebanon.

Fig. 7.26. The Aa-lavas of the Mrati-Touloul es Safa cover (β_7Q_6), Jebel Drouze.

the surface of the Holocene covers. They are related to the lava-supplying fault zones with northwestern trend (**Fig. 7.33**). For example, the centers of outflow of the Dmeir cover (β_1Q_4) are situated in the northern part of the large fault zone of Jebel Diret att-Touloul, which extends for some 30 km. Two rows of volcanoes are observed there, each containing up to 10 volcanic cones (Fig. 7.33 and **Fig. 7.34**).

These volcanoes show well-preserved cones with distinctly manifested craters. In addition to the main crater, they sometimes have 2-3 additional 'parasite' craters (Fig. 7.33 and **Fig. 7.35**).

Usually the volcanic cones are 70-100 m high, but sometimes they reach 200-250 m. The largest craters are not larger than 500 m in diameter and 100 m deep. The craters are cup-like depressions with a flat floor, covered by loamy formations of

Fig. 7.27. The column (pillar)-like structure of the Middle Quaternary basalts of the Western Qaratchok Darh.

deluvial-elluvial origin in the comparatively earliest Holocene volcanoes, and covered by lava blocks and volcanic bombs in later ones. A morphologically distinct vent is only seen in the craters of the youngest volcanoes, belonging to the 6th and 7th generations of covers.

The covers of the Recent basalts consist of a series of lava flows separated by horizons of volcanic breccias and ropy lavas. Each flow is about 2-5 m thick. For all the flows the internal structure

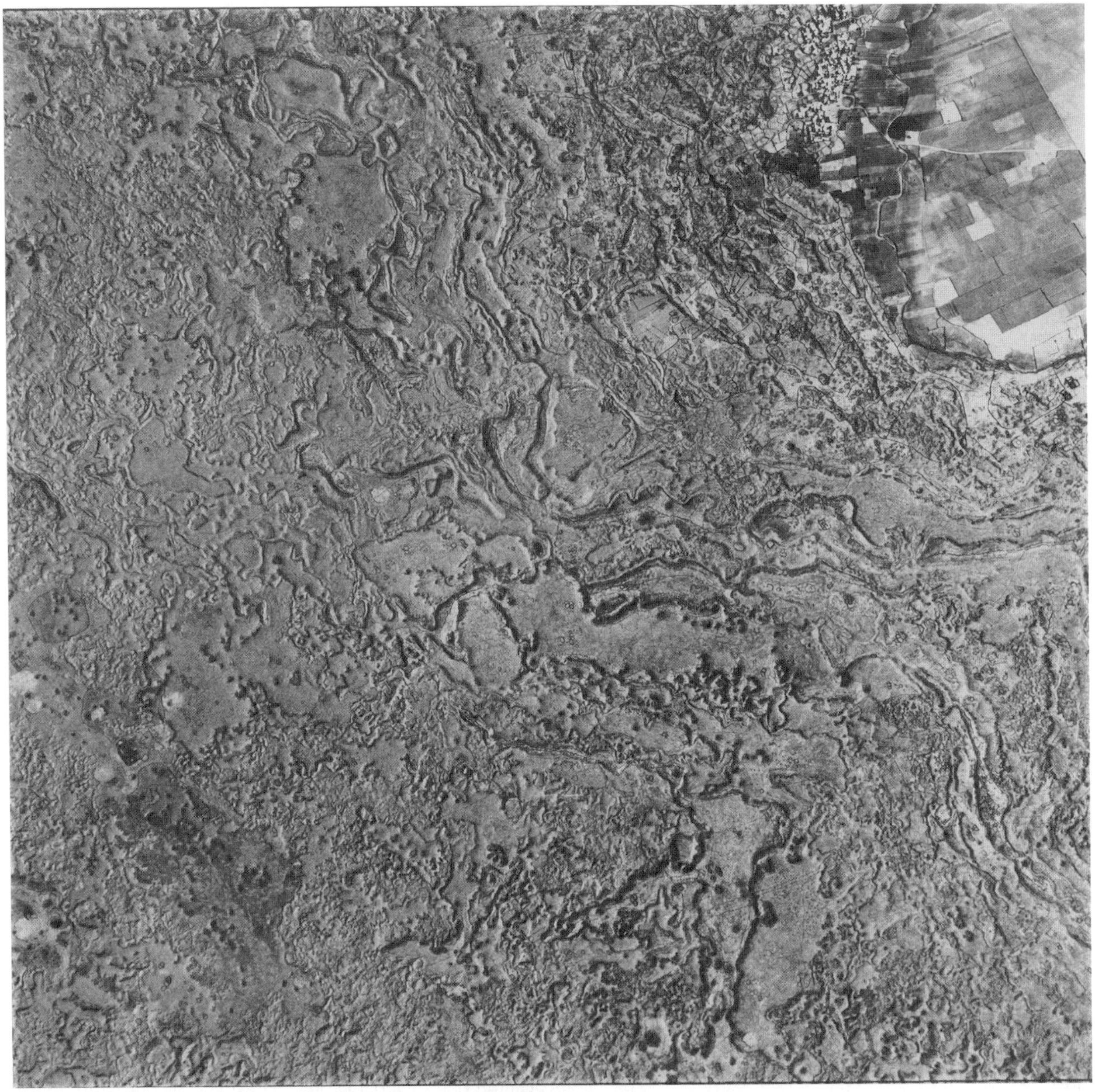

Fig. 7.28. Occurrence of the Recent basalts (β_4Q_4) of the Leja cover on the Lower Quaternary basalts (βQ_1). Elongated bars along the cover's periphery and subsidence funnels are clearly visible. The northern margin of the Hauran Plateau.

is the same: in the upper part there are ropy lavas or scoria crusts (15-30 cm), below them the rock is comprised of bubbled lavas grading down to smaller bubbles, and massive effusive rocks are found at the base of the flow (Fig. 7.35). The maximum thickness of the Holocene covers may reach more than 100 m, but usually they are only tens of meters thick.

In the Mesopotamian Basin Recent effusive rocks are developed in the area of Hassetche (Ponomarev and Bivshev, 1966). They form the Kowkab Massif, which consists of seven volcanic cones. These are grouped along two faults with a north-northeastern trend. The volcanoes are surrounded by three lava flows. The oldest is up to 12 km long; and its thickness decreases from 10 to 3 m from its source to its terminus. The largest cone is 10-15 m high and about 200 m in diameter. The middle flow is shorter (not more than 4 km) but it is very thick (50-60 m) above its source, decreasing to 5 m at the end of the flow. Five volcanoes have been observed on the surface of this flow. The upper flow is shorter (3 km), its thickness near the cone is about 50

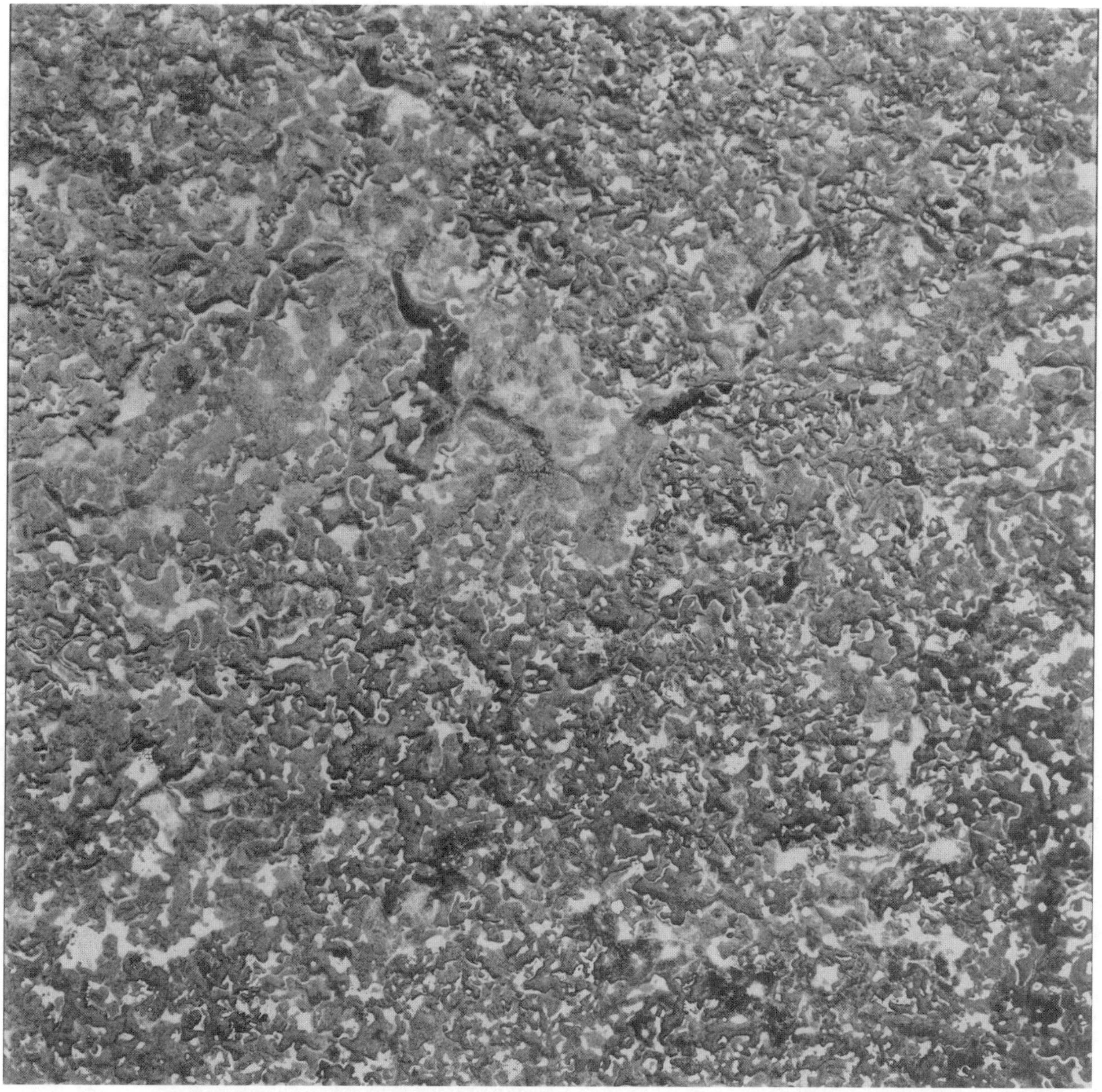

Fig. 7.29. The Safa cover (β_5Q_4) - exhibiting a closely crocheted-lace-like surface of the lava. Aerial photograph, 1:25,000 scale. The area is located to the north-east of Jebel Drouze.

m, decreasing to 15 m at its terminus. The rocks are ankaramites, basalts, picrites, scoria and tuff. In the lower flow ankaramites dominate, the middle one is built up of ankaramites and basalts, and the upper one by picrites.

In northwestern Syria the undifferentiated Quaternary basalts are known only within the Kara Su River valley. This valley is a graben, representing the continuation of the Al-Ghab graben to the north of Lake Amuk. Only the eastern boundary slope of the graben (a tectonic scarp) is in Syrian territory. Minor exposures of the Quaternary basalts are observed below the scarp on the left bank of the Kara Su River.

Two lava flows have been distinguished by Dubertret (1953) and Protasevich and Maksimov (1966). The oldest one is composed of amygdaloidal plagio-basalts and is characterized by a more weathered surface which nevertheless preserves traces of lava streams and features of lava cooling such as arched swells, scoria crusts, and cooling fractures. Typically, the structures of the younger cover are very well preserved, and lava fills the recent

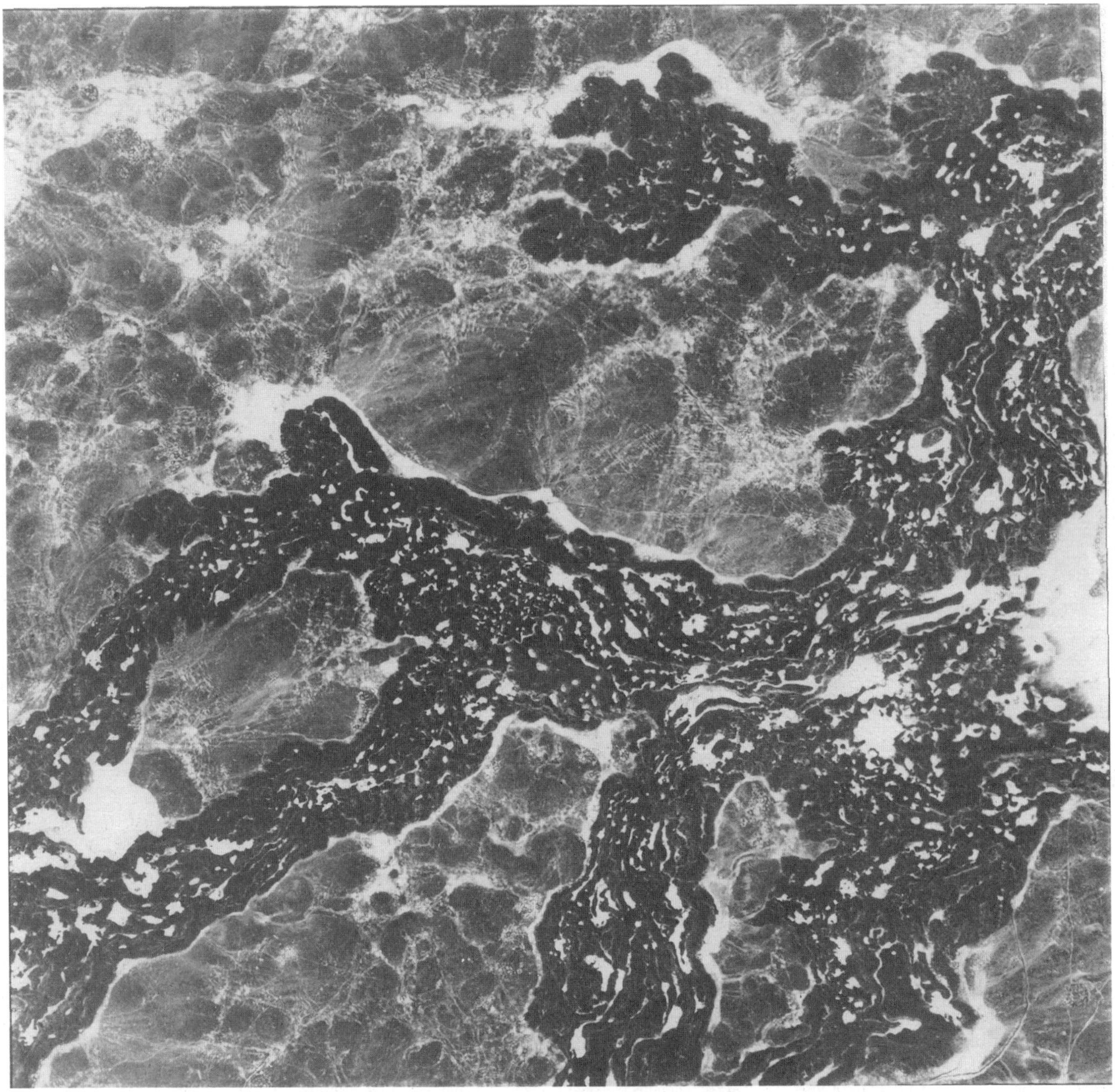

Fig. 7.30. The Safa cover. Occurrence of the 'wavy' lavas of the fifth-generation (β_5Q_4) on the basalts of Pliocene age (βN_2). The flow of young lavas drowning the valley is clearly visible. Aerial photograph, 1:25,000 scale.

relief forms. Beautifully preserved volcanic cones, representing the centers of lava outflow, are seen on the Turkish side of the Kara Su River. According to their mineral composition, the basalts belong to the alkaline type group.

The Neogene-Quaternary basalts of Syria, like those in other continental regions of tectonic magmatic activity, contain numerous xenolites of mantle material. Data for xenolites in the basalts of the volcanic fields of Saudi Arabia have been published previously (Camp et al., 1991; McCuire and Bohannon, 1989). Until recently, the xenoliths of Syria were not comprehensively studied, but this was remedied by Sharkov et al. (1986, 1987, 1992, 1993, 1996). Xenolites have been recovered from the Miocene basalts of central Syria (Jebel Manaa), from the Pliocene-Quaternary basalts of northern Syria (near Jisr Ash-Shaghour), and from the basalts

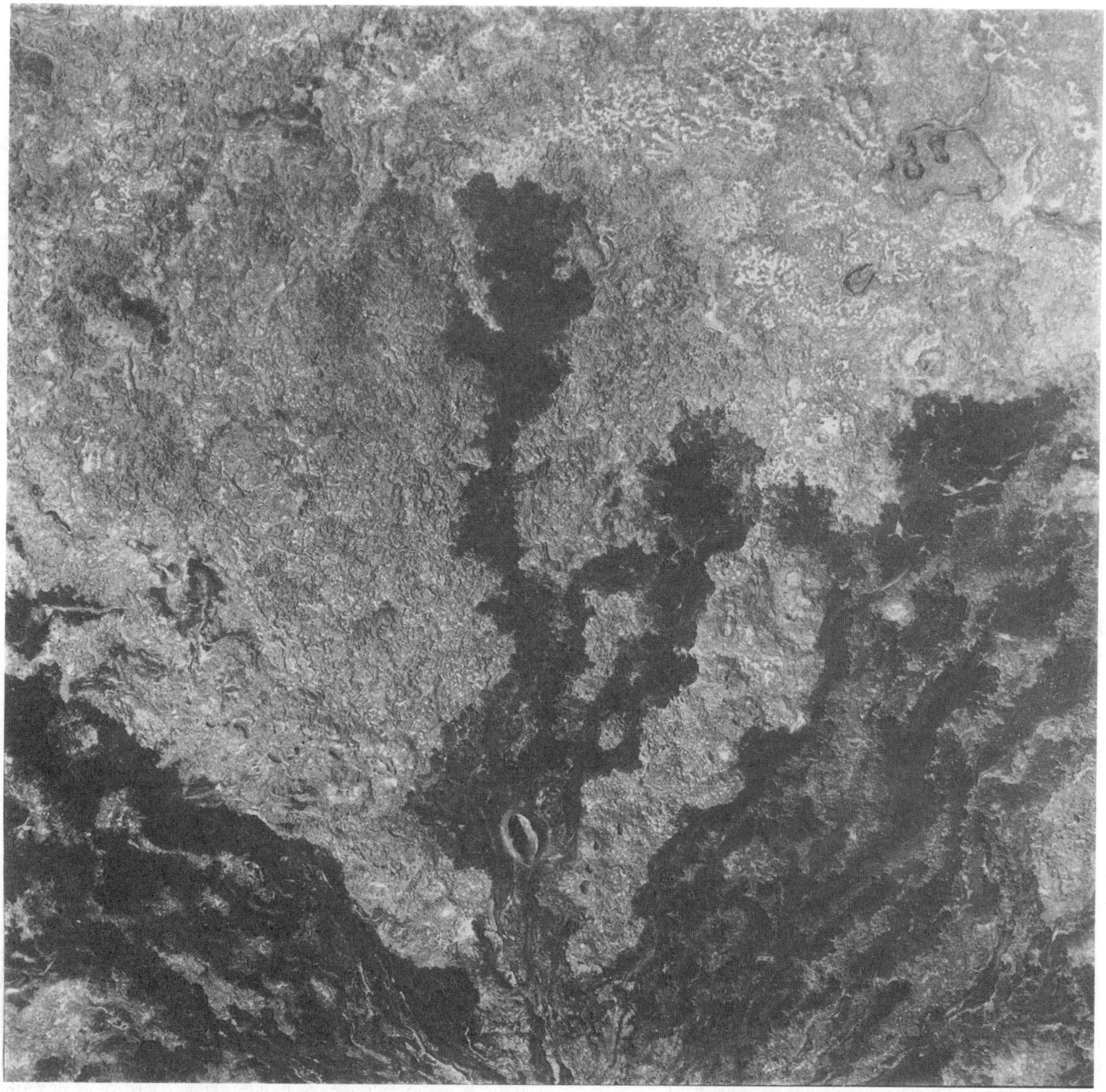

Fig. 7.31. The lava cover Mrati Touloul As-Safa (β_7Q_4) overlying the Tell Safa cover (β_6Q_4). There is a clear photogenic difference between surfaces of the covers. Aerial photograph, 1:25,000 scale.

of the Jebel Drouze highland in the southwest. The best known xenolites are those which Sharkov et al. (1992, 1996) recovered from the Tell Danun Quaternary volcano, which is situated in the northeastern part of the Jebel Drouze volcanic plateau, 30 km east of the town of Shahba.

The tuff-lavas and pyroclasts of moderately alkaline basalts which form volcanic cones contain numerous mantle xenolites which belong to three groups of rocks:

1) mostly spinel-peridotites and spinel-lherzolites;

2) garnet-spinel and spinel websterite (making up about 60% of the xenolites);

3) hornblendite, and hornblende-clinopyroxenite, and also megacrysts of kaersutite (up to 10-15 cm), ilmenite (1-2 cm), and kaersuitite-augite aggregates (about 30% of the xenoliths). The

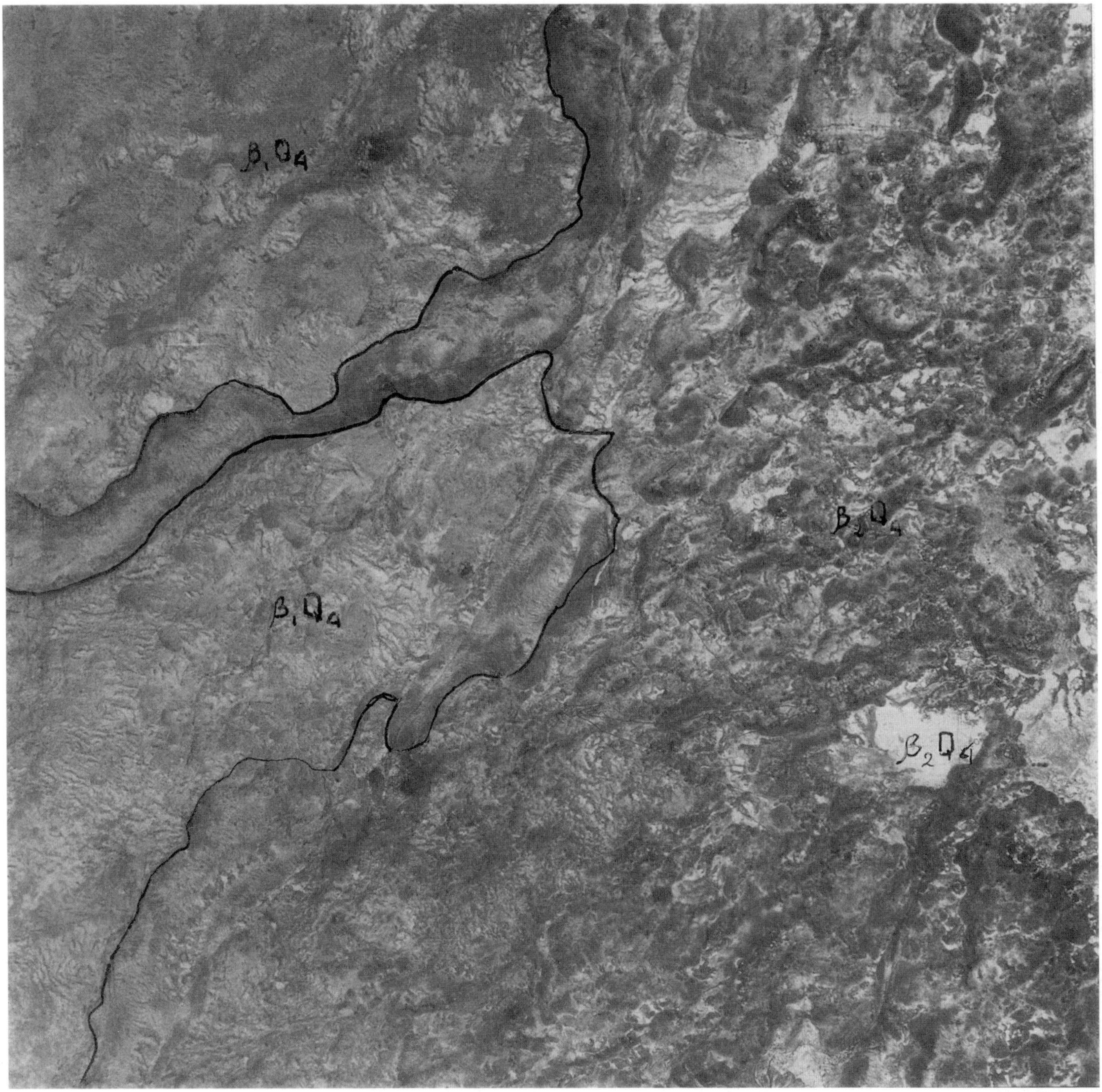

Fig. 7.32. The Medbah ed Drour (β_2Q_4) lava cover overlying the Ard el Hamra (β_1Q_4) cover. The different characters of the cover surfaces are clearly visible; the surface of the Ard el Hamra cover looks smoother.

remaining 10% of the xenolites are composed of spinel-gabbro, pyroxene granulites, and dolerites.

Petrologic-geochemical study of xenolites, including thermobarometry of garnet-spinel websterites, shows that xenolites comprise an interrelated association of rocks formed under different thermo-dynamic conditions: spinel-peridotites are a mantle matrix whose material underwent depletion during melting at the top of asthenosphere diapirs; the rocks of the second group were formed through interaction with mantle fluids in the fractured zones of the cooled upper parts of the diapirs. Mantle peridotites

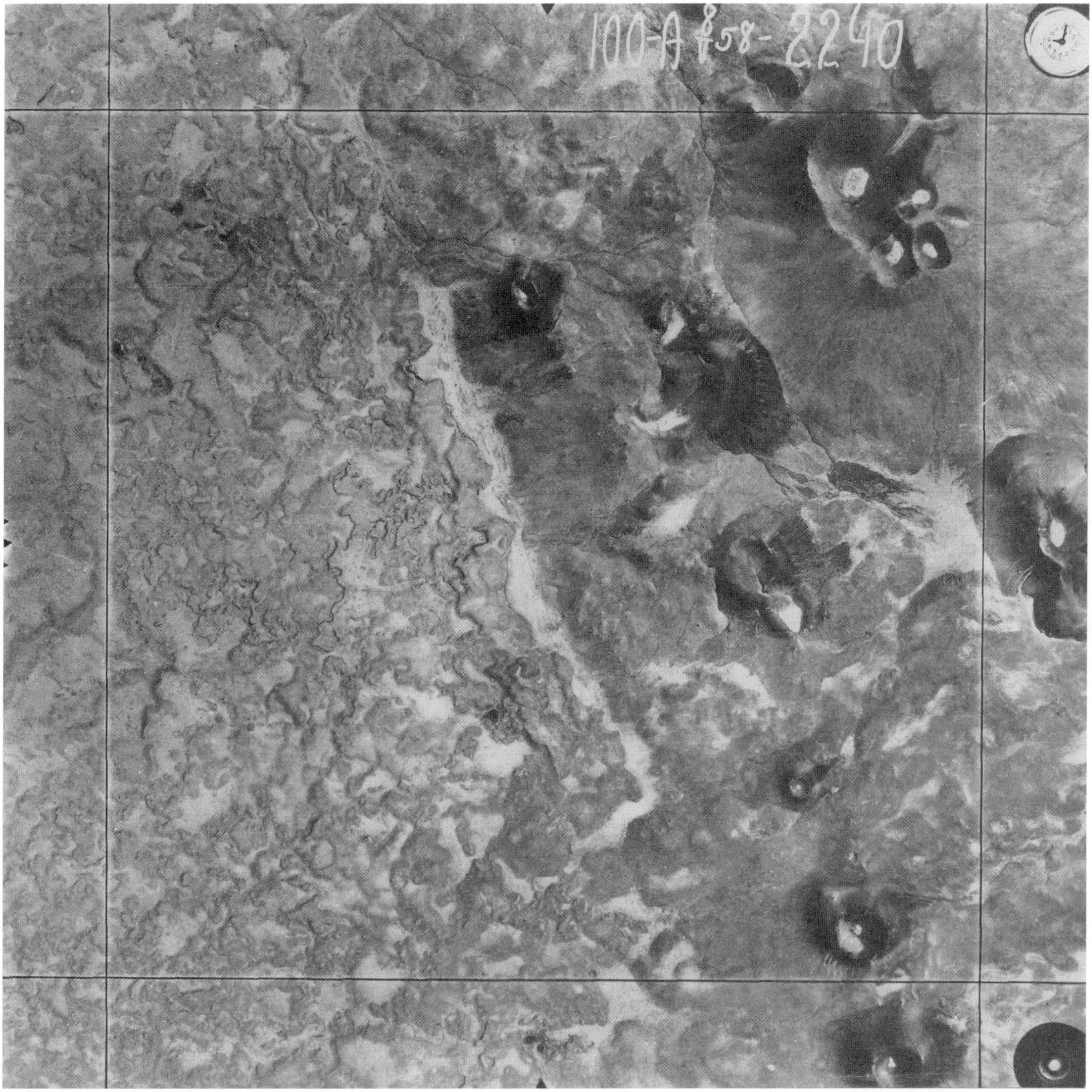

Fig. 7.33. The fault of the northwestern trend, traced by the chain of volcanoes. In the upper part of the photograph is the largest volcano (Oum Edhenn, elevation 932 m) of the Recent first-generation Dmeir cover (β_1Q_4).

produced basaltic magma which rose to form dikes which now consist of garnet-pyroxenites. Thermo-barometric measurements on the garnet-pyroxenites and spinel-peridotites showed that the temperature of crystallization of the former reached 946-1045°C and 8-10 kbar pressure, and in case of the latter - 755-1080°C and 10-20 kbar, which reflect depths of 24-30 km and 30-60 km, respectively. These depths determine the position of the top of the asthenosphere mantle beneath the Arabian platform.

Fig. 7.34. The view on the system of linearly oriented Touloul Lelle volcanic cones taken from the top of Tell el Hajar (the boundary of the Damascus and Sabaa Biar sheets).

Petrochemical Characteristics of the Mesozoic and Cenozoic Basalts

Data on the petrographic and petrochemical composition of Jurassic and Lower Cretaceous effusive rocks of Syria and Lebanon are practically absent. Possibly, they are similar to those which have been studied in Israel, where the following petrographic types have been identified: olivine-nepheline basalts, monzonites, trachy-basalts, basanites and others. Petrochemically all these rocks (which are especially rich in potassium) belong to the alkaline-rich sodium-potassium basaltic series (Freund et al., 1975), which is essentially different from the potassium-sodium Neogene-Quaternary association.

Although the data on the Jurassic-Lower Cretaceous volcanics of Israel and Lebanon is also limited, an important difference exists in the geodynamic conditions during their formation (**Fig. 7.36**). Thus, according to the data of Freund et al. (1975), the volcanics of the Ramon in southern Israel are very different from those in the north in the Galilee, as indicated by the K_2O/SiO_2 ratio. The former are mostly potassium (4-6%), while the latter have a much lower potassium content (1.6-2.6%). This difference is related to the geodynamic situation during their formation on the margin of the Mesozoic Tethys Ocean.

The petrochemical characteristics of the Neogene-Quaternary basalts in Syria are more reliable because they are based on numerous chemical analyses. The basalts belong to the potassium-sodium alkaline olivine-basalt association. The SiO_2 content varies from 41% to 47.5%, the Na_2O from 2.58% to 4.5%, the K_2O from 0.63% to 1.92%, and the sum of the alkalines ($Na_2O + K_2O$) varies from 3.47% to 6.12%. According to their chemistry, they are belong to three different groups: normal basalts, sub-alkaline basalts, and alkaline basalts (Krasnov et al., 1966; Razvalyaev, 1966a; Ponikarov et al., 1967) (**Figs. 7.37** and **7.38**).

The Normal Basalt Group (Table 7.1)

Basalts of this group are well-developed within the covers of the Middle Miocene, Pliocene, and Quaternary. They are represented by dolerites, anamesites, and infrequently by basalts. They are typically under-saturated with respect to SiO_2, especially in the Recent basalt covers.

According to their chemical composition, the rocks are close to the middle type of tholeiitic basalts. Their petrographic and petrochemical similarity, close spatial relationship, and the existence of transitions between the basalts of these sub-groups show that their facies are interrelated, and that they were formed during the earliest stages of the Miocene, Pliocene, and Quaternary volcanic cycles. These sub-groups are defined by their different contents of the dark-colored minerals and plagioclase. Their general composition is close to that of the primary parent magma. The contents of CaO, TiO2, Fe_2O_3 + FeO and alkalis vary insignificantly. At the same time, the contents of MgO and to some extent Al_2O_3, change in the following manner:

1) in melanocratic basalts the contents of these oxides are 9.83% and 16.22%, respectively,

2) in normal basalts - 5.7% and up to 19%, respectively,

3) in leucocratic basalts - 1.91% and 23%, respectively.

It means that in the melanocratic basalts there is an accumulation of dark colored minerals (olivine and pyroxene) which is accompanied by a decrease in plagioclase content. In contrast, in the leucocratic basalts the quantity of the dark colored minerals decreases and plagioclase is enriched.

Fig. 7.35. The structure of the lava flow of the Diret-att-Touloul cover (β_3Q_4). The upper part of the flow is built by vesicular basalt, the lower part is massive.

The Sub-Alkaline Basalt Group (Tables 7.2a, b)

This group of effusive rocks makes up of the upper parts of the Middle Miocene, Pliocene, Middle Quaternary, Upper Quaternary, and Recent (Holocene) covers. In the case of the Quaternary covers, it can be clearly seen that the changes in chemical composition occur in the second half of each volcanic cycle as the basalt composition is reflecting the evolution of the primary, parent magma. The rocks of this group are also characterized by a pronounced lack of silica. They can be divided into three sub-groups according to the degree of alkali saturation. These sub-groups are: slightly melanocratic, slightly leucocratic, and also sub-alkaline basalts themselves.

According to their chemistry, the rocks of the sub-alkaline basalt group occupy an intermediate position between the normal and typically alkaline basalts. In sections of the volcanic covers they have a definite position, reflecting the stages in volcanic activity and the evolution of the primary basaltic magma. The transitional

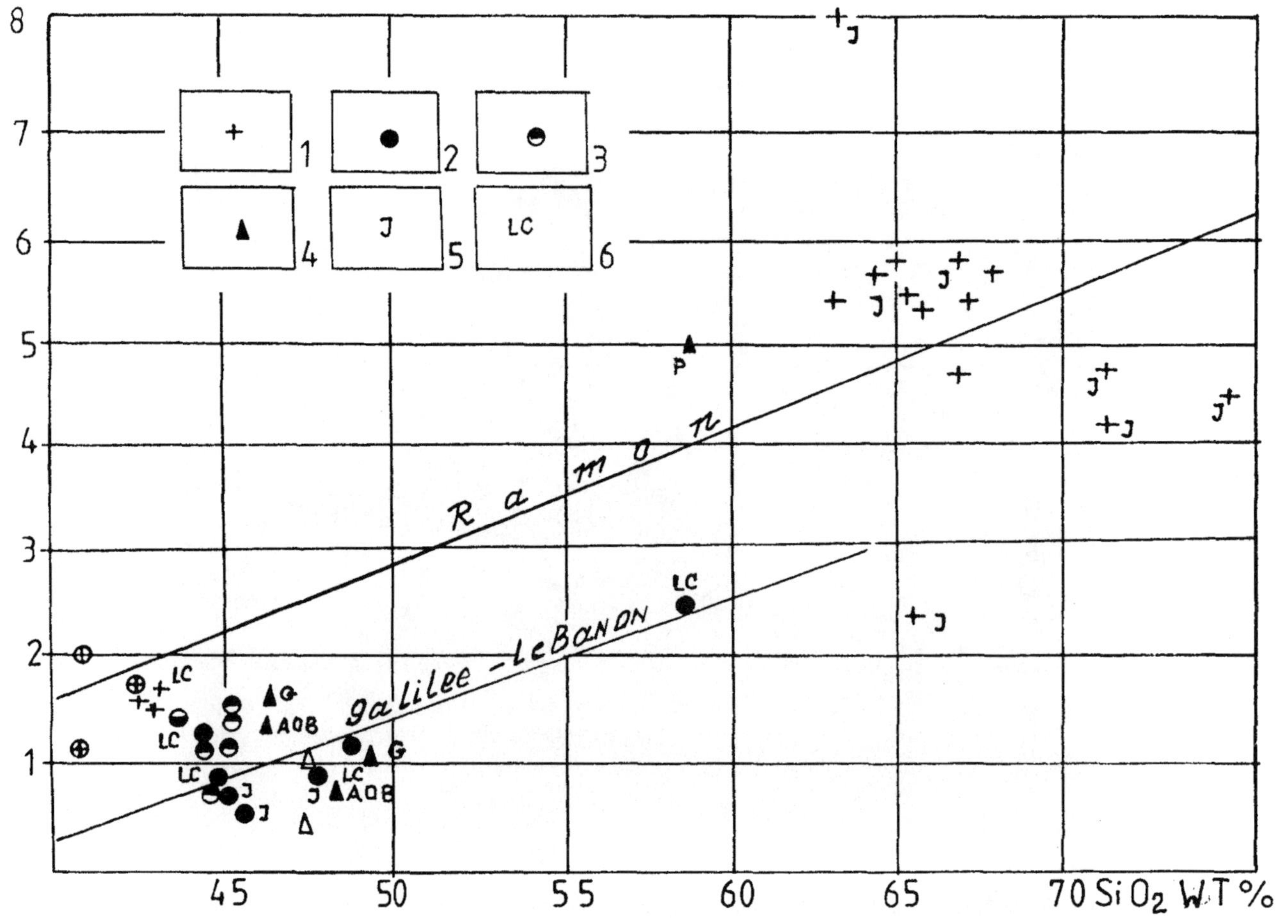

Fig. 7.36. The potassium-oxide – silica ratio-diagram for the Mesozoic volcanics of Israel and Lebanon (after Goldberg et al., 1975).
Legend: 1 – volcanics of the Ramon area; 2 – volcanics of Galilea and Lebanon; 3 – Recent basalts of southwestern Syria; 4 – basalts of Saudi Arabia; 5 – Jurassic volcanics; 6 – Lower Cretaceous volcanics; AOB – alkaline olivine basalts; P – phonolites; G – havayites.

character of the rocks of this group is manifested by a well-pronounced undersaturation in silica. The higher alkalinity was only recorded among the sub-alkaline basalts. In two other types of rocks within this group, which have a composition more typical of the normal basalts, their higher alkalinity is only expressed by the development of aegirine zones around the core of titan-augite, and by the presence of little needles of aegirine. It should be noted that these rocks are characterized by the high normative content of nepheline and orthoclase. Most of the melanocratics are magnesium-rich rocks of the first sub-group, and leucocratics are more alumina-rich rocks of the second and the third sub-groups.

The Alkaline Basalt Group (Tables 7.3a, b)

The rocks of this group are well represented in the covers of the Middle Miocene, Pliocene, Middle Quaternary, Upper Quaternary, and Recent (Holocene). They are recorded in the upper parts of these covers; indicating that they belong to the last stage of volcanic cycles. In other words, they were produced by the very last eruptions. The alkaline basalts form separate flows in between flows of sub-alkaline basalts. This is best observed in the upper part of the Pliocene basalt section in Wadi Raqqad on the Golan Plateau (Krasnov et al., 1966).

Petrographically, the rocks of this group include nepheline basalts, nepheline basanites, analcime basanites, teshenites, and others. All the types listed are similar and mostly differentiated according to the ratio of the main rock-forming minerals, and their structures. According to their chemical composition, they are characterized by further enrichment in alkali.

Thus the petrochemical patterns of the Neogene-Quaternary basalts of Syria are connected with a drop in silica and an enrichment in alkali. This chemical feature may relate to the evolution of the parent magma in the magma chamber or to fluctuations in the depth at which the magma is generated.

In most of the basaltic fields of Arabia the alkaline-olivine basalts and nodule-containing basanites predominate. Hawaiites and differentiated rocks exist to a lesser degree. The final product

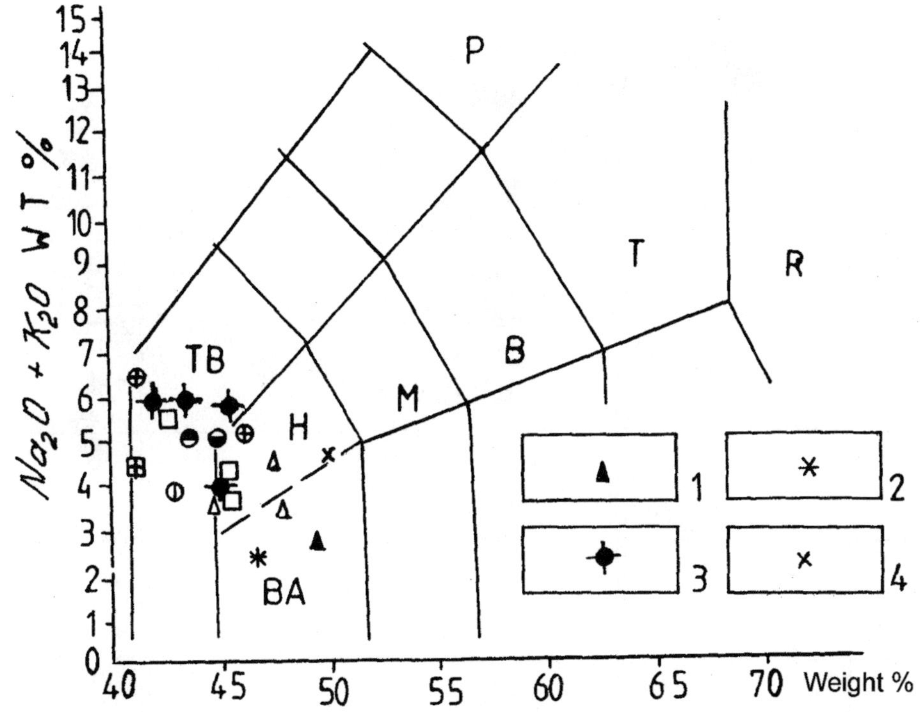

Petrochemical Groups of Basanites	Miocene	Pliocene	Pleistocene	Holocene
Normal	Δ	○	□	◓
Sub-Alkaline	Δ	⦶	◫	◒
Alkaline	▲	⊕	⊞	●

Fig. 7.37. The K_2O+Na_2O and SiO_2 ratio (after Le Bas et al., 1986). The dotted line – boundary of the alkaline and sub-alkaline fields (after Irvine and Baragar, 1971).

Legend: 1 – basalts of the axial part of the Red Sea (Coleman et al., 1977); 2 – the Miocene basalts of the coast of the Red Sea, Soudan (Razvalyaev, 1991); 3 – the Miocene basaltic dykes of the Sinai Peninsula (Baldridge et al., 1991); 4 – basalts of Saudi Arabia (Camp et al., 1991). Abbreviations: R – ryolites; P – phonolites; T – trachites; B – benmorites; M – moudjierites; H – havayites; TB – tephrit-basanites; BA – basanites.

of this process of differentiation is represented by phonolites. Some other magmatic fields are made up of sub-alkaline basalts where basalts without nodules are present. The latter include a small amount of alkaline-olivine basalts and hawaiites, but gradational to trachytes and less often to comendites (Camp et al., 1991).

On the diagram showing the alkali-to-silica ratio (Fig. 7.36) the basalts of Syria correspond to the higher alkalinity tephrite-basanite field. According to their petrochemical characteristics, they are more similar to the alkaline-olivine basalts and dikes of Saudi Arabia (Coleman, 1971; Camp et al., 1991), and differ significantly from the basalts of the axial trough of the Red Sea and the Sudanese coast (Razvalyaev, 1991).

From this diagram one can conclude that the general alkalinity of the basalts increases with time. The Pliocene basalts in the northeast of Syria are the most alkaline, but are not the youngest. This could be explained by magmas here derived from a deeper source, and not from the tectono-thermal system beneath the Red Sea which is a long distance away.

Taking into account the petrochemical similarity of the Syrian basalts and those of Saudi Arabia and the fact that they belong to the same alkaline olivine-basalt province, it can be concluded that the depths of their magma generation was about 80-100 km (Camp et al., 1991). They were generated by a parent magma with a low degree of partial melting of the mantle matrix. The normal alkaline rocks were probably formed in transitory chambers at the boundary of the mantle and the crust (at about 40 km) or higher. The parent magma was definitely differentiated by crystallization at this depth and later enriched by alkalis (Camp et al., 1991).

The petrochemical and petrographic similarity of the Neogene and Quaternary basalts of Syria and of the neighboring parts of Jordan, as well as the entire area of northern Arabia, indicates that the volcanic cycles followed one and the same pattern. The lavas that flowed at the beginning of each cycle produced a variegated spectrum of normal basalts. Later the lava outflows were impoverished in silica and were enriched in alkalis, which resulted in the formation of sub-alkaline basalts. In the last stages of each volcanic cycle the lavas became even poorer in silica and more enriched in alkalis, producing flows of alkaline basalts. This

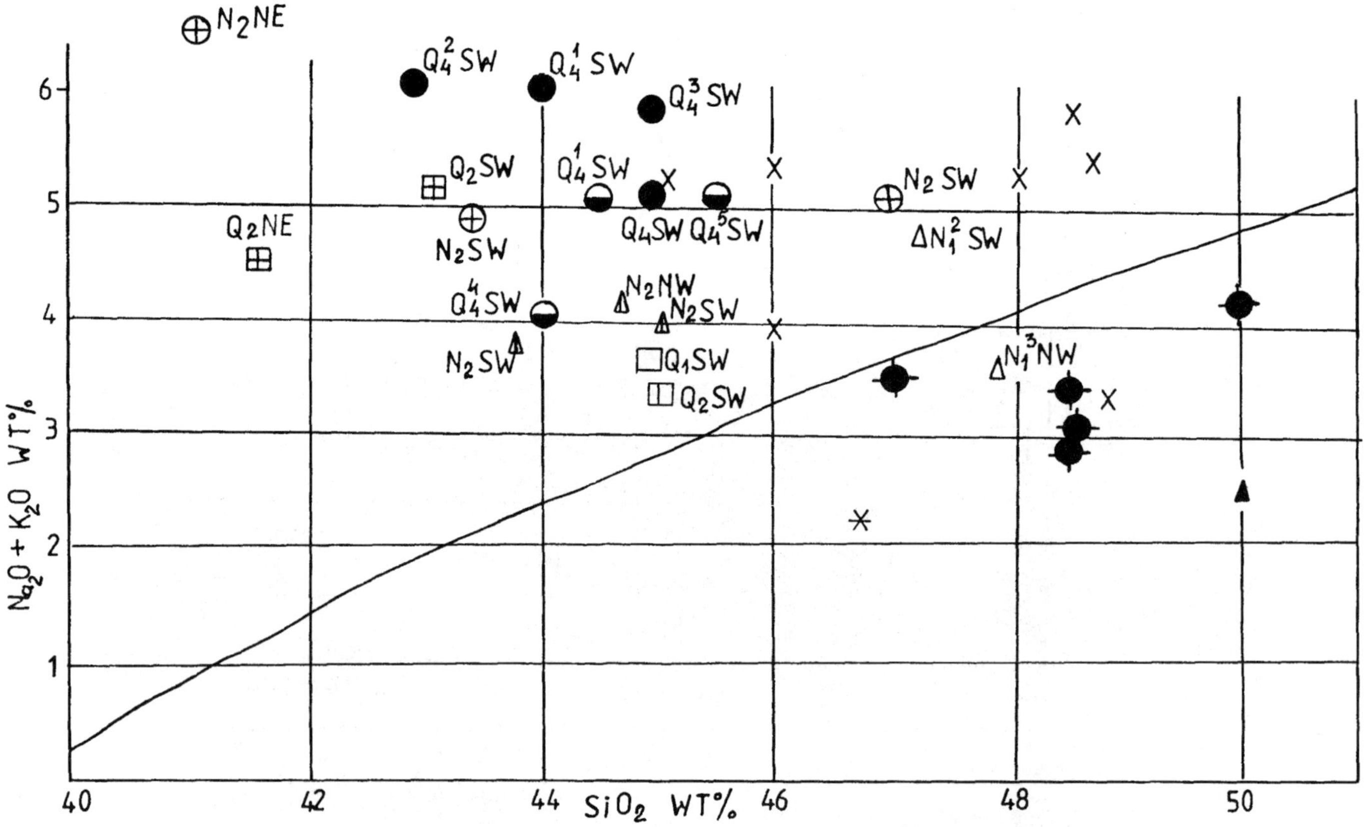

Fig. 7.38. The K_2O+Na_2O and SiO_2 ratio for the Miocene, Pliocene and Quaternary basalts of northeastern (NE) and south-western (SW) Syria. The line, separating alkaline olivine basalts (top) and tholeiites (bottom), is given after Irvine and Baragar (1971). The basalts of the axial trough of the Red Sea and of the dikes of the Sinai Peninsula are presented for comparison.
Legend: see Fig. 7.37.

volcanic evolution pattern is very typical for the Quaternary cycles and less evident for the Pliocene cycles.

The petrochemical tendency of the Neogene-Quaternary volcanism of Syria to increase alkalinity towards the end of the volcanic cycles, is well-founded, and based in general, on reliable data. It reflects the general geodynamic orientation of the Arabian plate and its tectono-magmatic activation. Tholeiitic and sub-alkaline series in other regions of intra-plate volcanism are usually separate entities, but in Syria they are connected via intermediate types, forming a single special continuous series. Nevertheless, a peculiar feature of the Neogene-Quaternary volcanism is found in the area of the Homs basaltic field (Sharkov et al, 1987). It is an intercalation (especially in the lower and middle parts of the basaltic plateau sections) of magma products from different depths – tholeiites, moderate alkaline basalts, and less often alkaline basalts. According to Sharkov et al. (1987), this finding can be related to the Syrian-Levantine transform fault. The greater tectonic mobility near the transform fault may have made the continuous inflow of alkaline melt from the deep chamber into a transitory one, with alternating flows of tholeiite basalts with increased alkalinity and titanium content, and potassium-sodium alkaline basalts.

Geochronology of the Neogene-Quaternary Basalts of Syria

During the first stages of the geological investigations in Syria, the age of the Late Cenozoic basalts was determined by their interrelations with the sedimentary rocks. However, for the Miocene, Pliocene, and Quaternary this method was successful only for northwestern Syria, where sediments of these ages are well-stratified and characterized by marine fauna. The internal regions of Syria (Palmyrides, Syrian Desert, the Anti-Lebanon, Homs basin, Hauran, and Jebel Drouze highland) are occupied by Late Cenozoic continental sediments which are practically barren of organic remains or which contain rare fossils. Their age (and the age of the associated basalts) was determined, with reservations, by indirect methods (the lithologic correlation, paleogeographic setting, stages of geological development). The situation regarding the Quaternary basalts is slightly different, as they are situated on the local river terraces or on peneplaned surfaces containing relicts of ancient cultures (flint tools).

This is why the isotopic (radiological) investigations were so important. These were carried out in the covers of basalt developed in the northwest and southwest of Syria (Giannerini et al., 1988;

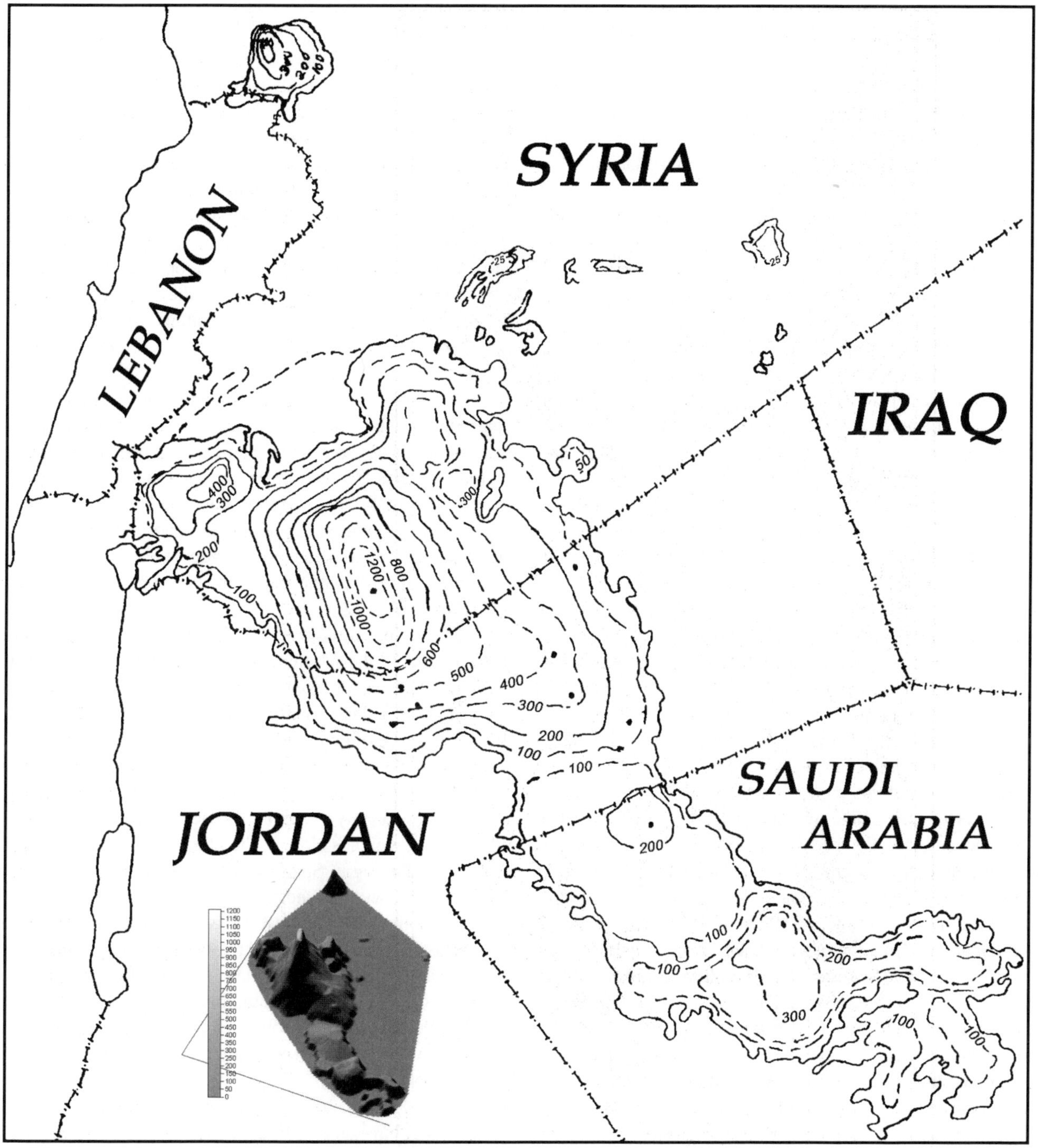

Fig. 7.39. Distribution of the Neogene-Quaternary basalts between the Rutba anteclise in the east and the Jordan uplift in the west and total thicknesses of basaltic covers.

Sharkov et al., 1994, 1998; Sharkov, 2000; Novikov et al., 1993), and in the neighboring vicinity of Israel on the eastern coast of Lake Tiberias (Michelson and Lipson-Benitah, 1986; Mor and Steinitz, 1982; Michelson and Mor, 1985). These investigations yield more exact ages for the basalts, their weathering crusts (*i.e.,* they dated the climatic changes), and for the associated continental sedimentary rocks. To our regret, it was not always possible to connect the isotopic age values to the field data from the 1:200,000 scale geological mapping of Syria.

Five stages of magmatic activity were identified (and four of them dated):

1) The Early Miocene basalts (26-16 Ma) were found in the north of the Hauran-Drouze plateaus (20.9 and 20.4 Ma), in the Damascus depression (20.4; 20,3; and 19.9 Ma), in the Al-Daww basin (24.7-26.0 Ma), and some are on the Shin Plateau (17.3 Ma).

2) A Middle Miocene age (16-11 Ma) was ascribed to basalts on the Aleppo Plateau (12.0 and 10.8 Ma), the Shin Plateau (12.8 Ma), the northern part of the Jebel Drouze highland (13.6 Ma), and to the basalt layers in the Hordos Formation on the eastern coast of Lake Tiberias, Israel, west of the Hauran Plateau (12.1 and 10.8 Ma,).

3) According to numerous dates, basalts of the Late Miocene (11-6 Ma)-Early Pliocene (6-4 Ma) are widely developed in the Hauran Plateau, the Jebel Drouze highland, and the Shin Plateau. Two periods of volcanic activity may mark the Late Miocene: on the Shin Plateau and the Aleppo Plateau (9.7 and 7.8 Ma), and southeast of Jebel Drouze (7.54 Ma);

The ages of the transitional layers in the boundary interval between the Miocene and Pliocene on the Shin Plateau are: 6.3; 6.2; 5.64; 5.63; 5.40; 5.23; 5.21; 5.19; 5.08; 5.07; 4.87; 4.79; and 4.76 Ma, and for the Hauran-Drouze Plateau: 5.4; 5.2; and 4.5 Ma.

4) Radiometric ages for the Late Pliocene-Pleistocene basalts were as follows: on the Hauran Plateau (the so-called cover basalts, 3.7;3.1; 2.19; 1.49; 1.07; 0.64; and 0.3 Ma), and the northern part of the Al-Ghab graben near Jisr Ash-Shaghour (1.95; 1.3; and 1.1 Ma).

5) The Holocene basalts could not be dated as they were below the reliable detection limit of the K/Ar method.

Sharkov et al. (1994, p. 448) wrote that "The periods of most intensive magmatic activity coincide in time with the main geodynamic events: the Early Miocene – with the opening of the Red Sea Rift; the Late Miocene-Early Pliocene – with the opening of the Gulf of Aqaba and the formation of the Levantine transform fault; and the Late Pliocene-Quaternary – with the general uplift of the region and the creation of the present relief". Sharkov et al. (1994) emphasized that the monotype basalt volcanism in Syria continued over a period of 21 Ma. This feature differentiates the Near East from other regions of typical trap-volcanism (*e.g.* the Deccan Plateau in India, Eastern Siberia in Russia), where huge amounts of basaltic lavas erupted over a very short (1-2 Ma) interval. In their opinion, "This attests to the existence of principle differences between the regions of the trap-magmatism and structures related to the opening of new oceans" (Sharkov et al., 1994, p. 448).

The Diatrems of Kimberlite-like Rocks and Carbonatites

Twenty-three diatremes were discovered on the Jebel An-Nusseiriyeh highland in northwestern Syria. They are composed of kimberlite-like rocks containing xenolites from great depths (Sharkov et al., 1987, 1992). These are ellipsoidal explosion-pipe funnels in the Jurassic and Lower Cretaceous limestones. The throat of the funnels is flat and covered by debris and deluvium. Their diameters are usually 500-1500 m and the depth is from 400 m to 500 m.

The largest of these pipes is the Nabi-Matta diatreme situated southwest of the town of Missiaf. It constitutes a lens-like body about 40 km^2 in area formed by pyroclastics of ultrabasic kimberlite-like matter 100-150 m thick. The diatreme exploded through the limestones of Aptian-Albian age. It is overlain by Cenomanian limestones. It follows that the age of the diatreme is close to the Albian (the late Early Cretaceous). The K/Ar isotope ages of the diatreme (120-123 Ma on average, *i.e.* Lower Aptian) are not in accord with this observation (Sharkov et al., 1987). However it should be noted that for the diatremes in this part of Syria, geologists have different opinions regarding the ages: in the case of this diatreme some think it is Cretaceous and some, Pliocene.

The xenolites are mostly found in the central part of the diatreme and they decrease in quantity towards its periphery, but various grains of garnet and kaersutite have been found throughout. Eclogite, granulite, pegmatoidal spinel-garnet and pyroxene-hornblende rocks, spinel gabbroids, megacrysts of titan-augite, kaersutite, pyrope-garnet, and garnet clinopyroxene growths were identified in the explosion pipe (Snyder et al., 1993). In some of the diatremes small diamond fragments have occasionally been found (Sharkov et al., 1992).

The xenolite rocks from the Nabi-Matta diatreme are similar to the xenolites of the sub-alkaline Neogene-Quaternary basalts from certain volcanoes, for example, the Tell Danun volcano (Sharkov et al., 1987, 1993, 1994). But the diatreme clearly differs from them by the presence of granulite and eclogite xenolites.

The rocks of the eclogite-granulite series are represented by large (up to 20-40 cm) sub-ellipsoidal xenolites, comprising 70% of the entire quantity of xenolites. They have a massive and gneiss-like texture, which is close to the granoblastic structure typical for schists. According to Sharkov et al. (1992), the latter attest to the metamorphic nature of the rock.

The composition of these rocks allows their subdivivision into non-feldspar 'eclogites' and plagioclase-garnet granulites. Taking into consideration their chemical composition, the rocks of the Nabi-Matta diatreme are distinguished from typical kimberlites by their higher alumina content and lower magnesium content. Therefore, they should more correctly be called 'kimberlite-like' (Sharkov et al., 1992).

The megacrysts of the dark-red garnet (pyrope) reach 10 cm in diameter, have an ellipsoidal form, and are covered by a shiny, glassy crust, indicating their partial melting.

Based on the xenolith assemblage in the Nabi Mata explosive center Sharkov et al. (1992) assumed that they represent materials from the lower crust and upper mantle of the southeastern margin of the Mesozoic Tethys.

The Neogene-Quaternary sub-alkaline basalts of the Tell Danun volcano, located approximately 10 km to the north-east of Shahba, contain xenoliths of spinel peridotites and garnet-spinel websterites ('the green chromium-diopside series'), as well as hornblendites and hornblende-bearing clinopyroxenites ('the black aluminum-titanium augite series'). Megacrysts of kaersutite and ilmenite are present in the host basalts together with xenoliths (Snyder et al., 1993; Sharkov et al., 1996). Garnet clinopyroxenites from the Tell Danun volcanic field are the dismembered substance of dikes from depths of 24-30 km within the lower crust and/or upper mantle beneath the Arabian plate (Snyder et al., 1993). Based on the thermobarometry results, xenoliths of spinel peridotites were transported from depths of 30-60 km and can be treated as the substance of depleted upper mantle (Sharkov et al., 1992, 1993, 1994, 1996). A remarkable geochemical property of spinel xenoliths of the Tell Danun volcano is their enrichment of LREE. This phenomenon can be related to the magmatic interaction between xenoliths and the alkaline basaltic magma which transported them. On the other hand, this enrichment in LREE may reflect the influence of mantle metasomatism (Snyder et al., 1993).

As with the other tectono-magmatically 'activized' regions of the platform (*i.e.* the sudden outburst of magmatic activity on what was until then a very stable craton), the investigation of the intra-plate sub-alkaline Neogene-Quaternary volcanism and the Lower Cretaceous diatremes of Syria with their xenolites of mantle matter is very important for understanding the compositional evolution of the mantle beneath the Arabian platform.

Nevertheless, there is also a principle difference in the xenolite composition which is expressed in the presence of eclogite xenolites in the pre-Neogene diatremes and in their complete absence in the Neogene-Quaternary sub-alkaline basalts. The latter have only spinel lherzolite xenolites.

The difference in composition between the diatreme xenolites and the Neogene-Quaternary basalts (volcanoes) may testify to the lateral and vertical irregularities in the lithosphere. The short distance (10-15 km) from the Nabi-Matta diatreme to the Jebel Saida volcano may attest to the degree of vertical irregularity in the upper mantle.

The eclogites of these diatremes represent material in the lower crust or the uppermost mantle which was formed during the Late Pre-Cambrian stage of development of the Arabian platform. Later, during the Neogene-Quaternary stage of 'activization' of the platform, they were brought to the surface by sub-alkaline basalts related to mantle diapirs. The mantle evolutionary process probably resulted in the ancient mostly eclogite Late Pre-Cambrian upper mantle (disgorged by the Lower Cretaceous diatremes) being replaced by lherzolite matter during the Neogene-Quaternary stage of the 'activization' of the Arabian platform. This process is related to the rise of mantle diapirs, which produced alkaline basalt volcanism and their subsequent horizontal spreading. Later on, the eclogite matter flowed down to the periphery of the diapirs, leaving a place in the center for the uplift of the lherzolite matter (Sharkov et al., 1992).

An excellent example of the development of the initial within-plate magmatism is the appearance of the Lower Pleistocene complex of carbonatite dikes, tuffs and lavas, which occurred in several areas of southern Syria (Mahfoud and Beck, 1995a, b). Their carbonates are high in alkalis (up to 28 wt%) and can be classified as alkalic-dolomitic carbonatites. The mineralogical and geochemical features of the rocks which comprise this complex, allow us to attribute them to two main geochemical groups. The first group, more distributed, is probably derived from the partial melting of a CO_2-rich source in the garnet peridotite zone. The rock has fine to microporphyritic textures, a trace of P_2O_5 and a disequilibrium reaction relationship with nepheline-bearing inclusions. The second less common group was probably formed by melting at shallower depths in the mantle. It has emulsion textures, no K, Sr, and Ba, and scattered xenocrysts of spinel lherzolite mineral (Mahfoud and Beck, 1995a, b). These scientists assumed that the carbonatite magmatism of southern Syria reflects the earliest stage of mantle melting beneath the Arabian plate connected with the opening of the Red Sea, the tectonic 'activization' of the Arabian plate, and the spreading and rifting processes to come.

The Relationship of Volcanism to the Tectonic Elements of the Arabian Platform

Almost all investigators traditionally relate the Syrian volcanism to the development of the West Arabian fault system. But in most cases this relationship has only an indirect character. The Neogene-Quaternary volcanism is controlled, first of all, by the northwestern trending faults, which are oblique to the Dead Sea faults and to the entire West Arabian fault system. This is clearly seen, especially in southwestern Syria on the Hauran Plateau, Jebel Drouze, and Jebel Diret-att-Touloul. The volcanism of northeastern Syria (between the Al-Furat and Nahr El-Khabour Rivers) is probably related to the faults of the Euphrates strike-slip zone.

The largest area of volcanism is located in southwestern Syria and continues into Jordan and Saudi Arabia (Hamad Basin Project, 1983). As already noted, the main faults controlling the volcanism also have a northwestern trend there. They are concentrated in three zones: the western, known by the name Hauranian; the central – Jebel Drouze; and the eastern – Diret-att-Touloulian.

It was primarily assumed that the basalts of Jebel Drouze and the Hauran Plateau are associated with the Drouze depression (Razvalyaev, 1964; Ponikarov et al., 1969). This would be correct if the structure of this region was discussed in its broadest aspects, *e.g.* between the Rutba anteclise on the east and the Jordan uplift on the west (**Fig. 7.39**). And in fact, when studied in more detail, especially using data from hydrogeological bore-holes (Figs. 7.4 and 7.5), it became clear that this assumption is justified. However, this may not really be the case since the surface of the underlying basalts doesn't show much downwarping where it transits the Jordan Uplift, and according to its degree of dislocation, it is comparable with the rest of the Arabian platform. The downwarp basalts pass into the eastern part of the Jordan uplift and its northern periclinal plunge.

The data on the Drouze depression attest to its complex structure. Drilling results show that within the underlying Paleogene deposits there are some comparatively downwarped and uplifted secondary features. For example, in the Damascus depression the Paleogene sediments lie at depths of more than 800 m (Fig. 7.5, profile IV). To the southwest and near the Jebab settlement they rise to the surface. The Jebab uplift can be traced southwestward for a distance of 20 km (Fig. 7.5, profiles III and IV).

The Jebel Drouze mountains formed by thick covers of Neogene-Quaternary basalts have a volcanogenic-accumulative nature. Theoretically we cannot exclude the existence of a downwarped surface of the underlying basalts in the central part of Jebel Drouze, since there was no drilling in this place.

A large region, which included southern Syria, Jordan, and a part of Saudi Arabia, has been studied in the framework of the Hamad Basin Project (1983). Most of the Jebel Drouze Plateau and its southern continuation are located within this area. The results of the project proved the role of the northwestern faults in controlling the magma flows. Faults of this trend in Syria determined the linear placement of the volcanic cones and basaltic dikes. They are clearly visible on the satellite images (Galaktionov, 1989). Thus, in the Keraq region, the northwest trending fault zone can be traced through all the central part of the Hauran Plateau for a distance of over 300 km. Along the length of this zone, the faults are accompanied by isolated basalt fields and basaltic and dolerite dikes. North of it, another shorter (20 x 100 km) zone can be traced. In the south of Jordan the northwestern trending faults are also accompanied by basalt and dolerite dikes intruded into the Paleozoic sandstones.

These faults have a trend similar to that of the Red Sea rift and related linear zones of dikes and grabens in Arabia and in the Sinai Peninsula (Mart, 1984; Mart and Rabinovitz, 1986; Mechie and El-Isa, 1988; Baldridge et al., 1991; Camp et al., 19991). Thus, the faults are genetically related to the extensional structures of the Red Sea rift zone.

As a rule the large sub-longitudinal faults of the West Arabian rift zone are not accompanied by volcanism. In Lebanon and Syria this is probably explained by their kinematics, since they are strike-slip faults developed under transpressional conditions.

To some extent the opening of the northwest-southeast-trending fractures is related to the strike-slip dislocations along the West Arabian (Levantine) faults. It is important to note the following:

1) The basalt flows in southwestern Syria began in Late Oligocene-Early Miocene (20-26 Ma), which means long before the strike-slip dislocation in the West Arabian system (about 12 Ma) according to Le Pichon and Gaulier (1988). In this case it is impossible to speak of any relationship.

2) In western Syria (Jebel Drouze, the Aleppo Plateau, Jebel Az-Zaouiye) the volcanic activity is clearly connected with the curved segments of the fault plane, which constrain the main displacement to the east (Yammuneh and Al-Ghab faults). In these segments there apparently exists a transpressional regime, which Kopp et al. (1994) have called 'an oblique motion', in which the mass of the Arabian plate appears to have 'flowed around' this kink-like obstacle. Simultaneously, in the wide zone of this 'flow around' deformational conditions have appeared which approximate a clear sinistral slip. As it was shown by Kopp et al. (1994), under such conditions, the northwest-southeastern trending fractures operated under tensional conditions and the northeast-southwestern faults under a compressional regime. This way, during the periods of strike-slip dislocations coinciding with the stages of the Red Sea opening (Middle-Late Miocene and Pliocene), the northwest-southeast trending faults could have opened and served as conduits for basaltic magma. As a result, the opening of the fractures with the Red Sea trend might begin in connection with the earliest stage of the Red Sea rift formation. Later on, this process was stimulated by specific conditions related to the evolution of the West Arabian transform system.

At the end of the Jurassic and beginning of the Cretaceous the Arabian Plate possibly underwent a general uplift accompanied by regression of the sea and marine erosion. Near the transform boundary with Mesogea this event was associated with basaltic volcanism, possibly related somehow to the movement along the transform zone.

The Early Cretaceous-Cenomanian magmatism around 90-120 Ma also can be related to the tectonic reconstruction taking place in the basins of Mesogea and the Tethys Ocean. During that period some important changes took place in the geodynamic system of the Tethys. On its northern active margin the zones of subduction were blocked. In contrast, new subduction zones appeared near the southern passive margin of the Tethys paleo-ocean and also in Mesogea. On the Arabian margin of the paleo-ocean this event was reflected in a general trans-tension, rift-genesis, regression, and local basalt volcanism. The reason for the intra-plate trans-tension could be the pull of the oceanic litho-spheric plate, dragged into the newly formed subduction zones.

The geodynamic nature of the basaltic volcanism in the Neogene and Quaternary was different. The intensive volcanism of this period was unquestionably related to the evolution of the African-Arabian rift system, and to the development of the Red Sea rift in particular. The initial stage of the Red Sea opening (Late Oligocene-Early Miocene, 22-30 Ma) was accompanied by basaltic volcanism within its eastern margin. Such localization could probably be explained by the asymmetric position of the asthenosphere nose in relation to the rift axis (Wernicke, 1985; Bohannon, 1989). In Syria the Lower Miocene volcanism is relatively limited. The basalts exist in the southwestern part of the country, in the northern part of Jebel Drouze (20-21 Ma), in the area of Damascus (19-20 Ma), in the Ad-Daww basin (24.7-26 Ma), on the Shin Plateau, and in the Homs basin (17 Ma). In northwestern Syria they are totally absent.

In southwestern and northwestern Syria the Late Miocene and Pliocene basaltic volcanism eruptions coincide with the two main phases of sinistral strike-slip motion along the West Arabian (Levantine) system, and accordingly, with the two main phases of the opening of the Red Sea and the Aden rifts. The movements began 12-13 Ma ago (Garfunkel, 1981; Le Pechon and Gaulier, 1988; Ben-Avraham and Grasso, 1991). This event coincided with the continuing volcanism in southwestern Syria and with the intensive volcanism in the northwest – the Shin Plateau (5-13 Ma), and to some extent in the Aleppo Plateau (7-12 Ma), and in the Aafrine area.

The second large phase of basaltic volcanism in Syria belongs to the Pliocene-Quaternary with its maximum in the Late Pliocene and Quaternary. This phase correlates with the second main phase of the movements along the Syrian-Levantine fault and with the second phase of the opening of the Red Sea and Aden Gulf rifts. The Miocene-Pliocene and Quaternary basalt flows are clearly connected to the northwest-oriented fractures. This relationship is especially obvious in southwestern Syria on the Hauran Plateau, where it is accentuated by the direction of the oriented chains of Quaternary volcanoes. In northwestern Syria, the northwest-trending fractures probably also acted as magma-supplying conduits. It is not difficult to imagine that under conditions of sinistral strike-slip motion along the Lebanese-Syrian faults, these fractures (possibly reactivated faults in the Precambrian basement) had to be of the tensional type, in other words, to have opened.

Conclusions

1) The Syrian volcanism occured during two epochs: Mesozoic (Early Jurassic-Early Cretaceous) and Neogene-Quaternary. The volcanism of the earlier epoch is related to the origin of the Neo-Tethys and Mesogea and with the development of the eastern Mediterranean continental margin, which probably had a transform character.

2) The Neogene-Quaternary volcanism is connected to the opening of the Red Sea rift and to the movements along the Levantine fault system. Five periods of volcanic activity can be discerned: Early Miocene (15-21 Ma), Middle Miocene (11-15 Ma), Late Miocene-Early Pliocene (4.5-11 Ma), Late Pliocene-Pleistocene (0.3-1.95 Ma), and Holocene.

3) The Neogene-Quaternary volcanism was controlled by the northwest-trending faults of the Red Sea trend.

4) The Neogene-Quaternary volcanics of Syria had a tendency to change over time from a tholeiitic to sub-alkaline and even alkaline composition and to belong to the single, Afro-Arabian alkaline olivine basalt province. The alkalinity changes in the petrochemical composition were dictated by the depth of magma formation and the grade of melting of the mantle matter.

REFERENCES CITED

Baldridge, W. S., Eyal, Y., Bartov, Y., Steinitz, G., and Eyal, M., 1991. Miocene magmatism of Sinai related to the opening of the Red Sea. Tectonophysics, vol. 197, p. 181-201.

Ben-Avraham, Z., and Grasso, M., 1991. Crustal structure variations and transcurrent faulting at the eastern and western margins of the Eastern Mediterranean. Tectonophysics, vol. 196, p. 269-277.

Bentor, Y. K., 1957. Cenozoic volcanism of Israel. XX sess. Intern. Geol. Congr., Mexico.

Blanckenhorn, M., 1914. Syrien, Arabien und Mesopotamien. Handbuch der regionalen Geologie, vol. 5, Heft 4, 159 p.

Bohannon, R. G., and Eittreim, S. L., 1991. Tectonic development of passive continental margins of the southern and central Red Sea with a comparison to Wilkes Land, Antarctica. Tectonophysics, 198(2-4), p. 129-154.

Burdon, D. J., 1959. Handbook of the geology of Jordan to accompany and explain the three sheets of the 1:250,000 geological map of 'Jordan east of the rift' by A. M. Quennel. Colchester (Benham and Co), 82 p.

Camp, V. E., Roobol, M. J., and Hooper, P. R., 1991. The Arabian continental alkali basalt province: Part II. Evolution of Harrats Khaybar, Ithnayn and Kura, Kingdom of Saudi Arabia. Geol. Soc. Amer. Bull., vol. 103, p. 363-391.

Coleman, R. G., 1971. Plate tectonic emplacement of upper mantle peridotites along continental edges. J. Geophys. Res., v. 76, p. 1212-1222.

Coleman, R. G., 1977. Ophiolites. Ancient Oceanic Lithosphere? Springer-Verlag, Berlin-Heidelberg-New York, 229 p.

Doss, B., 1886. Ueber die basaltischen Laven und Tuffe der Provinz Hauran und vom Diret et Tullul in Syrien. Tschermak's Miner. Petrogr. Mitt., N F, vol. 7, Wien, 461-535 p.

Dubertret, L., 1929. Études des régions volcaniques du Haouran, du Djebel Druze et du Diret el Touloul (Syrie). Rev. géogr. phys. et géol. dynam., vol. 2, N. 4, p. 275-321.

Dubertret, L., 1933. Les grandes nappes basaltiques syriennes: âge et relation avec la tectonique. C. R. Soc. Géol. France, vol. 11, p. 178-180.

Dubertret, L., 1937. Le Massif Alaouite. Notes et Mém. Syrie et Liban, vol. 2, p. 9-42.

Dubertret, L., 1940. Sur l'âge du volcanisme en Syrie et au Liban. C. R. Soc. Geol. France, vol. 6, p. 55-57.

Dubertret, L., 1953. Carte géologique au 200,000-e. Feuille d'Antioche (notice). Damascus, 67 p.

Dubertret, L., 1954. Basaltes et roches vertes du Liban, de Syrie et du Hatay (ancien Sandjak d'Alexandrette, Turquie). C. R. Congr. Géol. Internat., Alger 1952, XIX sess., fasc. 19, p. 29-36.

Dubertret, L., 1955. Carte géologique du Liban au 200,000-e (notice). Beyrouth, 74 p.

Dubertret, L., and Wetzel, R., 1949. Carte géologique au 200,000-e, feuille de Tripoli. Damascus, 62 p.

Faradzhev, V. A., 1966. The Geological map of Syria. Scale 1:200,000. Sheet I-37-XIV (Al Qaryatein). Explanatory notes. Technoexport, Moscow, 146 p.

Freund, R., Goldberg, M., Weissbrod, T., Druckman, Y., and Derin, B., 1975. The Triassic-Jurassic structure of Israel and its relation to the origin of the Eastern Mediterranean. Geol. Surv. Israel, Bull., v. 65, 26 p.

Galaktionov, A. B., 1989. Map of identification of space images of Syria. Scale 1: 500,000. Technoexport, Damascus, Syria. (Unpublished).

Galactionov, A. B, Demidov, V. A., and Katkova, N. S., 1966. The Geological map of Syria. Scale 1:200,000. Sheets J-37-VI, XII (Al-Qamishli, Sinjar) and J-38-I, VII (Qarashouk Dagh). Explanatory notes. Technoexport, Moscow, 83 p.

Garfunkel, Z., 1981. Internal structure of the Dead Sea leaky transform (rift) in relation to plate kinematics. Tectonophysics, vol. 80, p. 81-108.

Giannerini, G., Campredon, R., Feraud, G., and Abou Zakhem, B., 1988. Déformations intraplaques et volcanisme associé: example de la bordure NW de plaque au Cénozoique. Bull. Soc. Géol. France, N. 6, p. 938-947.

Hamad Basin Project (Series of maps), 1983. Damascus.

Irvine, T. N., and Baragar, W. R., 1971. A guide to the chemical classification of the common volcanic rocks. Canadian J. Earth Sciences, v. 8, p. 523-548.

Kazmin, V. G., 1991. On tectonic conditions of the Troodos and Mamonia complex formation. Geotectonica, N6, p. 104-116. (In Russian).

Kazmin, V. G., 2003. Tectonics of Syria, Chapter 8. This volume.

Kozlov, V. V., (assisted by Artemov, A. V., and Kalis, A. F.,), 1966. The Geological map of Syria. Scale 1:200,000. Sheets I-36-XVIII, I-37-XIII (Trablus, Homs). Explanatory notes. Technoexport, Moscow, 68 p.

Kozlov, V. V., Ponikarov, V. P., and Razvalyaev, A. V., 1967 The young volcanism of the northern slope of the Arabian part of the African platform. Trans. of the Moscow University, Ser. Geol., vol. 2, p. 42-51. (In Russian).

Krasnov, A. A., Kazmin, V. G., and Kulakov, V. V., 1966. The Geological map of Syria. Scale 1:200,000. Sheets I-36-VI, I-37-I, II (As-Suweida, Az-Zalaf). Explanatory notes. Technoexport, Moscow, 73 p.

Kurbanov, N. K., and Zaryanov, Yu. P., 1963. The morphological features of the Upper Quaternary volcanic eruption products from volcanoes of southwestern Syria. Trans. ZNIGRI, vol. 57, p. 73-96. (In Russian).

Lartet, L., 1877. Exploration géologique de la Mer Morte, de la Palestine et de l'Idumee. Paris (A. Bertrand), 326 p., geol. maps 1:1 000,000 and 1:300,000.

Le Pichon, X., and Gaulter, J. -M., 1988. The rotation of Arabia and the Levant fault system. Tectonophysics, vol. 153, p. 271-294.

Mahfoud, R. F., and Beck, I. N., 1995a. Composition, origin and classification of extrusive carbonatites in rifted southern Syria. Intern. Geol. Rev., vol. 37, p. 361-378.

Mahfoud, R. F., and Beck, I. N., 1995b. Petrogenesis and geodynamic significanse of ultramafic inclusions and basanites from southern Syria. Intern. Geol. Rev., vol. 37, p. 448-470.

Mart, Y., 1984. The tectonic regime of the South-Eastern Mediterranean continental margin. Marine Geology, vol. 55, p. 365-386.

Mart, Y., 1991. The Dead Sea Rift: from continental rift to incipient ocean. Tectonophysics, vol. 197, p. 155-179.

Mart, Y., and Rabinowitz, P. D., 1986. The northern Red Sea and the Dead Sea rift. Tectonophysics, vol. 124, p. 85-113.

McCuire, A. V., and Bohannon, R. G., 1989. Timing of mantle upwelling: evidence for a passive origin for the Red Sea Rift. J. Geophys Res., vol. 94, N. 132, p. 1677-1682.

Mechie, J., and El-Isa, Z. H., 1988. Upper lithosphaeric deformations in the Jordan Dead Sea transform regime. Tectonophysics, vol. 153, p. 153-159.

Medvedev, V. Ya., 1966a. The Geological map of Syria. Scale 1:200,000. Sheet I-37-XXII (Ar-Raqqa). Explanatory Notes. Technoexport, Moscow, 56 p.

Medvedev, V. Ya., 1966b. The Geological map of Syria. Scale 1:200,000. Sheets I-37-XXIII, XXIV (Deir Az-Zor, Al-Buvara). Explanatory Notes. Technoexport, Moscow, 42 p.

Michelson, H., and Mor, D., 1985. The Southern Golan Heights geological map (Gamla), Scale 1: 50,000. The Geological Survey of Israel.

Michelson, H., and Lipson-Benitah, S., 1986. The Litho- and Biostratigraphy of the Southern Golan Heights. Israel J. Earth Sci., vol. 35, p. 221-240.

Mikhailov, I. A., 1966. The Geological map of Syria. Scale 1:200,000. Sheet I-37-XX (Salamieh). Explanatory notes. Technoexport, Moscow, 64 p.

Mor, D., and Steinitz, G., 1982. K-Ar age of the Cover Basalt surrounding the Sea of Galilee. Geol. Surv. Israel Rep., ME/82/6.

Novikov, V. M., Sharkov, E. V., Chernishev, I. V., Devyatkin, E. V., Ivanenko, V. V., Karpenko, M. I., Hanna, S., and Akhtum, N., 1993. Geochronology of the weathering crusts of the plateau basalts of Syria and evolution of the paleoclimate of the region during the last 20 m. y. Stratigraphy and Geological Correlation, vol. 1, N. 6, p. 66-74. (In Russian).

Oufland, A. K., 1966. The Geological map of Syria. Scale 1:200.000. Sheet J-37-III (Jrablus). Explanatory Notes. Technoexport, Moscow, 56 p.

Ponikarov, V. P., Kazmin, V. G., Mikhailov, I. A., Razvalyaev, A. V., Krasheninnikov, V. A., Kozlov, V. V., Soulidi-Kondratiyev, E. D., Mikhailov, K. Ya., Kulakov, V. V., Faradzhev, V. A., Mirzayev, K. M., 1967. The Geology of Syria. Explanatory notes on the Geological map of Syria. Scale 1:500,000. Part I. Stratigraphy, igneous rocks and tectonics. Technoexport, Moscow, 229 p.

Ponikarov, V. P., Kazmin, V. G., Kozlov, V. V., Krasheninnikov, V. A., Michailov, I. A., Razvalyaev, A. V., Soulidi-Kondratyev, E. D., Oufland, A. K., and Faradzhev, V. A., 1969. Geology and Mineral resources of foreign countries. Syria. Trans. of NILZarubezhgeologia, vol. 18, Publ. House 'Nauka', Leningrad, 216 p. (In Russian).

Ponikarov, V. P., Soulidi-Kondratyev, E. D., Kozlov, V. V., and Kazmin, V. G., 1964. Tectonics of the northwestern part of the Arabian platform. Sovetskaya geologia, N. 1, p. 39-48. (In Russian).

Ponomarev, B. Ya., and Bivshev, A. S., 1966. The Geological map of Syria. Scale 1:200,000. Sheets J-37-V, XI (Al-Hassakeh). Explanatory Notes. Technoexport, Moscow, 66 p.

Protasevich, L. N., and Maksimov, A. A., 1966. The Geological map of Syria. Scale 1:200,000. Sheets J-37-I, II (Halab, Antakya). Explanatory notes. Technoexport, Moscow, 73 p.

Razvalyaev, A. V., 1964. Tectonics of southwestern Syria and adjacent parts of neighboring countries. Geologia and razvedka, N. 12, p. 14-26. (In Russian).

Razvalyaev, A. V., 1966a. The Geological map of Syria. Scale 1:200,000. Sheets I-37-VII, I-36-XII (Damascus, Beirout). Explanatory notes. Technoexport, Moscow, 124 p.

Razvalyaev, A. V., 1966b. The Geological map of Syria. Scale 1:200,000. Sheet I-37-VIII (Saba Abar). Explanatory notes. Technoexport, Moscow, 63 p.

Razvalyaev, A. V., 1971. Structural features and geological development of the west Arabian Rift system. Bull. of the Moscow Soc. of researches of Nature, vol. XLVI, N. 3, p. 28-39. (In Russian).

Razvalyaev, A. V., 1991. Continental rift formation and its prehistory. Oxford and IBH Publishing Co., P. V. T. LTD, New Dehli, Bombay, Calcutta

Sharkov, E. V., 2000. Mesozoic and Cenozoic basaltic magmatism. In: Outline of geology of Syria, Trans. of the Geological Institute of the Russian Acad. Sci., v. 526, p. 177-200. (In Russian).

Sharkov, E. V., Chernyshev, I. V., Devyatkin, E. D., Dodonov, A. E., Ivanenko, V. V., Karpenko, M. I., Leonov, Yu. G., Novikov, V. M., Hanna, S., and Khatib, K., 1994. Geochronoligy of the Late Cenozoic basalts of western Syria. Petrologia, vol. 2, N. 4, p. 439-448. (In Russian).

Sharkov, E. V., Chernyshev, I. V., Devyatkin, E. V., Dodonov, A. E., Ivanenko. V. V., Karpenko, M. I., Lebedev, V. A., Novikov, V. M., Hanna, S., and Khatib, K., 1998. New data on geochronology of the Late Cenozoic plateau-basalts of the northeastern territory of the Red Sea rift region (northern Syria). Rep. of the Russian Academy of Sciences, vol. 358, N. 1, p. 96-98. (In Russian).

Sharkov, E. V., and Hanna, S., 1987. Evolution of the upper mantle matter within the bounds of the intraplate basaltic magmatism regions (as examplified by western Syria). Rep. of the USSR Acad. Sci., vol. 297, N. 3, p. 684-686. (In Russian).

Sharkov, E. V., Jarmacani, I., Hanna, S., and Bagdasarian, G. P., 1987. The first K/Ar datings of the diatreme of the kimberlite-like rocks in the coastal region of Syria. Rep. of the USSR Acad. Sci., vol. 297, N. 3, p. 943-946. (In Russian).

Sharkov, E. V., Lazko, E. E., and Hanna, S., 1992. The deep-seated inclusions of the explosive center of Nabi-Matta (northwestern Syria). Geochimia, N. 9, p. 1241-1261. (In Russian).

Sharkov, E. V., Lazko, E. E., Fedosova, S. P., and Hanna, S., 1989. Deep-seated inclusions of the Quaternary volcano Tell Danun (southern Syria) in connection with the problem of intraplate basaltic magmatism. Geochimia, N. 11, p. 1609-1622. (In Russian).

Sharkov, E. V., Lazko, E. E., Fedosova, S. P., Hanna, S., Ali, S., Muravitzkaya, G. N., and Ivanova, V. K., 1989. Xenolites of clinopyroxene-hornblende rocks with the Ba-zeolite from diatremes of northwestern Syria. Trans. of EVMO, N. 1, p. 50-56. (In Russian).

Sharkov, E. V., Sindeev, A. S., Ali, S., Hanna, S., and Sharaff, M., 1987. Petrology of the Homs basaltic area (western Syria). Trans. of the USSR Acad. Sci., ser. geol., N. 7, p. 61-71. (In Russian).

Sharkov, E. V., Sneider, G. A., Taylor, L. A., Lazko, E. E., Dzherdzhe, E., and Hanna, S., 1996. Geochemical features of the asthenosphere beneath the Arabian plate according to the study of the mantle xenolites of the Quaternary Tell Danun volcano, the Syrian-Jordanian Plateau, southern Syria. Geochimia, N. 9, p. 819-835. (In Russian).

Shatsky, V. N., Kazmin, V. G., and Kulakov, V. V., 1966. The Geological map of Syria. Scale 1:200,000. Sheets I-37-XIX, I-36-XXIV (Al-Latheqiyeh, Hama). Explanatory notes. Technoexport, Moscow, 106 p.

Snyder, G. A., Taylor, L. A., Jerde, E., Sharkov, E. V., and Laz'ko, E. E., 1993. Petrogenesis of garnete, pyroxenite and spinel peridotite xenoliths of the Tell-Danun Alkali Basalt Volcano, Harrat As Shamah, Syria. Intern. Geol. Rev., vol. 35, p. 1104-1120.

Soulidi-Kondratyev, E. D., and Kozlov, V. V., 1963. Dormant volcanoes of the Syrian Desert. Priroda, N. 10. (In Russian).

Tyrrell, G. W., 1930. Pleistocene and Recent basalts of Transjordan and Palestine. Geol. Mag., vol. 67, N. 795.

Vaumas, E. de., 1956a. Sur le volcanisme du Rhab (Syrie). C. R. Acad. Sci. France, vol. 242, N. 5, p. 660-662.

Vaumas, E. de., 1956b. Sur le volcanisme du Djebel Zaouiye (Syrie). C. R. Acad. Sci. France, vol. 242, N. 5, p. 539-541.

Wernicke, B., 1985. Uniform-sense normal simple shear of the continental lithosphere. Canadian J. Earth Sci., 22,(1), p. 108-125.

APPENDIX: Chemical analyses of the basanitess of Syria (Ponikarov et al., 1969).

TABLE 7.1: Chemical analyses of the basanites of Syria

Group	Group of Normal Basanites												
Sub-Group	Melanocratic Varieties	Normal Basalts											Leucocratic Varieties
Regions	SW Syria	NW Syria	SW Syria	SW Syria	Palmyria	NW Syria	SW Syria					NW Syria	SW Syria
Sample	4-C	XII-3138	XII-3387	IX/527/3	IV-195/4	VII-35	VIII-83	I-2229a	I-2319	I-2507	I-2711-34	XII-361	I-3243
Age	β_1Q_1	βN_1^3	βN_1^2	βN_1^2	βQ1	βN1	βN_1^3	β_1Q_1	β_1Q_1	β_2Q_1	$\beta_2N_1^2$	βN_1h	β_1Q_1
SiO_2	44.72	47.69	47.48	46.8	49.76	47.2	48.07	43.51	46.33	45.82	47.63	47.93	45.36
TiO_2	2.4	1.46	1.65	2.05	1.74	2.0	2.60	2.49	2.53	2.76	1/57	2.25	2.69
Al_2O_3	16.22	13.72	13.8	15.85	15.84	17.45	17.30	19.78	18.58	18.56	18.15	15.59	23.0
Fe_2O_3	2.66	5.71	8.34	4.44	3.35	3.8	4.20	5.48	2.36	3.78	2.89	6.26	3.2
FeO	9.19	5.59	2.74	7.81	8.13	7.43	7.60	7.49	9.56	8.59	7.10	5.23	8.56
MnO	0.14	0.14	0.06	0.08	0.165	-.-	0.03	0.22	0.17	0.19	0.16	0.16	0.18
MgO	9.83	6.22	6.32	6.51	5.52	5.92	5.38	5.77	5.72	5.69	5.68	5.92	1.91
CaO	9.41	11.51	11.17	10.12	0.57	9.26	11.8	9.45	9.73	9.75	10.22	9.59	10.23
Na_2O	1.58	2.91	2.62	3.22	3.11	3.31	2.40	2.47	3.2	2.74	3.02	3.31	3.11
K_2O	1.36	0.8	0.57	0.67	1.15	1.10	1.19	0.95	0.94	1.06	0.6	1.61	1.13
P_2O_5	0.3	0.24	0.32	0.24	0.278	0.43	0.2	0.36	0.31	0.46	0.3	0.65	0.45
SO_3	-.-	0.065	0.17	-.-	0.115	-.-	Trace	0.1	0.09	0.11	0.35	0.086	0.2
Losses	1.14	3.45	4.15	1.25	1.14	1.96	0.5	1.11	0.49	1.12	1.7	0.77	0.73
Σ	98.95	100.4	99.39	99.04	99.88	99.86	100.82	99.98	101.47	100.36	99.54	99.36	100.75

TABLE 7.2a: Chemical analyses of the basanites of Syria

Group	Group of Sub-Alkaline Basanites											
Sub-Group	Slightly Melanocratic Basalts - Strongly Undersaturated in SiO_2			Slightly Leucocratic Basalts - Strongly Undersaturated in SiO_2								Sub-Alkaline Basalts
Region	NW Syria	SW Syria		NW Syria		SW Syria						
Sample	XII-179-7	IX-156	IX-611	VIII-854	VIII-173/3	2T	I-3005	I-3244	IX-408/2	IX-408/1	I-1	I-6
Age	βN_2	β_1Q_1	βN_2	βN_2	βN_2	βQ_2	βN_2	β_4Q_4	β_1Q_4	β_1Q_4	β_5Q_4	βN_2
SiO_2	43.82	45.0	46.88	44.65	45.55	45.5	44.71	44.02	44.45	44.3	45.52	43.45
TiO_2	1.81	1.31	1.73	2.83	2.07	1.8	1.35	2.46	2.38	2.66	2.73	2.47
Al_2O_3	14.46	14.19	15.5	17.39	16.81	17.73	16.15	18.92	19.9	21.46	19.58	17.36
Fe_2O_3	4.63	4.79	6.19	4.85	4.03	1.71	3.87	2.23	3.51	2.85	2.81	4.56
FeO	8.98	7.99	6.6	7.59	8.94	9.07	7.89	10.73	7.37	8.09	8.07	7.7
MnO	0.173	0.181	0.172	0.03	0.176	Traces	0.16	0.19	0.06	0.07	0.17	0.19
MgO	8.08	9.64	8.73	5.06	6.57	9.59	6.46	8.23	2.76	1.85	3.8	8.63
CaO	10.43	10.56	8.88	11.4	11.09	9.64	10.93	8.79	10.98	10.32	9.67	9.43
Na_2O	3.32	3.07	3.76	3.07	3.10	2.67	3.11	3.2	3.71	3.71	3.71	3.16
K_2O	1.1	1.170	1.2	1.08	1.1	0.85	1.0	0.91	1.5	1.13	1.3	1.35
P_2O_5	0.301	0.222	0.307	0.62	0.352	0.26	0.75	0.5	0.37	0.53	0.57	0.44
SO_3	0.033	0.192	0.017	0.55	0.026	-.-	0.34	0.18	-.-	-.-	0.65	0.21
Losses	2.84	2.04	0.34	1.52	0.7	1.20	2.42	0.02	2.09	2.16	1.5	0.72
Σ	99.97	100.4	100.38	100.64	100.57	100.02	99.14	100.38	99.08	99.13	100.08	99.7

TABLE 7.2b: Chemical analyses of the basanites of Syria (Continued)

Group	Group of Sub-Alkaline Basanites										
Sub-Group	Slightly Melanocratic Basalts - Strongly Undersaturated in SiO_2										
Region	NE Syria	SW Syria							NE Syria	Palmyra	NW Syria
Sample	V-1298/3	1-T	3-C	I-3001	I-3005a	I-3245	$I\text{-}3005_B$	I-3246	IV-193/2	VII-48	I-13-1
Age	βQ_2	βN_2	βN_2	β_5Q_4	β_5Q_4	β_1Q_4	β_2Q_4	β_3Q_4	βN_2	βN_1	βN_2
SiO_2	40.89	41.45	43.85	45.7	43.69	44.07	44.52	44.25	44.71	44.88	45.67
TiO_2	4.05	2.4	1.9	2.25	2.29	2.56	2.06	2.26	1.9	1.93	1.61
Al_2O_3	12.38	17.73	15.38	14.49	13.95	14.62	14.11	13.84	13.16	15.69	13.89
Fe_2O_3	4.77	2.26	5.53	3.22	3.37	4.28	2.54	3.51	3.02	2.73	4.06
FeO	9.18	9.22	6.1	8.92	9.29	8.31	9.45	8.62	10.19	8.98	7.71
MnO	-.-	Trace	0.21	0.15	0.17	0.17	0.16	0.16	0.194	-.-	0.174
MgO	10.68	9.76	8.58	9.6	10.09	8.17	9.49	10.11	11.31	11.45	9.68
CaO	10.82	10.64	10.48	9.48	10.91	10.94	10.31	10.85	9.72	9.40	10.85
Na_2O	2.78	2.45	2.33	3.06	2.65	3.10	2.65	2.63	2.66	3.21	3.15
K_2O	1.02	0.87	1.79	0.9	0.82	1.0	0.83	0.82	1.28	1.05	1.29
P_2O_5	0.73	0.96	0.43	0.36	0.35	0.42	0.38	0.33	0.342	0.44	0.293
SO_3	0.38	-.-	-.-	0.39	0.24	0.11	0.53	0.26	0.333	-.-	0.176
Losses	1.02	2.10	1.44	1.04	1.52	1.6	2.5	1.94	1.44	0.68	2.25
Σ	96.68	99.84	98.02	99.56	99.34	100.27	99.53	99.58	99.95	100.44	100.77

TABLE 7.3a: Chemical analyses of the basanites of Syria

Group	Group of Alkaline Basanites											
Sub-Group	Sub-Alkaline Basalts			Nepheline Basalts							Analcime Basalts	
Region	SW Syria		Palmyra	NW Syria	Palmyra	NE Syria					Palmyra	
Sample	I-2341	I-3238	VII-209	XI-766/1	VII-186	V-1298/2	V-1298/1	V-5258/3	IV-856	IV-855	VII-199	VI-3505
Age	βQ_2	$\beta_4 Q_4$	βN_2	$\mu\beta N_2$	βQ_2	βQ_2	βQ_2	βQ_2	βN_2	βN_2	βN_1	βQ_2
SiO_2	43.25	44.78	44.0	36.8	39.12	41.68	41.0	41.84	38.39	43.61	41.53	42.41
TiO_2	2.75	2.61	2.8	3.38	4.64	3.99	3.99	3.47	3.65	2.4	3.68	4.48
Al_2O_3	20.3	19.21	17.79	13.0	13.41	12.07	11.82	12.35	10.9	11.58	14.93	13.73
Fe_2O_3	2.74	3.58	0.87	4.84	3.77	4.1	4.26	3.0	15.23	12.22	0.63	2.36
FeO	9.04	8.7	9.93	8.5	7.43	9.89	9.96	11.07	-.-	-.-	9.78	9.78
MnO	0.18	0.18	0.22	0.16	0.19	-.-	-.-	-.-	-.-	-.-	0.19	0.21
MgO	4.46	5.57	7.0	9.58	8.48	10.25	10.36	10.75	11.8	8.99	11.11	10.29
CaO	10.09	8.87	10.09	13.35	13.66	11.22	11.28	10.87	11.13	12.08	9.99	10.42
Na_2O	3.72	3.61	3.8	2.82	3.60	3.22	3.22	3.6	4.81	4.21	3.4	3.5
K_2O	1.45	1.5	1.38	1.21	1.7	1.2	1.24	1.24	2.1	1.85	1.18	1.29
P_2O_5	0.91	0.53	0.42	2.17	0.42	0.84	0.82	0.76	0.83	0.31	0.42	0.4
SO_3	0.11	0.2	-.-	0.19	-.-	0.43	0.13	0.32	0.06	0.19	-.-	Traces
Losses	0.13	0.26	1.87	3.5	4.41	0.97	0.89	0.19	1.22	2.3	1.95	1.55
Σ	99.13	99.6	99.95	99.5	100.83	99.86	98.97	99.46	100.72	99.74	98.73	100.40

TABLE 7.3b: Chemical analyses of the basanites of Syria (Continued)

Group	Group of Alkaline Basanites										
Sub-Group	Analcine Basalts							Alkaline Basalts		Nepheline Basalts	
Region	NW Syria	Palmyra	SW Syria				NE Syria	SW Syria		SW Syria	
Sample	X-2394/12	VII-232	IX-151	IX-154	IX-161	IX-536/1	IV-857	IX-616	IX-613	I-3211	IX-414
Age	βN_2	βN_1	$\beta_1 Q_4$	$\beta_3 Q_4$	βN_2	βN_2	βN_2	$\beta_2 N_2$	βN_2	$\beta_2 Q_4$	$\beta_2 Q_4$
SiO_2	42.84	43.03	43.92	44.73	43.43	42.56	44.85	47.64	47.37	43.0	42.71
TiO_2	2.94	3.68	2.28	2.13	2.11	1.42	2.4	1.76	1.85	2.97	2.73
Al_2O_3	15.33	14.66	14.37	15.58	14.03	16.59	12.46	15.94	15.27	19.04	17.66
Fe_2O_3	3.94	1.81	3.21	3.42	4.4	3.76	12.33	2.64	3.99	3.42	6.17
FeO	8.94	9.15	9.39	8.72	8.52	8.32	-.-	9.15	8.1	8.09	6.44
MnO	0.07	0.20	0.171	0.175	0.175	0.1	-.-	0.163	Traces	0.16	0.09
MgO	7.11	11.27	9.57	7.89	8.44	8.51	9.22	7.71	8.14	6.71	4.9
CaO	9.24	9.54	8.96	10.75	11.72	11.6	11.36	8.78	8.75	9.02	9.91
Na_2O	3.99	3.7	3.76	3.76	3.5	3.82	3.92	4.00	4.0	3.92	4.72
K_2O	1.58	1.25	2.18	1.95	1.39	1.27	1.71	1.92	1.6	1.73	1.9
P_2O_5	1.11	0.36	0.623	0.549	0.399	0.21	0.4	0.595	0.598	0.84	0.69
SO_3	0.18	-.-	0.053	0.051	0.089	-.-	0.1	0.017	0.034	0.1	-.-
Losses	1.98	1.39	1.21	1.15	1.72	1.54	1.88	0.41	0.44	0.84	1.96
Σ	99.25	100.04	99.72	100.61	99.97	99.7	100.61	100.73	100.38	99.84	99.88

Chapter 8

Tectonics

Vladimir G. Kazmin
P. P. Shirshov Institute of Oceanology of the Russian Academy of Sciences, Nakhimovsky Prospect 36, 117 851 Moscow, Russia

INTRODUCTION

The first data on Syrian tectonics came from geological traverses by M. B. Blanckenhorn (1891, 1914, 1925, 1931, 1940 and others), L. Kober (1915) and E. Krenkel (1924, 1941, 1957). Blanckenhorn described the Syrian territory as a platform deformed by the vertical movement of basement blocks. Folding in Lebanon, Anti-Lebanon and Western Syria were also explained in the same way. Contrary to these views, L. Kober and E. Krenkel regarded these structures as frontal folds of the Alpine belt, formed in a compressional setting. In the early works of E. Krenkel, the Palmyrides ('Damascus arch'), Lebanon, and Anti-Lebanon were already jointly described as the 'Syrian Arc'.

Of special importance to the tectonics of Syria and Lebanon are the numerous works by L. Dubertret. He compiled the first geological maps of Syria and Lebanon (1942, 1945, and others). These precisely outline the fold belt of the Palmyrides, the anticlines of Lebanon and Anti-Lebanon which are separated by the Beqaa valley, the uplift of Jebel An-Nusseiriyeh, the Al-Ghab graben in Western Syria, and the anticlines of Jebel Abd El Aziz, Qamishly, and Qarashouk in the northeast of this country. In his early publication Dubertret (1932) shared the mobilistic concept of Lartet (1869), who suggested 160 km of left-lateral displacement along the Levantine fault system. Accordingly, Dubertret interpreted an assemblage of serpentinites, gabbro, basalts, and associated rocks in northwestern Syria ('roches vertes') as a nappe originating from the Amanous mountains in southeast Turkey. Later, when mapping had not revealed significant lateral displacement along the major faults in Lebanon, he rejected both the ideas of a great displacement along the Dead Sea faults, and of the allochtonous position of the 'roches vertes' (Dubertret, 1947, 1970). The latter were described as the product of differentiation of a basaltic melt formed in a gigantic 'bulb', erupted *in situ* on the sea floor at the platform margin (Dubertret, 1936, 1937, 1953), an interpretation also advocated by J. Aubouin (1959).

A new period in the studies of Syrian tectonics began in the middle of the 19th century in connection with prospecting for hydrocarbons. Geological mapping at 1:200,000 scale, and aeromagnetic and gravimetric surveys covered the country, conducted by Russian experts,. A number of deep prospection wells were drilled. A synthesis of the new tectonic data was presented in the 'Tectonic Map of Syria' (Ponikarov, 1964), in Explanatory Notes to the Geological Map of Syria (Ponikarov et al., 1967), and in the monograph 'Syria: Geology and Mineral Resources' (Ponikarov et al., 1969).

Two major positive structures were recognized, the Rutbah Uplift in the south, and the Aleppo High in the north of the country. The Palmyrides belt, separating these blocks, was interpreted as an intracratonic trough (aulacogene – according to A. Bogdanov, 1961), filled with Mesozoic-Cenozoic sediments and deformed in Neogene-Quaternary times. Though evidence of horizontal compression was noted (Faradzhev, 1966), vertical block movements were assumed to be the main cause of deformation. In western Syria the uplifts of the Jebel An-Nusseiriyeh and Jebel Az-Zawieh were interpreted as wings of a domal uplift with the Al-Ghab graben along its axis zone. (Kazmin et al., 1964). Later it appeared that the role of horizontal movements was greatly underestimated.

A similar situation occurs in the case of the Baer-Bassit ophiolites. The authors of the 1:50,000 survey conducted in 1959 (Kazmin and Kulakov, 1968) came to the conclusion that the ophiolites formed a gigantic nappe. However, the criticisms of leading Russian scientists were so great that this point of view was either excluded entirely from the reports, or left as a less probable alternative interpretation. Anyway, the geological maps, compiled in 1958-1961 by the author and his colleagues remain a major source of information on Syrian geology and tectonics.

In the 1970s a renewed effort toward oil and gas prospection began. About 100,000 km of multichannel seismic profiles were obtained between 1974 and 1990, revealing geological structures as deep as 10-15 km. Hundreds of prospection as well as several deep stratigraphic holes were drilled. The collected database was used by experts from Cornell University (USA) and the Syrian Petroleum Company for a number of publications on the deep structure of Syria (Chaimov et al., 1990; Best et al., 1990; McBride et al., 1990; Al-Saad et al., 1992; Chaimov et al., 1992; Sawaf et al., 1993; Chaimov et al., 1993).

The synthesis and analysis of the data concentrated mainly on two traverses across Central and Eastern Syria. It was revealed that the Mesozoic-Cenozoic folded sediments in the Palmyrides are detached from the older rocks, presumably along the Triassic evaporite horizon. A buried Al-Furat rift associated with the Al-Furat transcurrent fault zone was discovered, and the evolution of the Palmyrides aulacogene and other sedimentary basins described in some detail. A basement depth map was later produced by experts of the Syrian Petroleum Company (S. P. C., 1986).

A group of scientists from the Geological Institute of the Russian Academy of Sciences began work in Syria in the 1980s. A number of important results were obtained by this research program. These include the detailed study of the Baer-Bassit nappe

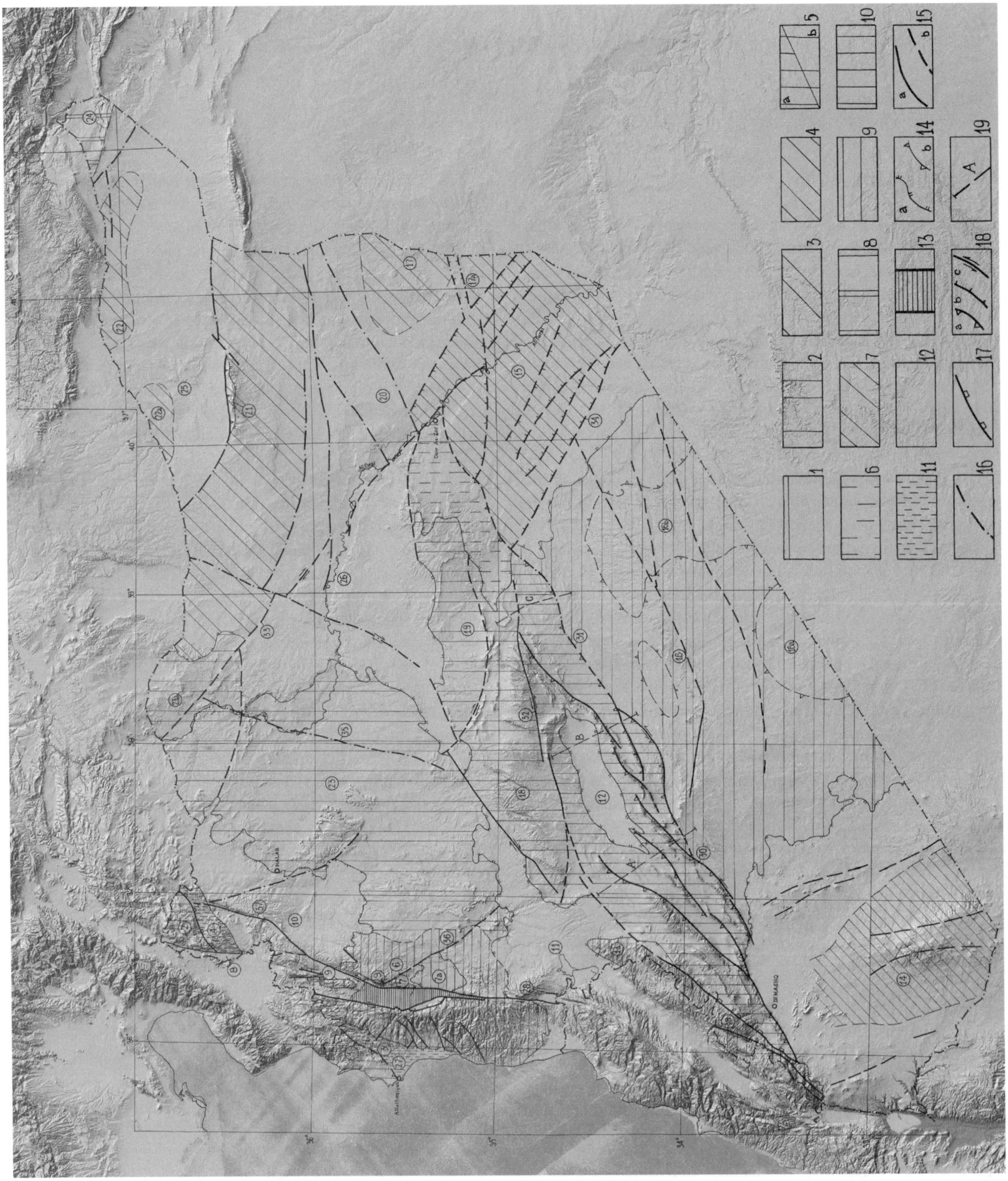

structure (Knipper et al., 1987; 1988), as well as the assessment of the present-day and the Pliocene-Quaternary displacements along the Al-Ghab fault which is the northern extension of the Levantine fault system (Trifonov et al., 1991). An original model for the structural evolution of Western Arabia, and, in particular, for the Palmyrides fold belt and the Al-Ghab graben was proposed (Kopp et al., 1994; Adjamyan et al., 1994; Kopp et al., 1999). According to these studies, the deformation in Western Arabia was caused by the restraining bend of the Yammuneh fault which produces a 'diffused' shear in the adjacent part of the platform (Kopp et al., 1994). The results of these Geological Institute studies were summarised in Kopp and Leonov (2000).

A. Galaktionov (1989a, b) made an important contribution in the late 1980s with his two maps showing his interpretation of the space imagery, and the neotectonics of Syria.

Though the available data provide a good basis for interpretation of the Syrian tectonics, a number of problems remain unsolved. These are:

a) The total amount and time of displacement along the Syrian segment of the Levantine fault system;

b) The magnitude of displacement on the low-angle detachment along the Triassic evaporites;

c) The morphology and evolution of the Al-Furat and Drouz depressions; and

d) The geodynamic setting of the Jebel An-Nusseiriyeh, Al-Ghab, and Jebel Zawieh structures in western Syria.

Major structures of Syria

Syria is situated in the northwest of the Arabian Peninsula. From Permian to Paleogene time, it first belonged to the passive margin of Gondwana, then to Africa, and finally it was deformed during the collision of Cenozoic Africa with Eurasia. Most of the deformation was localized in the intracratonic rift-related basin (aulacogene) of the Palmyrides, which divides Syria into northern and southern blocks (**Fig. 8.1**).

The southern block includes the weakly deformed northern slope of the Rutbah Uplift, which is characterized by thick Paleozoic and thinner Mesozoic sediments (McBride et al., 1990; Al-Saad et al., 1992). The core of this uplift, located within Iraq, consists of Early Paleozoic sediments, which occur at shallow depth below the Late Paleozoic-Triassic cover (Ponikarov, 1964).

The Rutbah Uplift is bound by the Drouz and Al-Furat depressions, to the southwest and northeast respectively. The age of the former is unknown, as it is covered by thick Neogene-Quaternary basaltic flows of Jebel Drouz and the Hauran plateau (Sharkov et al., 1998; Segev, 2000). The Al-Furat depression has a complex horst and graben structure, formed in the Permian or Triassic.

The northern block or Aleppo High has a different Paleozoic and Mesozoic history. Its Paleozoic-Lower Mesozoic cover is sharply reduced, while the Late Cretaceous sediments are the same thickness as in the Palmyrides (McBride et al., 1990; Al-Saad et al., 1992). Cenozoic deformation is more prominent here than in the Rutbah Uplift. Towards the Al-Furat River the Paleogene sediments of the Aleppo high plunge below undeformed Neogene sediments.

The Palmyrides fold belt is an inverted rift or aulacogene, which began to subside after the first rifting stage in the Late Carboniferous-Early Permian. Two other stages of rifting and subsidence occurred in the Triassic and latest Early Cretaceous. The inversion began in the Senonian and culminated in Neogene-Quaternary time. The Palmyrides plunge below undeformed Neogene sediments west of the Al-Furat transcurrent zone, and are not traced east of it. However, faults of the Palmyrides trend are inferred from geophysical data east of the Al-Furat at the margins of the Albeida and Derro highs (Sawaf et al., 1993).

A plain covered by undeformed Neogene-Quaternary sediments extends east of the Rutbah and Aleppo Highs. In the 'Tectonic Map of Syria' (Ponikarov, 1964) this area is named 'the Al-Furat depression'. To the southeast it merges with the 'Central Iraqi basin' (Lovelock, 1984), an outer undeformed zone of the Mesopotamian foredeep. In Syria, flat-lying sediments of the outer zone conceal a number of older structures: the Mesozoic-Cenozoic

⇦

Figure 8.1. Tectonic units of Syria. 1-3: uplifts of the platform basement. 1-Southern block; 2-Northern block; 3-uplifted blocks buried by the Neogene-Quaternary sediments; 4-Mesozoic rifts below the Neogene-Quaternary cover; 5-9 zones of the Neogene-Quaternary deformation. 5-6 - Palmyra fold belt; 5a - South Palmyra zone; 5b - North Palmyra zone; 6 - buried extension of the Palmyra belt; 7 - Sinjar zone; 8 - folded foreland (extension of the Zagros fold belt); 9 - structures related to the Levant fault system; 10 - frontal zone of the Alpine fold belt; 11 - early-late Alpine foredeep; 12 - Neogene-Quaternary platform cover; 13 - deepest part of the Al-Ghab depression; 14 - (a) limits of the Neogene-Quaternary basins; (b) limits of domal uplifts; 15 - mapped faults: a - traced; b - inferred; 16 - faults revealed by geophysical methods; 17 - flexures; 18 - types of faults: a - thrusts; b - normal faults; c - strike-slip faults; 19 - position of restored cross-sections with shortening measured.

Structures: 1 - Baer-Bassit block; 2 - Kurd-Dagh block; 3 - Nahr El- Kebir foredeep; 4 - Aafrine foredeep; 5 - Jebel An-Nusseiriyeh uplift; 6 - Jebel Az-Zawieh uplift; 7 - Al-Ghab depression; 7a - Asharneh depression; 8 - Karasu depression; 9 - Jebel Wastani anticline; 10 - Aafrine depression; 11 - Homs depression; 12 - Al Daww depression; 13 - Anti-Lebanon anticline; 14 - Drouz depression; 15 - Al-Furat depression; 16 - slope of the Rutbah Uplift; 16a - Tanf domal uplift; 16b - El Murabbaa domal uplift; 17 - Rawda Uplift; 18 - El-Balas block; 19 - Bishri anticline; 20 - Albaida-Derro zone; 21 - Jebel Abd El-Aziz anticline; 22 - Qamishli and 22a - Ras El-Ein swells (eastern termination of the Mardin uplift); 23 - Aleppo High; 23a - Jrablus block; 24 - Qarashouk anticline; 25 - slope of the Mardine-Qamishli uplift; 26 - buried extension of the Aleppo High.

Faults: 27 - Al-Latheqiyeh-Kilis; 28 - Yammuneh; 29 - Al-Ghab; 30 - South Palmyra; 31 - Sawwaneta-Heil; 32 - Jhar; 33 - Al-Furat transcurrent zone; 34 - El Madabe fault step; 35 - Maksar; 36 - Hama; 37 - Sheikh Barakat; 38 - Serrhaya.

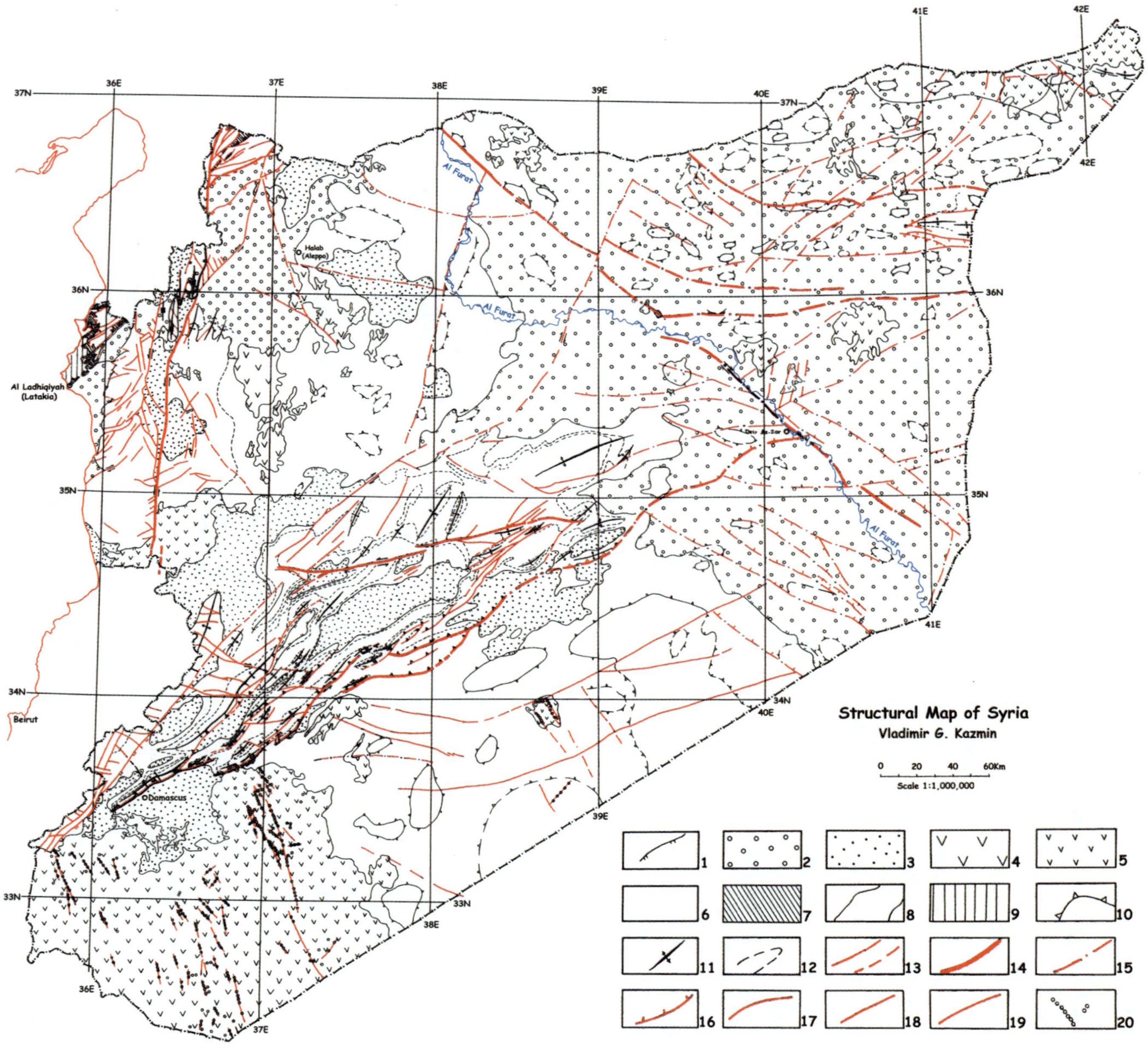

Figure 8.2. Structural map of Syria. 1 - outer margin of the pre-Neogene outcrops; 2 - Miocene-Quaternary cover; 3 - Pliocene depressions; 4 - Upper Miocene basalts; 5 - Pliocene basalts; 6 - outcrops of pre-Neogene platform sediments; 7-8 - allochthonous complexes; 7 - Ophiolite nappe; 8 - Tamima nappe; 9 - neoautochthone; 10 - limits of domal uplifts and brahyanticlines; 11 - linear folds; 12 - structural line; 13 - mapped faults (a - certain ; b - uncertain); 14 - major faults; 15 - faults revealed by geophysical methods; 16 - normal faults; 17 - nappes, thrusts; 18 - strike-slip faults; 19 - flexures; 20 - volcanic cones.

rift-related basin of the Al-Furat depression, the Rawda (or Khleissia) Uplift, the Albeida High, and the Derro High.

The inner folded zone of the Mesopotamian foredeep, described as a folded foreland, is exposed in the extreme northeast of Syria. Linear folds, characteristic of this zone, are represented here by the Qarashouk anticline. To the northwest the Mesopotamian foredeep shallows and terminates as it approaches the Aleppo High and the Mardin Uplift in southeastern Turkey (Lovelock, 1984; Fig. 8.1). The distal part of the Mardin Uplift, the Qamishli swell, extends into Syria.

A prominent structure in northeastern Syria is the Sinjar Trough, a Mesozoic rift-related basin situated between the Rawda and Mardin-Qamishli Uplifts (Lovelock, 1984; Sawaf et al., 1993). The basin was inverted in Neogene-Quaternary times but compression here was much milder than in the Palmyrides.

The structures of northwestern Syria are the asymmetric anticlinorium of the Jebel An-Nusseiriyeh, the Al-Ghab depression, and the Jebel Az-Zawieh uplift, which is itself a deformed margin of the Aleppo High, which evolved in a transpressional setting. All are closely related to displacement along the Levantine fault system. In this respect, they resemble the Lebanon-Al-Beqaa-Anti-Lebanon structural assemblage (Beydoun, 1977).

In the extreme northwest of the country a frontal zone of the Alpine belt is exposed in Baer-Bassit and Kurd-Dagh. This zone was overridden in the Maastrichtian by ophiolite nappes and later went through several episodes of moderate compression. As a result, the Nahr El-Kabir and Aafrin foredeeps were formed in front of the Alpine structures.

The southern platform block

The southern platform block includes the northern slope of the Rutbah Uplift as well as the Drouz and Al-Furat depressions. The northern limit of the block is the South Palmyra fault zone (Fig. 8.1).

The slope of the Rutbah Uplift consists of Paleogene sediments, which plunge to the west and east below Neogene-Quaternary deposits. The Paleozoic core of the uplift is found at shallow depth on the Iraqi side. The northward thickness and facies changes of the Paleozoic to Cenozoic sediments, point to the long evolution of the uplift (Ponikarov et al., 1967). According to drilling and seismic data, the Phanerozoic cover of the Rutbah Uplift is 6-7 km thick (McBride et al., 1990).

Detailed processing of the multichannel seismic data along a N-S traverse (Fig. 8.5) correlates the top of the basement with a reflector with boundary velocities of 6 km/s at about 8 km depth (Brew et al., 1997). This is 2-2.5 km deeper than was earlier assumed (Ponikarov, 1964; Leonov et al., 1989; S. P. C., 1986). It has therefore been suggested that the Lower Cambrian and Upper Precambrian sediments of Rutbah have a much greater thickness than was originally expected.

Devonian sediments are absent and an angular unconformity lies at the base of the Carboniferous-Permian sequence. This configuration may be attributed to the 'Hercynian orogeny' of Gvirtzman and Weissbrod (1984). Within the northern slope of Rutbah, the Triassic and Jurassic deposits pinch out, away from the Palmyrides. Another unconformity is found at the base of the ~1 km thick Cretaceous sequence. The Rutbah Uplift is described as an Early Paleozoic depocenter uplifted in the Devonian and Carboniferous and evolving as a positive structure throughout most of the Mesozoic and Cenozoic (Lovelock, 1984; McBride et al., 1990). The Paleogene layers on the Syrian side dip gently to the north at angles from 1° to 4°. In this context two structural trends are discernable. The northwesterly trending structures are broad antiforms outlined by the outcrops of the Paleocene-Lower Eocene sediments (Geological Map of Syria, 1964). Drilling on the At Tanf structure (Fig. 8.1) revealed the absence of Triassic deposits; Here Cretaceous sediments rest directly on Permian rocks. Thinning and pinching out of the Lower Mesozoic sediments is seen in the seismic profile running along another antiform farther to the east (McBride et al., 1990). Some fractures controlling the Quaternary volcanoes also strike NW-SE.

A chain of brachyanticlines, elongated in an northeasterly direction, runs parallel to the southern front of the Palmyrides (Soulidi-Kondratiev, 1966). At depth it appears as a basement swell with a gentle southern limb, dipping about 1°, and a steeper northern limb (McBride et al., 1990). Subvertical faults with visible displacements of 20 to 30 m have the same strike (**Fig. 8.2**). The great length of these faults, up to 200-250 km, suggests that they are deeply penetrating and may have much greater amplitude at depth. Originally these faults, on the northern slope of the Rutbah Uplift, were probably normal step-faults which became reverse faults during the Neogene-Quaternary compression.

The magnetic and gravity fields of the Rutbah Uplift are monotonous without prominent anomalies (Ponikarov et al., 1967; Best et al., 1990). In the western part of the uplift the magnetic anomalies are negative and vary from -30 to -40 gammas. In the east, on the slope to the Al-Furat depression, the values decrease gradually to -10 gammas.

Bouguer gravity anomalies vary from -40 to -50 mgal. In the western part of the uplift some vaguely defined north-trending relative gravity maxima which vary from -40 to 0 mgal (Rmah anomalies) are presumably associated with intrusions in the basement (Best et al., 1990).

Mesozoic rift depressions of the southern platform block.

The Rutbah Uplift is bound on both flanks by deep northwesterly trending basins: the Al-Furat depression in the east, and the Drouz depression in the west (Figs. 8.1. and 8.2). Reliable seismic and drilling information is only available for the Al-Furat depression (Lovelock, 1984; Sawaf et al., 1993; S. P. C., 1986; Brew et al., 1997).

The Al-Furat (or Euphrates) depression is about 90 km wide and is completely buried beneath Neogene-Quaternary sediments. A number of fault-bounded steps lower the basement surface from 5000 m at the Rutbah Uplift margin to a depth of 9000 m near the Euphrates (or Al Furat) fault (Brew et al., 1997). The depression is

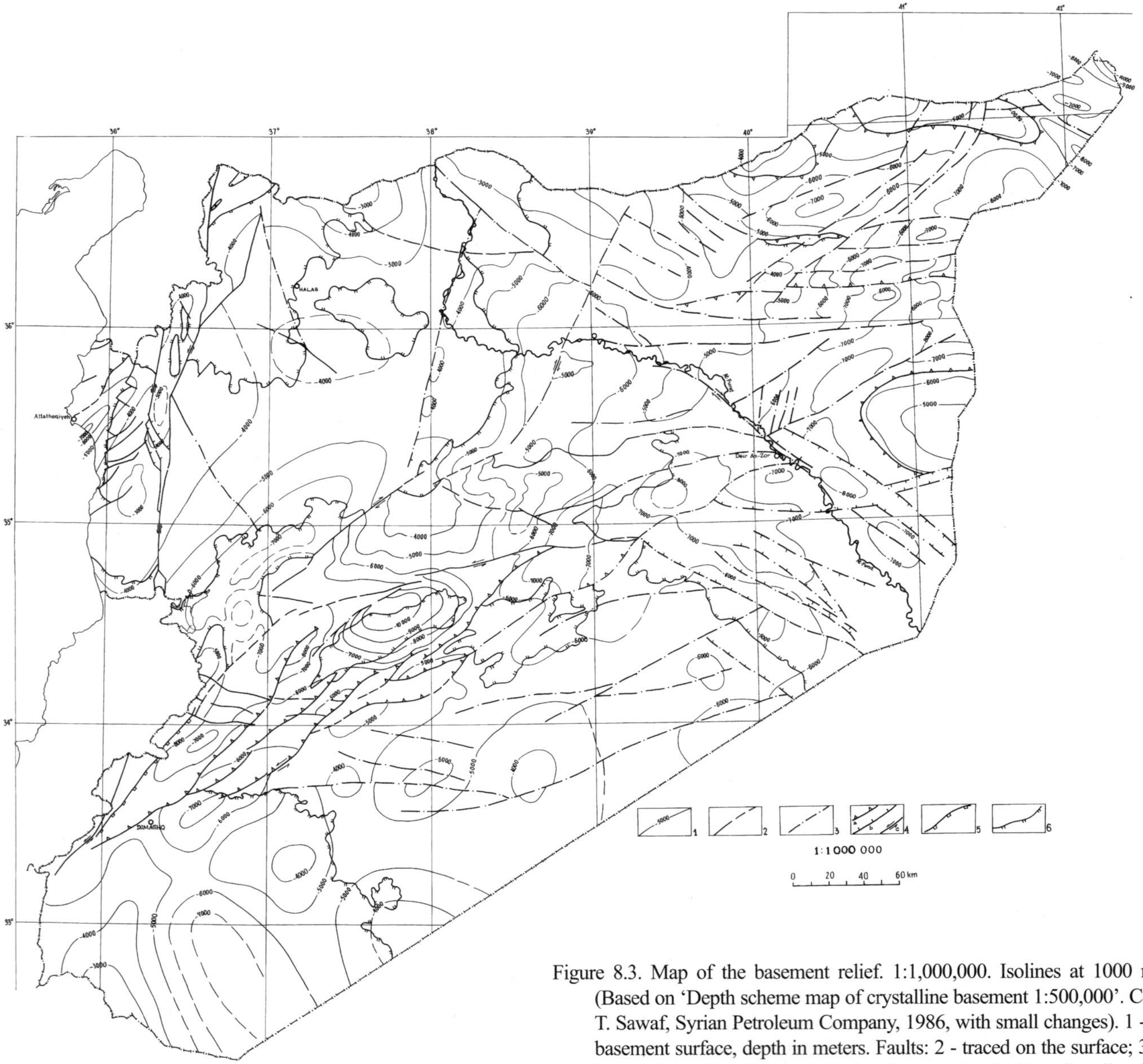

Figure 8.3. Map of the basement relief. 1:1,000,000. Isolines at 1000 m intervals (Based on 'Depth scheme map of crystalline basement 1:500,000'. Compiled by T. Sawaf, Syrian Petroleum Company, 1986, with small changes). 1 - isolines of basement surface, depth in meters. Faults: 2 - traced on the surface; 3 - revealed by geophysical methods; 4 - thrust (a), normal fault (b); strike-slipe fault (c); 5 - flexure; 6- limit of the uplifted block (contact with the Neogene cover).

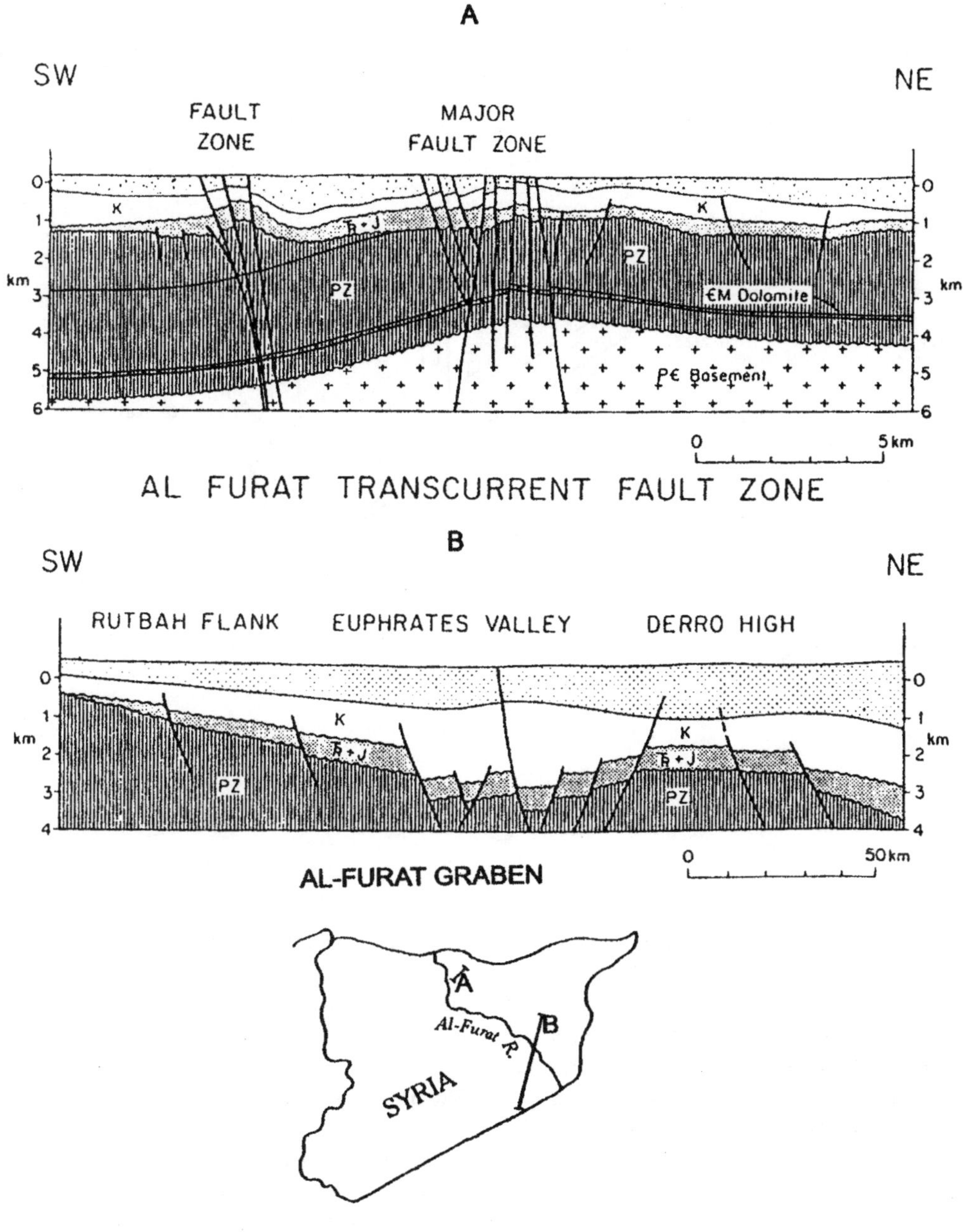

Figure 8.4. Schematic cross section of the Al-Furat transcurrent fault zone (a) and the Al-Furat depression (b) (After Lovelock, 1984).

bound on the east by the narrower and steeper slope of the Rawda (or Khleissia) High, where the basement is at about 4 km depth (S. P. C., 1986). It has been suggested that the Rawda High has a similar geological structure to the Rutbah Uplift (Lovelock, 1984). However in contrast to the Rutbah Uplift the gravity field of the Rawda High is positive (Ponikarov et al., 1967; Best et al., 1990), a property which remains unexplained.

On its southern end the Al-Furat depression joins the Anah graben almost at right angles, at a point coinciding with the east-west segment of the Al-Furat River. In the northeast it terminates against the South Palmyra fault zone (S. P. C., 1986; see Fig. 8.1. and **Fig. 8.3**).

A schematic profile across the innermost part of the depression (**Fig. 8.4b**) shows a complex horst and graben structure. According to P. Lovelock (1984), the older sediments in the Al-Furat depression are Triassic. However the Thayyem-103 drillhole in the northern part of the depression struck Permian beds at the bottom (Sawaf et al., 1993). It is therefore possible that subsidence

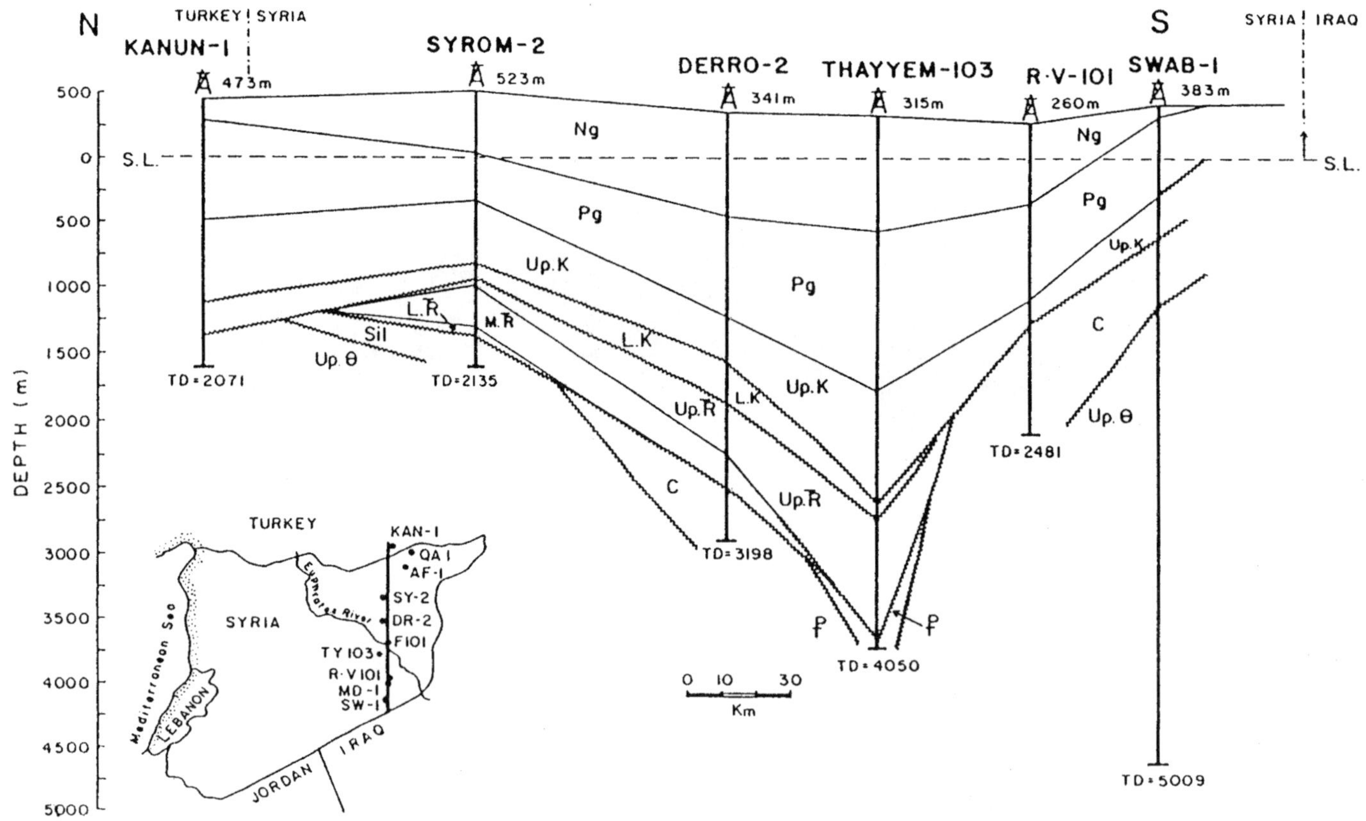

Figure 8.5. Well correlation along the East Syrian transect (After Sawaf et al., 1993).

here might have begun at the same time as in the Palmyrides. The second major period of subsidence began in the late Early Cretaceous, again simultaneously with the Palmyrides. About 2.5 km of sediments accumulated in the Al-Furat depression during the Late Cretaceous (**Fig. 8.5**). In contrast to the Palmyrides, the Al-Furat depression did not undergo the Cretaceous and Cenozoic inversions and continued to subside uninterruptedly throughout the Paleogene and Neogene.

The Drouz depression, situated between the Rutbah and Jordan Highs, is another northwesterly trending and possibly rift-related basin. On the surface the depression is covered by basaltic flows which reach a maximum thickness of about 1500 m at Jebel Drouz, a volcanic ridge rising up to 1600 m above sealevel in the axial zone of the depression. According to drilling data, the base of the basaltic pile beneath the ridge is at 100 m above sea level. *i.e.* 400 m deeper than at the margins of the basaltic plateau (Galaktionov, 1989a and **Fig. 8.6**). At the same time the depth of the basement in the Drouz depression is ~7 km or more (S. P. C., 1986; Brew et al., 1997). It means that the depression is filled by a thick sequence of Paleogene, Cretaceous and possibly older sediments (Ponikarov et al., 1967). There is a clear similiarity in basement depth and width (80-90 km) between the Drouz and Al-Furat depressions.

Basaltic eruptions at Jebel Drouz and the Hauran plateau were controlled by northwesternly trending fractures marked by chains of volcanic cones. As in other parts of Arabia, these fractures were most probably re-activated basement faults, which may control subsidence at the rifting stage.

The northern platform block

A large plateau of Paleogene rocks represents the northern platform block or the Aleppo High. A small outcrop of Cretaceous sediments is found in its central part (Figs. 8.1. and 8.2.). In the south, a system of faults or steep flexures separates the high from the Palmyrides fold belt. Eastwards, the Paleogene rocks plunge below the Neogene sediments while in the west the high is bound by the Levantine fault system. The deformation in Jebel Az-Zawieh, which is the part of the Aleppo High adjacent to this system, will be dealt with separately.

Ponikarov et al. (1967, 1969) state that in contrast to the Rutbah Uplift, the Aleppo High belongs to the 'unstable' part of the Arabian platform. However, seismic data (McBride et al., 1990) reveal a flat-lying Phanerozoic sequence rather similar to that of Rutbah but less thick. The so-called reflector 'D', interpreted as the top of the basement or the top of the carbonate-evaporite Infracambrian sequence, lies either at 4 km (S. P. C., 1986; and Fig.

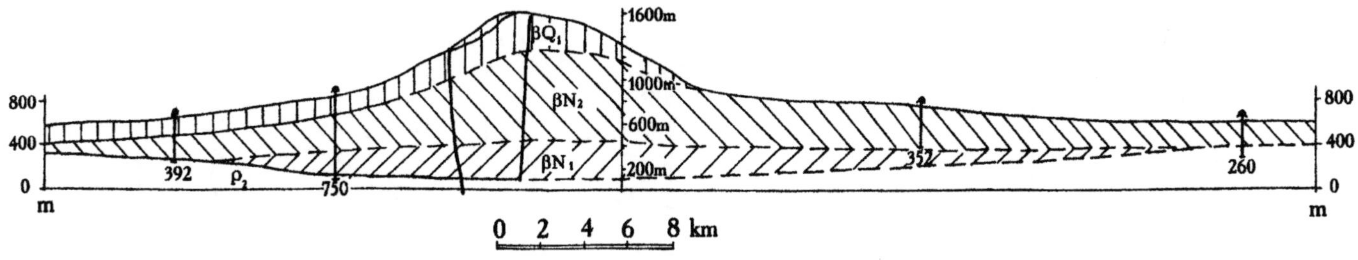

Figure 8.6. EN-WS cross section of the Jebel Araba (Jebel Drouz) volcanic ridge. (After Galaktionov, 1989a). N - Miocene basalts; N - Pliocene basalts; Q - Pleistocene basalts.

8.3.) or at 5.5-6 km (Brew et al., 1997), *i.e.* at a much shallower depth than in Rutbah Uplift. Subsurface correlation shows that the consequent northward wedging out of the upper and middle Paleozoic units (**Fig. 8.7**) is perhaps related to erosion at the end of the Paleozoic. In contrast to the Palmyrides, the Triassic sequence of the Aleppo High is greatly reduced in thickness, while the Jurassic sediments are missing. On the other hand, the Cretaceous deposits are similar in thickness to those of the Palmyrides, in striking contrast to the Rutbah section.

The most conspicuous features of the Aleppo High, as revealed in the gravity field (Ponikarov et al., 1967) and the aerial and space imagery, are extensive faults striking to the northeast and northwest. Two of the northeast trending faults (Fig. 8.2.) are probably strike-slips associated with the main faults of the Levantine system. Steps within the basement surface, whose eastern sides were uplifted, are associated with both faults (Fig. 8.3.).

Sinistral motion along the Maksar fault is suggested by a number of asymmetric anticlines formed at an acute angle to its eastern side (Oufland, 1966). Other evidence of strike-slip displacement comes from a seismic profile across the Maksar fault. It shows the typical flower structure usually associated with strike-slips. As seen in space photos (Galaktionov, 1989a), northeast-trending faults seem to branch from the northern boundary faults of the Palmyrides.

Controversy exists about the position of the Jrablus block (Fig. 8.1), east of the Al-Furat River. Large outrops of Paleogene and Cretaceous rocks and a shallow basement depth of 3000 m suggest that it is a part of the Aleppo High. Alternatively, this block is viewed as an uplifted part of the Sinjar zone (Oufland, 1966; Ponikarov et al., 1967). A characteristic of the Jrablus block is the great number of northwesterly trending folds which are especially well developed in a narrow zone along the Al-Furat fault (the 'Jrablus rampart' according to Oufland, 1966). Away from the fault, asymmetric linear folds whose steep limbs are adjacent to the fault give way to gentle symmetric anticlines.

The magnetic and gravity fields of the Aleppo High differ from those of the Rutbah Uplift. East of Aleppo a symmetric magnetic anomaly of up to 175 gammas amplitude is associated with the top of the basement high (Ponikarov et al., 1967). It coincides with a Bouguer anomaly of -20 mgal. The average Bouguer values vary between 0 and -20 mgal (Best et al., 1990). The Maksar and other northeast striking faults are very prominent in the gravity map, and are marked by zones of steep linear gradient.

Sharp differences in the geophysical fields and geological history suggest that the Aleppo High and Rutbah Uplift were individual Precambrian terranes later joined together along a Late Precambrian or Early Paleozoic suture, the latter being concealed beneath the Palmyride sediments (Best et al., 1990).

Palmyrides fold belt

The up to 90 km wide Palmyrides fold belt extends in a NE direction for about 400 km from the Damascus area (Ponikarov et al., 1967, 1969; Geological Map of Syria, 1964; Figs. 8.1. and 8.2). To the east, some 50 km west of the Al-Furat River, the Palmyrides folds and their corresponding ridges plunge beneath the Neogene-Quaternary cover. In the south, the South Palmyra fault zone consists of a series of steep north-dipping reversed faults and associated small amplitude overthrusts which separate the Palmyrides from the Rutbah Uplift (Kopp et al., 1994; Adjamyan et al., 1994) (**Fig. 8.8**). An *en echelon* arrangement of faults and fault-related asymmetric and overturned folds suggests a left-lateral component along the South Palmyra zone (Kopp et al., 1994). This is in agreement with one available focal mechanism solution (Ben-Menachem et al., 1976, 1981). Seismic data show that the South Palmyra fault zone extends to the northeast as far as the Al-Furat transcurrent fault. The buried extension of the South Palmyra zone is known as the Sawwaneta-Heil fault (Fig. 8.1).

Interpretation of multichannel seismic data suggests flattening of the faults at a shallow detachment zone on the Triassic evaporites (**Fig. 8.9**). Structural disharmony between the folded Mesozoic-Cenozoic sequence and the sub-salt Paleozoic deposits is well expressed in the Western Palmyrides, but disappears in the Eastern Palmyrides (McBride et al., 1990; Chaimov et al., 1990; Chaimov et al., 1992). Even in the Western Palmyrides the shortening does not exceed 20-25 km, which is much less than was earlier supposed (Ala and Moss, 1979).

North of the South Palmyra fault zone, the Pre-Mesozoic basement, at a depth of 5000 m beneath the Rutbah Uplift, drops to 10,000 m in the Southern Palmyrides Al-Daww depression (S. P. C., 1986; Brew et al., 1997; Fig. 8.3). The zone of the deep Southern Palmyrides Mesozoic-Cenozoic subsidence is separated

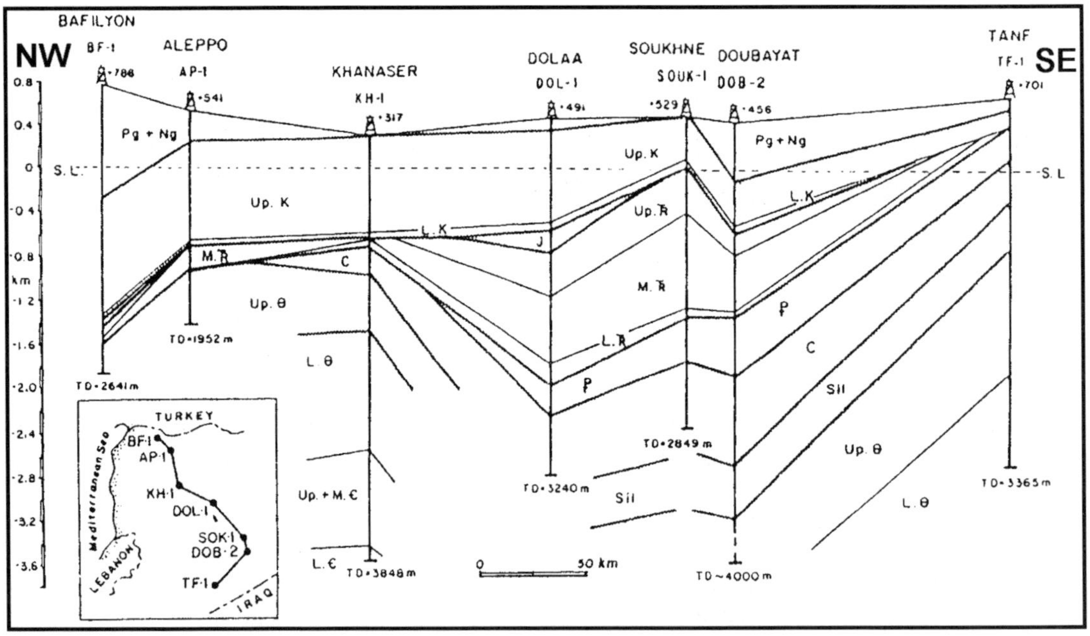

Figure 8.7. Well correlation section across Palmyrides. (After McBride et al., 1990).

from the Northern Palmyrides by the Jahr fault, which is expressed on the surface by a chain of discontinuous faults and flexures (Geological Map of Syria, 1964). The large vertical displacement on this fault is probably accompanied by a strike-slip component of unknown amplitude.

The single focal mechanism available for the Jahr fault is of sinistral-slip type (Ben-Menachem et al., 1976, 1981). It is possible that the fault represents the *en echelon* extension of the faults and flexures separating the Palmyrides from the Anti-Lebanon (Figs. 8.1. and 8.2).

In the Northern Palmyrides the pre-Mesozoic basement is at 4000 to 5000 m depth, which is similar to its depth in the adjacent Aleppo High. Thus, in contrast to its sharp southern boundary, the northern boundary of the Palmyrides is a rather gently dipping slope complicated by faults and flexures (McBride et al., 1990; Kopp et al., 1994) (Fig. 8.9).

Most authors describe the Palmyrides as a Late Paleozoic-Mesozoic rift or aulacogene (Ponikarov et al., 1967, 1969; McBride et al., 1990). Subsidence began in the Late Carboniferous or Early Permian, and intensified in the Middle Triassic (Fig. 8.7). As the distribution of thickness map shows, the late Paleozoic sedimentary basin was asymmetric, with steep northern and gentle southern margins, and in the Triassic the basin acquired a more symmetric shape. The next stage of rifting and subsidence began in the latest Early Cretaceous. It continued during the Late Cretaceous, when the Palmyra aulacogen became again asymmetric, this time with a steep southern margin (Fig. 8.7).

The present-day structure of the Palmyrides was formed during several phases of compression from the Late Senonian to the Quaternary. Depending upon the basement depth and thickness of sediments, the deformation in the Southern and Northern Palmyrides resulted in different structural patterns.

Southern Palmyrides

In the Southern Palmyrides, *i.e.* in the zone of maximum Paleozoic and Mesozoic subsidence, the belts of asymmetric and overturned linear folds are associated with major faults, and primarily with the South Palmyra fault zone. Fold axes strike northeast, oblique to the general trend of the Palmyrides. The southeastern fold limbs are usually truncated by reverse faults; the fault-related nature of the Palmyra folds is commonly accepted (Faradzhev, 1966; Soulidi-Kondratiev, 1966; Kopp et al., 1994) (**Fig. 8.10**).

Narrow belts of anticlinal ridge-forming folds are separated by wide synclines or depressions filled by Pliocene-Quaternary molasse (*e.g.* the Al-Daww depression).

For the Southern Palmyrides, Kopp et al. (1994) and Kopp and Leonov (2000) describe a combination of compressional structures. They include, besides folds, E-W trending dextral strike-slips with 100 to 500 m of displacement (Fig. 8.2). Similar dextral strike-slips are common in the Lebanon and Anti-Lebanon where they are related to anticlockwise block rotation in a transpressional setting (Ron, 1987). It is believed that the present day reverse and strike-slip faults are inverted normal and transverse faults which were formed at the rifting stage of the Palmyrides evolution (Chaimov et al., 1993).

Northern Palmyrides

In contrast to the Southern Palmyrides, the structure of the Northern Palmyrides is simple. It consists of a large anticlinorium

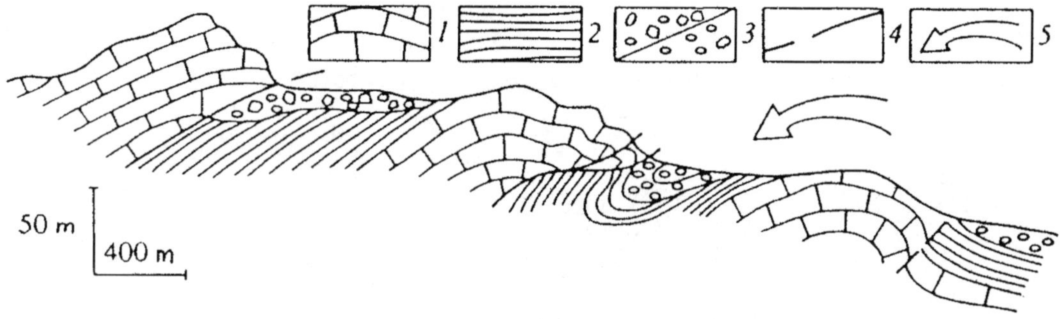

Figure 8.8. Types of overthrusts in the South Palmyra fault zone. (After Adjamyan et al., 1994). Sediments: 1 - Jurassic-Cretaceous; 2 - Paleogene-Miocene; 3 - Pliocene (a - olistostromes, b - conglomerates); 4 - overthrusts; 5 - direction of the Arabia's underthrusting.

formed by Cretaceous and Paleogene rocks, known as the El-Balas block in the west, and the Jebel Bishri anticline in the east. As pointed out by Razvalyaev (1966) and Kozlov et al. (1966), this anticlinorium may represent the continuation of the Anti-Lebanon, separated from the Southern Palmyrides by a deviation of the fold axis in the southern part the Homs depression. Drilling shows that below the Neogene sediments of this depression there is an anticline in the Paleogene and older rocks (the Awar anticline - Fig. 8.2) which forms a link between the Anti-Lebanon and the eastern closure of the El-Balas block (Kozlov et al., 1966).

The eastern part of the anticlinorium, known as the Jebel Bishri anticline, is a northeast plunging fold with a broad and flat arch and steep flexure-type limbs. The Jebel Bishri fold is separated from the El-Balas block by a northwest trending strike-slip fault with about 5 km of dextral displacement (Al-Saad et al., 1992). Broad open second-order folds are characteristic of this block and strike north-northeast, deviating significantly from the general northeast strike of the Southern Palmyrides. If the kinematics of the faults that limit the El-Balas block to the south and northeast are correctly interpreted, the block was pushed to the west, creating an east-west compression responsible for initiating the second-order folds.

Horizontal NW-SE shortening across the Palmyrides fold belt was estimated via three palinspastically restored cross sections (Chaimov et al., 1990). The western cross-section (A in Fig. 8.1) transects the widest southwestern part of the Southern Palmyrides and demonstrates a minimum shortening of about 20 km (or 20%). The central cross-section (B) shows only 3 km of shortening slightly west of Tudmor (Palmyra), where the Southern Palmyrides are only three quarters the width. Finally, along the eastern section crossing part of the Jebel Bishri anticline (C in Fig. 8.1.) the estimated shortening is 1 km. It is clear that most of the shortening occurs in the Southern Palmyrides and that it decreases dramatically from SW to NE, *i.e.* parallel to the narrowing South Palmyra zone of deep subsidence. In the Northern Palmyrides the shortening is apparently insignificant.

The extension of the Palmyrides to the northeast of the Al-Furat transcurrent zone is problematic. The top basement map (S. P. C., 1986) shows faults of Palmyrides trend extending for at least 100 km beyond the Al-Furat River and bounding a 30 km wide crustal block. Within the block are located two local basement highs, the Albeida and Derro (Lovelock, 1984). No folds or reverse-faults are mapped here. A conspicuous feature are the eruptions of Quaternary basalts, which are completely absent in the Palmyrides. Thus a transition from compressional to tensional setting is very probable across the transcurrent zone.

Sinjar Trough

The Sinjar Trough is an east-west trending zone about 50 km wide extending from northwestern Iraq to northeast Syria (Figs. 8.2 and 8.3). In its deepest part, in the east, the basement lies at a depth of more than 9000 m (Brew et al., 1997); to the west, in the area of the Tuwal El-Aba anticline and the Jrablus block, the basement shallows to 3000-4000 m. This basement depression is bounded by faults, expressed in the gravity field by steep gradients (Galaktionov and Demidov, 1966; S. P. C., 1986). A short segment of the northern boundary fault can be seen on the surface along the northern flank of the Jebel Abd El-Aziz anticline. It appears as a steep south-dipping reverse fault (Sawaf et al., 1993; and Site 21 in Fig. 8.1). Presumably the boundary faults were initially normal faults, converted to thrusts during the Late Miocene-Pliocene inversion. The most prominent structure in the eastern part of the trough is the long linear Sinjar anticline, while in the west, a very gentle 30 km wide swell, formed in Upper Miocene sediments, occupies the whole of its width. Besides the fault-related Jebel Abd El-Aziz anticline, second order folds are brachyanticlines . The largest of them is the Tuwal El-Aba anticline. Some linear folds only appear near the Syrian-Iraqi border (Galaktionov and Demidov, 1966). A number of E-W faults (possibly reverse-faults) and a northeast trending transverse fault zone were revealed by geophysical surveys (Figs. 8.1 and 8.3).

P. Lovelock (1984) was the first to suggest that the Sinjar Trough has much in common with the Palmyrides, although it may be somewhat younger. The main periods of subsidence correspond here to the Middle Triassic and Late Cretaceous (Lovelock, 1984; Sawaf et al., 1993).

The first indication of compression is evident in the Jebel Abd El-Aziz anticline, where Campanian-Maastrichtian beds include

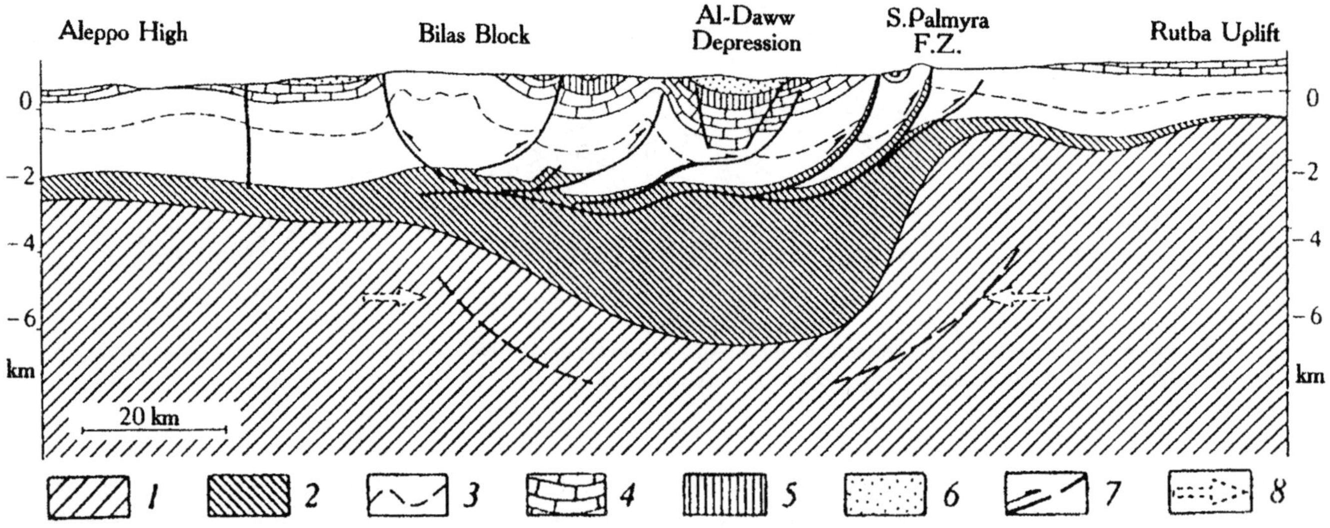

Figure 8.9. Tentative cross section of the Palmyrides fold belt. (After Kopp et al., 1994). 1 - crystalline basement and Paleozoic sediments; 2 - Triassic-Jurassic (including Triassic evaporites); 3 - Cretaceous; 4 - Paleogene; 5 - Miocene; 6 - Pliocene-Quaternary; 7 - faults (observed and hypothetical); 8 - direction of compression.

an olistostrome with blocks of Carboniferous rocks (Leonov et al., 1986). Apparently this anticline started to grow at the end of the Cretaceous. In comparison with the El-Bouab drill hole farther north, it has a thin Paleogene section (Ponomarew and Bivshev, 1966).

The Sinjar swell and its smaller folds and reversed faults were formed mainly during the Late Miocene-Pliocene inversion, which was weaker here than in the Palmyrides.

Qamishli and Ras El-Ein swells

The Qamishli and Ras El-Ein swells are situated at the extreme northeast of Syria and represent extensions of the major positive structure which is the Mardin domal uplift in southeast Turkey (Sites 22, 22a in Fig. 8.1). Precambrian metamorphics are exposed in the core of the Mardin dome. They have a thinner cover of incomplete Paleozoic and Mesozoic sedimentation, representing a long history of uplift and erosion (Temple and Perry, 1962; Ponikarov and Kazmin, 1965).

On the surface the gentle Qamishli and Ras El-Ein swells are vaguely defined by outcrops of Middle to Late Miocene sediments. At the same time, these swells are clearly expressed in the magnetic and gravity fields. Seismic profiles reveal basement at 4000 m depth, and the oldest sediments, penetrated by drillholes, are of Ordovician age. These are covered by Lower(?)-Middle Triassic deposits. The Upper Triassic and Jurassic are missing, while the Cretaceous and Cenozoic sediments attain a considerable thickness of 1800-2000 m (Ala and Moss, 1979; Sawaf et al., 1993).

It would appear that south of the Qamishli and Ras El-Ein swells the thickness of the Paleozoic and Mesozoic deposits increases (Fig. 8.5). The wide strip between these swells and the Sinjar Trough was referred to by different authors as the Sinjar foredeep (Galaktionov and Demidov, 1966), the North Khabour depression (Ponomarev and Bivshev, 1966), or the North Abd El-Aziz depression (Sawaf et al., 1993). In the Early Mesozoic, this strip was most probably uplifted relative to the subsiding Sinjar Trough, *i.e.* forming the slope of the Qamishli swell. In the Late Cretaceous both the slope and trough subsided at approximately the same rate. During the first stage of the Latest Cretaceous inversion in the Sinjar Trough (Ponikarov et al., 1967), an asymmetric depression was formed along its northern front and filled by a 6-7 km thick succession of Cenozoic sediments. Gravity data support this interpretation. The Sinjar Trough is limited on the north by a steep gradient zone (basement fault scarp?), while further north the negative gravity anomaly gradually increases to -80 and -40 mgal between the trough and the cores of the Qamishli and Ras El-Ein swells respectively. The sedimentary cover of the Qamishli swell is only slightly deformed by several small-amplitude E-W trending faults.

Folded foreland

The folded foreland is a direct continuation of the Zagros fold belt in northern Iraq. Within a small area in the extreme northeast of Syria, the basement drops to a depth of 10,000 m (Fig. 8.3). Simultaneously, the thickness of the Pliocene sediments increases to 2000 m. In the gravity map the southern limit of the deep basement zone corresponds with a steep negative gravity gradient with values of -50 to -80 mgal, that indicates a boundary fault or faults striking northwest. On the surface, the linear West and East Qarashouk anticlines are associated with the step in the basement.

The folded foreland was formed on the passive margin of Arabia which evolved at least since the Early Permian. With the onset of the Arabia-Eurasia collision in the Late Eocene, the

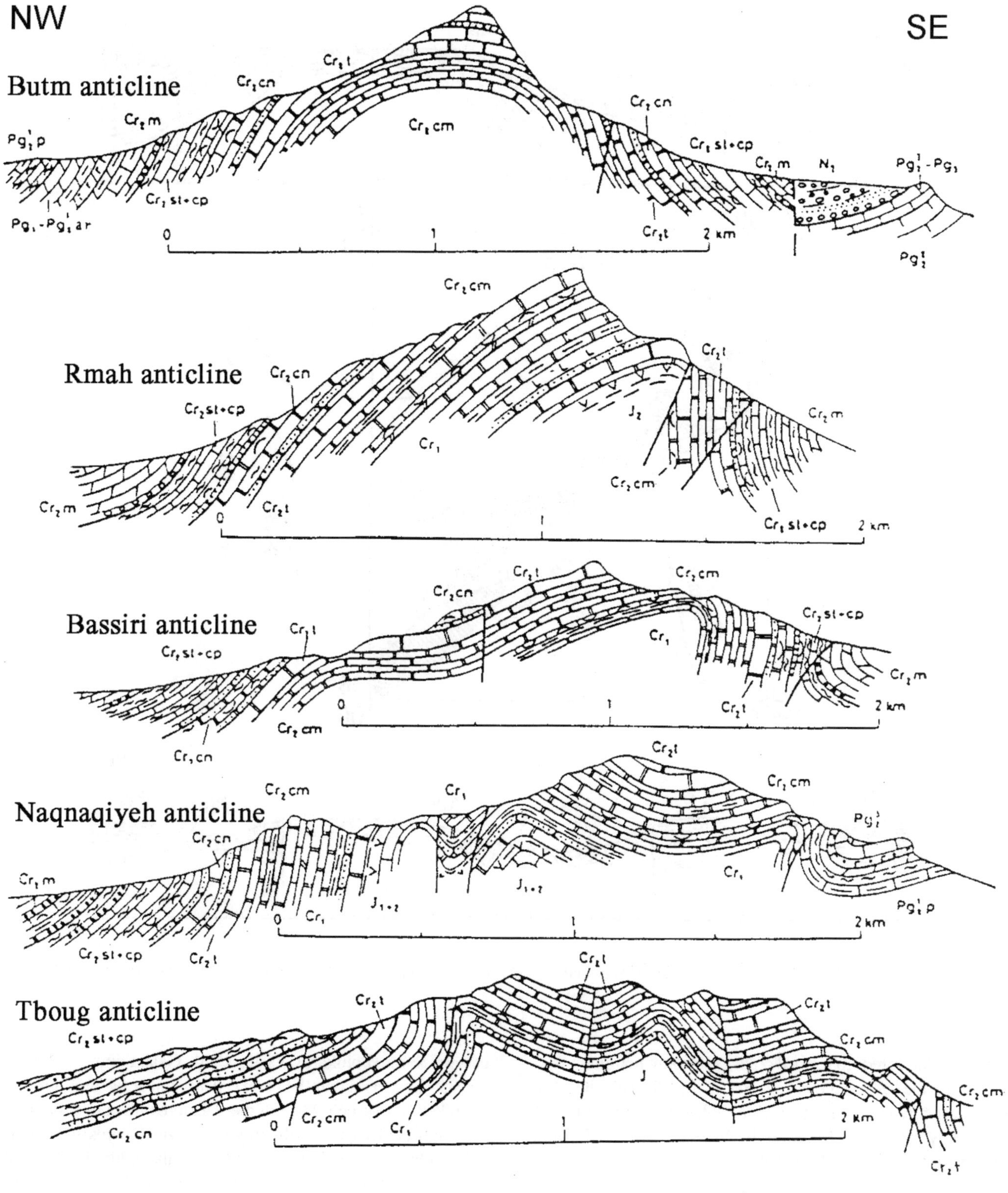

Figure 8.10. Folds associated with the South Palmyra fault zone, Tudmor (Palmyra) district. (After Faradzhev, 1966).

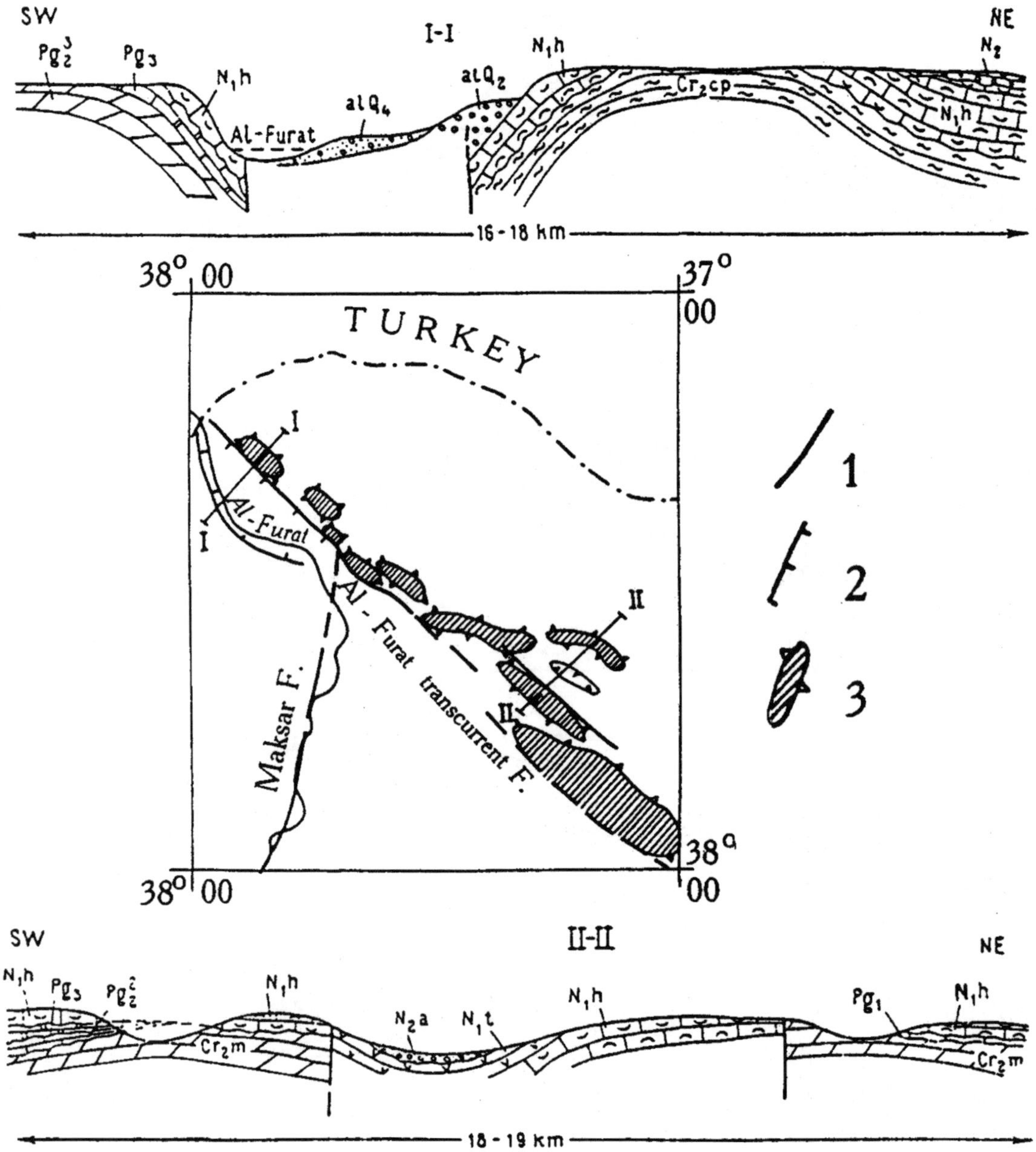

Figure 8.11. Northern segment of the Al-Furat transcurrent zone and cross sections. (After Oufland, 1966 with changes). 1 - major faults; 2 - normal faults; 3 - folds.

Mesopotamian foredeep formed on the subsiding margin.

During the latest stage of collision, in the Pliocene, the passive margin and foredeep sediments were folded by a system of linear anticlines which can be traced to the Mardin uplift in SE Turkey, where the foredeep narrows and dies out.

Al-Furat transcurrent zone

In aerial photographs and space imagery this zone can be clearly seen as a 640 km long NW-SE trending lineament crossing eastern Syria (Ponikarov et al., 1964; Galaktionov, 1989a, b). It consists of several parallel or *en echelon* linear features, grouped in a 20-25 km wide belt. Most of these features are purely topographic and mainly correspond to the straight marginal scarps of the Al-Furat valley. Only in the extreme northwest, in the uplifted Jrablus block, is the zone expressed by faults. Here, a northwest trending segment of the Al-Furat River coincides with the Qara Qozaq graben, or graben-syncline, which is filled by the riverine sediments (**Fig. 8.11**). A chain of flat, often box-shaped or asymmetric anticlines with steep (up to 20-25°) southwestern limbs is associated with the

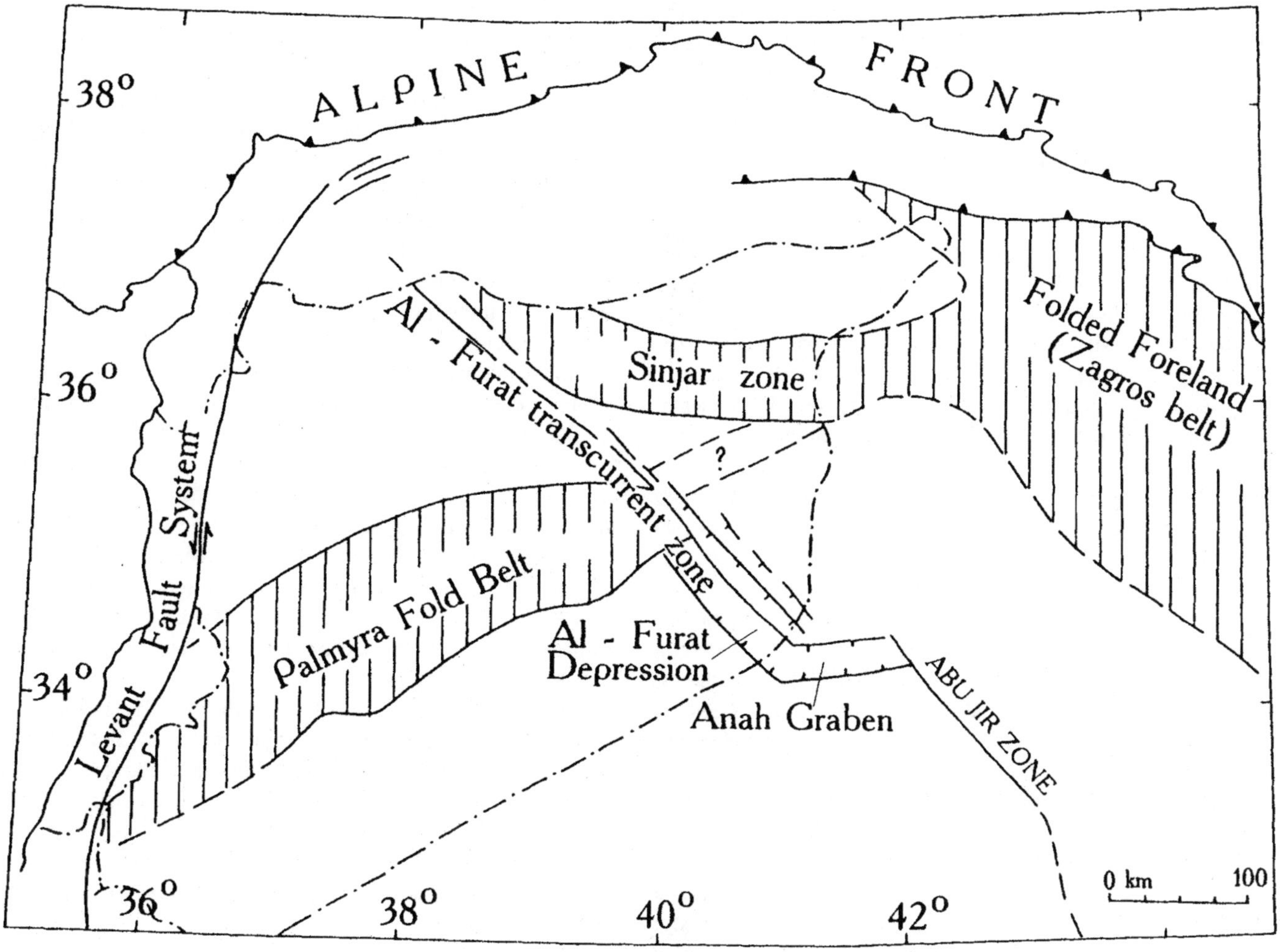

Figure 8.12. Relationships between the Palmyrides and the Sinjar zone. (Modified after Lovelock, 1984).

main fault on the northeastern margin of the graben. This deformed zone was previously described as 'Jrablus rampart' (Oufland, 1966). The Late Miocene sediments were involved in the folding while coarse clastics of Pliocene age fill the cores of the synclines. Thus, at least for the northern part of the transcurrent zone, a Late Miocene-Pliocene age of movement is established. A seismic profile shows that a 3 km wide zone of principal displacement consists of several closely spaced subvertical and branching faults, forming a typical 'flower structure' indicative of strike-slip motion (Lovelock, 1984; and Fig. 8.4A).

In southeastern Syria the Al-Furat zone is poorly expressed on the surface and is traced mainly by geophysical methods (S. P. C., 1986; Lovelock, 1984; Sawaf et al., 1993). Small fields of Quaternary basalts are possibly controlled by second-order transtensional fractures, associated with a major strike-slip fault. Some fractures in the Al-Furat alluvium were detected in space imagery (Galaktionov, 1989a, b), however no confirmation by ground observation is available. Near Deir Az-Zor, the Palmyrides fold belt is cut by the transcurrent zone, so that the Mesozoic basin and its related compressional structures are absent beyond the Al-Furat River. However some faults of the Palmyrides strike are recognized here (Fig. 8.3).

In the extreme southeast of Syria the Al-Furat transcurrent zone is identified by seismic surveys in the central, deepest part of the Al-Furat depression. As mentioned before, Neogene-Quaternary sediments fill the rift-related depression. These sediments are not at all or only slightly deformed by the Al-Furat transcurrent faults.

The sense of displacement along the transcurrent zone is not precisely known. South of the Syrian-Iraqi border the zone joins at a right angle the western end of the east-west trending Anah graben, which joins farther east, again at a right angle, the Abu Jir fault zone (**Fig. 8.12**). This structural pattern points to a left-lateral displacement along the Al-Furat–Abu Jir fault zone.

Nevertheless, the present-day compression in the Palmyrides and sinistral motion of the Rutbah and Jordan blocks along the Dead Sea fault are compatible with the dextral sense of the motion.

It is possible that the sinistral motion along the Al-Furat zone took place during the Late Paleozoic-Early Mesozoic opening of the Palmyrides, while the small-amplitude dextral displacement characterizes the Neogene-Quaternary compressional stage.

Frontal zone of the Alpine fold belt

This zone is exposed in two small areas in northwestern Syria, near the Turkish border: at Baer-Bassit, a coastal area north of Al- Latheqiyeh, and at Kurd Dagh, a mountaneous area north of Aafrine (Sites 1 and 2 in Fig. 8.1).

The Baer-Bassit block is bounded on the southeast by the Al-Latheqiyeh fault (Kazmin and Kulakov, 1965, 1968) an early Alpine nappe front modified by younger deformation. It is believed that the same fault forms the southeastern edge of the Kurd-Dagh block, separating it from the Aafrine foredeep. However it is impossible to trace the Al-Latheqiyeh fault directly to Kurd-Dagh because, for the most part, it is concealed by the young sediments of the Orontes River and Lake Amik.

The most conspicuous features of the Baer-Bassit and Kurd Dagh are the early Alpine nappes of Late Cretaceous ophiolites (90-85 Ma) and the volcanic-sedimentary sequence including rocks of Late Triassic to Cenomanian age (Kazmin and Kulakov, 1968; Parrot, 1977; Knipper et al., 1988). Most of Baer-Bassit is covered by allochtonous complexes, which in Kurd Dagh occur only as small erosional remnants.

In Baer-Bassit the autochthon appears in the core of the large Jebel Aqraa antiform (**Fig. 8.13A**). It is represented by platform-type carbonates of Jurassic to Early Maastrichtian age, which are identical to the platform sediments elsewhere in North Arabia (Kazmin and Kulakov, 1968). A south dipping fault-plane and a narrow zone of crushed volcanic and sedimentary rocks separate the core of the antiform from the tectonic sliver of the Qara Douran ridge, in which the carbonate section of Jebel Aqraa is repeated. In the paper by Knipper et al. (1988), the Qara Douran ridge is referred to as 'para-autochthone'. The contact of the latter with the allochthonous ophiolites is the steep south-dipping Saldirene arcuate fault, which Kazmin and Kulakov (1968) interpreted as a re-folded nappe surface.

The allochthonous complex of Baer-Bassit consists of two major nappes. The lower Tamima nappe is comprised of volcanic and sedimentary rocks of Triassic-Late Cretaceous age. The upper nappe is built of ophiolites which form three large massifs: the Bassit, Qeuzeul Dagh, and the Southeastern massif (Kazmin and Kulakov, 1968).

Ophiolites are represented by tectonized harzburgites, peridotite cummulates, layered gabbro, sheeted dykes, and pillow-lava. The normal succession is often disrupted by ultramafics and gabbro overthrust dykes and lavas. At the base of the ophiolites, greenschist to amphibolite facies metamorphics were formed during the early stage of the ophiolite slab motion from sediments and volcanics of the Tamima complex (Parrot, 1977; Silantyev, 1993). The latter represents a slice of oceanic crust and upper mantle, whose general structure is presented in **Fig. 8.13B**. Rocks of the Tamima nappe are exposed in strips and belts between the ophiolite massifs, and have an extremely complex and even chaotic inner structure. In early publications (Kazmin and Kulakov, 1968) this was compared with the 'Ankara melange' (Bailey and McCallian, 1953). The main components of the melange are siliceous mudstones, cherts, radiolarites, carbonate rocks of pelagic and shallow-water type, basic volcanics and sometimes terrigenous clastics. Blocks of gabbro and ultrabasites, presumably fragments of the ophiolite nappe, are ubiquitous. By analogy with the Mamonia and Hawashina complexes (Cyprus and Oman respectively), it was suggested that the Tamima rocks represent slope and rise deposits of the African-Arabian passive margin (Parrot, 1977).

In the process of thrusting the inner structure of the nappe pile became more and more complicated. Sediments and volcanics of the Tamima type occur structurally above the Bassit ophiolites in the core of the synform. These relationships most probably resulted from repeated thrusting that brought the rocks of the lower nappe above those of the upper one. However another interpretation should also be kept in mind, that the sediments and volcanics outcropping above the ophiolites may represent the original cover (layers 1 and 2) of the oceanic crust. In this case their lithological similarity to the Tamima rocks is fortuitous, and their age must be close to that of the ophiolites.

In contrast to Baer-Bassit, the allochthonous formations of Kurd Dagh are represented by small remnants of the Tamima nappe, preserved in synforms above the autochthonous Mesozoic sequence. The ophiolites form only small isolated blocks.

The youngest autochthon sediments in Jebel Aqraa and Kurd Dagh are of Campanian-Early Maastrichtian age. At the same time the oldest beds in the neo-autochthon sequence, which rest on ophiolites in the southern part of the Baer-Bassit, are Late Maastrichtian. It follows that the time of thrusting was Maastrichtian. If the Baer-Bassit ophiolites belong to the frontal part of the same nappe, as the ophiolites of Amanus in southeast Turkey, the amplitude of the nappe is not less than 150 km.

It is possible that the Nahr El-Kabir and Aafrin foredeeps already originated in front of the ophiolite nappes in the Late Maastrichtian, as indicated by the great thickness of the Paleogene deposits (up to 1700 m) (Ponikarov et al., 1967). In post-Maastrichtian time the autochthon, nappes, and neoautochthon were subject to several phases of deformation, which increased in intensity from the southeast to northwest, *i.e.* towards the Arabia-Eurasia suture. Folds in the Late Cretaceous-Paleogene sediments of the neoautochthon have axes parallel to the Lathaqiyeh-Kilis front, and limbs which dip at 20-35°. A good example of folds in the allochthon is the above mentioned synform in the Bassit Massif, with volcanics and sediments of the Tamima nappe in the core above ophiolites (Kazmin and Kulakov, 1968; Knipper et al., 1988). The dome-shaped antiform of Jebel Aqraa is the most impressive post-nappe structure. Here the autochthon rocks were deformed together with the parautochthon slivers and the main ophiolite nappe. Numerous northeast striking steep reverse faults and dip-slip faults with a strike-slip component also belong to post-nappe structures. Steep reverse faults of pre-Middle Miocene age

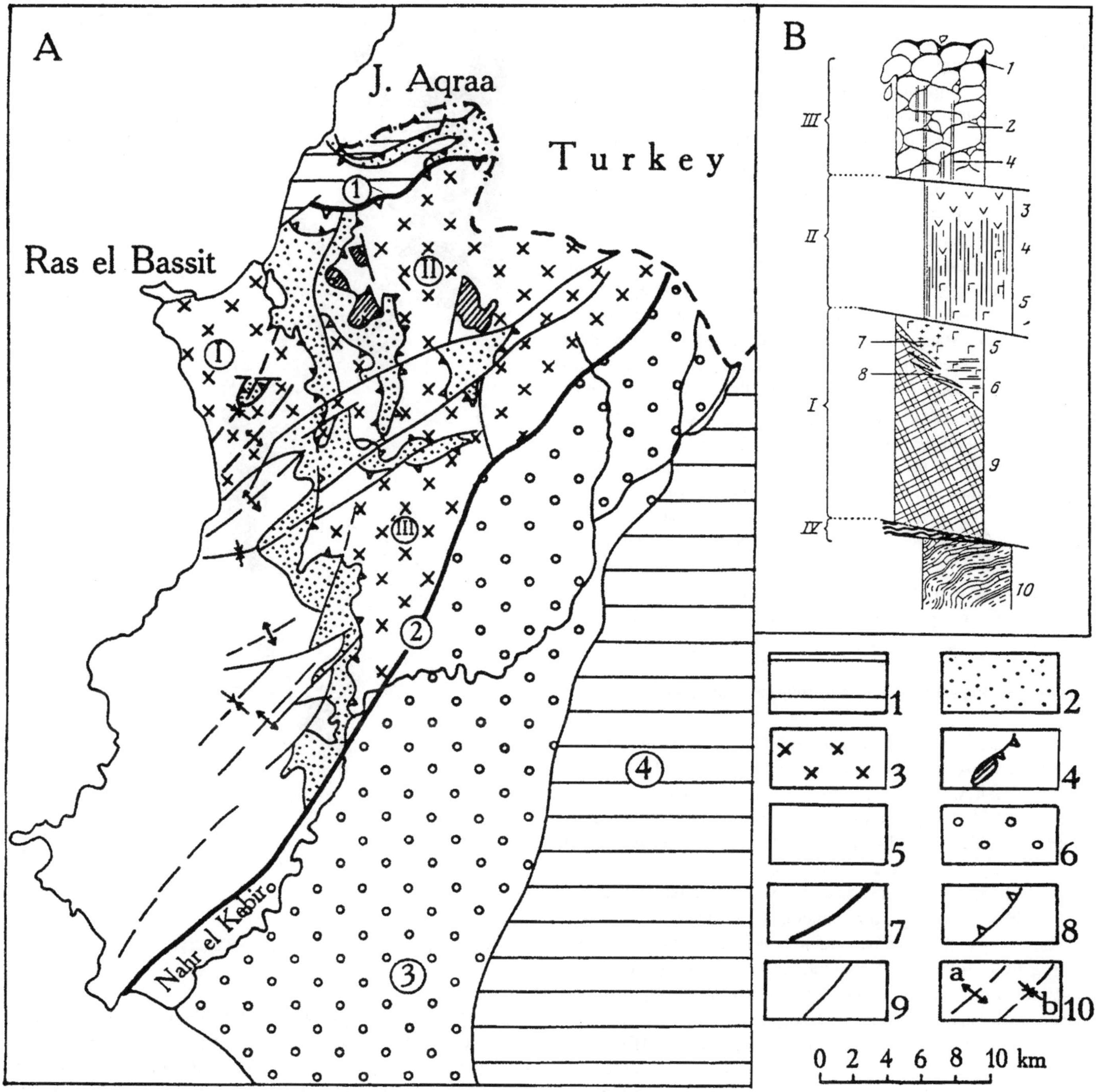

Figure 8.13.

A. Tectonic sketch of the Baer-Bassit Massif (after Kazmin and Kulakov, 1968; Knipper et al., 1988). 1 - sediments of the Arabian platform cover (autochthone); 2 - Tamima nappe; 3 - ophiolite nappe (I - Bassit Massif, II - Qeuzeul Darh Massif, III - Southeastern Massif); 4 - metamorphics at the base of the ophiolite nappe; 5 - neoautochthone; 6 - Neogene sediments; 7 - main faults; 8 - nappe front, overthrust; 9 - other faults; 10 - axes of the post-nappe folds. 1 - Saldirene fault; 2 - Al-Latheqiyeh-Kilis suture; 3 - Nahr El-Kebir foredeep; 4 - Jebel An-Nusseiriyeh.

B. Structure of the ophiolite nappe (after Knipper et al., 1988). Major rock-complexes forming separate tectonic slices: I - gabbro and ultramafics, II - diabases and gabbro, III - volcanites; IV - metamorphics. 1 - umbres; 2 - upper pillow-lava; 3 - lower pillow-lava; 4 - dikes, dike swarms, sheeted dike complex; 5 - massive gabbro; 6 - layered gabbro; 7 - plagiogranite; 8 - gabbro-pegmatite; 9 - serpentinized dunite and harzburgite; 10 - volcanics and sediments of the Tamima nappe.

are associated with the Latheqiyeh-Kilis nappe front (Trifonov et al., 1991).

The post-nappe deformation began at the end of the Middle Eocene to the beginning of the Late Eocene. The absence of Late Eocene sediments in Baer-Bassit is evidence of this. This event coincides with the onset of the Arabian-Eurasian collision (Dercourt et al., 1986, 1993). After that, deformation continued in the Oligocene and Neogene, reflecting the continent-continent collision. In the Early Miocene, in particular, fast subsidence of the Nahr El Kabir and Aafrine foredeeps began.

In Pliocene-Quaternary time, according to Knipper et al. (1988), northeast striking faults, including the Al-Lathekieh fault, were re-activated as sinistral shears. This re-activation coincides with the major stage of displacement along the Levantine fault system. It is important to note that the Al-Lathekieh fault continues into the eastern Mediterranean, following the southeast margin of the Latakia ridge (Kogan and Stenin, 1994; Vidal et al., 2000). The latter, in turn, joins the Cyprean trench in the west.

The faults of the Latakia and Larnaca ridges are steeply dipping sinistral strike-slips. Together with the Cyprean trench, they constitute the boundary between the African and Anatolian plates, the latter being expelled to the West (Vidal et al., 2000). If this interpretation is correct, the Al-Lathekieh fault and the Amanous fault, the extension of the Larnaca fault, should be regarded as the continuation of the East Anatolian fault, and must accommodate large sinistral displacement.

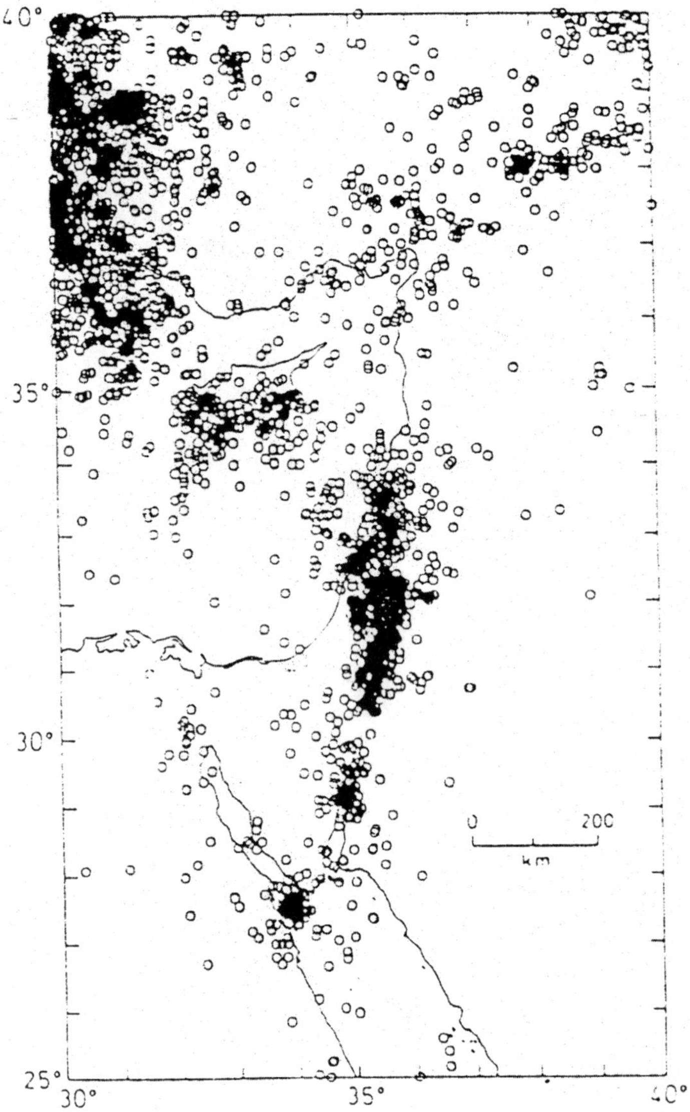

Figure 8.14a. Seismicity of the Levant fault zone (Girdler, 1990).

The Levantine fault system and related structures

The Levantine fault system consists of three segments. The Dead Sea segment, in the south, runs from the Gulf of Aqaba/Elat through Wadi Arava and Dead Sea grabens to the Hula depression at 33°-30'N. It consists of several *en echelon* right-stepping faults, with the total sinistral displacement of about 105 km (Quennell, 1958, 1984; Freund et al., 1970; Joffe and Garfunkel, 1987). Rhomb-shaped 'pull-apart' basins develop at fault overlaps. A belt of high seismicity corresponding to the Dead Sea segment continues up to the Hula depression, where it terminates abruptly (**Fig. 8.14 A**). In the plate tectonic scheme, the Dead Sea segment is interpreted as a transform boundary between the Arabian and Sinai plates, with rotation of Arabia relative to Sinai being around a pole at 32.7°N; 19.8°E (Joffe and Garfunkel, 1987) (**Fig.8.14B**).

North of the Hula depression the single shear zone of the Dead Sea segment splits into several fan-like faults within the Lebanon segment: these are the Roum, Yammuneh, Rachaya, Serrhaya, and the faults of the Palmyrides fold belt. For most of these, small scale sinistral displacement is either established or suspected. At the same time the seismicity becomes diffused with rare and usually weak earthquakes distributed among several major faults. Main displacement occurs along the Yammuneh fault running along the western side of the Litani River, and then along the Al-Beqaa valley (Dubertret, 1970; Hencock and Atiya, 1979). About 7 km of sinistral displacement was tentatively estimated by Hencock and Atiya (1979). They suggested that displacement is equal to the length of the small rhomb-shaped pull-apart depression associated with the Yammuneh fault. Another characteristic feature of the Lebanon segment is a fan-like system of folds that includes the Palmyrides, Anti-Lebanon, and the Lebanon. Following Freund et al., (1970), practically all authors relate the origin of the folds to the restraining bend of the Yammuneh fault to the east. This bend leads to transpression, significant horizontal shortening, and reduction of the sinistral displacement along the Levantine fault system. Though many mechanisms of crustal shortening in the Lebanon segment were suggested, such as folding, motion along deep detachment planes, or block and rotation around vertical axes, the total amount of shortening remains uncertain.

The northern or Syrian segment of the Levantine fault system includes the Al-Ghab and Karasu faults and associated structures:

the asymmetric anticlinorium of Jebel An-Nusseiriyeh, the Al-Ghab depression, and the Jebel Az-Zawieh uplift (Fig. 8.14).

The Al-Ghab fault substitutes for the Yammuneh fault near the Lebanon-Syria border. A possible link between the two faults is the small rhomb-shaped Al-Bukaya depression. North of Al-Bukaya the Al-Ghab fault is expressed by a 500 m high scarp on its western side. The margin of the Late Miocene-Pliocene basaltic plateau is displaced sinistrally by the fault for about 10 km (Shatsky et al., 1966). This is similar to the displacement along the Yammuneh fault. South of Missiaf, the main fault splits into two parallel branches 1.5 km apart. The 30 km long depression between them is filled by coarse alluvial and proluvial clastics, containing Pliocene fossils (Kozlov et al., 1966). Left-lateral displacements of the stream valleys were observed along faults on both sides of the depression (Trifonov et al., 1991). These authors estimate a total Late Pleistocene-Holocene displacement of 450 m and a mean slip rate of 0.5-0.6 cm per year. North of Tell Selkhab, the western boundary fault is again displaced to the west along a northwest striking fault (**Fig. 8.15**). Between the Al-Ghab and Jebel An-Nusseiriyeh fault, the 15 km wide Al-Ghab depression extends about 80 km north up to the town of Jisr Ash-Shaghour.

The principal Al-Ghab fault runs along the eastern margin of the depression. Immediately north of the branch point, a 30 m high scarp separates the deeper Al-Ghab depression from the higher Asharneh depression. This has an angular shape and is most probably an extensional basin filled with Pliocene-Quaternary alluvium and controlled by NW-SE and NE-SW trending faults (Shatsky et al., 1966). Further north, the hanging wall of the Al-Ghab fault is uplifted 600 m above the Al Ghab depression floor, forming the Jebel Az-Zawieh Massif which is the uplifted margin of the Aleppo plateau. Here a conspicuous crush zone consisting of several closely-spaced vertical fault-planes associated with the Al-Ghab fault.

North of the Asharneh depression, the Al-Ghab fault deviates 20° to the northeast and near 36°N it splits into two branches. The eastern branch, the Sheikh Barakat fault (**Fig. 8.16**), strikes northeast and can be followed up to the Aafrine foredeep, where it dies out (Protasevich and Maksimov, 1966). Trifonov et al. (1991) report the ~0.8 km displacement along this fault of a small dome-like structure.

The western branch of the Al-Ghab fault is traced up to the Syrian-Turkish border where it runs along the eastern 500 m high margin of the Karasu depression (Protasevich and Maksimov, 1966). In southeast Turkey the Karasu fault (or the faults associated with it) continues up to the latitude of Marash, where it splits into a number of northeast striking faults and then dies out (Geological Map of Turkey, 1989).

Jebel An-Nusseiriyeh forms the western margin of the Al-Ghab depression. This ridge, rising 1200 m above the floor of the depression, gradually drops northward. Structurally the ridge is an asymmetic horst-anticline with the western limb gently dipping towards the Mediterranean Sea and the eastern limb being a very steep flexure-type and fault truncated (**Fig. 8.17**). The Jebel An-Nusseirieh fault, bounding the horst-anticline on the east,

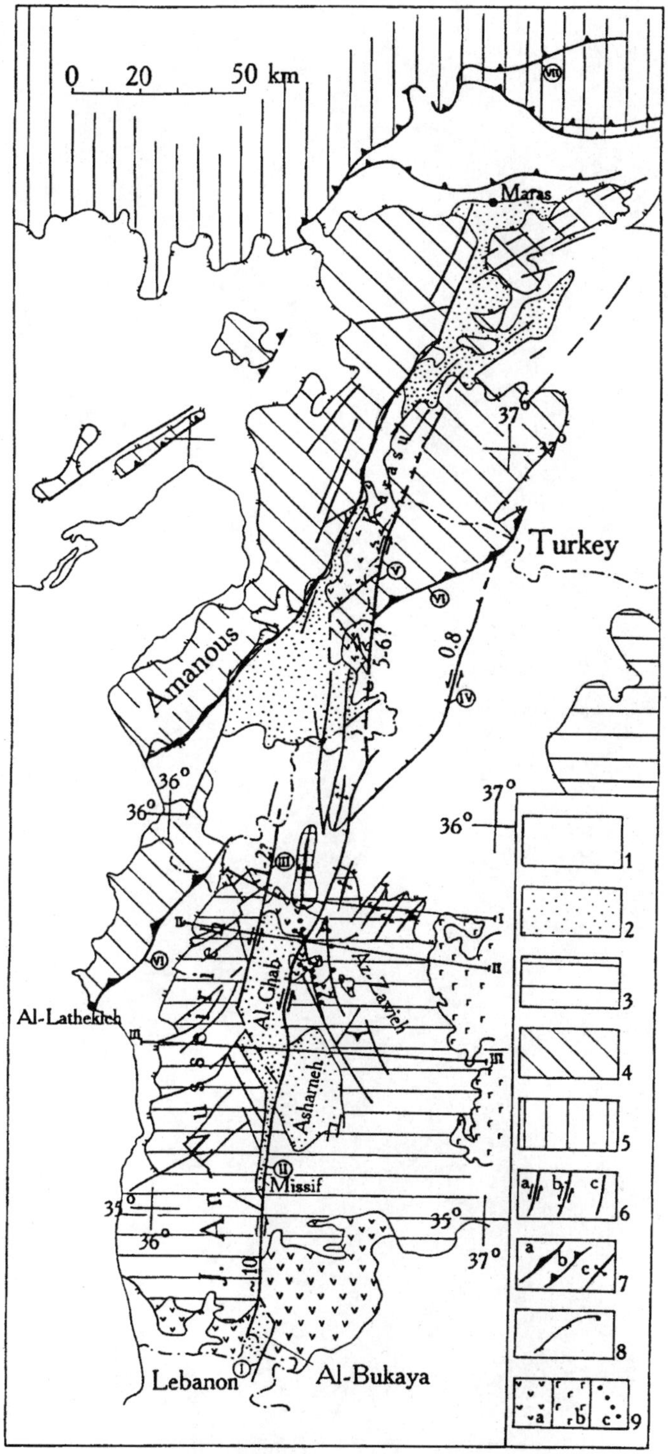

Figure 8.14B. Northern segment of the Levant fault zone. 1 - Neogene cover; 2 - Pliocene-Quaternary sediments associated with the Levant fault zone; 3 - pre-Neogene cover of the Arabian platform; 4 - frontal zone of the Alpine fold belt; 5 - Taurides fold belt; 6 - strike-slip faults (a), strike-slip faults with significant vertical amplitude (b), minor faults (c); 7 - early Alpine front (a), late Alpine suture (b), fold axis (c); 8 - limit of the pre-Neogene outcrop; 9 - volcanites: a - Pliocene; b - Upper Miocene; c - volcanic cones.

Figure 8.15. Southern end of the Al-Ghab depression. Northwesterly trending fault offsets the boundary fault of Al-Ghab to the west. Aerial photograph 1:20,000.

Figure 8.16. The Jebel Sheikh Barakat fault (an eastern branch of the Al-Ghab fault). Branching faults oblique to Sheih Barakat are clearly seen. Aerial photograph 1:20,000.

has not less than 1.5 km of vertical throw and a small left-lateral component evidenced by the displacement of watercourses. The horst-anticline plunges below young sediments not far from the Syrian-Turkish border. Simultaneously the Jebel An-Nusseiriyeh fault is also buried or dies out.

The Al-Ghab depression has some peculiar properties such as no normal step-faults at its margins, as are typical for extensional grabens. Both margins of the depression are formed by strike-slip faults with large vertical displacement. Near the faults, beds are bent and form steep flexures (Fig. 8.17). The deepest part of the depression is in the south, where 150 m deep drill-holes remained in the lacustrine marls. According to the mollusk studies of G. F.

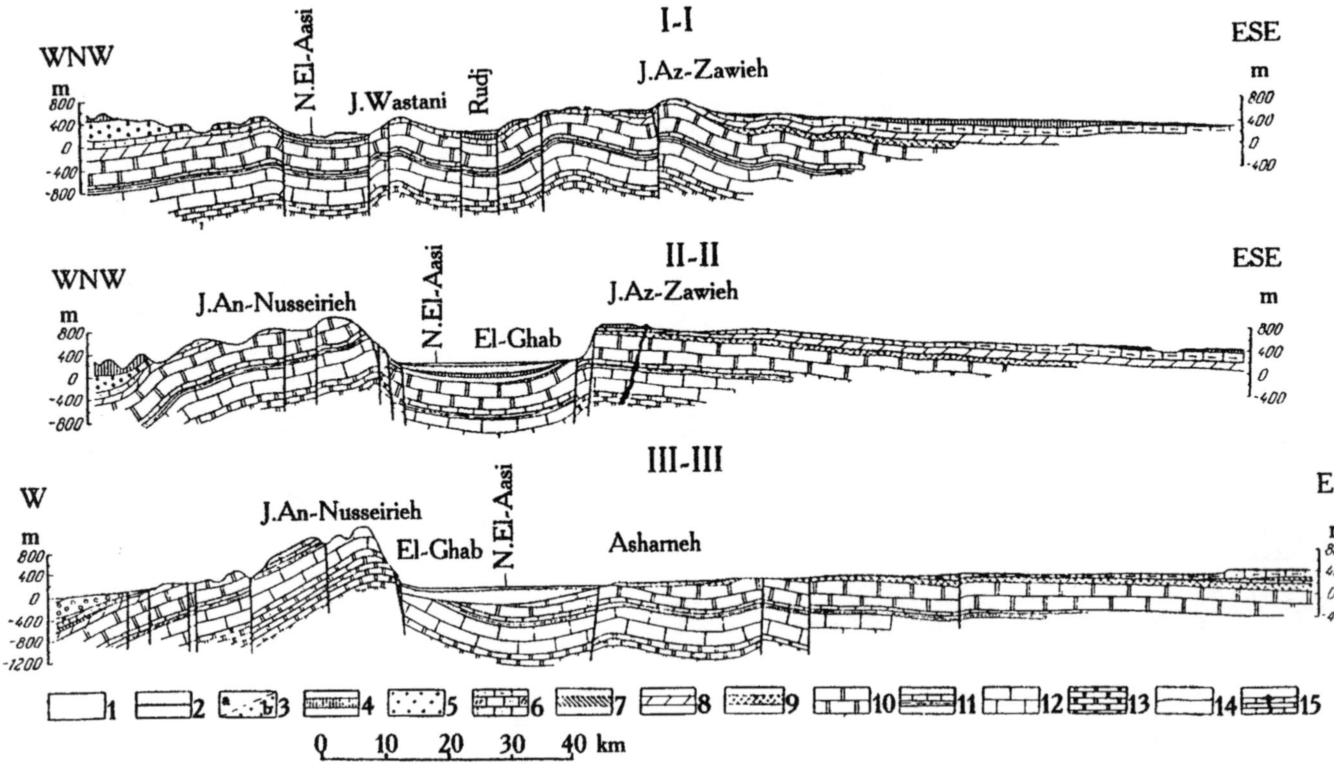

Figure 8.17. Cross sections of the Jebel An-Nusseiriyeh - Al-Ghab - Jebel Az-Zawieh. (After Kazmin et al., 1964). 1 - Quaternary; 2 - flood basalts; 3 - Pliocene: a - lacustrine; b - marine; 4 - Middle-Upper Miocene; 5 - Lower Miocene; 6 - Middle Eocene: a - nummulitic limestone, b - chalky limestone; 7 - Paleocene - Lower Eocene; 8 - Maastrichtian; 9 - Campanian; 10 - Cenomanian-Turonian; 11 - Aptian-Albian; 12 - Upper Jurassic; 13 - Middle Jurassic; 14 - faults; 15 - Pliocene volcanoes.

Lungersgausen, the marls have a Late Pliocene age (Shatsky et al., 1966). A gravity minimum also indicates deep subsidence in the southern portion of the depression (**Fig. 8.18**). Recent geophysical studies demonstrated that the thickness of the Pliocene-Quaternary sediments is about 900 m (Kopp et al., 1999).

Northwards the pre-Pliocene basement of the depression rises gradually, so that the Middle-Miocene and Eocene limestones emerge at the surfase near Jisr Ash-Shaghour. It appears that the Al-Ghab depression formed due to subsidence of a narrow block along a tensional fracture (or fractures) at its southern end (**Fig. 8.19**). As mentioned above, displacement to the west of the western (Jebel An-Nusseiriyeh) fault occurs along a northwest trending fracture (Fig. 8.15). Sinistral motion along the Jebel An-Nusseiriyeh fault creates extension along this fracture that leads to a dip-slip displacement. It follows that displacement along the Jebel An-Nusseiriyeh fault is accommodated by a normal fault or several faults at its southern end (Fig. 8.19) and cannot be significant.

An attempt was made to compare the Al-Ghab depression to the Gulf of Aqaba and the Dead Sea depression, where the structures formed in a transtensional framework (Kopp et al., 1994). For both of them a stress-partition model was suggested, according to which strike-slip motion concentrates along one of the marginal faults while on the opposite margin extension and normal step-faults dominate (Ben-Avraham and Zoback, 1992). The Al-Ghab depression has a different structure, indicating a general transpressional setting. In our interpretation, Jebel An-Nusseiriyeh, Al-Ghab, and Jebel Az-Zawieh form a structural assemblage, similar to the Lebanon–Al-Beqaa–Anti-Lebanon compressional structures of the Lebanon segment. It should be noted that Kopp et al., (1999) expressed somewhat similar views in their latest publication. Contrary to their earlier interpretation, they argue that the structures of northwest Syria are transpressional and that Al-Ghab is not a rift, but belongs to a specific type of depression, formed between branching strike-slip faults.

The transpressional regime in northwest Syria may have different origins. In the early global plate tectonic models the vector of the Arabian motion relative to Eurasia was defined as N002°E (Chase, 1978) or N359°E (Minster and Jordan, 1978). In the more up-to-date model (DeMets et al., 1990) this vector is N338°E, *i.e.* at a sharp angle to the Al-Ghab and Karasu faults. With the African Plate moving in a north-northeastern direction relative to Eurasia, transpression along the Levantine fault system can thus be expected..

Another and more evident reason is the deviation of the Al-Ghab fault plane to the east. It creates a 'restraining bend' effect, not as significant as in case of the Yammuneh fault but still

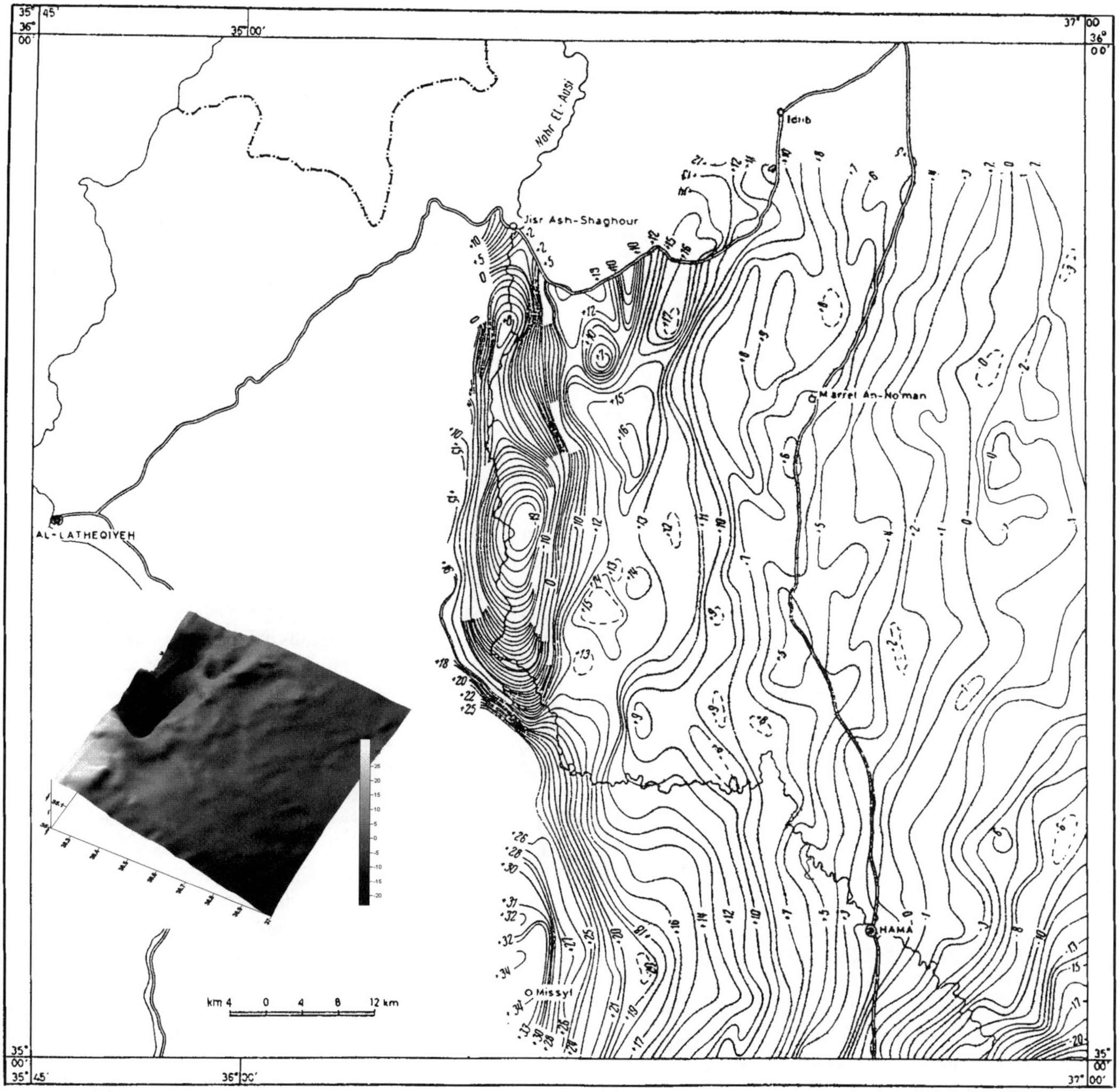

Figure 8.18. Bouquer anomaly map of the Al-Ghab depression. (Lobachevsky and Shcheglova, 1961). Isoanomals are 1 mgal. Inset shows 3-D representation of the digitized contours.

important. The Paleogene and Neogene rocks of Jebel Az-Zawieh are folded near the Al-Ghab fault bend, with the intensity of folding increasing towards the fault (Shatsky et al., 1966; and Figs. 8.14 and 8.17). Simultaneously the Al-Ghab depression splits into several graben synclines and horst-anticlines. The closure of the largest box-shaped anticline of Jebel Wastani is shown in the aerial photography (**Fig. 8.20**). Compressional N-S or NE-SW structures coexist with tensional fractures trending NW-SE, thus acting as feeders for the Pliocene basaltic eruptions (Fig. 8.14).

The 10 km Late Pliocene-Quaternary sinistral displacement noted for the southern segment of the Al-Ghab fault is partly compensated for by displacements along the secondary Jebel An-Nusseiriyeh and Jebel Sheikh Barakat strike-slip faults, and by folding. It is suggested that the displacement along the Karasu fault

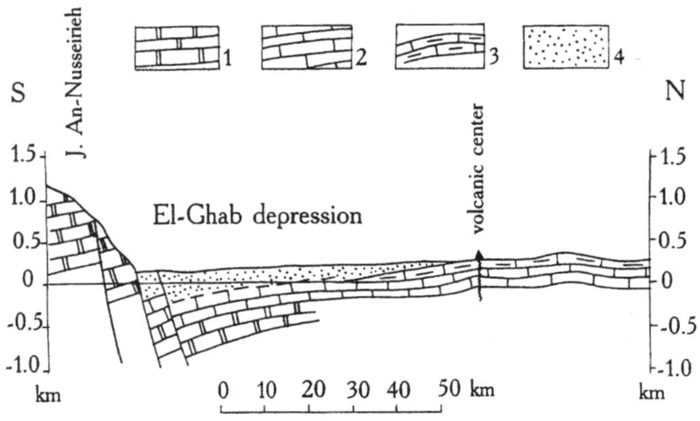

Figure 8.19. Cross section along the Al-Ghab depression. 1 - Jurassic limestones; 2 - Cretaceous-Paleogene limestones and marls; 3 - Miocene limestones; 4 - Pliocene-Quaternary laqustrine sediments.

is less than 10 km.

Trifonov et al. (1991) and Kopp et al. (1994) attempted to estimate the amount of sinistral displacement along the Al-Ghab-Karasu line, using the apparent displacement of the Al-Latheqieh-Kilis nappe front, which in their opinion is about 20 km. However this figure is not reliable. The southwestern part of the front (the Al-Latheqieh fault) plunges below young sediments 50 km southwest of the Al-Ghab-Karasu faults and cannot be traced directly to them (Fig. 8. 14). At the same time the nappe front is sinuous, so that it is impossible to project its extension as a straight line. The real displacement between the Baer-Bassit and Kurd Dagh segments of that front may, in fact, be much less than 20 km.

Structures related to the Syrian segment of the Levantine fault system are very young. It was noted that the Missiaf, Asharneh, and Al-Ghab depressions are not older than Pliocene, which is the age of their infilling sediments (Kazmin et al., 1964; Shatsky et al., 1966; Trifonov et al., 1991). Basaltic eruptions, controlled by fractures related to the Al-Ghab fault, are of Plio-Pleistocene age, and basalts erupted on the floor of the Karasu depression are Late Pleistocene to Holocene (Protasevich and Maksimov, 1966). Since the origin of these depressions and of the volcanism were closely related to lateral motion along the Al-Ghab and Karasu faults, this motion could not be older than Pliocene. Of course, older lateral movements cannot be excluded, but it is significant that no evidence for these has thus far been found.

The Anti-Lebanon anticlinorium, straddling the Syrian-Lebanon border is the primary structure of the Lebanon segment of the Levantine system. Only the northern and southeastern parts of the Anti-Lebanon are exposed in Syria.

The Anti-Lebanon is separated into two parts by the Serrhaya fault (Figs. 8.1 and 8.2). East of the fault, the Anti-Lebanon is an anticlinorium with a flat and wide arch with dips of 5° to 10°, and steep limbs dipping at 35-45° and sometimes even up to 90°. In the west it joins the Al-Beqaa graben-syncline and in the southeast with the Palmyrides fold belt. The anticlinal structure is best observed in the northern part. The anticlinorium extends to the Homs depression where its periclinal closure plunges below Neogene sediments (Kozlov et al., 1966). West of the Serrhaya fault, the southern block of the Anti-Lebanon is a horst-anticline. Its central part, the Hermon Massif, rises to 2884 m. On both sides the horst is bounded by faults which are branches of the Levantine fault system, with the Serrhaya on the east, and the Rechaya on the west.

As seen from the structural pattern, both faults are left-lateral shears. This is especially clear for the Serrhaya fault which is associated with the Zebdani depression, a small rhomb-shaped basin which is presumably of the 'pull-apart' type.

To the north, the depression continues as the long and narrow Serrhaya trough, filled with coarse conglomerates (Razvalyaev, 1966; Kurbanov and Zarjanov, 1968). The Zebdani depression and the Serrhaya trough are bounded on the east by the Serrhaya fault and by its parallel branch to the west. On a small scale the structure resembles the Missiaf and Al-Ghab depressions, associated with the Al-Ghab fault. In the vicinity of Beit Jann, asymmetric anticlines with steep southeastern limbs develop along the Serrhaya fault, indicating a transpressional set-up.

An estimation of the sinistral displacement along the Serrhaya fault is crucial to understanding the evolution of the Levantine fault system. According to Dubertret (1955 and others), the Rechaya and Serrhaya faults obliquely cross the Anti-Lebanon and die out within its confines. Remote sensing methods confirm Dubertret's conclusion (Galaktionov, 1989a). The Serrhaya fault reaches the eastern margin of the Al-Beqaa valley, but does not displace it. The possibility remains that the fault continues along this margin and then links up with the marginal faults of the northern Palmyrides (Walley, 1988). However this suggestion is not supported by geological mapping in the field (Dubertret, 1955; Kozlov et al., 1966).

It has been suggested that the initial sinistral displacement of about 30 km occurred along the Serrhaya fault at an early stage (Miocene) in the evolution of the Levantine fault system (Walley, 1988; Khair et al., 1993). This estimation is based on the length of the Zebdani depression and the Serrhaya trough, both interpreted as 'classic' pull-apart basins, for which a complete separation of crustal blocks is equal to their length. In this case a pronounced positive gravity anomaly could have been expected, but was never observed. As in the case of Missiaf and Al-Ghab, the actual lateral displacement along the Serrhaya fault did not exceed a few kilometers.

As mentioned above, the compressional origin of the Anti-Lebanon is related to the effect of the restraining band of the Yammuneh fault (Freund et al., 1970; Ron, 1987; Kopp et al., 1994). However the time of deformation has still not been precisely determined. According to Razvalyaev (1966), in the depressions along the southeastern limb of the Anti-Lebanon, Pliocene sediments rest on deeply eroded Cretaceous and Paleogene rocks. In the erosional cirque of Aarneh, Dubertret (1960) has mapped tilted Neogene conglomerates, resting on Late Triassic to Early Jurassic rocks. These observations point to a significant

Figure 8.20. Eastern limb and closure of the Jebel Wastani anticline with Middle Eocene limestones. The Rudj depression is in the east. Aerial photograph 1:200,000.

pre-Pliocene uplift (Razvalyaev, 1966). The formation of the Anti-Lebanon possibly began as early as the Eocene (Hencock and Atiya, 1979), *i.e.* simultaneously with the onset of folding in the Palmyrides. Since sediments of the Zebdani basin have not been dated, the time of displacement along the Serrhaya fault is not precisely defined. It may be coeval with the Late Pliocene-Quaternary motion along the Yammuneh and Al-Ghab faults or even older.

Main stages of evolution

This section deals with the geological history of Syria, starting from the latest Paleozoic. Data on earlier periods are very scarce. It has been suggested that the Syrian basement is composed of two or three large terranes, welded together in the course of the Late Precambrian-Early Cambrian accretion (Best et al., 1993; Sawaf et al., 1993; Brew et al., 1997). These hypothetical terranes are recognized on the basis of strong differences in their geophysical character and variations in the thickness of the Infra- to pre-Middle Cambrian lower part of the platform cover. Major differences in the gravity and magnetic fields between the Rutbah Uplift and the Aleppo High were already discussed above. The thickness of the Late-Precambrian-Early Cambrian sediments is more than 3 km on the Rutbah and less that 1 km on the Aleppo High. It is believed that these blocks were joined in the Late Precambrian-Early Cambrian along a suture coinciding with the present-day Palmyrides fold belt. The Khlessia (or Rawda) Uplift, with its very distinct positive gravity field contrasting with the above-mentioned blocks, may represent a third independent terrane.

After a major transgression, the ~200 m thick carbonate rocks of the Middle Cambrian Burj Formation were deposited. They were followed by the terrigenous clastics of the Upper Cambrian-Lower Silurian, varying in thickness from 4.5 km in the south to about 2.5 km in the north of Syria.

According to G. Stampfli (1996), from the end of the Ordovician to Early Devonian, a chain of microcontinents which included the Turan block, the Scythian platform, and the Intra-Alpine terranes, was rifted off from the Gondwana margin. Behind these blocks the Paleotethys began to open. The absence of Late Silurian and Devonian sediments in Syria may reflect thermal uplift of the continental margin during this rifting. Terrigenous sedimentation resumed in Syria only in the Carboniferous.

The Late Paleozoic stage. 1st phase of rifting (Fig. 8.21).

The most important event during this period was the rifting and creation of the Palmyrides aulacogene. A number of drillholes (Doubayat, Soukhne, Dolaa, Fig. 8.7) have penetrated up to 1300 m of mainly terrigenous sediments of Latest Carboniferous-Permian age. These are also traced in seismic profiles (McBride et al., 1990; Lovelock, 1984). The Carboniferous-Permian depocenter roughly coincided with the Palmyrides. These sediments pinch out gradually southward on the Rutbah Uplift, and are absent on the Aleppo High, (*op. cit.*). We may speculate that the initial rift was asymmetric, with the main fault scarp on its northern margin. Permian sediments in the Thayyem-103 borehole indicate that the Al-Furat rift depression was also subsiding.

Rifting in North Arabia coincided with the initial opening of the Neotethys, and the separation of Iran from Arabia occurred in the Early Permian (Dercourt et al., 1986, 1993; Kazmin, 1991a). According to Stampfli et al. (1991) and Stampfli and Pillevuit (1993), the Anatolian microplate was rifted from Africa at the same time, marking the birth of the eastern Mediterranean or Mesogea

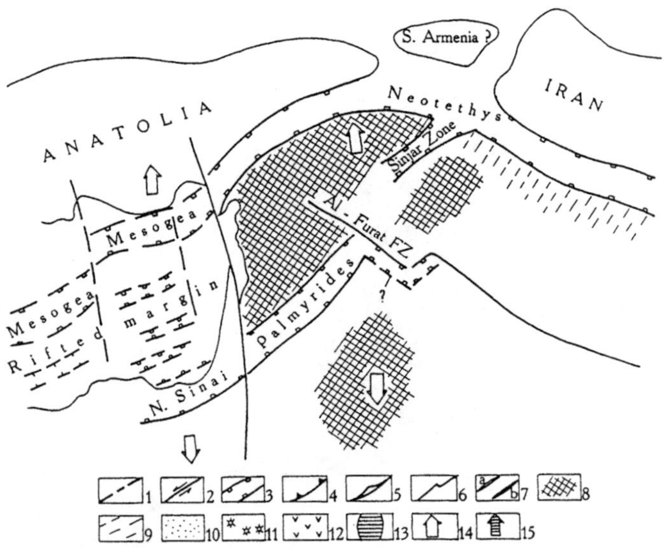

Figure 8.21. Paleotectonic scheme for the Early Permian time. 1st tensional stage. Symbols for Figs. 8.21 - 8.28. 1 - fault; 2 - strike-slip fault; 3 - normal fault; 4 - nappe front, overthrust; 5 - subduction; 6 - spreading center; 7 - folding: a - weak; b - strong; 8 - uplift; 9 - subsiding margin; 10 - foredeep; 11 - volcanic arc; 12 - basalts; 13 - fault-related large antiforms; 14 - stress direction; 15 - Arabia/Eurasia relative motion vector. COB - continent-ocean boundary; T - Troodos Massif; B - Bassit Massif.

basin. Referring to borehole sections on the African coast, these authors argued that the opening of Mesogea was preceded by only one rifting phase, and was immediately followed by spreading. Though the real history was probably more complicated (see below), Stampfli and colleagues are right in including the Anatolian microplate in 'Cimmeria', the chain of microcontinents rifted from Gondwana during the Permian (engör, 1979, 1984).

In the schematic reconstruction (**Fig. 8.21**) the following assumptions were made.

1) The Levantine fault system is shown as a plate boundary between Arabia and the eastern Mediterranean. Only a small sinistral displacement is known to have occurred along the Yammuneh and Al-Ghab faults (Hencock and Atiya, 1979; Trifonov et al., 1991; this chapter), and the shortening in the Palmyrides is modest (Chaimov et al., 1990). It is therefore provisionally accepted that the main pre-Pliocene displacement was taken up by the Roum fault and its hypothetical marine extension (Girdler, 1990).

2) A wide (400-500 km) passive margin, with attenuated continental crust, was formed in the eastern part of Mesogea (Phoenician basin). The reason for this interpretation is given below.

The cause of the Late Paleozoic rifting at the Gondwana margin remains unclear. A. M. C. engör (1990, 1991) suggested that rifting occurred in a backarc setting behind a north-facing subduction zone. However, a backarc mechanism is unlikely, since no suprasubduction magmatism of Carboniferous-Permian age is known on the microcontinents that rifted from Gondwana at that time (Iran, Afghanistan, Central Pamirs, Anatolia). In the Latest Paleozoic a number of compressional events took place on the Tethyan active margin, which cannot be accounted for by collision with microcontinents or island arcs. These events are known in the Caucasus (Belov, 1981), North Pamirs (Shvolman, 1977), and Kunlun (Pan Yusheng, 1996). Suprasubduction volcanism at the active margin was of limited scale, declining from the Early to Late Permian. It is suggested here that both phenomena were caused by ridge-trench collision, that temporarily hampered subduction. When subduction renewed, the active Paleotethyan margin and Gondwana were not separated by a spreading axis. Extension caused by slab-pull might in this situation have been translated to the Gondwana margin, creating rifting along a pre-existing weak zone, and finally leading to the formation of the new spreading axis of Neotethys.

***The Triassic stage. 2nd phase of extension* (Fig. 8.22).**

The major subsidence in the Palmyrides occurred in the Middle-Late Triassic. The thickness of Triassic sediments increases from 200 m on the margins of the Rutbah and Aleppo Highs to 1700 m in the Palmyrides (McBride et al., 1990). Presumably the depocenter was bounded by marginal normal faults which were later converted into reverse faults.

Subsidence in the western (Syrian) part of the Sinjar zone started in the Triassic, *i.e.* later than in the Palmyrides (Lovelock, 1984). It is possible, however, that the eastern and deeper part of the zone began to subside earlier. The Sinjar zone can be interpreted as a third failed arm of a triple junction in the Neotethyan rift system (Fig. 8.22).

The Al-Furat rift also subsided in Triassic times. In contrast to the Palmyrides, subsidence here was not active in the Middle, but in the Late Triassic (Sawaf et al., 1993; Fig. 8.7). It remains uncertain, whether in Syrian rifts there were two independent, Middle and Late Triassic impulses of rifting and subsidence. Simultaneous opening of the NE and NW trending rifts implies an approximately N-S (present-day coordinates) orientation of the main tensional stress and transtensional regime in the rifts (Fig. 8.22).

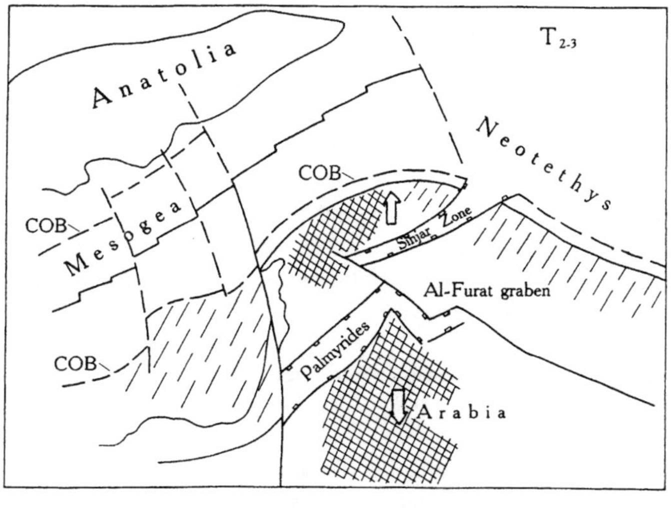

Figure 8.22. Paleotectonic scheme for the Middle-Late Triassic. 2nd tensional stage. Symbols in Fig. 8.21.

Triassic rifting in North Arabia was closely related to the evolution of the eastern Mediterranean or Mesogea. In contrast to the Neotethys, the Permian rifting in Mesogea most probably had not led to continental breakup but was followed by subsequent stages of Triassic rifting. In the allochthonous complexes of Baer-Bassit (Syria), Cyprus, and Antalya (Turkey) the lower nappes are composed of volcanic and sedimentary rocks, representing the passive margin and deep basin of Mesogea. In Baer-Bassit they are described as the Tamima nappe (Parrot and Whitechurch, 1978; Knipper et al., 1988), in Cyprus as the Mamonia complex (Lapierre, 1975; Robertson and Woodcock, 1979; Krylov et al., Chapters 4 and 5 of this volume), and in Antalya as the Kumulcha nappe (Robertson and Woodcock, 1980). The complicated imbricated structure of these complexes originated in accretionary prisms during subduction and the following obduction. Remarkably, in all these complexes, rocks older than Upper Triassic were never discovered, suggesting a Triassic age for the passive margin and deep basin. Therefore, transition to spreading in Mesogea hardly occurred before the Late Triassic and was preceded by one or more post-Permian phases of continental rifting. The Asher-Aarneh volcanics straddle the Triassic-Jurassic boundary. They extend from the Mediterranean coast along the Carmel escarpment, where they are over 2 km thick, then across Galilee to cross the Levantine Fault system into the Anti-Lebanon, where they are still 80 m thick in the cirque of Aarneh (Hermon), (Garfunkel and Derin, 1984; Hirsch, 1990; Mouty, 2000). This rift-related volcanism obviously accompanied the latest stage of Triassic rifting.

The same driving force of slab-pull of the sinking plate at the active margin was active in Mesogea until a new plate boundary (spreading axis) formed in the Late Triassic or Early Jurassic.

***The late Early Cretaceous stage. 3rd phase of extension* (Fig. 8.23).**

Triassic rifting was followed by a long period of emergence, development of a peneplain, and generation of a lateritic weathering crust which encompassed most of the Liassic. Triassic rifts were filled with sediments, but a topographic depression still might have remained in the Palmyrides.

Jurassic sediments in Syria are known from the margin of the eastern Mediterranean, in Jebel An-Nusseriyeh, and the Anti-Lebanon. They thin to about 200-300 m of Middle-Late Jurassic marine sediments in the shallow embayment of the Palmyrides.

A global sealevel drop in the Neocomian (Haq et al., 1987) was accompanied by the emergence and deposition of continental red beds over all of Syria. Basalts sporadically erupted in Jebel An-Nusseiriyeh and in two or three other places in the western Palmyrides.

A new transgression began in the Albian (Ponikarov et al., 1967). It followed a renewal of rifting in all the Syrian rifts (Ponikarov et al., 1967; Chaimov et al., 1992) and sporadic volcanic activity along the Levantine fault system, and in the adjacent portion of the Palmyrides (Faradzhev, 1966; Razvalyaev et al., Chapter 7 in this volume). Rifting was followed by post-rift subsidence, when up to 1 km of Upper Cretaceous sediments accumulated in the Palmyrides, Sinjar Trough and Al-Furat basin. Lovelock (1984) states that during this stage some marginal rift faults were initiated.

In the Al-Furat basin, not affected by the inversion, the post-rift subsidence continued uninterrupted throughout the Late Cretaceous and Cenozoic. About 3.5 km of sediments accumulated during this time (Sawaf et al., 1993; our Fig. 8.5). In the Palmyrides the thickness of the Late Cretaceous sediments drops sharply at the boundary with the Rutbah Uplift (Fig. 8.7). At the same time these sediments have almost the same thickness in the Palmyrides and on the Aleppo High. It seems that subsidence occurred mostly close to the southern marginal fault of the basin, *i.e.* the polarity of the rift structure changed in comparison with that of Late Paleozoic time.

In the Sinjar basin, the Late Cretaceous sediments thin considerably towards the Khlessia uplift in the south but slightly increase in thickness northward in the direction to the Qamishli Uplift (Fig. 8.5). As in the Palmyrides, the northern limit of the basin is poorly defined.

The Cretaceous rifting phase correlates well with a major reorganisation of the plate boundary. In the Aptian-Albian a trench system originated around the Arabian promontory of the African plate. A jump of the trench system to the south possibly resulted from the ridge-trench collision on the Tethyan active margin (Dercourt et al., 1986, 1993). There was no spreading axis between the newly formed trenches and the continental margin, so that the slab-pull was directly transmitted to the continental margin to produce extension and rifting.

A problem remains of the possible relationship between Mesozoic rifting in North Arabia and plume activity. Segev (2000) recently attempted a correlation of magmatic cycles with the fragmentation of Gondwana. He recognized two magmatic periods in the Levant, related to superplume activity, in the Jurassic (205-160 Ma) and the Cretaceous (138-183 Ma).

Jurassic magmatic rocks (basalts, gabbro) occur sporadically in the Levant in the near-coastal areas of Israel and Lebanon. No Jurassic rifting is known either in North Arabia or on the northern margin of Gondwana. Jurassic magmatism was certainly related to the opening of the Central Atlantic, but had no tectonic correlation on the southern Tethyan margin, including Arabia.

The Cretaceous 'superplume' was manifested in the Levant within the 'Levant-Nubian province (swell)'. Again sporadic magmatic activity occurred in the N-S elongated oval area parallel to the Mediterranean margin and the Levant fault system (*op. cit.* Fig.4). East of this zone, magmatism is practically absent. There was no spatial connection of the 'swell' with the North Arabian rifts but its structural link, with the transform margin of the eastern Mediterranean basin, is very clear. On a broader scale the question is: are any manifestations of intraplate magmatism necessarily associated with plume activity? Experimental data show that on a limited scale magmatism may accompany 'cold rifting' without plume interference, the melt being produced by decompression of the rising asthenosphere in the extended lithosphere (White and McKenzie, 1989, 1995). There was certainly no Mesozoic LIP (the Large Igneous Province of Coffin and Eldholm, 1991) in North Arabia indicating strong thermal perturbation and large-scale partial plume-related melting. Volcanic rocks were very modest in volume and were mostly alkaline basalts.

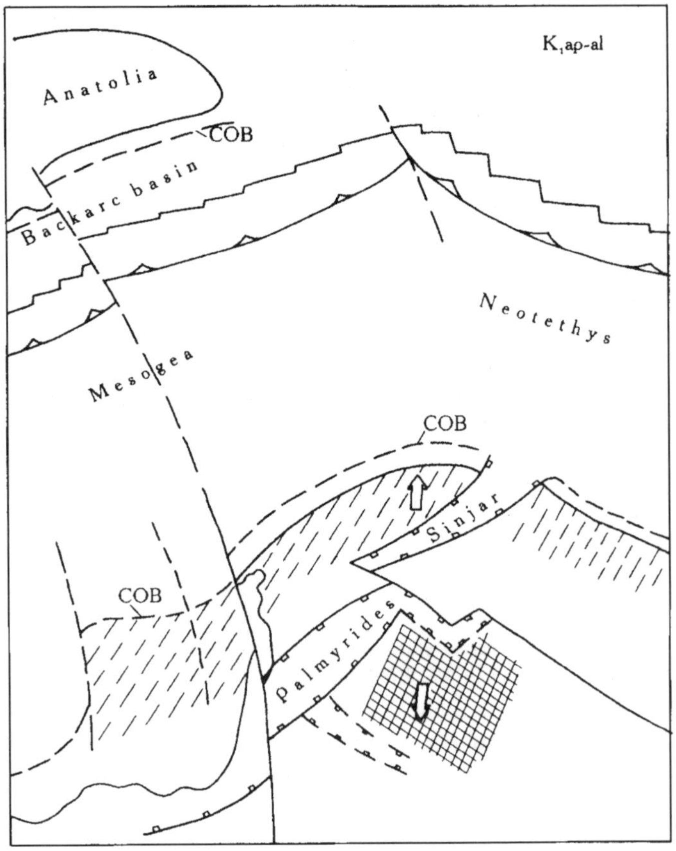

Figure 8.23. Paleotectonic scheme for the Aptian-Albian. 3-d tensional stage. Symbols in Fig. 8.21.

On the other hand, the North Arabian rifts were formed in a time of major plate boundary reorganizations, which could cause extension and rifting in the Arabian plate. As discussed above, this casual link was especially clear in late Early Cretaceous time.

Campanian-Early Maastrichtian stage. 1st compressional phase. (Fig. 8.24).

The earliest manifestation of compression occurred in the Early Senonian, when the platform margins in Jebel An-Nusseiriyeh and Kurd-Dagh were uplifted (Ponikarov et al., 1967). However the first major phase of compression in the Syrian basins was in the

Late Senonian-Early Maastrichtian. Structural differentiation was evident in the Palmyrides and Sinjar basins in the Campanian, with the growth of folds. Folding intensified in the Early Maastrichtian leading to considerable differences in thickness of the post-Maastrichtian sediments in anticlines and synclines (Ponikarov et al., 1967; Faradzhev, 1966; Soulidi-Kondratiev, 1966).

Inversion in the Sinjar basin was much weaker than in the Palmyrides. The basin still actively subsided in the Campanian and Early Maastrichtian; the thickness of the Maastrichtian sediments here reaches 1000 m. In the Jebel Abd El-Aziz anticline the Lower Maastrichtian sediments contain an olistostrome with Lower Cretaceous and Carboniferous olistoliths. Thus in the Early Maastrichtian the site of the future anticline was apparently a depression into which blocks of various lithologies were sliding. It has been suggested that the blocks came from the early Alpine nappe front which is about 180 km north of the Jebel Abd El-Aziz area (Leonov et al., 1986). The source was more likely closer, perhaps on the Mardin or Khlessia highs, where faulting and uplift might have taken place. In any case the Jebel Abd El-Aziz anticline began to grow in the Late Maastrichtian: the section of the Paleogene sediments in its core is incomplete and reduced in thickness.

The Campanian-Early Maastrichtian(?) folding was also active in the Syrian Arc of northern Sinai. The E-W and NE trending folds created at this time were covered with sharp angular unconformity by the Maastrichtian and Paleogene sediments (Hirsch, 1990). Thus folding in the intracratonic basins of Syria and on the passive African margin was simultaneous.

The correlation of this deformation with events in the Tethys is evident. At this time the passive margins of Africa-Arabia collided with a system of trenches, an event which culminated in the obduction of ophiolites (Dercourt et al., 1986, 1993). Metamorphic basement at the base of the ophiolitic slabs in Baer-Bassit and Amanous were formed in the Senonian from sediments and volcanics of the passive margin (Silantyev, 1993). In Baer-Bassit, obduction terminated in the Maastrichtian, *i.e.* collision and obduction were synchronous with deformation in the Syrian rift basins. Accretion of the Mamonia melange in the south-facing trench terminated in the Campanian and the nappe structure of Cyprus was finally formed in the Early Maastrichtian (Knipper and Sharaskin, 1995; Krylov et al., Chapters 4 and 5 of this volume). It has been suggested that these events were caused by the extensive collision of a wide African passive margin with a trench in Mesogea (Kazmin, 1991a) (Fig. 8.24). This way the temporal coincidence of events in the Baer-Bassit Massif, Cyprus, and Syrian Arc can be explained.

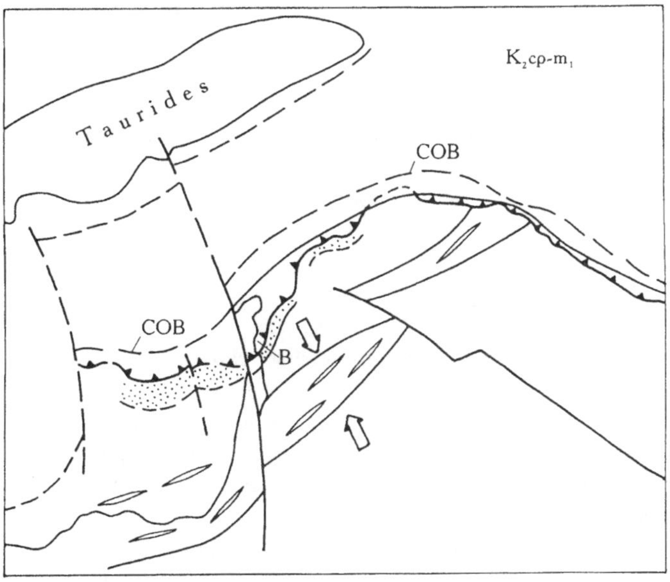

Figure 8.24. Paleotectonic scheme for the Campanian-Early Maastrichtian. 1st compressional stage. Symbols in Fig. 8.21.

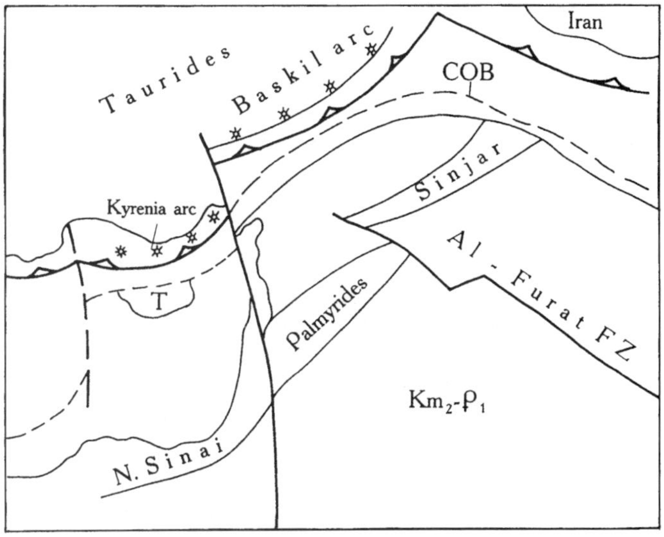

Figure 8.25. Paleotectonic scheme for the Late Maastrichtian-Paleocene. Symbols in Fig. 8.21.

Late Maastrichtian to Middle Eocene. Tectonic quiscence. **(Fig. 8.25)**

The period from the Late Maastrichtian to Middle Eocene was a time of tectonic quiescence and accumulation of a thick series of chalky limestones and marls all over Syria. The thickness of these sediments reaches 700-900 m in the Palmyrides and about 1200 m in the Al-Furat basin. The Nahr El-Kabir and Aafrin foredeeps were also rapidly subsiding.

The quiet tectonic period corresponds to the drift of Africa-Arabia towards the active Tethyan margin, which followed the ophiolite obduction. In particular, the oceanic space between Africa-Arabia and the Taurus carbonate platform (a backarc basin?) was closing (Kazmin, 1991a, b; Knipper and Sharaskin, 1995). The oceanic lithosphere of this basin was subducted below the Taurus platform, where the volcanic Kyrenia arc was formed (Baroz, 1980; our Figs. 8.14 to 8.25).

Late Eocene-Oligocene stage. 2nd phase of compression (Fig. 8.26).

The situation gradually began to change in the Middle Eocene, when the Jebel An-Nusseirieh and Anti-Lebanon anticlinoria began to grow. On these structures shallow water limestones replaced pelagic sediments of Middle Eocene age.

Compression and deformation intensified in the Late Eocene. In the Palmyrides, folding was renewed and the contrast between types and rates of sedimentation between the anticlines and synclines increased. The growth of the anticlines was accompanied by the widespread deposition of shallow marine, often terrigenous, sediments. Late Eocene sediments are missing from Jebel An-Nusseirieh which was uplifted. These processes reached a culmination at the end of the Oligocene when the sea finally retreated from most of Syria. Terrigenous clastics became ubiquitous in the Palmyrides. A marine basin still remained in the northeastern part of the country, in the Mesopothamian depression.

Deformation, uplift, and regression were consequences of the collision of Arabia with Eurasia. As demonstrated by reconstructions (Dercourt et al., 1986, 1993; Paleogeographic Atlas of Northern Eurasia, 1997), the Neotethys branch between Arabia and Iran closed about 55-50 Ma ago. A backarc basin (a remnant of Mesogea) lying between the African margin and Taurus carbonate platform was also closed at approximately the same time. The region entered a long period of continent-continent collision which sharply intensified in the late Oligocene (30 Ma ago) when the Gulf of Aden-Red Sea rifts were initiated.

Neogene-Quaternary stage. Formation of the present-day structure.

The present-day structure of Syria was formed during the last 12-13 Ma. The structural evolution was determined by the following:

a) A sharp increase in the rate of opening of the Gulf of Aden

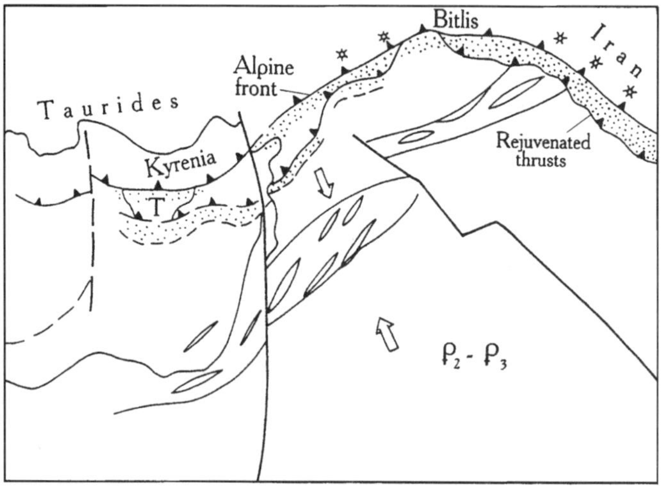

Figure 8.26. Paleotectonic scheme for the Middle-Late Eocene. Arabia-Eurasia collision and related compression. Symbols in Fig. 8.21.

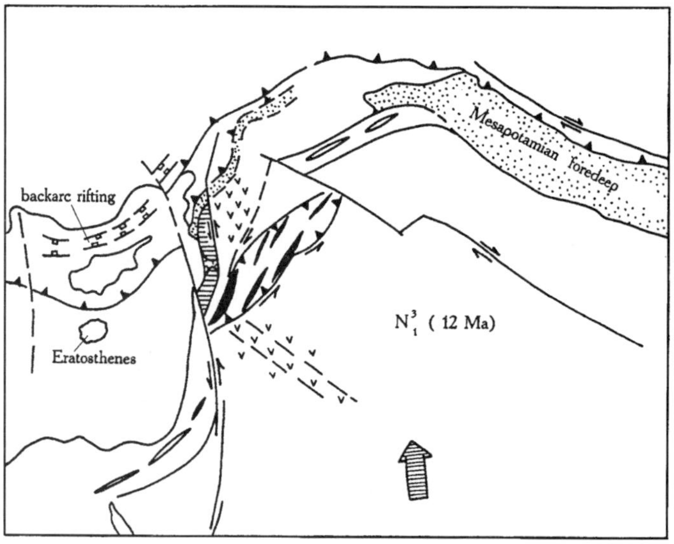

Figure 8.27. Paleotectonic scheme for the Middle-Late Miocene. The Miocene-Quaternary compression, 1st phase. Symbols in Fig. 8.21.

and the Red Sea (Le Pichon and Gaulier, 1988);

b) Separation of the Arabian plate from Sinai; and

c) A 105 km sinistral displacement along the Levantine transform (Quennell, 1958, 1984; Freund et al., 1970; Joffe and Garfunkel, 1987). An increase from 1 to 3 cm per year in the rate of convergence between Arabia and Eurasia (Savostin et al., 1986; DeMets et al., 1990) led to an intensification of the collision and extensive deformation of the Arabian foreland.

Two phases are usually distinguished in the motion of Arabia along the Levantine transform zone: a Middle-Late Miocene phase and a Pliocene-Quaternary phase with 65 and 40-45 km displacements respectively. Since only 8 to 10 km of the Late Pliocene-Quaternary displacement was observed along the Al-Ghab fault, it is provisionally accepted here that during the Miocene phase the principal motion occurred along the Roum fault which acted as a plate boundary at that time (Girdler, 1990). Consequently, in the Miocene the Yammuneh and Al-Ghab faults occupied the intraplate position (**Fig. 8.27**). The author is aware, of course, that so far no geological evidence has been obtained of a Late Tertiary motion on the Roum fault.

The scale of Miocene deformaton in Syria is difficult to estimate. In the Palmyrides ridges, corresponding to anticlines, were growing and supplying coarse clastics to the nearby synclinal depressions. In the South Palmyra zone, reverse faults formed as a result of inversion of normal boundary faults. No Miocene deformation has so far been described for the Sinjar basin (Galaktionov and Demidov, 1966; Ponomarev and Bivshev, 1966).

There is also no direct evidence of motion along the Al-Ghab and Karasu faults. Indirectly however, some motion can be inferred from the specific features of Late Miocene volcanism on the Aleppo plateau. Here basaltic eruptions were controlled

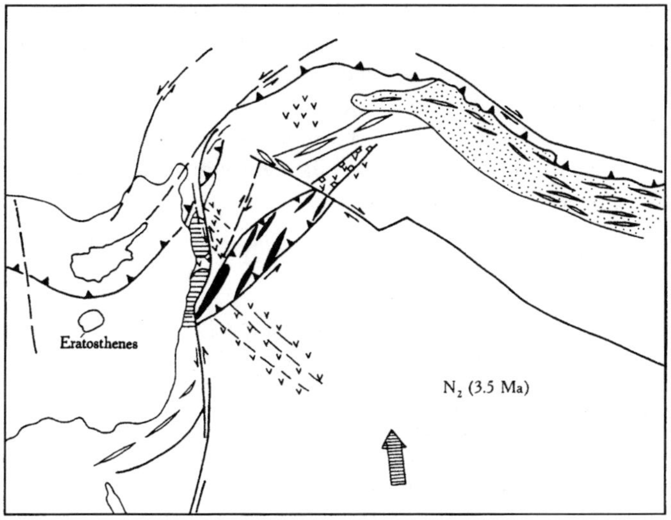

Figure 8.28. Paleotectonic scheme for the Pliocene-Quaternary. The Miocene-Quarternary compression, 2nd phase. Symbols in Fig. 8.21.

by NW-SE trending fractures, which probably join the Al-Ghab fault. Similar fractures served as magma conduits in the Drouz depression. According to Kopp et al. (1994) and Kopp and Leonov (2000), these fractures are an integral part of the structural assemblage created in the zone of influence of the restaining bend of the Yammuneh fault. Deformation of the Arabian plate in this zone belongs to the pure shear type and is distributed in an N-S strip about 300 km wide. Within this strip NW-SE and NE-SW trending structures are respectively transtensional and transpressional. Similar conditions could exist on the Aleppo plateau in conjunction with the restraining bend of the Al-Ghab fault.

The interpretation of Kopp and Leonov (2000) satisfactorily explains the structural pattern along the Levantine fault system. However, NW-SE trending fractures have controlled basaltic eruptions in the areas adjacent to the Red Sea since the Late Oligocene-Early Miocene, *i.e.* before the Levantine system was activated. There is, on the other hand, a good correlation between magmatic impulses and the opening stages of the Red Sea and Gulf of Aden. It appears that throughout the last 30 Ma of Arabia's motion, conditions were favorable for the re-opening of NW-SE trending Precambrian fractures in the transtensional regime.

The most conspicuous deformation in Syria took place in the Pliocene-Quaternary (**Fig. 8.28**). In the Palmyrides a system of ridges (anticlines) and mollasic basins (synclines) was finally formed. Horizontal shortening, due to folding and thrusting, reduced the western Palmyrides by at least 20 km (Chaimov et al., 1990). A shallow subhorizontal detachment separated the deformed Mesozoic-Cenozoic complex from the Paleozoic formations (Chaimov et al., 1993). There is little doubt that the Rutbah block underthrusts the Palmyrides, but the scale of the underthrusting remains unknown.

In western Syria 8 to 10 km of sinistral displacement along the Al-Ghab and Karasu faults was accompanied by sinking of the Missiaf and Al-Ghab depressions and elevation of the associated Jebel An-Nusseiriyeh and Jebel Az-Zawieh Uplifts. Similar to the Lebanon–Beqaa–Anti-Lebanon assemblage, the structures of northwestern Syria were formed in a transpressional set-up, enhanced by the restraining bend of the Al-Ghab fault. Sharp intensification of the deformation in the Palmyrides and along the Levantine fault system might have resulted from the change in position of the plate boundary, which, in the Pliocene, migrated to the Yummaneh and Al-Ghab faults.

East of the Al-Furat transcurrent zone, compression was weak. The Sinjar basin was inverted so as to form a wide and low swell. Linear folds are absent here, with the exception of the Jebel Abd El-Aziz anticline. Very young linear folds were formed only in the Mesopothamian foreland (*e.g.* the Qarashouk anticline).

The Quaternary basaltic eruptions east of the Al-Furat River, along the faults of the Palmyrides trend, point to a tensional setup. It means that the Al-Furat transcurrent zone still acts as a major structural divide.

A decrease in the horizontal shortening in the Palmyrides, from 20 km in the southwest to 1 km in the northeast (Chaimov et al., 1990) implies some clockwise rotation of the Rutbah block. In this case the dextral motion along the South Palmyra fault zone suggested by Quennell (1984) and Chaimov et al. (1993) is hardly possible. Ground observations and focal mechanisms of earthquakes (Ben-Menachem et al., 1976; Ben-Menachem and Aboodi, 1981) indicate sinistral motion that is in good agreement with the inferred Rutbah block movement. Clockwise rotation of the Rutbah block around a pole not far from the Al-Furat transcurrent zone would explain the small lateral displacement along this zone and transtension along the NW-SE trending faults in the Drouz depression.

Conclusions

1) The Mesozoic and Cenozoic intraplate structures in Syria evolved in response to plate boundary forces.

2) The Palmyrides and Sinjar rifts originated due to Permian and Triassic extension, connected to rifting and formation of new spreading axes in the Neotethys and Mesogea. Slab-pull at the active Tethyan margin is suggested as a possible cause for these processes.

The stress was transmitted to the passive margin simultaneously with mid-oceanic ridge subduction, *i.e.* at that time no plate boundary existed between the active and passive margins of the Tethys. As soon as new spreading centers were formed, the extension in Arabia ceased.

3) The Early Cretaceous (Aptian-Albian) rifting in Syria coincides with formation of the Peri-Arabian trench system. Again, the absence of a spreading axis between the newly-formed trenches and the margin of Arabia resulted in the slab-pull being transmitted directly to the margin, thus producing extensive rifting.

4) The first stage (Campanian-Early Maastrichtian) of compression in the Syrian rift basins, and in Sinai, coincides with the collision of the African-Arabian passive margin with the Peri-Arabian trench system.

5) The second stage of compression (Middle Eocene-Oligocene) took place as a consequence of the Arabian-Eurasian collision, and produced deformation in North Arabia as well as inversion in the Palmyrides and Sinjar rift basins. This process intensified dramatically 12-14 Ma ago, following the onset of the fast opening of the Red Sea and Gulf of Aden, which resulted in transform motion along the Levantine fault system. The restraining bends of the Yammuneh–Al-Ghab transform boundary played an important role in the formation of the transpressional structures of Lebanon and northwestern Syria.

REFERENCES CITED

Adjamyan, J., Kopp, M. L., Sigachev, S. P., Faciani, F., and Hafez, E., 1994. Structural indications of horizontal compression in southern Palmyrides (Syria). Geotektonika, No. 6, p. 54-63. (In Russian).

Al-Saad D., Sawaf, T., Gebran, A., Barazangi, M., Best, J. A., and Chaimov, T. A., 1992. Crustal structure of central Syria: the intracontinental Palmiride mountain belt. Tectonophysics, 207, p. 345-358.

Ala, M. A., and Moss, B. J., 1979. Comparative petroleum geology of southeast Turkey and northern Syria. J. Petrol. Geol., 1, p. 3-27.

Aubouin J., 1959. Essai sur l'évolution paleogéographique et le developpement tecto-orogenique d'un systeme geosynclinal: le secteur grec des Dinarides (Hellenides). Bull. Soc. géol. Fr. 6eme ser., 8, No. 7, p. 731-750.

Bailey, E. B., and McCallian, W. C., 1953. Serpentine lavas, the Ankara melange and the Anatolian thrust. Roy. Soc. Lond. Phil. Trans., 62, p. 403-442.

Baroz, F., 1980. Volcanism and continent-island arc collision in the Pentadactylos Range, Cyprus. Ophiolites. Proceeds. Intern. Ophiol. Symp. Cyprus, 1979. Nicosia, Geol. Surv. Depart., 1980, p. 73-85.

Belov, A. A., 1981. Tectonic evolution of the Alpine fold area in the Paleozoic. Nauka, Moscow, 212 p. (In Russian).

Ben-Avraham, Z., and Zoback, M. D., 1992. Transform-normal extension and asymmetric basins: An alternative to pull-apart models. Geology, 20, p. 423-426.

Ben-Menachem, A., and Aboodi, E., 1981. Micro- and macroseismicity of the Dead Sea rift and off-coast eastern Mediterranean. Tectonophysics, 80, p. 199-234.

Ben-Menachem, A., Nur, A., and Vered, M., 1976. Tectonics, seismicity and structure of the Afro-Eurasian junction - the breaking of an incoherent plate. Phys. Earth Planet. Interiors, 12, p. 1-50.

Beydoun, Z. R., 1977. The Levantine countries: the geology of Syria and Lebanon (maritime regions). In A. E. M. Nairn, W. H. Kanes and F. G. Stehli (eds.). The ocean basins and margins, v. 4A. The Eastern Mediterranean. New York, Plenum Press, p. 319-353.

Best, J. A., Barazangi, M., Al-Saad, D., Sawaf, T., and Gebran, A., 1990. Bouguer gravity trends and crustal structure of the Palmyride Mountain belt and surrounding northern Arabian platform in Syria. Geology, 18, p. 1235-1239.

Best, J. A., Barazangi, M., Al-Saad, D., Sawaf, T., and Gebran, A., 1993. Continental margin evolution of the Northern Arabian platform in Syria. Am. Assoc. Petrol. Geol. Bull., 77(2), p. 173-193.

Blankenhorn, M., 1891. Grundzuege der Geologie und physicalischen Geographie von Nordsyrien. Friedlaender, Berlin, 102 p., 2 maps.

Blankenhorn, M., 1914. Syrien, Arabien und Mesopotamien. Handbuch der regionalen Geologie. 5, fasc. 17, 159 p.

Blankenhorn, M., 1925. Der sogenannte syrische Bogen und erythraische Geosynclinale. Deutsch. geol. Ges., 77, p. 206-226.

Blankenhorn, M., 1931. Geologie Palastinas nach heuter Auffassung. Z. Deutsch. Palastina-Verein, 54, Leipzig, 50 p.

Blankenhorn, M., 1940. Syrien and Palaestina. Geol. Jahresber., Berlin, Borntraeger Verlag., 2, p. 351-361.

Bogdanov, A. A., 1961. Some problems of the European tectonics. Vestnik MGU, N 5, p. 1-18. (In Russian).

Brew, G. E., Litak, R. K., Seber, D., Barazangi, M., Al-Imam, A., and Sawaf, T., 1997. Basement depth and sedimentary velocity structure in the northern Arabian platform, eastern Syria. Geophys. J. Int., 128, p. 617-631.

Chaimov, A., Barazangi, M., Al-Saad, D., Sawaf, T., and Gebran, A., 1990. Crustal shortening in the Palmyride fold belt Syria and implications for movement along the Dead Sea fault system. Tectonics, 9(6), p. 1369-1386.

Chaimov, T. A., Barazangi, M., Al-Saad, D., Sawaf, T., and Gebran, A., 1992. Mesozoic and Cenozoic deformation inferred from seismic stratigraphy in the southwestern intracontinental Palmyride fold-thrust belt, Syria. Geol. Soc. Am. Bull., 104, p. 704-715.

Chaimov, T. A., Barazangi, M., Al-Saad, D., Sawaf, T., and Khaddour, M., 1993. Seismic fabric and 3-D structure of the southwestern intracontinental Palmyride fold belt. Am. Assoc. Petrol. Geol. Bull., 77(12), p. 2032-2047.

Chase, C. G., 1978. Plate kinematics: The Americas, East Africa and the rest of the World. Earth Planet. Sci. Lett., 37, p. 355-368.

Coffin, M. F., and Eldholm, O., (eds.), 1991. Large igneous provinces. JOI/USSAC Workshop Report. The Univ. of Texas, at Austin. Inst. for Geophysics. Rep. N114, 79 p.

De Mets, C., R. Gordon, G., Argus, D. F., and Stein, S., 1990. Current plate motions. Geophys. J. Int., 101, p. 425-478.

Dercourt, J., Zonenshain, L. P., Ricou, L. -E., Kazmin, V. G., Le Pichon, X., Knipper, A. L., Grandjacquet, C., Sbortshikov, I. M., Geyssant, J., Lepvrier, C., Pechersky, D. H., Boulin, J., Sibuet, J. -C., Savostin, L. A., Sorokhtin, O., Westphal, M., Bazhenov, M., Laurer, J. P., and Biju-Duval, B., 1986. Geological evolution of the Tethys belt from the Atlantic to the Pamir since Lias. Tectonophysics, 123, p. 241-315.

Dercourt, J., Ricou, L. E., and Vrielinck, B., 1993. Atlas Tethys, palaeo-environmental maps. Gauthier-Villars, Paris, 307 p.

Dubertret, L., 1932. Les formes structurales de la Syrie et de la Palestine: leur origine. C. R. hebd. Séances Acad. Sci. Paris.195, p. 65-67.

Dubertret, L., 1953. Géologie des roches vertes du nord-ouest de la Syrie et du Hatay (Turquie). Theses doct. Sci. natur. Fac. Sci. Univ. Paris, Museum nat. histoire natur. 226 p.

Dubertret, L., 1936. Stratigraphie des régions recovertes par les roches vertes. C. R. Acad. Sci., Paris, 203, p. 1173-1174.

Dubertret, L., 1937. Sur la constitution et la génese des roches vertes Syriennes. C. R. Acad. Sci., Paris. 204, p. 1663-1664.

Dubertret, L., 1942. Carte géologique du Moyen Orient au 1:2,000,000e. Beyrouth, 67 p.

Dubertret, L., 1945. Carte géologique de la Syrie et du Liban au 1:1,000,000 (3 ed). Beyrouth, 56 p.

Dubertret, L., 1947. Problèmes de la géologie du Levant. Bull. Soc. géol. Fr. 17, p. 3-31.

Dubertret, L., 1955. Carte géologique du Liban au 1:200,000e. Beyrouth, 74 p.

Dubertret, L., 1960. Feuille Hermon, Notice explicative. Rep. Libanaise, Min. Trav. Publ., Beyrouth, p. 1-46.

Dubertret, L., 1970. Review of structural geology of the Red Sea and surrounding areas. Philos. Trans. Roy. Soc. London. A267, p. 9-20.

Faradzhev, V. A., 1966. The Geological Map of Syria, scale 1:200,000, sheet J-37-IV. Explanatory notes. V. O. Technoexport, Moscow. 146 p.

Freund, R., Garfunkel, Z., Goldberg, M., Weissbrod, T., and Derin, B., 1970. The shear along the Dead Sea rift. Phil. Trans. Roy. Soc. London, A267, p. 107-130.

Galaktionov, A. B., and Demidov, V. A., 1966. Geological Map of Syria, Scale 1:200,000, sheets J-37 VI, XII. Explanatory notes. Technoexport, Moscow. 83 p.

Galaktionov, A. B., 1989a. Map of identification of space images (and Explanatory note). Scale 1:500,000. Syrian Petroleum Company, Damascus (Unpublished).

Galaktionov, A. B., 1989b. Neotectonic map. Scale 1:500,000. Syrian Petroleum Company, Damascus (Unpublished).

Garfunkel, Z., and Derin, B., 1984. Permian-Early Mesozoic tectonism and continental margin formation in Israel and its implication for the history of the eastern Mediterranean. In: J. E. Dixon and A. H. F. Robertson (eds.), The geological evolution of the eastern Mediterranean. Geol . Soc. London, Spec. Publ., 17, p. 187-201.

Geological Map of Syria, 1964. Scale 1:1,000,000 (V. P. Ponikarov and I. A. Mikhailov, eds.). V. O. Technoexport, Moscow.

Geological Map of Turkey, 1989. Scale 1:2,000,000. General Directorate of Mineral Research and Exploration. Ankara.

Girdler, R. W., 1990. The Dead Sea transform fault system. Tectonophysics, 180, p. 1-3.

Gwirtzman, G., and Weissbrod, T., 1984. The Hercynian geoanticline of Helez and the Late Palaeozoic history of the Levant. In J. E. Dixon and A. H. F. Robertson (eds.) The geological evolution of the eastern Mediterranean: Oxford, Blackwell Scientific Publications, p. 177-186.

Haq, B. U., Hardenbol, J., and Vail, P. R., 1987. Chronology of fluctuating sea levels since the Triassic (250 million years ago to present). Science, v. 235, p. 1156-1167.

Hencock, P. L., and Atiya, M. S., 1979. Tectonic significance of meso-fracture systems associated with the Lebanese segment of the Dead Sea transform fault. J. Struct. Geol., 1, p. 143-153.

Hirsch, F., 1990. Apreçu de l'histoire phanérozoique d'Israel. J. African Earth Sci., 11(1-2), p. 177-196.

Joffe, S., and Garfunkel, Z., 1987. Plate kinematics of circum Red Sea - a re-evaluation. Tectonophysics, 141, p. 5-22.

Kazmin, V. G., 1991a. Collision and rifting in the Tethys Ocean: geodynamic implications. Tectonophysics, 196, p. 371-384.

Kazmin, V. G., 1991b. Tectonic setting of the Mamonia and Troodos complexes formation (Cyprus). Geotectonika, No. 6, p. 104-114. (In Russian).

Kazmin, V. G., and Kulakov, V. V., 1965. Ophiolite formation of northwestern Syria. Geologiya i Razvedka, No. 2, p. 3-14. (In Russian).

Kazmin, V. G., and Kulakov, V. V., 1968. Geological Map of Syria. Scale 1:50,000. Technoexport. Moscow, 124 p.

Kazmin, V. G., Mikhailov, I. A., and Shatsky, V. N., 1964. Rift structures of northwestern Syria. Sovetskaya Geologiya, No. 6, p. 81-92. (In Russian).

Khair, K., Khawlie, M., Haddad, F., Barazangi, M., Seber, D., and Chaimov, T., 1993. Bouguer gravity and crustal structure of the Dead Sea transform fault and adjacent mountain belts in Lebanon. Geology, 21, p. 739-742.

Knipper, A. L., and Sharaskin, A. Ya., 1995. Correlation of tectonic events in the Mesozoic history of northeastern Mediterranean. Geotektonika, No. 1, p. 12-22. (In Russian).

Knipper, A. L., Kopaevich, L. F., and Rukie, M., 1987. Age and origin of ophealiites in the Baer-Bassit ophiolite massif. Izvestia Akad. Nauk SSSR, Seriya Geologicheskaya, No. 6, p. 124-129. (In Russian).

Knipper, A. L., Savel'ev, A. A., and Rukie, M., 1988. Ophiolite association of northwest Syria. Geotektonika, No. 1, p. 92-104. (In Russian).

Kober, L., 1915. Geologische Forschungen in Vorderasien 1. Zur tektonik des Libanon. Denkschr. Akad. Wiss. Wien, math.-naturwiss. Kl. 91, p. 379-427.

Kogan, L. E., and Stenin, A. M., 1994. Multichannel MOV-OGT (CDP seismic profiling during Cruise 5 of R/V Akademik Nikolaj Strakhov (Preliminary results). In: V. A. Krasheninnikov and J. K. Hall (eds.), Geological Structure of the Northeastern Mediterranean, Historical Productions-Hall, Jerusalem, p. 99-112.

Kopp, M. L., Leonov, Yu. G., and Adjamyan, J., 1994. Intraplate deformation in the west of the Arabian Plate as a result of the inhomogenous shear along the Levantine fault. Geotektonika, No. 3, p. 61-76. (In Russian).

Kopp, M. L., Adjamyan, J., Ilyas, K., Fakiani, F., and Hafez, A., 1999. Mechanism of the Al-Ghab (Syria) graben formation and propagation of the Levantine transform. Geotektonika, No. 5, p. 52-66. (In Russian).

Kopp, M. L., and Leonov, Yu. G., 2000. Tectonics. In: Yu. G. Leonov (ed.). Notes on the Syrian geology. Trudy Geologicheskogo Inst., RAS, No. 526, Moscow. (In Russian).

Kozlov, V. V., Artemov, A. V., and Kalis, A. F., 1966. Geological Map of Syria, Scale 1:200,000, sheets I-36-XVIII, I-37-XIII. Explanatory notes. V. O. Technoexport, Moscow, 68 p.

Krenkel, E., 1924. Der Syrische Bogen. Cbl. f. Mineral., No. 9, p. 274-281, No. 10, p. 301-313.

Krenkel, E., 1941. Syrabien (Arabien und Syrien). Deutsch. Geol. Ges. Z., 93(7-8), p. 357-376.

Krenkel, E., 1957. Geologie und Bodenschatze Afrikas. 2. Aufl.15 und 461 p. Geest and Porring, Leipzig.

Krylov, K. A., Kaleda, K. G., and Bragin, N. Yu., 2003. Chaotic rock complexes in south-western and central Cyprus. Chapter 4 in Krasheninnikov et al., (eds.), Geological Framework of the Levant, Volume I, Historical Productions-Hall, Jerusalem.

Krylov, K. A., Silantyev, S. A., and Krasheninnikov, V. A., 2003. The tectonic structure and evolution of south-western Cyprus. Chapter 5 in Krasheninnikov et al., (eds.), Geological Framework of the Levant, Volume I, Historical Productions-Hall, Jerusalem.

Kurbanov, N., and Zarjanov, Yu., 1968. The Geological Map of Syria. Scale 1: 50,000. Syrian Arab Republic. V. O. Techoexport, Moscow. 120 p.

Lapierre, H., 1975. Les formations sedimentaries et eruptives des nappes de Mamonia et leurs relations avec le massif de Troodos (Chypre occidentale). Mem. Soc. géol. Fr., 123, 123 p.

Lartet, L., 1869. La géologie de la Palestine. Ann. Sci. geol. 1, p. 1-116 and 149-329. Paris.

Leonov, Yu. G., Macarel, Ch., and Zaza, T., 1986. To the origin of the olistostroma in the core of the Abd El-Aziz anticline (Syria). Geotektonika, No. 2, p. 85-91. (In Russian).

Leonov, Yu. G., Sigachev, S. P., Otri, M., Yusef, A., Zaza, T., and Sawaf, T., 1989. New data on the Paleozoic complex of the platform cover of Syria. Geotectonics, 23, p. 538-542.

Le Pichon, X., and Gaulier, J. -M., 1988. The rotation of Arabia and the Levant fault system. Tectonophysics, 153, p. 271-294.

Lobachevsky, I. V., and Shcheglova, I. F., 1961. Report on the gravity survey in the Homs-Idlib area (Syrian Arab Republic) in 1960-1961. Dimashq, 1961 (unpublished). (In Russian).

Lovelock, P. E. R., 1984. A review of the tectonics of the northern Middle East region. Geol. Mag., 121(6), p. 577-587.

McBride, J. H., Barazangi, M., Best, J., Al-Saad, D., Sawaf, T., Al-Otri, M., and Gebran, A., 1990. Seismic reflection structure of intracratonic Palmyride fold-thrust belt and surrounding Arabian platform, Syria. Am. Assoc. Petrol. Geol. Bull., 74(3), p. 238-259.

Minster, J. B., and Jordan, T. H., 1978. Present-day plate motions. J. Geophys. Res., 83, p. 5331-5354.

Mouty, M., 2000. The Jurassic in Syria: an overview. Lithostratigraphic and biostratigraphic correlations with adjacent areas. In: S. Crasquin-Soleau and E. Barrier (eds.), Peri-Tethys Memoir 5: New data on Peri-Tethyan sedimentary basins. Mem. Mus. Hist. Nat., 186, p. 159-168.

Oufland, A. K., 1966. Geological Map of Syria, scale 1:200,000, sheet I-37-XXI. Explanatory notes. V. O. Technoexport, Moscow. 55 p.

Pan Yusheng (ed.)., 1996. Geological evolution of the Karakorum and Kunlun mountains. Seismological Press, Beijing, 288 p.

Parrot, J. F., 1977. Assemblage ophiolitique du Baer-Bassit et termes effusits et volcano sedimentairies. Travaux et documents de l'O.R.S.T.O.M. 72, Paris.

Parrot, J. F., and Whitechurch, H., 1978. Subductions anterieures au chariage Nord-Sud de la tethysienne: facteur de metamorphisme de series sedimentaires et volcaniques liées aux assemblages ophiolitiques syro-turcs en schists verts et amphibolites. Rev. Géogr. Phys. Géol. Dynam., 20(2), p. 153-170.

Paleogeographic Atlas of Northern Eurasia, 1997. L. M. Natapov and V. G. Kazmin (eds.). Inst. of Tectonics of Lithospheric Plates. Moscow.

Ponikarov, V. P., 1964. Tectonic Map of Syria. Scale 1:1,000,000. Technoexport, Moscow.

Ponikarov, V. P., Soulidi-Kondratiev, E. A., Kozlov, V. V., and Kazmin, V. G., 1964. Tectonics of the northwestern part of the Arabian Peninsula. Sovietskaya Geologiya, No. 1, p. 39-48. (In Russian).

Ponikarov, V. P., and Kazmin, V. G., 1965. Precambrian and Paleozoic in the northwestern part of the Arabian Peninsula. Sovietskaya Geologiya, No. 3, p. 89-99. (In Russian).

Ponikarov, V. P., Kazmin, V. G., Mikhailov, I. A., et al., 1967. The Geology of Syria. Explanatory Notes on the Geological Map of Syria, Scale 1:500,000. V. O. Technoexport, Moscow, 229 p.

Ponikarov, V. P., Kazmin, V. G., Kozlov, V. V., et al., 1969. Syria. Geology and Mineral resources. Leningrad, Nedra, 216 p. (In Russia).

Ponomarev, B. Ya., and Bivshev, A. S., 1966. Geological Map of Syria, scale 1:

200,000, sheet J-37-IV. Explanatory notes. V. O. Technoexport, Moscow. 66 p.

Protasevich. L. N., and Maksimov, A. A., 1966. The Geological Map of Syria, scale 1:200,000, sheets J-37-I, II (Halb, Antakya), Explanatory notes, V. O. Technoexport, Moscow, 73 p.

Quennell, A. M., 1958. The structure and geomorphic evolution of the Dead Sea rift. Quart. J. Geol. Soc. Lond., 114, p. 1-24.

Quennell, A. M., 1984. The Western Arabia rift system. In: The geological evolution of the Eastern Mediterranean. J. E. Dixon and A. H. F. Robertson, (eds.). Geol. Soc. Spec. Publ., Blackwell, London, p. 775-788.

Razvalyaev, A. V., 1966. Geological Map of Syria, scale 1:200,000, sheet I-37-VIII. Explanatory notes. V. O. Technoexport, Moscow. 124 p.

Krylov, K. A., Kazmin, V. G., and Galaktionov, A. B., 2003. Volcanism. Chapter 7 in Krasheninnikov et al., (eds.), Geological Framework of the Levant, Volume I, Historical Productions-Hall, Jerusalem.

Robertson, A. H. F., and Woodcock, N. H., 1979. Mamonia Complex, southwest Cyprus: Evolution, and emplacement of a Mesozoic continental margin. Geol. Soc. Am. Bull., 90, p. 651-665.

Robertson, A. H. F., and Woodcock, N. H., 1980. Tectonic setting of the Troodos massif in the East Mediterranean. Ophiolites. Proceeds. Intern. Ophiol. Symp. Cyprus, 1979. Nicosia: Geol. Surv. Depart. p. 36-49.

Ron, H., 1987. Deformation along the Yammuneh, the restraining bend of the Dead Sea transform: paleomagnetic data and kinematic impliations. Tectonics, 6(5), p. 653-666.

Savostin L. A., Sibuet, J. -C., Zonenshain, L. P., Le Pichon, X., and Roulet, M. J., 1986. Kinematic evolution of the Tethys belt from the Atlantic Ocean to the Pamirs since the Triassic. Tectonophysics, 123, p. 1-35.

Sawaf, T., Al-Saad, D., Gebran, A., Barazangi, M., Best, J. A., and Chaimov, T. A., 1993. Stratigraphy and structure of eastern Syria across the Euphrates depression. Tectonophysics, 220, p. 267-281.

Segev, A., 2000. Synchronous magmatic cycles during fragmentation of Gondwana: radiometric ages from Levant and other provinces. Tectonophysics, 325, p. 257-277.

Şengör, A. M. C., 1979. Mid-Mesozoic closure of Permo-Triassic Tethys and its implications. Nature, 279, p. 590-593.

Şengör, A. M. C., 1984. The Cimmeride orogenic system and the tectonics of Eurasia. Geol. Soc. Am. Spec. Paper, 195. 82 p.

Şengör, A. M. C., 1990. A new model for the late Palezoic-Mesozoic tectonic evolution of Iran and implication for Oman. Geol. Soc. London Spec. Publ., 49 , p. 797-831.

Şengör, A. M. C., Cin, A., Rowley, D. B., and Shangyou, N., 1991. Magmatic evolution of Tethysides: a guide to reconstruction of collage history. Palaeogeogr., Palaeoclimat., Palaeoecology, 87, p. 411-440.

Sharkov, E. V., Tshernyshev, I. V., Devyatkin, E. V., Dodonov, A. E., Ivaninko, V. V., Karpenko, M. I., Lebedev, V. A., Novikov, V. M., Hann, S., and Khatib, K., 1998. New data on geochronology of the Late Cenozoic plateau basalts on the north-eastern periphery of the Red Sea rift area (North Syria). Doklady Acad. Nauk, 338(1), p. 96-99. (In Russian).

Shatsky, V. N., Kazmin, V. G., and Kulakov, V. V., 1966. The Geological Map of Syria, Scale 1:200,000, Sheets J-37-XIX, J-36-XXIV. Explanatory Notes. Technoexport, Moscow, 106 p.

Shvolman, V. A., 1977. Tectonic evolution of the Pamirs in the Cretaceous and Paleogene. Nauka, Moscow, 159 p. (In Russian).

Silantyev, S. A., 1993. Metamorphic complexes of eastern Mediterranean: nature of the protolith and geodynamic regime of formation. Petrology, 1(4), p. 449-472. (In Russian).

Soulidi-Kondratiev, E. D., 1966. The Geological Map of Syria, Scale 1:200,000. Sheet 1-37-XV. Explanatory notes. Technoexport, Moscow. 104 p.

S. P. C. (Syrian Petroleum Company), 1986. Depth scheme map of the crystalline basement scale 1:500,000. (Compiled by T. Sawaf). Damascus (unpublished).

Stampfli, G., 1996. The Intra-Alpine terrain: a Paleotethyan remnant in the Alpine Variscides. Eclogae geol. Helv., 89, p. 13-42.

Stampfli, G., Marcoux, J., and Baud, A., 1991. Tethyan margins in space and time. In: Paleogeography and paleoceanography of Tethys (J. E. T. Channell, E. L. Winterer, and L. F. Jansa eds.), Palaeogeogr., Palaeoclim., Palaeoecol., 87, p. 373-410.

Stampfli, S. M., and Pillevuit, A., 1993. An alternative Permo-Triassic reconstruction of the kinematics of the Tethyan realm. In: Atlas Tethys Palaeoenvironmental Maps. Explanatory notes (J. Dercourt, L. -E. Ricou, and B. Vrielinck eds.), Gauthier-Villars, Paris, p. 55-62.

Temple, P. G., and Perry, L. J., 1962. Geology and oil occurrence, southeast Turkey. Bull. Am. Assoc. Petrol. Geol., 40(9), p. 1596-1612.

Trifonov, V. G., Trubikhin, V. M., Adjamyan, J., Jalled, Z., El-Mair, Yu., and Aied, A., 1991. The Levant fault zone in northwestern Syria. Geotektonika, No. 2, p. 63-75. (In Russian).

Vidal, N., Klaeschen, D., Kopf, A., Docherty, C., Von Huene, R., Krasheninnikov, V. A., 2000. Seismic images at the convergence zone from the south of Cyprus to the Syrian coast, eastern Mediterranean. Tectonophysics, 329, p. 157-170.

Walley, C. D., 1988. A braided strike-slip model for the northern continuation of the Dead Sea fault and its implications for the Levantine tectonics. Tectonophysics, 145, p. 63-72.

White, R. S., and McKenzie, D., 1989. Magmatism at rift zones: the generation of volcanic continental margins and flood basalts. J. Geophys. Res., 94(B6), p. 7685-7729.

White, R. S., and McKenzie, D., 1995. Mantle plumes and flood basalts. J. Geophys. Res., 100(B9), p. 17,543-17,585.

Opposite and Overleaf: High resolution satellite images of modern Damascus, capitol of Syria. North is to the left in this overlapping images.
Quoting from http://www.spaceimagingme.com/content/Gallery/Image_of_the_week/index.asp:
"Syria, located on the eastern coast of the Mediterranean Sea, extends between 32°19'N to 37°30'N and from 35°43'E to 42°25'E." "Damascus, the capital, is by far the largest and most important of the Syrian cities. Said to be the world's oldest inhabited city, Damascus today is a thriving modern capital. It lies in southwest Syria, between the Anti-Lebanon Mountains on the west and the Syrian Desert on the east. The Anti-Lebanon Mountains, seen in the picture, form the border between Syria and neighboring Lebanon. The city is on an oasis in a semiarid plain. The Barada River flows through Damascus and has proved to be it's lifeline for thousands of years."
"Damascus is an interesting mixture of the old and the new. Cultural attractions in Damascus include the University of Damascus, the National Library, museums, and theaters. The old city has many fine works of Islamic architecture, being home to more than 200 mosques and traditional one-storey Damascene houses."

Page 497: Ikonos Geo 5 meter black-and-white of Damascus, Syria, from August 2001. ©Spaceimaging.com.
Available on the web at http://www.spaceimagingme.com/content/Gallery/Image_of_the_week/index.asp.

Page 498: The city design of Damascus, Syria is featured in this four-meter resolution image originally in color. The image was taken March 30, 2001 by Space Imaging's IKONOS satellite. ©Spaceimaging.com. Available online at:
http://www.homsonline.com/SatelliteImages.htm with the image at
http://www.homsonline.com/SatelliteImages/Ikonos_Damascus-Syria.jpg.

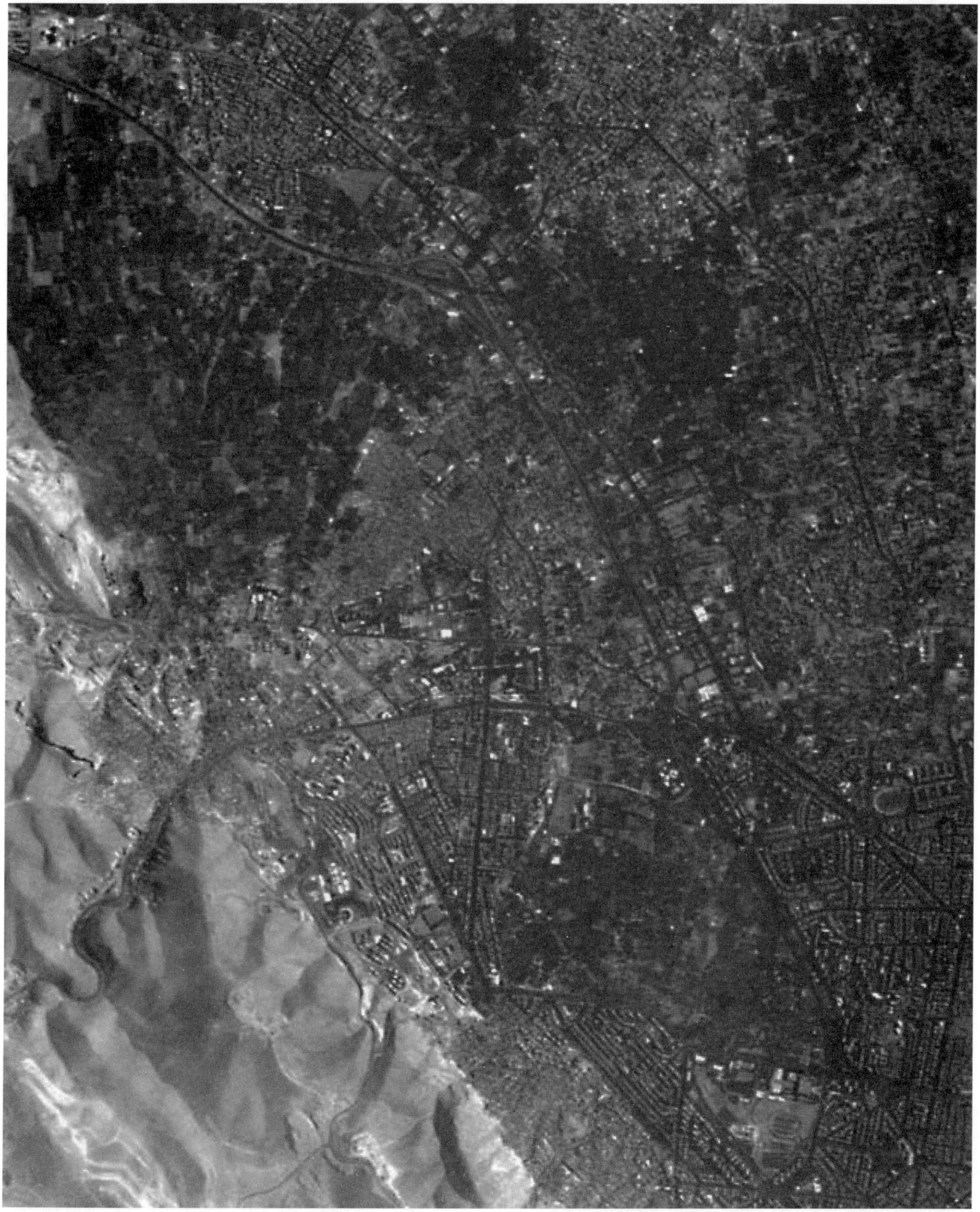

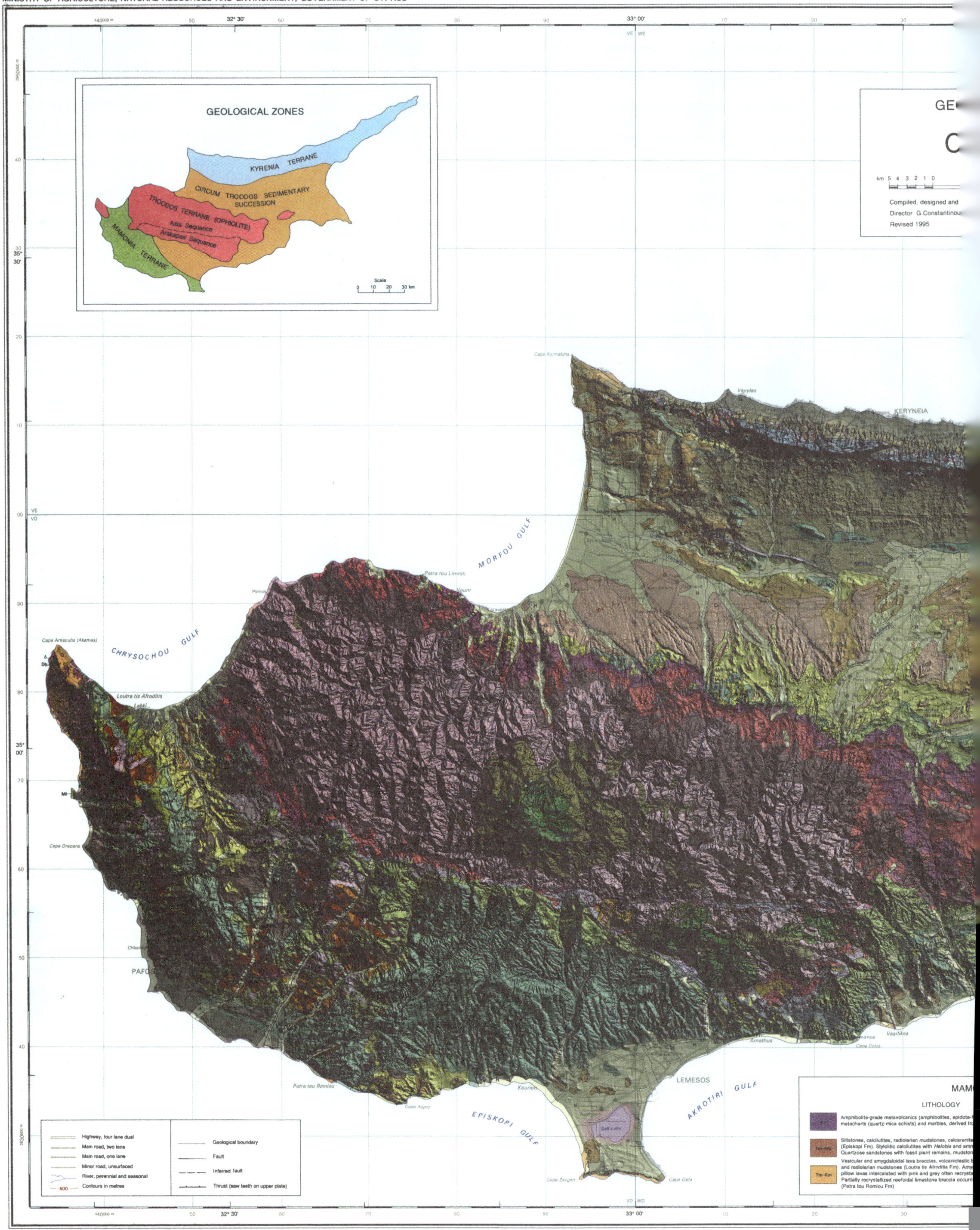

© State Copyright, 1995